Lecture Notes in Artificial Intelligence 4692

Edited by J. G. Carbonell and J. Siekmann

Subseries of Lecture Notes in Computer Science

Bruno Apolloni   Robert J. Howlett
Lakhmi Jain (Eds.)

Knowledge-Based Intelligent Information and Engineering Systems: KES 2007 - WIRN 2007

11th International Conference, KES 2007
XVII Italian Workshop on Neural Networks
Vietri sul Mare, Italy, September 12-14, 2007
Proceedings, Part I

Springer

Series Editors

Jaime G. Carbonell, Carnegie Mellon University, Pittsburgh, PA, USA
Jörg Siekmann, University of Saarland, Saarbrücken, Germany

Volume Editors

Bruno Apolloni
Dipartimento di Scienze dell'Informazione
Università degli Studi di Milano
20135 Milano, Italy
E-mail: apolloni@dsi.unimi.it

Robert J. Howlett
University of Brighton
Centre for SMART Systems, School of Engineering
Brighton, BN2 4GJ, UK
E-mail: R.J.Howlett@bton.ac.uk

Lakhmi Jain
University of South Australia
Knowledge-Based Intelligent Engineering Systems Centre
SA 5095, Australia
E-mail: lakhmi.jain@unisa.edu.au

Library of Congress Control Number: 2007934283

CR Subject Classification (1998): I.2, H.4, H.3, J.1, H.5, K.6, K.4

LNCS Sublibrary: SL 7 – Artificial Intelligence

ISSN 0302-9743
ISBN-10 3-540-74817-2 Springer Berlin Heidelberg New York
ISBN-13 978-3-540-74817-5 Springer Berlin Heidelberg New York

Springer is a part of Springer Science+Business Media

springer.com

Printed in Germany

Typesetting: Camera-ready by author, data conversion by Scientific Publishing Services, Chennai, India
Printed on acid-free paper SPIN: 12120499 06/3180 5 4 3 2 1 0

Preface

These three volumes are a collection of the contributions presented to the joint conferences of KES 2007, the 11th International Conference on Knowledge-Based and Intelligent Information and Engineering Systems, and the WIRN 2007, the 17th Italian Workshop on Neural Networks, held in Vietri sul Mare, Italy, in September 2007.

The formula by which KES conferences gather over 500 people each year from the four corners of the globe to discuss the topic of knowledge-based and intelligent information and engineering systems is: an open mind with rigor. Within the vast universe of the conferences centered around the keywords "information" and "computational intelligence," we encourage in our meetings the offering of new ideas and techniques to give solutions to the never-ending series of problems and challenges that our own intelligence poses.

As a precious attribute of the human brain, we will never be disturbed by the novelty, and possibly the provocation, of new mental paradigms and hazardous conjectures, especially if they are raised by fresh research teams. At the same time, we have riddled each contribution using the sieve of scientific quality, checking the rigor with which the ideas are illustrated, their understandability and the support of the theory or of the experimental evidence.

The structure of the conference reflects this philosophy. In addition to regular tracks on the main fields of the discipline, we invited scientists to propose sessions focused on topics of high interest. Their response was generous and based on all sources, we received some 1203 submissions. From this number we collected 11 general track sessions and 47 invited sessions to make a total of 409 papers after a severe referee screening, an acceptance rate of 34%. Thus the reader may have from these volumes an almost exhaustive overview of researcher's and practitioner's current work in the field of information extraction and intelligent systems.

WIRN 2007 was the annual meeting of the Italian Society for Neural Networks (SIREN). Joining the two conferences provided the double benefit of giving the meeting a qualified and interested local committee on the one hand, and putting the local scientific community in touch with an international cluster of researchers in similar fields, on the other. The efficiency and effectiveness of the result suggest that it would be beneficial to replicate this formula in the future.

We wish to express our sincere thanks to many people who worked hard for the success of the conference, to the authors who sent their contributions, and to the whole scientific community who contributed to intelligence and knowledge values being preserved and enhanced for future generations.

September 2007

Bruno Apolloni
Robert J. Howlett
Lakhmi C. Jain

Preface

[illegible]

Organization

KES 2007 was organized by KES International – Innovation in Knowledge-Based and Intelligent Engineering Systems. WIRN 2007 was organized by IIASS – Istituto Italiano Alti Studi Scientifici.

KES 2007 and WIRN 2007 Conference Chairs

General Chair: B. Apolloni (University of Milan, Italy)
Executive Chair: R. J. Howlett (University of Brighton, UK)
Honorary Invited Session Committee Chair: L. C. Jain (University of South Australia)

KES Conference Series

KES 2007 is part of the KES Conference Series
Conference Series Chairs, L. C. Jain and R. J. Howlett
KES Executive Chair, R. J. Howlett (University of Brighton, UK)
KES Founder, L. C. Jain (University of South Australia)

WIRN Conference Series

WIRN 2007 is part of the WIRN Conference Series
Conference Chairs: M. Marinaro (IIASS, Italy) and B. Apolloni (University of Milan, Italy).

Local Organizing Committee

S. Bassis, S. Gaito, D. Malchiodi, G. L. Galliani, L. Valerio, A. Zippo (University of Milan, Italy)
M. Marinaro, A. Esposito: IIASS

International Program Committee and KES 2007 Board of Reviewers

A. Abe
J. Abe
M. Abulaish
Y. Adachi
A. Adli
S. Akama
G. Alfonso
E. Al-Hasel
U. Amato
P. Angelov
D. Anguita
C. Angulo-Bahon

M. Anisetti
A. Antonella
B. Apolloni
J. Arima
S. Arima
M. Aritsugi
A. Azzini
N. Baba
I. Bae
S. Bae Cho
J. Bajo
B. Balachandran
S. Balzarotti
S. Bandini
B. Baruque
R. Bashar
S. Bassis
K. Basterretxea
R. Batres
L. Bedini
K. Belda
V. Bellandi
B. Berendt
A. Bertoni
P. Beullens
M. Bianchini
F. Biassoni
M. Bielikov
Y. Bin Kwon
L. Bobrowski
G. Bogdan
J. Bohm
A. Bonaldi
N. Borghese
A. Bouchachia
P. Bouquet
A. Brega
D. Brown
L. Brun
I. Buciu
H. Byun
C. Caiafa
P. Campadelli
V. Capasso
F. Capkovic
C. Carpineto
M. Ceccarelli
P. Ceravolo
B. Chabr
C. Chan
H. Chan Cho
J. Chan Lee
C. Chang
D. Chen
K. Chen
M. Chen
W. Chen
Y. Chen
G. Chetty
L. Chilwoo
W. Chou
J. Chung
S. Chung
A. Ciaramella
M. R. Ciceri
A. Colla
S. Colucci
A. Columbari
D. Cook
M. Corazza
E. Corchado
J. Corchado
R. Corchuelo
P. Cosi
A. Cosio
R. Cox
P. Crippa
M. Cristani
A. Cuzzocrea
C. d'Amato
E. Damiani
A. DAnjou
L. D'Apuzzo
P. Davidsson
C. de Campos
S. De Capitani di Vimercati
M. Degemmis
D. Deng
E. Di Claudio
E. Di Iorio
T. Di Noia
E. di Sciascio
D. Difabio
X. Ding
M. do Carmo Nicoletti
Y. Dourisboure
L. Dumitriu
R. Duro
A. Edman
A. Eleuteri
A. Esposito
F. Esposito
L. Eun-Sur
J. Fahimi
P. Falcoz
N. Fanizzi
M. Fansi
G. Fassano
J. Feng
A. Fernandez-Caballero
S. Ferraresi
S. Fiori
A. Formisano
F. Frati
T. Fuchino
C. Fugazza
S. Fujii
T. Fujinami
M. Fukumi
T. Fukumoto
H. Funaoi
C. Furlanello
A. Gabillon
B. Gabrys
S. Gaito
L. Galliani
G. Gao
K. Gao
M. Garcia-sebastian
P. Gastaldo
T. Gavrilova
D. Gendarmi
H. Ghodosi
F. Gianfelici

G. Gianini
P. Giorgini
S. Giove
W. Goh
S. Gon Kong
L. Gonzalez
E. Gouardères
G. Gouardères
M. Grana
M. Graña
K. Grant
D. Gu
H. Guo
T. Guy
K. HaengKon
M. Hagenbuchner
M. Haindl
A. Håkansson
B. Hammer
A. Hara
K. Harada
F. Harris
R. Hartung
S. Hasegawa
Y. Hashimoto
A. Hassanien
Y. Hayashi
X. He
M. Hemmje
M. Hiot Lim
K. Hiraishi
T. Hochin
S. Ho-Jun
X. Hong
S. Hori
A. Hotho
R. Howlett
P. Hraber
E. Hsiao
X. Huang
Y. Huang
F. Hussain
S. Hyun Kim
T. Ichikawa
T. Ichimura
K. Iizuka
N. Inuzuka
Y. Iribe
H. Ishibuchi
Y. Ishida
N. Ishii
H. Ito
J. Itou
Y. Iwahori
S. Iwashita
L. Jain
R. Jain
M. Jason
D. Jeng
M. Jeng
I. Jeon
J. Jiang
H. Joo Lee
S. Joon Yoo
J. Jung
S. Jung
K. Juszczyszyn
J. Kacprzyk
H. Kanai
T. Kanda
Y. Kang
M. Karny
W. Karwowski
R. Katarzyniak
N. Kato
S. Kato
P. Kazienko
L. Kenneth
A. Keskar
D. Keysers
B. Kim
D. Kim
H. Kim
I. Kim
S. Kim
Y. Kim
S. King
M. Kinnaert
D. Kitakoshi
P. Klan
T. Kojiri
T. Kokogawa
S. Kollias
H. Kosaka
A. Koukam
D. Król
N. Kubota
K. Kubota
S. Kunifuji
H. Kunimune
C. Kuroda
Y. Kurosawa
P. Kyu Rhee
K. Lam
K. Le
C. Lee
Y. Lee
F. Leporati
P. Leray
L. Lhotska
J. Li
L. Lin
P. Linh
H. Liu
Y. Liu
B. López
P. Lops
S. Luan
W. Ma
M. Maggini
L. Magnani
M. Majid
S. Makrogiannis
D. Malchiodi
J. O. Maldonado
D. Malerba
L. Mal-Rey
M. Mancini
S. Marinai
M. Marinaro
S. Marrara
G. Martinelli
R. Martone
F. Mason
F. Masulli

J. Matas
N. Matsuda
N. Matsui
H. Matsumoto
N. Matsumura
M. Matsushita
G. Mauri
Q. Meng
F. Menolascina
K. Mera
Y. Mi Kwon
F. Michaud
S. Miguet
H. Minami
H. Mineno
K. Misue
H. Mitsuhara
Y. Mitsukura
H. Miura
M. Miura
T. Mizuno
M. Mizuta
D. Mladenic
H. Mochizuki
Y. Mogami
M. Mohammadian
D. Monekosso
A. Montuori
I. Morgan
A. Morici
P. Motto Ros
N. Mukai
C. Mumford
J. Munemori
M. Muselli
M. Nachtegael
I. Nagy
T. Nakada
K. Nakamatsu
S. Nakamura
T. Nakamura
R. Nakano
T. Nakano
J. Nam Jung
Y. Nara
J. Nascimento
O. Nasraoui
D. Nauck
D. Ndedi Monekosso
M. Negoita
N. Nguyen
G. Nicosia
C. Niederée
A. Nijholt
T. Nishida
K. Nishimoto
T. Nishiura
H. Nobuhara
A. Nowak
M. Nowostawski
A. Nuernberger
Y. Ochi
S. Oeda
R. Oehlmann
L. Oestreicher
N. Ogata
Y. Ohsa
Y. Ohsawa
M. Okada
T. Okamoto
M. Ozden
V. Palade
F. Palmieri
D. Pan
M. Paprzycki
R. Parisi
T. Parisini
G. Park
Y. Park
F. Parra
E. Pasero
G. Pasi
W. Pedrycz
E. Pessa
T. Pham
L. Phong
F. Picasso
A. Pieczynska
L. Prevost
A. Ragone
G. Raiconi
G. Raimondo
J. Ramon
R. Ranawana
R. Rascuna
K. Rattan
L. Razmerita-Hockerts
M. Refice
P. Remagnino
M. Resta
L. Reyneri
A. Rohani
M. Ryoke
G. Ryung Uh
K. Saito
L. Saitta
M. Sakalli
E. Salerno
M. G. Sami
R. Sassi
M. Sato
Y. Sato
M. Sato-Ilic
A. Scarrelli
F. Scarselli
Z. Schindler
M. Schlegel
F. Schwenker
F. Scotti
G. Semeraro
C. Seng Chan
G. Sergiadis
R. Serra
S. Sessa
D. Shen
Y. Shiau
M. Shikida
B. Shizuki
V. Shkodirev
A. Sidhu
J. Smith
J. Sobecki
P. Somol
D. Soo Kim
F. Sorbello

Z. Sosnowski
A. Sperduti
A. Staiano
G. Stamou
R. Stecher
H. Stoermer
Y. Su Choi
T. Sugihara
K. Sugiyama
M. Suka
Z. Sun
I. Sun Choi
W. Sunayama
I. Tabakow
R. Tagliaferri
E. Takahagi
M. Takahashi
O. Takahashi
O. Takata
F. Takeda
H. Taki
H. Tamura
J. Tan
Y. Tanahashi
J. Tanaka
M. Tanaka-Yamawaki
P. Tann
Y. Tateiwa
C. Teeling
L. Tesař
H. Thai
C. Thanh Hoang
N. Thanh Nguyen
P. Tichavský
I. Ting
P. Tino
A. Tonazzini
D. Toshinori
D. Tran
E. Trentin
F. Trinidad
F. Trojani
K. Tsuda
Y. Tsuge
S. Tsumoto
N. Tsuyoshi
G. Tummarello
C. Turchetti
J. Tweedale
K. Umeda
A. Uncini
T. Ushiama
G. Valentini
I. Villaverde
S. Vitabile
I. Vlachos
T. Wadayama
D. Wan Kim
A. Wang
D. Wang
J. Wang
P. Wang
J. Wata
J. Watada
T. Watanabe
Y. Watanabe
Y. Wen
Y. Weo Lee
N. Wessiani
G. Wren
B. Wu
X. Wu
L. Xi
Y. Xiong
F. Xu
X. Xu
Y. Yabuuchi
T. Yamakami
Y. Yamashita
C. Yang
T. Yoshino
M. Young Sung
D. Yu
Z. Yu
T. Yuizono
M. Zalili
A. M. Zanaboni
A. Zeng
X. Zeng
B. Zhang
Y. Zhang
X. Zhou
G. Zhu
Y. Zhu
A. Zippo
I. Zoppis
R. Zunino

General Track Chairs

Generic Intelligent Systems Topics

Artificial Neural Networks and Connectionists Systems
Ryohei Nakano (Nagoya Institute of Technology, Japan)

Granular Computing
Detlef Nauck (BT, UK), Zensho Nakao (University of Ryukyus, Japan)

Machine Learning and Classical AI
Floriana Esposito (University of Bari, Italy)

Agent Systems
Ngoc Thanh Nguyen (Wroclaw University of Technology, Poland)

Knowledge-Based and Expert Systems
Anne Hakansson (Uppsala University, Sweden)

Miscellaneous Intelligent Algorithms
Honghai Liu (University of Portsmouth, UK)

Applications of Intelligent Systems

Intelligent Vision and Image Processing
Tuan Pham (James Cook University, Australia)

Knowledge Management and Ontologies
Guy Gouarderes (University of Bayonne, France), Gloria Wren (Loyola College in Maryland, USA), Lakhmi Jain (University of South Australia, Australia)

Web Intelligence, Text and Multimedia Mining and Retrieval
Andreas Nuernberger (University of Magdeburg, Germany)

Intelligent Signal Processing, Control and Robotics
Miroslav Karny (Czech Republic Academy of Science, Czech Republic)

Other Intelligent Systems Applications
Viacheslaw Shkodirev St. Petersburg State Poly. University, Russia)

Invited Session Chairs

Ambient Intelligence, *Cecilio Angulo-Bahon (Universitat Politecnica de Catalunya, Spain) Honghai Liu (University of Portsmouth, UK)*

Artificial Intelligence Applications in Digital Content, *Mu-Yen Chen (National Changhua University of Education, Taiwan), Hsiao-Ya Chiu (Yu-Da College of Business)*

Artificial Intelligence Applications in Security, *Emilio Corchado (University of Burgos, Spain) Rodolfo Zunino (Genoa University, Italy)*

Artificial Intelligence Methods for Information Processing (AIMIP 2007), *Lifeng Xi, Jifang Li, Kun Gao (Zhejiang Wanli University, Ningbo, China)*

Communicative Intelligence 2007, *Toyoaki Nishida (University of Kyoto, Japan) Ngoc Thanh Nguyen (Wroclaw University of Technology, Poland)*

Computation Intelligence for Image Processing and Pattern Recognition, *Yen-Wei Chen (Ritsumeikan University, Nojihigashi, Japan)*

Human Computer Intelligent Systems, *Takumi Ichimura, Kazuya Mera (Hiroshima City University, Japan)*

Hybrid Artificial Intelligence Systems Workshop (HAIS 2007-KES2007), *Juan M. Corchado (University of Salamanca, Spain) Emilio Corchado(University of Burgos, Spain)*

Innovations in Intelligent Data Analysis, *Mika Sato (University of Tsukuba, Japan) Lakhmi Jain (University of South Australia, Australia)*

Intelligent Agents and Their Applications, *Dharmendra Sharma, Wanli Ma (University of Canberra, Australia), Haeng Kon Kim (Catholic University of Daegu, Korea)*

Intelligent and Adaptive Systems in Economics, Finance and Management, *Marco Corazza (University Ca' Foscari, Venice) Norio Baba (Osaka Kyoiku University, Japan)*

Intelligent Automation Systems, *MuDer Jeng (National Taiwan Ocean University)*

Intelligent Control Theory and Applications, *Kazumi Nakamatsu (University of Hyogo, Japan) Scheng-Luen Chung (National Taiwan University of Science and Technology)*

Intelligent Data Processing in Process Systems and Plants, *Tetsuo Fuchino (Tokyo Institute of Technology, Japan) Yoshiyuki Yamashita (Tohoku University, Japan)*

Intelligent Mechanism for Knowledge Innovation, *Toyohide Watanabe (Nagoya University) Teketoshi Ushiama (Kyushu University)*

Intelligent Multimedia Solution and Security in the Next-Generation Mobile Information Systems (IMSS), *Dong Chun Lee (Howon University, Korea) Hyuncheol Kim (Namseoul University, Korea)*

Intelligent Techniques for Biometric-Based Authentication, *Ernesto Damiani, Antonia Azzini, Stefania Marrara (University of Milan, Italy)*

Logic-Based Intelligent Information Systems, *Kazumi Nakamatsu (University of Hyogo, Japan)*

Chance Discovery, *Akinori Abe (ATR Knowledge Science Laboratories, Japan) Yukio Ohsawa (University of Tokyo, Japan)*

Knowledge-Based Interface Systems I, *Naohiro Ishii (Aichi Institute of Technology, Japan) Yuji Iwahori (Chubu University, Japan)*

Knowledge-Based Interface Systems II, *Yoshinori Adachi (Chubu University, Japan) Nobuhiro Inuzuka (Nagoya Institute of Technology, Japan)*

Knowledge and Information Management in a Social Community, *Toyohide Watanabe (Nagoya University, Japan) Naoto Mukai (Tokyo Science University, Japan) Jun Feng (Hohai University, China)*

Knowledge and Ontological Engineering for Intelligent Information System Development (KOS), *Tatiana Gavrilova (St. Petersburg State Polytechnic University, Russia) Vyacheslav Shkodyrev (Polytechnic of St. Petersburg, Russia)*

Knowledge Engineering in Multi-Robot Systems, *Manuel Graa, Richard Duro (Universidad del Pais Vasco, Spain)*

Knowledge-Based Creativity Support Systems, *Susumu Kunifuji, Motoki Miura (JAIST, Japan) Kazuo Misue (Tukuba University, Japan)*

Knowledge-Based Multi-Criteria Decision Support, *Hsuan-Shih Lee (National Taiwan Ocean University)*

Knowleged-Based Systems for e-Business, *Kazuhiko Tsuda (University of Tsukuba, Japan) Masakazu Takahashi (Shimane University, Japan)*

Computational Learning Methods for Unsupervised Segmentation (CLeMUS), *Emanuele Salerno (Consiglio Nazionale delle Ricerche, Italy) Simon Wilson (Trinity College, Ireland)*

Computational Methods for Intelligent Neuro-Fuzzy Applications, *Gwi-Tae Park, Dongwon Kim (Korea University)*

Learning Automata and Soft Computing Techniques and Their Applications, *Norio Baba (Osaka Kyoiku University, Japan) Ann Nowe, Katja Verbeeck (Vrije Universiteit, Belgium)*

Learning from Uncertain Data, *Dario Malchiodi (University of Milan, Italy)*

Neural Information Processing for Data Mining, *Ryohei Nakano, Kazumi Saito (Nagoya Institute of Technology, Japan)*

Neural Networks: Advanced Applications, *Eros Pasero (University of Turin, Italy)*

Soft Computing Approach to Management Engineering, *Junzo Watada (Waseda University, Japan) Huey-Ming Lee (Chinese Culture University, Taiwan) Taki Kanda (Bunri University of Hospitality, Japan)*

Soft Computing in Electromagnetic Applications, *Raffaele Martone (University of Naples, Italy)*

Advanced Cooperative Work, *Jun Munemori, Takashi Yoshino (Wakayama University, Japan) Takaya Yuizono (JAIST, Japan)*

Behavior Support in Advanced Learning Collaborations, *Toyohide Watanabe, Tomoko Kojiri (Nagoya University, Japan)*

Context-Aware Adaptable Systems and Their Applications, *Phill Kyu Rhee (Inha University, Korea) Rezaul Bashar (Islamic University, Bangladesh)*

Engineered Applications of Semantic Web - SWEA, *Tommaso Di Noia, Eugenio di Sciascio (Politechnic of Bari, Italy) Giovanni Semeraro (University of Bari, Italy)*

Environment Support in Advanced Learning Collaborations, *Toyohide Watanabe, Tomoko Kojiri (Nagoya University, Japan)*

Immunity-Based Systems, *Yoshiteru Ishida (Toyohashi University of Technology, Japan) Giuseppe Nicosia (University of Catania, Italy)*

Interactive Visualization and Clustering, *Roberto Tagliaferri (University of Salerno, Italy)*

Multi-Agent Systems Design, Implementation and Applications, *Dharmendra Sharma, Bala M. Balachandran (University of Canberra, Australia)*

Multimedia Systems and Their Applications Focusing on Reliable and Flexible Delivery for Integrated Multimedia (Media 2007), *Yun Ji Na (Convergence Information Technology Research Center, Korea) Il Seok Ko (Dongguk University, Korea)*

Recommender Agents, *Dariusz Król, Janusz Sobecki (Wroclaw University of Technology, Poland)*

Skill Acquisition and Ubiquitous Human Computer Interaction, *Hirokazu Taki (Wakayama University, Japan) Satoshi Hori (Institute of Technologists, Japan)*

XML Security, *Stefania Marrara, Ernesto Damiani (University of Milan, Italy) Majirus Fansi, Alban Gabillon (University of Pau, France)*

Keynote Speakers

Jean-François Cardoso, École Nationale Supérieure des Télécommunications, France: *Independent Component Analysis: Concepts and Applications*

Stephanie Forrest, University of New Mexico, USA: *Self-Healing Systems and Autonomic Network Security*

Walter J. Freeman, University of California, Berkeley, USA: *Thermodynamic Model of Knowledge Retrieval in Brain Dynamics for Information Processing*

Mario Gerla, University of California, Los Angeles, USA: *Probing and Mining the Urban Environment Using the Vehicular Sensor Network*

Hans-Andrea Loeliger, ETH, Zurich, Switzerland: *The Factor Graph Approach to Model-Based Signal Processing*

Yoshiteru Ishida, Toyohashi University, Japan: *The Immune System Offered a Glimpse: What Makes Biological Systems Distinct from Artificial Ones*

Sponsoring Institutions

Università degli Studi di Milano

Seconda Università di Napoli

Comune di Vietri sul Mare

Comune di Salerno

Regione Campania

Ministero per le Riforme e le Innovazioni nella P.A.

Centro Regionale Information Communication Technology

Table of Contents – Part I

I General Tracks

Artificial Neural Networks and Connectionists Systems

Granular Computing

Machine Learning and Classical AI

Agent Systems

Knowledge Based and Expert Systems

Miscellaneous Intelligent Algorithms

Intelligent Vision and Image Processing

Knowledge Management and Ontologies

Web Intelligence, Text and Multimedia Mining and Retrieval

Intelligent Signal Processing, Control and Robotics

Other Intelligent Systems Applications

Table of Contents – Part II

I Intelligence Everywhere

Ambient Intelligence

Artificial Intelligence Applications in Digital Content

Artificial Intelligence Applications in Security

Artificial Intelligence Methods for Information Processing (AIMIP 2007)

Communicative Intelligence 2007

Computation Intelligence for Image Processing and Pattern Recognition

Human Computer Intelligent Systems

Hybrid Artificial Intelligence Systems Workshop (HAIS07-KES2007)

Innovations in Intelligent Data Analysis

Intelligent Agents and Their Applications

Intelligent and Adaptive Systems in Economics, Finance and Management

Intelligent Automation Systems

Intelligent Control Theory and Applications

Intelligent Data Processing in Process Systems and Plants

Intelligent Mechanism for Knowledge Innovation

Intelligent Techniques for Biometric Based Authentication

Logic Based Intelligent Information Systems

II Intelligent Knowledge

Chance Discovery

Knowledge - Based Interface Systems (I)

Knowledge - Based Interface Systems (II)

Knowledge and Information Management in Social Community

Knowledge and Ontological Engineering for (KOS)

Knowledge Engineering in Multi Robot Systems

Knowledge-Based Creativity Support Systems

Knowledge-Based Multi-criteria Decision Support

Knowleged Based Systems for e-Business

Table of Contents – Part III

I Intelligent Processing

Computational Learning Methods for Unsupervised Segmentation (CLeMUS)

Computational Methods for Intelligent Neuro-Fuzzy Applications

Learning Automata and Soft Computing Techniques and Their Applications

Learning from Uncertain Data

Soft Computing Approach to Management Engineering

Soft Computing in Electromagnetic Applications

II Intelligent Systems –

Advanced Cooperative Work

Behavior Support in Advanced Learning Collaboration

Context-Aware Adaptable Systems and Their Applications

Engineered Applications of Semantic Web - SWEA

Environment Support in Advanced Learning Collaboration

Immunity-Based Systems

Interactive Visualization and Clustering

Multi-agent Systems Design, Implementation and Applications

Multimedia Systems and Their Applications Focusing on Reliable and Flexible Delivery for Integrated Multimedia (Media07)

Recommender Agents

Skill Acquisition and Ubiquitous Human Computer Interaction

XML Security

A New Neural Network with Adaptive Activation Function for Classification of ECG Arrhythmias

Gülay Tezel[1] and Yüksel Özbay[2]

[1] Selcuk University, Computer Engineering, 42031, Konya, Turkiye
[2] Selcuk University, Electrical & Electronics Engineering, 42031, Konya, Turkiye
{gtezel,yozbay}@selcuk.edu.tr

Abstract. This study presents a comparative study of the classification accuracy of ECG signals using a well-known neural network architecture named multi-layered perceptron (MLP) with backpropagation training algorithm, and a new neural network with adaptive activation function (AAFNN) for classification of ECG arrhythmias. The ECG signals are taken from MIT-BIH ECG database, which are used to classify ten different arrhythmias for training. These are normal sinus rhythm, sinus bradycardia, ventricular tachycardia, sinus arrhythmia, atrial premature contraction, paced beat, right bundle branch block, left bundle branch block, atrial fibrillation and atrial flutter. For testing, the proposed structures were trained by backpropagation algorithm. Both of them tested using experimental ECG records of 10 patients (7 male and 3 female, average age is 33.8±16.4). The results show that neural network with adaptive activation function is more suitable for biomedical data like as ECG in the classification problems and training speed is much faster than neural network with fixed sigmoid activation function

Keywords: ANN, Adaptive activation function, classification, ECG, arrhythmia.

1 Introduction

Electrocardiography deals with the electrical activity of the heart. Monitored by placing sensors at the limb extremities of the subject, Electrocardiogram (ECG) is a record of the origin and the propagation of the electrical potential through cardiac muscles. It is considered a representative signal of cardiac physiology, useful in diagnosing cardiac disorders [1-2].

The state of cardiac heart is generally reflected in the shape of ECG waveform and heart rate. It may contain important pointers to the nature of diseases afflicting the heart. However, bio-signals being non-stationary signals, the reflection may occur at random in the time-scale (That is, the disease symptoms may not show up all the time, but would manifest at certain irregular intervals during the day). Therefore, for effective diagnostics, ECG pattern and heart rate variability may have to be observed over several hours. Thus the volume of the data being enormous, the study is tedious and time consuming. Naturally, the possibility of the analyst missing (or misreading) vital information is high. Therefore, computer-based analysis and classification of diseases can be very helpful in diagnostics [1]. Several algorithms have been developed in the

B. Apolloni et al. (Eds.): KES 2007/WIRN 2007, Part I, LNAI 4692, pp. 1–8, 2007.

literature for detection and classification of ECG beats. Most of them use either time or frequency domain representation of the ECG waveforms, on the basis of which many specific features are defined, allowing the recognition between the beats belonging to different classes. The most difficult problem faced by today's automatic ECG analysis is the large variation in the morphologies of ECG waveforms, not only of different patients or patient groups but also with in the same patient. The ECG waveforms may differ for the same patient to such extend that they are dissimilar to each other and at the same time they are similar for different types of beats. This is main reason that the beat classifier, performing well on the training data generalizes poorly, when presented with different patients ECG waveforms [2]. One of the methods of ECG beat recognition is neural network classification method [3-5].

Artificial Neural Network has played an important role in a wide variety of applications, such as pattern recognition and classification tasks. In traditional ANN model such as multi-layered perceptron network (MLP), each neuron computes the weighted sum of its inputs and applies to sum a non-linear function called activation functions [6,7]. In general the performance of MLP depends on the number of hidden layers, the number of hidden neurons, the learning algorithm and the activation function for each neuron [8]. MLP have the ability to perform tasks involving nonlinear relationships, in which all the neurons may perform the same type of activation function or different layers of neurons may realize different kinds of activation functions [6]. The commonly investigated activation functions in literature are sigmoid function, generalized sigmoid functions and the radial basis function, so on. These functions which all fixed and cannot be adjusted to adapt to different problems represent a relation between a single input, the weighted sum, and a single output, the neuron response. One common characteristic of these activation functions is that they are activation function is critical as the behavior performance of MLP depends on it [9-11]. So far there have been limited studies with emphasis on setting a few free parameters in the activation function. In Liu[12], real variables, node offset(c) and slope of the sigmoid function(s) in sigmoid activation function were adjusted during learning process. Yu at al., established an adaptive activation function for multilayer MLP to solve N-Parity and two spiral problems. Vecci at al.[13] and Solazzi and Uncini [7] studied with adaptive spline activation function neural networks. Xu and Zhang [8-11,14-16], studied Adaptive Higher Feed-Forward Neural Networks for financial analysis. Networks with such activation functions called AAFNN, seems to provide better performance than classical architectures with fixed activation function neurons.

In this paper, two AAFNN models with different adaptive activation functions with free parameters are proposed. A learning algorithm was derived for adjusting the free parameters as well as weights between neurons. We improved a new neural network with adaptive activation function with free parameters to classification of ECG arrhythmias.

2 Architecture of Neural Network with Adaptive Activation Function (AAFNN) and Learning Algorithm

The neural network with adaptive activation function (AAFNN) considered here has three layers (an input layer, one hidden layer and output layer) like as generally MLP.

The net input of hidden and output layers is weighted sum of its inputs. It is used no activation function in the input neurons of input layer. Sigmoid activation function with fixed parameters is used in the output neurons of the output layer. But adaptive activation functions with free parameters are used as the activation function in the hidden nodes of the hidden layer. It is used here two adaptive activation functions as defined in Eq.1 for AAF NN-1 model and Eq.2 for AAF NN-2 model. The structures in this study were implemented with MATLAB R2006a software package.

Adaptive activation function for the hidden neurons in the AAF NN-1 model is defined in Eq.1, adaptive activation function for the hidden neurons in the AAF NN-2 is defined Eq.2. The sigmoid function with fixed parameters in Eq.3 is used as activation function in hidden and output layers for the classical FFNN model. It is used only in the output layer of the AAF NN-1and AAF NN-2.

$$\psi_1(x) = \frac{a}{1+e^{-bx}} \tag{1}$$

$$\psi_2(x) = a_1 \mathrm{Sin}(b_1 x) + \frac{a_2}{1+e^{-b_2 x}} \tag{2}$$

$$\psi_3(x) = \frac{1}{1+e^{-x}} \tag{3}$$

Where a, b, a1, a2, b1, b2 are real variables which will be tuning during training as weights between neurons. There are two free parameters (a, b) in Eq.1 and four free parameters (a1,a2,b1 and b2) in the Eq.2.[8,10,11,14-17].

In our simulations, we used a learning algorithm which is not far from traditional backpropagation algorithm. These free parameters in adaptive activation functions as weights between neurons are adjusting with this learning algorithm that is based on steepest descent rule. In backpropagation algorithm, there are two phases: feed-forward and error backpropagation [6-11].

Firstly, all the weights and biases initialized to small real random values to the initial values in feed-forward phase [17,18]. The choice of initial weights will influence whether the net reaches a global minimum of the error and if so, how quickly it converges. Nguyen-Widrow initialization method gives much faster learning performance and depends on the number of input neurons and hidden neurons. Because of this Nguyen-Widrow method was used in this study [17].

After initializing, it is presented training pair (input vector and corresponding desired responses) to the network inputs. Each hidden unit sums its weighted signals in Eq.4, applies its selected activation function (ψ_1 or ψ_2) like as Eq.5 or Eq.6 to compute its output signal and applies fixed sigmoid activation function in Eq.3 in each output unit which sums its weighted signals to calculate the output signals of output in feed-forward phase. The input of ith neuron in the kth layer is defined as Eq.4.

$$I_{i,k}(u) = \sum_j \left\lfloor w_{i,j,k} o_{j,k-1}(u) \right\rfloor + \theta_{i,k} \tag{4}$$

$$o_{i,k}(u) = \psi_1(I_{i,k}(u)) = \frac{a}{1+e^{-b.I_{i,k}(u)}} \tag{5}$$

$$o_{i,k}(u) = \psi_2(I_{i,k}(u)) = a1_{i,k}\,\mathrm{Sin}(b1_{i,k}.I_{i,k}(u)) + \frac{a2_{i,k}}{1+e^{-b2_{i,k}.I_{i,k}(u)}} \tag{6}$$

Where j is neuron number in the layer (k-1). Eq.1 and Eq.2 are the value of output from i^{th} neuron in the kth layer for ψ_1 and ψ_2, respectively.

For an efficient learning algorithm, this method specifies how to reduce the mean squared error for all patterns through an adjustment of these free parameters simultaneously in backpropagation phase. The mean squared error function in Eq.7 is sum of the squared error between the actual network and he desired output for all input patterns [6-12,17,18]. It is suggested using gradient descent to perform steepest descent in which the adjustment of weight is proportional the first derivative of the output function in each neuron (Eq.8 and Eq.9). Similarly, the adjustment of free parameters in each activation functions is proportional the first derivative of the output function in each neuron (Eq.10 and Eq. 11). The network is training to minimize the error function by adjusting the weight and free parameters in the activation functions by using steepest descent rule expressed in Eq.(8-11).

$$E = \frac{1}{2}\sum_{j=1}^{m}(d_j(u) - o_{j,l})^2 \tag{7}$$

$$w^r_{i,j,k} = w^{r-1}_{i,j,k} + \beta\frac{\partial E}{\partial w_{i,j,k}} \tag{8}$$

$$\theta^r_{i,k} = \theta^{r-1}_{i,k} + \beta\frac{\partial E}{\partial \theta_{i,k}} \tag{9}$$

$$a^r_{i,k} = a^{r-1}_{i,k} + \beta\frac{\partial E}{\partial a_{i,k}} \tag{10}$$

$$b^r_{i,k} = b^{r-1}_{i,k} + \beta\frac{\partial E}{\partial b_{i,k}} \tag{11}$$

The other parameters (a1, a2, b1, b2) can be adjusted in the similar way for second activation function. Where $I_{i,k}(u)$ is the input of ith neuron in the kth layer , $w_{i,j,}$ is the weight between j^{th} neuron in the layer (k-1) and i^{th} neuron in the layer k, $o_{i,k}(u)$ is the value of output from i^{th} neuron in the kth layer, $\theta_{i,k}$ is the threshold value of ith neuron in the kth layer, β is learning rate, $d_j(u)$ is the desired value of jth output neuron, m is total number of neurons in the output layer, p is total number of neurons in the hidden layer, l is total number of network layers, r is the iteration number,

In this algorithm, the weights are updated after each training pattern is presented. An epoch is one cycle through the entire set of training vectors. At the end of the every epoch, free parameters (a,b for ψ_1 adaptive activation function as Eq.1 and a1, b1,a2,b2 for ψ_2 adaptive activation function as Eq.2) are adjusted as weights. After completing the training procedure of the neural network, the weights of AAF NN-1 and AAF NN-2 are frozen and ready for use in the testing mode [17, 18].

3 Structure and Training Data

Training data of ECG arrhythmias used in this study was taken from MIT-BIH ECG Arrhythmias Database. Selected types of arrhythmias were normal sinus rhythm (N; 15 segment), sinus bradycardia (Br; 15 segments), ventricular tachycardia (VT; 6 segments), sinus arrhythmia (SA; 15 segments), atrial premature contraction (APC; 6 segments), paced beat (P; 10 segments), right bundle branch block (R; 10 segments), left bundle branch block (L; 10 segments), atrial fibrillation (A.Fib; 10 segments) and atrial flutter (A.Fl; 9 segments). Training patterns had been sampled at 360 Hz, so we arranged them as 200 samples in the intervals of R-R for all arrhythmias, which are called as a segment. Training patterns were formed by mixed from the arrhythmias pre-processed by the order given above. The size of the training patterns was 106 segments*200 samples. The combined these training patterns were called as original training set [19].

In this paper, two different models, AAFNN -1 and AAF NN-2 with two adaptive activation functions are adapted for ECG Arrhythmia data set. $\psi_1(x)$ and $\psi_2(x)$ are used for activation functions of hidden neurons in AAF NN-1 and AAF NN-2, respectively. As a result of experiments, it was seen that the performance of the model with adaptive activation function in output layer is worse than the model with fixed sigmoid activation function in the output layer. For this reason, it was only used the adaptive activation function in the hidden layer. ECG data set, used for test process was selected from the study of Özbay et al [19]. We calculated training errors given in tables according to this study. It is used an algorithm for evaluation of test results. This algorithm was comprehensively explained in the study of Özbay et al [19].

4 Results

In order to compare training and test performance of the proposed structures (AAF NN-1 and AAH NN-2), the experimental results are discussed here. The proposed models, trained for 5000 iterations by using training set were tested using recorded from 10 patients. Table 1 describes the training error for training data set and test errors for test data set which is used for validation. Table 1 and Fig.1 show that the optimum number of hidden nodes was 17 with the highest classification accuracy of 100% for training data and 98% for test data with AAF NN-2. Our ECG test data contains 268 segments from ten patients in the bottom row, gives the total number of segments for each arrhythmia for ten patients. The right most column gives the

Table 1. Comparison of FFN, AAF NN-1 and AAF NN-2 on the performance of training and test on the ECG classification task

	MLP		AAFNN-1		AAFNN-2	
HN	TE	TestE	TE	TestE	TE	TestE
17	0,21	5	0,1	20,4	0,082	1,48
19	0,24	3,61	0,36	2,84	0,279	2,47
30	0,2	2,57	0,11	2,89	0,062	2,215
48	0,198	3,79	0,073	4,61	0,22	2,39

HN is the number of hidden neuron, TE is the training error and TestE is the test error.

classification error for each patient in Table 2. Table 2 shows that the best performance of test was obtained with the structure of AAF NN-2 and the number of Q11 misclassification set is minimum value for the structure of AAF NN-2.

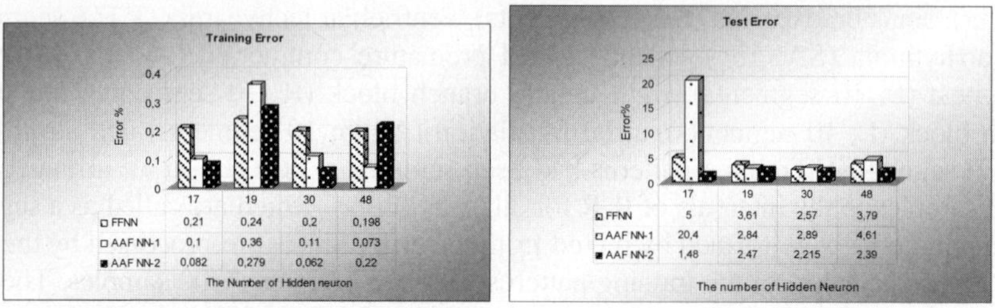

Fig. 1. Classification results for ECG problems: (a) the performance of Training error (b) the performance of test error

Table 2. The classification results by test data (NoS is the number of segment in data set)

(a) The test results of Traditional FFN

No	Sample	NoS	Q1 N	Q2 Br	Q3 T	Q4 S	Q5 Apc	Q6 P	Q7 R	Q8 L	Q9 Afib	Q10 Aflt	Q11 ?	Error %
1	3000	15	0	0	0	0	0	0	15	0	0	0	0	7,57
2	2400	12	0	0	0	0	0	0	12	0	0	0	0	7,29
3	6000	30	0	0	0	0	0	0	30	0	0	0	0	7,22
4	3400	17	0	16	0	0	0	0	0	0	0	0	1	3,72
5	4400	22	0	21	0	0	0	0	0	0	0	0	1	1,38
6	3000	15	15	0	0	0	0	0	0	0	0	0	0	0,12
7	6200	31	27	0	0	2	0	0	0	0	2	0	1	0,54
8	7000	35	0	0	0	35	0	0	0	0	0	0	10	6
9	6000	30	29	0	0	0	0	0	0	0	0	0	1	0,57
10	12100	61	1	0	0	0	0	0	0	0	3	0	57	5
Total	53600	268	72	37	0	37	0	0	57	0	5	0	71	3,94

(b) The test results of AAFNN-1

No	Sample	NoS	Q1 N	Q2 Br	Q3 T	Q4 S	Q5 Apc	Q6 P	Q7 R	Q8 L	Q9 Afib	Q10 Aflt	Q11 ?	Error %
1	3000	15	0	0	0	0	0	0	5	0	0	0	10	7,57
2	2400	12	0	0	0	0	0	0	6	0	0	0	6	7,29
3	6000	30	0	0	2	0	0	0	16	0	0	0	12	7,22
4	3400	17	0	7	0	0	0	0	0	0	0	0	3	3,72
5	4400	22	0	8	0	0	0	0	0	0	0	0	14	1,38
6	3000	15	15	0	0	0	0	0	0	0	0	0	0	0,12
7	6200	31	27	0	0	0	2	0	0	0	2	0	1	0,54
8	7000	35	0	0	0	15	0	0	0	0	0	0	10	6
9	6000	30	29	0	0	0	0	0	0	0	0	0	1	0,57
10	12100	61	1	0	0	0	0	0	0	0	3	0	57	2,04
Total	53600	268	72	15	2	15	2	0	27	0	5	0	114	3,65

Table 2. *(continued)*

(c) The test results of AAFNN-2

No	Sample	NoS	Q1	Q2	Q3	Q4	Q5	Q6	Q7	Q8	Q9	Q10	Q11	Error
			N	Br	T	S	Apc	P	R	L	Afib	Aflt	?	%
1	3000	15	0	0	0	0	0	0	15	0	0	0	0	2,34
2	2400	12	0	0	0	0	0	0	12	0	0	0	0	2,17
3	6000	30	0	0	0	0	0	0	30	0	0	0	0	2,17
4	3400	17	0	14	0	0	0	0	0	0	0	0	3	2,76
5	4400	22	0	20	0	0	0	0	0	0	1	0	1	2,45
6	3000	15	15	0	0	0	0	0	0	0	0	0	0	0,11
7	6200	31	27	0	0	2	0	0	0	0	1	0	1	0,65
8	7000	35	1	0	0	0	0	0	0	0	34	0	0	4,87
9	6000	30	29	0	0	0	0	0	0	0	0	0	1	0,71
10	12100	61	0	0	0	0	0	0	0	0	50	0	11	1,48
Total	53600	268	72	34	0	2	0	0	57	0	86	0	17	1,97

5 Conclusion

In this paper, we proposed two models, AAF NN-1 and AAF NN-2 with adaptive activation function with free parameters which are adjusting with this learning algorithm that is based on steepest descent rule as weights between neurons. It is different from works in the literature; that the number of neurons in the output layer of these adaptive models is more than one and that it was only used the adaptive activation function in the hidden layer. ECG data set were used to compare performance of MLP, AAFNN-1 and AAFNN-2 for classification problems. It was observed that the structures of AAFNN-1 and AAFNN-2 are faster and reduce network size and simulation error than the structure of MLP.

Acknowledgments. This work is supported by the Coordinatorship of Selcuk University's Scientific Research Projects.

References

1. Acharya, R., Bhat, P.S., Iyengar, S.S.: Classification of heart rate data using artificial neural network and fuzzy Eqivalence relation. The Journal of the Pattern Recognition Society (2002)
2. Osowski, S., Linh, T.H.: ECG beat recognition using fuzzy hybrid neural network. IEEE Transaction on Biomedical Engineering 48(11), 1265–1271 (2001)
3. Ozbay, Y., Karlik, B.: A recognition of ECG arrhythmias using artificial neural network, Proceedings-23rd Annual Conference- IEEE/EMBS. Istanbul, Turkey (2001)
4. Ozbay, Y.: Fast Recognition of ECG Arrhythmias, PhD Thesis, Institute of Natural and Applied Science. Selcuk University (1999)
5. Foo, S.Y., Stuart, G., Harvey, B., Meyer-Baese, A.: Neural network-based ECG pattern recognition. Elsevier Science Engineering Applications of Artificial Intelligence 15, 253–260 (2002)

6. Yu, C.C., Tang, Y.C., Liu, B.D.: An adaptive Activation Function for Multilayer Feedforward Neural Networks, Proceeding of IEEE TENCON'02 (2002)
7. Solazzi, M., Uncini, A.: Artificial Neural Networks with Adaptive Multidimensional Spline Functions. Neural networks 17, 247–260 (2000)
8. Xu, S., Zhang, M.: Justification of A Neuron-Adaptive Activation Function. In: IEEE, Proceeding of IEEE-INNS-ENNS International Joint Conference on Neural Networks, IJCNN 2000, vol. 3, pp. 465–470, 24–24 (July 2000)
9. Xu, S., Zhang, M.: Adaptive Higher-Order Feedforward Neural Networks IJCNN '99 IEEE International Joint Conference on Neural Networks (10-16 July), vol. 1, pp. 328 – 332 (1999)
10. Xu, S., Zhang, M.: A Novel Adaptive Activation Function. In: Proceedings IJCNN'01 International Conference on Neural Networks, vol. 4, pp. 2779–2782 (2001)
11. Xu, S., Zhang, M.: Data Mining- An Adaptive Neural Network Model for Financial Analysis, ICITA'05 (2005)
12. Liu, T.I.: On-line Sensing of Drill Wear Using Neural Network Approach. IEEE International Conference on Neural Networks, 2002, 690–694 (1993)
13. Vecci, L., Piazza, F., Uncini, A.: Learning and Approximation Capabilities of Adaptive Spline Activation Function Neural Networks. Neural Networks 11, 259–270 (1998)
14. Zhang, M., Xu, S., Fulcher, J.: Neuron-Adaptive Higher Order Neural-Network Models for Automated Financial Data Modeling. IEEE Transactions on Neural Networks 13(1) (2002)
15. Zhang, M., Xu, S., Fulcher, J.: Neuron-Adaptive Higher Order Neural-Network Group Models. In: International Joint Conference on Neural Networks IJCNN'99, vol. 1, pp. 337–374 (1999)
16. Xu, S., Zhang, M.: Aproximation to Continuous Functionals and Operators Using Adaptive Higher-Order Feedforward Neural Networks. In: International Joint Conference on Neural Networks IJCNN'99, vol. 1, pp. 337–374 (1999)
17. Fauset, L.: Fundamentals of Neural Networks:Architectures, Algorithms and Applications. Prentice Hall, Inc. A simon & Schuster Company (1994)
18. Haykin, S.: Neural Networks: A Comprehensive Foundation. Macmillan, New York (1994)
19. Özbay, Y., Ceylan, R., Karlik, B.: A fuzzy clustering neural network architecture for classification of ECG arrhythmias. Elsevier Science Computers in Biology and Medicine 36, 376–388 (2006)

A Simple and Effective Neural Model for the Classification of Structured Patterns

Edmondo Trentin and Ernesto Di Iorio

DII, Università degli Studi di Siena, V. Roma 56 Siena, Italy
{trentin,diiorio}@dii.unisi.it

Abstract. Learning from structured data (i.e. graphs) is a topic that has recently received the attention of the machine learning community, which proposed connectionist models such as recursive neural nets (RNN) and graph neural nets (GNN). In spite of their sound theoretical properties, RNNs and GNNs suffer some drawbacks that may limit their application. This paper outlines an alternative connectionist framework for learning discriminant functions over structured data. The approach, albeit preliminary, is simple and suitable to maximum-a-posteriori classification of broad families of graphs, and overcomes some limitations of RNNs and GNNs. The idea is to describe a graph as an algebraic relation, i.e. as a subset of the Cartesian product. The class-posterior probabilities given the relation are reduced to products of probabilistic quantities estimated using a multilayer perceptron. Experimental comparisons on tasks that were previously solved via RNNs and GNNs validate the approach.

Keywords: Structured pattern recognition, relational learning, graph neural network, bombastic neural network.

1 Introduction

In recent years, the machine learning community has manifested interest in the development of paradigms that are able to learn from data containing information on relations among different entities. These relations arise from the very nature of the task and they improve the description of the input domain, possibly strengthening the learning process. Four major instances of this scenario are the following: (1) relational learning, also known as inductive logic programming, where predicate descriptions are developed from examples and background knowledge (all in the form of logic programs); (2) probabilistic relational learning, where statistical inference is accomplished over the tables of a relational database; (3) graphical models, in which statistical dependencies in the form of conditional probabilities between pairs of random variables in the feature space are modeled via a graph structure; (4) machine learning over structured domains, i.e. feature spaces that have a graphical representation [12]. The present research focuses on the last scenario, and introduces a connectionist paradigm that learns maximum-a-posteriori discriminant functions from structured data. The neural

B. Apolloni et al. (Eds.): KES 2007/WIRN 2007, Part I, LNAI 4692, pp. 9–16, 2007.

network community developed a few, significant connectionist models for graph processing, including the application of RAAM [13] and Cascade Correlation [11] to tree-structured data. A few *ad hoc* connectionist architectures and training algorithms were proposed and thoroughly investigated, namely recursive neural networks (RNN) [14] and graph neural networks (GNN) [10]. They basically rely on the idea of unfolding the neural architecture over the (possibly labeled) graph structure, in a way similar to the popular backpropagation-through-time algorithm for recurrent neural nets. The RNN training algorithm [14] is suitable only for trees or directed acyclic graphs (preventing endless loops within the unfolding scheme), while GNNs can deal with cycles, since infinite recursion is smoothed down by reaching for steady states of the neural dynamics. In spite of their strong theoretical properties, RNNs and GNNs suffer from some intrinsic limitations that might be the rationale behind their limited real-world application: (i) RNN can process only acyclic structures which only seldom fit real-world scenarios; (ii) both RNN and GNN are complex machines, both from a formal and from a computational point of view. In particular, training over large graphs requires the unfolding of the network onto a very large and deep architecture, which poses numerical stability problems to the backpropagation of the gradients; (iii) above all, as pointed out also in [3], they suffer from a drawback which they share with the classic recurrent neural nets, namely the "long term dependencies" problem [2]. In terms of graphical structures this problem takes the form of graphs with long shortest-paths between certain pairs of nodes, e.g. high trees.

In the present research, we shall try to look at the problem of learning (discriminant functions) on structured domains from a different perspective and shall introduce a simple and effective attempt to overcome the above limitations. The proposed approach is suitable for directed or undirected, connected or unconnected, cyclic or acyclic graphs. Labels may be attached to nodes or edges as well. The approach simplifies significantly the formalism and the computational requirements and may be implemented by means of ordinarily available software simulators. In particular, it does not suffer from the long-term dependencies problem. The idea is to describe the graph as an algebraic binary relation, i.e., as a subset of the Cartesian product in the definition domain of the graph. The class-posterior probabilities given the graph can then be reduced to a product (joint probability) of probabilistic quantities which, in turn, can be estimated using a multilayer perceptron (MLP). This formulation is suitable for structured pattern classification within a maximum *a posteriori* Bayesian framework. Experimental comparisons with RNNs and GNNs on the Caltech benchmark dataset (Section 3) show that the approach is promising.

2 The Proposed Technique

A graph $\mathcal{G}$ is a pair $\mathcal{G} = (V, E)$ where V is an arbitrary set of nodes (or, vertices) over a given universe U, and $E \subseteq V \times V$ is the set of edges. We consider directed as well as undirected, connected and unconnected finite graphs ($\mathcal{G}$ is undirected iff $(a,b) \in E \leftrightarrow (b,a) \in E$), either cyclic or acyclic. From an algebraic point of view, the graph is a binary relation over U. More generally, given the arbitrary universes U_1 and U_2, we consider graphs as binary relations in the form of any subsets of the Cartesian product $U_1 \times U_2$, namely $E = \{\mathbf{x}_j = (a_j, b_j) \mid a_j \in U_1, b_J \in U_2, j = 1, \ldots, n\}$ for a proper cardinality n. All the binary relations (graphs) involved in the learning problem at hand (both in training and test) are assumed to be defined over the same domain ($U \times U$, or $U_1 \times U_2$). From now on, we rely on the assumption that the universe U (or, U_1 and U_2) is a (Lebesgue-) measurable space, in order to ensure that probability measures can actually be defined. The measurability of finite graphs defined over measurable domains (and with measurable labels) like countable sets or real vectors is shown in [11].

Labels may be attached to vertices or edges, assuming they are defined over a measurable space. For the vertices, we consider a labeling $\mathcal{L}$ in the form of d-dimensional vectors associated with nodes, namely $\mathcal{L}(\mathcal{G}) = \{\ell(v) \mid v \in V, \ell(v) \in \mathcal{R}^d\}$. Labels are accounted for by modifying the definition of $\mathbf{x}_j = (a_j, b_j) \in E$ slightly, taking $\mathbf{x}_j = (a_j, \ell(a_j), b_j, \ell(b_j))$. As regards the edge labels, for each $(a_j, b_j) \in E$ a label is allowed in the form $\ell_e(a_j, b_j) \in \mathcal{R}^{d_e}$, where d_e is the dimensionality of the continuous label domain. Then, $\mathbf{x}_j$ is extended as follows: $\mathbf{x}_j = (a_j, b_j, \ell_e(a_j, b_j))$ (if the graph has edge labels, but no node labels), or $\mathbf{x}_j = (a_j, \ell(a_j), b_j, \ell(b_j)), \ell_e(a_j, b_j))$ (if the graph has both).

Remark 1: the present framework requires that the nodes in the graph are individual elements of a well-defined universe. Consequently, it does not explicitly cover scenarios in which the nodes act only as "placeholders" in the specific graphical representation of the data. If this is the case, and the actual input features are completely encapsulated within label vectors, the previous definitions may be replaced by $\mathbf{x}_j = (\ell(a_j), \ell(b_j))$ for each pair $(a_j, b_j) \in E$. This may turn out to be effective in practical applications, but it is mathematically justified only iff each label identifies the corresponding node in an univocal manner.

Some examples of structures that fit the present framework are the following: 1) semantic networks, e.g. whose nodes are words from a given dictionary and the edges represent a semantic relation between words. 2) Subgraphs of the World Wide Web, where nodes are extracted form the universe of possible URLs, node labels are a representation of the information contained in the web page, and the edges are the hyperlinks. 3) Scene descriptions in syntactic pattern recognition, whenever nodes are extracted from the universe of terminal/nonterminal symbols and edges represent a relation (e.g., spatial) between symbols. Section 3 shows two more examples in the area of image processing.

Let $\omega_1, \ldots, \omega_c$ be a set of classes or states of nature. We assume that each graph belongs to one of the c classes. The posterior probability of i-th class given the graph is the class-posterior given the corresponding binary relation, namely $P(\omega_i \mid \{\mathbf{x}_1, \ldots, \mathbf{x}_n\})$, where the relation is a set of n pairs $\mathbf{x}_j = (a_j, b_j) \in E, j = 1, \ldots, n$, and each $\mathbf{x}_j$ is interpreted as a random vector whose characteristics and dimensionality depend on the nature of the universe U (or, of the universes U_1 and U_2). The assumption of dealing with measurable universes allows the adoption of probabilistic measures, and applying Bayes' theorem [8] we can write:

$$P(\omega_i \mid \{\mathbf{x}_1, \ldots, \mathbf{x}_n\}) = \frac{p(\{\mathbf{x}_1, \ldots, \mathbf{x}_n\} \mid \omega_i) P(\omega_i)}{p(\{\mathbf{x}_1, \ldots, \mathbf{x}_n\})} \tag{1}$$

where $P(.)$ denotes a probability measure, and $p(.)$ denotes a probability density function (pdf) which reduces to a probability if its support is discrete in nature. The quantity $p(\{\mathbf{x}_1, \ldots, \mathbf{x}_n\} \mid \omega_i)$ is a joint pdf that expresses the probabilistic distribution of the overall binary relation $\{\mathbf{x}_1, \ldots, \mathbf{x}_n\}$ over its domain according to the law $p(.)$. We assume that the pairs $\mathbf{x}_j, j = 1, \ldots, n$ (including the corresponding labels) are independently and identically distributed (iid) according to the class-conditional density $p(\mathbf{x}_j \mid \omega_i)$. In order to understand the meaning of $p(\mathbf{x}_j \mid \omega_i)$, it may be helpful to underline that it implicitly expresses three different, yet joint probabilistic quantities, all of them conditioned on ω_i: (1) the likelihood of observing any given pair of nodes (edge), (2) the probability distribution of node labels, and (3) the pdf of edge labels. In so doing, the probability of having an edge between two vertices is modeled jointly with the statistical properties of the nodes and of their labels.

The iid assumption is in line with classical and state-of-the-art literature on statistical pattern recognition [8] and on random graphs. In the ER random graph model [5] edges are iid according to a unique (e.g. uniform) probability distribution all over the graph. In the *small worlds* paradigm [15] iid edges are inserted during the rewiring process that generates the graph. Again, scale-free networks rely on a common probability law (eventually leading to a Power Law distribution of nodes connectivity) that characterizes the distribution of iid edges [1]. In the present framework, *hubs* might be modeled by values of $p(\mathbf{x}_j \mid \omega_i)$ peaked around the hubs themselves. Finally, [6] extended the scale-free paradigm and asserted that the likelihood of an edge between two nodes is related to the statistical properties (*fitness*) of the nodes, as we do.

Remark 2: the iid assumption does not imply any loss in terms of structural information. The structure is encapsulated within the binary relation, which does not depend on the probabilistic properties of the quantities involved in Equation 1.

Applying again Bayes' theorem with the iid assumption, we can write:

$$\begin{aligned} p(\{\mathbf{x}_1, \ldots, \mathbf{x}_n\} \mid \omega_i) &= \prod_{j=1}^{n} p(\mathbf{x}_j \mid \omega_i) \\ &= \prod_{j=1}^{n} \frac{P(\omega_i \mid \mathbf{x}_j) p(\mathbf{x}_j)}{P(\omega_i)}. \end{aligned} \tag{2}$$

Substituting Eq. 2 into Eq. 1 we obtain

$$P(\omega_i \mid \{\mathbf{x}_1, \ldots, \mathbf{x}_n\}) = \left\{ \prod_{j=1}^{n} \frac{P(\omega_i \mid \mathbf{x}_j) p(\mathbf{x}_j)}{P(\omega_i)} \right\} \frac{P(\omega_i)}{p(\{\mathbf{x}_1, \ldots, \mathbf{x}_n\})}$$
$$= \left\{ \prod_{j=1}^{n} \frac{P(\omega_i \mid \mathbf{x}_j)}{P(\omega_i)} \right\} P(\omega_i) \quad (3)$$

since $p(\{\mathbf{x}_1, \ldots, \mathbf{x}_n\}) = \prod_{j=1}^{n} p(\mathbf{x}_j)$, where $p(\mathbf{x}_j) = \sum_{k=1}^{c} P(\omega_k) p(\mathbf{x}_j \mid \omega_k)$.

Remark 3: since the pairs $\mathbf{x}_j$ are extracted from a well-defined universe and the joint probabilities (e.g. $p(\{\mathbf{x}_1, \ldots, \mathbf{x}_n\})$) are invariant w.r.t. arbitrary permutations of their arguments, there is no "graph matching" problem in the present framework. Representing the graph as a relation implies looking at the structure as a whole. This is a major difference w.r.t. other techniques that require a visit of the graph in a specific order, and that are faced with the problem of possible infinite recursion over cyclic structures.

In order to apply Eq. 3, we need to estimate $P(\omega_i)$ and $P(\omega_i \mid \mathbf{x}_j)$ for $i = 1, \ldots, c$ and $j = 1, \ldots, n$. If good estimates of these quantities are obtained, the maximum-a-posteriori decision rule expressed by Equation 3 is expected to yield the minimum Bayesian risk (i.e., minimum probability of classification error) [8]. The quantity $P(\omega_i)$ can be estimated from the relative frequencies of classes over the training sample, as usual. A MLP with c output units (one for each of the different classes) is then used to estimate $P(\omega_i \mid \mathbf{x}_j)$. The MLP is known to be a universal non-parametric probability model [4] and it may optimally approximate the Bayesian posterior-probability, once it is trained via Backpropagation (BP) on a supervised training set featuring class labels (i.e., 0/1 targets) [4]. The MLP outputs are then substituted in the right-hand-side of Eq. 3 which, eventually, yields $P(\omega_i \mid \mathcal{G})$. A standard MLP simulation software may be used, i.e. no implementation of complex, *ad hoc* algorithms is required. Note that a link-focused strategy is adopted, instead of the typical node-focused approach usually taken by RNNs and GNNs.

3 Experimental Results

We compare the technique with RNNs and GNNs in two image classification problems from the Caltech benchmark dataset [9]. The first experiment (as in [7]) is based on 4 classes, i.e. images of *bottles*, *camels*, *guitars*, and *houses*. For each class, a subset of 350 images was extracted from the Caltech dataset. Half of the images consists of positive examples of the class, while the others are negative examples, i.e. images randomly sampled from the other classes. The same data subsets as in [7] were used, each divided into training, validation and test sets (150, 50, and 150 images, respectively). Each image was represented as an undirected Region Adjacency Graph (RAG), obtained using the Mean Shift algorithm and the k-means color quantization procedure as in [7]. Since RNNs cannot deal with undirected graphs, application of the RNNs requires that the RAGs are transformed into directed acyclic graphs (DAG) *via* breadth-first visit and substitution of each undirected edge with a directed one. Each node of the RAG has a 23-dimensional vector label, while edge labels are 5-dimensional [7].

In [7], the Authors carry out experiments with three different ANN configurations, called "small", "medium", and "big". For a fair evaluation, we compare the performance of the present approach w.r.t. the RNN and the GNN configurations that, class-by-class, yielded the best results. The RNN model involves three different 2-layer ANNs, i.e. the following number of free parameters to be learnt: 376 (small), 575 (medium), and 893 (big), respectively. The GNN relies on 2 distinct ANNs, for a total of 466 (small), 659 (medium), and 1081 (big) free parameters. It should be born in mind that the space complexity of these models during training increases with the size of the graph being processed, since the unfolding strategy requires to create multiple instances of the encoding ANN (as many copies of the original ANN as the number of nodes in the graph). The results reported in [7] were obtained applying 1000 training epochs to all the above models, approximately corresponding to worst-case training times of up to 2 hours (RNN), and 6.5 hours (GNN), for the software implementations used in [7], on an Apple G5 (TM) biprocessor architecture with 4 GB RAM. As described in section 2, the present approach uses a standard MLP architecture, the complexity of which turned out to be small in the experiments. In this first experimental setup, 12 hidden units were used, for a total of 637 free parameters (connection weights and bias of hidden and output sigmoids). The architecture was determined via cross-validation, as well as the number of training epochs, ranging from 8 (*camels* class) to 50 (*bottles*). Roughly speaking, this is less than 2 minutes worst-case training time on a PC architecture with 1.0GHz processor and 256 MB RAM. Table 1 reports the results of the first experiment. Results are expressed in terms of recognition accuracy on a class-by-class basis (see [7]). Average of the accuracies and their standard deviation are reported in the last two columns of the Table. In spite of its simplicity and its computational speed, the present approach outperforms the RNNs, and it yields also a significant average improvement over the GNNs. Moreover, a much more stable behavior is obtained w.r.t. the class change, as the standard deviation of results shows.

The second experiment (again, using the same data and feature space as in [3] for the sake of fairness) is based on a different subset of the Caltech benchmark database. It contains images of *airplanes*, *motorbikes*, *faces* and *cars*.

Table 1. Recognition accuracies in the first experiment [7]

Models	*Bottles*	*Camels*	*Guitars*	*Houses*	Avg.	Std. Dev.
Present approach	82.49	**84.00**	**93.67**	**94.67**	**88.71**	**5.49**
GNN	**84.67**	74.67	70.67	84.67	77.84	**6.21**
RNN	70.66	65.33	62.67	81.33	69.33	**7.17**

Table 2. Second experiment [3], results in terms of (1− ROC eer)

Models	*Motorbikes*	*Cars*	*Airplanes*	*Faces*	Avg.	Std. Dev.
Present approach	96.82	**98.96**	98.96	**100**	**98.68**	1.15
RNN (GNN)	**97.91**	92.7	**100**	**100**	97.65	2.98

For each class, the data were partitioned into training (95 images), test (95 images), and validation (48 images) sets, equally distributed in positive and negative examples. Images are represented by the corresponding Multiresolution Trees (MRTs) [3]. Again, the average size of the MRT representation of images is 100 nodes and about 600 edges. Results are shown in Table 2 in terms of the average quantity (1− ROC equal error rate), as in [3], along with their average and standard deviation (last two columns). GNNs results are not reported, since GNNs reduce to RNNs whenever the input graphs are trees. The RNN results are taken from [3], choosing the best architectures described therein, i.e., an overall 182 free parameters model to be unfolded over each MRT during training.

For the proposed approach, 2-layer MLP architectures were applied, namely 6 hidden sigmoids plus an output sigmoid for the *motorbikes*, *airplanes* and *faces* classes (235 free parameters total, 10 training epochs), and 12 hidden units (i.e., 469 parameters, 50 training epochs) for the *cars* class. The topologies and the number of BP iterations were determined through cross-validation. Although the high scores yielded by the RNN/GNN leave little room to improvement, the present approach fits the scenario and compares favorably. Again, its average class recognition rate is higher than RNNs/GNNs, and its standard deviation is lower (i.e., its overall behavior is more robust to changes in the class at hand).

4 Conclusion

Connectionist approaches to the problem of learning in structured domains have been proposed so far in the form of recursive neural architectures that unfold over the input graph in a BP-through-time fashion. RNNs and GNNs may suffer from some limitations, mostly due to their burden and to the long-term dependencies problem. We proposed a preliminary, alternative viewpoint, namely the description of the input structure in terms of a binary algebraic relation, i.e. a set of pairs of input entities. This representation makes it easy to compute an approximate estimate of the posterior probability of classes in pattern recognition problems, relying on decomposition of the overall posterior into a product (joint probability) of probabilistic quantities estimated via MLP. We argue that the underlying iid assumption is a point of strength in several practical applications, since it reduces substantially the burden of the model without affecting its capability to take into consideration the whole structural information. The resulting machine is simple, computationally efficient and easy to implement (relying on standard BP). Its performance on image processing experiments from the Caltech benchmark database turned out to be promising, yielding an improvement over the results that can be found in literature for RNNs and GNNs (even dramatic w.r.t. RNNs), at a much lower computational cost. Although the presented framework was introduced for the classification of structured patterns, ongoing research is focused on an extension of the paradigm to: (a) the learning of more general input-output relations (i.e., any vectorial functions of the graph could be modeled, e.g. in regression tasks), as in RNNs and GNNs; and to (b) the computation of functions of individual nodes in the graph given the whole structure, e.g., the page-rank of a certain Web page given the graph describing a portion of the WWW (RNNs do not explicitly have this capability, while GNNs do).

References

1. Barabási, A.-L., Reka, A.: Emergence of scaling in random networks. Science 286, 509–512 (1999)
2. Bengio, Y., Simard, P., Frasconi, P.: Learning long-term dependencies with gradient descent is difficult (Special Issue on Recurrent Neural Networks, March 94). IEEE Transactions on Neural Networks 5(2), 157–166 (1994)
3. Bianchini, M., Maggini, M., Sarti, L.: Object recognition using multiresolution trees. Joint IAPR International Workshops SSPR 2006 and SPR 2006, pp. 331–339 (2006)
4. Bishop, C.M.: Neural Networks for Pattern Recognition. Oxford University Press, Oxford (1995)
5. Bollobs, B.: Random Graphs, 2nd edn. Cambridge University Press, Cambridge, UK (2001)
6. Caldarelli, G., Capocci, A., De Los Rios, P., Muoz, M.: Scale-free networks from varying vertex intrinsic fitness. Physical Review Letters 89(25), 258–702 (2002)
7. Di Massa, V., Monfardini, G., Sarti, L., Scarselli, F., Maggini, M., Gori, M.: A comparison between recursive neural networks and graph neural networks. World Congress on Computational Intelligence, pp. 778–785 (July 2006)
8. Duda, R.O., Hart, P.E.: Pattern Classification and Scene Analysis. Wiley, New York (1973)
9. Fergus, R., Perona, P., Zisserman, A.: A sparse object category model for efficient learning and exhaustive recognition. In: Proceedings of the IEEE Conference on Computer Vision and Pattern Recognition (CVPR), pp. 380–387. IEEE Computer Society Press, Los Alamitos (2005)
10. Gori, M., Monfardini, G., Scarselli, F.: A new model for learning in graph domains. In: Proc. of IJCNN-05 (August 2005)
11. Hammer, B., Micheli, A., Sperduti, A.: Universal approximation capability of cascade correlation for structures. Neural Computation 17(5), 1109–1159 (2005)
12. Hammer, B., Saunders, C., Sperduti, A.: Special issue on neural networks and kernel methods for structured domains. Neural Networks 18(8), 1015–1018 (2005)
13. Pollack, J.: Recursive distributed representations. Artificial Intelligence 46(1–2), 77–106 (1990)
14. Sperduti, A., Starita, A.: Supervised neural networks for the classification of structures. IEEE Transactions on Neural Networks 8(3), 714–735 (1997)
15. Watts, D., Strogatz, S.: Collective dynamics of small world networks. Nature 393, 440–442 (1998)

CSFNN Synapse and Neuron Design Using Current Mode Analog Circuitry

Burcu Erkmen and Tülay Yıldırım

Yildiz Technical University,
Department of Electronics and Communications Engineering
34349 Besiktas, Istanbul-Turkey
{bkapan,tulay}@yildiz.edu.tr

Abstract. In this paper, a neuron and synapse circuitry of Conic Section Neural Network (CSFNN) is presented. The proposed circuit has been designed to compute the Radial Basis Function (RBF) and Multilayer Perceptron (MLP) propagation rules on a single hardware to form a CSFNN neuron. Decision boundaries, hyper plane (for MLP) and hyper sphere (for RBF), are special cases of Conic Section Neural Networks depending on the data distribution of a given applications. Current mode analog hardware has been designed and the simulations of the neuron and synapse circuitry have been realized using Cadence with AMIS 0.5μm CMOS transistor model parameters. Simulation results show that the outputs of the circuits are very accurately matched with ideal curve. Open and closed decision boundaries have also been obtained using designed circuitry to demonstrate functionality of designed CSFNN neuron.

Keywords: Conic Section Function Neural Networks, Current Mode Analog Design, Neuron and Synapse Circuitry.

1 Introduction

Hardware realization of neural networks with their generalization capability is useful for numerous pattern recognition and signal processing applications. Neural network model require a lot of computing time to be simulated on a sequential machine resulting a great difficulty to investigate the behavior large neural networks and to verify their ability to solve problems. The neural system solves complicated problems by parallel operation of neurons. Performed in hardware, the operations performed by these circuits will take place in parallel, and in real-time [1]. As such, they will allow the neural network to converge at a higher speed than software-hased counterparts.

In literature, several architectures have been introduced for realization of artificial neural networks. Among the architectures, MLP and RBF are the two most popular neural network structures. Due to complementary properties of these networks several attempts have been performed to bring MLPs and RBFs under unified framework to make simultaneous use of advantages of both networks. Hybrid neural structures exist

B. Apolloni et al. (Eds.): KES 2007/WIRN 2007, Part I, LNAI 4692, pp. 17–25, 2007.

in literature. In [2], a hybrid Radial Basis Function-Multilayer Perceptron (RBF-MLP) network was used to improve performance. Dorffner [3] proposed Conic Section Functions Neural Networks (CSFNN) as a unified framework for MLP and RBF networks. Decision boundaries hyper plane (for MLP) and hyper sphere (for RBF) are special cases of Conic Section Neural Network.

Analog implementation [4] and digital implementation [5] of CSFNN neuron exist in literature. In [4], analog implementation of CSFNN neuron was realized with voltage mode operation using 2.4 micron parametres. In this paper, we propose current mode CSFNN synapse and neuron circuitry using submicron technologies. In current mode design, arithmetic operations are easily implemented and the frequency of operation is increased due to use of low-impedance internal nodes. Implemented circuit computes the weighted sum for MLP and the Euclidean distance for RBF with unified structure. The theory of CSFNN is overviewed in Section 2. In Section 3, synapse and neuron design with their sub-circuits are given. The simulation results are showed in Section 4. Decision boundaries of CSFNN neuron are demonstrated in Section 5. Finally conclusions are given in Section 6.

2 Conic Section Functions Neural Networks

The conic section function neural network (CSFNN), first described by Dorffner [3], is capable of making automatic decisions with respect to open (hyper plane) and closed (hyper sphere) decision regions and can use these regions whenever appropriate, depending on the data distribution of a given data. Both hyper plane and hyper sphere are special cases of the CSFNN. These are the decision boundaries of MLP and RBF, respectively. There would be intermediate types of decision boundaries such as ellipses, hyperbolas or parabolas in between those two cases which are also all valid for decision regions. Mathematically, the conic sections are formed from the intersection between a cone and a plane.

The neural computation is different in hidden neurons and output neurons in CSFNN. Hidden neurons realize the propagation rule of CSFNN and sigmoid activation function. The output neurons are inner product type. The following equations are obtained for n-dimensional input space for CSFNN neuron.

$$u_j^p(x) = \sum_{i=1}^{n} (x_i^p - c_{ij}) w_{ij} - \cos\omega_j \sqrt{\sum_{i=1}^{n} (x_{pi} - c_{ij})^2} \tag{1}$$

$$f_j^p(x) = \frac{2}{1 + e^{-2 \cdot u_j^p}} - 1 \tag{2}$$

Where x_{pi} refers to input vector for p. pattern, w_{ij} refers to the weights for each connection between the input and hidden layer, c_{ij} refers to center coordinates and ω_j refers to opening angles. i and j are the indices referring to the units in the input and hidden layer, respectively. This equation consists of two major parts analogous to the

MLP and the RBF. The equation simply turns into the propagation rule of an MLP network, which is the dot product (weighted sum) when ω is π/2. Second part of the equation gives the Euclidean distance between the inputs and the centers for an RBF network. Fig.1. illustrates the structure of a CSFNN.

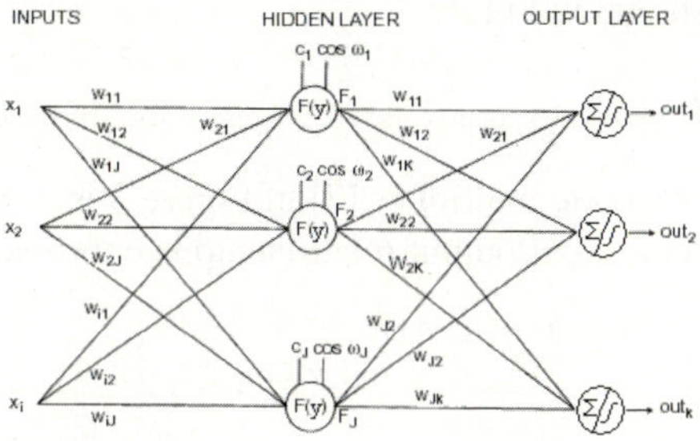

Fig. 1. Conic Section Function Neural Network structure

3 Synapse and Neuron Circuit Design

In this work, the synapse and neuron circuitry is designed using current mode analog circuits. Current mode signal processing offers several advantages when used in neural circuits. One of the most apparent advantages is that the summing of many signals is most readily accomplished when these signals are current. Arithmetic operations, such as addition, subtraction and scaling, are typically difficult to implement and it is often area- and power-consuming in a voltage-mode system. Other advantage is increased frequency of operation due to use of low-impedance internal nodes, and increased dynamic range of signals allowed when MOS transistors can be operated over a wide range of signals allowed when MOS transistors can be

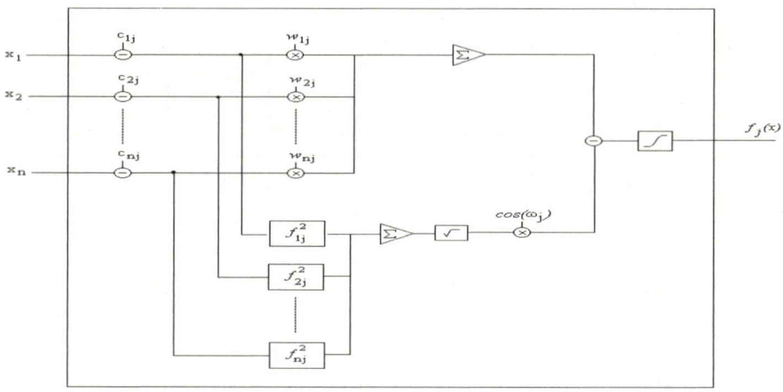

Fig. 2. Functional diagram of the synapse and neuron circuitry

operated over a wide range, from weak inversion to strong inversion [6]. The CSFNN synapse and neuron is composed of analog sub-circuits such as square root, squarer, multiplier and sigmoidal circuits. Cascode current mirrors are also used for repetition, multiplication with coefficient and for reversion the direction of currents. Each circuit has been examined in following sub-chapters. A functional diagram of the synapse and neuron circuitry is shown in Figure 2.

3.1 Multiplier Circuit

The four-quadrant current mode multiplier [7] in Figure 3 is used for multiplication in CSFNN neuron. The circuit input/output relationship is expressed as in Eq 3.

$$I_{out} = K \cdot I_x \cdot I_y \tag{3}$$

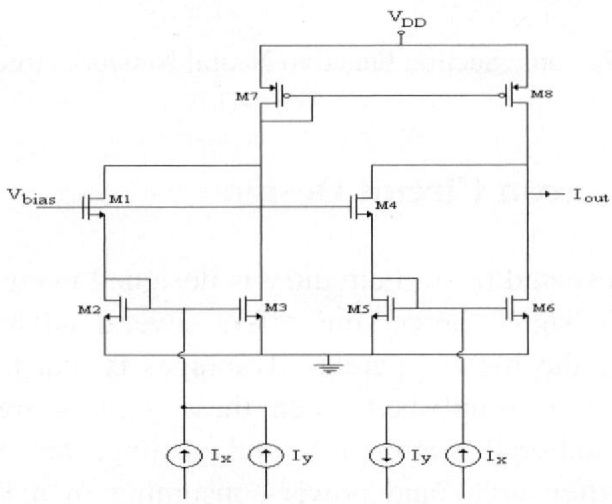

Fig. 3. Schematic diagram of the four-quadrant multiplier

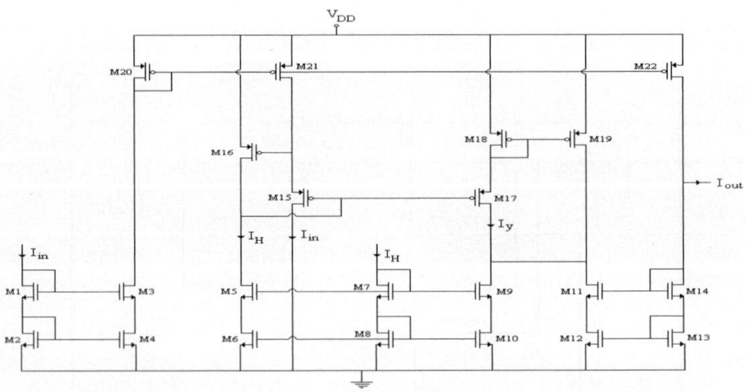

Fig. 4. Schematic diagram of the square-root circuit

3.2 Square-Rooter Circuit

The operation of square-rooter circuit in Figure 4 is to multiply the input currents I_{in} with bias current I_H and then to calculate the square-root of their product. [8]

$$I_{out} = 2 \cdot \sqrt{I_{in} \cdot I_H} \tag{4}$$

3.3 Squarer Circuit

The squarer circuit in Figure 5 is obtained with some modifications on the analog multiplier (Fig. 3). The input/output relationship is written as in Equation 5.

$$I_{out} = K \cdot I_{in}^2 \tag{5}$$

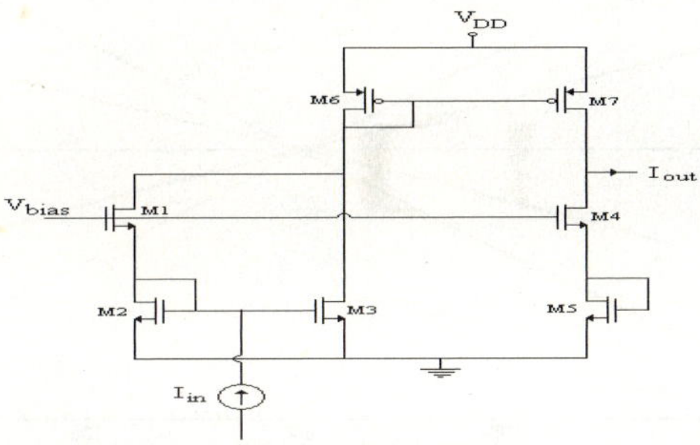

Fig. 5. Schematic diagram of the Squarer Circuit

3.4 Sigmoidal Circuit

The sigmoidal circuit [9] in Figure 6 is used to obtain activation function for CSFNN neuron. In Equation 6, α essentially controls the slop of the sigmoid function by changing the aspect ratios of transistors M_6 and M_7.

$$I_{out} = I_b \cdot \tanh(\alpha \cdot I_{in}) \tag{6}$$

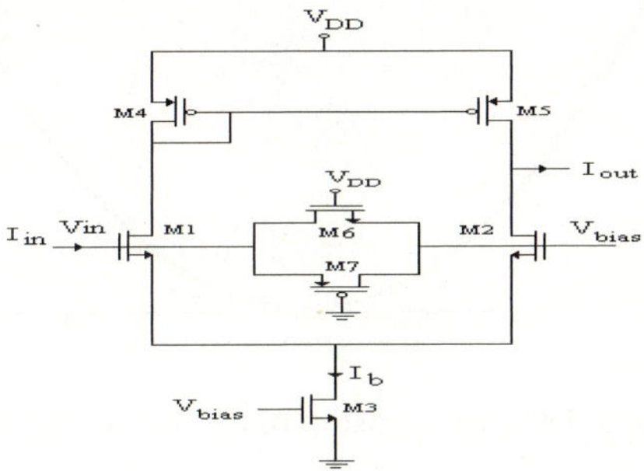

Fig. 6. Schematic diagram of the sigmoidal circuit

4 Simulation Results

Cadence software tool has been used to simulate the circuits performed with Spectre in Analog Artist environment. The simulations have been done to the neuron circuitry using Cadence with AMIS 0.5µm CMOS transistor model parameters. These circuits are all operated at a 5V supply voltage. The bias voltage V_{bias} = 2.5V is applied by external voltage source. The bias current source is set to I_H =1µA for square-root circuit. Results are very accurately matched with ideal curves.

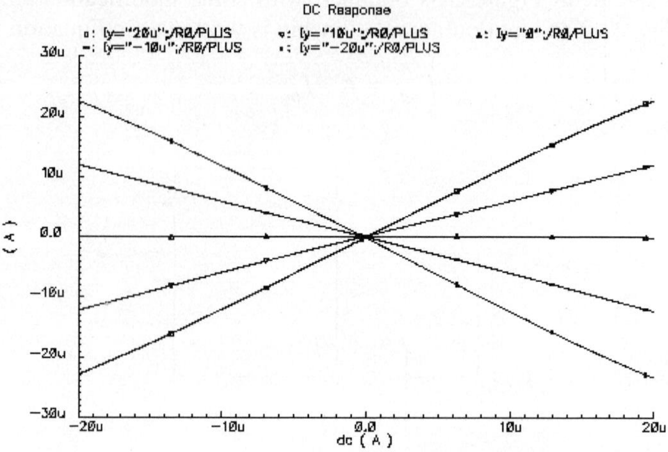

Fig. 7. DC characteristics of the multiplier circuit

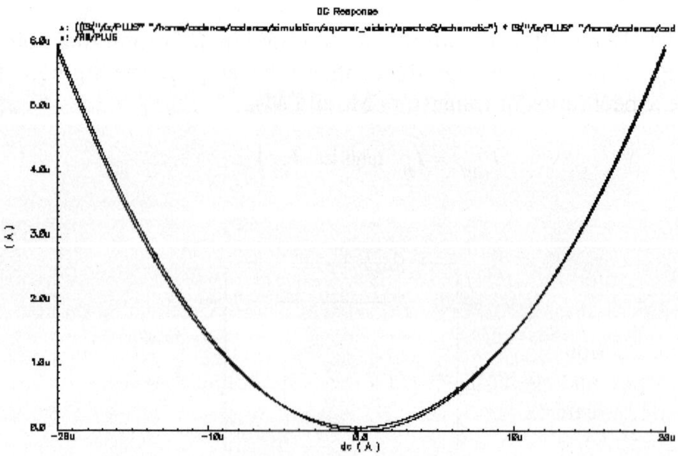

Fig. 8. DC characteristics of the squarer circuit

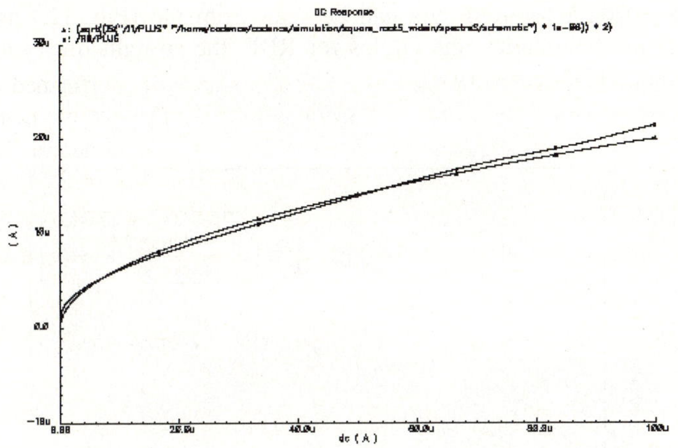

Fig. 9. DC characteristics of the square-root circuit

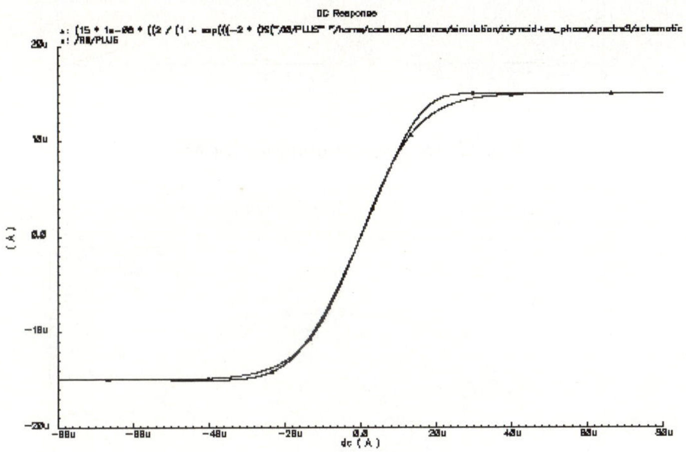

Fig. 10. DC characteristics of the sigmoidal circuit

5 Decision Boundaries for CSFNN Neurons

Decision boundaries, hyper plane (for MLP) and hyper sphere (for RBF), are special cases of Conic Section Neural Networks depending on the data distribution of a given applications. Open and closed decision boundaries have been obtained using designed circuitry to demonstrate functionality of CSFNN neuron. A fixed center value (Icen1=Icen2 =5μA) was chosen and the input of first synapse, Iin_1, was swept from 0μA ÷ 10 μA to draw the contours of decision regions for RBF and MLP. Then, the input of second synapse, Iin_2, was parameterized from 0μA ÷ 10 μA. By taking different output values and using the graph Iin_1 against to Iout, Iin_2 was plotted against to Iin_1 for different Iout contours. The opening angle was fixed to Ia = 0μA for

MLP and Ia = 20μA for RBF. As can be seen from the (Fig. 11.) and (Fig. 12), different decision boundaries, the circles for RBF, the straight lines for MLP, were obtained using CSFNN circuitry.

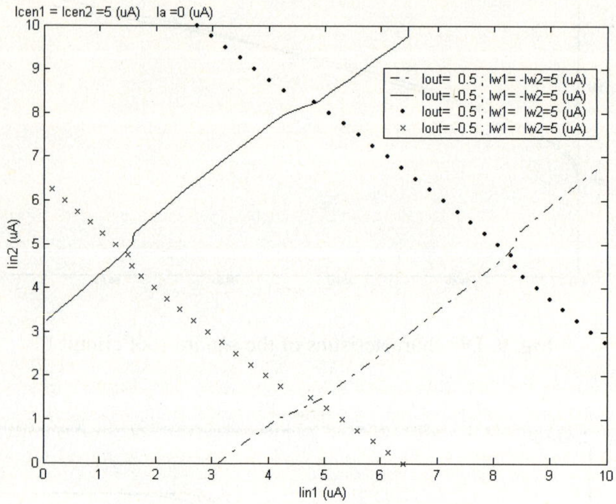

Fig. 11. Decision boundaries for MLP

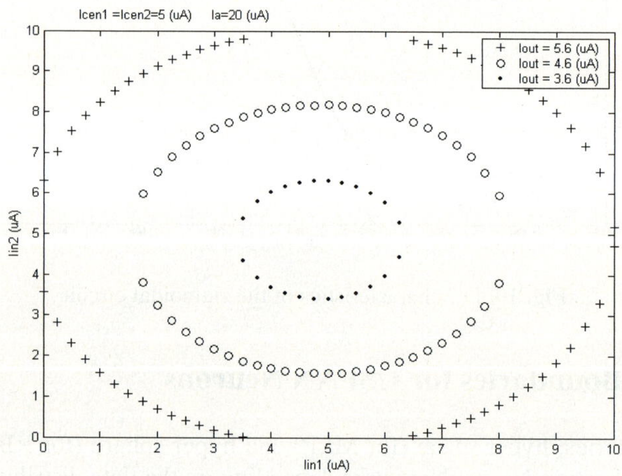

Fig. 12. Decision boundaries for RBF

6 Conclusion

In this work, CSFNN synapse and neuron circuitry was designed using current mode analog subcircuits. This implementation computes the Radial Basis Function (RBF) and Multilayer Perceptron (MLP) propagation rules with unified framework on a single hardware. Simulations of the sub-circuits have been realized by AMIS 0.5µm CMOS transistor model parameters using Cadence Software Tool. The simulation results of circuit outputs matched with ideal curves. Furthermore, open and closed decision boundaries have been obtained using designed circuitry to show the functionality of CSFNN neuron. In further work, the full CSFN network will be designed using implemented synapse and neuron circuitry to realize various classification problems on the hardware.

Acknowledgments. This research has been supported by TUBITAK- The Scientific and Technological Research Council of Turkey. Project Number : 104E133.

References

1. Sheu, B., J., Choi, J.: Neural Information Processing and VLSI, pp. 3–17. Kluwer Academic Publishers, USA (1995)
2. Chaiyaratana, N., Zalzala, A.M.S.: Evolving Hybrid RBF-MLP Networks Using Combined Genetic/Unsupervised/Supervised Learning. In: UKACC Int. Conf. on Control'98, Swansea, UK, vol. 1, pp. 330–335. IEE Publication 455 (1998)
3. Dorffner, G.: Unified frameworks for MLP and RBFNs: Introducing Conic Section Function Networks. Cybernetics and Systems 25, 511–554 (1994)
4. Yıldırım, T., Marsland, J.S.: An RBF/MLP Hybrid Neural Network Implemented In VLSI Hardware. In: Conf. Proc. of NEURAP'95 Neural Networks and Their Applications, Marseilles, France, pp. 156–160 (1996)
5. Esmaelzadeh, H., Farshbaf, H., Lucas, C., Fakhraie, S.M.: Digital Implementation For Conic Section Function Networks. Microelectronics, 2004. ICM 2004 Proceedings. The 16th International Conference, pp. 564–567 (2004)
6. Fakhraie, S.M., Smith, K.C.: VLSI-Compatible Implementations for Artificial Neural Networks, 1st edn. Springer, Heidelberg (1996)
7. El-Atta, M.A., Abou El-Ela, M.A., El Said, M.K.: Four-Quadrant Current Multiplier and Its Application as a Phase-Detector. Radio Science Conference (NRSC 2002). Proceedings of the Nineteenth National, pp. 502–508 (2002)
8. Liu, B.D., Chen, C.Y., Tsao, J.Y.: A Modular Current-Mode Classifier Circuit for Template Matching Application. IEEE Transactions on Circuits and Systems—II: Analog And Digital Signal Processing 47(2), 145–151 (2000)
9. El-Masry, E.I., Maundy, B.J., Yang, H.K.: Analog VLSI Current Mode Implementation of Artificial Neural Networks. Circuits and Systems, 1993, Proceedings of the 36th Midwest Symposium 2, 1275–1278 (1993)

Design of Neural Networks

Claudio Moraga

European Centre for Soft Computing, 33600 Mieres, Asturias, Spain
and Dept. Computer Science, University of Dortmund, 44221 Dortmund, Germany
mail@claudio-moraga.eu

Abstract. The paper offers a critical analysis of the procedure observed in many applications of neural networks. Given a problem to be solved, a favorite NN-architecture is chosen and its parameters tuned with some standard training algorithm, but without taking in consideration relevant features of the problem or possibly its interdisciplinary nature. Three relevant benchmark problems are discussed to illustrate the thesis that "*brute force solving is not the same as understanding*".

Keywords: Neural networks, pre-processing, problem-oriented design.

1 Introduction

Since the publication of "the" book of Rumelhart, and McClelland [22] the interest for developing and using neural networks for classification, approximation, and prediction tasks has grown quite impressively. Most nets applied for these tasks are feedforward nets with sigmoidal activation functions and tuned with some improved version (*e.g.* [13]) of the original gradient descend algorithm for which the name "Backpropagation" was coined, or are RFB-nets, which use non-monotonic activation functions, mostly Gaussian "bells". It is not quite clear whether the convincing power of the theorems on universal approximation of feedforward neural networks [9], [12] and RBF nets [11] or the effectivity of the training algorithms and the high speed of present PCs, which have contributed to favor a situation in which the "design" of a neural network reduces mainly to finding either per trial and error, or in the best case, by means of an evolutionary algorithm, the optimal number of hidden neurons. This is considered to be a negative trend. This paper analyses three benchmarks to show the results that may be obtained if instead of a "blind approach", knowledge –(possibly interdisciplinary)- of the designer, related to the problem domain and to fundamentals of neural networks, is taken into consideration and becomes an important component of the design process.

The selected benchmark problems are: The two spirals, the n-bit parity and the RESEX time series problems. The two spirals problem was posed by Alexis Wieland [26] in the late 80s as a challenge to the "triumphalists" in the NN-community. Two sets of 97 points each, are distributed along two concentric spirals. See Fig. 1. A neural network should be obtained, able to separate the two classes. Later on the problem was made tougher, by asking the neural network to separate *the spirals* and not only the sample points, i.e. the neural network was required to exhibit a very good

B. Apolloni et al. (Eds.): KES 2007/WIRN 2007, Part I, LNAI 4692, pp. 26–33, 2007.

generalization. Soon after the statement of the problem appeared the first solutions by Lang and Witbrock [14] and by Fahlman and Lebière [8]. Even though both solutions represented totally different architectures, both used 15 hidden neurons. Today, the two spirals problem is a typical homework in any neural networks course and probably most students *solve* the problem using the method criticized in the former paragraph. A 2-30-1 solution of recent times may be found in [17]. The second selected problem is the n-bit parity problem. A neural network with n binary inputs should be able to distinguish when the number of 1s in the input is even or odd and, following [24] it is possible to build such a neural networks with sigmoidal activation function using at most n/2 +1 hidden neurons if n is even or (n+1)/2, if n is odd. If shortcut connections are allowed, then a solution with only n/2 hidden neurons if n is even is possible [18]. A few solutions matching the bounds stated above may be found in the scientific literature. A solution without shortcuts and using only n/2 hidden neurons for the 8-bit parity problem is given in [2]. The RESEX time series [16] is based on the calling activity of the clients of a Canadian telephone company and is famous for a big outlier due to a special low-cost calling offer on Christmas time. This makes of this series a very tough prediction problem (not only) for a neural network.

2 From the Problem to the Neural Network

The two spirals problem, *as stated,* is indeed a difficult problem for a feedforward neural network using sigmoidal activation functions. However, experiences from signal processing, for instance, have shown that depending on the problem, it may be better to process the signal in the time domain or in the frequency domain. A rather obvious first choice for a different representation of the two spirals problem is to use polar coordinates. Every point with original coordinates (x, y) will be given the new coordinates (ρ, θ), where $\rho = (x^2 + y^2)^{1/2}$ and $\theta = \arctan(y/x) + \pi\cdot\mathrm{sign}(y)$. The computational complexity of these operations is relatively low and could be assigned as transfer functions of two auxiliary neurons with x and y as inputs. Using a new orthogonal coordinates system (ρ, θ), the representation of the problem in the new domain is illustrated at the right hand side of Fig. 1. The fact that the two spirals look now as a set of regularly alternating straight line segments within the $[-\pi, \pi]$ interval (each segment corresponding to a different class than that of its neighbor segments), is a consequence of the fact that A. Wieland must have chosen Archimedean spirals when he designed the problem. This kind of spirals has the property that the radius grows linearly with respect to the angle. From Fig. 1 (right) it may be seen that the slope of the line segments is $1/\pi$ and that, if $\theta = 0$, a symmetric square wave of period T = 2 could separate the alternating line segments corresponding to different classes (see Fig. 2). A symmetric square wave may be expressed as $\mathrm{sign}(\mathrm{sine}(\omega\rho))$, where ω denotes the angular frequency. The period T of a periodic function satisfies the condition $\omega T = 2\pi$, from where $\omega = 2\pi/T = 2\pi/2 = \pi$. Therefore the required square wave is $s = \mathrm{sign}(\mathrm{sine}(\pi\rho))$. If $\theta \neq 0$ a proper lag is required, which is proportional to the product of the value of θ and the slope of the line segment. Finally, the square wave accomplishing the separation of the two classes is given by $s = \mathrm{sign}(\mathrm{sine}(\pi\rho - \theta/\pi))$. This analysis leads to a neural network consisting of three neurons: the two auxiliary neurons mentioned above and a third one having as inputs ρ and θ, and as activation function, the periodic symmetric square wave *s*. All weights are equal to 1 and all biases are equal to 0. Furthermore, *no training is needed*. See Fig. 2. It may be argued that the proposed system is not really a neural network, however it satisfies the definition of a neural network: it is a distributed dedicated computing system, each neuron computes a single function of bounded complexity without requiring memory and the interconnection structure is as dense as possible by three neurons under the feedforward without shortcuts model.

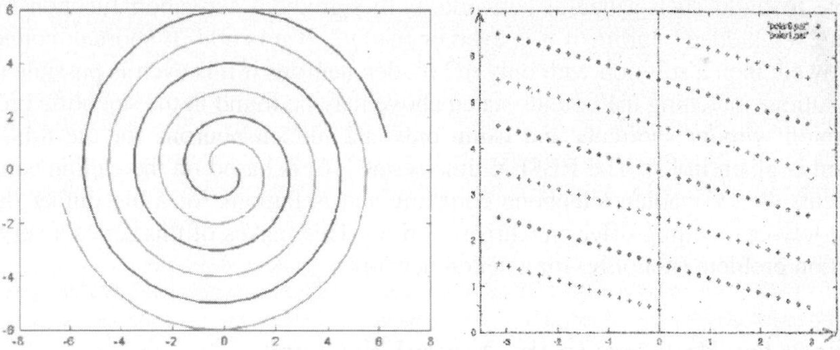

Fig. 1. The two spirals problem: left: representation in Cartesian coordinates. Right: representation in polar coordinates. (Vertical axis: radius horizontal axis, angle in radians).

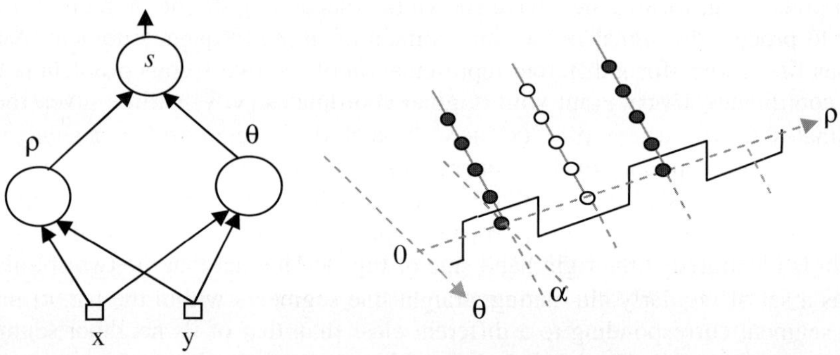

Fig. 2. Left: A minimal neural network *designed* to efficiently solve the two spirals problem. Right: Pseudo perspective of a partial view of the separation of the two unrolled spirals, where $\tan(\alpha) = 1/\pi$.

Solving the two spirals problem using polar coordinates was preliminary discussed in [19], but leading to a more complex solution, even though also without requiring learning the weights. The idea was later on also considered in [3], providing still a different final solution, however with emphasis in training the weights.

As mentioned in the introduction, important efforts and contributions have been done related to the n-bit parity problem. Most of them have focused on working with feedforward nets, eventually with shortcuts, using sigmoidal activation functions. An alternative approach was presented in [25] by introducing an "unusual" activation function quite different from a sigmoide to solve the parity problem with only two hidden neurons. In [15] the authors discuss the use of "product units" [7] instead of sigmoidal or Gaussian neurons. In these units, exponents are associated to the input variables. The output of the unit is the product of the exponentially weighted inputs. Obviously product units cannot directly process Boolean signals and the recoding to the set {1, -1} is a needed preprocessing. It becomes apparent that if the weights are chosen to be odd, a single product unit followed by proper decoding solves the parity problem. In [15] however the authors were interested in using the parity problem to test the ability of different training algorithms to "learn" appropriate weights. Unfortunately, these lines of research do not seem to have been continued. In what follows it is shown that is possible to design a problem-oriented activation function to solve the problem with only one neuron accepting the standard weighted sum of inputs, without needing a recoding of the inputs and decoding of the output and without training of the weights.

Let $A(x)$ denote the weighted sum of the inputs to a neuron. The standard sigmoidal activation function is given by $f(x) = [1 + \exp(-A(x))]^{-1}$. A parameterized extension was introduced in [10] as $f(x) = a[1 + \exp(-bA(x))]^{-1} + c$, where the parameters a,b and c were also adjusted to minimize the square error of performance. Notice that if $a = 1$ and $c = 0$, then $f(x)$ behaves as the classical sigmoide, except that the parameter b contributes to increase the speed of convergence by modifying all weights of a given neuron at the same time. If on the other hand, $a = 2$ and $c = -1$, then $f(x)$ behaves like a hyperbolic tangent function. A different view may be obtained by using $k^{-A(x)}$ instead of $\exp(-b(A(x))$ and by adjusting k instead of b. This leads to $f(x) = a[1 + k^{-A(x)}]^{-1}$. Notice that by choosing $a = ½$, $k = -1$, setting all weights to 1 and not taking the inverse value of the expression, the activation function turns to:

$$f(x) = ½\,[1 + (-1)^{-A(x)}] \qquad (1)$$

It is simple to see that whenever $A(x)$ is even, $(-1)^{-A(x)} = 1$ and $f(x) = 1$; meanwhile if $A(x)$ is odd, then $(-1)^{-A(x)} = -1$ and $f(x) = 0$. This means that *a single neuron* with this activation function, setting all weights to 1 and all biases to 0, is all that is needed to solve the n-parity problem *for any n* and, by the way, without training and without requiring the domain and codomain of the parity function to be {-1, 1}. It may be argued, that a single neuron is not a net. If this objection is accepted, the above results proves that no neural network is needed to solve the n-parity problem, but only a neuron *designed* to accomplish parallel counting modulo 2 and adding 1. The proposed solution was obtained by adapting a classical activation function to the requirements of the problem. It is however very closely related to the single neuron solution to the n-bit parity problem, presented in [1]. This other solution is formally based on principles of multiple-valued threshold logic over the field of complex numbers. The method requires using complex-valued weights, but up to n = 5 *all* binary functions may be realized with a single neuron.

The effectiveness of the training algorithms for neural networks and the similarities between neural networks and auto regressive systems (see *e.g.* [4]) possibly motivated the interest in using neural networks for prediction in the context of time series. The first intuitive approach to apply a neural network to predict one next event of a time series is to use the past history of the series to train a neural network using a sliding "time window". The samples within the window are the inputs to the neural network and the output is the predicted next event. Since during the training phase the next event is known, a prediction error may be calculated, which should be minimized by adjusting the weights of the network. The window moves one time step ahead and the process is repeated. This continues until all past samples are processed and the final prediction error is within tolerances. It is easy to see that the width of the time window is one additional parameter that should be adjusted. When the last block of samples is processed, the first actual prediction takes place. If the time window is moved one further step ahead, the just predicted signal will be within the time window, i.e., feedback is needed at the start of the prediction phase. (See Fig. 3, left). The above discussed scheme works particularly well in the case of short-term predictions (see *e.g.* [6]), since nonstationarity or seasonality of the time series may possibly be not noticeable. More sophisticated training algorithms have allowed however to use the above scheme together with support vector machines, even with chaotic series [21].

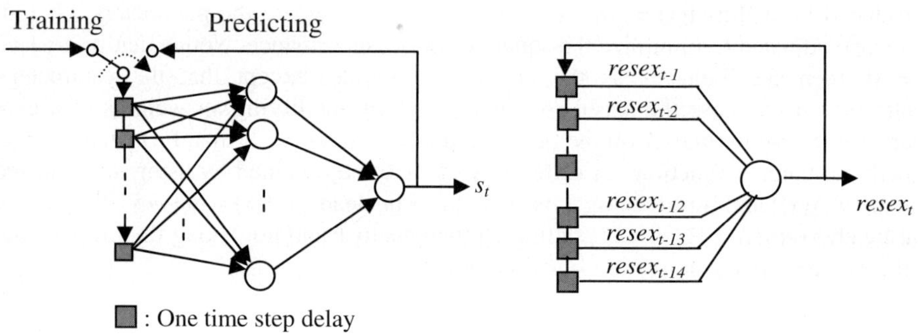

Fig. 3. Left: Time series prediction with a neural network (without taking in account data relevance). Right: An ARIMA-based one-neuron-net to predict the RESEX series.

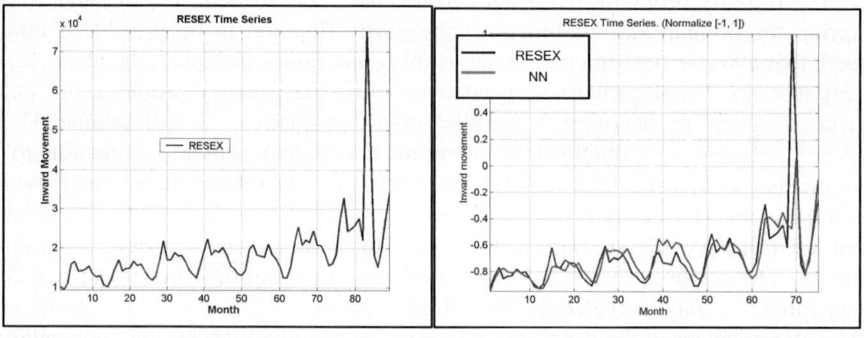

Fig. 4. Left: The RESEX time series. Right: A neural network based one step prediction.

Time series analysis is however a well established research area in Statistics, starting at least in 1976 with the publication of the seminal book of Box and Jenkins [5]. Statisticians have developed sound methods not only to find the appropriate width of the time window, but also to evaluate the relevance of the data within the window. Non relevant data within the time window adds a noise-like effect increasing the difficulty of the predicting process.

The RESEX time series is based on measurements of the internal connections of residential telephone extensions in a certain area of Canada during a period of 89 months [16]. The series is characterized by a clear yearly seasonality and a big outlier –(atypical value)- near the end of the series, corresponding to a low price Christmas promotion of the telephone company (see Fig. 4, left). Box and Jenkins [5] introduced a seasonal ARIMA model –(AutoRegressive model with Integrated Moving Average)- which, for this series, corresponds to ARIMA(2,0,0)(0,1,0)12. From this model it is possible to deduce that the relevant lags are { -1, -2, -12, -13, -14}. With this information, a *one* neuron net with sigmoidal activation function was able to predict the behaviour of the series on a one step basis [23]. See Fig. 4 (right hand side).

3 Conclusion

The discussed design of neural networks for three benchmark problems has shown that by including (eventually interdisciplinary) knowledge about the problem and about fundamentals of neural networks, both effectiveness and efficiency of the solution may be increased. Taking advantage of the methods developed long ago by statisticians it was possible to obtain a minimal solution to the prediction problem without facing an online noise filtering problem. The solution for the two spirals problem exhibits perfect generalization and unconstrained scalability: should the spirals be prolonged another round, then the *same* neural net would solve the problem. Similarly, the neuron designed to solve the parity problem, solves the problem for any n. Both the solution for the two spirals problem and the parity problem were obtained *without training*: the value of the free parameters could be deduced or calculated. This means that in this case, by trying to do a "knowledge-based design" of a neural network, *an analytical solution of the problem was obtained.*

The strategies discussed above are not new, but have often been absent when applying neural networks to solve a problem, with possibly the exception of Kernel-based methods (see *e.g.* [20]), which use idea of change of domain, as a basic component of their design strategy. A nonlinear problem is mapped into a linear domain, where a solution is most of the time simple, to later translate back the solution to the original domain.

People in pattern recognition are used to do first feature extraction, which is a search for relevant data, before starting to do classification. Statisticians can do that formally in the case of time series prediction. It is not unusual that engineers solve differential equations applying the Laplace transform. The "pocket calculator" of *my* generation was the slide-rule, where products and divisions were very efficiently done by adding and subtracting logarithms graphically represented on the rule!

Brute force should always be the very last resource. "*Brute force solving a problem is not the same as understanding the problem*".

References

1. Aizenberg, I.: Solving the parity n problem and other nonlinearly separable problems using a single universal binary neuron. Computational Intelligence. In: Reusch, B. (ed.) Theory and Applications, Springer, Berlin (2006)
2. Aizenberg, I., Moraga, C.: Multilayer feedforward neural network based on multi-valued neurons (MLMVN) and a backpropagation learning algorithm. Soft Computing 11(2), 169–183 (2007)
3. Alvarez-Sánchez, J.R.: Injecting knowledge into the solution of the two-spiral problem. Neural Computing and Applications 8, 265–272 (1999)
4. Allende, H., Moraga, C., Salas, R.: Artificial neural networks in forecasting: A comparative analysis. Kybernetika 38(6), 685–707 (2002)
5. Box, G.E., Jenkins, G.M.: Time series analysis: forecasting and control, Holden-Day, Oakland CA, USA (1976)
6. Chow, T.W.S., Leung, C.T.: Nonlinear autoregressive integrated neural network model for short-term load forecasting. IEE Proc. Gener. Transm. Distrib. 143(5), 500–506 (1996)
7. Durbin, R., Rumelhart, D.: Product units: A computationally powerful and biologically plausible extension to backpropagation networks. Neural Computation 1, 133–142 (1989)
8. Fahlman, S.E., Lebiere, C.: The Cascade Correlation Learning Architecture. In: Touretzky, S. (ed.) Advances in Neural Information Processing Systems, Morgan Kaufmann, San Francisco (1990)
9. Funahashi, K.I.: On the approximate realization of continuous mappings by neural networks. Neural Networks 2(3), 183–192 (1989)
10. Han, J., Moraga, C.: The Influence of the Sigmoid Function Parameters on the Speed of Backpropagation Learning. In: Sandoval, F., Mira, J.M. (eds.) From Natural to Artificial Neural Computation. LNCS, vol. 930, pp. 195–201. Springer, Heidelberg (1995)
11. Hartman, E.J., Keeler, J.D., Kowalski, J.M: Layered neural networks with Gaussian hidden units as universal approximators. Neural Computation 2, 210–215 (1990)
12. Hornik, K., Stinchcombe, M., White, H.: Multilayer feedforward neural networks are universal approximators. Neural Networks 2(5), 359–366 (1989)
13. Igel, C., Huesken, M: Improving the Rprop Learning Algorithm. In: Proc. 2nd Int. Symposium on Neural Computation, pp. 115–121. Academic Press, London (2000)
14. Lang, K.J., Witbrock, M.J.: Learning to tell two spirals apart, Proceedings of the 1988 Connectionist Models Summer School. Morgan Kaufmann, San Francisco (1988)
15. Leerink, L.R., Giles, C.L., Horne, B.G., Marwan, A.J.: Learning with product units. Advances in Neural Information Processing. NIPS-94, 537–544 (1994)
16. Martin, R.D., Smarov, A., Vandaele, W.: Robust methods for ARIMA models. In: Zellner, A. (ed.) Proc. Conf. applied time series analysis of economic data, ASA-Census-NBER, pp. 153–169 (1983)
17. Mizutani, E., Dreyfus, S.E.: MLP's hidden-node saturations and insensitivity to initial weights in two classification benchmark problems: parity and two spirals. In: Proc. IEEE Intl. Joint Conf. on Neural Networks, pp. 2831–2836. IEEE Computer Society Press, Los Alamitos (2002)
18. Minor, J.M.: Parity with two layer feedforward net. Neural Networks 6, 705–707 (1993)
19. Moraga, C., Han, J.: Problem Solving =/= Problem understanding. In: Proceedings XVI International Conference of the Chilean Computer Science Society, pp. 22–30. SCCC-Press, Santiago (1996)

20. Müller, K.-R., Mika, S., Rätsch, G., Tsuda, K., Schölkopf, B.: An introduction to Kernel-based learning algorithms. In: Hu, Y.H., Hwang, Y.-N. (eds.) Chapter 4 of Handbook of Neural Networks Signal Processing, CRC-Press, Boca Raton, USA (2002)
21. Müller, K.-R., Smola, A., Rätsch, G., Schölkopf, B., Kohlmorgen, J., Vapnik, V.: Using Support Vector Machines for Time Series Prediction. In: Gerstner, W., Hasler, M., Germond, A., Nicoud, J.-D. (eds.) ICANN 1997. LNCS, vol. 1327, pp. 999–1004. Springer, Heidelberg (1997)
22. Rumelhart, D.E., McClelland, J.L.: Parallel Distributed Processing: Explorations in the Microstructure of Cognition. MIT Press, Cambridge (1986)
23. Salas R.: Private communication, 2002 and 2007
24. Sontag, E.D.: Feedforward nets for interpolation and classification. Jr. Comput. Systems Science 45, 20–48 (1992)
25. Stork, D.G., Allen, J.D.: How to solve the N-bit parity problem with two hidden units. Neural Networks 5, 923–926 (1992)
26. Wieland, A.: Two spirals. CMU Repository of Neural Network Benchmarks (1988), http://www.bolz.cs.cmu.edu/benchmarks/two-spirals.html

Fast Fingerprints Classification Only Using the Directional Image

Vincenzo Conti[1], Davide Perconti[1], Salvatore Romano[1], G. Tona[1], Salvatore Vitabile[2], Salvatore Gaglio[1], and Filippo Sorbello[1]

[1] Dipartimento di Ingegneria Informatica
Università degli Studi di Palermo
Viale delle Scienze, Ed. 6, 90128 Palermo, Italy
{conti,gaglio,sorbello}@unipa.it
[2] Dipartimento di Biotecnologie Mediche e Medicina Legale,
Università degli Studi di Palermo
Via del Vespro, 90127 Palermo, Italy
vitabile@unipa.it

Abstract. The classification phase is an important step of an automatic fingerprint identification system, where the goal is to restrict only to a subset of the whole database the search time. The proposed system classifies fingerprint images in four classes using only directional image information. This approach, unlike the literature approaches, uses the acquired fingerprint image without enhancement phases application. The system extracts only directional image and uses three concurrent decisional modules to classify the fingerprint. The proposed system has a high classification speed and a very low computational cost. The experimental results show a classification rate of 87.27%.

Keywords: Fingerprint classification, c-means algorithm, bayesian network, neural network, decision network.

1 Introduction

In a real-time identification system it is required high response speed and, for intrinsic security reasons, a low false acceptance rate. The identification performed in a database divided in classes is faster, since it is reduced the number of necessary comparisons. This latency time can be reduced searching the template image in a database of the same class. In this work, a fingerprint classification system for a small-medium environment is presented.

The fingerprint is composed by ridges and valleys which forms unique geometric pattern in the skin [1]. Fingerprint is formed by a set of ridge lines which often runs parallel and they are characterized by end points and bifurcations. These points are called minutiae and they are fingerprint micro features. In the fingerprints there are other characteristics, called macro features, characterized by regions of the image where the ridge line flow is irregular. These macro features are called Delta and Core. The core point is the center of a circular edge pattern on a fingerprint image, and the

B. Apolloni et al. (Eds.): KES 2007/WIRN 2007, Part I, LNAI 4692, pp. 34–41, 2007.

delta point is the center of a triangular edge pattern [2]. This last are the most frequently used for fingerprint classification [3].

In the literature many classification approach have been proposed. Ballan, Evans and Sakarya [2], reduce distorsion and enhance contrast to compute the directional image. Successively, from directional image they have extracted the singularity points and have classified the fingerprints respect topological and numerical considerations on these points. Maio and Maltoni [3], have proposed a structural approach using relational graph. Others classification approaches are based on artificial intelligence technique, Kamijo uses a neural network [4], Mohamed and Nyongesa [5] use a fuzzy based technique.

In the proposed approach, no image processing is used before classification. Classification robustness is reached using three concurrent classification modules, each of one using its own features and paradigms.

This paper is organized as follows: in section 2 the proposed system is presented, in section 3 the experimental results are proposed and finally in the section 4 some conclusions are reported.

2 The Proposed System

The system classifies the fingerprint into four classes: Tented Arch, Whorl, Left Loop and Right Loop. The figure 1 shows the considered classes.

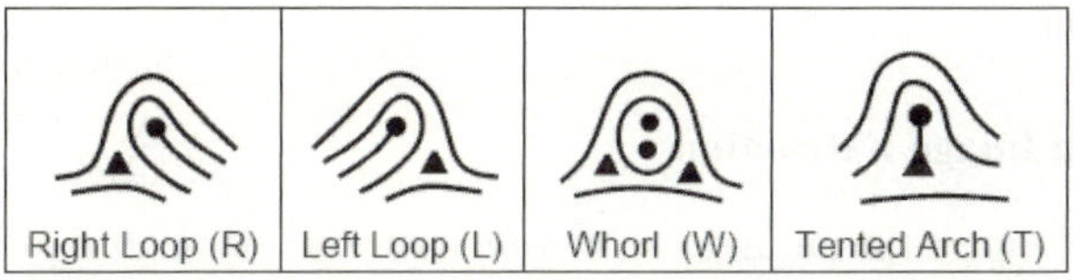

Fig. 1. The four fingerprint classes considered in this work: right loop, left loop, whorl and tented arch

The focus of this approach is to classify the fingerprints computing the directional image from original image without enhancement phases. The directional image is an image in which every element represents the local orientation of the ridges in the original gray-scale image. The directional image is computed in two steps: extraction the direction for each pixel and processing the output of step 1 assembling the pixels in 8x8 blocks and computing the predominant direction for each block.

The directional image extracted is processed by three concurrent modules: neural network, fuzzy c-means algorithm and Bayesian network approaches. The three classification results are processed by a decision network. The decision network classifies each fingerprint looking at modules classification results. The modules are trained using the same training set, see figure 2.

The Fuzzy c-means algorithm divides the space into four areas which represent the four classes. This approach classifies the template minimizing the Euclidian distance between the introduced template from the four classes center.

The Bayesian network approach computes the conditional probability of membership to the four classes.

The neural network model classifies using the training and the testing set.

The classification rate of the proposed system is based on the decision network. The decision network is a majority network, if in input at least two modules have classified the fingerprint in the same class then this class is considered correct, else the system gives a response of "indecision" (each module gives a different classification result), see figure 2.

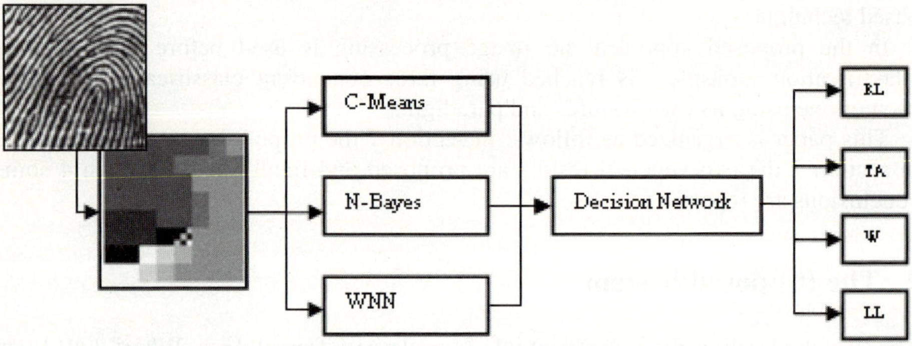

Fig. 2. The proposed classification system. The directional image is introduced and processed with the c-means, Bayesian and neural approaches. The decision network gives the final classification result.

2.1 Directional Image Extraction

The directional image is an image in which every element represents the local orientation of the ridges in the original gray-scale image. The noise present in typical ink-on-paper images or in images taken from different sensor, however, can affect the calculation of predominant directions inside noisy zones of the image that could result different from the predominant direction extracted in the neighbor zones.

In the directional image every pixel represents the local orientation of the ridges in the original image in scale of greys [2]. The phases of the algorithm are now described.

The direction $K(i,j)$ of the point (i, j) is defined through the following equation (1):

$$K\left(i,j\right) \equiv Min \sum_{K=1}^{L} \left[C\left(i_k, j_k\right) - C(i,j)\right] \quad (1)$$

where $C(i,j)$ and $C\left(i_k, j_k\right)$ indicate the level of grey to the point (i, j) while L is the number of select pixel in this calculation along a considered direction.

In the images used in this paper eight directions have been chosen, while a number of L=16 pixels along the direction (eight in a direction and eight in the opposite direction) have been considered. These values have been experimental determinate. As shown in figure 3, from 0° to 180° eight directions can be determinate hence in every block of 8x8 pixels it will be established the predominant direction.

In every block 8x8 the directions of the single constituent pixels are found, using the (1), the direction with greater frequency will be attributed to the considered block.

The possible presence of noise in the image used introduces a problem, in fact, the directions in noisy blocks could be very different in comparison to the direction assumed by their neighbours. This could cause mistakes in the characteristics extraction phase, therefore a smoothing has been applied on the directional image. This is achieved by calculating the directional histogram. The directional histogram is obtained from a comparison of direction of an area of 3x3 blocks. The final direction of the central block, the directional histogram, of the considered area was replaced by the majority of the neighbouring blocks. The reduced size of directional image decreases the complexity of the rest of the algorithm.

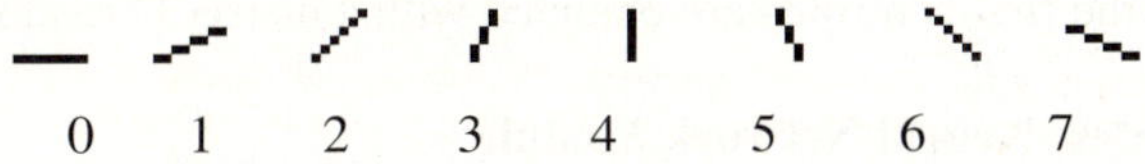

Fig. 3. Directional code for ridge orientation

2.2 The Fuzzy C-Means Module

Fuzzy C-means Clustering (FCM) is a clustering technique that employs fuzzy space partitioning. A data point can belong to all groups in the space with a different membership rate, usually between 0 and 1. FCM is an iterative algorithm and its aim is to find cluster centers (centroids) to minimize a dissimilarity function. To accommodate the introduction of fuzzy partitioning, the membership matrix U is randomly initialized according to formula (2):

$$\sum_{i=1}^{c} u_{ij} = 1, \forall j = 1,...,n \tag{2}$$

The dissimilarity function used in FCM is the following:

$$J(U, c_1, c_2, ..., c_c) = \sum_{i=1}^{c} J_i = \sum_{i=1}^{c}\sum_{j=1}^{n} u_{ij}^{\ m} d_{ij}^{\ 2} \tag{3}$$

where u_{ij} is a value in [0,1], c_i is the i_{th} centroid, d_{ij} is the Euclidian distance between i_{th} centroid(c_i) and j_{th} data point, m is a weighting exponent with values in $[1,\infty]$.

To reach a dissimilarity function minimum the following formulas, (4) and (5), are used:

$$c_i = \frac{\sum_{j=1}^{n} u_{ij}^{\ m} x_j}{\sum_{j=1}^{n} u_{ij}^{\ m}} \tag{4}$$

$$u_{ij} = \frac{1}{\sum_{k=1}^{c} \left(\frac{d_{ij}}{d_{kj}} \right)^{2/(m-1)}} \tag{5}$$

With more details, the proposed fuzzy c-means algorithm follows the algorithm in [12]. This algorithm is composed by following steps:

1. *the membership matrix (U) is randomly initialized respecting the constraints in (2);*
2. *centroids* (c_i) *are calculated using formula (4);*
3. *the dissimilarity function, between centroids and data points, is computed using formula (3)*
   a. *if its improvement is below a threshold the algorithm is stopped*
   b. *else, a new U using formula (5) is computed and the process is repeated from point 2.*

This FCM algorithm iteratively moves the centroids toward the "right" position within a data set. Nevertheless FCM not guarantees the convergence to an optimal solution because the first centroids are obtained with a matrix U randomly initialized.

2.3 The Weightless Neural Network Module

A Weightless Neural Network (WNN), known as “discriminator”, is based on a concept different from conventional neural networks. With WNNs the training data is stored in a memory, whereas with conventional neural networks the training data is used to adjust its weights. WNNs possess all the main advantageous characteristics of conventional neural networks including learning by examples, mapping capability, robust performance, parallel processing, and generalization capability. In addition, they do not suffer from the problems of local minima, are easy to train and update, and have a fast speed of operation [10] [14].

The network shown in Figure 4 can be used to distinguish k classes; it consists of k discriminators.

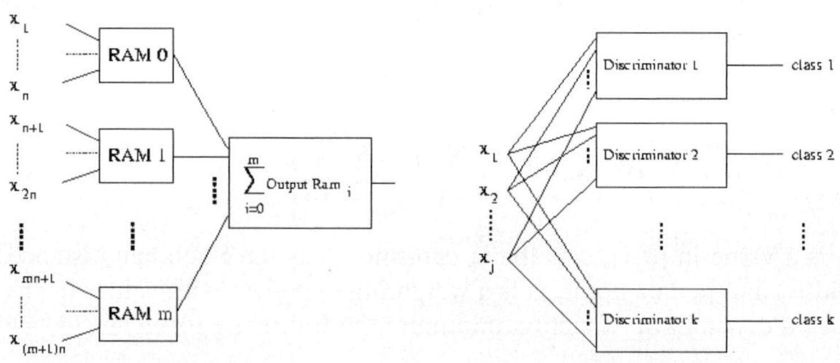

Fig. 4. The proposed weightless neural network schema. On the left: the training architecture, on the right the test architecture.

The discriminator is the device which performs the generalization. It consists of several vectors and one node which sum the outputs of the vectors in test mode. For each class of input patterns one discriminator is needed, which is trained with the input data of its class. In the test mode each discriminator responds with the number of matching subpatterns. The pattern is classified according to the discriminator with the highest response.

The proposed approach in this paper is the following: the input vector is divided into fixed number of parts; each part is connected to the inputs of a 1-Bit-Vector unit. The output of all the vectors inside each discriminator are summed up. In our architecture four discriminators are used to classify the input data.

2.4 The Naive Bayesian Network Module

A Naive Bayesian Network (NBN) is a probabilistic classifier based on applying Bayes theorem with strong independence assumptions. Every NBN classifier can be trained very efficiently in a supervised learning setting. A Bayesian network encodes the joint probability distribution of a set of n variables, {X1, . . ,Xn}, as a directed acyclic graph and a set of conditional probability distributions (CPDs). Each node corresponds to a variable and its CPD gives the probability of each state of the considered variable with every possible combination of states of its parents [9]. The structure of the network encodes the assertion that each node is conditionally independent of its non-descendants given its parents, Figure 5.

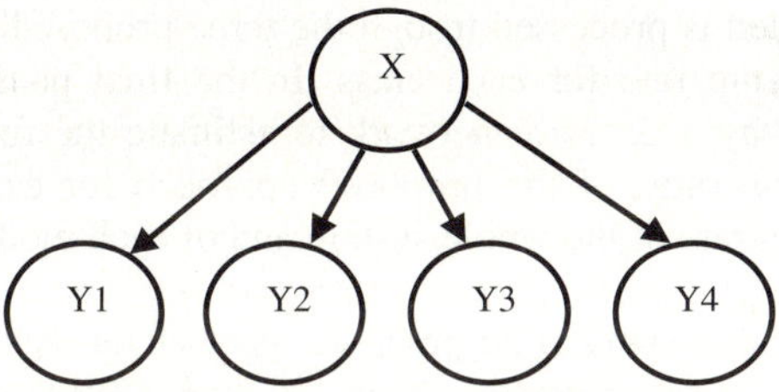

Fig. 5. The proposed Naive bayesian network schema

The main characteristics of this approach are the following. Suppose that the values of N input attributes, $X=\{x_1,x_2...x_N\}$, can be considered independent both unconditionally and conditionally respect a fixed class y. This means that the total probability of x can be written as the following product

$$P(x)=P(x_1)*P(x_2)*...*P(x_N) \quad (6)$$

and the relative probability of x within each class y

$$P(x|y)=P(x_1|y)*P(x_2|y)*...*P(x_N|y) \quad (7)$$

With these two probability relations the conditional probability is

$$P(y|x)=P(y)*P(x|y)/P(x)=P(y)*(P(x_i|y)/P(x_i)) \quad (8)$$

Equation (8) is the basis for the Naive Bayesian Classier [11]. The name naive is due to simplistic assumption that different input attributes are independent.

2.5 The Decision Network

The decision networks are used where uncertainties exist in the statistical environment. The proposed decision network has the ability to process its input data consisting of the classifier modules output and to produce a final decision regarding the fingerprint image correct class. The decision network is a majority network and the output result is "indecision" if the three modules give a different classification result, while the output result is a correct class if at least two modules have classified the fingerprint in the same class.

3 Experimental Results

In this section the experimental results, obtained with the proposed approach, are reported. All the modules were trained with the same data set. The training set is composed by 4 fingerprint images for each class type. The data set is composed of 55 fingerprint images divided in the four different classes: *right loop*, *left loop*, *whorl* and *tented arch*. The fingerprint images have been acquired through the sensor "Precise 100 MC" of the Precise Biometrics [13] and they have 200x200 pixels.

Three phases have been realized to learn the three architectures. In the first phase the directional image have been encoded in a 24x24 pixels matrix. This matrix represents the predominant directions of each block 8x8 pixels of the original image. In the second phase the image encoded is processed trough the three proposed concurrent modules to calculate the membership rate for each class. In the final phase, the results of each module are processed by a decision network to estimate the final class. The Table 1 shows the classification rates of the proposed approach for each class. The Table 2 shows the classification rate of the whole system and of each module.

Table 1. Classification rates of the proposed approach for each considered class

Module	NBN Correct/Total	WNN Correct/Total	FCM Correct/Total
Right Loop	11/15 (74%)	14/15 (94%)	14/15 (94%)
Left Loop	3/3 (100%)	3/3 (100%)	3/3 (100%)
Whorl	25/31 (81%)	24/31 (77%)	24/31 (77.42%)
Tented Arch	6/6 (100%)	4/6 (67%)	5/6 (83.33%)

Table 2. Classification rate of the whole system and of each module

	NBN	WNN	Fuzzy C-mean
Single Module	81.82 %	81.82 %	83.64 %
Total System		87.27 %	

4 Conclusion

This work presents an automatic method for fingerprint classification using only the directional image as input. With this approach, a high response speed has been obtained with a low false acceptance rate. The proposed system is composed by three concurrent architectures to classify fingerprint images in four classes. Successively, a decision network, classifies a fingerprint if at least two architectures are in agreement. The experimental result shows a classification rate of the whole system of 87.27 % over a database of 55 images.

References

1. Jain, A., Hong, L., Bolle, R.: On-Line Fingerprint Verification IEEE Trans. Pattern Analysis and Machine Intelligence 19(4), 302–314 (1997)
2. Ballan, M., Ayhan Sakarya, F.: A Fingerprint Classification Technique Using Directional images, Signal, System and Computer, 1997. Conference Record Of The Thirty-First Asilomar Conference on 1, 101–104 (1997)
3. Maltoni, D., Maio, D.: A structural Approach To Fingerprint Classification. In: Proceedings of the 13th Conference on, vol. 3, pp. 578–585, 25–29 (August 1996)
4. Kamijo.: Classifing Fingerprint Images Using Neural Network: Deriving The Classification State. IEEE International Conference on neural networks 3, 1932–1937 (1993)
5. Mohamed., Nyongesa.: Automatic Fingerprint Classification System Using Fuzzy Neural Techniques. In: Proceedings of the 2002 IEEE International Conference on Fuzzy System, 2002, vol. 1, pp. 358–362, 12–17 (May 2002)
6. Whang., Zhang., Whang.: Fingerprint Classification By Directional Fields. In: Proceedings. Fourth IEEE International Conference on Multimodal Interfaces, 2002, pp. 395–399, 14-16 (October 2002)
7. Maltoni., Maio.: Neural Network based minutiae filtering in fingerprints. In: Proceedings Fourteenth International Conference on Pattern Recognition, vol. 2, pp. 1654–1668, 16–20 (August 1998)
8. Jain., Prabhakar., Pankanti.: A filterbank-based representation for classification and matching of fingerprints, International Conference on Neural Networks, vol. 5, pp. 3284–3285, 10–16 (July 1999)
9. Tang., Pan., Li., Xu.: Fuzzy Naive Bayes classifier based on fuzzy clustering. IEEE International Conference on System, Man and Cybernetics, vol. 5, pp. 6, 6–9 (October 2002)
10. Mitchell, R.J., Bishop, J.M., Box, S.K., Hawker, J.F.: Comparison of methods for processing grey level data in weightless networks. In: Bisset, D. (ed.) Proc. of Weightless Neural Network Workshop WNNW95, Kent at Canterbury, UK, pp. 76–81 (1995)
11. Ceci, M., Appice, A., Malerba, D.: Mr-SBC: A multi-relational nave Bayes classifier. In: Lavrač, N., Gamberger, D., Todorovski, L., Blockeel, H. (eds.) PKDD 2003. LNCS (LNAI), vol. 2838, pp. 95–106. Springer, Heidelberg (2003)
12. Zhang, J.-S., Leung, Y.-W.: Improved Possibilistic C-Means Clustering Algorithms. IEEE Transaction on Fuzzy Systems, 12(2) (2004)
13. Precise 100 MC, Precise Biometrics, http://www.precisebiometrics.com
14. Rohwer, R., Morciniec, M.: A theoretical and experimental account of n-Tuple classifier performance. In: proc. of Neural Computation, (ISSN 0899-7667) vol. 8(3), pp. 629–642 (1996)

Geometric Algebra Rotors for Sub-symbolic Coding of Natural Language Sentences

Giovanni Pilato[1], Agnese Augello[2], Giorgio Vassallo[2], and Salvatore Gaglio[1,2]

[1] ICAR - Italian National Research Council,
Viale delle Scienze, Ed.11, 90128, Palermo, Italy
g.pilato@icar.cnr.it
[2] DINFO - University of Palermo,
Viale delle Scienze, Ed. 6, 90128 Palermo, Italy
augello@csai.unipa.it, {gvassallo,gaglio}@unipa.it

Abstract. A sub-symbolic encoding methodology for natural language sentences is presented. The procedure is based on the creation of an LSA-inspired semantic space and associates rotation operators derived from Geometric Algebra to word bigrams of the sentence. The operators are subsequently applied to an orthonormal standard basis of the created semantic space according to the order in which words appear in the sentence. The final rotated basis is then coded as a vector and its orthogonal part constitutes the sub-symbolic coding of the sentence. Preliminary experimental results for a classification task, compared with the traditional LSA methodology, show the effectiveness of the approach.

Keywords: Geometric Algebra Rotors, Latent Semantic Analysis.

1 Introduction

Natural language documents representation is one of the main issues of information retrieval, data mining, documents classification problems. Traditional methods are based on a *bag of words* approach and represent a document as vector of words occurrences. This kind of representation implies that the comparison between two documents is evaluated on the strength of the number of words that documents share, neglecting that similar concepts can be expressed using different terms. Another kind of approach is given by the Latent Semantic Analysis(LSA) paradigm[1], which allows to infer indirect similarity relations among words and documents by means of an induction-dimension optimization. This allows to overcome the issues due to the synonymy properties of natural language. LSA claims a human like generalization capability that is similar to induction processes obtained from associative, semantic, and neural network models[1]. Despite of its generalization properties, LSA neglects all information regarding the word order inside the document or the sentence. However this kind of information is very important for the semantic understanding of a phrase or discourse. As a matter of fact, according to Haugeland[2] the meaning of a sentence is determined by the meanings of its components, together with their mode

B. Apolloni et al. (Eds.): KES 2007/WIRN 2007, Part I, LNAI 4692, pp. 42–51, 2007.

of composition. A sentence like: "*The cat eats the mouse*" has a completely different semantic with respect to the sentence "*The mouse eats the cat*", although both sentences share exactly the same words.

In this work an algorithm for sub-symbolically coding natural language sentences is proposed. The aim is to obtain a representation of the sentence which takes into account both the semantics of the words composing it, and the structure of the sentence itself, intended as the order in which words appear in the phrase. This will lead to the properties that the replacement of one or more words with their corresponding synonyms should not alter too much the coding, which should be instead affected by a changing of the words order. Moreover an effective coding should be independent of the number of words belonging to the sentence and should be characterized by an enough high dimensionality in order to contain a large amount of information and to be easily classifiable from the other codings. We propose to use rotation trajectories of an orthogonal basis in a semantic space derived from a LSA-inspired methodology. The rotation is performed by means of semantic rotation operators derived from geometric algebra named rotors; the rotors are obtained analyzing the semantic context of the words composing the sentence. A sequence of rotors is associated to the sequence of bigrams in the sentence and is applied to the canonical basis represented by the unit matrix. The orthogonal component of the vector representing the final rotated basis will constitute the sub-symbolic coding of the phrase. In this way the resulting coding will not be influenced by the sentence length.

A classification task has been used as a testbench for the proposed algorithm employing a corpus of questions labeled according to a predefined set of categories[3][4]. The obtained results compared to the traditional LSA methodology show that the approach can be a valid step towards a better encoding of sentences.

2 Related Works

Bag of words based approaches do not take into account word order in sentences. A solution is to consider N-grams instead of bag of words features[5][6]. However this approach has the drawback of generating vectors of very high dimensionality, which are difficult to manage for large datasets. Liu et al.[7] propose a text representation model, named HOSVD based on higher order tensors to handle high dimensional features vectors. Another approach consists in the creation of state space models[8] used to capture the information given by the order of words appearing in the sentence.

A different solution is to parse natural language sentences and represent them as syntactic trees. The trees associated to the sentences can be used as features in a classification task, based on different algorithms, such as support vector machines[9] or maximum entropy and boosting models[10]. This kind of representation allows to obtain better classification results with respect to the traditional approaches; however it strongly depends on the parser used to get the syntactic trees and in particular on the corpus used to train the parser.

Some works try to introduce information about the word order in documents representations obtained through LSA methodology. In particular English Syntactically Enhanced LSA (SELSA)[11] generalizes LSA by considering a pair composed by a word along with the part-of-speech tag of its preceding word as a unit of information. In structured LSA (SLSA)[12], similarity between two sentences has been evaluated by averaging the similarity of sub-sentence structures like noun phrase, verb phrase, object phrase etc. Other approaches have been reported in[13][14][15].

3 Theoretical Background

3.1 Building of the Semantic/Conceptual Space

Latent semantic analysis methodology allows to obtain a semantic representation of words and documents through statistic analysis of a texts corpus. The strength of LSA is an induction-dimension optimization obtained through the truncated singular value decomposition (TSVD) that converts the initial representation of information into a condensed representation that captures indirect, higher-order associations between words[1].

In this work the LSA paradigm has been applied according to the approach reported in [16]. In particular a word-word co-occurrences matrix has been built, where its *(i, j)-th* entry represents the number of times a bigram composed of the *i-th* word followed by the *j-th* word appears in a documents corpus inside a window of a fixed number of words. It is important to point out that the dimension of the matrix is determined only by the number of words included in the vocabulary and it is independent of the number of documents. According to Agostaro et al.[16]the resulting matrix, which is not symmetrical, is preprocessed substituting each cell of the matrix with its square root in order to interpret the TSVD as a statistical estimator and to obtain the best rank k approximation of $\mathbf{A}$ with respect to the Hellinger's distance.

The result of the truncated SVD is the following:

$$\mathbf{A} \approx \mathbf{A}_k = \mathbf{U}_k \mathbf{\Sigma}_k \mathbf{V}_k^T. \tag{1}$$

where $\mathbf{U}_k$, $\mathbf{\Sigma}_k$ and $\mathbf{V}_k$ are matrices that provide compressed information about the left and the right context of the word. In particular the *i-th* row of $\mathbf{U}_k$, multiplied by the square root of the σ_{ii} element of $\mathbf{\Sigma}_k$ represents the *right context* of the *i-th* word, while the the *i-th* row of $\mathbf{V}_k$, multiplied by the square root of the σ_{ii} element of $\mathbf{\Sigma}_k$ represents the *left context* of the *i-th* word (see Fig.1).

Therefore in the generated semantic space it is possible to associate to each word two different vectors: $\mathbf{l}_i$ and $\mathbf{r}_i$, the former representing the left context and the latter representing the right context of the word.

To evaluate the distance between two generic vectors $\mathbf{v}_i$ and $\mathbf{v}_j$ belonging to this space maintaining the probabilistic interpretation, a similarity measure is defined as follows:

$$sim(\mathbf{v}_i, \mathbf{v}_j) = \begin{cases} \cos^2(\mathbf{v}_i, \mathbf{v}_j) & \text{if } \cos(\mathbf{v}_i, \mathbf{v}_j) \geq 0 \\ 0 & \text{otherwise} \end{cases}. \tag{2}$$

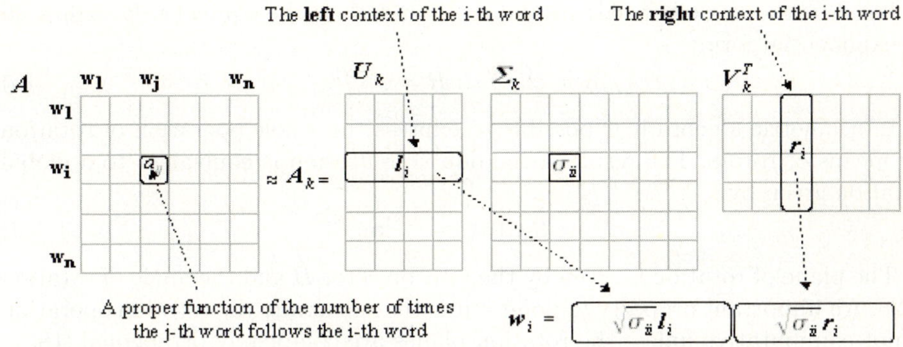

Fig. 1. Coding of words by means of Hellinger based TSVD applied to a word-word matrix A

3.2 Rotation Operators in the Geometric Algebra

Let **m** and **n** be two unit vectors, according to the geometric algebra, the rotation of a vector **a** in the $\mathbf{m} \wedge \mathbf{n}$ plane[17] is obtained by two subsequent reflections of **a** with respect to **m** and **n**(see Fig.2).

The result of the first reflection with respect to the versor **m** produces the vector **a'** given by $\mathbf{a'} = \mathbf{a} - 2(a \cdot \mathbf{m})\mathbf{m}$ where $(\mathbf{a} \cdot \mathbf{m})\mathbf{m}$ is the component of **a** parallel to **m**. In a similar manner the result of the second reflection is $\mathbf{a''} = \mathbf{a'} - 2(\mathbf{a'} \cdot \mathbf{n})\mathbf{n}$. Let N be the dimension of the vectors **m** and **n** the computational cost of the whole rotation is $O(N)$.

Let **nm** be the geometric product between the vectors **m** and **n** defined as [17]:

$$\mathbf{nm} = \mathbf{n} \cdot \mathbf{m} + \mathbf{n} \wedge \mathbf{m}. \tag{3}$$

A rotation operator rotor R and its reverse $\tilde{R}$ are defined by the following formulas:

$$R = \mathbf{nm} \;\; ; \;\; \tilde{R} = \mathbf{mn}. \tag{4}$$

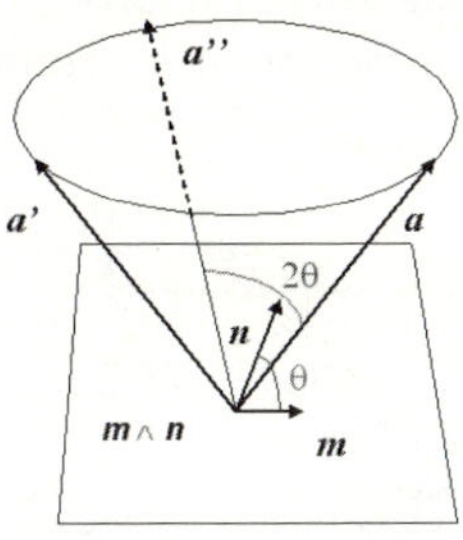

Fig. 2. Rotation of a vector through a double reflection in the $\mathbf{m} \wedge \mathbf{n}$ plane

Let θ be the angle between $\mathbf{m}$ and $\mathbf{n}$, R and $\tilde{R}$ can be equivalently expressed in exponential form:

$$R = e^{(-\hat{B}\theta)} \; ; \tilde{R} = e^{(\hat{B}\theta)}. \tag{5}$$

In geometric algebra it is possible to express the whole operation of rotation by means of rotors. The result of the double reflection is equivalent to a global rotation given by:

$$\mathbf{a}" \mapsto R\mathbf{a}\tilde{R} = e^{(-\hat{B}\theta)}\mathbf{a}e^{(\hat{B}\theta)}. \tag{6}$$

The plane of rotation is given by the unit bivector $\hat{B}$ and the angle of rotation is 2θ. An important property to point out is that generally the rotation operation is not commutative unless the rotation planes are completely orthogonal [18].

4 Geometric Algebra Rotors for Sub-symbolic Coding of Natural Language Sentences

The proposed approach consists in an unsupervised encoding procedure that tries to enhance the traditional LSA technique partially injecting information about the sentence structure into the coding. The temporal sequence of words appearing into the sentence generates a rotation trajectory of an orthogonal basis in a semantic/conceptual space created as described in Sect.3.1. Given a bigram composed by the words w_i and w_j, let $\mathbf{l}_i$ and $\mathbf{r}_i$ be the left and right contexts of the word w_i and $\mathbf{l}_j$ and $\mathbf{r}_j$ the left and right contexts of the word w_j (see Sect. 3.1), a rotor represented as the geometric product:

$$R'_{ij} = \mathbf{r}_i\mathbf{l}_j. \tag{7}$$

will be associated to the bigram. The rotor R'_{ij} rotates the basis of 2θ where θ is the angle between $\mathbf{r}_i$ and $\mathbf{l}_j$. A new rotor operator $R_{ij} = exp(-\hat{B}\frac{\theta}{2})$ for each R'_{ij} is obtained in order to rotate the basis of an angle θ.

The canonical basis of k dimensions represented by the identity matrix, is associated to the sentence that has to be coded. For a sentence of M words, M-1 bigrams are present, hence M-1 rotors are associated to the phrase(see Fig.3).

The sequence of these rotors will be applied to the original basis, transforming it M-1 times. At the end of the rotation process, the matrix representing the rotated basis is considered as a vector of k^2 components. The final coding will

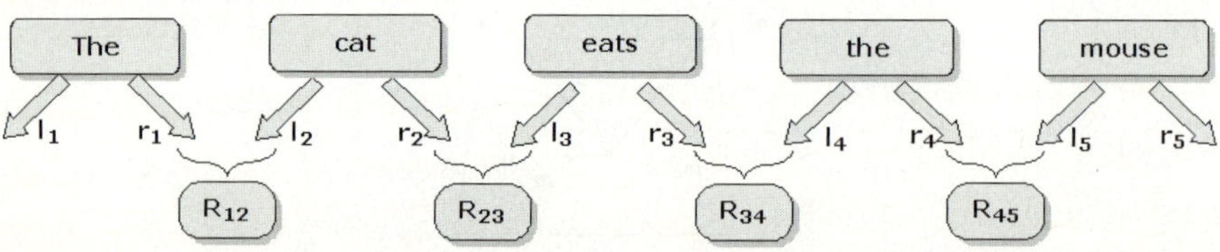

Fig. 3. Sequence of rotors associated to a sentence

be the orthogonal part of the vector related to the rotated basis, with respect to the original basis. This allows to obtain a coding which is independent of the sentence length. It is important to point out that cyclical coding should not appear if the dimension of the semantic space is higher than the number of rotations associated to the sentence. An analogy can be outlined between the proposed model and a state transition system. The initial state is represented by the k-dimensional orthogonal base given by the unit matrix, while the state transition is generated by the application of rotors associated to the bigrams in the sentence. The intermediate states are coded as rotated basis resulting from the application of the rotors. The final matrix represents a synthesis of the words sequence history within the sentence and corresponds to its sub-symbolic coding(see Fig.4).

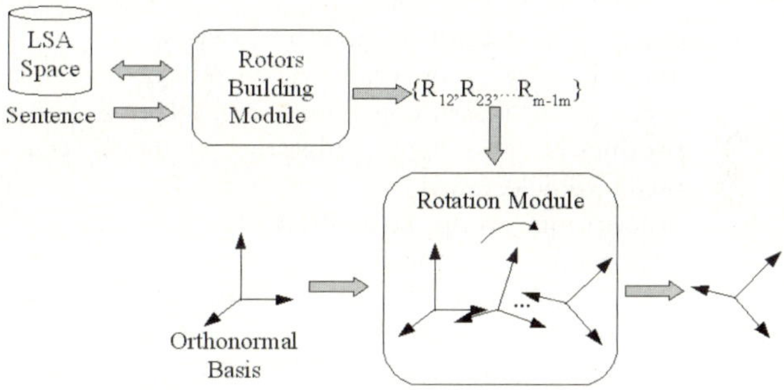

Fig. 4. Rotation of the orthonormal basis associated to a sentence

It is easy to show that this kind of coding satisfies the requisites reported in the introduction. The obtained representation takes into account the semantics of the words composing it because the rotors are defined in a semantic space generated by using LSA. The substitution of one or more words with their corresponding synonyms does not affect too much the coding because synonyms will have similar contexts. Besides, thanks to the non-commutability property of rotation, the final coding of the sentence will be a function of the words sequence into the sentence. The dimensionality of the coding can be considered enough high since each sentence is coded with a k^2 vector. Finally, the coding can be assumed independent of the number of words belonging to the sentence since we take into account only the orthogonal part of the vector obtained at the end of the procedure.

5 Experimental Results

5.1 Dataset

The proposed algorithm has been applied to a questions classification problem. A corpus of labeled questions has been employed. The questions are labeled by means of a hierarchical classifier, guided by a layered semantic hierarchy of expected answer types extracted from the TREC corpus[3][4]. There are two available class layers, a coarse-grained layer and a fine-grained layer, as shown in the taxonomy reported in Table1. The questions are splitted in two datasets: a set of training questions and a set of test questions. The test set is given by 500 questions belonging to the TREC10 corpus. All 5500 training questions are splitted in 5 training sets including respectively a number of 1000, 2000, 3000, 4000 and 5500 questions.

Table 1. Question Taxonomy in the UIUC corpus

Coarse Classes	Fine Classes
ABBR	abbreviation, expansion
DESC	definition, description, manner, reason
ENTY	animal, body, color, creation, currency, disease/medical, event, food, instrument, language, letter, other,plant, product,religion, sport, substance, symbol, technique, term, vehicle, word
HUM	description, group, individual, title
LOC	city, country, mountain, other, state
NUM	code, count, date, distance, money, order, percent, period, speed, temperature, size, weight

5.2 Classification Results

Two different semantic spaces have been created in order to compare the proposed approach with the traditional LSA methodology.

The first space has been obtained through the creation of a word-documents co-occurrences matrix of (9610 x 1066) dimension, followed by the classical TSVD with $k = 100$. The training and test sentences have been coded, according to the *folding-in* approach, as vectors calculated as weighted sum of the vectors associated to each word composing the sentence.

Table 2. Precision obtained on the coarse grained classified datasets

Algorithm	**1000**	**2000**	**3000**	**4000**	**5500**
LSA	69.6%	72.4%	75.6%	74.6%	75%
Proposed Approach	73.4%	78.2%	77.8%	80.2%	82.4%

Table 3. Precision obtained on the fine grained classified datasets

Algorithm	**1000**	**2000**	**3000**	**4000**	**5500**
LSA	57.2%	62%	62.8%	52.6%	66%
Proposed Approach	59.8%	67%	67.2%	68.2%	70.4%

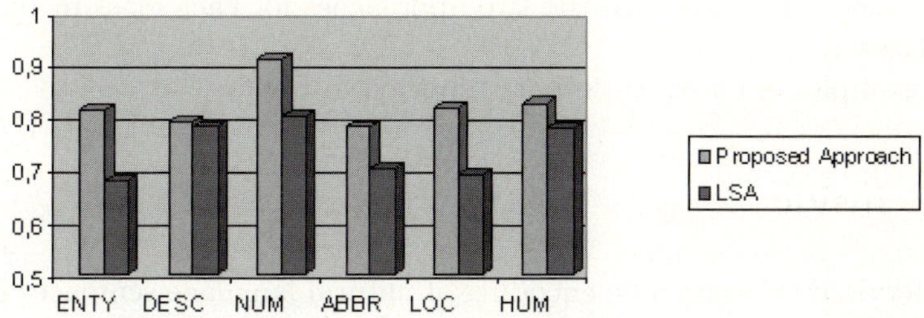

Fig. 5. Precision values obtained with LSA and proposed approach for the coarse grained classes (dataset: 5500)

Test Question	Selected Training Question	
	Proposed Approach	LSA
What is the abbreviation for Texas? **(Abb-abb)**	What is the abbreviation for micro? **(Abb-abb)**	CNN is the abbreviation for what? **(Abb-exp)**
Where is the Eiffel Tower? **(Loc)**	Where is the Danube? **(Loc)**	How do you get to the top of the Eiffel Tower? **(Desc)**
What is the percentage of water content in the human body? **(Num-perc)**	What percentage of the body is muscle ? **(Num-perc)**	The corpus callosum is in what part of the body? **(Enty-Body)**
What American composer wrote the music for "West Side Story "? **(Hum-Ind)**	What American poet wrote " Good fences make good neighbors "? **(Hum-Ind)**	What was the reason for the partition of the Anglican and Vatican churches? **(Desc-reason)**

Fig. 6. Some examples of sentences classified by the LSA and the proposed approach

The second space has been obtained building a word-word co-occurrences matrix of dimension (9610 x 9610). This matrix has been preprocessed in order to minimize the Hellinger distance[16] through the TSVD with $k = 100$. The training and test sentences have been coded as described in Sect.4. The size of the bigrams windows has been experimentally fixed to 8.

In both approaches all words, including stopwords have been considered in order to take into account the syntactic structure of the sentences.

The nearest neighbors algorithm has been used to perform the classification task. Let N_c the number of correctly classified questions, and N the total number of questions; for each class has been evaluated the precision measure given by:

$$Precision = \frac{N_c}{N}.$$

Preliminary results, evaluated for both the coarse-grained and the fine-grained datasets are shown in the following tables. Figure 5 shows a comparison of

Precision values obtained with the two approaches for each class in the coarse-grained dataset.

Some examples of classification are reported in Fig.6.

6 Conclusion

An unsupervised sub-symbolic encoding of natural language sentences has been presented. The procedure tries to overcome the limits of traditional Latent Semantic Analysis approach, which does not take into account the order in which words appear in a sentence. The proposed methodology starts from an LSA-based semantic space and associates to each bigram of a sentence a rotor defined according to the geometric algebra framework. The sequence of words in the phrase is encoded as a succession of rotors applied to an orthonormal basis. At the end of the process, the basis is interpreted as a vector and its orthogonal part constitutes the sub-symbolic coding of the sentence. The methodology has been tested on a classification task. In particular a corpus of questions labeled according to a predefined set of categories has been employed. The obtained results have been compared to the traditional LSA methodology approach and show the effectiveness of the proposed solution. Future works will regard a deeper exploration and validation of the methodology, with particular regard on the exploration of the meaning of the rotation planes defined by rotors as well as the rotation angle.

Acknowledgments. This work has been partially funded through the Programma di Rilevante Interesse Nazionale (PRIN) 2005, contract no. 2005103830-002, entitled: "Artificial Intelligence Techniques for Processing, Analysis, Preservation and Retrieval of Spoken Natural Language Archives".

References

1. Landauer Thomas, K., Dumais, S.T.: A Solution to Plato's Problem: The Latent Semantic Analysis Theory of the Acquisition, Induction, and Representation of Knowledge. Psychological Review 104(2), 211–240 (1997)
2. Haugeland, J.: Understanding Natural Language. The Journal of Philosophy. Seventy-Sixth Annual Meeting of the American Philosophical Association, Eastern Division 76(11), 619–632 (1979)
3. Li, X., Roth, D.: Learning Question Classifiers. In: Proceedings of the 19th International Conference on Computational Linguistics (COLING'02) (2002)
4. `http://l2r.cs.uiuc.edu/cogcomp/Data/QA/QC/`
5. Cavnar, W.B., Trenkle, J.M.: N-Gram-Based Text Categorization. In: Proceedings of the SDAIR-94, 3rd Annual Symposium on Document Analysis and Information Retrieval, pp. 161–169 (1994)
6. Croft, W.B., Lafferty, J.: Language Modeling for Information Retrieval. Kluwer Academic Publishers, Dordrecht (2003)

7. Liu, N., Zhang, B., Yan, J., Chen, Z., Liu, W., Bai, F., Chien, L.: Text Representation: From Vector to Tensor. In: Proceedings of the Fifth IEEE international Conference on Data Mining (2005). ICDM, pp. 725–728. IEEE Computer Society Press, Washington, DC (2005), http://dx.doi.org/10.1109/ICDM.2005.144
8. Madsen, R.E.: Modeling Text using State Space Model. Technical Report (2004), http://www2.imm.dtu.dk/pubdb/p.php?3998
9. Zhang, D., Lee, W.S.: Question classification using support vector machines. Research and development in information retrieval (2003)
10. Nguyen, M.L., Shimazu, A., Nguyen, T.T.: Subtree mining for question classification problem. Twentieth International Joint Conference on Artificial Intelligence (IJCAI), Hyderabad, India (January 6-12, 2007)
11. Kanejiya, D., Kumar, A., Prasad, S.: Automatic evaluation of students answers using syntactically enhanced LSA. In: Proceedings of the Human Language Technology Conference (HLT-NAACL 2003), Workshop on Building Educational Applications using NLP (2003)
12. Wiemer-Hastings, P., Zipitria, I.: Rules for Syntax, Vectors for Semantics. In: Proceedings of the 23rd Annual Conference of the Cognitive Science Society, Edinburgh (2001)
13. Dennis, S.: Introducing word order in an LSA framework. In: Landauer, T., McNamara, D., Dennis, S., Kintsch, W. (eds.) Handbook of Latent Semantic Analysis, Erlbaum (2006)
14. Doucet, A., Ahonen-Myka, H.: Non-Contiguous Word Sequences for Information Retrieval. Second ACL Workshop on Multiword Expressions: Integrating Processing(July 88–95, 2004)
15. Li, Y., McLean, D., Bandar, Z.A., O'Shea, J.D., Crockett, K.: Sentence Similarity Based on Semantic Nets and Corpus Statistics. IEEE Transactions on Knowledge and Data Engineering 18(8), 1138–1150 (2006)
16. Agostaro, F., Pilato, G., Vassallo, G., Gaglio, S.: A Sub-Symbolic Approach to Word Modelling for Domain Specific Speech Recognition. In: Proceedings of IEEE CAMP. International Workshop on Computer Architecture for Machine Perception, pp. 321–326 (2005)
17. Lounesto, P.: Clifford Algebra and Spinors. Cambridge University Press, Cambridge (1997)
18. Schoute, P.H.: Mehrdimensionale Geometrie. Leipzig: G.J.Gschensche Verlagsha 1 (Sammlung Schubert XXXV): Die linearen Rume, 1902. 2 (Sammlung Schubert XXXVI): Die Polytope (1905)

Neural Network Models for Abduction Problems Solving

Viorel Ariton[1] and Doinita Ariton[2]

[1] "Danubius" University, Lunca Siretului no. 3,
800416, Galati, Romania
variton@univ-danubius.ro
[2] "Dunarea de Jos" University, Domneasca no. 47,
800001, Galati, Romania
dariton@ugal.ro

Abstract. Due to its' connectionist nature, abductive reasoning may get neural network implementations that yet require structure adaptation to the abduction problems which Bylander and the team asserted. The paper proposes neural models for all known abduction problems, in a really unified manner, and with a sound and straightforward embedding in the existing neural network paradigms.

1 Introduction

Reasoning deals with discrete concepts and causal relations between them, apparently observed in the real world. Deductive reasoning asserts effects from causes – so, useful in control, abductive reasoning asserts causes from effects – useful in diagnosis.

In Fault Diagnosis, causes are faults and manifestations are effects; a fault evokes many manifestations and the same manifestation is evoked by many faults. The many-to-many relations between faults and manifestations easily lead to connectionist models, suited for ANN implementations. Many approaches use ANN for their recognition facilities but few refer to them suited for abduction problems solving.

Bylander et al. [3] reveal four categories of abduction problems in diagnosis:

- *independent* abduction problems - no interaction exists between causes;
- *monotonic* abduction problems - an effect appears at cumulative causes;
- *incompatibility* abduction problems – pair of causes are mutually exclusive;
- *cancellation* abduction problems – pair of causes cancel some effects, otherwise explained separately by one of them.

Ayeb et al. [2] have a sound approach in the neural network modelling of the abduction problems, also they introduce a fifth category:

- *open* abduction problems - when observations consist of three sets: present, absent and unknown observations.

The following approach is based on [1], where the abduction consists in sequentially applying *plausibility criteria* – to obtain the set of causes possibly evoked by the set of present effects, then *relevance criteria* – to obtain the minimum cardinality subset of causes (the "parsimonious principle" as in [5]) but also other

B. Apolloni et al. (Eds.): KES 2007/ WIRN 2007, Part I, LNAI 4692, pp. 52–59, 2007.

restrictions coming from the running context (e.g. frequency, reliability). In the ANN model, the plausibility criteria are the "excitatory" links from effects to causes, and the relevance criteria are competitions – each induce each by a given restriction.

The present paper only deals with the plausibility criteria, proposing neural network models for each abduction problem above, in a unified and simple approach.

2 New Connectionist Approaches for Abduction

According to [7], the Peng and Reggia connectionist approach hardly can cover the best explanation using competition-based networks – except for small and simple abduction problems. On the other side, Hebb's rule employed by Wang and Ayeb in [6] for learning the right set of connection weights is very sensitive to the initial conditions of the network. Goel and Ramanujam try in [4] to solve some of the abduction problems (regarding associativity and incompatibility interactions) using Hopfield architectures.

Ayeb et al. present in [2] specific a unified connectionist model for the abduction problems solving. They introduce excitatory and inhibitory links between effects and causes (each established individually), competition between weights, an additional layer for the additive interactions between causes – in case of monotonic abduction problems, competition between causes sharing the same effects. The connectionist approach is quite complicated, proposing different solutions, each specific to a certain abduction problem; the only unified approach is the use of the neural mechanisms: excitatory and inhibitory links, competition and new layers introduced between the effects and causes layers.

The present approach is much simpler and offers a real unified solution for all the abduction problems; each solution uses neural network models regarding the links between effects and causes and between causes, representing the plausibility as in [1]. The input function of the cause-neuron (on output layer) in a common ANN paradigm, get a neural structure with some "logical overload" that embeds the deep knowledge of the human diagnostician on specific effects and causes as from real context.

2.1 Plausibility Involves Logical Relations Between Causes and Effects

Connectionist models for the abduction problems should take into account that effects and causes enter logical pre-processing, e.g. a conjunction of effects and causes when evoking a fault – logical AND, or negation between causes – when they are mutually exclusive. Plausibility criteria refer to effects-to-causes and cause-to-cause links, as logical OR, AND, NOT that will affect the input function of the neuron.

The neuron is a processing element that performs numerical processing as below:

$$F_i = f(\sum_{j=1}^{|M|} w_{ij} \cdot M_j + \theta_i) \tag{1}$$

The cause/fault neuron F_i is fired according to the activation function f, which argument is the input function of the neuron – a cumulative action from each effect /

manifestation M_j. In terms of diagnosis, a manifestation M_j from the set $\boldsymbol{M}$ (with $|M|$ the cardinality) evokes in a specific measure (i.e. weight) w_{ji} the plausible fault F_i, while the last result from a cumulative effect of all other manifestations, eventually, surpassing the threshold θ_i sand obeying the activation function f. The activation values of F_i and M_j falls in [0, 1] so, they may get logical (qualitative) meanings; e.g. if M_j equals 0 then the effect is absent, if 1 – it is certain, while 0.5 is the doubt level.

Natively, the input function's processing is a kind of logical OR: each input M_j contributes (in a weighted degree) to the neuron's activation - see the argument of f in (1). How should it look the logical AND or the logical NOT in a similar approach?

To proceed to a solution, the input function is considered a separated processing, in the so called "site" – see SNNS [8]. Primarily, the site processing is the cumulative action in (1) – associated to logical OR; logical AND and NOT are performed according to the logical overloaded sites proposed in Fig. 1. so, acting also as gates.

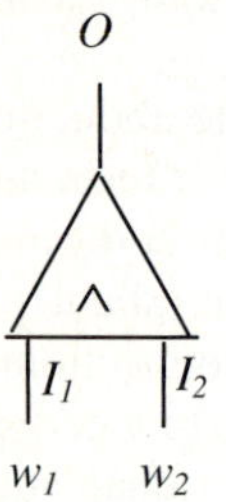

Conjunctive site (∧) AND

O	$I_1 \le w_1/2$	$I_1 > w_1/2$
$I_2 \le w_2/2$	0	0
$I_2 > w_2/2$	0	$I_1 + I_2$

Truth table for inputs I_1, I_2

a)

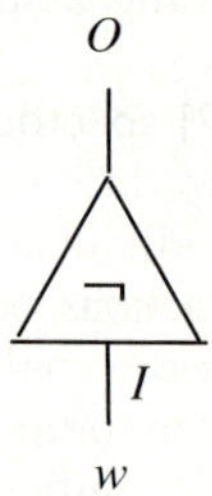

Negation site (¬) NOT

O	$I \le w/2$	$I > w/2$
	$w\text{-}I$	$w\text{-}I$

Truth table for input I

b)

Fig. 1. Logical AND (a) and logical NOT (b), as neural sites / input functions

The activation of a manifestation-neuron M_j gets a logical meaning depending on the range of numerical values it may fall: when in [0. 0.5] it is "not important", when in [0.5, 1] it is "important", but it passes through a weighted link to the input I, hence:

$$\textit{if } I > w/2 \textit{ then } I = \text{"important"} \textit{ else } I = \text{"not important"}. \quad (2)$$

So, the input function gets a "logical overload", and the site acts as a logical gate – see Fig.1 a for the conjunctive site (logical AND). Manifestation-neurons attack the inputs I_1 and I_2, and enter the fault neuron in a cumulative way only when both inputs are "important"; otherwise, the site is a blocking gate (the site output $O = 0$).

The two logical processing AND, NOT get truth tables as in Fig. 1., provided logical OR is the simple (native) cumulative processing for the site, as follows.

Disjunctive aggregation, performed by the "*disjunctive site*" through the default cumulative processing, i.e. all m inputs simply cumulate their activation I_j:

$$O = \sum_{j=1}^{m} I_j \quad (3)$$

Conjunctive aggregation, performed by the "*conjunction site*", which output O obeys the truth table from Fig. 1 a, following the rule:

$$\text{if } I_1 > w_1/2 \text{ AND } I_2 > w_2/2 \text{ then } O = I_1 + I_2 \text{ else } O = 0 \quad (4)$$

Negation, performed by the "*negation site*". The output O is obtained from the input I according to Eq. 5, and the truth table in Figure 1b:

$$O = w - I \quad (5)$$

The logical overload of the input function, using the proposed sites, makes possible the interaction of effects and (plausible) causes, for each abduction problem.

2.2 Neural Models of Plausibility for the Abduction Problems

The ANN model for an abduction problem consists in a structure of neural sites performing the logical aggregation specific for effects and causes in concern (as M and F neurons in Fig. 2). Each structure is placed in the target ANN architecture according to the deep knowledge of the human diagnostician on effects and causes.

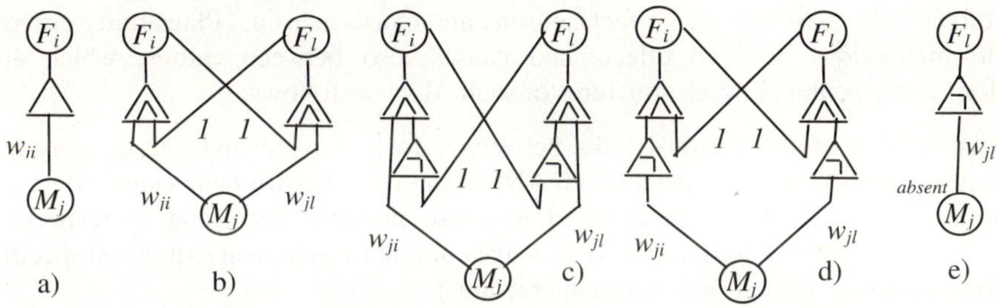

Fig. 2. Each abduction problem is solved by a neural structure of sites with logical overload

Each type of abduction problem is solved in Fig. 2. through a specific structure of neural sites, involving forward links from effects to causes and from causes to causes:

- For *independent* abduction problems – excitatory links apply directly from the effect M_j to the corresponding cause F_i (see Fig. 2 a). If there exists a conjunction grouping of effects to the cause, a conjunction site is provided at the input of the cause neuron. Note that by default, the neuron implements a disjunctive grouping of inputs (sum - Eq. 3), represented by the simple triangle.
- For *monotonic* abduction problems – the causes F_i and F_l both evoke the same effect M_j, hence they suffer conjunction with one-another and with the common effect through conjunction sites, as shown in Fig. 2 b, and expressed by the rule:

$$F_i \leftarrow F_l \text{ AND } M_j, \quad F_l \leftarrow F_i \text{ AND } M_j \quad (6)$$

- For *incompatibility* abduction problems – the pair F_i and F_l of causes are mutually exclusive (i.e. they are not both active in the same time), both evoking the same effect M_j. Each of them suffers conjunction with the negation of the other cause and with the common effect, as shown in Fig. 2 c, and expressed by the rule:

$$F_i \leftarrow \text{NOT } F_l \text{ AND } M_j, \quad F_l \leftarrow \text{NOT } F_i \text{ AND } M_j \quad (7)$$

- For *cancellation* abduction problems – the pair of causes F_i and F_l reduce the effect M_j when both occurred, although each of them evokes it separately. They suffer conjunctions as in Fig. 2 d, according to the following rule:

$$F_i \leftarrow F_l \ AND \ NOT \ M_j, \quad F_l \leftarrow F_i \ AND \ NOT \ M_j \tag{8}$$

- For *open* abduction problems – the main task is dealing with absent effects, so the cause F_i is activated if no effect M_j exists (Fig. 2 e), according to:

$$F_i \leftarrow NOT \ M_j \tag{9}$$

Links between cause-neurons in abduction problems of type b, c, d above, have all weights between cause neurons equal to 1 if they are symmetric (one to another), else they are set according to deep knowledge of the human expert.

2.3 Adding the Neural Models for Abduction to an ANN Paradigm

The neural models above may be used in any ANN paradigm in the diagnosis task, when direct links exist between effect-neurons and cause-neurons. Plausibility criteria refer to interactions between effects and causes also between causes, which are embedded in the neural network structures of such ANN, as follows:

- into weights of the forward links between effects evoked and causes – as the shallow knowledge obtained through ANN training on known pairs cause-effects;
- into neural sites structures attached to cause neurons, according to respective abduction problem – as deep the knowledge coming from human experts on specific effect-to-cause and cause-to-cause interactions;
- into threshold of the site – as deep knowledge from human experts (usually set to 0).

When building a neural network meant for diagnosis by abduction, one should be acquainted with the set of effects and causes – regarding their interactions as indicated by the deep knowledge. The chosen neural network paradigm is one with two layers – e.g. Adaline, which exhibits direct links between input and output neurons.

During the training phase, the gate functioning of the sites is disabled, i.e. the "logical overload" is not present and only the "classical" cumulative input function is running. The training procedure runs as usual, adapting weights of the links between (effect and cause) neurons so, the embedding of "shallow knowledge" takes place. Note that the interaction between effects and causes (specific to a given abduction problem) just happen that time and it is caught into the links between respective neurons of the ANN.

In the recall phase, the logical overload in sites will better reproduce the situation from the training phase (regarding the abduction problems) because, besides shallow knowledge, now the deep knowledge provided in the neural models of the abduction problems will contribute to the cause (output) neurons' activation.

3 Comparison to Other Approaches

The neural models above were added to the ANN Adaline and it was compared to [2] approach – the only one referring to all abduction problems. The approaches were used for the fault diagnosis in a simple hydraulic installation (see Fig. 3). First, some practical considerations then an experiment will be presented.

The installation in Fig. 3. comprises the Supply Unit (consisting in pump, tank and pressure valve), the Hydraulic Brake (control valve, brake's cylinder), and the Conveyor (control valve, self, the conveyor's cylinder). It present faults: 2 at the tank, 4 at the pump, 3 at the pressure valve, 2 at the pipes, 2×2 at the control valves, 2 at the damper (Drossel), 2×2 at the hydraulic cylinders.

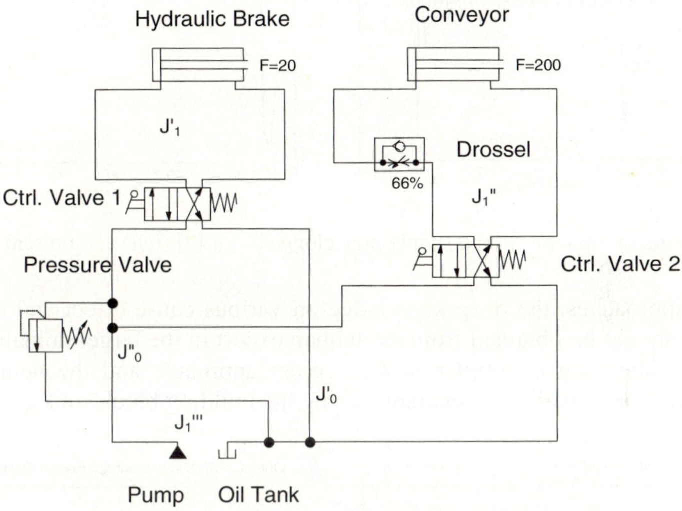

Fig. 3. Simple hydraulic installation under the fault diagnosis

The neural networks used for the experiment comprised the same number of neurons on the output layer for the 21 faults, and different numbers of neurons on the input layer: 48 observations which are also manifestations for the [2] approach, while in the present approach for a manifestations correspond 2 or 3 neurons, depending on the logical meaning attached: on/off or low/normal/high - respectively.

In the case of independent abduction problems, [2] introduces excitatory links from effects to causes and additional inhibitory links (competition) between causes sharing same effects, eventually freezing weights' values. In the present approach, competition is a relevance criterion (among others – beside the minimum cardinality). So, it offers a more flexible way on selecting the diagnostic. The minimum cardinality in [2] is always 1, while in the present approach multiple fault diagnosis is allowed.

For monotonic abduction problems, [2] introduces a third layer which combines the incompatibility abduction problems – because of the cause conjunction in the two cases. Here, some links skip the third layer (see the independent problems above), some enter the layer; so, the building of the network structure is non-homogenous and difficult. Moreover, the compromise between inhibitory and excitatory links for causes in conjunction, may lead to instability during the training (see Fig. 4 - left); Adaline ANN rapidly converges when effect-neurons attack specific cause-neurons.

The training procedure is different for different abduction problems in the [2] approach, and it involves also competition (between weights and cause neurons). In the present approach, the original ANN paradigm's training is kept for all neurons.

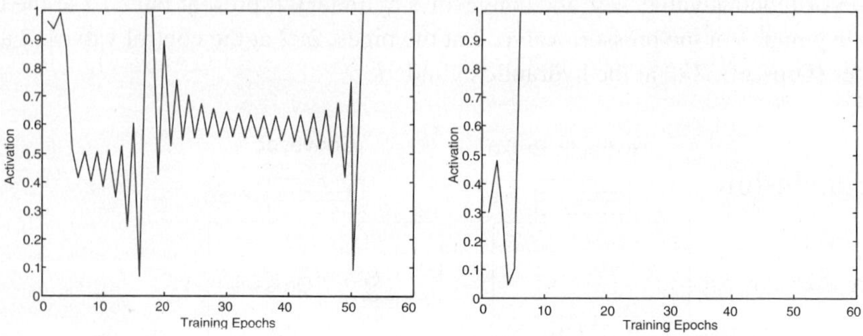

Fig. 4. Training of the fault 'Pump supply pipe clogged', for [2] (left) and present approaches

In both approaches, the deep knowledge on various cause-effect and cause-cause interactions should be obtained from the human expert in the target domain. The deep knowledge embedding is simpler in the present approach, and the neural network structure can be automatically generated, using the building blocks in Fig. 2.

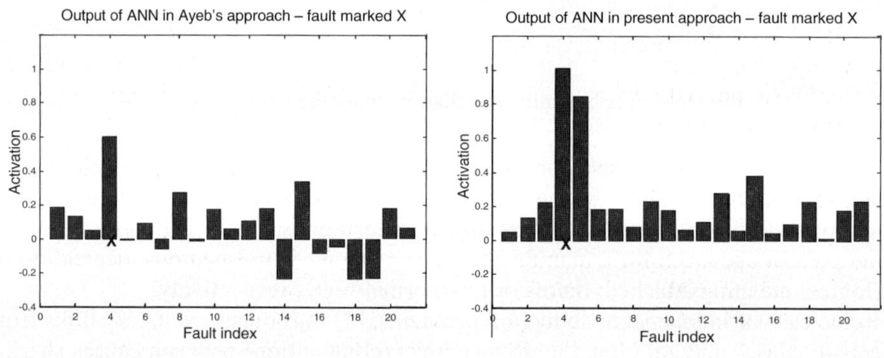

Fig. 5. Recognized fault 'Pump supply pipe clogged'(X), for [2] (left) and present approaches

In the recall phase the same patters of effects are applied to the inputs of the two neural networks for the 21 faults. In the case of the 4th fault ('Pump supply pipe clogged') the output of the two networks are depicted in Fig. 5. In the [2] approach (left), the activation of candidate faults and competition between them take place the same time, while in the present approach the 'plausibility criteria' (i.e. neural models for abduction) activate faults as in Fig. 5 right, then the 'relevance criteria' (i.e. multiple competitions) will later assert plausible and relevant causes (the 4th fault wins against 5th in Fig. 5).

The case study referred to a simulated behaviour of the target installation and included all types of abduction problems presented above. On the whole, all 21 faults get recognized after the recall phase – consisting in plausibility and relevance.

The reason for the good results is the way plausible faults get obtained, i.e. using the deep knowledge embedded in the proposed neural models for abduction problems. There are still faults, not included in the set of causes that may occur in the target hydraulic installation so, the open space of causes may induce errors in the diagnostic of the real installation.

4 Conclusion

The abductive reasoning may proceed by applying plausibility and relevance criteria, which may correspond – in a connectionist approach, to excitatory links between effects and causes, and to competition links between causes respectively. Plausibility criteria involve abduction problems that may occur in the frame of interactions between causes and effects, and they require a "logical overload" of the common artificial neural models (e.g. for causes in "conjunction" to an effect).

The paper proposes gate-sites for the input functions of the neurons – by means of a "logical overload" between inputs, and neural models suited for all known abduction problems – as structures of such gate-sites. The neural models are attached to the output neurons of a common ANN (two layer) paradigm – for example, ANN Adaline.

The neural models of abduction embed the deep knowledge of human diagnosticians on causes and effects interaction, so diagnostic is better than using only shallow knowledge embedded through a common ANN training. The present unified approach is sound and simple, and it may be used for various ANN paradigms – if direct links exist between effect and cause neurons.

References

1. Ariton, V., Ariton, D.: A General Approach for Diagnostic Problems Solving by Abduction. In: Proc. of IFAC-SAFEPROCESS, Budapest, Hungary, pp. 446–451 (2000)
2. Ayeb, B., Wang, S., Ge, J.: A Unified Model for Abduction-Based Reasoning. IEEE Trans. on Systems Man and Cybernetics - Part A: Systems and Humans 28(4), 408–424 (1998)
3. Bylander, T., Allemang, D., Tanner, M.C., Josephson, J.R.: The Computational Complexity of Abduction. Artificial Intelligence 49, 25–60 (1991)
4. Goel, A., Ramanujam, J.: A Neural Architecture for a Class of Abduction Problems. IEEE Transactions on Systems Man and Cybernetics 26(6), 854–860 (1996)
5. Peng, Y., Reggia, J.: Abductive Inference Models for Diagnostic Problem Solving. Springer, Heidelberg (1990)
6. Wang, S., Ayeb, B.: Diagnosis: Hypothetical Reasoning With A Competition-Based Neural Architecture. In: Proc. International Joint Conference on Neural Networks, vol. I, pp. 7–12 (1992)
7. Xu, Y., Zhang, C.: An improved Critical Diagnosis Reasoning Method. ICTAI, Toulouse, France, vol. 1, pp. 170–173 (1996)
8. Zell, A., Mache, N., Sommer, T., Korb, T.: SNNS – Neural Network Simulator, User Manual. University of Tuebingen (1991)

Online Training of Hierarchical RBF

Francesco Bellocchio[1], Stefano Ferrari[2], Vincenzo Piuri[2], and N. Alberto Borghese[1]

[1] Department of Computer Science, University of Milano, Italy
borghese@dsi.unimi.it

[2] Department of Information Technologies, University of Milano, Italy
{ferrari,piuri}@dti.unimi.it

Abstract. Efficient multi-scale manifold reconstruction from point clouds can be obtained through the Hierarchical Radial Basis Function (HRBF) network. An online training procedure for HRBF is here presented and applied to real-time surface reconstruction during a 3D scanning session. Results show that the online version compares well with the batch one.

Keywords: Online training, Hierarchical RBF, 3D scanner.

1 Introduction

Online learning is a widely diffused neural networks learning modality [1][2][3][4]. It is adopted in non stationary problems [5], and for real-time learning [6], such as to reconstruct a data manifold, while sampling from it. This second domain has interesting applications. For instance, the real-time reconstruction of the surface of an artifact, while it is being 3D scanned [7], would be of great help to drive the sampling procedure where the details are missing [8]. Up to now, methods based on splatting [9] have been mainly used for providing the perception of a continuous surface without giving its analytical description.

We propose here to reconstruct the 3D surface, as the output of Hierarchical Radial Basis Function (HRBF) network [10], introducing a new on-line training procedure, which can produce in real-time multi-scale reconstruction.

In Section 2 the batch version of the HRBF training procedure is reported, while the proposed online version is reported in Section 3. The algorithm has been implemented and challenged in real-time surface reconstruction problem. Results are reported in Section 4 and discussed in Section 5.

2 The HRBF Model

Let us assume that the manifold can be described as a $\mathbb{R}^D \to \mathbb{R}$ function. In this case, the input dataset is a height field: $\{(P_i, z_i) \,|\, z_i = S(P_i),\, P_i \in \mathbb{R}^D,\, 1 \leq i \leq N\}$, and the manifold will assume the explicit analytical shape: $z = S(P)$. The output of a HRBF network is obtained by adding the output of a pool

B. Apolloni et al. (Eds.): KES 2007/WIRN 2007, Part I, LNAI 4692, pp. 60–68, 2007.

of Radial Basis Function (RBF) networks, organized as a stack of hierarchical layers, each of which is characterized by a scale parameter, σ_l, with $\sigma_l > \sigma_l + 1$. If the units are equally spaced on a grid support and a normalized Gaussian function, $G(\cdot; \sigma) = \frac{1}{\sqrt{\pi\sigma^2}^D} \exp\left(-|| \cdot ||^2/\sigma^2\right)$, is taken as the basis function, the output of each layer can be written as a linear low-pass filter:

$$S(P) = \sum_{l=1}^{L} a_l(P; \sigma_l) = \sum_{l=1}^{L} \sum_{k=1}^{M_l} w_{l,\,k} G(||P - P_{l,\,k}||; \sigma_l) \tag{1}$$

where M_l is the number of Gaussian units of the l-th layer. The $G(\cdot)$ are equally spaced on a D-dimensional grid, which covers the input domain, that is the $\{P_{l,\,k}\}$s are positioned in the grid crossings of the l-th layer. The side of the grid, ΔP_l, is a function of σ_l: the smaller is σ_l, the shorter is ΔP_l, the denser are the Gaussians and the finer are the details which can be reconstructed.

The actual shape of the surface in (1) depends on a set of parameters: the number, $M = \sum_l M_l$, the scale ensemble, $\{\sigma_l\}$, the position, $\{P_{l,\,k}\}$, and the weights of the Gaussians, $\{w_{l,\,k}\}$. Each RBF grid, l, realizes a reconstruction of the surface up to a certain scale, determined by σ_l. Signal processing theory allows to set ΔP_l as $\sigma_l = 1.465\, \Delta P_l$ and to determine consequently M and the $\{P_{l,\,k}\}$ [10]. If only the l-th layer would be used, from the analogy between (1) and linear filtering theory, the weights $\{w_{l,\,k}\}$ can be as: $w_{l,\,k} = S(P_{l,\,k}) \cdot \Delta P_l^D$ [10]. As the data set usually does not include the $\{S(P_{l,\,k})\}$ (or they could be corrupted by noise), these values should be estimated.

A weighted average of the data points that lie in a neighborhood of $P_{l,\,k}$ can be used to estimate $S(P_{l,\,k})$. This neighborhood, called *receptive field*, $A(P_{l,\,k})$, can be chosen as a spherical region, with the radius proportional to ΔP_l. A possible weighting function, which is related to the Nadaraya-Watson estimator [11], is:

$$\tilde{S}(P_{l,\,k}) = \frac{n_{l,\,k}}{d_{l,\,k}} = \frac{\displaystyle\sum_{P_m \in A(P_{l,\,k})} S(P_m)\, e^{-||P_{l,\,k} - P_m||^2/\sigma_l^2}}{\displaystyle\sum_{P_m \in A(P_{l,\,k})} e^{-||P_{l,\,k} - P_m||^2/\sigma_l^2}} \tag{2}$$

Although a single layer with Gaussians of very small scale could reconstruct the finest details, this would produce an unnecessary dense packing of units in flat regions and an unreliable estimate of $\tilde{S}(P_{l,\,k})$ if too few points fall in $A(P_{l,\,k})$. A better solution is to adaptively allocate the Gaussian units, with an adequate scale in the different regions of the domain, by adding and configuring one layer at time, starting from the largest scale one, with $\sigma_l = 2\,\sigma_{l+1}$.

All the layers after the first one will be trained to approximate the residual, that is the difference between the original data and the actual output produced by the already configured layers. Hence, the residual, r_l, is computed as:

$$r_l(P_m) = r_{l-1}(P_m) - a_l(P_m) \quad \text{where} \quad r_0(P_m) = z_m \tag{3}$$

and it is used for estimating $\{w_{l,\,k}\}$ by substituting $S(P_m)$ in (2).

The L_1 norm of the local residual inside $A(P_{l,\,k})$, defined as:

$$R(P_{l,\,k}) = \frac{1}{|A(P_{l,\,k})|} \sum_{P_m \in A(P_{l,\,k})} |r_{l-1}(P_m)|. \tag{4}$$

may be used for evaluating the quality of the approximation of the Gaussian in $P_{l,\,k}$. This measure represents the *local residual error.*

When $R(P_{l,\,k})$ is over a given threshold, ϵ, the Gaussian is inserted: Gaussians at a smaller scales are inserted only in those regions where there are still some missing details. The introduction of new layers ends when the residual error is under threshold over the entire domain (uniform approximation).

As the Gaussian function decreases very fast to zero with the distance from its center, computational time can be saved by allowing each Gaussian to contribute to the residuals only for those points that belong to an appropriate neighborhood of the Gaussian center, $P_{l,\,k}$, called *influence region*, $I(P_{l,\,k})$.

This batch HRBF training procedure exploits the knowledge of the entire input dataset, and adopts local estimates to setup the network parameters with a fast configuration which can be parallelized, but that has to wait that all the data points are available.

3 Online Training Procedure

When the data set is not entirely known, but grows one point at a time, the schema described in Section 2 cannot be applied. In fact, let us assume that a HRBF has been already configured with a given data set, S_{old}. When a new point, P_{new} is sampled over the manifold, the estimate in (2) becomes out of date for all the units $(1,\,k)$ such that $P_{\text{new}} \in A(P_{1,k})$, and has to be estimated with the new data set, $S_{\text{old}} \cup P_{\text{new}}$. This modifies a_l inside the influence region of the updated units. As a consequence, the residual for the points that belong to this region changes, making out of date the weights for all those Gaussians of the second layer whose *receptive field* intersects with this region. This causes a chain-reaction that, at the end, may involve an important subset of the units of the HRBF network. Moreover, the need for a new layer can also occur.

If the computational power cannot sustain the updating of the network weights for every new input data, some approximations have to be accepted to obtain a real-time configuration.

The algorithm proposed here is based on updating the network parameters every Q points (with $Q << N$) to filter out possible outliers effects and collect a meaningful statistics. In the first phase, a few internal variables, associated to those Gaussians whose *receptive field* includes the collected points, are computed and the network weights are updated. In the second phase, the residual error, (4), is computed, and new Gaussians are inserted where it is over threshold. The two phases are iterated as far as data points are sampled.

3.1 Data Structures

For every layer, l, the input space is partitioned into squares, $\{C_{l,k}\}$, each centered in a different Gaussian center, $P_{l,k}$. The points, which belong to $C_{l,k}$ will be closer to $P_{l,k}$ than to any other Gaussian of the l-th layer. We therefore call $C_{l,k}$ the *close neighborhood* of the (l, k)-th Gaussian.

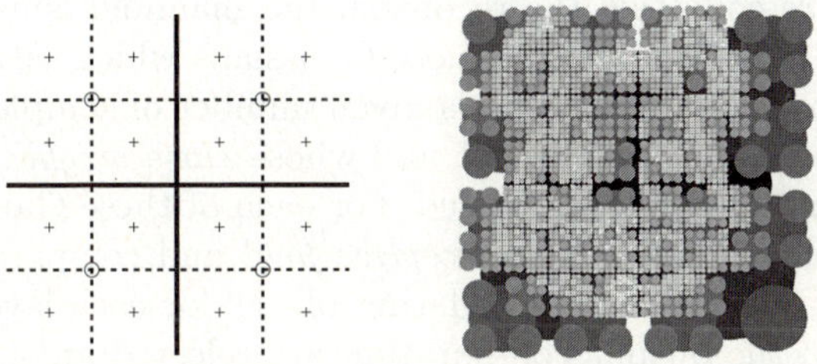

(a) (b)

Fig. 1. (a) The *close neighborhood* of each Gaussian ('o') is partitioned into four *close neighborhoods* of Gaussians of the next layer ('+'). (b) During the training, the partitioning is iterated adaptively. For sake of clarity, the *close neighborhood* of the Gaussians is depicted as a circle, having a diameter slightly smaller than the *close neighborhood* side. The circles changes their color from black to white depending on the layer (lower to higher) they belong to.

At the (l, k)-th Gaussian is associated a data structure containing: $P_{l,k}$, $\sigma_{l,k}$, the points inside $C_{l,k}$, its weight, $w_{l,k}$, the numerator, $n_{l,k}$, and the denominator, $d_{l,k}$, of (2). As $\Delta P_l = \frac{1}{2}\Delta P_{l-1}$, the *close neighborhood* of each Gaussian (*father*) of the l-th layer will be formed by the *close neighborhood* of the four corresponding Gaussians (*children*) of the $(l+1)$-th layer. This relationship, depicted in Fig. 1a, is used to organize in a quad-tree the Gaussians data: from each *father* Gaussian, it is easy to access to every *children* of the higher layers.

3.2 First Phase: Parameters Updating

When a new point, P_{new}, is acquired, the parameters $n_{1,k}$, $d_{1,k}$, and, hence, $w_{1,k}$ of the Gaussians of the first layer such that $P_{\text{new}} \in A(P_{1,k})$ are updated:

$$n_{1,k} := n_{1,k} + S(P_{\text{new}}) \cdot e^{-||P_{1,k} - P_{\text{new}}||^2/\sigma_1^2} \quad (5)$$

$$d_{1,k} := d_{1,k} + e^{-||P_{1,k} - P_{\text{new}}||^2/\sigma_1^2} \quad (6)$$

This procedure is iterated on the higher layers, considering for updating only the *children* of the updated Gaussians. According to (3), in the higher layers, $r_l(P_{\text{new}})$ is used instead of $S(P_{\text{new}})$. The computation of the r_l's is limited to P_{new}: the contribution of the old points to $n_{l,k}$ and $d_{l,k}$ is not modified.

Overall, the processing of P_{new} produces a modification of the network output in those regions covered by Gaussians whose *receptive field* contains P_{new}. However, only a small subset of the network units is involved, for each P_{new}.

After updating the weights, P_{new} is inserted into the data structure associated to the Gaussian such that $P_{\text{new}} \in C_{l,\,k}$.

3.3 Second Phase: Splitting

When Q points are collected, the reconstructed manifold is examined in correspondence of the *receptive fields* of those Gaussians which satisfies three conditions: they have no *children*, at least a given number of sampled points, K, was sampled inside its *close neighborhood*, and whose *close neighborhood* includes at least one of the last Q points processed. For each of these Gaussians, the residual error is computed inside their *receptive field* and compared with the error threshold, ϵ, as in (4). As for the updating phase, for each layer, the Gaussians that contribute to the output computation are selected among the *children* of the units considered in the previous layer. If the reconstruction is not satisfactory for the Gaussian $(l,\, k)$, four Gaussians, are inserted in the next layer, $l+1$, and the points collected by the Gaussian $(l,\, k)$ are distributed between the new Gaussians. The parameters of the new Gaussians ($n_{l,\,k}$, $d_{l,\,k}$, and $w_{l,\,k}$) are then set using the points contained in their *close neighborhood*.

3.4 Initialization

The only *a priori* information needed is the bounding box of the input space to be sampled. It is used to set up the first layer, which will be composed by only one Gaussian. Let B the maximum side length of the bounding box, σ_1 will be set as $\sigma_1 = 1.465\,B$, $P_{1,\,1}$ will be positioned at the center of the bounding box, and $C_{1,\,1}$, is defined as the square centered in $P_{1,\,1}$, having a side length of B.

The parameters of the online training algorithm, ϵ, Q, and K, may depend from the application, as they should be set proportional to the noise of the input data and to the computational power of the processing system.

4 Results

We applied the online HRBF model to 3D scanning, where the artifact's surface in Fig. 2a is reconstructed in real-time. The Autoscan system [7] is used here, which allows to sample for a longer time where more details are concentrated. The acquired 3D dataset is reported in Fig. 2b: it is composed of 32 000 points. Acquisition was stopped when the visual quality of the reconstructed surface (Fig. 2c-f) was judged sufficient and no significant improvement was observed adding new points (Fig.2e and Fig.2f). Alternatively the procedure could be stopped when no splitting occurs. In this example, $\epsilon = 0.8$, $Q = 100$, and $K = 9$.

As can be seen in Fig. 3a the number of units increases with the number of data points, improving the quality of the reconstructed model. However, comparing the number of allocated Gaussians (Table 1), with the maximum number

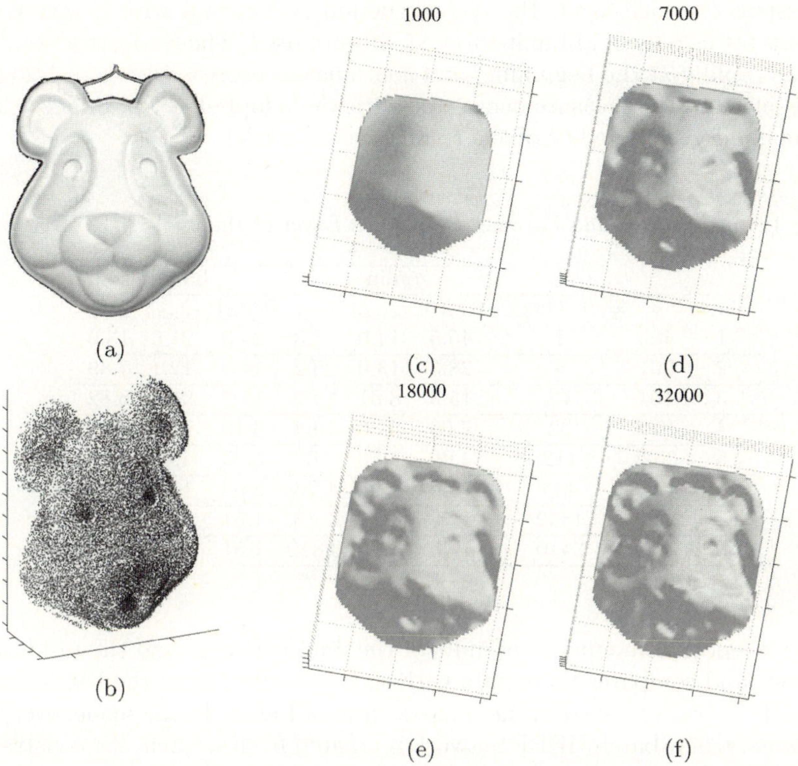

Fig. 2. A typical data set used to test the online training procedure. From the mask reported in (a), 32 000 points, which constitute the input data set (b), are sampled by the Autoscan system [7]. Notice the higher point density in the mouth and eyes regions. Panels (c), (d), (e), and (f) show the reconstruction after 1 000, 7 000, 18 000, and 32 000 points have been sampled.

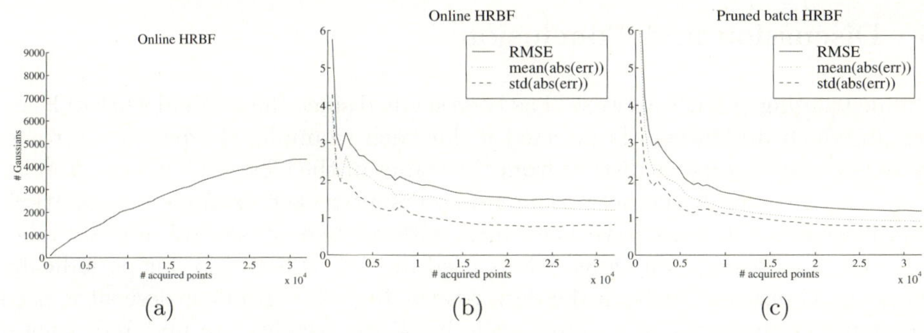

Fig. 3. Performance indexes of the batch HRBF model with respect to the number of processed points. The number of Gaussians that contribute to the reconstruction (i.e., those with an associated non-zero weight) for online HRBF (a). The batch HRBF has been pruned to match the same number of Gaussians. The error figures: RMSE, mean and standard deviation of the absolute error for the online, (b), and batch, (c), HRBF.

of Gaussians for each layer, the reconstruction is obtained with a very sparse Gaussians set as only 3,774 units over 87,381 are used. The reconstruction error decreases rapidly at the beginning, and continues to decrease down to 1.20 mm, which is close to the measurement error with the adopted setup, in accordance with the good visual quality of the reconstructed model.

Table 1. Performance and Parameters of Each Layer of the final HRBF Networks

			online			batch		
#layer	σ	#eff. Gauss.	RMSE	ϵ_{mean}	ϵ_{std}	RMSE	ϵ_{mean}	ϵ_{std}
1	403	1	45.5	14.0	43.3	25.3	21.6	13.0
2	201	4	28.3	13.0	25.2	14.5	12.1	7.89
3	101	12	15.3	8.67	12.7	11.7	9.56	6.82
4	50.4	32	7.09	4.30	5.64	9.16	7.47	5.31
5	25.2	112	4.09	2.70	3.07	5.24	4.02	3.36
6	12.6	405	2.60	1.94	1.73	2.68	1.97	1.81
7	6.29	1342	1.78	1.37	1.13	1.61	1.18	1.10
8	3.15	2410	1.45	1.20	0.810	1.51	1.15	0.986

Table 1 reports the number of units of the final networks and the residuals of each layer, and compares these data with those obtained from the batch version of a HRBF network applied to the same dataset of Fig. 2. To the scope, every 500 data points, a new batch HRBF network is created from scratch. We constrained the maximum number of layers and the scale parameters of the batch HRBF network to be equal to the corresponding parameters of the online version. The final error of the two HRBF models was quite comparable, 1.20 ± 0.810 mm versus 1.15 ± 0.986 mm. However, as expected, the batch version grown faster and converged faster to a smaller error than the on-line version (Fig 3(b)-(c)).

5 Discussion and Conclusion

Online training of RBF networks has been studied since the seminal work of Platt [12] in which a Gaussian is inserted if, for each example, the network error is large and the example is distant from the center of the closest Gaussian. A delta rule is used to adapt the networks parameters after each example presentation. The insertion criterions have been further developed by several authors (e.g., [13] and [3]). Besides, an aging rule is used in both [13] and [3] for periodically removing Gaussians far from the data points. In [13], a gradient descent is used to adapt the networks parameter, while in [3] the weights are updated using a recursive least square method. In [3], the dynamics of the RBF model trained by algorithms based on gradient descent is studied. Online HRBF shares with these methods the principle of inserting new Gaussians in regions poorly covered and where the local network error is over a given threshold. However, in the HRBF model only the weights are updated and it is operated locally with respect to the presented example. As no gradient descent based procedure is used in HRBF training, the results reported in [4] do not apply to the HRBF training algorithm.

The manifold height in $P_{l,k}$ (2) is estimated as the ratio between $n_{l,k}$ and $d_{l,k}$, which is obtained as a run-time sum of the values derived from each sampled point. However, this value is equal to that in the batch model only for the first layer as the residual is subject to changes as the points are sampled, and the contribution of the older points in the residual estimation may become unreliable as the training proceeds. However, due to the non-orthogonality of their basis functions, the HRBF model is able to compensate the error in one layer, with the approximation achieved by the next layer [14].

The choice of a suitable value for K may be critical as it can lead to an early splitting: if K is too small, the weight estimate of the new Gaussians may be unreliable because the contribution of the first points will not be changed. This effect could be mitigated by using an aging strategy for the estimated n and d. On the other hand, the more the points are processed, the more the weights of the lower layers become reliable. Hence, a late splitting involves a more accurated reconstruction. However, using a too large value of K, the Gaussians are inserted slowly, and the reconstruction details cannot became apparent.

Depending on the rate the example are available and by the computational power of the system, idle time may exploited for re-processing old points. This activity does not increase the input data points density and do not contribute to splitting, but should compensate the bad estimate operated in the early stages.

An online training procedure for the HRBF model is here presented, illustrated by an application to real-time surface reconstruction problem. The accuracy of the online procedure results comparable with that of the batch version. This online version can be used in all the domains of low dimensionality, where real-time manifolds approximation is required.

References

1. Saad, D. (ed.): On-Line Learning in Neural Networks. Cambr. Univ. Press, Cambridge (1998)
2. Bishop, C.M.: Neural Networks for Pattern Recognition. Oxford Univ. Press, New York (1995)
3. Alexandridis, A., Sarimveis, H., Bafas, G.: A new algorithm for online structure and parameter adaptation of RBF networks. Neural Networks 16(7), 1003–1017 (2003)
4. Freeman, J.A.S., D.S.: Online learning in radial basis function networks. Neural Computation 9(7), 1601–1622 (1997)
5. Rubaai, A., Kotaruand, R., Kankam, M.D.: Online training of parallel neural network estimators for control of induction motors. IEEE Trans. on Ind. Appl. 37(5), 1512–1521 (2001)
6. Fan, J., Dimitrova, N., Philomin, V.: Online face recognition system for videos based on modified probabilistic neural networks. In: Proc. of the 2004 Int. Conf. on Image Processing, 2004. ICIP '04, vol. 3, pp. 2019–2022 (2004)
7. Borghese, N.A., Ferrigno, G., Baroni, G., Ferrari, S., Savaré, R., Pedotti, A.: Autoscan: a flexible and portable 3D scanner. IEEE C. G. & A. 18(3), 38–41 (1998)

8. Rusinkiewicz, S., Hall-Holt, O., Levoy, M.: Real-time 3D model acquisition. In: Proc. of the 29th conf. on Comp. graph. and int. tech., pp. 438–446. ACM Press, New York (2002)
9. Rusinkiewicz, S., Levoy, M.: Qsplat: A multiresolution point rendering system for large meshes. In: Proc. of SIGGRAPH 2000, pp. 343–352 (2000), (ISBN 1-58113-208-5)
10. Ferrari, S., Frosio, I., Piuri, V., Borghese, N.A.: Automatic multiscale meshing through HRBF networks. IEEE Trans. on I. & M. 54(4), 1463–1470 (2005)
11. Girosi, F., Jones, M., Poggio, T.: Regularization theory and neural networks architectures. Neural Computation 7(2), 219–269 (1995)
12. Platt, J.: A resource-allocating network for function interpolation. Neural Computation 3(2), 213–225 (1991)
13. Fritzke, B.: Growing cell structures — A self-organizing network for unsupervised and supervised learning. Neural Networks 7(9), 1441–1460 (1994)
14. Ferrari, S., Borghese, N.A., Piuri, V.: Multiscale models for data processing: an experimental sensitivity analysis. IEEE Trans. on I. & M. 50(4), 995–1002 (2001)

Selecting Features by Learning Markov Blankets

Antonino Freno*

Dipartimento di Ingegneria dell'Informazione,
Università degli Studi di Siena
Via Roma 56, I-53100 Siena, Italy
freno@unisi.it

Abstract. In this paper I propose a novel feature selection technique based on Bayesian networks. The main idea is to exploit the conditional independencies entailed by Bayesian networks in order to discard features that are not directly relevant for classification tasks. An algorithm for learning Bayesian networks and its use in feature selection are illustrated. The advantages of this algorithm with respect to other ones are then discussed. Finally, experimental results are offered which confirm the reliability of the algorithm.

Keywords: Feature selection, Bayesian networks, learning.

1 Introduction

Typical supervised classification tasks involve representing the patterns in a dataset as vectors of features; algorithms that learn to classify patterns are then applied to such vectors. In complex domains, the number of features can be very large [1], and the vector size has an important impact on the performance of the algorithms, for at least two reasons. First, high-dimensional vectors increase both the amount of space and the time required for learning. Second, when the vectors also include irrelevant variables, the presence of such variables may degrade the accuracy of the learned classification model. Therefore, reducing the dimensionality of vectors may improve the performance of a learning algorithm in two respects. First, it can decrease the complexity of the learning process. Second, it can lead the learning algorithm to induce more reliable models of the domain at hand. One way of addressing dimensionality reduction in vectors of features is through *feature selection* [2].

Feature selection techniques can be classified into two broad families [3,4]: *filters* and *wrappers*. The distinction between filters and wrappers depends on the relationship they bear to the classification algorithm to be applied to the task at hand. Filters select features based on criteria which are independent of the particular learning algorithm to be applied to the data. Hence, the bias guiding

* I am grateful to Edmondo Trentin, Ernesto Di Iorio, and Marco Gori for providing suggestions and comments concerning this work.

B. Apolloni et al. (Eds.): KES 2007/WIRN 2007, Part I, LNAI 4692, pp. 69–76, 2007.

feature selection in filters is not related to the learning bias of the classifier. In wrappers, a learning algorithm is first chosen, and then the classification accuracy on different subsets of features is used as a heuristic for selecting the best subset.

Wrapper methods have both a practical shortcoming and a theoretical limitation [5]. On the one hand, they typically incur a high computational cost, since a classifier is required to be trained several times on the same task, with different subsets of features. On the other hand, they do not permit to *investigate* the problem of dimensionality reduction as such. Since features are selected only insofar as classification accuracy is increased, wrappers obscure the nature of the interaction between dimensionality reduction and learning accuracy. While the first limitation of wrapper techniques can be overcome by devising more efficient algorithms [2], the second limitation is more general, since it depends on the very nature of these techniques.

In this paper I propose a filter method. The basic idea consists in modeling the features, together with the label attached to the vector, as a set of (discrete) random variables, and then learning the joint probability distribution of the variables in the form of a Bayesian network (BN). Once a BN has been learned, the independencies entailed by the BN can be used to select a subset S of variables such that S makes the label independent of all other variables. The set S is referred to as a *Markov blanket* (MB) of the class variable. Therefore, within the framework of BNs, feature selection can be reduced to the task of extracting a MB for the class from the set containing all the original variables. I will call the algorithm 'Markov Blanket Filter' (MBF).

The paper is organized as follows: Section 2 reviews some fundamentals of BNs; Section 3 describes MBF; Section 4 discusses the advantages of MBF over a couple of similar algorithms [5,6]; Section 5 provides experimental results concerning MBF; Section 6 outlines a direction for future work.

2 Bayesian Networks as Filters

BNs [7,8] represent joint probability distributions underlying sets of random variables. BNs are made up of two components: a directed acyclic graph (DAG), and a set of conditional probability tables (CPTs). Each node in the DAG represents a variable, and for each node there is a CPT specifying the conditional probability distribution of the variable given (each possible combination of) the values of its parents in the DAG. The main advantage of using BNs is that, while the space complexity of a direct representation of a joint probability distribution grows exponentially with the number of variables, the space complexity of a BN grows only linearly with that number [9].

In order to derive a joint probability distribution from a BN, one has to make the Markov assumption, according to which each variable is independent of its non-descendants (in the DAG) given the values of its parents. Consider the random variables $X_1, \ldots, X_n$, and let PA_i denote the set of parents of X_i. The Markov assumption entails the following equality:

$$P(X_1, \ldots, X_n) = \prod_{i=1}^{n} P(X_i | PA_i) \tag{1}$$

Since the local distributions $P(X_i|PA_i)$ are provided by the CPTs, (1) specifies how to compute a joint probability distribution from a set of CPTs.

Let the notation $I(X, Y|Z)$ mean that X is independent of Y given Z. Then, if V denotes a set of random variables $X_1, \ldots, X_n$, a MB of X_i in V is any subset S of V such that $I(X_i, V - S - \{X_i\}|S)$. An interesting property of BNs is that, for any variable X, the set containing the parents of X, the children of X, and the parents of its children is a MB of X [7]. An example is given in Fig. 1.

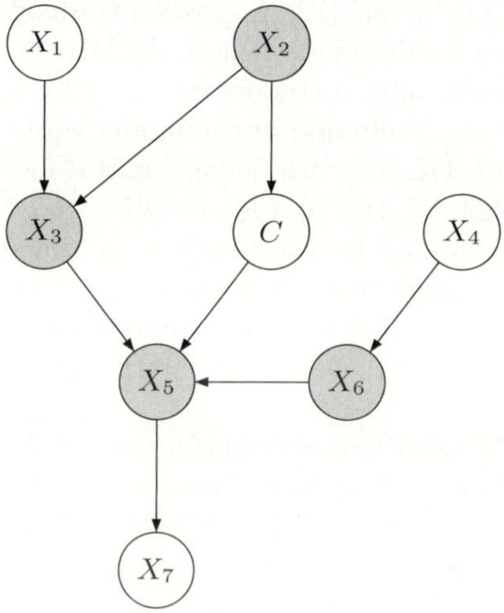

Fig. 1. The DAG of a BN representing the variables $X_1, \ldots, X_7, C$. The MB of C is denoted by shaded nodes. In this BN, $I(C, \{X_1, X_4, X_7\}|\{X_2, X_3, X_5, X_6\})$.

Now, consider the traditional task of supervised learning, where the training data consists of feature vectors $\mathbf{X} = (X_1, \ldots, X_n)$ such that the label C is known. Clearly, each feature X can be regarded as a random variable ranging over a fixed domain of possible values. The same is true for the label C, which can be treated as a random variable ranging over the set of possible labels $c_1, \ldots, c_m$. If we take as training data the set of vectors $\mathbf{X}' = (X_1, \ldots, X_n, C)$, we may reduce the problem of feature selection to the following one: what is the MB of C in $\{X_1, \ldots, X_n\}$? Since a BN implicitly specifies a MB for every variable contained in the DAG, the latter problem can be solved by learning a BN for the probability distribution underlying the data. In practice, any algorithm capable of learning the *structure* of BNs (not only the CPTs for a fixed DAG, but also the DAG itself) can be used as a feature selection algorithm.

3 A Feature Selection Algorithm

Structure learning in BNs has been investigated extensively in the last decade [10,8]. The learning algorithm we employ performs heuristic search in a space of candidate models [8], where each model is obtained by learning the CPTs for a given DAG. The search is aimed at finding the BN h that maximizes the heuristic function $score(h)$, defined as follows [11]:

$$score(h) = \log P(D|h) - \frac{k}{2} \log N \tag{2}$$

where D is the dataset, k is the number of parameters specified in the BN, and N is the size of the dataset. This heuristic, which is based on the Minimum Description Length (MDL) principle [12], imposes a tradeoff between the likelihood of a candidate model and its simplicity, where the latter is measured by the number k of independent probability estimates required for learning the CPTs. The idea is to maximize the log-likelihood of the model while keeping its complexity as low as possible. The MDL heuristic is very useful for approximating the full Bayesian score of a model, i.e. the sum $\log P(D|h) + \log P(h)$. In particular, the complexity penalty determined by the heuristic is aimed at approximating the prior probability of the candidate models when no estimate of priors is available. This heuristic is much more effective than a maximum likelihood heuristic that takes into account only $\log P(D|h)$, for the latter typically leads to models that overfit the dataset.

Computing the likelihood of h is straightforward. If $D = \{\mathbf{x}_1, \ldots, \mathbf{x}_m\}$, where $\mathbf{x}_j = (x_{1_j}, \ldots, x_{n_j})$, then (by assuming that the patterns in the dataset are mutually independent given h) we can derive the following equality:

$$P(D|h) = \prod_{j=1}^{m} \prod_{i=1}^{n} P(x_{i_j}|pa_{i_j}^h) \tag{3}$$

where pa_i^h denotes the values of the parents of X_i in h.

In order to find the model with the highest MDL score, we use a hill-climbing algorithm [8]. The algorithm starts from a BN h, and generates a set of 'neighbors' of h, where a neighbor is a BN obtained by operating on the DAG in h in one of three ways: (1) adding one arc, (2) removing one arc, or (3) reversing one arc. Once a new DAG has been constructed, a corresponding BN is obtained by learning the CPTs for that DAG. (Cyclic graphs are simply discarded). The neighbors of h are scored, and the highest-scoring neighbor h' is compared to h. If h' scores better than h, the whole cycle is iterated for h', and the process continues until no neighbor of the currently accepted BN improves on the score of that BN. Since the search algorithm is a greedy one, we run it several times, starting each time from a different, randomly generated DAG, so as to reduce the risk of getting stuck in local maxima. (In order to speed up the search, a tabu list can be used to record the states explored at each cycle, so as to prevent the algorithm from visiting the same states several times).

To address the feature selection task, we simply proceed as follows: we learn a BN h for the dataset by running the iterated hill-climbing algorithm, and then we pick out the features belonging to the MB of the class variable.

4 Discussion

The idea of applying the notion of MB to feature selection was first proposed in [5]. The algorithm formulated in [5] requires the user to specify in advance some important parameters, in particular the number of features that have to be eliminated. But since in typical applications the most sensible value for that number is unkown, this limitation either leads to suboptimal results (when the user is not able to choose a sensible number), or makes it necessary to run the algorithm several times (setting that number each time to a different value, until good results are obtained). On the other hand, MBF does not require the user to set any parameters at all when it is applied to new data.

In [6], feature selection is also dealt with by learning BNs. However, the learning algorithm employed in [6], called 'K2χ^2', has two serious drawbacks. K2χ^2 is based on K2 [13], which requires the user to specify two parameters: (1) an ordering over the variables such that, if variable X_i precedes variable X_j in the given ordering, an arc from X_j to X_i will not be allowed in the resulting BN; (2) an upper bound on the number of parents a node is allowed to have in the resulting BN. Concerning the first parameter, K2χ^2 orders the variables according to the strength of the dependence relationship between each one of them and the class variable, measuring this dependence through the χ^2 test. Thus, variables with higher χ^2 scores precede variables with lower scores. While it is shown in [6] that the resulting ordering works better than a random ordering, no connection is established between the χ^2 test and the semantics of the node ordering in BNs. This leaves room for some doubts as to how reliable the resulting ordering will be. Concerning the second parameter, no method is provided in [6] for specifying it. This means that the user will have to provide the algorithm with the right parameter setting each time it is applied to new data, and there seems to be no general criterion for going about this task. On the other hand, MBF requires neither to specify a variable ordering, nor to fix an upper bound on the number of parents allowed for each node, and this makes MBF both more general and more appealing for application to domains where no prior knowledge can be used to constrain the learning process.

5 Experimental Results

I tested MBF on four different datasets, concerning one synthetic and three real-world domains, which I describe next. The LED24 dataset [14] is a synthetic one, containing 3200 patterns. Each pattern is characterized by 24 boolean features, and it belongs to one of 10 possible classes. Only 7 features are relevant for predicting the class, and the class entirely determines the value of each relevant feature (except for a 0.1 probability that the value of the feature will be chosen randomly, introducing a certain noise). The values of the other 17 features are random. The Mushroom dataset [14] contains 8124 patterns. Each pattern describes a mushroom through 22 features. All features are nominally valued, and range over domains containing up to 12 possible values. The class variable has two possible values: either 'edible' or 'poisonous'. The Splice-junction dataset [14] consists of 3190 patterns. The patterns are characterized by 60 four-valued attributes, where each attribute stands for a position on a DNA sequence. The task is to recognize whether the pattern is an exon/intron boundary, an intron/exon boundary, or neither. The HS3D dataset has been extracted from the Homo Sapiens Splice Sites database [15]. I prepared the dataset by merging the Exon-Intron true splice sites data from that database with the first 1500 items of the Exon-Intron false splice sites data. In this way, I obtained a dataset containing 4296 patterns, belonging to two possible classes: either true or false exon/intron splice site. Each pattern is a vector of 140 features, describing a DNA sequence; as for the preceding Splice-junction dataset, each feature is a four-valued variable, standing for the nucleotide occupying the specific position in the sequence.

The strategy used to test MBF on each dataset is the following: (1) train some classifiers on the dataset, and record their classification accuracy; (2) run MBF on the data; (3) train the classifiers used in step 1 on the dataset obtained by eliminating from the original data the features discarded in step 2, and compare their classification accuracy to the accuracy recorded in step 1. Classification accuracy is measured using ten-fold cross-validation. The employed classification algorithms are the following (as supplied by the WEKA machine learning workbench [11]): Naive Bayes [16]; J4.8 (which is WEKA's implementation of the C4.5 algorithm [17]); IB1 [18]. These specific algorithms are chosen so as to make it possible to evaluate the impact of MBF across a variety of classification techniques (i.e. Bayesian classifiers, tree classifiers, and nearest-neighbor classifiers, respectively). Furthermore, the results of MBF are compared to the results of the CFS algorithm [19], which is a well-known and efficient feature selection method (also supplied by WEKA) and hence provides an interesting baseline for evaluating MBF.

The results are summarized in tables 1 and 2. Table 1 compares, for each dataset, the number of features selected by MBF to the number of features selected by CFS. Table 2 shows the classification accuracy of the three classifiers for each dataset, comparing their accuracy on the original datasets to the accuracy on the datasets filtered by MBF and by CFS respectively. Concerning the number of features selected by the two algorithms, MBF seems to perform better: summing over all four cases, CFS ends up selecting 5 features more than MBF. A strong difference emerges from the LED24 dataset, which is the only synthetical domain. Since we know in advance which features are relevant, this dataset provides a preliminary benchmark for MBF. The result is particularly encouraging, since the selected features for MBF are exactly the relevant ones, whereas CFS selects 8 irrelevant features in addition to the relevant ones. MBF almost always improves the classification accuracy, regardless of the chosen

Table 1. An overview of the datasets, with the last two columns reporting the number of features selected by MBF and CFS respectively

Dataset	# Features	# Labels	# Patterns	# MBF Features	# CFS Features
LED24	24	10	3200	7	15
Mushroom	22	2	8124	10	4
Splice-junction	60	3	3190	21	24
HS3D	140	2	4296	21	21

Table 2. Accuracy percentages of the classifiers on the original data and on the data filtered by MBF and CFS respectively

Dataset	Classification Accuracy								
	Naive Bayes			J4.8			IB1		
	Original	MBF	CFS	Original	MBF	CFS	Original	MBF	CFS
LED24	75.43%	75.37%	75.87%	73.46%	75.56%	74.46%	49.31%	64.53%	61.50%
Mushroom	95.82%	98.05%	98.52%	100%	100%	99.01%	100%	100%	93.89%
Splice-junction	95.29%	96.20%	96.08%	94.38%	94.48%	94.48%	75.92%	81.94%	80.37%
HS3D	83.44%	91.55%	90.17%	91.38%	91.55%	91.03%	76.65%	81.40%	80.60%

classifier: in only 1 out of 12 cases (i.e. Naive Bayes on the LED24 data) the accuracy decreases after filtering, and the decrease is quite small (0.06%). The accuracy after using CFS decreases instead in 3 cases (ranging from a 0.99% to a 6.11% decrease). In 9 cases, the classification accuracy achieved after filtering the data with MBF is higher than the one achieved by CFS; in 1 case the resulting accuracy is the same (J4.8 on the Splice-junction dataset); in the remaining 2 cases CFS performs better than MBF. The maximum increase in accuracy produced by filtering the data with MBF is 15.22%, while for CFS this value amounts to 12.19% (both values are recorded on LED24 with IB1).

6 Future Work

One interesting direction for future work is to devise more efficient algorithms for learning MBs. Since in feature selection we are properly interested in learning a MB for the class, rather than learning the *whole* BN containing the variables at hand, then we may try to constrain the search in the space of BNs so that the algorithm does not waste time optimizing irrelevant regions of the candidate model. That is, we may try to prevent the search algorithm from applying the search operators when the operations do not relate in any important way to the class variable. At least one problem will have to be solved in order for this strategy to work: we will need to make sure that the additional constraints on the application of the search operators do not raise significantly the risk of getting stuck in local maxima. Since structure learning in BNs is a computationally challenging task, achieving positive results in this direction would make MBF more appealing for application to domains involving very large numbers of features.

References

1. Duda, R.O., Hart, P.E., Stork, D.G.: Pattern Classification, 2nd edn. John Wiley & Sons, New York (NY) (2001)
2. Guyon, I., Elisseeff, A.: An introduction to variable and feature selection. Journal of Machine Learning Research 3, 1157–1182 (2003)
3. John, G.H., Kohavi, R., Pfleger, K.: Irrelevant features and the subset selection problem. In: Cohen, W.W., Hirsh, H. (eds.) Machine Learning: Proceedings of the Eleventh International Conference, pp. 121–129. Morgan Kaufmann, San Francisco (1994)
4. Kohavi, R., John, G.H.: Wrappers for feature subset selection. Artificial Intelligence 97, 273–324 (1997)
5. Koller, D., Sahami, M.: Toward optimal feature selection. In: Saitta, L. (ed.) Machine Learning: Proceedings of the Thirteenth International Conference, pp. 284–292. Morgan Kaufmann, San Francisco (1996)
6. Hruschka, E.R.J., Hruschka, E.R., Ebecken, N.F.F.: Feature selection by Bayesian networks. In: Tawfik, A.Y., D., G.S (eds.) Advances in Artificial Intelligence, pp. 370–379. Springer, Heidelberg (2004)
7. Pearl, J.: Probabilistic Reasoning in Intelligent Systems. Morgan Kaufmann, San Francisco (CA) (1988)
8. Neapolitan, R.E.: Learning Bayesian Networks. Prentice Hall, Upper Saddle River (NJ) (2004)
9. Russell, S., Norvig, P.: Artificial Intelligence, 2nd edn. Prentice Hall, Upper Saddle River (NJ) (2003)
10. Heckerman, D.: A tutorial on learning with Bayesian networks. Technical Report MSR-TR-95-06, Microsoft Research, Redmond (WA) (1996)
11. Witten, I.H., Frank, E.: Data Mining, 2nd edn. Morgan Kaufmann, San Francisco (CA) (2005)
12. Rissanen, J.: Stochastic complexity. Journal of the Royal Statistical Society. Series B 49, 223–239 (1987)
13. Cooper, G.F., Herskovits, E.: A Bayesian method for the induction of probabilistic networks from data. Machine Learning 9, 309–347 (1992)
14. Newman, D., Hettich, S., Blake, C., Merz, C.: UCI repository of machine learning databases (1998)
15. Pollastro, P., Rampone, S.: Homo sapiens splice sites dataset (2003)
16. Domingos, P., Pazzani, M.: On the optimality of the simple Bayesian classifier under zero-one loss. Machine Learning 29, 103–130 (1997)
17. Quinlan, J.R.: C4.5 Morgan Kaufmann, San Mateo (CA) (1993)
18. Aha, D.W., Kibler, D., Albert, M.K.: Instance-based learning algorithms. Machine Learning 6, 37–66 (1991)
19. Hall, M.A.: Correlation-based feature selection for discrete and numeric class machine learning. In: Proceedings of the Seventeenth International Conference on Machine Learning (ICML 2000), pp. 359–366. Morgan Kaufmann, San Francisco (2000)

ANFIS Based Emotions Recognision in Speech

Shubhangi Giripunje[1] and Narendra Bawane[2]

[1] Department of Electronics, G.H.R.C.E., Nagpur, India
[2] Department of Computer science and Engg., G.H.R.C.E., Nagpur, India
shubha_giripunje@yahoo.com, narenbawane@rediffmail.com

Abstract. In this paper, a speech based emotion classification method is presented. Five basic human emotions including anger, fear, happiness, sadness and neutral are investigated. This paper explores use of Adaptive Neuro-Fuzzy Inference System (ANFIS) to design a classifier that can discriminate between various emotions. The results found to be are significant, both in cognitive science and in speech technology. For emotion recognition, we selected statistics of the pitch like, first and second formants, and Energy and speaking rate as the base features. ANFIS based recognizer is created. Ensembles of such recognizer are used as an important part of decision support system for prioritizing voice messages and assigning a proper agent to response the message. The recognition of emotion in human speech has gained increasing attention in recent years due to the wide variety of applications that benefit from such technology such as emotional robot or Computer system.

1 Introduction

Automatic emotion recognition of speech can be viewed as a pattern recognition problem. The ability to express and recognize emotions or attitudes through the modulation of the intonation of the voice is fundamental to human communication. A new wave of interest has recently risen attracting both psychologists and artificial intelligence specialists. There are several reasons for this renewed interest such as: technological progress in recording, storing, and processing audio and visual information. A new field of research in AI known as affective computing has recently been identified [4]. As to research on recognizing emotions in speech, on one hand, psychologists have done many experiments and suggested theories. AI researchers had made contributions in the areas of emotional speech synthesis [1-3], recognition of emotions [5], and using agents for decoding and expressing emotions [6-8]. In current paper, an attempt has been made in the area of speech recognition. The proposed system can be used in application such as telephone call center and to develop emotional robot and computer system.

2 Emotions in Speech

Emotion recognition solutions depend on which emotions we want a machine to recognize and for what purpose. Emotion recognition has applications in talking toys, video and computer games, and call centers. Human emotional states can only be

B. Apolloni et al. (Eds.): KES 2007/ WIRN 2007, Part I, LNAI 4692, pp. 77–84, 2007.

identified indirectly. Different emotional states affect the speech production mechanism of a speaker in different ways, and lead to acoustical changes in their speech. Listeners can perceive these changes as being due to emotion. Generally, emotion refers to short-term states. Emotions have some mechanical effects on physiology, like heart rate modulation or dryness in the mouth, which in turn have effects on the intonation of the voice. This is why it is possible in principle to predict some emotional information from the prosody of a sentence. The various emotions in speech about their characteristics can be expressed as follows.

Anger: Anger generally seems to be characterized by increase in mean F0 (fundamental frequency), F0 (fundamental frequency) variability and mean energy. Further anger effects include increases in high frequency energy and downward directed F0 (fundamental frequency) contours.

Sadness: A decrease in mean F0 (fundamental frequency), F0 range and mean energy is usually found, as are downward directed F0 (fundamental frequency) contours. There is evidence that high frequency energy.

Happiness: Findings converge on increases in mean F0 (fundamental frequency), F0 (fundamental frequency) range, F0 (fundamental frequency) variability and mean energy. There is some evidence for an increase in high frequency energy.

Fear: The mean F0 is high but less than happiness, the rhythm is slow, with a high variance of phoneme durations, very few syllables are accented, the last syllable is accented, and the contours of syllables are rising.

Neutral: No *strong* expression of emotion, including weak or contrasting expressions, as well as no expression. Neutrality is assigned by default in this case, but acoustic and prosodic features, including moderate loudness, tempo, and inflection, give further evidence for the *neutral* label

To recognize emotions in speech, the existing approaches are K-nearest neighbors, and set of experts etc. In this paper, the possibility of use of ANFIS is explored to detect various types of emotions in speech. While designing classifiers for emotions, it is essential to learn how well people recognize emotions in speech, to find out which features of speech signal could be useful for emotion recognition, and explore different mathematical models for creating reliable recognizers.

3 Features Extraction Technique

All studies in this field point to the pitch (fundamental frequency) as the main vocal cue for emotion recognition. The other acoustic variables contributing to vocal emotion signaling are: vocal energy, frequency spectral features and formants. Usually only one or two first formants (F1, F2) and temporal features (speech rate and pausing) are considered. Another approach to feature extraction is to enrich the set of features by considering some derivative features such as LPC (linear predictive coding) parameters of signal or features of the smoothed pitch contour and its derivatives. For our study authors estimated the acoustical variables like fundamental frequency F0 (fundamental frequency), energy, speaking rate, first three formants (F1, F2, and F3) and their bandwidths (BW1, BW2, and BW3), and calculated some descriptive statistics for them. Then ranking has been done for the statistics using feature selection techniques, and picked a set of most "important" features. All other parameters, which have been calculated, are mean, standard deviation, minimum, maximum, and range. The pitch of an utterance offers a both meaningful and reasonably reliably detectable representation. The scheme outlined above yielded *F0 (fundamental frequency)* values consistent with the perceived pitch. In most analyses, pitch values were used only as static description of the system, e.g. in the form of average values, deviation, or percentile ranges. The relationships to other variables and the resulting perceived impression could readily be investigated by analysis of variance, correlation, multiple regression, perceptions, etc

F0 (Fundamental Frequency) Contour

F0 (fundamental frequency) contours on the other hand do just this. The contour seems to hold part of the information necessary to classify the recording according to other impressions, as for instance angry or happy to approximate all other points with pieces of sinusoidal. F0 contour of a wave has even been reported that it suffices to record the position and size of the contour maxima (in addition to information about the onset of the vowel) and the slope of both adjacent parts on the curve. Variation of amplitude versus time for Neutral and sad emotion is revealed in figs. 1 and 2 respectively.

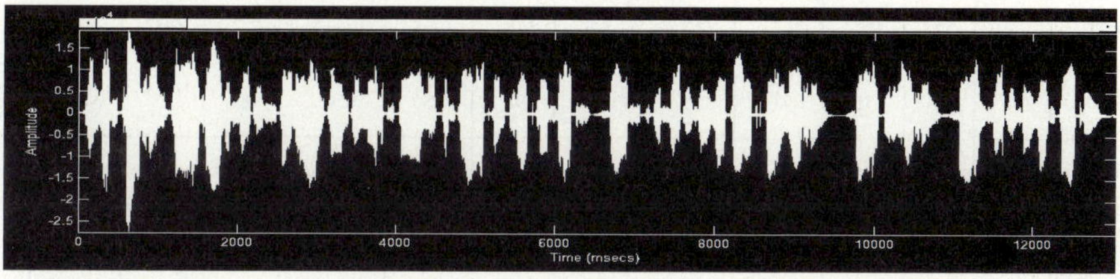

Fig. 1. Wave File for Emotion 'Neutral'

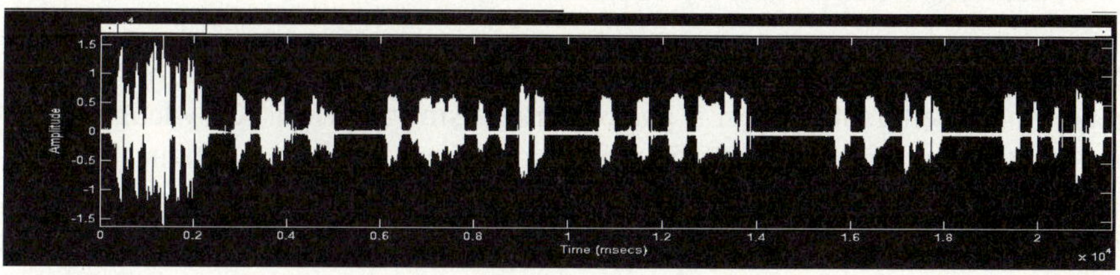

Fig. 2. Wave File for Emotion 'Sad'

Formant

Formant frequencies (F1, F2 and F3) are computed by solving for the roots of the LPC polynomial. Heuristic rules are used to ensure formant continuity between frames. Thus for feature extraction method, formants are determined selected speech segment or the formant track of the whole wav file.

All studies in the field point to the pitch (fundamental frequency F0) as the main vocal cue for emotion recognition. The other acoustic variables contributing to vocal emotion signaling are : vocal energy, frequency spectral features, formants, and temporal features (speech rate and pausing). Another approach to feature extraction is to enrich the set of features by considering some derivative features such as LPC (linear predictive coding) parameters of signal or features of the smoothed pitch contour and its derivatives . For our study we calculated some descriptive statistics for the following acoustical variables: fundamental frequency F0, energy, speaking rate, first three formants (F1, F2, and F3) Then we ranked them using feature selection techniques, and picked a set of most "important" features. For all other parameters we calculated the following statistics: mean, standard deviation, minimum, maximum, and range as shown in Table 1.

Table 1. Feature parameters for different emotions

Feature parameters / Emotions	Fo avg	Fo std	F1 max	F1 mean	F1 median	F1 std	F2 min	F2 mean	F2 median	F2 std	F3 min	F3 mean	F3 std
Normal	120	66	3149	637	350	624	535	1968	1826	801	1317	3056	496
Anger	174	97	3054	708	429	646	515	1985	1862	771	1454	3009	537
Glad	136	80	2514	693	355	678	498	2086	1956	784	1755	3183	417
Fear	213	115	2904	815	472	705	656	2063	1930	790	1340	2995	538
Sad	123	65	3107	778	267	781	518	2179	2082	756	1911	3196	452

4 Predicting Emotional Speech Using ANFIS

ANFIS is the fuzzy-logic based paradigm that grasps the learning abilities of ANN to enhance the intelligent system's performance using a priori Knowledge. Using a given input/output data set, ANFIS constructs a fuzzy inference system (FIS) whose membership function parameters are tuned (adjusted) using either a back-propagation algorithm alone, or in combination with a least squares type of method. This allows your fuzzy systems to learn from the data they are modeling.

These techniques provide a method for the fuzzy modeling procedure to learn information about a data set, in order to compute the membership function parameters that best allow the associated fuzzy inference system to track the given input/output data. This learning method works similarly to that of neural networks.

It is worth noting that the Layer-1 of ANFIS structure consists of membership functions described by the generalized bell function

$$\mu(X)=(1+((X - c)/a)2b)-1 \quad (1)$$

where a, b and c are adaptable parameters. Layer-2 implements the fuzzy AND operator, while Layer-3 acts to scale the firing strengths. The output of the Layer-4 is comprised of a linear combination of the inputs multiplied by the norm alized firing strength w:

$$Y = w\,(pX + r) \quad (2)$$

where p and r are adaptable parameters. Layer-5 is a simple summation of the outputs of Layer-4. The adjustment of modifiable parameters is a two-step process. First, information is propagated forward in the network until Layer-4 where the parameters are identified by a least-squares estimator. Then the parameters in Layer-2 are modified using gradient descent. The only user specified information is the number of membership functions in the universe of discourse for each input and output as training information.

ANFIS uses back propagation learning to learn the parameters related to membership functions and least mean square estimation to determine the consequent parameters. Every step in the learning procedure includes two parts.

The input patterns are propagated, and the optimal consequent parameters are estimated by an iterative least mean square procedure. The premise parameters are assumed fixed for the current cycle through the training set. The pattern is propagated again, and in this epoch, back propagation is used to modify the premise parameters while the consequent parameters remain fixed.

To use ANFIS for classification problem, the designer needs to perform the following steps:

1. Design a Sugeno FIS appropriate for the classification problem.
2. Hands optimize the FIS, given actual input classification data.
3. Set up training and testing matrices.

The training and testing matrices will be composed of inputs and the desired classify-cation corresponding to those inputs.

4. Run the ANFIS algorithm on the training data.
5. Test the results using the testing data.

The parameters associated with the membership functions will change through the learning process. The computation of these parameters (or their adjustment) is facilitated by a gradient vector, which provides a measure of how well the fuzzy inference system is modeling the input/output data for a given set of parameters. Once the gradient vector is obtained, any of several optimization routines could be applied to adjust parameters that will reduce some error measure (usually defined by the sum of the squared differences between actual and desired response).

ANFIS uses either back propagation or a combination of least squares estimation and back propagation for membership function parameter estimation.

Authors have estimated the acoustical variables like fundamental frequency F0, energy, speaking rate, first three formants (F1, F2, and F3) and their bandwidths (BW1, BW2, and BW3) for calculating descriptive statistics of them. Then they ranked the statistics using feature selection techniques, and picked a set of most "important" features.

ANFIS based emotion recognition system (ANFIS1) is developed using Matlab fuzzy logic toolbox. Rule viewer of the designed ANFIS1 system and surface view (for input in1 and in13) is shown in figures 3 and 4. ANFIS1 is having 13 inputs (shown in Table 1) and only one output. It contains only 5 rules as shown in Rule viewer.

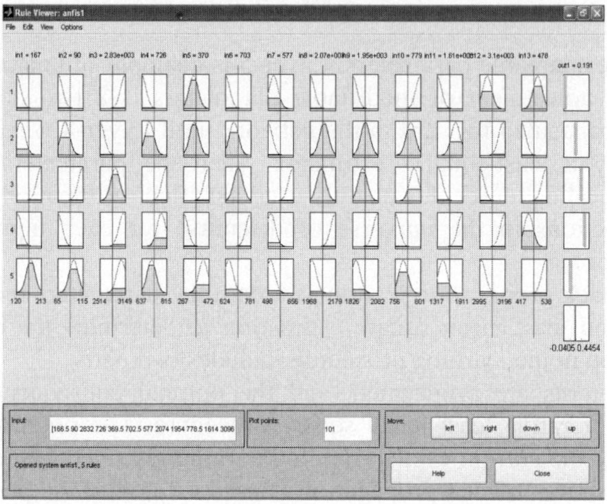

Fig. 3. Rule view

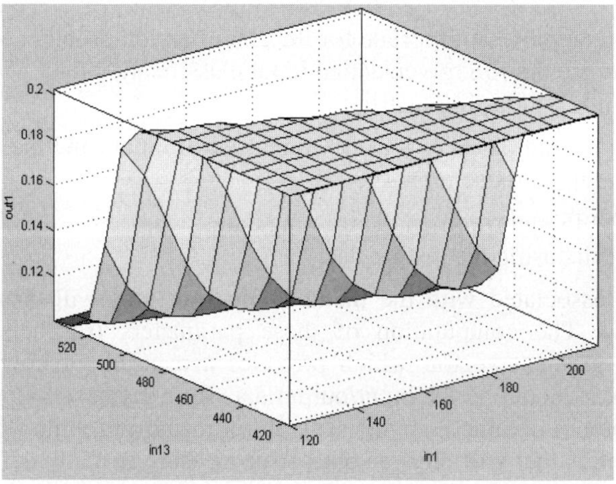

Fig. 4. Surface view of ANFIS1

5 Classification

Each speech file is a continuous recording of several emotions from one speaker. A splitter component that takes these recordings and the associated transcripts and emits separate utterances and their transcripts. Each utterance file represents one utterance by one actor expressing one emotion. These are short utterances of 3-4 words each 16-bit PCM, 22.05 KHz, and one channel.

The subjects heard these sentences:

I'm almost finished.
I saw your name in the paper.
I thought you really meant it.
I'm going to the city.
Look at that picture.

We compared the performance of classifiers by using Neural Network and ANFIS. In NN system, it gives quite satisfactory emotion detection performance, yet demonstrates a significant increase in versatility through its propensity for language independence. The ANFIS-based classifier has the advantages over other NN classifiers that frame length normalization is not necessary and temporal dynamics of the base features can be reflected by using the state transition probability. The experimental result for emotion 'fear' is shown in fig.5.

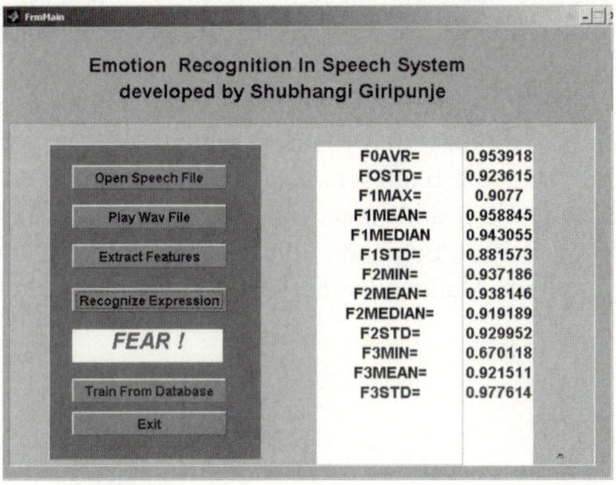

Fig. 5. Final result for emotion 'fear'

6 Conclusion

ANFIS is designed as a classifier that can discriminate between different emotions in speech. Adaptability, Noise tolerance are the inherent features of such classifier. Some important statistics of the pitch like first and second formants, Energy and Speaking rate are used as relevant features using feature selection techniques. The developed ANFIS is of very simple structure and contains only 5 rules and thus can be easily implemented. The areas of applicability are a number of other areas where a broad, but shallow computer representation of emotion and personality might be useful.Other appropriate uses include military applications where training to deal with stressful situations *must* include a model of personality and emotion and social interaction, as well as role; understanding of political personalities. and social simulations. Emotion recognition using a ANFIS is explored in the context of automated call center applications in this paper. Its provides improved performance compared with other classifiers such as Neural network.. Also, as the number of input features decreases, the performance with respect to classification error has reduced. This explains that there is clearly need to apply signal processing techniques to examine the optimal input features in given applications.

References

1. Giripunje, S., Panat, A.: Speech recognition for emotion using Neural network: A design approach. In: Negoita, M.G., Howlett, R.J., Jain, L.C. (eds.) KES 2004. LNCS (LNAI), vol. 3213, Springer, Heidelberg (2004)
2. Zhou, G., et al.: Nonlinear feature based classification of speech under stress. IEEE Trans. Speech, Audio Proc. 9, 201–216 (2001)
3. Erickson, D., Abramson, A., Maekawa, K., Kaburagi, T.: Articulatory Characteristics of Emotional Utterances in Spoken English. In: Proceedings of the ICSLP (2000)
4. Picard, R.: Affective computing. MIT Press, Cambridge (1997)
5. Dellaert, F., Polzin, T., Waibel, A.: Recognizing emotions in speech. ICSLP 96 (1996)
6. Tosa, N., Nakatsu, R.: Life-like communication agent - emotion sensing character MIC and feeling session character MUSE. In: Proc. of IEEE conference on Multimedia pp. 12–19 (1996)
7. Banse, R., Scherer, K.R.: Acoustic profiles in vocal emotion expression. Journal of Personality and Social Psychology 70, 614–636 (1996)
8. Cohn, J.F., Katz, G.S.: Bimodal Expressions of Emotion by Face and Voice. In: Workshop on Face/Gesture Recognition and their Applications, The Sixth ACM International Multimedia Conference, Bristol, England (1998)
9. User manual, Fuzzy toolbox, Mathworks Inc.

Binary Particle Swarm Optimization for Black-Scholes Option Pricing

Sangwook Lee[1], Jusang Lee[2], D. Shim[3], and Moongu Jeon[1,⋆]

[1] School of Information & Mechatronics, GIST, Gwangju 500-712, Republic of Korea
{yashin96,mgjeon}@gist.ac.kr
http://acl.gist.ac.kr
[2] Consulting Division, Samsung SDS, Republic of Korea
jussang.lee@samsung.com
[3] Dept. of Traffic Engineering, Kwandong University, 522 Naegok-Dong,
Gangrung-Si, Gangwon-Do 201-701, Republic of Korea
sdyng@kwandong.ac.kr

Abstract. The Black-Scholes option pricing (BS) model is a landmark in contingent claim theory and has been widely accepted in financial markets. However, it has a difficulty in the use of the model, because the volatility which is a nonlinear function of the other parameters must be estimated. The more accurately the volatility is estimated, the more accurate estimates of theoretical option prices will be. Thus, we propose a new model based on particle swarm optimization (PSO), which finds more precise theoretical values of options with estimates of the implied volatility than genetic algorithms (GAs).

Keywords: Black-Scholes option pricing, binary particle swarm optimization, bit change mutation.

1 Introduction

Recently, there have been enormous research activities regarding developments of derivative securities. Option markets are the most popular shares of financial institutions, and modern option pricing is often considered the most complex mathematically in all applied areas of finance. To explore the markets with improved investment yields, researchers including bankers, investors, and economists have paid considerable attention on option pricing. Modeling and predicting option prices have been very important to practitioners.

BS model [1] is the most widely used option pricing model, which can find a closed form solution for the price of a European call and European put with assumption that the price of the underlying stock follows a lognormal random variable. The time to expiration, the price of the underlying security, the exercise price, the risk-free rate of return, and the volatility of the security return are the vital attributes to be dealt with. The volatility can only be estimated, and

⋆ Corresponding author.

B. Apolloni et al. (Eds.): KES 2007/WIRN 2007, Part I, LNAI 4692, pp. 85–92, 2007.

its estimation itself is the interesting research subject [2,3,4,5]. The objective is to determine a 'good' estimate of the volatility that can literally be used to find a precise option price. Once a theoretical price is obtained, investors can search for mispriced options and collect arbitrage profits from subsequent trades.

The difficulty is that the volatility is a complex nonlinear function of the other related parameters, which requires robust numerical search techniques. In one of the earliest studies, Manaster and Koehler [2] analyzed it with the Newton-Raphson method, but it is too sensitive to the starting value for the search. Although hill climbing algorithms can be used to find a global optimum for nonlinear functions, the convergence may also hung up on a local optimum. Bruce K. [6] demonstrated GA which can more effectively find accurate values of the parameters than calculus-based search methods.

PSO is a recently invented optimizer that is very easy to understand and implement. It is one of evolutionary computation techniques developed by Kennedy and Eberhart in 1995 [7], and similar in some ways to GAs and other evolutionary algorithms. However, it requires less computational book-keeping and encapsulates fewer coding lines [8]. In this paper, we apply PSO to BS model and compare its performance to that of GAs.

The paper is organized as follows. In the next section, the BS model and PSO are briefly reviewed. The section 3 introduces a bit change mutation for binary PSO. In section 4, three kinds of PSO model and GAs are applied to BS model based on daily transaction real data with the Korea Composite Stock Price Index (KOSPI) 200. Finally, we make some concluding remarks in section 5.

2 Overview

2.1 Black-Scholes Formula and Volatility

In this paper, we deal with the European call only. The call price, c in the BS model is computed as follows:

$$c = S \cdot \mathbf{N}(d_1) - X \cdot e^{r_f T} \cdot \mathbf{N}(d_2), \tag{1}$$

$$d_1 = (\ln(S/X) + r_f T) / \left(\sigma T^{1/2}\right) + 0.5\sigma T^{1/2}, \tag{2}$$

$$d_2 = d_1 - \sigma T^{1/2}, \tag{3}$$

where S is the stock price, X is the strike price, T is the time remaining until expiration expressed as a percent of a year, r_f is the current continuously compounded risk-free interest rate, σ is the implied volatility of stock price, and $\mathbf{N}(\cdot)$ is the standard normal cumulative distribution function.

The Black-Scholes model assumes that the option can be exercised only at expiration. It requires that both the risk-free rate and the volatility of the underlying stock price remain constant over the period of analysis. The model also assumes that the underlying stock does not pay dividends; adjustments can be made to correct for such distributions. For example, the present value of estimated dividends can be deducted from the stock price in the model.

The implied volatility is the most critical parameter for option pricing; option prices are very sensitive to changes in volatility which cannot be directly observed but must be estimated by computational techniques.

The implied volatility outperforms past volatility in forecasting future volatility and even subsumes the information content of past volatility in some of their specifications. Volatilities attribute different results according to their use of longer time series and non-overlapping data. While implied volatility is commonly used the argument that this is the "best" estimate is somewhat circular. Skilled options traders will not rely solely on implied volatility but will look behind the estimates to see whether or not they are higher or lower than we would expect from historical and current volatilities, and hence whether options are more expensive or cheaper than perhaps they should be. It's a slight over simplification, but basically implied volatility gives the price of an option; historical volatility gives an indication of its value.

2.2 Particle Swarm Optimization

The PSO algorithm was first introduced by Kennedy and Eberhart as an optimization technique for continuous problems in 1995 [7]. It is a kind of evolutionary computation technique motivated by the behavior of organisms such as fish schooling and bird flocking. Because of its simple and easy implementation, PSO has been widely used in optimization problems which might be solved by other optimization techniques. Robinson [9] explained the PSO with a swarm of bees in a field. Their goal is to find the location in the field with the highest density of flowers. Without any knowledge of a priori, the bees are attracted both to the best location (group best) of highest concentration found by the entire swarm and the best location (personal). Eventually, after being attracted to areas of high flower concentration, all the bees swarm around the best location (group best). Fig. 1 describes Robinson's explanation.

PSO algorithm is initialized with a population of individuals randomly placed on the search space and searching for optimal solution by updating individual generations. In each iteration, the velocity and the position of each particle are updated according to its previous best position ($p_{best,n}$) and the best position of group ($g_{best,n}$). In the original continuous version, each particle's velocity and position are adjusted by the following formula [7,12].

$$\begin{aligned} v_n(t+1) = wv_n(t) + c_1R_1\left(p_{best,n} - x_n(t)\right) \\ +c_2R_2\left(g_{best,n} - x_n(t)\right), \end{aligned} \tag{4}$$

$$x_n(t+1) = x_n(t) + v_n(t+1), \tag{5}$$

where n is the index of particle in the swarm, t represents the iteration number, $v_n(t)$ is the velocity vector of the nth particle, and $x_n(t)$ is the position vector. c_1 and c_2 are the acceleration constants which are positive numbers, R_1 and R_2 are random numbers uniformly distributed between 0 and 1, and w is called the inertia weight.

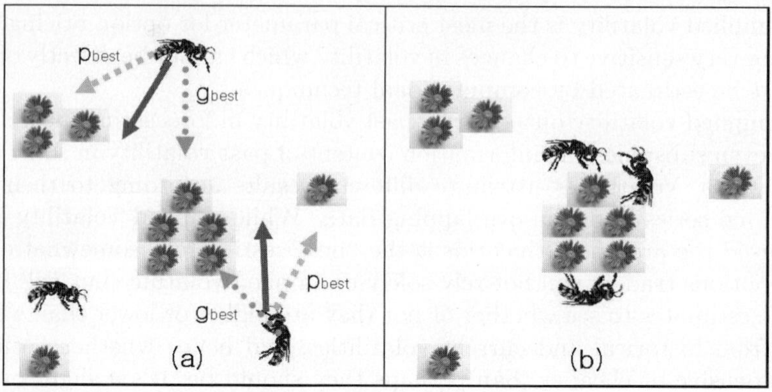

Fig. 1. (a) Bees are attracted both to the best location (group best) by the entire swarm and the best location (personal). (b) Eventually, all the bees swarm around the best location (group best).

2.3 Binary Particle Swarm Optimization

The binary particle swarm optimization (BPSO) algorithm was introduced by Kennedy and Eberhart to allow the PSO algorithm to operate in binary problem spaces [13]. It uses the concept of velocity as a probability that a bit (position) takes on one or zero. In BPSO, the Eq.(4) of updating a velocity remains unchanged, but the Eq.(5) of updating a position is re-defined by the following rule [13]:

$$x_n(t+1) = \begin{cases} 0 & \text{if } rand() \geq S(v_n(t+1)) \\ 1 & \text{if } rand() < S(v_n(t+1)) \end{cases} \tag{6}$$

where $S(\cdot)$ is a sigmoid function for transforming the velocity to the probability constrained to the interval [0.0, 1.0] as follows

$$S(v_n(t+1)) = \frac{1}{1+e^{-v_n(t+1)}}, \tag{7}$$

and *rand*() is a quasirandom number selected from a uniform distribution in [0.0, 1.0].

BPSO is susceptible to sigmoid function saturation, which occurs when velocity values are either too large or too small. In such cases the probability of a change in bit value approaches zero, thereby limiting exploration. For a velocity of 0, the sigmoid function returns a probability of 0.5, implying that there is a 50% chance for the bit to flip. Velocity clamping will delay the sigmoid function saturation from occurring.umber selected from a uniform distribution in [0.0, 1.0].

3 Bit Change Mutation for BPSO

When the BPSO was introduced, it was shown that the method performed quite well on problem with a unique global optimum, but it has trouble jumping out

Algorithm 1. *Original BPSO (BPSO1)*

```
begin
    t=0;                                        {t is the generation index}
    initialize particles x_n(t)
    evaluate x_n(t);
    while (terminationcondition ≠ true) do
        v_n(t) = update v_n(t);                 {by equation (4)}
        x_n(t) = update x_n(t);                 {by equation (6)}
        evaluate x_n(t);
        t = t + 1;
    end while
end
```

Algorithm 2. *BPSO with Bit Change Mutation (BPSO2)*

```
begin
    t=0;                                        {t is the generation index}
    initialize particles x_n(t)
    evaluate x_n(t);
    while (terminationcondition ≠ true) do
        v_n(t) = update v_n(t);                 {by equation (4)}
        v_n(t) = mutate v_n(t);                 {by equation (7)}
        x_n(t) = update x_n(t);                 {by equation (6)}
        evaluate x_n(t);
        t = t + 1;
    end while
end
```

of good local optima[13]. In BPSO, the current position value is used not in the position update rule but in the velocity update function. So, the position value of the next iteration is decided by only the corresponding velocity value. Because of this fact, it seems that a velocity, not a position, is a particle in BPSO. This implies that even though the final solution space is the binary position domain, the actual search space is the continuous velocity domain.

When BPSO starts iteration to find a optimum, the velocity tends to go into v_{max} or -v_{max} by the velocity update according as the corresponding target position is one or zero, respectively. If a velocity converges near v_{max} or -v_{max}, it is hard to change the corresponding position with a small change of velocity. Because of this fact, it is hard to escape from a good local optimum in BPSO. In order to go out from this undesired position, the large movement of velocity, which is unconcerned with two current best positions (p_{best} and g_{best}), is required. For accomplishing above objective, we suggest the insertion of the following operation between velocity update (Eq.(4)) and position update(Eq.(6)) in BPSO process.

$$\textbf{if}(rand() < r_{mute}) \ \ \textbf{then} \ v_n(t+1) = -v_n(t+1), \tag{8}$$

where r_{mute} is a probability which decides if the proposed operation is conducted or not. If this operation is executed when velocity is near v_{max} or -v_{max}, the position will be changed from one to zero or from zero to one, respectively. This operation works similarly in BPSO with a bit change mutation in GAs [11]. Thus, we call this operation a bit change mutation [10]. The pseudo codes for BPSO without and with a bit change mutation are described as follows.

4 Experiments

In this tests three variants of PSO model; continuous PSO (CPSO), discrete binary PSO (BPSO1), and discrete binary PSO with mutation (BPSO2) are applied to BS model, and then PSO with best performance is compared to GAs. For these tests a case study was carried out based on daily transaction data of the KOSPI 200 options which are obtained from the resorted format of the Samsung Finance Networks Systems. 356 end prices of each transaction from 27 February 2005 to 10 March 2005, (involving the call) are used. The risk-free rates are represented by 3-month certificate of deposit (CD) rate and are obtained from the Bank of Korea. Note that the main problem of BS model is finding an estimate of σ, $\hat{\sigma}$ with which the estimated price of the call is the same as the actual price of the call. Thus, fitness function is |estimated call value - actual call value|, which should be minimized.

For representing the solution $\sigma (0 < \sigma < 1)$, the 30 dimensional binary code is used in BPSO1, BPSO2, and GAs while 1 dimensional real value code is used in CPSO. In all tests, the number of evaluations is $2 \cdot 10^4$ (PSO: population 20, number of iteration 1,000 and GAs: population 100, number of iteration 200). All tests are repeated ten times and results are presented as average values.

In GAs, we use the uniform crossover with 0.6 probability and bit change mutation with 0.05/bit probability, but we cross-checked the results of this parameter set with other genetic operators and parameter sets. In PSO as well as GAs, parameter selection is very important for finding a global optimum. Especially, the inertia weights control the balance between the global and local exploration abilities of PSO [14]. A larger inertia weight w facilitates global exploration while a smaller inertia weight tends to facilitate local exploration in CPSO [14]. However, in BPSO, the inertia weight facilitates reversely [13].

We tested various inertia weights with fixing $c_1 = c_2 = 2$ in PSO experiments. In CPSO with $v_{max} = 0.3$, the best performance was shown by the case that inertia weight w decreases from 1.0 to 0.4 during the first 700 iterations and stays constant at 0.4 for the remaining 300 iterations. In BPSO1 with $v_{max} = 6.0$ and BPSO2 with $v_{max} = 6.0$, $r_{mute} = 0.05$, the best performance was shown by the case that inertia weight w increases from 0.1 to 0.7 during the first 700 iterations and stays constant at 0.7 for the remaining 300 iterations. A good performance with these parameter sets is the same tendency as [14].

Table 1 shows the comparison of performances among three PSO models. Clearly, we can find that BPSO with mutation (BPSO2) demonstrates the best

Table 1. BS fitness comparison among PSOs

	Mean	Std.Dev.	Min	Max
Diff CPSO	0.174021	0.361269	0	2.013315
Diff BPSO1	0.040357	0.160138	0	1.752168
Diff BPSO2	0.037884	0.157915	0	1.752161

Table 2. BS volatility estimate comparison between BPSO2 and GAs

	Mean	Std.Dev.	Min	Max
GA-$\hat{\sigma}$	0.253784	0.122157	0.033451	0.994912
BPSO2-$\hat{\sigma}$	0.238807	0.121745	0.033253	0.994922
Actual Call	10.05762	9.036544	0.01	35.2
GA-Call	10.09754	9.88325	0.01	35.19993
BPSO2-Call	10.09541	9.87302	0.01	35.19998
Diff GA	0.039127	0.158643	0	1.752163
Diff BPSO2	0.037884	0.157915	0	1.752161

performance. The reason for very poor results of CPSO is that the dimension of this CPSO is 1. That means the particles can move only on a line so that the search of vectorial direction between $p_{best,n}$ and $g_{best,n}$ is meaningless. The better performance of BPSO2 than BPSO1 must be from the bit change mutation [10] because all parameter sets are identical except the mutation.

The comparison results between BPSO2 and GAs are shown in Table 2. Note that GA-$\hat{\sigma}$ is the implied volatility using GAs; BPSO2-$\hat{\sigma}$ is the implied volatility using BPSO2; GA-Call is the BS call value with GA-$\hat{\sigma}$; BPSO2-Call is the BS call value with BPSO2-$\hat{\sigma}$; Diff GA is the difference between the actual call value and GA-Call; and Diff BPSO2 is the difference between the actual call and BPSO2-Call. As can be seen, BPSO2 outperforms GAs. The call values estimated by BPSO2 is more closer actual values than those by GAs. Moreover, all standard deviation by BPSO2 is smaller than those by GAs. This indicates that BPSO2 is more stable than GAs.

5 Conclusion

The usefulness of BPSO with the bit change mutation in the option pricing and in numerical parameter optimization has been demonstrated. The implied volatilities found with PSO produced option prices much closer to theoretical values than GAs. The PSO shares the ability of the GA to handle arbitrary nonlinear functions, but with much simpler implementation it clearly demonstrates high potential for widespread use in Finance. Now is a good time to think about how PSO can be utilized in facilitating economics from a fundamental perspective, because PSO is able to explore large numbers of different possibilities.

Acknowledgments

This research was supported by the UCN Project, the MIC 21C Frontier R&D Program in Korea, and the Center for Distributed Sensor Networks at GIST.

References

1. Black, F., Scholes, M.: The pricing of options and corporate liabilities. Journal of Political Economy 81, 637–653 (1973)
2. Manaster, S., Koehler, G.: The calculation of implied variances from the Black-Scholes model: a note. The Journal of Finance 37, 227–230 (1982)
3. Krausz, J.: Option parameter analysis and market efficiency tests. A simultaneous solution approach, Applied Economics 17, 885–897 (1985)
4. Corrado, C.J., Miller, T.W.: Optimal volatility estimates based on Black-Scholes implied standard deviations, Working Paper, University of Missouri (1993)
5. Chu, S., Freund, S.: Volatility estimation for stock index options. a GARCH approach, Quarterly Review of Economics and Finance 36, 431–450 (1996)
6. Bruce, K.: Black-Scholes option pricing via genetic algorithms. Applied Economics Letters 7, 129–132 (2000)
7. Kennedy, J., Eberhart, R.C.: Particle swarm optimization. In: Proc. IEEE Intl. Conf. on Neural Networks, IEEE Service Center, Piscataway, NJ, IV (1995) (1942–1948)
8. Boeringer, D.W., Werner, D.H.: Particle swarm optimization versus genetic algorithms for phased array synthesis. IEEE Transaction on Antennas and Propagation 52(3), 771–779 (2004)
9. Robinson, J., Rahmat-Sami, Y.: Particle swarm optimization in electromagnetics. IEEE Transactions on Antennas and Propagation 52(2), 397–407 (2004)
10. Lee, S., Park, H., Jeon, M.: Binary particle swarm optimization with bit change mutation, IEICE Fundamentals, in press (2007)
11. Wang, G., Goodman, D.E., William, F.P.: Simultaneous multi-level evolution, Submitted to Evolutionary Computation (1996)
12. Shi, Y., Eberhart, R.C.: A modified particle swarm optimizer. In: IEEE International Conf. on Evolutionary Computation, Anchorage, Alaska, USA, pp. 69–72 (1998)
13. Kennedy, J., Eberhart, R.C.: A discrete binary version of the particle swarm algorithm. In: Proc. 1997 Conf. Systems, Man, Cybernetics, Piscataway, NJ, pp. 4104–4108 (1997)
14. Shi, Y., Eberhart, R.C.: Parameter selection in particle swarm optimization. In: Proc. Seventh Annual Conf. on Evolutionary Programming, pp. 591–601 (1998)

Design of Very High-Speed Integer Fuzzy Controller Without Multiplications by Using VHDL

Sang-Gu Lee[1], Michio Miyazaki[2], and Jin-Il Kim[3]

[1] Dept. of Computer Engineering, Hannam University
133, Ojeong-dong, Daedeok-gu, Daejeon, Korea
sglee@hannam.ac.kr
[2] Dept. of Electronics, Kanto Gakuin University
1-50-1 Mutsuura-Higashi, Kanazawa-ku, Yokohama, Kanagawa 236-8501, Japan
miyazaki@kanto-gakuin.ac.jp
[3] Liberal Arts Education Center, Paichai University
439-6 Doma-2Dong, Seo-Gu, Daejeon, Korea
comclass@pcu.ac.kr

Abstract. This paper proposes an algorithm to map real values of the fuzzy membership functions in the consequent part onto an integer grid, as well as a method of eliminating the unnecessary operations of the leading and trailing zero items in the defuzzification stage. This allows a center of gravity method to be implemented with only integer additions and one integer division. The proposed system by using VHDL shows approximately an order of magnitude increase in speed as compared with conventional methods. This system can be applied to real-time high-speed intelligent systems such as robot arm control or fuzzy clustering for large data set.

Keywords: Fuzzy control, Defuzzification, Integer operation, COG, VHDL.

1 Introduction

Fuzzy logic models are recently successfully being used in many applications, including medical diagnosis, data mining, pattern recognition and robotics [3]. For them, a large number of floating-point operations in each fuzzy membership function are performed. Most conventional fuzzy systems, despite having fast computing hardware, use much time by using floating point operations in the condition part, in the consequent part, and in the defuzzification stage. Generally, in fuzzy computing, computations require the most time in the consequent part and the defuzzification stage rather than the condition part, so specially designed hardware modules and fast algorithms using only integer are required to process data rapidly in these stages. In cases involving complex and large volumes of computations, as found in typical fuzzy systems, the difference of speed can dominate overall system performance.

To overcome these problems, this paper uses Bresenham's scan line algorithm [1] to map the real values to integer grid. A method of eliminating the unnecessary operations of the continuous leading and trailing zero items in the defuzzification stage is also proposed. This allows a center of gravity method to be implemented with only integer additions and one integer division. The proposed system is applied to a truck backer-

B. Apolloni et al. (Eds.): KES 2007/ WIRN 2007, Part I, LNAI 4692, pp. 93–100, 2007.

upper control system for performance evaluation. The proposed system is an order of magnitude faster than the conventional methods using floating-point operations and introduces only a minimal error. This system can be applied to the real-time high-speed intelligent systems that can tolerate the minimal error introduced.

2 Very High-Speed Fuzzy System Using Integer Operations

In fuzzy inferencing, if fuzzy inputs are given as singleton values, it takes little times to find α, the degree of fulfillment, in each fuzzy rule. However, in the consequent part, it takes much more time than in the condition part due to the many real-valued operations in [0, 1] corresponding to the universe of discourse. Also, in the defuzzification stage, in order to find COG, many floating-point multiplications, additions, and one division are required.

In this paper, we propose a novel method to use only integer operations in the consequent part and the defuzzification stage. Here, we use an integer pixel grid in the consequent part composed of (M × N) pixels. After computing the α value, it is multiplied by N and converted to an integer by rounding. Let *Round*(α × N) be β. The integer β value is transferred to the consequent as a modified degree of fulfillment.

2.1 Integer Mapping Algorithm

Integer mapping of the fuzzy membership function is the first step in the proposed algorithm. Removing floating-point operations produces significant improvement in speed as compared to conventional methods. In this proposed method, we have a 512 × 32 integer pixel array for easy adapting to hardware implementation. In this grid, a segment of a membership function can be represented by connecting the integer pixels from the starting point pixel to the ending point pixel. Using mid-point technique [1], it is very efficient to compute the next *y* pixel value at given *x* pixel using only integer additions. This procedure is as follows.

Let $F(x,y) = ax + by + c = 0$. If $dy = y_1 - y_0$, and $dx = x_1 - x_0$, then $y = \frac{dy}{dx}x + B$, where B is the y-intercept, (x_0,y_0) is the start point, and (x_1,y_1) is the end point.

$$\therefore F(x,y) = dy \cdot x - dx \cdot y + Bdx = 0,$$

where $a = dy$, $b = -dx$, and $c = Bdx$.

To choose the next pixel, the first midpoint is $(x_p + 1, y_p + \frac{1}{2})$. We need only to compute the value of $F(x_p + 1, y_p + \frac{1}{2})$, and to test its sign. If the sign is positive, we select point $(x + 1, y + 1)$ as a next pixel. If the sign is negative, we select the point $(x + 1, y)$. If d is the decision variable for this choice, then

$$d_{start} = F(x_p + 1, y_p + \frac{1}{2}) = a(x_p + 1) + b(y_p + \frac{1}{2}) + c \;= a + \frac{b}{2} \;\longrightarrow\; 2dy - dx$$

If $d < 0$, then the next pixel is $(x_p + 1, y_p)$, so we select 'E'(East).
If $d \geq 0$, then the next pixel is $(x_p + 1, y_p + 1)$, so we select 'NE' (North East).

If 'E' is selected, the next mid-point is incremented by 1 in the x direction. To find the next pixel, we have to compute the amount of variation, ΔE, and update the decision variable d.

$$d_{new} = F(x_p + 2, y_p + \frac{1}{2}) = a(x_p + 2) + b(y_p + \frac{1}{2}) + c = 2a + \frac{b}{2} \rightarrow 4dy - dx$$

$$d_{old} = F(x_p + 1, y_p + \frac{1}{2}) = a(x_p + 1) + b(y_p + \frac{1}{2}) + c = a + \frac{b}{2} \rightarrow 2dy - dx$$

$$\Delta E = d_{new} - d_{old} = 2dy$$

If 'NE' is selected, as the similar way, $\Delta NE = d_{new} - d_{old} = 2(dy - dx)$.

All of the pixels can be found using integer operations in this procedure. Fig. 1 shows the algorithm for integer line mapping with start point (x_1,y_1) and end point (x_2,y_2). Here, β is the modified degree of fulfillment and has an integer value in the range $0 \le \beta \le 31$. The value defuzz(xa) is the value of the membership function in the consequent part when the value of the integer pixel in the x-axis is xa.

```
dx←x2-x1 , dy←y2-y1 , d←2dy-dx
xa←x1 ,  ya←y1, a← β
defuzz(xa)←ya
while  (ya < a+1) do
  begin
   xa←xa+1
     begin
       if (d<0)  d←d+2dy
       else
         ya←ya+1
         d←d+2(dy-dx)
       end
     defuzz(xa)←ya
  end
```

Fig. 1. Integer mapping algorithm (left_line)

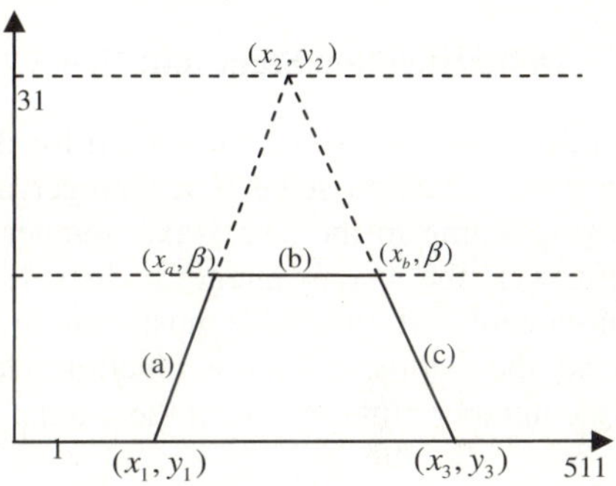

Fig. 2. Representation of the consequent part

Next, we describe the mapping procedures of the line (b) and (c), the remaining portions of the membership function, as shown in Fig. 2.

Because the required part in the defuzzification stage is in trapezoidal form, the part from integer β+1 to 31 in the y-axis point does not need mapping. Therefore, we must have mappings until the value of y is β. In order to map the membership function in Fig. 2, we first process line segment (a), then (c), and finally (b), following the order (a)→(c)→(b). Following this order has the advantage of calculating the endpoints of segment (b) before it is processed.Now we describe the integer mapping in line (c). It is similar to the process used to map line (a), except x is decremented for each iteration.

$$d_{start} = F(x_p - 1, y_p + \frac{1}{2}) = a(x_p - 1) + b(y_p + \frac{1}{2}) + c = -a + \frac{b}{2} \rightarrow -2dy - dx$$

If $d < 0$, we select 'NW'. If $d \ge 0$, we select 'W'.

If 'NW' is selected, the next mid-point is decremented by 1 in the x direction and is incremented by 1 in the y direction. To find the next pixel, we compute the amount of variation, ΔNW, and update the decision variable d.

$$\Delta NW = d_{new} - d_{old} = -2dy - 2dx = -2(dy + dx)$$

If 'W' is selected, as the similar way, $\Delta W = d_{new} - d_{old} = -2dy$

After mapping lines (a) and (c) to the integer pixels, the final task is the mapping of line (b) to the grid. To map line (b), we must set $y = \beta$ for all pixels from point (x_a,β) to (x_b,β). These two points are the ending points previously computed in lines (a) and (c). Therefore, all y values are set to β in the interval of the x-axis from x_a to x_b.

This integer mapping method offers a significant advantage over the methods that use floating-point operations. However, this does require a tradeoff. In this algorithm, a quantization error between the real y value and its equivalent integer pixel y value exists because this method selects the nearer integer pixel among two neighboring pixels. This quantization error is inversely proportional to the number of y-axis integer pixels. The root mean square value of the quantization error decreases as the number of y-axis integer pixels increases. In spite of this drawback, the increase in speed justifies the use of this procedure for application that can tolerate the small error it introduces. In practice, the size of the y-axis pixels can be adjusted according to the application domain.

2.2 Defuzzification Stage and New COG Method Without Multiplications

In this paper, we use COG method for defuzzification. The required data structure in the defuzzification stage is an integer array that can store the y-axis integer values corresponding to the 512 x-axis integer pixels in the consequent part. In this array, defuzz(x), the y-axis integer values from 0 to 31 are stored. On computing the consequent part, this array must contain the maximum values in all fuzzy rules. Fig. 3 shows the algorithm for this function. Here, n is the number of the fuzzy rules, and m is the number of the pixels in the x-axis.

```
integer array defuzz[1:m] , max[1:m]←0
 for i:=1 step 1 until n do
  for j:=1 step 1 until m do
  begin
   operation of integer pixel mapping
   return value ← defuzz(1:m)
  if max( j ) < defuzz( j )
  then max( j ) ← defuzz ( j )
  end
```

Fig. 3. Update algorithm of Max value

```
for i := upper step = –1 until lower
 begin
   temp ← defuzz( i ) + temp
   sum ← sum + temp
 end
COG ← sum / temp
COG ← COG + (lower – 1)
```

Fig. 4. The proposed algorithm for COG operations

After computing the consequent part, in general, many values of *defuzz(x)* have values of zero continuously from either of the array. It is not necessary to compute any operations in that interval less than lower or greater than upper in the defuzzification process. To overcome this problem, we check x_1 and x_3 for each fuzzy rule, setting *lower* to the minimum of all x_1 and *upper* to the maximum of all x_3.

In this paper, we also propose an algorithm to eliminate the computation times for the multiplications. We use only integer additions to calculate the nominator of the COG operation. One integer division operation is still needed to compute the final COG. Fig. 4 shows the algorithm to compute the nominator of the COG function. The final "temp" value in Fig. 4 becomes the denominator. In this method, for n non-zero items, only $2 \times n$ additions and 1 division are required.

3 Hardware Structure of the Proposed Fuzzy Processor

3.1 Entire Structure and Operations

The structure of the proposed one-chip fuzzy processor [5] is shown in Fig. 5. It contains modules of rule control interface, integer mapping, defuzzification, and memory part. The ROM stores all conditions to be processed from the conditional part in the rule interface.

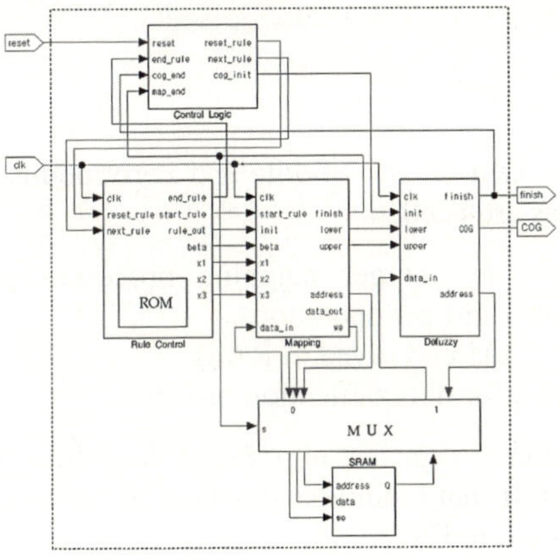

Fig. 5. Block diagram of proposed processor

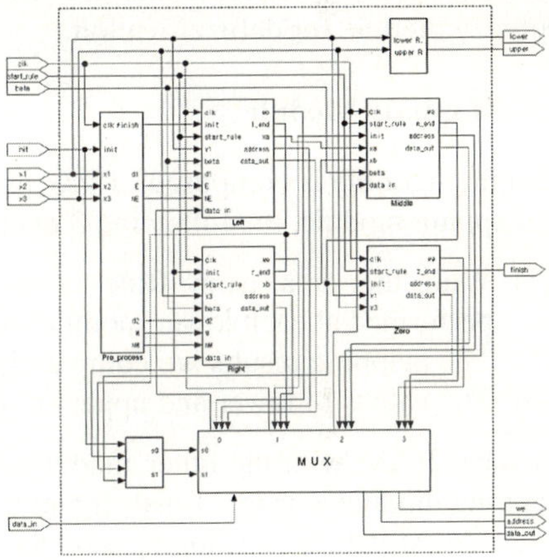

Fig. 6. Mapping module

The operation of the fuzzy processor is activated from reset signal. First, in the rule control module, the points of (x1, 0), (x2, 31), (x3, 0) in the ROM are transferred to mapping modules. In mapping module, y-coordinates (ranged from 0 to 31) are computed according to the x-coordinates (0~511) and this module compares that value and the content of the addressed RAM so that the larger one is saved to RAM. If start_rule = '1', then the computed y value is saved to RAM instead of referencing to the contents in RAM, and all '0' data is saved to the addresses (0～x1) and (x3～511), respectively. After saving, this module signals to the rule control module in order to receive the next data. After receiving the final data, the control signal is sent to the Defuzzy module. In Defuzzy module, after reading the contents in the RAM, additions and one division operation are performed, and compute and output the COG.

3.2 Rule Control Module

Data format for Rule control module is shown in Fig. 7 to process in the conclusion part by using data in conditional part. This data format is composed of 9-bit × 4.

	8	7	6	5	4 3 2 1 0
LSB	start	end	×	×	beta
	x1				
	x2				
MSB	x3				

Fig. 7. Data format for the rule control module

Here, start_rule and end_rule are control signals for controlling each module. The values of x1, x2, x3 and β are transferred to mapping module for computing in the conclusion part. In the case of end_rule = '1', Mapping module is finished. So all the final results in this module are saved to RAM and the control is transferred to Defuzzy module for defuzzification.

3.3 Mapping Module

Mapping module is composed of Pre_process, Left, Right, Middle and Zero modules. In mapping module, the following 3 operations are needed.

(1) With the data from Rule control module, integer mapping processes are performed at each lower module using only integer operations.
(2) The mapping results are compared with RAM and saved to RAM.
(3) The values of lower and upper are updated for defuzzification process.

To share RAM with the other modules, the finish signal is introduced. So while the mapping module is using, finish signal is reset. If that module is finished, finish signal is set. The structure of Mapping module is shown in Fig. 6.

Pre_process module inputs the values of x1, x2 and x3 and output the values of d1, E, NE, d2, W and NW for initial values of integer mapping operations. **Left module** receives the values of x1, beta, d1, N and NE for mapping the (a) part of Fig. 2. This module computes the y coordinate according to x coordinate from x1 in step 1. In the case of rule_start = '0', the larger one between the computed value and stored value in the RAM is saved to RAM. In the case of rule_start = '1', the computed value is saved to RAM. Here, if computed y value is equal to the beta, because corresponding x value is xa, the value of xa is outputted and control is transferred to Right module.

Right module receives the values of x3, beta, d2, W and NW for mapping the (c) part of Fig. 2. This module computes the y coordinate according to x coordinate from x3 in step -1. In the case of rule_start = '0', the larger one between the computed value and stored value in the RAM is saved to RAM. In the case of rule_start = '1', the computed value is saved to RAM. Here, if computed y value is equal to the beta, because corresponding x value is xb, the value of xb is outputted and control is transferred to Middle module.

Middle module receives the values of xa (from Left), xb (from Right) and beta for mapping the (b) part of Fig. 2. This module computes the y coordinate according to x coordinate from xa in step 1. In the case of rule_start = '0', the larger one between the value in the RAM and beta is saved to RAM. In the case of rule_start = '1', beta value is saved to RAM. Here, after repeating the same operation until x is equal to xb, the control is transferred to Zero module. In **zero module,** for the initialization of RAM, in the case of start_rule = '1', '0's are saved to the addresses from 0 to x1 and from x3 to 511, respectively, and output the finish signal. In the case of start_rule = '0', no operations are performed and output the finish signal.

3.4 Defuzzy Module

In Defuzzy module, COG is computed by referencing RAM where the final data in the conclusion part are saved. Defuzzy module is composed of Cog_add and Cog_divide modules.

4 Performance Analysis

To analyze the performance of the proposed method, we simulated it using both high-level programs and VHDL synthesis. First, algorithms for both methods were coded in C/C++ and execution speeds were timed. By using only integer operations, the proposed algorithm executed approximately 12.75 times faster than the traditional algorithm that uses floating point operations.

We also developed a VHDL implementation of the proposed algorithm and compared its performance to that of the traditional algorithm. For this implementation, the proposed algorithm reduced the time needed to calculate system outputs by an order of magnitude [5]. Both the software and VHDL simulation results are consistent with the speedup achieved for the truck backer-upper system [6]. Table 1 shows the comparison results of each fuzzy hardware system. FLASP and the proposed system have flexible bit-width for each fuzzy variable, while the other systems in this table support only fixed bit lengths. In the defuzzification procedure, the FLASP system and

Table 1. Comparison results by each fuzzy hardware system

	FLASP [3]	**KAFA [4]**	**Aranguren's [2]**	***Proposed system***
Max freq.	15 MHz	10 MHz	12 MHz	40 MHz
Bitwidth of variables	variable	fixed	Fixed (8 bits)	variable
Fuzzy operator	Min/Max	various	Min/Max	Min/Max
Defuzzification	Pipelining	Parallel processing	Pipelining (COA)	No multiplications (COG)
Parallel structure	yes	no	yes	yes
HW/SW codesign	yes	no	yes	yes
Membership function computing	LUT	Floating point operations	LUT	Only Integer Additions

Aranguren's controller [2] use a pipelining technique. In the computation of each membership function, we note that the existing algorithms use either lookup tables or floating point operations. Only the proposed algorithm uses integer operations.

Finally, we note that it is possible to improve the performance of the proposed algorithm by utilizing parallel processing techniques. For example, the functions for line segments (a) and (c) in Fig. 2 are independent and can be calculated in parallel. In addition, the points on line segment (b) in that figure are also independent of each other and do not have to be calculated sequentially. For membership functions that are symmetric, it is also possible to calculate only the points on line segment (a) and then make use of symmetry to generate the points on (c) without significant computations.

5 Conclusion

In this paper, we have proposed the followings.

(1) integer pixel mapping using only integer operations in the consequent part.
(2) eliminating the zero items in the consequent part for fast defuzzification
(3) novel COG method without multiplications

The proposed method was simulated for the truck backer-upper control system to verify its performance as compared to the conventional systems. As the results of the simulation, the proposed system is 12.75 times faster than the conventional method for the classic truck backer-upper problem and an order of magnitude faster on general fuzzy control problems. This system can also be applied to build powerful architectures for control applications, such as robotic control, with time-critical sensor integration or fuzzy clustering for very large data set.

Acknowledgment

This work was supported by a grant from Security Engineering Research Center of Korea Ministry of Commerce, Industry and Energy.

References

[1] Angel, E.S.: Interactive Computer Graphics, 3rd edn. Addison-Wesley, Boston, MA, USA (2002)

[2] Aranguren, G., et al.: Hardware implementation of a pipeline fuzzy controller. Fuzzy Sets and Systems 128, 61–79 (2002)

[3] Ibrahim, A.M.: Fuzzy Logic for Embedded Systems Applications. Elsevier, Amsterdam (2004)

[4] Kim, Y.D.: High speed flexible fuzzy hardware for fuzzy information processing. IEEE Trans. Systems, Man, and Cybernetics – Part A 27(1), 45–56 (1997)

[5] Lee, S.G.: High-speed Integer Fuzzy Controller without Multiplications. International Journal of Fuzzy Logic and Intelligent Systems 6(3), 223–231 (2006)

[6] Li, T-H S., Chang, S.-J.: Autonomous Fuzzy Parking Control of a Car-Like Mobile Robot. IEEE Trans. on Systems, Man and Cybernetics - Part A 33(4), 451–465 (2003)

Extended Fuzzy C-Means Clustering in GIS Environment for Hot Spot Events

Ferdinando Di Martino[1], Vincenzo Loia[1], and Salvatore Sessa[2]

[1] Università degli Studi di Salerno, Dipartimento di Matematica e Informatica, Via Ponte Don Melillo, 84084 Fisciano (Salerno), Italy
fdimarti@unina.it, loia@unisa.it

[2] Università degli Studi di Napoli Federico II, Dipartimento di Costruzioni e Metodi Matematici in Architettura, Via Monteoliveto 3, 80134 Napoli, Italy
sessa@unina.it

Abstract. The Extended Fuzzy C-Means (EFCM) algorithm in a Geographic Information System (GIS) is used for identifying the volume clusters as Hot Spot areas, being the data events geo-referenced as points on the geographic map. We have implemented EFCM with the usage of the software tools ESRI/ARCGIS and ESRI/ARCVIEW 3.x and moreover we have made a comparison with the classical Fuzzy C-Means (FCM) algorithm. The application concerns a specific problem of maintenance, executed in the years 2001-2005, over the buildings constructed before 1960 in the city of Cava de' Tirreni, located in the district of Salerno (Italy).

Keywords: Fuzzy C-Means, EFCM, GIS, Hot Spot Event, Spatial Analysis.

1 Introduction

A Hot Spot is generally defined as an area containing dense clusters of localized events (for example, criminal incidents). Identifications of Hot Spots on the geographic map is usually made by geo-referencing as points the events happened in a certain period. One of the most known approaches for the detection of Hot Spots is to use a cluster technique, which is an effective method for determining areas with elevated concentrations of localized events. The most used clustering techniques in Spatial Hot Spot Analysis are the Fuzzy C-Means (FCM) method [1,2,3] and the density based estimation techniques [7]: the FCM method identifies cluster prototype as point on the map and determines the membership degree of a point event to a cluster; the density estimation techniques identify clusters by searching for dense concentrations of events. In [5] the authors present the Extended Fuzzy C-Means (EFCM) as generalization of the classical FCM algorithm. There the point cluster prototypes are extended to hypervolumes whose size is determined automatically from the data and further the authors show clearly the advantage of EFCM with respect to FCM in the determination of the exact number of clusters. We have experimented the EFCM algorithm for determining cluster volume prototypes as hyperspheres, implementing it in the tools GIS ESRI/ARCMAP and ESRI/ARCVIEW 3.x. The main goal of this paper is to verify the use of these volume prototypes in Hot Spot Spatial Analysis in

B. Apolloni et al. (Eds.): KES 2007/WIRN 2007, Part I, LNAI 4692, pp. 101–107, 2007.

the context of a application concerning the specific problem of buildings maintenance in the city of Cava de' Tirreni, located in the district of Salerno (Italy).

2 The EFCM Algorithm

The FCM clustering method [1, 2, 3] creates a fuzzy partition of the data set in C clusters. Formally, let $\mathbf{X}=\{\mathbf{x}_1,...,\mathbf{x}_N\}\subset R^n$ be the data set composed by N elements (feature vectors) in R^n. The FCM is an iterative algorithm based on the minimization of the following objective function:

$$\mathrm{J}(\mathbf{X},\mathbf{U},\mathbf{V}) = \sum_{i=1}^{C}\sum_{j=1}^{N} u_{ij}^{m} d_{ij}^{2} \tag{1}$$

where $\mathbf{U} = [u_{ij}]$, being u_{ij} the membership degree of the j-th vector $\mathbf{x_j}$ to the i-th cluster with i = 1,…,C, $\mathbf{V} = \{\mathbf{v}_1,...,\mathbf{v}_C\} \subset R^n$ is the set of the centers of the C clusters (point prototypes), d_{ij} is the distance between the center $\mathbf{v_i}$ of the i-th cluster and the j-th vector $\mathbf{x_j}$ calculated with the Euclidean metric, $m \in [1,+\infty)$ is the "fuzzifier" parameter which affects the membership values and determines the degree of fuzziness of the partition. The algorithm FCM is efficient, but its main defect is that an expert must define a priori the number C of prototypes or obtain C by using a specific validity index. To overcome this defect, many authors developed methods to calculate a specific validity index C for the cluster number as maximum or minimum of a suitable objective function. Some known validity indexes are the Partition Coefficient [8], Partition Entropy [1, 2], Fukuyama and Sugeno Coefficient [4], Xie and Beni Coefficient [9]. A comparison of many cluster validity indexes is described in [10]. In [5] the authors have proposed the EFCM algorithm based on the concept of volume prototype $\mathbf{V}\subseteq R^n$ defined as a convex and compact surface, that is as subspace of the clustering space and its geometric shape depends on the metric adopted. Generally the volume prototypes are hyperellipsoids in R^n [6]: in particular, they are hyperspheres with centers $\mathbf{v_i}$ under the Euclidean metric; we calculate the distance d_{ij} between the feacture vector $\mathbf{x_j}$ and the volume prototype $\mathbf{V_i}$ and assign the maximum membership degree of $\mathbf{x_j}$ to $\mathbf{V_i}$ if $d_{ij} \le r_i$, being r_i a radius assigned to $\mathbf{V_i}$ estimated by calculating firstly the covariance matrix $\mathbf{P_i}$ of the i-th cluster defined as

$$\mathbf{P}_\mathrm{i} = \frac{\sum_{j=1}^{N} u_{ij}^{m}(\mathbf{x}_\mathrm{j} - \mathbf{v}_\mathrm{i})(\mathbf{x}_\mathrm{j} - \mathbf{v}_\mathrm{i})^T}{\sum_{j=1}^{N} u_{ij}^{m}}, \tag{2}$$

secondly by obtaining det(Pi) which gives its volume and finally by setting

$$r_i = \sqrt[n]{\det(P_i)^{1/n}} . \tag{3}$$

In clustering algorithms with volume prototypes the objective function to be minimized is defined as

$$J = \sum_{i=1}^{C} \sum_{j=1}^{N} u_{ij}^{m} (d_{ij}^{2} - r_i^{2}) \tag{4}$$

By using the Lagrange multiplicators, the minimization of (4) implies that

$$u_{ij} = \frac{1}{\sum_{k=1}^{C} \left(\frac{d_{ij}^{2} \max\left(0, 1 - \frac{r_i^2}{d_{ij}^2}\right)}{d_{kj}^{2} \max\left(0, 1 - \frac{r_k^2}{d_{kj}^2}\right)} \right)^{1/(m-1)}} = \frac{1}{\sum_{k=1}^{C} \left(\frac{d_{ij}^{2} w'_{ij}}{d_{kj}^{2} w'_{kj}} \right)^{1/(m-1)}} \tag{5}$$

The terms $d_{kj}^{2} w'_{kj} = \max\left(0, d_{kj}^{2} - r_k^{2}\right) = d_{kj}^{'2}$ can be interpreted as a squared distance of $\mathbf{x_j}$ from the k-th volume prototype $\mathbf{V_k}$. This term is null if the j-th feature vector $\mathbf{x_j}$ is in the hypersphere $\mathbf{V_k}$. For each j-th feature vector $\mathbf{x_j}$, we consider the value φ_j equal to the number of clusters for which $d'_{kj} = 0$ with k = 1,…,C; the related membership value u_{ij} can be written as

$$u_{ij} = \frac{1}{\sum_{k=1}^{C} \left(\frac{d'_{ij}}{d'_{kj}}\right)^{2/(m-1)}} \qquad \text{if } \varphi_j = 0$$

$$u_{ij} = \begin{cases} 0 & \text{if } d'_{ij} > 0 \\ \dfrac{1}{\varphi_j} & \text{if } d'_{ij} = 0 \end{cases} \qquad \text{if } \varphi_j > 0 \tag{6}$$

The features inside of the hypersphere of radius r_i centered in the point $\mathbf{v_i}$ do not influence the other clusters because the membership degree to other clusters is zero. This approach allows to overcome the problem found in the FCM algorithm relative to the infuence of zones with high data points density over zones with lower data points density. To avoid the negative effect of diminishing the objective function when a meaningful number of features is placed in a cluster (due to the value zero of the distance from this cluster), in [5] the authors increase the radius r_i at each iteration by a factor $\beta^{(l)}/C^{(l)}$, where $C^{(l)}$ is the number of clusters at l-th iteration and $\beta^{(l)}$ is a parameter defined recursively as

$$\begin{aligned} &\beta^{(0)} = 1 \\ &\beta^{(l)} = \min(C^{(l-1)}, \beta^{(l-1)} + 1) \end{aligned} \tag{7}$$

In order to determine the number of clusters, the EFCM method adopts the following formula as similarity measure between the i-th and k-th clusters:

$$S_{ik} = \max(I_{ik}, I_{ki}) \tag{8}$$

where

$$I_{ik} = \frac{\sum_{j=1}^{N} \min(u_{ij}, u_{kj})}{\sum_{j=1}^{N} u_{ij}} \tag{9}$$

The merging between the two i-th and k-th clusters, that have maximum similarity $S_{ik} = S_{\max}^{(l)}$, is done when at the l-th iteration we have $\left| S_{\max}^{(l)} - S_{\max}^{(l-1)} \right| \le \varepsilon$, where $S_{\max}^{(l)} \ge \alpha^{(l)}$, being the parameter $\alpha^{(l)} \in [0,1]$ depending from the density, the dimension of the clusters and the value of the index m. In [5], it is proposed to determine the value of the threshold $\alpha^{(l)}$ at the l-th iteration with the following adaptive formula:

$$\alpha^{(l)} = \frac{1}{C^{(l-1)} - 1} \tag{9}$$

The EFCM algorithm determines in each iteration the two indexes i* and k* with the greatest similarity similarity value; if it is greater than of $\alpha^{(l)}$, then i*-th and k*-th clusters are merged by setting

$$\begin{cases} u_{i^*j}^{(l)} = u_{i^*k^*}^{(l)} + u_{k^*j}^{(l)} & \forall j \in \{1,\ldots,N\} \\ C^{(l)} = C^{(l-1)} + 1 \end{cases} \tag{10}$$

and by removing the k*-th row from the matrix **U**. In other words, the steps of the EFCM algorithm can be summarized in the following way:

(1) The user assigns the initial number $C^{(0)}$ of clusters, the fuzzifier parameter $m > 1$ and the stop iteration parameter $\varepsilon > 0$. Moreover, the initial value of the similarity $S_{\max}^{(0)} = 1$ and the parameter $\beta^{(0)} = 1$ are fixed .

(2) The membership degree of the j-th feature to the i-th cluster $u_{ij}^{(0)}$ for j=1,…,N and i =1,…,$C^{(0)}$ are assigned randomly.

(3) The centers of the clusters $\mathbf{v}_i$ are calculated by using the following formula

$$\mathbf{v_i} = \frac{\sum_{j=1}^{N} u_{ij}^m \mathbf{x_j}}{\sum_{j=1}^{N} u_{ij}^m}. \tag{12}$$

(4) The radii of the clusters are calculated by using (3).

(5) The elements u_{ij} of the matrix **U** of the membership degree of the j-th feature vector to the i-th cluster are calculated by using (6).
(6) The elements s_{ik} of the similarity matrix S between the i-th and k-th clusters are calculated by using (8) and the i*-th and k*-th clusters are determined in such a way $S_{i^*k^*}$ assumes the maximum value.
(7) If $|S^{(l)}_{i^*k^*}$ - $S^{(l-1)}_{i^*k^*}| < \varepsilon$ and $S^{(l)}_{i^*k^*} > \alpha^{(l)} = 1/(C^{(l-1)} - 1)$, then the i*-th and k*-th clusters are merged via formula (11).
(8) If $\|\mathbf{U}^{(s)} - \mathbf{U}^{(s-1)}\| < \varepsilon$ at the s-th iteration, then stop otherwise go to 3) for the (s+1)-th iteration.

3 Implementation and Results

We have implemented a spatial clustering algorithm, based on the EFCM method, by using the tools GIS ESRI/ARCMAP and ESRI/ARCVIEW 3.x. This algorithm allows to choose features of a punctual spatial vectorial layer and determines the hyper-spherical cluster volumes. We used the GIS tool and EFCM in Hot Spot Spatial Analysis in order to study the resulting cluster volumes. In order to achieve this aim, as input data we used the punctual features constituted by the centroids of the buildings of Cava de' Tirreni constructed before 1960, in which a maintenance operation in the years 2001÷2005 was made. We used the buildings layer of the last known fotogrammetry (in scale 1:2000) by selecting the buildings with at least a manteinance

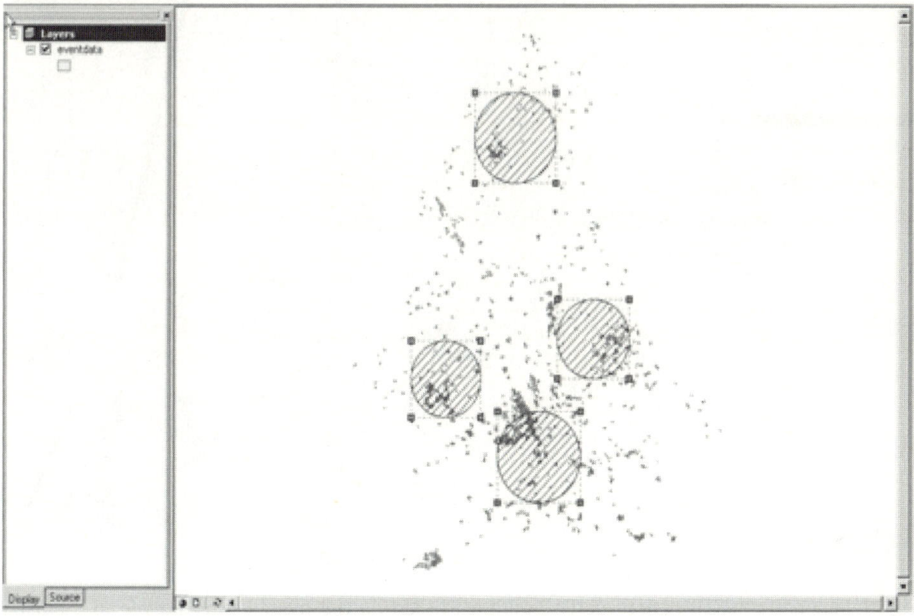

F

Fig. 1. The resulting cluster volumes

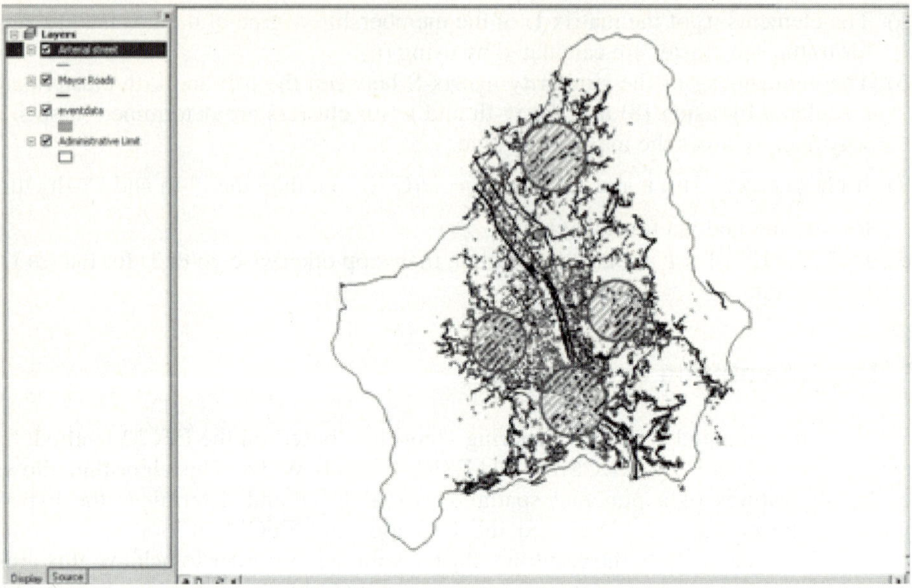

Fig. 2. The resulting geographic map with the roads and buildings

Fig. 3. Thematic classification of the buildings based on the membership degree to a cluster volume

operation and by deriving the related centroids. Since the average of the dimensions of a building is remarkably inferior with respect to the area under study, it was reasonable to approximate every polygonal feature building with its centroid. We have used the EFCM algorithm in the tool GIS and it has produced four clusters, situated in correspondence with different old zones as shown in Fig. 1. We have also displayed the roads and the buildings on the geographic map in Fig. 2.

One of the advantages of the use of the tool GIS is to obtain the cluster volumes as circles on the geographic map. In the areas covered from four cluster volumes there are many buildings constructed before 1960 and at least a maintenance operation was realized in the years 2001-2005 to many of these buildings.

The cluster volumes can represent circular buffer areas, that is circular areas of impact regarding a phenomenon localized in the center of this buffer, e.g. the epicenter of an earthquake. We can utilize the tool GIS to display thematic geographic maps based on the membership degree to a cluster. Such value can be interpreted as an impact degree of the feature to the phenomenon localized in the circular cluster volume. Fig. 3 shows a geographic map of the theme "Buildings" obtained by creating five intervals of equal width based on the membership degree to a cluster volume.

References

1. Bezdek, J.C.: Cluster Validity with Fuzzy Sets. IEEE Journal of Cybernetics 8(3), 58–73 (1974)
2. Bezdek, J.C.: Numerical Taxonomy with Fuzzy Sets. Journal of Math. Biol. 1, 57–71 (1974)
3. Bezdek, J.C.: Pattern Recognition with Fuzzy Objective Function Algorithms. Plenum Press, New York (1981)
4. Fukuyama, Y., Sugeno, M.: A New Method of Choosing the Number of Clusters for the Fuzzy C-Means Method. In: Proceedings of Fifth Fuzzy Systems Symposium, Kobe, Japan, pp. 247–250 (1989)
5. Kaymak, U., Setnes, M.: Fuzzy Clustering with Volume Prototype and Adaptive Cluster Merging. IEEE Trans. on Fuzzy Systems 10(6), 705–712 (2002)
6. Krishnapuram, R., Kim, J.: Clustering Algorithms Based on Volume Criteria, IEEE Trans. IEEE Trans. on Fuzzy Systems 8(2), 228–236 (2000)
7. Silverman, B.W.: Density Estimation for Statistics and Data Analysis. Chapman & Hall, New York (1986)
8. Trauvert, E.: On the Meaning of Dunn's Partition Coefficient for Fuzzy Clusters. Fuzzy Sets and Systems 25, 217–242 (1988)
9. Xie, X.L., Beni, I.G.: A Validity Measure for Fuzzy Clustering. IEEE Trans. Pattern Analysis Machine Intell 13, 841–847 (1991)
10. Wu, K.L., Yang, M.S.: A Fuzzy Validity Index for Fuzzy Clustering. Pattern Recognition Letters 26, 1275–1291 (2005)

Fuzzy Fusion in Multimodal Biometric Systems

Vincenzo Conti[1], Giovanni Milici[1], Patrizia Ribino[1], Filippo Sorbello[1], and Salvatore Vitabile[2]

[1] Dipartimento di Ingegneria Informatica
Università degli Studi di Palermo
Viale delle Scienze, Ed. 6, 90128 Palermo, Italy
{conti,milici,sorbello}@unipa.it
[2] Dipartimento di Biotecnologie Mediche e Medicina Legale,
Università degli Studi di Palermo
Via del Vespro, 90127 Palermo, Italy
vitabile@unipa.it

Abstract. Multimodal authentication systems represent an emerging trend for information security. These systems could replace conventional mono-modal biometric methods using two or more features for robust biometric authentication tasks. They employ unique combinations of measurable physical characteristics: fingerprint, facial features, iris of the eye, voice print, hand geometry, vein patterns, and so on. Since these traits are hardly imitable by other persons, the aim of these multibiometric systems is to achieve a high reliability to determine or verify person's identity. In this paper a multimodal biometric system using two different fingerprints is proposed. The matching module integrates fuzzy logic methods for matching score fusion. Experimental trials using both decision level fusion and matching score level fusion were performed. Experimental results show an improvement of 6.7% using the matching score level fusion rather then a mono-modal authentication system.

Keywords: Fuzzy fusion, fingerprint, afas, matching score level fusion.

1 Introduction

A biometric system regards a pattern recognition system acquiring biometric data from a user, extracting features from it, and comparing extracted features with stored template sets.

Traditionally biometric systems, operating on a single biometric feature, have many limitations [14]:

- *troubles on data sensor*: sensor captured data are often affected from "noise" due to the environmental conditions (insufficient light, powder etc) or to user physiological conditions (cold, cut fingers etc);
- *distinctiveness ability*: all biometric features have not the same distinctiveness (for example, hand geometry biometric systems are less selective than those based on fingerprints);
- *not universality*: every biometric feature is universal, but some people not owned them for some physical invalidity (lack of a finger or the voice).

B. Apolloni et al. (Eds.): KES 2007/WIRN 2007, Part I, LNAI 4692, pp. 108–115, 2007.

Multimodal biometric systems [6][7][8] are a new approach to overcome these problems. These systems promise significant improvements over biometric system using a single characteristic, in terms of higher accuracy and most resistance to spoofing.

In literature different approaches for multimodal biometric systems have been proposed. These systems are based on different biometric features or introduce different fusion algorithm of these features.

In [8] Jain *et al.* have proposed a multimodal biometric system which integrates face-fingerprints-voice recognition modules to realize personal identification system. The authors have used a statistical decision theory based on Neyman-Pearson rule to classify a given observation.

In [9] Lau *et al.* have presented a multimodal biometric system combining face-voice-fingerprints. The authors used a fuzzy logic based approach, in order to consider the effect of external conditions on the system. With more details they have implemented fuzzy logic module to calculate the weights for each recognition subsystem to realize the weight sum rule.

In [10] Zewail *et al.* have presented an identity verification system based on soft and hard biometrics. The authors proposed a fingerprint-iris based system integrating extra soft biometric information, as the colour of the iris, to enhance its performance.

In this paper a multimodal biometric system, combining two fingerprint authentication systems, is proposed. With the proposed approach some biometric monomodal authentication systems limitations has been reduced. The used fusion techniques are based on fuzzy logic considering fingerprints image quality as in particular, two fusion methods have been tested: the first one implements a decision level fusion, the second one implements a matching score fusion. The proposed approach uses the images quality as goodness index, as the authors in [9], but nothing weight is obtained from fuzzy rules to characterize the system phases. The developed approach can be adopted with general multimodal authentication systems involving different biometric features.

The paper is organized as follows. Section II deals with taxonomy of multimodal systems; section III introduces the proposed system; in the section IV the experimental results are reported; and finally in section V some conclusions are described.

2 Taxonomy of Multimodal Systems

The multimodal biometric systems can be classified considering the biometrics source number and the biometrics samples number [7][11]. Four different models of multimodal systems can be distinguished:

- the MSSS (MultiSample-SingleSource) model: many samples of one single biometric feature are used to recognize a user;
- the SSMS (SingleSample-MultiSource) model: only one sample of many biometric features is used;
- the MSMS (MultiSample-MultiSource) model: many sensors for every biometric feature are used;
- the multi-algorithms model: different matching algorithms are used in the recognition of one single sample of one biometric feature.

The proposed multimodal approach implements the SSMS model.

3 The Proposed Multimodal Biometric System

3.1 Standard Fingerprint Authentication

In general, an automatic fingerprint authentication system [1][2][12], commonly called AFAS (Automatic Fingerprint Authentication System), consists of three main processing phases [3]: *image acquisition*, *feature extraction*, and *matching*. In the first phase, a sensor captures a fingerprint image, successively a *features vector* (containing information about minutiae) will be extract, and finally some indexes of similarity are calculated to verify the matching score between the input image and the stored template. In such system a scalar number called *matching score* is the considered measure of identity correctness. Generally: if the matching score is smaller than a fixed threshold the user will be refused otherwise the user will be accepted.

3.2 Fusion in the Proposed Multimodal Authentication System

A typical multimodal authentication system is composed of two or more parallel mono-modal systems and a fusion module. Fusion module takes two or more data and combines them in order to obtain the authentication result: impostor or genuine user.

In this work a new module estimating input image quality have been added to a traditional AFAS. Image quality is estimated through the analysis of two specific image parameters: the number of *erased segments* (*ES*) and the number of the *candidate minutiae* (*CM*). The *eliminated segments* are portions of fingerprint too much noisy, so no useful information can be found inside them. The adopted matching algorithm is based on a structural matching executed on similar geometric forms between input fingerprint and the stored template. In particular triangle forms, produced by triplet of minutiae, have been used to perform the matching. We have called the produced score by a single AFAS "*recognition degree*".

The fusion strategies are divided into two main categories [6]: *pre-mapping fusion* (before matching) and *post-mapping fusion* (after matching). The first strategy deals with the *sensor data fusion level* and *feature vector fusion level*. These techniques are not used because they give many implementation problems [6]. The second strategy is realized through the *decision level fusion*, based on some algorithms which combine single decisions for each component system, or through the *matching score level fusion*, which combines the matching scores of each component system. Figure 1 shows the blocks scheme of a recognition process with many fusion levels.

In this paper the decision level fusion and the matching level fusion were implemented and evaluated using two fingerprints based subsystems as sample. The whole recognition system also incorporates the knowledge about the image quality. As started before, image quality is based on both *ES* and *CM* features. Experimental trials aimed to find a global image Goodness Index (*GI*) in term of *ES* and *CM* values were performed. Figure 2 shows the *ES* and *CM* ranges with relative membership function values and the membership function range of the *GI*. These ranges are experimentally obtained.

The Goodness Index *GI=f(ES, CM)* is obtained applying "*moment defuzzify function*" [16]. The *moment defuzzify function* returns a defuzzified floating point value, which represents the defuzzified fuzzy set using the centre of gravity (or first moment of inertia) function. The centroid method is a good method to use in many processes since it tends to smooth out the fuzzy region [16]. Figure 3 shows three different fingerprint images with a different *GI* value. The image on the left is a fingerprint of good quality, *GI=82.01*, on the center a fingerprint of medium quality, *GI=55.01*, and on the right a fingerprint of bad quality, *GI=0*.

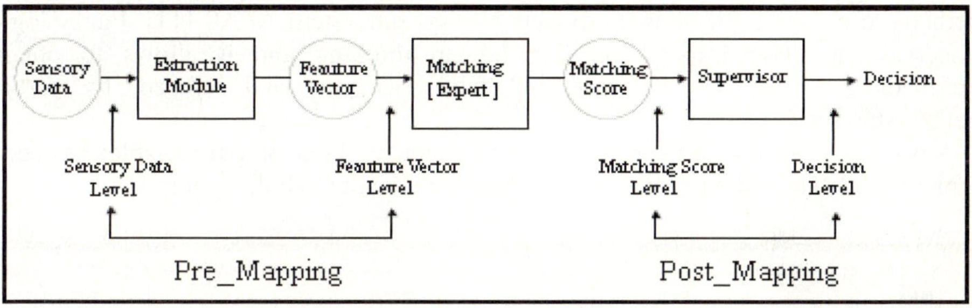

Fig. 1. Distinction of the different fusion levels: on the left the pre_mapping fusion, on the right the post_mapping fusion (used in this work)

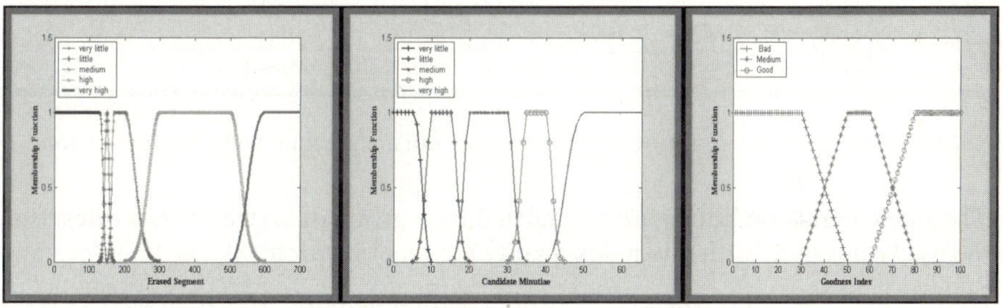

Fig. 2. *ES*, *CM* and *GI* ranges. On the left *ES* values, on the center *CM* values and on the right *GI* values.

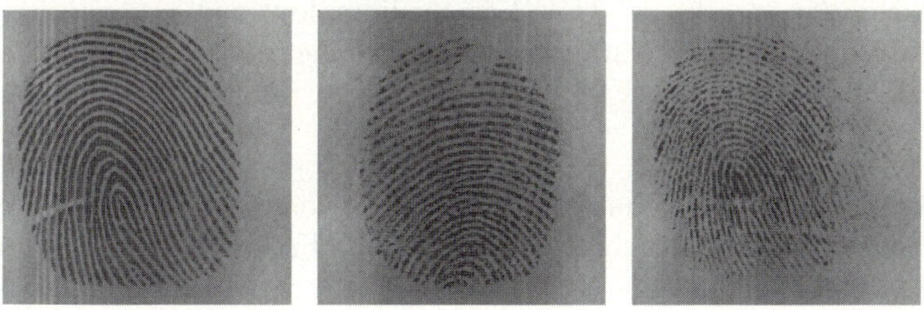

Fig. 3. Two typical fingerprints of different quality. On the left a fingerprint image of good quality (*GI=82.01*), on the center a fingerprint image of medium quality (*GI=55.01*), while on the right a fingerprint image of bad quality (*GI=0*).

3.3 Fuzzy Fusion

Decision Level Fusion

The authentication system is composed of two modules AFAS to realize a multimodal authentication system combining the obtained scores from the single systems, Figure 4. With more details, the proposed authentication system is composed of "AFAS1" subsystem that performs index fingerprint processing and recognition, and of "AFAS2" subsystem that performs the middle finger's fingerprint recognition and processing.

The proposed and implemented fusion module uses the fuzzy logic principles and methods to combine the scores products by two subsystem AFAS [11]. Fuzzy logic processes imprecise information like human thinking and it allows to obtain intermediate values between true and false, accepted and refused, by partial membership set.

A fuzzy inference system composed of two input and one output variables has been implemented. The output represents the decision taken by whole system.

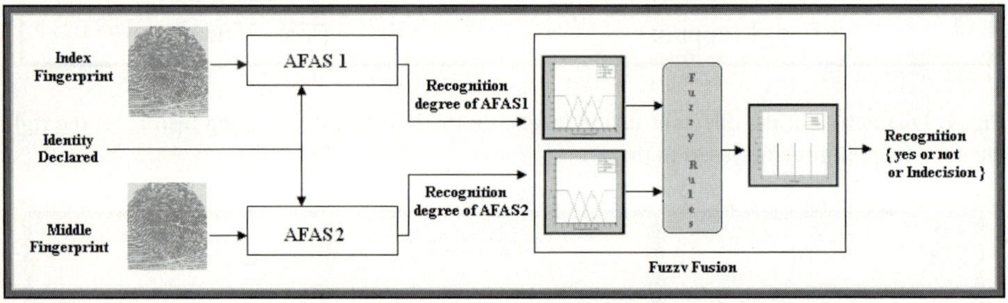

Fig. 4. The general scheme of the proposed system architecture with decision fusion

The fuzzy logic conditions are formulated by a group of sixteen fuzzy rules. Some of the used rules used with this index quality are the following:

1. if *recognition degree of AFAS1* is high and *recognition degree of AFAS2* is high then final decision is *genuine user.*
2. if *recognition acknowledgment degree of AFAS1* is low and *recognition degree of AFAS2* is low then final decision is *impostor user.*
3. if *recognition degree of AFAS1* is medium low and *recognition degree of AFAS2* is medium low then final decision is *indecision.*
   etc…

Matching Score Level Fusion

The proposed system architecture is composed of two AFAS modules and the fusion is realized combining the matching score of both AFAS and the quality measure of fingerprint images.

The fusion has been obtained by a fuzzy system with four input and one output variables. The output variable represents the decision of the whole system. The fuzzy system uses the knowledge base built with above fuzzy rules. Each rule has the following common guidelines:

- if the input images have good quality and fingerprints are very similar to the stored fingerprints then certainly the user is who he/she claims to be.
- if the input images have good quality and fingerprints are not very similar to the stored fingerprints then certainly the user is an impostor".
- if AFAS1 works with an input bad quality image and AFAS2 works with an input good quality image then the AFAS2 decision will be more discriminant than AFAS1.

The figure 5 shows the scheme of the proposed architecture with matching score level fusion.

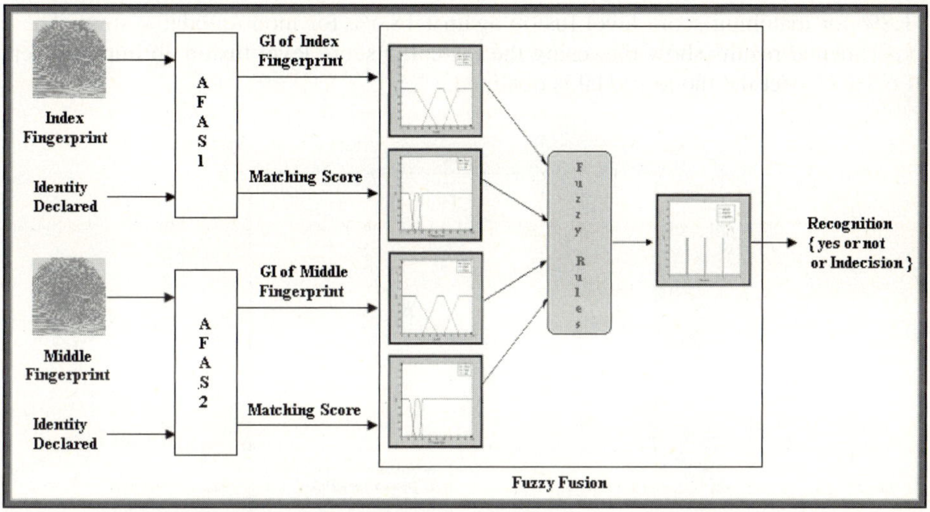

Fig. 5. General scheme of the proposed architecture with matching score level fusion

4 Experimental Results

The purpose of an authentication system is to decide if a person is who he/she says to be. In other words, the system can make two types of errors: a False Rejection Rate (FRR, when a client is rejected) and a False Acceptance Rate (FAR, when an impostor is accepted) [13]. The performances of an authentication system are given in terms of error rates computed during the test phase. An optimal identity verification have FAR=0 and FRR=0.

The overall performance of an identity verification system is well characterized by the *Receiver Operating Characteristic* (ROC) curve [5][13], which represents the FAR as a function of FRR or its complementary GAR=1-FRR (Genuine Acceptance Rate). The Equal Error Rate (EER: i.e. when FAR=FRR), is often used as the only performance measure of an identity verification method, although this measure gives just one point of the ROC.

The algorithms are described in previous sections are used to merge the results coming from two modalities. The fuzzy rules have been deduced by analysis of the experimental trials. To evaluate algorithm performance the DB3 database of the

FVC2002 [15] has been used. The database is composed of eight samples coming from one hundred users. The 800 images have 300*300 pixel dimension. The experimental results have been performed using the method of FVC competition to calculate FAR and FRR indexes [15].

The results show that the multimodal system performances are better than mono-modal system. In addition, the tests show that the matching score level fusion gives better result than the decision level fusion ones.

Figure 6 shows the performance of the proposed approach on DB3 using the ROC curves. In this figure the ROC curve of the mono-modal system without filter [4] is depicted. The Equal Error Rate is about 13,7% for decision level fusion and about 11,8% for matching score level fusion against 18,5% for mono-modal system. These experimental results show that using the matching score level fusion an improvement of 6.7% respect the mono-modal is obtained.

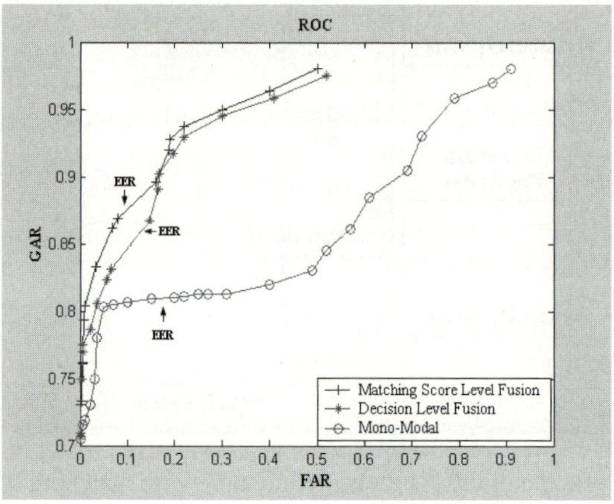

Fig. 6. The ROC curves about our approach on FVC DB

5 Conclusion

The multimodal methods for biometric authentication are used to overcome some limitations of a single modal system. In this work a multimodal biometric system which combines two different fingerprints to realize a personal authentication has been proposed. In particular, two type of fusion based on fuzzy logic techniques have been developed and tested. The experimental results show that the matching score level fusion gives good result. However, the system needs to be tested on a large database in a real operating environment.

References

1. Conti, V., Pilato, G., Vitabile, S., Sorbello, F.: A Robust System for Fingerprints Identification, Knowledge-Based Intelligent Information Engineering System & Allied Technologies, Crema, pp. 1162–1166 (16-18 September, 2002)

2. Conti, V., Pilato, G., Vitabile, S., Sorbello, F.: Verification of Ink-on-paper Fingerprints by Using Image Processing Techniques and a New Matching Operator. AI*IA Siena pp. 594 – 601, (10-13 September 2002)
3. Conti, V., Milici, G., Vitabile, S., Sorbello, F.: A New Directional Morphological Filter in a Automated Fingerprint Identification System. In: accepted to CISI'06 Conferenza Italiana Sistemi Intelligenti, Ancona (Italia), pp. 17–29 (September 2006)
4. Milici, G., Raia, G., Vitabile, S., Sorbello, F.: Fingerprint Image Enhancement Using Morphological Filter. In: IEEE EUROCON 2005 - The 8th International Conference on Computer as a tool, Belgrade, Serbia & Montenegro, 21-24 November, pp. 967–970 (2005)
5. Jain, A.K., Ross, A., Prabhakar, S.: An Introduction to Biometric Recognition. In: IEEE Transactions on Circuits and Systems for Video Technology, vol. 14(1) (January 2004)
6. Bubeck, U.M.: Multibiometric Authentication – An Overview of Recent Developments, San Diego University (spring 2003), http://www.thuktun.org/cs574/papers/multibiometrics.pdf
7. Poh, N., Bengio, S., Korczak, J.: A multi-sample multi-source model for biometric authentication. In: Proc. IEEE 12th Workshop on Neural Networks for Signal Processing, pp. 375–384 (2002), http://citeseer.ist.psu.edu/thian02multisample.html
8. Jain, A.K., Hong, L., Kulkarni, Y.: A Multimodal Biometric System Using Fingerprint, Face and Speech. In: Conference on Audio-Video based Biometric Person Authentication (1999)
9. Lau, C.W., Ma, B., Meng, H.M., Moon, Y.S., Yam, Y.: Fuzzy Logic Decision Fusion in a Multimodal Biometric System. In: Proceedings of the 8th International Conference on Spoken Language Processing (ICSLP), Korea (October 2004)
10. Zewail, R., Elsafi, A., Saeb, M., Hamdy, N.: Soft and Hard Biometrics Fusion for Improved Identity Verification. In: the 47th IEEE International Midwest Symposium on Circuits and Systems (2004)
11. Prabhakar, S., Jain, A.K.: Decision-level Fusion in Biometric Verification. Pattern Recognition 35(4), 861–874 (2002)
12. Jain, A.K., Hong, L., Bolle, R.: On-line fingerprint verification. IEEE Trans. Pattern Analysis and Machine Intelligence 19(4), 302–313 (1997)
13. Dahel, K., Xiao, Q.: Accuracy Performance Analysis of Multimodal Biometrics, Proceedings of the, IEEE, Workshop on Information Assurance (2003)
14. Verlinde, P., Chollet, G., Acheroy, M.: Multi-Modal Identity Verification Using Expert Fusion. Information Fusion 1(1), 17–33 (2000)
15. http://bias.csr.unibo.it/fvc2002/
16. Zadeh, L.A.: Fuzzy sets. Information and Control 8, 338–353 (1965)

Parameter Determination of Induction Machines by Hybrid Genetic Algorithms

Mümtaz Mutluer[1], Osman Bilgin[1], and Mehmet Çunkaş[2]

[1] Department of Electrical and Electronics Engineering
Selçuk University, Konya, 42031, Turkey
[2] Department of Electronic and Computer Education,
Selçuk University, Konya, 42075, Turkey
mmutluer@selcuk.edu.tr, mcunkas@yahoo.com

Abstract. In general, a genetic algorithm combined with other algorithms (e.g. tabu search, simulated annealing, etc.) is well known to be a powerful approach. In this paper, an efficient hybrid approach containing local search and genetic algorithms is presented. The purpose of the using local search mechanisms is to provide better the solution quality and to increase the convergence speed. It is demonstrated that the performance of the proposed algorithms is significantly better than the conventional genetic algorithm methods.

Keywords: Induction machine, hybrid genetic algorithm, parameter determination.

1 Introduction

Artificial intelligent methods have been increasingly employed to determine electrical parameters of induction machine recently [1],[2]. Genetic algorithms (GAs) are widely accepted for its simplicity, ability to converge with minimal input from the user with a short computational time [3],[4]. Genetic algorithm developed by Holland is a very important stochastic search algorithm for solving optimization problems during the last two decades. Generally, the GA consists of operators such as selection, crossover, and mutation. A local search or metaheuristic algorithm is incorporated into these genetic operators to create a more powerful algorithm. Such an algorithm is called a "hybrid GA" (HGA). In general, the HGA is based on applying the genetic operators to a set of local optima found by local search algorithms. This combined algorithm can keep the advantages and avoid the disadvantages of both algorithms. Owing to the existence of Simulated Annealing (SA), this combined algorithm has better fine-tuning ability to search for the global optimum solutions and has stronger hill-climbing ability for escaping from local minima than the standard GA [9],[10],[11]. These methods are sometimes not able to find the optimal solution for difficult constrained non-linear optimization problems.

In this paper, a Hybrid Genetic Algorithm (HGA), combined with GAs and local search algorithm, is proposed for identifying the induction motor parameters. It is designed to improve the fine-tuning ability, the hill-climbing ability, and the speed of convergence to the global optimum solutions or near global optimum solutions.

B. Apolloni et al. (Eds.): KES 2007/WIRN 2007, Part I, LNAI 4692, pp. 116–124, 2007.

Furthermore, it could contribute to enriching of the search space. An appropriate combination of the two algorithms has therefore accelerated the convergence speed and minimized the possibility of being trapped in a local optimum. Several samples are carried out on a standard test functions and the results are compared with conventional genetic algorithms. The effectiveness and potential of the proposed approach is demonstrated on determining of induction motor parameters.

2 Hybrid Genetic Algorithms

It is assumed that the reader is familiar with the conventional GA. More detailed overviews about GA in given in Ref [8]. In this section, the proposed algorithm is introduced. The flowchart of HGA optimization procedure is shown in Fig. 1.

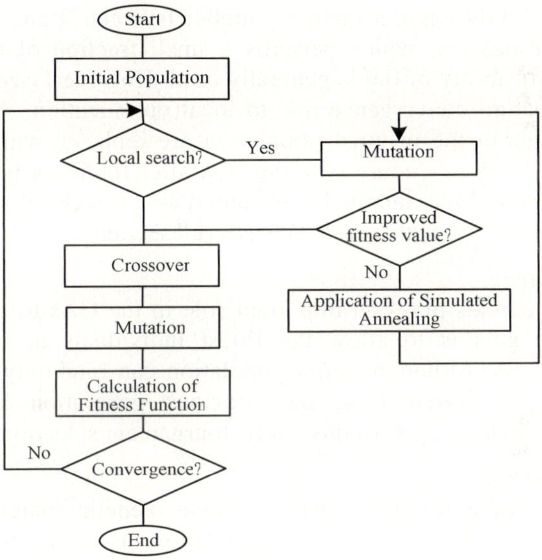

Fig. 1. Flowchart of hybrid genetic algorithm

The HGAs consist of two parts which one of them is global search algorithm and the other is local search algorithm. The global search is fundamental part of hybrid genetic algorithms. The HGAs straightly converge to optimum solution by the local search assistance. There are two main types in hybrid GAs application: Lamarckian and Baldwinian GAs. The Lamarckian GAs update the chromosome after the local optimizations, while the Baldwinian GAs do not update the chromosome after the local optimizations, and the local optimizations are only used as a fitness evaluation. Generally, the Lamarckian GAs have been more widely accepted and the Baldwinian GAs' use has been somewhat restricted [5],[6]. In this paper, we used the concepts of the Lamarckian GAs. The proposed algorithm is explained as below:

i) Initial population
GAs start by randomly generating a population of N individuals, each of length *l*. it is worth noting that each individual represents a possible solution to the problem. First of all, size of population and numbers of a chromosome bits must be determined, due to the fact that these values influences the accuracy and speedy of genetic algorithms. Number bits of a chromosome depend on the boundary values. Commonly, the population size ranges from 50 to 300 and number of a chromosome bits extends from 10 to 30. Population size 100 and 15 bits are used to code each design variable.

ii) Local search
The disadvantages of local search algorithm require many function calls and longer computational time. Moreover, excessive use of local optimization can degrade the diversity of the population. Therefore, the proposed local search algorithm is not applied in every iteration of the process. In this procedure, a general mutation method is applied only when a random number between 0 and 1 is bigger than a predetermined probability, which perturbs a small fraction of the offspring. The predetermined probability of 0.9 is generally used. However, larger values are used to alleviate premature convergence due to local optimization. After the mutation, the worst solutions in the original population are replaced with newly generated ones. The local search procedure is terminated after finding a better solution from among all population. More detailed explanation about the local search procedure is given in the following ***"Local search Procedure"*** section.

iii) Selection strategy
The selection of parents plays an important role in the GAs to produce successful generations. The goal is to allow the fittest individuals to be selected in the reproduction process. Although initial population can randomly be generated, the next generation is chosen from the previous generation members using a probabilistic selection process. In this study, tournament selection is used.

iv) Crossover operation
Crossover is a mixing operator that combines genetic material from selected parents. Among all the members of population, two distinct individuals are randomly selected as indicated in Fig. 2, afterwards chosen members are crossed over a random point of each member. Variety of population is finally obtained and good features of old population are transmitted to new population. In this case, single-point crossover and 0.8 crossover rate are used to perform crossover operation.

v) Mutation operation
This is a common genetic manipulation operator and it involves the random alteration of genes during the process of copying a chromosome from one generation to the next. Raising the ratio of mutations increases the algorithm's freedom to search outside of the current region of parameter space. In this case, bit mutation and 0.01 mutation rate are used. An example of how mutation works is displayed in Fig. 3.

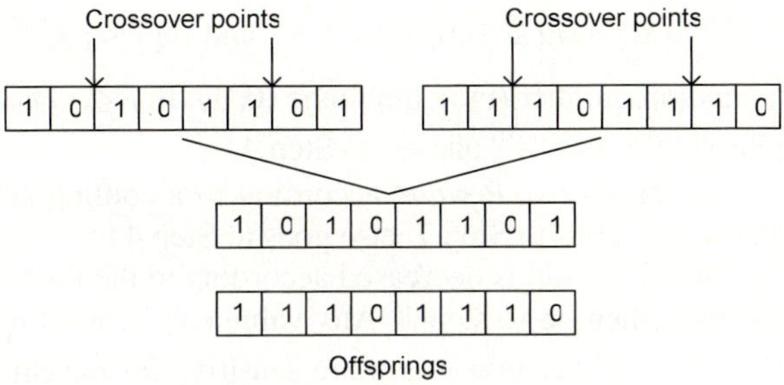

Fig. 2. Usage of a single point crossover between two chromosomes

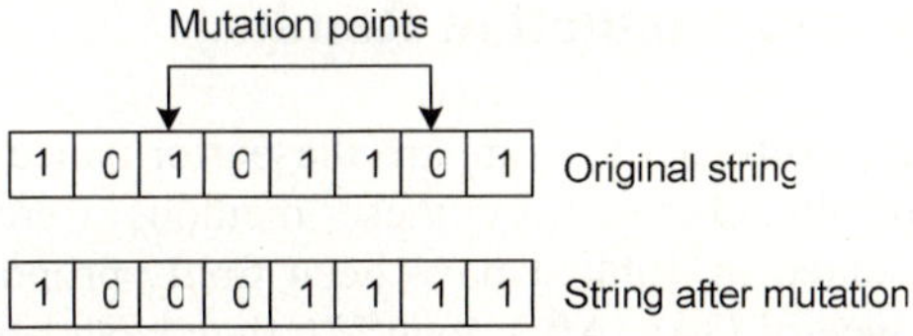

Fig. 3. An example of mutation using binary representation

vi) Fitness function
The individual fitness is a metric to assess the performance of population members. The fitness of each individual in this situation is calculated by the fitness function given in Equation (4).

vii) Reinsert offspring
At this point a new population is achieved as follows; all individuals are evaluated as described in Step vi and the subsequent steps are repeated. The procedure is completed when test convergence value in Fig. 1 is reached, i.e. a prefixed number of generations is reached.

Local search Procedure: The proposed local search algorithm is as follows:

Step 1) Some genes are selected for mutation and the number of selected genes is defined as follows:

$$N_c = N_p x P$$

where N_c is the number of genes that change their states, N_p is the total bit number of population, and P is a probability between 0 and 1.

Step 2) Here, some improvement of the fitness value is detected. When better solution among the current population is found, the local search procedure terminate, else go to Step 3.

Step 3) In this step, Simulated Annealing is applied to check the fitness value after the mutation. Set temperature to P (local mutation rate in this case), Population size N (the same initial population size), and set .. to 0.

Step 4) Calculate a fitness value (i.e. energy) of neighboring point, $E(x^{t+1}$, this point is randomly created.

Step 5) If $\Delta E=E(X^{(t+1)})-E(X^{(t)})<0$, set $t=t+1$ and replace $X^{(t+1)}$ with $X^{(t)}$; or else create a random number (r) in the range (0, 1). If $r \le \exp(-\Delta E / P)$ set $t=t+1$, and replace $X^{(t+1)}$ with $X^{(t)}$;and go to Step 4.

Step 6) If t mod N is zero, P *value* according to a cooling schedule is lowered and then process goes to Step 7 else goes to Step 4

Step 7) The mutation ratio is decreased according to the following procedure: $P(new) = P(old)x0.985$, then go to step 1. Any value can be used instead of 0.985. In this paper, 0.985 is selected because it is more sensitive for the Simulated Annealing processes.

3 Test Functions and Simulation Results

In this section, the test problems to compare the performance of GAs and HGA are described and the results obtained by these methods were reviewed. Four test functions (F_1-F_4) presented in Table 1 have been firstly proposed by De Jong [7] to measure the performance of GAs. After Jong [7], they have been extensively used by GA researchers and other algorithm communities. In the present study, the test functions which are convex (F_1), non-convex (F_2), discontinuous (F_3) and multimodal (F_4) are used.

Table 1. Test functions

Functions	Parameter Limits	Optimum solutions
$F_1 = \sum_{i=1}^{3} x_i^2$	$-5.12 \le x_i \le 5.12$	$x_i = 0;\ i = 1,2,3;\ F_1 = 0$
$F_2 = 100(x_1^2 - x_2)^2 + (1 - x_1)^2$	$-2.048 \le x_i \le 2.048$	$x_i = 1;\ i = 1,2;\ F_2 = 0$
$F_3 = \sum_{i=1}^{5} \lVert x_i \rVert$	$-5.12 \le x_i \le 5.12$	$x_i = -5.12;\ i = 1..5;\ F_3 = -30$
$F_4 = 0.002 + \sum_{j=1}^{25} \frac{1}{j + \sum_{i=1}^{2}(x_i - a_{ij})^6}$	$-65.536 \le x_i \le 65.536$	$x_i = -32;\ i = 1,2;\ F_4 = 0$

The average number of generations for finding the optimum solutions shown in Table 1 is presented in Fig. 4. The performance of the HGA to that of other well known GA is also compared in Fig. 4. These algorithms were run 50 times for each test function to achieve average results. For each run, the initial population was randomly created by means of using different seed numbers.

When the results produced by the HGA for all the functions are evaluated together, it is seen in Fig 4 that the convergence speed of the HGA is significantly better than that of GAs. From the results, it can be concluded that the HGA is usually able to find optimum solutions for the functions with a less number of generations than conventional GAs. The significant improvement is observed in function F2, while results found for F1 are the same in both algorithms. HGA is generally producing the best results for the same functions than the GA in resulting smaller average of number of generations.

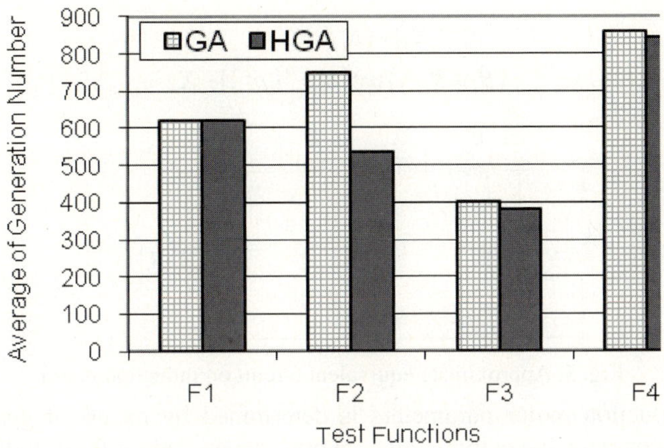

Fig. 4. Average number of generation for finding the optimal solution

4 Parameter Identification

The identification of suitable motor parameters can be carried out using limited information from motor that can easily be retrieved. This information consists of motor nameplate data along with common data provided with a motor data sheet or measurements can be taken with common test equipment. The identification of induction motor parameter, as shown in Fig.5 is performed using the HGA. The full-load, locked-rotor, and breakdown torque equations are formed for optimization problem. Each equation consists of three or more machine parameters. The fitness function [1] can be written as follows:

$$f_1 = \frac{U_p^2(R_r/s)}{\omega_s[(R_s + R_r/s)^2 + X_l^2]} - T_{fl} \tag{1}$$

$$f_2 = \frac{U_p^2 R_r}{\omega_s[(R_s + R_r)^2 + X_l^2]} - T_{lr} \tag{2}$$

$$f_3 = \frac{U_p^2}{2\omega_s[R_s + \sqrt{R_s^2 + X_l^2}]} - T_{bd} \tag{3}$$

$$fitness\_function = \frac{1}{f_1^2 + f_2^2 + f_3^2} \tag{4}$$

Where T_{fl} is the full-load torque, T_{lr} is the lock-rotor torque, and T_{bd} is the breakdown torque. Parameters of induction machine are the stator resistance R_s, the rotor resistance R_r, the combination of stator and rotor reactance X_l, and the phase voltage U_p. Magnetizing reactance X_m is calculated by using four parameters obtained from HGA, as follows:

$$X_m = \frac{X_l^2 + (R_s + R_r/s)^2}{(R_s + R_r/s)\tan(\cos^{-1}(pf)) - X_l} \tag{5}$$

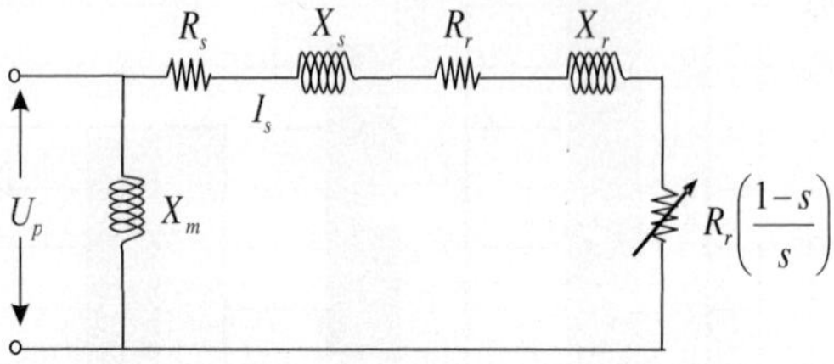

Fig. 5. Approximate equivalent circuit on induction motor

Size of induction motor parameters is determined by means of given parameter values. If parameters have more bits, hybrid genetic algorithms will slow down. Therefore, optimum length of each parameters and sensitivity values are selected 15 bits and 10000, respectively. As shown in Table 2, three induction motors with known equivalent circuit parameters are used to test the identified parameters as in the case of Pillay's study [1].

Table 2. Actual induction motor parameters

Hp	Volt	Speed[rpm]	Frequ.[Hz]	Rs[Ω]	R_r[Ω]	X_l[Ω]	X_m[Ω]
5	230	1750	60	0.434	0.303	2.511	24.61
50	460	1705	60	0.087	0.228	0.604	13.08
500	2300	1773	60	0.262	0.187	2.412	54.02

5 Results

The parameter values obtained from GA and HGA are shown in Table 3 and 4. The estimated parameters are compared with the actual motor parameters and then the errors are calculated. The results obtained from HGA are better than that from GA. It is concluded that the errors for the parameter R_s, R_r, X_l and X_m are in acceptable level for industrial application.

Table 3. Genetic algorithm results

	Motor 1--5 Hp	Motor2--50 Hp	Motor3--500 Hp
R_s (est)	**0.428**	**0.082**	**0.258**
R_s (act)	0.434	0.087	0.262
% error	**1.379**	**4.807**	**1.296**
R_r (est)	0.303	0.229	0.187
R_r (act)	0.303	0.228	0.187
% error	**0.123**	**0.546**	**0.069**
X_l (est)	2.517	0.608	2.415
X_l (act)	2.511	0.604	2.412
% error	**0.241**	**0.668**	**0.151**
X_m (est)	24.650	13.150	54.092
X_m (act)	24.61	13.08	54.02
% error	**0.163**	**0.540**	**0.134**

Table 4. Hybrid genetic algorithm results

	Motor1--5 Hp	**Motor2--50 Hp**	**Motor3--500 Hp**
R_s (est)	0.433	0.088	0.259
R_s (act)	0.434	0.087	0.262
% error	**0.001**	**1.506**	**0.783**
R_r (est)	0.303	0.227	0.187
R_r (act)	0.303	0.228	0.187
% error	**0.003**	**0.096**	**0.036**
X_l (est)	2.511	0.602	2.414
X_l (act)	2.511	0.604	2.412
% error	**0.001**	**0.210**	**0.088**
X_m (est)	24.610	13.063	54.062
X_m (act)	24.61	13.08	54.02
% error	**0.001**	**0.128**	**0.077**

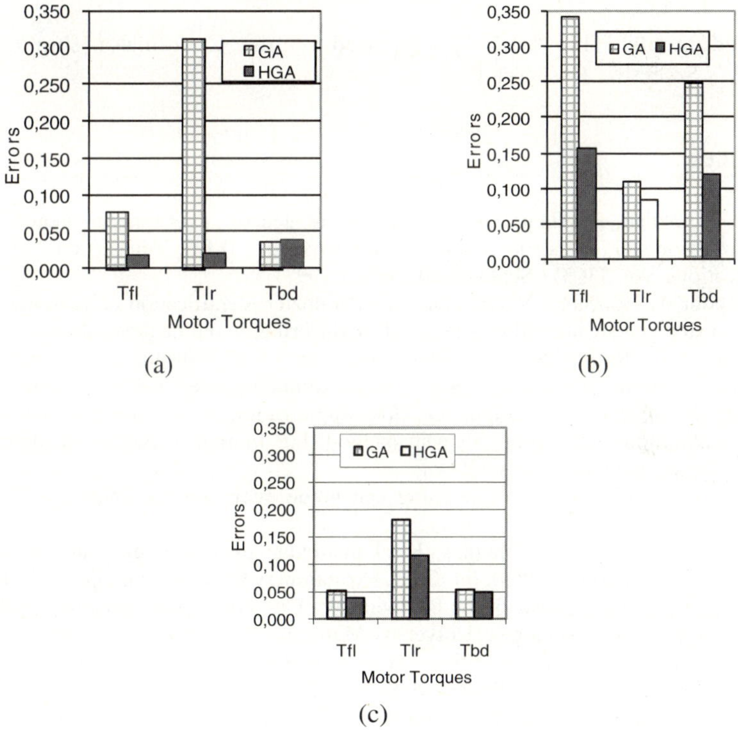

Fig. 6. Torque errors for (a) 5hp (b) 50hp (c) 500hp (Tfl: full-load, Tlr: starting, and Tbd: breakdown torques)

Larger errors were produced in R_1 for 50hp and 500hp motors in HGA procedure. However, Fig.6 shows that the errors in R_1 do not affect the torque calculations significantly. In that case, while there were 1.506 % errors in R_1 as larger errors, the maximum error produced in the estimation of any torque is 0.155%. This is normally a situation, since the error function was defined to minimize the torque, not electrical parameters. The results in Fig.6 show that the estimated torque error is at most 1.506%, and is smaller than GA results. Therefore, the results obtained by HGA are more realistic to parameter identification.

6 Conclusion

In this paper, a hybrid GA for identifying induction motor parameters was introduced. To improve conventional GA, a novel local optimization algorithm was proposed and hybridized with the existing GA. This proposed algorithm was then successfully applied to the test problems and the motor parameter identification, and proved to be a very promising optimization algorithm for using in optimization problems. Furthermore, the simulation results have shown that the HGA minimizes the torque error, and provides an accurate determination of electrical parameters.

Acknowledgments. This work is supported by the Coordinator ship of Selçuk University's Scientific Research Project.

References

1. Pillay, P., Nolan, R., Haque, T.: Application of genetic algorithms to motor parameter determination for transient torque calculations. In: IEEE Transactions on Industry Applications, vol. 33(5) (September/October, 1997)
2. Rahimpour, E., Rashtchi, V., Pesaran, M.: Parameter identification of deep-bar induction motors using genetic algorithm. Electrical Engineering, in online press (2006)
3. Thilagar, S.H, Rao, G.S.: Parameter estimation of three-winding transformers using genetic algorithm. Engineering Appl. of Artificial intelligence 15, 429–437 (2002)
4. Çunkaş, M., Akkaya, R., Bilgin, O: Cost optimization of submersible motors using a genetic algorithm and a finite element method. Int. Journal of Advanced Manufacturing Technologies, In Online Press (2006)
5. Hinton, G., Nowlan, S.: How learning can guide evolution. Complex Sys. 1, 495–502 (1987)
6. Whitley, D., Gordon, V., Mathias, K.: Lamarckian evolution, the Baldwin effect and function optimization. In: Proc. Int. Conf. Evolutionary Computation, pp. 6–15 (1994)
7. De Jong, K.A.: An Analysis of the Behavior of a Class of Genetic Adaptive Systems. Phd Thesis, University of Michigan (University Micro_lms No. 76-9381) (1975)
8. Goldberg, D.E.: Genetic Algorithms in Search, Optimization, and Machine Learning. Addison-Wesley Publishing, London, UK (1989)
9. Jeong, I.K., Lee, J.: Adaptive simulated annealing genetic algorithm for system identification. Eng Appl Artif Intel. 9, 523–532 (1996)
10. Tan, K.C., Li, Y., Murray-Smith, D.J., Sharman, K.: System identification and linearization using genetic algorithms with simulated annealing. In: Proc. IEEE genetic algorithms in engineering system: innovations and applications, conf. Public, vol. 414, pp. 164–169 (1995)
11. Adler, D.: Genetic algorithm and simulated annealing: a marriage proposal. In: Proceeding of the IEEE international conference on neural network, pp. 1104–1109 (1993)

Prediction of E.Coli Promoter Gene Sequences Using a Hybrid Combination Based on Feature Selection, Fuzzy Weighted Pre-processing, and Decision Tree Classifier

Bayram Akdemir, Kemal Polat, and Salih Güneş

Selcuk University, Dept. of Electrical & Electronics Engineering,
42075, Konya, Turkey
{bayakdemir,kpolat,sgunes}@selcuk.edu.tr

Abstract. In this paper, we have investigated the real-world task of recognizing biological concepts in DNA sequences. Recognizing promoters in strings that represent nucleotides (one of A, G, T, or C) has been performed using a hybrid approach based on combining feature selection (FS), fuzzy weighted pre-processing, and C4.5 decision tree classifier (DCS). Dimensionality of E.coli Promoter Gene Sequences dataset has 57 attributes and 106 samples including 53 promoters and 53 non-promoters. The proposed approach consists of three stages. Firstly, we have used the FS process to reduce the dimensionality of E.coli Promoter Gene Sequences dataset that has 57 attributes. So the dimensionality of this dataset has been reduced to 4 attributes by means of FS process. Secondly, fuzzy weighted pre-processing has been used to weight E.coli Promoter Gene Sequences dataset that has 4 attributes in interval of [0,1]. Finally, C4.5 decision tree classifier algorithm has been run to estimation the E.coli Promoter Gene Sequences. In order to show the performance of the proposed system, we have used the predicton accuracy and 10-fold cross validation. 93.33% classification accuracy has been obtained by the proposed system using 10-fold cross validation. This success shows that the proposed system is a robust and effective system in the prediction of E.coli Promoter Gene Sequences.

1 Introduction

The estimation of E.coli promoter gene sequences is important issue in molecular biology field. Here, we first have explained the promoters in molecular biology. Then a novel approach based on combining feature selection process, fuzzy weighted pre-processing, and C4.5 decision tree has been explained.

Promoters are DNA (Deoxyribonucleic acid) sequences which affect the frequency and location of transcription initiation though interaction with RNA polymerase. Two conserved regions about 35 and 10 base pairs (bp) upstream from the transcription start (-35 and -10 regions, respectively) were identified by comparison of relatively few promoters. More extensive compilations and comparisons of prompters for genes of E.coli and its phage and plasmids supported and extended the concept of a "consensus" promoter sequence: a-35 (TTGACA) and -10 (TATAAT) region separated by 17 bp with transcription initiating at a purine about 7 bp downstream from the '3' end of the -10 region. While the -35 and -10 regions show the greatest

B. Apolloni et al. (Eds.): KES 2007/WIRN 2007, Part I, LNAI 4692, pp. 125–131, 2007.

conservation across promoters and are also the sites of nearly all mutations which affect transcriptional strength, other bases flanking the -35 and -10 regions, in addition to the start point also occur at greater than random frequencies and sometimes affect promoter activity. In addition, variation in spacing between the -35 and -10 regions plays a role in promoter strength [1].

Promoter compilations and analysis have led to computer programs which predict the location of promoter sequences on the basis of homology either to the consensus sequence or to a reference list of promoters. Such programs are of practical significance in searching new sequences; thus promoter compilations are important beyond proving data regarding promoter structure. However, current compilations are based on sequences aligned by eye in attempts to maximize homology to a consensus sequence. Unfortunately, sequences closer to the consensus sequence may be missed thus weakening the homology between promoters and consequently reducing the predictive power of algorithms. Although promoter elements evidence that pin-point bases which interact with RNA (Ribonucleic acid) polymerase, such data is unavailable for most genes [1].

There are few studies related with predicting of E.coli Promoter Gene Sequences in literature. Among these, Ranawana et al. proposed a neural network based multi-classifier system for gene identification in DNA sequences and obtained the values of precisions 0.9515, 0.9636, 0.9818 using Majority voting, LAP, and LOP2 classifier fusion approaches [8]. Mahadevan et al. predicted the E.coli promoter structures using neural networks and obtained 98% successful in promoter recognition and 90.2% successful in non-promoter recognition when tested on 5000 randomly generated sequences [9].

In this study, we have proposed a novel hybrid combination based on three parts to predict the E.coli promoter gene sequences. Firstly, we have used the FS process to reduce the dimensionality of E.coli Promoter Gene Sequences dataset that has 57 attributes (one of A, G, T or C). So the dimensionality of this dataset has been reduced to 4 attributes by means of FS process. Second, we have weighted E.coli Promoter Gene Sequences dataset that has 4 features using fuzzy weighted pre-processing. Finally, C4.5 decision tree classifier algorithm has been run to predict the E.coli Promoter Gene Sequences. We obtained 93.33% prediction accuracy from the experiments made on the E.coli promoter gene sequences dataset taken from machine learning database.

2 E.Coli Promoter Gene Sequences Dataset

In the present study, the real-world task of recognizing biological concepts in DNA sequences has been investigated. In particular, the task is to recognize promoters in strings that represent nucleotides (one of A, G, T, or C). A promoter is genetic region which initiates the first step in the expression of an adjacent gene (transcription) [2].

Table 1 presents the initial domain theory used in the promoter recognition task. The first rule says that the promoter includes two subcategories: a contact and a conformation region. The second rule states that a contact involves two regions, while subsequent rules define alternative ways these regions can appear [2].

Dimensionality of E.coli Promoter Gene Sequences dataset has 57 sequential DNA nucleotides and 106 samples including 53 promoters and 53 non-promoters. A special notation is used to simplify specifying locations in the DNA sequence. The biological literature counts locations relative to the site where transcription begins. Fifty nucleotides before and six following this location constitute an example. When a rule's antecedents refer to input features, they first state the starting location, and then list the sequence that must follow. In these specifications, "x" indicates that any nucleotide will suffice. Hence, the first rule for conformation says that there must an "a" 45 nucleotides before the site where transcription begins. Another "a" must be at -44, then any two nucleotides can appear, and finally there must be a "t" at location -41 [2, 3].

Table 1. A Domain Theory for Promoters

Promoter	-contact, conformation.
contact	- minus_35, minus_10.
minus_35	-@_37 "cttgac".
minus_35	-@_36 "ttgxca".
minus_35	-@_36 "ttgaca".
minus_35	-@_36 "ttgac".
minus_10	-@_14 "tataat".
minus_10	-@_13 "taxaxt".
minus_10	-@_13 "tataat".
minus_10	-@_12 "taxxxt".
conformation	-@_45 "aaxxa".
conformation	-@_45 "axxxa", @_4 "t", @_28 "txxxtxaaxxtx".
conformation	@_49 "axxxxt", @_1 "a", @_27 "txxxxaxxtxtg".
conformation	@_47 "caaxttxac", @_22 "gxxxtxc", @_8 "gcgccxcc".

3 The Proposed Method

The proposed method consists of three main parts: Feature selection process, Fuzzy weighted pre-processing and C4.5 DCS. First, in the feature selection stage, dimensionality of E.coli Promoter Gene Sequences that has 57 features has been reduced to 4 features using FS process. Second, in the pre-processing stage, we have weighted E.coli Promoter Gene Sequences dataset that has 4 features using fuzzy weighted pre-processing. Finally, in the classification stage, C4.5 decision tree classifier has been used to classify subjects as promoters or non- promoters. We explain the details of the feature selection, preprocessing and classification steps in the following subsections.

3.1 Feature Selection Process

The number of features (attributes) and number of instances in the raw dataset can be enormously large. This enormity may cause serious problems to many data mining systems. Feature selection is one of the long existing methods that deal with these problems. Its objective is to select a minimal subset of features according to some reasonable criteria so that the original task can be achieved equally well, if not better. By choosing a minimal subset of features, irrelevant and redundant features are removed according to the criterion. Simpler data can lead to more concise results and their better comprehensibility. Since feature selection can only deal with discrete features (attributes), you need to run Discretion first if there are continuous features in the dataset. FS process with C4.5 decision tree rules cannot directly use a data file with continuous attributes in mining. To solve this problem, a discrimination system has to be used. You might pass your data file with continuous attributes to this system, which will discrete the continuous attributes and output a file that contains the discredited data. Upon return, discredited data can then be used in the mining [4].

Feature selection (FS) was done according to the feature distributions over the E.coli promoter gene sequences dataset. After it was detected for classes to be mixed, it was concluded with the consultations of experts that these features were not critical in classification. So the number of features was reduced to 4 by removing these features with the use of C4.5 decision tree rules. We have used the 6th, 15th, 16th, and 17th attributes in prediction of E.coli promoter gene sequences by means of FS process.

3.2 Fuzzy Weighted Pre-processing

In the fuzzy weighted pre-processing, we extract new feature values for each feature. To do this, we first apply two membership functions, known as the input and output membership functions, and we then use weighting procedure to the given data. The membership functions are selected as triangular membership functions as shown in Fig. 1(a) and Fig. 2(b) respectively.

The formation of these membership functions is realized as follows: First, the arithmetic mean values of each feature for corresponding samples are calculated using Eq. (1):

$$m_i = \frac{1}{N}\sum_{k=1}^{N} x_{k,i} \tag{1}$$

In this expression, $x_{k,i}$ represents the ith feature value of sample x_k, k=1,2,…N.

After the calculation of the sample means for each feature, the input membership function is formed by triangles as shown in Fig 1(a). The supports of these triangles

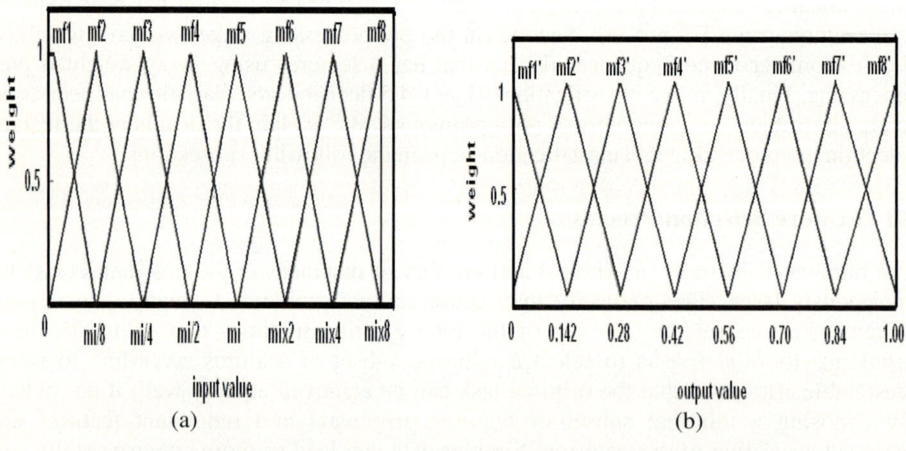

Fig. 1. (a) Input Membership Function, (b) Output Membership Function

are determined by $m_i/8$, $m_i/4$, $m_i/2$, m_i, $2m_i$, $4m_i$, $8m_i$. The lines of the input membership functions are named as mf1, mf2,…mf8. To form the output membership function, the interval [0,1] is divided into 8 equal parts and the corresponding lines are named as mf1', mf2',…mf8' as shown in Fig 1(b). Note here that these input and output membership functions are formed for each feature so that there will exist different input-output membership function configurations for each feature since the sample means of each feature differs.

After determining the input and output membership functions for each feature, the weighted procedure can start. In this procedure, a feature value $x_{k,i}$ is assumed to be in the x axis of input membership function and y axis values for the points at which this value cuts the input membership functions are determined. For example, if a feature value is between 0 and $m_i/8$, then this point will cut both lines mf1 and mf2. The y values at these intersection points, say y_1 and y_2, are known as membership values (μ) and they will then be used in a fuzzy rule base in the following manner: First, the input membership value, $\mu(i)$, is determined by using the above mentioned intersection points as follows:

$$\mu(i)=\mu_{A\cap B}(x_{k,i}) = \mathrm{MIN}\,(\mu_A(x_{k,i}), \mu_B(x_{k,i})),\ x\in X \qquad (2)$$

In this expression, $\mu_A(x_{k,i})$ and $\mu_B(x_{k,i})$ membership values correspond to the intersection points. The rule base for our system is used as presented in Table 1. After $\mu(i)$ value is determined using Eq. (2) for the $x_{k,i}$ feature value, the output weight value is determined by using output membership functions and the rules in Table 1. To do this, first, the input membership value, $\mu(i)$, is presented to the output membership function to determine the corresponding weighted value of the original feature value. This membership value is then taken as a point in y-axis of the output membership functions and, the intersection points, which are cut by this membership value, are determined. As the output membership functions show there will be more than one intersection points. Which of them will be used is decided by the rules given in Table 2.

Table 2. Fuzzy rule base for our system

k value	*Rules*
1	If Input_value cuts mf(k) and mf(k+1) then Output_value=(mf(k)'(y)+mf(k+1)'(y))/2
2	if Input_value cuts mf(k) and mf(k+1) then Output_value= mf(k)'(y)+mf(k+1)'(y))/2
3	if Input_value cuts mf(k) and mf(k+1) then Output_value= mf(k)'(y)+mf(k+1)'(y))/2
4	if Input_value cuts mf(k) and mf(k+1) then Output_value= mf(k)'(y)+mf(k+1)'(y))/2
5	if Input_value cuts mf(k) and mf(k+1) then Output_value= mf(k)'(y)+mf(k+1)'(y))/2
6	if Input_value cuts mf(k) and mf(k+1) then Output_value= mf(k)'(y)+mf(k+1)'(y))/2
7	if Input_value cuts mf(k) and mf(k+1) then Output_value= mf(k)'(y)+mf(k+1)'(y))/2

For example, if the input feature value cuts mf1 and mf2 lines in input membership functions then the output value for this feature will be the arithmetic mean of two points that μ(i) cuts mf1' and mf2' at the output membership functions [5].

3.3 C4.5 Decision Tree Classifier

C4.5 Decision tree learning is one of the most widely used and practical methods for inductive inference. It is a method for approximating discrete-valued functions that is robust to noisy data and capable of learning disjunctive expressions [6, 7].

C4.5 Decision tree learning is a method for approximating discrete-valued functions, in which the learned function is represented by a decision tree. Learned trees can also be-represented as sets of if-then rules to improve human readability. These learning methods are among the most popular of inductive inference algorithms and have been successfully applied to a broad range of tasks from learning to diagnose medical cases to learning to assess credit risk of loan applicants. C4.5 Decision tree learning is a heuristic, one-step lookahead (hill climbing), non-backtracking search through the space of all possible decision trees [6, 7].

4 The Experimental Results and Discussion

We have used two methods for performance evaluation of E.coli Promoter Gene Sequences prediction. These methods are prediction accuracy and 10-fold cross validation.

The proposed approach consists of three parts: feature selection process, fuzzy weighted pre-processing, and C4.5 DCS. Here, we have used four different methods to predict of E.coli Promoter Gene Sequences. These methods are C4.5 DCS,

Table 3. Obtained Prediction Accuracies for recognizing E.coli promoter gene sequences by other methods in literature using 10-fold cross validation

Author	*Method*	Number of Attributes	Prediction Accuracy (%)
Geoffrey G. Towell et al. [2]	*Standard Backprogation*	57	92.45
Geoffrey G. Towell et al. [2]	*O' Neill*	57	88.67
Geoffrey G. Towell et al. [2]	*Nearest Neighbor*	57	87.73
Geoffrey G. Towell et al. [2]	*ID3*	57	82.07
Our study (2007)	C4.5 Decision tree classifier (DCS)	57	70.08
Our study (2007)	Combining fuzzy weighted pre-processing- C4.5 DCS	57	88.88
Our study (2007)	Combining feature selection- C4.5 DCS	4	76.82
Our study (2007)	The proposed method	4	93.33

combining fuzzy weighted pre-processing- C4.5 DCS, combining feature selection-C4.5 DCS, and combining feature selection-fuzzy weighted pre-processing- C4.5 DCS. The obtained prediction accuracies are 70.08%, 88.88%, 76.82%, and 93.33% using above methods for 10-fold cross validation. Table 3 presents the obtained results. These results show that the proposed method is a robust and effective hybrid approach in prediction of E.coli Promoter Gene Sequences.

5 Conclusions and Future Work

In this paper, a novel hybrid system based on feature selection, fuzzy weighted pre-processing, and C4.5 decision tree classifier has been applied to the task of predicting E.coli promoter gene sequences and the most accurate learning methods have been evaluated. Experiments have been conducted on the E.coli promoter gene sequences dataset to decide whether any sequence belong to promoters or not. While only C4.5 decision tree classifier obtained 70.08% prediction accuracy, the proposed method obtained 93.33% prediction accuracy in predicting E.coli promoter gene sequences for 10-fold cross validation. The result which is obtained by proposed method has higher prediction accuracy in the reported studies at literature for recognizing E.coli promoter gene sequences. The results strongly suggest that proposed method can aid in the predicting E.coli promoter gene sequences. It is hoped that more interesting results will follow on further exploration of data.

Acknowledgments. This study has been supported by Scientific Research Project of Selcuk University (Project No: 05401069).

References

1. Harley, C., Reynolds, R.: Analysis of E.coli promoters sequences. Nucl. Acids Res. 15, 2343–2361 (1987)
2. Geoffrey, G.T., Jude, W.S., Michiel, O.N.: Refinement of Approximate Domain Theories by Knowledge-Based Neural Networks. In: Proc. of the eight national conf. on Artificial Intelligence, pp. 861–866 (1990)
3. UCI Machine Learning Repository (last arrived: January, 2007), http://www.ics.uci.edu/~mlearn/MLRepository.html
4. Polat, K., Şahan, S, Kodaz, H., Günes, S.: A New Classification Method for Breast Cancer Diagnosis: Feature Selection Artificial Immune Recognition System. In: Wang, L., Chen, K., Ong, Y.S. (eds.) ICNC 2005. LNCS, vol. 3611, pp. 830–838. Springer, Heidelberg (2005)
5. Polat, K., Şahan, S., Güneş, S.A: A New Method to Medical Diagnosis: Artificial Immune Recognition System (AIRS) with Fuzzy Weighted Pre-processing and Application to ECG Arrhythmia. Expert Systems with Applications 31(2), 264–269 (2006)
6. Mitchell, M.T.: Machine Learning. McGraw-Hill, Singapore (1997)
7. Quinlan, J.R.: Induction of decision trees, Machine Learning, vol. 1, pp. 81–106 (1986)
8. Ranawana, R., Palade, V.: A Neural Network Based Multi-Classifier System for Gene Identification in DNA Sequences. Neural Computing and Applications 14(2), 122–131 (2005)
9. Mahadevan, I., Ghosh, I.: Analysis of E.coli promoter structures using neural networks. Nucleic Acids Res. 22(11), 2158–2165 (1994)

A Hybrid Symbolic-Statistical Approach to Modeling Metabolic Networks

Marenglen Biba, Stefano Ferilli, Nicola Di Mauro, and Teresa M.A. Basile

Department of Computer Science, University of Bari
Via E. Orabona, 4 - 70125 Bari, Italy
{biba,ferilli,ndm,basile}di.uniba.it

Abstract. Biological systems consist of many components and interactions between them. In Systems Biology the principal problem is modeling complex biological systems and reconstructing interactions between their building blocks. Symbolic machine learning approaches have the power to model structured domains and relations among objects. However biological domains require uncertainty handling due to their hidden complex nature. Statistical machine learning approaches have the potential to model uncertainty in a robust manner. In this paper we apply a hybrid symbolic-statistical framework to modeling metabolic pathways and show through experiments that complex phenomenon such as biochemical reactions in cell's metabolic networks can be modeled and simulated in the proposed framework.

Keywords: Systems biology, metabolic networks, symbolic-statistical frameworks, machine learning, statistical relational learning.

1 Introduction

Biological systems' behavior is determined by complex interactions between their building components. In Systems Biology [1] the main problem is to uncover and model how function and behavior of the biological machinery are implemented through these interactions. Since biological circuits are hard to model and simulate, many efforts [2] have been made to develop computational models that can handle their intrinsic complexity. In this paper we focus on a particular problem of Systems Biology that concerns the modeling of metabolic pathways. A metabolic pathway is a sequence of chemical reactions occurring within the cell. These reactions are catalyzed by enzymes which are particular proteins that convert metabolites (input molecules) in other molecules that represent the products of the reaction. These products can be stored in the cell under certain forms or can cause the initiation of another metabolic pathway. A metabolic network of a cell is formed by the metabolic pathways occurring in the cell.

Since a reaction can happen if the input molecules are available to the catalytic enzyme, a modeling framework must be able to model relations among entities. Symbolic approaches such as logic-based techniques have the potential to model relations in structural complex domains and there have been a growing number of

B. Apolloni et al. (Eds.): KES 2007/WIRN 2007, Part I, LNAI 4692, pp. 132–139, 2007.

biological applications of these methods [3]. First-order logic representations have also the advantage that models are easily comprehensible to humans. Moreover, since most part of biological systems performs its activity remaining hidden to the human modeler, machine learning techniques can play an important role in discovering latent phenomena. However, symbolic-only approaches suffer from the incapability of handling uncertainty. In models built with symbolic-only approaches, the learned rules are deterministic and do not incorporate any kind of mechanism for uncertainty modeling. On the other side, biological systems intrinsically behave in a uncertain fashion with many interactions probable to happen. Since cell's life is determined by the most probable interactions, handling uncertainty is crucial when the cell's machinery must be modeled. Statistical approaches based on the probability theory represent a valuable mechanism to govern uncertainty. However, observations of biological systems rarely reflect exactly what happens inside them. Therefore, estimation techniques are precious in order to model what we cannot observe. Statistical machine learning methods have the ability to learn probability distributions from observations and hence are suitable for modeling biological systems. On the other side, statistical-only approaches rarely are able to reason about relations and/or interactions among biological circuits as symbolic approaches do. Hence, there is strong motivation on developing and applying hybrid approaches to modeling biological systems.

The paper is organized as follows. Section 2 describes the problem of modeling the aromatic amino acid pathway of yeast and the necessity for symbolic-statistical machine learning. Section 3 describes the hybrid framework PRISM. Section 4 describes modeling of the problem in the framework PRISM. Section 5 presents experiments and Section 6 concludes and presents future work.

2 Metabolic Networks

Metabolic networks can be represented as graphs where nodes represent reactions and there are two kinds of arcs, those entering the node and labeled with the input molecules and those exiting the node and labeled with the products of the reaction. Fig. 1. shows part of the aromatic pathway for Yeast presented in [4]. The node represents the reactions made possible by the enzymes 2.5.1.19, 4.6.1.4 and 5.4.99.5 (KEGG identifiers are used for enzymes and reactions). A possible reaction is that of converting the metabolites C00074 and C00008 into the products C00009 and C01269. Then the metabolite C01269 takes part into another reaction made possible by the enzyme 4.6.1.4 and so on. Such interactions can be easily expressed using a first-order logic representation. For example the interactions can be expressed by the following predicates:

```
enzyme(2.5.1.19, reaction_2_5_1_19, [C00074,C00008], [C00009,C01269]).
enzyme(4.6.1.4, reaction_4_6_1_4, [C01269], [C00009,C00251]).
enzyme(5.4.99.5, reaction_5_4_99_5, [C00251], [C00254]).
```

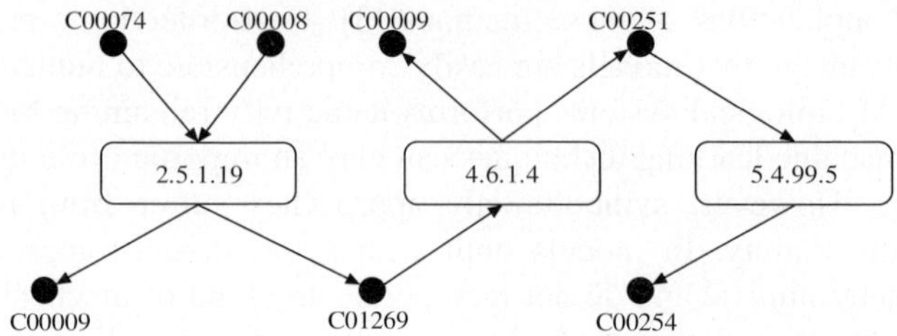

Fig. 1. Part of the aromatic pathway for Yeast

However, this representation does not incorporate any further information about the reactions. For example, there can be competing reactions if two enzymes elaborate the same input metabolites and the occurring of any of the reactions determines a certain sequence of successive reactions instead of another. Hence, it is important to know which reaction among the two is more probable to happen. Moreover, input metabolites not always are available. Their absence can cause a certain reaction not to occur and give rise to another sequence in the metabolic pathway. Therefore, it is crucial to know how probable a certain reaction is. This situation can be modeled by attaching to each reaction the probability that it happens. This requires a first-order representation framework that can handle for each predicate that expresses a reaction the probability that the predicate is true.

The simple incorporation of probabilities is not enough to model complex metabolic networks. The probabilities of the reactions depend on many factors, such as initial quantity of input metabolites, changes in the physical-chemical environment surrounding the cell and many more. For this reason it is a hard task to observe all the states of the biological machinery and try to assign probabilities to reactions. Therefore there is a need for machine learning methods that can learn distribution of probabilities from observations.

In order to model metabolic networks, two tasks must be performed. First, a relational model that describes the reactions must be built. This can be constructed manually by the expert or learned automatically by machine learning multi-relational methods. Once the model has been built, the second task is the assignment of probabilities. In this paper we do not deal with model building. We use the metabolic pathway built in [4] and model this pathway in a hybrid symbolic-statistical framework in order to automatically learn the probabilities of the reactions. A logic program that describes this metabolic pathway has been given in [5]. We extend this logic program in order to statistically model the pathway in the framework PRISM. After modeling the pathway in PRISM we perform learning of probabilities and show through experiments the feasibility of accurately learning reactions probabilities from metabolomics data using PRISM.

3 The Symbolic-Statistical Framework PRISM

PRISM (PRogramming In Statistical Modelling) [6] is a symbolic-statistical modeling language that integrates logic programming with learning algorithms for probabilistic programs. PRISM programs are not only just a probabilistic extension of logic programs but are also able to learn from examples through the EM (Expectation-Maximization) algorithm which is built-in in the language. The parameter learning algorithm [7], provided by the language, is a new EM algorithm called graphical EM algorithm that when combined with the tabulated search has the same time complexity as existing EM algorithms, i.e. the Baum-Welch algorithm for HMMs (Hidden Markov Models), the Inside-Outside algorithm for PCFGs (Probabilistic Context-Free Grammars), and the one for singly connected BNs (Bayesian Networks) that have been developed independently in each research field. Since PRISM programs can be arbitrarily complex, all the formalisms such as HMMs, PCFGs and BNs can be described by these programs.

PRISM programs are defined as logic programs with a probability distribution given to facts that is called basic distribution. Formally a PRISM program is $P = F \cup R$ where R is a set of logical rules working behind the observations and F is a set of facts that models observations' uncertainty with a probability distribution. Through the built-in graphical EM algorithm the parameters (probabilities) of F are learned and through the rules this learned probability distribution over the facts induces a probability distribution over the observations. As an example, we present a hidden markov model with two states slightly modified from that in [7]:

```
values(init,[s0,s1]).          % State  initialization
values(out(_),[a,b]).          % Symbol emission
values(tr(_),[s0,s1]).         % State transition
hmm(L):-                       % To observe a string L:
str_length(N),                 % Get the string length as N
msw(init,S),          % Choose an initial state randomly
hmm(1,N,S,L).         % Start stochastic transition (loop)
hmm(T,N,_,[]):- T>N,!.         % Stop the loop
hmm(T,N,S,[Ob|Y]) :-           % Loop: state S,time  T
msw(out(S),Ob),                % Output Ob at the state S
msw(tr(S),Next),               % Transit from S to Next.
T1 is T+1,                     % Count up time
hmm(T1,N,Next,Y).              % Go next (recursion)
str_length(10).                % String length is 10
set_params :- set_sw(init, [0.9,0.1]), set_sw(tr(s0),
[0.2,0.8]), set_sw(tr(s1), [0.8,0.2]),
set_sw(out(s0),[0.5,0.5]), set_sw(out(s1),[0.6,0.4]).
```

The most appealing feature of PRISM is that it allows the users to use random switches to make probabilistic choices. A random switch has a name, a space of possible outcomes, and a probability distribution. In the program above, msw(init,S) probabilistically determines the initial state from which to start by tossing a coin. The predicate set_sw(init, [0.9,0.1]), states that the probability of starting from state s0 is 0.9 and from s1 is 0.1. The predicate *learn* in PRISM is used to learn from examples (a set of strings) the parameters (probabilities of init, out and tr) so that the ML (Maximum-Likelihood) is reached. For example, the learned parameters from a set of examples can be: switch init: s0 (0.6570), s1 (0.3429); switch out(s0): a (0.3257), b (0.6742); switch out(s1): a (0.7048), b (0.2951); switch tr(s0): s0 (0.2844), s1 (0.7155); switch tr(s1): s0 (0.5703), s1 (0.4296).␣After learning these ML parameters, we can calculate the probability of a certain observation using the predicate *prob*:

prob(hmm([a,a,a,a,a,b,b,b,b,b]) = 0.000117528. This way, we are able to define a probability distribution over the strings that we observe. Therefore from the basic distribution we have induced a probability distribution over the observations.

4 PRISM Modeling of Aromatic Amino Acid Pathway of Yeast

The logic foundation of PRISM facilitates the construction of a representation of the metabolic pathway described in the previous section. Predicates that describe reactions remain unchanged from a language representation point of view. What we need to statistically model the metabolic pathway is the extension with random switches of the logic program that describes the pathway. We define for every reaction a random switch with its relative space outcome. For example, in the following we describe the random switches for the reactions in Fig. 1.

values(switch_rea_2_5_1_19,[rea_2_5_1_19(yes, yes, yes, yes),rea_2_5_1_19(yes, yes, no, no)]).
values(switch_ea_4_6_1_4,[rea_4_6_1_4(yes, yes, yes),rea_4_6_1_4(yes, no, no)]).
values(switch_rea_5_4_99_5,[rea_5_4_99_5(yes, yes),rea_5_4_99_5(yes, no)]).

For each of the three reactions there is a random switch that can take one of the stated values at a certain time. For example, the value *rea_2_5_1_19(yes, yes, yes, yes)* means that at a certain moment the metabolites C00074 and C00008 are present and the reaction occurs producing C00009 and C00251. While the other value *rea_2_5_1_19(yes, yes, no, no)* means that the input metabolites are present but the reaction did not occur, thus the products C00009 and C00251 are not produced. Below we report the PRISM program for modeling the pathway in Figure 1. (The complete PRISM code for the whole metabolic pathway can be requested to the authors).

```
enzyme('2.5.1.19',rea_2_5_1_19,[C00074,C03175],[C00009,
C01269]).
enzyme('4.6.1.4',rea_4_6_1_4,[C01269],[C00009,C00251]).
enzyme('5.4.99.5',rea_5_4_99_5,[C00251],[C00254] ).
can_produce(Metabolites,Products) :-
can_produce(Metabolites,[],Products).
can_produce(Metabolites,Stalled,Products) :-
(possible_reaction(Metabolites,Stalled,Name,Inputs,Outp
uts,Rest) ->
reaction_call(Reaction,Inputs,Outputs,Call),
rand_sw(Call,Value),
((Value == rea_2_5_1_19(yes,yes,yes,yes);
Value == rea_4_6_1_4(yes,yes,yes);
Value == rea_5_4_99_5(yes,yes))  ->
can_produce(Rest,Stalled,Products )
;
can_produce(Metabolites,[Reaction|Stalled],Product));
Products = Metabolites).
```

```
rand_sw(ReactAndArgs,Value):-
ReactAndArgs=..[Pred|Args],
(Pred == rea_2_5_1_19 ->msw(switch_rea_2_5_1_19,Value);
(Pred == rea_4_6_1_4 ->msw(switch_rea_4_6_1_4,Value);
(Pred == rea_5_4_99_5 -> msw(switch_rea_5_4_99_5,Value)
;
true))). % do nothing
```

In the following, we trace the execution of the program. The top goal to prove that represents the observations in PRISM is *can_produce(Metabolites,_,Products)*. It will succeed if there is a pathway that leads from *Metabolites* to *Products*, in other words if there is a sequence of random choices (according to a probability distribution) that makes possible to prove the top goal. The predicate *possible_reaction* controls among the first three clauses of the program, if there is a possible reaction with *Metabolites* in input. Suppose that at a certain moment *Metabolites* = [C00074,C00008] and thus the reaction can happen. The variables *Inputs* and *Outputs* are bounded respectively to [C00074,C00008] and [C00009,C01269]. The predicate *reaction_call* constructs the body of the reaction that is the predicate *Call* which is in the form: *rea_2_5_1_19 (_,_,_,_)*. This means that the next predicate *rand_sw* will perform a random choice for the switch. This random choice which is made by the built-in predicate *msw(switch_rea_2_5_1_19,Value)* of PRISM, determines the next step of the execution, since *Value* can be either *rea_2_5_1_19(yes, yes, yes, yes)* or *rea_2_5_1_19(yes, yes, no, no)*. In the first case it means the reaction has been probabilistically chosen to happen and the next step in the execution of the program which corresponds to the next reaction in the metabolic pathway is the call *can_produce(Rest, Stalled, Products)*. In the second case, the random choice *rea_2_5_1_19(yes, yes, no, no)* means that probabilistically the reaction did not occur and the sequence of the execution will be another, determined by the call *can_produce(Metabolites, [Reaction|Stalled],Products)*.

In order to learn the probabilities of the reactions we need a set of observations of the form *can_produce(Metabolites,_,Products)*. These observations that represent metabolomic data, are being intensively collected through available high throughput instruments and stored in metabolomics databases. In the next section, we show that from these observations, PRISM is able to accurately learn reaction probabilities.

5 Experiments

The scope of the experiments is to show empirically that on a medium-sized metabolic pathway the learning of the probability distributions from metabolomics data is feasible in PRISM. In order to assess the accuracy of learning the probabilities of the reactions we adopt the following method. A probability distribution P_1 ,..., P_M is initially assigned to the clauses of the logic program so that each reaction has a probability attached. We call these M parameters the true parameters. Then we sample from this probability distribution *S* samples (observations) by launching the top goal *can_produce(Metabolites,_,Products)*. Once that we have these samples, we replace the probabilities by uniformly distributed ones. At this point the built-in predicate *learn* of PRISM is called in order to learn from the samples. PRISM learns M new parameters

$P_1^{'}, \ldots, P_M^{'}$, that represent the learned reaction probabilities from the observations. In order to assess the accuracy of the learned $P_i^{'}$ towards P_i we use the RMSE (Root Mean Square Error) for each experiment with *S* samples.

$$\text{RMSE} = \sqrt{\left(\sum_{i=1}^{M} \frac{(P_i - P_i^{'})^2}{M} \right)}$$

We performed experiments on two types of networks. In the first there are not alternative branches in the metabolic pathway. It means that starting from any node in the network there are not multiple paths to reach another node in the network. While in the second network we add an alternative path. For each network, we have performed different experiments with a growing number *S* of samples in order to evaluate how the number of samples affects the accuracy and the learning time. For each *S* we have performed 10 experiments in order to assess the standard deviation of RMSE for different experiments with the same number of samples.

Table 1. Experiments on the 2 networks

S – Number of samples	Mean of RMSE on 10 experiments		Standard Deviation. of RMSE on 10 experiments		Mean learning time on 10 experiments (seconds)	
	Network 1	Network 2	Network 1	Network 2	Network 1	Network 2
100	0,14860	0,18080	0,00013	0,000021	0,031	0,078
200	0,13377	0,14723	0,00001	0,000041	0,078	0,094
400	0,09909	0,11796	$1{,}5 * 10^{-7}$	0,000308	0,079	0,156
600	0,08263	0,10471	0,00001	0,000458	0,094	0,182
800	0,07766	0,08317	$7{,}2 * 10^{-7}$	$2{,}2 * 10^{-7}$	0,098	0,141
1000	0,07200	0,07708	0,00006	$8{,}9 * 10^{-7}$	0,104	0,172
2000	0,06683	0,07027	0,000014	0,000686	0,118	0,194
4000	0,06442	0,06672	0,00001	$2{,}9 * 10^{-7}$	0,140	0,204
6000	0,05667	0,05768	0,000018	0,000351	0,156	0,219
8000	0,05279	0,05306	0,000106	$5{,}3 * 10^{-7}$	0,182	0,266
10000	0,05164	0,05231	0,000037	0,000481	0,203	0,281

As Table 1 shows, we get better results in terms of accuracy as *S* grows and the learning time is very low considering that the two networks are of medium size where Network 1 and Network 2 contain respectively 21 and 25 reactions. As RMSE decreases we note a slight increase of the learning time. Comparing the two networks, we can see that on the second network RMSE and the learning time are greater than on the first network. This is due to more nodes to explore during learning as the same node can be reached in different ways. However, the experiments show that given metabolomics data, learning accurately reaction probabilities in PRISM is feasible.

In a related work [5], SLPs (Stochastic Logic Programs) [8] were applied to the same problem. The advantage of our approach stands in the parameter learning phase. Parameter estimation in SLPs [9] requires the intractable computation of a normalizing constant. In [9] it is shown that the approach of simply enumerating refutations in the SLD-tree is tractable only for small problems because it requires the exploration of the entire SLD-tree of the top goal. Moreover, for parameter learning of SLPs there have not yet been developed tabulation techniques such as in PRISM where tabulated search greatly increases efficiency [7]. However, structure learning for SLPs has been dealt with in [10] (in [9] the structure is supposed to be learned by another method and it only applies the parameter estimation algorithm to the given structure), while structure learning for PRISM programs has not been attempted.

6 Conclusion

We have applied the hybrid symbolic-statistical framework PRISM to a problem of modeling metabolic pathways and have shown through experiments the feasibility of learning reaction probabilities from metabolomics data for a medium-sized network. To the best of our knowledge this is the first application of the framework PRISM to a problem in Systems Biology. Very good probability estimation accuracy and learning times validate the hybrid approach to a problem where both relations and uncertainty must be handled.

As future work, we intend to investigate larger networks and the problem of model building from observations. We believe PRISM fast learning algorithm will help in exploring larger metabolic networks in reasonable times.

References

1. Kitano, H.: Foundations of Systems Biology. MIT Press, Redmond, Washington (2001)
2. Kriete, A., Eils, R.: Computational Systems Biology. Elsevier - Academic Press, Amsterdam (2005)
3. Page, D., Craven, M.: Biological Applications of Multi-Relational Data Mining. Appears in SIGKDD Explorations, special issue on Multi-Relational Data Mining (2003)
4. Bryant, C.H., Muggleton, S.H., Oliver, S.G., Kell, D.B., Reiser, P., King, R.D.: Combining inductive logic programming, active learning and robotics to discover the function of genes. Electronic Transactions in Artificial Intelligence 5-B1(012), 1–36 (2001)
5. Angelopoulos, N., Muggleton, S.H.: Machine learning metabolic pathway descriptions using a probabilistic relational representation. Electronic Transactions in Artificial Intelligence 6 (2002)
6. Sato, T., Kameya, Y., PRISM,: PRISM: A symbolic-statistical modeling language. In: Proceedings of the 15th International Joint Conference on Artificial Intelligence, pp. 1330–1335 (1997)
7. Sato, T., Kameya, Y.: Parameter learning of logic programs for symbolic-statistical modeling. Journal of Artificial Intelligence Research 15, 391–454 (2001)
8. Muggleton, S.H.: Stochastic logic programs. In: de Raedt, L. (ed.) Advances in Inductive Logic Programming, pp. 254–264. IOS Press, Amsterdam (1996)
9. Cussens, J.: Parameter estimation in stochastic logic programs. Machine Learning 44(3), 245–271 (2001)
10. Muggleton, S.H.: Learning structure and parameters of stochastic logic programs. In: Proceedings of the 10th International Conference on Inductive Logic Programming, Springer, Berlin (2002)

Boosting Support Vector Machines Using Multiple Dissimilarities

Ángela Blanco and Manuel Martín-Merino

Universidad Pontificia de Salamanca
C/Compañía 5, 37002, Salamanca, Spain
{ablancogo,mmartinmac}@upsa.es

Abstract. Support Vector Machines (SVM) are powerful machine learning techniques that are able to deal with high dimensional and noisy data. They have been successfully applied to a wide range of problems and particularly to the analysis of gene expression data.

However SVM algorithms rely usually on the use of the Euclidean distance that often fails to reflect the object proximities. Several versions of the SVM have been proposed that incorporate non Euclidean dissimilarities. Nevertheless, different dissimilarities reflect complementary features of the data and no one can be considered superior to the others. In this paper, we present an ensemble of SVM classifiers that reduces the misclassification error combining different dissimilarities. The method proposed has been applied to identify cancerous tissues using Microarray gene expression data with remarkable results.

Keywords: Machine Learning, Support Vector Machines, Dissimilarity Based Classifiers, Gene Expression Data Analysis, DNA Microarrays.

1 Introduction

Support Vector Machines (SVM) are powerful non-linear techniques that are able to handle high dimensional and noisy data [17]. They have been proposed under a strong theoretical foundation and exhibit a high generalization ability.

An interesting application of the SVM is the identification of cancerous tissues using the gene expression levels. However, common SVM algorithms rely on the use of the Euclidean distance which fails often to reflect the proximities among the cellular samples [5,10,13]. Several versions of the SVM have been proposed in the literature that incorporate non-euclidean dissimilarities [15]. However, no dissimilarity outperforms the others because each one reflects just different features of the data.

In this paper, we propose an ensemble of classifiers based on multiple dissimilarities. It well known that combining non-optimal classifiers can help to reduce particularly the variance of the predictor [11,16]. In order to achieve this goal, different versions of the classifier are usually built by sampling the patterns or the features [3]. Nevertheless, in our application, this kind of sampling techniques increases the bias of individual classifiers and thus, the ensemble of classifiers

B. Apolloni et al. (Eds.): KES 2007/WIRN 2007, Part I, LNAI 4692, pp. 140–147, 2007.

often fails to reduce the error [16]. To overcome this problem, we propose to build a diversity of classifiers considering that each dissimilarity reflects different features of the data. To this aim, the dissimilarities are first embedded into an Euclidean space where a SVM is adjusted for each measure. Next, the classifiers are aggregated using a voting strategy [11]. The method proposed has been applied to the prediction of different kinds of cancer using the gene expression levels with remarkable results.

This paper is organized as follows. Section 2 discusses the problem of distances in the context of gene expression data analysis. Section 3 introduces our method to combine classifiers based on dissimilarities. Section 4 illustrates the performance of the algorithm in the challenging problem of gene expression data analysis. Finally, section 5 gets conclusions and outlines future research trends.

2 Dissimilarities for Gene Expression Data Analysis

An important step in the design of a classifier is the choice of a proper dissimilarity that reflects the proximities among the objects. However, the choice of a good dissimilarity for the problem at hand is not an easy task. Each measure reflects different features of the dataset and no dissimilarity outperforms the others in a wide range of problems. In this section, we comment shortly the main differences among several dissimilarities applied to the analysis of gene expression. For a deeper description and definitions see [5,10,6].

The Euclidean distance evaluates if the gene expression levels differ significantly across different samples. When the experimental conditions change from one sample to another the cosine dissimilarity reflects better the proximities between the sample profiles. This dissimilarity will become small when the rate between the gene expression levels is similar for the samples considered. It differs significantly from the Euclidean distance when the data is not normalized.

The correlation measure evaluates if the expression levels of genes change similarly in sample profiles. Correlation based measures tend to group together samples whose expression levels are linearly related. The correlation differs significantly from the cosine if the means of the sample profiles are not zero. This measure is distorted by outliers. The Spearman rank dissimilarity avoids this problem by computing a correlation between the ranks of the gene expression levels. An alternative measure that helps to overcome the problem of outliers is the kendall-τ index. The Kendall's τ is related to the Mutual Information probabilistic measure [6]. Finally, the Kullback-Leibler divergence evaluates the distance between the probability distribution of the gene expression levels for samples.

Due to the large number of genes, the sample profiles are codified in high dimensional and noisy spaces. In this case, the dissimilarities mentioned above are affected by the 'curse of dimensionality' [1,12]. Hence, most of the dissimilarities become almost constant and the differences among dissimilarities are lost [9]. To avoid this problem, the number of features is reduced aggressively before computing the dissimilarities.

3 Combining Classifiers Based on Dissimilarities

The SVM is a powerful machine learning technique that is able to work with high dimensional and noisy data [17]. However, the original SVM algorithm is not able to work directly from a dissimilarity matrix. To overcome this problem, we follow the approach of [15]. First, each dissimilarity is embedded into an Euclidean space such that the inter-pattern distances reflect approximately the original dissimilarities. Next, the test points are embedded via a linear algebra operation and finally the SVM is adjusted and evaluated.

Let $D \in \mathbb{R}^{n \times n}$ be a dissimilarity matrix made up of the object proximities. A configuration in a low dimensional Euclidean space can be found via a metric multidimensional scaling algorithm (MDS) [4] such that the original dissimilarities are approximately preserved. Let $X = [\boldsymbol{x}_1 \dots \boldsymbol{x}_n]^T$ be the matrix of the object coordinates for the training patterns. Define $B = XX^T$ as the matrix of inner products which is related to the dissimilarity matrix via the following equation:

$$B = -\frac{1}{2} J D^{(2)} J \,, \tag{1}$$

where $D^{(2)} = (\delta_{ij}^2)$ is the matrix of the square of dissimilarities, $J = I - \frac{1}{n}\mathbf{1}\mathbf{1}^T \in \mathbb{R}^{n \times n}$ is the centering matrix and I is the identity matrix. If B is positive semi-definite, the object coordinates in the low dimensional space $\mathbb{R}^k$ can be found through a singular value decomposition [4,7]:

$$X_k = V_k \Lambda_k^{1/2} \,, \tag{2}$$

where $V_k \in \mathbb{R}^{n \times k}$ is an orthogonal matrix with columns the first k eigen-vectors of B and $\Lambda_k \in \mathbb{R}^{k \times k}$ is a diagonal matrix with the corresponding eigenvalues. Several dissimilarities introduced in section 2 generate inner product matrices B non positive semi-definite. The negative values are usually small in our application and therefore can be neglected.

Once the training patterns have been embedded into a low dimensional space, the test pattern can be added to this space via a linear projection [15]. Next we detail briefly the process.

Let $X_k \in \mathbb{R}^{n \times k}$ be the object configuration found for the training patterns in $\mathbb{R}^k$ and $X_n = [\boldsymbol{x}_1 \dots \boldsymbol{x}_s]^T \in \mathbb{R}^{s \times k}$ the matrix of the object coordinates sought for the test patterns. Let $D_n^{(2)} \in \mathbb{R}^{s \times n}$ be the matrix of square dissimilarities between the s test patterns and the n training patterns that have been already projected. The matrix $B_n \in \mathbb{R}^{s \times n}$ of inner products among the test and training patterns can be found as:

$$B_n = -\frac{1}{2}(D_n^{(2)} J - U D^{(2)} J) \,, \tag{3}$$

where $J \in \mathbb{R}^{n \times n}$ is the centering matrix and $U = \frac{1}{n}\mathbf{1}^T\mathbf{1} \in \mathbb{R}^{s \times n}$. Since the matrix of inner products verifies

$$B_n = X_n X_k^T \tag{4}$$

then, X_n can be found as the least mean-square error solution to (4), that is:

$$X_n = B_n X_k (X_k^T X_k)^{-1} , \tag{5}$$

Given that $X_k^T X_k = \Lambda_k$ and considering that $X_k = V_k \Lambda_k^{1/2}$ the coordinates for the test points can be obtained as:

$$X_n = B_n V_k \Lambda_k^{-1/2} , \tag{6}$$

which can be easily evaluated through simple linear algebraic operations.

Next we introduce the method proposed to combine classifiers based on different dissimilarities.

Our method is based on the evidence that different dissimilarities reflect different features of the dataset (see section 2). Therefore, classifiers based on different measures will missclassify a different set of patterns. Figure 1 shows for instance that bold patterns are assigned to the wrong class by only one classifier but using a voting strategy the patterns will be assigned to the right class.

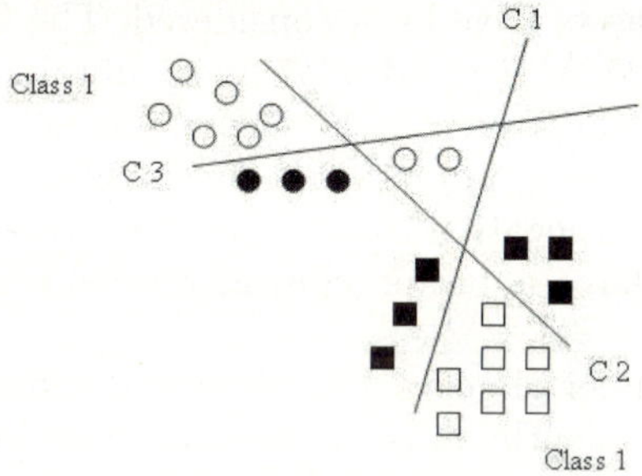

Fig. 1. Aggregation of classifiers using a voting strategy. Bold patterns are missclassified by a single hyperplane but not by the combination.

Hence, our combination algorithm proceeds as follows: First, a set of dissimilarities are computed. Each dissimilarity is embedded into an Euclidean space via the method explained in this section. Next, we train a SVM for each dissimilarity computed. Thus, it is expected that misclassification errors will change from one classifier to another. So the combination of classifiers by a voting strategy will help to reduce the misclassification errors.

A related technique to combine classifiers is the Bagging [3,2]. This method generates a diversity of classifiers that are trained using several bootstrap samples. Next, the classifiers are aggregated using a voting strategy. Nevertheless there are three important differences between bagging and the method proposed in this section.

First, our method generates the diversity of classifiers by considering different dissimilarities and thus using the whole sample. Bagging trains each classifier using around 63% of the training set. In our application the size of the training set is very small and neglecting part of the patterns may increase the bias of each classifier. It has been suggested in the literature that bagging doesn't help to reduce the bias [16] and so, the aggregation of classifiers will hardly reduce the misclassification error. A second advantage of our method is that it is able to work directly with a dissimilarity matrix. Finally, the combination of several dissimilarities avoids the problem of choosing a particular dissimilarity for the application we are dealing with. This is a difficult and time consuming task.

Notice that the algorithm proposed earlier can be easily applied to other classifiers such as the k-nearest neighbor algorithm that are based on dissimilarities. k-NN has a larger variance than the SVM and so, it is expected that the ensemble of classifiers will reduce even more the misclassification error.

4 Experimental Results

In this section, the ensemble of classifiers proposed is applied to the identification of cancerous tissues using Microarray gene expression data.

Two benchmark datasets have been considered. The first one consisted of 72 samples (47 ALL and 25 AML) and 6817 genes obtained from acute leukemia patients at the time of diagnosis [8]. The second dataset consisted of 49 samples and 7129 genes from breast tumors [18], 25 classified as positive to estrogen receptors (ER+) and 24 negative to estrogen receptors (ER-). Those positive to estrogen receptors have a better clinical outcome and require a different treatment.

Due to the large number of genes, samples are codified in a high dimensional and noisy space. Consequently, most of dissimilarities defined in section 2 will be correlated [9,12]. To avoid this problem, the number of genes has been aggressively reduced using the standard F-statistics [6].

The dissimilarities have been computed without normalizing the variables because this may increase the correlation among them. Once the dissimilarities have been embedded in a Euclidean space, the variables are normalized to unit variance and zero mean. This preprocessing improves the SVM accuracy and the speed of convergence.

The C regularization parameter for the SVM has been set up by ten fold-crossvalidation [14]. We have considered linear kernels in all experiments, because the small size of the training set in our application favors the overfitting of the data. Consequently error rates are smaller for linear kernels than for non linear one.

Regarding to the ensemble of classifiers, an important issue is the dimensionality in which the dissimilarity matrix is embedded. To this aim, a metric Multidimensional Scaling Algorithm is first run. The number of eigenvectors considered is determined by the curve induced by the eigenvalues. For the dataset considered in this paper about 85% of the variance is captured by the first eleven eigenvalues. Therefore they preserve the main structure of the data.

The combination strategy proposed in this paper has been also applied to the k-nearest neighbor classifier. An important parameter in this algorithm is the number of neighbors which has been estimated by cross-validation.

The classifiers have been evaluated from two different points of view: on the one hand we have computed the misclassification errors. But in our application, false negative and false positives errors have unequal relevance. For instance, in breast cancer, false negative errors corresponds to tumors positive to estrogen receptors that have been classified as negative to estrogen receptors. This will lead a wrong treatment with very dangerous consequences to the patient. Therefore, false negative errors are much more important than false positive errors.

Table 1. Experimental results for the ensemble of SVM classifiers. Classifiers based solely on a single dissimilarity and Bagging have been taken as reference.

	% Error		% False negative	
Method	Breast	Leukemia	Breast	Leukemia
Euclidean	10.2%	6.9%	4%	6.94%
Cosine	14.2%	1.38%	4%	1.38%
Correlation	14.2%	2.7%	6.1%	2.7%
χ^2	12.2%	1.38%	4%	1.38%
Manhattan	12.2%	5.5%	4%	4.16%
Spearman	16.3%	8.3%	6.1%	5.5%
Kendall-Tau	18.3%	8.3%	6.1%	5.5%
Kullback-Leibler	16.3%	30.5%	12.2%	19.4%
Bagging	6.1%	1.38%	2%	1.28%
Random genes	4.2%	4.16%	2.04%	4.16%
Combination	8.1%	1.38%	2%	1.38%

Table 1 shows the experimental results for the ensemble of classifiers using the SVM. The method proposed has been compared with bagging introduced in section 3 and a variant of bagging that generate the classifiers by sampling the genes. Finally, the classifiers based on a single dissimilarity have been taken as a reference.

From the analysis of table 1, the following conclusions can be drawn:

- The error for the Euclidean distance depends on the dataset considered, breast cancer or leukemia. For instance, the misclassification error and false negative error are large for Leukemia. On the other hand, the combination of dissimilarities improves significantly the Euclidean distance which is usually considered by most of SVM algorithms.
- The algorithm based on the combination of dissimilarities improves the best single-distance which is χ^2. Notice that for breast cancer false negative errors are significantly reduced.
- The combination of dissimilarities performs similarly to bagging sampling the patterns or the genes. However,we remark that our method is able to work directly from the dissimilarity matrix.

Table 2. Experimental results for the ensemble of k-NN classifiers. Classifiers based solely on a single dissimilarity and Bagging have been taken as reference.

	% Error		% False negative	
Method	Breast	Leukemia	Breast	Leukemia
Euclidean	14.2 %	6.94%	6.1%	4.1%
Cosine	16.3 %	2.77%	8.1%	2.77%
Correlation	14.2 %	4.16%	8.1 %	4.16%
χ^2	10.2%	2.77%	2.0%	2.77 %
Manhattan	8.1 %	2.7%	2.0%	2.7%
Spearman	10.2 %	2.77 %	4.0 %	2.77%
Kendall-tau	8.1 %	2.77 %	2.0 %	2.77 %
Kullback	51 %	76 %	46.9 %	11.1 %
Bagging	14.2 %	6.9 %	6.1 %	6.9 %
Combination	8.1 %	1.38 %	2.0 %	1.38 %

Table 2 shows the experimental results for the ensemble of k-NNs. The primary conclusions are the following:

- The combination of dissimilarities improves the best classifier based on a single dissimilarity and particularly for Leukemia data.
- The Euclidean distance performs very poorly and it is significantly improved by the combination of dissimilarities.
- The combination of dissimilarities outperforms clearly the bagging algorithm.
- Bagging errors are larger for k-nn classifiers than for SVM classifiers. This can be justified because SVM is more robust when a subset of patterns is neglected due to bootstrap sampling. The combination of dissimilarities does not suffer from this drawback.

5 Conclusions and Future Research Trends

In this paper, we have proposed an ensemble of classifiers based on a diversity of dissimilarities. Our approach aims to reduce the misclassification error of classifiers based solely on a single distance working directly from a dissimilarity matrix. The algorithm has been applied to the classification of cancerous tissues using gene expression data.

The experimental results suggest that the method proposed improves both, misclassification errors and false negative errors. We also report that our algorithm outperforms classifiers based on a single dissimilarity. A widely used combination strategy such as Bagging is improved particularly for k-NN classifiers.

As future research trends, we will try to increase the diversity of classifiers by random sampling the patterns for each dissimilarity.

References

1. Aggarwal, C.C.: Re-designing distance functions and distance-based applications for high dimensional applications. In: Proc. of SIGMOD-PODS, vol. 1, pp. 13–18 (2001)
2. Bauer, E., Kohavi, R.: An empirical comparison of voting classification algorithms: Bagging, boosting, and variants. Machine Learning 36, 105–139 (1999)
3. Breiman, L.: Bagging predictors. Machine Learning 24, 123–140 (1996)
4. Cox, T., Cox, M.: Multidimensional Scaling, 2nd edn. Chapman & Hall/CRC Press, Boca Raton, USA (2001)
5. Drăghici, S.: Data Analysis Tools for DNA Microarrays. Chapman & Hall/CRC Press, New York (2003)
6. Gentleman, R., Carey, V., Huber, W., Irizarry, R., Dudoit, S.: Bioinformatics and Computational Biology Solutions Using R and Bioconductor. Springer, Berlin (2006)
7. Golub, G.H., Loan, C.F.V.: Matrix Computations, 3rd edn. Johns Hopkins University press, Baltimore, Maryland, USA (1996)
8. Golub, T., Slonim, D., Tamayo, P.: Molecular classification of cancer: Class discovery and class prediction by gene expression monitoring. Science 286(15), 531–537 (1999)
9. Hinneburg, C.C.A.A., Keim, D.A.: What is the nearest neighbor in high dimensional spaces? In: Eiter, T., Libkin, L. (eds.) ICDT 2005. LNCS, vol. 3363, pp. 506–515. Springer, Heidelberg (2004)
10. Jiang, D., Tang, C., Zhang, A.: Cluster analysis for gene expression data: A survey. IEEE Transactions on Knowledge and Data Engineering 16(11) (November 2004)
11. Kittler, J., Hatef, M., Duin, R., Matas, J.: On combining classifiers. IEEE Transactions on Neural Networks 20(3), 228–239 (1998)
12. Martín-Merino, M., Muñoz, A.: A new Sammon algorithm for sparse data visualization. In: International Conference on Pattern Recognition (ICPR), vol. 1, pp. 477–481. IEEE Press, Cambridge (UK) (2004)
13. Martín-Merino, M., Muñoz, A.: Self organizing map and Sammon mapping for asymmetric proximities. Neurocomputing 63, 171–192 (2005)
14. Molinaro, A., Simon, R., Pfeiffer, R.: Prediction error estimation: a comparison of resampling methods. Bioinformatics 21(15), 3301–3307 (2005)
15. Pekalska, E., Paclick, P., Duin, R.: A generalized kernel approach to dissimilarity-based classification. Journal of Machine Learning Research 2, 175–211 (2001)
16. Valentini, G., Dietterich, T.: Bias-variance analysis of support vector machines for the development of svm-based ensemble methods. Journal of Machine Learning Research 5, 725–775 (2004)
17. Vapnik, V.: Statistical Learning Theory. John Wiley & Sons, New York (1998)
18. West, M., Blanchette, C., Dressman, H., Huang, E., Ishida, S., Spang, R., Zuzan, H., Olson, J., Marks, J., Nevins, J.: Predicting the clinical status of human breast cancer by using gene expression profiles. PNAS, 98(20) (September 2001)

Inductive Concept Retrieval and Query Answering with Semantic Knowledge Bases Through Kernel Methods

Nicola Fanizzi and Claudia d'Amato

Dipartimento di Informatica, Università degli Studi di Bari
Campus Universitario, Via Orabona 4, 70125 Bari, Italy
{fanizzi,claudia.damato}@di.uniba.it

Abstract. This work deals with the application of kernel methods to structured relational settings such as semantic knowledge bases expressed in Description Logics. Our method integrates a novel kernel function for the $\mathcal{ALC}$ logic in a *support vector machine* that could be set up to work with these representations. In particular, we present experiments where our method is applied to the tasks of concept retrieval and query answering on existing ontologies.

Keywords: Inductive Concept Retrieval, Query Answering, Kernel Methods, Kernel Function, Description Logics, Semantic Web.

1 Learning in Multi-relational Settings

Many application domains, spanning from computational biology and chemistry to natural language processing, require operating on structured data representations. A new emerging domain is represented by the Semantic Web (SW) [1] where knowledge intensive manipulations on complex relational descriptions are foreseen to be performed by machines. In this context, *Description Logics* (DLs) [2] have been adopted as the core technology for ontology languages, such as OWL. This family of languages is endowed with well-founded semantics and reasoning services (see Sect. 2). Unfortunately, machine learning through logic-based methods is inherently intractable in multi-relational settings, unless language bias is imposed to constrain the representation. Yet, for the sake of tractability, only very simple DL languages have been considered so far.

Kernel methods [3] are a family of efficient statistical learning algorithms, including the *support vector machines* (SVMs), that have been effectively applied to a variety of tasks, recently also in domains that typically require structured representations [4,5]. They can be very efficient because they map, by means of a kernel function, the original feature space of the considered data set into a high-dimensional space, where the learning task is simplified. However, such a mapping is not explicitly performed (*kernel trick*): it requires a sound definition of a positive definite kernel function on the feature space; the validity of such a function ensures that the embedding into a new space exists, so that it corresponds to the inner product in this space [3].

In this work, we exploit a kernel function for DLs representations, specifically for the $\mathcal{ALC}$ logic [6]. It encodes a notion of similarity of individuals in this representation, based on both structural and semantic aspects of the reference representation (see

B. Apolloni et al. (Eds.): KES 2007/WIRN 2007, Part I, LNAI 4692, pp. 148–155, 2007.

Sect. 3). By means of the resulting SVM, many tasks based on inductive classification can be tackled. Particularly, we demonstrate how to perform important inferences on semantic knowledge bases, namely concept retrieval and query answering. These tasks are generally grounded on merely deductive procedures which easily fail in case of (partially) inconsistent or incomplete knowledge. We show how the methods performs comparably well w.r.t. a standard deductive reasoner, allowing the suggestion of new knowledge that was not previously logically derivable. Indeed, the method was implemented and experimentally tested on artificial and real ontologies drawn from standard repositories as illustrated in Sect. 5.

2 Reference Representation Space

We will recall the basics of $\mathcal{ALC}$ (see [2] for a thorough reference). Such a logic is not trivial as it is endowed with the basic constructors employed by the standard ontology languages and deductive reasoning is quite computationally expensive [7].

Descriptions are inductively defined starting with a set N_C of *primitive concept* names and a set N_R of *primitive roles*. Complex descriptions are built using primitive concepts and roles and the language constructors. The semantics of the descriptions is defined by an *interpretation* $\mathcal{I} = (\Delta^{\mathcal{I}}, \cdot^{\mathcal{I}})$, where $\Delta^{\mathcal{I}}$ is a non-empty set, the *domain* of the interpretation, and $\cdot^{\mathcal{I}}$ is the *interpretation function* that maps each $A \in N_C$ to a set $A^{\mathcal{I}} \subseteq \Delta^{\mathcal{I}}$ and each $R \in N_R$ to $R^{\mathcal{I}} \subseteq \Delta^{\mathcal{I}} \times \Delta^{\mathcal{I}}$. The *top* concept $\top$ is interpreted as the whole domain $\Delta^{\mathcal{I}}$, while the *bottom* concept $\bot$ corresponds to $\emptyset$. Complex descriptions can be built in $\mathcal{ALC}$ using the following constructors. *Full negation*: given any description C, it is denoted $\neg C$ and amounts to $\Delta^{\mathcal{I}} \setminus C^{\mathcal{I}}$. *Concept conjunction*, denoted by $C_1 \sqcap C_2$, yields an extension $C_1^{\mathcal{I}} \cap C_2^{\mathcal{I}}$ and, dually, *concept disjunction*, denoted $C_1 \sqcup C_2$, yields the union $C_1^{\mathcal{I}} \cup C_2^{\mathcal{I}}$. Finally, the *existential restriction*, denoted $\exists R.C$, is interpreted as the set $\{x \in \Delta^{\mathcal{I}} \mid \exists y \in \Delta^{\mathcal{I}}((x, y) \in R^{\mathcal{I}} \wedge y \in C^{\mathcal{I}})\}$ and the *value restriction* $\forall R.C$, has the extension $\{x \in \Delta^{\mathcal{I}} \mid \forall y \in \Delta^{\mathcal{I}}((x, y) \in R^{\mathcal{I}} \rightarrow y \in C^{\mathcal{I}})\}$.

The main inference is *subsumption* between concepts based on their semantics: given two descriptions C and D, C *subsumes* D, denoted by $C \sqsupseteq D$, iff for every interpretation $\mathcal{I}$ it holds that $C^{\mathcal{I}} \supseteq D^{\mathcal{I}}$. When $C \sqsupseteq D$ and $D \sqsupseteq C$ then they are equivalent, denoted with $C \equiv D$.

A *knowledge base* $\mathcal{K} = \langle \mathcal{T}, \mathcal{A} \rangle$ contains a *TBox* $\mathcal{T}$ and an *ABox* $\mathcal{A}$. $\mathcal{T}$ is the set of definitions $C \equiv D$, meaning $C^{\mathcal{I}} = D^{\mathcal{I}}$, where C is the concept name and D is its description. $\mathcal{A}$ contains assertions on the world state, e.g. $C(a)$ and $R(a, b)$, meaning that $a^{\mathcal{I}} \in C^{\mathcal{I}}$ and $(a^{\mathcal{I}}, b^{\mathcal{I}}) \in R^{\mathcal{I}}$.

A related inference is *instance checking*, that is deciding whether an individual is an instance of a concept [7,2]. Conversely, it may be necessary to find the concepts which an individual belongs to (*realization problem*), especially the most specific one:

Definition 1 (most specific concept). *Given an ABox $\mathcal{A}$ and an individual a, the* most specific concept *of a w.r.t. $\mathcal{A}$ is the concept C, denoted $\mathsf{MSC}_{\mathcal{A}}(a)$, such that $\mathcal{A} \models C(a)$ and for any other concept D such that $\mathcal{A} \models D(a)$, it holds that $C \sqsubseteq D$.*

In some cases, the MSC may not be expressed by a finite description [2], yet it may be approximated. Generally approximations up to a certain depth k are considered, denoted MSC^k. We will generically indicate a maximal depth approximation with MSC^*.

Another inference is *retrieval* which consists in finding the extension of a given concept C, namely, all individuals a such that $\mathcal{K} \models C(a)$.

Many semantically equivalent (yet syntactically different) descriptions can be given for the same concept. Nevertheless, equivalent concepts can be reduced to a normal form by means of rewriting rules that preserve their equivalence [2]. Some notations are necessary to define the $\mathcal{ALC}$ normal form. $\mathsf{prim}(C)$ is the set of all the primitive concepts (and their negations) occurring at the top-level of C; $\mathsf{val}_R(C) = C_1 \sqcap \cdots \sqcap C_n$ if there exists a value restriction $\forall R.(C_1 \sqcap \cdots \sqcap C_n)$ on the top-level of C, otherwise $\mathsf{val}_R(C) = \top$; $\mathsf{ex}_R(C)$ is the set of the descriptions C' appearing in existential restrictions $\exists R.C'$ at the top-level conjunction of C. The normal form is defined as follows:

Definition 2 **($\mathcal{ALC}$ normal form).** *A description C is in $\mathcal{ALC}$* normal form *iff $C \equiv \bot$ or $C \equiv \top$ or if $C = C_1 \sqcup \cdots \sqcup C_n$ with*

$$C_i = \bigsqcap_{P \in \mathsf{prim}(C_i)} P \sqcap \bigsqcap_{R \in N_R} \left(\forall R.\mathsf{val}_R(C_i) \sqcap \bigsqcap_{E \in \mathsf{ex}_R(C_i)} \exists R.E \right)$$

where, for all $i = 1, \ldots, n$, $C_i \not\equiv \bot$ and, for any $R \in N_R$, $\mathsf{val}_R(C_i)$ and every sub-description in $\mathsf{ex}_R(C_i)$ are in normal form.

3 Kernel Functions

In the kernel methods, the learning algorithm (inductive bias) and the choice of the kernel function (language bias) are almost completely independent. Thus, an efficient algorithm for attribute-value instance spaces can be converted into one suitable for structured spaces (e.g. trees, graphs) by merely replacing the kernel function with a suitable one. This motivates the increasing interest addressed to the SVMs and other kernel methods [3] that reproduce learning in high-dimensional spaces while working like in a vectorial representation.

Kernels are endowed with the closure property w.r.t. many operations. In particular this class is closed w.r.t. convolution [8]: such kernels can deal with compounds by decomposing them into their parts, provided that valid kernels have already been defined for them. Other works have continued this line of research introducing kernels for strings, trees, graphs and other discrete structures [4]. In particular, [5] shows how to define generic kernels based on type construction where types are defined in a declarative way. While these kernels were defined as depending on specific structures, a more flexible method is building kernels as parametrized on a uniform representation. Cumby and Roth [9] propose the syntax-driven definition of kernels based on a simple DL representation, the *Feature Description Language*. They show that the feature space blow-up is mitigated by the adoption of efficiently computable kernels. These functions transform the initial representation of the instances into the related active features, thus allowing learning the classifier directly from the structured data.

Grounded on [5], a (family of) valid kernel for the space X of $\mathcal{ALC}$ descriptions has been proposed [6]. Recurring to the convolution kernels [8], the normal form is used to decompose complex descriptions level-wise into sub-descriptions as follows:

Definition 3 **($\mathcal{ALC}$ kernel).** *Given an interpretation $\mathcal{I}$, the $\mathcal{ALC}$* kernel *based on $\mathcal{I}$ is the function $k_{\mathcal{I}} : X \times X \mapsto \mathbb{R}$ inductively defined as follows. Let two descriptions in normal form be $D_1 = \bigsqcup_{i=1}^{n} C_i^1$ and $D_2 = \bigsqcup_{j=1}^{m} C_j^2$, then:*

disjunctive descriptions: $k_{\mathcal{I}}(D_1, D_2) = \lambda \sum_{i=1}^{n} \sum_{j=1}^{m} k_{\mathcal{I}}(C_i^1, C_j^2)$ *with* $\lambda \in]0, 1]$

conjunctive descriptions:

$$k_{\mathcal{I}}(C^1, C^2) = \prod_{\substack{P_1 \in \mathsf{prim}(C^1) \\ P_2 \in \mathsf{prim}(C^2)}} k_{\mathcal{I}}(P_1, P_2) \cdot \prod_{R \in N_R} k_{\mathcal{I}}(\mathsf{val}_R(C^1), \mathsf{val}_R(C^2)) \cdot \prod_{R \in N_R} \sum_{\substack{C_i^1 \in \mathsf{ex}_R(C^1) \\ C_j^2 \in \mathsf{ex}_R(C^2)}} k_{\mathcal{I}}(C_i^1, C_j^2)$$

primitive concepts: $k_{\mathcal{I}}(P_1, P_2) = k_{\mathrm{set}}(P_1^{\mathcal{I}}, P_2^{\mathcal{I}}) = |P_1^{\mathcal{I}} \cap P_2^{\mathcal{I}}|$

where k_{set} is the kernel for set structures defined in [5]. This case includes also the negation of primitive concepts using: $(\neg P)^{\mathcal{I}} = \Delta^{\mathcal{I}} \setminus P^{\mathcal{I}}$.

This kernel computes the similarity between disjunctive as the sum of the cross-similarities between any couple of disjuncts from either description (λ is employed to downweight the similarity of the sub-descriptions on the grounds of the level where they occur). The conjunctive kernel computes the similarity between two input descriptions, distinguishing among primitive concepts, those referred in the value restrictions and those referred in the existential restrictions. These similarity values are multiplied reflecting the fact that all the restrictions have to be satisfied at a conjunctive level. The similarity between primitive concepts is measured in terms of the intersection of their extension.

The kernel can be extended to the case of individuals $a, b \in \mathsf{Ind}(\mathcal{A})$ simply by taking into account the approximations of their MSCs: $k_{\mathcal{I}}(a, b) = k_{\mathcal{I}}(\mathsf{MSC}^*(a), \mathsf{MSC}^*(b))$.

The application of the kernel function to most expressive DL is not trivial. DLs allowing normal form concept definitions can only be considered. Moreover, for each constructor not included in the $\mathcal{ALC}$ logic, a kernel definition has to be provided.

4 Concept Retrieval by Means of Kernel Methods

SVMs are classifiers, that, exploiting a kernel function, map the training data into a higher dimensional feature space where they can be classified using a linear classifier. The SVM, as any other kernel method, can be applied to whatever knowledge representation, provided a kernel function suitable for the chosen representation. Hence, a SVM can be applied to an $\mathcal{ALC}$ knowledge base, considering the kernel function in Def. 3. In this paper, the SVM is used to solve the following classification problem:

Definition 4 **(Problem Definition).** *Given a knowledge base $KB = (\mathcal{T}, \mathcal{A})$, let Ind($\mathcal{A}$) be the set of all individuals in $\mathcal{A}$ and $C = \{C_1, \ldots, C_s\}$ the set of all concepts (both* *primitive and defined) in $\mathcal{T}$. The problem to solve is: considered an individual $a \in \mathsf{Ind}(\mathcal{A})$ determine the set of concepts $\{C_1, \ldots, C_t\} \subseteq C$ to which a belongs to.*

In the general setting of SVMs, the classes for the classification are disjoint. This is not generally verified in the SW context, where an individual can be instance of more than one concept. To solve this problem, a new answering procedure is proposed. It is based on the decomposition of the multi-class problem into smaller binary classification problems (one per class). Therefore, a simple binary value set ($V = \{-1, +1\}$) can be employed, where $(+1)$ indicates that an example x_i occurs in the ABox w.r.t. the considered concept C_j (namely $C_j(x_i) \in \mathcal{A}$); (-1) indicates the absence of the assertion in the ABox. As an alternative, it can be considered $+1$ when $C_j(x_i)$ can be inferred from the knowledge base, and -1 otherwise.

Another issue has to be considered. In the general classification setting an implicit assumption of *Closed World* is made. On the contrary, in the SW context the *Open World Assumption* (OWA) is generally made. To deal with the OWA, the absence of information on whether a certain instance x_i belongs to the extension of concept C_j should not be interpreted negatively, as seen before, rather, it should count as neutral information. Thus, another value set has to be considered, namely $V = \{+1, -1, 0\}$, where the three values denote, respectively, assertion occurrence ($C_j(x_i) \in \mathcal{A}$), occurrence of the opposite assertion ($\neg C_j(x) \in \mathcal{A}$) and assertion absence in $\mathcal{A}$. Occurrences can be easily computed with a lookup in the ABox. Moreover, as in the previous case, a more complex procedure may be devised by substituting the notion of occurrence (absence) of assertions in (from) the ABox with the one of derivability from the whole KB, i.e. $\mathcal{K} \vdash C_j(x_i)$ ($\mathcal{K} \not\vdash C_j(x_i)$), $\mathcal{K} \not\vdash C_j(x_i)$ *and* $\mathcal{K} \not\vdash \neg C_j(x_i)$, respectively.

Hence, considered the query instance x_q, for every concept $C_j \in C$ the classifier will return $+1$ if x_q is an instance of C_j, -1 if x_q is an instance of $\neg C_j$, and 0 otherwise. The classification is performed on the ground of a set of training examples from which such information can be derived.

The classification results can be used to improve concept retrieval service. By classifying the individuals in the Abox w.r.t. all concepts, concept retrieval is performed exploiting an inductive approach. As will be experimentally shown in the following, the classifier, besides of having a comparable behavior w.r.t. a standard reasoner, is also able to induce new knowledge that is not logically derivable. Moreover it can be employed for the query answering task by determining, as illustrated above, the extension of a new query concept built from concepts and roles in the considered ontology.

5 Experimental Evaluation

In order to solve the classification problem presented in the previous section and assess the validity of the $\mathcal{ALC}$ kernel function (see Def. 3), a SVM from the LIBSVM library[1] has been considered. The instance classification has been performed on nine different ontologies represented in OWL: FAMILY and UNIVERSITY handmade ontologies, FSM, SURFACE-WATER-MODEL, NEWTESTAMENTNAMES, SCIENCE, PEOPLE, NEWSPAPER and WINES ontologies from the Protégé library[2]. Although they are represented in

[1] `http://www.csie.ntu.edu.tw/~cjlin/libsvm`

[2] See the webpage: `http://protege.stanford.edu/plugins/owl/owl-library`

languages that are different from $\mathcal{ALC}$, constructors that are not allowed by $\mathcal{ALC}$ are simply discarded, in order to apply the kernel function.

The classification method was applied to all the individuals in each ontology; namely, the individuals were checked to assess if they were instances of the concepts in the ontology through the SVM. The performance was evaluated comparing its responses to those returned by a standard reasoner[3] used as baseline. Specifically, for each individual in the ontology the MSC is computed and enlisted in the set of training (or test) examples. Each example is classified applying the SVM and the $\mathcal{ALC}$ kernel function with $\lambda = 1$ (see Def. 3). The experiment has been repeated twice, adopting the leave-one-out cross validation procedure for ontologies with less then 50 individuals, and the ten-fold cross validation procedure for the other ontologies. For each concept in the ontology, the following parameters have been measured for the evaluation: *match rate* computed as the number of cases of individuals that got exactly the same classification by both classifiers with respect to the overall number of individuals; *omission error rate* computed as the amount of unlabeled individuals (namely the method could not determine whether it was an instance or not) while it was to be classified as an instance of that concept; *commission error rate* computed as the amount of individuals (analogically) labeled as instances of a concept, while they (logically) belong to that concept or vice-versa; *induction rate* computed as the amount of individuals that were found to belong to a concept or its negation, while this information is not logically derivable from the knowledge base. The average rates obtained over all the concepts in each ontology are reported, jointly with their range. By looking at Tab. 1, reporting the experimental outcomes, it is important to note that, for every ontology, the commission error is quite low. This means that the classifier did not make critical mistakes, i.e. cases when an individual is deemed as an instance of a concept while it really is an instance of another disjoint concept. Particularly, the commission error rate is not null in case of UNIVERSITY and FSM ontologies and consequently also the match rate is the lowest. It is worthwhile to note that these ontologies have the lowest number of individuals for concepts. Specifically, the number of concepts is almost similar to the number of individuals, this may represent a situation in which there is not enough information for separating the feature space and then produce a correct classification. However, also in this condition, the commission error is quite low, the matching rate is considerably high and the classifier is able to induce new knowledge (induction rate not null).

In general, looking at Tab. 1 it is possible to note that the match rate increases with the increase of the number of individuals in the considered ontology with a consequent strong decrease of the commission error rate that is close to 0 in such cases. Almost always the classifier is able to induce new knowledge. Anyway it presents also a conservative behavior, indeed the omission error rate is very often not null. To decrease the tendency to a conservative behavior of the classifier, a threshold could be introduced for the consideration of the "unknown" (namely labeled with 0) training examples.

Another experiment has been done, to test the method as a means for performing inductive concept retrieval w.r.t. new query concepts built from a considered ontology. The method has been applied to perform a number of retrieval problems applied to the considered ontologies using $\lambda = 1$ for the kernel function. The experiment was quite

[3] PELLET: `http://pellet.owldl.com`

Table 1. Results (average and range) of the experiments with $\lambda = 1$

ONTOLOGY		*match rate*	*induction rate*	*omis. err. rate*	*comm. err. rate*
PEOPLE	*avg.*	**0.866**	**0.054**	**0.08**	**0.00**
	range	0.66 - 0.99	0.00 - 0.32	0.00 - 0.22	0.00 - 0.03
UNIVERSITY	*avg.*	**0.789**	**0.114**	**0.018**	**0.079**
	range	0.63 - 1.00	0.00 - 0.21	0.00 - 0.21	0.00 - 0.26
FSM	*avg.*	**0.917**	**0.007**	**0.00**	**0.076**
	range	0.70 - 1.00	0.00 - 0.10	0.00 - 0.00	0.00 - 0.30
FAMILY	*avg.*	**0.619**	**0.032**	**0.349**	**0.00**
	range	0.39 - 0.89	0.00 - 0.41	0.00 - 0.62	0.00 - 0.00
NEWSPAPER	*avg.*	**0.903**	**0.00**	**0.097**	**0.00**
	range	0.74 - 0.99	0.00 - 0.00	0.02 - 0.26	0.00 - 0.00
WINES	*avg.*	**0.956**	**0.004**	**0.04**	**0.00**
	range	0.65 - 1.00	0.00 - 0.27	0.01 - 0.34	0.00 - 0.00
SCIENCE	*avg.*	**0.942**	**0.007**	**0.051**	**0.00**
	range	0.80 - 1.00	0.00 - 0.04	0.00 - 0.20	0.00 - 0.00
S.-W.-M.	*avg.*	**0.871**	**0.067**	**0.062**	**0.00**
	range	0.57 - 0.98	0.00 - 0.42	0.00 - 0.40	0.00 - 0.00
N.T.N.	*avg.*	**0.925**	**0.026**	**0.048**	**0.001**
	range	0.66 - 0.99	0.00 - 0.32	0.00 - 0.22	0.00 - 0.03

Table 2. Results (average) of the querying experiments

ONTOLOGY	*match rate*	*ind. rate*	*omis. err. rate*	*comm. err. rate*
PEOPLE	0.886	0.040	0.074	0.0
UNIVERSITY	0.72	0.16	0.009	0.111
FSM	0.878	0.009	0.0	0.114
FAMILY	0.663	0.045	0.292	0.0
NEWSPAPER	0.779	0.0	0.221	0.0
WINES	0.943	0.0	0.057	0.0
SCIENCE	0.978	0.005	0.016	0.0
S.-W.-M.	0.804	0.134	0.062	0.0
NTN	0.906	0.022	0.072	0.0

intensive involving the classification of all the individuals in each ontology; namely, the individuals were checked through the inductive procedure to assess whether they were retrieved as instances of a query concept. Therefore, 15 queries were randomly generated by means of conjunctions/disjunctions of primitive and/or defined concepts of each ontology. As for the previous experiment, the leave-one-out procedure was performed in case of ontologies with less than 50 individuals and a ten-fold cross validation was performed for the others. The outcomes are reported in Tab. 2, from which it is possible to observe that the behavior of the classifier mainly remains the same as in the experiment whose outcomes are reported in Tab. 1.

Summarizing, the $\mathcal{ALC}$ kernel function can be effectively used, jointly with a SVM, to perform inductive concept retrieval, guaranteeing almost null commission error and interestingly the ability to induce new knowledge. The performance of the classifier increases with the increase of the number of individuals populating the considered ontology that have to be preferable homogeneously spread w.r.t. the concept in the ontology.

6 Conclusions and Future Work

In this work we have tested a kernel function for $\mathcal{ALC}$ descriptions integrated with a SVM in a (multi)relational learning setting. The resulting classifier has been used to improve concept retrieval and query answering tasks in the ontological setting. It has been experimentally shown that its performance is not only comparable to the one of a standard reasoner, but it is is also able to induce new knowledge, which is not logically derivable. Particularly, an increase in prediction accuracy was observed when the instances are homogeneously spread.

The realized classifier can be exploited for predicting/suggesting missing information about individuals, thus completing large ontologies. Specifically, it can be used to semi-automatize the population of an ABox. Indeed, the new assertions can be suggested to the knowledge engineer that has only to validate their inclusion. This constitutes a new approach in the SW context, since the efficiency of the statistical and numerical approaches and the effectiveness of a symbolic representation have been combined.

The main weakness of the approach is on its scalability towards more complex DLs. While computing MSC approximations might be feasible, it may be more difficult focusing on a normal form when comparing descriptions. Indeed, as long as the expressivity increases, the gap between syntactic structure semantics of the descriptions becomes more evident. As a next step, we can foresee the investigation of defining kernels for more expressive languages w.r.t. $\mathcal{ALC}$, e.g. languages enriched with (qualified) number restrictions and inverse roles [2].

References

1. Berners-Lee, T., Hendler, J., Lassila, O.: The semantic web. Scientific American 284(5), 34–43 (2001)
2. Baader, F., Calvanese, D., McGuinness, D., Nardi, D., Patel-Schneider, P.: The Description Logic Handbook. Cambridge University Press, Cambridge (2003)
3. Schölkopf, B., Smola, A.J.: Learning with Kernels. MIT Press, Cambridge (2002)
4. Gärtner, T.: A survey of kernels for structured data. SIGKDD Explorations 5, 49–58 (2003)
5. Gärtner, T., Lloyd, J., Flach, P.: Kernels and distances for structured data. Machine Learning 57(3), 205–232 (2004)
6. Fanizzi, N., d'Amato, C.: A declarative kernel for $\mathcal{ALC}$ concept descriptions. In: Esposito, F., Raś, Z.W., Malerba, D. (eds.) In Proceedings of the 16th International Symposium on Methodologies for Intelligent Systems. LNCS, vol. 4203, pp. 322–331. Springer, Heidelberg (2006)
7. Donini, F.M., Lenzerini, M., Nardi, D., Schaerf, A.: Deduction in concept languages: From subsumption to instance checking. Journal of Logic and Computation 4(4), 423–452 (1994)
8. Haussler, D.: Convolution kernels on discrete structures. Technical Report UCSC-CRL-99-10, Department of Computer Science, University of California – Santa Cruz (1999)
9. Cumby, C.M., Roth, D.: On kernel methods for relational learning. In: Fawcett, T., Mishra, N. (eds.) Proceedings of the 20th International Conference on Machine Learning, ICML2003, pp. 107–114. AAAI Press, Stanford, California, USA (2003)

Sub-symbolic Mapping of Cyc Microtheories in Data-Driven "Conceptual" Spaces

Giovanni Pilato[1], Agnese Augello[2], Mario Scriminaci[2], Giorgio Vassallo[2], and Salvatore Gaglio[1,2]

[1] ICAR - Italian National Research Council,
Viale delle Scienze, Ed.11, 90128, Palermo, Italy
g.pilato@icar.cnr.it
[2] DINFO – University of Palermo,
Viale delle Scienze, Ed. 6, 90128 Palermo, Italy
{augello,scriminaci}@csai.unipa.it, {gvassallo,gaglio}@unipa.it

Abstract. The presented work aims to combine statistical and cognitive-oriented approaches with symbolic ones so that a conceptual similarity relationship layer can be added to a Cyc KB microtheory. Given a specific microtheory, a LSA-inspired conceptual space is inferred from a corpus of texts created using both *ad hoc* extracted pages from the Wikipedia repository and the built-in comments about the concepts of the specific Cyc microtheory. Each concept is projected in the conceptual space and the desired layer of sub-symbolic relationships between concepts is created. This procedure can help a user in finding the concepts that are "sub-symbolically conceptually related" to a new concept that he wants to insert in the microtheory. Experimental results involving two Cyc microtheories are also reported.

Keywords: Data-Driven Conceptual Spaces, Ontologies, Cyc.

1 Introduction

Ontologies generally describe individuals, classes, attributes and relations [7], [9]. In the last years, the Cycorp Inc company has developed the Cyc commonsense knowledge base (KB) [13] which has a very large ontology constituted by over one hundred thousands atomic terms axiomatized by a set of over one million assertions, rules or commonsense ideas formulated in *n*-th order predicate calculus. The Cyc KB, at present, is the largest and most complete general knowledge base, equipped with a good performing inference engine. In Cyc the knowledge base is composed by microtheories (Mt), that are particular collections of concepts and facts in a specific domain. Cyc is suitable for automated logical inference to support knowledge-based reasoning applications. It also supports interoperability among software applications, it is extensible, provides a common vocabulary, and is suitable for mapping to/from other ontologies.

In recent years an interest towards methodologies for automatically learning ontologies from text corpora has grown [5], [6], [7], [8], [10]. At the same time there has been a great deal of research, which leads to the so-called hybrid symbolic/

B. Apolloni et al. (Eds.): KES 2007/WIRN 2007, Part I, LNAI 4692, pp. 156–163, 2007.

sub-symbolic systems. Many attempts have been made to integrate connectionist and symbolic methodologies, most of the time in solving learning problems [11]; only few works have been done that face the problem of linking a conceptual level to a symbolic one and they usually concern robotics [12].

Latent Semantic Analysis (LSA) paradigm [3], [4] is one of the most useful sub-symbolic techniques to represent the latent relations between the words belonging to a large collection of documents. According to this methodology, words are represented as vectors in a large-dimensional semantic space, while the semantic similarity between words can be calculated as the geometric distance between their representative vectors. The work presented in [2] shows how the semantic space created employing LSA technique can be thought as a "conceptual space", which is entirely data driven, since it is built by processing a matrix automatically derived from the analysis of a text corpus. In this space the concepts of an ontolology are represented projecting their verbal description. Moreover it is not necessary to introduce any hierarchical structure, since orthonormality of basis vectors ensures independency among the vectors which generates the conceptual space.

This paper focuses on a technique to enhance the Cyc commonsense KB with a sub-symbolic conceptual similarity relationship layer, combining statistical and cognitive-oriented approaches with symbolic ones.

Given a specific microtheory, a conceptual space is properly inferred from a corpus of text built using both ad hoc extracted pages from the Wikipedia [14] repository, and the comments on the concepts already present in the specific microtheory. Each concept is projected in this space and a layer of sub-symbolic semantic relationships between concepts is automatically created. The approach allows to overcome the limitations of classic rule based knowledge bases thanks to the added associative properties provided by a data driven, automatically constructed, "conceptual" space which has the same psychological basis claimed by the LSA [4].

The procedure can help a user in: *a)* finding concepts already stored in the KB by applying an associative sub-symbolic path in the ontology which automatically arises from the descriptions of the concepts and *b)* properly inserting new concepts in the ontology finding immediately the concepts which are "sub-symbolically conceptually related" to the new ones.

In the remainder of the paper the whole procedure is explained and experimental results, regarding two Cyc microtheories (*AcademicOrganizationMt* and *BiologyMt*) are illustrated. Conclusions and future works are reported in the last paragraph.

2 LSA-Based Data Driven "Conceptual" Space

In what follows an interpretation of the LSA framework, which leads to a data-driven "conceptual" space creation, is briefly recalled [1], [2], [3], [4]. Let N be the number of documents of a text corpus, and let M be the number of unique words present in the corpus. Then let $\mathbf{A}=\{a_{ij}\}$ be the M×N matrix whose (i,j)-th entry is the count of the occurrences of the i-th word in the j-th paragraph. According to the Singular Value Decomposition theorem, $\mathbf{A}$ can be decomposed in the product $\mathbf{A}=\mathbf{U}\mathbf{\Sigma}\mathbf{V}^T$, where $\mathbf{U}$ is a column-orthonormal $M\times N$ matrix, $\mathbf{V}$ is a column-orthonormal N×N matrix and $\mathbf{\Sigma}$ is a $N\times N$ diagonal matrix, whose elements are called singular values of $\mathbf{A}$.

It can be supposed, without loss of generality that **A**'s singular values are ranked in decreasing order. Let R be a positive integer with $R < N$, and let $\mathbf{U}_R$ be the $M \times R$ matrix obtained from **U** by suppressing the last $N-R$ columns, $\mathbf{\Sigma}_R$ the matrix obtained from $\mathbf{\Sigma}$ by suppressing the last $N-R$ rows and the last $N-R$ columns and $\mathbf{V}_R$ be the $N \times R$ matrix obtained from **V** by suppressing the last $N-R$ columns. Then $\mathbf{A}_R = \mathbf{U}_R\mathbf{\Sigma}_R\mathbf{V}_R^T$ is a $M \times N$ matrix of rank R obtained from the matrix **A** through Truncated Singular Value Decomposition (TSVD). $\mathbf{A}_R$ is the best rank-R approximation of the matrix **A** (among the M × N matrices) with respect to the Frobenius metric. The i–th row of the matrix $\mathbf{U}_R$ may be considered as representative of the i–th word, while the j-th row of the matrix $\mathbf{V}_R$ may be considered as representative of the j -th document. However, if we normalize the matrix **A**, dividing each element by the sum of all its elements, **A** can be considered as a sample set. If we subsequently calculate a matrix $\mathbf{B}=\{b^{(R)}_{ij}\}$ whose component is the square root of a_{ij}, it can be shown that performing the TSVD on **B** is equivalent to evaluate the best rank R approximation $\mathbf{B}_R=\{b^{(R)}_{ij}\}$ to **B** with respect to the *Hellinger* distance, defined by:

$$d_H(B, B_R) = \sqrt{\sum_{i=1}^{M}\sum_{j=1}^{N}\left(\sqrt{b_{ij}} - \sqrt{b_{ij}^{(R)}}\right)} \; . \tag{1}$$

This allows interpreting the TSVD as a sufficient statistical estimator [2]. Therefore the singular vectors of **B** can be seen as probability distributions (besides **B**'s singular vectors all square to 1).

To evaluate the distance between two vectors $\mathbf{v}_i$ and $\mathbf{v}_j$ belonging to this space which is coherent with this probabilistic interpretation, a similarity measure is defined as follows:

$$sim(v_i, v_j) = \begin{cases} \cos^2(v_i, v_j) & \text{if } \cos(v_i, v_j) \geq 0 \\ 0 & \text{otherwise} \end{cases} \; . \tag{2}$$

The two matrices **U** and **V** obtained after decomposition process reflect a breakdown of the original relationships into linearly-independent vectors [2]. These independent R dimensions of the R^R space can be tagged in order to interpret this space as a "conceptual" space. Since these vectors are orthogonal, they can be regarded as principal axes, and so they can be regarded as axes, which represent the "fundamental" concepts residing in the data driven space generated by the LSA.

3 Sub-symbolic Mapping of Cyc Microtheories

The proposed procedure consists of two phases: conceptual space creation and new concepts introduction and it is illustrated in Fig. 1.

3.1 Conceptual Space Creation

For the conceptual space creation, all constants belonging to the selected microtheory are searched through the associated *VocabularyMt*. All assertions are also analyzed and all the links in the ontology are stored for validation purpose only.

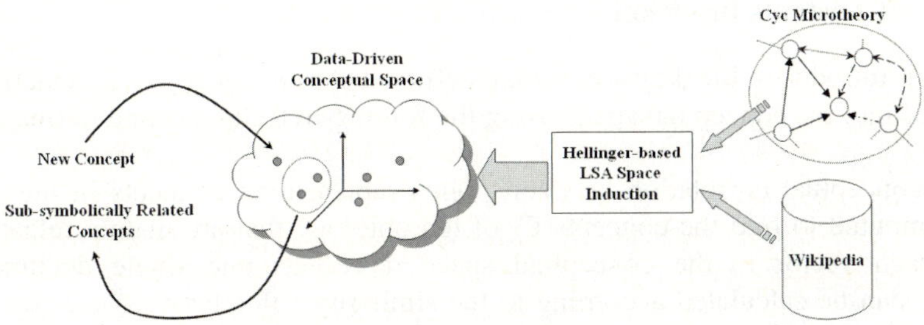

Fig. 1. An overview of the sub-symbolic mapping process

A semantic space is built using the LSA-inspired technique illustrated in section 2, which minimizes the Hellinger distance. A large meaningful text corpus, which is also coherent with the topic characterizing the selected microtheory, is needed. The collection of these documents is a critical phase of this step, because the quality of the corpus determines the effectiveness of the semantic space creation. Therefore it is chosen to use the English version of the Wikipedia [14] repository, which nowadays is one of the most complete semi-structured free documents repository. We have used the internal search engine of Wikipedia for retrieving documents that are pertinent to the topic of the microtheory using the names of the concepts as keywords where a relevance threshold has been experimentally fixed to 50%.

Each retrieved page is then filtered in order to remove non-textual and non-informative content such as HTML code, images, scripting and so on. Each page has been divided in paragraphs; therefore, for every concept there is a variable number of texts, each one corresponding to an extracted paragraph. A large variability of document number can exist for different concepts depending on articles of Wikipedia: the more an argument is widespread, the more links related to the main articles are found. The set of documents used to build the semantic space has been extended using also the Cyc comments of each concept in the selected microtheory.

After the retrieval and the preprocessing of the documents, the semantic vector space is created. A vocabulary is built including all words belonging to the documents corpus, excluding words that do not carry any informative content. A co-occurrence word-document matrix **A** is then created, whose *(i,j)-th* element is the sample probability of finding the *i-th* word in the *j-th* document. Each row is associated to a word in the vocabulary, while each column is associated to a paragraph or a Cyc concept comment. A matrix $\mathbf{B}=\{(a_{ij})^{1/2}\}$ is then calculated from **A** and subsequently decomposed according to TSVD procedure into the product of three matrices **U**, **Σ**, **V** through the minimization of the Hellinger distance. A dimensionality reduction is then operated leading to an Hellinger based LSA conceptual space in which all concepts of a given microtheory can be projected into it by coding their Cyc definitions and their related documents; i.e. each concept will be identified by a set of vectors each one related to the comment already present in the Cyc knowledge base or to a Wikipedia paragraph directly referred to the concept.

3.2 New Constants Insertion

The user introduces the keywords that define the new concept C_x, which is then projected into the conceptual space using the folding-in technique and normalized [1], [2], [3].

The conceptual correlation between points representing concepts in the space is then computed to find the concepts C_k of the ontology that are mainly related to C_x. Since each vector in the conceptual space represents one single document, the measure can be calculated according to the similarity value between the two vectors defined by eq.2. Given two concepts C_i and C_j and the sets $D_i = \{\mathbf{d}_{ik}\}$ and $D_j = \{\mathbf{d}_{jm}\}$ of their vectors mapped in the conceptual space and associated to the documents describing the concepts C_i and C_j, with $k=1...N_i$ and $m=1...N_j$ (where N_i and N_j are the number of documents associated to the concepts C_i and C_j respectively), the closeness between C_i and C_j is evaluated according to the following formula:

$$closeness(C_i, C_j) = \max\{sim(d_{ik}, d_{jm})\}, \quad \forall d_{jk} \in D_j \;\; \forall d_{ik} \in D_i . \tag{3}$$

Hence, a set CR of concepts C_i sub-symbolically conceptually related to the new concept C_x introduced by the user is calculated according to this formula:

$$CR = \{C_i \,|\, closeness(C_x, C_i) \geq T\}. \tag{4}$$

where T is a threshold value that can be interactively determined. The elements of CR are eventually shown to the user together with their associated *closeness* values.

4 Implementation

In order to evaluate the proposed system, two different microtheories of Cyc [13] have been analyzed: a smaller one in order to validate the proposed technique on the entire set of microtheory concepts and a greater one in order to carry out a test on one practical scene. The chosen application domains are:

- the american academic structure, described by *AcademicOrganizationMt* microtheory. In this microtheory there are 31 strongly connected elements: in the worst case, each one is far four steps from all the others. For these reasons it represents a good example of analyzable Cyc microtheory, even if it is limited in the number of elements;
- the *BiologyMt*, it is one of the Cores Theories, the more important specific domains microtheories in Cyc. It contains 1040 elements and it describes the biological world: there are animal and vegetable kingdoms taxonomies, biological behaviors, etc.

Both of them have been analyzed in detail and a semantic space has been built in order to represent in a sub-symbolic way the concepts described by the microtheory. For the first one six new constants have been inserted in it, and for the second one we have simulated the insertion of "new" concepts with constants that are already present in the microtheory.

4.1 AcademicOrganizationMt

The *AcademicOrganizationMt* microtheory describes the american academic structure and the main relations between several scholastic institutions through a set of collections. Abstract concepts are also present, like properties (*hasAlumni*) and actions (*DoingAHomeworkAssignment*).

In order to obtain all the constants belonging to the selected microtheory the *AcademicOrganizationVocabularyMt* has been used with the particular assertion: (*#$definingMt? X #$AcademicOrganizationVocabularyMt*).

A typical query to the inferential engine is:

- question: (*?X #$AgricultureDeparment ?Y*);
- answer: (*genls AcademicDepartment*),
  (*isa AcademicDepartmentTypeBySubject*),
  (*conceptuallyRelated Agriculture*).

as it can be seen from this example, different results for one generic query are obtained and each one can reference to different constants and predicates. Besides the found constants could not belong to the selected microtheory; therefore external constants have been also included in the analysis of microtheory. Such choice brought to an increase of analyzed concepts number; in this particular case 134 analyzed elements are reached from 31. For the examined microtheory 46 predicates have been found and it has been created a matrix for each one. A vector space has then been built retrieving Wikipedia articles about the domain. A matrix has been constructed and subsequently decomposed into the three matrices $\mathbf{\Sigma}_R$, $\mathbf{V}_R$ and $\mathbf{U}_R$ according to the TSVD technique with R=100.

New Constant Insertion. Using the property of the created semantic space, it is possible to suggest relations between constants already present in the microtheory and a new inserted constant by an external user. In order to verify the semantic relations between ontology concepts and the new constant, it has been estimated the similarity measure given by eq.2 between the vectors related to the microtheory concepts and the vector coding the new constant inserted by the user.

In order to validate the quality of the results, some *appropriate*, *less appropriate* and *not appropriate* constants to the chosen domain have been estimated. The concept, described by keywords, has been sub-symbolically compared to the corpus of reported documents. The comparison threshold has been fixed to T=0.5 (see eq. 4).

Pertaining Constants. Three new constants pertaining to the chosen domain have been inserted: *MathematicsDepartment*, *PrivateUniversity* and *PublicUniversity*. The results are shown in Table 1. For *MathematicsDepartment* concepts that represent departments and university concept are found. The retrieved relations are pertinent, but some concepts like *HistoryDepartment* and *AnthropologyDepartment* are not found. For *PrivateUniversity* concept appropriate relations with *College*, *University* and *UniversitySystem* are found. It is worthwhile to highlight the relation with the *PublicUniversity* concept, which is a new constant inserted by the user. For *PublicUniversity* pertinent relations with the constants *University* and *UniversitySystem* are pointed out.

Table 1. Concepts found associated to three new "pertaining constants". Relations are found also with newly inserted constants and not only with the concepts already present in the microtheory (e.g. *PrivateUniversity* and *PublicUniversity*).

MathematicsDepartment		*PrivateUniversity*		*PublicUniversity*	
0.80	BiologyDepartment	0.72	University	0.75	University
0.73	PhysicsDepartment	0.68	PublicUniversity	0.68	PrivateUniversity
0.70	AgricultureDepartment	0.61	College	0.65	UniversitySystem
0.65	University	0.59	UniversitySystem		

Less Pertaining Constants. A test with a less appropriate constant, *Campus*, has been carried out. It has weak connections with the chosen domain because it belongs to the university world but not to the academic structure. One single link has been found with *UniversitySystem* with a score of 0.79, the reason can be found in the fact that in many documents referring to *UniversitySystem* there are names of various university campuses.

Not Pertaining Constants. Two tests with not pertaining constants have been carried out: *Bedroom* and *Telephone*. Such constants have been chosen in order to verify two possible situations: the former has been chosen because it appears rarely in the domain documents; while the latter is very frequent in the retrieved text corpus.

For *Bedroom* no links with other constants have been found, while for *Telephone* a semantic link *UniversitySystem* with a score of 0.79 has been found, but it is not correct. That can be explained by many documents related to the UniversitySystem concept: there are references to the telephone numbers of some university.

4.2 BiologyMt

Previous results show a good precision of the proposed system. In order to validate such results a big Cyc microtheory has been analyzed: the *BiologyMt*. For this microtheory 7304 documents have been recovered and a conceptual space has been constructed. The used technique is the same one applied to the previous microtheory. The insertion of some already present constants in the ontology has been simulated and the found relations have been evaluated. Fourteen constants have been randomly chosen and for each of them many semantic relations are found. Those relations have been compared with the three biggest Cyc predicates: *isa*, *genls* and *conceptuallyRelated*. Subsequently a manual comparison has been conducted between the ontology structure and the found sub-symbolic relations. Semantic relations have been chosen with equal or higher closeness value than 0.9, since there are more than one thousand constants in the microtheory. The percentages of correct relations are high (80% for "*ConceptuallyRelated*", 92% for "*genls*", and 69% for "*isa*" relationships).

5 Conclusion and Future Works

Ontology learning from text is a very challenging task, however learning semantic concepts and relations instead of manually creating them could lead to correctness problems. For this reason it is virtually impossible to completely automatically induce ontologies from raw data. Nevertheless the proposed approach, based on the automatic induction of "conceptual" spaces, can help in adding a sub-symbolic relation between concepts of existing ontologies and help users in extending them. Experimental results are encouraging and further research will regard the coding of the kind and the directionality of relations in the ontology.

References

1. Agostaro, F., Augello, A., Pilato, G., Vassallo, G., Gaglio, S.: A Conversational Agent Based on a Conceptual Interpretation of a Data Driven Semantic Space. In: Bandini, S., Manzoni, S. (eds.) AI*IA 2005: Advances in Artificial Intelligence. LNCS (LNAI), vol. 3673, pp. 381–392. Springer, Heidelberg (2005)
2. Agostaro, F., Pilato, G., Vassallo, G., Gaglio, S.: A Subsymbolic Approach to Word Modelling for Domain Specific Speech Recognition. In: Proc. of IEEE CAMP05 International Workshop on Computer Architecture for Machine Perception, Terrasini - Palermo, pp. 321–326 (July 4-6, 2005)
3. Landauer, T.K., Foltz, P.W., Laham, D.: Introduction to Latent Semantic Analysis. Discourse Processes 25, 259–284 (1998)
4. Dumais, S.T., Landauer, T.K.: A solution to Plato's problem: the latent semantic analysis theory of acquisition, induction and representation of knowledge. Psychological Review (1997.2)
5. Navigli, R., Velardi, P.: Learning domain ontologies form document warehouses and dedicated web sites. Computational Linguistics, 50 (2004)
6. Sanderson, M., Croft, B.: Deriving concept hierarchies from text. In Research and Development in Information Retrieval, 206–213 (1999)
7. Biemann, C.: Ontology learning from text: a survey of methods. In LDV-Forum 2005 – Band 20(2), 75–93 (2005)
8. Buitelaar, P., Cimiano, P., Magnini, B.: Ontology Learning form Text: an Overview. In: Frontiers in Artificial Intelligence and Applications, vol. 123, IOS Press, Amsterdam (2005)
9. Suguraman, V., Storey, V.C.: Ontologies for conceptual modeling: their creation, use and management. In Data & Knowledge Engineering 42, 251–271 (2002)
10. Maedche, A., Staab, S.: Discovering conceptual relations from text. In: Proc of the 14th European Conference on Artificial Intelligence (ECAI 2000) (2000)
11. Ultsch, A.: The Integration of Connectionist Models with Knowledge based Systems: Hybrid Systems. Proc. of the IEEE SMC 98 International Conference, Oktober, San Diego, 11(14), 1530–1535 (1998)
12. Chella, A., Frixione, M., Gaglio, S.: An Architecture for Autonomous Agents Exploiting Conceptual Representations. Robotics and Autonomous Systems 25, 231–240 (1998)
13. http://research.cyc.com
14. http://www.wikipedia.org

A Belief-Desire Framework for Goal Revision

Célia da Costa Pereira and Andrea G.B. Tettamanzi

Università degli Studi di Milano
Dipartimento di Tecnologie dell'Informazione
Via Bramante 65, I-26013 Crema (CR), Italy
pereira@dti.unimi.it, andrea.tettamanzi@unimi.it

Abstract. A rational agent revises its goals when new information becomes available or its "desires" (e.g., tasks it is supposed to carry out) change. In this paper, we propose a logical framework, compatible with BDI theory, to represent changes in the mental state of an agent depending on the acquisition of new information and/or on the arising of new desires. Based on these changes, we estabilish fundamental postulates that the function which generates the goal set must obey, given the assumption of agent rationality.

Keywords: Beliefs, Desires, Goals, Goal change and Revision, Agent Systems.

1 Introduction

Although there has been much discussion on belief revision, goal revision has not received much attention. Most of the works on goal *change* found in the literature do not build on results on belief revision. That is the case of [2], in which the authors propose a formal representation for goals as rational desires and introduce and formalize dynamic goal hierarchies, but do not formalize explicitly beliefs and plans; or of [10], in which the authors propose an explicit representation of goals suited for conflict resolution based on a preference ordering of sets of goals. A more recent approach is [8], which models a multi-agent system in which an agent adopts a goal if requested to do so and the new goal is not conflicting with existing goals. This approach is based on goal persistence, i.e., an agent maintains its goals unless explicitly requested to drop them by the originating agent. The main lack of these approaches is that agents do not use their own knowledge for revising goals.

A static approach to the problem of how goals arise that has been proposed within planning is over-subscription planning [9,3]. In over-subscription planning, the problem of identifying the best subset of goals, given resource constraints, is addressed. The work presented in [5,4] is very much in that line, except that it attempts to provide a model of rationality, which is a slightly different focus. The approach consists in constructing dynamically the goal set to be pursued by a rational agent, by considering changes in its mental state with a very simple formalism intended as a first step. This paper extends and adapts

B. Apolloni et al. (Eds.): KES 2007/WIRN 2007, Part I, LNAI 4692, pp. 164–171, 2007.

such work to a formalism devised specifically within the BDI framework [11], a model of agency based on three primitive modalities: beliefs, desires, and intentions [7], of which we consider below the former two only.

2 Preliminaries

In this section, we present the formalism which will be used throughout the paper. Such formalism is inspired by the one used in [11]. However, unlike [11], the objective of our formalism is to analyze, not to develop, agent systems. Precisely, our agent must single out the *best* set of goals to be given as an input to a traditional planner component. That is because the intentions of the agent are not considered. We merely consider beliefs (the agent has about the world states), desires (or motivations) and relations (desire-adopting rules) wich define how the desire base will change with the acquisition of new beliefs and/or new desires. This work is very much in line with the work carried out in [9] on *oversubscription planning problems*, in which the main objective is to find the maximal set of desires to be reached in a given period and with a limited quantity of resources, and in [6].

2.1 Beliefs, Desires, and Goals

The basic components of our language are *beliefs* and *desires*. Beliefs are represented by means of a *belief base*. A belief base consists of a consistent set of propositional formulas which describe the information the agent has about the world and internal information. Desires are represented by means of a desire base. A desire base consists of a set of propositional formulas which represent the situations the agent would like to achieve. However, unlike the belief base, a desire base may be inconsistent, i.e., $\{\phi, \neg\phi\}$ may be a desire base. Goals, on the other hand, are represented by consistent desire bases.

Belief Base and Desire Base. Let $\mathcal{L}$ be a propositional language with a typical formula ϕ and the connectives $\wedge$ and $\neg$ with the usual meaning. The agent belief base, denoted by σ, is a subset of $\mathcal{L}$, i.e., $\sigma \subseteq \mathcal{L}$. Similarly, the agent's desire base is denoted by γ, where $\gamma \subseteq \mathcal{L}$.

Definition 1 (Belief and Desire Formulas). *Let ϕ be a formula of $\mathcal{L}$. An element, β, of the set of belief formulas $\mathcal{L}_B$ and an element κ of the set of desire formulas $\mathcal{L}_D$ are defined as follows:*

$$\beta ::= \top | \mathbf{B}\phi | \neg\mathbf{B}\phi | \beta_1 \wedge \beta_2,$$
$$\kappa ::= \top | \mathbf{D}\phi | \neg\mathbf{D}\phi | \kappa_1 \wedge \kappa_2.$$

Note that the modal operators **B** *and* **D** *cannot be nested.*

Definition 2 (Desire-Adoption Rules). *The set of* desire-adoption rules $\mathcal{R}_D$ *is defined as follows:*

$$\mathcal{R}_D = \{\beta, \kappa \Rightarrow_D^+ \phi \mid \beta \in \mathcal{L}_B, \kappa \in \mathcal{L}_D, \phi \in \mathcal{L}\}. \quad (1)$$

The antecedent of a desire-adoption rule consists of a belief condition β and a desire condition κ; the consequent is a propositional formula ϕ. Intuitively, this means that if the belief and the desire conditions in the antecedent hold, the formula in the consequent is automatically adopted as a desire.

Given a desire adoption rule R, we shall denote lhs(R) the antecedent of R, and rhs(R) the consequent of R. Furthermore, if S is a set of rules, we define $\text{rhs}(S) = \{\text{rhs}(R) : R \in S\}$.

2.2 Mental State Representation

We assume that an agent is equipped with three bases:

- belief base $\sigma \subseteq \mathcal{L}$;
- desire base: $\gamma \subseteq \mathcal{L}$;
- desire-adopting rule base $\mathcal{R}_D$;

The state of an agent is completely described by a triple $\mathcal{S} = \langle \sigma, \gamma, \mathcal{R}_D \rangle$.

The belief base, σ, represents the agent's beliefs about the world, $\mathcal{R}_D$ contains the rules which generate desires from beliefs and other (more basic) desires, and the desire base, γ, contains all desires which may be deduced from the agent's beliefs and the agents's desire-adopting rule base.

The semantics we adopt for the belief and desire formulas are inspired by the semantics of belief and "goal" formulas proposed in [11].

Semantics of Belief Formulas. Let $\phi \in \mathcal{L}$ and $\mathcal{S} = \langle \sigma, \gamma, \mathcal{R}_D \rangle$ be the mental state of an agent. Let $\beta_1, \beta_2 \in \mathcal{L}_B$. The semantics of belief formulas is given as

$$
\begin{aligned}
&\mathcal{S} \models_{\mathcal{L}_B} \top, \\
&\mathcal{S} \models_{\mathcal{L}_B} \mathbf{B}\phi \Leftrightarrow \sigma \models \phi, \\
&\mathcal{S} \models_{\mathcal{L}_B} \neg\mathbf{B}\phi \Leftrightarrow \mathcal{S} \not\models_{\mathcal{L}_B} \mathbf{B}\phi, \\
&\mathcal{S} \models_{\mathcal{L}_B} \beta_1 \wedge \beta_2 \Leftrightarrow \mathcal{S} \models_{\mathcal{L}_B} \beta_1 \quad \text{and} \quad \mathcal{S} \models_{\mathcal{L}_B} \beta_2.
\end{aligned}
$$

Semantics of Desire Formulas. Let $\phi \in \mathcal{L}$ and $\mathcal{S} = \langle \sigma, \gamma, \mathcal{R}_D \rangle$ be the mental state of an agent. Let $\kappa_1, \kappa_2 \in \mathcal{L}_D$. The semantics of desire formulas is given as

$$
\begin{aligned}
&\mathcal{S} \models_{\mathcal{L}_D} \top, \\
&\mathcal{S} \models_{\mathcal{L}_D} \mathbf{D}\phi \Leftrightarrow \exists \gamma' \subseteq \gamma : (\gamma' \not\models \bot \quad \text{and} \quad \gamma' \models \phi), \\
&\mathcal{S} \models_{\mathcal{L}_D} \neg\mathbf{D}\phi \Leftrightarrow \mathcal{S} \not\models_{\mathcal{L}_D} \mathbf{D}\phi, \\
&\mathcal{S} \models_{\mathcal{L}_D} \kappa_1 \wedge \kappa_2 \Leftrightarrow \mathcal{S} \models_{\mathcal{L}_D} \kappa_1 \quad \text{and} \quad \mathcal{S} \models_{\mathcal{L}_D} \kappa_2.
\end{aligned}
$$

Definition 3 (Active Desire Adoption Rule). *Let $R \in \mathcal{R}_D$ be a desire adoption rule with* lhs(R) $= \langle \beta, \kappa \rangle$. *$R$ is said active iff $\mathcal{S} \models_a$* lhs(R), *i.e.,*

$$\mathcal{S} \models_a \text{lhs}(R) \Leftrightarrow (\mathcal{S} \models_{\mathcal{L}_B} \beta) \wedge (\mathcal{S} \models_{\mathcal{L}_D} \kappa). \tag{2}$$

Semantics of Desire Adoption Rules. Let $\phi \in \mathcal{L}$ and let $\mathcal{S}$ be the mental state of an agent.

$$\phi \in \gamma \Leftrightarrow \exists R \in \mathcal{R}_D : \text{rhs}(R) = \phi \wedge \mathcal{S} \models_a \text{lhs}(R).$$

Such a desire is said to be a *justified* desire.

Definition 4 (Candidate Goals). *A candidate goal set is a subset of the desire base which is consistent, i.e., it is a set of consistent justified desires.*

The main point about goals is that we expect a rational agent to try and manipulate its surrounding environment to fulfill them. In general, considering a planning problem $\mathcal{P}$ to solve, not all goals can be fulfilled. For example, if one of the goals considered, ϕ, is not satisfied and there is no an action in the description of $\mathcal{P}$, with ϕ in the list of its effects, ϕ will never be fulfilled. We assume we dispose of a function $\mathcal{F}_{\mathcal{P}} : 2^{\mathcal{L}} \times 2^{\mathcal{L}} \rightarrow \{\bot, \top\}$ wich, given a belief base σ and a goal set γ, returns $\top$ if γ is feasible for $\mathcal{P}$, and $\bot$ otherwise.

A candidate goal set γ is said to be feasible for a planning problem $\mathcal{P}$ if and only if $\mathcal{F}_{\mathcal{P}}(\sigma, \gamma) = \top$.

3 Changes in the State of an Agent

The acquisition of a new belief in state $\mathcal{S}$ may cause a change in the belief base σ and this may also cause a change in the desire set γ with the retraction of existing desires and/or the assertion of new desires. A desire ϕ is retracted from the desire set γ if and only if ϕ becomes not justified, i.e., all active desire-adoption rules such that $\text{rhs}(R) = \phi$ become inactive. A desire ϕ is asserted into a desire set γ if and only if the new information activates a desire adoption rule R with $\text{rhs}(R) = \phi$.

3.1 Changes Caused by a New Belief

The next definition introduces a notation to refer to the set of rules that become active, resp. inactive, after the acquisition of new information β in a given state $\mathcal{S} = \langle \sigma, \gamma, \mathcal{R}_D \rangle$. Let $*$ be the well known AGM operator for belief revision [1] and $\mathcal{S}' = \langle \sigma * \beta, \gamma, \mathcal{R}_D \rangle$ be the new resulting state.

Definition 5 (Rule Activated/Deactivated by a Belief). *We define the subsets* $\text{Act}^{\mathcal{S}}_{\beta}$ *of* $\mathcal{R}_D$ *composed by the rules which becames activated because of* β *as follows:*

$$\text{Act}^{\mathcal{S}}_{\beta} = \{R : (\mathcal{S} \not\models_a \text{lhs}(R)) \wedge (\mathcal{S}' \models_a \text{lhs}(R))\}. \tag{3}$$

$\text{Act}^{\mathcal{S}}_{\beta}$ *contains rules which are directly or indirectly activated by* β. *In the same way, we define the subset of* $\mathcal{R}_D$, $\text{Deact}^{\mathcal{S}}_{\beta}$, *containing the rules which become directly or indirectly deactivated because of* β.

Two considerations must be taken into account:

1. By definition of the revision operator $*$, $\mathcal{S}' \models_{\mathcal{L}_B} \beta$, thus all desire-adoption rules $R \in \mathrm{Act}_\beta^{\mathcal{S}}$ become active and all new desires $\phi = \mathrm{rhs}(R)$ are asserted into the desire set γ.
2. If, before the arrival of β, $\mathcal{S} \models_{\mathcal{L}_B} \neg\beta$, then all active desire-adoption rules R, such that $\neg\beta \in \mathrm{lhs}(R)$, become inactive and, if there is not an active desire-adoption rule R', such that $\mathrm{rhs}(R') = \mathrm{rhs}(R)$, then the desire $\phi = \mathrm{rhs}(R)$ is retracted from the desire set γ.

We can summarize the above considerations into one desire-updating formula which tells how the desire set γ of a rational agent in state $\mathcal{S}$ should change in response to the acquisition of a new belief β. Let $A_\beta^{\mathcal{S}}$ be the set of desires acquired because of the new belief β:

$$A_\beta^{\mathcal{S}} = \mathrm{rhs}(\mathrm{Act}_\beta^{\mathcal{S}}). \tag{4}$$

Let $L_\beta^{\mathcal{S}}$ be the set of desires lost because of the acquisition of the new belief β:

$$\begin{aligned} L_\beta^{\mathcal{S}} = \{\phi : \phi \in \mathrm{rhs}(\mathrm{Deact}_\beta^{\mathcal{S}}) \wedge \\ \neg\exists R\,(\mathcal{S} \models_a \mathrm{lhs}(R) \wedge R \notin \mathrm{Deact}_\beta^{\mathcal{S}} \wedge \mathrm{rhs}(R) = \phi)\}. \end{aligned} \tag{5}$$

Let $\oplus$ be our operator for desire updating, and γ the base of agent's desires. According to the above considerations, we have:

$$\gamma \oplus \beta = (\gamma \cup A_\beta^{\mathcal{S}}) \setminus L_\beta^{\mathcal{S}}. \tag{6}$$

It is easy to verify that $A_\beta^{\mathcal{S}} \cap L_\beta^{\mathcal{S}} = \emptyset$, for all state $\mathcal{S}$.

3.2 Changes Caused by a New Desire

In this work, for the sake of simplicity, we consider that a new desire ϕ may only be represented by a desire adoption rule R with an empty left hand side and such that $\mathrm{rhs}(R) = \phi$. Because the desire base may be inconsistent, the new acquired desire ϕ and the desires in the antecedent of the rules activated because of ϕ are automatically asserted into γ.

Let $\phi \in \mathcal{L}$ be a new desire arising in state $\mathcal{S} = \langle \sigma, \gamma, \mathcal{R}_D \rangle$. Let $\mathcal{S}' = \langle \sigma, \gamma \oplus \phi, \mathcal{R}_D \rangle$ be the resulting mental state.

Definition 6 (Rule Activated by a New Desire). *We define the subsets* $\mathrm{Act}_\phi^{\mathcal{S}}$ *of* $\mathcal{R}_D$ *composed by the rules which become activated because of* ϕ *as follows:*

$$\mathrm{Act}_\phi^{\mathcal{S}} = \{R : (\mathcal{S} \not\models_a \mathrm{lhs}(R)) \wedge (\mathcal{S}' \models_a \mathrm{lhs}(R))\}. \tag{7}$$

$\mathrm{Act}_\phi^{\mathcal{S}}$ contains rules which are directly or indirectly activated because of ϕ.

In the same way, we may define the subset of $\mathcal{R}_D$, $\mathrm{Deact}_\phi^{\mathcal{S}}$ containing the rules which become directly or indirectly deactivated because of ϕ.

Let $\mathcal{S}$ be the state of the agent, and $A_\phi^{\mathcal{S}} = \{\mathrm{rhs}(R) : R \in \mathrm{Act}_\phi^{\mathcal{S}}\}$ be the set of desires acquired with the arising of ϕ in state $\mathcal{S}$. Let $\otimes$ be the operator for updating the desire-adoption rule base.

How does $\mathcal{S}$ change with the arising of the new desire ϕ?

1. The desire-generating rule $\top, \top \Rightarrow_D^+ \phi$ is added to $\mathcal{R}_D$,
2. ϕ is added to γ,
3. All desire-generating rules R in $\text{Act}_\phi^\mathcal{S}$ become activated, and all desires appearing in the right hand side of these rules are also added to γ.

Therefore,

$$\gamma \oplus \phi = \gamma \cup \{\phi\} \cup A_\phi^\mathcal{S}, \tag{8}$$

$$\mathcal{R}_D \otimes \phi = \mathcal{R}_D \cup \{\top, \top \Rightarrow_D^+ \phi\}. \tag{9}$$

In general, a rational agent will try to choose a consistent set of goals which, first of all, is feasible and, secondly, gives the greatest possible pay-off.

3.3 Comparing Candidate Goals and Sets of Candidate Goals

An agent may have many sets of feasible candidate goals. However, it is essential to be able to represent the fact that not all goals have the same importance or urgence for a rational agent. A natural choice for representing the importance of goals would be to use their expected pay-off.

A pay-off function for goals is a function $f : \mathcal{G} \to \mathbb{R}$ which associates a real value, a pay-off, to all goals.

One problem with pay-offs is that we are not always able, in general, to attach a precise numerical value to goals. An alternative approach would be to establish a (partial or total) ordering among goals. In either case, we can define preference between desires as follows.

Preference between Candidate Goals. A goal ϕ is at least as preferred as ϕ', denoted $\phi \succeq \phi'$ iff the agent desires ϕ at least as much as it desires ϕ'.

The $\succeq$ relation, which is reflexive and transitive, can be extended from candidate goals to sets of candidate goals.

Definition 7 (Preference between Sets of Candidate Goals). *A candidate goal set γ_1 is at least as preferred as γ_2, denoted $\gamma_1 \succeq \gamma_2$:*

- *If pay-offs are defined, iff*

$$\sum_{\phi \in \gamma_1} f(\phi) \geq \sum_{\phi' \in \gamma_2} f(\phi'); \tag{10}$$

- *Otherwise, let $\gamma_1' = \gamma_1 \setminus \gamma_2$ and $\gamma_2' = \gamma_2 \setminus \gamma_1$. Iff one of the following two conditions is satisfied:*
  1. *$\forall \phi_2 \in \gamma_2'$, $\exists \phi_1 \in \gamma_1'$, s.t. $\phi_1 \succeq \phi_2$;*
  2. *$\forall \phi_1 \in \gamma_1'$, ($\exists \phi_2 \in \gamma_2'$, such that $\phi_1 \succeq \phi_2$ and $\neg\exists\, \phi_3 \in \gamma_2'$ such that $\phi_3 \succ \phi_1$)*

It can be proved that the above preference relation is reflexive and transitive.

In case neither pay-offs nor a preference relation on cadidate goals are available, it is still possible to define a preference relation on sets of candidate goals as the partial ordering $\gamma_1 \succeq \gamma_2 \equiv \gamma_2 \subseteq \gamma_1$, which is quite reasonable under the assumption that, if no goal is preferred over another, all goals must be worth the same; therefore, the more goals an agent can fulfill, the more "satisfied" (whatever this means) it would be.

4 Revising Goal Sets

The main point about goals is that we expect a rational agent to try and manipulate its surrounding environment to fulfill them. Therefore, a rational agent will select a particular set of feasible candidate goals to realize.

4.1 Postulates for Goal Revision

In general, given a set of desires γ, there are many possible subsets $\gamma' \subseteq \gamma$ of feasible candidate goals. However, a rational agent in state $\mathcal{S} = \langle \sigma, \gamma, \mathcal{R}_D \rangle$ will elect as the set of goals it is pursuing one precise goal set γ^*, which depends on $\mathcal{S}$.

Let us call G the function which maps a state $\mathcal{S}$ into the goal set elected by a rational agent in state $\mathcal{S}$: $\gamma^* = G(\mathcal{S})$. This goal election function G must obey two fundamental postulates:

- **(G1)** $\forall \mathcal{S}, G(\mathcal{S})$ is a feasible goal set;
- **(G2)** $\forall \mathcal{S}$, if $\gamma' \subseteq \gamma$ is a feasible goal set, then $G(\mathcal{S}) \succeq \gamma'$, i.e., a rational agent always selects the most preferable feasible candidate goal set.

In [4], three alternatives, G_u, $G_{\succeq}$, and $G_{\subseteq}$, for the definition of the goal set election function have been proposed. These definitions are applicable, respectively, to the case whereby pay-offs are defined, to the weaker case in which the total ordering $\succeq$ of desires is available, and to the weakest case in which only a partial ordering or no ordering at all of desires is available.

5 Conclusion

Previous work on goal selection for a rational agent has been recast within a logical framework devised for cognitive agent programming and inspired by BDI theory. Formulating the goal selection problem within such framework makes it available to a larger community and provides a framework for goal revision for various kinds of rational and cognitive agents.

An important point of the framework developed above is that the two aspects of how goals are selected by an agent and how the selected goals are achieved can be conceptually separated: this means, the goal selection mechanics are independent of the planning process or algorithm, although interactions between these two aspects are not ruled out. This is a requirement for an agent design where the cognitive and planning modules are clearly distinguished.

References

1. Alchourrón, C.E., Gärdenfors, P., Makinson, D.: On the logic of theory change: Partial meet contraction and revision functions. J. Symb. Log. 50(2), 510–530 (1985)
2. Bell, J., Huang, Z.: Dynamic goal hierarchies. In: Foo, N.Y., Göbel, R. (eds.) PRICAI 1996. LNCS, vol. 1114, pp. 88–103. Springer, Heidelberg (1997)
3. Benton, J., Do, M.B., Kambhampati, S.: Over-subscription planning with numeric goals. In: Kaelbling, L.P., Saffiotti, A. (eds.) IJCAI-05, Proceedings of the Nineteenth International Joint Conference on Artificial Intelligence, Edinburgh, Scotland, UK, July 30–August 5, pp. 1207–1213. Professional Book Center (2005)
4. da Costa Pereira, C., Tettamanzi, A.: Towards a framework for goal revision. In: Pierre-Yvese Schobbens, W.V., Schwanen, G. (eds.) BNAIC-06, Proceedings of the 18th Belgium-Netherlands Conference on Artificial Intelligence, Namur, Belgium, October 5-6, 2006, pp. 99–106. University of Namur (2006)
5. da Costa Pereira, C., Tettamanzi, A., Amgoud, L.: Goal revision for a rational agent. In: Brewka, G., Coradeschi, S., Perini, A., Traverso, P. (eds.) ECAI 2006, Proceedings of the 17th European Conference on Artificial Intelligence, Riva del Garda, Italy, A, August 29–September 1, pp. 747–748. IOS Press, Amsterdam (2006)
6. Hulstijn, J., Broersen, J., Dastani, M., van der Torre, L.: Goal generation in the boid architecture. Cognitive Science Quarterly Journal 2(3–4), 428–447 (2002)
7. Rao, A.S., Georgeff, M.P.: Modeling rational agents within a BDI-architecture. In: Proceedings of the 2nd International Conference on Principles of Knowledge Representation and Reasoning (KR'91) (1991)
8. Shapiro, S., Lespérance, Y., Levesque, H.J.: Goal change. In: Kaelbling, L.P., Saffiotti, A. (eds.) IJCAI-05, Proceedings of the Nineteenth International Joint Conference on Artificial Intelligence, Edinburgh, Scotland, UK, July 30–August 5, pp. 582–588. Professional Book Center (2005)
9. Smith, D.E.: Choosing objectives in over-subscription planning. In: Zilberstein, S., Koehler, J., Koenig, S. (eds.) Proceedings of the Fourteenth International Conference on Automated Planning and Scheduling (ICAPS 2004), Whistler, British Columbia, Canada, June 3–7, pp. 393–401. AAAI, Stanford, California, USA (2004)
10. Thangarajah, J., Padgham, L., Harland, J.: Representation and reasoning for goals in bdi agents. In: CRPITS '02: Proceedings of the twenty-fifth Australasian conference on Computer science, Darlinghurst, Australia, pp. 259–265. Australian Computer Society, Inc (2002)
11. van Riemsdijk, M.B.: Cognitive Agent Programming: A Semantic Approach. PhD thesis, Ludwig-Maximilians-Universität München (2006)

An Investigation of Agent-Based Hybrid Approach to Solve Flowshop and Job-Shop Scheduling Problems

Joanna Jędrzejowicz[1] and Piotr Jędrzejowicz[2]

[1] Institute of Computer Science, Gdańsk University,
Wita Stwosza 57, 80-952 Gdańsk, Poland
jj@inf.univ.gda.pl

[2] Department of Information Systems, Gdynia Maritime University,
Morska 83, 81-225 Gdynia, Poland
pj@am.gdynia.pl

Abstract. The paper investigates a possibility of combining the population learning algorithm and the A-Team concept with a view to increase quality of results and efficiency of computations. To implement the idea a middleware environment called JABAT is used. The proposed approach is validated experimentally using benchmark datasets containing instances of the two well-known combinatorial optimization problems: flow shop and job shop scheduling.

Keywords: Population learning algorithm, A-Team, flow shop and job shop scheduling.

1 Introduction

Recently, a number of agent-based approaches have been proposed to solve different types of optimization problems [14], [1], [13]. One of the successful approaches to agent-based optimization is the concept of A-Teams. According to [17] an A-Team is a problem solving architecture in which the agents are autonomous and co-operate by modifying one another's trial solutions. Various implementations of the A-Team concept seem particularly well suited to support population-based hybrid approaches to solving optimization problems.

The paper investigates a possibility of combining the population learning algorithm (PLA) introduced in [6] and the A-Team concept with a view to increase quality of results and efficiency of computations. PLA is a hybrid approach where different improvement procedures including random and local search techniques, greedy and construction algorithms etc., are sequentially applied to a population representing solutions to the problem at hand. Unlike some other population based methods, in the PLA population of individuals representing solutions is not constant but decreases after each computation stage. Moreover, at later computation stages more computationally complex improvement procedures are used.

B. Apolloni et al. (Eds.): KES 2007/WIRN 2007, Part I, LNAI 4692, pp. 172–179, 2007.

The authors have already successfully applied the population learning algorithm to solve some difficult scheduling problems [7], [8]. The proposed approach is validated experimentally using benchmark datasets containing instances of the two well-known combinatorial optimization problems: flow shop and job shop scheduling. The paper is organized as follows. Section 2 provides some details on the middleware environment used and presents search strategy used by the proposed teams of agents. Section 3 contains problem formulation and a detailed description of the optimization agents developed for solving permutation flow shop and job shop scheduling problems, respectively. Section 4 presents the results of the computational experiment carried out. Conclusions include an evaluation of the approach and suggestions for further research.

2 JABAT Middleware Environment

To implement the proposed approach a middleware environment referred to as JABAT (JADE-based A-Team) is used. The JADE-based A-Team environment (JABAT) described in a more detailed manner in [10] supports the construction of the dedicated A-Team architectures used for solving variety of computationally hard optimization problems. JADE, on the other hand, is an enabling technology, for the development and run-time execution of peer-to-peer applications which are based on the agents paradigm [3]. JADE allows each agent to dynamically discover other agents and to communicate with them according to the peer-to-peer paradigm. The main functionality of JABAT is searching for the optimum solution of a given problem instance through employing a variety of agents representing solution improvement algorithms. The search involves a sequence of the following steps:

- Generating an initial population of solutions.
- Applying solution improvement algorithms which draw individuals from the common memory and store them back after attempted improvement, using some user defined replacement strategy.
- Continuing reading-improving-replacing cycle until a stopping criterion is met.

To perform the above two classes of agents are used. The first class includes *OptiAgents*, which are implementations of the improvement algorithms. The second class includes *SolutionManagers*, which are agents responsible for maintenance and updating of individuals in the common memory. All agents act in parallel. Each *OptiAgent* is representing a single improvement algorithm (simulated annealing, tabu search, genetic algorithm, local search heuristics etc.). An *OptiAgent* has two basic behaviors defined. The first is sending around messages on readiness for action including the required number of individuals (Solutions). The second is activated upon receiving a message from some *SolutionManager* containing the problem instance description and the required number of individuals. This behavior involves improving fitness of individuals and resending the improved ones to a sender. A *SolutionManager* is brought to life for each problem instance. Its behavior involves sending individuals to *OptiAgents* and updating the common memory. Main assumption behind the proposed approach is its independence from a problem definition and solution algorithms. Hence, main classes *Task* and *Solution* upon which agents act, have been defined at a rather general level. Interfaces of both classes include function *ontology()*, which returns JADE's ontology designed for classes *Task* and *Solution*, respectively. Ontology in JADE is a class enabling definition of the vocabulary and semantics for the content of message exchange between agents. More precisely, an ontology defines how the class is transformed into the text message exchanged between agents and how the text message is used to construct the class (here either *Task* or *Solution*). In JABAT the *SolutionManager* is responsible for executing the population-based search for the best solution. Hence, the *SolutionManager* not only manages interactions between optimization agents and the common memory but also assures that these are in accordance with the paradigm of the adopted population-based method. In this paper a variant of A-Team called JABAT-PLA is used. The approach is based on the population learning algorithm with the search strategy outlined in the following pseudo-code:

begin
Initialize the common memory by using a random mechanism
(or other user defined) to produce P individuals
(here feasible solutions of the problem at hand);
Set within a parallel and distributed environment $n \times m$ *agents,*
where n *is a number of improvement procedures employed*
and m *is a number of agents of each kind;*
Set number of stages corresponding to the number of improvement
procedures employed.
Order stages according to the complexity of the corresponding improvement
procedure.
for *stages* 1 **to** n **do**
  Activate all available agents executing an improvement procedure
  of the current stage.
  repeat *for each agent*
    Draw randomly the required number of individuals from P
    and copy them into working memory;
    Improve individuals in the working memory by executing
    the improvement procedure;
    Update common memory;
  until *stopping criterion is met*
  Decrease number of individuals in the common memory
end for
Select best individual from P as a solution
end

3 Problem Definition and Optimization Agents Employed

3.1 Permutation Flow-Shop Scheduling Problem

In the permutation flowshop scheduling problem (PFSP) there is a set of n jobs. Each of n jobs has to be processed on m machines $1 \ldots m$ in this order. The processing time of job i on machine j is p_{ij} where p_{ij} are fixed and nonnegative. At any time, each job can be processed on at most one machine, and each machine can process at most one job. The jobs are available at time 0 and the processing of a job may not be interrupted. In the PFSP the job order is the same on every machine. The objective is to find a job sequence minimizing schedule makespan (i.e., completion time of the last job).

To solve an instance of the PFSP the proposed JABAT-based system uses sequentially four kinds of *OptiAgents*: Cross-entropy algorithm, Evolutionary algorithm with cross-over and mutation, Tabu search, Simulated annealing.

Cross-entropy algorithm [4] is broken into two phases: generating random job sequences and updating the parameters at each iteration using probabilities $prob(i, j)$ of job j following job i, for any pair of different jobs. The two phases are repeated until no changes in probabilities are observed. Evolutionary algorithm based *OptiAgent* (E-b-Opti) acts upon the population P, which is read from the common memory and transmitted to each active E-b Opti by the *SolutionManager*. After having received a population the E-b-Opti performs *no-gen* iteration steps with *mut-perc* as the percentage of individuals undergoing mutation. In each iteration the agent selects randomly two individuals x1, x2 from the population P, performs cross-over in a randomly chosen point, adds the better result of cross-over to the working population new-P. After *no-gen* steps the agent replaces half of the worst individuals from P by the best individuals from new-P and mutates *mut-perc* individuals from the thus obtained population. After stopping criterion has been met the agent resends population P to the *SolutionManager*. Each tabu search based *OptiAgent* receives an individual (here a single problem solution) from the *SolutionManager* and attempts to improve it through local search. Such search consists of a sequence of moves. The move $N(x)$ is understood as relocating a random job x in a permutation under the improvement to another randomly selected location within this permutation. The process is controlled using two kind of memories - the short term (STM) and the long term (LTM) one as proposed in [5]. Both memories are managed according to the reverse elimination method. After stopping criterion has been met the agent resends the improved individual to the *SolutionManager*. Simulated annealing *OptiAgents* try to improve individuals through local search based on the simulated annealing metaheuristic [11]. The neighborhood includes all possible pairwise exchanges of jobs within a schedule.

3.2 Job-Shop Scheduling Problem

An instance of the job-shop scheduling problem (JSSP) consists of a set of n jobs and m machines. Each job consists of a sequence of n activities so there are $n \times m$

activities in total. Each activity has a duration and requires a single machine for its entire duration. The activities within a single job all require a different machine. An activity must be scheduled before every activity following it in its job. Two activities cannot be scheduled at the same time if they both require the same machine. The objective is to find a schedule that minimizes the overall completion time of all the activities. In this paper a permutation version of the job-shop scheduling problem is used. That is, given an instance of the job-shop scheduling problem, a solution is a permutation of jobs for each machine defining in a unique manner a sequence of activities to be processed on this machine. For a problem consisting of a set of n jobs and m machines a solution is a set of m permutations of n elements each. A feasible solution obeys all the problem constraints including precedence constraints. To solve an instance of the JSSP the proposed JABAT-based system uses sequentially three kinds of *OptiAgents*: Cross-entropy algorithm, Tabu search, Simulated annealing. In the first step random solutions are generated such, that for each machine the jobs are scheduled according to the order of activities. This guarantees a feasible schedule. Cross-entropy algorithm is used to improve these solutions. Tabu search *OptiAgents* try to improve the allocated individuals through local search procedure similarly as in case of the PFSP. The only difference is that the local search is carried over a collection of permutations (one for each machine) and that there are two neighborhood structures used. Respectively, there are two kinds of moves - relocating an activity within a single permutation and exchanging activities within a single permutation. Only moves producing feasible solutions are allowed. Simulated annealing *OptiAgents* try to improve the allocated individuals through local search based on the simulated annealing metaheuristic [11]. The neighborhood structure is based on the critical path of the solution. The transition operator exchanges pairs of adjacent critical operations.

4 Computational Experiment Results

To validate the proposed approach computational experiment has been carried out. It involved a number of instances from the OR-LIBRARY benchmark datasets. The experiment has been designed to assure comparability with some other recent approaches including the PLA-Team proposed by the authors in [9]. PLA-Team differs from the JABAT-PLA in two respects. It uses both - different optimization agents and different strategy of searching for the best solution. To solve instances of both scheduling problems identical hardware configuration has been used. It included a network of 3 PC computers with 2.4 GHz processors and 1 GB RAM. For evaluating different approaches the average deviation from the currently known upper bound is used.

4.1 Flowshop Scheduling

JABAT-PLA has been run to solve all 120 benchmark instances from the OR-LIBRARY and the data from a total of 10 independent runs have been averaged.

Table 1. The average deviation from the currently known upper bound (%)

instance	NEHT	GA	HGA	SAOP	PLA-team	JABAT-PLA
20 × 5	3.35	0.29	0.20	1.47	0.00	0.00
20 × 10	5.02	0.95	0.55	2.57	0.32	0.23
20 × 20	3.73	0.56	0.39	2.22	0.26	0.16
50 × 5	0.84	0.07	0.06	0.52	0.03	0.01
50 × 10	5.12	1.91	1.72	3.65	0.71	0.65
50 × 20	6.20	3.05	2.64	4.97	1.62	1.04
100 × 5	0.46	0.10	0.08	0.42	0.01	0.00
100 × 10	2.13	0.84	0.70	1.73	0.52	0.43
100 × 20	5.11	3.12	2.75	4.90	1.35	1.28
200 × 10	1.43	0.54	0.50	1.33	0.41	0.37
200 × 20	4.37	2.88	2.59	4.40	1.04	0.98
500 × 20	2.24	1.65	1.56	3.48	1.09	1.01
Total	3.33	1.33	1.15	2.64	0.61	0.51

As a stopping criteria all compared algorithms have been allocated execution time of 30 seconds for instances with 500 jobs, 12 seconds for instances with 200 jobs, 6 seconds for instances with 100 jobs, 3 seconds for instances with 50 jobs and 1.2 seconds for instances with 20 jobs. During computations there have been 3 copies, each on different platform, of each kind of agent active. The common memory has had 400 individuals, reduced to 200 after the cross entropy stage, and further reduced to 100 after the evolutionary algorithm stage, and to 50 individuals after the tabu search stage. In Table 1 the results obtained by JADE-PLA are compared with the PLA-Team results reported in [9] and [15]. The algorithms reported in [15], include the NEH heuristics with enhancements - NEHT, the genetic algorithm - GA, the hybrid genetic algorithm - HGA and the simulated annealing algorithm - SAOP. From the experiment results it can be easily observed that JABAT-PLA should be considered as a useful and competitive tool for the permutation flowshop scheduling.

4.2 Job-Shop Scheduling

The computational experiment has been designed with a view to compare the performance of the JABAT-PLA with other approaches including agent-based and distributed algorithms. The results obtained by the JABAT-PLA have been compared with the A-Team results reported in [1] on a set of 10 × 10 instances from the OR-LIBRARY. All results have been averaged over 10 independent runs. During computations there have been 3 copies, each on different platform, of each kind of agent active. The common memory has had 100 individuals reduced to 50 after the cross entropy stage, and further reduced to 25 after the tabu search. In all runs solutions generated by the PLA-Team have been equal to optimal results for Abz5, Abz6, La16, La17, La18, La19, La20 and Orb1 instances. Only in case of Ft10 average deviation from the optimal solution has been 0,11%. The A-Team of [1] has produced average deviation from the optimal solution at the average level above 1% and 3% in case of Ft10.

A further experiment aimed at comparing the JABAT-PLA performance with the following state of the art algorithms: rescheduling based simulated annealing (SAT), tabu search algorithm (KTM) reported in [16], the hybrid genetic and simulated annealing algorithm (KOL) reported in [12], and parallel modular simulated annealing (MSA) reported in [2]. The results including average deviation from the best known result (%) and computation time in seconds, averaged over 10 independent runs, are shown in Table 2.

Table 2. JABAT-PLA versus other algorithms

	SAT		KTM		KOL		MSA		JABAT-PLA	
	Dev.	Time	Dev.	Time	Dev.	Time	Dev.	Time	Dev.	Time
Abz7	4.55	5991	x	x	x	x	3.08	1445	2.90	123
Abz8	9.40	590	x	x	x	x	7.71	1902	1.95	145
Abz9	7.71	5328	0.38	1720	0.48	594	0.23	838	0.94	89
La21	1.53	1516	0.75	1170	0.58	509	0.49	570	1.06	38
La24	1.50	1422	0.36	1182	0.20	644	0.08	1035	0.03	101
La25	1.29	1605	1.00	919	0.76	3650	0.84	982	0.72	87
La27	2.28	3761	3.83	3042	3.47	4496	4.65	1147	2.64	108
La29	5.88	4028	1.21	3044	0.54	5049	1.56	1143	1.30	133
La38	1.87	3004	0.96	6692	0.59	4544	0.59	1894	0.84	149
La40	1.72	2812	0.38	1720	0.48	594	0.23	838	0.13	91

From the experiment results it can be easily observed that JABAT-PLA should be considered as an efficient and useful tool for the job shop scheduling, in particular from the computation time point of view. It also produces excellent results in case of smaller problem instances.

5 Conclusion

The proposed JABAT-PLA architecture has several advantages inherited from both - JABAT middleware and the population learning algorithm. Among them one should mention: (1) Ability to simplify the development of the distributed A-Teams composed of autonomous entities that need to communicate and collaborate in order to achieve the working of the entire system. (2) Ability to achieve a synergetic effect through integrating within one framework different optimization approaches and algorithms and using them to produce results of a good quality in a competitive time. (3) Ability to use efficiently available computational resources including different hardware platforms and different software components.

Future research should lead towards developing a library of optimization agents and a user interface allowing for easily composing own PLA implementations from the available components.

References

1. Aydin, M.E., Fogarty, T.C.: Teams of Autonomous Agents for Job-Shop Scheduling Problems. An Experimental Study, Journal of Intelligent Manufacturing 15(4), 455–462 (2004)
2. Aydin, M.E., Fogarty, T.C.: A Simulated Annealing Algorithm for Multi-agent Systems: a Job-Shop Scheduling Application. Journal of Intelligent Manufacturing 15(6), 805–814 (2004)
3. Bellifemine, F., Caire, G., Poggi, A., Rimossa, G.: JADE. A White Paper 3(3), 6–20 (2003)
4. de Boer, P.T., Kroese, D.P, Mannor, S., Rubinstien, R.Y: A Tutorial on the Cross-Entropy Method. Annals of Operations Research 134(1), 19–67 (2005)
5. Glover, F.: Heuristics for Integer Programming Using Surrogate Constraints. Decision Sciences, 8(1), 156–166 (1977)
6. Jędrzejowicz, P.: Social Learning Algorithm as a Tool for Solving Some Difficult Scheduling Problems. Foundation of Computing and Decision Sciences 24, 51–66 (1999)
7. Jędrzejowicz, J., Jędrzejowicz, P.: PLA-Based Permutation Scheduling. Foundations of Computing and Decision Sciences 28(3), 159–177 (2003)
8. Jędrzejowicz, J., Jędrzejowicz, P.: New Upper Bounds for the Permutation Flowshop Scheduling problem. In: Ali, M., Esposito, F. (eds.) IEA/AIE 2005. LNCS (LNAI), vol. 3533, pp. 232–235. Springer, Heidelberg (2005)
9. Jędrzejowicz, J., Jędrzejowicz, P.: Agent-Based Approach to Solving Difficult Scheduling Problems. In: Ali, M., Dapoigny, R. (eds.) IEA/AIE 2006. LNCS (LNAI), vol. 4031, pp. 24–33. Springer, Heidelberg (2006)
10. Jędrzejowicz, P., Wierzbowska, I.: JADE-Based A-Team Environment. In: Alexandrov, V.N., van Albada, G.D., Sloot, P.M.A., Dongarra, J.J. (eds.) ICCS 2006. LNCS, vol. 3993, pp. 719–726. Springer, Heidelberg (2006)
11. Kirkpatrick, S.: Optimization by Simulated Annealing. Science 220, 671–680 (1983)
12. Kolonko, M.: Some New Results on Simulated Annealing Applied to Job Shop Scheduling Problem. European Journal of Operational Research 113, 123–136 (1999)
13. Marinescu, D.C., Boloni, L.: A Component-Based Architecture for Problem Solving Environments. Mathematics and Computers in Simulation 54, 279–293 (2000)
14. Parunak, H.V.D.: Agents in Overalls: Experiences and Issues in the Development and Deployment of Industrial Agent-Based Systems, Intern. J. of Cooperative Information Systems 9(3), 209–228 (2000)
15. Ruiz, R., Maroto, C., Alcaraz, J.: New Genetic Algorithms for the Permutation Flowshop Scheduling Problems. In: Proc. The Fifth Metaheuristic International Conference, Kyoto, 63–1–63–8 (2003)
16. Satake, T., Morikawa, K., Takahashi, K., Nakamura, N.: Simulated Annealing Approach for Minimising the Makespan of the General Job-Shop. International Journal of Production Economics, pp. 60–61, 515–522 (1999)
17. Talukdar, S., Baerentzen, L., Gove, A., de Souza, P.: Asynchronous Teams: Cooperation Schemes for Autonomous, Computer-Based Agents, Technical Report EDRC 18–59–96, Carnegie Mellon University, Pittsburgh (1996)

Calculating Optimal Decision Using Meta-level Agents for Multi-Agents in Networks

Anne Håkansson[1] and Ronald Hartung[2]

[1] Department of Information Science, Computer Science,
Uppsala University,
Box 513, SE-751 20,
Uppsala, Sweden
Anne.Hakansson@dis.uu.se
[2] Department of Computer Science,
Franklin University,
201 S. Grant Avenue,
Columbus, Ohio 43215, USA
Hartung@franklin.edu

Abstract. In spatial graphs with a vast number of nodes, it is difficult to compute a solution to graph optimisation problems. We propose using meta-level agents for multi-agents in a network to calculate an optimal decision. The network contains nodes and arcs wherein the agents are information carriers between the nodes and, since there is one agent per arc, the agents are statically located. These agents, operating at a ground level, communicate with a comprehensive agent, operating at a meta-level. The agents at the meta-level hold information computed by the ground-level agents, but also include ground-level agents' special conditions. As an example, we apply the work to the travelling salesman problem and use a map, with cities and roads, constituting the network where the information about the roads is carried in the meta-level agents. For multi-agents in maps, we use parallel computing.

Keywords: Intelligent Agents, Multi-Agent Systems, Meta-Agents, Undirected Graphs.

1 Introduction

Spatial networks can, in general, be used to describe any network in which links or arcs connect nodes. In physical spaces, spatial networks can be derived from maps, using features like the road segments. Spatial networks can also represent complex social networks (Internet) and computer networks. To work with these different networks, there have been proposals of using a range of intelligent agents [13].

A challenge with network problems is finding and extracting information in the network within an acceptable time bound. As solutions, there have been attempts at applying effective search strategies, e.g., heuristics, A*, Best-first search, breadth-first, depth-first search, Dijkstra's algorithm, Kruskal's algorithm, the nearest neighbour algorithm, and Prim's algorithm [3; 10]. However, searching in enormous

B. Apolloni et al. (Eds.): KES 2007/WIRN 2007, Part I, LNAI 4692, pp. 180–188, 2007.

networks still remains a challenge with many of the interesting problems being NP-Hard. The NP-Hard problems include Clique, Independent set, Vertex cover, travelling salesman problem (TSP), Hamiltonian cycle, Graph partition, Vertex cover, Edge cover, and Graph isomorphism [2; 12]. In this paper we focus on TSP, and as a solution we suggest using meta-level agents over multiple agents in a network. The searching starts with a breadth-first approach and subsequently applies weighting costs for reducing the number of arcs to traverse.

Tools can be used for the route-finding problem in, e.g., computer networks, operations planning, and travel planning. Moreover, tools can be used for the touring problems, and visiting every city once (Hamiltonian path), or finding the shortest path for travelling salesman problem [3]. But they can also be used for the information overflow problem associated with Internet. Predicting optimal solutions in spatial networks remains a computationally hard problem. Although many attempts have been made, the problem of finding the paths through the large spatial space still remains. As a solution, we suggest calculating optimal decisions using multi-agents in networks. The optimal solution is held by a higher-level agent comprising the multi-agents involved in the solution and holds information about the computed travel time. These higher-level agents are called meta-level agents.

The characteristics of intelligent agents include operating under autonomous control, perceiving the environment, persisting over time, adapting to change and taking on another agent's goal [10, 14]. The agents, in multi-agent systems [15], can have these characteristics. The agents rely on prior knowledge with static characteristics, such as the nodes it connects and the distance. However, putting them in a real-world example will make them autonomous because the agents need to operate without inventions. The agents perceive their environment by considering the constraints and obstacles and subsequently act under the conditions that affect the agents' performance. To begin acting, the starting point for an agent is either data inserted by the user or a goal achieved by another agent. From this data or goal, the agents perform their task. At the lowest level, each agent performs one specific task and will unceasingly complete this task in the current system. However, if the constraints and obstacles in the task change, the agent needs to adjust to those changes. The agents continuously check the dynamic characteristics, and if needed, they adapt to their environment. Since optimal solutions usually include several agents, the system needs to take on other agents' goals. And as Roth [11] proposed, an agent can transfer commitments to another agent, our intelligent agents at the ground-level transfer environmental conditions in the network through the use of meta-level agents. The ground-level agents are information carriers between nodes in a network taking on the other agents? goals through the meta-level agents.

As software, we use a logic system for the ground-level multi-agents and meta-level agents. Each ground-level agent has knowledge about the route and moves by following the road, until it reaches its goal. Messages about time and the agents, involved in the optimal solution, are passed from the ground-level agents to the meta-level agents. From the messages, the meta-agents can collect the environmental conditions for every ground-level agent. By applying multi-agents in maps, the ground-level agents compute independently from other agents. With independency between the agents, we can use parallel computing and for maps we use simulated model of geographical information systems (GIS). The parallel computing is at the agents' level where the geographical information systems keep the agents updated with environmental information.

2 Related Work

The "Ant colony system" is a distributed algorithm that has been applied to the travelling salesman problem [4; 5]. The system uses a set of cooperating agents to find good solutions for the TSP. The agents cooperate using an indirect form of communication mediated by a pheromone deposited on the edges of a travelling salesman problem graph while building solutions [5]. Moreover, they use several agents per arc. In our work, we use agents in the arcs of the network, but we do not use the pheromone deposit on the edges. Instead we use information about the current circumstances. Take, for example, distance, constraints in speed due to topographic environment and temporary obstacles, such as current road conditions. The result of executing the agents is the time to accomplish their tasks, i.e., the travel time. Moreover, at each arc there is one agent working as an information carrier between the nodes. The basic principles of the "Ant colony" algorithm and its method of design and has been implemented in a multi-agent system [7]. Multi-agent structures can be described as an object or functional approach with an egalitarian subordination structure, a variable coupling strategy and an emergent constitution [6; 1]. The agents we use in our multi-agent system work locally in a spatial arrangement. Action of the agents is local modification. Moreover, the agents communicate locally and are perceptive of their local area via a GIS backend system.

Meta-reasoning for agents is when meta-level agents are controlling the agents' ability to trade off its resources between actions at an object level with those at a ground level. The meta-level control allows the agents to adapt their object-level computation. Moreover, the meta-level agents are horizontally modular [6]. For example, it has been shown that meta-level control with bounded computational overhead can allow complex agents to solve problems. The complex agents can solve problems more efficiently than current approaches in open dynamic multi-agent environments [9]. The meta-agents can be used to provide the control over agents and can stop the execution of some agents when further execution is not needed [9]. The meta-level agents collect data at a global level of knowledge, plan and schedule, and coordinate by using inter-agent negotiation. In our work, we do not apply meta-level control of the agents. Instead we use meta-agents to hold information collected by the agents at object level (ground level). The meta-agents are initiated from the work of the object-level agents and then used as a meta-level agent for the routes. The meta-level agents are then used to make leaps in the network ignoring some nodes without any loss of significant information. However, changes in the environment cause the meta-level agent to update in order cope with the new information as generated by the ground-level agents.

3 Multi-Agents in the Network

The multi-agent system has several ground-level agents operating between nodes in a network. In the network, each agent acts as an information carrier and picks up all the information it apprehend during its execution, i.e., while moving from one node to another node. These two nodes are the only information the agent has from the beginning. During the execution, the agent works with the conditions in the environment. These conditions affect the agent's time for execution. This time is the most important feature, since this is what we use in the network and it affects the other agents.

In the network, an event will cause an agent to operate unless the agent has fulfilled the termination condition. Before executing the network, the agent does not know which node is the starting node and thus is symmetric. Which of the nodes becomes the starting node and the termination node for the particular agent depends on the user's input to the network or the other agents. The starting node either is the user input inserted to the system, i.e., the initial event, or an agent invoking another agent. The invocation occurs when an agent reaches its end node, which then becomes the start node for the agents associated with that node.

To collect information in the multi-agent system, the agents associated with the starting point execute the task of carrying information along the arc where the agents are working. These agents perceive their environment by knowing the starting and ending nodes together with the constraints in the arcs as well as obstacles. The constraints are permanent impacts in the environment and the obstacles are temporary problems, in which both constraints and obstacles become resistances to travelling in the arcs. Constraints are static information covering distance between nodes, allowed speed, topography and nature of the environment such as number of lanes. Obstacles are dynamic information about the conditions changing over a short period of time, like weather, road quality and field of vision but also degree of passable roads such as road constructions. The constraints and obstacles are not limited to these enumerated above and can be expanded with the information needed in the agents. Because of changes in the environment, the agents have to adapt to these changes while calculating the cost to move across the arc.

The time it takes to move between the nodes, distance multiplied with speed, is the initial cost for the agent. During the execution, the agents work under the circumstances input by constraints and obstacles, which affect the time it takes for the agent to travel. Each constraint and obstacle is weighted as an additional cost to the agent, and the degree of influence affects the size of the extra cost. Small influences give minor extra costs while greater influences increase the cost gradually. For the multi-agents in the network, each condition has a scale ranging from 0-5 multiplied with the factor of 2.5. This corresponds to reduce the speed from 90 km/h to 70 km/h. Worst-case condition makes the agents to increase their costs with 50, which reduces the speed with about 60%. This time is the key factor for calculating the optimal solution. After the execution, each agent keeps track of its travel time.

In the multi-agent system, each agent starts from a node, works through the arc, and ends up at a node. At this ending node, other agents are waiting for execution taking on the first agent's goal. As mentioned above, the result from the first agent is the computed time of transit. Hence, the waiting agent cannot start executing before the first has reached the node with information. The information about the time will be carried further on in the network with the other agents. However, these agents also collect information about obstacles during their execution and the last agent round up the information from the agents operating from the starting node.

Since the multi-agents are working simultaneously from the starting node, they can work in parallel. As long as agents can work independently of each other, they will continue to execute until a termination node is reached by the system. However, reaching the termination node may not end the execution. If there are some agents still running at a lower cost than the agent that reached the terminating node, those agents will continue to compute until they either reach the terminating node or their costs exceeds the costs of the agent that had reached the final node.

A limitation of the multi-agents in the system is that the agents are not allowed to invoke other agents at a node already visited. Thus, if one agent already has been activated, this agent must not be called again for the same computation. Almost certainly, the agent is in a shorter path through the network. To know if the agent has been activated, each node has an execution flag, which holds the information about whether it has been visited by another agent.

When agents reach a node already visited, the agents that have the highest costs do not have any further possibility to execute and have to terminate. These can be removed from the calculation process without affecting the execution. The agents that run into dead-end roads also need to terminate directly. This speeds up the computational time because the system ignores these agents.

3.1 Example of a Network with Agents

To illustrate the multi-agents in a network, we provide an example of a graph symbolising a map with nodes and arcs. In the map, the nodes are cities and the arcs are the roads between the cities. In this example, the cities are denoted with the characters S and A-F where the S is the departure city and F is the destination city, see Figure 1. In the multi-agent system, these nodes have the real names of the cities. Between the cities, each road (arc) has an agent asserted to it.

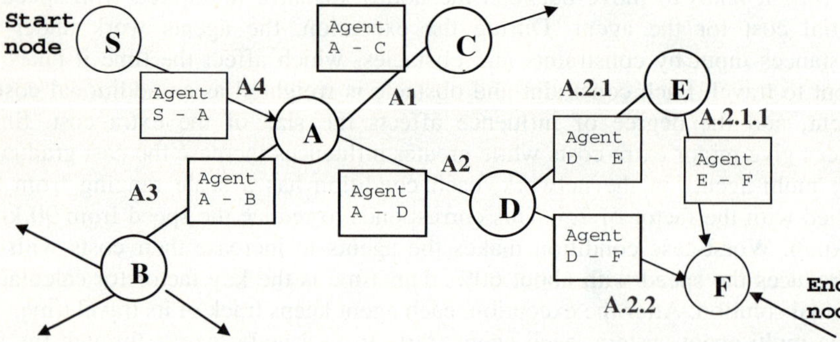

Fig. 1. An example of multi-agents in a network

Each agent holds the information about the cities it works between, for example the agent S-A and agent A-D. In the figure, the roads also have character and number attached to distinguish them, e.g., A4, A2, and A2.1, corresponding to the real road denotations. These are used to collect and calculate the time for executing the agents. The agent S-A carries information between the start city (S) and the A city and agent A-D carries information between A and D city. The agent A-D and agent D-F carry information between the A and D and D and the end city (F), respectively. There is an alternative path through the network. It is the route from the starting city (S) to end city (F) through the A city, D city and E city. In this case, there are three agents involved instead of two agents. Although several agents are used, the route might be the optimal solution between the cities.

The information carried between the agents can be more or less relevant to the result of optimal solution in the graph optimisation problems. The shortest computation time to reach a goal is, of course, the optimal solution. However, it is interesting to know reason for selecting the particular route. Therefore, the agents need to keep information about the temporary conditions, such as road constructions, that probably affect computing time for the routes. There can also be more radical changes in the constraints, like changing the number of lanes. The amount of information to be kept can be decided for each computation through the network. However, a lot of information can affect the time for computation, which is the reason for introduce agents at a meta-level.

The system needs to keep track of the multiple agents in a network while finding the optimal solution because of its inherent heavy computing load. Therefore, we apply meta-level agents on top of ground-level agents, which preserve the result from executing the agents in the network. These meta-agents collect the other agents' computational results and save the results in order to finally present the optimal solution to the user. Also the meta-level agents can accumulate local information. That is, the agent can look for optimal paths between nodes in a local neighbourhood. This can be constructed as a side affect of other computations. When the meta-agents come through the agent, the agent can offer more interesting information to the meta-agent.

4 Agents at a Meta-level

The meta-agents are created from ground-level agents after performing the calculation of the time it takes to execute the intended task. Hence, the meta-agents can be used to follow the multi-agents work through the network [8]. Meta-agents keep track of time and information about the conditions, passed as messages between the agents, i.e., ground-level and meta-level, in the network. Usually, several ground-level agents are involved in computing an assigned task. These agents' information, carried by each agent message, is used to build the meta-agent. The meta-agents execute in the context of the all the agents in the arcs. At the top-level, the meta-agent holds the goal of the computation. It will have the logic to determine when a solution has been achieved and to present the solution as the result of the computation. This logic is a set of conditions and constraints that select the best paths to help schedule the computation and determine when a solution is reached. A simple approach to scheduling can be based on the shortest path generated at any time.

For the example mentioned above, the system produces one meta-agent. The meta-level agent comprises the ground-level agents involved in finding the optimal solution while moving from S to F, see Figure 2. The meta-agent is located at the top in the figure. This agent incorporates the information about the departure and destination and the time for executing each ground-level agent. The meta-agent also has the information about the nodes (cities) and conditions that the ground-level agents have collected during their execution, which is taken into account in the optimal solution. The ground-level agent to the left (agent S-A) has information about the initial start node, the road number and the stop node. After executing, the ground-level agents also have collected information about the constraints and obstacles, which have been translated into costs. The cost is sent as a message of time to the meta-level agent, (S 2.32). The ground-level agent to the right (agent E-F) has information about the activated node, road number and the end node.

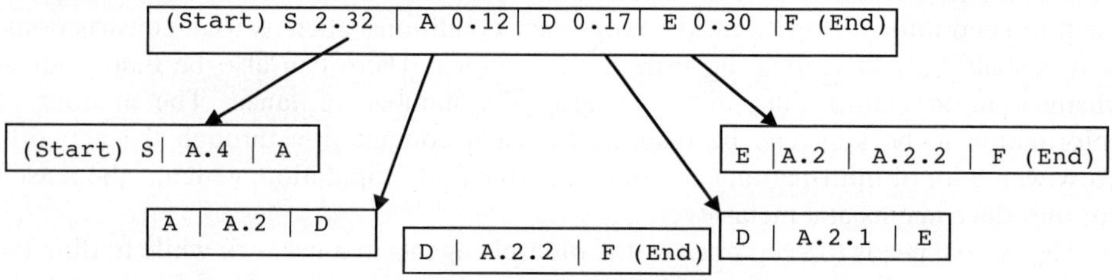

Fig. 2. Meta-level agents in the network

To the left in Figure 2, the first agent (agent S-A) is activated by the user. This agent reach the A node which causes the agent A-D, agent D-E, agent D-F and agent E-F to start running. The agent D-F is not captured in meta-level agent because it required higher cost than the other path, D-E and E-F. Additionally, for each agent, the cost for executing between the arcs is hold by the meta-agent. The cost for S-A agent is 2.32 hour, A-D agent is 12 min, D-E agent 17 min and E-F agent 30 min.

The selection of the optimal solution is made by the meta-agents. The meta-agent enters each node of corresponding agents in the network and performs the calculation for the agents. It has access to all information contained in the ground-level agents, including all collected derived conclusions. The meta-agent simply adds the agents' performance in time, taken into account the static information and the dynamic information.

It is possible to search for multiple solutions in the network. For example, a practical problem is to set up a set of routes for multiple deliveries to nodes in the network. This is a useful problem for airlines, trucking and other carriers. In this case, several meta-agents can be started simultaneously to search the network for solution. For each computation in parallel, several meta-agents may be created since the first agent to reach the end node might not be the fastest. Then, the meta-agents must be tested against each other.

5 Software for the Agents in the Network

As software, we use a logic system for the ground-level agents and meta-level agents. The network is a small map of a part of Sweden and part of United States of America. Between the nodes there are several arcs, which are corresponding to the reality. Each ground-level agent has knowledge about the route and follows the road, until it reaches its goal. Messages about time are passed from the ground-level agents to the meta-level agents, which are stored in the database of the system. The meta-level agent can keep track of the dead end roads and make the ground-level agent to avoid running in to it again.

For applying multi-agents in maps, we use parallel computing in simulated geographical information systems. The parallel computing is at the ground-agents' level. To fully utilise the parallel facility, we duplicate the agents for each node that has two or more computing agents. The number of computing agents decides the number of duplicates. In this way, the system can continue to execute all the agents without needing to explore one at the time.

By applying meta-level agents, we can calculate the optimal path for between two nodes. This facility is used for calculating the optimal path through the graph, producing several meta-agents and comparing the cost for the meta-agents. The current system gives alternative paths between nodes, especially when the agents are visiting several cities on their way to the goal. The costs are used as a weighting facility when choosing the optimal road.

6 Conclusions and Further Work

In this paper, we have presented an approach of using meta-level agents for multi-agents in networks to suggest a solution to the optimal solution problem in graphs. The intelligent agents follow the arcs and compute the time by acting upon the arc information, which are the constraints and obstacles that the specific arc has about the environment. The computation time becomes the significant factor to find the optimal solution. To keep track of the computational time of the ground-level multi-agents, the network also uses meta-level agents to which the ground-level agents pass the computed time as messages. Beside time, the meta-level agents can also hold environmental information from the ground-level agents. The use of meta-level agents for multi-agents in networks support parallel computing of independent ground-level agents. The parallel agents work as long as ground-level agents are independent of other agents and have not reached a node that is already reached by another agent. However, when the agent has been activated from a node, this node turned the execution flag to "visited". This facility needs a lot of testing before deciding that it works in all situations.

Moreover, to apply multi-agent in reality, we need to develop a parallel computing system using a GIS interface. The parallel computing handles the multi-agents, and the geographical information systems provide the agents with updated environmental information. The system also needs testing to check the extent to which the GIS interface supports the agents.

One appealing aspect of meta-agent approach is the ease of scheduling the multi-agents for execution. One simple approach is to partition the agents onto processors. The agents on a processor remain idle until something affects them and causes them to run. This can work in a loosely coupled parallel processing system. A second approach is to have a single scheduling queue on a central processor that can assign work to processors as available. Clearly, there are trade-offs involved, especially with the amount of information to be moved between processors. Also, we have the question of whether or not the work is evenly distributed between agents or whether there are some agents that are executed more frequently than others.

References

1. Attoui, A.: Real-Time and Multi-Agent Systems, 1st edn. Springer, Heidelberg (2000)
2. Berman, K., Paul, J.: Algorithms: Sequential, Parallel, and Distributed. Course Technology; 1 th edition, ISBN-10: 0534420575 (2004)
3. Cormen, T., Leiserson, C., Rivest, L., Stein, C.: Introduction to Algorithms, 2nd edn. MIT Press and McGraw-Hill, Cambridge (2001)
4. Dorigo, M., Maniezzo, V., Colorni, A.: Ant System: Optimization by a Colony of Cooperating Agents. IEEE Transactions on Systems, Man, and Cybernetics, Part B 26(1), 29–41 (1996)
5. Dorigo, M., Gambardella, L.M.: Ant colony system: a cooperative learning approach to the traveling salesman problem. IEEE Transactions on Evolutionary Computation 1(1), 53–66 (1997)
6. Ferber, J.: Multi-Agent Systems. Addison Wesley Co, London, UK (2002)
7. He, J.-m., Min, R., Wang, Y.-y: Implementation of Ant Colony Algorithm Based-On Multi-agent System. In: Lu, X., Zhao, W. (eds.) ICCNMC 2005. LNCS, vol. 3619, pp. 1234–1242. Springer, Heidelberg (2005)
8. Håkansson, A., Hartung, R.L.: Using Meta-Agents for Multi-Agents in Networks (ICAI'07). In: The 2007 International Conference on Artificial Intelligence, WORLDCOMP'07, Las Vegas, USA (June 25th -28th, 2007)
9. Raja, A., Lesser, V.: A Framework for Meta-level Control in Multi-Agent Systems. Autonomous Agents and Multi-Agent Systems. Springer, Heidelberg (2007)
10. Russell, S., Norvig, P.: Artificial Intelligence - a Modern Approach, pp. 32–752. Prentice Hall, Englewood Cliffs (2003)
11. Roth, V.: Mutual protection of co–operating agents. In: Vitek, J., Jensen, C. (eds.) Secure Internet Programming. LNCS, vol. 1603, pp. 275–285. Springer, Heidelberg (1999)
12. Skiena, S.: The Algorithm Design Manual, 1st edn. Springer, Heidelberg (1998)
13. Turban, E., Aronson, J., Liang, T.-P.: Decision Support Systems and Intelligent Systems, 7th edn. Pearson, London (2005)
14. Wooldridge, M., Jennings, N.: Intelligent agents: Theory and practice. Knowledge Engineering Review 10(2) (1995)
15. Wooldridge, M.: An Introduction to MultiAgent Systems. John Wiley & Sons Ltd, Chichester (2002)

Determining Consensus with Dependencies of Set Attributes Using Symmetric Difference*

Michał Zgrzywa

Institute of Information and Engineering, Wroclaw University of Technology, Poland
michal.zgrzywa@pwr.wroc.pl

Abstract. In this paper the author considers some problems related to attribute dependencies in consensus determining. These problems concern the dependencies of attributes representing the content of conflicts, which cause that one may not treat the attributes independently in consensus determining. It is assumed that attribute values are represented by sets. In this paper the author presents conditions guaranteeing determining a correct consensus despite treating the attributes independently. Next, the algorithm for calculating the proper consensus in the cases when these limitations are not met is presented. Finally, the differences between proper consensus and consensus proposals calculated with treating attributes independently are considered.

Keywords: Consensus theory, Conflict, Dependency, Set, Distributed system.

1 Introduction

Conflict resolution is one of the most important aspects in distributed systems and multi-agent systems. The resources of conflicts in these kinds of systems come from the autonomy feature of their sites (nodes). This feature means that each site of a distributed or multi-agent system processes a task independently. There are several reasons to organize a system in such an architecture [4]. First of all, information collected in the system is easier to obtain – some sites may be nearer to the user or not as busy as others. Although the reliability of such systems is better – the failure of one node may be compensated by using others. Finally, the trustworthiness of the system may be increased when several agents are investigating the same issue. Unfortunately, there may arise such a situation that for the same task, different sites may generate different solutions. Thus, one deals with a conflict.

In distributed and multi-agent systems three origins of conflicts can be found: insufficient resources, differences of data models and differences of data semantic [7]. Consensus models, among others, seem to be useful in semantic conflict solving [9]. The oldest consensus model was worked out by such authors as Condorcet, Arrow and Kemeny [1]. This model serves to solve such conflicts in which the content may be represented by orders or rankings. Models of Barthelemy and Janowitz [2],

* This work was supported by Polish Ministry of Science and Higher Education grant No. N516 033 31/3447.

B. Apolloni et al. (Eds.): KES 2007/WIRN 2007, Part I, LNAI 4692, pp. 189–197, 2007.

Barthelemy and Leclerc [3] and Day [5] enable to solve such conflicts for which the structures of the conflict contents are n-trees, semillatices, partitions etc. The common characteristic of these models is that they are one-attribute, which means that conflicts are considered only referring to one feature. Multi-feature conflicts have not been investigated.

In the work [6] the author presents a consensus model, in which multi-attribute conflicts may be represented. Furthermore, in this model attributes are multi-valued, which means that for representing an opinion on some issue an agent may use not only one elementary value (such as +, –, or 0) [7] but a set of elementary values. This model enables to process multi-feature conflicts, but attributes are mainly treated as independent. However, in many practical conflict situations some attributes are dependent on others. For example, in a meteorological system attribute *Wind_power* (with values: *weak*, *medium*, *strong*) is dependent on attribute *Wind_speed*, the values of which are measured in unit *m/s*. This dependency follows that if the value of the first attribute is known, then the value of the second attribute is also known. It is natural that if a conflict includes these attributes then in the consensus the dependency should also take place. The question is: Is it enough to determine the consensus for the conflict referring to attribute *Wind_speed*? And if not, how should one calculate the proper consensus that fulfills the dependency?

In this paper we consider the answers for questions about calculating proper consensus for conflicts with dependencies. For this aim we assume some dependencies between attributes and show their influence on consensus determining.

2 The Outline of Consensus Model

The consensus model which enables processing multi-attribute and multi-valued conflicts has been discussed in detail in work [6]. In this section we present only some of its elements with extensions needed for the consideration of attribute dependencies. We assume that a real world situation is commonly considered by a set of agents (or nodes) that are placed in different sites of a distributed system. The interest of the agents consists of events which occur (or have to occur) in this world. The task of the agents is based on determining the values of attributes describing these events and reporting them to some central unit. If several agents consider the same event then they may generate different descriptions (which consist of, for example, scenarios, timestamps etc.) for this event. Thus we say that a conflict takes place.

For representing ontologies of potential conflicts we use a finite set $\boldsymbol{A}$ of attributes and a set $\boldsymbol{V}$ of attribute elementary values, where $\boldsymbol{V} = \bigcup_{a\in A} V_a$ (V_a is the domain of attribute a). Let $\Pi(V_a)$ denote the power set of V_a and $\Pi(V_B) = \bigcup_{b\in B}\Pi(V_b)$. Let $B\subseteq \boldsymbol{A}$, a tuple r_B of type B is consist of all values of some function f_r: $B \rightarrow \Pi(V_B)$ where $f_r(b) \subseteq V_b$ for each $b\in B$. Empty tuple is denoted by symbol ϕ. The set of all tuples of type B is denoted by $TYPE(B)$. The conflict ontology is defined as $<\boldsymbol{A}, \boldsymbol{X}, \boldsymbol{P}, \boldsymbol{F}>$, where:

- $\boldsymbol{A}$ is a finite set of attributes, which includes a special attribute *Agent*; any value of attribute a where $a\neq Agent$ is an element of V_a; values of attribute *Agent* are singletons which identify the agents;

- $\boldsymbol{X}=\{\Pi(V_a): a\in \boldsymbol{A}\}$ is a finite set of conflict carriers;
- $\boldsymbol{P}$ is a finite set of relations on carriers from $\boldsymbol{X}$, each relation $P\in \boldsymbol{P}$ is of some type T_P (for $T_P\subseteq \boldsymbol{A}$ and $Agent \in T_P$). Relations belonging to set $\boldsymbol{P}$ are classified into groups of two, identified by symbols "+" and "−" as the upper index to the relation names. For example, if R is the name of a group, then relation R^+ is called the positive relation and R^- is the negative relation. Positive relations contain tuples representing such descriptions which are possible for events. Negative relations, on the other hand, contain tuples representing such descriptions which are not expected for events. When there is only a positive relation, the upper index may be omitted.
- Finally, $\boldsymbol{F}$ is a set of function dependencies between sets of attributes.

The structures of the conflict carriers are defined by means of a distance function between tuples of the same type. In this chapter we will use distance functions that measure the distance between 2 sets X and Y ($X,Y\subseteq V_a$ for $a\in \boldsymbol{A}$) as the minimal costs of the operation which transforms set X into set Y. The symbol δ will be used for distance functions.

A consensus is considered within a conflict situation, which is defined as a pair $s = <\{P^+,P^-\}, (A,B)>$ where $A,B\subseteq \boldsymbol{A}$, $A\cap B=\varnothing$, and $r_A\neq\phi$ holds for any tuple $r\in P^+\cup P^-$ (P^+,P^- are relations of $TYPE(\{Agent\}\cup A\cup B)$). The first element of a conflict situation (i.e. set of relations $\{P^+,P^-\}$) includes the domain from which consensus should be chosen, and the second element (i.e. 2-tuple (A,B)) presents the structure of consensus. For a subject e (as a tuple of type A, included in P^+ or P^-) there should be assigned only one tuple of type B. A conflict situation yields a set $Subject(s)$ of conflict subjects which are represented by tuples of type A. For each subject e two conflict profiles, i.e. $profile(e)^+$ and $profile(e)^-$, as relations of $TYPE(\{Agent\}\cup B)$ may be determined. Profile $profile(e)^+$ contains the positive opinions of the agents on the subject e, while profile $profile(e)^-$ contains agents' negative opinions on this subject.

Definition 1. *Consensus on a subject $e\in Subject(s)$ is a 2-tuple $(C(s,e)^+,C(s,e)^-)$ of 2 tuples of type $A\cup B$ which fulfill the following conditions:*

a) $C(s,e)^+_A=C(s,e)^-_A=e$ *and* $C(s,e)^+_B\cap C(s,e)^-_B=\phi$,

b) The sums $\sum_{r\in profile(e)^+}\delta(r_B,C(s,e)^+_B)$ *and* $\sum_{r\in profile(e)^-}\delta(r_B,C(s,e)^-_B)$ *are minimal.*

Any tuples $C(s,e)^+$ and $C(s,e)^-$ satisfying the conditions of Definition 1 are called consensuses of profiles $profile(e)^+$ and $profile(e)^-$, respectively.

Example 1. Let us consider the meteorological system from the beginning of the first section. The ontology of that conflict is the quadruple: <{Agent, Region, Temperature}, {Π(Silesia, Great Poland, Little Poland, Pomerania), Π(-30,-29,…,34,35)}, {Weather$^+$}, ϕ>. We can distinguish one conflict situation: <{Weather$^+$,ϕ}, {Region}→ {Temperature}>. Suppose that a meteorological station is not always precise in its forecast. In such case it proposes a set of possible values of temperature. Information about the conflict for the subject Silesia is gathered below.

Table 1. The relation Weather$^+$ with sets of values

Agent	Region	Temperature
station$_1$	Silesia	{24, 25, 26}
station$_2$	Silesia	{25, 26}
station$_3$	Silesia	{25, 26, 27}

We need a different distance function for attribute values that are sets. We will use function $\delta_{Sym\text{-}Dif}$ defined as: $\delta_{Sym-Dif}(set_1, set_2) = | set_1 \div set_2 |$ (number of elements in a symmetric difference of two sets). After calculating all the distances from possible values to the whole profile we will find out that our consensus is the set {25, 26}.

Function $\delta_{Sym\text{-}Dif}$ is very popular and widely used for comparing sets. Unfortunately, it also has some drawbacks. It should not be used when it is possible to calculate the distances between elements of sets. $\delta_{Sym\text{-}Dif}$ function assumes that any two elements may be only identical or not identical – it does not concern how different they are. The author considered more set-comparing functions in work [11].

3 Some Aspects of Attribute Dependencies

In Definition 1, condition b) is the most important. It requires the tuples $C(s,e)^+_B$ and $C(s,e)^-_B$ to be determined in such a way thus the sums $\sum_{r \in profile(e)^+} \partial(r_B, C(s,e)^+_B)$ and $\sum_{r \in profile(e)^-} \partial(r_B, C(s,e)^-_B)$ are minimal. These tuples could be calculated in the following way: for each attribute $b \in B$ one can determine sets $C(s,e)^+_b$ and $C(s,e)^-_b$, which minimize sums $\sum_{r \in profile(e)^+} \partial(r_b, C(s,e)^+_b)$ and $\sum_{r \in profile(e)^-} \partial(r_b, C(s,e)^-_b)$ respectively. This way is an effective one, but it is correct only if the attributes from set B are independent ($\boldsymbol{F}=\phi$). In this section we consider consensus choice assuming that some attributes from set B are dependent on some others. The definition of attribute dependency given below is consistent with those given in the information system model [8]:

Definition 2. *Attribute b is dependent on attribute a if and only if there exists a surjective function* $f^a_b: V_a \rightarrow V_b$ *for which in conflict ontology* <**A**,**X**,**P**,**F**> ($f^a_b \in \boldsymbol{F}$) *for each relation* $P \in \boldsymbol{P}$ *of type* T_P *and* $a,b \in T_P$ *formula* $(\forall r \in P)(r_b = \bigcup_{x \in r_a} \{f^a_b(x)\})$ *is true.*

The dependency of attribute b on attribute a means that in the real world if for some object the value of a is known then the value of b is also known. In practice, owing to this property for determining the values of attribute b it is enough to know the value of attribute a. Instead of $\bigcup_{x \in Y} \{f^a_b(x)\}$ we can write more concisely $f^a_b(Y)$.

Consider now a conflict situation $s=<\{P^+,P^-\},A \rightarrow B>$, in which attribute b is dependent on attribute a where $a,b \in B$. Let $profile(e)^+$ be the positive profile for given conflict subject $e \in Subject(s)$. The problem relies on determining consensus for this profile. We can solve this problem using two approaches:

1. Notice that $profile(e)^+$ is a relation of type $B \cup \{Agent\}$. There exists a function from set $TYPE(B \cup \{Agent\})$ to set $TYPE(B \cup \{Agent\} \setminus \{b\})$ such that for each profile $profile(e)^+$ one can assign exactly one set $profile'(e)^+ = \{r_{B \cup \{Agent\} \setminus \{b\}}: r \in profile(e)^+\}$. Set $profile'(e)^+$ can be treated as a profile for subject e in the following conflict situation $s' = <\{P^+, P^-\}, A \to B \setminus \{b\}>$.

Notice that the difference between profiles $profile(e)^+$ and $profile'(e)^+$ relies only on the lack of attribute b and its values in profile $profile(e)^+$. Thus one can expect that the consensus $C(s,e)^+$ for profile $profile(e)^+$ can be determined from the consensus $C(s,e)'^+$ for profile $profile(e)'^+$ after adding to tuple $C(s,e)'^+$ attribute b and its value which is equal to $f_b^a(C(s,e)'^+{}_a)$. In a similar way one can determine the consensus for profile $profile(e)^-$.

2. In the second approach attributes a and b are treated independently. This means that they play the same role in consensus determining for profiles $profile(e)^+$ and $profile(e)^-$.

The consensus for profiles $profile(e)^+$ and $profile(e)^-$ are defined as follows:

Definition 3. *The consensus for subject* $e \in Subject(s)$ *considered in situation* $s=<\{P^+,P^-\},A \to B>$ *is a tuple* $(C(s,e)^+, C(s,e)^-)$ *of type* $A \cup B$*, which satisfy the following conditions:*

a) $C(s,e)^+{}_A = C(s,e)^-{}_A = e$ *and* $C(s,e)^+{}_B \cap C(s,e)^-{}_B = \phi$,

b) $C(s,e)^+{}_b = f_b^a(C(s,e)^+{}_a)$ *and* $C(s,e)^-{}_b = f_b^a(C(s,e)^-{}_a)$,

c) *The sums* $\sum_{r \in profile(e)^+} \partial(r_B, C(s,e)^+_B)$ *and* $\sum_{r \in profile(e)^-} \partial(r_B, C(s,e)^-_B)$ *are minimal.*

We are interested in the cases when conditions b) and c) of Definition 3 can be satisfied simultaneously. Unfortunately it is not true that if set $C(s,e)^+{}_a$ is a consensus for profile $profile(e)^+{}_a$ (as the projection of profile $profile(e)^+$ on attribute a) then set $f_b^a(C(s,e)^+{}_a)$ will be a consensus for profile $profile(e)^+{}_b$ (as the projection of profile $profile(e)^+$ on attribute). The limitations guaranteeing determining a correct consensus for single-element values of attributes were considered in previous work [10]. In [11] the author found these limitations for set attributes for a few difference functions.

In this paper we will focus only on $\delta_{Sym\text{-}Dif}$ function. We will consider three questions: What are the limitations guaranteeing determining a correct consensus? What is the algorithm of calculating the proper consensus in cases when these limitations are not met will? And how different is the proper consensus from consensus proposals calculated with treating attributes independently?

4 Conditions Sufficient for Treating Attributes Independently

In this section the conditions sufficient for treating attributes independently will be shown. First, the following regularity has to be noticed.

Theorem 1. *If function $\delta_{Sym\text{-}Dif}$ is used for measuring distances between agents' propositions for attributes a and b and there is a dependency $f_b^{\,a}$ then* $f_b^a(C(s,e)'_a) \subseteq C(s,e)'_b$.

In other words, if a value is a consensus for attribute a then its image is also a consensus for attribute b. However, consensus for attribute b may also include some elements that are not images of consensus elements for attribute a.

Proof. Let us suppose that a_0 is an element of consensus for attribute a. This means that the sum of all the distances from the consensus (including a_0) to all the agents' propositions (called profile) is less then the sum of all the distances from the consensus without a_0 to the profile. As function $\delta_{Sym\text{-}Dif}$ is used, we know that the sum of distances between consensus and the profile will increase by 1 for each proposition that does not contain a_0. This leads us to the conclusion that the number of propositions in the profile containing a_0 must be bigger than the number of propositions without a_0. This means that also $f_b^{\,a}(a_0)$ will be included in at least half of the propositions in the profile for attribute b, which is enough to claim that $f_b^{\,a}(a_0)$ will be included in consensus for b.

Additionally, the case when a_0 is included in exactly half of the proposals must be considered. In such a situation, either set with a_0 or set without a_0 could be a proper consensus. This may cause a problematic situation when consensus for a includes a_0 and consensus for b does not include $f_b^{\,a}(a_0)$. To avoid such problems, during consensus determining one should always use one rule concerning elements included in exactly half of the profile.

As was shown, dependency function transforms all the elements of consensus for a into elements of consensus for b. But how can we determine the whole consensus for attribute b? The following theorem may be useful.

Theorem 2. *If function $\delta_{Sym\text{-}Dif}$ is used for measuring distances between agents' propositions for attributes a and b and there is a differential dependency $f_b^{\,a}$ then* $f_b^a(C(s,e)'_a) = C(s,e)'_b$.

Proof. If function $f_b^{\,a}$ is differential then $(a_1 \neq a_2) \Rightarrow (f_b^{\,a}(a_1) \neq f_b^{\,a}(a_2))$. In such a case it is certain that each element b_i of consensus for b has exactly one element a_i in the domain of a that can be transformed to it. As b_i is a consensus then it is included in at least half of the profile. Thus also a_i is included in at least half of the profile, which means that a_i is a consensus for a.

5 Calculating Consensus in Case of Attribute Dependency

As was shown in previous section, when dependency function is differential, we can treat attributes independently. But how should we determine consensus when this condition is not met? Let us introduce the following example.

Example 2. Let us assume that our system is observed by agents: $agent_1$, … , $agent_5$. Their knowledge is gathered in relation P and is described by two set attributes: a and b. Attribute b depends on attribute a in the following way: $f_b^{\,a}(a_1)=b_1$, $f_b^{\,a}(a_2)=b_2$,

Table 2. Relation P (conflict)

Agent	*a*	*b*
$agent_1$	$\{a_2,a_3\}$	$\{b_2,b_3\}$
$agent_2$	$\{a_3\}$	$\{b_3\}$
$agent_3$	$\{a_2,a_4\}$	$\{b_2,b_4\}$
$agent_4$	$\{a_2,a_5,a_7\}$	$\{b_2,b_3,b_4\}$
$agent_5$	$\{a_6,a_7\}$	$\{b_3,b_4\}$

$f_b^a(a_3)=b_3$, $f_b^a(a_4)=b_4$, $f_b^a(a_5)=b_4$, $f_b^a(a_6)=b_4$, $f_b^a(a_7)=b_3$. Now, a conflict takes place in our system (Table 2).

How can we find the best solution for a conflict? Algorithm 1 calculates (with a polynomial complexity $O(n^2)$) the correct and optimal consensus in such a situation.

Algorithm 1.
Input: Profile X with n proposals consisting of two set attributes a and b, dependency f_b^a.
Output: Consensus C for profile X.

1. Create sets $elements_a$ and $elements_b$ which contains all the values of attribute a and b proposed in the profile.
2. If $elements_a$ is empty then go to step 8.
3. Pick one element – value v_a – from set $elements_a$.
4. Count the occurrences of v_a in agents' proposals: $occ(v_a)$ and store it.
5. If $occ(v_a) \geq n/2$ then add v_a to C_a, add $f_b^a(v_a)$ to C_b and remove $f_b^a(v_a)$ from set $elements_b$.
6. Remove v_a from set $elements_a$.
7. Go to step 2.
8. If $elements_b$ is empty then go to step 16.
9. Pick one element – value v_b – from set $elements_b$.
10. Count the occurrences of v_b in agents' proposals: $occ(v_b)$.
11. If $occ(v_b) < n/2$ then go to step 14.
12. From the elements of new set $\{v_a: f_b^a(v_a) = v_b\}$ choose element v_a, which has the most occurrences in agents' proposals ($occ(v_a)$ was calculated and stored earlier for every possible value v_a).
13. If $(n/2 - occ(v_a)) \leq (occ(v_b) - n/2)$ then add v_a to C_a and v_b to C_b.
14. Remove v_b from set $elements_b$.
15. Go to step 8.
16. Return C.

Now, we will use Algorithm 1 in Example 2. First, we create two sets: $elements_a = \{a_2,a_3,a_4,a_5,a_6,a_7\}$ and $elements_b = \{b_2,b_3,b_4\}$. There is only one value which occurs in at least half of the profile: a_2, so we add 2-tuple (a_2,b_2) to calculated consensus. We move further. Set $elements_b$ now includes two elements: $\{b_3,b_4\}$. Both values occur in at least half of the profile. First we will consider value b_3. We are looking for such a value v_a ($f_b^a(v_a) = b_3$) that has the greatest number of occurrences in agents' proposals. In this case we can use either a_3 or a_7 (2 occurrences). Thus, the condition $(n/2 - occ(v_a)) \leq (occ(v_b) - n/2)$ will become $(5/2 - 2) \leq (3 - 5/2)$ which is true. We add

2-tuple (a_3,b_3) (or 2-tuple (a_7,b_3)) to calculated consensus. Next we will consider value b_4. This time three values have the greatest number of occurrences in the profile: a_4, a_5 and a_6 (1 occurrence). The condition becomes $(5/2 - 1) \leq (3 - 5/2)$ which is not true, so we omit b_4. The algorithm ends by returning the 2-tuple $\{a_2,a_3\}$ and $\{b_2,b_3\}$ (or the 2-tuple $\{a_2,a_7\}$ and $\{b_2,b_3\}$). The distance from such calculated consensus to the profile is 16 (10 for attribute *a* and 6 for attribute *b*) which is the best solution.

Additionally, the construction of Algorithm 1 leads to the following theorem.

Theorem 3. *If function $\delta_{Sym\text{-}Dif}$ is used for measuring distances between agents' propositions for attributes a and b and there is a dependency f_b^a then:*

a) $f_b^a(C(s,e)'_a) \subseteq C(s,e)_b \subseteq C(s,e)'_b$,

b) $C(s,e)'_a \subseteq C(s,e)_a \subseteq \{v_a : f_b^a(v_a) \in C(s,e)'_b\}$.

6 Conclusion

In this paper we described how dependencies of set attributes influence the possibilities of consensus determining. Assuming that $\delta_{Sym\text{-}Dif}$ distance function is used, the limitations of dependency functions were shown, guaranteeing determining a correct consensus despite treating attributes independently. Using such functions provides the following profits. First of all, they enable determining a consensus for only a part of the attributes (the rest may be calculated using dependency functions). Secondly, they prevent determining an incorrect consensus, which does not fulfill some of the dependencies of attributes. Also the algorithm of consensus determining (with polynomial complexity) was shown, which may be used when the limitations are not met. Additionally, the differences between proper consensus and consensus proposals calculated with treating attributes independently were considered.

The presented theorems do not solve all of the problems of this area. The following issues, among others, need to be considered:

- what other limitations are necessary when many attributes may depend on many attributes?
- how to calculate correct consensus when other distance functions are used?
- how to calculate correct consensus for element structures different from sets?

Work on these subjects is being continued. The results should enable construction of effective algorithms which will aid conflict resolution in distributed systems.

References

1. Arrow, K.J.: Social Choice and Individual Values. Wiley, New York (1963)
2. Barthelemy, J.P., Janowitz, M.F.: A Formal Theory of Consensus, SIAM J. SIAM J. Discrete Math 4, 305–322 (1991)
3. Barthelemy, J.P., Leclerc, B.: The Median Procedure for Partitions. DIMACS Series in Discrete Mathematics and Theoretical Computer Science 19, 3–33 (1995)

4. Coulouris, G., Dollimore, J., Kindberg, T.: Distributed systems, Concepts and design. Addison-Wesley, London, UK (1996)
5. Day, W.H.E.: Consensus Methods as Tools for Data Analysis. In: Bock, H.H. (ed.) Classification and Related Methods for Data Analysis, North-Holland, pp. 312–324 (1988)
6. Nguyen, N.T.: Methods for Consensus Choice and their Applications in Conflict Resolving in Distributed Systems. Wroclaw University of Technology Press (in polish) (2002)
7. Pawlak, Z.: An Inquiry into Anatomy of Conflicts. Information Sciences 108 (1998)
8. Skowron, A., Rauszer, C.: The Discernibility Matrices and Functions in Information Systems. In: Słowiński, E. (ed.) Intelligent Decision Support, Handbook of Applications and Advances of the Rough Sets Theory, pp. 331–362. Kluwer Academic Publishers, Dordrecht (1992)
9. Tessier, C., Chaudron, L., Müller, H.J.: Conflicting agents: conflict management in multi-agent systems. Kluwer Academic Publishers, Boston (2001)
10. Zgrzywa, M., Nguyen, N.T.: Estimating and Calculating Consensus with Simple Dependencies of Attributes. In: Zgrzywa, M., Nguyen, N.T. (eds.) Proceedings of CORES 2005, Rydzyna, Poland, Advances in soft computing, pp. 319–328. Springer, Heidelberg (2005)
11. Zgrzywa, M.: Determining Consensus with Dependencies of Multi-element Attributes. In: Katarzyniak, R. (ed.) Ontologies and Soft Methods in Knowledge Management, Advanced Knowledge International, Australia, pp. 119–136 (2005)

Field-Based Coordination of Mobile Intelligent Agents: An Evolutionary Game Theoretic Analysis

Krunoslav Trzec[1] and Ignac Lovrek[2]

[1] Ericsson Nikola Tesla, R&D Centre,
Krapinska 45, HR-10000 Zagreb, Croatia
krunoslav.trzec@ericsson.com
[2] University of Zagreb,
Faculty of Electrical Engineering and Computing,
Department of Telecommunications
Unska 3, HR-10000 Zagreb, Croatia
ignac.lovrek@fer.hr

Abstract. The paper deals with field-based coordination of agent team in which the continental divide game is applied as a coordination mechanism. The agent team consists of self-interested mobile intelligent agents whose behaviour is modelled using coordination policies based on adaptive learning algorithms. Three types of learning algorithms have been used: three parameter Roth-Erev algorithm, stateless Q-learning algorithm, and experience-weighted attraction algorithm. The coordination policies are analyzed by replicator dynamics from evolutionary game theory. A case study describing performance evaluation of coordination policies according to the analysis is considered.

1 Introduction

An intelligent software agent is an autonomous program which acts on behalf of its user. Efficient coordination is essential if agents are to achieve their goals in a team. The need for such coordination occurs in agent-based provisioning of context-aware services that require multilateral negotiation and/or mutual concession of situational resources, which we named group-oriented context-aware services. Context in such services is influenced by preferences the users in a group are interested in. Users' preferences are represented by self-interested agents (personal assistants) in a ubiquitous (pervasive) network environment. We believe it is important for the agents to be able to use a variety of coordination policies in order to successfully provide group-oriented context-aware services, so that they can apply a policy that has an optimal outcome for a coordination mechanism at hand. Each agent in a team should adopt a policy that converges as quickly as possible (i.e. the most efficiently) towards an optimal outcome for a group of users.

However, the coordination policy that determines agent's actions in a team is usually imposed upon the agent at design time. This means that in many cases the policy that is hard-coded may not be a suitable choice for a given coordination mechanism. To circumvent this problem, an evolutionary game theoretic performance evaluation

B. Apolloni et al. (Eds.): KES 2007/WIRN 2007, Part I, LNAI 4692, pp. 198–205, 2007.

of coordination policies is applied that may help us to provide agents with a policy that is well suited for given coordination mechanism, or to build agents that can dynamically switch to a suitable policy using, for example, rule-based approach.

Self-interested agents have to be able to efficiently coordinate their activities to achieve goals together. By exploiting some nature-inspired coordination mechanisms, it is possible to reproduce natural phenomena of adaptive self-organization. The field-based coordination [1] represents such coordination models that takes its inspiration from the physical world, in particular from the way masses in our universe move and globally self-organize according to the contextual information represented by gravitational fields. Such coordination aims at supporting agents' activities by providing, through the concept of "computational fields", an abstraction that promotes uncoupled and adaptive interactions, as well as provide agents with simple, yet expressive, contextual information. Consequently, field-based coordination of personal assistants in a ubiquitous network environment may be a promising approach for agent-based provisioning of group-oriented context-aware services.

The rest of the paper is organized as follows: Section 2 deals with field-based coordination and elaborates the properties of the used field. The applied coordination policies that are based on different types of adaptive learning algorithms are explained in Section 3. Section 4 gives insight into an evolutionary game theoretic analysis of coordination policies. A case study describing performance evaluation of coordination policies according to the analysis is given in Section 5, while Section 6 concludes the paper.

2 Field-Based Coordination Mechanism

Computational fields propagated in a ubiquitous network environment can give mobile intelligent agents some clue about the context they are interested in. By following this idea, we have experimented with a group-oriented context-aware service which enables that team of personal assistants meet together in a location that is as close as possible to a context-dependent location of interest. Initially, the personal assistants had been arbitrarily dispersed in the network at different locations. In order to become a team and act in a coordinated manner, they followed the field intensity generated by a context manager agent by iteratively exploring new locations in the network. The context manager was aware of personal assistants' movements. On the other hand, each personal assistant was only able to perceive the intensity of computational field on a visited location. The field intensity actually represented the reward (i.e. payoff) the agent received by visiting the location.

The configuration of field intensity encoded in distributed way context-dependent location information. In particular, it reflected two context-dependent locations interesting for the users in a group. Depending on agents' intentions (that were captured by median of agents' initial actions in a team), the field generated by the context manager at appropriate time intervals (iterations) influenced movements of personal assistants in the team towards a meeting location that is as close as possible to the context-dependent location which attracted the majority of them.

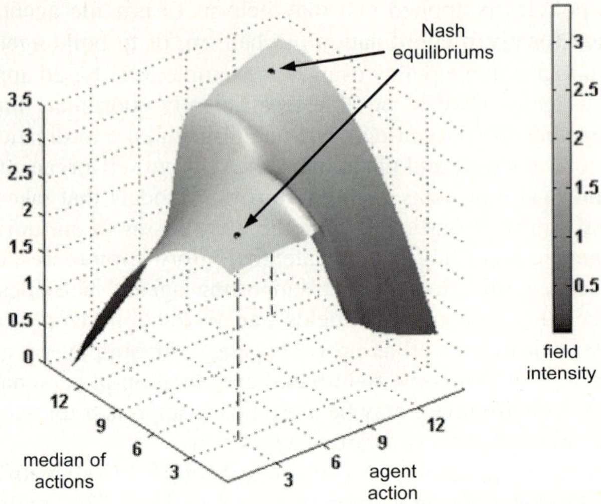

Fig. 1. Configuration of computational field for the continental divide game

We have been conducting simulations with mobile intelligent agents that are autonomous and self-interested. Therefore, the field intensity configuration was specified bearing in mind game theoretic point of view. In particular, intensity of the field, shown in Fig. 1, was specified according to the payoffs in the continental divide (CD) game. This game, studied by Van Huyck et al. [2], is a coordination game [3] with two pure Nash equilibriums which in our case represent the most desirable interaction outcomes among the agents in a team. An agent payoff in the game depends on both its own action and the median of actions of all other agents in a team. Depending on the median of initial actions, the game payoffs influence self-interested learning agents to meet together at a location that is as close as possible to a context-dependent location of interest (reflected by one of the two pure Nash equilibriums in Fig. 1). In our simulations we wanted from the personal assistants to learn to meet together as quickly as possible in a location that is as close as possible to the sources of context-dependent information. Although agents always met together (by using available coordination policies) the question we had to answer was which policy performed the best given initial agents' intentions in a situation interesting for their users.

3 Agent Coordination Policies

A coordination policy dictates reaction of intelligent mobile agent to intensity of computational field in the ubiquitous network environment. In particular, according to field intensity and applied coordination policy, the agent chooses an action that determines its next target location. In our simulation settings, the selection of agent action was guided by an adaptive learning algorithm that makes use of past experience and can be characterized by some initial values and free parameters, as well as by two types of rules: a decision rule which describes how actions are taken given the available information, and an updating rule that can be in terms of beliefs, propensities or, in general, attractions assigned to each of the actions. Taking into consideration a way attractions are updated, adaptive learning algorithms can be divided in three groups: reinforcement algorithms, belief-based algorithms, and algorithms that combine reinforcement and belief-based learning [4].

The first coordination policy, denoted by RE, used three parameter Roth-Erev myopic reinforcement learning algorithm [5] which forms the basis of a model of human behaviour in competitive games. The applied Roth-Erev algorithm is characterized by three free parameters: the strength of initial propensities s_0 that influences the rate of change of action probabilities (i.e. the speed of learning), the experimentation parameter ε, and the forgetting (recency) parameter χ. The parameters ε and χ facilitate responsiveness of reinforcement learning to a changing environment (i.e. behaviours of other agents). The former takes into account not only actions that were successful in the past, but also more often reinforce actions similar to (i.e. near) successful one, while the later asserts that recent experience plays a larger role than past experience. Moreover, the Roth-Erev algorithm obeys power law of practice: learning curves tend to be steep, and then flatten out. In other words, as the "weight of history" becomes greater, it becomes harder to change an action that has been performing well. The power law of practice is incorporated in the learning algorithm through the use of cumulative propensities assigned to available actions.

The second coordination policy, denoted by Q, was based on the stateless Q-learning algorithm with Boltzmann exploration [6] that represents a form of temporal difference reinforcement learning in which agents learn an evaluation function over its actions. The evaluation function determines the maximum expected reward (discounted and cumulative) the agent can obtain by applying an action. The algorithm can be successfully employed even when the learner has no prior knowledge of how its actions affect its environment. It is characterized by the learning rate α, the discount rate γ, and the temperature parameter T which determines an exploration/exploitation rate and usually obeys an annealing (or cooling) scheme.

The third coordination policy, denoted by EWA, was based on the experience-weighted attraction learning algorithm [7] that represents a hybrid of reinforcement and belief-based learning. It is characterized by the imagination parameter δ, the change parameter ϕ, the exploration/exploitation parameter κ, and the initial experience weight N_0 that can be interpreted as the strength of initial attractions, relative to incremental changes in attractions due to experience and payoffs. A key feature of the experience-weighted attraction learning is that attractions are not only updated when an action is taken, but the model weights hypothetical payoffs that forgone (i.e. non-chosen) actions would have earned by the imagination parameter δ. Therefore, the parameter δ can be interpreted as a kind of responsiveness to forgone payoffs. A higher δ means players move more strongly, in a statistical sense, towards "ex post best responses". The change parameter ϕ denotes decay rate which reflects a combination of forgetting and "motion detection", i.e. the degree to which players realize that other players are adapting, so that old observations are obsolete and should be ignored. The lower ϕ means that agents decay old observations more quickly and are responsive to the most recent observations. The exploration/exploitation parameter κ determines the growth rate of attractions, which reflects how quickly players lock in to an action. Consequently, the parameter κ roughly captures the distinction between "exploration" (low κ) and "exploitation" (high κ) what is known by locking in to a good action.

4 Evolutionary Selection Process of Coordination Policies

In order to compare learning performances of applied coordination policies we were performing analysis in which the personal assistants had not been completely committed to just one way of behaving. Rather, coordination policies had been available to them simultaneously. The selection of coordination policies was depending on social learning (e.g. imitation) of the agent. The evolutionary change that "population of coordination policies" in the agent undergoes may be analogous to biological evolution. In particular, we have borne in mind the analogy between social learning and biological evolution governed by replicator dynamics [8]. Although different in discrete time, it is shown that social learning and biological evolution in normal-form games exhibit identical, or related behaviour, once a continuous time limit is satisfied.

The replicator dynamics (RD) considers players playing the same one-shot normal form game repeatedly in discrete time. Since the agents played the continental divide game that consists of several iterations, one-shot normal form game that is equivalent to the CD game had to be obtained. Therefore, we have adopted the methodology proposed by Walsh et al. [9], which transforms the game and its available actions (i.e. game theoretic pure strategies) into a one-shot game with a limited number of coordination policies. In the obtained one-shot normal form game coordination policies were treated as pure strategies. The payoff of a coordination policy was calculated as total payoff at the end of the CD game and denoted the speed of convergence (efficiency) towards a location where the agents met together.

It was supposed that all agents had the same set of coordination policies to play, as well as that receive the same payoffs. In other words, a one-shot normal form game, obtained by the transformation of the CD game, was modelled as a symmetric game in which the size of payoff table, that specifies the expected payoff to each agent in the game when playing a mixed strategy (i.e. probability distribution of coordination policies), was significantly reduced due to the fact that in the symmetric game the payoff table is built by considering the number of agents playing each policy (rather than considering which policy each of the agents is playing).

In the evolutionary game theoretic approach based on replicator dynamics, at each point in time, all agents in the symmetric game are characterized by the same probability distribution over a set of available coordination policies, which is represented by a vector $x(t) = (x_1(t), \dots , x_i(t), \dots , x_m(t))$, where $x_i(t)$ is a probability that an agent chooses coordination policy e^i. Since all probabilities x_i are non-negative and sum up to one, the vector x belongs to the unit simplex in m-dimensional Euclidian space. Moreover, as we dealt with a sufficiently large population of agents from which a smaller group of agents was randomly selected at each time step to play the game, the probability distribution x was treated as the game theoretic mixed strategy, i.e. as a continuous variable. The replicator dynamics was used to model the evolution of x with time as follows

$$\frac{dx_i}{dt} = \left(u(e^i,t) - \langle u(t) \rangle\right) x_i, \tag{1}$$

where $u(e^i, t)$ is the average payoff to coordination policy e^i when all agents play mixed strategy x, and $\langle u(t) \rangle$ is the mean population payoff when all agents play x. For each game and each policy, the individual payoffs of agents using policy e^i (obtained from the payoff table) were averaged. As it can be deduced, the replicator dynamics models an evolutionary process in which agents select a coordination policy that appears to be more successful, with a probability proportional to the expected payoff.

The RD can show the mixed strategy trajectories and how they converge to an equilibrium in its phase space, although they do not necessarily settle at a rest point. An equilibrium to which trajectories converge and settle is known as an attractor, otherwise the equilibrium is unstable and represents a saddle point. The region within all trajectories converge to a particular equilibrium is known as the basin of attraction. It is used for measuring a probability of convergence towards an attractor. The attractors, at which a larger range of initial mixed strategies will end up, are equilibriums that are more likely to be reached (assuming a uniform initial distribution of mixed strategies). The replicator dynamics has the property that a Nash equilibrium of the game represents its stationary point. Moreover, when trajectories converge to a Nash equilibrium of the game, the equilibrium is asymptotically stable (i.e. being robust to local perturbations) and represents an attractor, which as a result of evolutionary force, influence agent's behaviour in an environment.

5 Case Study: Performance Evaluation of Coordination Policies

Evolutionary game theoretic performance evaluation of coordination policies, when mobile intelligent agents are coordinated in the ubiquitous network environment according to the computational field defined by the payoff matrix of a CD game, is chosen as a case study, taken from the broader research of agents and their application in a new generation networks [10, 11]. In order to calculate coordination policies' payoffs, each entry in the payoff table of the transformed CD game was computed by averaging the payoff of each coordination policy across 2000 simulations of continental divide game. At the beginning of a simulation, initial action probabilities in all coordination policies were set equal. Each CD game consisted of 40 iterations. This number of iterations ensured that the agents always met together during the game. In each iteration, the agents chose among 14 actions (marked by integers from 1 to 14) that represented available locations to the agents in the ubiquitous network environment.

The experimentation parameter ε, the forgetting parameter χ, and the strength of initial propensities s_0 in the RE policy were set to 0.1, 0.5, and 9, respectively. In the Q policy, the learning rate α and the discount rate γ were set to 0.5 and 0.9, respectively. The temperature parameter T was set to obey the geometric decrease cooling scheme $T_{j+1} = \beta T_j$ at iteration j+1 of the game, where the cooling rate β and the initial temperature T_0 were set to 0.9 and 10, respectively. In the EWA policy, the learning algorithm used the logit decision rule with the sensitivity to attractions λ set to 1.5. The initial attractions were set to 0, while the parameters N_0, δ, ϕ, and κ were set to 1, 0.2, 0.2, and 0.8, respectively.

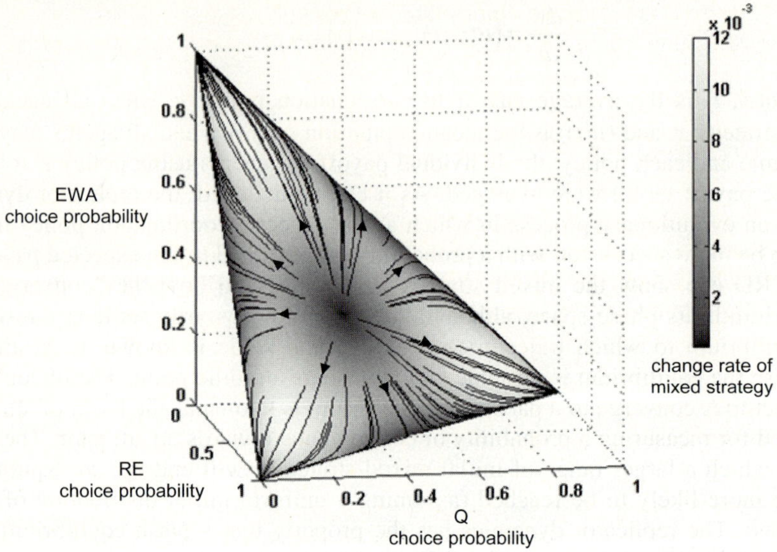

Fig. 2. Evolutionary selection of coordination policies for 7 mobile intelligent agents in a team

The performance evaluation of coordination policies, that captured the influence of evolutionary force to personal assistant's behaviour, was analyzed by the use of RD. A team of seven agents had been considered. The team size was application-specific requirement. Each agent was randomly assigned a mixed strategy $x(x_1, x_2, x_3)$ where x_1, x_2, and x_3 denote the choice probabilities for RE, Q, and EWA policies, respectively. An initial mixed strategy was progressively adjusted as a result of the dynamics described by Eq. 1. In Fig. 2, which shows phase portrait (plane) of the replicator dynamics, the mixed strategies are indicated by points of the simplex. The simplex contains the trajectories generated from 2500 randomly sampled initial mixed strategies. The direction of trajectories indicates that the vertices of simplex act as attractors, i.e. as asymptotically stable symmetric Nash equilibriums [8]. It was calculated that the EWA policy performed slightly better (i.e. earned approx. 1 % higher payoff) than the Q and RE policies which had almost the same performance. The shading, which is proportional to $|dx/dt|$, denotes the rate of change of the mixed strategy. The number of trajectories leading to a vertex indicates its basin of attraction. Fig. 2 shows that the EWA policy has the largest basin of attraction (with influence of approx. 36 %), while the Q and RE policies have almost the same basins of attraction (each with influence of approx. 32 %). Consequently, the EWA policy is slightly better choice for all agents in the team compared to the Q and RE policies.

6 Conclusion

The results of the evolutionary game theoretic analysis have shown that all applied coordination policies ensured the mobile intelligent agents in a team similar learning performances in finding a meeting location of interest. In order to further make insight into agent-based provisioning of group-oriented context-aware services, future work will include an information theoretic analysis of field-based coordination of self-interested mobile intelligent agents.

Acknowledgements

This work was carried out within the research projects "New Architectures and Protocols in Converged Telecommunication Networks" and "Content Delivery and Mobility of Users and Services in New Generation Networks", supported by the Ministry of Science, Education and Sports of the Republic of Croatia.

References

1. Mamei, M., Zambonelli, F.: Field-Based Coordination for Pervasive Multiagent Systems. Springer, Berlin (2006)
2. Van Huyck, J.B., Cook, J.P., Battalio, R.C.: Adaptive Behavior and Coordination Failure. Journal of Economic Behavior and Organization 32, 483–503 (1997)
3. Cooper, R.W.: Coordination Games. Cambridge University Press, Cambridge (1999)
4. Salmon, T.C.: An Evaluation of Econometric Models of Adaptive Learning. Econometrica 6, 1597–1628 (2001)
5. Erev, I., Roth, A.E.: Predicting How People Play Games: Reinforcement Learning in Experimental Games with Unique, Mixed Strategy Equilibria. American Economic Review 4, 848–881 (1998)
6. Sutton, R.S., Barto, A.G.: Reinforcement Learning: An Introduction. MIT Press, Cambridge (1998)
7. Camerer, C., Ho, T.-H.: Experience-Weighted Attraction Learning in Normal Form Games. Econometrica 4, 827–874 (1999)
8. Weibull, J.W.: Evolutionary Game Theory. MIT Press, Cambridge (1997)
9. Walsh, W.E., Das, R., Tesauro, G., Kephart, J.O.: Analyzing Complex Strategic Interactions in Multi-Agent Systems. In: Proceeding of the AAAI 2002 Workshop on Game Theoretic and Decision Theoretic Agents, Edmonton, Canada, pp. 109–118 (2002)
10. Trzec, K., Lovrek, I., Mikac, B.: Agent Behaviour in Double Auction Electronic Market for Communication Resources. In: Negoita, M.G., Howlett, R.J., Jain, L.C. (eds.) KES 2006. LNCS (LNAI), vol. 4251, pp. 318–325. Springer, Heidelberg (2006)
11. Lovrek, I., Sinkovic, V.: Mobility Management for Personal Agents in the All-Mobile Network. In: Negoita, M.G., Howlett, R.J., Jain, L.C. (eds.) KES 2004. LNCS (LNAI), vol. 3213, pp. 1143–1149. Springer, Heidelberg (2004)

Hybrid Filtering Methods Applied in Web-Based Movie Recommendation System

Ngoc Thanh Nguyen[1], Maciej Rakowski[2], Michal Rusin[1], Janusz Sobecki[2], and Lakhmi C. Jain[3]

[1] Institute of Information Science and Engineering, Wroclaw Univ. of Technology, Poland
[2] Institute of Applied Informatics, Wroclaw Univ. of Technology, Poland
[3] School of Electrical and Information Engineering, Univ. of South Australia, Australia
{thanh,maciej.rakowski,michal.rusin,sobecki}@pwr.wroc.pl
Lakhmi.Jain@uniisa.edu.au

Abstract. In this paper web-based movie recommendation system using hybrid filtering methods is presented. The recommender systems deliver one of the methods for increasing the web-based systems attractiveness and usability. We can distinguish three basic filtering methods that are applied in recommender systems: demographic, content-based, and collaborative. The combination of these approaches that is called hybrid method.

Keywords: Hybrid filtering, web-based systems, movie recommendation.

1 Introduction

The success of the today's web-based information systems relies on the delivery of customized information for their users. The systems with this functionality are often called recommender systems [6].

One of the most popular applications of recommendation systems is movie or video recommendation. We can find on the Web numerous commercial and noncommercial such as: Hollywood Video, jimmys.tv, MovieLens, Everyone's a Critic and Film Affinity. There are also quite many research projects that applies different collaborative and hybrid recommendation methods in this area, such as: MovieMagician [4], application of Artificial Immune System [1] and application of neural networks [2].

In this paper we will present a hybrid method that applies both fuzzy reasoning and consensus methods for recommendation of movies. This method differs from many other movie recommendation systems because applies not only collaborative and content-based reasoning but also demographic recommendation.

This paper is organized as follows: in the next paragraph the general architecture of the hybrid recommendation system is presented, and user model and movie models are presented. In the fourth paragraph we will present the utilization of the user model in hybrid recommendation concerning demographic stereotype reasoning, collaborative filtering using consensus methods and content based recommendation.

B. Apolloni et al. (Eds.): KES 2007/WIRN 2007, Part I, LNAI 4692, pp. 206–213, 2007.

2 General Architecture of the Hybrid Recommendation of Movies

We can distinguish three basic types of filtering: demographic (DF), content-based (CBF) and collaborative (CF). DF is using stereotype reasoning [5] and is based on the information stored in the user profile that contains different demographic features [6]. According to [5], stereotype reasoning is a classification problem that is aimed at generating initial predictions about the user that is mainly based on demographic data and is mainly used in the initial steps of the collaborative user interface recommendations [7]. DF has however two basic disadvantages [6]: for many users may be too general; does not provide any adaptation to user interests changing over time.

The content-based filtering takes descriptions of the content of the previously evaluated items to learn the relationship between a single user and the description of the new items [6]. It is believed that application of content-based filtering enables personalized and effective recommendations for particular users, however has some disadvantages: content-based approaches depends on so called objective description of the recommended items; it tends to overspecialize its recommendations; content-based approach is based only on the particular user relevance evaluations, but users usually are very reluctant to give them explicit, so usually other implicit, possibly less adequate, methods must be used.

Finally the CF that is most often used in many movie recommendation applications, is able to deliver recommendations based on the relevance feedback from other similar users. Its main advantages over the content-based architecture are the following [6]: we can obtain subjective data about items; CF is able to offer novel items; CF utilizes item ratings of other users to find the best fitting one. CF has however two main disadvantages [10]: early-rater problem occurs when a user is one of the first from his or her neighborhood to enter the rating for the item and sparsity problem that is caused by only few ratings for even popular items. In [6] also other collaborative filtering disadvantages have been encountered: lack of transparency in the process of prediction and finally the user's personal dislike may be overcome by the number of other similar users' opinions.

The disadvantages of CBF and CF could be overcome by applying the hybrid approach (HA). In the works [7, 11], where the concept and implementation of the collaborative user interface adaptation using consensus method is presented, the disadvantage of the insufficient number of the similar users at the early stages of the system operation was overcome by application of the DF. However in this architecture the CBF is not implemented, but individual preferences of the interface settings that are selected manually by the user and stored in the user profile and used in every system session.

The Movie is a system designed for movie recommendation used in video and DVD rental shop. Movies are recommended for the users and may be ranked by them after watching on returning to the rental shop. In *The Movie* two system elements, i.e. movies and user interface layout , are recommended by means of HA.

At the beginning of working with the system each new user is obliged to fill out the registration form and enter some demographic and movie preference data together with rating several movies. All these data are stored in the user profile, however not straightforward but after application of some fuzzy logic interference rules [3].

Then according to the combination of DF and CF that is described in [10] appropriate user interface layout id recommended. The recommendation concerns the following user interface elements: background image, type of buttons, soundtrack and its loudness, background transparency, text highlight color and frame color. However the presentation recommendation may be considered as less important as the content recommendation it is very important in overall system usability performances.

The movie recommendation is based on DF, CBF and CF. DF uses fuzzy inference rules to initialize the user profile values of movie preferences of favorite genres. The values of favorite genres and features of interest are further modified by movie ranking that is made by each user during the whole process of using the system. These assessments are also used to modify the movie profile interest values. Then we find several most similar movies to the user interest that are exhibited by the user profile. Considering that "The Movies" is applied in the video and DVD rental shop we may not recommended the movies that the user has watched before. This type of recommendation is based on the content-based approach. However as in many other movie recommendation systems we also apply the collaborative approach by finding the similar users and recommend the movies that they ranked as being interesting. To select the recommended movies we first select the group of similar users and then using consensus [7] methods we select the most highly ranked movies. Then these two recommendation methods are combined by selecting several most highly ranked movies rejecting already watched movies.

3 User and Movie Models

User model is the central element of all recommendation systems and contains the knowledge about the individual preferences which determine his or her behavior in the system [8]. The user model contains the composition of beliefs concerning users demographic data, their preferences, knowledge and attributes from the particular domain [5]. We should also consider the area of application of the recommended items: web-pages, documents, links, movies, products etc. In defining the recommender system architecture we need to define the user model, the item model (in our case the movies).

In *The Movie*, movies are the recommended items. Their content may be modified during the system operation so their representation and initialization is very important in the process of recommendation. In this section user and movie models representation and initialization will be described.

3.1 User Model Representation

In the user model we can distinguish user data and usage data. The user data contains five elements: the demographic data, preferred movie genres, preferred movie features, ranked movies and interface profile settings. The user model is represented as a tuple that is a function $p: A \rightarrow V$ where A is a set of attributes and V is a set of their values and $(\forall a \in A)(p(a) \in V_a)$, where V_a is a set of values of attribute a.

The demographic data contains the following attributes: education, age, gender, children and place of living. Values of all these attributes are elementary. The set of the values of the attributes are following:

- $V_{education}$={elementary school, secondary school, vocational school, vocational secondary school, technical secondary school, post-secondary school, university, technical university},
- V_{age}= INT (the set of integers)
- V_{genre} = {male, female},
- $V_{children}$ = {yes, no},
- $V_{place\_of\_living}$ = {village, small_town, city}.

The preferred movie genres contain the following attributes: drama, action, science-fiction, comedy, horror, documentary, sport, war. All these attributes have the equal set of values {0, 1} that represent how much the user is interested in the specific genre (0 – not at all, 1- very much).

The preferred movie features contain the following attributes: plot, special effects, realism, soundtrack, actors play and action. All these attributes have the equal set of values {0,1} that represent how much the user is interested in the specific movie feature (0 – not at all, 1- very much).

The usage data contains the ranked movies that are represented by the list of tuples containing movie identifier together with the rank given by the user. The user ranks each movie feature giving one of the following ranks: worthless, very poor, poor, doesn't mater, good, very good, outstanding, which are represented by integer values from -3 to 3 accordingly.

The usage data contains also information on the user interface settings that are represented by the following attributes: background image, type of buttons, soundtrack and its loudness, background transparency, text color and frame color. Their values are some identifiers that represent actual settings implemented in "The Movie".

3.2 Movie Model Representation

The movie model contains all the data concerning each particular movie and is also represented by a tuple. In our system the movie description was automatically uploaded from the Internet Movie Database (IMDb). The movie model contains most of the fields that are present in the movie description in IMDb, such as: id, titles, directors, year, credits, music, actors, etc., together with two elements: movie genres and movie features.

The movie genres are directly downloaded from the IMDb and represented as a value of the attribute genre, where $V_{movie\_genre} = 2^{\text{Movie\_genres}}$, where Movie_genres = {drama, action, science-fiction, comedy, horror, documentary, sport, war}. The movie features are represented in the same way as this element in the user model and contain the following attributes: plot, special effects, realism, soundtrack, actors play and action.

3.3 User and Movie Models Initialization and Modification

The process of the initialization is very important in the user modeling. The model may be empty in the beginning, entered manually or automatically. In *The Movie* different elements of the user and movie models are initialized and modified in different ways.

The demographic data of the user model are initialized manually by each user. The user model movie genres are first initialized automatically by means of the set of fuzzy inference rules [3], then their values are modified first in the initial process of ranking of selected movies and then during the process of ranking the rented movies. The user model movie features are initially entered by the user and later, they are modified while user is ranking movies.

The general intuition of the movie genres and features part of the user model modification is as follows: the higher rank of the movie increases the value of the user interest of the movie genre or its feature and the lower rank decreases these values. Of course these changes are quite small and need at least several specific rankings to change the user profile attribute values significantly and they have lower and upper bounds, 0 and 1 accordingly. The precise equation that modifies movie genres features are presented in the work [9]. We applied square function to ensure non-linear changes of feature's interest (rates close to neutral are less more important than extreme ones) and learning coefficient that was experimentally determined as a value from the interval [0.005, 0.015] for genres and [0.001, 0.005] for features.

4 User and Movie Profiles Exploitation

The user model exploitation in the recommendation systems defines precisely the filtering methods used in the system. *The Movie* uses HA, where DF is used on the initialization stage of the movie genres and features of the user profile and was described above. CF and CBF are used to recommend movies.

4.1 Collaborative Filtering (CF)

The basic intuition behind this method is that people are interested in movies that other similar people liked to see. To determine the collaborative recommendation first we must define how we determine similar system users and then how we select the movies these users liked the most. In the literature to find similar user quite often the Pearson coefficient is used [2]:

$$r(x,y)=\frac{\sum_{m\in movie}(rate_{x,m}-\overline{rate_x})(rate_{y,m}-\overline{rate_y})}{\sqrt{\sum_{m\in movie}(rate_{x,m}-\overline{rate_x})^2\ \sum_{m\in movie}((rate_{y,m}-\overline{rate_y})^2}}$$

where $rate_{x,m}$ is the mean of ranks (for all six features) given by the user x for them movie m, and $\overline{rate_x}$ is the mean value of $rate_{x,m}$ for all movies evaluated by user x. The value of the correlation $r(x,y)$ is a real number that lies within [-1, 1]. By defining the given threshold τ, we can determine the most similar users, then using consensus methods we can find the movies they like the most. We used the algorithm for determining of consensus for ratings presented in the work [7]. Given m binary matrixes $\mathbf{A}^{(k)}$ (for k=1,...,m), of dimension $n\times n$ representing rankings of n values, where m represents the number of similar users and n is the number of rated movies and $a(i,j)$=1 if rate of the movie i is greater than j and else equals 0. The result is matrix $\mathbf{C}$ of dimension $1\times n$ representing consensus.

```
Procedure
Var
  B: array[1..n, 1..n] of real;
  E: array[1..n] of {0,1};
BEGIN
    For i:=1 to n do for j:=1 to n do B[i,j]:=0;
    For l:=1 to m do
      For i:=1 to n do
      For j:=1 to n do
        B[i,j]:= B[i,j]+A^(l)[i,j];
      For  i:=1 to n do
      For j:=1 to n do
      Begin
        If B[i,j] < (m/2) then B[i,j]:=0;
        If B[i,j] ≥ (m/2) then B[i,j]:=1;
      End;
   For i:=1 to n do E[i]:=0;
   For i:=1 to n do
      Begin
        k:=0;
        For j:=1 to n do k:=k+B[i,j];
        k:=n–k +1;
        If E[k]=0 then E[k]:=i
        Else
         Begin
            l:=k+1;
            While E[l]>0 do l:=l+1;
             E[l]:=i;
         End;
      End;
    C:=E
END.
```

As a result we receive the list of movies, together with their ranks, which received the highest ranks among the similar users.

4.2 Content-Base Filtering (CBF) and Hybrid Approach (HA)

CF the most popular among most of movie recommendation systems, has some disadvantages presented above, i.e. early-rater and sparsity problems. To overcome these problems we also applied CBF in *The Movie* system. Here CBF is based on the user interests represented in the user profile by movie favorite genres and features attributes. CBF has many in common with the domain of Information Retrieval. The process of determining the movies user may like the most is made with the following formula:

$$\underset{mv \in Movies}{\forall} mv_{rating} = 1{,}5 * \frac{1}{n}\sum_{i=1}^{n} umg_i + \frac{1}{6}\sum_{k=1}^{6} uf_k * mf_k ,$$

where $umg_i \in user_{genres} \cap movie_{genres}$, n is the number of elements in $user_{genres} \cap movie_{genres}$, uf_h, mf_k is the value of user's interest in feature k and the rate of k movie feature. The system records twenty movies with the best ratings.

Having two lists of recommended movies that were determined by CF and CBF we must now combine them. There may be many methods for doing this: presenting one from CF and the second from CBF, and so on; presenting the all the movies with the best ratings, however sometimes it is difficult to find the best consensus among these two filtering methods. In "The Movie" we applied the first method.

5 Implementation, Experimental Results and Future Works

The Movies application has been implemented with Flash Macromedia for the user interface layer, MySQL for the database layer and PHP for business logic layer. As a source for movie descriptions we used IMDB, so it was necessary to implement special application to download several thousands complete descriptions of the movies, with list of actors, user comments, and additional information like runtime, color or certificates which movie gets. All this data has been saved to the system and served as base for verification of implemented HA.

The user after registration that consists of delivering both demographic data and favorite movie genres and features is asked to rank several movies. Then the system recommends the user interface for the user that may be further modified. The main goal of the system is movie recommendation. We selected that each time we recommend 20 movies according to CBF and CF that define the HA. So far, we managed to make experiments only with several users, however we conducted standard usability studies with five typical users, giving them five typical tasks. The results have shown, verified by the questionnaires that they appreciate both recommended movies and user interface.

In this paper we presented application of hybrid methods for movie and user interface recommendation in *The Movie* system. The hybrid recommendation methods are usually combination of several other methods so it is very difficult to describe them in details in the conference paper. Unfortunately here we were able to give only some general descriptions of the methods used.

However the system usability was tested, we have not determined the efficiency of the movie recommendation, especially in comparison with other methods. We have tested some parts of the system, i.e. DF, CF and CBF using MovieLens dataset [12]. The results were promising, however not significantly better than those described in the literature. Unfortunately we were not able to test the full version because the lack of the complete data we require to run the system.

References

1. Chen, Q., Aickelin, U.: Movie Recommendation Systems using an artificial immune system. Poster Proceedings of ACDM, Bristol, UK. Engineers House (2004)
2. Christakou, C., Stafylopatis, A.: A Hybrid Movie Recommender System Based on Neural Networks. In: Proc. Fifth Int. Conf. on Intelligent Systems Design and Applications, pp. 500–505 (2005)

3. Elkan, C.: The Paradoxical Success of Fuzzy Logic. IEEE Expert, 3–8, 9–46 (August 1994) (First version in AAAI'93 proceedings, pp. 698-703) (1994)
4. Grant, S., McCalla, G.: A hybrid approach to making recommendations and its application to the movie domain. In: Proc. 2001 Canadian AI Conference, pp. 257–266 (2001)
5. Kobsa, A., Koenemann, J., Pohl, W.: Personalized Hypermedia Presentation Techniques for Improving Online Customer Relationships. Knowledge Eng. Rev. 16(2), 111–155 (2001)
6. Montaner, M., Lopez, B., de la Rosa, J.P.: A Taxonomy of Recommender Agents on the Internet. Artificial Intelligence Review 19, 285–330 (2003)
7. Nguyen, N.T., Sobecki, J.: Using Consensus Methods to Construct Adaptive Interfaces in Multimodal Web-based Systems. Universal Access in Inf. Society 2(4), 342–358 (2003)
8. Papatheodorou, C.: Machine Learning in User Modeling. Machine Learning and Its Applications, 286–294 (2001)
9. Rakowski, M., Rusin, M., Sobecki, J.: Hybrid recommendation applied in web-based movie information system. In: Multimedia and network information systems. Proceedings, Wrocław, September 21-22, / ed. by A. Zgrzywa. Wrocław: Oficyna Wydaw. PWroc, 2006. s. 361–369 (2006)
10. Sarwar, B., Konstan, J., Borchers, A., Herlocker, J., Miller, B., Riedl, J.: Using Filtering Agents to Improve Prediction Quality in the GroupLens Research Collaborative Filtering System. In: CSCW'98, Seattle Washington USA, pp. 1–10 (1998)
11. Sobecki, J., Weihberg, M.: Consensus-based Adaptive User Interface Implementation in the Product Promotion. To be published as a chapter in book "Design for a more inclusive world, London. Springer, Heidelberg (2004)
12. Data sets, downloaded in (April 2007), http://www.grouplens.org/taxonomy/term/14

Network Simulation in a Fragmented Mobile Agent Network

Mario Kusek, Gordan Jezic, Kresimir Jurasovic, and Vjekoslav Sinkovic

University of Zagreb
Faculty of Electrical Engineering and Computing
Unska 3, HR-10000, Zagreb, Croatia
{mario.kusek,gordan.jezic,kresimir.jurasovic,vjekoslav.sinkovic}@fer.hr

Abstract. This paper deals with the simulation of a multi–agent system based on the Fragmented Mobile Agent Network model. The model consists of agent teams performing remote software management operations and network elements that connect processing nodes and allow agent mobility. A case study considering a scenario in which multi-operation teamwork agents install new software in a network is included. An analysis of simulation results based on operation execution in a simulated large-scale network with different fragment sizes, network sizes and node/link capabilities is elaborated.

Keywords: Mobile agent network, coordination, large–scale network, simulation.

1 Introduction

The New Generation Network (NGN) consists of different types of networks, nodes and terminals aimed at providing the appropriate environment for advanced services emerging from the convergence of Internet information services and traditional telecom services (e.g. telephony). In the NGN concept, where many nodes are distributed over heterogeneous networks, enabling service flexibility and software portability is crucial. Thus, the need for remote software management (software operations and maintenance) is increasing. We have developed a system, called the Multi–Agent Remote Maintenance Shell (MA–RMS), which may be considered an environment for supporting remote software maintenance operations at various nodes. MA-RMS will be explained in detail in section 4.

In this paper we extend a model of a Mobile-Agent Network (MAN), presented in [1] to include network elements. Multi–agent system (MAS) in the MAN is divided into fragments (F–MAN) and consists of agent teams performing remote software management operations. The paper focuses on the effects of introducing a network infrastructure and the impact of fragmented teams aimed at improving execution time. The paper is organized as follows: Section 2 deals with team organization in a fragmented mobile agent network, particularly with network description. Network simulation is elaborated upon in Section 3. A case study describing software installation in a large-scale network, along with a simulation analysis, is given in Section 4, and Section 5 concludes the paper.

B. Apolloni et al. (Eds.): KES 2007/WIRN 2007, Part I, LNAI 4692, pp. 214–221, 2007.

1.1 Related Work

Multi-agent simulation has been a topic of several research projects. The Multi-Agent System Simulator (MASS) is the result of one such project. It focuses on validating coordination's and adaptive qualities in an unpredictable environment [2]. This simulator does not consider environments with mobile agents that use computer networks. Paper [3] concentrates on how to simulate agents in a distributed environment and they are using network only to run a distributed simulation. Implementing a computer network in this environment would be too complicated. Event based simulation with a completely connected network has been developed by the authors in the paper [4]. Unfortunately this simulator does not conform to our MAN model since the duration of one operation cannot be modeled using Distilled StateCharts based approach. Since we did not find any simulator capable of simulating our model of MAN we have decided to develop our own simulator that meets our requirements.

2 Model of Fragmented Mobile Agent Network

The Fragmented Mobile Agent Network (F-MAN) is used for modeling network organization and agent coordination in an agent team. The F-MAN is represented by a triple $\{SA, FS, N\}$, where SA is a multi–agent system organized as a team and divided into subteams $SA = \{SA_1, SA_2, \ldots, SA_j, \ldots, SA_n\}$, FS represents a set of network fragments $FS = \{f_1, f_2, \ldots, f_j, \ldots, f_n\}$ on which subteams perform services, and N is a network that connects processing nodes and allows agent mobility. Each subteam SA_j performs operations on a specified fragment, $f_j \Rightarrow SA_j$. Each fragment f_j includes a set of processing nodes $f_j = \{S_{j1}, S_{j2}, \ldots, S_{jq}\}$ (Fig. 1).

Each processing node S_i has a unique $address_i$ from the set of addresses, $address = \{address_1, address_2, \ldots, address_i, \ldots, address_m\}$. An agent is defined by a triple, $agent_k = \{name_k, address_k, task_k\}$, where $name_k$ defines the agent's unique identification, $address_k \in address$ represents the list of nodes to be visited by the agent and $task_k$ denotes the functionality it provides in the form of $task_k = \{s_1, s_2, \ldots, s_i, \ldots, s_p\}$ representing a set of assigned elementary operations s_i. When hosted by node $Si \in address_k$, $agent_k$ performs operation $s_i \in task_k$. If an operation requires specific data, the agent carries this data during migration [5].

A network N is represented by an undirected graph, $N = (S, E)$ which denotes network connections and assures agent mobility. The set of processing nodes is denoted as $S = \{S_1, S_2, \ldots, S_i, \ldots, S_m\}$. E represents the set of edges e_{ij} between S_i and S_j implying that nodes S_i and S_j are connected. The communication time c_{ij} between tasks v_i and v_j is associated with edge (link) e_{ij} which connects these nodes. This way, a delay is incorporated into the communication channel. The following three types of network elements, with corresponding capacities, are defined: processing nodes, switches, and links.

A subteam includes one subteam coordination agent (SCA) and subteam members which share a subteam coordination plan (SCP). An initial request

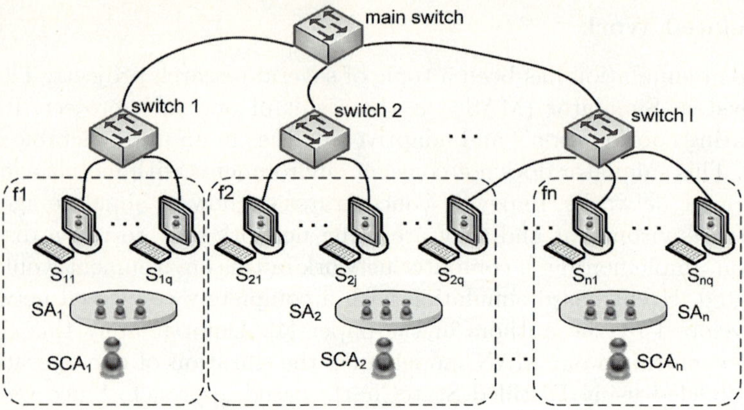

Fig. 1. A Fragmented Mobile Agent Network

is first submitted to a management agent (MA). The MA then defines fragments and divides the initial request into fragment requests which are sent to the SCAs of each fragment, i.e. agents in $CA = \{SCA_1, SCA_2, \ldots, SCA_j, \ldots, SCA_n\}$. The number of SCAs is equal to the number of the fragments which in turn depends on the network size. After receiving a request, each SCA creates a subteam and a shared plan, distributes operations among them and starts their executions. An SCA must be able to, on the basis of a user request and operations, create a shared SCP, form a subteam of member agents, and send them to perform the corresponding operations.

An SCP defines the complexity of subteam agents (i.e. the number of elementary operations per agent) and the size of a subteam. We chose the scenario with an SCP that organizes agents within subteams according to the results from [1]. The selected SCP allows a subteam agent to execute multiple operations on only one node in the fragment.

3 Network Simulation

The last part of the triple, which represents the F–MAN, is the network element N. This element represents the physical network that agents use while migrating to the location of their target nodes. Since the network introduces delays caused by limited network capacity and processing times at network elements (e.g. switches), it was crucial to create a model of the network in the simulator.

A component common to all the network elements in the simulator is the component entity. This entity can be regarded as a black box with a set of connectors. Each connector (marked with symbol C_i where i is the connector number) represents an input/output of the component. Connectors connect different components with logical links (LL_i). Logical links only connect entities in the network and do not introduce any link delay. There are three implementations of component: the link, switch and processing node entities.

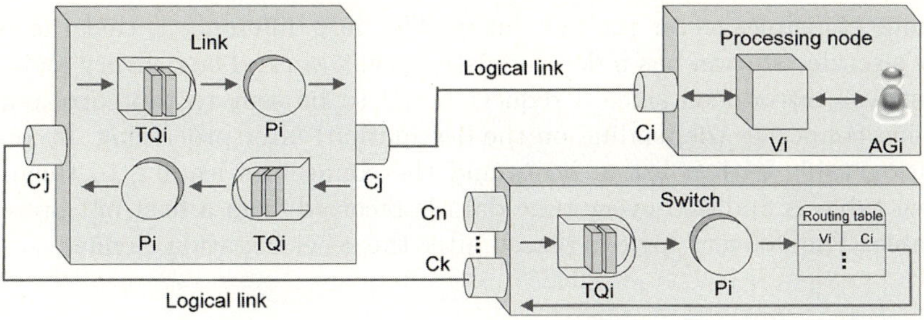

Fig. 2. Network elements structure

In Fig. 2 an example of a network is shown in which there is one link, one switch and one processing node. The processing node is connected via its C_i connector to the link's C_j connector with a logical link. Furthermore, the link's C_j' connector is connected with a logical link to the C_k connector of the switch. The switch entity can have more than one connector allowing connections with multiple processing nodes or switches.

Processing node (S_i) represents a network node from the F–MAN model. It contains two elements: a network host (V_i) and an agent node (AG_i). The network host offers communication functions to the agent node. The agent node represents the agent platform running on the processing node.

Link entities represent full-duplex physical links which connect nodes and switches in the network. Each link is limited by its network capacity which causes delay when sending data over the link. In accordance with queuing theory [6], a link can be divided into two components: a queue (TQ_i) and a service station (P_i). The queue is used to store processing requests that cannot be processed at that particular time since the service station is already processing some other request. In the network model, a processing request is data regarding the agent sent during the process of agent migration or the content of an ACL message. The service station represents an Ethernet card used to send data through the network. The process of sending data over a link is performed in the following manner: first the link receives a processing request from a component connected to it through a connector. After receiving the request, it is stored in the queue. The service station then takes the request from the queue and sends the data to the destination component through the corresponding connector. The time needed to send the data is defined as follows: $t_{si} = b_i/C$, where t_{si} is the service time for request i, b_i is the size of the data being sent for request i and C is the link capacity. The processes of receiving and processing requests are performed in parallel, separately for upload and download link directions. In the network model we assume that the queue is infinite employing the first-come-first-served queuing discipline. In our model there is only one service station at each link.

The switch entity represents a network switch used to transfer data between fragments. The switch is composed of three components: a queue, a service station and delivery logic. The queue and the service station are modeled using the same principles as for the link entity. The only difference is that the switch entity's service station has a deterministic service time. The delivery logic component was introduced since a request needs to be sent to the corresponding outgoing connector (depending on the destination) after processing. It contains a routing table with a list of hosts and the connectors leading to them. The routing table is updated every time data is received from a host not present in the table. The delivery logic is placed after the service station element.

4 Case Study: Simulation of Software Installation Using MA–RMS

The MA–RMS [7] is an agent based framework used for remote control over software on remote locations. It's a distributed system comprised of two main components: the RMS Console and the RMS Maintenance Environment. The RMS Console is the centralized client part of the RMS which offers a GUI enabling the user to define management operations (requests) on remote nodes. The RMS Maintenance Environment, also called the RMS Core, is the server part of RMS which must be preinstalled on remote network nodes in order for them to be managed. All software operations are performed by mobile agents, using appropriate coordination strategies [8,9], created at the RMS Console.

In our experiment, we simulated the installation of new software. The software installed by RMS consists of two parts: an Application Testbed and an Application Version. The Application Testbed is an interface between the application and the RMS Core, while the Application Version provides the actual functionality of the application. The operations required to install new software are as follows: migration of the application testbed and version, testbed installation, version installation, configuration and starting the application.

4.1 Simulation Results

We simulated the execution of software management operations in a large-scale mobile agent network with 101 network nodes. The nodes were divided into different numbers of network fragments ranging from 1 to 100. One node, denoted as S_0 is dedicated to the RMS Console and it is connected to the main switch (Fig. 1). The remaining nodes are nodes running remote systems and are connected to the switches in the second layer. The agent team performance evaluation is based on the following assumptions:

- Each agent belongs to only one subteam and agents can not change subteams;
- All subteams are organized with the same SCP (one agent for one node);
- The size of an agent is 10KB, a loaded agent (with software) is 5MB and a message is 10KB;
- All links have the same capacity;
- The network is organized in 2 layers of switches (Fig. 1);
- The network is initialized (the routing table is set);
- The number of switches in the 2nd layer is changed from 0 to 50 switches;

- The delay of a switch is 20 μs (per message or per agent);
- Operation execution time depends on the node capacity and is the same for all operations on all nodes. Time is measured in nanoseconds (ns).

The performance analysis is based on the time execution assessment defined in [5]. Figure 3 shows how the total execution time depends on the number of switches and the number of fragments in the network. The x-axis shows the number of fragments, while the y-axis shows the number of switches. The z-axis represents the total execution time in nanoseconds. The graph indicates that the number of switches does not have much influence over the total execution time. However, increasing the capacity of the links in the network significantly decreases execution time. For a very small number of fragments, the total execution time is high. This is also true for cases with a large number of fragments. Namely, fragments are executed in parallel. If there are only a small number of them, most of the work is still executed consecutively within each fragment. With an increase in the number of fragments, and hence with increased parallelization, the total execution time decreases, eventually reaching its minimal value. Further increasing the number of fragments causes an increase in execution time due to the overhead generated by subteam coordination agents (SCA).

The graph in (Fig. 4), plots the number of switches on the x-axis and the node delay (i.e. node capacity) on the y-axis. The node delay is the time needed to execute one operation. The z-axis shows the minimal total execution time. This minimal total execution time is found by changing the number of fragments. The graph shows that while the number of switches does not affect it, increasing the node delay causes a linear increase in the minimal total execution time.

Next, we examine for which fragment number the total execution time is minimal. This is shown on the graphs in Fig. 5. The graphs show that the number

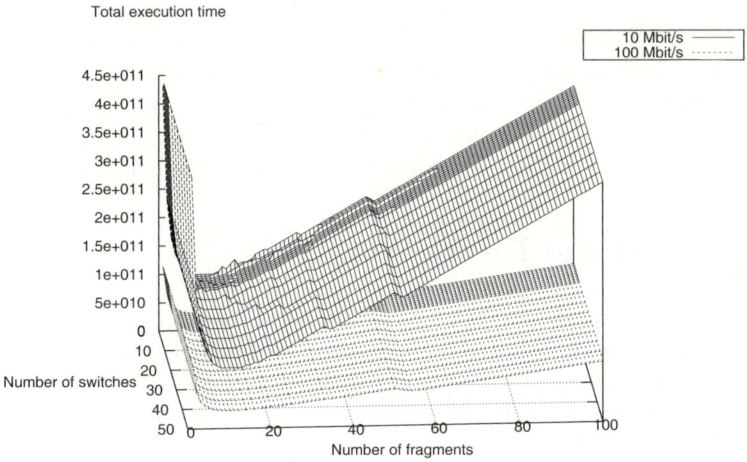

Fig. 3. Execution graph for 10Mbps and 100Mbps

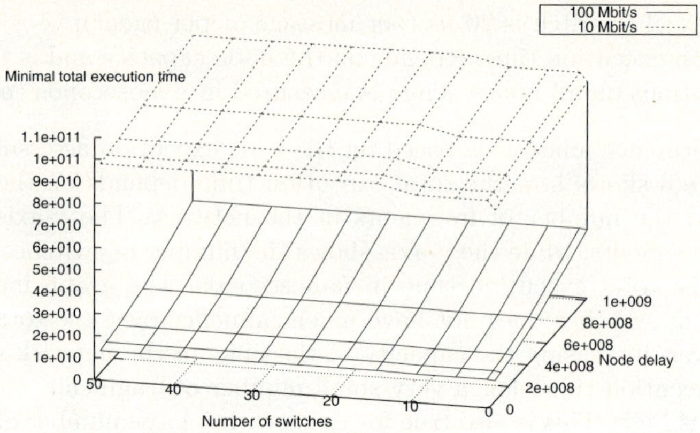

Fig. 4. Minimal total execution time graph

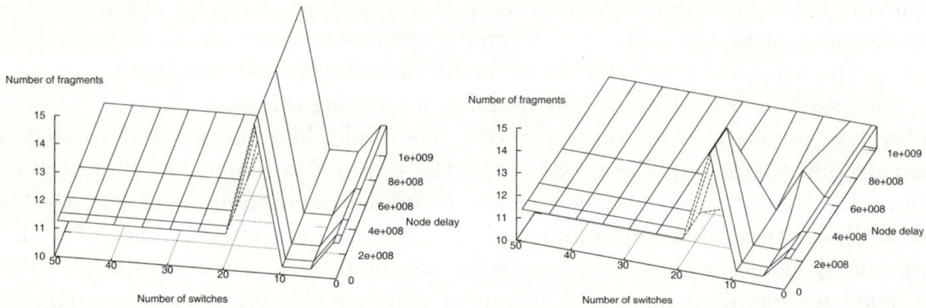

a) 10 Mbps b) 100 Mbps

Fig. 5. Minimal total execution time graph

of fragments for which the minimal total execution time is achieved, ranges from 11 to 15. This depends mostly on the number of switches. For the 100Mbps network, we can see that for slower nodes (those with higher node delay), the minimal total execution time is achieved for optimal fragment number 11.

5 Conclusion and Future Work

In this paper, we discussed a MAS based on the F-MAN model. Our focus was on network simulation. From the simulation results it can be concluded that introducing fragments improves results with respect to agent systems without fragments. Although increasing the network bandwidth causes a decrease in the total execution time, the number of switches in the lower layer was shown to have little influence. The main benefit is achieved by changing the number of fragments. Thus, the best results were obtained for fragment numbers ranging from 11 to 15, depending on the number of switches. Future work on the simulator will include further investigation of the F-MAN on various network topologies by introducing routers and changing the link capacities of certain links in the network. The influence of agent distribution strategies will also be investigated.

Acknowledgments. This work was carried out within the research project 036-0362027-1639 "Content Delivery and Mobility of Users and Services in New Generation Networks", supported by the Ministry of Science, Education and Sports of the Republic of Croatia.

References

1. Jezic, G., Kusek, M., Sinkovic, V.: Teamwork coordination in large-scale mobile agent networks. In: Gabrys, B., Howlett, R.J., Jain, L.C. (eds.) KES 2006. LNCS (LNAI), vol. 4251, pp. 236–243. Springer, Heidelberg (2006)
2. Horling, B., Lesser, V., Vincent, R.: Multi-Agent System Simulation Framework. 16th IMACS World Congress 2000 on Scientific Computation, Applied Mathematics and Simulation (2000)
3. Logan, B., Theodoropolous, G.: The distributed simulation of multi–agent systems. Proceedings of the IEEE 89(2), 174–185 (2001)
4. Fortino, G., Garro, A., Russo, W.: A discrete-event simulation framework for the validation of agent-based and multi–agent systems. In: Corradini, F., Paoli, F.D., Merelli, E., Omicini, A. (eds.) WOA, Pitagora Editrice Bologna, pp. 75–84 (2005)
5. Kusek, M., Lovrek, I., Sinkovic, V.: Agent team coordination in the mobile agent network. In: Khosla, R., Howlett, R.J., Jain, L.C. (eds.) KES 2005. LNCS (LNAI), vol. 3681, pp. 240–245. Springer, Heidelberg (2005)
6. Kleinrock, L.: Queueing Systems, vol. 1. John Wiley & Sons, Chichester (1975)
7. Jezic, G., Kusek, M., Marenic, T., Lovrek, I., Desic, S., Trzec, K., Dellas, B.: Grid service management by using remote maintenance shell. In: Jeckle, M., Kowalczyk, R., Braun, P. (eds.) GSEM 2004. LNCS, vol. 3270, pp. 136–149. Springer, Heidelberg (2004)
8. Paul, Xu.Y., Liao, E., Lai, J., Sycara, K.: Scaling teamwork to very large teams. In: Kudenko, D., Kazakov, D., Alonso, E. (eds.) Adaptive Agents and Multi-Agent Systems II. LNCS (LNAI), vol. 3394, Springer, Heidelberg (2005)
9. Tambe, M.: Agent architectures for flexible, practical teamwork. In: Proceedings of the 14th National Conference on Artificial Intelligence, pp. 22–28. AAAI, Stanford, California, USA (1997)

RSS-Based Blog Agents for Educational Applications

Euy-Kyung Hwang, Yang-Sae Moon*, Hea-Suk Kim, Jinho Kim, and Sang-Min Rhee

Department of Computer Science, Kangwon National University
192-1, Hyoja2-Dong, Chunchon, Kangwon 200-701, Korea
bonabia@nate.com, {ysmoon,hskim,jhkim,smrhee}@kangwon.ac.kr

Abstract. In recent years, blogs are widely used in many Web applications to easily share information between individuals or to effectively promote products in business marketing. In this paper we propose a novel notion of *blog agents* to easily exploit blogs in educational applications. We first investigate problems of current educational blogs. We then explain how we can solve those problems by using the blog agents. We also show that, if exploiting the blog agents, we can easily design the blogs for homework or consultation management. Using XML-based RSS (Really Simple Syndication) we finally implement *RSS-based blog agents* for the homework or consultation management blogs. We believe that our RSS-based blog agents will be widely used in many educational applications.

1 Introduction

Blogs are widely used as a representative personal media service through the Internet [4]. By providing commentary, news, personal diaries, and essays, many netizens exploit the blog to implement their own applications such as personal publication, online community, and personal broadcasting. Commercial sites providing the blog service include Windows Live Spaces [14], Google Blogger [7], and Cyworld [5]. Many futurologists are expecting that, like e-mails, blogs will be one of the most important communication methods in the near future [13].

There have been many attempts to exploit blogs in educational applications. We can find several educational blogs [1,2,3] in which teachers assign homework or provide lecture materials to their students. Most of these educational blogs are operated as follows: students visit their teacher's blog, and obtain their homework or lecture materials assigned by the teacher. In these blogs, however, content movement among bloggers can be seen as one-way communication rather than two-way communication, i.e., from a teacher to students, but not vice versa. Thus, only a little active students participate in educational blogs.

In this paper we investigate a major problem in exploiting blogs for educational applications, especially for homework management. The problem incurs

* Corresponding author.

B. Apolloni et al. (Eds.): KES 2007/WIRN 2007, Part I, LNAI 4692, pp. 222–229, 2007.

as the following reasons: 1) students have to visit their teacher's blog *very frequently* to see if there are new assignments or updated deadlines; and 2) the teacher also has to visit *many* students' blogs *frequently* to see students' reports and to evaluate their reports. Thus, we conclude that the biggest obstacle in using blogs for homework management is the frequent visiting to many blogs.

To solve those problem, we present a novel notion of *blog agents* that automatically deliver contents among educational blogs. An agent is an autonomous process that performs some (pre)defined jobs for its user or owner [11,12]. In a similar way, our blog agent enables automatic delivery of students' or teachers' contents among appropriate blogs maintained by students or teachers. We then show that, if exploiting the blog agents, we can reduce the number of visits to other people's blogs, and use the educational applications much easier.

Finally, we use RSS (Really Simple Syndication) [8] to implement our blog agents. RSS is a simple XML-based system that allows users to subscribe to their favorite Web sites [10]. In this paper we apply the *RSS-based blog agents* to educational blogs in homework or consultation management. We believe that our RSS-based blog agents will be widely used in many educational applications.

2 Related Work

A lot of netizens maintain their own blogs as the representative personal media service [13]. The term "blog" is a contraction of "Web log," and this blog makes many netizens implement their own applications such as personal publication, online community, and personal broadcasting. Currently, many portal companies including Windows Live Spaces [14], Google Blogger [7], and Cyworld [5] provide blog service as their most important business item.

There have been many efforts to exploit blogs in educational applications. Figure 1 shows an example blog for English education [1], which is maintained by an English teacher in an elementary school. The teacher tries to help students learn English grammar and writing by posting lecture materials and assignments in the blog. The working mechanism is as follows: 1) the teacher posts homework (or quiz) to the teacher's own blog; 2) each student posts the corresponding report to the student's own blog; and 3) the teacher visits each student's blog to evaluate the report or to leave some comments. In each student's point of view, teacher's visit to his/her blog will be very helpful to improve English skills. In the teacher's point of view, however, it will be terribly difficult to visit all the students' blogs.

Besides [1], there have been several attempts to use blogs in educational applications [2,3]. Blog [2] gives lessons in composition to improve students' writing skills. Blog [3] provides an educational course for Mathematics by posting frequently asked questions and their answers. These blogs, however, have two major problems: 1) frequent visiting to many blogs and 2) one-way communication. Due to these problems, only a little teachers maintain educational blogs, and only a few active students participate in current educational blogs.

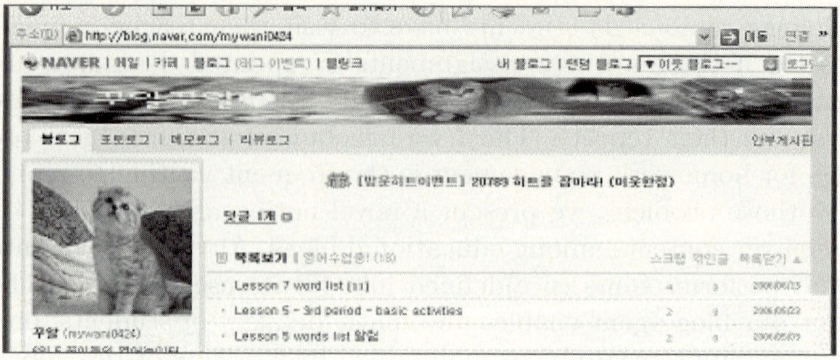

Fig. 1. An example blog for English education (in Korean)

3 RSS-Based Blog Agents for Educational Applications

3.1 Problems on Existing Educational Blogs

Figure 2 shows the brief operation mechanism of blogs for homework management. The followings explain each step in Figure 2.

① A teacher assigns homework in the teacher's own blog.
② Each student obtains the homework from the teacher's blog.
③ Each student posts the corresponding report to the student's own blog.
④ The teacher evaluates the report by visiting each student's blog.

In Figure 2, we note that both teachers and students have to frequently access several blogs for homework management. We explain the problem in details: 1) in Step ②, each student has to consume much time in frequent visiting the teacher's blog to obtain the homework; and 2) in Step ④, the teacher also has to consume much time in frequent visiting many students' blogs to evaluate their reports. We think that this frequent visiting to many blogs is the biggest obstacle in exploiting blogs in educational applications. To overcome this obstacle, we therefore need a new concept of automatic content delivery mechanism, which we call *blog agents*. To reduce the number of visits to other people's blogs, our blog agents automatically deliver contents among students' and teachers' blogs.

3.2 Design of Blog Agents

We redraw Figure 2 as Figure 3 by adopting the concept of automatic delivery in homework management. The followings explain each step in Figure 3.

① A teacher assigns homework in the teacher's own blog.
② Each student obtains homework from the *student's* own blog.
③ Each student posts the corresponding report to the student's own blog.
④ The teacher evaluates the report in the *teacher's* own blog.

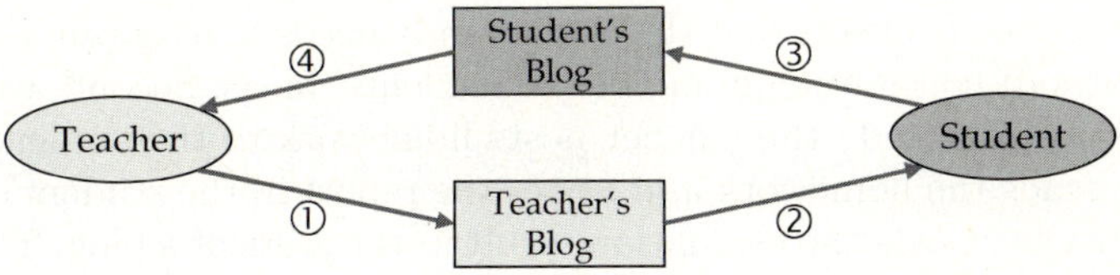

Fig. 2. The current blog mechanism for homework management

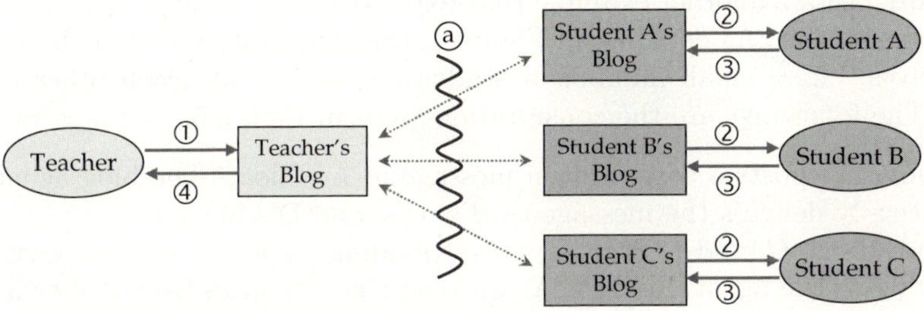

Fig. 3. Concept of a blog agent for homework management

To use automatic delivery steps, we need an appropriate medium or tool, that is, ⓐ in Figure 3, that delivers the contents from each student's blog to the teacher's blog, and vice versa. The blog agent performs this role. We redraw Figure 3 as Figure 4 by using an explicit blog agent. As shown in Figure 4, the blog agent delivers homework that the teacher posts to the teacher's own blog to many students' blogs. Similarly, it also automatically delivers the reports that students posted to their own blogs to the teacher's blog.

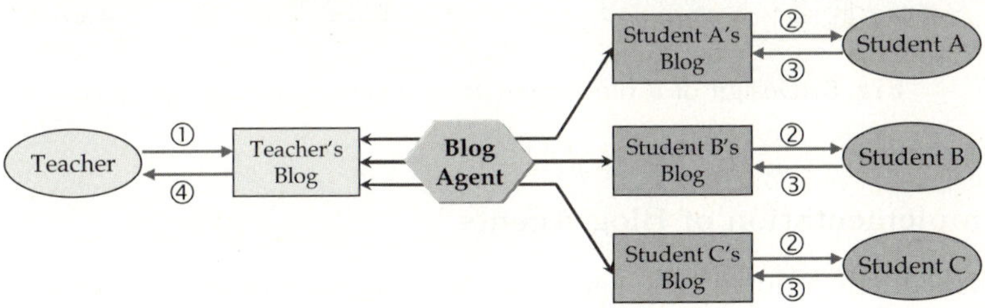

Fig. 4. Design of a blog agent for homework management

Using a blog agent as in Figure 4, we can reduce the number of visits to other people's blogs in both students' and teachers' points of view. First, the teacher might think as follows: 1) the teacher leaves homework in the teacher's blog; 2) each student reads the homework and posts the corresponding report to the teacher's blog; and 3) the teacher reads and evaluates the report in the teacher's own blog. It means that the teacher only needs to maintain the teacher's own blog without considering a number of students' blogs. Second, each student might think as follows: 1) the teacher posts homework to the student's blog; 2) the student reads the homework and leaves the report in the student's own blog; and 3) the teacher posts the evaluation result in the student's blog. That is, each student also only needs to maintain the student's own blog without considering the teacher's blog. Likewise, the proposed blog agent makes users, i.e., teachers and students, easily use blogs in homework management.

Figure 5 shows another example that exploits a blog agent in group consultation management. As depicted in Figure 5, the consultation is done by students themselves, i.e., a small number of students communicate each other through blogs. The followings are the consultation steps in Figure 5.

① Student A posts a consultation message in A's blog. (The blog agent automatically delivers the message to B's, C's, and D's blogs.)
② Students B, C, and D post the corresponding responses to their own blogs.
③ Owing to the agent, Student A can read the responses from A's own blog.

This blog agent-based consultation has the following advantages. First, it might help students form a small positive group whose every member has a role of consultant for other members. Second, students can share the consulting results only with their private members. Third, troubled students can quickly get consulting feedbacks from their friends due to automatic delivery of messages.

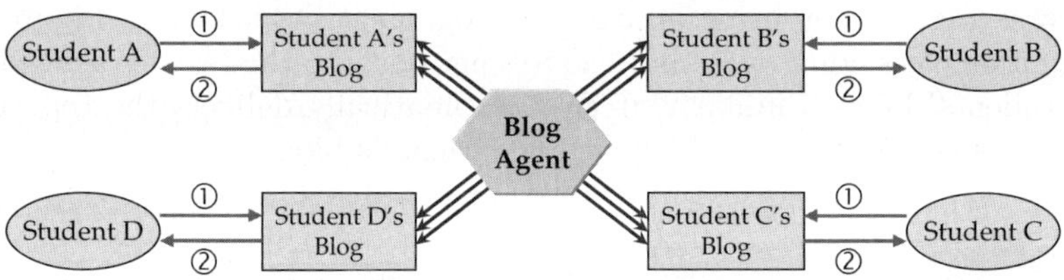

Fig. 5. Design of a blog agent for consultation management

3.3 Implementation of Blog Agents

To realize an educational blog agent, we first investigate several Web programs and services. Table 1 shows summary of these programs and services used in the current Web environment. First, a *Trackback* [9] is simply an acknowledgment. This acknowledgment is sent via a network signal (ping) from Site A (originator) to Site B (receiver). In general, these Trackbacks are used primarily to facilitate communication between blogs. Second, a *Web crawler* (also known as a *Web spider* or *Web robot*) [6] is a program or automated script which browses the World Wide Web in an automated manner. These Web crawlers can be used in automating maintenance tasks on a Web site, such as checking links or validating HTML codes. Third, RSS is a simple XML-based system that allows users to

Table 1. Summary of agents and their functions

Functions	Trackback	Web crawler	RSS (Reader)
Content browsing	×	○	○
Content management	×	×	○
Automatic delivery	○	○	○
Delivery of update information	×	×	○

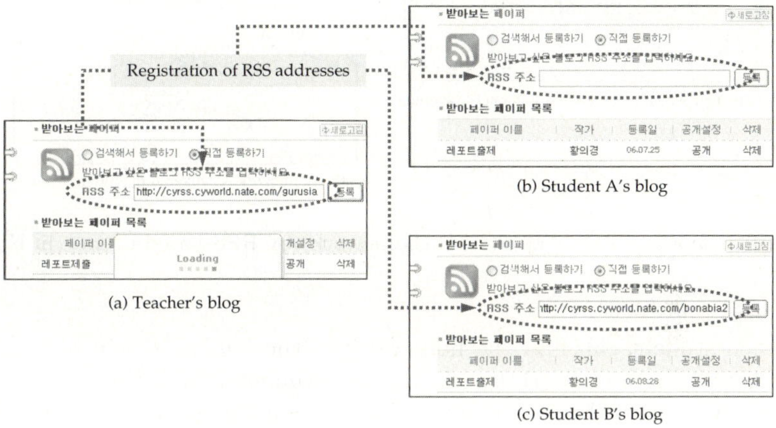

(a) Teacher's blog

(b) Student A's blog

(c) Student B's blog

Fig. 6. Registration of RSS addresses for homework management (in Korean)

subscribe to their favorite Web sites [10]. Among these programs and services, we select RSS to implement our blog agents since, as shown in Table 1, it satisfies all of our requirements for the agent, content browsing, content management, automatic delivery, and delivery of update information.

Likewise, in this paper we have implemented blog agents using XML-based RSS [8,10], and applied the agents to homework management and consultation management explained in Section 3.2. As the first implementation example, we explain homework management in detail. First, we register students' RSS addresses to the teacher's blog, and the teacher's RSS address to each student's blog. This registration process can be thought as implementing the blog agent, and through this registration process, the contents can be delivered automatically among blogs.

We actually implemented educational blog sites for a teacher and students in a middle school class. Figures 6 and 7 show an example of these blog sites for homework management. Figure 6(a) shows an example of registering Students A's and B's RSS addresses to the teacher's blog. Figures 6(b) and 6(c) show examples of registering the teacher's RSS address to A's and B's blogs, respectively. By this registration process, the contents will be automatically delivered among blogs. Figure 7 shows an example of delivering the homework from the teacher's blog to students' blog whose RSS addresses are registered in Figure 6.

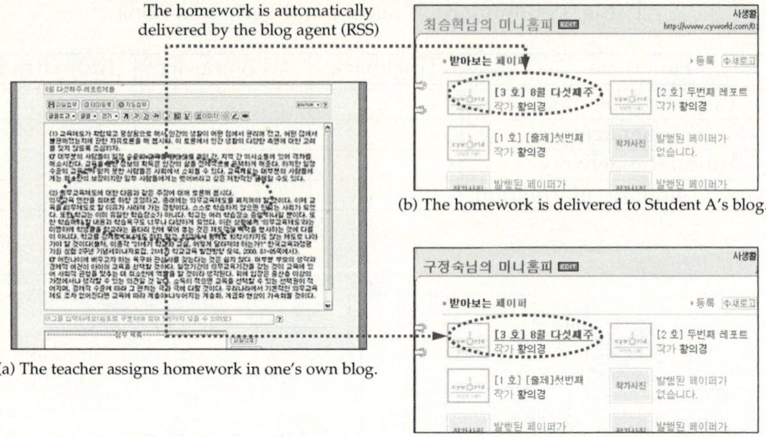

(a) The teacher assigns homework in one's own blog.

(b) The homework is delivered to Student A's blog.

(c) The homework is delivered to Student B's blog.

Fig. 7. Automatic delivery of content (homework) by RSS-based agents (in Korean)

Thus, each student can find the homework in the student's own blog as if the teacher leaves the homework by visiting the student's blog. Next, the students post their reports to their own blogs. Since we register students' RSS addresses to the teacher's blog, the reports are also automatically delivered to the teacher's blog. Thus, the teacher can evaluate each student's report in the teacher's own blog as if the student submits the report by visiting the teacher's blog.

We have also implemented consultation management by registering RSS addresses among students who are joined in the same consulting group. It can be done as follows: if students A, B, and C want to join a consulting group, we register 1) A's and B's RSS addresses to C's blog, 2) B's and C's addresses to A's blog, and 3) C's and A's addresses to B's blog. By this simple registration process, we can complete the implementation of our blog agents for the consultation management. We omit implementation details due to space limitation.

4 Conclusion

In this paper we have presented a blog agent-based approach for educational applications. This paper can be summarized as follows. First, we have pointed out that the major problem of educational blogs is frequent visiting to many blogs. Second, to solve this problem, we have presented the novel notion of *blog agents* that automatically deliver the contents among blog sites. Third, by exploiting the blog agents, we have designed the blog sites for homework or consultation management. Fourth, we have implemented the *RSS-based blog agents* for the educational blogs. These results indicate that our work will provide a practical framework that makes it easy to exploit blogs in many educational applications.

Acknowledgements

This work was supported by the Ministry of Science and Technology (MOST)/ Korea Science and Engineering Foundation (KOSEF) through the Advanced Information Technology Research Center (AITrc).

References

1. A Blog Site for English Education, `http://blog.naver.com/mywani0424/`
2. A Blog Site for Writing Skills, `http://blog.naver.com/necewarm/`
3. A Blog Site for Mathmatics Education, `http://blog.empas.com/es00ksop1004/`
4. Blood, R.: How Blogging Software Reshapes the Online Community. Communications of the ACM 47(12), 53–55 (2004)
5. Cyworld Mini-Homepage, `http://www.cyworld.com`
6. Dikaiakos, M.D., Stassopoulou, A., Papageorgiou, L.: An Investigation of Web Crawler Behavior: Characterization and Metrics. Computer Communications 28(8), 880–897 (2005)
7. Google Blogger, `http://www.blogger.com`
8. Hansen, F.A., et al.: RSS as a Distribution Medium for Geo-Spatial Hypermedia. In: Proc. of the 16th ACM Conf. on Hypertext and Hypermedia, pp. 254–256 (2005)
9. Kimura, M., Saito, K., Kazama, K., Sato, S.: Detecting Search Engine Spam from a Trackback Network in Blogspace. In: Proc. of the 9th Int'l Conf. on Knowledge-Based Intelligent Information and Engineering Systems, pp. 723–729 (September, 2005)
10. Lyndersay, S.: Windows and RSS: Beyond Blogging. In: In Int'l Conf. on Management of Data, ACM SIGMOD, p. 723 (2006)
11. Maes, P.: Agents that Reduce Work and Information Overload. Communications of the ACM 37(7), 31–40 (1994)
12. Srbljinovic, A., Skunnca, O.: An Introduction to Agent Based Modelling and Simulation of Social Processes. Interdisciplinary Description of Complex Systems 1, 1–8 (2003)
13. Wagner, F.: Blog Perspectives. Metainformatics Symposium , 212–219 (2004)
14. Windows Live Spaces, `http://spaces.msn.com`

Soft Computing Approach to Contextual Determination of Grounding Sets for Simple Modalities

Radosław Piotr Katarzyniak[1], Ngoc Thanh Nguyen[1], and Lakhmi C. Jain[2]

[1] Institute of Information Science and Engineering, Wrocław University of Technology, Wybrzeże Wyspiańskiego 27, 50-350 Wrocław, Poland
{Radosław.Katarzyniak,Thanh}@pwr.wroc.pl
[2] School of Electrical and Information Engineering, University of South Australia
Lakhmi.Jain@uniisa.edu.au

Abstract. Four strategies for computing the grounding sets are suggested. An original model for grounding of simple modalities is briefly outlined and the need for its contextualization is discussed. References are made to works in which soft computing methods are presented to make effective implementation of these strategies possible for the case of software agents.

Keywords: Modality, Grounding theory, Modal logic, Software agent.

1 Introduction

The language grounding problem belongs to the class of the main research issues considered in the field of artificial intelligence and cognitive sciences [1,15,16,19]. Grounding deals with referring symbols of interpreted languages to actual worlds. In case of the natural language it defines the way in which particular sentences are to be referred to surroundings described by these sentences. It is quite obvious that this process is highly subjective and strongly depends on agents that ground languages in external worlds. In the field of artificial intelligence the language grounding has already been considered for at least a few artificial languages [14].

In [2-7] an original theory of grounding a modal language of communication has been proposed. It has been given for a particular class of artificial agents able to observe external world and store the results in internal knowledge bases. The modal language has been given as a set of modal formulas with commonsense interpretation assigned. Simple modalities are a subset of this language. Their list and commonsense interpretation are given in Table 1. The above mentioned theory of grounding states detailed requirements for a proper way in which interpreted modal formulas should be connected to the world in order to use them as the adequate world's descriptions. The theory consists of multiple theorems that overview properties of grounding process considered for simple modalities [3, 7], modal conjunctions [4, 5, 6, 7] and modal alternatives [7].

One of the main assumptions accepted in the theory for grounding reflects the ability of natural cognitive agents to fill gaps experienced in autonomously created models for actual worlds with some mental ′patterns′ extracted from previous empirical

B. Apolloni et al. (Eds.): KES 2007/WIRN 2007, Part I, LNAI 4692, pp. 230–237, 2007.

experiences. Due to the fact that the stored empirical material can be very rich, the cognitive agents are often forced to create alternative models for unknown parts of their worlds. Additionally these alternative models can be assigned different strength with which they influence individual beliefs of the agents. All these aspects of stored experiences influence the way cognitive agents refer their modal languages to external worlds.

Table 1. Commonsense semantics for simple modalities

Formula	Commonsense interpretation
$p(o)$	Object *o* exhibits property *P*.
$\neg p(o)$	Object *o* does not exhibit property *P*.
$Pos(p(o))$	It is possible that object *o* exhibits property *P*.
$Bel(p(o))$	I believe that object *o* exhibits property *P*.
$Know(p(o))$	I know that object *o* exhibits property *P*.
$Pos(\neg p(o))$	It is possible that object *o* does not exhibit property *P*.
$Bel(\neg p(o))$	I believe that object *o* does not exhibit property *P*.
$Know(\neg p(o))$	I know that object *o* does not exhibit property *P*.

In [2-7] the basic case of the theory for grounding has been developed. In this approach it has been assumed that in case of an agent's impossibility to observe the state of a property in an object o the agent reduces a related lack of knowledge with previous experiences. The stored empirical material helps the agent to ground private beliefs on the current state in the collected data.

It is assumed in the theory of grounding that at each time point t the agent is equipped with the set $KS(t)$ of multiple observations PB collected by this agent up to the time point t. These observations in which an object o was perceived as exhibiting the property P constitutes the only content of a set $A^1(t)$. This set consists of everything that shapes the agent's opinions on the way the property P 'realizes' in the object o. In consequence this set can be used to ground modalities $Know(p(o))$, $Bel(p(\mathrm{o}))$ and $Pos(p(\mathrm{o}))$ when the actual state of P in the object o is not accessible. Similarly all observations in which the object o was observed as an entity without the property P are the only members of the set $A^2(t)$. The role of this set is similar to the role of $A^2(t)$, too. However it is applied to ground all simple modalities that are built from negated atom $\neg p(\mathrm{o})$.

In this case of theory for grounding both sets $A^1(t)$ and $A^2(t)$ are determined without consideration of any context in which they are created. In particular it is not taken into account that at least some parts of the model for actual world's state are given by the latest observations of this world. Such approach to computing the content of $A^1(t)$ and $A^2(t)$ assigns the same degree of importance to all members of $A^1(t)$ and $A^2(t)$. However this importance can be graded by a reference to the actual similarity between each previous observation and the latest one. This remark results in a conclusion that a context-dependent model for computing the sets $A^1(t)$ and $A^2(t)$

could be developed to make the grounding theory better suited to model natural processes of language grounding.

2 The Model for Non-contextual Grounding Simple Modalities

Let us assume that objects $\Omega=\{o_1,o_2,...o_N\}$ in the world can exhibit properties from the set $\Im=\{P_1,P_2,...P_K\}$. The agent observes this world, develops subjective representations of realized observations and stores them in an internal knowledge base. Observations are represented by the so called base profiles [8]. A base profile related to a time point $t\in T$ is a representation of an individual observation that was autonomously realized and encapsulated by the agent in its internal knowledge base. The structure of the base profile is given as follows:

$$BP(t) = <\Omega,P^+_1(t),P^-_1(t),P^+_2(t),P^-_2(t),...P^+_K(t),P^-_K(t)> \quad (1)$$

where:

a) t denotes a time point to which the base profile $BP(t)$ is related,
b) $P^+_i(t)\subseteq\Omega$ and for each object $o\in\Omega$ the condition $o\in P^+_i(t)$ holds if and only if the agent perceived o as exhibiting the property P_i at the time point t.
c) $P^-_i(t)\subseteq\Omega$ and for each object $o\in\Omega$ the condition $o\in P^-_i(t)$ holds if and only if the agent perceived o as not exhibiting the property P_i at the time point t.

The state of knowledge $KS(t)$ related to a time point $t\in T$ is defined by the overall collection of collected base profiles and is given as follows:

$$KS(t) =\{BP(l): l\in T \text{ and } l\leq^{TM}t\}. \quad (2)$$

Let us assume that the following sets are defined:

$$A^1(t,o)=\{BP(l): l\leq^{TM}t \text{ and } BP(l)\in KS(t) \text{ and } o\in P^+(l)\}, \quad (3)$$

$$A^2(t,o)=\{BP(l): l\leq^{TM}t \text{ and } BP(l)\in KS(t) \text{ and } o\in P^-(l)\}, \quad (4)$$

where $KS(t)$ has been defined above. The sets $A^1(t)$ and $A^2(t)$ are called the grounding sets and induce two higher level knowledge structures called mental models $m^a{}_1$ and $m^a{}_2$. They are related to the object o and all observed states of the property P in this object o. In consequence they support beliefs of the agent developed as regards to formulas based on the atom $p(o)$ and negated atom $\neg p(o)$, respectively.

The influence of both grounding sets on beliefs created by the agent for property P and object o is assumed to be dependent on the cardinality of $A^i(t)$, i=1,2. Let the symbols $G^A{}_i$ be introduced to represent their cardinality:

$$G^A{}_i=card(A^i(t)). \quad (5)$$

The main role in implementing the grounding is however given to the concept of relative grounding values defined below. For both formulas $p(o)$ and $\neg p(o)$ they can be computed at each time according to the following formulas:

$$\lambda(t, p_i(o)) = \frac{G_1^A}{G_1^A + G_2^A}, \tag{6}$$

$$\lambda(t, \neg p_i(o)) = \frac{G_2^A}{G_1^A + G_2^A}. \tag{7}$$

It is assumed in the cited theory of grounding [2-7] that these values are used to choose an appropriate modal operator to extend both formulas $p(o)$ and $\neg p(o)$ in order to make them well-grounded. In order to make this evaluation possible the agent is equipped with a system of modality threshold ($\lambda_{minPos}, \lambda_{maxPos}, \lambda_{minBel}, \lambda_{maxBel}$) satisfying the following inequalities:

$$0 < \lambda_{minPos} < \lambda_{maxPos} \leq \lambda_{minBel} < \lambda_{maxBel} \leq 1. \tag{8}$$

To complete the model for grounding a concept of knowledge disribution $DA(t)=\{RA^1(t),\ TA^1(t),\ RA^2(t),\ TA^2(t)\}$ is introduced in the cited theory. This distribution describes a distribution of both sets $A^1(t)$ and $A^2(t)$ over the working and permanent memory of the agent. It is given as follows:

$$\begin{aligned} RA^i(t) &= PR(t) \cap A^i(t), \\ TA^i(t) &= PT(t) \cap A^i(t), \\ RA^i(t) \cap TA^i(t) &= \emptyset, \\ RA^i(t) \cup TA^i(t) &= A^i(t). \end{aligned} \tag{9}$$

The original application of grounding sets follows directly from definitions for the so called epistemic satisfaction relation. This relation describes conditions which have to be fulfilled by the internal state knowledge state to make a simple modality well-grounded:

Let a time point $t \in T$, a distribution $DA(t)$ and a system of modality thresholds $0 < \lambda_{minPos} < \lambda_{maxPos} \leq \lambda_{minBel} < \lambda_{maxBel} \leq 1$ be given. For each property $P \in \Im$ and each object $o \in \Omega$ the following is assumed:

- The epistemic satisfaction relation $DA(t) \models_G Pos(p(o))$ holds if and only if
  $$o \in \Omega \backslash (P^+(t) \cup P^-(t)),\ RA^1(t) \neq \emptyset \text{ and } \lambda_{minPos} \leq \lambda(t,p(o)) < \lambda_{maxPos}. \tag{10}$$
- The epistemic satisfaction relation $DA(t) \models_G Bel(p(o))$ holds if and only if
  $$o \in \Omega / (P^+(t) \cup P^-(t)),\ RA^1(t) \neq \emptyset \text{ and } \lambda_{minBel} \leq \lambda(t,p(o)) < \lambda_{maxBel}. \tag{11}$$
- The epistemic satisfaction relations $DA(t) \models_G Know(p(o))$ and $DA(t) \models_G p(o)$ hold if and only if

$$o \in P^+(t) \tag{12}$$

or

$$o \in \Omega / (P^+(t) \cup P^-(t)) \text{ and } RA^1(t) \neq \emptyset \text{ and } \lambda(t,p(o)) = 1. \tag{13}$$

- The epistemic satisfaction relation $DA(t) \models_G Pos(\neg p(o))$ holds if and only if
  $o \in \Omega/(P^+(t) \cup P^-(t))$, $RA^2(t) \neq \varnothing$ and $\lambda_{minPos} \leq \lambda(t, \neg p(o)) < \lambda_{maxPos}$. (14)
- The epistemic satisfaction relation $DA(t) \models_G Bel(\neg p(o))$ holds if and only if
  $o \in \Omega/(P^+(t) \cup P^-(t))$, $RA^2(t) \neq \varnothing$ and $\lambda_{minBel} \leq \lambda(t, \neg p(o)) < \lambda_{maxBel}$. (15)
- The epistemic satisfaction relations $DA(t) \models_G Know(\neg p(o))$ and $DA(t) \models_G \neg p(o)$ hold if and only if

$$o \in P^-(t)) \quad (16)$$

or

$$o \in \Omega/(P^+(t) \cup P^-(t)),\ RA^2(t) \neq \varnothing \text{ and } \lambda(t, \neg p(o)) = 1. \quad (17)$$

3 Strategies for Contextual Determination of Grounding Sets

According to the above definitions the grounding sets are to be computed without any consideration of the latest experiences stored in the base profile $PB(t)$ at each time point t. The individual contribution of each base profile $BP' \in A^1(t) \cup A^2(t)$ to the relative strength of grounding can be however influenced by additional factors:

1. the degree of similarity (distance) between BP' and the latest base profile $BP(t)$,
2. the generalized distance between a set $K \subseteq A^i(t)$, i=1,2 and the latest base profile $BP(t)$ provided that $BP' \in K$ and K consists of all stored based profiles close to BP' according to a distance (similarity) measure,
3. the cardinality or consistency of the above cluster K.

Below four strategies for using these factors in grounding of simple modalities are suggested. These strategies assume the reference to the following concepts of soft computing:

- a distance function d: $KS(t) \times KS(t) \rightarrow \Re$ [11,12] ,
- a similarity function sim: $KS(t) \times KS(t) \rightarrow [0,1]$ [11,12],
- a consensus function $cons$: $2^{KS(t)} \rightarrow KS(t)$ [8,11,12],
- a consistency measure ∂: $2^{KS(t)} \rightarrow KS(t)$ [9,12],
- a fuzzy cardinality functions $fuzz\text{-}card$: $2^{F(KS(t))} \rightarrow \Re$ [17].

3.1 Strategy A

This strategy is the simplest one. Namely for each i=1,2 it reduces the content of the grounding set $A^i(t)$ by removing all its base profiles with their distance to a current base profile $PB(t)$ exceeding a predefined level of distance d_{max}. This strategy has already been partly used to compute grounding sets for conjunctional modal responses to an information query, provided that a responding agent has got limited knowledge of a current state of its world [9]. A general outline of this strategy can be given as follows:

Input: $P\in\Im$, $o\in\Omega$, a knowledge state $KS(t)$.
Output: $G^{A}{}_{i}$ $i = 1,2$.
Step 1. Compute $A^{1}(t)$ and $A^{2}(t)$.
Step 2. Compute $modA^{i}(t)=\{PB'\in A^{i} : d(PB(t), PB')\le d_{max}\}$.
Step 3. Compute $G^{A}{}_{i}=card(mod(A^{i}(t)))$.

The values $\lambda(t, p(o))$ and $\lambda(t,\neg p(o))$ are computed according to formulas (6-7).

3.2 Strategy B

This strategy assumes additional classification of all content in $A^{1}(t)$ and $A^{2}(t)$. This way additional analysis is carried out to determine tendencies in the actual world as they have been subjectively experienced by the agent. Clusters are analyzed to determine their distance to a current base profile. This evaluation requires previous determination of consensus base profiles representing each cluster. If the distance between a consensus base profile to a current base profile exceeds a predefined value d_{max}, then the related cluster is rejected from an appropriate grounding set. A general outline of this strategy can be given as follows:

Input: $P\in\Im$, $o\in\Omega$, a knowledge state $KS(t)$, d_{max}.
Output: $G^{A}{}_{i}$ i=1,2.
Step 1. Compute $A^{1}(t)$ and $A^{2}(t)$.
Step 2. Apply a clustering method to compute classifications $K^{1}=\{K^{1}{}_{1}, K^{1}{}_{2}, \ldots K^{1}{}_{n}\}$ and $K^{2}=\{K^{2}{}_{1}, K^{2}{}_{2}, \ldots K^{2}{}_{m}\}$ for sets $A^{1}(t)$ and $A^{2}(t)$, respectively.
Step 3. For each i=1,2 and each class $K'\in K^{i}$ compute the consensus $cons(K')$. If $d(cons(K'), PB(t))>d_{max}$, then remove all members of K' from the set $A^{i}(t)$.
Step 4. For i=1,2 compute $G^{A}{}_{i}=card(mod(A^{i}(t)))$.

Obviously the values $\lambda(t, p(o))$ and $\lambda(t,\neg p(o))$ are given as in (6-7).

3.3 Strategy C

This strategy assumes fuzzification of sets $A^{1}(t)$ and $A^{2}(t)$. A possible reformulation relies on the assumption that the membership of each base profiles to a grounding sets depends on a similarity function. The higher the similarity of a base profile to a current base profile the stronger its influence on current beliefs developed by the agent in relation to an unknown state of property P in the object o.

Input: $P\in\Im$, $o\in\Omega$, a knowledge state $KS(t)$.
Output: $G^{A}{}_{i}$ i=1,2.
Step 1. Compute $A^{i}(t)$, i=1,2.
Step 2. Compute $FA^{i}(t)=\{(BP,sim(BP,BP(t))): PB\in A^{i}(t)\}$, $i = 1,2$.
Step 3. Compute $G^{A}{}_{i}=fuzz\text{-}card(mod(A^{i}(t)))$, $i = 1,2$.

Both values $\lambda(t, p(o))$ and $\lambda(t,\neg p(o))$ are given by formulas (6-7).

3.4 Strategy D

It is assumed for this strategy that the membership values for all base profiles in $A^{1}(t)$ and $A^{2}(t)$ relies on base profiles' similarities and other numeric characteristics of stored empirical data. In this proposal it is suggested that for each base profile in $A^1(t)$ and $A^2(t)$ its grounding strength is influenced by consistency of its closest neighbour. The closest neighbourhood is defined as a cluster to which each analyzed base profile belongs. As it has been already been proved the consistency of clusters for which consensus is computed, is an important factor. Details related to models for consistency have been originally developed in [10,13]. the general assumption acdepted for the strategy D is that the higher the consistency of the closest neighbourhood of a base profile in $A^1(t)$ and $A^2(t)$ the higher the influence of this base profile on beliefs developed by the agent. An outline of an apprpriate strategy for this type of grounding contextualization can be given as follows:

Input: $P\in\Im$, $o\in\Omega$, a knowledge state $KS(t)$, d_{max}.
Output: $G^A{}_i$ i = 1, 2.
Step 1. Compute $A^1(t)$ and $A^2(t)$.
Step 2. Compute classifications $K^1=\{K^1{}_1, K^1{}_2, \ldots K^1{}_n\}$ and $K^2=\{K^2{}_1, K^2{}_2, \ldots K^2{}_m\}$.
Step 3. Compute fuzzy grounding sets $FA^i(t)=\{(BP, fuzz(BP)): PB\in A^i(t)\}$, i=1,2, where for each i=1,2 and $BP\in K\in K^i$:

1. $\partial(K)$ denotes the consistency of cluster K,
2. $sim(BP,cons(K))$ denotes a similarity between a current context $BP(t)$ and consensus $cons(K)$ derived from K, and
3. $fuzz(BP)=\begin{cases}\partial(K)\cdot sim(BP(t)),cons(K)) & d(BP(t),cons(K))\le d_{max}\\ 0 & \text{otherwise}\end{cases}$ is this value with which the base profile BP grounds the related formula.

Step 7. Compute $G^A{}_i$=$fuzz\text{-}card(mod(A^i(t)))$, i = 1, 2.

Obviously now values $G^A{}_1$ and $G^A{}_2$ can be used to determine relative strenght of grounding p(o) and $\neg p$(o), respectively.

4 Final Remarks

In this paper four strategies for computing the context-related grounding for the case of simple modalities have been outlined. It has been assumed in all of them that the grounding of each modal language depends not only on the overall experience stored in the agent but should be modified by the latest experiences. It has been also suggested that effective implementation of modified grounding should be based on appropriate soft computing methods. There are multiple possibilities to implement these strategies in real agent systems. However, all of them should be defined by particular choice of similarity, distance, consistency or cardinality measures.

Acknowledgment

This work was partially supported by the Polish Ministry of Science and Higher Education under the grant no N516 013 32/1733 (2007-2008).

References

1. Harnad, S.: The symbol grounding problem. Physica D 42, 335–346
2. Katarzyniak, R.: The language grounding problem and its relation to the internal structure of cognitive agents. Journal of Universal Computer Science 11(2), 357–374 (2005)
3. Katarzyniak, R.: On some properties of grounding simple modalities, Systems Science. Systems Science 31(3), 59–86 (2005)
4. Katarzyniak, R.: On some properties of grounding non-uniform sets of modal conjunctions, Int. Journal of Applied Mathematics and Computer Science 16(3), 399–412 (2006)
5. Katarzyniak, R.: On some properties of grounding uniform sets of modal conjunctions. Journal of Intelligent & Fuzzy Systems 17(3), 209–218 (2006)
6. Katarzyniak, R.: Some notes on grounding singletons of modal conjunctions. Systems Science 32(1), 45–55 (2005)
7. Katarzyniak R.: Grounding modal language of communication in agent systems, EXIT Warsaw (to be published in Polish) (2007)
8. Katarzyniak, R., Nguyen, T.N.: Reconciling inconsistent profiles of agent's knowledge states in distributed multiagent systems using consensus methods. Systems Science 26(4), 93–119 (2000)
9. Katarzyniak, R., Pieczyńska-Kuchtiak, A.: Grounding and extracting modal responses in cognitive agents: 'and' query and states of incomplete knowledge. Int. J. Applied Mathematics and Computer Science 14(2), 249–263 (2004)
10. Nguyen, N.T.: Susceptibility to Consensus of Conflict Profiles. In: Terano, T., Nishida, T., Namatame, A., Tsumoto, S., Ohsawa, Y., Washio, T. (eds.) New Frontiers in Artificial Intelligence. LNCS (LNAI), vol. 2253, pp. 327–332. Springer, Heidelberg (2001)
11. Nguyen, N.T.: Using distance functions to solve representation choice problem. Fundamenta Informaticae 48(4), 295–314
12. Nguyen, N.T.: Consensus system for conflict solving in distributed systems. Information Sciences 147(1-4), 91–122 (2002)
13. Nguyen, N.T., Małowiecki, M.: Consistency Measures for Conflict Profiles. In: Peters, J.F., Skowron, A., Grzymała-Busse, J.W., Kostek, B., Świniarski, R.W., Szczuka, M. (eds.) Transactions on Rough Sets I. LNCS, vol. 3100, pp. 169–186. Springer, Heidelberg (2004)
14. Roy, D., Reiter, E.: Connecting language to the world. Artificial Intelligence 167(1-2), 1–12 (2005)
15. Vogt, P.: Anchoring of semiotic symbols. Robotics and Autonomous Systems 43, 109–120 (2003)
16. Vogt, P.: The Physical Grounding Symbol Problem. Cognitive Systems Research 3, 429–457 (2002)
17. Wygralak, M.: Questions of cardinality of finite fuzzy sets. Fuzzy Sets and Systems 102, 185–210 (1999)
18. Ziemke, T., Rethinking Grounding, W., Riegler, A., Peschl, M., von Stein, A.: Understanding Representation in the Cognitive Sciences, pp. 177–190. Plenum Press, New York (1999)

The Statistical Verification of Rough Classification Algorithms*

Adrianna Kozierkiewicz and Ngoc Thanh Nguyen

Institute of Information Science and Engineering,
Wroclaw University of Technology, Poland
adrianna.kozierkiewicz@student.pwr.wroc.pl, thanh@pwr.wroc.pl

Abstract. In this paper some heuristic algorithms for rough classification are presented and verified by using statistical tests. The results of these tests allow finding our which algorithm is more effective and should be applied in E-learning systems for redefining the classification criterion for the set of learners.

Keywords: Rough classification, E-learning systems, statistical analysis.

1 Introduction

We live in an information society which has been developed especially in many countries where teleinformatic technologies reached very fast growth rate. This fact is characterized by the wide access to information and communication technologies, open networks and the ability to data processing. There is no doubt that the key to building the new type of society is education. New forms of education such as E-learning play more and more important role in our life. Distance education systems should guarantee effective learning based on preferences, abilities and learning styles of learner. These systems should be able to adapt the content and structure of knowledge and learning style for an individual. In traditional learning, where only one teacher works with different learners, there are no possibilities of recommendation of an individual learning scenario for a single learner. In E-learning systems this kind of personalization is easier to achieve. The method for this possibility is introduced in this paper.

Learner preferences which are used for personalization are stored in a user profile that is created during the registration process. It contains two types of data: user data and usage data [7]. The user data consists of learner login, name, age, sex, educational level, IQ and other information referring to his preferable learning styles such as: perception, receiving, processing and understanding. The usage data contains of scenarios with a learner successfully has finished the course. Some of user data attributes (most often selected by experts) create the criterion of classification of learners, and a new learner is classified to a class on the basis of his user data. For this learner an opening scenario is determined on the basis of scenarios of learners who belong to the

* This work was partially supported by the Polish Ministry of Science and Higher Education under the grant no N516 013 32/1733 (2007-2008).

B. Apolloni et al. (Eds.): KES 2007/WIRN 2007, Part I, LNAI 4692, pp. 238–245, 2007.

same class. The opening scenario is being modified during the learning process. The final successful scenario is added to the usage data and stored in the user profile of this leaner. It is assumed that the attributes selected by the experts should be verified by the learner results. Therefore, after a period of functioning the classification criterion is re-created by the so called rough classification methods which is based on clustering the scenarios belong to learner usage data. On the basis of this clustering the method determines the set of attributes which create the new criterion for learner classification [8, 13]. The cycle of the system life is described in work [12] and can be in brief presented as follows: The experts are asked for suggesting attributes from user data for learner classification. Each new learner is classified on the basis of values of these attributes. Owing to learning processes the usage data of learners are gathered. We may expect that the passed scenarios will not be similar even for the learners of the same class. If so, the situation means that it is needed to have a new criterion for classification. Using the passed scenarios we may perform clustering for the set of learners generating new partition Q of learners' set. If the actual learners partition (classification) differs meaningfully from partition Q then using the rough classification method we can determine the new criterion for learner classification. Next this new criterion is used for new learners.

We assume that in an E-learning system there is a set of learners $U=\{u_1,u_2,...,u_K\}$. Each learner is described by a set of user data attributes $A=\{a_1,a_2,...,a_N\}$. Set A generates partition P, i.e. a set of nonempty and disjoint classes which are subsets of U, whose union is equal U. In the learning process the user profile is completed with usage data, that is opening and final scenario. Based on passed scenarios the clustering process is done and gives partition Q. The rough classification task is to find a subset $B \subseteq A$, where the distance between Q and partition P_B generated by the attribute set B is the smallest according to certain distance function δ, where

$$\delta(P,Q)=\frac{1}{2}\sum_{i,j=1}^{K}|\, p_{ij}-q_{ij}\,|,$$

$$p_{ij}=\begin{cases}0 \text{ if } x_i,x_j \text{ in different class of } P\\ 1 \text{ otherewise}\end{cases},$$

and

$$q_{ij}=\begin{cases}0 \text{ if } x_i,x_j \text{ in different class of } Q\\ 1 \text{ otherwise}\end{cases}$$

In this paper such described problem is called RC-Problem and is formulated as follows: *For given partition Q of set U one should determine set $B \subseteq A$ such that $\delta(P_B,Q)$ is minimal.*

In this paper we present some heuristic algorithms for this problem and the results of verification process using statistical tests.

2 Related Works

2.1 Personalized System

There are many intelligent tutoring systems present on the market. In 1995-1996 web-based systems appeared. ELM-ART [2], which is used to help people learn LISP programming language, is one of them. One can think of it as an on-line course book. However, certain features make it intelligent and adaptive system, which personalizes content of pages based on learner's level and his previous results. NetCoach [18] is a learning system which adapts to learner's knowledge, goals and preferences. Mechanisms implemented in the system help find an optimal learning path. It is based on a user model which contains interactions with the system. ActiveMATH [10] is a system which supports math classes by generating interactive courses based on learner's abilities. SQL-Tutor [11] is a SQL teaching environment. As well as NetCoach, it uses user model containing a list of tests which were taken by a particular learner. The difficulty level changes together with learner's knowledge. German Tutor [5] is another intelligent system which helps learning foreign languages. It stores all test taken and results which are used to plan additional exercises. VC-Prolog [16] is another adaptive system that helps learning Prolog programming language. EDUCE System [6] is an interactive system which allows learner to pick from four different subjects. Depending on option chosen, content pages are being personalized.

Much more intelligent system that interacts and adapts to user have been classified and described in Brusilovsky's works [3, 4]. The theory of personalization of web-based learning systems can be found in [1, 7, 8].

2.2 Rough Classification

Rough sets theory has been defined by Pawlak in 1982 [14]. The definition of information system and rough classification [15] was a breakthrough for data-mining, knowledge discovery, classification and decision support systems. The rough sets theory is used in medical, pharmacology, engineer, banking and industry applications. Throughout the years many computing support systems based on rough sets have been developed including ROSETTA, RSES2, ROSECON, TEMPLATES, ARES [9]. Another aspect of rough classification was described in [12], where rough classification problem was formulated and a solution was suggested. The rough classification was used in an adaptive system to choose the most essential attributes and their values in order to group similar users. There are no articles considering rough classification in learning system, so far.

3 Solution of the RC-Problem

The introduced problem of rough classification is a NP-completed problem. Exact methods are not used in practice due to long operation time. The large computational complexity of algorithms causes that in real applications to solve this problem heuristic algorithms are applied. In this paper three methods are introduced.

3.1 Genetic Algorithms

Genetic algorithms belong to general heuristic methods and are based on mechanism of natural selection and heredity. This method is universal therefore several elements should be designed before using it. First of them is to choose coding elements of solution space. For the considering problem binary coding is assumed. The value of *i* gene of chromosome *c* defines which attributes belong to set *B*:

$$c(i)=\begin{cases}1 \ \textit{if}\ a_i \in B\\ 0 \ \textit{otherwise}\end{cases}$$

Fitness function is: $\delta(P_c,Q)+\sum_{i=1}^{N} c(i)$, where P_c is a partition after decoding chromosome *c*. A scheme of classical genetic algorithms [17] is presented below:

Given: *U* - set of user, *A* - set of attributes, partition *Q*

Result: Set *B,* where $\delta(P_B,Q)$ is minimal

BEGIN
1. Let t:=0
2. Initialize base population $P(t)$
3. Estimate fitness of population $P(t)$
4. $P_R(t)$=reproduction $P(t)$
5. $P_K(t)$=crossing $P_R(t)$
6. $P_M(t)$=mutation $P_K(t)$
7. t:=t+1
8. $P(t)= P_M(t)$
9. Does condition STOP satisfy? If not GOTO 3.
10. Show the best chromosome from population.

END.

This algorithm is characterized with simplicity and fast working time.

3.2 Random Algorithm with Decomposition

The algorithm searches for the space of solutions in a random way. The decomposition is based on separating *N* subspaces of solutions where *card*(*B*)=*d.* In the next iteration *d* different attributes are randomized, added to set *B* and quality of solution $\delta(P_B,Q)$ is checked.

Given: *U*- set of user, *A* - set of attributes, *N* - cardinality of set *A*, *M* - number of iterations, partition *Q*

Result: Set *B,* where $\delta(P_B,Q)$ is minimal

BEGIN
1. Randomize attribute a_z and let $d_{\min}=\delta(P_{a_z},Q)$; $B_{\min}=a_z$, $z\in\{1,\dots,N\}$
2. Let i:=1; j:=0;
3. Randomize i attributes and add them into set B
4. If $\delta(P_B,Q)<d_{\min}$ then $d_{\min}=\delta(P_B,Q)$; $B_{\min}=B$

5. $j:=j+1$
6. If $j<M$ GOTO 3
7. $i:=i+1$
8. If $i<N$ GOTO 3

END.

The advantage of this algorithm is simplicity and small computational complexity $O(n^2)$. The defect is randomization.

3.3 Algorithm with Undistinguishable Matrix

In this subsection algorithm with undistinguishable matrix is presented. To show this method a few auxiliary definitions are quoted.

Definition 1. *If* $S=(U,A,V,f)$ *is information system where* $U=\{u_1,u_2,...,u_K\}$ *and* $A=\{a_1,a_2,...,a_N\}$ *then the undistinguishable matrix of information system S is defined as:* $MN(S)=\{a\in A: f(u_i,a)=f(u_j,a)\ dla\ i,j=1,...,K\}$.

Definition 2. *If* $S=(U,A,V,f)$ *information system then the undistinguishable vector for partition Q is equal:*

$$W(S)=MN(S)[i,j],\ \ if\ i,j\in belong\ to\ the\ same\ class\ Q\ \ for\ i\neq j$$

Given: U - set of user, A - set of attributes, partition Q

Result: Set B, where $\delta(P_B,Q)$ is minimal

BEGIN
1. $B=\varnothing$
2. Assign undistinguishable matrix $MN(S)$
3. Create $W(S)$ using undistinguishable matrix $MN(S)$
4. For each $a_i\in A, i\in 1,...,N$ calculate $f(a_i)$, that is frequency of appearances of element a_i in sets of $W(S)$
5. For each $a_i\in A, i\in 1,...,N$ if $f(a_i)>\frac{\sum_{i=1}^{N} f(a_i)}{N}$ then $B=B\cup a_i$
6. $C:=B$
7. For each $b\in B$ if $\delta(P_{B-b},Q)\le\delta(P_B,Q)$ then $B=B\backslash\{b\}$
8. If $B\neq C$ GOTO 6

END.

The idea of this algorithm is based on a construction of solution when partition Q is known. The small computational complexity is an advantage of this method.

4 Statistical Verification

Algorithms described in chapter 3, exact algorithm based on brute force and the third algorithm for RC-3 problem [12] is implemented in Java where the environment for simulation is created. K-means are used as a clustering method. Ranking selection, one point crossover and STOP after M iterations are assumed. All algorithms are statistically researched on a random population of 50.

Assumptions:

$$M=30;\ K=30;\ p_k=0.8;\ p_m=0.01;$$

Statistically tested feature:

$$\alpha = d(P_{exact}, Q) - d(P_{tested}, Q)$$

Hypothesis 1

Algorithm 3rd for RC-3 problem [12] satisfies following assumption: $H_0 : \overline{\alpha} = 2$ or alternate assumption: $H_1 : \overline{\alpha} < 2$ where $\overline{\alpha}$ - an average value of statistically tested feature.

Proof

There are no information considering standard deviation and the tested sample is small, that is why for testing the null hypothesis the Learner's t-distribution is assumed:

$t = \frac{\overline{\alpha} - \mu}{S}\sqrt{p-1}$ with (p-1) degrees of freedom, where p-sample size, μ -assumed average value, S-standard deviation. The significance level of 0.05 is assumed.

$$t = \frac{1.08 - 2}{2.88}\sqrt{50-1} = -\frac{6.44}{2.88} = -2.24\ ;\ t_{kr} = t_{(0.05;49)} = -1.68$$

The tested statistical value is smaller than the critical value, that is why the null hypothesis is rejected and alternate hypothesis is assumed.

Conclusion 1. *The average distance between the 3rd algorithm for the RC-3 problem and the exact algorithm is not greater than 2, which shows that no more than 3.33% of population could be classified better.*

Hypothesis 2

Genetic algorithm satisfies following assumption: $H_0 : \overline{\alpha} = 4.5$ or alternative assumption: $H_1 : \overline{\alpha} < 4.5$, where $\overline{\alpha}$ - an average value of statistically tested feature.

Proof

As it was done before, the Learner's t-distribution is assumed:

$t = \frac{\overline{\alpha} - \mu}{S}\sqrt{p-1}$ with (p–1) degrees of freedom, where p-sample size, μ -assumed average value, S-standard deviation. The significance level of 0.05 is assumed. We have $t = \frac{3.16 - 4.5}{3.72}\sqrt{50-1} = -\frac{9.38}{3.72} = -2.52$; $t_{kr} = t_{(0.05;49)} = -1.68$; $t_{kr} > t$, that is why the null hypothesis is rejected and alternate hypothesis H_1 is assumed.

Conclusion 2. *The average distance between the genetic algorithm and the exact algorithm is not greater than 4.5, which shows that no more than 15% of population could be classified better.*

Hypothesis 3
Algorithm 3rd for RC-3 problem [12] and random algorithm satisfies following assumptions: $H_0 : \overline{\alpha}_{algorithm3} = \overline{\alpha}_{random}$ or alternate assumption:

$$H_1 : \overline{\alpha}_{algorithm3} > \overline{\alpha}_{random},\quad H_1 : \overline{\alpha}_{algorithm3} < \overline{\alpha}_{random}$$

where $\overline{\alpha}$ - an average value of statistically tested feature.

Proof
For testing the null hypothesis the following statistic is assumed:

$$t = \frac{|\overline{\alpha}_{algorithm3} - \overline{\alpha}_{random}|}{\sqrt{\dfrac{S^2_{algorithm3} + S^2_{random}}{p-1}}}$$

with $(p-1)$ degrees of freedom, where p-sample size, μ -assumed average value, S-standard deviation. The significance level of 0,05 is assumed. We have

$$t = \frac{1.08 - 0.52}{\sqrt{\dfrac{8.28 + 2.58}{49}}} = \frac{0.56}{0.47} = 1.19\,;$$

$$t_{kr} = t_{(0.05;49)} = 2.01\,;\ t < t_{kr},$$

that is why the null hypothesis cannot be rejected.

Conclusion 3. *The average distance between the* 3rd *algorithm for the* RC-3 *problem and the exact algorithm as well as the average distance between the random algorithm and the exact algorithm are not significantly different.*

5 Conclusions and Further Work

In the paper three algorithms for solving the rough classification problem are introduced. They are compared with the 3rd algorithm described in [12] and exact algorithm. The statistical tests show that the 3rd algorithm is not significantly different than the random algorithm. However, it is much better than genetic algorithm.

The 3rd algorithm seems to be a good heuristic in comparison with the exact algorithm. Only 3.33% learners could be classified better. Considering the amount of time needed for the brute force method, it is a good approximation. A system implementation for remote education using the rough classification methods is planned. Next step would be testing described algorithms on a real data. Introduced methods can be used in intelligent tutoring systems to increase efficiency and personalization. Being able to adapt the knowledge to different learner's preferences (age, education level, IQ level etc.) can lead to more efficient education process.

References

1. Brusilovsky, P.: Methods and techniques for adaptive hypermedia. User Modeling and User-Adapted Interaction 6(2-3), 87–129 (1996)
2. Brusilovsky, P., Schwarz, E., Weber, G.: ELM-ART: An intelligent tutoring system on World Wide Web. In: Lesgold, A., Frasson, C., Gauthier, G. (eds.) ITS 1996. LNCS, vol. 1086, pp. 261–269. Springer, Heidelberg (1996)
3. Brusilovsky, P.: Adaptive and Intelligent Technologies for Web-based Education. In: Rollinger, C., Peylo, C. (eds.) Künstliche Intelligenz, Special Issue on Intelligent Systems and Teleteaching, vol. 4, pp. 19–25 (1999)
4. Brusilovsky, P.: Adaptive Educational Hypermedia. In: A summary of an invited talk at PEG'01 conference. In: Proceedings of Tenth International PEG conference, Tampere, Finland, June 23-26, pp. 8–12 (2001)
5. Heift, T., Nicholson, D.: Web Delivery of Adaptive and Interactive Language Tutoring. International Journal of Artificial Intelligence in Education 12(4), 310–325 (2001)
6. Kelly, D., Tangney, B.: Predicting Learning Characteristics in a Multiple Intelligence Based Tutoring System. Intelligent Tutoring Systems, 678–688 (2004)
7. Kobsa, A., Koenemann, J., Pohl, W.: Personalized Hypermedia Presentation Techniques for Improving Online Customer Relationships, The Knowledge Engineering Review, vol. 16(2), pp. 111–155. Cambridge University Press, Cambridge (2001)
8. Kukla, E., Nguyen, N.T., Daniłowicz, C., Sobecki, J., Lenar, M.: A model conception for optimal scenario determination in an intelligent learning system. ITSE - International Journal of Interactive Technology and Smart Education 1(3), 171–184 (2004)
9. Matusiewicz, Z.: Computer technology in teaching fuzzy sets, Varia Informatica. Obliczenia matematyczne. red: Grzegórski, S., MiŁosz, M., Muryjas, P.(eds.) PTI Lublin (2006)
10. Melis, E., et al.: ActiveMath: A Generic and Adaptive Web-Based Learning Environment. International Journal of Artificial Intelligence in Education 12, 385–407 (2001)
11. Mitrovic, A., Hausler, K.: Porting SQL-Tutor to the Web, K. In: Gauthier, G., VanLehn, K., Frasson, C. (eds.) ITS 2000. LNCS, vol. 1839, pp. 37–44. Springer, Heidelberg (2000)
12. Nguyen, N.T.: Advanced Methods for Inconsistent Knowledge Management. Springer, London (2007)
13. Nguyen, N.T., Sobecki, J.: Determination of User Interfaces in Adaptive Systems Using a Rough Classification based Method. New Generation Computing 24(4), 377–402 (2006)
14. Pawlak, Z.: Information systems- theoretical foundation, Wydawnictwo Naukowo-Techniczne, Warszawa (1983)
15. Pawlak, Z.: Rough classification. Int. J. Human-Computer Studies 51, 369–383 (1999)
16. Peylo Ch, T.T., Rollinger C, G.H.: A web-based intelligent educational system for PROLOG. In: Gauthier, G., VanLehn, K., Frasson, C. (eds.) ITS 2000. LNCS, vol. 1839, Springer, Heidelberg (2000)
17. Rutkowska, D., Piliński, M., Rutkowski, L.: Neural Networks, Genetic Algorithms and Fuzzy Systems. Wydawnictwo Naukowe PWN, Warszawa (1999)
18. Weber, G., Kuhl H.Ch., Weibelzahl, S.: Developing Adaptive Internet Based Courses with the Authoring System NetCoach. In: Reich, S., Tzagarakis, M., de Bra, P. (eds.) Hypermedia: Openness, Structural Awareness, and Adaptivity, pp. 226–238. Springer, Berlin (2001)

Toward a Novel Multi-modal HCI: Fusion Architecture Using Confidence Score and Fuzzy Value

Jung-Hyun Kim[1], Jeh-Seon Youn[2], and Kwang-Seok Hong[1]

[1] School of Information and Communication Engineering, Sungkyunkwan University, 300, Chunchun-dong, Jangan-gu, Suwon, KyungKi-do, 440-746, Korea
kjh0328@skku.edu, kshong@skku.ac.kr
http://hci.skku.ac.kr
[2] HCILAB Co., LTD., 235-2, Guro-Dong, Guro-Gu, Seoul, 152-740, Korea
sunhci@hotmail.com

Abstract. This paper describes a wireless PDA-based Multi-Modal Interaction Architecture (hereinafter, MMIA) fusing 3 sensory channels, which are a detachable stylus that can be used on the touch screen by coupling embedded speech and KSSL recognizer. The MMIA fuses and recognizes 15 word-based instruction models that are represented by stylus (or touching), speech and KSSL (Korean Standard Sign Language), and then translates the recognition result into synthetic speech and visual illustration, for multi-modal PDA handling and interaction. The experimental results, each average recognition rate for the MMIA fusing 3 sensory channels was 95.73% in clean environments, and 93.29% was shown in noisy environments.

1 Introduction

Many original PDAs (Personal Digital Assistants) featured touch screens and virtual keyboards that a keyboard is shown on the touch screen, having only a few buttons usually reserved for shortcuts to often used programs, for user interaction [1], [2], [3]. To name only a few examples, they proved to be a viable aid for visually impaired users, an alternative to WIMP (Windows, Icon, Menu, and Pointer based interfaces) interfaces in mobile computing, an entertaining extension of computer games [4]. However, the next generation of HCI technology for more advanced and personalized PC system such as wearable computer and PDA based on wireless network and wearable computing, may require and allow new interfaces and interaction techniques such as tactile interfaces with haptic feedback methods, and gesture interfaces based on hand gestures, to serve different kinds of users. Namely, for perceptual experience and behavior to benefit from the simultaneous stimulation of multiple sensory modalities that are concerned with the (five) senses, fusion and fission technologies of the information from these modalities are very important and positively necessary. In other words, neo-HCI technology for the next generation PC requires the (five) senses information processing technology and user interface technology that are concerned with various input/output devices as well as a touch screen and a virtual keyboard that is shown on the touch screen, for more efficient and intelligent HCI than the old one. Consequently, we implement a wireless PDA-based MMIA using a detachable stylus

B. Apolloni et al. (Eds.): KES 2007/WIRN 2007, Part I, LNAI 4692, pp. 246–253, 2007.

by coupling embedded speech and KSSL recognizer, and suggest an improved synchronization method between multi-modalities for the simultaneous and sequential multi-modality.

2 A Wireless PDA-Based MMIA Fusing 3 Sensory Channels

The PDA that is used in this study is based on Intel XScale PXA270 operating at 624 MHz. Capacities of the main memory and storage memory are 64 MB and 128 MB correspondingly, the storage memory has 85 MB allocated for user data. What concerns wireless interfaces, it has Wi-Fi, Bluetooth and IrDA. This PDA operates under Windows Mobile 2003 SE.

2.1 Touch; A Touch Screen PDA-Based Mmulti-Touch

A touch screen PDA-based interaction is then done by tapping, double-tapping, or touching the screen to activate buttons or menu choices, and dragging the stylus on a touch screen. Particularly, a resistive touch screen technology is a reliable and affordable technology that is widely used by individuals and in less demanding workplace applications, and is pressure sensitive so it responds to any input device, including finger, gloved hand, or pen stylus. Resistive touch screen rely on a touch overlay, which is composed of a flexible top layer and a rigid bottom layer separated by insulating dots, attached to a touch screen controller. The inside surface of each of the two layers is coated with a transparent metal oxide coating (ITO) that facilitates a gradient across each layer when voltage is applied. Pressing the flexible top sheet creates electrical contact between the resistive layers, producing a switch closing in the circuit. The control electronics alternate voltage between the layers and pass the resulting X and Y touch coordinates to the touch screen controller. The touch screen controller data is then passed on to the PDA operating system for processing [5], [6].

2.2 Speech; Embedded Speech Recognizer

We implement a wireless PDA-based embedded Vocabulary-Independent Speech Recognizer (VISR) composing standard models such as CV, VCCV, VC. Where, standard models are consisted of C (Consonant) and V (Vowel) in phonology of the Korean language, for example, standard model CV is represented by "Consonant + Vowel". In embedded VISR, a sampling rate is 8 KHz, and a frame length is 10ms; we allow the 50% overlapping. To extract feature vectors at each frame from given speech data, we applied the MFCC (Mel Frequency Cepstral Coefficient-whole the 39th feature vectors: MFCC the 13th, Delta and Delta-Delta). As recognition and search algorithm, we used a Continuous Hidden Markov Model (CHMM) and the Viterbi algorithm. A HMM is an extended stochastic model from Markov chain to find a sequence of states with observation value, a probabilistic function of a state, and the parameters of a HMM are given in Eq. (1).

$$\lambda = \{A, B, \pi\} \tag{1}$$

Where, $A = \{a_{ij}\}$ is the set of transition probability from state i to state j, $B = \{b_j(k)\}$ is the set of observation probability given the current state j, and $\pi = \pi_i$ is the set of initial state probability. However, in CHMM, $b_j(\boldsymbol{x})$ in Eq. (2) is substituted for $B = \{b_j(k)\}$ in Eq. (1).

$$b_j(x) = \sum_{K=1}^{M} c_{jk} b_{jk}(x) = \sum_{K=1}^{M} c_{jk} N(x, \mu_{jk}, \sum\nolimits_{jk}) \tag{2}$$

Where, $\boldsymbol{x}$ is the observation vector being modeled, C_{jk} is the mixture coefficient for the k-th mixture in state j, and N is the Gaussian pdf with mean vector μ_{jk} and covariance matrix $\sum_{jk}$ for the k-th mixture component in state j. In addition, embedded VISR finds a confidence score that indicates the posterior probability of an utterance being correctly matched to a word model, and then it decides on the priority order between the recognition results. According as confidence scores in embedded VISR ranged from approximately −30 to near +70, they are normalized by a positive number between 0 and 100, and in case normalized confidence score is *over than 53,* embedded VISR returns speech recognition result to user. The flowchart of embedded VISR is shown in Fig. 2 (in section 2.4).

2.3 Gesture; Embedded KSSL Recognizer

For a wireless PDA-based embedded KSSL recognizer, we constructed the 15 word-based instruction models, according to "Korean Standard Sign Language Tutor [7]", and we utilized 5DT company's wireless data gloves and Fastrak® to capture the KSSL hand and motion gesture [8]. In addition, for a clustering method to extract feature vectors from captured dynamic KSSL data, and to construct distributed computing-based KSSL instruction models, we suggest an improved RDBMS (Relational Data-Base Management System) clustering module [9]. As KSSL recognition algorithm, we applied trapezoidal shaped membership functions and the fuzzy max-min composition that is defined in Eq. (3).

$$\textit{For}\ (x, y) \in A \times B,\ (y, z) \in B \times C,$$

$$\mu_{S \bullet R}(x, z) = \underset{y}{Max}\ [Min\ (\mu_R(x, y), \mu_S(y, z))] \tag{3}$$

- *R and S : fuzzy relations → R ⊆ A × B, S ⊆ B × C, S•R □A×C*
- *A, B and C : defined fuzzy numbers-set*
- *S•R = SR : The composition of two relations*

Also, embedded KSSL recognizer calculates and produces a fuzzy value that is the posterior probability of a gesture being correctly matched to a KSSL recognition model from the user's dynamic KSSL via a fuzzy reasoning and composition process, and then decides and recognizes various KSSL according to produced fuzzy value [9], [10]. The flowchart of KSSL recognizer is shown in Fig. 2 (in section 2.4) together with the flowchart of the MMIA.

2.4 The Fusion Scheme for a Wireless PDA-Based MMIA

A wireless PDA-based fusion and fission schemes between speech-touch-gesture modalities consists of six major steps: 1) the user inputs prescribed instruction models using speech, KSSL and touch to a wireless PDA-based MMIA, simultaneously, 2) inputted 3 sensory information are recognized and recorded by a touch controller (that is a PDA application), the embedded VISR and KSSL recognizer, and then recognition results are saved to the MMDS (Multi-Modal Database Server; The MMDS is a responsible database for synchronizing between modalities, 3) at this point, the user's touch, speech and KSSL data are synchronized depending on the internal SQL logic in the MMDS, and then 4) while suggested the MMIA runs a validity check function on touch, ASR and gesture recognition results with prescribed instruction models, the NAT (Noise Analysis Tool) analyzes a noise, and 5) the MMIA applies weights into an individual embedded recognizer by improved fusion and fission rules, 6) finally,

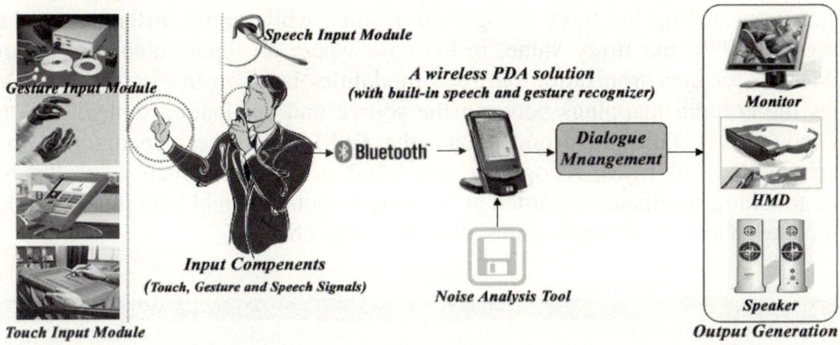

Fig. 1. Fusion architecture fusing 3 sensory channels

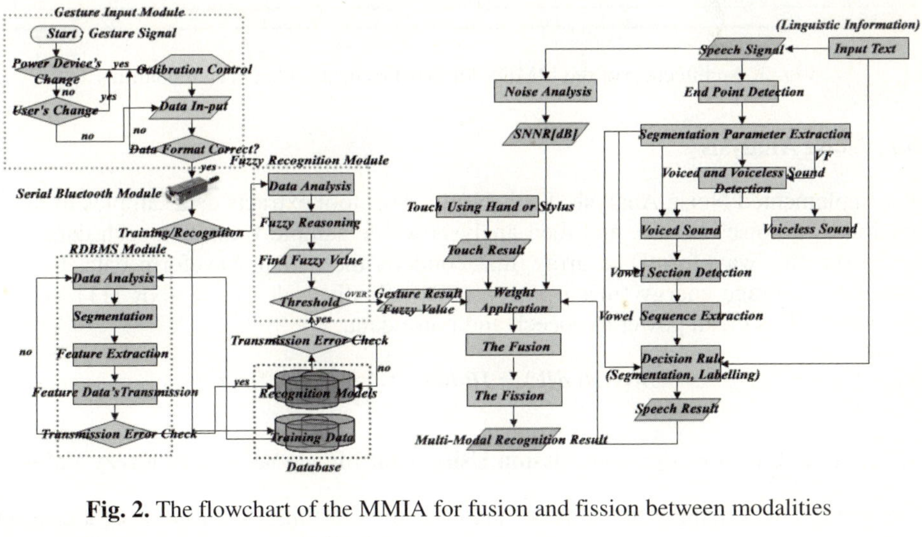

Fig. 2. The flowchart of the MMIA for fusion and fission between modalities

user's intention is provided through TTS and visualization. The suggested architecture and flowchart of MMIA are shown in Fig. 1 and 2.

3 Improved Fusion and Fission Rules for the MMIA

3.1 A Synchronization Between Touch-Audio-Gesture Modalities

This paper solves the asynchronous control problems between touch, speech and gesture signals using a SQL-logic and word-unit input method based on the database (the MMDS in section 2.4). In other words, after individual speech, KSSL recognizer and touch controller recognize instruction models, they transmit recognition results into the MMDS. However, the transmission time of recognition results has some time delay because of asynchronous communication between three input signals. As a result, touch, speech and KSSL recognition results are recorded sequentially to the MMDS, and the MMDS is kept in standby mode via internal SQL-logic in case one was not input among the three recognition results, while apply different weight according to SNNR and fuzzy value, in the case where all input values are recorded. The MMDS for synchronization between modalities is given in Fig. 3. The MMDS defines the column mappings between the source and destination tables. The 'Gesture' and 'Voice' fields save and update the KSSL and speech recognition results from the embedded KSSL recognizer and VISR, as well as touch controller-based double-touching results are recorded at 'Double-Touching' field, and then 'Checking' field updates a synchronization states after a validity check.

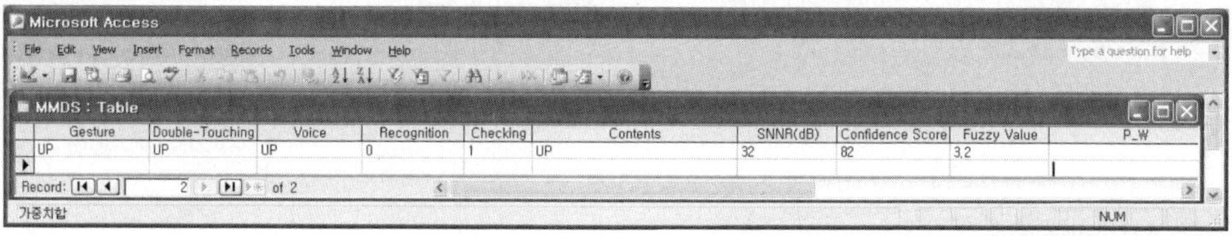

Fig. 3. Architecture of the MMDS for synchronization between modalities

3.2 Noise Analysis

We implemented Noise Analysis Tool (NAT). This tool extracts data samples in mute section as a noise domain, and then analysis whole sample frames; speech data section from the 'waveformData array' that contains the actual waveform data, then it calculates average energy (mean power; [dB]), finally, calculate SNNR [11] by Eq. (4), where, P is mean power of speech and noise data.

$$SNNR(dB) = 10 \log_{10} \frac{P_{signal+noise}}{P_{noise}} \tag{4}$$

3.3 SNNR-Based Fusion and Fission Using Confidence Score and Fuzzy Value

The process of weighting in this paper involves emphasizing some aspects of a data set that has a probability value - giving them 'more weight' in the final recognition result. So, we used normalized confidence score in embedded VISR (in section 2.2) as a speech probability value, and a gesture probability value utilized the fuzzy value that is found from embedded KSSL recognizer(in section 2.3), as well as a touch probability value applied average touch success rates by double-touching using a finger. The fusion and fission processing between modalities consists of three major steps: 1) as touch, speech and KSSL probability value, it calls the normalized confidence score *(S)* depending on the defined SNNR, average touch success rates (*T*), and the normalized fuzzy value *(G)* for KSSL recognition, from the MMDS, and then applies defined weights (W_S, W_T and W_G) in called-individual probability values (*S, G* and *T*), 2) and then calculate the expected value (*P_W*) after weighting via by Eq. (5), 3) return recognition result by *P_W* value to user. Where, the maximum of the normalized fuzzy value is 3.5, and the minimum critical value is 3.2 (in our study). Accordingly, the normalized fuzzy value (*G*) in the embedded KSSL recognizer is defined Eq. (6). In addition, because each confidence scores in defined SNNR find very various values, an average confidence score (*CS_Ave*) is inevitably necessary to give standard by which to calculate speech probability (*S*) with Eq. (7). Therefore, we achieved repeatedly speech recognition experiments 10 times with the 10 test recognition models, for every 5 reagents in the given each *SNNR*, and then an average confidence score (*CS_Ave*) was found with Table 1.

Table 1. In case *CS_Current* is 90 and *FV_Current is 3.2,* a weight value depending on SNNR and each probabilities. Note, because speech and KSSL probability *(S and G)* are changed according to *CS_Current* and *FV_Current*, *P_W* is changed justly.

SNNR Critical value	Weight values							
	Speech			KSSL		Touch		*P_W*
	W_S	S	*CS_Ave*	W_G	*G*	W_T	*T*	
more than 40 [dB]	0.96	0.873	0.970	0.01	0.914	0.03	0.879	0.874
35 [dB] ≤ SNNR < 40 [dB]	0.95	0.858	0.953	0.02	0.914	0.03	0.879	0.860
30 [dB] ≤ SNNR < 35 [dB]	0.91	0.841	0.934	0.06	0.914	0.03	0.879	0.847
25 [dB] ≤ SNNR < 30 [dB]	0.88	0.831	0.923	0.09	0.914	0.03	0.879	0.840
20 [dB] ≤ SNNR < 25 [dB]	0.85	0.825	0.917	0.12	0.914	0.03	0.879	0.837
15 [dB] ≤ SNNR < 20 [dB]	*Feed-Back*							
10 [dB] ≤ SNNR < 15 [dB]	0.05	0.223	0.248	0.90	0.914	0.05	0.879	0.878
5 [dB] ≤ SNNR < 10 [dB]	0.03	0.154	0.171	0.87	0.914	0.10	0.879	0.888
0 [dB] ≤ SNNR < 5 [dB]	0.02	0.115	0.128	0.83	0.914	0.15	0.879	0.893
less than 0 [dB]	0.01	0.036	0.040	0.80	0.914	0.19	0.879	0.899

As a result, if *P_W* value is greater than or equal to 0.837 and SNNR is over than or equal to 25[dB], the MMIA fissions and returns recognition result of embedded VISR, and it fissions the embedded KSSL recognizer in case *P_W* value is greater than or equal to 0.878 and SNNR is less than or equal to 5[dB], as well as if *P_W* value is greater than or equal to 0.837 and less than 0.878 and SNNR is greater than or equal to 10[dB] and less than 25[dB], it fissions touch recognition mode. Note, in case SNNR critical value for weight decision is ambiguous; 15[dB] ≤ SNNR < 20[dB] in Table 1, according as the MMIA requests re-input (touch, speech and KSSL) to user via the feed-back function for clear a declaration of intention, more improved language processing is available.

$$P\_W = \sum(W_S \times S\ ,\ W_G \times G\ ,\ W_T \times T\ ) \tag{5}$$

- *P_W: The expected value after weight application*
- *W_S: Defined weight for speech recognition mode*
- *W_G: Defined weight for KSSL recognition mode*
- *W_T: Defined weight for touch recognition mode*
- *S: Speech probability (Normalized confidence score, depending on each SNNR)*
- *G: KSSL probability (Normalized fuzzy value from embedded KSSL recognizer)*
- *T: Touch probability (Average touch success rates by double-touching is **0.874**)*

$$G = \frac{FV\_Current}{FV\_Max} = \frac{FV\_Current}{3.5} \tag{6}$$

- *FV_Current: The fuzzy value from the current gesture(KSSL)*
- *FV_Max = 3.5: The maximum of the fuzzy value for KSSL recognition*

$$S = CS\_Current \times CS\_Ave \tag{7}$$

- *CS_Current: A confidence score for current utterance in the given SNNR*
- *CS_Ave: Average confidence score, by speech recognition experiments in defined each SNNR*

4 Experiments and Results

The experimental set-up is as follows. The distance between the KSSL input module and a wireless PDA with a built-in KSSL recognizer approximates radius 10M's ellipse form. In KSSL gesture, touch and speech, we move the wireless data gloves and the motion tracker to the prescribed position. Also, while the user inputs KSSL using data gloves and a motion tracker, speaks using the blue-tooth headset in a telephone terminal, and inputs a touch modality by double-tapping using a finger on the touch screen. For every 10 reagents, we repeat this action 10 times in noisy and clean

Table 2. The average recognition results for the 15 word-based instruction models

Evaluation / Reagents	Uni-modal instruction Interface						The MMIA	
	KSSL (%)		Speech (%)		Double-Touch (%)		KSSL + Speech + Touch (%)	
	Noise	Clean	Noise	Clean	Noise	Clean	Noise (%)	Clean (%)
Reagent 1	92.8	92.7	62.4	98.3	88.6	88.5	93.1	97.3
Reagent 2	93.2	93.3	61.3	95.1	89.3	89.3	92.9	96.9
Reagent 3	94.1	94.2	53.7	94.7	87.6	87.7	94.3	94.7
Reagent 4	92.6	92.4	64.1	95.4	91.8	91.8	92.7	95.1
Reagent 5	93.1	92.9	57.3	96.2	89.6	89.6	93.5	96.8
Reagent 6	91.8	92.1	56.1	96.1	90.9	90.8	92.3	96.1
Reagent 7	91.7	91.8	61.2	93.2	87.7	87.7	92.5	98.3
Reagent 8	93.9	93.7	52.9	93.9	91.3	91.4	93.7	94.4
Reagent 9	92.8	93.0	58.7	94.3	88.5	88.5	94.2	93.2
Reagent10	93.1	93.1	54.9	95.8	87.1	87.2	93.7	94.5
Average	92.91	92.92	58.26	95.30	89.24	89.25	93.29	95.73

- *As a noisy environment, the average SNNR using actual waveform data is recorded in laboratory space, including the music and the mechanical noise, was about 13.59[dB].*
- *The average SNNR using actual waveform data that remove noise elements for a clean environment was about 38.37[dB].*
- *Note, if the SNNR changes by experimental conditions such as the music and the mechanical noise, because weight also is changed, experiment result can be changed.*

environments. Experimental results, the uni-modal and the MMIA's average recognition rate using a wireless PDA, in noisy and clean environment for the 15 word-based instruction models, are shown in Table 2 respectively. These experiments were achieved on below experimental conditions and weights.

5 Conclusion

Wearable and ubiquitous computing-based multi-modal user interfaces have implications for accessibility. This study combines natural language and artificial intelligence techniques to allow human computer interaction with an intuitive mix of speech, touch, and gesture (KSSL) based on a wireless PDA. The MMIA's average recognition rates for the 15 word-based instruction models were 95.73% in clean environments (e.g. office space), as well as 93.29% were shown in noisy environments. In addition, the MMIA supports two major types of multi-modality for language recognition based on embedded and ubiquitous computing. Finally, we clarify that this study is a fundamental study for more natural the (five) senses information processing, and for an advanced multi modal recognizer integrating the five senses such as sight, smell, and taste.

Acknowledgement

This research was supported by MIC, Korea under ITRC IITA-2006-(C1090-0603-0046).

References

1. Myers, B.A.: A brief history of human-computer interaction technology, Interactions. ACM Press 5(2), 44–54 (1998)
2. Jacko, J.A., et al.: Handbook for Human Computer Interaction, Mahwah, Lawrence Erlbaum & Associates (2003)
3. Wooldridge, M.J., Jennings, N.R.: Intelligent agents: Theory and practice, Know. Eng. Review 10(2), 115–152 (1995)
4. Buxton,B.: Multi-Touch Systems that I Have Known and Loved, Microsoft Research(2007)
5. Fuchs, P.M.: Architecture of Multi-modal Dialogue System. In: Sojka, P., Kopeček, I., Pala, K. (eds.) TSD 2000. LNCS (LNAI), vol. 1902, pp. 433–438. Springer, Heidelberg (2000)
6. Owen, R., et al.: When it Gets More Difficult, Use BothHhands - Exploring Bimanual Curve Manipulation. Proc. Graphics Interface, GI'05, 17–24 (2005)
7. Kim, S.-G.: Korean Standard Sign Language Tutor, 1st edn. Osung Publishing Company, Seoul (2000)
8. Kim, J.-H., et al.: Hand Gesture Recognition System using Fuzzy Algorithm and RDBMS for Post PC. In: Wang, L., Jin, Y. (eds.) FSKD 2005. LNCS (LNAI), vol. 3614, pp. 170–175. Springer, Heidelberg (2005)
9. Kim, J.-H., et al.: An Implementation of KSSL Recognizer for HCI Based on Post Wearable PC and Wireless Networks. In: Gabrys, B., Howlett, R.J., Jain, L.C. (eds.) KES 2006. LNCS (LNAI), vol. 4251, Springer, Heidelberg (2006)
10. kandasamy, W.B.V.: Smaranda Fuzzy Algebra. American Research Press, Seattle (2003)
11. Stein, J.Y.: Digital Signal Processing: A Computer Science Perspective. John Wiley & Sons, Chichester (2000)

Using Uncertainties as Basis for Evaluating Plans

Christofer Waldenström

Uppsala university,
Institution of information science,
Dept. Computer science
christofer.waldenstrom@dis.uu.se

Abstract. Decision-making requires coping with uncertainties. Technical decision-support has traditionally focused on trying to represent these uncertainties and perform reasoning under uncertain conditions using probability theory or one of its relatives. An alternative method for handling uncertainty is to study what decision-makers actually do when encountering uncertainty and try to develop a system that supports or utilizes those activities. This paper suggests an agent based system that gives the decision-maker an opportunity to evaluate what impact different assumptions have on a plan. The system tracks uncertainties encountered during planning and use these uncertainties to generate a set of scenarios. The scenarios are then tested for workability in a cooperative process between the planners and the system. The purpose of such evaluation is to make sure the plan is workable in a worst-case scenario and to identify the uncertainty(s) that has most impact on the mission.

Keywords: Decision making, uncertainty, assumption, scenario.

1 Introduction

Much planning involves coping with uncertainty [1] [2] [3]. Technical decision-support has traditionally focused on trying to represent these uncertainties and perform reasoning using probability theory or one of its relatives, such as fuzzy set theory, Dempster-Shafer theory and certainty factors [4]. An alternative method for handling uncertainty during planning is to study what decision-makers actually do when encountering uncertainty, and try to develop a system that supports or utilizes those activities. Studies of how decision-makers cope with uncertainty have been performed and they show that decision-makers distinguish between three types of uncertainty: inadequate understanding, incomplete information and undifferentiated alternatives[3] [5]. Uncertainty can be defined as "a sense of doubt that blocks or delays action"[3]. This definition of uncertainty will be used throughout this paper. This paper describes an approach to utilize the uncertainties caused by incomplete information for post-planning evaluation of the plan. The purpose of such evaluation is to (1) make sure the plan is workable in a worst-case scenario and (2) identify the uncertainty(s) that has most impact on the mission. This paper implements (1) using an agent-based approach.

B. Apolloni et al. (Eds.): KES 2007/WIRN 2007, Part I, LNAI 4692, pp. 254–261, 2007.

2 Overview of the Procedure

During planning, whenever an uncertainty is encountered, the cause, its relation(s) to the plan and the action(s) taken to overcome the uncertainty are noted and transferred to a dedicated uncertainty record. Tracking the uncertainties in this way requires extra effort from the planners, but since uncertainties are a major cause of perceived difficulties the extra effort may be justified [6]. When the planning is finished, the plan together with the list of uncertainties is used as the basis for scenario generation. After the scenarios have been generated, an evaluation of the plan takes place with focus on the following questions:

1. Does the plan look workable in light of worst-case assumptions?
2. Which uncertainties have the greatest impact on the plan?

As an illustration, we apply the procedure of tracking uncertainties to the military domain[1]. Figure 1 shows the output of a military planning process. At the bottom is the mission statement with a sketch to the left. The mission statement can be on any level of detail (here displayed as a rough outlined course of action). To the right is the list of uncertainties together with alternatives for each uncertainty.

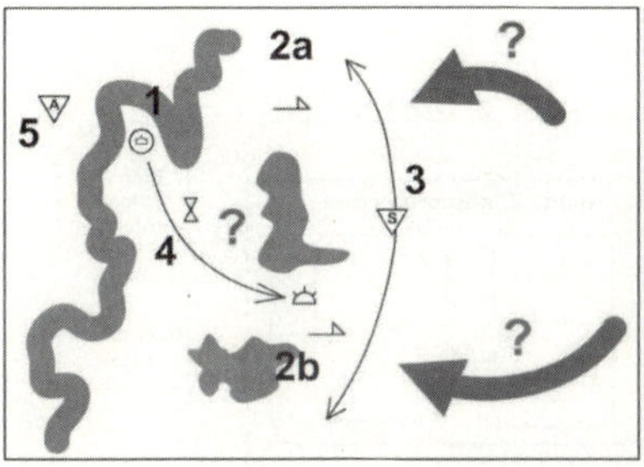

Uncertainty 1	Strength of enemy surface attack
Alt 1	2 Cruisers, 4 Corvettes
Alt 2	2 Cruisers, 2 Corvettes
Alt 3	4 Corvettes
Uncertainty 2	**Enemy course of action**
Alt 1	From north-east
Alt 2	From south-east
Alt 3	Already in mission area
Uncertainty 3	**Own course of action**
Alt 1	SA1, SA2 together with AA
Alt 2	SA1, SA2 north area, AA south area

Mission: *Task Group 01 lays before 070800 out minefield 8123*
Resources: *Minelayer ML; surface attack SA1, SA2; reconnaissance aircraft RA; attack air-craft AA; close protection group CP.*
Implementation: *(1) ML lays out minefield 8123 protected by CP. (4) CP performs free anti submarine warfare before ML departure, and resumes escort duty when ML leaves archipelago. (2a) SA1 and (2b) SA2 conducts attack missions to prevent enemy naval units from attacking minelayer. (3) RA supports SA1, SA2 with surface surveillance. (5) AA supports SA1, SA2 with attack missions. AA in high alert from 070100.*

Fig. 1. A mission statement for a naval mine laying mission together with list of uncertainties

[1] Since the uncertainties tracked here actually relates to the decision making during planning any other situation that involves decision making under uncertainty (or risk) could have been chosen. For examples of this extensive field, see [7] [8] [9].

3 Representing the Actors and the Environment

Uncertainties may relate to any aspect of a task. This implies that all relevant aspects of a task must be modelled in order to perform an evaluation. For example, from the sketch in Fig. 1, one can extract two central aspects: Actors and environment. The actors perform actions in order to solve missions. Actors are friendly units, enemy units and other units present in the mission area. The actors are located in the environment, which poses constraints on the actions of the actors.

3.1 The Actors

Each actor in the real world is implemented as a software agent in the system (see Fig. 2). From here on all references of an agent refers to software agents in the system. The agent consists of three components: (1) sensors, (2) effectors and (3) a manager. The sensors are used to detect agent surroundings; the environment and other agents. The performance of the sensors is affected by the environment as well as other agents[2]. The data from the sensors is relayed to the manager which also controls the use of the sensors. The agent exerts influence on the environment and other agents through its effectors. The effectors are controlled by the manager, which is also given feedback from the effectors of their performance (adapted from [10])[3].

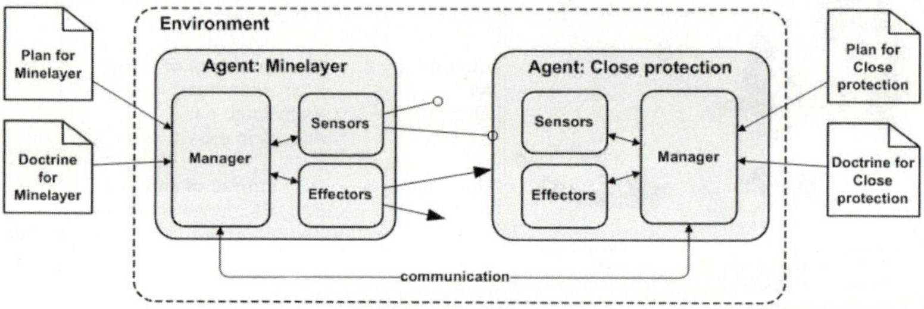

Fig. 2. The agents and their main components

The manager is responsible for deciding the agents' actions given its mission and the preconditions at hand. The agent's mission is extracted from the overall plan and patterns of how to perform different missions are described in doctrines.

---

[2] Sensor ranges can be affected both by environmental factors such as rain, fog and temperature. The sensor can also be affected by other agents through the use of active counter measures such as jamming or chaff.

[3] If the agent instead represents a company, the sensors would represent the people or units within the company responsible for gathering and analysing information about the market and competitors. The effectors would represent the people or units responsible for affecting the market and/or competitors such as marketing and sales departments. The manager would of cause represent the company's managerial group.

The manager also communicates with other agents in order to share data about the environment, about other agents and to give or receive task to/from other agents (for more elaborate models see [11] [1]). A plan is a description of intended actions. Since actors performs actions, the concept of a plan relates to all modelled actors. A plan is a complex linkage of series of assigned actions (tasks) [12]. The tasks can be linked according to chain of command (e.g. tasks given to an institution may be decomposed into tasks for subordinate departments) or according to functional decomposition (e.g. the task of "brewing a cup of coffee" can be decomposed into "pour coffee into machine" and "brew coffee"). A plan also has a temporal dimension. This dimension states that action A can not start before Action B has completed.

When representing the plan, a slightly modified excerpt from JC3IEDM [4]is used [12]. JC3IEDM is a world wide standard used by NATO and several other countries. The central concept in this model is that of an *Action*. The purpose of performing an action is to reach a desired *End state*. *Divergence* describes a difference between the current state and the end state. It is a perceived divergence that triggers the execution of an action. An action can be related to another action via the *Action association*. It is the action association that describes the relations between the different tasks in the plan[5]. Each action is conducted by a *resource* and each action consumes and/or produces resources. What is consumed is described by *Consumption* and what is produced is described by *Action effect*. The action effect takes care of a divergence in order to reach the desired end state. Each action is formalized by a *task* (i.e. expressed in an order). The task is in turn given to/received by an *Actor*. The actor can be related to other actors through the *Actor relation*. A special kind of actor relation is that of *Actor C2* which describes the chain of command. All actions and all resources adhere to *Rules*. *Decision* is a type of rule that regulates actions. The plan contains all information needed to load an agent. Each actor in the plan is modelled as single agent. A data structure representing the agent could look like:

```
agent(Name, Manager, Sensors, Effectors)
```

Name is the name of the agent. The *Manager* represents the manager of the agent. The manager contains the plan, rules of how to perform the tasks and a communication schema. *Sensors* is a list representing the agents sensors. *Effectors* is a list representing the agents effectors.

3.2 The Environment

The environment is also modelled as agents. The environmental agents can be stationary (ground, water) or moving around (rain, fog). The difference between

[4] JC3IEDM = Joint Command Control and Communications Information Exchange Data Model.

[5] *Action association* can either (1) be sub-typed into *temporal association*, *hierarchal association* and *functional association*, or (2) an attribute on Action association can be used to distinguish between the different relations [13].

an actor agent and an environmental agent is that the environmental agent is much more limited when it comes to controlling its own behaviour[6]. An environmental agent can be modelled as follows:

```
environment_agent(ID, Type, ListOfNeighbours).
environment_type(name, course, speed, height, density).
```

For the environment agent, the *ID* is a unique identifier of the agent. *Type* denotes which type of environment the agent represents (land, air or water). *ListOfNeighbours* describes other environmental agents adjacent to the agent. For the environment type agent, *Name* describes the type (e.g. "land"), *height* denotes the height of the agent, *course* and *speed* applies if the agent is moving (e.g. weather) and *density* denotes how much effect the agent has on other agents effectors and sensors.

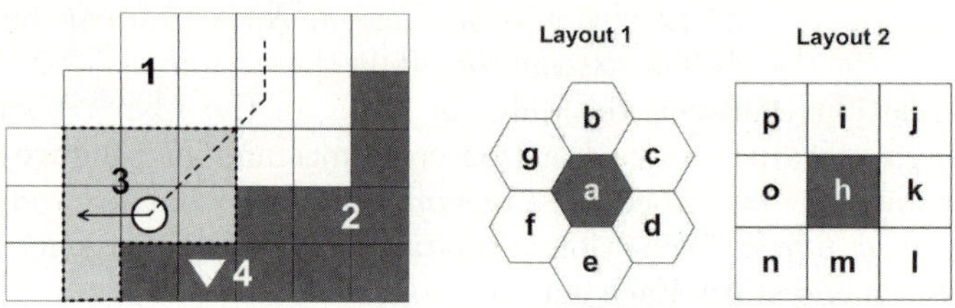

Fig. 3. Example of agent moving and environment layouts

As an illustration different layouts of environmental agents are displayed in Fig. 3. The environmental agents affect the performance of the actor agents when they move through the environment. The sea agents (1) and the land agents (2) have different effects on the sensor range of the moving agent and therefore the moving agent can not detect the agent on land(4). The sensor range is represented by the dotted line around the grey boxes (3).

4 Generating and Evaluating Scenarios

When the planning is finished, the planning team hand over the plan and the list of uncertainties to the evaluation team. The planning team, together with the evaluation team elaborate all of the assumptions made so that they can be formalised as a single instance and represented in the plan structure. A scenario for evaluation is generated according to following:

[6] Some environmental agents must have the capability to follow a plan e.g. the development of weather over a period of time might be modeled as a plan and given to an agent.

Let P denote the plan.
Let $U = \{u_1, u_2, \ldots, u_i\}$, $u_i \in P$ denote the set of uncertainties in P
Let $A_i = \{a_{i1}, a_{i2}, \ldots, a_{ij}\}$ denote the set of assumptions made for u_i

A scenario S is defined as:

$$S = \{s_1, s_2, \ldots, s_k\}, \; k \in [1, i], \; s_k \in A_k \tag{1}$$

The total number of assumptions elaborated is: $A_1 + A_2 + \cdots + A_i$
The total number of scenarios evaluated is: $A_1 \times A_2 \times \cdots \times A_i$

Looking at the uncertainties from a scenario point of view, there are two types of uncertainties: (1) uncertainties external to the own plan (external uncertainties, $U_{external}$) and (2) uncertainties internal to the own plan (internal uncertainties, $U_{internal}$). The external uncertainties refer to uncertainties related to the environment and other actors. The internal uncertainties refer to uncertainties related to own courses of actions (OCA).

$$\begin{aligned} U_{internal} &= \{ui_1, ui_2, \ldots, ui_i\} \text{ so that } ui_i \in U \text{ and } ui_i \text{ refers to } OCA \\ U_{external} &= \{ue_1, ue_2, \ldots, ue_i\} \text{ so that } ue_i \in U \text{ and } ue_i \notin U_{internal} \end{aligned} \tag{2}$$

Using these definitions, $U_{internal}$ defines the set of OCA and $U_{external}$ defines the set of surroundings in which each OCAs should be evaluated. To illustrate how one scenario is evaluated, we will use the worst-case scenario as an example.

The worst-case surrounding E is defined as:

$$\begin{gathered} E = \{s_1, \ldots, s_k\}, \; k \in [1, i], \; s_k \in A_k, \; s_k \in U_{external} \\ \text{and } s_k = \text{the worst assumption in } A_k \end{gathered}$$

An OCA I is defined as:

$$I = \{s_1, \ldots, s_l\}, \; l \in [1, i], \; s_l \in A_l \text{ and } s_l \in U_{internal}$$

A worst-case scenario S is defined as:

$$S = E \cup I \tag{3}$$

Looking at Fig. 1 following example can illustrate the process of generating the worst-case scenarios. The uncertainties are grouped as follows:

$U_{internal}$ = {Own course of action}
$U_{external}$ = {Strength of enemy surface attack, Enemy course of action}

Assume that the planners have identified Alt. 1(2 cruisers, 4 corvettes) as the most dangerous assumption in uncertainty 1 and Alt. 3(Already in mission area) as the most dangerous assumption in uncertainty 2. The worst-case surrounding E and, own courses of actions I are defined as:

E = {2 cruisers and 4 corvettes, already in mission area}
I = {SA1 & SA2 together with AA, SA1 & SA2 north area; AA south area}

Using (3) the worst-case scenarios S are generated as:

$$S_1 = \{\text{2 cruisers and 4 corvettes, already in mission area, SA1 \& SA2 together with AA}\}$$
$$S_2 = \{\text{2 cruisers and 4 corvettes, already in mission area, SA1 \& SA2 north area; AA south area}\}$$

The question to be answered when evaluating the worst-case scenarios is: "Does the plan look workable in light of the worst-case assumptions". To identify if segments in the plan look fragile under these circumstances, one has to examine if the plan "leads to descriptions or prescriptions that are unworkable" (this corresponds to *workability testing* in Silverman's framework for critiquing a body of knowledge [14]).

The proposed solution adds two modules to the agent based system: an *agent monitoring* module responsible for monitoring selected agents, and a *result compilation* module responsible for analysing the gathered data and compile a result.

Before running the simulation the user must decide the following: (1) what aspect of the scenario (plan) should be evaluated and (2) which agents in the scenario should be evaluated. The *evaluation aspect* can be anything the user is interested in evaluating: movement times, detection probabilities or, if the domain is commerce, maybe cash-flow. The *agents to monitor* define the set of agents for which the evaluation aspect should be monitored. Optionally the user can specify an *acceptance level* on an evaluation aspect. The acceptance level describes a threshold or interval of acceptable performance of an agent in that evaluation aspect. If the evaluation aspect is time, an acceptance level could be defined as "it must take no longer than two hours for agent A to complete task B".

During the simulation the agent monitoring module collects data from the agents monitored, according to the evaluation aspects. When the simulation is finished the collected data is transferred to the result compilation module. This module constructs a graph for each agent representing each evaluation aspect of that agent for that simulation run. These graphs can either be presented as they are (e.g. it took agent A 2.3 hours to complete task B) or they could be subjected to the acceptance levels (e.g. agent A did not perform satisfactory on task B). When subjected to acceptance levels, the system automatically alerts the user of when and where an agent shows insufficient performance on an evaluation aspect.

5 Conclusions and Further Work

To evaluate if a plan is workable in light of the worst-case assumptions a set of worst-case scenarios has to be created and evaluated. Evaluation of those scenarios is based on simulations that verify that the plan does not violate any physical constraints or any domain heuristics. This workability testing requires active participation of users; it is a cooperative process[14].

A key assumption in this paper is that it is possible to generate scenarios based on the uncertainties encountered during planning. Empirical studies have to be performed to examine if that is possible. If these studies look promising, the next step is to study if the different assumptions can be represented in such way so a set of scenarios can be generated by automatic permutation of those assumptions. This paper also puts forward a question: "How is the uncertainty that has the most impact on the plan identified?" In order to answer this question, empirical studies must be conducted to capture the strategies decision-makers use when they identify these uncertainties. The strategies can then be used as basis for development of algorithms. Evaluation of the algorithms is carried out through experiments where their performance is compared with that of an expert. The final evaluation of the system is a series of experiments that test if the proposed system actually improves the quality of the plans.

References

1. van Creveld, M.: Command in War. Harvard University Press, Cambridge, MA (1985)
2. of the Navy, D.: MCDP 6, Command and Control. Headquarters United States Marine Corps, Washington, D.C. 20380-1775 (1996)
3. Lipshitz, R., Strauss, O.: Coping with uncertainty: A naturalistic decision-making analysis. Organizational Behaviour and Human Decision Processes 69, 149–163 (1997)
4. Turban, E., Aronson, J., Liang, T.P.: Decision Support Systems and Intelligent Systems, 7th edn. Pearson Prentice Hall, Englewood Cliffs (2005)
5. Rydmark, J., Kuylenstierna, J.: Hantering av osäkerhet i beslutsprocessen (managing uncertainty during decision making). Case study report, FHS Skr 21 105: 61305:Supp. 2. Swedish National Defence College, Stockholm (2003)
6. Waldenström, C.: An initial study of what is difficult in sense-making. Swedish National Defence College, Dept. of War Studies, Stockholm (2006)
7. Haistie, R., Dawes, R.: The Psychology of Judgement and Decision Making. Sage Publications, Thousand Oaks, CA (2001)
8. Jackson, N., Carter, P.: Organisationsbeteende i nytt perspektiv (Rethinking Organisational behaviour). Liber Ekonomi ISBN 91-4706366-1 (2002)
9. Kahneman, D., Slovic, P., Tversky, A.: Judgement under uncertainty: Heuristics and biases. Cambrige University Press, Cambrige (1982)
10. Macal, C., North, C.: Tutorial on agent based modelling a simulation. In: Proceedings of the 2005 Winter Simulation Conference (2005)
11. Brehmer, B.: One loop to rule them all. In: Proceedings 11th ICCRTS (2006)
12. (MNMB), M.N.M.B.: The Joint C3 Information Exchange Data Model Overview. 3.1 (2006)
13. Elmasri, R., Navathe, S.: Fundamentals of Database Systems, 2nd edn. Addison-Wesley, London, UK (1994)
14. Silverman, B.: Survey of expert critiquing systems: Practical and theoretical frontiers. Communications of the ACM 35, 106–127 (1992)

A Knowledge Sorting and Matrix Representation Approach for Developing Knowledge-Based Product Design Systems

ZhiMing Rao[1] and Chun-Hsien Chen[2]

School of Mechanical & Aerospace Engineering,
Nanyang Technological University, 50 Nanyang Avenue, Singapore 639798
{raoz0001,mchchen}@ntu.edu.sg

Abstract. Knowledge acquisition plays a very important role in constructing a knowledge-based system. As the most time-consuming phase, it has been recognized as the bottleneck in developing a knowledge-based system (KBS). This paper presents a knowledge sorting and matrix representation approach to facilitate the effectiveness of knowledge acquisition in building a KBS. The proposed knowledge acquisition approach, which is based on the relationship sorting and matrix representation for different knowledge elements in product design, comprises four major consecutive steps for extracting rules. The procedure in each step is elaborated. A case study on solid circular power transmission shaft design is employed to illustrate how the proposed approach works.

1 Introduction

It is well-known that knowledge acquisition is the most time-consuming phase and hence the bottleneck in constructing a knowledge-based system (KBS) [1-2]. Generally, a knowledge acquisition process covers the activities of collecting domain knowledge and transforming it into a form that can be processed by a computer. The barrier in knowledge acquisition is threefold: (1) difficulty in soliciting knowledge from domain expert(s) due to inexperienced knowledge engineer(s) used; (2) difficulty in extracting useful knowledge from the solicited domain knowledge owing to the limitation and bias of human cognition; and (3) difficulty in conveying knowledge from a human to a machine because of different knowledge representation format.

A number of researchers have suggested various measures to tackle the problems encountered in knowledge acquisition. As summarized by Chen & Occeña [3], there are two common approaches for knowledge acquisition: (1) acquire knowledge directly from the domain expert(s); and (2) acquire knowledge through the historical records, i.e. by rule induction. A popular tool for rule induction is ID3 algorithm, which is a machine learning technique. A number of studies on ID3 algorithm for rule induction can be found in the literature [1, 4]. From the aspect of facilitating the efficiency of knowledge acquisition, Chen & Occeña [3] proposed a so-called knowledge sorting process. The process capitalizes on the relationships between attributes and factors, dependent and independent variables, and interrelationships between attributes. It consists of three consecutive stages, viz. identification of knowledge sources,

B. Apolloni et al. (Eds.): KES 2007/WIRN 2007, Part I, LNAI 4692, pp. 262–269, 2007.

generation of taxonomic trees, and organization of acquired knowledge, to extract rules from sorted knowledge. Yan et al. [5] established a QFD (quality function deployment) -enabled product conceptualization system, in which design knowledge is elicited using laddering technique and organized using RCE (restricted coulomb energy) neural network. In addition, some data mining methodologies have been postulated for knowledge discovery and association rule mining [6-7].

However, there are two major omissions in existing work: first, the quantitative characteristics of information and the relationships between the various design knowledge elements, which are essential for rule extraction and of critical importance in constructing knowledge-based systems, have not been well addressed; and second, it is imperative to provide an integrated and complete knowledge acquisition process with well-organized knowledge representation scheme to facilitate the effectiveness of rule generation. For this purpose, a knowledge acquisition approach using knowledge sorting techniques and matrix representation for constructing knowledge-based systems is proposed in this work.

2 Overview of the MRM Approach

A KBS is a computer program used to solve problems requiring expert knowledge in some application domains. In order to construct such a system, knowledge is elicited from domain expert(s) or related literature. The proposed knowledge acquisition approach begins by identifying the appropriate knowledge sources, which may include

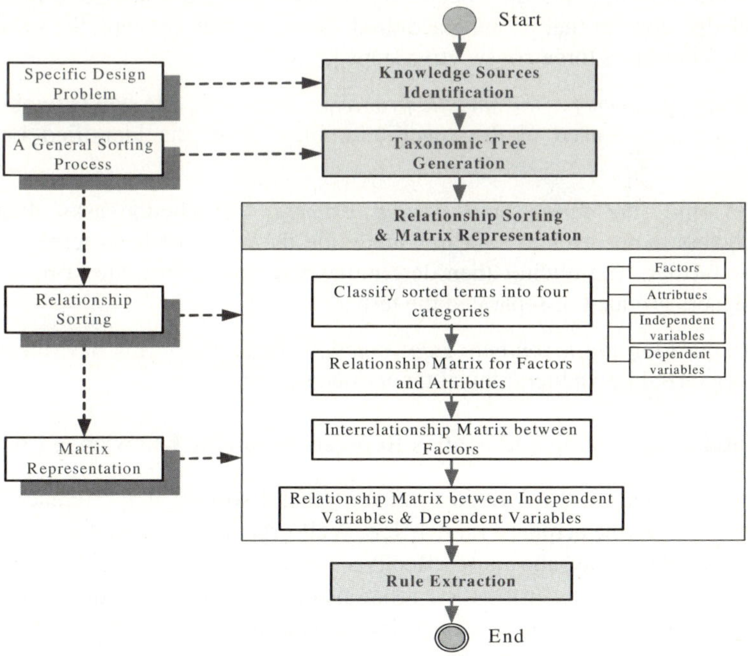

Fig. 1. Overall framework of the proposed approach

domain expert(s), historical records, related literature, etc. After the relevant knowledge is acquired from different knowledge sources, it is preliminarily organized in a taxonomic manner with a general knowledge sorting process, which is used to facilitate knowledge organization. Based on the taxonomic tree generated, a series of relationship sorting and matrix representation operations for the categorized knowledge elements will be performed to extract the rules. The proposed approach involves four major consecutive steps, namely, knowledge sources identification, taxonomic tree generation, relationship sorting and matrix representation and rule extraction. Figure 1 shows the overall procedure of the proposed approach.

3 Description of the Proposed Approach

3.1 Knowledge Sources Identification

The knowledge sources required to build a KBS may include domain experts, historical records, and related literature, etc. In constructing a specific KBS, knowledge engineers should identify the appropriate knowledge sources required and plan for strategies to acquire knowledge from these sources.

3.2 Taxonomic Tree Generation (General Sorting)

In order to organize the acquired knowledge into a well-structured taxonomic manner, a general sorting process is employed to facilitate the process. The sorting process aims to learn and organize the vocabulary (categories) and knowledge hierarchy of the knowledge domain that is being acquired. Based on this concept, the general sorting process comprises three consecutive steps [3, 8-9].

Step 1. Obtaining items for the specific problem. The aim of this step is to obtain a list of terms (or items) used in the specific problem from the identified knowledge sources.

Step 2. Sorting the terms with certain criteria. An elicitor asks domain experts/designers themselves to sort and subdivide the terms until no terms can be divided any further. The elicitor then determines the reasons or criteria by which experts/designers grouped or separated the terms.

Step 3. Generating the taxonomic tree structure. Once the terms have been sorted, they are organized into a hierarchical taxonomic tree structure.

3.3 Relationship Sorting and Matrix Representation for Knowledge Elements

With the taxonomic tree(s) derived from the general sorting as a reference, the next stage is to organize the acquired knowledge. Well-organized knowledge will make the extraction of rules easier and more effective. For this purpose, a new knowledge organization process aiming to sort the relationships between the acquired knowledge elements was developed. The process is based on a fundamental understanding of the specific product design problem obtained from general sorting. Specifically, there are four steps involved in this process. Firstly, the sorted terms are classified into four categories: factors, attributes, independent variables and dependent variables. Subsequently, three matrices are employed to sort the relationships between the knowledge elements in the various categories. In rule extraction, the knowledge elements such as factors, attributes, independent and dependent variables in the relationships can be regarded as the keywords to form the IF and THEN clauses in the rules. Consequently, rules can be generated with the IF and THEN clauses based on the relationships defined in the matrices. The four steps of this stage are presented as follows.

Step 1: Classify the sorted terms into factors, attributes, independent variables and dependent variables. The factors are the specifications and/or characteristics or components of the specific product. In the product design domain, factors pertain to what the designer typically controls or specifies. The attributes are those features that are either affect or affected by the product specifications and/or characteristics. The independent variables include all the attributes and those specifications and characteristics that will affect the performance of the specific product. The dependent variables are those factors that can be used to judge the performance of the specific product. Based on these definitions, the domain expert(s) select and classify the sorted terms into the four different categories.

Step 2: Identify the relationship between factors and attributes using matrix representation. In the product design stage, both factors and attributes are relevant considerations and the relationships in between should be identified for subsequent rule extraction. The results are represented with a quantified matrix, i.e. $\mathbf{M_{F\text{-}A}}$. Consult the domain expert(s) and mark each grid of the matrix with "1", "-1" and "0" to indicate the factor-attribute pair relationships. In general, "1" denotes where factors (rows) affect the attributes (columns); "-1" denotes factors are affected by the attributes; and "0" represents there is no relationship for a specified factor-attribute pair.

Step 3: Discover the interrelationship between factors using a matrix. With the same factors as in Step 2, identify the interrelationships between factors and present the results in a square matrix, i.e. $\mathbf{M_{F\text{-}F}}$. Ask the domain expert(s) and mark each grid of the matrix with "1" or "0" for each factor-factor pair. Here, value "1" indicates that the two factors are related whereas value "0" denotes there is no relationship. The summation of values of all the grids in each row or column in the matrix suggests the importance of each factor.

Step 4: Establish the relationship between independent and dependent variables using a matrix. As aforementioned, the independent variables include all the attributes in Step 2 and those specifications and characteristics that will affect the performance of the specific product. The dependent variables are those factors which can be used to judge the performance of the specific product. In the product design domain, performance is a key evaluation criterion. This step aims at figuring out the relationship between dependent and independent variables. In so doing, consult the domain expert(s) and mark each grid in a matrix, i.e. $\mathbf{M_{ID\text{-}D}}$, to express the relationship of each independent-dependent variable pair. In this case, the values 1, 5 and 9 denote the strength of the relationships between the variables, which forms a basis to determine the confidence levels of the extract rules. The value "0" is used to denote that there is no relationship for a specified independent-dependent variable pair.

4 A Case Study

The following case study regarding the development of a KBS for solid circular power transmission shaft design [10] is used to demonstrate the proposed knowledge acquisition approach. The knowledge acquisition process follows the procedure presented in Section 3.

Initially, shaft designers can acquire knowledge from senior shaft designers or individual working experience. Besides, shaft design knowledge can also be acquired from such indirect sources as shaft design books [10], shaft design catalogues, etc. As a result, 65 terms used in reference to shafts were acquired from the domain expert (a professional shaft designer), literature and trade journals. These terms were then sorted and organized according to six major criteria, namely types of machine element on shaft, types of geometric discontinuity on shaft, factor variables used in shaft design, types of shaft material strength, kinds of shaft surface condition and types of design stress. From the general sorting and the list of criteria, a sample taxonomic tree for solid circular power transmission shaft was generated as shown in Figure 2.

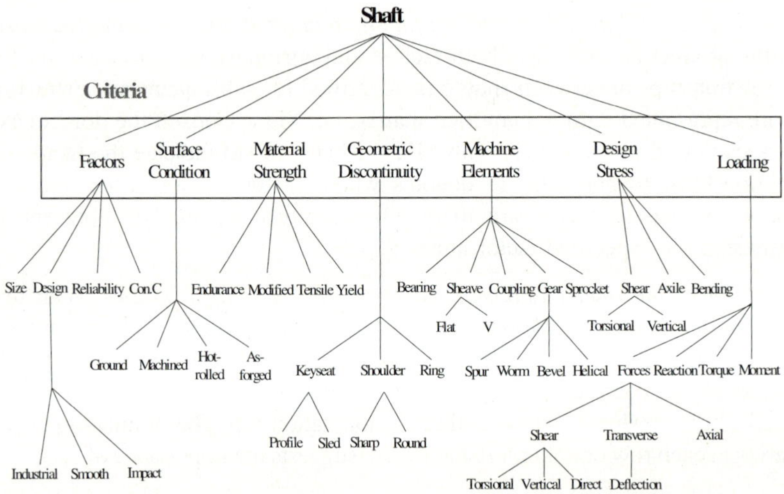

Fig. 2. Taxonomic tree generated from general sorting

Based on the generated taxonomic tree, the terms were categorized into factors, attributes, independent variables and dependent variables of the shaft design by the domain expert. The classification is in accordance with the definitions described in Section 3.3. Subsequently, consulted the domain expert and marked the relationships between each pair of factor-attribute, factor-factor, and independent-dependent variables. The results of the relationships between various knowledge elements are presented in Matrix $\mathbf{M}_{\mathbf{F\text{-}A}}$ (Table 1), Matrix $\mathbf{M}_{\mathbf{F\text{-}F}}$ (Table 2), and Matrix $\mathbf{M}_{\mathbf{ID\text{-}D}}$ (Table 3), respectively. All the relationships shown in the three matrices were then verified by the domain expert.

Table 1. Relationship matrix ($\mathbf{M}_{F\text{-}A}$) between factors and attributes (1: Factors affect Attributes, -1: Factors affected by Attributes, 0: No relationship)

	(A1) Loading Condition	(A2) Desired Reliability	(A3) Bend. Moment	(A4) Shear. Force	(A5) Rotational speed	(A6) Polar Section Modu.	(A7) Rect. Section Modu.	(A8) Trans. Torque	(A9) Shaft Geom.
(F1) Stress Concentra.	-1	0	-1	-1	0	0	0	-1	-1
(F2) Design Factor(*N*)	-1	0	0	0	0	0	0	0	0
(F3)Reliability Factor	-1	-1	0	0	0	0	0	0	0
(F4) Size Factor (*Cs*)	0	0	0	0	0	0	0	0	-1
(F5)Shaft Material	0	1	0	0	0	0	0	0	0
(F6)Shaft Surf. Finish	-1	1	0	0	0	0	0	0	0
(F7) Min Shaft Diam.	0	-1	-1	-1	0	1	1	-1	-1
(F8) Ini.. Estimate of Shaft Diam.	0	0	0	0	0	1	1	0	1
(F9) Shaft Length	-1	0	-1	0	0	0	0	0	1

Table 2. Interrelationship matrix ($\mathbf{M}_{F\text{-}F}$) between factors

	(F1) Stress Concentra	(F2) Design Factor(*N*)	(F3) Reliability Factor	(F4) Size Factor	(F5) Shaft Material	(F6)Shaft Surf. Finish	(F7) Min Shaft Diam.	(F8) Ini.. Estimate of Shaft Diam.	(F9) Shaft Length
(F1) Stress Concentra.	0	0	0	0	0	0	1	1	0
(F2) Design Factor(*N*)	0	0	0	0	0	0	0	0	0
(F3)Reliability Factor	0	0	0	0	0	0	0	0	0
(F4) Size Factor	0	0	0	0	0	0	1	1	0
(F5)Shaft Material	0	0	0	0	0	1	0	0	0
(F6)Shaft Surf. Finish	0	0	0	0	1	0	0	0	0
(F7) Min Shaft Diam.	1	0	0	1	0	0	0	1	0
(F8) Ini.. Estimate of Shaft Diam.	1	0	0	1	0	0	1	0	0
(F9) Shaft Length	0	0	0	0	0	0	0	0	0

Based on Tables 1 – 3, a KBS, SHAFTES, was constructed using an expert system shell, Exsys®, on a PC. In constructing SHAFTES, two functions most useful to shaft designers were identified: (1) determination of the minimum required shaft diameter at various points along the shaft; and (2) provision of appropriate recommendations for shaft design. Rules were extracted from the organized knowledge according to each function. For the first function, the rules regarding determining the minimum required shaft diameter were extracted from Tables 1($\mathbf{M}_{F\text{-}A}$) and 2 ($\mathbf{M}_{F\text{-}F}$). For instance, suppose that the computed minimum required shaft diameter is greater than the initially estimated shaft diameter, then, from Table 2, it can be found that the computed minimum shaft diameter (*DM*) (F7) and the initially estimated shaft diameter (*DE*) (F8) are affected by the size factor (C_s) (F4). Thus, these three factors can be identified as the keywords to form the IF and THEN clauses. Based on the knowledge of the relationship between the computed minimum required shaft diameter, the initially estimated shaft diameter and the size factor, the following rule can be generated.

Rule: *If* minimum required shaft diameter [*DM*] > initially estimated shaft diameter [*DE*] *Then* repeat the shaft diameter computation with a smaller size factor C_s value to obtain a corrected shaft diameter.

Table 3. Relationship matrix (M_{ID-D}) between independent and dependent variables (1: Weak, 5: Medium, 9: Strong, 0: No relationship)

	(D1) Tensile Strength	(D2) Yield Strength	(D3) Endur. Strength	(D4) Modified Endur. Strength	(D5) Torsional Shear. Stress	(D6) Vertical Shear. Stress	(D7) Bend. Stress	(D8) Fatigue Failure	(D9) Shaft Deflect.
(A1) Loading Condition	0	0	0	0	5	5	5	0	9
(A2)Desired Reliability	1	1	0	9	0	0	0	5	0
(A3)Bend. Moment	0	0	0	0	0	0	9	0	5
(A4)Shear. Force	0	0	0	0	0	9	0	0	5
(A5) Rotational speed	0	0	0	0	1	0	0	0	0
(A6) Polar Section Modu.	0	0	1	0	9	0	0	0	0
(A7) Rect. Section Modu.	0	0	1	0	0	9	0	0	0
(A8)Trans. Torque	0	0	0	0	9	0	0	0	0
(A9) Shaft Geom.	0	0	0	0	1	1	1	9	0
(S1) Shaft Material	9	9	9	9	0	0	0	1	0
(S2) Min Shaft Diam.	0	0	0	0	9	9	9	5	5
(S3)Shaft Surf. Finish	0	0	9	0	0	0	0	9	0
(S4)Ini..Estimate of Shaft Diam.	0	0	0	0	9	9	9	0	5
(S5) Shaft Length	0	0	0	0	0	0	9	9	9
(S6) Design Factor(*N*)	0	0	0	0	0	9	0	0	0
(S7) Stress Concentra	0	0	0	0	0	0	0	5	0

For the second function, two relationship matrices $\mathbf{M_{F\text{-}A}}$ and $\mathbf{M_{ID\text{-}D}}$ (Tables 1 and 3) are used for rule extraction. Table 3 points out the relationship between fatigue failure (D8) with shaft geometry (A9), shaft surface finish (F6) and shaft length (F9). Also, Table 1 reveals if there is any relationship among shaft geometry, shaft surface finish and shaft length. Using the dependent variable D8 and independent variables A9, F6 and F9 as the keywords, the following rule can be generated.

Rule: *If* the task is to provide appropriate design recommendations and ideas for shaft design *And* the design objective involves reducing the probability of fatigue failure *Then* improve fatigue strength by eliminating steps (diameter changes) *And* improve fatigue strength by specifying an improved shaft surface finish *And* make shaft length as short as possible.

As a result, 198 rules were extracted in constructing the knowledge base of SHAFTES. The system was tested using Exsys® Developer's validation function and the validation procedure proposed by Ignizio [11] for consistency and completeness. The performance of the system has also been verified by shaft designers to see if the results conformed to what was expected. Besides, 14 undergraduates who took engineering design course were invited for a usability test to evaluate the system. The results were promising and revealed the potential of the system.

5 Conclusion

In this work, a knowledge sorting and matrix representation approach is proposed to improve the process of knowledge acquisition and facilitate the effectiveness of rule extraction. The proposed approach begins by identifying the appropriate knowledge sources, followed by acquiring relevant knowledge and organizing the acquired knowledge in a taxonomic manner using a general knowledge sorting process. Based on the taxonomic trees generated, a series of relationship sorting and matrix representation operations are performed to extract the rules used to develop a KBS. In the approach, the relationships between the various knowledge elements were sorted and represented with quantified matrices. This enables both the quantitative characteristics of information and relationships in the acquired knowledge be identified. Furthermore, the approach paves a new avenue to establish an integrated and complete knowledge acquisition process with a well-organized knowledge representation scheme. The capability of the proposed approach was illustrated with a case study on developing a KBS for designing solid circular power transmission shaft. The results of the case study are promising and reveal the effectiveness of the proposed knowledge acquisition process.

References

1. Shao, X., Zhang, G., Li, P., Chen, Y.: Application of ID3 algorithm in knowledge acquisi tion for tolerance design. Journal of Materials Processing Technology, 117, 66–74 (2001)
2. Rao, Z.M., Chen, C.-H.: A matrix representation and mapping approach to knowledge acquisition for product design. In: Khosla, R., Howlett, R.J., Jain, L.C. (eds.) KES 2005. LNCS (LNAI), vol. 3681, pp. 311–317. Springer, Heidelberg (2005)
3. Chen, C.-H., Occeña, L.G.: Knowledge sorting process for a product design expert system. Expert Systems, 16, 170–182 (1999)
4. Wu, C.H., Kao, S.C.: Induction-based approach to rule generation using membership function. International Journal of Computer Integrated Manufacturing, 15, 86–96 (2002)
5. Yan, W., Khoo, L.P., Chen, C.-H.: A QFD-enabled product conceptualisation approach via design knowledge hierarchy and RCE neural network. Knowledge-Based Systems, 18, 279–293 (2005)
6. Tsay, Y.-J., Chang-Chien, Y.-W.: An efficient cluster and decomposition algorithm for mining association rules. Information Sciences, 160, 161–171 (2004)
7. Lee, Y.-C., Hong, T.-P., Lin, W.-Y.: Mining association rules with multiple minimum supports using maximum constraints. International Journal of Approximate Reasoning, 40, 44–54 (2005)
8. Rugg, G., McGeorge, P.: Sorting techniques: a tutorial paper on card sorts, picture sorts, and item sorts. Expert Systems, 14, 80–92 (1997)
9. Yan, W., Pritchard, M.C., Chen, C.-H., Khoo, L.P.: A study of bidding-oriented collaborative product conceptualization. International Journal of Computer Integrated Manufacturing, 19, 161–177 (2006)
10. Mott, R.L.: Machine Elements in Mechanical Design, 4th edn. Pearson Education, Inc., Upper Saddle River (2004)
11. Ignizio, J.P.: Introduction to Expert Systems-The Development and Implementation of Ruled-based Expert Systems. McGraw-Hill, New York (1991)

Automated Testing for Knowledge Based Systems

Ronald Hartung[1] and Anne Håkansson[2]

[1] Department of Computer Science,
Franklin University,
201 S. Grant Avenue,
Columbus, Ohio 43215, USA
Hartung@franklin.edu
[2] Department of Information Science, Computer Science,
Uppsala University,
Box 513, SE-751 20,
Uppsala, Sweden
Anne.Hakansson@dis.uu.se

Abstract. Building and modifying knowledge-based systems requires testing of the knowledge for quality assurance, such as verification and validation. This is especially important when reverse engineering is applied to a system that needs to be remodeled or renewed. However, the modification of a knowledge-based system is a difficult process. Commonly, the documentation is poor, and the original domain expertise is lacking. Therefore, testing must be applied on existing knowledge to be able to verify the changed knowledge. To this objective we apply an automated test generation system to verify the operation of the modified system.

Keywords: Automated testing, knowledge-based systems, re-engineering, UML, rule-based systems, visualization, verification.

1 Introduction

During the development, extension, or modification of a medium or large knowledge based system, testing is needed to verify and validate the knowledge. Verifying and validating a knowledge base includes checking correctness, consistency, and completeness of the knowledge. Since these systems work in diffuse domains, the requirements are seldom clearly specified. Therefore, the testing needs to be performed with a domain expert commenting on the test results or tested against the specified requirements, (i.e., if there exist any).

When making profound changes in a knowledge-based system or applying reverse engineering[1] (reengineering) to the knowledge base, testing becomes both necessary and complicated. In these cases, the knowledge is often modified to accomplish new tasks or to suit a new system. Since a small change can affect considerable parts of the knowledge base and it is difficult to discover the side effects, the knowledge base

[1] Reverse engineering is used to extract the knowledge in unused or old systems to be able to reuse the knowledge in another system.

B. Apolloni et al. (Eds.): KES 2007/WIRN 2007, Part I, LNAI 4692, pp. 270–278, 2007.

needs to be retested. In older systems, the testing might not have an existing reference against which to verify and validate the knowledge. The domain expertise may not be available, and the requirements vague and poor. Moreover, the knowledge engineer, who performed the knowledge acquisition, may not be available. In these cases, testing must be based on the previous existing knowledge.

The testing problem, mentioned above, has been emphasized in a reengineering tool, used for collecting the knowledge, as well as, the reasoning strategies in a knowledge-based system [4]. The tool uses an extraction process to find the knowledge in a system and put it into different Unified Modeling Language-like diagrams, such as sequence diagrams. Since knowledge-based systems vary in the knowledge representation used for implementation, the extraction needs to be parameterized to cover multiple systems. While rule-based systems are most common, the format of the rules varies widely. Moreover, some of the systems rely on programming code, for example interpreters written in a programming language, such as Prolog or CLIPS. One decision is to determine how much of the logic of the underlying implementation, e.g. Prolog, to extract for testing the significant parts of the system.

The reengineering tool produces diagrams to help the user visualize and understand the knowledge-based system prior to modifying the system [4]. The tool can also be used to generate test cases for the knowledge-base system. The test cases are run on the modified system and the failures found during the testing must be inspected. Some failures are errors introduced by the user's modifications and can rather easily be detected. However, some failures are not errors, but results of modifying the system and belong to the new or modified knowledge. In this case, the user needs to inspect the result of running the test cases and verify the particular result. Then user needs to change the test cases to match a desired outcome (result). The diagrams are a helpful visualization during this process. As the user modifies the knowledge base, the user needs to create new test cases to verify the intended changes.

2 Related Work

Software testing has a long history and a well developed theory [7]. Theories have been applied in all kinds of systems and good discussions of testing and verification have been made for knowledge-based systems [9]. There are two approaches for testing knowledge-based systems: one analyzes the knowledge base for internal structural problems; the other facilitates testing of the knowledge. Of particular note is the work of Kiper [8] since this work assists the developer in testing the knowledge. This approach is similar to our approach, in that the analysis used will result in same test structure as the reengineering tool used as the base for this work. The extraction process is different and the reengineering tool used provides a diagrammatic view of the knowledge and the test cases. This is especially useful when the knowledge-based system is old and lacks the original staff or documentation to base the modification work.

However, a challenge in knowledge-based systems is detecting coverage, since the execution sequence is declarative, and most of the developed testing methods are based on imperative languages. A testing technique has been developed for rule-based systems [1] but our technique does give an approach to generating test cases that provide coverage.

3 Testing Knowledge in Knowledge Based Systems

Testing is the process of executing the system with the intention of finding errors or bugs in the software. For testing knowledge-based systems, the knowledge is checked to see whether it meets the domain expert's expectations and the system requirements. Testing is especially important when the rules are changed because a modification may involve many rules. The interactions between rules are hard to detect and to correct, which is why the test cases must ensure that any changes affecting the existing knowledge are detected.

Test cases can be adopted in knowledge-based systems to ensure the correctness of the knowledge base. A test case specifies how to test the implementation of a particular requirement or a design decision and to determine the criteria used to define success. We use test cases to test the individual units of facts and rules contained within the knowledge base.

In our approach, we use an automated framework for the testing. The framework is built on object-oriented testing tool but applied to rules and our test cases are automatically constructed from the reengineering tool. The construction process produces individual test cases on top of the rules to make the testing for knowledge-based systems a feasible process.

4 Design for a Test Driver

A test driver is required for automated testing of the knowledge-based system. The test driver we choose is the JUnit framework. In the design of a test driver, the most important factor is the interface used by the knowledge-based system. The classic expert systems use an interface (possibly a GUI) that is designed to communicate with the user, generally with text input and output. A newer application of rule-based expert systems is found in the middle tier business logic of enterprise systems [10,11]. In these systems, the interface can be an object accessed through method calls in a variety of programming languages. Another possibility for logic in the business tier is via a service interface in service-oriented architecture systems. Our approach can accommodate any of these interfaces. Some, for example business tier logic, will require adding code to the knowledge-based system to provide access for testing.

A common approach to unit testing is found in the JUnit toolset [2]. JUnit is an automated testing system that is easy to use. There are other toolsets, such as Unit libraries for other languages, in which our testing tool can be implemented, but our work employed Java. The JUnit features of interest for our tool are outlined in the following paragraphs.

There are two main parts to the JUnit framework: test cases and testing tools. Test cases are derived by extending the base class TestCase from the JUnit library. Methods in the class with names starting with "test" are assumed to be the test cases. The second part is a testing tool. The tool runs a set of test cases and reports the failures. The real power of this approach is two fold. It is extremely easy to run an entire suite of tests, thus, encouraging the developer to test frequently. Additionally, it collects the test cases in a maintainable systematic structure, thus, encouraging the developers to save and reuse them.

The TestCase class API has a method for initializing the conditions for the test case, called setUp, and one to clean up afterward, called tearDown. Our design uses the setUp method. Some knowledge base system may need a tearDown routine if they have stored state that persists across invocations. However, the systems we have run do not require tearDown, since restarting the system is sufficient for running the test cases.

The other part of the TestCase class API is the methods that facilitate the creation of test cases. These are methods that have names beginning with "assert". These routines test a result returned from the system and depending on the result, signals success or failure of the test. The set of routines cover a wide set of data types enabling a flexible mechanism for checking results. Our work, so far, has used the string data type since the knowledge bases return results and prompts (for questions) as strings.

The generator builds the test cases. There is one part that must be written by the user, and that is an interface class for the knowledge-based system. This class has a defined standard interface that the generator uses in constructing the test cases. When the test cases run, they will use the interface class to access the knowledge-based system under test. Hereafter, the knowledge-based system is referred to as the "system under test".

A setup method is needed to start and initialize the system. The Java Runtime class provides a way to start a process that can contain the running system under test. The setup method creates a Runtime object running the system. The runtime object can access the input and output streams for the system. For many systems, the interface used simple console I/O and the streams are all that is required.

The other part of the interface class is a routine to read in prompts that are output from the system under test, and a routine to supply input to the system. For many systems, a read and write routine applied to the I/O streams is used. An additional complication can occur if a sophisticated GUI is used by the knowledge-based system. A GUI requires some adapting in the expert system to provide suitable access for the test driver, which is generally easy to achieve.

There are two other common types of systems. One system type is an object encapsulating a rule system, which may be used to implement business rules in a multi-tier enterprise system. This requires the use of method calls to access the object. This is directly implementable in JUnit without user parameterization. A similar situation will occur in the other common type, a rules engine encapsulated into a network service. This requires the user to parameterize the network messages used to communicate with the service.

5 Constructing Test Case

The construction of test cases is based on reengineering data automatically derived from an existing knowledge-based system. The reengineering tool collects knowledge

[4] on which we apply test cases. The test case generator is a fully automated algorithm. The algorithm works with production rules as the one described in the figure below, Figure 1. The test case generator uses the internal data extracted by the reengineering tool and diagram is for the visualization of the test cases.

```
rule(32, 'recent_largeobject', 'recent_large_text', 1000):-
  check('maps_largeobject', 'modern_maps_small_area3', 200),
  fact('environment', 'Tropical climate (A)'),
  reply('map_tools','Yes'),
  not reply('map_tools', 'Modern topographical maps').
```

Fig. 1. An example of a production rule

The above figure shows a rule from an existing expert system used in tropical environment giving advice about dam sites and hydropower buildings [6]. The system contains about 40 questions, 200 rules and 30 conclusions handling environmental impact assessment for building constructions. The reengineering tool extracts the rules into an internal form. Then the system constructs possible proofs, or reasoning sequences, that the systems can generate. The proofs are used for constructing diagrams like the one shown figure 2. This is a UML sequence diagram.

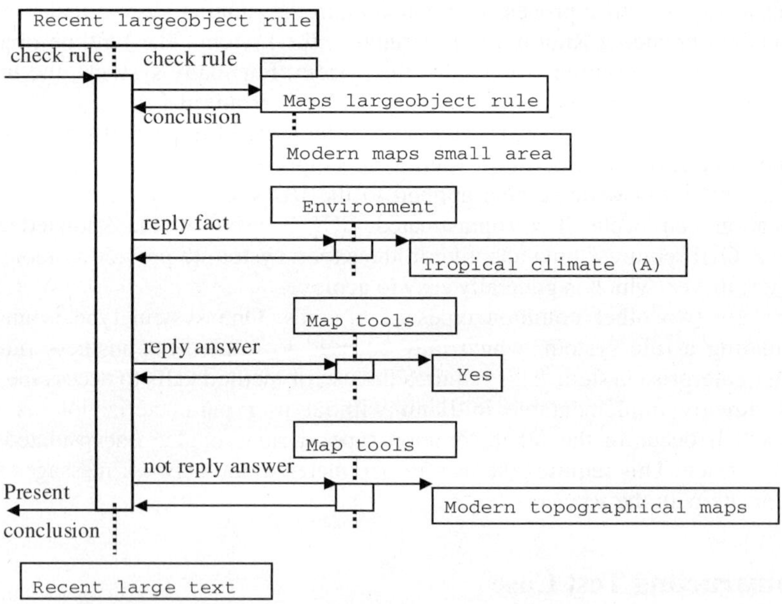

Fig. 2. A sequence diagram for the recent large object rule

The diagram, in the figure, corresponds to a proof or steps of a proof that occurs in the rule-based system. The diagrams offer a useful display to the user to help them understand the reasoning of the knowledge-based system. The diagram is a visual projection of the internal representation. The internal representation together with an algorithm is able to generate the test cases.

The other piece of information required by the generator is the input and output functions or predicates used in the knowledge-based system. These are defined for the generator by a regular expression pattern and a translation into a proper method call to the standard interface used by the test case. The generator uses the regular expression to find the input and output and create the proper calls to the methods in the interface class.

To illustrate this parameterization, a useful example is the expert system shell described by Ivan Bratko [3]. This shell is an interpreter for rules written in Prolog. The rules are placed in a file separate from the shell and are invoked by the system. This differs from the example used above in the predicate that generates the I/O request. The rules trigger I/O by the use of a predicate "askable". An approach to reengineering the knowledge-based system, described in Bratko, is to include the Prolog expert system shell and apply reengineering down to Prolog I/O primitives. A second approach is to have the reengineering tool detect the predicate "askable".

The algorithm, that generates the test case, takes each reasoning sequence and follows the sequence of the rule firings. This is best understood by looking at figure 2, although the generator used the internal representation. The diagram is traversed to determine a sequence of method calls for the test case body. There are two cases, one for prompts (questions) and one for inputs to the knowledge-based system.

For each prompt in the sequence, a read method is called to receive the prompt from the system under test. The prompt string received is tested for the correct value. The reengineering tool determines the value of the prompt string.

For each input after the prompt, a write method is called to send a value to the system. The value sent is selected to make sure the sequence will continue. That is, the value matched the needed condition for the predicate in the rule.

The last step in the body of the test case is to read the conclusion from the system and verify that it is correct according to the result predicted by the reengineering tool.

In the example in figure 2, the check and reply fact predicates do internal operations to verify a rule or a fact in the knowledge-store. These do not need to generate any code in the body of the test case. The reply and not-reply predicates prompt the user and then test the result. These two generate the prompt and input value to the system under test. The conclusion is printed by the system and it read and verified.

The example shows a single rule application. However, the generator must construct complete reasoning sequences for this rule. These sequences are then translated into test cases.

6 Complexity Analysis

The feasibility of the testing approach depends on two factors, one of which is an automated approach for building and running the test cases. The preceding sections of this paper demonstrate this. This section focuses on the second issue, which is generating a bounded and practical set of test cases to provide adequate coverage. This analysis is dependent on some reasonable basic assumptions. It is possible that some systems may violate these assumptions, but we have not seen an example of such a system yet.

The first observation is that the number of rules is finite. Therefore, the number of possible proofs in the system is finite. However, the number of possible inputs need not be finite. But the rules use predicates, and the possible set of input values for any predicate can be partitioned into two sets, one for true and one for false. Therefore, all possible input values are not necessary. Just a representative true producing input and a false producing input are required for any given predicate.

However, the number of possible proofs to test can be impractically large. A simple argument for this is to look at all possible proofs by simple combinatoric methods. That is, look at all possible sequences of rules running in length from 1 to the total number of rules in the knowledge base. However, this is not representative of typical expert systems we have observed. Knowledge tends to be structured into partitions and the interconnections between rules tend to be limited. While we do not have proof for this assertion, observation of existing knowledge bases makes this appear reasonable. A more useful predicator of the number of test cases is the number of possible conclusions the system can generate. While this is a lower bound on the number of test cases needed, it is generally the case that the number of required test cases is bounded by O(number of conclusions).

Techniques from classical software testing can be applied to construct a useful set of test cases. Knowledge-based systems are declarative in nature and, therefore, do not have control flow in the sense of imperative programs. The reengineering extraction process builds a graph of connections between the rules. The conclusions become the leaves in this graph. Proofs can be located by tracing paths from conclusions back to starting nodes. This is similar to Path Testing (Jorgensen 2002). For rule systems, this is easier than in imperative programs, since loops do not exist. When the extraction is done, we can use graph theory to predict the number of possible test cases. A simple approach is to use a breadth first approach to build the test cases and set a limit on the total number, to prevent the generation process from going out of control.

7 Testing Process

The process starts by invoking the generated test case set. For each failed test case, the user must determine if the failure is valid. A valid failure is when the knowledge is changed and the unmodified system is no longer correct. Otherwise, the failure indicates that a change was made that has a bad interaction with the old knowledge. For the valid failure, we can remove the old test case. However, in the actual error case, the system needs to be debugged to find the cause of the error. The diagram associated with the test case can be located to help the developer see what the failure was. The reengineering tool can also be used to find the actual reasoning that occurred to cause the failure, by applying the tool to extract a new diagram of the actual reasoning. This is a powerful use of the diagrams generated by the tools.

The next phase is to add in the new test cases to define reasoning of the new knowledge that has been added. These test cases should all pass or else we have an error to fix. Since these test cases are being added, the test cases required will be understood by the knowledge engineer and written by the engineer.

The other part of the tool suite, which is used during the modification of the knowledge base, is the automated testing of consistency, completeness, conflicts between rules and redundant rules. This verification and validation testing is based on the work in the KANAL system [5]. This work helps detect problems in the system that test cases alone may fail to locate.

8 Conclusion

Automated test generation is a viable technique to assist in reengineering a knowledge-based system. Sequence diagrams illustrate the proofs with inputs from and outputs to the users. Starting from the inputs, the proofs are 'walked through', i.e., a process involving the content of the rules, and during this process, test cases are generated. All the proofs are processed to build a complete set of test cases. One can argue about the correctness of the test cases, but the assumption has to be made that the knowledge base is correct when we apply reengineering to it. If the knowledge base is correct, the set of test cases should be complete for the existing knowledge base. In the extreme, some systems may exist that produce an intractably large number of test cases, but this is uncommon in our experience.

The collected test cases are implemented in an automatic testing framework. Each execution of a test case is noted in a test log that, therefore, will contain information of all the test cases that have been executed and the errors obtained. The errors are the outcome of the execution of the test cases and the result of comparing the test case on the modified rules with the original rules.

By using these test cases, the rules are detected and verified either as being correct or as producing errors. This leads to a solid verification process ensuring that the modifications to the knowledge base do not interact negatively with the existing knowledge. While new test cases need to be generated manually to test the new knowledge, this process gives a solid framework to capture those as well. In addition, the original work covers testing the knowledge base for consistency, completeness, and correctness of the knowledge base.

The test driver needs to be applied to more knowledge-based systems. So far, the driver had only been applied in two different rule based systems, but we will apply it to other rule based systems. The test bed we can use has more than 20 systems. However, we need to apply the driver to other kinds of programming techniques as well. We also have systems using decision tables in the test bed. [If you want to read a bit about it – see http://www.cs.ucy.ac.cy/compulog/wisdomon.htm].

References

1. Barr, V.: Applications of Rule-Based Coverage Measures to Expert System Evaluation. Journal of Knowledge-Based Systems 12, 27–35 (1999)
2. Kent, B.: JUint Pocket Guide. Published by O'Reilly (2004)

3. Bratko, I.: Prolog Programming for Artificial Intelligence, 3rd edn. Addison Wesley, London, UK (2000)
4. Håkansson, A., Hartung, R.: Reengineering for Knowledge in Knowledge Based Systems. In: Gabrys, B., Howlett, R.J., Jain, L.C. (eds.) KES 2006. LNCS (LNAI), vol. 4253, Springer, Heidelberg (2006)
5. Håkansson, A., Widmark, A.: The KANAL System - a Tool for Knowledge Acquisition. Masters Thesis. Computing Science, Uppsala (1996)
6. Håkansson, A.: An Expert System for the Environment Impact Assessment Method. Research Report 2004:1. Department of Information Science, Division of Computer Science, University of Uppsala, Sweden, (2004)
7. Jorgensen, P.: Software Testing A Craftmans Approach, 2nd edn. CRC Press, Boca Raton, USA (2002)
8. Kiper, J.D.: Structural testing of rule-based expert systems. ACM Transactions on Software Engineering and Methodology 1(2), 168–187 (1992)
9. Preece, A.: Validation of Knowledge-Based Systems: The State-of-the-Art in North America. Journal of Communication and Cognition-Artificial Intelligence (1994)
10. Common, I.B.M.: Rules (2007-01-15) (2002), http://www.alphaworks.ibm.com/tech/commonrules
11. Owen, J.: Bring business logic to light. Java World(2003) (2007-01-20) (2003), http://www.javaworld.com/javaworld/jw-09- jw-0919-iw-jrules.html

Building Maintainable Knowledge Bases with Knowledge Objects

John Debenham

Faculty of IT, University of Technology, Sydney
PO Box 123, Broadway 2007, Australia
debenham@it.uts.edu.au

Abstract. Knowledge 'objects' are constructors for building chunks of knowledge that enable the hidden links in the knowledge to be identified. A single operation for objects enables these hidden links to be removed from the knowledge thus simplifying maintenance. Analysis of the maintenance links shows that they are of four different types. The density of the maintenance links is reduced by transforming that set into an equivalent set. In this way the knowledge base maintenance problem is analysed and simplified. A side benefit of knowledge items as a formalism is that they contain knowledge constraints that protect the knowledge from unforeseen modification.

Keywords: Knowledge-based systems, maintenance.

1 Introduction

Each chunk of knowledge is represented as a single "item" [1]. However items can share the same essential wisdom [2]. The integrity of items represented in this model is maintained by following maintenance links — the structure of these links is simplified by a decomposition process. The expressive power of a unified representation must be able to describe *at least* the data, information and knowledge things. The unified representation described here also contains two classes of constraints that apply equally to knowledge, information and to data. These constraints provide safeguards against invalid maintenance operations. In [3] these constraints are generalised to fuzzy acceptability measures of knowledge base integrity. Item and object join has been extended to apply to those measures [4].

Approaches to the maintenance of declarative conceptual models are principally of two types [5]. First, approaches that take a model 'as is' and then try to control the maintenance process [6] [7]. Second, approaches that *engineer* a model so that it is in a form that is inherently easy to maintain [8] [9]. The approach described here is of the second type because maintenance is driven by a maintenance link structure that is simplified by transforming the model. The majority of conceptual models for knowledge-based systems treat the "rule base" component separately from the "database" component. This enables well established design methodologies to be employed, but the use of two separate models means that the interrelationship between the things in these two models cannot be represented, integrated and manipulated naturally within the model [10]. Further, neither of these two separate models is able to address completely the validity of the whole knowledge base.

B. Apolloni et al. (Eds.): KES 2007/WIRN 2007, Part I, LNAI 4692, pp. 279–286, 2007.

2 A Unified Representation

Items represent *all* data, information and knowledge things in an application [10]. An *item* is a named triple $A[S_A, V_A, C_A]$ with *item name* A, S_A is the *item semantics* of A, V_A is the *item value constraints* of A and C_A is the *item set constraints* of A. The item semantics, S_A, is a λ-calculus expression that recognises the members of the value set of item A. The expression for an item's semantics may contain the semantics of other items $\{A_1, ..., A_n\}$ which are called that item's *components*.

$$\lambda x_1 x_2 y_1 y_2 z_1 z_2 \cdot [(S_{part/sale\text{-}price}(x_1, x_2) \wedge S_{part/cost\text{-}price}(y_1, y_2) \wedge S_{part/mark\text{-}up}(z_1, z_2)) \wedge (((x_1 = y_1) \wedge (x_1 = z_1)) \rightarrow (x_2 = z_2 \times y_2))] \cdot$$

This expression asserts that the pair (x_1, x_2) satisfies the semantics of the item *part/sale-price*, that (y_1, y_2) satisfies the semantics of the item *part/cost-price*, that (z_1, z_2) satisfies the semantics of the item *part/mark-up*, and that:

$$(((x_1 = y_1) \wedge (x_1 = z_1)) \rightarrow (x_2 = z_2 \times y_2))$$

holds. This last component of the semantics expression is the substantive part. The important feature of the λ-calculus form is that it represents all of the if-then interpretations of a single knowledge chunk.

In general the item semantics is an expression of the form:

$$\lambda y_1^1 \dots y_m^1 \dots y_{m_n}^n \cdot [S_{A_1}(y_1^1, \dots, y_{m_1}^1) \wedge \dots \wedge [S_{A_n}(y_1^n, \dots, y_{m_n}^n) \wedge J(y_1^1, \dots, y_m^1, \dots, y_{m_n}^n)] \cdot$$

where J is a first-order predicate. The item value constraints, V_A, is a λ-calculus expression:

$$\lambda y_1^1 \dots y_m^1 \dots y_{m_n}^n \cdot [V_{A_1}(y_1^1, \dots, y_{m_1}^1) \wedge \dots \wedge [V_{A_n}(y_1^n, \dots, y_{m_n}^n) \wedge K(y_1^1, \dots, y_m^1, \dots, y_{m_n}^n)] \cdot$$

where K is a first-order predicate, that should be satisfied by the members of the value set of item A as they change in time. So if a tuple satisfies S_A then it should satisfy V_A [11]. The expression for an item's value constraints contains the value constraints of that item's components. The item set constraints, C_A, is an expression of the form:

$$C_{A_1} \wedge C_{A_2} \wedge \dots \wedge C_{A_n} \wedge (L)_A$$

where L is a logical combination of:

- Card lies in some numerical range;
- $\text{Uni}(A_i)$ for some i, $1 \leq i \leq n$, and
- $\text{Can}(A_i, X)$ for some i, $1 \leq i \leq n$, where X is a non-empty subset of $\{A_1, \dots, A_n\} - \{A_i\}$

subscripted with the name of the item A, "$\text{Uni}(a)$" means that "all members of the value set of item a must be in this association". "$\text{Can}(A, X)$" means that "the value set of the set of items X functionally determines the value set of item A". "Card" means "the number of things in the value set". The subscripts indicate the item's components to which that set constraint applies.

To make the inherent structure of items clear, 'objects' are introduced as item building operators. The use of objects to build items enables the hidden links in the knowledge to be identified. A single operation for objects enables these hidden links to be removed from the knowledge thus simplifying maintenance.

An n-adic *object* is an operator that maps n given items into another item for some value of n. Further, the definition of each object will presume that the set of items to which that object may be applied are of a specific "type". Examples of item *type* include D^m for m-adic data items, I^m for m-adic information items and K^m for m-adic knowledge items. Items may also have unspecified, or free, type which is denoted by X^m. The formal definition of an object is similar to that of an item. An *object* named **A** is a typed triple $\mathbf{A}[E,F,G]$ where E is a typed expression called the *semantics* of A, F is a typed expression called the *value constraints* of A, and G is a typed expression called the *set constraints* of A. For example, the *part/cost-price* item can be built from the items *part* and *cost-price* using the **costs** operator:

$$\mathit{part/cost\text{-}price} = \mathbf{costs}(\mathit{part}, \mathit{cost\text{-}price})$$

$$\mathbf{costs}[\lambda P : \mathrm{X}^1 Q : \mathrm{X}^1 \cdot \lambda xy \cdot [S_P(x) \wedge S_Q(y) \wedge costs(x,y)] \cdot \cdot,$$

$$\lambda P : \mathrm{X}^1 Q : \mathrm{X}^1 \cdot \lambda xy \cdot [V_P(x) \wedge V_Q(y) \wedge ((1000 < x < 1999) \rightarrow (y \leq 300))] \cdot \cdot,$$

$$\lambda P : \mathrm{X}^1 Q : \mathrm{X}^1 \cdot [C_P \wedge C_Q \wedge (\mathrm{Uni}(P) \wedge \mathrm{Can}(Q,P))_{\nu(\mathbf{costs},P,Q)}] \cdot]$$

where $\nu(\mathbf{costs}, P, Q)$ is the name of the item $\mathbf{costs}(P, Q)$.

Data objects provide a representation of sub-typing. Data objects are used in the conceptual model to derive individual data items from the universal item U, where $U = U[\lambda x \cdot x : \mathrm{U} \cdot, \lambda x \cdot \top \cdot, \emptyset]$, "U" is the "universe of discourse" and $\top$ is the constant "true" expression. The data object **part** is:

$$\mathbf{part}[\lambda P : \mathrm{X}^1 \cdot \lambda x \cdot [S_P(x) \wedge \text{with-part}(x)] \cdot \cdot,$$

$$\lambda P : \mathrm{X}^1 \cdot \lambda x \cdot [V_P(x) \wedge (x \geq 0)] \cdot \cdot, \lambda P : \mathrm{X}^1 \cdot [C_P \wedge (\mathrm{Card} \geq 1)_{\nu(part,P)}] \cdot]$$

If the object **part** is applied to the universal item U it then generates the item *part* by: *part* = **part**U.

Declarative rules are quite clumsy when represented as items; objects provide a more compact representation. For example, consider the *[part/sale-price, part/cost-price, part/mark-up]* knowledge item which represents the chunk of knowledge [K1] "The sale price of a part is the cost price of that part marked up by the markup rate for that part". This item can be built by applying a knowledge object **mark-up-rule** of argument type $(\mathrm{I}^2, \mathrm{I}^2, \mathrm{I}^2)$ to the items *part/sale-price*, *part/cost-price* and *part/mark-up*. That is:

[part/sale-price,part/cost-price, part/mark-up] =

mark-up-rule(*part/sale-price*, *part/cost-price*, *part/mark-up*)

Item join provides the basis for item decomposition [10]. Given items A and B, the item with name $A \otimes_E B$ is called the *join* of A and B on E, where E is a set of components common to both A and B. Consider: $A[S_A, V_A, C_A]$ and $B[S_B, V_B, C_B]$. Suppose that S_A has n variables, that is A is an n-adic item. Suppose that S_B has m variables, that is B is an m-adic item. Some of the components of A and B may be identical. Suppose that k pairs of components of A and B that are identical are identified, where $k \geq 0$. Let E be an ordered set of components where each is one of these identical pairs of components of both A and B. E may be empty. To ensure that the definition is well defined the order of the components in the set E is the same as order in which they occur as components of A. Suppose the semantics expressions of the components from item A (or item B) that are in the set E are expressed in terms of a total of p variables. Let A^* be an n-adic item that is identical to item A except for the order of its variables. The last p variables in A^* are those variables in A that belong to the components of A in the set E. Let B^* be an m-adic item that is identical to item B except for the order of its variables. The first p variables in B^* are those variables in B that belong to the components of B in the set E. Let π be a permutation that turns the ordered set of variables of A^* into the ordered set of variables of A. Let π' be a permutation that turns the ordered set of variables of B^* into the ordered set of variables of B. Suppose that x is an $(n-p)$-tuple of free variables, y is a p-tuple of free variables and z is an $(m-p)$-tuple of free variables. Then the item with name $A \otimes_E B$ is the *join* of A and B on E, it is defined to be:

$$(A \otimes_E B)[\lambda xyz \cdot [S_A(\pi(x,y)) \wedge S_B(\pi'(y,z))]^{\cdot},$$
$$\lambda xyz[V_A(\pi(x,y)) \wedge V_B(\pi'(y,z))]^{\cdot}, C_{A\otimes_E B}]$$

where $C_{A\otimes_E B}$ is defined as follows. Suppose that C_A is an expression of the form $c_A \wedge G$ where c is that part of C_A that carries the subscript 'A' and G is that part of C_A that carries subscripts other than 'A'. Likewise suppose that C_B is an expression of the form $d_B \wedge H$. Then:

$$C_{A\otimes_E B} \triangleq (c \wedge d)_{A\otimes_E B} \wedge (G \wedge H)$$

The set E is a set of identical pairs of components of A and B. If E is the set of all identical pairs of components of A and B then $A \otimes_E B$ may be written as $A \otimes B$.

Using the method of composition , knowledge items, information items and data items may be joined with one another regardless of type. For example, the knowledge item:

$$\textit{[cost-price, tax]}[\lambda xy \cdot [S_{\textit{cost-price}}(x) \wedge S_{tax}(y) \wedge x = y \times 0.05]^{\cdot},$$
$$\lambda xy \cdot [V_{\textit{cost-price}}(x) \wedge V_{tax}(y) \wedge x < y]^{\cdot}, C_{\textit{[cost-price, tax]}}]$$

can be joined with the information item *part/cost-price* on the set {*cost-price*} to give the information item *part/cost-price/tax*. In other words:

$$\textit{[cost-price, tax]} \otimes_{\{\textit{cost-price}\}} \textit{part/cost-price} =$$
$$\textit{part/cost-price/tax}[\lambda xyz \cdot [S_{part}(x) \wedge S_{\textit{cost-price}}(x) \wedge S_{tax}(y) \wedge$$
$$\text{costs}(x,y) \wedge z = y \times 0.05]^{\cdot},$$
$$\lambda xyz \cdot [V_{part}(x) \wedge V_{\textit{cost-price}}(x) \wedge V_{tax}(y) \wedge$$
$$((1000 < x < 1999) \rightarrow (0 < y \leq 300)) \wedge (z < y)]^{\cdot}, C_{\textit{part/cost-price/tax}}]$$

Using the item join operator, items may be joined together to form more complex items. The $\otimes$ operator also forms the basis of a theory of decomposition in which each item is replaced by a set of simpler items. An item I is *decomposable*; into the set of items $D = \{I_1, I_2, ..., I_n\}$ if: I_i has non-trivial semantics for all i, $I = I_1 \otimes I_2 \otimes \cdots \otimes I_n$, where each join is *monotonic*; that is, each term in this composition contributes at least one component to I. If item I is decomposable then it will not necessarily have a unique decomposition. The join operator for objects is defined in a similar way to item join and is also denoted by $\otimes$. When $\otimes$ is used to join two objects it is subscripted with a set of pairs of positive integers that identify the component pairs in the first and second argument of $\otimes$ that are being joined. The method of decomposition is: "Given a conceptual model discard any items and objects which are decomposable". For example, this method requires that the item *part/cost-price/tax* should be discarded in favour of the two items *[cost-price,tax]* and *part/cost-price*.

3 Analysis of the Conceptual Model

The items in the conceptual model are constructed by applying a set of object operators to U. A *maintenance link* joins two items in the conceptual model if modification of one item means that the other item must be checked for correctness, and maybe modified, so that the consistency of the conceptual model is preserved [12]. The number of maintenance links can be very large. So maintenance links can only form the basis of a practical approach to knowledge base maintenance if there is some way of reducing their density on the conceptual model.

RESULT
Sub-item links may be reduced to sub-type links between data items.

Demonstration:
Given two items A and B, where both are n-adic items with semantics S_A and S_B respectively, if π is permutation such that:

$$(\forall x_1 x_2 \ldots x_n)[S_A(x_1, x_2, \ldots, x_n) \leftarrow S_B(\pi(x_1, x_2, \ldots, x_n))]$$

then item B is a *sub-item* of item A. These two items should be joined with a maintenance link. If A and B are both data items then B is a sub-type of A. Suppose that:

$$X = \mathbf{E}D; \quad \text{where} \quad D = \mathbf{C}AB \tag{1}$$

for items X, D, A and B and objects $\mathbf{E}$ and $\mathbf{C}$. Item X is a sub-item of item D. Object $\mathbf{E}$ has the effect of extracting a sub-set of the value set of item D to form the value set of item X. Item D is formed from items A and B using object $\mathbf{C}$. Introduce two new objects $\mathbf{F}$ and $\mathbf{J}$. Suppose that object $\mathbf{F}$ when applied to item A extracts the same subset of item A's value set as $\mathbf{E}$ extracted from the "left-side" (ie. the "A-side") of D. Likewise $\mathbf{J}$ extracts the same subset of B's value set as $\mathbf{E}$ extracted from D. Then:

$$X = \mathbf{C}GK; \quad \text{where} \quad G = \mathbf{F}A \quad \text{and} \quad K = \mathbf{J}B \tag{2}$$

so G is a sub-item of A, and K is a sub-item of B. The form (2) differs from (1) in that the sub-item maintenance links have been moved one layer closer to the data item layer, and object $\mathbf{C}$ has moved one layer away from the data item layer. Using this method repeatedly sub-item maintenance links between non-data items are reduced to sub-type links between data items. □

RESULT
There are four kinds of maintenance link in a conceptual model built using the uniform knowledge representation.

Demonstration:
Consider two items A and B, and suppose that their semantics S_A and S_B have the form:

$$S_A = \lambda y_1^1 \ldots y_{m_1}^1 \cdots y_{m_p}^p \cdot [S_{A_1}(y_1^1, \ldots, y_{m_1}^1) \wedge \cdots \wedge S_{A_p}(y_1^p, \ldots, y_{m_p}^p) \wedge J(y_1^1, \ldots, y_{m_1}^1, \ldots, y_{m_p}^p)] \cdot$$

$$S_B = \lambda y_1^1 \ldots y_{n_1}^1 \cdots y_{n_q}^q \cdot [S_{B_1}(y_1^1, \ldots, y_{n_1}^1) \wedge \cdots \wedge S_{B_q}(y_1^q, \ldots, y_{n_q}^q) \wedge K(y_1^1, \ldots, y_{n_1}^1, \ldots, y_{n_q}^q)] \cdot$$

S_A contains $(p+1)$ terms and S_B contains $(q+1)$ terms. Let μ be a maximal sub-expression of $S_{A \otimes B}$ such that:

$$\text{both} \quad S_A \rightarrow \mu \quad \text{and} \quad S_B \rightarrow \mu \tag{3}$$

where μ has the form:

$$\lambda y_1^1 \ldots y_{d_1}^1 \cdots y_{d_r}^r \cdot [S_{C_1}(y_1^1, \ldots, y_{d_1}^1) \wedge \cdots \wedge S_{C_r}(y_1^r, \ldots, y_{d_r}^r) \wedge \quad L(y_1^1, \ldots, y_{d_1}^1, \ldots, y_{d_r}^r)] \cdot$$

If μ is empty, ie. 'false', then the semantics of A and B are independent. If μ is non-empty then the semantics of A and B have something in common and A and B should be joined with a maintenance link.

Now examine μ to see why A and B should be joined. If μ is non-empty then there are three cases. First, if:

$$S_A \leftrightarrow S_B \leftrightarrow \mu \tag{4}$$

then items A and B are equivalent and should be joined with an *equivalence link*. Second if (4) does not hold and:

$$\text{either} \quad S_A \leftrightarrow \mu \quad \text{or} \quad S_B \leftrightarrow \mu \tag{5}$$

then either A is a sub-item of B, or B is a sub-item of A and these two items should be joined with a *sub-item link*. Third, if (4) and (5) do not hold then if Δ is a minimal sub-expression of S_A such that $\Delta \rightarrow \mu$. Then:

$$\text{either} \quad S_{A_j}(y_1^j, \ldots, y_{m_j}^j) \in \Delta \quad \text{for some} \quad j \tag{6}$$

$$\text{or} \quad J(y_1^1, \ldots, y_{m_j}^1, \ldots, y_{m_p}^p) \in \Delta \tag{7}$$

Both (6) and (7) may hold. If (6) holds then items A and B share one or more component items to which they should each be joined with a *component link*. If (7) holds then items A and B may be constructed with two object operators whose respective semantics are logically dependent. Suppose that item A was constructed by object operator **C** then the semantics of **C** will imply:

$$\Phi = \lambda Q_1 : X_1^{i_1} Q_2 : X_2^{i_2} \ldots Q_j : X_j^{i_j} \cdot \lambda y_1^1 \ldots y_{d_1}^1 \ldots y_{d_r}^r \cdot [S_{P_1}(y_1^1, \ldots, y_{d_1}^1) \wedge \ldots$$
$$\wedge S_{P_r}(y_1^r, \ldots, y_{d_r}^r) \wedge L(y_1^1, \ldots, y_{d_1}^1, \ldots, y_{d_r}^r)] \cdot$$

where the Q_i's take care of any possible duplication in the P_j's. Let **E** be the object $\mathbf{E}[\Phi, \top, \emptyset]$ then **C** is a sub-object of **E**; that is, there exists a non-tautological object **F** such that:

$$\mathbf{C} \approx_W \mathbf{E} \otimes_M \mathbf{F} \tag{8}$$

for some set M and where the join is not necessarily monotonic. Items A and B are weakly equivalent, written $A \approx_W B$, if there exists a permutation π such that:

$$(\forall x_1 x_2 \ldots x_n)[S_A(x_1, x_2, \ldots, x_n) \leftrightarrow S_B(\pi(x_1, x_2, \ldots, x_n))]$$

where the x_i are the n_i variables associated with the i'th component of A. If A is a sub-item of B and if B is a sub-item of A then items A and B are weakly equivalent.

If (8) holds then the maintenance links are of three different kinds. If the join in (8) is monotonic then (8) states that **C** may be decomposed into **E** and **F**. If the join in (8) is not monotonic then (8) states that either $\mathbf{C} \approx_W \mathbf{E}$ or $\mathbf{C} \approx_W \mathbf{F}$. So, if the join in (8) is not monotonic then either **E** will be weakly equivalent to **C**, or **C** will be a sub-object of **E**. □

It has been shown above that sub-item links between non-data items may be reduced to sub-type links between data items. So if:

- all equivalent objects have been removed by re-naming, and
- sub-item links between non-data items have been reduced to sub-type links between data items

then the maintenance links will be between nodes marked with:

- a data item that is a sub-type of the data item marked on another node, these are called the *sub-type links*;
- an item and the nodes marked with that item's components, these are called the *component links*, and
- an item constructed by a decomposable object and nodes constructed with that object's decomposition, these are called the *duplicate links*.

If the objects employed to construct the conceptual model have been decomposed then the only maintenance links remaining will be the sub-type links and the component links. The sub-type links and the component links cannot be removed from the conceptual model.

Unfortunately, decomposable objects, and so too duplicate links, may be hard to detect. Suppose that objects **A** and **B** are decomposable as follows:

$$\mathbf{A} \approx_W \mathbf{E} \otimes_M \mathbf{F} \qquad \mathbf{B} \approx_W \mathbf{E} \otimes_M \mathbf{G}$$

Then objects **A** and **B** should both be linked to object **E**. If the decompositions of **A** and **B** have not been identified then object **E** may not have been identified and the implicit link between objects **A** and **B** may not be identified.

4 Conclusion

A model of knowledge objects has been described at a high-level abstraction by using a unified conceptual model. Five principles have been given to identify decomposable objects. Another kind of maintenance link is removed by reducing sub-item relationships to sub-type relationships. In this way the maintenance problem is simplified. All items, including knowledge items, have a set of constraints. The constraints of a knowledge item apply to any program that implements that item. So item constraints provide a mechanism that further protects against integrity violation during maintenance.

References

1. Debenham, J.: Why use a unified knowledge representation? In: Monostori, L., Váncza, J., Ali, M. (eds.) IEA/AIE 2001. LNCS (LNAI), vol. 2070, Springer, Heidelberg (2001)
2. Iglezakis, I., Reinartz, T., Roth-Berghofer, T.: Maintenance Memories: Beyond Concepts and Techniques for Case Base Maintenance. In: Advances in Case-Based Reasoning, pp. 227–241. Springer, Heidelberg (2004)
3. Debenham, J.: Fuzzy degrees of knowledge integrity. In: proceedings 9'th International Conference on Information Processing and Management of Uncertainty in Knowledge-Based Systems IPMU-2002, pp. 1391–1398 (2002)
4. Debenham, J.: A rigorous approach to knowledge base maintenance. In: Proceedings Sixteenth International Conference on Industrial and Engineering Applications of Artificial Intelligence and Expert Systems IEA/AIE-2003, Loughborough, UK, pp. 219–228 (2003)
5. Katsuno, K., Mendelzon, A.: On the difference between updating a knowledge base and revising it. In: proceedings Second International Conference on Principles of Knowledge Representation and Reasoning KR'91, Morgan Kaufmann, San Francisco (1991)
6. Barr, V.: Applying reliability engineering to expert systems. In: Proceedings 12'th International FLAIRS conference, pp. 494–498 (1999)
7. Chen-Burger, Y.-H.: Sharing and Checking Organisation Knowledge. In: Knowledge Management and Organizational Memories, Kluwer Academic, Dordrecht (2002)
8. Jantke, K., Herrmann, J.: Lattices of knowledge in intelligent systems validation. In: Proceedings 12'th International FLAIRS conference, Florida, US, pp. 499–505 (1999)
9. Darwiche, A.: Compiling knowledge into decomposable negation normal form. In: Proceedings International Joint Conference on Artificial Intelligence IJCAI'99, Stockholm, Sweden, Stockholm, Sweden, pp. 284–289 (1999)
10. Debenham, J.: Knowledge Engineering — Unifying Knowledge Base and Database Design. Springer, Heidelberg (1998)
11. Johnson, G., Santos, E.: Generalizing knowledge representation rules for acquiring and validating uncertain knowledge. In: Proceedings 13'th International FLAIRS conference, Florida, US, pp. 186–191 (2000)
12. Ramirez, J., de Antonio, A.: Semantic verification of rule-based systems with arithmetic constraints. In: Ibrahim, M., Küng, J., Revell, N. (eds.) DEXA 2000. LNCS, vol. 1873, pp. 437–446. Springer, Heidelberg (2000)

Influenza Forecast: Case-Based Reasoning or Statistics?

Rainer Schmidt and Tina Waligora

Institute for Medical Informatics and Biometry, University of Rostock, Germany
rainer.schmidt@medizin.uni-rostock.de

Abstract. Influenza is the last of the classic plagues of the past, which still has to be brought under control. It causes a lot of costs: prolonged stays in hospitals and especially many days of unfitness for work. Therefore many of the most developed countries have started to create influenza surveillance systems. Mostly statistical methods are applied to predict influenza epidemics. However, the results are rather moderate, because influenza waves occur in irregular cycles. We have developed a method that combines Case-Based Reasoning with temporal abstraction. Here we compare experimental results of our method and of statistical methods.

Keywords: Case-Based Reasoning, Statistics, Influenza.

1 Introduction

Since influenza results in many costs, e.g. for delayed stays in hospital and especially for an increased number of unfitness for work, many of the most developed countries have started to generate influenza surveillance systems (e.g. US: www.flustar.com, France [1], and Japan [2]). The idea is to predict influenza waves or even epidemics as early as possible and to indicate appropriate actions like starting vaccination campaigns or advising high-risk groups to stay at home. Mostly statistical methods are applied to predict influenza epidemics. However, the results are rather moderate, because influenza waves occur in irregular cycles and as Farrington pointed out statistical methods have difficulties to cope with infectious diseases characterized by an irregular cyclic behavior [3]. So, we developed a method that combines CBR with temporal abstraction. Before we explain our method and subsequently present comparative results of our method and of statistical methods, we firstly discuss the question which data should be used.

1.1 Data

It is well known that a couple of factors may be responsible for influenza outbreaks. One of them is the weather. It is often assumed that a strong winter increases the spread of influenza. However, studies could reveal only an extremely small relation between temperatures and spread of influenza [4]. Dowell even suggests that the occurrence of influenza in winter is not related to the temperature but to the annual light/dark pattern [5]. Other influences are the mutations of the virus and influenza outbreaks in foreign countries, even as far as Hongkong. Unfortunately, no exact

B. Apolloni et al. (Eds.): KES 2007/WIRN 2007, Part I, LNAI 4692, pp. 287–294, 2007.

knowledge about these influences is available. So far, only seasonal behaviour of influenza is well known from observations [3, 5]. Therefore all surveillance systems focus on observed numbers of infected people, especially on their increase.

Considerations about how to forecast influenza begin with questions about which data should be used and which data are available. The answer varies with the country and its health organisation. In some countries with rather private health systems like Germany, research groups interested in predicting influenza have started to develop surveillance nets based on voluntary participation of general practitioners. They felt that the data collected and provided by official health centres were insufficient, because they are usually available with a delay of two or even three weeks. These surveillance nets are based on general practitioners who give once a week some sort of standardised reports. It needs a huge effort to initiate and organise such nets and rural areas are very often not adequately represented, because it is difficult to find doctors willing to participate. Since the reports are always subjective, misjudgements and data interpretation errors may occur, which may lead to false assessments – especially in areas with low density of participating doctors. An alternative means to use official data from health centres. In Germany, these data are more objective, because they contain reports and laboratory results of all occurrences of notifiable diseases. Unfortunately, because of the hierarchical and bureaucratic organisation of the health centres, the availability of these data is delayed for at least two weeks. In countries with more public health systems like Japan [2], the situation seems to be better.

However, we have chosen another alternative. Since 1997 we receive data for our federal state Mecklenburg-Western Pomerania from the main health insurance scheme. These data are sick certificates of employees and of people who receive unemployment benefit. Fortunately we get the data daily. Of course there is a short delay between doctors writing the certificates and the insurance scheme receiving them by mail from their policyholders. Our method does not recur on the days when the certificates have been issued by doctors, but on the daily data sets received by the insurance scheme. Since daily fluctuations occur by chance, influenza surveillance systems usually use weekly aggregated data. The disadvantage of using insurance data is their superficiality, because the certificates usually contain just the first diagnoses, which might be refined or even changed later on. However, for influenza this is just a minor problem, because the symptoms of influenza, acute bronchitis, etc. are so similar that most surveillance groups use a superficial category anyway, namely all acute respiratory diseases to infer influenza.

2 Prognostic Methods

All influenza surveillance systems make use of developments in the past. Most of them have tried statistical methods. The usual idea is to compute average values and standard courses based on weekly incidences of former influenza seasons (from October till March) and to analyse deviations from a statistic normal situation.

Influenza waves usually occur only once a season, but they start at different time points and have extremely different intensities. Since Farrington pointed out that statistical methods are inappropriate for diseases like influenza that are characterised by irregular cyclic temporal spreads [3], we have developed a method that uses former influenza seasons more explicitly. We apply the Case-Based Reasoning idea: that means to determine the most similar former courses of weekly incidences and to use them to decide whether a warning is appropriate. Viboud [6] from the group that is responsible for the French Surveillance net has developed a method that is very similar to our one. However, both methods differ in their intentions. Viboud attempts to predict incidences few weeks in advance, while we are interested in more practical results, namely in the computation of appropriate warnings.

2.1 Case-Based Reasoning to Forecast Influenza

Inspired by our former program for the prognosis of kidney function courses [7], we have developed a method to decide about the appropriateness of warnings against approaching influenza waves (figure 1).

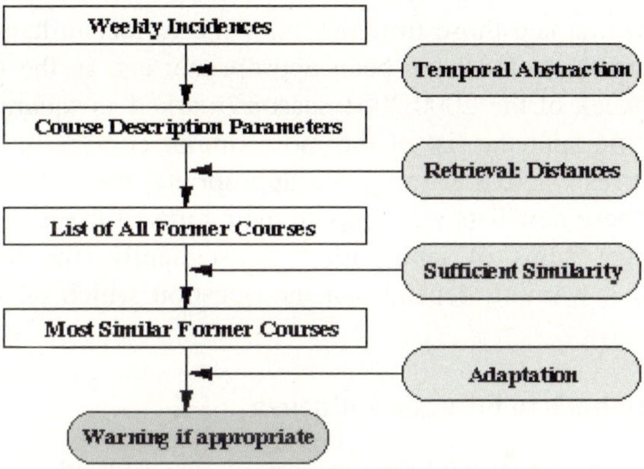

Fig. 1. Steps to forecast influenza by applying Case-Based Reasoning

Every influenza season consists of 26 weeks (from October till March). Since we consider weekly incidences, seasons are represented as sequences of 26 numeric values. Each week it has to be decided anew whether a warning is appropriate or not. For this decision, only the recent development is important. So, we always consider sequences of the four most recent weeks. When an influenza season is finished, it is separated into 23 four-week courses; all of them are stored as cases in the case base.

The first step of our method is a temporal abstraction of a sequence of four weekly incidences into three trend descriptions that assess the changes from last week to this week, from last but one week to this week and so forth. Secondly, these three assessments and the four weekly incidences are used to determine similarities between a current query course and all four-week courses stored in the case base. Our intention for using these two sorts of parameters is to ensure that a query course and an appropriate similar course are on the same level (similar weekly incidences) and that they have similar changes on time (similar assessments). More details about these first two steps of our method can be found in [8]. The result of computing distances is a very long list of all former four-week courses sorted according to their distances in respect to the query course. For the decision whether a warning is appropriate, this list is not really helpful, because most of the former courses are rather dissimilar to the query course. So, the next step means to find the most similar ones. We decided to filter the most similar cases by applying two explicit similarity conditions. First, the difference concerning the sum of the three trend assessments between a query course and a similar course must be below a threshold X. This condition guarantees similar changes on time. And secondly, the difference concerning the incidences of the current weeks must be below a threshold Y. This second condition guarantees an equal level of the current week of a similar case and the current week of the query course. We have learned good settings for the threshold parameters X and Y by taking in turn one season out of the case base and comparing the results when varying the settings. The result of this third step usually is a small list containing only the most similar former courses. As in compositional adaptation [9] we take the solutions of a couple of similar cases into account, namely of all courses in this small list. In retrospect, we have marked those time points of the former influenza seasons where we believed a warning would have been appropriate; e.g. in the 4th week of 2001, which is the 17th week of the 2000/2001 season (marked as square in fig.2). For the decision to warn, we split the list of the most similar courses in two lists. One list contains those courses where a warning was appropriate; the second list gets the other ones. For both of these new lists we compute their sums of the reciprocal distances of their courses to get sums of similarities. Subsequently, the decision about the appropriateness of a warning depends on the question which of these two sums is bigger.

2.2 Statistical Methods to Forecast Influenza

Under the assumption of binomial distribution, we have tested whether the observed weekly count of infected people is significant. With the following formula the probability of exactly k insured people being infected is calculated:

$$P(X=k)=\binom{n}{k} p^k (1-p)^{n-k} \; for\; k=0,\ldots,n$$

where k is the number of observed insured people being infected, n is the number of all insured people and

$$p = \frac{Number\ of\ infected\ people\ on\ average}{Number\ of\ insured\ people}$$

Here, "on average" means the average number of infected people per week concerning the specific months. Concerning the whole time period of three years or just the influenza seasons would be too vague, because influenza mainly occurs in some months, and concerning a specific calendar week might be randomly.

If the sum of the probabilities of k people being infected, k+1 people being infected etc. is 5% or more, an influenza wave can be assumed:

$$P(X \geq i) = \sum_{k=1}^{n} P(X = k) \, for \, i = 0, \ldots, n$$

3 Experimental Results

We have performed some experiments to compare both methods. However, we used different data. For doctors it is extremely difficult to distinguish between a real influenza infection and other acute respiratory infections, especially at the first diagnosis. So, by most influenza surveillance systems influenza is usually inferred from the counts of all acute respiratory infections. For both experiments, we did not used more up-to-date data, because in Mecklenburg-Western Pomerania the last influenza wave occurred in early 2001 and the course of the 2001/2002 season is typical for the following ones.

3.1 Case-Based Reasoning Method

For the Case-Based Reasoning method we used acute respiratory infections (ICD9: 460 to 487 and ICD10: J00 to J99, except very few chronic diseases) for the influenza seasons from 1997 to 2002. Firstly, we marked those time points where we, in retrospect, believed a warning would have been appropriate (the three squares in figure 2). Later on we assumed that these warnings might be a bit late. So, we attempted earlier desired warnings (the three circles in figure 2). For our first tests, we used the five seasons shown in figure 2 with the desired warnings depicted as squares. In turn we used one season as query course. Furthermore, we wanted to discover how much the results are improved by the number of seasons stored in the case base. So, for every query season we varied the case base, and we did not only put the seasons in chronological order into the case base, but attempted every combination. That means, for each query season we made four attempts with one season in the case base, six attempts with two seasons etc.

The results are shown in table 1. Sensitivity means proportion of computed warnings to desired warnings; specificity means proportion of computed "non-warnings" to desired "non-warnings".

Table 1. Sensitivity and specificity of our first experiment

	Sensitivity	Specificity
1 season in case base	50 %	100 %
2 seasons in case base	83 %	100 %
3 seasons in case base	100 %	100 %
4 seasons in case base	100 %	100 %

Since we assumed that the desired warnings of our first experiment might be a bit late, we tried earlier ones in a second experiment, in figure 2 depicted as circles. We made the same experiment again (table 2).

Table 2. Sensitivity and specificity of our second experiment: with earlier warnings

	Sensitivity	Specificity
1 season in case base	45 %	95 %
2 seasons in case base	69 %	96,7 %
3 seasons in case base	80 %	97,2 %
4 seasons in case base	80 %	96,1 %

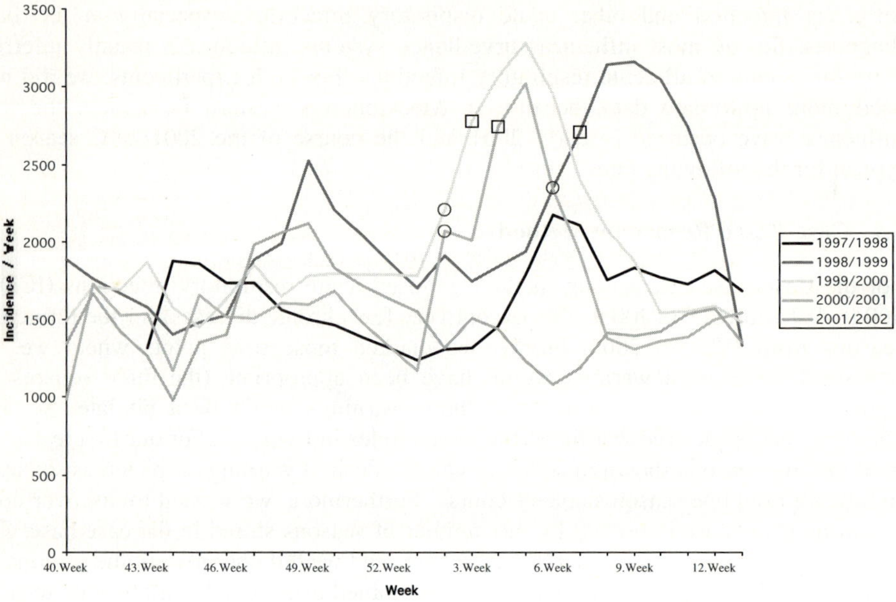

Fig. 2. Influenza seasons of Mecklenburg-Western Pomerania; ICD10: J00-J99, ICD9: 460-487

Now it is more difficult to compute the new desired warnings. However, the problems are mainly caused by the peak in the 49th week of 1998, which is the 10th week of the 1998/1999 season. Since the incidences of this peak are higher than the incidences of the desired earlier warnings and the developments are similar too, consequently a warning is computed. Only in retrospect it becomes clear that this was not the beginning of an influenza wave. And since this peak is marked as not worth for a warning, it prevents our program from computing desired warnings for other seasons. However, this is not so much a problem of the method, but rather a question of the availability of appropriate data. Since we use health insurance data, we do not have access to laboratory results, which often indicate causes. In fact, concerning the analysis of such data by the Robert-Koch Institute [11], the peak in the 49th weak of 1998 was probably (but this was never definitely proved) the result of a pathogen (respiratory syntactical virus) that causes similar symptoms as influenza. Unfortunately, such data from health centres even the Robert-Koch Institute gets only delayed (about two weeks).

3.2 Statistical Tests

For the statistical tests we did not consider all acute respiratory infected people but only those with real influenza infections (ICD10: J10 and J11). Since the counts of ICD9 and ICD10 did not fit together, we used only data for 2000, 2001, and 2002.

Though the considered infections differ, the increase of seasons 2000 and 2001 occurs obviously at the same moment. The fact that influenza waves start in two following seasons in the same week is poor chance. Sometimes influenza waves start much later, see e.g. the 1998/1999 season (figure 2). Parts of the results, namely for the first four calendar weeks, are shown in table 3. A significant count of infected people is indicated, when the value is below 5%. For 2000 the beginning of an influenza wave in the second calendar week is well discovered. For 2001 the obviously desired result is a start of an influenza wave in the second week too, but the statistical test discovers the start not until the fourth week. Of course, one problem is that we considered just three seasons. However, the results illuminate a problem of statistical methods applied to diseases with cyclical behaviour. In two of the three considered seasons influenza waves occurred which increase the average value very much. So, instead of using a general average value only those weeks without influenza waves should be considered as "normal" and should be used for computing average values. However, sometimes it is difficult to decide whether a week is "normal" and for some calendar weeks there may be just very few "normal" ones.

Table 3. Results of tests under assumption of binomial distribution. NIL means that no computation was necessary, because the observed count is even below the average value.

	Counts 2000	%	Counts 2001	%	Counts 2002	%
1.Week	184	41	163	Nil	195	11
2.Week	366	0,1	246	Nil (32)	151	Nil
3.Week	489	0,1	340	23	154	Nil
4.Week	559	0,1	620	0,1	102	Nil

4 Conclusion

In contrast to most medical diagnostic problems, we cannot ask experts about the correctness of the computed warnings. Instead, nobody knows in which week a first warning should be computed. However, the simultaneous increase in both sorts of data indicates that for the Case-Based Reasoning method the earlier warnings of the second experiment are probably ideal moments for first warnings against approaching influenza waves. So far, it is difficult to assess the quality of our Case-Based Reasoning method. However, the results are at least as good as by statistical tests that assume binomial distribution. Fortunately for the people but unfortunately for our research, there has not occurred an influenza wave since 2002 in Mecklenburg-Western Pomerania. Furthermore, we believe that the results of influenza surveillance depend more on the data than on the method. This does not only mean the a priori quality of the data and the speed of their availability, but additionally the quality for discriminating risky situations. The a priori quality of our health insurance data is rather poor, especially the diagnoses are often superficial, but there is only a very short delay concerning their availability. Official data from German health centres are more profound, but for bureaucratic reasons there availability is delayed for too long.

References

1. Prou, M., Long, A., Wilson, M., Jacquez, G., Wackernagel, H., Carrat, F.: Exploratory Temporal-Spatial Analysis of Influenza Epidemics in France. In: Flahault, A., Viboud, C., Toubiana, L., Valleron, A.-J. (eds.) Abstracts of the 3rd International Workshop on Geography and Medicine, Paris, vol. 17 (2001)
2. Shindo, N., et al.: Distribution of the Influenza Warning Map by Internet. In: Flahault, A., Viboud, C., Toubiana, L., Valleron, A.-J. (eds.) Abstracts of the 3rd International Workshop on Geography and Medicine, Paris, vol. 16 (2001)
3. Farrington, C.P., Beale, A.D.: The Detection of Outbreaks of Infectious Diseases. In: Gierl, L., et al. (eds.) International Workshop on Geomedical Systems, pp. 97–117. Teubner-Verlag, Stuttgart (1997)
4. Rusticucci, M., Bettolli, M.L., De Los Angeles Harris, M., Martinez, L., Podesta, O.: Influenza outbreaks and weather conditions in Argentina. In: Toubiana, L., Viboud, C., Flahault, A., Valleron, A.-J. (eds.) Geography and Health, Inserm, Paris, pp. 125–140 (2003)
5. Dowell, S.F.: Seasonal Variation in host susceptibility and cycles of certain infectious diseases. Emerg Inf Dis 7(3), 369–374 (2001)
6. Viboud, C., et al.: Forecasting the spatio-temporal spread of influenza epidemics by the method of analogues. In: Abstracts of 22nd Annual Conference of the International Society of Clinical Biostatistics, Stockholm, 71th edn. (2001)
7. Schmidt, R., Gierl, L.: Prognoses for Multiparametric Time Courses. In: Brause, R.W., Hanisch, E. (eds.) Medical Data Analysis, pp. 23–33. Springer, Heidelberg (2000)
8. Schmidt, R., Gierl, L.: A prognostic model for temporal courses that combines temporal abstraction and case-based reasoning. Int J Medical Informatics 74, 307–315 (2005)
9. Wilke, W., Smyth, B., Cunningham, P.: Using Configuration Techniques for Adaptation. In: Lenz, M. (ed.) Case-Based Reasoning Technology, From Foundations to Applications, pp. 139–168. Springer, Heidelberg (1998)

Knowledge Based Industrial Maintenance Using Portable Devices and Augmented Reality

Carlos Toro[1], Cesar Sanín[2], Javier Vaquero[1], Jorge Posada[1], and Edward Szczerbicki[2]

[1] VICOMTech Research Centre, Mikeletegi Pasealekua 57
20009 Donostia, Spain
{ctoro,jvaquero,jposada}@vicomtech.es
[2] Faculty of Engineering and Built Environment, University of Newcastle
University Drive, Callaghan, NSW 2308, Australia

Abstract. In this paper we present a framework and a system implementation for the exploitation of embedded knowledge in the domain of industrial maintenance in a mobile context, using Augmented Reality techniques. We base our approach in the SOUPA group of ontologies (Standard Ontology for Ubiquitous and Pervasive Applications). Our approach extends SOUPA with two new ontologies (*i*) the Set of Experience Knowledge Structure, used to model the user's experience and (*ii*) the AR ontology which models an Augmented Reality environment that is used to enhance the maintenance experience through virtual elements. As test case, we implemented our approach in different portable devices with video input capabilities such as UMPCs, PDAs and Tablet PCs.

Keywords: Ontologies, Industrial Maintenance, Knowledge Based Systems.

1 Introduction

Nowadays, engineering systems are based on evolving paradigms; the knowledge and the user experience take a big role in today's applications as we have now the computational capability of modeling such paradigms. From an industrial point of view, there are several processes involved in the product life cycle, being one of extreme interest the maintenance of the product itself; a set of actions which is known as Industrial Maintenance (IM). IM can be defined as the combination of all technical and administrative actions, including supervision actions, intended to retain an entity in, or restore it to, a state in which it can perform a required function [6]. According to Honkanen [5], machines have to be maintained in order to increase reliability and thereby avoid production disturbances. It is, therefore, assumed that the purpose of a single maintenance action is to increase reliability. There are several techniques and theories that can be applied to the Industrial Maintenance field. The explanation of those techniques is outside the scope of this paper, but an interested reader is referred to the work of Honkanen [5]. From an application point of view, different research projects have been presented by the scientific community involving the implementation of Virtual and Augmented Reality (VR/AR) to extend the user's understanding and, in general, his experience during the maintenance work ([3],[7]).

B. Apolloni et al. (Eds.): KES 2007/WIRN 2007, Part I, LNAI 4692, pp. 295–302, 2007.

To our acquaintance, most of these approaches however, miss the potential of using knowledge based theories in the domain that might enhance the user's experience. This user, in our case, is the maintenance worker in a typical industrial facility whose special needs include mobility, fast response and immediate access to the relevant data, like specifications, historical records, etc. We show in this paper that the use of semantics and augmented reality techniques provide additional support to the maintenance tasks, by improving the user understanding of the elements being maintained. The enhancement comes when the knowledge and the user experience related to the maintenance system is embedded in the AR environment as an important aid for the user, giving him a sense of immersion. This paper is organized as follows: In chapter 2, an overview of the conceptual basis is presented. In chapter 3, we introduce our knowledge based industrial maintenance system. In chapter 4, we show an application scenario, explaining briefly some highlights, and in chapter 5, we present our conclusions and future work.

2 Conceptual Basis and Background

In this chapter we introduce some concepts relevant to this paper. Our intention is not to give a wide description of the topics involved, but to give a short overview in order to offer a self-contained paper.

2.1 Knowledge Modeling Using Ontologies

Following is Tom Gruber's widespread accepted definition of what ontology is in the Computer Science domain: ontology is the explicit specification of a conceptualization; a description of the concepts and relationships in a domain [4]. In the context of Artificial Intelligence (AI), we can describe the ontology of a program by defining a set of representational terms. In such ontology, definitions associate names of entities in the universe of discourse with human-readable text describing what the names mean, and formal axioms that constrain the interpretation and well-formed use of these terms. Computer programs can use ontologies for a variety of purposes including inductive reasoning, classification, and problem solving techniques, as well as communication and sharing of information among different systems. In addition, emerging semantic web systems use ontologies for a better interaction and understanding between different agent web-based systems. Ontologies can be modelled using several languages, being the most widely used RDF and recently OWL (Ontology Web Language) [11]. Ontology modelling can deliver interesting benefits as it allows inferring semantically new derived queries. These queries relate concepts that are not explicitly specified by the user, but nevertheless relevant to the query. Modern inference engines and reasoners like Pellet and Racer deliver a highly specialized, yet efficient way to perform such queries via a JAVA compliant API [11]. In the literature, data handling by Ontology-based technology is reported by researchers in different fields [11]. Furthermore, user modelling, task and experience are also possible scenarios for the exploitation of semantic data by Ontology based technology as it was addressed for example in the European IST-Project WIDE [10].

2.2 SOUPA (Standard Ontology for Ubiquitous and Pervasive Applications)

SOUPA is a shared ontology expressed using OWL [2], designed to model and support pervasive computing applications (ambient intelligence). It consists of two distinctive but related set of ontologies, called SOUPA Core and SOUPA Extension as can be seen on figure 1.

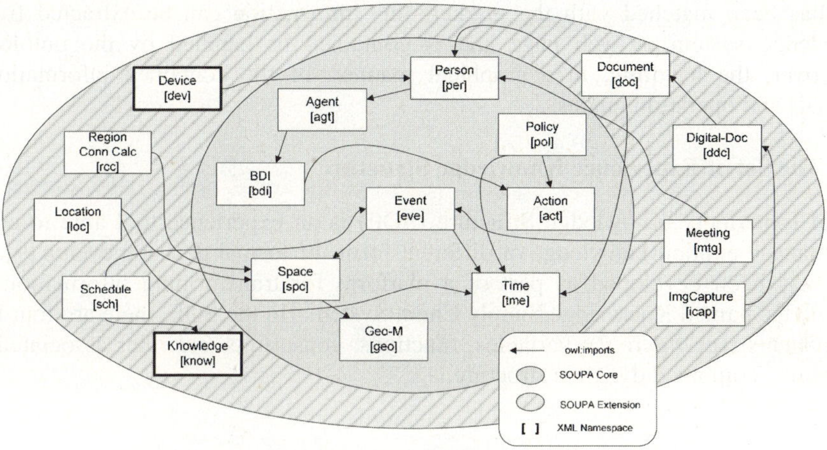

Fig. 1. The SOUPA group of ontologies (Core + Extension) [2]

The set of the SOUPA Core ontologies attempts to define generic vocabularies that are universal for different pervasive computing applications, vocabularies for expressing concepts associated with person, agent, belief-desire-intention (BDI), action, policy, time, space and event. The set of SOUPA Extension ontologies define additional vocabularies for supporting specific types of applications and provide examples for future ontology extensions. The Core and Extension of SOUPA have been applied in several research projects like the Context Broker Architecture (CoBrA) [2] which is agent based architecture for supporting context-aware systems in smart spaces. Central to this architecture is an intelligent agent called context broker (majordomo) that maintains a shared model of context on behalf of a community of agents, services, and devices in the space while providing privacy protections for the users. We intend to show in this paper that by extending SOUPA and using the CoBrA architecture is possible to implement a context aware (ubiquitous) system that helps a user in an Industrial Maintenance environment.

2.3 Virtual Reality and Augmented Reality in the Industrial Maintenance Environment

Virtual Reality (VR) is a virtual environment where the user is immersed in a virtual world in which the elements hold the physical resemblance to possible objects belonging to the real world. Milgram and Kishino created a *Virtuality Continuum* [8] to classify mixed worlds, in where we find Augmented Reality (AR) as a system that, according to Azuma [1], combines real and virtual objects in a real environment enhanced with computer generated graphics, running interactively, in real time, and aligns real and virtual objects with each other. In this work we use Augmented Reality in order to enhance the user's experience (maintenance engineer) by giving him contextualized information extracted from an ontology reasoning kernel. Augmented Reality in an industrial maintenance environment provides, for example, user experience enhancements like the viewing of information related to the object in sight that has been matched with the marker; this information can be extracted from a knowledge system or web page and is contained or handled by the ontologies. Moreover, the handling, in a graphical manner, of the contained information in manuals and operation guides.

2.4 The Set of Experience Knowledge Structure

Set of Experience Knowledge Structure (SOE) is an experience tool able to collect and manage explicit knowledge of different forms of formal decision events [9]. The SOE has been developed as part of a platform for transforming information into knowledge named Knowledge Supply Chain System. In the SOE, there are four basic components (see figure 2): variables, functions, constraints and rules associated and stored in a combined dynamic structure.

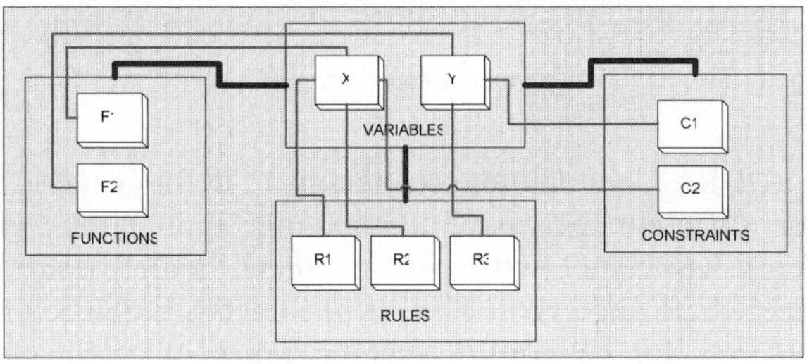

Fig. 2. The Set of Experience Knowledge Structure

The combination of the four components of the SOE offers distinctiveness due to the elements of the structure that are connected among themselves, imitating part of a long strand of DNA. A SOE produces a value of decision, called efficiency. Besides, it is possible to group sets of experience by category, that is, by objectives. These groups could store a "strategy" for such category of decisions. Furthermore, an OWL ontology modelling the SOE was presented by Sanín et al. [9], and it was added to SOUPA with the purpose of extending its functionality in a way that formal decision events can be suggested to the Maintenance User during his/her work. Those stored sets of experience are produced by past decisions made by maintainers with the same or similar elements. E.g. the user is checking during her/his duties a butterfly valve. Since it is a fluid restringing element, possible decisions taken in the past on such valve, or a similar element (ball valve), could be suggested by the system.

3 The UDKE System

In this chapter, we introduce our Knowledge based Industrial Maintenance system using portable devices and Augmented Reality. We call our architecture UDKE (User, Device, Knowledge and Experience). UDKE provides a possible conceptual model of a maintenance system that combines knowledge, user experience and AR techniques. The schema is divided in layers which are depicted in figure 3.

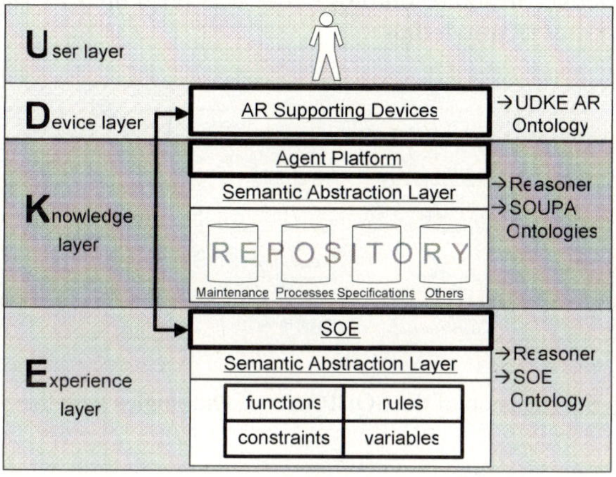

Fig. 3. The UDKE System

The *User layer* contains only the user, handling the particular profile, session, etc. The *Device layer* contains the modeling of the different devices used to capture the environment (a camera, a PDA, a pocket PC or a Tablet PC, etc) introduced as an extension to SOUPA that we call the UDKE AR ontology. In the figure 4, such extension is depicted (extending from the SOUPA class **dev:device**)

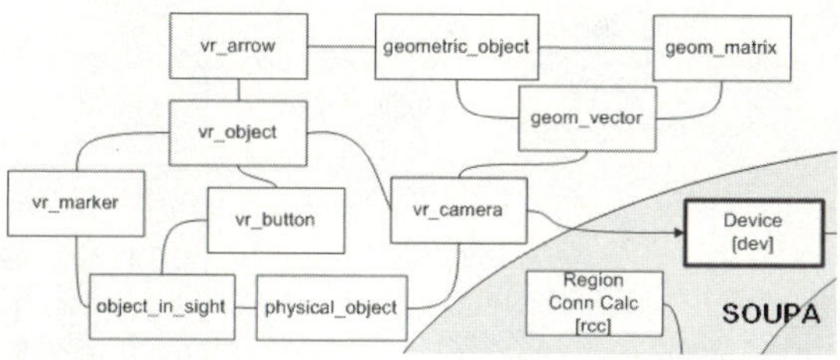

Fig. 4. The UDKE AR extension to SOUPA

The *Knowledge layer* contains the agent platform whose main objective is to interact via majordomo messages with the Semantic abstraction engine. It contains the SOUPA Core and the SOUPA Extension ontologies as well as a reasoner system that is in charge of performing the semantic queries. The Ontologies feed their instances from different repositories relevant to the maintenance domain (historical data, programmed stops, cycles, etc). In the *Experience Layer*, the SOE is enclosed as an OWL ontology with different data bases that feed the reasoning system with functions, rules, constrains and variables used to specify new decisional events or even to contain past decisional events taken over similar elements (where similar refers to the object in sight element or in other words the element in which the maintenance engineer is considering at a given moment). In the next figure, we show the extension place of the SOUPA set of Ontologies in which we derived the Set of Experience Knowledge Structure ontology that was presented by Sanin et al. [9], the extension class is **know:Knowledge**.

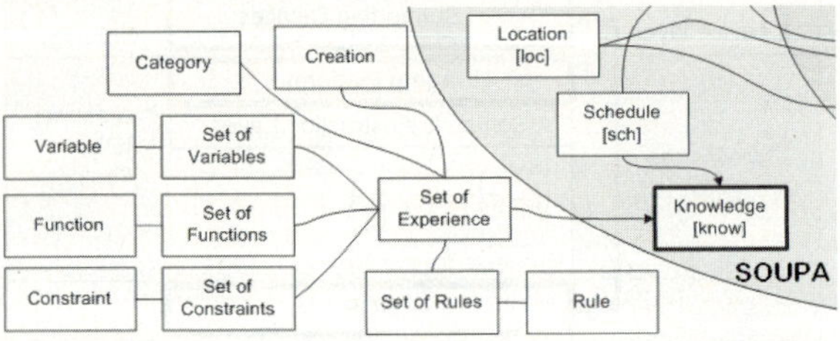

Fig. 5. Extension of the SOUPA set of Ontologies with the SOE

4 Application Scenario

At present time, the system is being tested in a real maintenance environment. Rather than describing the actual experiments being held, we would like to present in this article a possible application scenario leaving the experiment results to be exposed in

Fig. 6. AR Enhanced user view using the UDKE Platform

a future publication. In our scenario, the user during his maintenance patrol uses a portable device (PDA, UMPC or Tablet PC) with a camera connected. For every object to be maintained there exist a VR marker (following the sensor concept in ambient intelligence). Every marker is an unequivocal grayscale pattern that can be easily printed in white paper by a regular PC printer. When the camera recognizes a marker, a matching element to be maintained is identified according to the context (user, task, priority) and a set of information is extracted from the repositories. The output video stream of the camera is mixed with 3D objects and other relevant information and is displayed to the user in his portable device screen. As can be seen in figure 6, the user is in front of an element (in this case a fire extinguisher) and when the system recognizes the matching marker, the user receives in the screen information such as the name of the element, the next programmed change, the maintenance procedure etc. All information is obtained from the repositories in the Knowledge layer and is maintained by the Experience layer.

4.1 Implementation Issues

The system was tested using different portable devices, our implementation uses JAVA as the core language for the prototype implementation. The graphic library and the AR engine used were GL4Java and JARToolkit library respectively. All the ontology modeling was done in Protégé and the API used was the Protégé OWL API. The agent platform used in our implementation was JADE, and for reasoning purposes over the ontologies we chose RACER. When a marker is detected, the system calculates the matrices necessary to place the augmented information via JARToolkit calls. Following the application flow, the Agent platform begin its work starting a majordomo service whose main function is to serve as an intermediary between the user and the rest of the architecture. The majordomo handles the events in the knowledge layer data bases through reasoning over the SOUPA ontologies. The majordomo also handles the Experience layer through reasoning over the SOE ontology in order to obtain knowledge from past experiences or similar devices being maintained. Once that all that information is obtained/inferred and possible experience are acquired from the SOE using the reasoning system, a last step is performed by returning such information to the device (UMPC, Pocket PC, etc) and displayed (streamed) it in its graphical output.

5 Conclusions and Acknowledgements

In this paper, we presented a framework and a system implementation for the exploitation of embedded knowledge in the domain of industrial maintenance in a mobile context, using AR techniques. An extension to SOUPA was presented with two ontologies: the Set of Experience Knowledge Structure (presented by Sanin et al. [9]) and a new ontology for handling AR objects. The framework was implemented in different portable devices. For our future work, we plan to test the system in a real facility in where a maintenance system is already implemented in order to test the system's feasibility. Also, we plan to extend the SOE ontology from a decisional point of view by including Genetic Algorithms. We want to thank the Basque

Government for the partial financing of the project under the SAIOTEK 2005-2006 call (SEMTEK project). A special mention is given to the Faculty of Engineering and Built Environment of the University of Newcastle (NSW – Australia) for the priceless ideas they shared with us related to the use of the Set of Experience Knowledge Structure in our domain.

References

1. Azuma, R.T.: A Survey of Augmented Reality. Presence: Teleoperators and Virtual Environment, vol. 6(4) (1997)
2. Chen, H., H., Perich, F., Finin, T., Joshi, A.: SOUPA: Standard Ontology for Ubiquitous and Pervasive Applications. - MOBIQUITOUS. Cambridge, USA (2004)
3. Friedich, W.: ARVIKA – Augmented Reality for Development, Production and Service. In: Proc. of International Symposium on Mixed and Augmented Reality (ISMAR 2002), pp. 3–4 (2002)
4. Gruber, T.R.: Toward principles for the design of ontologies used for knowledge sharing. International Journal of Human-Computer Studies 43(5-6), 907–928 (1995)
5. Honkanen, T.: Modelling Industrial Maintenance Systems and the Effects of Automatic Condition Monitoring. PhD Thesis, Helsinki University of Technology - Finland (2004)
6. IEC 60050-191: International Electrotechnical Vocabulary – ch. 191: Dependability and quality of service, International Electrotechnical Commission (1990)
7. Makri, A., Weidenhausen, J., Eschler, P., Stricker, D., Machui, O., Fernandes, C., Maria, S., Voss, G., Ioannidis, N.: ULTRA Light Augmented Reality Mobile System. In: 4th IEEE and ACM International Symposium on Mixed and Augmented Reality, Austria (2005)
8. Milgram, P., Kishino, F.: A Taxonomy of Mixed Reality Visual Displays. IEICE Transactions Information Systems E77-D, 1321–1329 (1994)
9. Sanín, C., Szczerbicki, E., Toro, C.: An OWL Ontology of Set of Experience Knowledge Structure. In: Journal of Universal Computer Science, (in press) (February, 2007)
10. Sevilmis, N., Stork, A., Smithers, T., Posada, J., Pianciamore, M., Castro, R., Jimenez, I., Marcos, G., et al.: Knowledge Sharing by Information Retrieval in the Semantic Web. Lecture Notes of Computer Science, vol. 32, p. 471. Springer, Heidelberg (2005)
11. Toro, C., Posada, J., Termenón, M., Oyarzun, J., Falcón, J.: Knowledge Based Tools to Support the Structural Design Process. In: Gabrys, B., Howlett, R.J., Jain, L.C. (eds.) KES 2006. LNCS (LNAI), vol. 4251, Springer, Heidelberg (2006)

Modelling a Team of Radiologists for Lung Nodule Detection in CT Scans

Michela Antonelli, Marco Cococcioni, Graziano Frosini, Beatrice Lazzerini, and Francesco Marcelloni

Dipartimento di Ingegneria dell'Informazione:
Elettronica, Informatica, Telecomunicazioni
University of Pisa, Via Diotisalvi 2, 56122 Pisa, Italy
{michela.antonelli,m.cococcioni,g.frosini,b.lazzerini,
f.marcelloni}@iet.unipi.it

Abstract. This paper describes a system for automatic detection of pulmonary nodules in lung CT (Computed Tomography) images. After modelling the activity of a single radiologist as two subsequent phases, namely, the regions of interest (ROIs) detection phase and the nodule detection phase, we built a system which emulates a team of radiologists. This is achieved by providing a further phase of collaboration and opinion exchange among the experts at the end of each of the previous phases. We also present experimental results, based on the ROC convex hull method, which show how the team of radiologists obtains better performance than the single best radiologist in both phases. In particular, we achieved a sensitivity of 92.48% against a specificity of about 83.54% in the nodule detection phase.

Keywords: Computer-Aided Detection, ROC convex hull, Multi-Classifier Systems.

1 Introduction

In the last few years lung cancer has been the most common fatal malignancy in both men and women. Each year more than 50% of the new diagnosed cases are metastasis [1]. Due to the late appearance of the symptoms, lung cancer is usually diagnosed when it has developed to such an advanced stage that it cannot be operated on.

Since pulmonary nodules are typically asymptomatic, screening programs can help detect the disease earlier so as to increase the survival rate. People with a particular risk factor for cancer (typically smokers older than fifty) may undergo periodic screening programs with low-dose radiation CT (LDCT). Using LDCT scans, nodules can be detected earlier, i.e., at a smaller size (2-3 mm), than those observed with chest radiography [2]. On the other end, many more images are generated. As a consequence, Computer Aided Detection (CAD) systems have received increasing attention to assist radiologists for more accurate and quicker diagnosis [3].

In [15], Gurcan *et al.* describe a CAD system for lung nodule detection in CT scans using a rule-based classifier. First they segment the lung lobes and extract the ROIs using k-means clustering with pixel grey-level as feature. Then they apply the

B. Apolloni et al. (Eds.): KES 2007/WIRN 2007, Part I, LNAI 4692, pp. 303–310, 2007.

rule-based system to classify the region of interest as either nodule or blood vessel. The rules take into account morphological features of lung nodules, namely, size and shape, and the nodule location. To further reduce the number of false positives a linear discriminant analysis is exploited, based on grey-level, area and volume of ROIs. This method is applied to CT scans with slice thickness of 5 mm and a total of 63 nodules ranging from 4 mm to 28 mm. The sensitivity achieved is 84% with a specificity of 1.74 false positives per slice. *Specificity* is the proportion of subjects without the disease identified as negative by the system, relative to all the subjects without the disease, *sensitivity* is the proportion of subjects identified as positive by the system relative to all the subjects with the disease.

In [16], Armato III *et al.* propose a similar method: they segment the lung lobes applying multiple grey-level thresholds and use a rule-based approach considering morphological and grey-level features to distinguish between nodules and blood vessels. The method is applied to CT scans with slice thickness of 5 mm. The authors obtain a sensitivity of 70% with an average of 1.5 false positives per slice.

In this work, we have tried to give a novel contribution in the direction of the computerized detection of lung nodules looking at the problem from a non-standard perspective. More precisely, we have not modelled just a radiologist, but a team of radiologists.

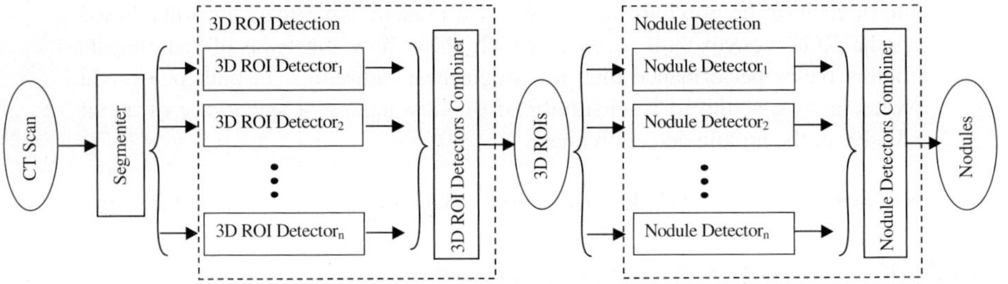

Fig. 1. Model of the CAD system

First of all, there is a common phase for all the radiologists: the CT images are segmented in order to extract the lung lobes so as to reduce the data to analyse. Then two distinct and subsequent phases are considered, namely, a first phase aimed at detecting the ROIs within the CT scan, and a second phase of classification of ROIs into nodules and non-nodules (i.e., bronchi and blood vessels).

Of course, at the end of both phases, it may be useful to let the radiologists of the team exchange their opinions; to this aim two additional modules are included in order to combine, respectively, the single outputs of the different ROI detection activities, and the single outputs of the different nodule detection activities performed by the radiologists.

We applied the CAD system to 30 multi-slice CT scans (all with medical report), each consisting, on average, of 300 slices, each slice being a 512×512 pixel matrix. The slice thickness is 1.25 mm and the pixel size is 0.6 mm.

The system has been evaluated considering its specificity and sensitivity. Of course, a screening program should have high sensitivity in order to avoid false negatives (i.e., missed nodules) without reducing the specificity so as to avoid false positives (i.e., non-nodules recognized as nodules). In the experiments, we achieved a sensitivity of 92.48% against a specificity of about 83.54%. These results are appreciable also compared with those found in the literature, in particular if we consider the fact that we work on CT slices with thickness of 1.25 mm so as to allow detection of nodules of diameter starting from 3 mm.

2 System Architecture

In this section we describe the architecture of the CAD system we have developed. The system, whose basic structure is illustrated in Fig. 1, emulates a team of radiologists. The system includes a first module, called Segmenter, which performs a phase, common to all the radiologists, aimed at segmenting the CT images in order to extract the lung lobes, thus reducing the data to analyse in the following phases. Then the CAD system is composed of two subsystems, named the 3D ROI Detection subsystem and the Nodule Detection subsystem. Both subsystems are implemented by a set of independent modules (3D ROI Detectors and Nodule Detectors, respectively), each representing a single radiologist which performs the pertinent phase. Different radiologists are obtained by applying different techniques, which provide different outputs for each phase. Therefore, two further modules, namely, the 3D ROI Detectors Combiner and the Nodule Detectors Combiners, respectively, are added in the corresponding subsystem to integrate the different radiologists' opinions.

3 The Segmenter

We chose to implement the Segmenter so as to let the CAD system itself be easily extended and used for the follow-up of possible nodules detected in the screening phase. Therefore, following the approach found in [4], the Segmenter is an expert system, which detects the pulmonary lobes within the CT images guided by an anatomical model of the thorax. We adopted the simple thorax model shown in Fig. 2.

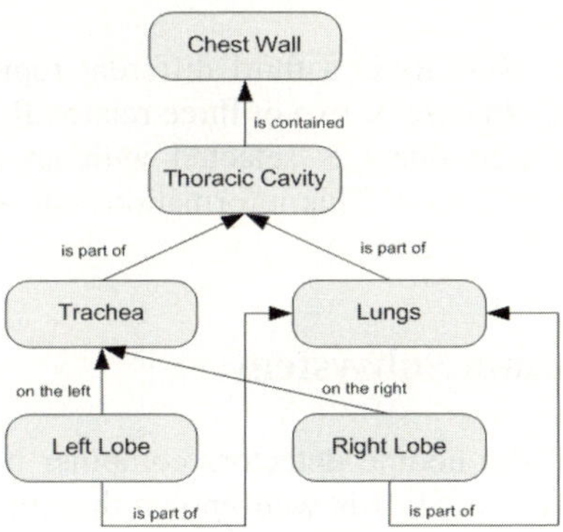

Fig. 2. The thorax model

A semantic network is used, whose nodes are the anatomical structures of interest and the directed arcs are spatial relations that allow identifying a given anatomical structure with respect to others previously found. In practice, one moves according to a specific order within the volume that makes the scan: first the chest wall is singled out, then, inside the chest, the thoracic cavity is looked for, and so on. From an implementation point of view, this means that the search for an anatomical structure is performed by searching, inside the area suggested by the anatomical model, the agglomerate of voxels whose characteristics best match the properties that describe that structure in the model. The first time a patient undergoes a CT exam the general model is adopted; this model is then specialized so as to fit the patient's specific anatomy and be used in following exams.

4 The 3D ROI Detection Subsystem

We have implemented three different 3D ROI detectors, namely, a clustering algorithm, an algorithm that builds 3D agglomerates, and a more classical region growing algorithm. More precisely, in the first case, the Robust Fuzzy c-Means algorithm (RFCM) [5] is applied to the volume occupied by the two lungs in order to find two clusters containing, respectively, the clear pixels, which correspond to nodules and blood vessels, and the dark pixels, which correspond to bronchi and air. The sets of connected voxels within the first cluster represent the 3D ROIs. In the second case, 3D ROIs are detected by simply searching for connected voxels whose grey level is within a fixed range. Finally, in the third case, we use a classical region growing algorithm, which starts from a voxel (seed point) automatically selected according to its grey level and adds voxels whose grey level differs from the seed point of a fixed value experimentally chosen.

The task of the 3D ROI detectors combiner is to provide the "union" of the ROIs detected by the single detectors. Actually, it realizes when ROIs detected by different detectors do correspond to the same real ROI. More precisely, two detected ROIs R_1 and R_2 are, in fact, two different representations of the same real ROI R_r if $\frac{R_1 \cap R_2}{R_1 \cup R_2} \geq \vartheta$, with $0.5 \leq \vartheta \leq 1$. In the experiments, we used $\vartheta = 0.8$. We call R_1 and R_2 *related* ROIs. Another ROI R_3 is a third different representation of R_r if R_3 is related with either R_1 or R_2. In case of two or three related ROIs, a single ROI (chosen at random among the related ones) is selected with an indication of how many detectors have actually identified it. This information will be appropriately exploited in future developments of the system.

5 The Nodule Detection Subsystem

The nodule detectors and the nodule detectors combiner have been developed as a multi-classifier system (MCS) [6]. It is well-known that the appropriate combination of the outputs of different classifiers in a MCS can achieve better performance than the single best classifier. In this paper we adopt a two-level architecture, in which the first-level classifiers are different classification algorithms that work on the same set of features extracted from the single 3D ROIs (i.e., average gray level, standard deviation, sphericity, inverse of elongation [7]). The five first-level classifiers are a linear classifier (ldc), a quadratic classifier (qdc), a logistic classifier (loglc), a decision tree (treec) and a radial basis function neural network (rbnc). At the second level we use seven classifiers without training (maximum, minimum, average, median, product, AND-like OWA (Ordered Weighted Average) [8], OR-like OWA [8]), and four classifiers with training (decision template [9], Dempster-Shafer [10], fuzzy integral [10], and Naïve-Bayes [11]).

In the medical diagnostic field the performance of a classifier is typically assessed based on the ROC (Receiver Operating Characteristic) analysis. Let $TPR = TP/(TP + FN)$ be the true positive rate (i.e., sensitivity) and $FPR = FP/(TN + FP)$ be the false positive rate (i.e., complement to 1 of the specificity), where *TP* = true positive (nodule correctly classified), *TN* = true negative (non-nodule correctly classified), *FP* = false positive (non-nodule classified as nodule), *FN* = false negative (nodule classified as non-nodule). The ROC curve is obtained building a chart with *TPR* on the ordinate axis, *FPR* on the abscissa axis and a set of pairs ($FPR(\gamma), TPR(\gamma)$), each corresponding to a specific value of the threshold γ (which discriminates negative from positive outputs). In particular, a binary classifier is represented by a single point. One point in the ROC space is better than another if it is located more north-west (higher *TPR*, lower *FPR*, or both). The ROC curve shows the behaviour of a classifier without any reference to the class distribution or the classification error cost. This means that the ROC analysis can be appropriate to choose the best global classifier only if there is a classifier that dominates all the others over the whole ROC space. On the other hand, we are concerned with a two-class problem: the "class of positives" and the "class of negatives", with unbalanced a-priori probabilities, and false positive error and false negative error which are very different from each other. A classification cost can be approximated as follows [12]

$$cost = FPR \cdot P(n) \cdot c(Y,n) + FNR \cdot P(p) \cdot c(N,p) \tag{1}$$

where $P(p)$ e $P(n) = 1 - P(p)$ are the a-priori probabilities of a positive example and a negative example, respectively, $C(Y,n)$ e $c(N,p)$ are, respectively, the false positive cost and the false negative cost, and $FNR = 1 - TPR$ is the false negative rate. In order to solve the problem of choosing the best global classifier, we use the technique of the ROC convex hull (i.e., the convex hull of the ROC curves of more classifiers) [13],[14]. This technique allows choosing the locally optimum classifier, that is, the classifier that is optimum with respect to given class distribution and classification error cost. Recalling the expression of the expected classification cost by the classifier represented by the point (FPR, TPR) in the ROC space, once fixed the values for $P(n)$, $c(Y,n)$, $P(p)$ e $c(N,p)$, we obtain a family of parallel lines (called *iso-cost* lines) with slope $\dfrac{P(n)c(Y,n)}{P(p)c(N,p)}$. For a given ROC curve, the point that minimizes the classification cost is the tangent point between the ROC curve and the family of parallel lines (local optimum condition). From an operation point of view, in order to compare a set of classifiers based on the ROC convex hull method, the following steps must be performed: i) the ROC curves of all the classifiers are drawn on the same chart; ii) the convex hull of these curves is added, and iii) an iso-cost line is drawn, which corresponds to the specific class distribution and erroneous classification error cost, tangent to the ROC convex hull. Then we verify which classifier the tangent point belongs to and for which value of γ. This classifier (with the found value of γ) is the optimum.

6 Experiments and Results

The proposed CAD system was evaluated on 30 CT scans (with medical report), containing 64 nodules in total. For the first phase of ROI detection, we just considered four distinct groups of consecutive slices (belonging to the same scan), each group containing a nodule, and we computed how many ROIs were correctly recognized, against a manual ROI identification.

The 3D ROI detectors combiner obtained better true positive rate (95.21%) than RFCM (90.75%), region growing (87.33%) and 3D agglomerates (68.49%). This means that the three detectors can detect different 3D ROIs, thus allowing the combiner to increase the number of true positives. Further, the combiner found a number of false positives lower than the sum of false positives generated by each of the three 3D detectors.

For the subsequent phase of nodule detection we used a data set consisting of 1984 3D ROIs, including 64 nodules, corresponding to $\cong 3.2\%$ of the total, and 1920 blood vessels, corresponding to $\cong 96.8\%$ of the total. We performed a 10-fold cross validation. In order to balance positive and negative examples in the training set, we adopted the fuzzy c-means (FCM) algorithm to reduce by a factor 30 the negative examples of each of the ten folds used for training. In particular, we executed FCM on the negative examples fixing the number of clusters to 64 (the number of nodules). Then, we used the prototypes computed by FCM in the training set instead of the negative examples. The ratio 1:30 was, however, kept on the test set.

After training the detectors and (possibly training and) applying the combiners, for each detector and for each combiner the average of the ROC curves obtained on the test set of the ten trials forming the cross-validation were computed.

Figure 3 shows the average ROC curves of the detectors and the combiners, and the ROC convex hull. From Fig. 4, which shows the detectors and the combiners that contribute to the ROC convex hull, we can deduce that the combiners give a contribution to the ROC convex hull; actually, in the north-west area of the ROC plane there are only combiners (fuzzy integral, Dempster-Shafer and OR-like OWA). From Fig. 6, which shows the overall ROC convex hull for both detectors and combiners, we can see how significant the contribution to the overall ROC convex hull by the combiners is.

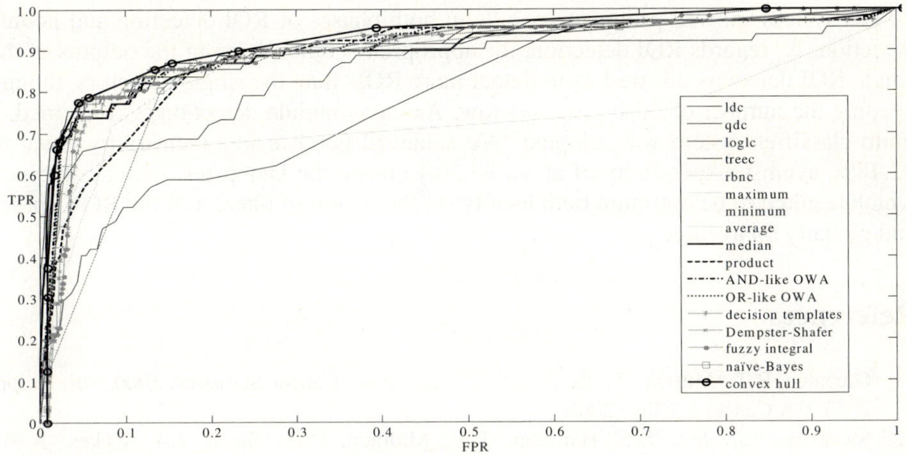

Fig. 3. ROC curves pertinent to both the detectors and the combiners, and ROC convex hull

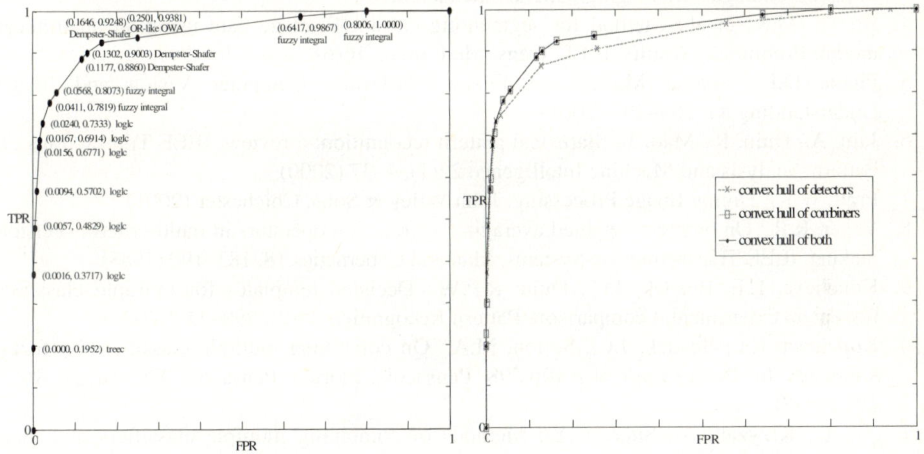

(a) (b)

Fig. 4. (a) Detectors and combiners which contribute to the ROC convex hull; (b) Overall ROC convex hull of the detectors and the combiners

7 Conclusion

We developed a CAD system for lung CT images able to automatically perform early detection of lung cancer. The system emulates a team of radiologists. To this aim, we adopted a model of the radiologist's typical activity consisting of two distinct and following phases, and precisely, the search for ROIs within the CT images and the classification of such regions into nodules and non-nodules. At the end of each phase, a further phase of collaboration and radiologists' opinion fusion is provided.

The system showed promising results in both phases of ROI detection and nodule detection. As regards ROI detection, the appropriate combination of the outputs of the single ROI detectors allowed us to detect more ROIs than the single detectors, though keeping the number of false positives low. As far as nodule detection is concerned, a multi-classifier system was adopted. We achieved good results (sensitivity equal to 92.48%, against a specificity of about 83.54%) using the Dempster-Shafer combiner, which resulted to be optimum both locally (in the region of interest of the ROC space) and globally on average.

References

1. Greenlee, R.T., Murray, T., Bolden, S., Wingo, P.A.: Cancer Statistics, 2000, vol. 50, pp. 7–33. CA Cancer J. Clin. (2000)
2. Swensen, S.J., Jett, J.R., Hartman, T.E., Midthun, D.E., Sloan, J.A., Sykes, A-M., Aughenbaugh, G.L., Clemens, M.A.: Lung Cancer Screening with CT: Mayo Clinic Experience. Radiology 226(3), 756–761 (2003)
3. Armato III, S.G.: CAD Dissects Growing Volume of Data from Lung CT exams, http://www.diagnosticimaging.com/advancedct2003/
4. Brown, M.S., et al.: Method for segmenting chest CT image data using an anatomical model: Preliminary results. IEEE Trans. Med. Imag. 16(6), 828–839 (1997)
5. Pham, D.L.: Spatial Models for Fuzzy Clustering. Computer Vision and Image Understanding 84, 285–297 (2001)
6. Jain, A., Duin, R., Mao, J.: Statistical pattern recognition: a review. IEEE Transactions on Pattern Analysis and Machine Intelligence 22(1), 4–37 (2000)
7. Pratt, W.K.: Digital Image Processing. John Wiley & Sons, Chichester (2001)
8. Yager, R.R.: On ordered weighted averaging aggregation operators in multi-criteria decision making. IEEE Transactions on Systems, Man and Cybernetics 18, 183–190 (1988)
9. Kuncheva, L.I., Bezdek, J.C., Duin, R.P.W.: Decision templates for multiple classifier fusion: an experimental comparison. Pattern Recognition 34(2), 299–314 (2001)
10. Kuncheva, L.I., Bezdek, J.C., Sutton, M.A.: On combining multiple classifiers by fuzzy templates. In: Proceedings of Nafips'98, Pensacola, Florida, Pensacola, Florida, pp. 193–197 (1998)
11. Xu, L., Krzyzak, A., Suen, C.Y.: Methods of combining multiple classifiers and their application to handwritten recognition. IEEE Transactions on Systems, Man, and Cybernetics 22, 418–435 (1992)
12. Adams, N.M., Hand, D.J.: Comparing classifiers when the misallocation costs are uncertain. Pattern Recognition 32(7), 1139–1147 (1997)
13. Provost, F., Fawcett, T.: Analysis and visualization of classifier performance: comparison under imprecise class and cost distributions. In: Proceedings of Third International Conference on Knowledge Discovery and Data Mining, Menlo Park, CA, pp. 43–48. AAAI Press, Stanford, California, USA (1997)
14. Provost, F., Fawcett, T.: Robust classification for imprecise environments. Machine Learning 42(3), 203–231 (2001)
15. Gurcan, M.N., Sahiner, B., Petrick, N., Chan, H-P., Kazerooni, E.A., Cascade, P.N., Hadjiiski, L.: Lung nodule detection on thoracic computed tomography images: Preliminary evaluation of a computer-aided diagnosis system. Med. Phys. 29(11), 2552–2558 (2002)
16. Armato III, S.G., Giger, M.L., MacMahon, H.: Automated detection of lung nodules in CT scans: preliminary results. Med Phys. 28(8), 1552–1561 (2001)

Parallel Computations for Logic-Algebraic Based Expert Systems

Leszek Borzemski and Mariusz Fraś

Wroclaw University of Technology, Institute of Information Science and Engineering,
Wybrzeze Wyspianskiego 27, 50-370 Wroclaw, Poland
{leszek.borzemski,mariusz.fras}@pwr.wroc.pl

Abstract. Task allocation is a very important issue in the area of knowledge-based parallel and distributed systems. The problems formulated for a wide class of expert systems with knowledge representation in the form of facts can be solved applying the logic-algebraic (LA) method. The recursive procedure applied in LA algorithm defines elementary tasks to be processed at different computation stages. The paper presents parallel processing architecture for LA method. Effectively distributed basic task of proper granularity is defined and three load balancing task allocation algorithms are proposed. The performance of the algorithms is evaluated through experiments with the use of randomly generated instances of different LA task profiles.

Keywords: Logic-algebraic method, knowledge-based system, parallel processing.

1 Introduction

A great variety of expert systems are increasingly used in commercial applications. New requirements for such systems enforce improvement of computational procedures and address performance as a key issue. One of possible approaches to speed up computations in expert systems is to develop parallel execution of expert tasks.

Parallel processing methods and algorithms have been widely studied in the literature. At first, in the area of artificial intelligence (AI) considerable piece of work focused on rule-based production systems. Typical parallel processing issues were exploited such as discovering of parallelism, proper granularity of tasks and task allocation i.e. load balancing problem [e.g. 7, 9]. In the successive years the researches have concentrated on applying parallel solutions for different areas of AI-based systems such as constraint programming [10] or generally to logic programming languages [6], and on adapting AI solutions for parallel and distributed computing [e.g. 5, 8, 11].

For a wide class of control, identification and recognition systems an appropriate object expert system (i.e. the system with knowledge representation in the form of facts describing object under consideration) is developed [3]. Two basic problems for such systems are formulated: analysis problem and decision-making problem. To solve these problems a logic-algebraic method based on logic algebra fundamentals is developed [4]. The LA method was applied for different problems such as decision-making in complex operation systems or pattern recognition, but its further use demands increasing its performance.

B. Apolloni et al. (Eds.): KES 2007/WIRN 2007, Part I, LNAI 4692, pp. 311–318, 2007.

In this paper the parallel processing architecture for the logic-algebraic method is introduced. Section 2 presents the background of LA approach. The subsequent sections describe the structure of parallel processing, task allocation algorithms and results of simulation experiments performed to evaluate performance of computations.

2 Logic-Algebraic Method

Knowledge representation in the form of facts can be described as a set of simple formulae α and sequence of logical facts $F(\alpha)$ composed with subsequences of α and logical operators [3]. Simple formula $\alpha_n, n \in \overline{1,N}$ is a property of the object under consideration in the form of logical sentence (for instance "the temperature t is above the threshold t_0", "operation O_3 follows operation O_6", etc.). The facts $F_k, k \in \overline{1,K}$ describe known relationships of the simple formulae. For the given object the sequence of input formulae α_x (i.e. formulae considering input properties of the object) and input fact $F_x(\alpha_x)$, as well as sequence of output formulae α_y and output fact $F_y(\alpha_y)$ are distinguished. Two basic problems are considered for such knowledge representation: the analysis problem: for the given $F_x(\alpha_x)$ find the output fact $F_y(\alpha_y)$ such that implication $F_x \Rightarrow F_y$ is true, and the decision-making problem: for the given $F_y(\alpha_y)$ find the input fact $F_x(a_x)$. For these problems the logic-algebraic method defines equivalent problems [4]: the appropriate sets S_a={a: $F(a)$=1}, S_x={a: $F_x(a_x)$=1} and S_y={a: $F_y(a_y)$=1}, where a is 0-1 sequence of logical values of the simple formulae (lvsf for short) to be determined according to proper logic algebraic equations. The sets are calculated using recursive procedure and the partitioning of set of facts into subsets: $\overline{F}_1(\overline{a}_0,\overline{a}_1)$, ..., $\overline{F}_j(\overline{a}_{j\text{-}1},\overline{a}_j)$, ..., $\overline{F}_J(\overline{a}_{J\text{-}1},\overline{a}_J)$, where: $\overline{F}_j$ - subset of facts (conjunction) containing proper subsequence $\overline{a}_{j-1}$ of lvsf, $\overline{a}_j$ - subsequences of remaining lvsf in $\overline{F}_j$, $\overline{a}_0 = a_y$ (or $\overline{a}_0 = (a_y, a_x)$ for decision-making). The recursive procedure determines at j-th decomposition stage (j=J, J-1, ...,1) appropriate subsets $\overline{S}_{j-1}$ of sequences $\overline{a}_{j-1}$ according to the formula:

$$\overline{S}_{j-1} = \left\{ \overline{a}_{j-1} \in S_{j-1} : \bigvee_{\overline{a}_j \in \overline{S}_j} \left[\overline{F}_j(\overline{a}_{j-1}, \overline{a}_j) = 1 \right] \right\}, \tag{1}$$

where: S_{j-1} - the set of all possible sequences $\overline{a}_{j-1}$, $\overline{S}_J = S_J$, $\overline{S}_0$ - the ultimate solution for the expert problem from which the set S_y or S_x is obtained. The formula (1) describes the basic processing to be done. The whole computational task Z (the primary task) is decomposed into J subtasks $Z_j, j \in \overline{1,J}$, each referring to calculations to be performed at the corresponding decomposition stage (Fig. 1). The task Z_j consists in *elementary tasks* $Z_j^{(e)}$, $e \in \overline{1,\gamma_j}$ (γ_j – the number of tested

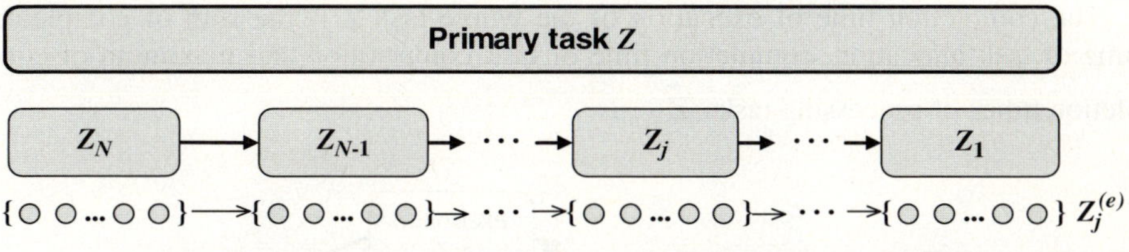

Fig. 1. General scheme of computations for logic-algebraic method

sequences), each calculating the value of the fact $\overline{F}_j(\overline{a}_{j-1}, \overline{a}_j)$ for given sequence of logical values.

It is possible to introduce a parallel processing of elementary tasks in different ways. However, there are usually too many such tasks to manage them efficiently. Moreover, elementary tasks at each stage are generated in an unknown pattern.

3 Parallel Computations Architecture

After the decomposition additional processing methods can be applied to improve the speed of computations. Considering characteristics of elementary tasks we proposed to perform parallel computations at each stage separately [1]. To reduce the possible task assignment overhead we define the *basic task* $Z_{j,i}(\overline{a}^{(i)}_{j-1}, \overline{a}_j)$ as follows:

for specified $\overline{a}^{(i)}_{j-1} \in S_{j-1}$ the condition $\bigvee_{\overline{a}_j \in \overline{S}_j} \left[\overline{F}_j(\overline{a}^{(i)}_{j-1}, \overline{a}_j) = 1\right]$ is tested for $\overline{a}_j \in \overline{S}_j$ computed at previous stage, and calculations terminate when the condition is true.

Basic tasks are independent and define portion of work that can be effectively assigned to processors without significant system overhead, using identifier w_i of basic task $Z_{j,i}$ equal to value of tested sequence $\overline{a}^{(i)}_{j-1}$ that is stated as follows:

$$w_i = \sum_{n_j=1}^{b_j} a_{n_j} \cdot 2^{n_j - 1}, \; i \in \overline{0, \beta_j - 1} \tag{2}$$

,

where: a_{n_j} - the value of n_j-th position in sequence $\overline{a}^{(i)}_{j-1}$, $a_{n_j} \in \{0,1\}$, $n_j \in \overline{1, b_j}$, b_j – the length of sequence $\overline{a}_{j-1}$, β_j - the number of sequences of set S_{j-1}.

The general computational procedure at j-th stage (see Fig. 2) partitions task Z_j into P_j subtasks $\widetilde{Z}_{j,p}$, $p \in \overline{1, P_j}$ (being the sum of basic tasks $Z_{j,i}$), executes subtasks $\widetilde{Z}_{j,p}$ assigned onto P_j processors in parallel and composes the task Z_{j+1} for the next stage of computation (i.e. distributes calculated sequences $\overline{a}_{j-1}$ to all processors).

The completion time of execution of the whole task Z is the sum of completion time of task allocation, completion time of task composition and maximum of completion times of processing tasks $\tilde{Z}_{j,p}$.

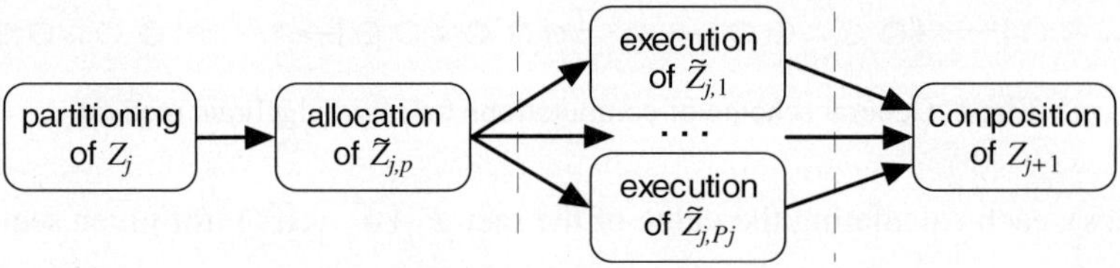

Fig. 2. The scheme of task execution at the j-th stage

Hereafter Q-processor homogenous environment is considered. We assume that the number of basic tasks is greater than the number of processors and P_j=Q. We don't investigate task composition problem which is explored in other papers.

4 Task Allocation Algorithms

The basic allocation strategy is to assign to each processor the equal number of basic tasks to be processed at j-th stage by partitioning the sorted sequence of task identifiers w_i of basic tasks $Z_{j,i}$ on Q continuous subsequences:

$$\tilde{Z}_{j,p} = \{Z_{j,i} : w_i \in \langle w'_p, w''_p \rangle\}, \tag{3}$$

where: $\tilde{Z}_{j,p}$ – tasks executed on processor p, $w'_p = \left\lceil (p-1) \cdot \frac{\beta_j}{Q} \right\rceil$,

$w''_p = \left\lceil p \cdot \frac{\beta_j}{Q} \right\rceil - 1$, w'_p, w''_p – boundary task identifiers for processor p, p=1,…,Q.

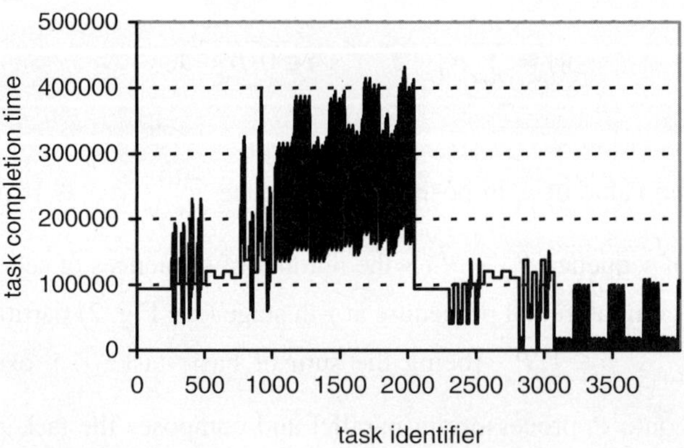

Fig. 3. An example of execution times of basic tasks at one computation stage

This allocation strategy is v efficient, however does not guarantee balanced processors' load [2]. Characteristic feature of computations for most LA tasks is that the basic tasks of shorter and longer completion time are grouped irregularly as shown in Fig. 3. The values of axis of ordinates correspond to internal counter simulation system and are proportional to execution time.

4.1 The Shuffle Algorithm RT

The shuffle algorithm RT allocates basic tasks to each processor from the whole range of task identifiers by random selection (Fig. 4) using random shuffle procedure:

1. Generate table T of sorted task identifiers w_i: T[v]=w_i : w_i=v , $v \in \overline{0, \beta_j - 1}$.
2. Shuffle randomly elements of T.
3. Assign tasks of consecutive identifiers to processor p:

$$\tilde{Z}_{j,p} = \{Z_{j,i} : w_i = \mathrm{T}[v] \wedge v \in \langle v'_p, v''_p \rangle\}, \tag{4}$$

where: $v'_p = \left\lceil (p-1) \cdot \frac{\beta_j}{Q} \right\rceil$, $v''_p = \left\lceil p \cdot \frac{\beta_j}{Q} \right\rceil - 1$ – boundary indexes specifying task identifiers in table T for processor p, p=1,…,Q.

The above procedure is performed on each processor simultaneously.

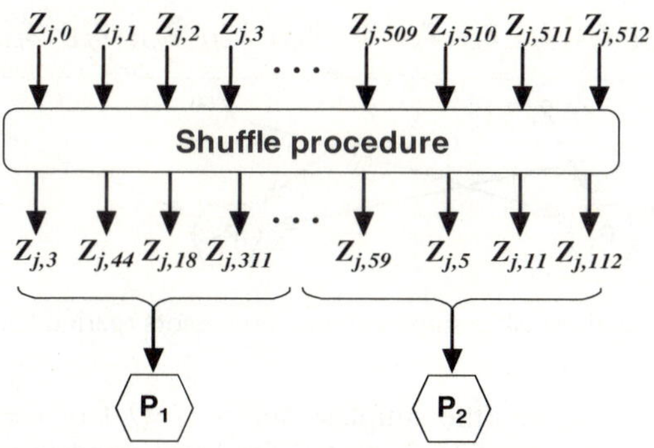

Fig. 4. Random assignment of 512 basic tasks on 2 processors

To satisfy requirement that all basic tasks must be executed and different tasks must be assigned to each processor, the shuffle algorithm has to generate the same order of identifiers on each processor. It is guaranteed by the use of linear congruence generator and Fisher-Yates shuffle algorithm.

Under some assumptions computational complexity of algorithm RT can be evaluated as:

$$f_{RT}(n) = O(J \cdot 2^{\frac{n}{J+1}}), \tag{5}$$

where: n – number of simple formulae, J – number of decomposition stages.

4.2 The Selective Algorithm RS

The selective algorithm RS specifies analytical formula of distribution of basic tasks to each processor. Let c be the sequence: $c_g = 2 \cdot g + 1,\ g \in \overline{0, G-1}$. The set $\tilde{Z}_j$ is divided into H subsets $\tilde{Z}_j^{(h)}$ as follows:

$$\tilde{Z}_j^{(h)} = \{Z_{j,i} : d_h \le w_i < d_h + c_{(h \bmod G)} \ \wedge\ w_i < \beta_j\}, \tag{6}$$

where: $h \in \overline{0, H-1}$, $d_0 = 0$, $d_h = \sum_{i=0}^{h-1} c_{(i \bmod G)}$ for $h \in \overline{1, H-1}$.

The consecutive sets have periodically changing odd number of basic tasks. Period length equals G and must not be multiple nor aliquot of the number of processors Q. The tasks assignment on processor p is given by:

$$\tilde{Z}_{j,p} = \bigcup_{(p-1)=(h \bmod Q)} \tilde{Z}_j^{(h)} \ . \tag{7}$$

The above procedure allocates to processor particular subsets of basic tasks alternately as shown in Fig. 5 for two-processor system example.

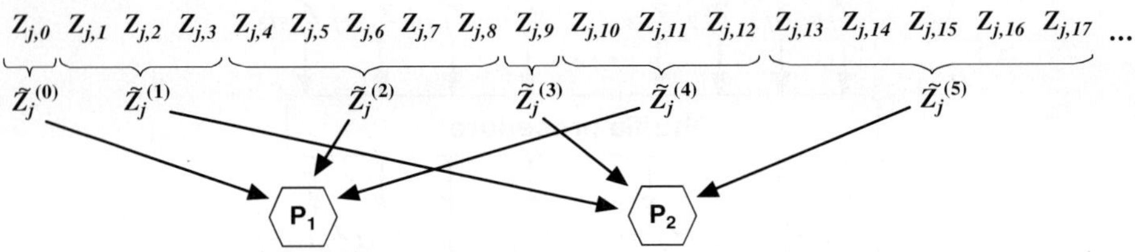

Fig. 5. Selective task assignment on 2 processors (period length G=3)

Under certain assumptions (the simplest one is G=Q-1 or G=Q+1) all processors are assigned similar number of basic tasks (for large number of task nearly equal) quasi-irregularly selected from whole range of task identifiers.

The computational complexity of algorithm RT can be evaluated as:

$$f_{RS}(n) = \frac{f_{RT}(n)}{G \cdot Q} \ . \tag{8}$$

5 Simulation Experiments

The effectiveness of the algorithms was evaluated with developed system SymLog [2] that is the expert system as well as the event-driven simulation system for studying execution issues of LA algorithms. The experiments were performed for randomly generated LA tasks producing different workload. The group of LA tasks of given profile had the fixed number of computation stages *J*, number of facts *LF* on one stage and number of simple formulae *LL* defining one stage. The series of experiments were performed for profiles that varied one of these parameters.

Fig. 6 and 7 present the speedup of computations σ versus the number of processors *Q* for LA tasks of *LL*=11 and *LL*=15 respectively. The figures show that the larger tasks are computed the better is performance, especially for the RT and RS algorithms. The effectiveness of the algorithm RS decreases when too small LA tasks are computed on too many processors. Fig. 8 shows that for larger number of computed basic tasks the performance of algorithm RS fast increases. The number of facts on each computation stage doesn't affect the speedup significantly (Fig. 9).

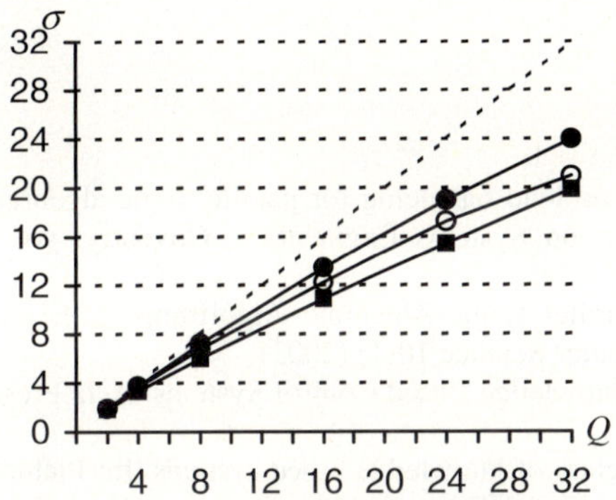

Fig. 6. The speedup σ versus the number of processors *Q* for algorithm R, RT and RS – task size: *LL*=10, *LF*=9, *J*=5

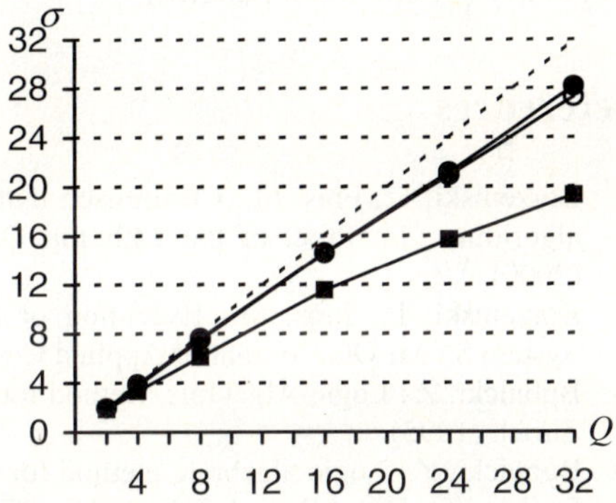

Fig. 7. The speedup σ versus the number of processors *Q* for algorithm R, RT and RS – task size: *LL*=15, *LF*=9, *J*=5

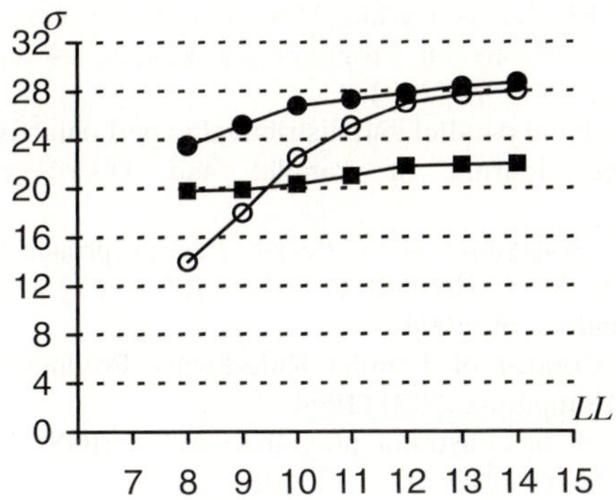

Fig. 8. The speedup σ versus the number of formulae *LL* for algorithm R, RT and RS; for *Q*=32, *LF*=10, *J*=3

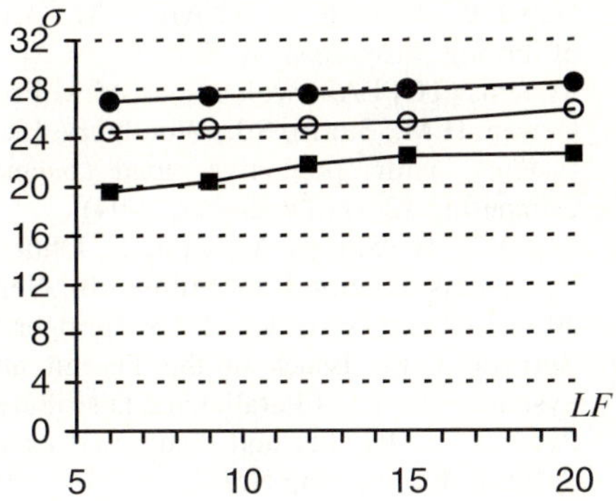

Fig. 9. The speedup σ versus the number of facts *LF* for algorithm R, RT and RS; for *Q*=32, *LL*=11, *J*=3

6 Conclusion

The paper presents parallel processing architecture and algorithms developed for logic-algebraic method that is applied in control, identification and recognition expert systems. Presented task allocation algorithms balance workload of processors and not involve interprocessor communication overhead. Performed simulation experiments show that significant performance improvement is achieved when introduced processing scheme is used. The best speedup assure the algorithm RT, however the algorithm RS for larger LA tasks is recommended, as its computational complexity is lower. Better effectiveness of algorithm RS can be probably achieved when advanced tuning of parameter G will be applied.

Presented algorithms can be used not only for described LA tasks but also for all tasks which have similar features, especially when their workload is grouped irregularly (in unknown pattern).

References

1. Borzemski, L., Fraś, M.: Customised dynamic load balancing for parallel logic-algebraic algorithms. In: Proc. of the 14th Int. Conf. on Systems Engineering, Coventry, vol. 1 (2000)
2. Borzemski, L., Fraś, M.: Execution of Parallel Logic-Algebraic Algorithms in Expert System SYMLOG. Journal of Applied Computer Science 10(1) (2002)
3. Bubnicki, Z.: Logic-Algebraic Method for Knowledge Based Control Systems. Acta Press, Zurich (1995)
4. Bubnicki, Z.: Logic-algebraic method for a class of knowledge based systems. In: Pichler, F., Moreno-Diaz, R. (eds.) EUROCAST 1997. LNCS, vol. 1333, Springer, Heidelberg (1997)
5. Choi, S.-Y., Cha, J.-S., Shin, M.-C.: Artificial Intelligent Application to Service Restoration Considering Load Balancing in Distribution Networks. In: Gabrys, B., Howlett, R.J., Jain, L.C. (eds.) KES 2006. LNCS (LNAI), vol. 4253, Springer, Heidelberg (2006)
6. Gupta, G., Pontelli, E., Khayri, A.M., Ali, C.M., Hermenegildo, M.V.: Parallel Execution of Prolog Programs: A Survey. ACM Transactions on Programming Languages and Systems (TOPLAS), vol. 23(4). ACM Press, New York (2001)
7. Dewan, H.M., Stolfo, S.J., Woodbury, L.: Scalable parallel and distributed expert database systems with predictive load balancing. Journal of Parallel and Distributed Computing 22(3) (September, 1994)
8. Ishikawa, H., Shioya, Y., Omi, T., Ohta, M., Katayama, K.: A Peer-to-Peer Approach to Parallel Association Rule Mining. In: Negoita, M.G., Howlett, R.J., Jain, L.C. (eds.) KES 2004. LNCS (LNAI), vol. 3213, Springer, Heidelberg (2004)
9. Neiman, D.E.: Issues in the Design and Control of Parallel Rule-Firing Production Systems. Journal of Parallel and Distributed Computing 23(3) (1994)
10. Pontelli, E.: Parallel and distributed execution of constraint programs. ACM SIGSOFT Software Engineering Notes, vol. 25(1). ACM Press, New York (2000)
11. Srinivasa, R., Reynolds, P.F., Williams, C.: Parallel Rule-Based Systems on Isotach Networks. In: Int. Conf. on Parallel and Distributed Computing and Systems, IASTED, Cambridge, Massachusetts (1999)

Process Control of an Event Filter Farm for a Particle Physics Experiment Based on Expert System Technology

Kristina Marasović[1], Bojana Dalbelo-Bašić[2], and Vuko Brigljević[3]

[1] University of Split, Faculty of Natural Sciences, Mathematics, and Kinesiology, Teslina 12, 21000 Split, Croatia
[2] University of Zagreb, Faculty of Electrical Engineering and Computing, Unska 3, 10000 Zagreb, Croatia
[3] Ruđer Bošković Institute, Bijenička cesta 54, P.O.Box 180, 10002 Zagreb, Croatia
Kristina.Marasovic@pmfst.hr, Bojana.Dalbelo@fer.hr, Vuko.Brigljevic@irb.hr

Abstract. Modern particle physics experiments observing collisions of particle beams generate large amounts of data. Complex trigger and data acquisition systems are built to select on-line the most interesting events and write them to persistent storage. The final stage of this selection process nowadays often happens on large computer farms. The stable and reliable operation of such event filter farms is critical for the success of these experiments. In this paper, the current status and plans in developing a Problem Solver based on expert system technology, which could be applied for maintaining reliability and uninterrupted operation of the Event Filter Farm, is described. The proposed Problem Solver has been tested with an Event Filter Farm prototype based on the architecture of the CMS experiment. A performance analysis of the Problem Solver integrated in the existing control system is given.

Keywords: Process control, expert systems, fault detection and diagnosis, recovery.

1 Introduction

As the complexity of a technical system grows, control system design becomes more difficult. With increasing complexity, there are more possibilities for failures. To maintain reliability and uninterrupted operation, control systems are often extended with advanced problem solving capabilities. Tasks such as monitoring, fault detection and diagnosis, and recovery may be required. The fault detection and diagnosis can be generally considered as a process of monitoring the operating conditions of the controlled system to detect and isolate any abnormal behavior, referred to as a fault. The terms fault diagnosis and fault isolation are often used synonymously to refer to a classification of the detected fault. Over the years, a number of different fault detection and diagnosis approaches have been proposed by mainly two distinct communities: the Fault Detection and Isolation (FDI) community, based on the fields of automatic control and statistical decision theory; and the Diagnosis (DX) community based on the fields of computer science and artificial intelligence (AI) [1].

B. Apolloni et al. (Eds.): KES 2007/WIRN 2007, Part I, LNAI 4692, pp. 319–326, 2007.

While the FDI community focuses its research on fault detection by processing quantitative data using an algorithmic solution, the DX community has introduced a new concept, namely, symbolic and qualitative data processing where more emphasis is placed upon fault diagnosis as a reasoning process.

FDI techniques and methodologies are more established than their DX counterparts since they were first introduced into the problem solving area. FDI approaches are mainly based on a dynamic quantitative model of the controlled system, usually represented as a set of differential algebraic equations, where fault detection and isolation is carried out in two steps. First the measured input and output of the controlled system are compared to the modeled one. The resulting difference, called residual, is an indicator of a potential fault of the system. In the second step, the residual is examined to establish whether a fault has occurred. If this is the case, diagnosis can be carried out, often using statistical methods for decision making [2]. In cases where no a priori knowledge about the process exists, data-driven approaches based on univariate or multivariate statistical analysis are used [3].

With increasing complexity, high dimensionality, and process nonlinearity, these conventional FDI approaches can no more practically and effectively satisfy the system requirements [2]. They do not provide the knowledge about the relationships between different system components which is essential for undertaking the task of fault diagnosis. Also, they cannot handle inexact or incomplete information [2]. Usage of AI technologies such as soft computing and expert systems can make the fault diagnosis task more reliable and efficient. Soft computing techniques are mainly used in developing hybrid intelligent systems. In a neuro-fuzzy system, rules are generated and optimized by a neural network, whereas in a fuzzy-genetic system this is done by a genetic algorithm. In reference [4], nine articles addressing this topic are described. Another set of articles describing how AI techniques have been used for fault diagnosis in electronic systems can be found in [5] and [6].

The FDI and the DX community each have their own terminology and their own set of conferences, which could result in conflicting understanding of the same issue. Efforts are being made to bring them together, such as the one from the Monet Network [1]. As a result, today's fault detection and diagnosis systems are mainly based on synergistic approaches from both communities and thus provide the best suited method for a given type of problem. Although many synergistic methods have been proposed and evaluated as successfully applicable in many industrial processes, there are not enough paper reports about their real acceptance. On the one hand, analytical model-based approaches for fault detection are usually combined with a fuzzy system for fault classification as in [7] or with a neural network as in [8]. On the other hand, data-driven approaches based on statistical analysis are also combined with AI approaches as in [9]. Additional information can be found in [1].

The *Problem Solver* (PS) of the *Event Filter Farm* (EvFF) prototype, described in this paper, is developed using statistical approach in combination with expert system (ES) technology. In the following section, an overview of the EvFF towards problem solving research is given. The architecture of the PS is described in the third section followed by a performance analysis of this successfully integrated system into the existing control environment of the EvFF. Finally, conclusion with remarks on future work is given.

2 Process Control Description of the CMS Event Filter Farm

The *Large Hadron Collider* (LHC), currently under construction at CERN[1] and due to start operating in 2008, is a large circular accelerator that will accelerate bunches of protons in two counter-rotating beams and collide them at four points. Large detector systems such as CMS (*Compact Muon Solenoid*) [10] are being built to observe the collisions. To collect the data from all subparts of the detector and to filter out those that will be written to persistent storage for off-line analysis, a middleware-based distributed *Trigger and Data Acquisition System* (TriDAS) is being designed [11]. The on-line filtering process will be performed in two stages. The last stage known as *High-Level Trigger* (HLT) [11] will be carried out in the EvFF consisting of about 1000 dual-CPU PCs. To configure, control and monitor the components of the Data Acquisition (DAQ) system, a *Run Control and Monitoring System* (RCMS) has been designed. Within the RCMS, the DAQ components are operated in a hierarchical structure and controlled by a hierarchy of *Function Manager* (FM), each modeled as a finite state machine. Configuration and control messages are being relayed from the top FM down to each DAQ node in parallel. The FMs are also provided with monitor and state information of the DAQ nodes. The EvFF will be controlled by its own function manager, the *Farm Manager* (FFM).The RCMS and FMs are implemented as Java-based Web services hosted on Tomcat servers [12] where SOAP/HTTP is used as communication protocol.

Operating the EvFF must ensure dead time free processing of large amount of data, thus, requiring 24-hours continuous status monitoring of each process node called *Filter Unit* (FU), and fast detection and problem solving. Ideally, problems should be recognized before they deteriorate the system performance.

Setting up the EvFF for data taking and filtering includes several stages. First, the EvFF needs to be *initialized* where bootstrap and configuration information for all relevant processes are downloaded by the FFMs from the appropriate servers and distributed in parallel to FUs. Afterwards, the farm is ready to be *configured* where binaries of the HLT applications with relevant libraries are uploaded to FUs into memory and connections are established for data taking. Finally, the EvFF can be *started* where data, referred to as events, are read-out and HLT algorithms are run to accept or reject events for storage. During operation, different parameters of each FU process such as CPU load, memory usage, number of processed events, fraction of accepted events, and processed events rate can be monitored for action taking in case problems occur. Problems could range from process or machine crashes to any deviation of the observed process parameters from normal operating conditions. The recovery procedure may involve modifying running parameters, reconfiguring and restarting one or more FUs, or even the whole EvFF. Problem reporting to the central information service of the experiment is also involved.

The process control of the EvFF, as in other particle physics experiments, is exclusively performed by a human operator, placing high demands difficult to accomplish. Automating crucial tasks could address this issue. So far, there has been very little or no experience in applying AI methods to this problem domain which requires the PS implementation to meet several demands. It should embed the

[1] European Organization for Nuclear Research, the world's largest particle physics centre.

operator's expertise, and be flexible enough for knowledge updating since no operational data are available before the first LHC run, thus, not all problems could be foreseen. It should meet real-time constraints of a controlled system which is non-linear, complex, distributed, and large-scaled. The PS should also cope with available computing resources for running continuously over extended period of time, handling and analyzing a vast amount of operational data. It should be modular and scalable to account for the complexity and the evolution of the hardware and software environments of the CMS experiment over time. Therefore, a hybrid solution based on ES technology and statistical data analysis is an appropriate choice.

Over time, a number of EvFF prototypes for small-scaled systems have been developed for test purposes. The prototype tested with the PS includes the EvFF whose input has been simulated using randomly generated data. All received events are being accepted and no output has been written to persistent storage.

3 Problem Solver Architecture

The PS has been implemented by using the Java technology which makes the PS easy to embed into the FFM where problem solving could be also propagated to higher levels in the control hierarchy. As indicated in Fig. 1, the PS incorporates several functions to help the human operator automating some problem solving tasks.

The *Monitor Data Collection Handler* retrieves collection of data about various process parameters from all FU nodes through HTTP. The data collection is then parsed into a data structure reusable by other PS components.

To analyze the data collection for the purpose of fault detection and diagnosis, a *Jess Handler* based on statistics and Jess (Java Expert System Shell) [13] is used. The responsibility of the *Jess Fact Handler* is to supply and update the Jess working memory with facts referred here to as sensors. Sensors are implemented as Java Beans whose properties describe the current problem solving environment. For each FU process, a *FU Sensor* is created whose properties reflect the current status of the observed process parameters as well as the state of the FU itself. In lack of response from a certain FU, its state is set to faulty. Assuming normally distributed data, a *Statistics Sensor* calculates tolerance intervals with statistically derived limits for each observed process parameter where interval coverage can be set to custom percent. For example, to calculate the tolerance interval for the processed events rate of a FU, the rate is examined across all FU nodes at a time and the limits can be set to 1.96 standard deviations of the mean, this way, covering 95% of the population. The Jess rule base implements following knowledge:

1. Any observed parameter whose value exceeds the associated tolerance interval is considered as an outlier, and warning messages are sent to the concerned parties.
2. When the state of a certain FU is faulty, two potential causes of this problem are considered. First, a connection with the FU node could not be established because some services on FU went down or the FU crashed, and secondly, the requested monitored data were lost during transportation. In the first case, the FU needs to be restarted and brought to the last recorded running state. In the second case, no problem is reported if the data loss is not repeated.

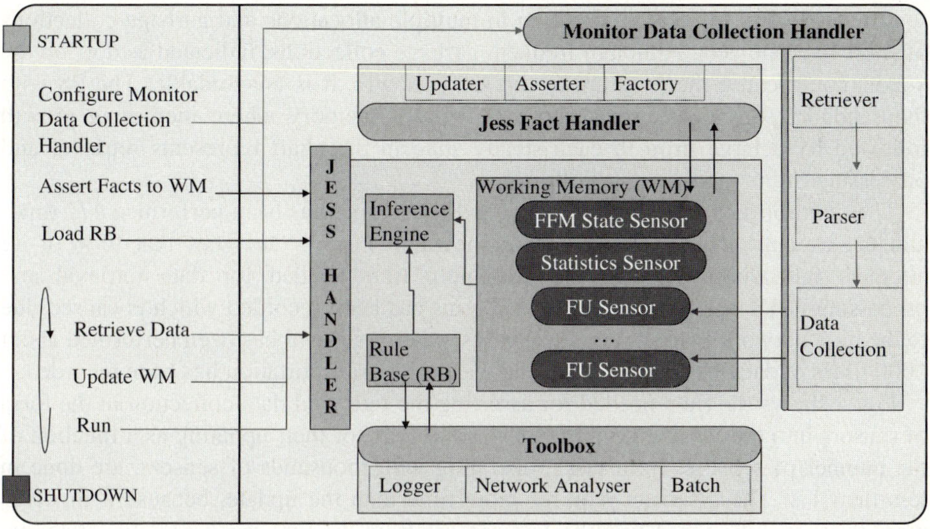

Fig. 1. Problem Solver Architecture Based on Expert System Technology

Execution of those problem solving tasks is carried out by utilizing the *Toolbox*. The *Logger* uses the log4j logging service for Java applications [14] to send warning messages via TCP/IP. Parties interested in receiving those remote messages should run a log4j log viewer such as Chainsaw [14]. The *Network Analyzer* is used for discovering the cause of the reported faulty state of a certain FU. Among others, it uses TCP sockets to establish a connection to a FU service port to test its reachability. When a FU needs to be restarted and brought back to the last recorded running state, the *Batch* tool capable of sending process commands to a FU in the form of SOAP/HTTP messages is being used.

Once started, the PS operates in the following cycles until it is shutdown. It retrieves collection of data from all FU nodes, updates the Jess working memory with these new data and triggers the running of the Jess inference engine.

4 Performance Results and Analysis

The PS has been tested for efficiency, robustness and reliability. Because of the limited amount of available resources, the PS has been tested on a small-scaled EvFF consisting of 7 FUs. To analyze the Jess decision making over a large-scaled EvFF, a component for sensor multiplication has been developed. Each sensor of a monitored FU has been cloned multiple times giving the appearance to the Jess of an EvFF consisting of thousands of FUs.

Fig. 2 shows the heap memory usage of the Tomcat server hosting RCMS and PS. From the total available memory of 2 GB, 256 MB have shown to be enough for the Java Virtual Machine. The startup procedure of the EvFF, during which the PS is also started, as well as each transitional state in the process control, is accompanied by multiple steep spikes in the chart. This is because a lot of SOAP messages are being sent by a FFM to FU nodes, resulting in multiple allocations and garbage collections of short-lived objects. Although frequent garbage collections, indicated as dots on the x-axis, could cause the program to run more slowly, it is unavoidable. The PS with thousands of sensors occupies about 10 MB of memory where increasing growth followed by a large drop in each steady state in the chart represents retrieval and processing of the monitor data collection.

The monitor data collection retrieval and processing has been performed 472 times and the execution time distribution is shown in Fig. 3. One sees that 95% of all attempts fall within 70 ms. The minimum time needed for data retrieval and processing is 17 ms. A maximum of 9201 ms has been recorded which is caused due to the high network load. In other test cases, where retrieval has been performed about 6000 times, a data loss between 1% and 4% during transportation has been recorded.

Fig. 4 shows the time needed for asserting the retrieved data collection in the form of sensors into the Jess working memory as well as for their updating as a function of the number of sensors. Both tasks, operating with thousands of sensors, are done in less than 1 s. The assertion requires more time than the update, because it involves more tasks such as generating the Jess sensors from the first received data collection, and asserting them into the Jess working memory. Assertion is performed only once, during the PS startup, whereas updating is performed periodically whenever a data collection is received.

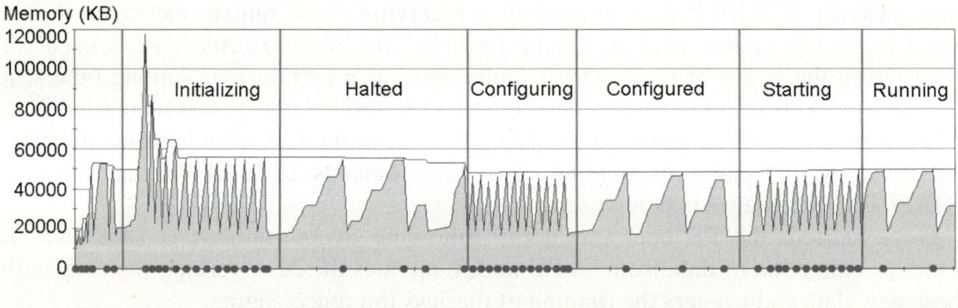

Fig. 2. Heap Usage Chart of the Tomcat Server Hosting RCMS & PS

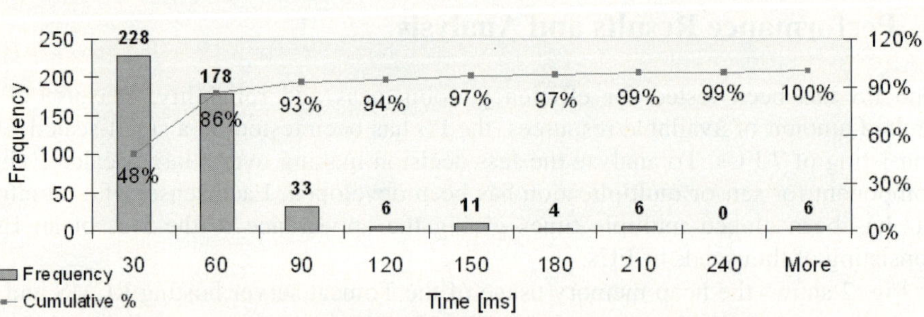

Fig. 3. Monitor Data Collection Retrieval and Processing

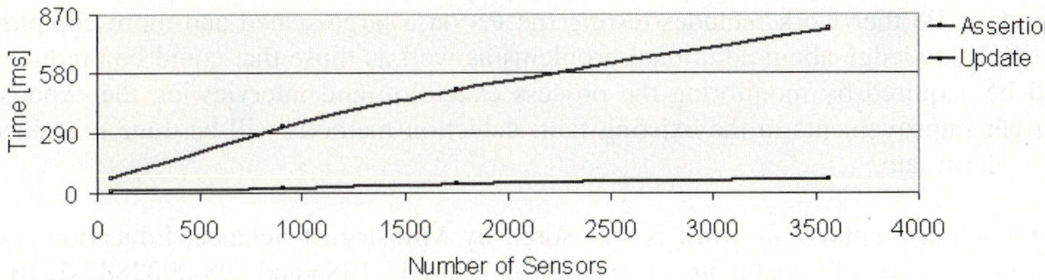

Fig. 4. Assertion and Update Time of PS Sensors as a Function of the Number of Sensors

The PS has been tested for failures and actions have been performed as expected. Since the execution time of each problem solving task depends on the rule's action part, the performance of the Jess engine has been tested with a special developed rule base. Thus, thousands of sensors were compared against 11 rules resulting in an agenda containing more than ten thousands activated rules with no actions on their right-hand side. The results in Fig. 5 show that the Jess engine has a very fast pattern matching algorithm whose execution takes less than 630 ms.

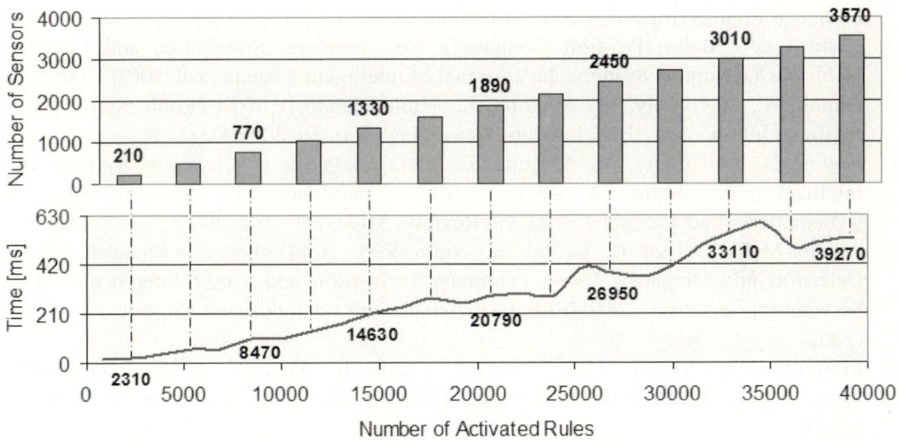

Fig. 5. Runtime Overhead of the Jess Inference Engine in Relation to the Number of Sensors

5 Conclusion

The results of this study show that the proposed PS meets the demands placed on the process control of the EvFF. Tested on a small-scaled EvFF, the PS with a large quantity of observed process parameters in its working memory does not deteriorate the system performance. The retrieval and processing time of the observed parameters is under 1s, and depends on the network load. Inference which includes pattern matching with thousands of facts and forming an agenda with more than ten thousands of rules is not a time consuming task. Problem solving tasks have been successfully performed whereby several, not all, problems where considered for test purposes. Further work includes testing the PS on a large-scaled and more complex EvFF. Knowledge about additional problems as well as those that could be predicted will be acquired by monitoring the process execution and interviewing the experts. Further improvements in the existing fault detection methods will be done to reduce false alarm rate.

Acknowledgments. This work is sponsored by Ministry of Science, Education and Sports, Republic of Croatia, under grants 036-1300646-1986 and 098-0982887-2878.

References

1. Biswas, G., Cordier, M.-O., Lunze, J., Trave-Massuyes, L., Staroswiecki, M.: Diagnosis of Complex Systems: Bridging the Methodologies of the FDI and DX Communities. IEEE Trans. Syst. Man Cybern. B Cybern. 34(5), 2159–2162 (2004)
2. Venkatasubramanian, V., Rengaswamy, R., Yin, K., Kavuri, S.N.: A Review of Process Fault Detection and Diagnosis, Part I: Quantitative Model-Based Methods. Computers & Chemical Engineering 27(3), 293–311 (2003)
3. Venkatasubramanian, V., Rengaswamy, R., Kavuri, S.N, Yin, K.: A Review of Process Fault Detection and Diagnosis, Part III: Process History Based Methods. Computers & Chemical Engineering 27(3), 327–346 (2003)
4. Castillo, O., Melin, P.: Soft Computing for Modeling, Simulation, and Control of Nonlinear Dynamical Systems, Int'l Journal of Intelligent Systems, vol. 20(2) (2005)
5. Fenton, B., McGinnity, M., Maguire, L.: Fault Diagnosis of Electronic Systems Using Artificial Intelligence. IEEE Instrum. Meas. Mag. 5(3), 16–20 (2002)
6. Fenton, B., McGinnity, M., Maguire, L.: Fault Diagnosis of Electronic Systems Using Intelligent Techniques: A Review. IEEE Transactions on Systems, Man, and Cybernetics—Part C: Applications and Reviews 31(3), 269–281 (2001)
7. Abidin, M.S.Z., Yusof, R., Kahlid, M., Amin, S.M.: Application of a Model-Based Fault Detection and Diagnosis Using Parameter Estimation and Fuzzy Inference to a DC-Servomotor. In: Proc. 2002 IEEE Int'l Symposium on Intelligent Control, pp. 783–788 (2002)
8. Murphey, Y.L., Masrur, M.A., Chen, Z., Zhang, B.: Model-Based Fault Diagnosis in Electric Drives Using Machine Learning. IEEE/ASME Trans. Mechatron. 11(3), 290–303 (2006)
9. Uluyol, O., Kim, K., Nwadiogbu, E.O.: Synergistic Use of Soft Computing Technologies for Fault Detection in Gas Turbine Engines. IEEE Transactions on Systems, Man and Cybernetics - Part C: Application and Reviews 36(4), 476–484 (2006)
10. CMS Collaboration. The Compact Muon Solenoid, technical proposal no. 7, CERN/LHCC 94-38 (1995)
11. CMS Collaboration, CMS The TriDAS Project Technical Design Report. Data Acquisition and High-Level Trigger, CERN/LHCC 2002-26, vol. 2 (2002)
12. Apache Tomcat (2006), [Online] (31 August, 2006), Available from, http://tomcat.apache.org
13. Jess and the Jess Logo is a registered trademark of Sandia National Laboratories. Jess source code, binary code and all Jess documentation associated with Jess code is owned and under copyright by Sandia Corporation. To license and download Jess, visit the Jess website at, http://www.jessrules.com and contact Craig Smith at casmith@sandia.gov
14. Log4j (2006), [Online] (31 August, 2006), Available from, http://logging.apache.org

The CTCN Temporal Model for Representing Knowledge in the Sleep Apnea Syndrome Diagnostic Task

Ángel Fernández-Leal and Vicente Moret-Bonillo

Laboratory for Research and Development in Artificial Intelligence (LIDIA), Department of Computer Science, University of A Coruña, A Coruña, Spain

Abstract. We present a prototype for implementing a framework for handling temporal information using the CTCN (Causal Temporal Constraint Networks) model as the representational schema. This prototype was developed for application to the diagnosis of the sleep apnea syndrome (SAS). The temporal information in SAS diagnosis is structured—according to its characteristics and temporal granularity—in terms of different interpretation contexts that are connected to each other via an inference mechanism.

Keywords: Temporal knowledge, Sleep Apnea Syndrome, CTCN, TCSP, medical diagnosis.

1 Introduction

The representation and handling of temporal information can be viewed as a temporal constraint satisfaction problem (TCSP), in which temporal constraints constitute a set of problems arising from general constraint satisfaction problems (CSP). The variables represent temporal entities such as instants or intervals, and the constraints represent permissible temporal relationships between temporal entities.

The most interesting TCSP approaches are hybrid models that endeavour to deal with quantitative and qualitative constraints in conjunction. The most representative of these models [1][2][3] fail to deal adequately, however, with causal relationships and inexact knowledge. In an attempt to address this drawback, we propose a new computable model for representing temporal information and for handling causality effectively: the general causal temporal constraint network (CTCN) [4][5][6].

A causal temporal constraint network (CTCN) represents a set of variables and the corresponding imprecision measurements $\{(O_1,M_1), ..., (O_n,M_n)\}$, each representing a temporal object (point or interval) and a set of unitary and binary constraints. When a variable represents a temporal point, its domain is the set of real numbers $\mathbb{R}$, and when a variable represents a temporal interval, its domain is the set of ordered pairs of real numbers denominated $\{(a , b) \mid a , b \in \mathbb{R}, a < b\}$.

Constraints may be quantitative, qualitative or causal. Each qualitative constraint is represented by a set of relationships R, and each quantitative constraint is represented by a set of intervals I.

Causal constraints label qualitative constraints in the form (r_i , CC), where each r_i represents any one of 13 basic temporal relationships that may exist between two

B. Apolloni et al. (Eds.): KES 2007/WIRN 2007, Part I, LNAI 4692, pp. 327–334, 2007.

objects, according to the Allen model [7], $r_i \in$ { e, b, bi, m, mi, s, si, f, fi, d, di, o, oi}, and CC represents any of 6 causal temporal restrictions that may exists between two objects, CC $\in$ {there cannot be (0), there may be (P), there must be (1), there cannot be causal (0C), there may be causal (PC), there must be causal (1C)}.

It is also assumed that unitary quantitative constraints are represented by equivalent binary constraints. A set of internal constraints relates each interval I = [I-, I+] with its extremes I^- {start} and I^+ {finish}.

A network of constraints is associated with a directed graph of constraints, where the nodes represent variables and an arc $i \rightarrow j$ indicates a constraint C_{ij} between the variables O_i and O_j. The arc is labelled by means of a set of intervals (when the constraint is quantitative). In the event that there are intervals that denote latency intervals in the causal or pseudocausal relationships, these will be preceded by an "L".

When the relationship is qualitative the arc is labelled by means of a set of pairs (r_i , CC), except for the start (s) and finish (f) relationships for the start and finish instants of an event with a duration greater than zero.

Unlike what happens with general temporal constraint networks, the causal constraint C_{ij} does not necessarily have to coincide with the inverse causal constraint C_{ji}, and so the constraint graph should indicate arcs $i \rightarrow j$ and $j \rightarrow i$.

Moreover, the absence of an arc between two temporal objects indicates the absence of constraints.

The novelty of CTCN is that it broadens the representation of causality by including the heuristic knowledge possessed by experts operating in the particular application domain that is to be modelled. This ensures that causal relationships are handled efficiently. The model also broadens the representation of imprecision, by adding the corresponding information items to the intrinsic representation of temporal relationships.

Based on the CTCN model, we have designed a working framework aimed at taking advantage of the improvements offered by the CTCN model in the design of computational systems for more efficient clinical diagnosis. Using this working model, expert systems can be constructed—for specific clinical application domains—that are capable of handling both temporal and causal relationships between temporal information entities and incorporating formal models for handling inexact knowledge [8].

In this research we present a prototype for implementing the work carried out in representing and handling temporal information in the diagnosis of the sleep apnea syndrome (SAS). This prototype is called TASAS (Temporal Analysis of Sleep Apnea Syndrome) [4].

2 Materials and Methods

In order to define construction of the domain for SAS, we used the same terminology used for defining the CTCN model [5][6] and for defining the framework for the treatment of temporal information based on CTCN [4][8].

2.1 Declarative Knowledge

Natural facts: Natural facts involve a degree of imprecision associated with the observation of occurrences. There are two types: *pneumological facts* and *electrophysiological facts*

Pneumological facts: The aim of analysing pneumological signals in SAS diagnosis is to evaluate the patient's respiratory function on the basis of three signals: a) respiratory flow, b) oxygen saturation (SaO_2), and c) respiratory effort.

Respiratory flow. Processing respiratory flow signals involves identifying four possible situations: a) normal respiratory flow, d) slightly reduced respiratory flow, c) clearly reduced respiratory flow, and d) completely reduced respiratory flow.

Oxygen saturation. Apneic events are associated with a fall in the SaO_2 signal (desaturation) and a hyperventilatory response (resaturation) that causes this signal to rise rapidly. Thus, two types of facts are considered for oxygen saturation: a) oxygen desaturation, and b) oxygen resaturation.

Respiratory effort. Classification of various types of apneic events is achieved through monitoring the presence or absence of: a) thoracic effort, and b) abdominal effort.

Electrophysiological facts: Electrophysiological facts are of three types: a) position, b) electromyogram (EMG), and c) hypnogram.

Position. a) supine, b) lateral (right or left), c) prone, and d) other (unclassified).

Electromyogram (EMG). The electromyogram (EMG) indicates the patient's muscular activity during polysomnographic monitoring. Four properties have been defined to describe EMG behaviour: a) muscle tone, b) tone level, c) phasic movement, and d) evolution. These properties allow us to consider the following as EMG events: a) relaxed EMG (low or medium muscle tone), and b) contracted EMG (low muscle tone).

Hypnogram. The hypnogram records periods of wakefulness, sleep stages and arousal, classified as follows: a) wakefulness, b) arousal, and c) sleep. Sleep can be classified as: a) non-REM sleep (light sleep, i.e. Stage 1 or 2, or deep sleep, i.e. Stage 3 or 4), and b) REM sleep (tonic REM or phasic REM).

Inferred facts: Inferred facts in SAS are apneic events, which can be classified as follows: a) apnea, and b) hypopnea, which in turn can be classified as: a) obstructive, b) obstructive REM-phase, c) obstructive REM-phase in the supine position, d) central, and e) mixed.

2.2 Contexts and Concepts

Contexts in SAS: The context identifying apneic facts is considered the natural context. Considered as inference contexts are the classification context for apneic facts and the results context.

Concepts in SAS: Concepts (which can be considered as atemporal information, meaning that its value does not change during polysomnographic analysis) are as follows: a) the patient's sleeping habits, and b) the patient's basic data. The patient's basic data includes name, sex, age, whether they are obese, and the time and date of the polysomnograph.

2.3 Procedural Knowledge

In this section we look at the types of rules applied to SAS diagnosis. Applying the CTCN model to the representation of temporal knowledge and to the rules for temporally correlating symbolic information will be seen to be simple and intuitive.

Fact-context occurrence rules: In the context for identifying apneic events, all occurrences of natural facts to be considered in SAS are analysed and the occurrence of apneas and hypopneas are identified. In the classification context for apneic facts, occurrence of the following facts is analysed: a) apneas, b) hypopneas, c) respiratory effort, and d) position. In the results context, the occurrence of classified apneic facts is observed, as follows: obstructive apnea, REM-phase obstructive apnea, REM-phase obstructive apnea in the supine position, central apnea, REM central apnea, mixed apnea, REM-phase mixed apnea, obstructive hypopnea, REM-phase obstructive hypopnea, REM-phase obstructive hypopnea in the supine position, central hypopnea, REM-phase central hypopnea, mixed hypopnea, and REM-phase mixed hypopnea.

Contextual temporal restriction rules: These rules allow restrictions to be established that are associated with the temporal relationships which arise among the facts of the three contexts defined for SAS using the CTCN model.

Figure 1 shows an example of a CTCN graph for a contextual temporal restriction rule, depicting apnea and thoracic respiratory effort facts in the classification context for apneic facts.

Fact-reference interval and fact-inference interval occurrence rules: For each context, facts are defined which can be observed in each reference interval, for example in the classification context for apneic events.

If: (1) an APNEA event occurs
and: (2) abdominal effort exists
and: (3) thoracic effort exists
Then: an OBSTRUCTIVE APNEA reference interval is confirmed.

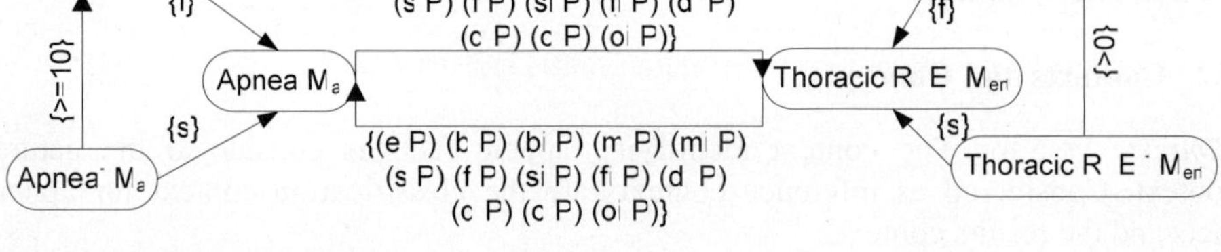

Fig. 1. CTCN graph of temporal relationships between an apnea and thoracic respiratory effort

Similarly defined for each reference interval are the facts that can be observed in each inference interval.

Temporal relationship-reference interval and temporal relationship-inference interval occurrence rules: These rules allow restrictions to be established that are associated with temporal relationships that arise between reference interval facts in the contexts defined for SAS using the CTCN model.

Figure 2 shows an example of a CTCN graph for a temporal relationship-reference interval occurrence rule for classification of an obstructive apnea described as follows:

If: (1) an apnea event occurs
and: (2) abdominal effort occurs during {d}
and: (3) thoracic effort occurs during {d}
Then: an obstructive apnea reference interval is confirmed.

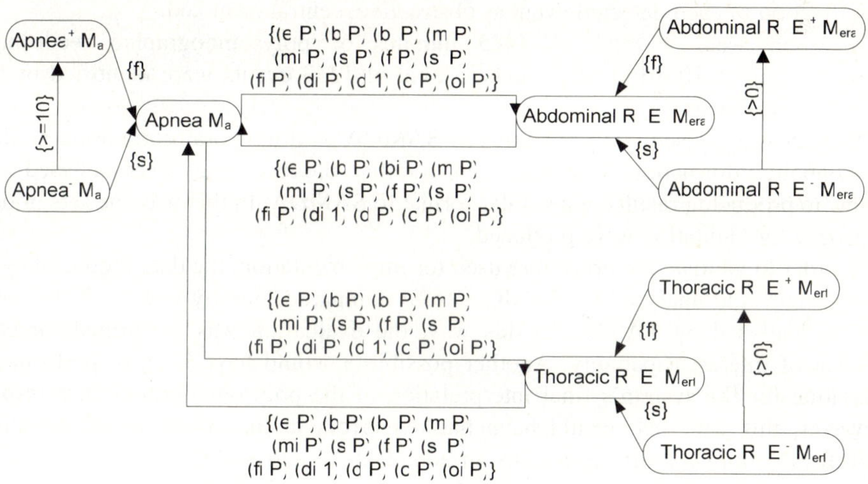

Fig. 2. CTCN graph for a temporal relationship-reference interval occurrence rule for apnea classification

Similarly, for each reference interval, restrictions are defined that are associated with temporal relationships which may occur between the inference interval facts in the reference intervals defined for SAS using the CTCN model.

Inference rules: Inference rules are defined so that, based on each defined inference interval, new events relevant to the diagnosis can be inferred. For example:

If: (1) an inference interval for obstructive apnea is confirmed
and: (2) an apnea event occurs
and: (3) the associated uncertainty is less than a value α
Then: a new obstructive apnea event with an associated uncertainty β is inferred.

Once the declarative and descriptive elements and the SAS concepts and contexts were defined, the method was implemented for validation purposes. The validation results are described below.

3 Consistency Analysis

In validating the procedure on which the implementation was based, a major consideration was the consistency analysis of the results produced by SAMOA [9], a system already validated against SAS clinical diagnosis experts.

In order to perform the validation, SAMOA was used to analyse data from recordings made by the Hospital Juan Canalejo of A Coruña (Spain) and produced by the commercial α-Somnostar® system [10]. The following two elements were identified as of interest in regard to testing TASAS results consistency: (a) identification of apneic events (apneas and hypopneas) for each record, and (b) classification of each detected event as obstructive, central or mixed.

Analysed were a total of 4485 minutes of polysomnographic recordings, corresponding to 10 different patients; a total of 1243 events were identified by the collaborating expert.

TASAS was subsequently included in SAMOA as a temporal correlation module for symbolic information. To ensure comparability, the recordings were analysed with a view to producing results in a similar format to SAMOA. In this way the sets of data necessary for validation were produced.

In order to validate the procedure used for implementation, the data produced by α-Somnostar® and analysed by SAMOA were compared, one by one, with the same data as analysed by TASAS. In this way, the evaluation was performed for each element of interest separately. Another possibility would have been to perform the validation for the system's final interpretation of the polysomnographic test results. However, this approach would have masked errors in the apneic event detection, identification and classification tasks, in both TASAS and SAMOA (it being possible to obtain an equivalent diagnosis on the basis of different inputs).

The results obtained were divided into two groups: (a) identification of apneic events, and (b) classification of apneic events.

3.1 Identification of Apneic Events

Evaluated were apneic events identified as apneas or hypopneas. This identification was very much conditioned by the detection of possible apneic events by SAMOA; this conditioning also applied to classification.

The results obtained for the set of patients evaluated in the validation with respect to the identification of apneas and hypopneas were: agreement index 0.941, within-one agreement index 0.981, and kappa index 0.899. Table 1 shows the values obtained for the relationships defined by means of the agreement indices and analysed by taking the opinions of SAMOA as the standard reference.

3.2 Classification of Apneic Events

Validation of the data on the classification of the apneic events was performed for all events detected by SAMOA [9] and TASAS.

The results obtained for the set of patients evaluated in the validation with respect to the classification of obstructive, central and mixed apneas and hypopneas were: agreement index 0.887, within-one agreement index 0.966, and kappa index 0.827.

Table 2 shows the values obtained for the relationships defined by means of the agreement indices and analysed by taking the results of SAMOA as the standard reference.

Table 1. Performance measurements (S-Sensitivity, SP-Specificity, PPV-Positive Predictive Value, NPV-Negative Predictive Value, P-Precision, J-Jackard, θ-Area under Roc) for the identification of apneas and hypopneas (A-Apnea, H-Hypopnea, NE-No Event)

		S	SP	PPV	NPV	P	J	θ
TASAS	NE	0.835	0.987	0.996	0.934	0.942	0.811	0.911
	A	0.988	0.927	0.942	0.985	0.961	0.932	0.958
	H	0.979	0.980	0.900	0.996	0.980	0.883	0.980

Table 2. Performance measurements for the classification of apneas and hypopneas as obstructive, central or mixed (AO-Obstructive Apnea, AC-Central Apnea, AM-Mixed Apnea, HO-Obstructive Hypopnea, HC-Central Hypopnea, HM-Mixed Hypopnea, NE-No Event)

		S	E	PPV	NPV	P	J	θ
TASAS	NE	0.835	0.987	0.966	0.934	0.942	0.811	0.911
	AO	0.913	0.936	0.934	0.915	0.924	0.858	0.924
	AC	1.000	0.957	0.145	1.000	0.957	0.145	0.979
	AM	0.837	0.998	0.953	0.993	0.992	0.804	0.918
	HO	0.911	0.983	0.901	0.985	0.973	0.828	0.947
	HC	1.000	0.987	0.333	1.000	0.987	0.333	0.994
	HM	0.600	0.999	0.750	0.998	0.998	0.500	0.800

4 Conclusion

The inclusion of a temporal reasoning module in an SAS diagnostic system—according to the methodology described and using the CTCN model—has a number of advantages.

The inclusion of this kind of module enables a formalisation, in clinical equipment, of both objective causality (commonly acknowledged as public or semi-public knowledge) and subjective causality (heuristic or private knowledge). This is done by using a single schema both for representing the temporal relationships that arise between two events in a specific domain and for specifying temporal patterns that enable the inference of new knowledge. In addition, the imprecision associated with a specific event is reduced and imprecision conditions are established for inferring new knowledge.

Moreover, this approach enables apneic events to be identified according to the instants when they start and end, rather than according to the number of the epoch (the 30-second interval on which polysomnographic studies are based) marking the commencement of the event. This improves the perception of events, particularly when several events occur in a short period of time.

5 Future Work

The results obtained for the validation of the system indicate the need to improve certain aspects of the system and to incorporate new functions. These include:

a) parametrization of TASAS by several experts and post-parametrization results validation.

b) studies on the applicability of the CTCN model and of the working framework for handling temporal information to other knowledge domains.

c) studies on the handling and propagation of inexact knowledge through the inferential process of the developed working framework.

Acknowledgments. This work has been partially funded by Project XUGA PGIDIT05TIC0510501PR of the Xunta de Galicia (Spain).

References

1. Kautz, H., Ladkin, P.B.: Integrating metric and qualitative temporal reasoning. In: Proceedings of AAAI-91, Anaheim, CA, pp. 241–246 (1991)
2. Meiri, I.: Combining qualitative and quantitative constraints in temporal reasoning. Artificial Intelligence 87(1-2), 295–342 (1996)
3. Navarrete, I., Marín, R., Palma, J.: Redes de restricciones entre puntos y duraciones. In: Garijo, F.J., Riquelme, J.-C., Toro, M. (eds.) IBERAMIA 2002. LNCS (LNAI), vol. 2527, pp. 15–24. Springer, Heidelberg (2002)
4. Fernández-Leal, A.: Aspectos temporales de la representación de conocimiento en el síndrome de apneas del sueño. A dissertation for the requirements for the degree of Doctor of Philosophy in Information and Computer Science. University of A Coruña (2006)
5. Fernández-Leal, A., Moret-Bonillo, V.: A computable model for representing temporal knowledge and managing causality. MELECON, Electronic Proceedings. ISBN: 1-4244-0088-0. Library of Congress 2005938112. IEEE Catalog Number 06CH37756C (2006)
6. Fernańdez-Leal, A., Moret-Bonillo, V.: A framework for representing temporal knowledge in the intelligent diagnosis of the sleep apnea syndrome. IFMBE Proceedings, vol. 6, MEDICON 2004. ISSN: 1727-1983. ISBN: 88-7780-308-8 (2004)
7. Allen, J.: Towards a general theory of action and time. Artificial Intelligence 23(2), 123–154 (1984)
8. Fernández-Leal, A., Moret-Bonillo, V.: A framework for handling temporal knowledge in clinical diagnosis problems. ESBME 2006. Electronic Proceedings (2006)
9. Cabrero-Canosa, M., Hernández-Pereira, E., Moret-Bonillo, V.: Intelligent Diagnosis of Sleep Apnea Syndrome. IEEE Engineering in Medicine and Biology 23(2), 72–81 (2004)
10. Burgess, R.C.: Computerized polysomnographic analysis systems. Journal of Clinical Neurophysiology 7(1), 145–154 (1990)

Alternative Methods of Wave Motion Modelling

Lukasz Korus

Wroclaw University of Technology
Institute of Informatics, Automatics and Robotics
Wybrzeze Wyspianskiego 27, 50-370 Wroclaw, Poland
lukasz.korus@pwr.wroc.pl
http://www.pwr.wroc.pl

Abstract. The paper presents two optional methods for wave motion modelling. The first one is the classic Finite Difference Method (FDM). It is grounded on transforming Partial Differential Equation (PDE) to difference equation and resolving the latter on a discrete grid. The second method bases itself on the theory of Cellular Automata (CA), and more precisely on a model defined as Lattice Gas Automata (LGA). Modelling algorithms and simulation of simple wave motion using the two aforementioned methods are shown. The drawbacks and advantages of both techniques are discussed and their potential application to modelling the acoustic wave in urban agglomerations is outlined.

Keywords: Wave equation, finite difference method, cellular automata, lattice gas automata.

1 Introduction

Wave motion (wave) is a propagation of a disturbance in a certain medium. Numerous physical phenomena are of wavy nature. According to the medium wherein a wave motion takes place, we can discriminate acoustic waves, electromagnetic waves, matter waves, gravitational waves, etc. As the wave motion is a common phenomenon, of great importance was to develop relevant effective simulation methods. The Finite Difference Method (FDM) is the classic, well formulated and mature method used for this purpose. It consists in transformation of Partial Differential Equation (PDE) for the wave motion to a difference analogue and resolving it on a discrete, spatial grid. The paper presents a wave equation, a simple difference analogue and simulation. An alternative technique is based on the theory of Cellular Automata (CA) [1]. Macro-scale physical phenomena are not modelled in CA with a direct way. They result from interactions between numerous particles moving within a discrete grid. Hence, models of this type are referred to in the bibliography as Lattice Gas Automata (LGA) [1] [2] [3] [4]. The LGA technique assumes that sophisticated phenomena shall not be modelled by complex mathematical equations. Instead, simple rules for elementary components shall be created and the complexity shall be allowed to emerge due to interactions between those components. The model of wave motion based

B. Apolloni et al. (Eds.): KES 2007/WIRN 2007, Part I, LNAI 4692, pp. 335–345, 2007.

on modified LGA method is given further on in the body of the paper and the results of simulation based on this model are also provided. The properties of both methods and their potential applicability to modelling the acoustic wave propagation in conurbations were discussed.

2 Wave Equation

A differential equation of wave motion may by given by:

$$\frac{\partial^2 \xi}{\partial x^2} - \frac{1}{v^2}\frac{\partial^2 \xi}{\partial t^2} = 0, \tag{1}$$

where x is the position on the axis and v is the wave velocity. This equation is met by each twice-differentiable function of the form $\xi = f(x - vt)$ and also $\xi = f(x+vt)$. Thus, the formula (1) refers to a wave of any shape moving in both directions along the axis x. The wave equation (1) may be simply generalized to a two-dimensional form:

$$\frac{\partial^2 \xi}{\partial x^2} + \frac{\partial^2 \xi}{\partial y^2} - \frac{1}{v^2}\frac{\partial^2 \xi}{\partial t^2} = 0, \tag{2}$$

which may be written in a more compact form as:

$$\nabla^2 \xi - \frac{1}{v^2}\frac{\partial^2 \xi}{\partial t^2} = 0, \tag{3}$$

where ∇^2 stands for two-dimensional Laplace operator:

$$\nabla^2 \xi \equiv \frac{\partial^2 \xi}{\partial x^2} + \frac{\partial^2 \xi}{\partial y^2}. \tag{4}$$

3 Finite Difference Method (FDM)

3.1 Difference Analogues for Wave Equation

Using the notion of difference quotient, the difference analogues for individual factors of wave equation [5] [6] [7] will be as follows:

$$\left.\frac{\partial^2 \xi}{\partial x^2}\right|_{i,j} = \frac{\xi^n_{i+1,j} - 2\xi^n_{i,j} + \xi^n_{i-1,j}}{\triangle x^2}, \tag{5}$$

$$\left.\frac{\partial^2 \xi}{\partial y^2}\right|_{i,j} = \frac{\xi^n_{i,j+1} - 2\xi^n_{i,j} + \xi^n_{i,j-1}}{\triangle y^2}, \tag{6}$$

$$\left.\frac{\partial^2 \xi}{\partial t^2}\right|_{i,j} = \frac{\xi^{n+1}_{i,j} - 2\xi^n_{i,j} + \xi^{n-1}_{i,j}}{\triangle t^2}, \tag{7}$$

where: $\triangle x$ - elementary (atomic) distance between two nodes of Euler grid denoted as i and $i+1$ along the axis x, $\triangle y$ - elementary (atomic) distance between two nodes of Euler grid denoted as j and $j+1$ along the axis y, $\triangle t$ - elementary (atomic) distance between two time moments denoted as n and $n+1$. Entering the analogues (5) and (6) into (4) and assuming the square grid $\triangle x = \triangle y$, the following central difference analogue of second-order-of-accuracy was found for Laplace operator:

$$\nabla^2 \xi \,\Big|_{i,j} = \frac{1}{\triangle x^2}\left(\xi^n_{i+1,j} + \xi^n_{i-1,j} + \xi^n_{i,j+1} + \xi^n_{i,j-1} - 4\xi^n_{i,j}\right). \tag{8}$$

When analogues (7) and (8) are used in wave equation (3), and when the central term is placed on the left-hand side of the equation, the following difference analogue is reached:

$$\xi^{n+1}_{i,j} = v^2 \frac{\triangle t^2}{\triangle x^2}\left(\xi^n_{i+1,j} + \xi^n_{i-1,j} + \xi^n_{i,j+1} + \xi^n_{i,j-1} - 4\xi^n_{i,j}\right) + 2\xi^n_{i,j} - \xi^{n-1}_{i,j}. \tag{9}$$

This is the difference analogue of the wave equation given by (3).

The wave equation (3) may be also supplemented with a term causing wave amplitude to decrease along the distance from the source of disturbance. Such term is represented by friction force given by:

$$F_t = -b\frac{\partial \xi}{\partial t}, \tag{10}$$

where b the friction factor.

The wave equation supplemented with friction force becomes now:

$$\nabla^2 \xi - b\frac{\partial \xi}{\partial t} - \frac{1}{v^2}\frac{\partial^2 \xi}{\partial t^2} = 0. \tag{11}$$

The difference analogue for additional term will be as follows:

$$\frac{\partial \xi}{\partial t}\Bigg|_{i,j} = \frac{\xi^{n+1}_{i,j} - \xi^n_{i,j}}{\triangle t}. \tag{12}$$

Substituting the analogues (7), (8), and (12) into the wave equation including attenuation (11), and placing the central term on the left-hand side of the equation, the following difference analogue is attained:

$$\begin{aligned}\xi^{n+1}_{i,j} = {} & \frac{\triangle t^2}{\triangle x^2\left(b\triangle t + \frac{1}{v^2}\right)}\left(\xi^n_{i+1,j} + \xi^n_{i-1,j} + \xi^n_{i,j+1} + \xi^n_{i,j-1} - 4\xi^n_{i,j}\right) + \\ & + \frac{b\triangle t}{b\triangle t + \frac{1}{v^2}}\xi^n_{i,j} + \frac{2}{v^2 b\triangle t + 1}\xi^n_{i,j} + \frac{1}{v^2 b\triangle t + 1}\xi^{n-1}_{i,j}.\end{aligned} \tag{13}$$

3.2 Initial Conditions

In order to run simulation, initial conditions shall be set for the Euler grid. The form of these conditions is dependent on the phenomenon subjecting to modelling. For example, in case of acoustic wave propagation, the function ξ will be identical with the pressure p. The Gaussian function was taken as the initial conditions for the needs of simulations:

$$f_p(x, y) = 0.5e^{-0.5(x^2+y^2)}. \tag{14}$$

3.3 Boundary Conditions

As the simulation area will be limited and some obstacles may be present therein, boundary conditions need to be determined [5] [6]. One out of two standard boundary conditions may be assumed: fixed end or free end ones. The first ones consist in entering 0 in place of an element outside the area, so: $\xi^n_{i,j} = 0$ when $i = 0$ or $i = I + 1$ or $j = 0$ or $j = J + 1$, the processing area having been a square discrete grid dimensioned $i = 1 \dots I$ and $j = 1 \dots J$. Free end border conditions consist in introducing the previous or next element in place of the out-of-area element, so: $\xi^n_{i+1,j} = \xi^n_{i-1,j}$ when $i = I$ or $\xi^n_{i,j+1} = \xi^n_{i,j-1}$ when $j = J$ or $\xi^n_{i-1,j} = \xi^n_{i+1,j}$ when $i = 1$ or $\xi^n_{i,j-1} = \xi^n_{i,j+1}$ when $j = 1$.

3.4 Simulation in Obstacle-Free Space

Simulation was made on Euler grid dimensioned $i \in (-10, 10)$ and $j \in (-10, 10)$. The minimum distance between two grid points was assumed to be $\triangle x = \triangle y = 0.5$ The attenuation factor was fixed as $b = 2$, and the velocity as $v = 0.1$. The fixed end initial conditions were assumed. Simulation was run through 160 iterations. After each eighty iterations, the algorithm was stopped to find the current state of the system. Figure 1 shows the system states in successive time moments. The Gaussian function was used as the initial conditions (14).

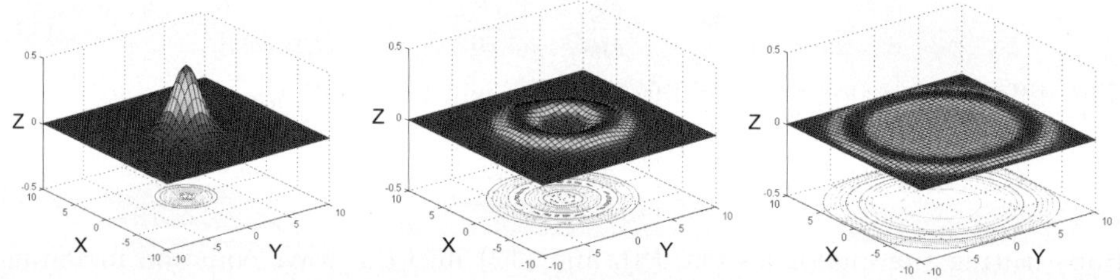

Fig. 1. Grid conditions at the start of simulation, after 80 and after 160 iterations, respectively

3.5 Simulation in Space Including an Obstacle

Simulation was run in similar conditions as previously. The difference depends on positioning an obstacle within simulation area and on modifying the boundary conditions to those of free end. Also the initial conditions were modified. The Gaussian function was moved with respect to the axis and its initial amplitude was increased:

$$f_p(x, y) = 0.8e^{-0.5((x+4)^2+y^2)}. \tag{15}$$

The attenuation factor was set to $b = 0.5$. The state of grid is shown at starting, after 120 and after 160 iterations, respectively (Figure 2).

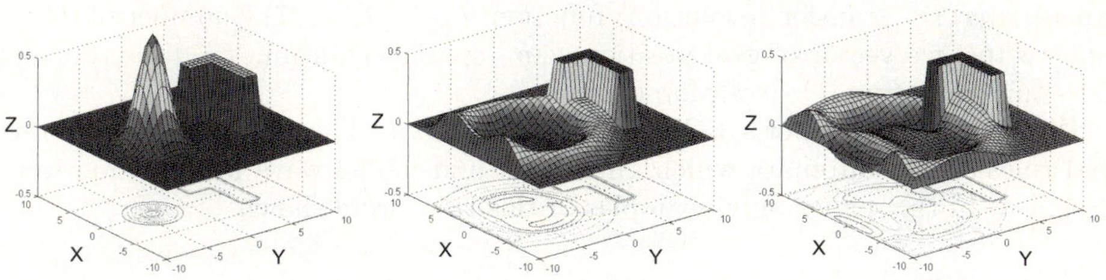

Fig. 2. Grid state at starting point, after 120 and after 160 iterations, respectively

4 Lattice Gas Automata

The Lattice Gas Automata (LGA) [2] [3] [4] are the Cellular Automata (CA) [1] wherein the notion of elementary particle wandering over a discrete grid was introduced. Cellular Automata represent a regular grid of identical components which take on discrete states out of the fixed set identical for the whole grid. The state of a given component is determined from the states of surrounding cells in the previous time moment. The components evolve to the rhythm of discrete time slices. In LGA model, the cell state determines the directions of particles exiting from a specific location. The transfer function of $n_i(\boldsymbol{r}, t)$ may be specified for the LGA [9]. It specifies the number of particles exiting the location $\boldsymbol{r}$ at time t with a specific velocity along the direction $\boldsymbol{c_i}$, where $i = 0, 1, \ldots, 7$ - determines eight directions in the space. This function can take two values only, 0 or 1. Let's assume that $\triangle t$ is the elementary time step, and $\triangle r$ is the elementary step in the space. Eight possible velocities in the space may be determined by the formula:

$$\boldsymbol{v_i} = \frac{\triangle r}{\triangle t}\boldsymbol{c_i}. \tag{16}$$

When no interaction between particles exists, the automat evolution equation becomes:

$$n_i(\boldsymbol{r} + \triangle r\boldsymbol{c_i}, t + \triangle t) = n_i(\boldsymbol{r}, t). \tag{17}$$

In this case, the particle leaving the position $\boldsymbol{r}$ moves further on in the direction $\boldsymbol{c_i}$, and in the next time step, it leaves the location $\boldsymbol{r} + \triangle r\boldsymbol{c_i}$ in the same direction. When a collision is introduced, particles may deviate from their primary directions to another ones [9]. If, for example, the variables n_i and n_{i+3} in a cell $\boldsymbol{r}$ are set to 1, there is a collision and the particle moving with the velocity $\boldsymbol{v_i}$ changes its velocity to $\boldsymbol{v_{i-1}}$ or $\boldsymbol{v_{i+1}}$. Operations on indexes i take place in the set $(0, 1, \ldots, 7)$. In order to describe the configuration wherein a collision takes place, collision factors were introduced. These parameters, determined for individual situations, are put together to create the resulting collision operator $\Omega_i(n_i(\boldsymbol{r}, t))$. The complete transfer function specifying the LGA dynamics may be given as:

$$n_i(\boldsymbol{r} + \triangle r\boldsymbol{c_i}, t + \triangle t) = n_i(\boldsymbol{r}, t) + \Omega_i(n_i(\boldsymbol{r}, t)), \tag{18}$$

where: $n_i(\boldsymbol{r}, t)$ - transfer (evolution) function, $i \in (0, 1, \ldots, 7)$ - number of direction vector, $\boldsymbol{r}$ - vector of grid position, $\triangle r$ - spatial quantum, t - discrete time, $\triangle t$ - time quantum, Ω - collision operator.

In this section author‘s LGA model was shown. The boundary conditions and rules of the automata which implement and model wave propagation were designed. In this paper only examples of this rules were shown.

4.1 Grid Parameters

A square grid was used in LGA system implementation. As eight directions are used in each cell, automat particles make an impression as if they move within triangle grid. Selecting such type of grid and the number of directions is a compromise between getting the highest possible degrees of freedom and the complexity of the model. The square grid (two-dimensional matrix) is a natural way of data saving in computer memory. In well-grounded cases, non-square grids are used; then, the location of each node is re-indexed respectively in computer memory. So, in the model being implemented, the square grid and eight variables in each automat cell related to eight vector senses, the particles are moving along, were used as shown in Figure 3.

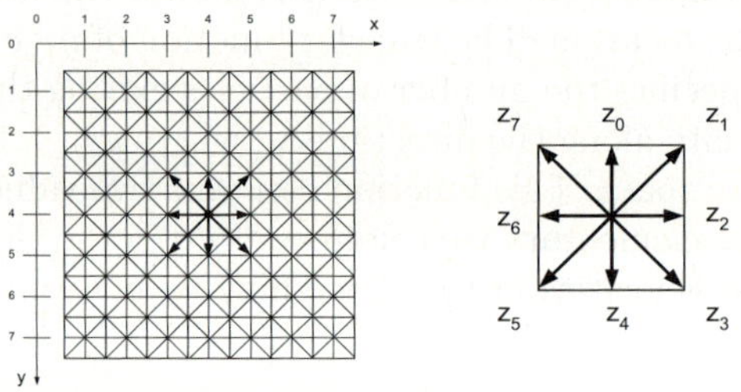

Fig. 3. Grid and directions of a single cell in LGA system.

Figure 3, in its right-hand side, shows eight vector senses which the particles may move along. Of importance is also the way of particle presentation on the grid in simulation program. In a specific time moment, all existing particles in each cell over the whole grid are displayed. A given direction vector, which represents the particle moving along it, is fixed in the position resulting from the indexes of computer table, whereas its end is located in the centre of surrounding component. In this way, an illusion of smooth particle movements within simulation space was attained.

4.2 Neighbourhood

A successive important parameter of LGA system is the cell neighbourhood used as a basis to calculate the cell value in the next time step [1]. The Moore neighbourhood was chosen in the implemented model. Such neighbourhood is a natural one for this case as it is in line with direction vector senses in each component of the automat. Figure 4 illustrates the Moore neighbourhood with notations used for LGA system.

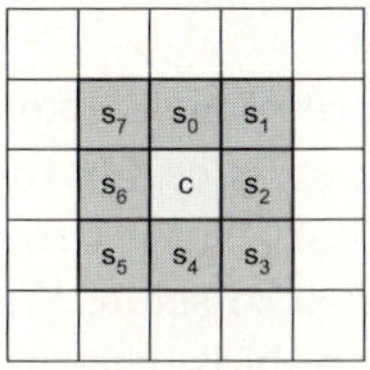

Fig. 4. Moore neighbourhood for LGA system.

Where: s_i - cell included in the neighbourhood, c - cell currently processed.

The directions in the cell under processing will be calculated in sequence according to fixed rules allowing for directions and senses of vectors moving in the neighbourhood of the particles.

4.3 Boundary Conditions

The purpose of creating the model is to simulate a wave motion. Thus, boundary conditions will be related to the rules describing particle interactions with room walls [8]. These rules must describe all possible cases which may exist on limited edges of the space over which the simulation is carried out. Figure 5 shows an example of event which may happen, states of cells in successive time moments and rules used to describe it. Operations of iterations for indexes run modulo 8 due to the number of direction vector senses. As the automat evolution is described using evolution equation, there is no need to specify the rules with result from rotation of the situation presented. Component's states for individual directions are calculated sequentially considering rotation of the configuration. Such method causes the number of rules which shall be specified to be much less than the number of rules actually affecting the particles.

Where: $D_i^k \in \{0, 1\}$ - k-th collision operator ($D_i^k = 1$ - collision in specific direction takes place), $w_i \in \{0, 1\}$ - wall ($w_i = 1$ wall in specific direction exists),

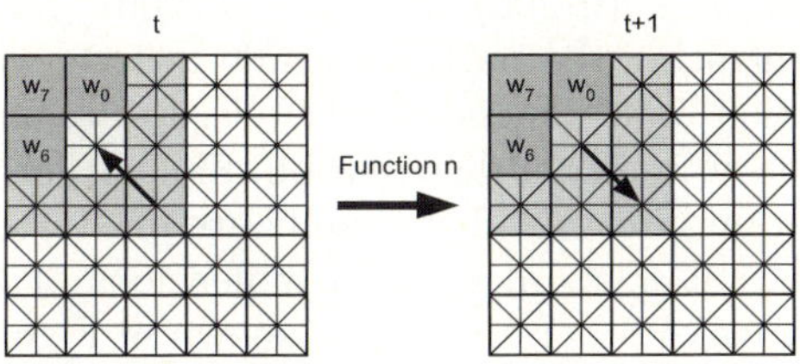

Fig. 5. Particle collision with wall corner. $D_i^1 = (w_{i+3}) \cap (s_{i,i+4}) \cap (w_{i+4}) \cap (w_{i+5})$.

$s_{i,j} \in \{0, 1\}$ - particle in position i of the neighbourhood is moving in direction j ($s_{i,j} = 1$ - respective particle exists in the grid), $i, j = (0, 1, \ldots, 7)$ - numbers of senses of direction vectors.

4.4 Rules of the Automata

The rules for particle collisions are defined like for the boundary conditions [8]. These operators are of great importance for the properties of the whole model (the system of large number of particles). Proper definitions of rules in micro-scale lead to desirable effects in macro-scale. While designing the operators, due considerations shall be given not only to the properties of the process itself, but also to the conditions in which it exists. A simple example may be the gravitation while designing automat to simulate sand spillage, or wind in matter (e.g. snow) transport model. Several rules for collisions of two or more particles were defined in the model designed.

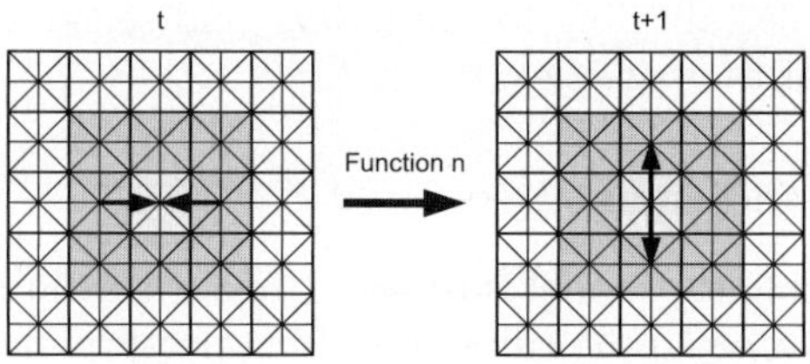

Fig. 6. Orthogonal collisions of two particles. $D_i^2 = (1-s_i)(1-s_{i+1})(s_{i+2})(1-s_{i+3})(1-s_{i+4})(1-s_{i+5})(s_{i+6})(1-s_{i+7})$.

Where: $s_i \in \{0, 1\}$ - particle in the i-th position of neighbourhood moving along the direction $[(i + 4) mod 8]$.

Figure 6 shows an exemplary rule and collision operator which defines this rule. Defined boundary and collision operators are summed up and substituted into the formula (18).

4.5 Simulation

Two simulation runs were carried out. The first case refers to simulation for the grid dimensioned: $x \in (0, 200)$ and $y \in (0, 200)$. Initial conditions were established by concentrating the particles inside a circle with radius $r = 10$ in the centre of simulation area. Outside the circle, the remaining particles were sampled at random from uniform distribution. Simulation was run through 70 iterations at 10 iteration intervals. The results are illustrated in Figure 7.

The second simulation was run for grid dimensions of $x \in (0, 70)$ and $y \in (0, 70)$. Thirty iterations of program run are shown in Figure 8.

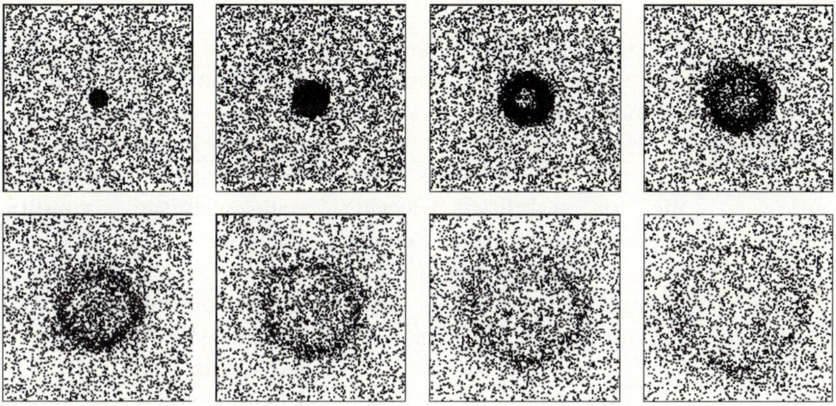

Fig. 7. Grid states from 0 to 70 iterations at 10 iteration interval

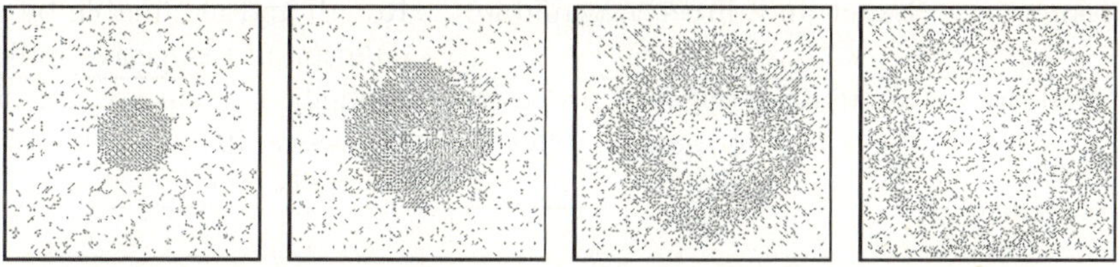

Fig. 8. Grid states from 0 to 30 iterations at 10 iteration interval

5 Comparasion of Methods

The FDM and LGA methods are created on different ideas. Each one features with its drawbacks and advantages. The FDM has longer history, hence it is better developed and used more frequently. The LGA method also subjects lately to intense development. New algorithms, expansions of this method, like Lattice Boltzman Methods (LBM) [9] have been appeared lately. It seems however that application areas of these two methods coincide to a low degree only. The FDMs have their well established position in modelling Partial Differential Equations used to describe physical phenomena (fluid mechanics, electromagnetism, etc.). The LGA methods are applicable wherever mathematical description of system behaviour is not known, but the rules of interaction between small components (particles) can be selected intuitively. Phenomena observed in macro-scale are the result of interactions between components in micro-scale. Cellular Automata are also the more efficient technique, especially for dedicated equipment (parallel calculations) [10]. For this reason, LGA and LBM are used to run simulations and examination of fluid mechanics especially for turbulent range [3] [9]. Basic properties of both methods are summarized in the table below:

	FDM	LGA
Idea	Solution of partial differential equations by determining difference analogues.	LGA are able to show collective behaviours. Simple rules in micro-scale - complex behaviours in macro-scale.
Model	Creating the analogues. Testing for stability. Selecting the parameters. Initial conditions. Boundary conditions.	Designing the rules in micro-scale. Initial conditions. Boundary conditions.
Simulation	Complex equations - time-consuming calculations.	Simple operations on tables. Dedicated equipment (parallel calculations).
Problem	It is necessary to know mathematical equations.	Phenomenon to be modelled may by known cursorily only. Rules may be selected empirically.
Solution	Exact and directly available.	Rough and attainable by calculation of respective indices.
Applications	Simulating physical processes described by partial differential equations.	Modelling complex physical processes in case the other methods failed.

6 Summary

The paper attempts to make a comparison between two alternative methods and to select proper technique to resolve a specific problem. The task stated was to simulate the propagation of acoustic wave in urban agglomerations. Obviously, the classic Finite Difference Method meets all requirements related to resolving this problem. The trials with Lattice Gas Automata were aimed at verification whether a simple square lattice and several rules allow modelling spherical wave motion for large number of particles. It was found that simple model of this type features very interesting and desirable properties. This model may be developed by introducing an additional dimension so as to represent 3D reality. It also seems reasonable to work on further improving the rules, especially those related to various boundary conditions.

References

1. Stephen, W.: New Kind of Science. Wolfram Media Inc (2002)
2. Hardy, J., Pazzis, O., Pomeau, Y.: Molecular Dynamics of a classical lattice gas: Transport properites and time correlation functions. Physical Review A 13(5), 1949 (1976)
3. Frish, U., Hasslacher, B., Pomeau, Y.: Lattice-Gas Automata for the Navier-Stokes Equation. Physical Review Letters 56(14), 1505 (1986)
4. Weimar, J.: Simulation with Cellular Automata. Los Verlag, Berlin (1997)

5. Ryszard, G.: Fundaments of Fluid Mechanics, PWN, Warszawa (1998)
6. Morawski, T., Gwarek, W.: Electromagnetic Field and Waves, WNT, Warszawa (1978-1998)
7. Lukasz, K.: Simulate Wave Equation with Finite Difference Method, Report, Wroclaw University of Technology, Wroclaw (2006)
8. Lukasz, K.: Simulate Wave Equation with Lattice Gas Automata, Report, Wroclaw University of Technology, Wroclaw (2006)
9. Chopard, B., Luthi, P., Masselot, A., Dupuis, A.: Cellular Automata and Lattice Boltzmann Techniques: An Approach to Model and Simulate Complex Systems. Advances in Complex System 5(2) (2002)
10. Margolus, N.: CAM-8: a computer architecture based on cellular automata, MIT Laboratory For Computer Science, Cambridge Massachusetts (1993)

Conceptual Enrichment of Locations Pointed Out by the User

Ana Alves[1,2], Raquel Hervás[2], Francisco C. Pereira[3], Pablo Gervás[2], and Carlos Bento[3]

[1] Instituto Superior de Engenharia de Coimbra, Portugal
aalves@isec.pt
[2] Universidad Complutense de Madrid, Spain
raquelhb@fdi.ucm.es, pgervas@sip.ucm.es
[3] CISUC, Department of Informatics Engineering, University of Coimbra, Portugal
{ana,camara,bento}@dei.uc.pt

Abstract. The growing availability of GPS and localization technologies opens up new challenges regarding the use of position information for Location Based Services. As a side effect, several research questions arise, one of which we consider in this paper: "What is the meaning of place?". Here, we present two ongoing projects that we believe can contribute to the answer. The first regards the semantic enrichment of place, starting from a simple description of the place towards a concept network; the second focuses the natural language description of this network, as a final output to the user. We present these two projects, focus on their intersection and give an example of the end results.

Keywords: Position-to-Place, Location Modeling, Information Extraction, WordNet, Ontologies, Semantic Network, Text Generation.

1 Introduction

A major question regarding places it is how we describe them. Considering as an example the British Library we have: "Position (LatitudeX, LongitudeY)", "Building at Euston Road, 96", "British Imperial Library", "A place for reading", "A library", etc. Clearly, given only the exact global position of a place (LatitudeX, LongitudeX), we do not necessarily know something of it or understand its meaning as a place. Humans do not think in longitude or latitude terms. Furthermore, the functional purpose or category of a place may be easily recognized by humans, but it can not be used by an automatic reasoning system without applying external sources of knowledge. This problem, nicely presented by Hightower [1], is being tackled by many (see Related Work - section 2).

Nowadays, freely available commonsense Ontologies, like WordNet [2], are helpful tools to deduce semantic meaning of several concepts, including places. This can lead to infer other concepts related to a place category through semantic relations between them. Normally, these concepts and their semantic relationships are attached to a generic perspective, thus not representing any

B. Apolloni et al. (Eds.): KES 2007/WIRN 2007, Part I, LNAI 4692, pp. 346–353, 2007.

instance of them. For example, the concept of "library" can be generically described as "a building that houses a collection of books and other materials" (in WordNet [2]), but if we talk about a specific library (e.g. U.S. National Library of Medicine), further exploration is needed to grab a more precise meaning of that place, which is commonly not possible using commonsense Ontologies.

On the other hand, the World Wide Web (Web) contains an enormous amount of information with great redundancy. Following its growth, many Information Retrieval (IR) algorithms has been developed to perform efficient search and to index massive Web databases. Until now, the vast majority of this information is represented as texts and lacks structure and semantics. Taking advantage of these facts, we propose an approach to our first question considering work well known in Artificial Intelligence (AI) areas: Geographical IR, Natural Language Processing (NLP) and Ontologies. Geographical IR algorithms and NLP (see Related Work - section 2) can help us to find, to extract relevant knowledge and to categorize specific information about a place name and geographical reference, mostly present in user written text on the Web; while Ontologies will allow the assessment of relatedness of new concepts given a place category.

All these techniques can provide a good conceptual framework, susceptible of being processed automatically in the context of particular applications, related to a given location. However, the very same characteristics that make such material machine-tractable tend to make it obscure to understand by everyday users. A traveller looking up a particular restaurant on his PDA while wondering along the street does not necessarily want an XML file that encodes all available information about it. Instead, he would probably be much happier if he could obtain a simple textual description of the most salient facts about it. If available information about his past preferences or experiences were available to the system, and it could be in some way applied to customise such a textual description appropriately, the user experience would be much better. This is where Natural Language Generation techniques can play a significant role in extracting the most benefit from the wealth of information provided by enrichment methods.

The present paper describes a two stage approach to the practical treatment of locations in service-oriented applications. During the first stage, available locations are enriched by resorting to various techniques. During the second stage, textual descriptions of the locations are automatically generated from the resulting enriched conceptual representation of the location.

2 Related Work

An inspiring paper [1] by Jeffrey Hightower argues that location must have more associated information than simply the absolute position in a global coordinate system. We may split this problem into two perspectives: **identify which places are "meaningful"**, [3,4] which groups of GPS readings are build by a clustering algorithm and then attach its user-defined meaning; and **focus on a set of "meaningful locations" constructed by a community** [5,6] and decide, for each context, which correct meaning to use. In the last case, which we are particulary more interested, the most common type of representation of these locations is by a tuple (LatitudeX, LongitudeX, NameX) also known as Point of Interest (POI). In such environments, ambiguitous names can be given to the same place (e.g vegetarian restaurant or Green Menu), or different locations may share an identical place name (e.g. "vito's pizza" in Coimbra or Dublin).

Common Sense Ontologies (such as WordNet, OpenCyc [7] and others) are collections of trivial and semantic knowledge that allow the extension of the computational reasoning process. Their generic concepts and relationships that we specifically find for a known category of places (restaurant, museum, hospital, cinema, pharmacy, etc.) can be used to build a *CommonSense Place Ontology* comprising not only semantic related concepts to a given category but all concepts referred by descriptive definitions (or glossaries). This ontology will be discussed in more detail in section 3. With the growth of the World Wide Web, some researchers have already considered the convergence of web-related technologies and ubiquitous computing. A good example of this was the CoolTown project [8]. Its main goal was to support web presence for people, places and things. They used Universal Resource Identifiers (URIs) for addressing, physical URI beaconing and sensing of URIs for discovery, and localized web servers for directories in order to create a location-aware ubiquitous system to support nomadic users. In the same line, Ubiquitous Web [9] was envisioned as a pervasive web infrastructure in which all physical objects are socially tagged and accessible by URIs, providing information and services that enrich users experiences in their physical context as the web does in the cyberspace.

It is widely accepted that the majority of on-line information is composed of unrestricted user-written texts. Information Extraction (IE) is a research subtopic in IR devoted to extract useful information from a body of text. Within this field, Term Extraction and Name Entity Recognition are Natural Language Processing (NLP) tasks. The general process of text generation takes place in several stages, during which the conceptual input is progressively refined by adding information that will shape the final text [10]. During the initial stages the concepts and messages that will appear in the final content are decided (*content determination*) and these messages are organised into a specific order and structure (*discourse planning*), and particular ways of describing each concept where it appears in the discourse plan are selected (*referring expression generation*). This results in a version of the discourse plan where the contents, the structure of the discourse, and the level of detail of each concept are already fixed. The *lexicalization* stage that follows decides which specific words and phrases should be chosen to express the domain concepts and relations which appear in the messages. A final stage of *surface realization* assembles all the relevant pieces into linguistically and typographically correct text.

3 Semantic Enrichment of Place

The problem of "position to place" is a well known challenge within the area of Ubiquitous Computing and relates deeply with the connection humans have with places, their functionality and meaning. Attached to a tag name, even when a category is included, a place needs a richer semantic representation in our perspective. This knowledge can be used for whatever processes that demand semantics of place (e.g. understanding POIs while in navigation; searching for a place that has specific characteristics; route planning using locations with specific functionalities; inferring users activity, etc.). We name this process as Semantic Enrichment of Place and it consists of using available Common Sense Ontologies and Web information to build a collection of generic and instance facts about these places.

3.1 Generic Place Model Ontology

It represents a collection of commonsense and generic information about well-known place categories, like restaurants, cinemas, museums, hotels, hospitals, etc. At a first stage, this information is manually extracted from WordNet by Semantic Augmentation (see subsection 3.2 for more detailed description), but as the system is used, it is dynamically fed by new examples, and thus enriched by specific extracted information of these instances. Information extracted from WordNet has a different weight in instance classification than the information that comes from the Web, since the former is more generic and applied to all individuals pertaining to the same place category.

3.2 Categorization of New Places

Given collected POI from Web databases, we conceive a semantic enrichment system for places comprising by the following three sequential tasks:

1. *Geographical Web Search* is responsible for finding Web pages using only POI data as keywords: place name and geographical address. This search is presently made by the freely available Yahoo API. As we described in Related Work section 2, Geo-Annotated Web Pages are not available yet and are not the central goal of this work. For the moment, we are using a simple heuristic that uses the geographical reference as another keyword in the search. We think that works like Geotumba[11] will contribute to achive better precision for this process. At the end of this module the N more relevant pages are selected.
2. *Concept Extraction* from Web pages found earlier is made at this point. This processing includes POS tagging and word sense disambiguation using available NLP tools. On completion of these two tasks for each web page, we are able to extract the most relevant terms that will be used in classification task. Since WordNet does not cover widely proper nouns yet, we are only considering common nouns as valid concepts. In our work, we attempt to propose a different approach to Location Geo-Recognition where we are not only interested in classifying a place name, but also capture the meaning of it (represented by its related concepts).

3. *Place Classification* will be obtained from the semantic distance between the concepts extracted before and Place Models, computing their semantic similarity measure [12]. This approach is not new [13] and we try to enhance its performance by doing off-line pre-processing of the related concept semantic similarities to perform faster categorization of new examples. We'll use the Generic Place Model Ontology to classify new examples. Thus, on one hand we have different place types and, on the other hand, we have distinct concept lists originated from the respective web pages. We believe that at this point we can conclude with some degree of confidence that the most scored category will eliminate ambiguity and noise from Geographical Web Search. Naturally, some problems will arise, like *to distinguish weighting between generic and instance concepts in the categorization task*, or *to have multiple possible categories to classify a new example.* We believe some well-known Machine Learning (ML) strategies can tackle to these questions.

4 Generating Textual Descriptions from the Enriched Conceptual Representation

The input for this stage consists in a enriched conceptual description of the POI to be described. This description is represented as a list of facts, where each facts contains one or two concepts and a relation between them. These relations are defined during the enrichment process of the POI, and they extend from the WordNet sense assigned to a concept to the relation between a concept and one of its hypernyms.

Table 1. Example of Input and Generated Text

WN_sense(non-smoking, smokeless.1). *WN_does_not_contain(range_of_wines).* *WN_sense(coffee, coffee.1).* *WN_direct_hype(beverage.1,hot_chocolate.1).* *WN_direct_hype(beer.1,draught_beer.1).* *WN_direct_hype(beverage.1,beer.1).*	*"Boddington Arms serves coffees such as cappuccino and espresso, beers such as draught beer and Guinness, and tea and hot chocolate.* *It admits credit card and debit card.* *[...]* *More information: Range of wines, Wheelchair access,..."*

As an example, a piece of the conceptual description of the POI corresponding to the pub Boddington Arms (coming from the previous stage, described in section 3.2) is shown in Table 1.

The process carried out over this representation follows the basic steps in NLG explained in Section 2. The generation of text is a knowledge-intensive process. Not only the information obtained from the enrichment is used, but also additional knowledge about the concepts that may be required for the generator at any stage of the process. In addition, a vocabulary for the concepts in the domain is necessary to express into words the description of the POI.

Content Determination. From the whole conceptual information available for the POI only a part corresponds to the specific data that describes it: the initial words obtained for the POI before the enrichment. The rest is additional information obtained during the enrichment, from synonyms to related concepts, also necessary during the generation process but not specifically descriptive for the POI. This includes words that are found in WordNet and contribute to the enrichment - which appear in a *WN_sense* relation assigned to the corresponding WordNet sense -, but also words not useful during the enrichment because there is no information related to them that can be used - they appear in a *WN_does_not_contain* relation. They are not completely discarded, but attached at the end of the text as extra information. The results of this filtering process can be aggregated according to conceptual criteria. Using relations as hypernymy concepts are organised into small taxonomies of related concepts. For instance, in the example above, a taxonomy of "beverages" is involved, with "coffee" and "beer" as subclasses, and "draught_beer" as subclass of "beer".

Discourse Planning. In order to determine the structure of the discourse needed to convey this information, we have considered that related concepts must be explained together in the text. As a first approximation, the discourse consists of a list of descriptions of sets of concepts, without relations between them. In the example, a possible plan for the discourse would be to describe the menu first, then the drinks, after that the accommodation facilities, and finally the payment options. Each of them would be a small taxonomy of related concepts. This basic solution could be refined by finding relations between the taxonomies of concepts in such a way that each one can introduce the next one. For the moment we have not explored this option.

Referring Expression Generation. In the textual description of the POI, various elements would be mentioned one or more times: beverage, Boddington Arms, restaurant,... Different natural language expressions can be used to describe them at each occurrence in the text: "a drink", "the beverage", "it",... The referring expression generation stage decides at each occurrence how to refer to a given element, ensuring that enough information is given to be able to uniquely identify the intended referent in the current context, while avoiding redundant or otherwise unnecessary modifiers. For the moment only the name of the concepts and the corresponding pronouns have been used.

Lexicalization. In this stage the structure of the text is already determined by the Discourse Planning, and the form of the concepts by the Referring Expression Generation. The words to use for each concept or relation are chosen between the different options stored in the vocabulary. For each entry the vocabulary may contain more than one possible word, so some heuristic is needed to choose between different options. For instance, the concept "non-smoking" could be expressed using "smokeless" or "smoke-free". In this approximation there is only one word assigned to each concept or relation stored in the vocabulary.

Surface Realization. Surface realization is a task very closely related to linguistic constraints, and reasonable remote from the specific restrictions of particular domains. For this reason, there are several available implementations for reusable modules that handle this task. We have developed a simple implementation that receives as input an abstraction corresponding to the content that needs to be realised as a sentence and unifies it with a grammar of the language in which the output is desired.

5 Conclusion

In this paper, we focussed on the question "What is the meaning of place?", which is becoming salient in many works on Location Based Services (LBS). We followed an approach that has some tradition in symbolic AI - to derive the "meaning" from a network of interrelated concepts. In this case, we showed an approach that is being developed to automatically build the network of concepts (that define a place), and finally described a method for generating a textual interpretation for this network.

The conceptual enrichment of an initial fragment of information can be greatly improved if the final goal for which the resulting information is intended is taken into account. The development described in this paper considers conceptual enrichment for two different tasks: information management - extending available POIs with enough information for the system to be capable of recognising their nature and improving the services provided to the user based on this knowledge; language generation - constructing a textual description of the POI that is at the same time useful and easy to read for the user. The results of the present work indicate that the enrichment required for enabling machine-processing of the POI - searching for associations with related concepts, as in query expansion during information retrieval - is not necessarily a positive contribution to the task of generating a textual description of the POI. In practical terms, it may be useful for the system to know that a POI corresponding to a restaurant has associations with concepts such as waiters, kitchens, bills,...; but to include such elementary information about the concept of restaurant in a textual description of the POI provided to a user would be redundant. In that case, the only information worth transmitting is whatever is exclusive to this particular POI in contrast with other POIs, and this should not include generic truths about restaurants (for instance).

On the other hand, whereas for simple identification of the nature of a POI considering a single level of hypernymy within the WordNet hierarchy might be enough - and in fact adding more might cloud the enriched result with too much clutter, when addressing tasks of discourse planning it may be more interesting to consider the distance to the closest common ancestor, sidestepping most of the actual information in the concepts found on the way.

The solution presented here, by operating in two separate stages, provides a good balance between enrichment for information management and enrichment for language generation. The language generation stage has been designed to perform an initial filtering to identify those items of information that are indeed most relevant as identifiers of a particular POI. To compensate for the possible loss of information resulting from this operation, this module has access to additional knowledge bases, either in the form of WordNet itself or in the form of OWL encoded ontologies, that it can search to find the information it requires to carry out its task in an informed manner. This procedure guarantees both the existence of a suitable enriched representation of POIs for internal treatment by the information system, and acceptable text outputs as descriptions of the POIs that may be both informative and easy to read by the user.

Acknowledgements

This work is partially supported by the Spanish Ministry of Education and Science project TIN2006-14433-C02-01.

References

1. Hightower, J.: From position to place. In: Proc. of the Workshop on Location-Aware Computing (part of the Ubiquitous Computing Conference), pp. 10–12 (2003)
2. Miller, G., Beckwith, R., Fellbaum, C., Gross, D., Miller, K.: Introduction to wordnet: An on-line lexical database. Int J Lexicography 3, 235–244 (1990)
3. Ashbrook, D., Starner, T.: Learning significant locations and predicting user movement with gps. In: Proc. of 6 IEEE Intl. Symp. on Wearable Computers (2002)
4. Zhou, C., Ludford, P., Frankowski, D., Terveen, L.: An experiment in discovering personally meaningful places from location data. In: extended abstracts on Human factors in Computing Systems, pp. 2029–2032. ACM Press, New York (2005)
5. Alves, A., Figueiredo, N., Gomes, H., Pereira, F., Marcelino, M., Bento, C.: The ejaki project: a quality of service regulator for citizens. In: Proc. of the Intl. Conference on Portable Information Devices (2007)
6. Tagzania: Last Acess on (June 15, 2007), `http://www.tagzania.com/`
7. Opencyc: Last Acess on (June 15, 2007), `http://www.opencyc.org`
8. Kindberg, T., Barton, J., Morgan, J.e.a.: People, places, things: Web presence for the real world. In: Proc. of WMCSA2000 (2000)
9. Vazquez, J., J., A., D.L., I.: The ubiquitous web as a model to lead our environments to their full potential. In: W3C Workshop on the Ubiquitous Web. Position paper (2006)
10. Reiter, E., Dale, R.: Building Natural Language Generation Systems. Cambridge University Press, Cambridge (2000)
11. Cardoso, N., Martins, B., Andrade, L., Silva, M.J.: Geotumba. Technical report, The XLDB Group at GeoCLEF (2005)
12. Seco, N., Veale, T., Hayes, J.: An intrinsic information content metric for semantic similarity in wordnet. In: ECAI, pp. 1089–1090 (2004)
13. Corley, C., Mihalcea, R.: Measuring the semantic similarity of texts. In: Proc. of the ACL Workshop on Empirical Modeling of Semantic Equivalence and Entailment, Association for Computational Linguistics, pp. 13–18 (2005)

Design of Urban Growth Probability Model by Using Spatial Association Rules

Seonghwi Cho[1], Sungeon Hong[2], Jungyeop Kim[3], and Soohong Park[3]

[1] IMU Korea Ltd, Seoul, Korea
imcho1@nate.com
[2] Dept. of Welfare·Land Information, Cheongju University, Cheongju, Korea
hongsu2005@cju.ac.kr
[3] Dept. of Geoinformatic Engineering, Inha University, Incheon, Korea
jyfloo@gmail.com, shpark@inha.ac.kr

Abstract. AOI(Attribute-Oriented Induction) is one of the data mining techniques with which phenomena of complicated spatial relationships can be expressed as specific rules by summarizing spatial or non-spatial data in accordance with super ordinate concepts. In addition to such AOI technique, if GIS(Geographic Information System) that has an advantage to solve space-related problems is combined, they can be used to expect the issues of urban growth. Combining the AOI technique and GIS, the study draws out spatial association rules focusing on a physical urban growth, and those rules are applied to urban growth models during the period from the 1960's to the 1990's. The results and analysis of the urban growth modes combined with data mining are compared with those of Clarke Keith's UGM(Urban Growth Model).

1 Introduction

Among many space-related problems, there have been many researches on analysis and expectations of an increasing population and its influences including enlarged urban in various kinds of fields such as human and social studies and engineering(Kang and Park, 2000; Jung, 2001). Especially, studies through physical factors affecting urban growth or geographical information have been done, but most of them are followed by a method to establish factors in accordance with specifically pre-defined patterns. This method can be general-purpose, but it has a limitation to reflect specific features of a concerned area and its models are very complicated.

The study is designed to find out spatial association rules of features hidden in a specific space by using GIS's data saving, management, and analysis functions. GIS's excellent spatial analysis function and its simultaneous processing function of spatial and information of attributes can be good devices to extract spatial association rules. Through this the study extracts out spatial association rules which are the core of urban growth probability models and verifies those rules by realizing the models. Also, their efficiency is examined by comparing and analyzing the UGM results.

The selected research area was Suwon city, Gyeonggi-do(121,093,951.3 m^2, as of January 1, 2003) among big cities located in the metropolitan area and by using various kinds of geographical information built for about 30 years from the 1960's to

B. Apolloni et al. (Eds.): KES 2007/WIRN 2007, Part I, LNAI 4692, pp. 354–362, 2007.

the 1990's spatial association rules were drawn out and applied to urban growth probability models.

The extraction of spatial association rules was performed through two approaching methods. First, by using an analyzing function provided by the GIS software pre-processing procedure was done, and then again a final range of rules were extracted by the data mining method. Those extracted rules were written in algorithms applicable to the models.

Urban growth probability models were embodied by utilizing the cellular automata systems (CAS) as a simulator based on cellular automata (CA). CAS provides a cellular language, which is called Cellang, and helps reflecting written algorithms (Park, 1997). The simulated results by using the models were compared and analyzed with those of Clarke Keith's UGM.

2 Building Experimental Data

With regard to the researched area, useful information including topographic maps on various reduced scales, thematic maps, and digital topographic maps were collected and analyzed for their attributes. The consistency in various forms of spatial data collected on the studied area during the concerned period (approximately more than 30 years) was maintained and GIS database was built in a serial form. As for inputting topical spatial data, passive digitizing, scanning, and vectorizing methods were used side by side for topographic maps printed on paper and thematic maps, and conversing process was used for digital topographic maps and other thematic maps.

The main contents of the built GIS database are as shown in Table 1 and the researched periods are the years of 1960, 1970, 1980, and 1990, four periods in total.

Table 1. Experimental GIS Database

	Main data	Sub-data	Time
Urban	Topographical map	Land use map	1960-1990
Road	Topographical map	Road network map,	1960-1990
Exclusive region	Topographical map	Channel network map	1960-1990
	Greenbelt	Topographical map (Greenbelt)	1970-1990
Slope	Numerical map	Numerical data	1990

At the initial stage of building database there were more items for data than now, but they have been integrated in conformation with purposes and methods of the study.

Agriculture and forestry areas and other areas were combined as non-urban areas and as for traffic networks two-lane and four-lane roads affecting substantial urbanization were used, while railroads and express ways were excluded. In regard of channel networks polygon types were considered only and line-type data were excluded because they were not in a cell range of 100m×100m. The data on development-limited districts were built from those of the 1970's, as these regions were set to prevent big cities from proliferating unrestrictedly from the 1970's.

Gradient maps were used by extracting contours and layers only from digital topographic maps, interpolating them in TIN(Triangulated Irregular Network), and then transforming into DEM(Digital Elevation Model) or a percent form. The format of the built database was in mdb, and then transformed as shape → grid → ASCII data, because the ASCII format is easy to approach from other applications or apply to spatial operations by piling up each layer.

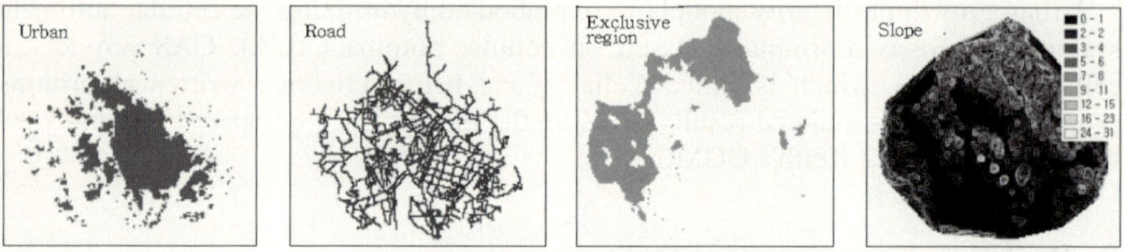

Fig. 1. Suwon data - 1990s

3 Extraction Procedure of Spatial Association Rules

3.1 Pre-processing Procedure for Extracting Rules

To find rules related to urban growth, newly established urban cells for each period were extracted by piling up layers of urban over the four periods on the preferential basis. As the new urban cells for each period belonged to non-urban areas during the previous period, relationships with surrounding areas were analyzed by piling up urban and road layers of a concerned year and its previous year.

Non-urban areas are transformed to cities due to the influence of existing urban and road cells around the areas. Generally in CA, surrounding cells affecting a concerned cell are called neighborhood, and Moore's concept of neighborhood is conceptually very similar to mask or window of spatial filters used in GIS(Park, 1997).

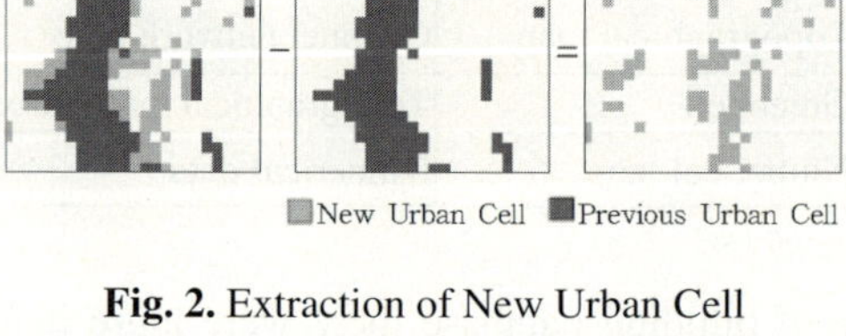

Fig. 2. Extraction of New Urban Cell

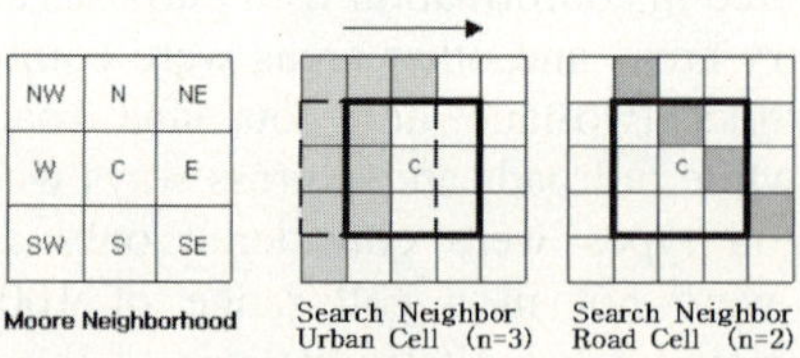

Fig. 3. Neighborhood Analysis using Moving Window

In many cases, irregular and various types of boundary establishment are required rather than regular establishment of surrounding areas, but in this study the same type and size of window was applied in order to extract a consistent pattern.

Analyzing surrounding urban and roads results in creation of each layer and by piling up excluded regions and gradient layers prospective cells for developing as a urban are chosen. Also by means of integrating selected cells for each period into one cell, time and spatial data on concerned areas are generalized as a group or a table form.

3.2 Extracting Spatial Association Rules by Using Data Mining Method

One of the important functions of GIS is to recognize a new pattern (Kim, 2000). As GIS has its information of attributes as relational database, it can be applied with attribute oriented induction (AOI), which is one of the data mining techniques. As a bottom-up summarizing method among information classifying methods, AOI repeats removing unnecessary characteristics until each attribute is generalized to a desired level and integrating tuple, and summarizes attributes or rules in a series of tuple that the information has.

Because AOI summarizes spatial or non-spatial data in accordance with generalized super ordinate concepts, urban phenomena with complicated spatial relationships can be compressed as a few rules. The process of information compression inevitably results in loss of information, but this can be a relatively efficient method given the fact that it is difficult in reality to build a model satisfying all the requirements and collect an enormous amount of information for such purpose.

Initially pre-process procedure was implemented at the concerned area and prospect cells for urban growth or spatial information predicating spatial association rules related to urban growth is classified. The study focused on physical urban growth, considering adjacent urban, roads, excluded areas, and gradients as input data.

Each input data is subject to generalization as an attribute. Gradients are divided as a limited condition before being applied with AOI. First of all, as the analysis on statistics of gradients shows that most values are in the range of standard deviation, 3σ(about 99.74%), the limited condition is decided under 11.

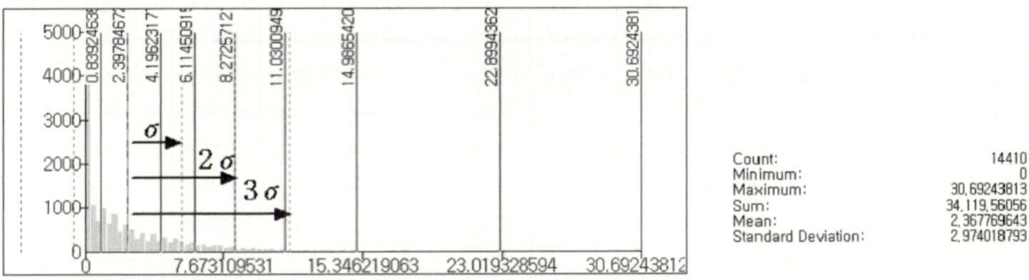

Fig. 4. Statistics of Slope

With regard to application of AOI, concerned attributes for generalization are defined as the numbers of surrounding urban and road cells and excluded areas. If excluded areas are chosen as attributes for removal and then tuple are removed from

Table 2. Support&Confidence of the Ur/Ro(Urban cells/Road Cells)

Ur/Ro	0	1	2	3	4	5	6	7	8
0	0.12456	0.01617	0.01806	0.03977	0.02709	0.01049	0.00451	0.00189	0.00160
1	0.05812	0.00772	0.00670	0.02185	0.01864	0.00582	0.00422	0.00218	0.00014
2	0.04924	0.00728	0.00932	0.01544	0.01588	0.00641	0.00451	0.00131	0.00087
3	0.03452	0.00553	0.00743	0.01398	0.01223	0.00713	0.00349	0.00291	0.00102
4	0.01879	0.00466	0.00582	0.01165	0.01311	0.00582	0.00539	0.00247	0.00145
5	0.01427	0.00378	0.00437	0.00859	0.01034	0.00757	0.00539	0.00291	0.00189
6	0.00859	0.00320	0.00204	0.00509	0.00713	0.00539	0.00495	0.00320	0.00204
7	0.00699	0.00218	0.00174	0.00670	0.00728	0.00597	0.00786	0.00364	0.00131
8	0.02695	0.01573	0.01646	0.03904	0.03991	0.03278	0.02811	0.01267	0.00641

Support = Each Association/All Association

Ur/Ro	0	1	2	3	4	5	6	7	8
0	0.51014	0.06622	0.07398	0.16288	0.11097	0.04295	0.01849	0.00775	0.00656
1	0.46341	0.06155	0.05342	0.17421	0.14866	0.04645	0.03368	0.01742	0.00116
2	0.44649	0.06605	0.08454	0.14002	0.14398	0.05812	0.04095	0.01188	0.00792
3	0.39108	0.06270	0.08415	0.15841	0.13861	0.08085	0.03960	0.03300	0.01155
4	0.27157	0.06736	0.08421	0.16842	0.18947	0.08421	0.07789	0.03578	0.02105
5	0.24137	0.06403	0.07389	0.14532	0.17487	0.12807	0.09113	0.04926	0.03202
6	0.20629	0.07692	0.04895	0.12237	0.17132	0.12937	0.11888	0.07692	0.04895
7	0.16000	0.05000	0.04000	0.15333	0.16666	0.13666	0.18000	0.08333	0.03000
8	0.12358	0.07214	0.07548	0.17902	0.18303	0.15030	0.12892	0.05811	0.02939

Confidence = Each Association/Neighbor Urban Cell

the regions, those tuple that don't belong to the exclusive areas are left and the attributes of exclusive areas are removed. On the assumption that roads and cities are interrelated, the two attributes, in other words the number value of surrounding urban cells and that of surrounding roads are 9(0-8) and 9(0-8) respectively, which make 81 combinations in total. By using the numbers of tuple in each combination, support and confidence are estimated. To remove meaningless values or rules it is necessary to identify a threshold by users. Thus, a threshold value for support is set over 0.01. After setting a threshold value for confidence at 0.10 among the filtered combinations, 13 combinations are selected.

As for the rest of the combinations based on urban the road values in the same row are applied with a formula of $\sum(\text{support}*\text{confidence})/\sum\text{confidence}$ and then 9 values are gained(Table 2). By integrating combinations of surrounding urban cells and roads and limited conditions using excluded areas and gradients, the study extracted the final spatial association rules and suggested the algorithms as shown in Table 3.

Table 3. Algorithm using Spatial Association Rules

```
If target cell = non-urban cell then
 If slope >=0 & slope <= 11  then
  If not exclusive region then

  Rule 1 :
     (urban 3 / road 0) then Int(0.391089109 * 100) >= random value 1   then  target cell → ruban cell
  Rule 2 :
     (urban 3 / road 3) then Int(0.158415842 * 100) >= random value 2   then  target cell → ruban cell
         .........
  Rule n :
     (urban 8 / road 6) then Int(0.128924516 * 100) >= random value n   then  target cell → ruban cell

  Else Rules :
       ∑(confidence * support) / ∑support >= random value   then  target cell → ruban cell
```

4 Design and Analysis of Urban Growth Probability Model

4.1 Design of Model

To verify the extracted spatial association rules, a simulation was implemented through realizing models. The study used CAS, a CA based simulator, and realized urban growth probability models. As CAS provides a cellular language, called Cellang, it can be applied to spatial association rules and realize them. In order to realize models the calibration stage in which an optimal rule can be found by repeating a number of times was repeated. To find rules reflecting changes over the four periods from the 1960's to the 1990's the best, one round of simulation was considered as one year and to minimize the influence of random numbers an average of accuracy by repeating 10 times was used.

4.2 Analysis and Comparison of Results

Among various types of urban growth, the study focused on physical urban growth. As a leading example of such types, UGM was compared with the results of the study.

Because UGM's information and program sources on models are open, it was possible to analyze and compare them. To assess accuracy of the models in which spatial association rules were used, Lee-Sallee index was introduced in the study (Clarke et al., 1997; Kang and Park, 2000).

$$Lee - Sallee\ index = \frac{A \cap B}{A \cup B}$$

Lee-Sallee index is estimated by using the number of cells which are matching between images of the urban and those of the simulated urban at a standard point of time. UGM also used Lee-Sallee index for the calibration of models, and the calibration stage was repeated three times. Generally, in case of UGM, 10 coefficient combinations with the highest Lee-Sallee indexes are decided at the last calibration stage. At this time, the top 10 Lee-Sallee indexes and the result values from the calibration stage of the study were compared. To make the round figures of the compared values identical, it was repeated 10 times.

Table 4. The Result of Comparison Our Experiment with UGM

	Our Experiment	UGM
Lee-Sallee Index	0.40441 - 0.41228	0.37077 - 0.36998

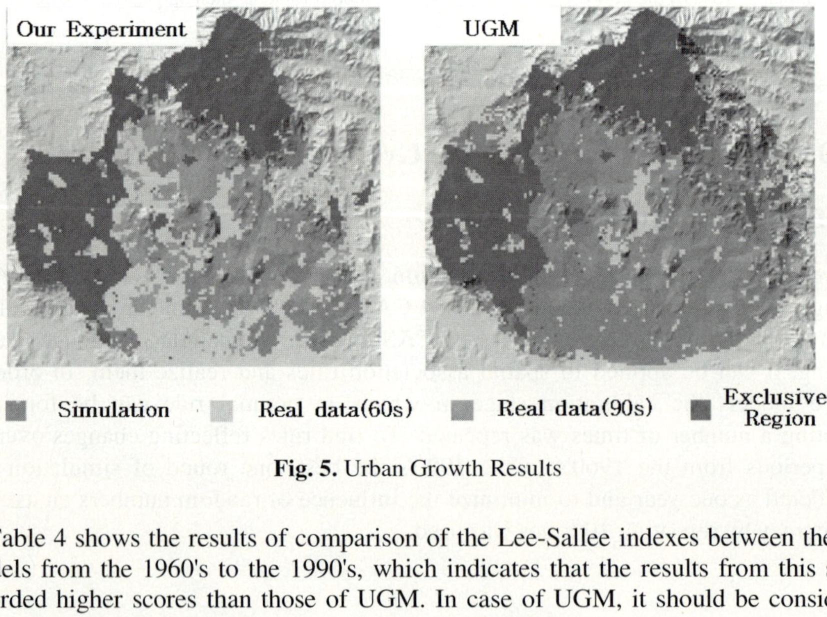

Fig. 5. Urban Growth Results

Table 4 shows the results of comparison of the Lee-Sallee indexes between the two models from the 1960's to the 1990's, which indicates that the results from this study recorded higher scores than those of UGM. In case of UGM, it should be considered that the both procedures of setting a range of attributes at each calibration stage and of setting a threshold at the generalization stage for information of attributes in the study require users' judgment. However, compared with UGM that compares results by using combinations of various attributes, relatively simple rules are applied to urban growth probability models. Also the result for accuracy showed relatively high values.

5 Conclusion

The study is designed to generalize types of urban growth by spatial analyzing functions of GIS and the data mining technique and extract spatial association rules.

To do so, from the 1960's to the 1990's the time and spatial database was built for each 10-year period and the data were integrated to be suitable for the study. To extract spatial association rules from the built data, GIS's spatial analyzing function and AOI, a data mining technique were applied and then a series of rules were drawn out. By utilizing the algorithms of the extracted rules, the analyzing stages of urban growth were simulated by CAS and a CA simulator.

The results of the simulation were compared with UGM, which is in coherence with this study regarding the physical urban growth. The comparison of accuracy between the two models by using Lee-Sallee index resulted in 0.40441 - 0.41228, which are relatively higher numbers than UGM. Compared with the existing CA models, finding spatial association rules inherent in data and utilizing those for modeling through GIS's functions of data processing and spatial analysis are better from the perspectives of time and efficiency. In the study, only a few geographical factors related to urban growth were applied. If other various factors including social and economic factors are applied, more reliable rules can be extracted.

On the other hand, because as a supervising classification technique the knowledge extraction technique which was applied for extraction of spatial association rules requires users' judgment at the stage of deciding rules, subjective factors can be intervened. Unlike other models, in order to find rules inherent in data, it is necessary to have measures to make up for such limitations. Another problem is that it is difficult to apply exterior variables which can calibrate the models from outside. As this can be a limited condition in making an active model, there should be complementary measures and subsequent studies in the near future.

References

1. Han, J., Kamber, M.: Data Mining Concepts and Techniques. Morgan Kaufmann Publishers, San Francisco (2001)
2. Jung, J.J.: Development of Cellular Automata Model for the Urban Growth. Seoul University (2001)
3. Clarke, K.C., Hoppen, S., Gaydos, L.: A Self-modifying Cellular Automata Model of Historical Urbanization in the San Francisco Bay Area. EPB 24, 247–261 (1997)
4. Kim, K.H.: GIS Introduction, Deayoungsa (2000)
5. Park, S.H., Joo, Y.G., Shin, Y.H.: Design and Development of a Spatio-Temporal GIS Database for Urban Growth Modeling and Prediction. The Korean Association of Professional Geographers 36(4), 313–326 (2002)

6. Park, S.H.: Design and Implementation of an Integrated CA-GIS System. Geographic Information System Association of Korea 5(1), 99–113 (1997)
7. Lu, W, Han, J., Ooi, B.C.: Discovery of General Knowledge in Large Spatial Databases. In: Proc. Far East Workshop on Geographic Information Systems, Singapore, pp. 275–289 (1993)
8. Kang, Y.O., Park, S.H.: A Study on the Urban Growth Forecasting for the Seoul Metropolitan Area. Journal of the Korean Geographical Society 35(4), 621–639 (2000)
9. www.ncgia.ucsb.edu/projects/gig/index.html
10. www.vbi.vt.edu/ dana/ca/cellular.shtml

Detecting Individual Activities from Video in a Smart Home

Oliver Brdiczka, Patrick Reignier, and James L. Crowley

INRIA Rhône-Alpes, Montbonnot, France
{brdiczka,reignier,crowley}@inrialpes.fr

Abstract. This paper addresses the detection of activities of individuals in a smart home environment. Our system is based on a robust video tracker that creates and tracks targets using a wide-angle camera. The system uses target position, size and orientation as input for interpretation. Interpretation produces activity labels such as "walking", "standing", "sitting", "interacting with table", or "sleeping" for each target. Bayesian Classifier and Support Vector Machines (SVMs) are compared for learning and recognizing previously defined individual activities. These methods are evaluated on recorded data sets. A novel Hybrid Classifier is then proposed. This classifier combines generative Bayesian methods and discriminative SVMs. Bayesian methods are used to detect previously unseen activities, while the SVMs are shown to provide high discriminative power for recognizing examples of learned activity classes. The evaluation results of the Hybrid classifier for the recorded data sets show that the combination of generative and discriminative classification methods outperforms the individual methods when identifying unseen activities.

1 Introduction

This paper describes a system for detecting individual activities in a smart home environment. The objective is to detect both predefined and unseen activities. The proposed system is based on a visual tracking process that creates and tracks moving targets using a wide-angle camera. Extracted target position, size and orientation are the input for framewise activity recognition for each target.

This paper makes two contributions. First, a Bayesian Classifier and Support Vector Machines (SVMs) are compared for learning and recognizing basic individual activities ("walking", "standing", "sitting", "interacting with table", "sleeping") from visual target properties. Both methods are tested and evaluated on data sets recorded in a laboratory mockup of a smart home environment. Secondly, a novel Hybrid Classifier is proposed for identifying previously unseen activities. Bayesian methods are used to create a training data model. The probability with regard to this model determines whether or not a predefined activity class can be attributed. If yes, SVMs are used to determine the learned activity class. If not, a wrong detection or a new activity class (to be learned) is identified. The proposed Hybrid Classifier has been tested and evaluated on the recorded data sets.

B. Apolloni et al. (Eds.): KES 2007/WIRN 2007, Part I, LNAI 4692, pp. 363–370, 2007.

2 Approach

In the following, we present an approach for activity detection from video. First, our smart home environment and the robust video tracking system are briefly depicted. Then, the activity labels and the recorded data sets are described. Finally, the Bayesian Classifier, Support Vector Machines and Hybrid Classifier are explained and the results for the data sets are presented.

2.1 Smart Home Environment

The experiments described in this paper are performed in a laboratory mockup of a living room environment in a smart home. The environment contains a small table surrounded by three armchairs and one couch (Fig. 1 left). Microphone arrays and video cameras are mounted on all walls in the environment. In this paper we concentrate on the use of a single wide-angle video camera mounted in a corner of the smart room (Fig. 1 middle) opposite the couch.

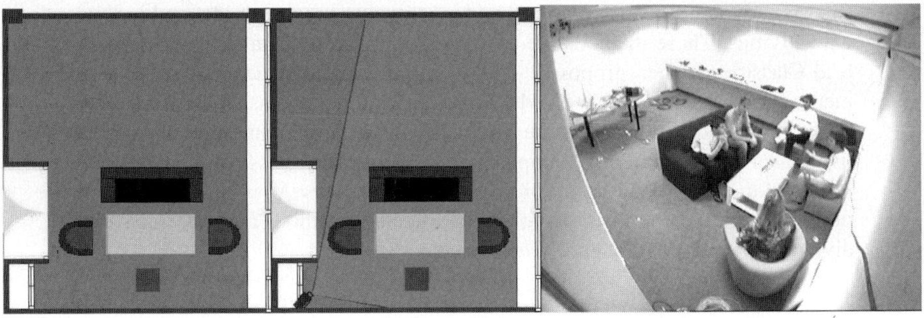

Fig. 1. Map of our Smart Room (left), wide-angle camera view indicated in gray (middle), wide-angle camera image (right)

The wide-angle camera observes the environment (Fig. 1) with a frame rate between 15 and 20 images per second. A real-time robust tracking system detects and tracks targets in the video images.

2.2 The Video Tracking System

In our smart environment, a real-time robust video tracking system [3] [11] is used to detect and track moving users in the environment. Targets can be detected by energy measurements based on background subtraction or intensity normalized color histograms. The video tracking system returns a vector of properties for each video frame. Each vector contains the position, size and orientation of one target detected and tracked by the system. The returned properties for each target are top position (x, y) of the bounding ellipse, the radius of the first and second axis of the ellipse and the angle describing the orientation of the ellipse (Fig. 2). Additional features including velocity, speed or energy can also be determined from the target tracking process.

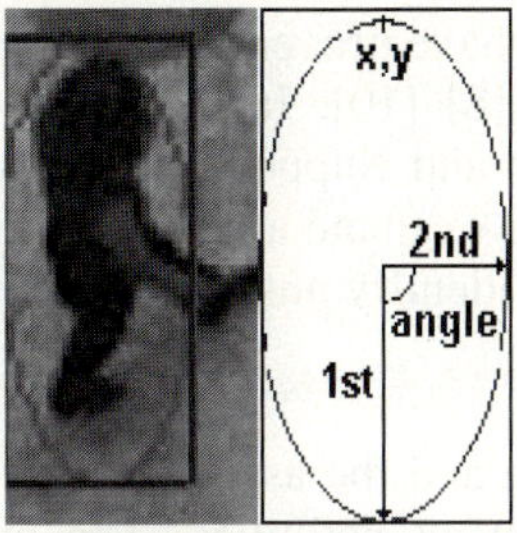

Fig. 2. Target properties estimated by the robust tracker

2.3 Individual Activities and Data Sets

Five categories of elementary activities are recognized: "walking", "standing", "sitting", "interaction with table" and "sleeping".

In order to develop and evaluate the detection process, we recorded 8 short video sequences in the environment. During these sequences, one or several individuals did different elementary activities in the smart room. The number of frames and the distribution of different activities played during the sequences are indicated in Table 1.

The activities played by the individuals in the video sequences have been hand labeled for use in learning and evaluation. The labeling process assigns an activity label to each target detected by the robust tracking system for each frame. The labeler had the possibility of assigning a "no activity" label if a detected target did not appear to do any of the five elementary activities. Thus, each of the 8 data sets contains a list of target properties (x, y, first radius, second radius, angle) and the associated activity label.

Table 1. Frame numbers of the video sequences and distribution of activities (in per cent)

Video Sequence	No. Frames
1	1352
2	6186
3	4446
4	4684
5	4027
6	4477
7	3067
8	3147
Total	31386

Class	% in data sets
Walking	0,18
Standing	0,09
Sitting	0,44
Inter. Table	0,19
Sleeping	0,10

2.4 Learning and Recognizing Individual Activities

By using machine learning methods, our system is to find a connection between the sensed information (target properties per frame) and the individual activities as perceived and labeled by the person who has provided the hand labeling. We are focusing particularly on Bayesian methods, because they are well adapted to deal with erroneous sensor data and they have proven to be useful in many application domains, in particular computer vision [8] [10]. In the following, we will first present and evaluate a Bayesian Classifier and Support Vector Machines on the recorded data sets. Then, we will propose and evaluate a novel Hybrid Classifier combing Bayesian methods and SVMs in order to identify unseen activity classes.

2.4.1 Bayesian Classifier

On the basis of the sensor data and the associated activity labels, we seek to learn a probabilistic classifier for relevant activities. The proposed Bayesian Classifier is similar to classifiers proposed in [7] [10]. The classification is done framewise, i.e. the classifier takes the target properties of one frame as input and generates the activity prediction for the frame as output.

We seek to determine the activity a_{MAP} with the maximum a posteriori (MAP) probability, given the target property set T (equation (1)).

$$a_{MAP} = \arg\max_{a} P(a \mid T) \tag{1}$$

$$P(a \mid \mathrm{T}) = \frac{P(T \mid a)P(a)}{P(T)} \tag{2}$$

We apply Bayes theorem (2) and we further assume that the prior probabilities *P(a)* for the activities are equal for each frame. As the constant denominator can be eliminated because of the argmax, we get equation (3).

$$a_{MAP} = \arg\max_{a} P(T \mid a) \tag{3}$$

We model *P(T|a)* for each activity as multidimensional Gaussian mixture distribution estimated by running EM algorithm [1] on the learning data. The initial number of Gaussians in the mixture is set to a high value (128); Gaussians with too weak contribution to the mixture are successively eliminated.

We evaluated the classifier on the video sequence recordings (Table 1) using 8-fold cross-validation. Each sequence has been used for testing once, while learning the model with the 7 remaining sequences. The overall results for the Bayesian Classifier can be seen in the left column of Table 2. We evaluated three different target property sets *T*. The first set was the position X, Y in the image. The results are good showing that the position in the environment is discriminating for individual activities. Position is, however, very dependent on environment configuration, e.g. couch and chair localization. Therefore, the second target set was (1st, 2nd, angle), which only contains information on the form of the ellipse and not its position. The results are quite similar to those obtained for the position. The combination of the first and second target property sets (X, Y, angle, 1st, 2nd) gives the best results.

2.4.2 Support Vector Machines

In order to further improve recognition results, we use Support Vector Machines (SVMs) as a classifier. SVMs [2] [5] classify data through determination of a set of support vectors, through minimization of the average error. The support vectors are members of the set of training inputs that outline a hyperplane in feature space. This l-dimensional hyperplane, where l is the number of features of the input vectors, defines the boundary between the different classes. The classification task is simply to determine on which side of the hyperplane the testing vectors reside.

The training vectors can be mapped into a higher (maybe infinite) dimensional space by the function ϕ. The SVM finds a separating hyperplane with the maximal margin in this higher dimensional space. $K(x_i, x_j) = \phi(x_i)^T \phi(x_j)$ is used as a kernel function. For multi-class classification, a "one-against-one" classification for each of the k classes can be performed. The classification of the testing data is accomplished by a voting strategy, where the winner of each binary comparison increments a counter. The class with the highest counter value after all classes have been compared is selected.

We evaluated the classifier on the video sequence recordings (Table 1) using 8-fold cross-validation. A radial basis function kernel with $C=11.0$ and $\gamma=11.0$ showed good results for our training data. The LIBSVM library [4] has been used for implementation and evaluation.

The overall results of SVM are shown in the right column of Table 2. Both SVM and the Bayesian Classification are applied framewise. That is, the target properties for each frame are used to produce an activity label, independent of values in other frames. Because the SVM is a discriminative method, it optimizes classification between the given/trained classes, outperforming the Bayesian Classifier. However, SVM does not learn the structure for a given data set, but only borders and margins between classes. As a result, with the SVM it is difficult or impossible to reject unseen test data ("garbage") or to discover new classes of activity.

Table 2. Recognition rates for Bayesian Classifier and SVMs

		Bayesian Classifier	SVMs
X,Y	Mean	0,7696	0,7855
	Std. dev.	0,0469	0,0398
1st, 2nd, angle	Mean	0,7691	0,7811
	Std. dev.	0,0393	0,0469
X,Y,1st,2nd,angle	Mean	0,8150	0,8610
	Std. dev.	0,0146	0,0276

2.4.3 Hybrid Classifier

SVMs are a discriminative classification method that outperforms the generative Bayesian Classifier for particular data sets. However, SVMs do not provide reliable information about whether or not a new data item may be coherent with the training data sets. Although there are probabilistic SVMs [9], the generated probabilities only refer to the distribution within the trained classes. Unseen data such as wrong target detections or new activity classes cannot be identified. These data will be attributed to one of the existing classes. The Bayesian Classifier is a generative classification method that generates a model for the training data, providing a possible probability output for each new data item. A Hybrid Classifier combines the strong points of each method: the probabilistic output of the Bayesian Classifier and the discriminative power of the SVMs. First approaches for such a classifier have been applied to text-independent speaker identification [6]. The focus, however, was on classification of trained speakers; unseen classes/data have not been considered.

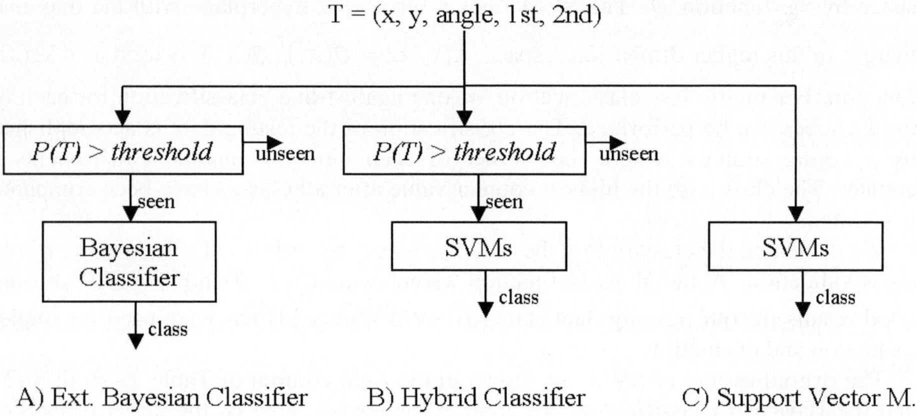

Fig. 3. Extended Bayesian Classifier, Hybrid Classifier and Support Vector Machines

In the following, we propose a Hybrid Classifier combining Bayesian methods for identifying unseen data and SVMs for classifying seen data. We will compare the method with an extended Bayesian Classifier and classical SVMs. The architecture of the classifiers can be seen in Fig. 3. For testing and evaluation, we will limit ourselves to the complete target property set (X, Y, angle, 1st, 2nd).

In section Bayesian Classifier, we used equation (3) to determine the class of a new data item. We modeled *P(T|a)* for each activity as multidimensional Gaussian mixture distribution estimated by EM. We have extended this by modeling additionally *P(T)* as multidimensional Gaussian mixture distribution estimated by EM. *P(T)* makes it possible to estimate the probability for a new data item to be generated from the training data set model. By using a threshold on this probability value, we can determine whether the new data item is part of the learned classes or whether it is unseen data (e.g. wrong detections or new class). The threshold can be automatically estimated from the training data sets (based on minimal probability of data items of the classes).

The Hybrid Classifier (Fig. 3 B) combines the estimation of *P(T)* (generative model) with SVMs trained on the classes. If a data item is determined to be seen data, the SVMs determine the class of this item. For evaluation, we compare the Hybrid Classifier with an extended Bayesian Classifier (Fig. 3 A) and classical SVMs (Fig. 3 C). The extended Bayesian Classifier combines the estimation of *P(T)* with a classical Bayesian Classifier. We want to show that the Hybrid Classifier outperforms both a purely Bayesian Classifier and a purely SVMs.

We evaluated the three different classifiers on the video sequence recordings (Table 1) using 8-fold cross-validation. In order to test the classifiers on unseen data, we excluded each class once from the training data sets. This resulted in *5*8 = 40* test runs. The obtained overall results for the classifiers are depicted in Table 3. The Hybrid Classifier outperforms the extended Bayesian Classifier and the SVMs for the complete data sets.

Table 3. Overall recognition rates for Bayesian Classifier, Hybrid Classifier and SVMs

	Bayesian Classifier	Hybrid Classifier	SVMs
Mean	0,7523	0,7786	0,7101
Std. dev.	0,0550	0,0639	0,0840

Table 4 shows the TP rate, FP rate, precision, recall and F-measure of the activity classes that have been excluded from training for the Hybrid Classifier. These results are identical for the extended Bayesian Classifier because the detection of the unseen classes by the probability values of *P(T)* is common for both classifiers. As the classical SVMs are not trained to detect the unseen classes, the TP rate, FP rate, precision, recall and F-measure are zero for SVMs. The detection results for the unseen activities "standing" and "interacting with table" are mediocre. From an activity point of view, both classes overlap with more frequent classes ("walking" and "sitting" respectively), which explains detection errors. A distinct activity class like "sleeping" is, however, very well recognized as unseen. The overall rates indicate that the approach can be used to identify unseen activity classes.

Table 4. TP rate, FP rate, precision, recall and F-measure of the unseen activity classes for the Hybrid Classifier ("walking"(0), "standing"(1), "sitting"(2), "interacting with table"(3), "sleeping"(4))

Class	% in data sets	TP rate	FP rate	Precision	Recall	F-measure
0	0,18	0,7374	0,1356	0,6481	0,7374	0,6763
1	0,09	0,0108	0,001	0,3938	0,0108	0,0208
2	0,45	0,7467	0,2677	0,6576	0,7467	0,6713
3	0,19	0,5336	0,1217	0,6845	0,5336	0,5867
4	0,10	0,8476	0,0631	0,6557	0,8476	0,723
Total	1,00	0,5752	0,1178	0,6079	0,5752	0,5356

3 Conclusion

We presented a method for visually detecting activities in a smart home environment. This method is based on a robust tracking system that creates and tracks targets in wide-angle camera images of the scene. A Bayesian Classifier and Support Vector Machines are used for classification. Both methods have been applied to the extracted target properties (x,y,1st radius, 2nd radius, angle) in order to learn and detect individual target activity classes "walking", "standing", "sitting", "interacting with table", "sleeping". The evaluation of both classifiers on recorded data sets showed good results. In order to detect unseen activity classes, a Hybrid Classifier has then been proposed combining generative Bayesian methods and discriminative SVMs. The overall detection results for unseen classes in the recorded data sets are good. The Hybrid

Classifier outperformed the Bayesian Classifier and the SVMs, showing that the proposed combination of generative and discriminative methods is beneficial.

References

1. Bilmes, J.A.: A Gentle Tutorial of the EM Algorithm and its Application to Parameter Estimation for Gaussian Mixture and Hidden Markov Models, Technical Report ICSI-TR-97-021. University of Berkeley (1998)
2. Boser, B., Guyon, I., Vapnik, V.: A training algorithm for optimal margin classifiers. In: Proceedings of the Fifth Annual Workshop on Computational Learning Theory (1992)
3. Caporossi, A., Hall, D., Reignier, P., Crowley, J.L.: Robust visual tracking from dynamic control of processing. In: Proceedings of International Workshop on Performance Evaluation for Tracking and Surveillance, pp. 23–32 (2004)
4. Chang, C.-C., Lin, C.-J.: LIBSVM, a library for support vector machines. Software (2001), available at http://www.csie.ntu.edu.tw/ cjlin/libsvm
5. Cortes, C., Vapnik, V.: Support-vector network. Machine Learning 20, 273–297 (1995)
6. Fine, S., Navratil, J., Gopinath, R.: A hybrid GMM/SVM approach to speaker identification. In: Proceedings of IEEE International Conference on Acoustics, Speech, and Signal Processing (ICASSP) (2001)
7. Muehlenbrock, M., Brdiczka, O., Snowdon, D., Meunier, J.-L.: Learning to Detect User Activity and Availability from a Variety of Sensor Data. In: Proceedings of Second IEEE International Conference on Pervasive Computing and Communications (2004)
8. Oliver, N., Rosario, B., Pentland, A.: A Bayesian Computer Vision System for Modeling Human Interactions, IEEE Trans. Pattern Analysis and Machine Intelligence 22(8), 831–843 (2000)
9. Platt, J.C.: Probabilities for SV Machines. In: Smola, A., Bartlett, P., Schölkopf, B., Schuurmans, D. (eds.) Advances in Large Margin Classifiers. ch. 5, pp. 61–74. MIT Press, Cambridge (1999)
10. Ribeiro, P., Santos-Victor, J.: Human activity recognition from Video: modeling, feature selection and classification architecture. In: Proceedings of International Workshop on Human Activity Recognition and Modelling (2005)
11. Zhou, S., Chellappa, R., Moghaddam, B.: Visual tracking and recognition using appearance-adaptive models in particle filters. IEEE Transactions on Image Processing 11, 1434–1456 (2004)

Harmony Search Algorithm for Solving Sudoku

Zong Woo Geem

Johns Hopkins University, Environmental Planning and Management Program,
729 Fallsgrove Drive #6133, Rockville, Maryland 20850, USA
geem@jhu.edu

Abstract. Harmony search (HS) algorithm was applied to solving Sudoku puzzle. The HS is an evolutionary algorithm which mimics musicians' behaviors such as random play, memory-based play, and pitch-adjusted play when they perform improvisation. Sudoku puzzles in this study were formulated as an optimization problem with number-uniqueness penalties. HS could successfully solve the optimization problem after 285 function evaluations, taking 9 seconds. Also, sensitivity analysis of HS parameters was performed to obtain a better idea of algorithm parameter values.

Keywords: Sudoku puzzle, harmony search, combinatorial optimization.

1 Introduction

Sudoku, which is Japanese term meaning "singular number," has gathered popularity in Japan, the UK, and the USA. The Sudoku puzzle consists of 9 × 9 grid and 3 × 3 blocks for all the 81 cells. Each puzzle, which has a unique solution, has some cells that have already been filled in. The objective of the puzzle is to fill in the remaining cells with the numbers 1 through 9 so that the following three rules are satisfied:

- Each horizontal row should contain the numbers 1 - 9, without repeating any.
- Each vertical column should contain the numbers 1 - 9, without repeating any.
- Each 3 × 3 block should contain the numbers 1 - 9, without repeating any.

In recent years researchers have started to apply various methods such as graph theory [1], artificial intelligence [2], and genetic algorithm [3] to solve the Sudoku puzzle. Eppstein [1] used the transformation from a directed or undirected graph to an unlabeled digraph to solve the puzzle. Although it was successful to the undirected case, the method is not successful to a directed one because the latter is NP-complete [4]. Caine and Cohen [2] proposed an artificial intelligent model named MITS (Mixed Initiative Tutoring System for Sudoku), in which the tutor takes the initiative to interact when the student lacks knowledge and makes moves that have low utility. Nicolau and Ryan [3] developed a system named GAuGE (Genetic Algorithm using Grammatical Evolution) for Sudoku, which uses a position independent representation.

B. Apolloni et al. (Eds.): KES 2007/WIRN 2007, Part I, LNAI 4692, pp. 371–378, 2007.

Each phenotype variable is encoded as a genotype string along with an associated phenotype position to learn linear relationships between variables.

Recently, a musicians' behavior-inspired evolutionary algorithm, harmony search (HS), has been developed [5] and applied to various optimization problems such as structural design [6], water network design [7], dam scheduling [8], traffic routing [9], satellite heat pipe design [10], oceanic structure mooring [11], hydrologic parameter calibration [12], and music composition [13].

From its success in various applications, HS in this study tackles the board game Sudoku, which can be formulated as an optimization problem with minimal violations of the above-mentioned three rules.

2 Harmony Search Model

The objective of the Sudoku problem is to fill in the cells with the numbers 1 through 9 only once while satisfying the above-mentioned three rules. In other words, the problem can be formulated as an optimization problem as follows:

$$\text{Minimize } Z = \sum_{i=1}^{9}\left|\sum_{j=1}^{9} x_{ij} - 45\right| + \sum_{j=1}^{9}\left|\sum_{i=1}^{9} x_{ij} - 45\right| + \sum_{k=1}^{9}\left|\sum_{(l,m)\in B_k} x_{lm} - 45\right| \quad (1)$$

where x_{ij} = cell at row i and column j, which has integer value from 1 to 9; and B_k = set of coordinates for block k.

The first term in Equation 1 represents the penalty function for each horizontal row; the second term for each vertical column; and the third term for each block. It should be noted that, although the sum of each row, each column, or each block equals 45, it does not guarantee that the numbers 1 through 9 are used exactly once. However, any violation of the uniqueness affects other row, column, or block which contains the wrong value jointly.

To this penalty-included optimization problem, HS was applied, which originally came from the behavioral phenomenon of musicians when they together perform improvisation [5]. HS basically mimics musician's behaviors such as memory consideration, pitch adjustment, and random consideration, but it also includes problem-specific features for some applications.

For the first step of the HS algorithm, solution vectors are randomly generated as many as HMS (harmony memory size), then they are stored in HM (harmony memory) as follows:

$$\begin{bmatrix} x_{11}^1 & x_{12}^1 & \cdots & x_{19}^1 \\ x_{21}^1 & x_{22}^1 & \cdots & x_{29}^1 \\ \vdots & \cdots & \cdots & \cdots \\ x_{91}^1 & x_{92}^1 & \cdots & x_{99}^1 \end{bmatrix} \Rightarrow Z(\mathbf{x}^1) \quad (2a)$$

$$\begin{bmatrix} x_{11}^2 & x_{12}^2 & \cdots & x_{19}^2 \\ x_{21}^2 & x_{22}^2 & \cdots & x_{29}^2 \\ \vdots & \cdots & \cdots & \cdots \\ x_{91}^2 & x_{92}^2 & \cdots & x_{99}^2 \end{bmatrix} \Rightarrow Z(\mathbf{x}^2) \tag{2b}$$

$$\cdots$$

$$\begin{bmatrix} x_{11}^{HMS} & x_{12}^{HMS} & \cdots & x_{19}^{HMS} \\ x_{21}^{HMS} & x_{22}^{HMS} & \cdots & x_{29}^{HMS} \\ \vdots & \cdots & \cdots & \cdots \\ x_{91}^{HMS} & x_{92}^{HMS} & \cdots & x_{99}^{HMS} \end{bmatrix} \Rightarrow Z(\mathbf{x}^{HMS}) \tag{2c}$$

where x_{ij}^n = cell at row i and column j in n[th] vector stored in HM; and $Z(\mathbf{x}^n)$ = function value for n[th] vector in HM.

For the next step, a new harmony in Equation 3 is improvised using one of the following three mechanisms: random selection, memory consideration, and pitch adjustment.

$$\mathbf{x}^{NEW} = \begin{bmatrix} x_{11}^{NEW} & x_{12}^{NEW} & \cdots & x_{19}^{NEW} \\ x_{21}^{NEW} & x_{22}^{NEW} & \cdots & x_{29}^{NEW} \\ \vdots & \cdots & \cdots & \cdots \\ x_{91}^{NEW} & x_{92}^{NEW} & \cdots & x_{99}^{NEW} \end{bmatrix} \tag{3}$$

Random Selection. For x_{ij}^{NEW}, random value is chosen out of value range ($1 \le x_{ij}^{NEW} \le 9$) with a probability of (1-HMCR). HMCR ($0 \le$ HMCR ≤ 1) stands for harmony memory considering rate.

$$x_{ij}^{NEW} \leftarrow x_{ij}, \quad x_{ij} \in \{1, 2, \ldots, 9\} \quad \text{w.p.} \quad \textit{(1 - HMCR)} \tag{4}$$

Memory Consideration. Instead of the random selection, the value can be chosen from any values stored in HM with a probability of HMCR.

$$x_{ij}^{NEW} \leftarrow x_{ij}, \quad x_{ij} \in \{x_{ij}^1, x_{ij}^2, \ldots, x_{ij}^{HMS}\} \quad \text{w.p.} \quad \textit{HMCR} \tag{5}$$

Pitch Adjustment. Once one pitch is obtained in memory consideration rather than random selection, the obtained value may further move to neighboring values with a probability of HMCR × PAR while the original value obtained in memory consideration does not move with a probability of HMCR × (1-PAR). PAR ($0 \le$ PAR ≤ 1) stands for pitch adjusting rate. Here, x_{ij}^{NEW} in the right hand side is the value originally obtained in memory consideration; and Δ is the amount of increment (Δ equals one if x_{ij}^{NEW} is not upper limit (9) or lower limit (1). Otherwise, Δ equals zero).

$$x_{ij}^{NEW} \leftarrow \begin{cases} x_{ij}^{NEW} + \Delta & \text{w.p.} \quad HMCR \times PAR \times 0.5 \\ x_{ij}^{NEW} - \Delta & \text{w.p.} \quad HMCR \times PAR \times 0.5 \\ x_{ij}^{NEW} & \text{w.p.} \quad HMCR \times (1 - PAR) \end{cases} \tag{6}$$

If the new harmony vector $\mathbf{x}^{NEW}$ is better than the worst harmony in the HM in terms of objective function value, $Z(\mathbf{x}^{NEW})$, the new harmony is included in the HM and the existing worst harmony is excluded from the HM.

If the HS model reaches MaxImp (maximum number of improvisations), computation is terminated. Otherwise, another new harmony is improvised by considering one of three mechanisms.

3 Applications

The HS model was applied to the Sudoku puzzle proposed by Nicolau and Ryan [3] as shown in Figure 1.

	5		3		6			7
				8	5		2	4
	9	8	4	2		6		3
9		1			3	2		6
	3						1	
5		7	2	6		9		8
4		5		9		3	8	
	1		5	7				2
8			1		4		7	

Fig. 1. Example of Sudoku Puzzle

The HS model found the optimal solution without any violation after 285 function evaluations using HMS = 50, HMCR = 0.7, and PAR = 0.1. Figure 2 shows the history of reaching global optimum.

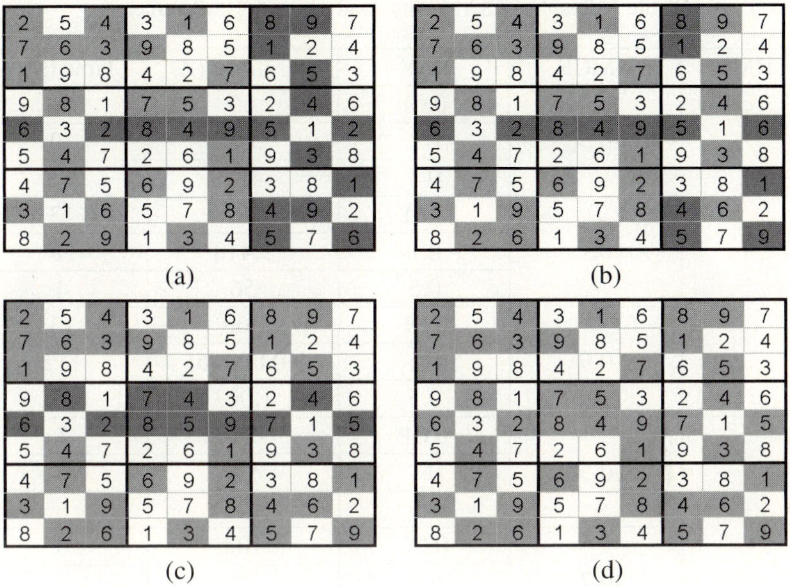

Fig. 2. Intermediate and Final Solutions of Test Sudoku Puzzle

While the green-colored cell (light-dark color in black & white) in Figure 2 means that there is no violation, the magenta-colored cell (dark color in black & white) indicates that there is at least one violation horizontally, vertically, or block-wise. Figure 2 (a) is the solution at 13 improvisations, which has a penalty of 21; Figure 2 (b) is the solution at 121 improvisations, which has a penalty of 5; Figure 2 (c) is the solution at 231 improvisations, which has a penalty of 2; and Figure 2 (d) is the solution at 285 improvisations, which has a penalty of 0.

This HS model further performed sensitivity analysis of algorithm parameters (HMS = {1, 2, 10, 50}, HMCR = {0.5, 0.7, 0.9}, PAR = {0.01, 0.1, 0.5}). Table 1 shows the analysis results. When only one vector is considered in the HM (HMS = 1), like simulated annealing or tabu search algorithm, the HS found global optimum except in one case (HMCR = 0.9, PAR = 0.1, Z = 6); When two vectors are considered in the HM (HMS = 2), partially similar to genetic algorithm, the HS also found global optimum except in two cases (HMCR = 0.7, PAR = 0.01, Z = 15; HMCR = 0.7, PAR = 0.1, Z = 27). However, when more than two vectors were considered in the HM (HMS = 10 or 50), there was no rule violation for the Sudoku example.

The HS computation was performed on Intel Celeron 1.8GHz CPU. The computing time ranged 4 - 38 seconds for HMS = 1 in order to arrive at the global optimum; for HMS = 2, it ranged 3 - 20 seconds; for HMS = 10, it ranged 3 - 8 seconds; and for HMS = 50, it ranged 7 - 12 seconds.

The HS model developed in this study was further applied to another Sudoku problem classified as "hard" as shown in Figure 3 [14]. When applied to the problem, the HS model was entrapped in one of local optima with a penalty of 14 after 1,064 function evaluations as shown in Figure 4.

Table 1. Results of Sensitivity Analysis with HS Parameters

HMS	HMCR	PAR	Iterations (Z)	Time (sec)
1	0.5	0.01	66	5
		0.1	337	10
		0.5	422	11
	0.7	0.01	287	13
		0.1	3,413	38
		0.5	56	4
	0.9	0.01	260	13
		0.1	**10,000** (6)	112
		0.5	1,003	19
2	0.5	0.01	31	3
		0.1	94	6
		0.5	175	6
	0.7	0.01	102	6
		0.1	77	6
		0.5	99	7
	0.9	0.01	**10,000** (15)	98
		0.1	**10,000** (27)	135
		0.5	1,325	20
10	0.5	0.01	49	3
		0.1	280	8
		0.5	188	5
	0.7	0.01	56	4
		0.1	146	5
		0.5	259	8
	0.9	0.01	180	5
		0.1	217	8
		0.5	350	8
50	0.5	0.01	147	9
		0.1	372	10
		0.5	649	12
	0.7	0.01	165	7
		0.1	285	9
		0.5	453	12
	0.9	0.01	87	7
		0.1	329	10
		0.5	352	11

3				1	7		5	
		1				8		4
			5	6				1
9		2						
			6		3			
						1		2
1				8	2			
7		8				2		
	5		7	4				6

Fig. 3. Another Sudoku Example (Hard Level)

3	8	4	2	1	7	6	5	9
6	3	1	9	2	5	8	7	4
4	9	7	5	6	8	3	2	1
9	1	2	4	3	6	5	8	7
5	2	9	6	7	3	4	1	8
8	6	3	8	5	4	1	9	2
1	7	6	3	8	2	9	4	5
7	4	8	1	9	1	2	6	3
2	5	6	7	4	9	6	3	6

Fig. 4. Local Optimum for Hard Example

4 Conclusion

The HS, musicians' behavior-inspired evolutionary algorithm, challenged the Sudoku puzzle with 40 given values in the literature, and could successfully find the unique global solution. The total searching space for this case is $9^{41} = 1.33 \times 10^{39}$ if integer programming formulation is considered. The proposed HS model found the global optimum without any row, column or block violation after 285 function evaluations, taking 9 seconds on Intel Celeron 1.8 GHz Processor.

When sensitivity analysis of algorithm parameters was performed, the HS could reach the global optimum 33 times out of 36 runs, taking 3 - 38 seconds (median for 33 successful cases is 8 seconds).

However, it failed to find the global optimum for hard level case with 26 given values, which has the searching space of $9^{55} = 3.04 \times 10^{52}$. The HS model was instead entrapped in one of local optima with the penalty of 14 after 1,064 function evaluations.

For study in the future, the HS model should consider additional problem-specific heuristics in order to efficiently solve a harder puzzle.

References

1. Eppstein, D.: Nonrepetitive Paths and Cycles in Graphs with Application to Sudoku. ACM Computing Research Repository. cs.DS/0507053 (2005)
2. Caine, A., Cohen, R.: A Mixed-Initiative Intelligent Tutoring System for Sudoku. In: Lamontagne, L., Marchand, M. (eds.) Canadian AI 2006. LNCS (LNAI), vol. 4013, pp. 550–561. Springer, Heidelberg (2006)
3. Nicolau, M., Ryan, C.: Solving Sudoku with the GAuGE System. In: Collet, P., Tomassini, M., Ebner, M., Gustafson, S., Ekárt, A. (eds.) EuroGP 2006. LNCS, vol. 3905, pp. 213–224. Springer, Heidelberg (2006)
4. Yato, T., Seta, T.: Complexity and Completeness of Finding Another Solution and its Application to Puzzles. IEICE Transactions on Fundamentals of Electronics, Communications and Computer Sciences 86, 1052–1060 (2003)
5. Geem, Z.W., Kim, J.H., Loganathan, G.V.: A New Heuristic Optimization Algorithm: Harmony Search. Simulation. 76(2), 60–68 (2001)
6. Lee, K.S., Geem, Z.W.: A New Structural Optimization Method Based on the Harmony Search Algorithm. Computers and Structures. 82(9-10), 781–798 (2004)
7. Geem, Z.W.: Optimal Cost Design of Water Distribution Networks using Harmony Search. Engineering Optimization. 38(3), 259–280 (2006)
8. Geem, Z.W.: Optimal Scheduling of Multiple Dam System Using Harmony Search Algorithm. In: Lecture Notes in Computer Science, vol. 4507, pp. 316–323 (2007)
9. Geem, Z.W., Lee, K.S., Park, Y.: Application of Harmony Search to Vehicle Routing. American Journal of Applied Sciences. 2(12), 1552–1557 (2005)
10. Geem, Z.W., Hwangbo, H.: Application of Harmony Search to Multi-Objective Optimization for Satellite Heat Pipe Design. In: Proceedings of 2006 US-Korea Conference on Science, Technology, & Entrepreneurship (UKC 2006). CD-ROM (2006)
11. Ryu, S., Duggal, A.S., Heyl, C.N., Geem, Z.W.: Mooring Cost Optimization via Harmony Search. In: Proceedings of the 26th International Conference on Offshore Mechanics and Arctic Engineering (OMAE 2007), ASME. CD-ROM (2007)
12. Kim, J.H., Geem, Z.W., Kim, E.S.: Parameter Estimation of the Nonlinear Muskingum Model Using Harmony Search. Journal of the American Water Resources Association. 37(5), 1131–1138 (2001)
13. Geem, Z.W., Choi, J.–Y.: Music Composition Using Harmony Search Algorithm. In: Lecture Notes in Computer Science, vol. 4448, pp. 593–600 (2007)
14. Web Sudoku (January 19, 2007), http://www.websudoku.com/

Path Prediction of Moving Objects on Road Networks Through Analyzing Past Trajectories

Sang-Wook Kim[1], Jung-Im Won[1], Jong-Dae Kim[1], Miyoung Shin[2], Junghoon Lee[3], and Hanil Kim[3]

[1] School of Information and Communications, Hanyang University, Korea
{wook,jiwon}@hanyang.ac.kr, jdkim@agape.hanyang.ac.kr
[2] School of Electrical Engineering and Computer Science
Kyungpook National University, Korea
shinmy@knu.ac.kr
[3] Dept. of Computer Science and Statistics
Cheju National University, Korea
{jhlee,hikim}@cheju.ac.kr

Abstract. This paper addresses a series of techniques for predicting a future path of an object moving on a road network. Most prior methods for future prediction mainly focus on the objects moving over Euclidean space. A variety of applications such as telematics, however, require us to handle the objects that move over road networks. In this paper, we propose a novel method for predicting a future path of an object in an efficient way by analyzing past trajectories whose changing pattern is similar to that of a current trajectory of a query object. For this purpose, we devise a new function for measuring a similarity between trajectories by considering the characteristics of road networks. By using this function, we search for candidate trajectories whose subtrajectories are similar to a given query trajectory by accessing past trajectories stored in moving object databases. Then, we predict a future path of a query object by analyzing the moving paths along with a current position to a destination of candidate trajectories. Also, we suggest a method that improves the accuracy of path prediction by grouping those moving paths whose differences are not significant.

1 Introduction

Recently, with the wide spread of portable mobile devices and the advance in wireless communication technologies, various *location−based services* have been provided[1]. In such services, the locations of moving objects as well as the behavior patterns of their users can be traced and understood by analyzing the trajectories of moving objects. The *trajectory*[1, 2, 3, 4, 5], which is a moving path of an object, can be described as a series of line segments in 3-dimensional space (x, y, t) that reflects the characteristics of the space (x, y) and the time (t). User queries on moving objects are classified into two categories: *historical queries* for retrieving past positions of moving objects and *future queries* for predicting their future positions. The historical query is further subdivided into

B. Apolloni et al. (Eds.): KES 2007/WIRN 2007, Part I, LNAI 4692, pp. 379–389, 2007.

three types of queries: a range query, a trajectory query, and a complex query[4]. The range query retrieves the moving objects residing within a given query range. The trajectory query retrieves the past trajectories of moving objects in a given time interval. The complex query, which is a combination of the former two, retrieves the trajectories of moving objects within a given query range and a time interval. The future query predicts upcoming positions of moving objects based on their current locations, moving speeds, and moving directions[1, 6, 7]. This paper focuses on the future queries.

For efficient processing of future queries, various index structures including the VCI-tree[6], the TPR-tree[1], and the TPR*-tree[7] have been proposed. They index all the possible objects in 2-dimensional Euclidean space (x, y) and predict their future positions under the assumption that its moving directions and speeds are constant. Among these, the TPR*-tree is the most popular because it enables us to quickly retrieve the CBRs(conservative bounding rectangle) that contain the current positions, directions, and speeds of moving objects in the R*-tree structure[8].

In real applications, however, most objects move on road networks rather than Euclidean space. The directions and speeds of objects tend to be dependent on the road network condition at a specific time[11, 12, 13, 14]. Accordingly, existing methods predict future positions of moving objects by taking their current positions, directions, and speeds into consideration. Thus, they are not appropriate for road network environments.

In this paper, we propose a novel method for predicting future paths of objects moving over road networks. There have been many research efforts on database technologies such as location tracking, similar trajectory searching, data generation, and road network indexing[15, 16]. To the best of our knowledge, however, the method for predicting future paths of moving objects on road networks has not been suggested yet.

In the proposed method, when a moving path of an object from a start location to a current location is given as a query trajectory, we search for candidate trajectories that contain subtrajectories similar to a given query trajectory by investigating past trajectories stored in moving object databases. Then, we predict a future path of a query object by analyzing moving paths along with the current position to the destination of candidate trajectories thus retrieved. Also, we suggest a method that improves the accuracy of path prediction by grouping moving paths that have just small differences.

This paper is organized as follows. The motivation and problem definition are given in Section 2. In Section 3, we propose a method for similar trajectory searching and path predicting of moving objects on road networks. In Section 4, the paper is summarized and concluded.

2 Motivation and Problem Definition

A *trajectory* T_i of an object moving on a road network is composed of (userId, moId, tId, $<rseg_1, rseg_2, ..., rseg_k>$), where userId is a user identifier, moId is an identifier of a moving object, and tId is a trajectory identifier. The $rseg_j$(1 $\leq$ j $\leq$ k) is a road segment, i.e., a part of a trajectory T_i, and is represented as ($rsId_j$, $rsLen_j$). $rsId_j$ is an identifier of the road segment, and $rsLen_j$ is the length of the road segment. In our research, trajectories are kept separate for each user because different objects even moving towards the same destination may take different paths according to the user's driving preference. Also, a road segment is considered as a sequence of roads between intersections[9] because the persons' path selections take place only at the intersections.

The road segment taken by a moving object on a road network is stored as trajectories in moving object databases. At times, some of them show iterative or similar patterns, which mostly reflect the user's driving preference. Some examples of the trajectories taken by user *A* on a road network is shown in Table 1.

Table 1. Example of trajectories generated on a road network

Starting	Tranjectories	Destination	Departure time
Cancer Center Hospital	Dongboo express way/Yongbi bridge/ Seongdong bridge/Hanyang Univ.	Hanyang Univ.	9:00
Cancer Center Hospital	Dongboo express way/Hankuk Univ./ Seongdong bridge/Hanyang Univ.	Hanyang Univ.	10:00
Seoul National Univ.	Hangang bridge/Gangbyeon express way/ Yongbi bridge/Hanyang Univ.	Hanyang Univ.	8:20
Kangnam Station	Hannam bridge/Olympic express way/ Sangdo tunnel/Sillim station	Sillim Station	7:30

Under these circumstances, it is possible to have some queries for predicting future paths of a moving object as follows. "Which path is likely to be taken to the destination *Hanyang University* by user *A* who has just arrived at *Yongbi bridge* from *Cancer Center Hospital* via *Dongboo express way*?" For this problem, we suggest a method to process future queries for path prediction described above in large moving object databases, under the assumption that such information as drivers, starting positions, moving paths to a current position, and destinations are given by users in advance.

Logistics transportation of a home delivery service company is a good example of real applications for this method. A typical procedure of logistics transportation is as follows: A transportation vehicle of a home delivery service company moves towards the place to pick up the delivery parcel, starting from a current location. A transportation vehicle's driver usually takes his/her own preferred path to the destination. A new pick-up request for parcels may occur while the transportation vehicle is moving. In such a situation, if we can predict future paths of each vehicle, it would be possible to allocate the new delivery parcel to the transportation vehicle of the highest possibility to move towards the place where the parcel can be picked-up, leading to an efficient logistics transportation system that minimizes moving paths of transportation vehicles.

Fig. 1 shows such an example. A_1, A_2, and A_3 denote the current locations of transportation vehicles MO_1, MO_2, and MO_3, respectively, moving on a road network. The solid line indicates a moving path of each vehicle up to the current location. At this point, if a new pick-up request for parcels occurs at location P, future paths of each vehicle can be predicted as marked with the dashed lines in Fig. 1. As a result, we can improve transportation efficiency by allocating the new parcels to the vehicle MO_1 that is most likely to move to location P.

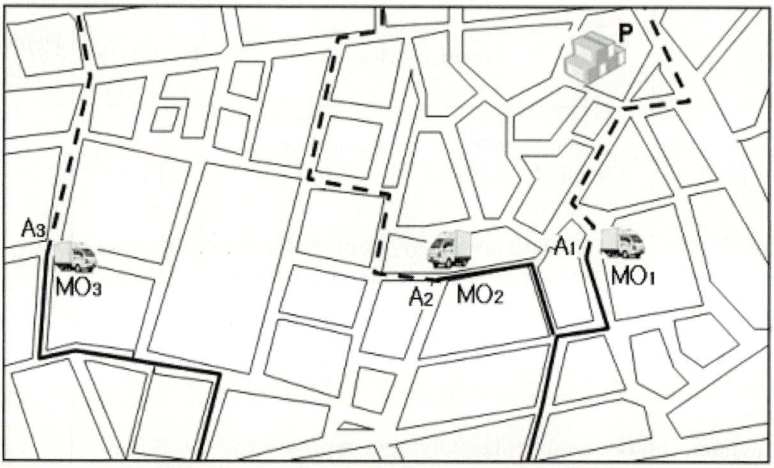

Fig. 1. Example of predicting future paths of the transportation vehicles

3 Proposed Method

3.1 Basic Strategy

The process of predicting future paths of moving objects consists of two steps. Given a moving path of an object from a start location to a current location as a query trajectory to the server, the first step is to search for candidate trajectories in databases whose changing patterns are similar to a query trajectory. Then, the second step is to predict moving paths of the object after the current location by computing the probabilities of the candidate trajectories to reach the destination based on their frequencies. The accuracy for path prediction can be further improved by grouping similar trajectories and computing the overall frequency of all the moving paths within a group.

Let us explain the procedure of predicting a future path of an object using the example shown in Fig. 2. Assume that an object started from location ⓢ, and got to location ⓒ by way of P_1 and P_3, moving towards the final destination ⓓ. Our interest is in predicting which path will be taken by the moving object from location ⓒ to ⓓ by analyzing its past trajectories. First, we retrieve the trajectories similar to the trajectory ⓢ to ⓒ from the past trajectories stored in moving object databases. As a result, we found 60 past trajectories T_{p1} which are completely equal to the trajectory ⓢ to ⓒ passing through P_1 and P_3, and 40 similar past trajectories T_{p2} passing through P_1 and P_2.

After this, for the retrieved trajectories T_{p1} and T_{p2}, their paths from ⓒ were investigated. The trajectory T_{f1} to destination ⓓ passing through P_4 and P_6 was found 68 times, the trajectory T_{f2} passing through P_4 and P_5 was found 2 times, the trajectory T_{f3} passing through P_7 and P_9 was found 26 times, and the trajectory T_{f4} passing through P_7 and P_9 was found 4 times. Thus, the future path of a moving object from ⓒ to ⓓ could be predicted by selecting the trajectory T_{f1} of the highest frequency among the above, which takes the route of P_4 and P_6 to destination ⓓ.

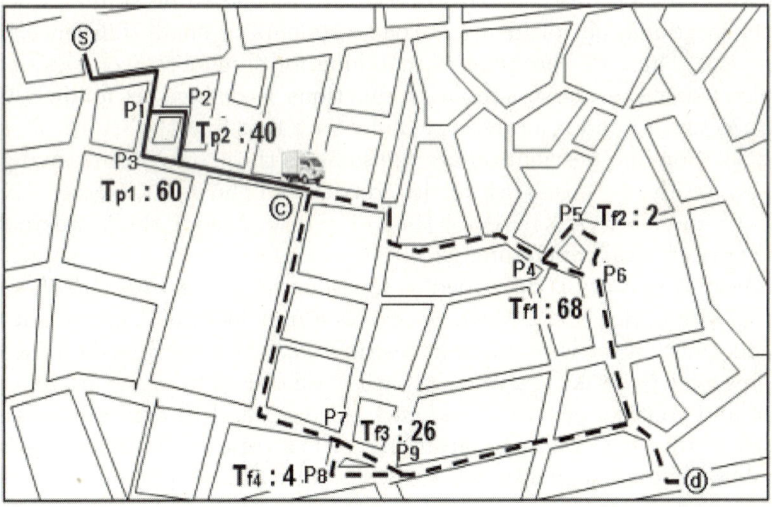

Fig. 2. Example of predicting future paths of a moving object

In this figure, however, it should be noted that the trajectory T_{f1} is almost same to T_{f2} and the trajectory T_{f3} to T_{f4}, which have just small differences from each other. Accordingly, instead of computing the frequencies, each of which is for the trajectories T_{f1}, T_{f2}, T_{f3}, and T_{f4}, we divide them into two groups $G_1=\{T_{f1}, T_{f2}\}$ and $G_2=\{T_{f3}, T_{f4}\}$, just computing the frequencies of each group G_1 ($T_{f1} + T_{f2} = 70$) and G_2 ($T_{f3} + T_{f4} = 30$).

3.2 Similar Trajectory Search

To perform a similar trajectory search on moving object databases, we consider only the trajectories whose starting and ending positions are consistent with those of a given moving object. From such trajectories, we search for past trajectories whose subtrajectories are similar to a given query trajectory. To define the similarity between subtrajectories on a road network, we devise a new function that computes the similarity between trajectories in a way different from a model based on the distance among trajectories in Euclidean space[10] or a model based on the characterization of temporal/spatial network constraints. This function measures the similarity between a query trajectory Q and a past trajectory T by using (1) DSN(Dissimilarity based on Segment Number) and (2) DSL(Dissimilarity based on Segment Length) below.

$$DSN(Q,T) = \frac{\#\ of\ Different\ Segments\ between\ Q\ and\ T}{Total\ \#\ of\ Segments\ in\ Q + Total\ \#\ of\ Segments\ in\ T} \tag{1}$$

$$DSL(Q,T) = \frac{\sum Lengths\ of\ Different\ Segments\ between\ Q\ and\ T}{\sum Lengths\ of\ Segments\ in\ Q + \sum Lengths\ of\ Segments\ in\ T} \tag{2}$$

With this method, any two trajectories are judged to be similar when there are either large number of identical road segments or small difference between their lengths. Thus, the problem of searching for similar trajectories of a moving object using the above similarity functions is defined as follows. Given a query trajectory Q and tolerances $\epsilon 1$ and $\epsilon 2$, we first find the trajectories whose starting position and destination are the same as those of Q from moving object databases, and from those trajectories, retrieve all the subtrajectories X whose DSN(Q, X) and DSL(Q, X) are less than $\epsilon 1$ and $\epsilon 2$. Among them, we finally pick only the trajectories T including X.

The query trajectory is described as Q=($userId$, $moId$, $start$, $dest$, <$qrseg_1$, $qrseg_2$, ..., $qrseg_k$>, $\epsilon 1$, $\epsilon 2$). Here, $userId$ is a user identifier, $moId$ is a moving object identifier, $start$ is a starting position, $dest$ is a destination of a moving object, $qrseg_i(1 \leq \mathrm{i} \leq \mathrm{k})$ is a moving path of an object from a start to a current position, and consists of ($rsId_i$, $rsLen_i$). Also, $\epsilon 1$ and $\epsilon 2$ denote tolerances for similar trajectory searching and grouping. $\epsilon 1$ is a ratio of the difference between the numbers of road segments while $\epsilon 2$ is a ratio of the difference between the lengths of road segments.

The reason to use a ratio is that those similarity tolerances can be applied for various objects having different characteristics. In other words, we need to apply them not only for computing the similarity between a query trajectory Q, consisting of k road segments from a start to a current position, and a past trajectory, but also for grouping the trajectories from a current position to the destination, consisting of l road segments. For example, if there are a small number of road segments from a start to a current position in a query trajectory Q, like 20 road segments, it would be useful to search for similar trajectories with a tolerance $\epsilon 1$=1. However, in predicting a future path of an object, if there are a large number of road segments, like 200 segments, from a current position to a destination, it would be meaningless to perform the grouping using the same tolerance $\epsilon 1=1$.

To make the implementation simple, we employ a sequential-scan based method for processing similar trajectory queries. First, since each user manages trajectory data in his/her own database and tends to iteratively follow the same moving path according to user's driving preference, there are usually small amount of trajectory data stored in the database. Second, subtrajectories to be compared by using tolerances $\epsilon 1$ and $\epsilon 2$ are not of long length. Third, even when given tolerances $\epsilon 1$ and $\epsilon 2$ are large, the lengths of original trajectories stored in a database are not long because a set of road segments up to the intersection is considered as a single road segment. Thus, even though a sequential-scan based method usually runs in O(n), it seems to be a reasonable choice because the number of trajectories to be retrieved is small and each trajectory is short in length.

3.3 Future Path Prediction

In order to predict a moving path of a given query Q to a destination with n candidate trajectories obtained from similar trajectory search, we may compute their moving frequencies from the current position to the destination and then to select such a trajectory that has the highest frequency as a future path of a query Q. As shown in Fig. 2, however, computing the frequencies of all the trajectories that have small differences from one another may decrease the accuracy of path predictions. Thus, to solve this problem, we propose a method of grouping similar trajectories retrieved.

For this purpose, we measure the similarity between moving paths of a query Q to a destination from a current position by using DSN and DSL. Next, we

Table 2. Examples of moving paths for candidate trajectories

Trajectories	Moving Paths
T_{f1}	<(R01,5),(R02,2),(R03,7),(R04,1),(R05,1),(R06,1),(R09,3),(R10,2)>
T_{f2}	<(R01,5),(R02,2),(R03,7),(R04,1),(R05,1),(R07,2),(R08,2),(R09,3),(R10,2)>
T_{f3}	<(R11,3),(R12,3),(R13,2),(R14,4),(R15,4),(R18,5),(R19,7)>
T_{f4}	<(R11,3),(R12,3),(R13,2),(R14,4),(R16,1),(R17,2),(R18,5),(R19,7)>

Table 3. Similarity values between candidate trajectories

Trajectories	DSN	DSL
T_{f1}, T_{f2}	0.176	0.106
T_{f1}, T_{f3}	1	1
T_{f1}, T_{f4}	1	1
T_{f2}, T_{f3}	1	1
T_{f2}, T_{f4}	1	1
T_{f3}, T_{f4}	0.2	0.127

group the trajectories whose DSN and DSL are both below tolerances $\epsilon 1$ and $\epsilon 2$. Thus, for predicting the future paths of moving objects, we use the frequencies obtained by adding up all the frequencies of trajectories in each group, rather than using individual frequencies of moving paths. Here, it may occur that a single trajectory belongs to several different groups at the same time. In such a case, if the frequency of a moving path is n and the number of groups including the moving path is m, the corresponding frequency of a moving path can be recalculated as $\lfloor n/m \rfloor$, which is used to compute a moving frequency of a group.

Table 2 shows moving paths of trajectories T_{f1}, T_{f2}, T_{f3}, and T_{f4} for the object of Fig. 2. Table 3 shows similarity values of the moving paths required for grouping, and from these, we can obtain $G_1 = \{T_{f1}, T_{f2}\}$ and $G_2=\{T_{f3}, T_{f4}\}$.

Algorithm PathPredict ($User\_Id$, P_{start}, P_{dest}, $P_{current}$, $Traj_{query}$)

$User\_Id$: User identifier
P_{start}: Start position
P_{dest}: Destination
$P_{current}$: Current position
$Traj_{query}$: Query trajectory from a start position to a current position

1. Among past trajectories in a database, select the trajectories that include $P_{current}$ and also contain user identifier, starting position, and destination corresponding to $User\_Id$, P_{start}, and P_{dest}, respectively. Then, put them into $TrajSet$.
2. **FOR** all the subtrajectories $TrajPrefix$ from P_{start} to $P_{current}$ in each trajectory $Traj_{data}$ in $TrajSet$
   IF (DSN($TrajPrefix$, $Traj_{query}$) $< \epsilon 1$) AND
   (DSL($TrajPrefix$, $Traj_{query}$) $< \epsilon 2$)
      Put $Traj_{data}$ into $TargetTrajSet$.
3. Classify the trajectories in $TargetTrajSet$ into $SimTrajGroup_i (1 \leq i < k)$ in such a way that for any two trajectories $Traj_{i,x}$ and $Traj_{i,y}$ in $SimTrajGroup_i$, their subtrajectories from $P_{current}$ to P_{dest} denoted by $TrajPostfix_{i,x}$, and $TrajPostfix_{i,y}$, respectively, should satisfy the condition that
   (DSN($TrajPostfix_{i,x}$, $TrajPostfix_{i,y}$) $< \epsilon 1$) AND
   (DSL($TrajPostfix_{i,x}$, $TrajPostfix_{i,y}$) $< \epsilon 2$).
4. **FOR** each group of trajectories, $SimTrajGroup_i (1 \leq i < k)$,
   Its prediction probability is assigned as $|SimTrajGroup_i| / |TargetTrajSet|$.

Algorithm 1. Predicting future paths

Algorithm 1 below shows the overall procedure of the proposed method for predicting future paths.

Fig. 3 shows an example of similar trajectory search. For user A, the starting position is SP_1, the destination is EP_1, and the moving path from the start to the current location is (R0-R1-R2-R3). Using these values, candidate trajectories whose the starting position and the destination are SP_1 and EP_1, respectively, are retrieved from all the past trajectories of user A. As shown in this figure, from the retrieved candidate trajectories, trajectories T1, T2, T3, and T4 were obtained, part of which are similar to a given path (R0-R1-R2-R3), and the corresponding frequencies were 15, 2, 5, and 6, respectively. Also, for those trajectories T1, T2, T3, and T4, the grouping is performed to build G1={T1, T2}, G2={T3}, G3={T4} and corresponding frequencies. The moving frequencies of G1, G2, G3 obtained from Fig. 3 are 17, 5, and 6, respectively. That is, the probability of moving by following T1 or T2 is 60.7%, the probability of moving by following T3 is 17.8%, and the probability of moving by following T4 is 21.4%. In the end, the predicted list of moving paths to the destination is provided to users in the decreasing order of moving probability.

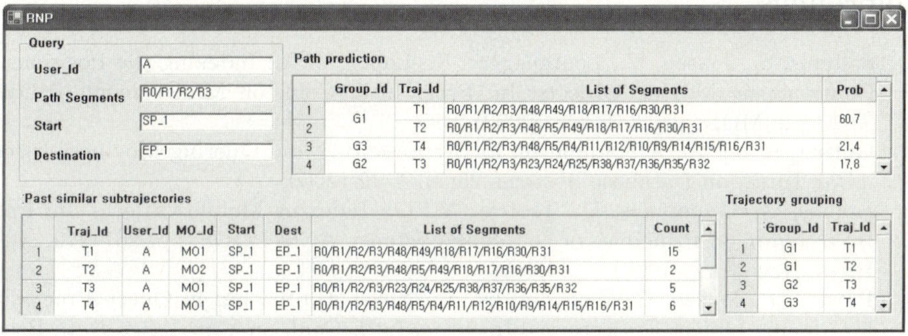

Fig. 3. Example of similar trajectory search

4 Conclusion

In this paper, we addressed the problem of predicting future paths of moving objects on a road network. For prediction of future paths, the proposed method takes an approach to retrieve the trajectories whose moving patterns are similar to that of a query trajectory from past trajectories of moving objects stored in databases. For this purpose, we represented a trajectory as a series of road segments which reflect the characteristics of a road network, proposed a novel similarity function which employs the number and the length of road segments, and proposed the method of predicting future paths based on these. Furthermore, to improve its prediction accuracy, moving paths with just small differences are grouped together and treated as one.

We did not conduct performance evaluation because the real-life trajectory data are not available due to privacy reasons. As future work, we will evaluate the performance of the proposed method for similar trajectory search and future path prediction to verify its accuracy and efficiency when the problem is resolved. In addition, the proposed method is applicable only for a special case satisfying the assumption that a driver, a moving path of an object to the current position, a starting position, and a destination are given. Hence, we plan to extend this approach so as to make it applicable for more general cases.

Acknowledgments

This research was supported by the MIC(Ministry of Information and Communication), Korea, under the ITRC(Information Technology Reserch Center) support program of supervised by the IITA(Institute of Information Technology Assessment)(IITA-2005-C1090-0502-0009) via Cheju National University. Sang-Wook Kim would like to thank Jung-Hee Seo, Suk-Yeon Hwang, Joo-Young Kim, and Joo-Sung Kim for their encouragement and support.

References

[1] Saltenis, S., Jensen, C., Leutenegger, S., Lopez, M.A.: Indexing the Positions of Continuously Moving Objects. In: Proc. Int'l. Conf. on Management of Data, ACM SIGMOD, pp. 331–342 (2000)
[2] Guting, R., et al.: A Foundation for Representing and Quering Moving Objects. ACM Trans. on Database Systems 25(1), 1–42 (2000)
[3] Kollios, G., Gunopulos, D., Tsotras, V.: On Indexing Mobile Objects. In: Proc. Int'l. Symp. on Principles of Database Systems, ACM PODS, pp. 261–272. ACM Press, New York (1999)
[4] Pfoser, D., Jensen, C., Theodoridis, Y.: Novel Approaches in Query Processing for Moving Object Trajectories. In: Proc. Int'l. Conf. on Very Large Data Bases, VLDB, pp. 395–406 (2000)
[5] Pitoura, E., Samaras, G.: Locating Objects in Mobile Computing. IEEE Trans. on Knowledge and Data Engineering 13(4), 571–592 (2000)
[6] Nascimento, M., Silva, J.: Towards Historical R-trees. In: Proc. ACM Symp. on Applied Computing, ACM SAC, pp. 235–240 (1998)
[7] Tao, Y., Papadisas, D., Sun, J.: The TPR*-tree: An Optimized Spatio-Temporal Access Method for Predictive Queries. In: Proc. Int'l. Conf. on Very Large Data Bases, VLDB, pp. 790–801 (2003)
[8] Beckmann, N., Kriegel, H., Schneider, R., Seeger, B.: The R*-tree: An Efficient and Robust Access Method for Points and Rectangles. In: Proc. Int'l. Conf. on Management of Data, ACM SIGMOD, pp. 322–331. ACM Press, New York (1990)
[9] Kim, K.-S., et al.: Fast Indexing and Updating Method for Moving Objects on Road Networks. In: Proc. IEEE Int'l. Conf. on Web Information Systems Engineering, pp. 34–42 (2003)
[10] Yanagisawa, Y., Akahani, J., Satoh, T.: Shape-Based Similarity Query for Trajectory of Mobile Objects. In: Proc. Int'l. Conf. on Mobile Data Management, pp. 63–77 (2003)
[11] Almeida, V., Güting, R.: Indexing the Trajectories of Moving Objects in Networks. Geoinformatica 9(1), 33–60 (2005)

[12] Benetis, R., et al.: Nearest Neighbor and Reverse Nearest Neighbor Queries for Moving Objects. In: Proc. Int'l. Conf. on Database Engineering Applications Symposium, IDEAS, pp. 44–53 (2002)
[13] Weqhe, N.V., et al.: Representation of Moving Objects along a Road Network. In: Proc. Int'l. Conf. on Geoinformatics (2004)
[14] Vazirgiannis, M., Wolfson, O.: A Spatiotemporal Model and Language for Moving Objects on Road Networks. In: Jensen, C.S., Schneider, M., Seeger, B., Tsotras, V.J. (eds.) SSTD 2001. LNCS, vol. 2121, pp. 20–35. Springer, Heidelberg (2001)
[15] Brinkhoff, T.: A Framework for Generating Network-based Moving Objects. GeoInformatica 6(2), 153–180 (2002)
[16] Hu, H., Lee, D., Lee, V.: Distance Indexing on Road Networks. In: Proc. Int'l. Conf. on Very Large Data Bases, VLDB, pp. 894–905 (2006)

Performance Analysis of WAP in Bluetooth Ad-Hoc Network System

Il-Young Moon

School of Internet Media Engineering,
Korea University of Technology and Education, Republic of Korea
iymoon@kut.ac.kr

Abstract. In this paper, it has analyzed performance enhancement for WAP (Wireless Application Protocol) in Bluetooth network system using a multi-slot segmentation scheme. In order for SAR to improve the transfer capability, the transmission of messages have been simulated using a fragmentation scheme that begins with the total package and incremental fragmentation for each layer using the WTP (Wireless Transaction Protocol) to define the resultant packet size and the level of fragmentation for each proceeding layer. The data is divided into individual packets at the baseband level. This scheme decreases transmission time of L2CAP (Logical Link Control And Adaptation Protocol) baseband packets by sending packets that span multiple slots. From the results, it was able to obtain packet transmission time and optimal WTP packet size for WAP in Bluetooth network system.

Keywords: Bluetooth, Ad-hoc Network, WAP.

1 Introduction

Until recently, as data communication services have become more broadly available, there is growing interest to provide services that take advantage of data capabilities as well. WAP (Wireless Application Protocol) was shaped to create a standards-based structure in which value-added data services can be deployed, ensuring some degree of interoperability [1]. As a result, WAP and HTML offers an interoperable presentation platform for end-user interfaces [2],[3]. Applications that will be increasingly popular in the future, such as man-machine and machine-machine interfaces, will drive WAP in Bluetooth environment to a new level of popularity. In many ways, Bluetooth can be used like other wireless networks with regard to WAP, supplying a bearer for transporting data between WAP clients and its closest WAP server. In the existing protocol, when a WAP client transports data to a WAP server, no more than 1 slot baseband packet in a WAP client transmits to the WAP server. This technique however, increases transmission time of baseband packets because only 1 slot baseband packet is sent by L2CAP at a time [4]. In contrast, a multi-slot segmentation scheme can be used to improve transport ability for WAP in Bluetooth network system. This scheme decreases transmission time for L2CAP baseband packets by sending packets spanning multiple slots. As illustrated in the simulation, it can achieve a greater efficiency for WAP packet transmission time using a multi-slot configuration

B. Apolloni et al. (Eds.): KES 2007/WIRN 2007, Part I, LNAI 4692, pp. 390–396, 2007.

as opposed to a typical 1-slot packet method. From this, it is able to extrapolate the optimal packet size for WTP of WAP in Bluetooth environment.

2 The Protocol Stack of WAP in Bluetooth Network

Bluetooth can be used to replace traditional medium and act as a bearer as specified by the WAP architecture. Bluetooth baseband describes the specification of the digital signal processing part of the hardware. A TDD (Time Division Duplex) scheme is used to resolve conflict over the wireless link, where each slot is 625ms long. A baseband packet normally occupies 1 slot, but can be extended to cover 3 or 5 slots. Above the baseband layer is the datalink layer, where both the LMP (Link Manager Protocol) and L2CAP are found. The LMP assumes the responsibility of managing connection states, enforcing equality among slaves and other management tasks. L2CAP supports a higher-level protocol for multiplexing and packet SAR, and also conveys quality of service information. For WAP connections over Bluetooth, when WAP clients are dynamically 'listening' for existing Bluetooth devices, the presence of a WAP server is detected using Bluetooth's service discovery protocol [5]. If the WAP client detects that communication has been lost with the WAP proxy/gateway, they may optionally decide to restart communication by repeating the above process.

The WAP protocol stack uses WDP (Wireless Datagram Protocol) at the lowest level for situations using WAP in Bluetooth network. This layer implements the bearer adaptation and is defined for a variety of bearers. WAP over Bluetooth differs from plain WAP in several aspects. For example, in the traditional WAP scenario, the terminal generally establishes connection to the server [6],[7]. In WAP over Bluetooth, however, the server itself has the ability to sense a nearby terminal and initiate a connection (connection established). Another difference lies in the coverage area for Bluetooth, which has a considerably shorter range that the WAP counterpart. Finally, bandwidth is higher in Bluetooth systems and a license free band is used. Figure 1 is a protocol stack of WAP in Bluetooth network system.

3 The Multi-Slot Scheme for WAP in Bluetooth Network

SAR(Segmentation and Reassembly) reduces overheads by spreading the packets used by higher layer protocols over several packets, covering 1, 3 or 5 slots in Bluetooth network [8]. It is defined slot limit as the maximum number of slots that cross the packet. The slot limit could be less than 5 due to a very high bit error rate in the wireless channel. This factor is passed by the LMP to the L2CAP through a signaling packet. The multi-slot segmentation scheme decreases transmission time of L2CAP packets by sending packets spanning multiple slots. This scheme is summarized in the following steps.

If slot_limit=5, divide 5 slot packets in the L2CAP packet. If the data remaining to be fragmented shall require more than 3 slot packets and less than slot packets, the data should be sent as 5 slot packets according to the L2CAP. Likewise, if the data to be sent exceeds 1 slot packet and is less than 3, the data should be transmitted as 3 slot packets. For smaller segments that are greater than 1 slot packet and less than 3 slot packets, the 3 slot configuration is defined by the protocol. Figure 2 depicts a flowchart of the multi-slot segmentation scheme by above steps.

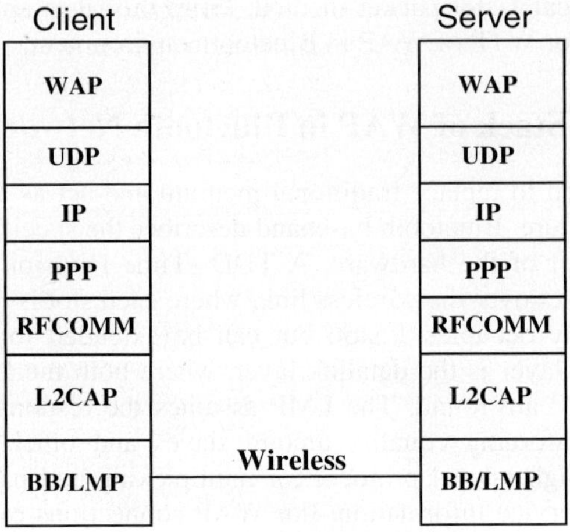

Fig. 1. The Protocol Stack of WAP in Bluetooth Network

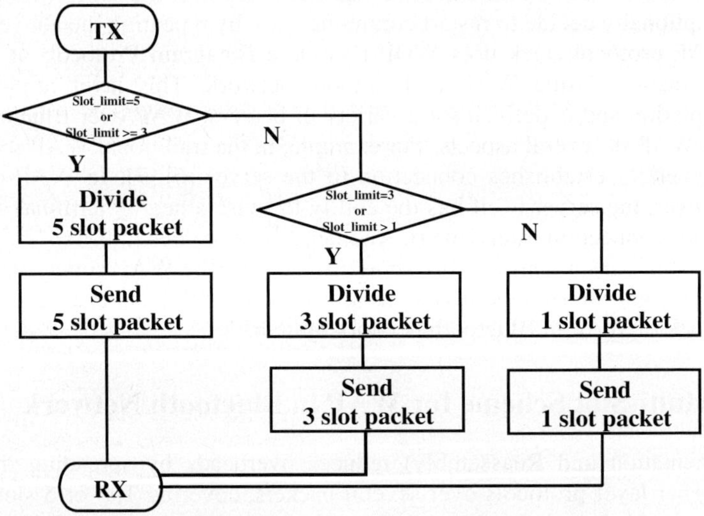

Fig. 2. The Multi-Slot Segmentation

4 GFSK Signal Model in Bluetooth Network System

GFSK (Gaussian Frequency Shift Keying) signal for Bluetooth can be written as

$$S(t) = \mathrm{Re}\left\{ \sqrt{\frac{2E}{T}} \exp\left(j2\pi \left\{ f_c t + h \int_{-\infty}^{t} g(t)dt \right\} \right) \right\}, \tag{1}$$

where, E is energy, T is time(period), fc is carrier frequency, h is modulation index, and $g(t)$ is the transfer function of Gaussian low-pass filter, and expressed as

$$g(t) = \sum_{K=-\infty}^{\infty} a_k v(t - kT) \quad , \tag{2}$$

where, ak = 1, -1,

$$v(t) = \frac{1}{2}\{erf(-\lambda B_b T) + erf(\lambda B_b (t+T))\} \tag{3}$$

where, $\lambda = \sqrt{2/\ln 2\pi}$, $B_b T = 0.5$, and $erf(t) = \int_0^t \frac{2}{\sqrt{\pi}} e^{-t^2} dt$

When the composite received signal consists of a large number of plane waves, the received complex envelope $g(t) = g_I(t) + g_Q(t)$ can be treated as a wide-sense stationary complex Gaussian random process. Some types of scattering environments have a specular or line-of-sight component. In this case, $g_I(t)$and $g_Q(t)$ are random Gaussian processes with non-zero means.

In order to simulate the BER performance for Bluetooth piconet, AWGN channel model is used to this paper. Figure 3 is BER performance of GFSK Bluetooth network system in AWGN channel model.

In addition, to achieve transmission time using multi-slot scheme for WAP in Bluetooth network, T_{MSG}, is defined as

$$\begin{aligned} T_{MSG} &= (K-1)T_{PKT}(q) + T_{PKT}(r) \\ &= (K-1)\frac{S_{TIME} \times q}{p} + \frac{S_{TIME} \times r}{p} \end{aligned} \quad , \tag{4}$$

where, $\lambda = \sqrt{2/\ln 2\pi}$, $B_b T = 0.5$, and $erf(t) = \int_0^t \frac{2}{\sqrt{\pi}} e^{-t^2} dt$

where, K is the number of total message packet, q and r is the number of time slot to be compute, $T_{PKT}(q)$ and $T_{PKT}(r)$ is a transmission time of WAP packet to be fragment q and r, S_{TIME} is slot time and p is a probability of data frame to be successfully transfer. Figure 4 depicts a process of segment transmission for WAP packet for the above equation.

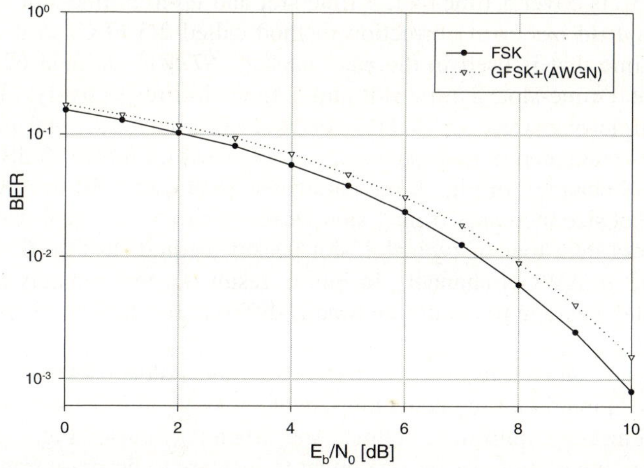

Fig. 3. BER performance of GFSK in Bluetooth network system

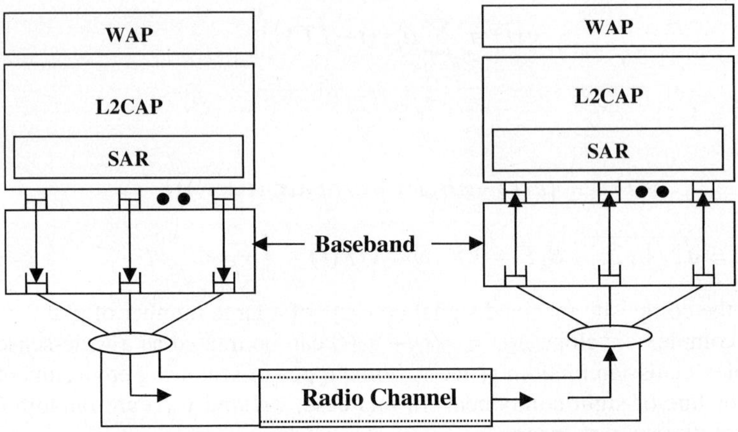

Fig. 4. System architecture of WAP in Bluetooth network

In this paper, simulation model of a Bluetooth network for WAP consists of a transmitter, wireless channel, and receiver. To find the transmission time of WAP, it must transmit the total message by first segmenting it into data packets. It values for the BER of the payload part in the receiver part has been calculated. In order to simulate the BER performance, an independent, static AWGN channel were assumed for every packet in a Bluetooth piconet network.

5 Multi-Slot Transmission Time in Bluetooth Network System

In wireless channel, it gained transmission time of packet and analyze BER performance of Bluetooth piconet network. A kind of packet used in this simulation is DM1, DM3, DM5 packet that carry data information only. DM stands for data medium rate. These DM packets cover 1 time-slot, 3 time-slot and up to 5 time-slot. Furthermore, DM packet payload has error correction method called 2/3 FEC, so it could acquire transmission time that is used in the each packet. *STIME* defined 625ms, 1875ms and 3125ms at 1 time-slot, 3 time-slot and 5 time-slot, respectively. For achieving transmission time of packet for WAP over Bluetooth network, total message transmission time is simulated at total packet size (5000 byte), *Eb/No* = 3 dB and *Eb/No* = 6 dB in AWGN channel. In Fig. 5, the parameter *Eb/No* set 3 dB in AWGN channel. When the packet size increases from 1 slot packet size to 5 slots packet size, transmission time is less than that of typical 1 slot packet method. In Fig. 6, the parameter *Eb/No* set 6 dB in AWGN channel. In Fig. 6, result is approximately the same with Fig. 5. But, total message transmission time is different by changing *Eb/No* in AWGN channel model.

From the result, when it used 3 slots packet size and 5 slots packet size rather than 1 slot packet size, a multi-slot segmentation scheme of WAP in Bluetooth network system decreases total message transmission time. And, when it evaluated Fig. 5 and 6, it found out that wireless transaction packet size ought to increase to decrease transmission time in wireless channel. Besides, considering BER in wireless channel, it is obtained appropriate wireless transaction packet size. Also, in case of optimal wireless transaction packet size is about 700 byte in AWGN model channel, the WAP packet transmission time considering trade-off between total message transmission time and wireless transaction packet size is about 210 ms (1 slot packet size), 120 ms and 115 ms (3 slot packet and 5 slot packet size) in AWGN model channel.

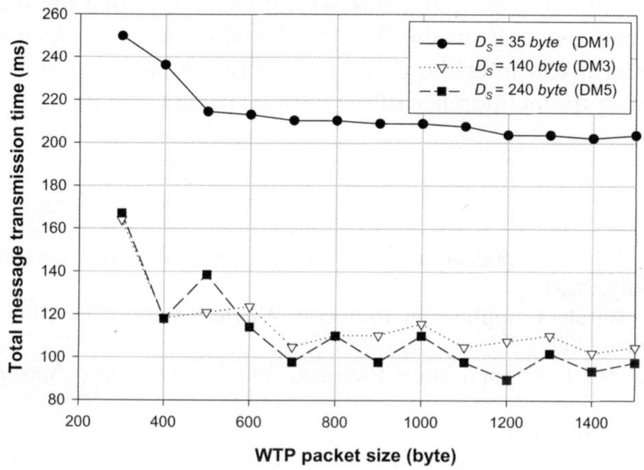

Fig. 5. Total transmission time of WAP in Bluetooth network system. (E_b/N_o = 3 dB, AWGN)

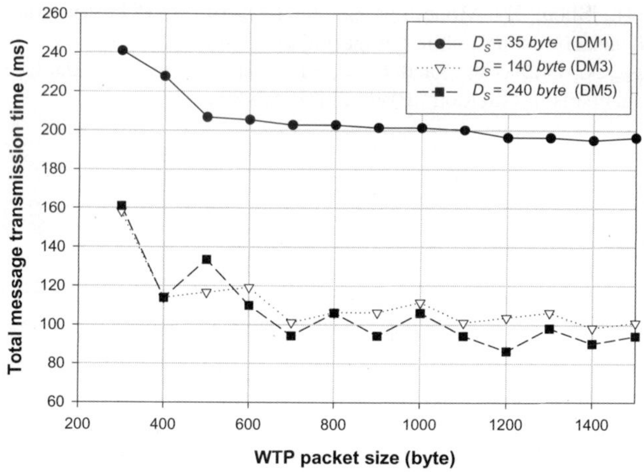

Fig. 6. Total transmission time of WAP in Bluetooth network system. (E_b/N_o = 6 dB, AWGN)

6 Conclusion

This paper has simulated WAP packet transmission times using multi-slot scheme. In order for SAR to progress the transfer capability, the whole messages are fragmented in WTP layer and segmented further as it passes through each layer towards the baseband layer, where the actual packets are sent sequentially. It is analyzed the WAP packet transmission time by changing *Eb/No* in AWGN model channels using DM1, DM3, and DM5 packets that carry a data payload. From the result, it can see that the multi-slot scheme of WAP in Bluetooth network system decreases the total message transmission time by using a multi-slot packet size as opposed to the single slot packet size transmission approach. Moreover, it can gather that the transmission time in wireless channel decreases as the WTP packet size increases. As a result, based on the data collected, it can infer the correlation between packet size and the transmission time, allowing for an inference of the optimal packet size in the WTP layer.

References

1. WAP forum, Wireless Application Protocol: Wireless Transaction Protocol Specification, Version 10 (July, 2001)
2. WAP forum, Wireless Application Protocol: Wireless Datagram Protocol Specification, Version 14 (June, 2001)
3. WAP forum, Wireless Application Protocol: WAP Architecture Specification, Version 12(July, 2001)
4. http://www.bluetooth.com
5. Hartwig, S., Rautenberg, T., Simmer, M., Temovic, D., van Bebber, A.: WAP over Bluetooth: Technology and applications. In: ICCE-2001. WAP forum, Wireless Application Protocol: Wireless Profiled TCP, version 31, pp. 12–13 (March, 2001)
6. Park, H.S., Heo, K.W.: Performance evaluation of WAP-WTP. The journal of the Korean institute of communication sciences 26(1A), 67–76 (2001)
7. Rutagemwa, H., Shen, X.: Modeling and Analysis of WAP Performance over Wireless Link. IEEE Transactions on mobile computing 2(3), 221–232 (2003)
8. Das, A., Ghose, A., Gupta, V., Razdan, A., Saran, H., Shorey, R.: Adaptive link-level error recovery mechanisms in bluetooth. PWC-2000 , 85–89 (2000)

Performance Evaluation of Embedded Garbage Collectors in CVM Environment

Chang-Il Cha[1], Sang-Wook Kim[1], Ji-Woong Chang[2], and Miyoung Shin[3]

[1] Department of Information and Communications, Hanyang University
{charose,wook}@agape.hanyang.ac.kr
[2] Department of Game and Multimedia Engineering, Korea Polytechnic University
jwchang@kpu.ac.kr
[3] School of Electrical Engineering and Computer Science, Kyungpook National University
shinmy@mail.knu.ac.kr

Abstract. Garbage collection in the Java virtual machine is a core function that relieves application programmers of difficulties related to memory management. In this paper, we evaluate the performance of GenGC and GenRGC, garbage collectors for the embedded Java virtual machine, CVM. To compare the performance of GenGC and GenRGC, we first evaluate the execution time of garbage collection and the delay time caused by garbage collection. Second, for more detailed performance analysis of GenRGC, we evaluate the execution time of garbage collection and the delay time caused by garbage collection while changing the sizes of a block and a frame. Third, we analyze the size of storage space required for performing GenRGC, and show GenRGC to be suitable for embedded environment with a limited amount of memory. Since CVM is the most representative one of embedded Java virtual machines, this performance study is quite meaningful in that we can predict the performance of garbage collectors in real application environments more accurately.

Keywords: Java, Java virtual machine, garbage collection, CVM.

1 Introduction

The garbage collection is a task of automatically collecting such memory objects that are no longer used and making them reusable [4]. To perform the garbage collection, it is necessary to distinguish the currently used objects from the objects which are no longer used by application programs. In this paper, we call *live objects* the objects that are currently used and *dead objects* the objects that are no longer used.

In the CVM, which is an embedded Java virtual machine developed by Sun Microsystems, a generational garbage collector [6] is used, and we call it *GenGC* in this paper. Since GenGC collects garbage for large regions at a time, however, time delay is too large to satisfy real-time requirements for embedded environments. GenRGC improves the problem of GenGC [2].

In this paper, we evaluate and analyze the performance of GenGC and GenRGC via experiments in CVM environment. Although some performance studies on various

B. Apolloni et al. (Eds.): KES 2007/WIRN 2007, Part I, LNAI 4692, pp. 397–404, 2007.

garbage collectors have been done earlier, most of them were done on simulation environments, not on actual Java virtual machines. Since CVM is the most popular one of embedded Java virtual machines, this performance study is quite meaningful in that we can predict the performance of garbage collectors in real application environments more accurately.

The remainder of this paper is organized as follows. In Section 2, GenGC and GenRGC are presented as related works. In Section 3, to evaluate the performance of GenGC and GenRGC in a comparative way, various experiments are performed and the results are discussed. Finally, in Section 4, this paper is concluded.

2 Related Works

2.1 GenGC

GenGC is based on generational garbage collection[7] which is known as being effective in application programs where most of objects live shortly[1]. The generational garbage collection divides the heap into more then two generational regions. GenGC divides the heap asymmetrically in to young generational region and old generational region. The former is relatively smaller than the later.

New objects are usually allocated to the young generational region. When the memory becomes short due to many new object allocations and overcrowding of the young generational region, garbage collection is attempted on the young generational region to acquire enough memory. We call it *young generational garbage collection.* The overhead of the young generational garbage collection is small since it is performed on a small part of a heap, the young generational region.

Live objects which are alive for a long time on the young generational region are moved to the old generational region. When we are no longer able to acquire memory by young generational garbage collection, the garbage collection is performed on the old generational region. We call it *old generational garbage collection.* The old generational garbage collection needs to identify live objects over large heap area, the old generational region, requiring a big overhead.

To minimize the execution frequency of the old generational garbage collection, when an old generational region becomes full, GenGC performs the young generational garbage collection instead of the old generational garbage collection. In this paper, we call such policy of GenGC *the delayed old generational garbage collection strategy.* This strategy reduces the execution frequency of the old generational garbage collection, but incurs the young generational garbage collection frequently.

Since garbage collection in GenGC is performed on a part of a heap, it is impossible to identify all the live objects. To solve this problem, the write-barrier, a mechanism that detects write attempts to a certain memory area, can be employed [8].

2.2 GenRGC

In GenGC, the old generational garbage collection is performed on the whole area of old generational region, so the time delay becomes large. Thus, this is not suitable for the embedded environment where real-time response is required.

GenRGC improves the problem of GenGC by dividing an old generational region into multiple equal-sized blocks[2]. The blocks become the units for object allocation. A certain number of blocks form a frame, which is a unit of garbage collection. Here, the number of such blocks is called a *frame size*.

The aged objects in the young generational region are promoted to the old generational region. The old generational garbage collection, unlike GenGC adopting the delayed old generational garbage collection strategy, is initiated when there is not enough space for moving the aged objects from the young generational region. At this time, the old generational garbage collection is performed on a frame.

Like GenGC, GenRGC employs the write-barrier to identify all the live objects efficiently. Furthermore, to resolve the problem of segmenting the old generational region into multiple frames, it uses the two-step write-barrier[2]. Using the write-barrier, GenRGC reduces the time delay of garbage collection by distributing the overhead required for tracking objects in a whole heap. However, the two-step write-barrier is employed which needs some additional space.

3 Performance Evaluation

GenRGC as well as GenGC were implemented and ported into CVM. Also, we employed SpecJVM98[5] as benchmark programs for our performance evaluation. SpecJVM98 consists of eight application programs that include a parser(_228_jack), a puzzle solver(_202_jess), a ray tracer(_205_raytrace), a ray tracer running in multi-threaded environment(_227_mtrt), an MPEG-3 decoder(_222_mpegaudio), a Java compiler(_213_javac), a simple database management system(_209_db), and a file compressing and decompressing system(_201_compress). Those programs are widely accepted for performance evaluation of Java garbage collectors.

Experiment 1. Comparisons Between GenGC and GenRGC

Experiment 1-1. Execution time of entire garbage collection with varying heap sizes
This experiment measures the execution time of entire garbage collection with varying heap sizes for a fixed size of a young generational region. Table 1 shows parameter settings in our experiments. The size of a young generational region is set to be 256KB for _222_jack and _202_jess, which use small memory while it is set to be 1MB for other programs. The block size is set to be from 128KB at minimum to the size which is large enough for largest objects in each benchmark program to be allocated. The base heap size is the minimum of heap sizes initiating the old generational garbage collection.

Table 1. Parameter settings for execution of benchmark programs

program	size of a young generational region(KB)	block size(KB)	base heap size(KB)
_228_jack	256	128	1,408
_202_jess	256	128	1,408
_227_mtrt	1,024	256	7,168
_213_javac	1,024	128	11,776
_209_db	1,024	2,048	8,704
_201_compress	1,024	4,096	8,704

Fig. 1 shows the trend of the execution time of entire garbage collection when each program is executed with varying heap sizes, just like 1, 1.5, 2, 2.5, 3 times the base heap size. The execution times of GenRGC are smaller than those of GenGC in almost all the experiments. This is because the time for identifying the live objects decreases in GenRGC since the old generational garbage collection performs on a small part of heap, a frame.

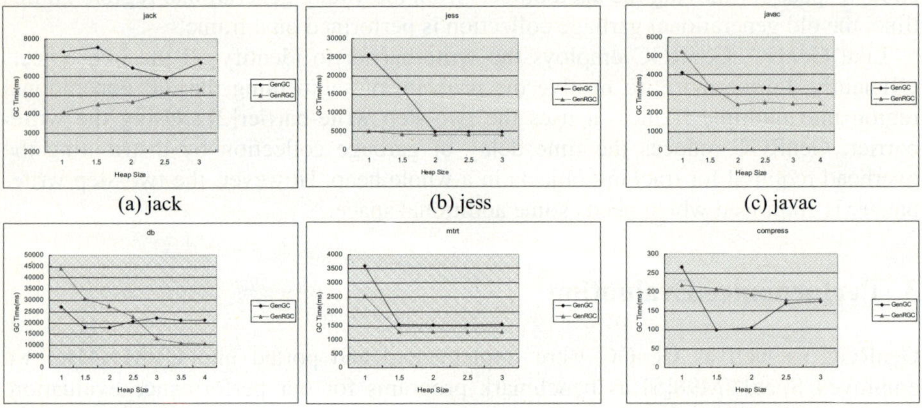

(a) jack (b) jess (c) javac

(d) db (e) mtrt (f) compress

Fig. 1. Execution time of entire garbage collection with varying heap sizes

The execution time of entire garbage collection generally tends to decrease as the heap size increases. This is because the execution frequency of garbage collection decreases, if the heap size increases.

In case of javac and db, when the heap size is small, GenGC shows better performance than GenRGC, but the performance is reversed for large heap size. This phenomenon occurs because the objects in the two programs live a long time. When the heap size is very small, the garbage collection only on a single frame leads to not enough space in GenRGC, then contiguous garbage collection occurs on other frames. However, such problem occurs just in case of abnormally small heap size.

We also have performed the other experiment that measures the execution time of applications with varying heap sizes for a fixed size of a young generational region. The execution time of applications shows similar trends to the execution time of entire garbage collection shown in Fig. 1. This means that the execution time of garbage collection affects overall execution time of applications largely.

Experiment 1-2. Execution time of garbage collection with varying sizes of young generational region
In this experiment, the heap size is fixed to be 32MB [1], the size of a young generational region is set to be from 32KB to 16MB by increasing it twice at a time. Fig. 2 shows the execution time of entire garbage collection with varying sizes of

[1] The heap size was set to be large enough to include the young and old generational region.

young generational region. Overall, as the size of young generational region increases, the execution time of entire garbage collection tends to decrease. This is because the execution frequency of the young generational garbage collection decreases owing to the easiness of object allocation for large size of young generational region. On the other hand, since the size of an area on which garbage collection should perform increases, the execution time of each young generational garbage collection also increases.

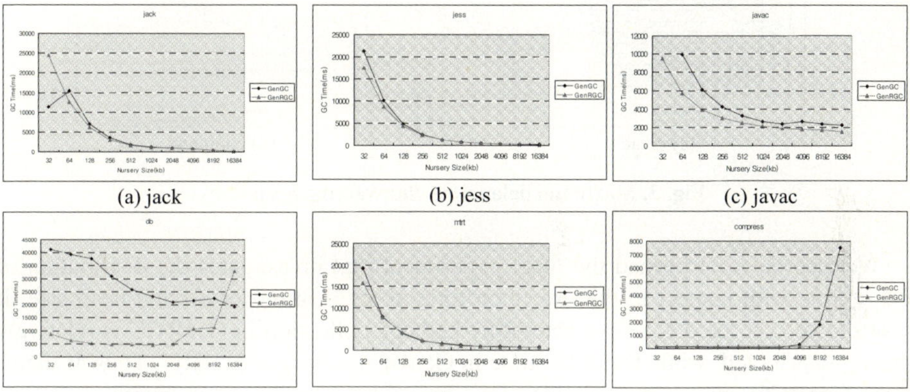

(a) jack (b) jess (c) javac

(d) db (e) mtrt (f) compress

Fig. 2. Execution time of entire garbage collection with varying sizes of young generational region

In case of db, GenRGC increases the execution time of entire garbage collection drastically when the size of young generational region is larger than 2,048KB. However, in case of compress, GenGC increases the execution time of entire garbage collection drastically when the size of young generational region is larger than 4096KB. These distortions occur due to the old generational garbage collection strategy.

Experiment 1-3 Maximum delay time with varying heap sizes

In this experiment, the size of young generational region and a block size are set to be the same as in Table 1, and while a frame size is fixed to be 1, maximum delay times of each method are measured with increasing the basic size of a heap by 1, 1.5, 2, 2.5, and 3 times. Then, the distribution of delay time with respect to the execution time of application is investigated for a fixed heap size.

Fig. 3 shows maximum delay time caused by a single execution of old generational garbage collection when each program is executed with varying heap sizes. GenGC shows the increase of the maximum delay time caused by the old generational garbage collection with the increase of a heap size, while GenRGC shows almost its consistency. This is because GenRGC always performs the garbage collection gradually for the fixed size of area.

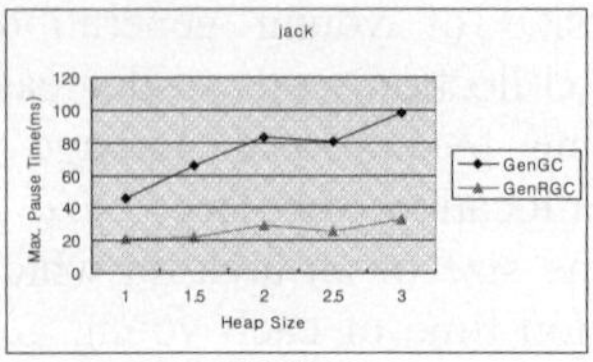

(a) jack

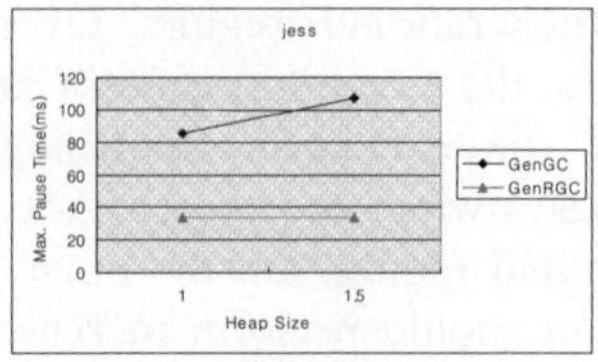

(b) jess

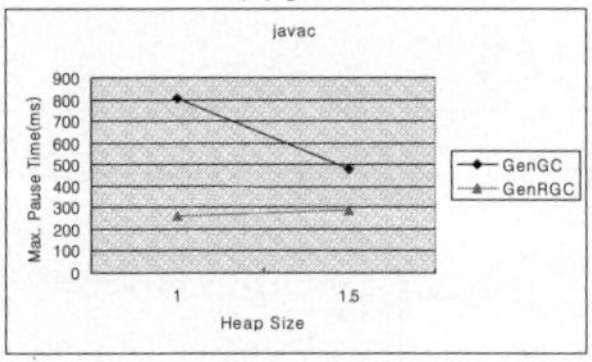

(c) javac

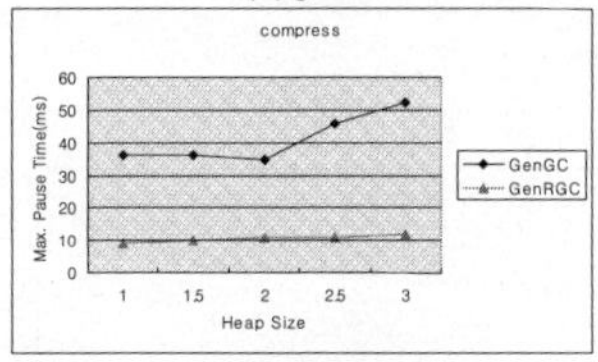

(d) compress

Fig. 3. Maximum delay time with varying heap sizes

We also have performed the other experiment that investigates the distribution of delay time with respect to the execution time of application for a fixed heap size. It was shown that the delay time caused by garbage collection in GenRGC is relatively evenly distributed, while the delay time in GenGC is changed greatly[2]. Thus, it seems that GenRGC is more suitable for embedded environments with real-time requirements.

Experiment 2. Performance evaluation of GenRGC

Experiment 2-1 Execution time of garbage collection with varying block and frame sizes

In this experiment, the size of young generational region was set to be the same as in Experiment 1-1. The execution time of garbage collection in GenRGC was evaluated by increasing a block size from 12KB to 1MB by 2 times and a frame size from 1 to 8 by 2 times, while heap size is fixed as a base size. We used all the programs used in Experiment 1, except db and compress[3].

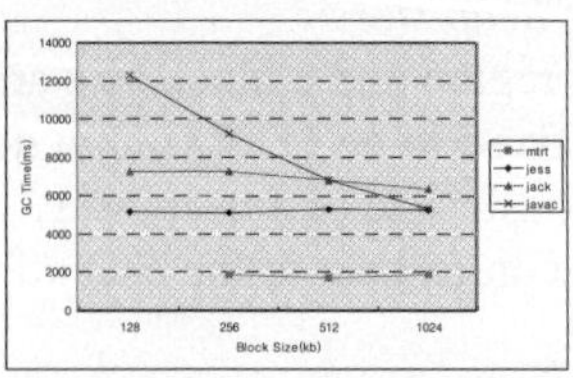

(a) Execution time of garbage collection with varying block sizes.

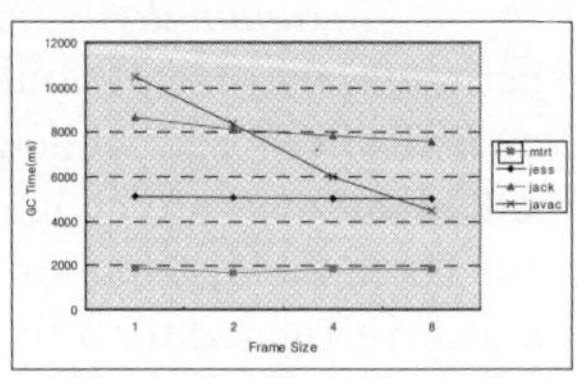

(b) Execution time of garbage collection with varying frame sizes.

Fig. 4. Execution time of garbage collection with varying block and frame sizes

2 Due to the limit of the paper length, we omit the detailed results here.

3 db and compress generate the objects of 1MB size or larger which require the block size to be at least 1MB or more. It is not suitable for embedded environments.

Fig. 4(a) shows the execution time of entire garbage collection with varying block sizes in GenRGC. Fig. 4(b) shows the execution time of entire garbage collection with varying frame sizes. In both figures, the execution time of entire garbage collection tends to be decreased with the increase of block sizes and frame sizes. This is because the execution frequency of entire garbage collection decreases when block sizes and frame sizes are increased.

Experiment 2-2. Maximum delay time with varying block sizes
Fig. 5 shows maximum delay time by the old generational garbage collection, when the heap is fixed as a base size and the block size is varied. In the figure, the maximum delay time increases with the increase of block size. This is because the area on which garbage collection should perform increases with the increase of block sizes. In this experiment, it is shown that the maximum delay time can be controlled by the change of block sizes. This means that the real-time requirements of applications can be satisfied by taking an appropriate size of blocks in GenRGC.

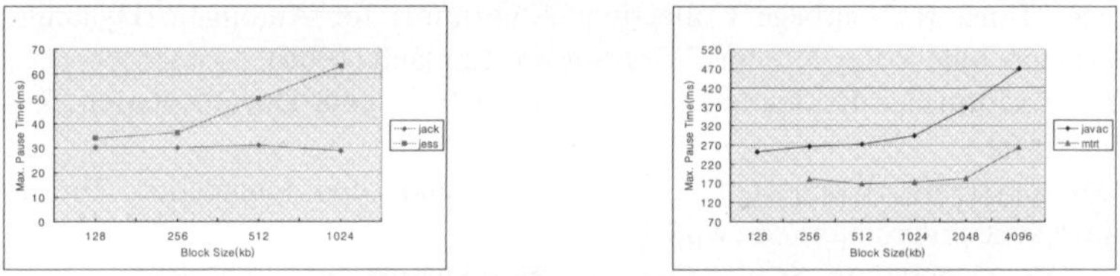

(a) jack and jess

(b) javac and mtrt

Fig. 5. Maximum delay time with varying block sizes

4 Conclusion

In this paper, via experiments, we evaluated the performance of GenGC and GenRGC on a real CVM environment. In the first experiment, the execution time and the delay time of garbage collection were compared between GenGC and GenRGC with varying heap sizes and the sizes of young generational region. In most experiments, GenRGC showed better performance than GenGC, having smaller execution time of garbage collection. Also, the maximum delay time of GenRGC was smaller than that of GenGC, with being evenly distributed, which is saying that GenRGC is more suitable for embedded environments and real-time requirements.

In the second experiment, for the detailed analyses of the performance of GenRGC, we measured the execution time of garbage collection and the maximum delay time with varying block and frame sizes of heaps. As results, with the increase of block sizes and frame sizes, the execution time of garbage collection was decreased while the maximum delay time was increased. This means that the execution time of garbage collection and the maximum delay time can be controlled by taking appropriate values of block sizes and frame sizes for applications. Also, we measured the storage space required for the use of GenRGC. As a result, the additional storage space required by GenRGC was at most 3% of a whole heap, showing that GenRGC can work under the embedded environments with a limited memory.

Acknowledgment

This research was supported by the MIC(Ministry of Information and Communication), Korea, under the ITRC support program of supervised by the IITA(IITA-2005-C1090-0502-0009) via Cheju National University.

References

1. Blackburn, S., Cheng, P., McKinley, K.: Myths and Reality: The Performance Impact of Garbage Collection. In: Proc. Int'l. Conf. on Measurement and Modeling of Computer Systems, SIGMETRICS, pp. 25–36 (2004)
2. Cha, C., et al.: Garbage Collection in an Embedded Java Virtual Machine. In: Gabrys, B., Howlett, R.J., Jain, L.C. (eds.) KES 2006. LNCS (LNAI), vol. 4251, Springer, Heidelberg (2006)
3. Chen, G., et al.: Tuning Garbage Collection in an Embedded Java Environment. In: Proc. Int'l. Symp. on High-Performance Computer Architecture, HPCA, pp. 92–103 (2002)
4. Jones, R., Lins, R.: Garbage Collection: Algorithms for Automatic Dynamic Memory Management. John Wiley & Sons, West Sussex, England (1996)
5. Standard Performance Evaluation Corporation, SPECjvm98 Documentation, Release 1.04 Edition (2001)
6. Sun Microsystems, Connected Device(CDC) and the Foundation Profile (2006), http://java.suncom/products/cdc/wp/CDCwp.pdf
7. Ungar, D.: Generation Scavenging: A Non-Disruptive High Performance Storage Reclamation Algorithm. ACM SIGPLAN Notices 19(5), 157–167 (1984)
8. Zorn, B.: Barrier Methods for Garbage Collection, Technical Report CU-CS-494-90, University of Colorado (1990)

Time Discretisation Applied to Anomaly Detection in a Marine Engine

Ian Morgan, Honghai Liu, George Turnbull, and David Brown

Institute of Industrial Research, The University of Portsmouth,
Portsmouth, PO1 3QL, England, UK
{ian.morgan,honghai.liu,george.turnbull,david.j.brown}@port.ac.uk

Abstract. An introduction to the problems associated with anomaly detection in a marine engine, explaining the benefits that the SAX representation brings to the field. Despite limitations in accuracy of the SAX representation in comparison with the normalised time series, we conclude that because of the reduction in data points that should be processed SAX should be considered further as a valid and efficient representation. Finally, a continuation of the work to make the approach more viable in the real world is briefly noted based upon Markov Chaining and Support Vector Machines.

Keywords: Time series, discretisation, anomaly detection, symbolic aggregate approximation.

1 Introduction

Marine engines are large mechanisms requiring constant maintenance during operation and due to the nature of the industry incurs massive expense if a ship is delayed through missed deadlines, recovery costs and working hours, seen recently with the incident involving the container ship MSC Napoli. Furthermore, the cost of lubricants can be seen to be just as expensive as maintenance of the entire engine [1] and hence the reduction of wear is of paramount importance. There is therefore much interest in predicting events in a ship's mechanism before they occur, and in some cases, events such as a necessary cylinder change could have been identified upto two weeks in advance, allowing the ship to continue normal operation if it had been previously detected as such [2].

Any implementation must function in conjunction with the SEA-Mate architecture[1] installed upon the *Sine Maersk* which manually receives aperiodic oil samples from over fifty locations aboard ship. These samples are scanned for the concentration of nine elements; iron, calcium, sulphur, copper, zinc, lead, nickel, chromium and vanadium, whereupon a graphical display of the trends can be accessed. It is these trends that are available for the algorithm to analyse, though

[1] Trademark of A.P. Moller and Rivertek Ltd. A unit aboard the ship where oil samples are plugged into and concentrations of elements are analysed using ferrography and spectrometry.

B. Apolloni et al. (Eds.): KES 2007/WIRN 2007, Part I, LNAI 4692, pp. 405–412, 2007.

it should be noted that samples may be taken as little as once or twice per week, and hence approaches used in, for example, anomaly detection in gas turbines [3] or flight data [4] are not entirely appropriate due to the regularity of samples taken.

Elemental analysis of oil is utilised to observe possible events within the engine; increased levels of iron, lead or zinc in the scavenge sample for example suggests scuffing of the piston [5]. Much previous work in the elemental analysis of marine engines has focussed upon expert or rule based systems [5, 1]. Where all of the necessary data is available, this approach is sufficient to model a finite number of anomalies, though where data is sparse, anomalies that are not present in the knowledge base may be ignored. The SEA-Mate architecture does not measure a number of important elements, for example viscosity, water content, aluminium or silicon; all possible indicators of external oil contamination. Viscosity is especially important as it will change the lubrication properties of the oil, possibly resulting in increased wear or higher heat generation due to the formation of sludge [5]. Consequently it should be trends and not absolute values that are considered, due to the imprecise nature of the collected data.

Taking into account the points mentioned above, a framework is proposed based upon the work described in [6], [7] and [8], where a good explanation of the underlying principles is provided. In short, the approach taken in this paper consists of discretisation of a time series into a collection of states which can be used to form a vector of elements for each cross-section along a time series, hereafter referred to as a node. In this paper, the efficacy of time series discretisation proposed by [8] is investigated by observing the classification accuracy of a standardised algorithm, LIBSVM [9] upon two datasets; the normalised feed and the descretised feed, which both encapsulate the same information.

2 Preliminaries

Although a good overview of time series discretisation is given in [6] and [8] a brief overview of the method and implementation will be given here.

In both data mining and investigation of a time series, it is necessary to find a compromise between computational complexity and accuracy, where an adequate summarisation of the time series reduces the data in such a way that the classification of an algorithm is not significantly reduced for the sake of efficiency. This of course is a subjective measure, and hence the designer should decide which they are more willing to sacrifice, as the two are in many approaches mutually exclusive. There are a number of methods of representation that are currently in use, including the Discrete Fourier Transform [10], Discrete Wavelet Transform [11] and Symbolic Aggregate approXimation (SAX) [8], the representation focussed upon in this paper.

2.1 Piecewise Aggregate approXimation

SAX as described in [8] is a technique to reduce the numerosity of the data, allowing traditional data mining techniques such as Markov Chains or Suffix trees

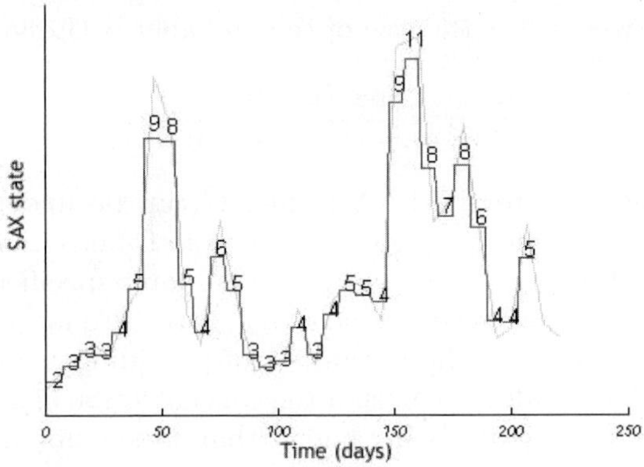

ss

Fig. 1. An example of the measured concentration of iron from cylinder 1 plotted against the equivalent SAX representation with associated state labels. Light grey represents the normalised data feed, and dark grey the PAA frames. The model discussed in this paper uses 12 cylinders with 7 elements per cylinder.

to be applied, which require discrete states (Figure 1). The reduction in data size is achieved using an approach known as Piecewise Aggregate Approximation (PAA) reducing the numerosity of the data by dividing the time series into equally sized frames, where all the datapoints within the frame are aggregated so that a single value can be extracted. Though this is a relatively simple approach, Lin and Keogh et al. [8] claim that this has "been shown to rival more sophisticated dimensionality reduction techniques like Fourier transforms and wavelets". PAA has been demonstrated to 'lower bound', or closely match the original time series however it should also be noted that equally sized frames may filter out significant information, such as peaks [12].

$$\bar{c}_i = \frac{w}{n} \sum_{j=\frac{n}{w}(i-1)+1}^{\frac{n}{w}i} c_j \tag{1}$$

More formally, this can be defined as a time series $C = c_1, \ldots, c_j$, where $\bar{C}$ is a collection of frames $\bar{c}_1, \ldots, \bar{c}_i$ and $w = \left|\bar{C}\right|$, reducing the time series from n dimensions to w dimensions. The indices are to ensure that the correct frame is being processed in relation to the individual datapoint (Equation 1).

2.2 Symbolic Aggregate approXimation

SAX applies states to the PAA representation, the result of which is referred to as a *word* or *string* [8]. As can be seen in Figure 1, the alphabet upon which the states are based in this case is numerical though their exact forms are unimportant and merely act as a unique distinguishing label.

A variation used in this paper upon the original implementation was to select state labels or *breakpoints* based upon the input data by dividing the range of the input data into equi-probable regions and using these same state labels at the testing stage where $n =$ the size of the alphabet α (Equation 2).

$$\alpha = \frac{\max(c) - \min(c)}{n} \tag{2}$$

The application of α to the PAA frames $\bar{C}$ can be understood as the new state $\bar{c}_i'$ is created when the aggregated value from frame $\bar{c}_i$ has a label j applied to it if the value of $\bar{c}_i$ is greater than the breakpoint specified at α_{j-1} and less than the breakpoint at α_j, where $\alpha = \alpha_1, \alpha_2, \ldots \alpha_n$ (Equation 3). In relation to the target domain, this has the advantage of providing a comparison for each element over the 12 cylinders, so even if the concentration of a particular element is low, it can still be noted as being higher than in neighbouring cylinders.

$$\bar{c}_i' = j, \alpha_{j-1} \leq \bar{c}_i < \alpha_j \tag{3}$$

As can be seen, SAX is a lightweight and reasonably intuitive approach for discretisation of a time series, however because of the specificities of the domain, it is necessary to compare a dataset that has been normalised with one that has been discretised using the SAX representation.

2.3 LIBSVM and Support Vector Machines

LIBSVM [9] is a C Support Vector Machine library that is run from the command line. It has been used in a number of other studies as a freely available standardised algorithm, and as in this case, run with the default tuning parameters [13,14]. Refer to [9] or [14] for a more in depth explanation of the package. The Support Vector Machine (SVM) stems from research into statistics and machine learning, and provides an approach which is similar in effect to neural networks, however one which is very different in its implementation. The current formulation of the SVM is based upon pre-processing the data into a high dimensional feature space and then calculating the number of *support vectors*[2]. The approach used in this paper is the classification of separable binary data, where $y = \{1, -1\}$. This is accomplished by utilising the principle of a *Maximal Margin Hyperplane* to separate the two classes. In effect, maximising the margin between the classes improves the generalisation ability of the SVM, and is related to the use of a weight decay factor in neural networks [15]. The support vectors are then calculated to be points with a non-zero lagrangian multiplier α_i, $i \in \{1, 2, \ldots, l\}$ where $l =$ number of training patterns.

3 Experimental Evaluation

The time series is provided as an NxM matrix, where N is equivalent to the number of elements scanned and M are the number of samples taken. In this paper, the

[2] Support vectors are datapoints which 'support' the margin, and hence where the name of the approach originates.

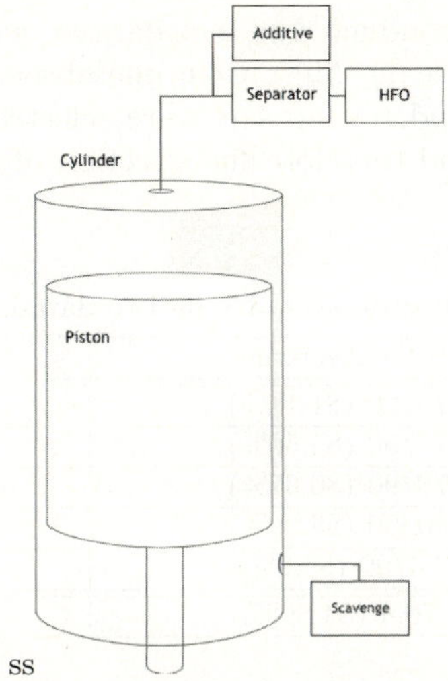

Fig. 2. A simplified diagram of the sampling points on a cylinder, where scavenge is the lubrication oil sampled subsequent to combustion. High levels of iron would indicate significant wear on the cylinder wall, which should be compared to the separator input prior to the combustion process ensuring the high iron content originated from the cylinder.

focus is placed upon seven of the measured elements, excluding calcium and sulphur as these are primarily found in heavy fuel oil (HFO) rather than lubricating system or cylinder oil (MESO and MECO). The MECO samples are taken from the scavenge outlet on each of the twelve cylinders, as can be seen in Figure 2.

It is necessary to note here a number of preprocessing steps taken on the time series. The values were initially normalised against engine specifications for the normal, maximum concentration expected. If during ship operation, the concentration rises above specification this does not necessarily indicate an anomaly and may be a result of higher engine load, operating in heavy weather, or at engine startup or shutdown [2]. This ensures that all the elements are compared equivalently, where a concentration between 0 and 1 represents expected levels.

Assuming the timescale is 250 days, the existing data points are interpolated across this scale to create a daily sample for all feeds. If two samples were taken on the same day, these are averaged and a single value extracted. This is not ideal, as the interpolation over 250 days of around 50 datapoints makes many assumptions as to the presence of events. Location of anomalies is significant, so a user can be alerted to a detected anomaly in a sub-mechanism of the ship, therefore these steps are taken separately for each feed.

Each node, or vector of elements, is given a target value of 1 or -1 depending on whether any one of the elements has an unexpected concentration. This is a relatively naïve approach to classification, however is a simple estimator of target value that can be used for comparison. The parameters for the SAX representation were kept constant at $n = 10$ and $w = 30$. A training and testing set were then selected randomly with a 4:1 ratio on all observations.

Five observations were conducted on each dataset, with the average percentage extracted from each (Table 1). More than one observation was conducted per dataset as the training and testing sets were selected randomly, though from the same universal set, and therefore the selection of points will differ between observations.

Table 1. SVM classification accuracy on two datasets over 5 observations

	SAX Accuracy	Raw Accuracy
	73/90 (81.11%)	630/663 (95.02%)
	77/90 (85.56%)	626/663 (94.4%)
	78/90 (86.67%)	624/663 (94.12%)
	80/90 (88.69%)	628/663 (94.72%)
	74/90 (82.2%)	624/663 (94.12%)
Mean	76.4 (84.89%)	626.4 (94.48%)

Unsurprisingly, as can be seen in Table 1 the accuracy of classification is higher with the raw dataset than with the SAX representation.

From these results it can be assumed that the SAX representation is not a completely accurate summary of the raw time series, however the reduction of accuracy compared to the reduction in data points is, in this domain, a sufficiently small compromise that can be accepted. Furthermore, the reduction of data points is important in classification, for example, a hypothesis that becomes too complicated, like a decision tree with a leaf for every training pattern, can overfit the training data and will make poor predictions on unseen input [16]. It should however be noted that SVMs demonstrate that it is possible to generalize well despite, in some cases, 'infinite' capacity; in the majority of cases "SVM generalization performance (i.e. error rates on test sets) either matches or is significantly better than that of competing methods" [17], and hence rarely suffer from overfitting.

The number of training patterns may also affect the classification accuracy of the SVM, as in the case of the SAX time series the number of patterns available to the model is four times less than that of the normalised data set.

The method used in this paper of identifying anomalies in a time series would not work in a real system and was used for comparison only. Therefore, a possible extension to this work will now be noted.

It is unrealistic to consider nodes as independent entities, or even as bigrams as in a standard Markov Chain and therefore a method of locating subpatterns should be found that is efficient. There are many methods of locating subpatterns within a time series, promising techniques include Probabilistic or Prediction suffix trees [8, 6, 7, 18]. A simpler approach can be considered within this domain however, that of considering only upward trends and treating each upward trend as a new subpattern, whereas a downward trend suggests that a man-made intervention has occurred, and hence is not significant[3]. Therefore, transition probabilities between individual elements can be used as attribute values; low transition probabilities are tagged as -1, as these patterns are unlikely from the *a priori* time series and high probabilities are tagged as +1. Refer to [19] for a combination of the two algorithms.

4 Concluding Remarks

This paper presents the SAX representation to the field of anomaly detection in marine engines, where a robust and fuzzy approach is required to ignore uncertainties in the time series. It has been demonstrated that the discretisation of the time series reduces accuracy of classification in comparison to a normalised time series, however we conclude that this is a suitably small compromise to the scale of the reduction that takes place; coupled with the advantages of being able to apply traditional data mining approaches to the discretised series. Finally, we have suggested an approach to enable further work in combination with SAX which demonstrates some promising properties for continuation in this particular field, though this should also be compared to more well known approaches such as DWT and DFT to appreciate the distinctions between representations, and has scope for further work.

Acknowledgements

The authors would like to thank the OACG, Steven Wilson (Rivertek-Industrial Ltd.) and Terry Robinson (Teedro Ltd.) for their collaboration in this work.

References

1. Dragsted, J., Bergeson, O.: Influence of low cylinder consumption on operating cost for 2-stroke engines. International council on combustion engines, CIMAC Congress, Kyoto 9 (2004)
2. Wilson, S.: System functional specification for elemental analysis system. Technical report. Rivertek Industrial Ltd (2005)
3. Palade, V., Patton, R., Uppal, F., Quevedo, J., Daley, S.: Fault diagnosis of an industrial gas turbine using neuro-fuzzy methods. In: Proceedings of the 15th IFAC World Congress, pp. 2477–2482 (2002)
4. Yan, W., Goebel, K., Li, C.: Flight regime mapping for aircraft engine fault diagnosis. In: Proceedings of the 58th Meeting of the Society of Mechanical Failures Prevention Technology, pp. 153–164 (2004)
5. Macian, V., Tormos, B., Sala, A., Ramirez, J.: Fuzzy logic-based expert system for diesel engine oil analysis diagnosis. Insight 8, 1–8 (2006)
6. Keogh, E., Chakrabarti, K., Pazzani, M., Mehrotra, S.: Dimensionality reduction for fast similarity search in large time series databases. Knowledge and Information Systems 3(3), 263–286 (2001)

[3] This could include the introduction of an alkali to reduce the acidity of the fuel, or manual intervention has occurred.

7. Keogh, E., Lonardi, S., Chiu, B.: Finding surprising patterns in a time series database in linear time and space. In: Proceedings of ACM Knowledge Discovery and Data Mining, pp. 550–556 (2002)
8. Lin, J., Keogh, E., Lonardi, S., Chiu, B.: A symbolic representation of time series, with implications for streaming algorithms. ACM Workshop on Research Issues in Data Mining and Knowledge Discovery, 2–11 (2003)
9. Chang, C.-C., Lin, C.-J.: LIBSVM: a library for support vector machines, Software (2001), available at http://www.csie.ntu.edu.tw/~cjlin/libsvm
10. Agrawal, R., Faloutsos, C., Swami, A.: Efficient similarity search in sequence databases, pp. 69–84 (1993)
11. Chan, K.p., Fu, A.W.-C.: Efficient time series matching by wavelets. ICDE (1999)
12. Lkhagva, B., Suzuki, Y., Kawagoe, K.: New time series data representation ESAX for financial applications. ICDE Workshops (2006)
13. Manevitz, L., Yousef, M.: One-class svms for document classification. Journal of Machine Learning Research 2, 139–154 (2001)
14. Lovell, B., Walder, C.: Support vector machines for business applications. Business Applications and Computational Intelligence, 267–290 (2006)
15. Suykens, A., Gestel, T., Brabanter, J., Moor, B., Vandewalle, J.: Least Squares Support Vector Machines, K.U. Leuven, Belgium (2002)
16. Cristianini, N., Shawe-Taylor, J.: An introduction to Support Vector Machines and other Kernel-based learning methods. Cambridge University Press, Cambridge (2000)
17. Burges, C.: A tutorial on support vector machines. Knowledge Discovery and Data Mining, 1–43 (1998)
18. Largeron-Leteno, C.: Prediction suffix trees for supervised classification of sequences. Pattern Recognition Letters 24, 3153–3164 (2003)
19. Altun, Y., Tsochantaridis, I., Hofmann, T.: Hidden markov support vector machines. In: Proceedings of the Twentieth International Conference on Machine Learning (ICML) (2003)

Using Weak Prior Information on Structures to Learn Bayesian Networks

Massimiliano Mascherini[1] and Federico M. Stefanini[2]

[1] European Commission, Joint Research Centre
Via E. Fermi 1, 21020 Ispra(VA), Italy
massimiliano.mascherini@jrc.it
[2] Dipartimento di Statistica "G.Parenti", Universita' di Firenze
Viale Morgagni 59, 50134, Florence, Italy
stefanini@ds.unifi.it

Abstract. Most of the approaches developed in the literature to elicit the a-priori distribution on Directed Acyclic Graphs (DAGs) require a full specification of graphs. Nevertheless, expert's prior knowledge about conditional independence relations may be weak, making the elicitation task troublesome. Moreover, the detailed specification of prior distributions for structural learning is NP-Hard, making the elicitation of large networks impractical. This is the case, for example, of gene expression analysis, in which a small degree of graph connectivity is a priori plausible and where substantial information may regard dozens against thousands of nodes. In this paper we propose an elicitation procedure for DAGs which exploits prior knowledge on network topology, and that is suited to large Bayesian Networks. Then, we develop a new quasi-Bayesian score function, the *P*-metric, to perform structural learning following a score-and-search approach.

Keywords: Prior information, structural learning, Bayesian Networks.

1 Introduction

Bayesian Networks (BNs), [1], are a widespread tool in many areas of artificial intelligence and automated reasoning because they perform probabilistic inference through very efficient algorithms. However, the problem of searching the BN that best depicts the dependence relations entailed in a database of cases it is hard to solve. Structural learning exploits algorithms which typically combine expert's knowledge with the information gathered from a database.

The complete specification of a prior distribution on the topology of a Bayesian Network (BN) is NP-Hard [2]. Most of the approaches in the literature require a complete specification of a prior probability distribution on the space of Directed Acyclic Graphs (DAGs). Nevertheless, there are problem domains in which such complete elicitation is difficult or infeasible, due to the lack of enough information to completely specify one network.

In this paper we develop a method to elicit partial beliefs about network structure without requiring the a-priori complete specification of structures. Elicited beliefs are refined by means of dissimilarity measures on network's topology. In order to perform

B. Apolloni et al. (Eds.): KES 2007/WIRN 2007, Part I, LNAI 4692, pp. 413–420, 2007.

structural learning in a score-and-search framework, we propose a new score function to evaluate causal Bayesian Networks: the P-metric. It is a quasi-Bayesian score obtained by modifying the Bayesian Dirichlet Equivalent metric, [3]. The peculiarity of a likelihood equivalent metric is to assign the same likelihood value to structures entailing the same conditional independence assertions. The P-metric is not likelihood equivalent and it exploits prior information to discriminate among causal structures within equivalence classes.

The paper is organized as follows: in section 2 after a general description of earlier approaches to elicit prior information on structures we detail our approach. Then a new elicitation procedure using the P-metric is presented in section 3. Numerical results from the analysis of some Machine Learning benchmark datasets are presented in section 4. Finally, in section 5, we present conclusions and issues to be addressed by further research.

2 From Prior Information to Score Functions

The elicitation of prior beliefs on network's structure has been not much considered in the literature. A straightforward elicitation of prior beliefs on complex structures is performed element-by-element assigning (subjective) probability values to graphs defined on a given set V of nodes. The enumerative approach is infeasible out of networks with a very small set of nodes because the space of DAGs has superexponential cardinality while increasing the number of nodes in V.

A simpler approach puts a uniform prior distribution on a subset H of all possible DAGs, [4], therefore some structures are a-priori excluded from the scoring procedure. Bounds on the number of parents/children are established to set hard constraints on elements in H.

Two more elaborated approaches have been proposed by [5, 6] to define a prior distribution on the space of BN structures. Both of them require a complete specification of beliefs over the network making their implementation not very practical in large networks. In fact the elicitation of expert's prior information element-by-element is performed through the assignment of (subjective) probability values to all possible arrows of a Bayesian Network, as in [5], but it becomes very difficult due to the superexponential cardinality of the space of structures for an increasing number of nodes. In large networks, a coherent and complete specification of a prior distribution on the space of networks seems very difficult, [6].

In general, the expert's prior information on a large problem domain may be strong but partial, for example it may deal with the orientation of some edges over hundreds (thousands), or with global network traits like the size of the graph. In gene expression analysis, for example, a small degree of graph connectivity is a priori expected and substantial knowledge may regard the partial order of ten against thousands genes. In order to fully exploit the a-priori structural information both local and global features have to be taken into account.

In our approach the expert is expected to express: (1) beliefs over some, but not all, possible edges of the network; (2) beliefs over some features of the network topology, like the expected number of node parents or the degree of network connectivity.

Given these assumptions, we propose to elicit the a-priori belief on the structure of a candidate network B_s by a score function $S_{prior}(B_s)$ capturing local and global network features. The score component $S_p^\delta(B_s)$ refers to edges elicited one at a time. The second score component, $S_p^\tau(B_s)$, describes global network features, related to DAG connectivity.

2.1 Encoding Local Features

The score component $S_p^\delta(B_s)$ encodes expert's belief on the presence of oriented edges, each one marginally considered.

DAG's structure is specified by the subset $E \subset V \times V$. We conventionally indicate a pair of nodes (v_i, v_j) in the canonical order $i < j$, and we use deponent $i \cdot j$ to refer to the edge between nodes v_i and v_j. A structure is more parsimoniously represented by a collection $\mathcal{M}$ of $F \leq n(n-1)/2$ variables $\mathcal{M} = \{m_1, \ldots m_f, \ldots, m_F\}$ each one taking values on $\chi = \{-1, 0, 1\}$ for each pair of nodes $(v_i, v_j), i < j$, in V. Values in the range χ respectively indicate: an arrow $i \leftarrow j$, no arrow, an arrow $i \rightarrow j$. Expert's belief takes the form of a set of probability distributions $\{p(x_{m_f} \mid \xi) : m_f \in \mathcal{M}\}$.

The distributions are now coded as vectors of probability values in the following way $P_{i \cdot j}^T = (p_{i \cdot j, -1}, p_{i \cdot j, 0}, p_{i \cdot j, +1})$ so that $\mathbf{1}^T P_{i \cdot j} = 1$. Following the approach proposed in [7], for each couple of nodes i and j, Connectivity vectors $C_{i \cdot j}$ are introduced to indicate the value taken by variables in a candidate structure. It follows that $\mathbf{1}^T C_{i \cdot j} = 1$. The probability value associated to the oriented edge for a pair $i \cdot j$ is $C_{i \cdot j}^T P_{i \cdot j}$.

The above construction leads to the specification of a probability distribution on the set of directed graphs $\mathcal{G}_{DG}$ in which the candidate directed graph B_D has a prior probability value equal to:

$$P(B_D \mid \xi) = \prod_{\{i \cdot j\}} C_{i \cdot j}^T P_{i \cdot j}$$

The above factorization refers to our prior judgment about the existence of a causal link between v_i and v_j without considering other nodes.

The space of DAGs is contained in the space of Directed Graphs, $\mathcal{G}_D \subseteq \mathcal{G}_{DG}$, therefore the above construction also induces a probability distribution over DAGs contained in the space of directed graphs, $B_s \in \mathcal{G}_{DG}$:

$$P(B_s \mid \xi) \propto I_{DAG}(B_s) \cdot \prod_{\{i \cdot j\}} C_{i \cdot j}^T P_{i \cdot j} \tag{1}$$

with $I_{DAG}(B_s)$ taking value one if B_s is a DAG, zero otherwise. The proportionally is due to an omitted constant depending on directed graphs which are not DAGs because of cycles. We remark that there is no difficulty in calculating the value of the normalization constant but the huge cardinality of spaces may be unworkable.

We define the score $S_\delta(B_s)$ of a candidate Bayesian Networks using (1):

$$S_\delta(B_s) = \log\left(\frac{P(B_s)}{P(\{\emptyset\})}\right) \tag{2}$$

with $P(\{\emptyset\})$ the probability assigned to the Bayesian Network in which E is empty (graphs without edges). A remarkable property of the score $S_\delta(B_s)$ in equation 2 regards the possibility of calculating scores by just considering the pair of nodes for which the expert defined a distribution.

2.2 Encoding Global Features

Partial prior beliefs on network topology may take the form of an expected degree of connectivity, for example if the expert has clues about the expected number of parents/children per node. In gene expression analysis, the regulation of one gene is expected to depend on few other genes, although cases of regulation over many different metabolic pathways are known. The score component $S_p^\tau(B_s)$ captures this class of beliefs about the topology of a candidate network.

In a constructional approach the topology of a n-nodes network B_s is encoded into a $n \times n$ connectivity matrix C_s [7], whose element i, j is one iff $v_i \in pa(v_j)$, zero otherwise. Matrix C_s is one-to-one with E, therefore it contains the whole structural information. Variables $x_{g_f}(B_s)$, $f = 1, 2, \ldots$ can be built to capture global network features like the mean cardinality of parent sets, the DAG size or the number of v-structures, i.e. the same set of collapsing edges, appearing on a directed path.

For simplicity, we consider here variables $\{x_{g_1}, \ldots, x_{g_n}\}$ defined to count the number of parents for each $v_i \in V$:

$$x_{g_i} = \sum_j C_{i,j} = \sum_{v_i \in V} \mid pa(v_i) \mid \tag{3}$$

The approach adopted to depict prior beliefs about network topology is based on a reference distribution $\mathcal{Q}_{pa}$ representing expert's belief about the fraction of total nodes bearing a given number of parents, $(0, 1, \ldots)$ and on the distribution $\mathcal{P}_{pa,s}$ of relative frequencies calculated on the candidate network s. The support of $\mathcal{P}_{pa}$ is $\chi = \{0, 1, 2, \ldots, n-1\}$. Whenever the elicitation of the probability distribution on the canonical sample space of the auxiliary variable x_{g_f} is beyond expert's ability, a partitioning of χ into a coarser grid of values is performed before elicitation.

The distribution $\mathcal{P}_{pa,s}$ is compared to $\mathcal{Q}_{pa}$ and the degree of dissimilarity enters in the score function. The Kullback-Leibler distance, [8], is here adopted to assess the degree of dissimilarity among the above distributions.Note that the Kullback-Leiber distance is not symmetrical and is equal to 0 if and only if $\mathcal{Q}_{pa} \equiv \mathcal{P}_{pa,s}$. A small value of KL distance means that the candidate network has a structure close to the a-priori belief as regards the connectivity. The score component $S_\tau(B_s)$ is defined as a function of the Kullback-Leibler distance:

$$S_\tau(B_s) = (-KL(\mathcal{P}_{pa} \| \mathcal{Q}_{pa})) \tag{4}$$

2.3 Score Function and Calibration

Given the quantities in equations 2 and 4, the proposed score function is a convex combination of two other functions:

$$S_{prior}(B_s) = \alpha S_p^\delta(B_s) + (1-\alpha) S_p^\tau(B_s) \tag{5}$$

with $0 \leq \alpha \leq 1$. By substitution, we have:

$$S_{prior}(B_s) = \alpha \log\left(\frac{P(B_s)}{P(\{\emptyset\})}\right) + (1-\alpha)\left(-KL(\mathcal{P}_{pa} \| \mathcal{Q}_{pa})\right) \tag{6}$$

The role of α is to balance the strength of the components due to edge orientation and the strength due to network topology. A value $\alpha = 1$ is suited to the lack of specific prior beliefs on network topology. The most a-priori probable structure is the structure that maximizes (6). The logarithmic score is convenient for computational reasons:

$$S_{prior}(B_s) = \log\left[\left(\frac{P(B_s)}{P(\{\emptyset\})}\right)^{\alpha} \cdot e^{(1-\alpha)(-KL(\mathcal{P}_{pa}\|\mathcal{Q}_{pa}))}\right] \tag{7}$$

3 The *P*-Metric

Structural learning of BNs may be performed using the score function (6) in a Bayesian-inspired metric, called *P-metric*, which mixes prior beliefs and experimental information following [6]. The Bayesian Dirichlet with Equivalence metric, (BDe), is peculiar in assigning the same likelihood value to structures which are likelihood equivalent, i.e. DAGs encoding the same assertions on conditional independence relations. The equivalence is obtained by estimating the parameters through a prior procedure in which Dirichlet hyperparameters are defined using the notion of equivalent sample size. We propose the *P*-metric below to assess the score of a candidate structure B_s, given a complete database of cases $\mathcal{D}$:

$$S_{P\text{-}metric}(B_s) = S_{prior}(B_s)^{\beta_z} \cdot P_{BDe}(D \mid B_s, \theta) \tag{8}$$

that may be rewritten as:

$$log\left(S_{P\text{-}metric}(B_s)\right) = \beta_z \cdot \log(S_{prior}(B_s)) + ll_{BDe}(D \mid B_s, \theta) \tag{9}$$

Being based on the BDe function, the *P*-metric inherits all the assumptions described in [6]. The role of the parameter β_z is to calibrate the strength of the prior score with respect to the likelihood function. The value of β_z depends on the size of the problem domain and on the sample size of cases as well as on the elicited belief. Even if heuristics to set β_z are still under investigation, here we propose to set β_z as a function of the score prior and the likelihood computed for the empty structure:

$$\beta_z = z \cdot \frac{ll_{BDe}(D \mid \{\emptyset\}, \theta)}{\log\left(S_p(\{\emptyset\})\right)}$$

with $0 \leq z \leq 1$. Clearly when $z = 0$ then $\beta_z = 0$ and the *P*-metric is equal to the BDe metric when uniform prior distribution over structures is assumed.

The *P*-metric makes easy to quantify beliefs taking the form of both global network features and (marginal) causal assertions on pairs of variables. The joint use of the score prior $S_p(B_s)$ and of the BDe likelihood enables the detection of score differences in causally distinct structures, that would be otherwise collapsed into the same equivalence class by using a uniform prior distribution over structures.

Numerical explorations on benchmark case studies suggest that the *P*-metric is a valuable tool for large and structured domains, like gene expression studies. Note that the proposed approach is one step beyond the use of hard constraints, which may cause a loss of information and even biased elicitation.

4 Results

We implement the *P*-metric in the package MASTINO, [9], coded in the *R* environment. MASTINO is a suite of *R* functions, built on the top of the package DEAL, [10], which includes several algorithms to learn Bayesian Networks and conditionally Gaussian networks from data. The package MASTINO can be download for free from the website `http://statind.jrc.it/mastino`.

We numerically investigated the proposed metric by means of two benchmark datasets which are often referred to in the machine learning literature. One is the famous ASIA network, proposed by [11] and the other is a subnetwork from the Hepatic Glucose Homeostasis network proposed by [12] that depicts a model for the genetic network controlling glucose metabolism in perinatal hepatocytes. These are two discrete networks, which handle 8 and 20 variables for a total of 8 and 33 arc respectively. The adoption of a simplified version of the HGH network is justified by the computational problems arisen with the *R* environment. For the two benchmark datasets, we ran the learning algorithm over three different sample of: 500, 1500, 3000 observations and we tested the *P*-metric for different combinations of parameters $z \in \beta_z$ and α. Results were compared to those from the BDe metric implemented in DEAL, where a uniform distribution over structures is assumed.

The results obtained for both the benchmark networks are quite encouraging and for all the sample the *P*-metric strongly improve the overall performance of the BDe metric implemented in DEAL.

In the ASIA network simulation we encoded local features supposing a partial weak prior belief (quantified in a probability value equal to 0.6) on the absence or the orientation of arcs in 5 different pairs of nodes. The prior belief encoded was coherent with the real network. As regards the network topology (global features), we supposed that 80% of network nodes has at most one parent. In all the cases considered, the best network found by the *P*-metric correctly identifies all the arcs of the ASIA networks and adding one incorrect arc, see table 2; in the best case obtained with DEAL, see table 1, just two arcs are correctly identified, six arcs are identified but with wrong orientation, and nineteen incorrect arcs are added. Results about the calibrating parameters suggest that

Table 1. The ASIA network,learned by DEAL

Sample	Total Arcs	Correct Arcs	Wrong Directed	Incorrect	Missing Arcs
500	27	2/8	6	19	0
1500	26	1/8	7	18	0
3000	26	1/8	7	18	0

Table 2. The ASIA network learned by *P*-metric

Sample	z	α	Total Arcs	Correct Arcs	Wrong Directed	Incorrect	Missing Arcs
500	0.05	0.2	12	8/8	4	0	0
500	0.50	0.5	9	8/8	1	0	0
1500	0.05	0.2	11	8/8	3	0	0
1500	0.50	0.5	9	8/8	1	0	0
3000	0.05	0.2	9	8/8	1	0	0
3000	0.50	0.5	9	8/8	1	0	0

Table 3. The HGH network, learned by DEAL (out of memory error invoked after 49(*), 40(**) and 19(***) iterations)

Sample	Total Arcs	Correct Arcs	Wrong Directed	Incorrect	Missing Arcs
500*	48	1/33	18	29	14
1500**	40	1/33	18	19	14
3000***	19	0/33	7	12	26

Table 4. The HGH network, [12], learned by *P*-metric (out of memory error invoked after 51(*) iterations)

Sample	z	α	Total Arcs	Correct Arcs	Wrong Directed	Incorrect	Missing Arcs
500*	0.05	0.5	50	23/33	1	26	9
500*	0.50	0.2	37	23/33	1	13	9
1500	0.05	0.5	45	22/33	1	22	10
1500	0.50	0.2	39	23/33	1	15	9
3000	0.05	0.5	37	21/33	1	13	11
3000	0.50	0.2	35	23/33	1	11	9

by increasing the sample size the best network is obtained even with smaller values of z. Small values of α seems to improve the overall performance of the search.

In the study of the HGH network we included prior information take the form of a partial order among few variables and high structural sparsity. The results obtained with different combinations of calibrating parameters are shown on table 4. Although the search of the best BN using DEAL is in the best case stopped after 49 iterations due to the "Out of Memory" message, it is clear that our proposed metric performed quite well. The use of prior information indeed improved the performance of structural learning. The limited number of correct arcs discovered, see table 1 and 3, casts some shadows on the BDe algorithm for discrete BNs implemented in the DEAL package.

5 Conclusion

In this paper we defined a new quasi-bayesian score function, called *P*-metric, to score networks representing causal relations among variables. The metric component dealing with structural information takes account of marginal causal beliefs on arcs and global network features without requiring the elicitation of a complete network, [5, 3]. The second component is based on the BDe metric, thus it exploits its peculiarities well known in the literature. The BDe metric does not distinguish structures entailing the same conditional independence assertions, but our score function makes possible to discriminate structures belonging to the same likelihood equivalence class at the price of loosing score equivalence property: the *P*-metric is suited to learn causal networks, [3].

The *P*-metric has been tested under two different Machine Learning benchmark datasets and compared against the metric implemented in the DEAL package. Successful numerical findings suggest that the *P*-metric could be very useful in large problem domains with associated substantial and partial information. Unfortunately, computational constraints forbade wide numerical testing in large networks using the R environment. Further code improvement is needed, especially an implementation under C++ or Java, in order to perform extensive numerical testings including the sensitivity analysis of calibration parameters with large networks.

References

1. Jensen, F.V.: An introduction to Bayesian Networks. Springer, Heidelberg New York (1996)
2. Chickering, D.M.: Learning Bayesian Networks is NP-Complete. In: Proceedings on Artificial Intelligence and Statistics, pp. 121–130 (1995)
3. Heckerman, D., Geiger, D., Chickering, D.M.: Learning Bayesian Network: A combination of knowledge and statistical data. In: Proceedings of 10th Conf. Uncertainty in Artificial Intelligence, pp. 293–301 (1994)
4. Heckerman, D., Meek, C., Cooper, G.: A Bayesian Approach to Causal Discovery. Technical Report MSR-TR-97-05. Microsoft Corporation, Redmond, WA (1997)
5. Buntine, W.L.: Theory of Refinement on Bayesian Networks. In: Proceedings of 7th Conference on Uncertainty in Artificial Intelligence, pp. 52–60 (1991)
6. Chickering, D.M., Geiger, D., Heckerman, D.: Learning Bayesian Network: A combination of knowledge and statistical data. Tecnical Report MSR-TR-94-17, Microsoft Research, Advanced Technology Division (1994)
7. Larrañaga, P., Poza, M.: Learning of Bayesian Networks by Genetic Algorithms: A Performance Analysis of Control Parameters IEEE. Journal on Pattern Analysis and Machine Intelligence 18(9), 912–926 (1996)
8. Kullback, S., Leibler, R.A.M.: On Information and Sufficiency. Annals of Mathematical Statistics 22, 79–86 (1951)
9. Mascherini, M., Mastino: A Suite of R Functions to learn Bayesian Networks from data. UseR! International Conference of R Users, Vienna(Austria) (2006)
10. Bøttcher, S.G., Dethlefsen, C.: DEAL: A package for Learning Bayesian Networks. Journal of Statistical Software. 8(20), 1–40 (2003)
11. Lauritzen, S.L., Spiegehalter, D.J.: Local Computation with probabilities on graphical structures and their application to expert system. Journal of the Royal Statistical Society - B Series 50(2), 157–192 (1988)
12. Le, P.P., Bahl, A., Ungar, L.H.: Using prior knowledge to improve genetic network reconstruction from microarray data. InSilico Biology 27(4) (2004)

3D α-Expansion and Graph Cut Algorithms for Automatic Liver Segmentation from CT Images

Elena Casiraghi, Gabriele Lombardi, Stella Pratissoli, and Simone Rizzi

Universita degli Studi di Milano, Computer Science Department,
Via Comelico 39, 20135 Milano, Italy

Abstract. Abdominal CT images have been widely studied in the recent years as they are becoming an invaluable mean for abdominal organ investigation. In the field of medical image processing, some of the current interests are the automatic diagnosis of liver pathologies and its 3D volume rendering. The first and fundamental step in all these studies is the automatic liver segmentation, that is still an open problem. In this paper we describe an automatic method to segment the liver from abdominal CT data, by combining an α-expansion and a graph cut algorithm. When evaluated on the data of 40 patients, by comparing the automatically detected liver volumes to the liver boundaries manually traced by three experts, the method achieves a symmetric volume difference of 94%.

Keywords: Computed Tomography, Liver Segmentation, energy minimization, α-expansion, graph-cut algorithm.

1 Introduction

Computed tomography (CT) images are nowadays the standard instruments for diagnosis of liver pathologies (e.g. cirrhosis, liver cancer, fulminant hepatic failure) for they provide accurate anatomical information about the visualized structures, thanks to their high Signal-to-Noise ratio and good spatial resolution. This motivates the great deal of research work, in the digital image processing field, aimed at the development of computerized methods for the automatic detection of liver pathologies [16,10], and the 3D liver volume measurement [13] and rendering [6], which have been shown to be helpful for surgical planning prior to living donor liver transplantation or to hepatic resection.

Whatever the aim of the system, the first and fundamental step is always the liver volume segmentation, that is usually done by expert radiologists who either manually trace the liver contour on each slice of the CT data, or employ semi-automated techniques [15]. Since both manual and semi-automatic procedures require the user interaction time, and they are affected by his/her errors and biases, a lot of research work has been devoted to the development of fully automatic liver segmentation techniques. Nevertheless the problem is still open [3] due to several factors. First of all, neighboring organs (e.g. liver, spleen and stomach) might have similar gray levels, since the gray tones in CT images are

B. Apolloni et al. (Eds.): KES 2007/WIRN 2007, Part I, LNAI 4692, pp. 421–428, 2007.

related to their (sometimes similar) tissue densities. Besides, the same organ may exhibit different gray levels both in the same patient, due to the administration of contrast media, and in different ones, for varying machine setups. As a result, methods relying on simple thresholding [4,11,12], where thresholds are set based on a priori knowledge or statistical analysis of manually segmented samples, are likely to fail when processing patients whose liver gray level characteristics are not captured by the analyzed sample.

Moreover, due to the partial volume effects resulting from spatial averaging, patient movement, beam hardening, and reconstruction artifacts, the acquired images have low contrast and blurred edges. Consequently, methods employing simple gray level dependent edge detectors (e.g. Sobel, Roberts) do not produce satisfactory results [9].

In addition, the liver presents significant anatomical variation in different image slices of the same patient; even at the same slice position, its shape may vary widely from patient to patient. This fact makes model fitting techniques [5,18], statistical shape models [8], and probabilistic atlases [14,17] not easy to be used, since they require a huge amount of training examples to capture as much shape variability as possible. Furthermore, when dealing with complex shapes, these techniques might require too much computation time before a good match between the model and the image data is obtained.

In [2] we reviewed the most relevant automatic liver segmentation works, and we noted that a comparison among them would not be meaningful due to the lack of a common dataset with its gold standard, i.e. a commonly accepted manual segmentation, and a unique measure of the discrepancy between the automatic and the manual segmentation. Besides, the private datasets employed by most authors are too small (less than 10 patients).

In this paper we propose our liver segmentation method (section 3), that has been evaluated on a set of 40 abdominal CT data (section 2) and achieves results comparable to the intra and inter-personal variability of manual segmentation by experts (section 4).

2 Materials

Our dataset is composed of 40 abdominal contrast enhanced CT images of the third phase. They have been acquired at the Niguarda Ca' Granda hospital in Milan, with a Siemens multi-detector spiral CT, after the injection of 2 Ml/Kg contrast material. The images, stored into a PACS system in DICOM format, have been exported by AGFA IMPAX software as a set of 2D axial slices in JPG format. For each patient a set of 80 axial slices with a 3 mm interval is acquired; each slice has a 1024×1024 pixel size, and a 0.165×0.165 mm pixel resolution. To expedite the computation they have been reduced to 256×256 pixels, and a 3×3 median filter has been subsequently applied to remove impulsive noise. The 3D coordinate system used in this work has the Z axis parallel to the body axis and oriented from the topmost to the bottommost slice, while the X and Y axis are oriented respectively along the width (from left to right) and the

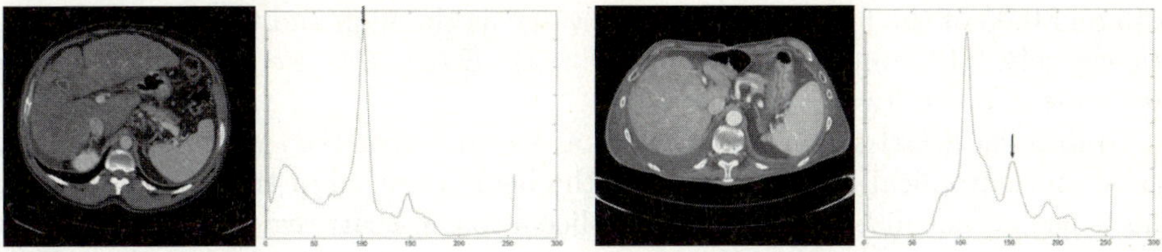

Fig. 1. Axial slices of two patients, and the gray level histograms of the whole patients' volumes

height (from top to bottom) of the 2D axial slices. While 'axial' slices are those obtained as cross sections on the Z axis, 'sagittal' and 'coronal' slices are those obtained as cross sections along the X and Y axis, respectively.

Our dataset contains patients with normal, fatty, cirrhotic, overextended livers, and livers with cancer, so that we have to take into account a big anatomical and gray level variability in the data. As an example of this wide variability, figure 1 shows two axial slices of two patients taken at the same vertical position on the Z axis, with the gray level histograms, H_{Vol}, of their whole volumes (the arrows point to the liver peak).

3 Liver Segmentation

After the definition of a processing area, the 'body box', strictly containing the abdominal structures, a binary edge map is computed by detecting significant edges in the axial, coronal and sagittal slices. The edges are at first used to segment the heart volume, whose identification helps to localize the liver, and to extract a liver data sample (subsection 3.1). This sample is used to automatically estimate, for each patient, a liver gray level range; this is the input of the α-expansion algorithm described in subsection 3.2.

3.1 Image Preprocessing, Edge Detection, and Heart Segmentation

At first we extract from the CT images the 3D volume strictly including the patient's body. Since the darkest gray levels correspond to air voxels, the first peak in the histogram H_{Vol} is searched for, and the gray level value corresponding to the first local minimum at its right side, is used to threshold the CT data. The patient's body is contained in the 3D 'body box' that includes the biggest 3D connected component in the thresholded result (see figure 2). All the following computation steps will be applied only to this 'body box' (simply referred as CT image, or CT volume).

To produce a binary edge map, we apply to each axial, coronal, and sagittal slice the first order derivative of a gaussian function (with $\sigma = 0.5$, to detect details in the image) evaluated in eight directions. To keep only the significant edge pixels in each direction, the result is thresholded with hysteresis, by using

0.15 and 0.05 of the maximum gradient value as the high and the low threshold, respectively. The computed binary edge map, $Edge_{3D}$, is used to segment both the heart and the liver.

To find the heart we initially define a 2D bounding box, B_H (see figure 2), based on anatomical knowledge about the heart position in the patient's body; this is used to localize in the first axial slice a coarse heart region, H_1, as follows: 1) convolve the image with a 2D gaussian filter (with $\sigma = 2$), 2) select the 10% of the pixels with the highest gray levels, 3) select H_1 by finding, in the thresholded image, the biggest connected region that intersects B_H.

A similar procedure is applied to each following axial slice, i, where the heart region H_i is identified by selecting in the thresholded image the biggest region that intersects H_{i-1}, detected in the previous slice. This process is repeated until the selected region H_i is less than $0.3 \times area(H_1)$.

The heart regions detected in successive slices form an initial 3D heart volume, V_H, that is further refined by a 3D region growing algorithm. It takes as seed points the voxels on the surface of V_H, and considers the 6-connected 3D neighborhood of each seed. Each analyzed voxel, v, is included into the heart volume, and it is used as a new seed, if: (i) it has not been considered yet; (ii) it is not an edge point in the $Edge_{3D}$ map; (iii) its gray level $g(v)$ is such that $||g(v) - \nu|| < c\ \sigma$, where c is a constant set to 2.0, ν and σ are the mean and the standard deviation of the gray levels in V_H. The region growing stops either when it finds no more voxels that can be added to the heart volume, or when a maximum number of 100 iterations has been reached.

3.2 Liver Gray Levels Estimation and Liver Segmentation

To obtain a reliable liver gray level estimate, that is crucial as it affects the number of voxels that are wrongly segmented as liver by the following algorithm, we process each patient separately, to account of the gray level variability among different CT volumes. Furthermore, we overcome problems due to the intra-patient gray level variability by automatically extracting, from the patient's volume, a significant liver sample set (by exploiting anatomical knowledge about liver size and shape). More precisely, we define a 3D box located below the heart volume, that surely contains the liver tissue; the height of this box along the Z axis is 20 voxels, while its position and dimensions in the horizontal X-Y plane are related to the body axial slice dimensions, as shown in figure 2. In the same figure the gray level histogram of the defined sample is plotted with a dashed line; it always shows a unique peak, corresponding to a narrow range of liver gray levels, that is used to correctly identify the liver peak, and its corresponding gray level G_{Liv}, in the histogram of the whole volume, H_{Vol} (solid line in figure 2). A proper liver gray level range $[MinG, MaxG]$ is defined by finding, at the left and at the right side of G_{Liv}, the nearest local minima.

The estimated liver gray levels are used as input of the α-expansion algorithm that interprets segmentation as a 3D labeling problem; the labels are assigned according to both gray levels and spatial relationships between neighboring voxels in the 3D 6-connected neighborhood $Neigh$. In particular, the image is

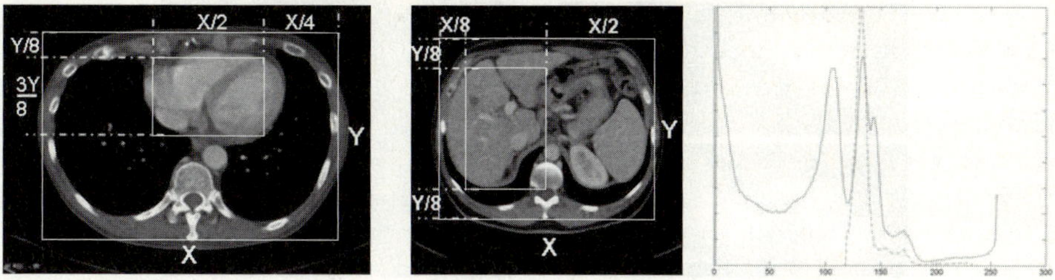

Fig. 2. Left:the axial section of a patient's 'body box', and the two dimensional heart bounding box, B_H; in the image we show the relationship among B_H and the two dimensions of the 'body box' in the X-Y plane. Center: an axial section of the 3D liver bounding box used to extract the liver sample, and its relationships to the axial section of the 'body box'. Right: the gray level histogram of the whole volume H_{Vol} (solid line), and the gray level histogram of the liver sample (dashed line).

partitioned into 5 disjoint classes corresponding to liver, bones and kidneys, spleen, stomach and organs with similar gray levels, and background. This partitioning (labeling) can be achieved by minimizing the following energy function:

$$E(L) = \sum_{i=1..V} E_1(L(i)) + \sum_{i,j \in Neigh} E_2(L(i), L(j))$$

where L is the labeling function, V is the number of voxels, $E_1(L(i))$ sets the cost of assigning the label $L(i)$ to the voxel i depending on its gray level, and $E_2(L(i), L(j))$ impose spatial smoothness as it defines the cost of assigning the labels $L(i)$ and $L(j)$ to the voxels i, j in $Neigh$.

To minimize $E(L)$ we use the α-expansion algorithm described in [1], whose input is the transformed CT volume, $CT_{Trasf}(i) = |g(i) - G|$, where $G = \sum_{k=MinG}^{MaxG} k \; H_{Vol}(k) / \sum_{k=MinG}^{MaxG} H_{Vol}(k)$ is the mean of the estimated liver gray levels and $g(i)$ is the gray level of the voxel i. Since we use 5 classes, CT_{Trasf} is then scaled and rounded to the range $[1, .., 5]$; this produces an initial labeling that is a first approximation of the solution.

Next, for each label $\alpha = [1, .., 5]$, chosen in a random order, an α-expansion step is applied to solve a two-classes partitioning problem. This is done by minimizing $E(B)$ (see equation above) via graph-cut algorithm [7], where B is the binary assignment that identifies the voxels that must be labeled as α, and setting the energy terms to:

$$E_1(B(i); CT_{Trasf}(i), L(i), \alpha) = \begin{cases} |CT_{Trasf}(i) - L(i)| & \text{for } B(i) = 0 \\ |CT_{Trasf}(i) - \alpha| & \text{for } B(i) = 1 \end{cases}$$

$$E_2(B(i), B(j); L(i), L(j), \alpha) = \begin{cases} |L(i) - L(j)| & \text{for } B(i) = 0 \text{ and } B(j) = 0 \\ |L(i) - \alpha| & \text{for } B(i) = 0 \text{ and } B(j) = 1 \\ |\alpha - L(j)| & \text{for } B(i) = 1 \text{ and } B(j) = 0 \\ 0 & \text{for } B(i) = 1 \text{ and } B(j) = 1 \end{cases}$$

In the original α-expansion algorithm the process of expanding the whole set of labels in a random order is iterated until a user defined convergence criterion is

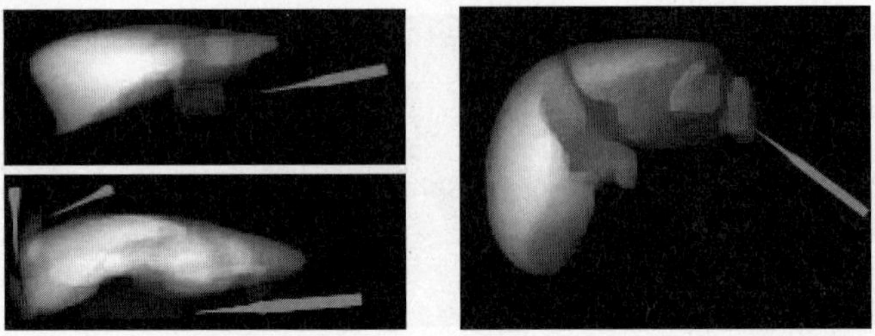

Fig. 3. First column: SUM_Y and SUM_X images; Second column: SUM_Z image

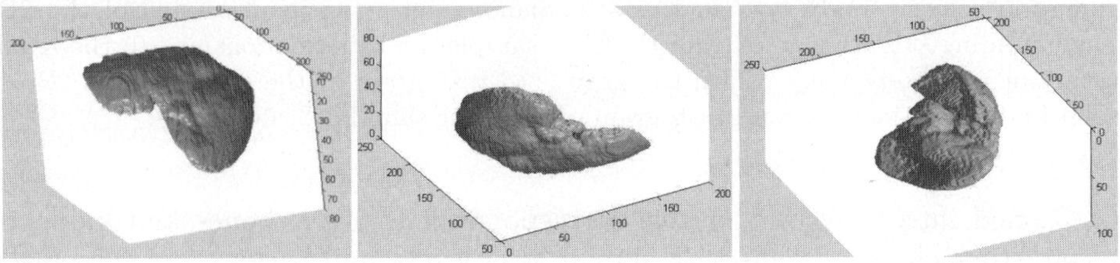

Fig. 4. 3D views of some results

matched; using our initialization, instead, one iteration is enough to obtain the final volume partitioning.

Once the CT image has been partitioned into 5 classes, the liver volume, *Liv*, is selected by taking the biggest labeled volume among those corresponding to the lowest values in CT_{Trasf} (i.e. those labeled 1).

At this stage, *Liv* might contain parts of neighboring organs, such as heart, stomach, portal vein, and spleen. The next steps of the algorithm have been developed to remove these unwanted parts, and to smooth the boundaries.

At first, we remove those voxels contained also in the segmented heart (see section 3.1). Then, we create three 2D images by projecting $Liv(x, y, z)$ onto the Y-Z, X-Z, and X-Y plane as follows:

$$SUM_X(y,z)=\sum_{i=1}^{N} Liv(i,y,z),\ SUM_Y(x,z)=\sum_{i=1}^{M} Liv(x,i,z),\ SUM_Z(x,y)=\sum_{i=1}^{P} Liv(x,y,i)$$

where N, M, P are the sizes of the CT data on the X,Y, and Z axis, respectively. As shown in figure 3, voxels belonging to not-liver organs can be identified because their projections have the lowest values in either one of the three SUM images. The wrongly segmented voxels are then removed by applying the following steps to the three images separately; without loss of generality we will refer to SUM_X only: 1) in the SUM_X image, find the 2D coordinates, (y_{del}, z_{del}), of the pixels whose value is less than $\frac{max(SUM_X)}{10}$; 2) delete from the liver volume, $Liv(x, y, z)$, all the voxels v with coordinates (x, y_{del}, z_{del}), where $x = [1, .., N]$; 3) select from the resulting liver volume the 3D biggest connected component. To smooth the boundaries, we finally perform a 3D morphological opening operation with a digital sphere with radius of 1.5 voxels.

This last step may remove, along with unwanted parts, also some liver voxel; to recover these regions we apply a refinement process. In the literature [3,11], this is usually done by complex techniques (eg.: snakes, level set methods), that might take much computational time and need a cost function to be defined; besides they are applied separately to each slice, neglecting the 3D relationships among neighboring slices. To overcome all these limitations we apply the 3D region growing algorithm described in section 3.1, where ν and σ are computed based on the liver volume, c is set to 3, and the maximum number of iterations is set to 100. Our 3D region growing method is simple, very fast, and considers inter-slice and intra-slice relationships. 3D views of some results are shown in figure 4; the segmentation system described in this paper takes about 50 seconds when running on a Pentium IV, 3,2GHZ/775; α-expansion and graph cut algorithms are implemented in C++, while the other steps are implemented in Matlab.

4 Results and Future Works

The segmentation method has been evaluated by comparing the automatically detected liver volumes, V_{Aut}, to the ground truth, V_{Man}, manually traced by three experts, to compensate human errors and biases. The employed measure of discrepancy is the 'symmetric volume overlap' (SVO), presented and used in [8]; it is a symmetric measure which accounts of both over segmentation and under segmentation errors. It is defined as:

$$SVO = \frac{|V_{Aut} \cap V_{Man}|}{\frac{1}{2}(|V_{Aut}| + |V_{Man}|)}$$

The method achieves a mean SVO of 94%. The good quality of the result is proved by the fact that it is comparable to both the mean intra-personal (96%) and inter-personal variation (95%). These two measures were evaluated on 10 patients, by computing SVO between the two liver volumes of the same patient, produced respectively by the same expert in two different times, and by two different experts. Besides, our results are comparable to those obtained in [8], where the author achieves a mean SVO of about 95%. Indeed, the author himself specifies that his dataset contains normal livers only with not so complex shapes, hence this comparison might not be fair.

Future works will be aimed at improving the system performance by integrating the edge information in the energy function, $E(L)$, used by the α-expansion algorithm. To enhance the anatomical hepatic information provided by the system, we will also focus on the segmentation of the hepatic vascular system. In addition, to achieve a more complete patient description, we are currently testing the α-expansion algorithm for the segmentation of the spleen; the final purpose is the automatic segmentation of all the abdominal organs.

References

1. Boykov, Y., et al.: Fast approximate energy minimization via graph cut. IEEE Transactions on Pattern Analysis and Machine Intelligence 23(11), 1222–1239 (2001)

2. Campadelli, P., Casiraghi, E.: Liver segmentation from ct scans: A survey. In: Proceedings of Computational Intelligence methods for Bioinformatics and Biostatistics (CIBB 2007), Portofino, Italy (July 7-10, 2007)
3. Foruzan, A.H., et al.: Automated segmentation of liver from 3d ct images. International Journal of Computer Assisted Radiology and Surgery 1(7), 71–73 (2006)
4. Gao, L., et al.: Automatic liver segmentation technique for three-dimensional visualization of ct data. Radiology 201, 359–364 (1996)
5. Gao, L., et al.: Abdominal image segmentation using three-dimensional deformable models. Investigative Radiology 33(6), 348–355 (1998)
6. Harms, J., et al.: Transplantation Proceedings. 37, 1059–1062 (2005)
7. Kolmogorov, V., Zabih, R.: What energy functions can be minimized via graph cuts? IEEE Trans. on Pattern Analysis and Machine Intelligence 26(2), 147–159 (2004)
8. Lamecker, H., et al.: Segmentation of the liver using a 3d statistical shape model. ZIB-Report 04-09, 04-09:1–25 (April, 2004)
9. Lee, C.-C., et al.: Identifying multiple abdominal organs from ct image series usinmg a multimodule contextual neural network and spatial fuzzy rules. IEEE Transaction on Information Technology in Biomedicine 7, 208–217 (2003)
10. Lee, C.-C., et al.: Classification of liver diseases from ct images using bp-cmac neural network. In: Proceedings of 9th International Workshop on Cellular Neural Networks and Their Applications, pp. 118–121 (2005)
11. Lim, S.-J., et al.: Automatic liver segmentation for volume measurement in ct images. Journal of Visual Communication and Image Representation 17(4), 860–875 (2006)
12. Liu, F., et al.: Liver segmentation for ct images using gvf snake. Medical Physics 32(12), 3699–3706 (2005)
13. Nakayama, Y., et al.: Automated hepatic volumetry for living related liver transplantation at multisection ct. Radiology 240(3), 743–748 (2006)
14. Park, H., et al.: Construction of an abdominal probabilistic atlas and its application in segmentation. IEEE Transactions on Medical Imaging 22(4), 483–492 (2003)
15. Schenk, A., et al.: Efficient semiautomatic segmentation of 3d objects in medical images. In: Delp, S.L., DiGoia, A.M., Jaramaz, B. (eds.) MICCAI 2000. LNCS, vol. 1935, pp. 186–195. Springer, Heidelberg (2000)
16. Shimizu, A., et al.: Preliminary report of cad system competition for liver cancer extraction from 3d ct imaging and fusion of the cads. International Journal of Computer Assisted Radiology and Surgery 1, 525–526 (2005)
17. Shimizu, A., et al.: Multi-organ segmentation in three dimensional abdominal ct images. International Journal of Computer Assisted Radiology and Surgery (CARS 2006) 1(7), 76–78 (2006)
18. Soler, L., et al.: Fully automatic anatomical, pathological, and functional segmentation from ct scans for hepatic surgery. Computed Aided Surgery 6(3), 131–142 (2001)

A Study on the Gesture Recognition Based on the Particle Filter

Hyung Kwan Kim[1], Yang Weon Lee[2], and Chil Woo Lee[3]

[1] Department of Computer engineering, Chonnam University, Yongbongdong, Gwangju, South Korea
freebeast@hanmail.net

[2] Department of Information and Communication Engineering, Honam University, Seobongdong, Gwangsangu, Gwangju, South Korea
ywlee@honam.ac.kr

[3] Department of Computer Engineering, Chonnam University, Yongbongdong, Gwangju, South Korea
cwlee@chonnam.ac.kr

Abstract. The recognition of human gestures in image sequences is an important and challenging problem that enables a host of human-computer interaction applications. This paper describes a gesture recognition algorithm based on the particle filters, namely CONDENSATION. The particle filter is more efficient than any other tracking algorithm because the tracking mechanism follows Bayesian estimation rule of conditional probability propagation. We used two models for the evaluation of particle filter and apply the MATLAB for the preprocessing of the image sequence. But we implement the particle filter using the C++ to get the high speed processing. In the experimental results, it is demonstrated that the proposed algorithm prove to be robust in the cluttered environment.

1 Introduction

Gesture is one interesting subspace of human motion. For the purposes of this paper, we define gesture to be motions of the body that are intended to communicate to another agent. Recently human gesture has received much interest in computer vision field for the applications such as human interface, robot, medicine, animation, video database, intelligent surveillance and virtual reality.

In this paper, we focused into the development of human gesture recognition using particle filter. Particle filter[1] is based on the Bayesian conditional probability such as *prior* distribution and *posterior* distribution. First of all, we expanded the existing algorithm[2] to derive the CONDENSATION-based particle filter for human gesture recognition. Also, we adopt the two hand motion model to confirm the algorithm performance such as leftover and paddle.

MATLAB package is used to preprocess the raw image data and tracking algorithm is implemented by the C++ language. The overall scheme for the gesture recognition system is shown in Figure1.

This paper consist of as follows: Following the introduction, CONDENSATION algorithm and its related model is described in Section 2 and the motion

B. Apolloni et al. (Eds.): KES 2007/WIRN 2007, Part I, LNAI 4692, pp. 429–438, 2007.

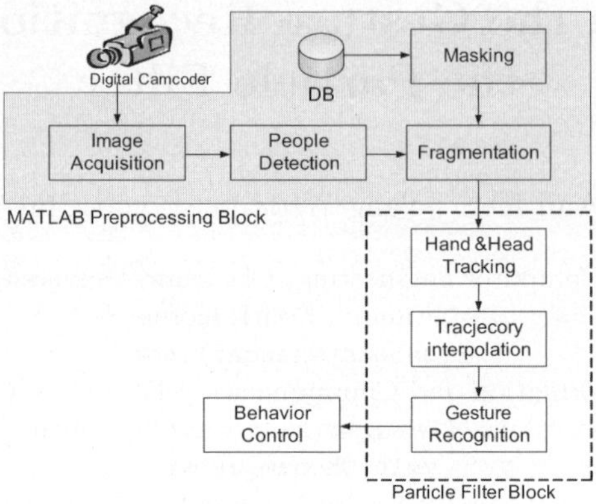

Fig. 1. Overall operation block diagram of recognition system

extraction process for the proposed algorithm test is explained in Section 3. In Section 4, the result of experiment for the proposed algorithm is described and finally, conclusion is followed.

2 Condensation Algorithm

2.1 Condensation Algorithm

The particle filter approach to track motion, also known as the condensation algorithm [1] and Monte Carlo localisation [?], uses a large number of particles to explore the state space. Each particle represents a hypothesised target location in state space. Initially the particles are uniformly randomly distributed across the state space, and each subsequent frame the algorithm cycles through the steps illustrated in Figure 2:

1. Deterministic drift: particles are moved according to a deterministic motion model (a damped constant velocity motion model was used).
2. Update probability density function (PDF): Determine the probability for every new particle location.
3. Resample particles: 90with replacement, such that the probability of choosing a particular sample is equal to the PDF at that point; the remaining 10throughout the state space.
4. Diffuse particles: particles are moved a small distance in state space under Brownian motion.

This results in particles congregating in regions of high probability and dispersing from other regions, thus the particle density indicates the most likely target states. See [3] for a comprehensive discussion of this method. The key strengths of the particle filter approach to localisation and tracking are its scalability (computational requirement varies linearly with the number of particles), and its ability to deal with multiple hypotheses (and thus more readily recover from tracking errors). However, the particle filter was applied here for several additional reasons:

- it provides an efficient means of searching for a target in a multi-dimensional state space.
- reduces the search problem to a verification problem, ie. is a given hypothesis face-like according to the sensor information?
- allows fusion of cues running at different frequencies.

The last point is especially important for a system operating multiple cues with limited computational resources, as it facilitates running some cues slower than frame rate (with minimal computational expense) and incorporating the result from these cues when they become available. If a cue takes n frames to return a result, by the time the cue is ready, the particles will have moved from where they were n frames ago. To facilitate such cues the system keeps a record of every particle's history over a specified number of frames k. The cue value determined for a particle nk frames ago can then be assigned to the children of that particle in the current frame, thus propagating forward the cues response to the current frame. Conversely, probabilities associated with particles that were not propagated are discarded.

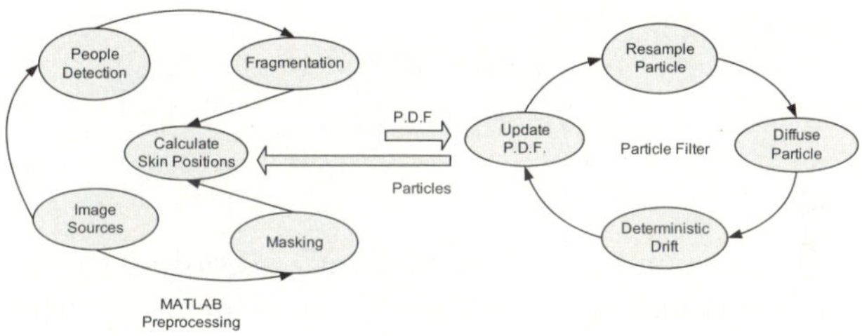

Fig. 2. Particle Filter Calculation Process

2.2 Application of Condensation for the Gesture Recognition

In order to apply the Condensation Algorithm to gesture recognition, we extend the methods described by Black and Jepson [2]. Specifically, a state at time t is described as a parameter vector: $s_t = (\mu, \phi^i, \alpha^i, \rho^i)$ where: μ is the integer index of the predictive model, ϕ^i indicates the current position in the model, α^i refers to an amplitudal scaling factor and ρ^i is a scale factor in the time dimension. Note that i indicates which hand's motion trajectory this ϕ^*, α^*, or ρ^* refers to left and right hand where $i \in \{l, r\}$. My models contain data about the motion trajectory of both the left hand and the right hand; by allowing two sets of parameters, I allow the motion trajectory of the left hand to be scaled and shifted separately from the motion trajectory of the right hand (so, for example,ϕ^l refers to the current position in the model for the left hand's trajectory, while ϕ^r refers to the position in the model for the right hand's trajectory). In summary, there are 7 parameters that describe each state.

Initialization. The sample set is initialized with N samples distributed over possible starting states and each assigned a weight of $\frac{1}{N}$. Specifically, the initial parameters are picked uniformly according to:

$$\begin{aligned} \mu &\in [1, \mu_{max}] \\ \phi^i &= \frac{1-\sqrt{y}}{\sqrt{y}}, y \in [0,1] \\ \alpha^i &= [\alpha_{min}, \alpha_{max}] \\ \rho^i &\in [\rho_{min}, \rho_{max}] \end{aligned} \tag{1}$$

Prediction. In the prediction step, each parameter of a randomly sampled s_t is used to s_{t+1}determine based on the parameters of that particular s_t. Each old state, s_t, is randomly chosen from the sample set, based on the weight of each sample. That is, the weight of each sample determines the probability of its being chosen. This is done efficiently by creating a cumulative probability table, choosing a uniform random number on $[0, 1]$, and then using binary search to pull out a sample (see Isard and Blake for details[1]). The following equations are used to choose the new state:

$$\begin{aligned} \mu_{t+1} &= \mu_t \\ \phi^i_{t+1} &= \phi^i_t + \rho^i_t + N(\sigma_\phi) \\ \alpha^i_{t+1} &= \alpha^i_t + N(\sigma_\alpha) \\ \rho_{t+1} &= \rho^i_t + N(\sigma_\rho) \end{aligned} \tag{2}$$

where $N(\sigma_*)$ refers to a number chosen randomly according to the normal distribution with standard deviation σ_*. This adds an element of uncertainty to each prediction, which keeps the sample set diffuse enough to deal with noisy data. For a given drawn sample, predictions are generated until all of the parameters are within the accepted range. If, after, a set number of attempts it is still impossible to generate a valid prediction, a new sample is created according to the initialization procedure above. In addition, 10 percent of all samples in the new sample set are initialized randomly as in the initialization step above (with the exception that rather than having the phase parameter biased towards zero, it is biased towards the number of observations that have been made thus far). This ensures that local maxima can't completely take over the curve; new hypotheses are always given a chance to dominate.

Updating. After the Prediction step above, there exists a new set of N predicted samples which need to be assigned weights. The weight of each sample is a measure of its likelihood given the observed data $Z_t = (z_t, z_{t_1}, \cdots)$. We define $Z_{t,i} = (z_{t,i}, z_{(t-1),i}, \cdots)$ as a sequence of observations for the ith coefficient over time; specifically, let $Z_{(t,1)}, Z_{(t,2)}, Z_{(t,3)}.Z_{(t,4)}$ be the sequence of observations of the horizontal velocity of the left hand, the vertical velocity of the left hand, the horizontal velocity of the right hand, and the vertical velocity of the right hand respectively. Extending Black and Jepson [2], we then calculate the weight by the following equation:

$$p(z_t|s_t) = \prod_{i=1}^{4} p(Z_{t,i}|s_t) \tag{3}$$

where $p(z_{t,i}|s_t) = \frac{1}{\sqrt{2\pi}} exp \frac{-\sum_{j=0}^{\omega-1}(z_{(t-j),i} - \alpha^* m^{\mu}_{(\phi-\rho^* j),i})^2}{2(\omega-1)}$ and where ω is the size of a temporal window that spans back in time. Note that ϕ^*, α^* and ρ^* refer to the appropriate parameters of the model for the blob in question and that $\alpha^* m^{(\mu)}_{(\phi^*-\rho^* j),i}$ refers to the value given to the ith coefficient of the model μ interpolated at time $\phi^* - \rho^* j$ and scaled by α^*.

Classification. With this algorithm in place, all that remains is actually classifying the video sequence as one of the two signs. Since the whole idea of Condensation is that the most likely hypothesis will dominate by the end, I chose to use the criterion of which model was deemed most likely at the end of the video sequence to determine the class of the entire video sequence. Determining the probability assigned to each model is a simple matter of summing the weights of each sample in the sample set at a given moment whose state refers to the model in question. The following graphs plot the likelihood of each model over time for an instance of each sign (the first is a sign that is classified as model 1, the second a sign that is classified as model 2).

3 Gesture Model and Image Preprocessing

We adopt the two gesture model to verify the proposed particle filter. As shown in Figure 3, gesture 1 means leftover and gesture 2 means paddle.

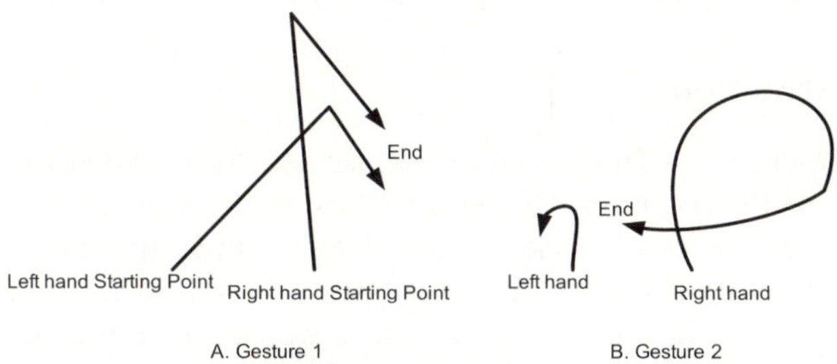

Fig. 3. Two gesture model

3.1 Raw Image Preprocessing

The image sequences were filmed using a Sony DCR Camcoder. They were manually aligned and then converted into sequences of TIFs to be processed in MATLAB. Each TIF was 320x240 pixels, 24bit color. The lighting and background

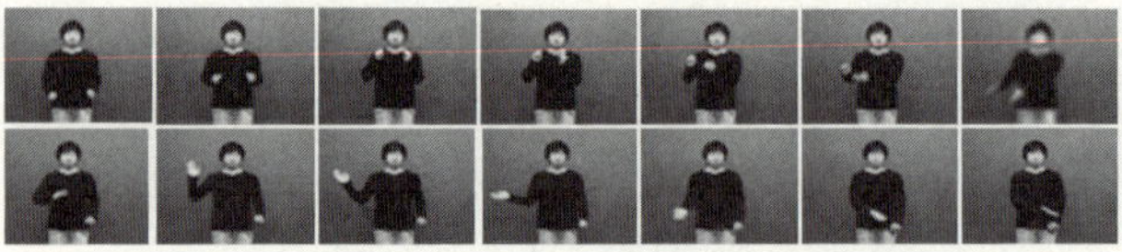

Fig. 4. Gesture Images of the Two Models

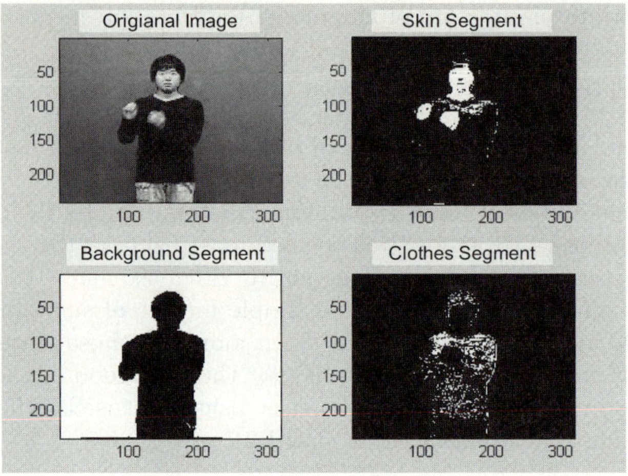

Fig. 5. Output of Segmentation

in each sequence is held constant; the background is not cluttered. The focus of my project was not to solve the tracking problem, hence I wanted the hands to be relatively easy to track. I collected 7 film sequences of each sign(see Figure 4).

3.2 Skin Extraction

In order to segment out skin-colored pixels, we used the color segment routine we developed in MATLAB. Every image in every each sequence was divided into the following regions: skin, background, clothes, and outliers. First of all, we set up the mask using the gaussian distribution based on mean and covariance value which is stored in the database. Then we segment the images into four section above mentioned regions. So, we get the the segment of skin as shown in Figure 5.

3.3 Finding Skin-Colored Blobs

We then calculated the centroid of the three largest skin colored 'blobs' in each image. Blobs were calculated by processing the skin pixel mask generated in the previous step. A blob is defined to be a connected region of 1's in the mask. Finding blobs turned out to be a bit more difficult than we had originally thought.

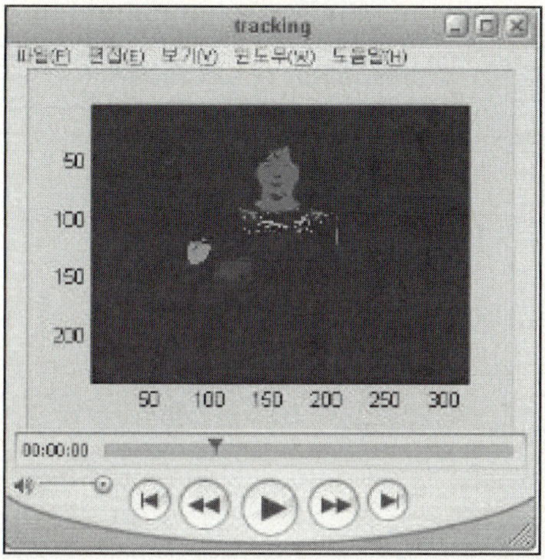

Fig. 6. Tracking result using centroid calculation

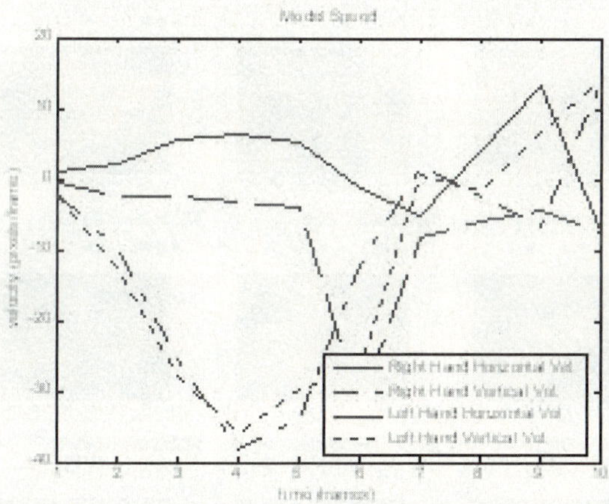

Fig. 7. Velocity of Model I

Our first implementation was a straightforward recursive algorithm which scans the top down from left to right until it comes across a skin pixel which has yet to be assigned to a blob. It then recursively checks each of that pixel's neighbors to see if they too are skin pixels. If they are, it assigns them to the same blob and recurses. On such large images, this quickly led to stack overflow and huge inefficiency in MATLAB. The working algorithm we eventually came up with is an iterative one that scans the skin pixel mask from left to right top down. When it comes across a skin pixel that has yet to be assigned to a blob, it first checks pixels neighbors (to the left and above) to see if they are in a blob. If they

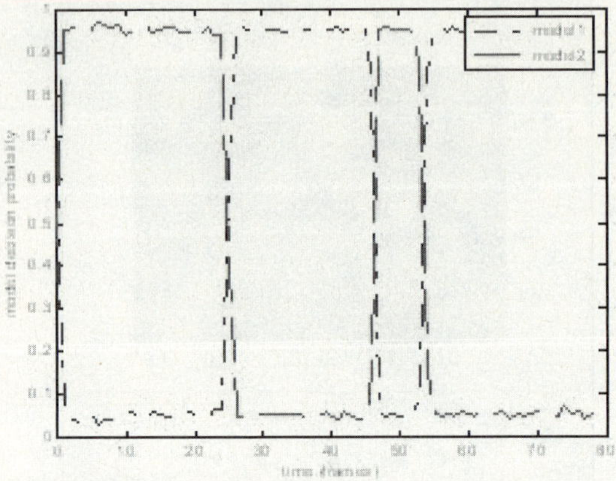

Fig. 8. Probability of Model 1 and 2

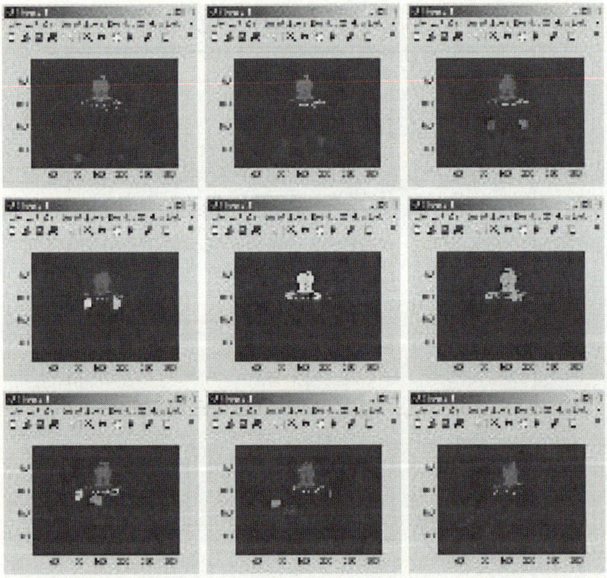

Fig. 9. The Tracking process of particle filter for the model 1(From left to right, top to down)

aren't, it creates a new blob and adds the newly found pixel to the blob. If any of the neighbors are in a blob, it assigns the pixel to the neighbor's blob. However, two non-adjacent neighbors might be in different blobs, so these blobs must be merged into a single blob. Finally, the algorithm searches for the 3 largest blobs and calculates each of their respective centroid.

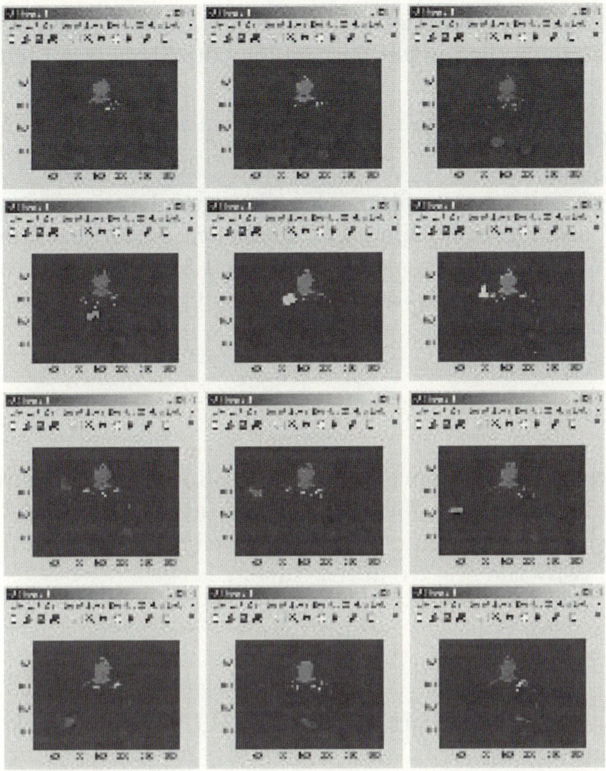

Fig. 10. The tracking process of particle filter for the model 2

3.4 Calculating the Blobs' Motion Trajectories over Time

At this point, tracking the trajectories of the blobs over time was fairly simple. For a given video sequence, we made a list of the position of the centroid for each of the 3 largest blobs in each frame. Then, we examined the first frame in the sequence and determined which centroid was farthest to the left and which was farthest to the right. The one on the left corresponds to the right hand of signer, the one to the right corresponds to the left hand of the signer. Then, for each successive frame, we simply determined which centroid was closest to each of the previous left centroid and called this the new left centroid; we did the same for the blob on the right. Once the two blobs were labelled, we calculated the horizontal and vertical velocity of both blobs across the two frames using [(change in position)/time]. We recorded these values for each sequential frame pair in the sequence. The example of the tracking is shown in Figure 6.

3.5 Creating the Motion Models

We then created models of the hand motions involved in each sign. Specifically, for each frame in the sign, we used 5 training instances to calculate the average horizontal and vertical velocities of both hands in that particular frame. The following graphs show the models derived for both signs (see Figure 7 and 8).

4 Experiment Result

To test the proposed particle filter scheme, we used two gesture model which is shown in Figure 3 in this paper. The coefficient of particle filter are $\mu_{max} = 2, \alpha_{min} = 0.5, \alpha_{max} = 1.5, \rho_{min} = 0.5, \rho_{max} = 1.5$ to maintain the 50Also, the other parameters are settled by $\sigma_{\phi} = \sigma_{\alpha} = \sigma_{\rho} = 0.1$. The variable of ω equation 3 is 10.

5 Conclusion

In this paper, we have developed the particle filter for the gesture recognition. This scheme is important in providing a computationally feasible alternative to classify the gesture in real time. We have proved that given an image, particle filter scheme classify the gesture in real time.

Acknowledgements

This work was supported in part by MIC and IITA through IT Leading R & D Support Project.

References

1. ISard, M., Blake, A.: CONDENSATION-conditional density propagation for visual tracking. International Journal of Computer Vision 29(1), 5–28 (1998)
2. Black, M.J., Jepson, A.D.: A Probabilistic Framework for Matching Temporal Trajectories: Condensation-based Recognition of Gestures and Expressions. Proceedings 5th European Conf. Computer Vision 1, 909–924 (1998)
3. Isard, M., Blake, A.: A mixed-state condensation tracker with automatic model-switching. In: Proceedings 6th International Conference of Computer Vision, pp. 107–112 (1998)
4. Lee, Y.W.: Adaptive Data Association for Multi-target Tracking using relaxation. In: Eisinger, N., Małuszyński, J. (eds.) Reasoning Web. LNCS, vol. 3564, pp. 552–561. Springer, Heidelberg (2005)
5. Lee, Y.W., Seo, J.H., Lee, J.G.: A Study on the TWS Tracking Filter for Multi-Target Tracking. Journal of KIEE 41(4), 411–421 (2004)

Analysis and Recognition of Touching Cell Images Based on Morphological Structures

Donggang Yu[1,2], Tuan D. Pham[1,2], and Xiaobo Zhou[3]

[1] Bioinformatics Applications Research Centre
[2] School of Mathematics, Physics and Information Technology,
James Cook University Townsville, QLD 4811, Australia
[3] HCNR Centre for Bioinformatics
Harvard Medical School, Boston, MA 02215, USA

Abstract. Automated analysis of molecular images has increasingly become an important research in computational life science. In this paper we present new morphological algorithms for the segmentation of touching cell images, which is essential for the task of cell screening. The proposed algorithms are useful for finding different models of touching images and image reconstruction.

Keywords: Cell screening, touching cell, morphological structure, segmentation and reconstruction, shape analysis.

1 Introduction

Automated cell-cycle screening using fluorescence microscopic images is very useful for biologists to understand complex processes of cell division under new drug treatment [1,2,3,4]. The most difficult task of such analysis [5,6,7] is finding the images of cell at different stages which can be presented by nuclear size and shape changes during mitosis. A key problem for identifying the size and shape of the cell nuclei is that they are touching each other. Therefore, it must be useful if we can detect touching cell nuclei so that they can be separated and reconstructed. This is the motivation of this paper. For example, the images of two frames are shown in Fig. 1. We can see that the sizes and shapes of some cells cannot be found because these cell images touch each other. Therefore, we have to find which cell images are touched, how many are touching cells, where these separation points, and how touching cell images are separated and reconstructed. This paper attempts to explore this issue. The rest of this paper is organized as follows. Section 2 presents the preprocessing of cell images. Section 3 presents the structural points of the touching cell images. The morphological structures of the touching models are discussed in Section 4. Separation points and reconstruction of touching cell images are determined in Section 5. Examples of the performance of the algorithms using real cell images are illustrated in each of these sections. Finally, a conclusion is given.

B. Apolloni et al. (Eds.): KES 2007/WIRN 2007, Part I, LNAI 4692, pp. 439–446, 2007.

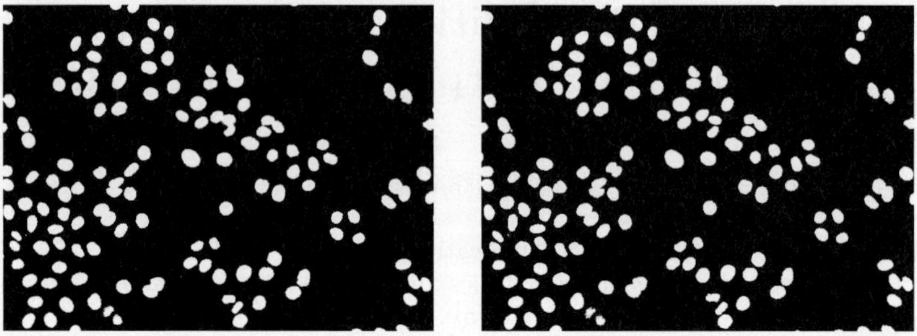

(1) (2)

Fig. 1. Binary images of two frames in one cell-cycle screening

2 Preprocessing of Touching Cell Images

The description of image contour plays an important role for the shape analysis and recognition of image. line segment, critical points and their convexity and concavity are useful features to analyze the shape of image contour. Many methods and algorithms are developed for the description of contours in the past [8,9,10,11]. We propose the morphological structure method to analyze and recognize contour shapes.

There are nine groups of touching cell images in Fig. 1(1). Three groups of touching cell images in Fig. 1(1) are shown in Fig. 2. We describe the algorithms as follows. Let the starting point of an binary image be the upper-left corner. Freeman code is used, and the contours are 8-connected. The direction of contour following is counter clockwise. The chain code set of contour k is represented as:

$$C_k = \{c_0, c_1 ... c_i, ... c_{n-1}, c_n\} \tag{1}$$

where i is the index of the contour pixels. The difference code, d_i, is defined as:

$$d_i = c_{i+1} - c_i. \tag{2}$$

In smooth followed contours, $|d_i|$ equals 0 or 1 [11]. The smoothed contour can be converted to a set of lines which consist of ordered pixels. Suppose that the direction chain code set of the smoothed contour is

$$\{c_l^{ln}[i] \quad (i = 0, ...(n_l^{ln} - 1))\}, \tag{3}$$

where ln is the ln-th line of a smoothed contour and n_l^{ln} is the number of points of the ln-th line. A linearized line has the following property: [11].

If

$$d_{ij} = c_l^{ln}[i] - c_l^{ln}[j] \quad (i = 0, ...k - 1), (j = 0, ...k - 1), \tag{4}$$

then

$$| d_{ij} | \leq 1 \quad (i = 0, ...k - 1), (j = 0, ...k - 1). \tag{5}$$

Therefore, a linearized line contains only two elements whose chain codes meet the above equation. Two element codes of the linearized line are represented by *cdir*1 and *cdir*2 respectively[11]. The smooth following and linearization results of images in Fig. 2 can be shown in Fig. 3 based on algorithms of smooth following, and linearization, where the spurious points in contours are removed and character "Y" is the first point of each linearized line.

(1) (2) (3)

Fig. 2. Binary images of three groups of touching cell images taken from Fig. 1

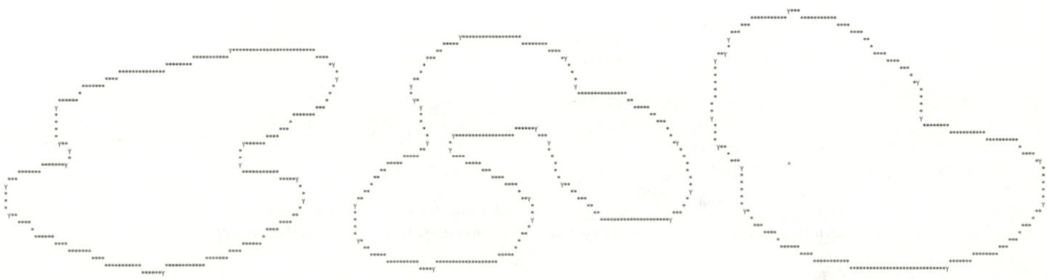

(1) (2) (3)

Fig. 3. The results of smooth following and linearization for the images in Fig. 2

3 Structural Points of Touching Cell Images

The structural points are some special points which can be used to represent convex or concave change in the direction of chain codes between two neighboring lines along the contour. Their definition and detection are based on the structure patterns of element codes of two lines. Assume that $line[ln]$ is the current line and that $line[ln-1]$ is the previous line.

Definition 1. The convex point in the direction of code 4 (represented with the character "$\wedge$"): If the element codes 3, 4 and 5 occur successively as a group of neighborhood linearized lines, then one convex point can be found as follows; if *cdir*1 of $line[ln]$ is code 4, *cdir*2 is code 5 and the direction chain code of the last pixel of $line[ln-1]$ is code 3, then the first pixel of the current line $line[ln]$ is a convex point which is represented with "$\wedge$".

Definition 2. The concave point in the direction of code 4 (represented with the character "m"): If the element codes 5, 4 and 3 occur successively as a group of neighborhood linearized lines, then one concave point can be found as follows; if $cdir1$ of $line[ln]$ is code 4, $cdir2$ is code 3 and the direction chain code of the last pixel of $line[ln-1]$ is code 5, then the first pixel of the current line, $line[ln]$, is a concave point which is represented with "m".

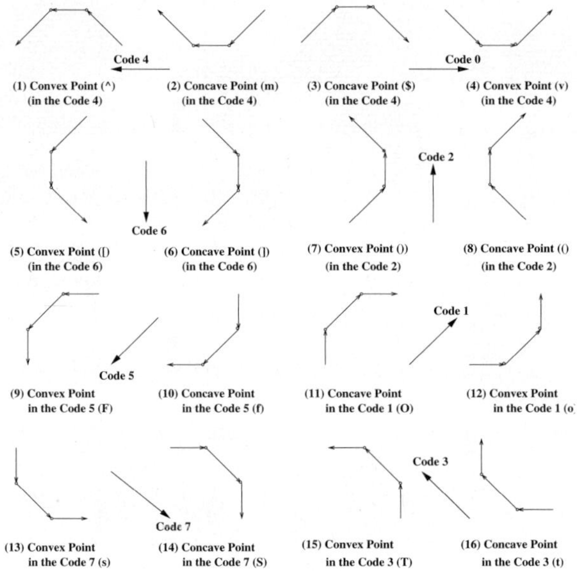

Fig. 4. Structural patterns of structural points

Similar to Definitions 1-2, other structural points can be defined and found. These points are convex points "v", "[', ")", "F", "o", "T", "s", and concave points "$", "]", "(", "f", "O", "t" and "S" which are shown in Fig. 5 respectively. These structural points describe the convex or concave change in different chain code directions along the contour, and they can therefore be used to represent the morphological structure of contour regions.

The series of structural points of touching cell images in Fig. 3 can be found and shown in Fig. 5 based on the above algorithm.

4 Morphology Structures of Touching Cell Images

We can see that there are some concave structural points on the contours of the images in Fig. 5. Based on the definition of structural points, one concave point means a concave change in the direction of one chain code on the contour. Based on the prior knowledge, the cell shape of cell-cycle screening images can approximate as an ellipse before it is divided. Therefore, if two or more cells are

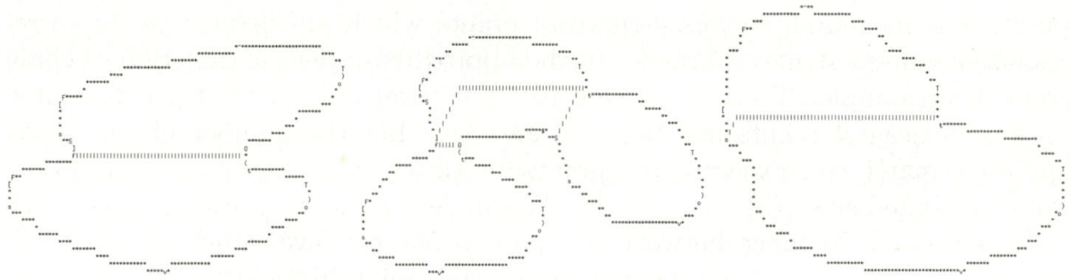
(1) (2) (3)

Fig. 5. The extracting structural points of the images in Fig. 3

touched, there are is one concave structural point at least on its outer contour. Also, its size is larger than that of one cell image as touching cell image consists of two or more cells. Let a series of concave structural points on the outer contour of touching cell images is

$$S_{cc} = \{s_{cc}(0), s_{cc}(1)...s_{cc}(i), ...s_{cc}(n-1), s_{cc}(n)\} \quad (6)$$

where $s_{cc}(i)$ is the structural point number of the i-th concave structural point on the contour, and there are n concave structural points on the contour. It is clear that $s_{cc}(i) < s_{cc}(i+1)$. In fact, one concave change on the contour may consists of several closest concave structural points. For example, if there exists $s_{cc}(i+1) - s_{cc}(i) = 1$ and $s_{cc}(i+2) - s_{cc}(i+1) = 1$, then that means one concave change consists of three concave structural points, $s_{cc}(i)$, $s_{cc}(i+1)$ and $s_{cc}(i+2)$. In this case, these three concave structural points should be merged into one group of concave structural points. After the above merging processing for S_{cc}, a series of groups of concave structural points (S_{cg}),

$$S_{cg} = \{s_{cg}(0), s_{cg}(1)...s_{cg}(i), ...s_{cg}(k-1), s_{cg}(k)\} \quad (7)$$

can be found, where k is the number of groups and $k \leq n$. For example, nine concave structural points in Fig. 5(2) are merged into three groups of concave structural points. The morphological patterns of touching cell images can be determined based on the number of groups of concave structural points. If $k = 1$ or $k = 2$, two cells are touched. If $k = 3$, three cells are touched. If $k = 4$, four cells are touched.

5 Separation Points and Reconstruction of Touching Cell Images

Separation Points of Touching Cell Images: The method of searching separation points can be described as follows.

Case 1 ($k = 1$): If $k = 1$, there is one group of concave points, $s_{cg}(0)$. Suppose $s_{cg}(0)$ contains p concave points, $s_{cg0}(0), ...s_{cg0}(p-1)p < 4$. For each concave point, find its match convex structural points which are defined as its corresponding convex structural points in the approximate reverse direction of chain code. For example, if $s_{cg0}(0)$ is concave structural point "∧", then its match convex structural points are "s", "v" and "o". Let the number of the corresponding match convex structural points for all $s_{cg0}(0), ...s_{cg0}(p)$ be q, and they are represented as $s_{cv}(0), ...s_{cv}(q-1)$. We can determine separation points which make minimum distance between one pair of one concave structural points in $\{s_{cg0}(0), ...s_{cg0}(p)\}$ and one convex structural point in $\{s_{cv}(0), ...s_{cv}(q-1)\}$. That is

$$\{s_{cg0}(m), s_{cc}(n)\} = mini\{|s_{cg0}(i), s_{cc}(j)|i < p, j < q\}, \tag{8}$$

where $s_{cg0}(m)$ and $s_{cc}(n)$ are selected separation points.

Case 2 ($k = 2$): If $k = 2$, there are two groups of concave points, $s_{cg}(0)$ and $s_{cg}(1)$. Suppose the number of concave structural points in $s_{cg}(0)$ is p_0, and in $s_{cg}(1)$ is p_1 respectively. In this case, we can determine separation points which make minimum distance between one pair of one concave structural point in $\{s_{cg0}(0), ...s_{cg0}(p_0)\}$ and one in $\{s_{cg1}(0), ...s_{cg1}(p_1)\}$. That is

$$\{s_{cg0}(m), s_{cg1}(n)\} = mini\{|s_{cg0}(i), s_{cg1}(j)|i < p_0, j < p_1\}, \tag{9}$$

where $s_{cg0}(m)$ and $s_{cg1}(n)$ are selected separation points.

Case 3 ($k > 2$): If $k > 2$, there are more than two groups of concave points, $s_{cg}(0)$ ... $s_{cg}(l)$ $l > 2$. In this case, we can determine each pair of separation points which make minimum distance between each pair of one concave structural point in $\{s_{cgx}(0), ...s_{cgx}(p_x)\}$ and one in $\{s_{cgy}(0), ...s_{cg1}(p_y)\}$, where $\{s_{cgx}(0), ...s_{cgx}(p_x)\}$ and $\{s_{cgy}(0), ...s_{cg1}(p_y)\}$ are neighboring groups of concave structural points. That is

$$\{s_{cgx}(m), s_{cgy}(n)\} = mini\{|s_{cgx}(i), s_{cgy}(j)|i < p_x, j < p_y\}, \tag{10}$$

where $s_{cgx}(m)$ and $s_{cgy}(n)$ are selected separation points. For example, if $k = 3$, there are three pairs of groups of concave structural points, $s_{cg}(0)$ and $s_{cg}(1)$, $s_{cg}(1)$ and $s_{cg}(2)$, and $s_{cg}(2)$ and $s_{cg}(0)$ respectively.

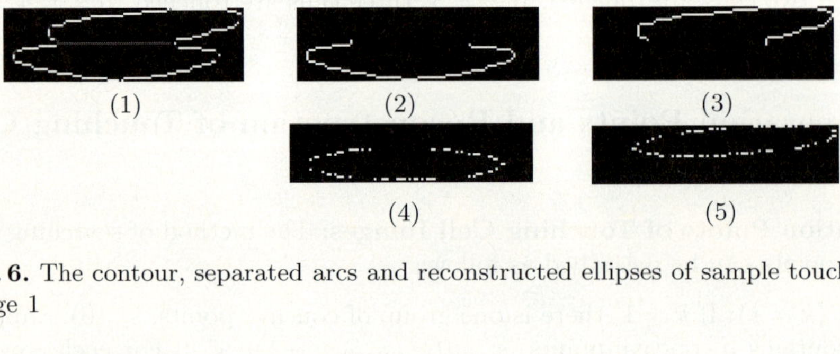

Fig. 6. The contour, separated arcs and reconstructed ellipses of sample touching cell image 1

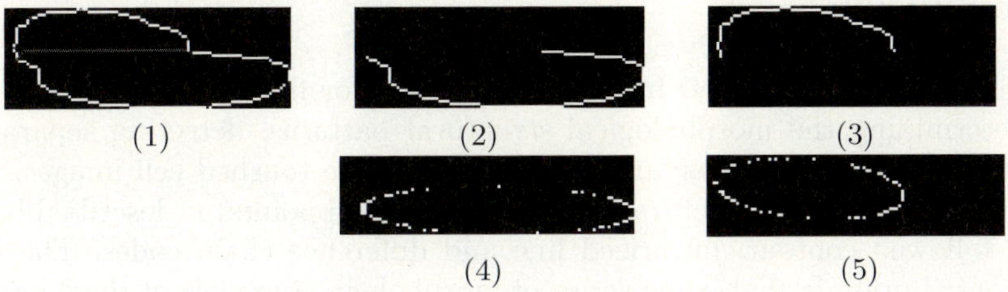

Fig. 7. The contour, separated arcs and reconstructed ellipses of sample touching cell image 2

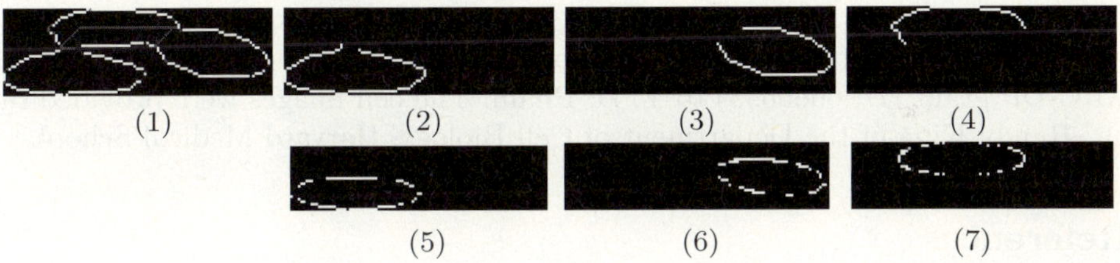

Fig. 8. The contour, separated arcs and reconstructed ellipses of sample touching cell image 3

Based on the above algorithm, the touching cell image in Fig. 5(1) is Case 2, in Fig. 5(2) Case 3, and in Fig. 5(3) Case 1. Based on the above algorithm, we can find all separation points of images in Figs. 5(1-4). We can find related separation lines (see Figs. 5(1-4)) and the coordinate data of related arcs which are shown in Figs. 6(2,3), 7(2,3) and 8(2,3,4) based on these separation points and series of points of the contour. These contours of touching cell images are shown in Figs. 6(1), 7(1) and 8(1).

Reconstruction of Touching Cell Images: We have found the coordinate data of all related arcs which are separated based on the above algorithm. As all cell shapes are approximately as an ellipse, touching cell images can be reconstructed by using the data of these separated arcs. The reconstruction method is direct least square fitting of ellipses [12]. The the reconstructed cell images are shown in Figs. 6(4,5), 7(4,5) and 8(5,6,7) based on the coordinate data of separated arcs respectively. The ellipse in Fig. 6(4) is the reconstruction result of the separated contour of touched cell image in in Fig. 6(2), and the ellipse in Fig. 6(5) is that in Fig. 6(3). Similarly, the ellipses in Figs. 8(5,6,7) are the reconstruction results of the separated contours of touched cell images in Figs. 8(2,3,4) respectively. The series of reconstruction results starts from left-upper point in anti-clock direction. The reconstructed touched cell images can help to determine which cell phase the cell when compare some features of the cell at current time with those at previous time and next time.

6 Conclusion

An efficient and new method has been developed for finding touching cell images, determining the morphological structural patterns, detecting separation points of touched cell images, and reconstructing the touched cell images. The algorithm of extracting structural features (structural points) is described based smooth followed contour, linearized line and difference chain codes. The best useful contribution is that some series of morphological models of touching cell images are developed and touched cell images are reconstructed based on our algorithm. Our method is efficient and novel in the sense that morphological structure models of touching cell images are constructed, and these models simulate artificial intelligence.

Acknowledgement. This work was supported by the Australia Research Council ARC-DP grant (DP0665598) to T. D. Pham. The cell images were provided by Dr. Randy King of the Department of Cell Biology, Harvard Medical School.

References

1. Fox, S.: Accommodating cells in HTS. Drug Discovery World 5, 21–30 (2003)
2. Feng, Y.: Practicing cell morphology based screen. European Pharmaceutical Review 7, 7–11 (2002)
3. Dunkle, R.: Role of image informatics in accelerating drug discovery and development. Drug Discovery World 5, 75–82 (2003)
4. Yarrow, J.C., et al.: Phenotypic screening of small molecule libraries by high throughput cell imaging. Comb Chem High Throughput Screen 6, 279–286 (2003)
5. Chen, X., Zhou, X., Wong, S.T.C.: Automated segmentation, classification, and tracking cancer cell nuclei in time-lapse microscopy. IEEE Trans. on Biomedical Engineering, in press
6. Pham, T.D., Tran, D., Zhou, X., Wong, S.T.C.: An automated procedure for cell-phase imaging identification. In: Proc. AI-2005 Workshop on Learning Algorithms for Pattern Recognition, pp. 52–29 (2005)
7. Pham, T.D., Tran, D.T., Zhou, X., Wong, S.T.C.: Classification of cell phases in time-lapse images by vector quantization and Markov models. In: Greer, E.V. (ed.) Neural Stem Cel l Research, Nova Science, New York (2006)
8. Moktarian, F., Mackworth, A.K.: Theory of Multiscale Curvature-Based Shape Representation for Planer Curvature Angles. IEEE Trans. Pattern Analysis Mach. Intell. 14(8), 789–805 (1992)
9. Fu, A.M.N., Yan, H., Huang, K.A: Curvature Angle Bend Function Based Method to Characterize Contour Shapes. Patt. Recog. 30(10), 1661–1671 (1997)
10. Sonka, M., Hlavac, V., Boyle, R.: Image Processing, Analysis and Machine Vision. Chapman & Hall Computing, Cambridge (1993)
11. Yu, D., Yan, H.: An efficient algorithm for smoothing binary image contours. Pro. of ICPR'96 2, 403–407 (1996)
12. Fitzgibbon, A., Pilu, M., Fisher, R.B.: Direct Least Square Fitting of Ellipse. Term Analysis and Machine Intelligence 21, 476–480 (1999)

Comparison of Accumulative Computation with Traditional Optical Flow

Antonio Fernández-Caballero, Rafael Pérez-Jiménez, Miguel A. Fernández, and María T. López

Departmento de Sistemas Informáticos, Universidad de Castilla-La Mancha
Escuela Politécnica Superior de Albacete, Albacete, Spain
caballer@dsi.uclm.es

Abstract. Segmentation from optical flow calculation is nowadays a well-known technique for further labeling and tracking of moving objects in video streams. A likely classification of algorithms to obtain optical flow based on the intensity of the pixels in an image is in (a) differential or gradient-based methods and (b) block correlation or block matching methods. In this article, we are going to carry out a qualitative comparison of three well-known algorithms (two differential ones and a correlation one). We will do so by means of the optical flow obtaining method based on accumulated image differences known as accumulative computation.

Keywords: Optical flow, Accumulative computation method, Image difference.

1 Introduction

One of the most interesting and productive techniques in the field of image sequence motion analysis is the technique known as optical flow [7]. Indeed, segmentation from optical flow calculation is nowadays a well-known technique for further labeling and tracking [10],[15],[17],[11] of moving objects in video streams, as motion is a major information source for segmenting objects perceived in dynamic scenes. Optical flow can be defined as the apparent displacement of the pixels in the image when there is relative motion between the camera and the objects under focus. Another possible definition is considering optical flow as the 2-D motion field obtained from the projection of the velocities of the three dimensional pixels, corresponding to the surfaces of a scene, onto the sensor's visual plane [8].

A possible algorithm classification to obtain optical flow [2] based on pixel intensity in the image would be (a) differential methods and (b) block correlation methods (matching). In this article, we are going to carry out a qualitative comparison of three well-known algorithms with our optical flow obtaining method, known as accumulative computation [6],[16]. Our method presents a new way of looking at optical flow and describes it as a measure of the time elapsed since the last significant change in the brightness level of each pixel in the image [12].

B. Apolloni et al. (Eds.): KES 2007/WIRN 2007, Part I, LNAI 4692, pp. 447–454, 2007.

Differential Methods are also called gradient-based methods. These techniques calculate the flow from the space-time derivatives of the intensities in the image, through the expression known as the brightness constancy equation of the optical flow computation. This method has become the most frequent approximation used in computer vision applications because of its swiftness and its good velocity estimation.

Horn and Schunck [9] propose a method based on first order derivatives and add a smoothness condition on the flow vectors to the general conditions. They assume that object motion in a sequence will be rigid and approximately constant, that a pixel's neighborhood in said objects will have similar velocity, therefore, changing smoothly over space and time. Nevertheless, this condition is not very realistic in many cases and it yields bad results [14] since the images' flow has a lack of continuity, especially in the boundaries between different objects. Therefore the results obtained in these areas will not be correct. Poor results are also obtained in the sequences where there are multiple objects, each having different motion.

Barron, Fleet and Beauchemin [2] suggest a modification where a Gaussian space-time pre-smoothing is done to the images and where the derivatives are calculated using the differential method with a coefficient mask. On the other hand, they introduce a gradient thresholding method in algorithm implementation and decide whether a velocity will be accepted or rejected. This decision is taken based on the gradient module; when it does not exceed the threshold value, the velocity in said pixel will be rejected. A great number of inaccurate results can be eliminated this way.

Lucas and Kanade's algorithm [13] is similar to Horn and Schunck's. Horn and Schunck use a global approach, whereas Lucas and Kanade use a local approach to an environment. The algorithm devised by Lucas and Kanade adds a flow smoothing constraint in local neighborhoods to the intensity conservation restraint. The method expects the velocities to be constant in a relatively small environment basing this on the fact that it is logical to expect pixels from the same object to have identical velocities.

Block Correlation Methods, also known as a block matching-based method, assume that the distribution of the intensity for the region which surrounds the pixel, whose motion is to be evaluated, is maintained. Thus, for each pixel whose flow is to be computed at a certain time, a window of pixels which surrounds that pixel is created. The purpose, in the following time, is to look for the maximum correspondence between said window and a set of windows of equal resolution within a neighborhood defined by a higher window, called a search a window in the following time.

Anandan's algorithm [1] fits into the matching methods. It proposes that, in a discrete case, the sum of the squared differences (SSD) is closely related to the correlation coefficient. To attain sub-pixel accuracy and to avoid problems due to aperture or great displacements, Anandan used a hierarchical scheme based on Gaussian or Laplacian pyramids, estimating velocity from the lowest to the highest resolution level. This way, sub-pixel displacements are estimated in two different phases. Anandan also proposes smoothing the resulting velocities, since it is expected for velocities present in a sequence to be fairly homogenous. In the final algorithm, a matching and smoothing of the resulting velocities is carried out for each level in the pyramid created, from the lowest to the highest resolution level.

2 Optical Flow Through Accumulative Computation

Accumulative computation is based on the allocation of charge levels assigned to every image pixel related to the history of a studied feature of the pixel. The general formula which represents the charge in an image pixel, due to accumulative computation [5],[4] is:

$$Ch[x,y,t] = \begin{cases} \min(Ch[x,y,t-\Delta t] + C, Ch_{max}), & \text{if "}\textit{property is fulfilled}\text{"} \\ \max(Ch[x,y,t-\Delta t] - D, Ch_{min}), & \text{otherwise} \end{cases} \tag{1}$$

In the LSR mode of operation (length-speed ratio) [3], $C = C_{Mov}$ is called a charge increase value. The idea behind is that if there is no motion in pixel (x,y), which is estimated as a change in the grey level between two consecutive times, charge value $Ch[x,y,t]$ increases up to a maximum value Ch_{max}. And if there is motion, there is a complete discharge (a minimum value Ch_{min} is assigned). In general, Ch_{max} and Ch_{min} take values of 255 and 0, respectively. Notice that charge value $Ch[x,y,t]$ represents a measure of the time elapsed since the last significant change in the image pixel's (x,y) brightness.

$$Ch[x,y,t] = \begin{cases} Ch_{min}, & \text{if "}\textit{motion is detected in } (x,y) \textit{ in } t\text{"} \\ \min(Ch[x,y,t-1] + C_{Mov}, Ch_{max}), & \\ \qquad \text{otherwise} \end{cases} \tag{2}$$

Once the image's charge map is obtained for the current time t, the optical flow considered as the velocity estimated from the stored charge values is obtained as detailed next. (1) $Ch[x,y,t] = Ch_{min}$: Motion is detected in pixel (x,y) in t. The map's value is the minimum charge value. (2) $Ch[x,y,t] = Ch_{min} + k \cdot C < Ch_{max}$: Motion in pixel (x,y) is not detected in t. Motion was last detected in $t - k \cdot \Delta t$. After k increments, the maximum charge has not yet been reached. (3) $Ch[x,y,t] = Ch_{max}$: Motion is not detected in pixel (x,y) in t. It is not known when motion was last detected. The map's value is the maximum charge value.

It is important to point out that the velocity obtained by these means is not the velocity of an object pixel, which is occupied by pixel (x,y) in time t, but the velocity of an object pixel responsible for motion detection when it went by pixel (x,y) $k = \dfrac{Ch[x,y,t] - Ch_{min}}{C_{Mov}}$ units of time ago. Therefore, a given charge has the same value in all pixels where motion was detected at the same time.

Now then, velocity is calculated in axis x, v_x, as well as in axis y, v_y. To calculate velocity in x, the charge value in (x,y), which an object is currently crossing, is compared to the charge value of another coordinate in the same image row $(x+l, y)$, where the same object is crossing. At best, that is when both values are different to Ch_{max}, the time elapsed from the last motion detection in (x, y) to the time when motion is detected in $t - k_{(x+l,y)} \cdot \Delta t$ en $(x + l, y)$ can be calculated as:

$$\begin{aligned} &Ch[x, y, t] - Ch[x + l, y, t] = \\ &= (Ch_{min} + k_{(x,y)} \cdot C_{Mov}) - (Ch_{min} + k_{(x+l,y)} \cdot C_{Mov}) = \\ &= (k_{(x,y)} - k_{(x+l,y)}) \cdot C_{Mov} \end{aligned} \tag{3}$$

Obviously, this cannot be calculated if either of the values is equal to Ch_{max}, since it is not known how many time intervals have elapsed since the last motion detection. Therefore, for valid charge values, we have:

$$\Delta t = \frac{(k_{(x,y)} - k_{(x+l,y)}) \cdot C_{Mov}}{C_{Mov}} = k_{(x,y)} - k_{(x+l,y)} \tag{4}$$

From equations (3) and (4):

$$\Delta t = \frac{Ch[x, y, t] - Ch[x + l, y, t]}{C_{Mov}} \tag{5}$$

Since $v_x[x, y, t] = \frac{\delta x}{\delta t} = \frac{l}{\Delta t}$, we finally have:

$$v_x[x, y, t] = \frac{C_{Mov} \cdot l}{Ch[x, y, t] - Ch[x + l, y, t]} \tag{6}$$

Velocity is calculated in the same way in y from values stored as charges:

$$v_y[x, y, t] = \frac{C_{Mov} \cdot l}{Ch[x, y, t] - Ch[x, y + l, t]} \tag{7}$$

3 Data and Results

Once the methods have been described, we go on to present the results obtained in the qualitative comparison of the different algorithms: (a) Barron, Fleet and Beauchemin, (b) Lucas and Kanade, (c) Anandan and (d) accumulative computation. For this experimental level comparison, different image sequences have been selected for each algorithm. The results show in a qualitative manner those pixels where some velocity different from zero is obtained.

Yosemite Sequence. This is a complex case in the synthetic sequence bank used in numerous benchmarks. It shows a virtual flight over the Yosemite valley. The clouds on the upper right of the image move at a velocity of 2 pixels/frame from left to right. The rest of the flow is divergent, with velocities of up to 5 pixels/frame in the lower left corner. This is an interesting sequence since it displays different types of motion, slightly different boundaries and it can resemble a real situation. In Fig. 1, we see the result of applying each of the four

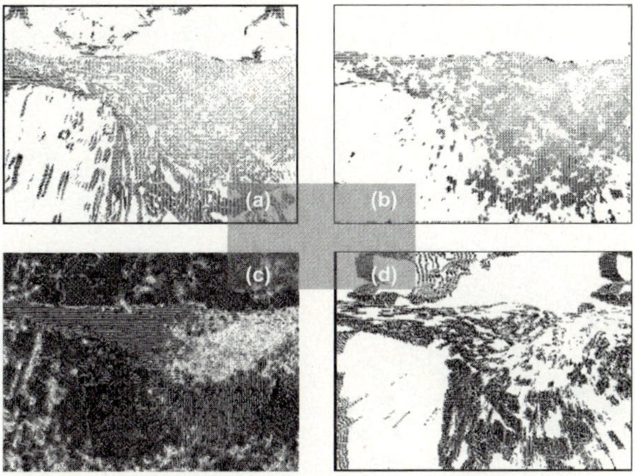

Fig. 1. Results obtained in the Yosemite sequence. (a) Barron, Fleet and Beauchemin's (BFB) method. (b) Lucas and Kanade's (LK) method. (c) Anandan's (A) method. (d) Accumulative computation (AC) method.

methods to the Yosemite sequence. In the first place, we are struck by Anandan's method's poor performance. It detects much more (and inaccurate) flow than other methods. We can also verify that both Barron's and the accumulative computation methods are able to detect cloud motion as opposed to Lucas and Kanade's which cannot.

Hamburg Taxi Sequence. This sequence is a classic in the computer vision field. There are four objects in motion: (1) the white taxi turning the corner, (2) a dark car in the lower left corner, moving from left to right, (3) a van, also dark, moving from right to left, and, (4) a pedestrian, who is fairly far away from the camera, in the upper left corner. In the foreground and slightly to the right, we see tree branches. The approximate velocity for each object is: 1.0, 3.0, 3.0 and 0.3 pixels/frame, respectively. The fields obtained for the Taxi sequence (Fig. 2) in general show all the displacements mentioned in its description, with the exception of the pedestrian's movement which can only be obtained with accumulative computation-based algorithm. This method also "outlines" objects better than others. In every case there is a lot of noise in the scene. We are also struck by the fact that the vehicles are not excessively well segmented (this would belong to an advanced level analysis). The vehicle closest to the right hand side is detected the worst because it is partially hidden by part of a tree.

Rubik's Cube Sequence. Another well-known sequence is this Rubik's cube rotating counter-clockwise. The velocity field caused by the cube's rotation is less than 2 pixels/frame. The surface on which the cube is placed has a motion between 1.2 and 1.4 pixels/frame. Good results are obtained, in general, in the Rubik's cube sequence (Fig. 3), obtaining the velocity of the cube's sides as well

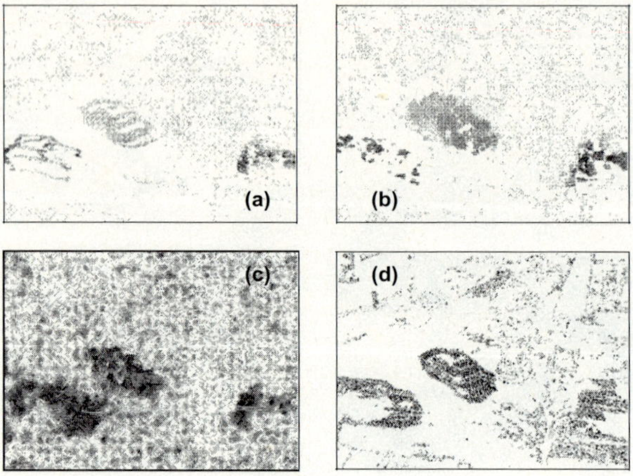

Fig. 2. Results obtained in the Hamburg Taxi sequence. (a) BFB method. (b) LK method. (c) A method. (d) AC method.

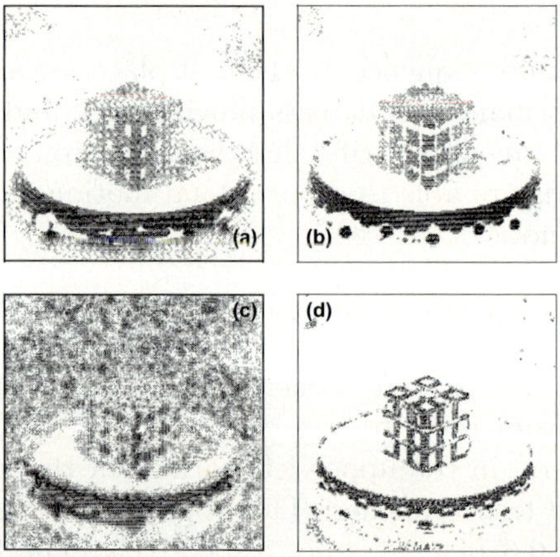

Fig. 3. Results obtained for the Rubik's Cube sequence. (a) BFB method. (b) LK method. (c) A method. (d) AC method.

as that of the rotary base. The cube's shadow motion is detected in Barron-Fleet-Beauchemin's and Anandan's algorithms and it is best filtered with Lucas and Kanade's method and with the accumulative computation-based method. The latter algorithms offer the best results for this sequence, qualitatively speaking. We see at a first glance that the accumulative computation algorithm eliminates the most noise from the scene. Again, Anandan offers poor results.

SRI Trees Sequence. This time, the camera moves from right to left, parallel to the plane in front of the group of trees. This is a complex sequence since it has a great number of occlusions, as well as low resolution. The velocities are

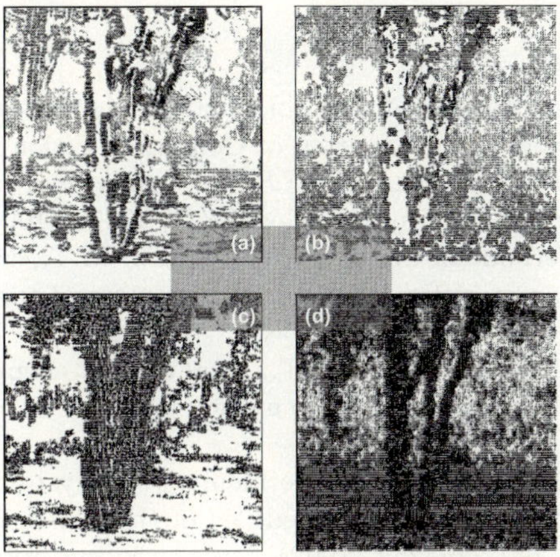

Fig. 4. Results obtained for the SRI Tress sequence. (a) BFB method. (b) LK method. (c) A method. (d) AC method.

greater than 2 pixels/frame. The SRI Trees sequence is very complex. Barron et al's algorithm performs better than the rest since it outlines the trees (Fig. 4). Other methods seem to be inefficient when working with movable cameras on static scenes. This is essentially so in the accumulative computation method.

4 Conclusion

In this work, we have presented a qualitative comparison of the different traditional optical flow computation methods with our new accumulative computation technique. Other methods have high computational costs as opposed to our accumulative computation method, based on simple additions and subtractions.

In this paper, accumulative computation is based on the allocation of charge levels assigned to every image pixel related to the history of motion presence detection of the pixel. Our accumulative computation method is new in the sense that it calculates the optical flow as a measure of the elapsed time since the last significant change in the brightness level for each pixel in the image.

In the results obtained in the segmentation of the shape of figures due to the motion inherent to he camera capture, we see that for most of the sequences tested, specifically Yosemite and Hamburg Taxi, the accumulative computation method offers similar or better quality than the other methods. We are currently working to offer a quantitative comparison of the results obtained, with regard to execution time and success rate in the results.

Acknowledgements

This work is supported in part by the Spanish CICYT TIN2004-07661-C02-02 grant and the Junta de Comunidades de Castilla-La Mancha PBI06-0099 grant.

References

1. Anandan, P.: A computational framework and an algorithm for the measurement of visual motion. International Journal of Computer Vision 2, 283–310 (1989)
2. Barron, J.L., Fleet, D.J., Beauchemin, S.S.: Performance of optical flow techniques. International Journal of Computer Vision 12(1), 43–77 (1994)
3. Fernández, M.A., Fernández-Caballero, A., López, M.T., Mira, J.: Length-speed ratio (LSR) as a characteristic for moving elements real-time classification. Real-Time Imaging 9, 49–59 (2003)
4. Fernandez, M.A., Mira, J., Lopez, M.T., et al.: Local accumulation of persistent activity at synaptic level: application to motion analysis. From Natural to Artificial Neural Computation, 137–143 (1995)
5. Fernández, M.A., Mira, J.: Permanence memory - A system for real time motion analysis in image sequences. In: Proceedings of the IAPR Workshop on Machine Vision Applications, pp. 249–252 (1992)
6. Fernández-Caballero, A., Fernández, M.A., Mira, J., Delgado, A.E.: Spatio-temporal shape building from image sequences using lateral interaction in accumulative computation. Pattern Recognition 36(5), 1131–1142 (2003)
7. Gibson, J.J.: The Perception of the Visual World. Houghton Mifflin (1950)
8. Horn, B.K.P.: Robot Vision. MIT Press, Cambridge (1986)
9. Horn, B.K.P., Schunck, B.G.: Determining optical flow. Artificial Intelligence 17, 185–204 (1981)
10. Liang, K.H., Tjahjadi, T.: Multiresolution segmentation of optical flow fields for object tracking. Applied Signal Processing 4(3), 179–187 (1998)
11. Lodato, C., Lopes, S.: An optical flow based segmentation method for objects extraction. Transactions on Enginnering, Comuting and Technlogy 12, 41–46
12. López, M.T., Fernández-Caballero, A., Fernández, M.A., Mira, J., Delgado, A.E.: Motion features to enhance scene segmentation in active visual attention. Pattern Recognition Letters 27(5), 469–478 (2005)
13. Lucas, B., Kanade, T.: An iterative image registration technique with an application to stereo vision. In: Proceedings of the DARPA IU Workshop, pp. 121–130 (1981)
14. Lucena, M.: Uso del flujo óptico en algoritmos probabilísticos de seguimiento. Tesis Doctoral, Departamento de Informática. Universidad de Jaén (2003)
15. Macan, T., Loncaric, S.: Hybrid optical flow and segmentation technique for LV motion detection. Proceedings of SPIE 4321, 475–482 (2001)
16. Mira, J., Delgado, A.E., Fernández-Caballero, A., Fernández, M.A.: Knowledge modelling for the motion detection task: The algorithmic lateral inhibition method. Expert Systems with Applications 2, 169–185 (2004)
17. Zitnick, C.L., Jojic, N., Kang, S.B.: Consistent segmentation for optical flow estimation. In: Proceedings of the Tenth IEEE International Conference on Computer Vision, vol. II, pp. 1308–1315 (2005)

Face Recognition Based on 2D and 3D Features

Stefano Arca, Raffaella Lanzarotti, and Giuseppe Lipori

Dipartimento di Scienze dell'Informazione
Università degli Studi di Milano
Via Comelico, 39/41 20135 Milano, Italy
{arca,lanzarotti,lipori}@dsi.unimi.it

Abstract. This paper presents a completly automated face recognition system integrating both two dimensional (texture) and three dimensional (shape) features. We introduce a novel fusion strategy that allows to automatically select, for each face, the most relevant features from each modality. The performance is evaluated on the largest public data corpus for face recognition currently available, the Face Recognition Grand Challenge version 2.0.

1 Introduction

A general statement describing the face recognition (FR) problem can be formulated as follows: given a stored database of face representations, identify subjects represented in input probes (2D and/or 3D images). This definition can then be specialized to describe either the *identification* or the *verification* problem. The former requires as input a face image, and the system determines the subject identity on the basis of a database of known individuals; in the latter situation the system has to confirm or reject the claimed identity of the input face.

FR systems are attractive for their non-intrusive nature and in the last three decades a great research effort has been devoted to tackle this problem [1]. Face recognition algorithms using 2D intensity or color images were the first to be investigated; we recall the methods based on subspaces [2,3](PCA, LDA, ICA), based on classifiers such as Neural Network or SVMs [4,5], and the Elastic Bunch Graph Matching technique [6]. Almost all of these methods achieve good performance in constrained environments; however, they encounter difficulties in handling large amounts of facial variations due to head pose, lighting conditions and facial expressions.

To overcome these limitations in recent years a great deal of research work has been devoted to the development of 3D face recognition algorithms that identify subjects from the 3D shape of a person's face. Indeed, while 2D (color or gray level) images provide a more precise description of the facial features (eyes, nose, mouth), 3D face models bring shape information that is not affected by pose, lighting conditions or makeup. Several 3D [7,8,9] and 2D+3D [10,11,12] face recognition systems have been surveyed by Bowyer et al. in [13]; most of them work on range images, that is images where the pixel values reflect the distance of the sensor from the object. It is a common thought that integration of both

B. Apolloni et al. (Eds.): KES 2007/WIRN 2007, Part I, LNAI 4692, pp. 455–462, 2007.

texture and shape information may lead to increased recognition rate, since such an approach should exploit the benefits of both 2D and 3D information and could make it possible to overcome their respective shortcomings. Another reason that makes the multimodal approach promising is that 3D and 2D information are likely to be weakly correlated, so that their combination should improve the system performance.

In order to better understand which are the most promising methods, a common and challenging dataset should be referred for tests and comparisons. The Face Recognition Grand Challenge [14] is a significant attempt to address this issue; indeed it provides a large amount of data and contains face images presenting both illumination and expression variations.

Examples of multimodal face recognition systems tested on the FRGC v2.0 are the works proposed by Maurer [15] and by Husken [16]. They report good performance for the verification problem, whilst in [17] Mian et al. show very promising results in the identification scenario.

Here we propose a fully automated multimodal system for face identification, that extends our previous work [18], dealing with 2D images of low quality and with 3D range data. 2D features are extracted examining the texture around a set of fiducial points, while 3D features characterize the shape in correspondence to a set of facial profiles. The system has been tested on the FRGC v2.0 database addressing in particular the Experiment 3, where the performance is measured referring to a gallery and a test set composed of 3D images (texture and shape).

The paper is organized as follows: Section 2 describes the method used to extract the 2D features; Section 3 introduces the 3D face representation; Section 4 presents the fusion recognition algorithm; finally in Section 5 and 6 the experimental results are showed and discussed.

2 2D Features Extraction

In [18] we presented a fully automated component-based face recognition system that does not need any training session and works on color images of good quality. It first localizes the face in the image and precisely determines the eye center positions by means of an eye locator. These points are used to normalize the image and to initialize the method that localizes the facial components (eyes, eyebrows, nose, and mouth) and precisely extracts 24 facial fiducial points: the eyebrow and chin vertices, the tip, the lateral extremes and the vertical mid-point of the nose, the eye and lip corners, their upper and lower mid-points, the mid-point between the two eyes, and four points on the cheeks. In order to generalize the system to deal with images presenting strong shadows (as the ones contained in the FRGC v2.0 database), the steps that need to be modified are the ones that extract the fiducial points of the mouth and the nose.

In [18] the mouth subimage and corners were determined on the basis of color information; here we extract the mouth subimage by exploiting a statistics on the face geometry that allows to estimate the position and the dimensions of the mouth knowing the eye positions. On the extracted mouth subimage we determine the mouth corners taking the extremes of the lip-cut characterized by both low gray level values and high horizontal derivative values. The upper and lower mid-points of the mouth are then determined applying the snakes as described in [18].

For the nose tip we exploit the 3D information available with the range data, and we determine it simply by considering the nearest point to the camera. Its projection onto the texture image gives a very precise localization of the 2D nose tip fiducial point.

Besides, in order to enrich the face description, we consider two additional points on each eyebrow; they are determined by considering the points on the parabolas describing the eyebrows [18], corresponding to the abscissas of the eye corners. Moreover we do not consider the chin fiducial point being most of the time inaccurate. The final set of fiducial points is thus composed of the 27 points illustrated in Figure 1-left.

Once the fiducial points have been extracted, we characterize each of them by convolving the portion of gray image around it with a bank of 40 Gabor kernels (5 scales, 8 orientations), as described in [18], obtaining a vector (*Jet*) of 40 real coefficients for each fiducial point. Each face is then represented by a vector V_{2D} of (40×27) real coefficients. In order to compare pairs of corresponding Jets we introduce the similarity measure *Sim2D* between Jets:

$$Sim2D(J^1, J^2) = \frac{\langle J^1, J^2 \rangle}{\sqrt{\|J^1\|^2 \cdot \|J^2\|^2}} \tag{1}$$

3 3D Features Extraction

The range images provided by the FRGC v2.0 consist of texture-range data pairs consistently registered: each pixel on the texture image is associated to its 3D point in the range data, making straightforward the determination of the 3D coordinates associated to any point in the 2D image. Given a range image we represent a face by means of a set of fifteen 3D facial profiles. Each profile is determined by considering the set of 3D points which correspond to a segment connecting or passing through certain fiducial points automatically determined in the texture image. The set of profiles is composed of three vertical and three horizontal profiles on the nose, three profiles on the area between the eyes, two profiles on the eyebrows and four profiles on the cheeks (see Figure 1-right).

Due to acquisition errors, the set P of the 3D points $\{p_i = (x_i, y_i, z_i)\}$ composing a facial profile might contain some outliers; in order to remove them we eliminate those points which are too distant from the center of mass of P. Once removed the outliers, we apply the Principal Component Analysis to the remaining points and we project the set of 3D points of the profile P on the first two principal directions. The set P_{2D} of the projected data $\{p^i_{2D} = (x^i_{2D}, y^i_{2D})\}$ is shown in Figure 2. In order to express all the profiles in the same reference frame, each set of points P_{2D} is translated and rotated to bring its first point in the origin, and its last point on the abscissa axes. Once each profile has been

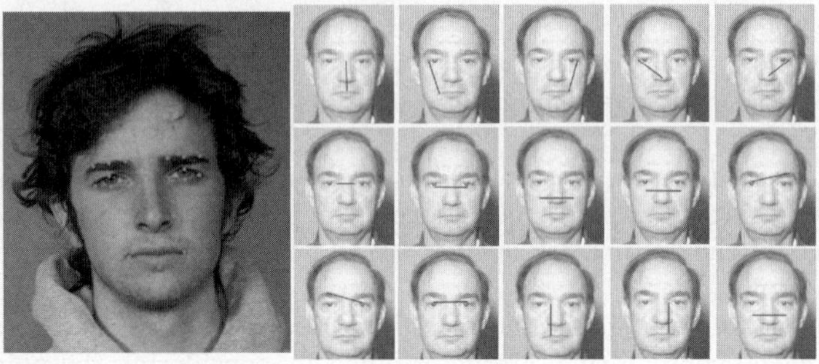

Fig. 1. Left: Set of the 2D fiducial points; Right: Set of segments considered to generate the 3D profiles

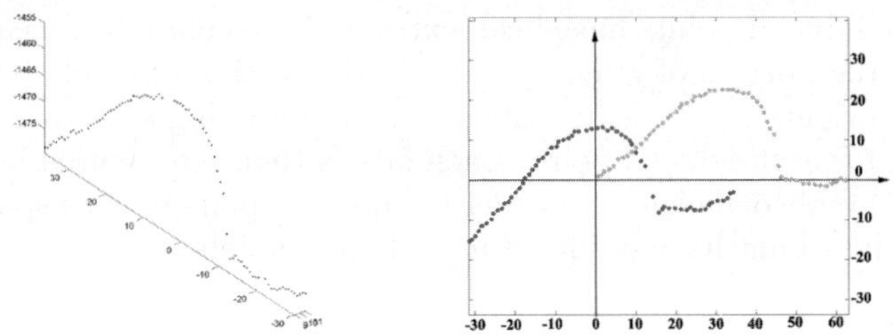

Fig. 2. Left: Set P of the 3D points composing the profile. Right: Projection of the profile along the two principal directions determined by the PCA (set P_{2D}) before and after the transformation.

expressed in the common reference frame, it is approximated in a least-squares sense, by a 15th order polynomial. and uniformly sampled (sampling step 0.25) obtaining the set SP of samples of a profile P.

Each face is then characterized by a feature vector V_{3D} containing in each row the samples of its fifteen profiles (3D features).

3.1 Similarity Measure Between 3D Profiles

In order to compare two profiles, we first maximize the overlapping between them by shifting the one with less samples on the other one and determining the translation $\bar{t}$ which minimizes the Sum of Squared Differences (SSD):

$$\bar{t} = argmin_t \{SSD\left[SP1(i-t), SP2(i)\right]\} \tag{2}$$

where t varies in the set $[-N/2, M + N/2]$ while N and M are the number of samples of $SP1$ and $SP2$ respectively. Denoting with *diff* the value of the SSD computed for the translation $\bar{t}$, normalized in the range $[0, 1]$, the similarity measure *Sim3D* between the two profiles is computed as:

$$Sim3D(SP1, SP2) = \begin{cases} 1 & if\ |L(SP1) - L(SP2)| \geq 30 \\ min\{1, diff\} & otherwise \end{cases} \quad (3)$$

where $L(SP)$ is the length of the sampled profile SP[1]. $Sim3D$ is defined in the range $[0, 1]$ and gives low values for those pairs of profiles which have approximatively the same shape (0 for the identity pair) whilst those pairs which have very different length are penalized.

4 Recognition

Once the 2D points characterization and the 3D profiles have been calculated, each face is represented by means of a vector V_{2D3D} composed of 42 features obtained appending V_{3D} to V_{2D}. The first 27 rows correspond to the 2D features while the last 15 are the 3D features. We notice that most of the multimodal recognition systems perform the fusion of 2D and 3D data combining the 2D and 3D similarity measures after they have been independently evaluated, here the 2D and 3D information are integrated at an early stage. In particular to recognize a test range image t, we compute for each range image i in a referring gallery G a *Score*, representing the closeness of i to the test image t; the face in t is recognized as the one in the gallery which obtained the highest score.

We proceed as follows:

- for each image $i \in G$ and each features $k = 1, .., 42$, compute the similarity measure between pairs of corresponding features:

$$S^{i,k} = \begin{cases} Sim2D(V_{2D3D}(t,k), V_{2D3D}(i,k))\ if\ k \leq 27 \\ Sim3D(V_{2D3D}(t,k), V_{2D3D}(i,k))\ if\ k > 27 \end{cases} \quad (4)$$

  where $V_{2D3D}(t, k)$ and $V_{2D3D}(i, k)$ are the kth feature of the test and the gallery image respectively.
- for each feature k, order the values $\{S^{i,k}\}$ in descending order, and assign to each of them a weight $w^{i,k}$ as a function of its ordered position $p^{i,k}$. The weight $w^{i,k}$ is determined as:

$$w^{i,k} = c \cdot [\ln(x + y) - \ln(x + p^{i,k})], \quad (5)$$

  where $y = \frac{|G|}{4}$, $x = e^{-\frac{1}{2}}$, and c is a normalization factor.
- for each gallery image i, consider the set, *BestFeatures*, of the 22 features[2] which have the highest weights, and determine the score:

$$\text{Score}(i) = \sum_{k \in BestFeatures} w^{i,k} S^{i,k}. \quad (6)$$

[1] The length of a profile is the sum of the Euclidan distance between its consecutive samples.

[2] The cardinality of the set *BestFeatures* has been set to half the total number of features (42) plus 1, that is 22. This number is a trade-off between the necessity to maintain enough information and to discard the less unreliable features contribution.

This technique allows to discard wrong matches on single points or single profiles: if either some fiducial points or some profiles are not precisely determined either in the test or in the gallery images, they will have low similarity measures so that they will not belong to the set *BestFeatures*, and they will not be used for the recognition. Moreover this method allows to automatically select the most suitable features to be used for the recognition of each face. For example if the information provided by the 2D features is not highly discriminative due to a bad illumination, it is likely that the 3D features (insensibible to illumination variations) will be used for the recognition. In this way the fusion allows to take the most relevant features from each modality.

5 Experiments

The experiments have been carried out on the FRGC version 2.0 [14] database. The set of the range data provided by this distribution is composed of 4007 3D faces of 466 subjects with their corresponding texture maps with resolution of 480 × 640 pixels. We consider for the experiments the subset composed of the 2902 images with neutral expression, and we process them as described in sections 2 and 3 in order to extract both the 2D and 3D features. Since the precision of the localized fiducial points is strongly dependent on the eye centers position, we decided to discard the images where the eye-localization error was larger than 10% of the interocular distance. This process eliminates 132 images (4.5%) leading to a final set of 2770 images of 441 subjects.

We set up three experiments in order to analyze the effect of the fusion strategy (2D+3D) with respect to the behavior of the system working either on 2D or 3D features only. To this end we built a gallery composed of 441 randomly chosen range images (one per subject), while the remaining 2329 were used to construct the test set. Table 1 shows the recognition results where the performance is evaluated according to the Cumulative Match Characteristic (CMC) metric presented in [19] and defined as $Pr(r) = |C(r)| \,/\, |T| \cdot 100$ where $C(r)$ is the set of images in the test set T that are recognized at rank r or better.

Observing the results in Table 1 we notice that, as expected, the multimodal system behaves better than those working either on 2D (+2.8%) or 3D (+15%) images only. Moreover we observe that the performance of the system working on 2D images is consistently higher than that obtained when only the 3D information is considered for the recognition. This fact highlights that the discriminative power of the 2D fiducial points is higher than that provided by the 3D profiles.

Table 1. Recognition Performance

Modality	$Pr(1)$	$Pr(2)$	$Pr(3)$	$Pr(4)$	$Pr(5)$
2D	92.6	94.4	95.3	95.7	96.2
3D	80.4	84.8	86.9	88.5	89.4
2D+3D	95.4	96.6	97.3	97.7	98.1

A direct comparison can be made with the recently presented (not yet published) method of Mian et al. [17]; it is, at our knowledge, the only multimodal face recognition technique presented in literature that tackles the identification problem on the FRGC version 2.0. In that work the authors propose an efficient rejection classifier based on a new Spherical Face Representation (SFR) for 3D faces, and the SIFT descriptor for the texture. They report an identification rate of 99% for tests on faces with neutral expression, that is 3.6% higher than our performance. Nevertheless we believe that better results could be achieved by our algorithm if the mouth fiducial points were localized more precisely. This could be done by devising a technique that would represent the mouth contour more suitably than the snakes. Besides, in order to increase the overall identification rate, we believe that additional 3D features, less correlated to the 2D fiducial points, could be extracted and used together with the 3D profiles.

6 Conclusion

This work presents a fully automated algorithm for face recognition that is based on the extraction of 2D and 3D features. The main characteristics of the algorithm regard: the local nature of the information used for face description, the integration of 2D and 3D features directly within the matching criterion, and the automatic discard of unreliable features. The combination of all these aspects allows to achieve a good robustness, as indicated by the results obtained on the FRGC v.2.0 database.

Regarding future works, we intend to improve the precision of both 2D and 3D features extraction: concerning the former, we observe that the algorithm would benefit from a more robust localization of the eye components, as well as from a more precise individuation of the mouth fiducial points; for what concerns 3D features, we plan to deepen the study of the profiles in order to make more profit from the availability of range data.

References

1. Zhao, W., Chellappa, R., Phillips, P., Rosenfeld, A.: Face recognition: A literature survey. ACM, Computing Surveys 35, 399–458 (2003)
2. Turker, M., Pentland, A.: Face recognition using Eigenfaces. Journal of cognitive neuroscience 3 (1991)
3. Shakhnarovich, G., Moghaddam, B.: Face recognition in subspaces. In: Handbook of Face Recognition, Springer, Heidelberg (2004)
4. Haddadnia, J., Ahmadi, M.: N-feature neural network human face recognition. Image and Vision Computing 22, 1071–1082 (2004)
5. Heisele, B., Ho, P., Poggio, T.: Face recognition with support vector machines: global vesus component-based approach. Proceedings IEEE Int'l Conf. Computer Vision 2, 688–694 (2001)
6. Wiskott, L., Fellous, J., Kruger, N., von der Malsburg, C.: Face recognition by elastic bunch graph matching. In: L.J., et al. (eds.) Intelligent biometric techniques in fingerprints and face recognition, pp. 355–396. CRC Press, Boca Raton, USA (1999)

7. Xu, C., Wang, Y., Tan, T., Quan, L.: Automatic 3D face recognition combining global geometric features with local shape variation information. International Conference on Vision Interface (2004)
8. Chua, C., Han, F., Ho, Y.: 3D human face recognition using point signature. International Conference on Automated Face and Gesture Recognition AFGR, 233–237 (2000)
9. Medioni, G., Waupotitsch, R.: Face recognition and modelling in 3D. IEEE International Workshop on Analysis and Modelling of Faces and Gestures AMFG, 232–233 (2003)
10. Chan, S., Wong, Y., Daniel, J.: Dense stereo correspondence based on recursive adaptive size multi-windowing. In: Proc. of Image and Vision Computing NZ, Palmerston North, pp. 256–259 (2003)
11. Wang, Y., Jain, A., Tan, T.: Face verification based on bagging RBF. In: Proc. IEEE Conf. on Biometrics (2006)
12. Lu, X., Colbry, D., Jain, A.: Matching 2.5d face scans to 3D models. IEEE Transactions on Pattern Analysis and Machine Intelligence 28(1), 31–43 (2006)
13. Bowyer, K., Chang, K., Flynn, P.: A survey of approaches and challenges in 3D and multi-modal 3D+2D face recognition. Computer Vision and Image Understanding 101, 1–15 (2006)
14. Phillips, J., Flynn, P., Scruggs, T., Bowyer, K.: Overview of the face recognition grand challeng. IEEE International Conference on Computer Vision and Pattern Recognition CVPR , 321–326 (2005)
15. Maurer, T., Guigonis, D., Maslov, I., Pesenti, B., Tsaregorodtsev, A., West, D., Medioni, G.: Performance of geometrix $activeid^{TM}$ 3D face recognition engine on the frgc data. IEEE Workshop on Face Recognition Grand Challenge Experiment FRGC (2005)
16. Husken, M., Brauckmann, M., Gehlen, S., von der Malsburg, C.: Strategies and benefits of fusion of 2D and 3D face recognition. IEEE Workshop on Face Recognition Grand Challenge Experiment FRGC (2005)
17. Mian, A., Bennamoun, M., Owens, R.: An efficient multimodal 2D-3D hybrid approach to automatic face recognition. IEEE Transactions on Pattern Analysis and Machine Intelligence (to be published) (2007)
18. Arca, S., Campadelli, P., Lanzarotti, R.: A face recognition system based on automatically determined facial fiducial points. Pattern Recognition 39, 432–443 (2006)
19. Grother, P., Micheals, R., Phillips, P.: Face recognition vendor test 2002 performance metrics. In: Proc. Int'l. Conf. Audio- and Video-based Biometric Person Authentication, pp. 937–945 (2003)

Generalization of a Recognition Algorithm Based on the Karhunen-Loève Transform

Francesco Gianfelici, Claudio Turchetti, Paolo Crippa, and Viviana Battistelli

DEIT – Dipartimento di Elettronica, Intelligenza Artificiale e Telecomunicazioni
Università Politecnica delle Marche, I-60131, Ancona, Italy
{f.gianfelici,turchetti,pcrippa}@deit.univpm.it
http://www.deit.univpm.it

Abstract. This paper presents a generalization of a recognition algorithm that is able to classify non-deterministic signals generated by a set of Stochastic Processes (SPs), the number of which may be arbitrarily chosen. This generalized recognizer exploits the nondeterministic trajectories generated by the Karhunen-Loève Transform (KLT) with no additional constraints or explicit limitations, and without the probability density function (pdf) estimation. Several experimentations were performed on SPs generated as solutions of non-linear differential equations with parameters and initial conditions being random variables. The results show a recognition rate which is close to 100%, thus demonstrating the validity of the generalized algorithm.

Keywords: Karhunen-Loève transform, recognition algorithm, signal classification, stochastic processes.

1 Introduction

Classification algorithms with high learning capability, low computational-complexity, and efficient decision rules are highly desirable in the recognition of non-deterministic signals. Moreover, it is worth noting as new approaches, optimized for limited *ensembles*, are *key factors* for the evolution of the learning theory [1], as clearly stated in [2].

During recent decades a large number of recognizers based on a *probabilistic setting*, such as the Hidden-Markov Model (HMM) [3], the Vector-Quantization (VQ), and the Dynamic Time Warping (DTW), have been developed [4]. It is well-known that the above techniques are affected by the following limitations: *a)* the high computational complexity of the probability density function (pdf), *b)* the low recognition performance in unsupervised cases, *c)* the large number of constraints on signal features, and/or the assumptions on system properties, *d)* the long elaboration time of the training phase, *e)* and the large number of signals used in the training phase.

The large interest in the recognition algorithms and the absence of a rigorous closed-form solution of this problem actually represent the starting point for the formalization and development of specific approaches and suitable techniques.

B. Apolloni et al. (Eds.): KES 2007/WIRN 2007, Part I, LNAI 4692, pp. 463–470, 2007.

A novel technique based on a *non-probabilistic setting* suited for the classification of signals generated by two Stochastic Processes (SPs) has been proposed in [5], and its effectiveness in the secure communications has been recently developed in [6]. In this technique each SP is modelled in terms of *canonical representations* as a linear combination of eigenfunctions of the correlation function [7], and the training phase extracts a collection of parameters by means of an *ad hoc* mathematical formulation. This formulation defines a group of eigenspaces that are associated with the realizations of the SPs, and the recognition algorithm extracts the projections of the realizations over the eigenfunctions of all eigenspaces. The redundancy in the training procedure is the key point of this technique since it guarantees the extraction of all parameter combinations by which the largest collection of nondeterministic *trajectories* can be determined.

The recognition phase analyzes the proximity-measures between the trajectories and the projections of a signal which has to be recognized, over all the eigenfunctions calculated in the training phase. In fact, the decision technology on which the recognition procedure is strongly based, is achieved using a non-probabilistic setting that takes into account both the principal and the minimal components.

In this work an effective and reliable generalization of the above algorithm for a generic *set* of SPs, the number of which may be arbitrarily chosen, was proposed. Methodologically, this approach is directly formalized by means of a novel recognizer that generalizes the previous algorithm with no additional constraints or explicit limitations, and without the estimation of the probability density function.

In order to evaluate the performance, the generalized recognizer was applied to several classes of SPs generated by stochastic differential equations that are able to effectively represent a large number of real systems, such as oscillating integrated circuits (ICs) affected by random device variations, or secure communication systems. The experimental results show a high recognition performance that is close to 100% with a limited *ensemble* of signals used in the training phase, thus demonstrating the validity of this generalized algorithm.

2 Stochastic Process Representation with the KLT

A signal *ensemble* in a *stochastic setting* can be represented with the Karhunen-Loève Transform (KLT) that is one of the most powerful frameworks in the SP theory thanks to its capabilities in the modelling of real phenomena with no limitations in terms of signal kind, applicative domain, and so on.

In order to introduce this fundamental result, let us consider a discrete-time finite-length SP $\{\boldsymbol{\xi}[n], n = 0, \ldots, L-1\}$ so that for every n the random variable $\boldsymbol{\xi}[n]$, defined on a fixed probability space, satisfies the condition $E\{\boldsymbol{\xi}[n]\} = 0$, where $E\{\cdot\}$ represents the expectation of a Random Variable (RV). The correlation matrix of the SP $\boldsymbol{\xi}$, is defined as:

$$R_{\xi\xi} = E\left\{\boldsymbol{\xi}\,\boldsymbol{\xi}^T\right\} \tag{1}$$

where $R_{\xi\xi} \in \mathbb{R}^{L\times L}$. Defining the matrix U whose columns are the orthonormal eigenvectors of $R_{\xi\xi}$, we obtain

$$U^T U = I, \tag{2}$$

$$R_{\xi\xi} U = U\Lambda, \qquad \Lambda = \operatorname*{diag}_{i=1,\ldots,L} \lambda_i \tag{3}$$

where $U \in \mathbb{R}^{L\times L}$, $I \in \mathbb{R}^{L\times L}$, and $\Lambda \in \mathbb{R}^{L\times L}$. The discrete Karhunen-Loève Transform (DKLT) of the SP $\boldsymbol{\xi}$ is written as

$$\boldsymbol{\xi} = U\,\mathbf{a} \tag{4}$$

where the vector $\mathbf{a} \in \mathbb{R}^L$ of the KLT-coefficients is

$$\mathbf{a} = U^T \boldsymbol{\xi}. \tag{5}$$

In general the exact knowledge of the correlation matrix is unachievable, so it is common practice to rely on approximations computed using an estimation technique based on sets of known realizations. By denoting the realizations of $\boldsymbol{\xi}$ collected during the estimation stage with $\xi^{(i)} \in \mathbb{R}^L$, $i = 1, \ldots, N$, the estimation for the correlation matrix can be computed in a convenient way by defining the matrix $C \in \mathbb{R}^{L\times N}$ containing the N realizations of length L in its columns, that is $C = \left[\xi^{(1)}, \ldots, \xi^{(N)}\right]$. The matrix C will be referred to as the data matrix. A commonly adopted estimator for the correlation matrix is:

$$R_{\xi\xi} \approx \frac{1}{N}\sum_{i=1}^{N} \xi^{(i)}\xi^{(i)^T} = \frac{1}{N} C\,C^T \tag{6}$$

that is the summation of the second moments between all possible pairs of components of the vectors $\xi^{(j)}$.

3 Stochastic Process Recognizer Based on the KLT Representation

3.1 Preliminary Definition

In order to give a mathematical formulation of the *generalized* recognizer, let us refer to a set of M real-valued zero-mean stochastic processes $\boldsymbol{\xi}_{(1)}, \ldots, \boldsymbol{\xi}_{(M)}$, and let $\xi_{(j)}^{(i)} \in \mathbb{R}^L$, $i = 1, \ldots, N$, $j = 1, \ldots, M$ be the i-th realization of the j-th stochastic process. Thus for each SP $\boldsymbol{\xi}_{(j)}$, the totality of realizations belongs to a set Ω_j, $\xi_{(j)}^{(i)} \in \Omega_j$ for any i. To perform a classification of the realizations $\xi_{(j)}^{(i)}$, it is useful to define the training set:

$$\Lambda_j = \left\{\xi_{(j)}^{(1)}, \ldots, \xi_{(j)}^{(N)}\right\} \tag{7}$$

with N being the number of realizations collected during the training stage. Obviously, for every j, it results $\Lambda_j \subset \Omega_j$. Let $\zeta \in \mathbb{R}^L$ be the signal to be recognized, thus the following property holds: $\zeta \in \Omega$, and $\zeta \notin \Lambda$, where $\Omega = \bigcup_{j=1}^{M} \Omega_j$, and $\Lambda = \bigcup_{j=1}^{M} \Lambda_j$. Both Λ, and ζ are input elements of the recognition algorithm.

The overall recognizer is made up of the succession of the training and recognition algorithms. The first algorithm extracts *ad hoc* parameters P_Φ using the input Λ, while the second one establishes whether ζ belongs to one of the M SPs by means of the extracted parameters P_Φ. The output of the algorithm represents the class, which the signal ζ is recognized as belonging to.

In the following sub-sections the generalization of the training and recognition algorithms of [5] will be given.

3.2 Training Algorithm

Given a set of M stochastic processes $\boldsymbol{\xi}_{(1)}, \ldots, \boldsymbol{\xi}_{(M)}$ where $\xi_{(j)}^{(i)}$ with $i = 1, \ldots, N$, $j = 1, \ldots, M$, is the i-th realization of the j-th SP. Each $\xi_{(j)}^{(i)}$ can be represented in matrix form as

$$\xi_{(j)} = \left[\xi_{(j)}^{(1)}, \ldots, \xi_{(j)}^{(N)}\right] \qquad j = 1, \ldots, M \tag{8}$$

where $\xi_{(j)} \in \mathbb{R}^{L \times N}$. Let $R_{(j)}$, $j = 1, \ldots, M$, be the autocorrelation matrices of $\xi_{(j)}$, therefore the bases $\Phi_{(j)} \in \mathbb{R}^{L \times N}$ of the SPs are defined in terms of the eigenvectors of the corresponding eigenproblems, namely

$$R_{(j)}\Phi_{(j)} = \Lambda_{(j)}\Phi_{(j)}, \qquad j = 1, \ldots, M \tag{9}$$

where $\Lambda_{(j)}$ are diagonal matrices containing the corresponding eigenvalues. By projecting all the realizations of $\xi_{(j)}$, $j = 1, \ldots, M$ onto the bases $\Phi_{(s)}$, $s = 1, \ldots, M$, we have the following matrices

$$A_{(s)(j)} = \Phi_{(s)}^T \xi_{(j)}, \qquad j, s = 1, \ldots, M \tag{10}$$

that can be rewritten as

$$P_\Phi = \begin{bmatrix} P_{(1)} \\ P_{(2)} \\ \cdots \\ P_{(M)} \end{bmatrix} = \begin{bmatrix} A_{(1)(1)} & A_{(1)(2)} & \cdots & A_{(1)(M)} \\ A_{(2)(1)} & A_{(2)(2)} & \cdots & A_{(2)(M)} \\ \cdots & \cdots & \cdots & \cdots \\ A_{(M)(1)} & A_{(M)(2)} & \cdots & A_{(M)(M)} \end{bmatrix} \tag{11}$$

where $P_\Phi \in \mathbb{R}^{MN \times MN}$ is the non-symmetric matrix of extracted features. Eq. (11) is equivalent to the following set of M equations

$$P_{(f)} = \left[\Phi_{(f)}^T \xi_{(1)} \; \Phi_{(f)}^T \xi_{(2)} \cdots \Phi_{(f)}^T \xi_{(M)}\right], \qquad f = 1, \ldots, M \tag{12}$$

with $P_{(f)} \in \mathbb{R}^{N \times MN}$.

3.3 Testing Algorithm

Letting ζ be the realization that has to be recognized as belonging to one the SPs $\boldsymbol{\xi}_{(1)}, \ldots, \boldsymbol{\xi}_{(M)}$, the first step in the algorithm is to calculate the projections of ζ onto $\Phi_{(1)}, \ldots, \Phi_{(M)}$ as:

$$l_{(f)} = \Phi_{(f)}^T \zeta \qquad f = 1, \ldots, M \tag{13}$$

where $l_{(f)} \in \mathbb{R}^N$.

As a second step let us define a transformation $\mathcal{T} : \mathbb{R}^{N \times K} \to \mathbb{R}^{N \times 2K}$, acting on the columns of an $N \times K$ matrix as:

$$\mathcal{T} v = \begin{bmatrix} v_1 & v_1 \\ v_1 & v_2 \\ \vdots & \vdots \\ v_1 & v_N \end{bmatrix} = \begin{bmatrix} v^{(1)} \\ v^{(2)} \\ \vdots \\ v^{(N)} \end{bmatrix}, \tag{14}$$

where $v = [v_1 \ v_2 \ \cdots \ v_N]^T$ is a generic column vector and $v^{(1)}, \ldots, v^{(N)}$ are elements of $\mathbb{R}^2$. Applying $\mathcal{T}$ to $l_{(f)}$ and $P_{(f)}$, we obtain:

$$l_{(f)} \to \mathcal{T} l_{(f)} = \begin{bmatrix} l_{(f)}^{(1)} \\ l_{(f)}^{(2)} \\ \vdots \\ l_{(f)}^{(N)} \end{bmatrix}, \qquad f = 1, \ldots, M \tag{15}$$

and

$$P_{(f)} \to \mathcal{T} P_{(f)} = \begin{bmatrix} P_{f1}^{(1)} & P_{f1}^{(2)} & \ldots & P_{f1}^{(MN)} \\ P_{f2}^{(1)} & P_{f2}^{(2)} & \ldots & P_{f2}^{(MN)} \\ \cdots & \cdots & \cdots & \cdots \\ P_{fN}^{(1)} & P_{fN}^{(2)} & \ldots & P_{fN}^{(MN)} \end{bmatrix}, \qquad f = 1, \ldots, M \tag{16}$$

with $\mathcal{T} l_{(f)} \in \mathbb{R}^{N \times 2}$, and $\mathcal{T} P_{(f)} \in \mathbb{R}^{N \times 2MN}$.

Thus we compute the matrices $D_{(f)} \in \mathbb{R}^{N \times MN}$ with $f = 1, \ldots, M$ whose generic ik-th elements are

$$[D_{(f)}]_{ik} = dist\left(l_{(f)}^{(i)}, P_{fi}^{(k)}\right) \qquad \begin{array}{l} i = 1, \ldots, N \\ k = 1, \ldots, 2N \end{array} \tag{17}$$

where *dist* is the Euclidean-distance between vector pairs. Let us define another transformation $\mathcal{S} : \mathbb{R}^{N \times MN} \to \mathbb{R}^{N \times MN}$ such that, when applied to a matrix Q, results in a novel matrix $\widetilde{Q} = \mathcal{S} Q$ with same dimensions, whose elements are:

$$[\widetilde{Q}]_{ik} = \begin{cases} 1\,, & [Q]_{ik} = \min_l [Q]_{il} \\ 0\,, & \text{elsewhere} \end{cases}. \tag{18}$$

In such a way the minimum distance in the rows of matrices $D_{(f)}$ is determined. It is easy to note that $\mathcal{S}$ represents the *decision* procedure of the recognizer in every row, inasmuch the 1's positions are determined as: $\boldsymbol{\xi}_{(1)}$ if they are placed in first N elements, $\boldsymbol{\xi}_{(2)}$ if they are placed in second N elements, and so on. Therefore we can define M vectors: $c_{(f)} = [c_{(f)}^{(1)}, c_{(f)}^{(2)}, \ldots, c_{(f)}^{(MN)}]$ whose elements are:

$$c_{(f)}^{(k)} = \sum_{i=1}^{N} [\widetilde{D}_{(f)}]_{ik} \qquad f = 1, \ldots, M \tag{19}$$

where

$$\widetilde{D}_{(f)} = \mathcal{S} D_{(f)} \tag{20}$$

and $c_{(f)} \in \mathbb{R}^{MN}$. The terms $c_{(f)}^{(k)}$ can be rewritten as elements of a novel matrix $\Pi \in \mathbb{R}^{M \times M}$ as:

$$\Pi = \begin{bmatrix} \sum_{k=1}^{N} c_1^{(k)} & \sum_{k=N+1}^{2N} c_1^{(k)} & \ldots & \sum_{k=Z}^{MN} c_1^{(k)} \\ \sum_{k=1}^{N} c_2^{(k)} & \sum_{k=N+1}^{2N} c_2^{(k)} & \ldots & \sum_{k=Z}^{MN} c_2^{(k)} \\ \ldots & \ldots & \ldots & \ldots \\ \sum_{k=1}^{N} c_M^{(k)} & \sum_{k=N+1}^{2N} c_M^{(k)} & \ldots & \sum_{k=Z}^{MN} c_M^{(k)} \end{bmatrix} \tag{21}$$

where $Z = (M-1)N - 1$. The elements of Π can be summed by columns, thus obtaining following numbers:

$$\mu_{(f)} = \sum_{h=1}^{M} [\Pi]_{fh}, \qquad f = 1, \ldots, M \tag{22}$$

Table 1. Results of recognition (M = 2, Testing Set = 1000)

Stochastic Processes	N	Sens.	Recogn. Perf.
(i) Van der Pol *vs.* Exp. Cos. SP	10	0	100.0%
(i) Van der Pol *vs.* Exp. Cos. SP	15	0	100.0%
(i) Van der Pol *vs.* Exp. Cos. SP	20	0	100.0%
(i) Van der Pol *vs.* Exp. Cos. SP	30	REF-*(i)*	100.0%
(ii) Pw. Dyn. Sys. *vs.* Exp. Cos. SP	10	0.0210	98.6%
(ii) Pw. Dyn. Sys. *vs.* Exp. Cos. SP	15	0.0300	98.5%
(ii) Pw. Dyn. Sys. *vs.* Exp. Cos. SP	20	0	100.0%
(ii) Pw. Dyn. Sys. *vs.* Exp. Cos. SP	30	REF-*(ii)*	100.0%
(iii) Van der Pol *vs.* Bessel	10	0	100.0%
(iii) Van der Pol *vs.* Bessel	15	0	100.0%
(iii) Van der Pol *vs.* Bessel	20	0	100.0%
(iii) Van der Pol *vs.* Bessel	30	REF-*(iii)*	100.0%
(iv) Duffing *vs.* Exp. Cos. SP	10	0	100.0%
(iv) Duffing *vs.* Exp. Cos. SP	15	0	100.0%
(iv) Duffing *vs.* Exp. Cos. SP	20	0	100.0%
(iv) Duffing *vs.* Exp. Cos. SP	30	REF-*(iv)*	100.0%
(v) Duffing *vs.* Van der Pol	10	-0.0138	99.0%
(v) Duffing *vs.* Van der Pol	15	-0.0285	99.5%
(v) Duffing *vs.* Van der Pol	20	0.0635	96.0%
(v) Duffing *vs.* Van der Pol	30	REF-*(v)*	98.1%

that represent the *likelihood-scores* of ζ respect to $\boldsymbol{\xi}_{(1)}, \ldots, \boldsymbol{\xi}_{(M)}$. Finally the recognition of ζ is performed as follows:

$$\zeta \in \boldsymbol{\xi}_{(f)} \qquad \text{if} \qquad \mu_{(f)} = \max[\mu_1, \ldots, \mu_M]. \tag{23}$$

The main benefit of this generalization is that the estimation of the probability density function, required in other recognition techniques, is not needed.

4 Experimental Results

In order to evaluate the performance, the generalized recognizer was applied to several classes of stochastic processes generated by stochastic differential equations, which are able to effectively represent a large number of real systems such as oscillating ICs affected by random device variations or secure communication systems. The SPs used in the experiments were generated as solutions of non-linear differential equations with randomly varying parameters and initial

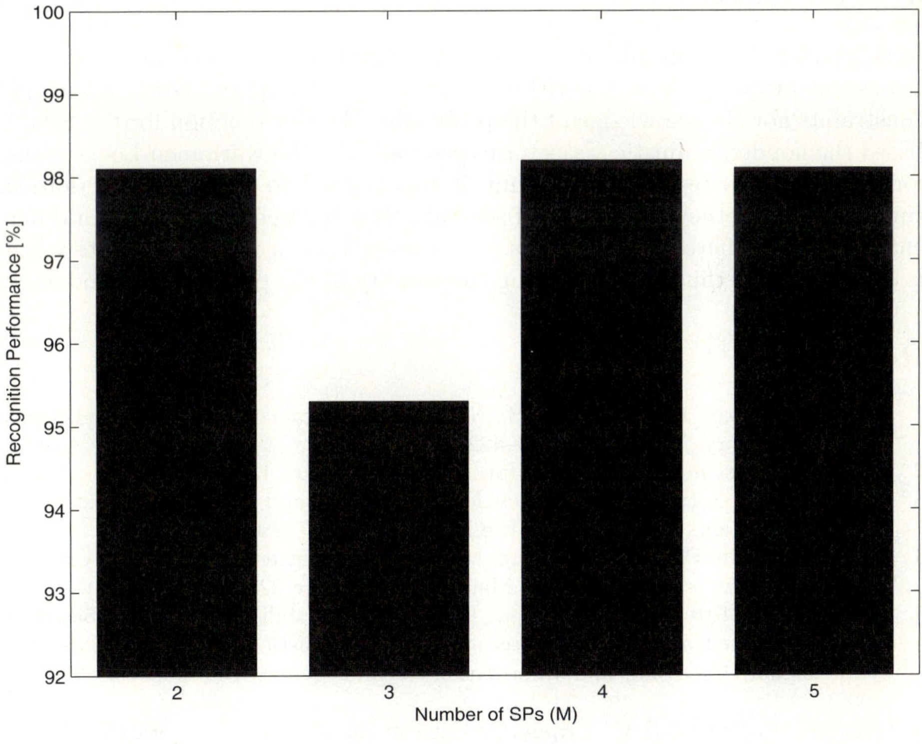

Fig. 1. Recognition performance as a function of the number M of SPs. For $M = 2$: Van der Pol SP, and Duffing SP; for $M = 3$: Van der Pol SP, Duffing SP, and Exp. Sin. SP; for $M = 4$: Van der Pol SP, Duffing SP, Exp. Sin. SP, and Bessel SP; for $M = 5$: Van der Pol SP, Duffing SP, Exp. Sin. SP, Bessel SP, and Pw. Dyn. Sys. SP.

conditions. No restrictions in terms of statistical distribution of the RVs were *a priori* established, and *a posteriori* results did not show unbalanced performance measures for several SPs. Examples of SPs used in the computer-aided simulations, are solutions of Van der Pol, Duffing, and Bessel equations having RVs as parameters and exponential-cosine SP.

In order to show the validity of the recognizer, Tab.1 reports the performance rate for $M = 2$.

Figure 1 shows the recognition performance for a number of stochastic processes $2 \leq M \leq 5$ where $N = 30$ realizations of each SP are considered and the testing set is made up of 1000 signals. For $M = 2$ Van der Pol SP, and Duffing SP; for $M = 3$ Van der Pol SP, Duffing SP, and Exp. Sin. SP; for $M = 4$ Van der Pol SP, Duffing SP, Exp. Sin. SP, and Bessel SP; for $M = 5$ Van der Pol SP, Duffing SP, Exp. Sin. SP, Bessel SP, and Pw. Dyn. Sys. SP were considered. In all cases the recognition performance was quite constant and the results show a recognition rate which is close to 100%, thus demonstrating the validity of the generalized algorithm.

5 Conclusion

In this paper a recognition algorithm for arbitrarily large *sets* of stochastic processes was proposed. This algorithm, which does not require neither additional constraints nor the knowledge of the probability density function for the SPs, is based the nondeterministic trajectories generated by the Karhunen-Loève transform. In order to test this algorithm, it was applied to recognize functions of time (signals) as the realizations of several SPs generated as solutions of different stochastic differential equations. The results show a recognition rate which is close to 100%, thus demonstrating the validity of the proposed methodology.

References

1. Poggio, T., Rifkin, R., Mukherjee, S., Niyogi, P.: General conditions for predictivity in learning theory. Nature 428, 419–422 (2004)
2. Tomasi, C.: Past performance and future results. Nature 428, 378 (2004)
3. Rabiner, L.R.: A tutorial on hidden Markov models and selected applications in speech recognition. Proceedings of the IEEE 77(2), 257–286 (1989)
4. Jain, A.K., Duin, P.R.W., Jianchang, M.: Statistical pattern recognition: A review. IEEE Trans. Pattern Analysis and Machine Intelligence 22(1), 4–37 (2000)
5. Gianfelici, F., Turchetti, C., Crippa, P.: A non probabilistic algorithm based on Karhunen-Loéve transform for the recognition of stochastic signals. IEEE Proc. Int. Symp. Signal Processing and Information Technology (ISSPIT 2006) 1, 385–390 (2006)
6. Gianfelici, F., Turchetti, C., Crippa, P.: Efficient classification of chaotic signals with application to secure communications. IEEE Proc. Int. Conf. Acustics, Speech and Signal Processing (ICASSP 2007) 3, 1073–1076 (2007)
7. Dougherty, E.R.: Random Processes for Image and Signal Processing. SPIE—IEEE Series on Imaging Science and Engineering. Bellingham (1998)

Intelligent Monitoring System for Driver's Alertness (A Vision Based Approach)

Rashmi Parsai[1] and Preeti Bajaj[2]

[1] Research Associate, ECE Dept, G.H.Raisoni College of Engineering, Nagpur, India
parsai_rashmi023@yahoo.co.in
[2] Professor & Head, ECE Dept, G.H.Raisoni College of Engineering, Nagpur, India
preetib123@yahoo.com

Abstract. International statistics shows that a large number of road accidents are caused by driver fatigue. Therefore, a system that can detect oncoming driver fatigue and issue timely warning could help prevent many accidents, and consequently save money and reduce personal suffering. The authors have made an attempt to design a system that uses security camera that points directly towards the driver's face and monitors the driver's eyes in order to detect fatigue. If the fatigue is detected a warning signal is issued to alert the driver. The authors have used the skin color based algorithm to detect the face of the driver. Once the face area is found, the eyes are found by computing the horizontal averages in the area. Once the eyes are located, measuring the distances between the intensity changes in the eye area determine whether the eyes are open or closed. A large distance corresponds to eye closure and small distance corresponds to eye open. If the eyes are found closed for 5 consecutive frames, the system draws the conclusion that the driver is falling asleep and issues a warning signal. The algorithm is proposed, implemented, tested, and found working satisfactorily.

1 Introduction

The ever- increasing number of traffic accidents all over the world is due to diminished driver's vigilance level. Drivers with a diminished vigilance level suffer from a marked decline in their perception; recognition and vehicle control abilities and therefore pose a serous danger to their own lives and the lives of the other people. For this reason, developing systems that actively monitors the driver's level of vigilance and alerting the driver of any insecure driving condition is essential for accident prevention.

Many efforts have been reported in the literature [1-4] for developing an active safety system for reducing the number of automobiles accidents due to reduced vigilance. Driver's drowsiness can be detected by sensing of physiological characteristics, sensing of driver operation, sensing of vehicle response, monitoring the response of driver. Among these methods, the techniques based on human physiological phenomena are the most accurate. This technique is implemented in two

B. Apolloni et al. (Eds.): KES 2007/WIRN 2007, Part I, LNAI 4692, pp. 471–477, 2007.

ways: measuring changes in physiological signals, such as brain waves, heart rate, and eye blinking; and measuring physical changes such as sagging posture, leaning of the driver's head and the open/closed states of the eyes. The first technique, while most accurate, is not realistic, since sensing electrodes would have to be attached directly onto the driver's body, and hence be annoying and distracting to the driver. The second technique is well suited for real world driving conditions since it can be non-intrusive by using video cameras to detect changes. Driver operation and vehicle behavior can be implemented by monitoring the steering wheel movement, accelerator or brake patterns, vehicle speed, lateral acceleration, and lateral displacement. These too are non-intrusive ways of detecting drowsiness, but are limited to vehicle type and driver condition. The final technique for detecting drowsiness is by monitoring the response of the driver. This involves periodically requesting the driver to send a response to the system to indicate alertness. The problem with this technique is that it will eventually become tiresome and annoying to the driver.

The proposed system relies on the eyelid movement, visual cue to detect the fatigued state of the driver. Micro sleeps are the short periods of sleep lasting 3 to 4 seconds are the good indicator of the fatigued state. Thus by continuously monitoring the eyes of the driver one can detect the state of the driver.

2 System Overview

A flowchart of the designed system is shown in Fig.1. After inputting a facial image, the skin color based algorithm is applied to detect the face in the image. The top and sides of the face are detected to narrow down the area of where the eyes exist. Moving down from the top of the face, horizontal averages (average intensity value for each x coordinate) of the face area are calculated. Large changes in the averages are used to define the eye area. Using the horizontal average values of both, sides of the face the open or closed state of the eyes are detected. If the eyes are found closed for 5 consecutive frames, the system draws the conclusion that the driver is falling asleep and issues a warning signal.

3 Face Detection

A lot of research has been done in the area of human face detection [5-6]. The authors have used skin filter method to detect the face [7]. The face detection is performed in three steps. The first step is to classify each pixel in the given image as a skin pixel or a non-skin pixel. The second step is to identify different skin regions in the skin-detected image by using connectivity analysis. The last step is to decide whether each of the skin regions identified is a face or not. They are the height to width ratio of the skin region and the percentage of skin in the rectangle defined by the height and width.

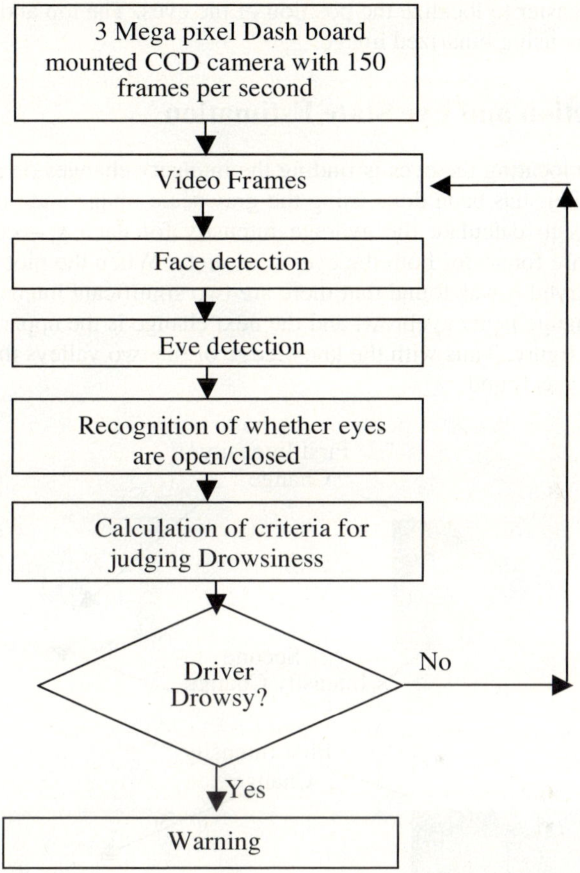

Fig. 1. System Flowchart

With the help of binarization the pixel, not belonging to the face are all made 0 and those belonging to the face are made 1. The result of binarization is shown in the Fig.2.

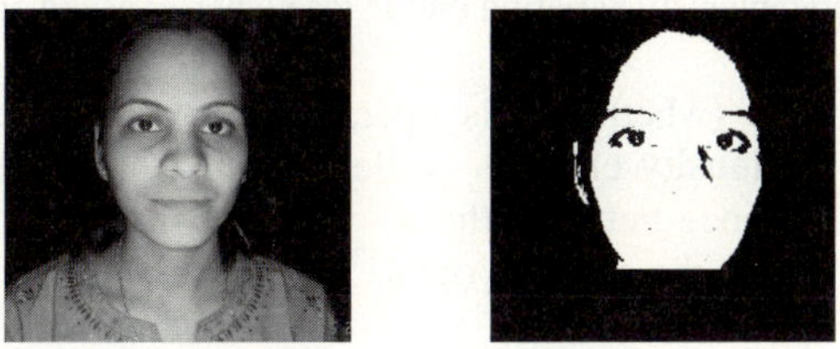

(a) (b)

Fig. 2. Face Detection (a) Original image (b) Binarized image

The next step is determining the top and side of the driver's face. This is important since finding the outline of the face narrows down the region in which the eyes are, which makes it easier to localize the position of the eyes. The top and the edges of the face are found by using binarized image.

4 Eye Detection and Eye State Estimation

The next step in locating the eyes is finding the intensity changes on the face as given by Parmar [8]. This has been done using the gray scale image and not the rgb image. The first step is to calculate the average intensity for each x – coordinate. These average values are found for both the eyes separately. When the plot of these average values was observed it was found that there are two significant intensity changes. The first intensity change is the eyebrow, and the next change is the upper edge of the eye, as shown in the figure. Thus with the knowledge of the two valleys the position of the eyes in the face was found.

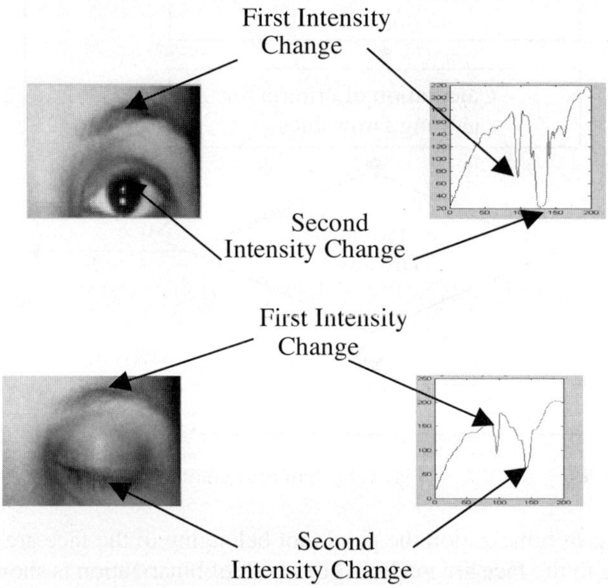

Fig. 3. Average intensity variation on the face when eyes are open and close

The state of the eyes (whether it is open or closed) is determined by distance between the first two intensity changes (valleys) found in the above step. When the eyes are closed, the distance between the x – coordinates of the intensity changes is larger if compared to when the eyes are open.

5 Drowsiness Detection

Each frame is observed to check whether the eyes are closed or open. Thus, the eye blink frequency is determined and if the eye blinks increases beyond the normal limit, the alarm is activated. Also if the eyes are found closed for consecutive 5 to 6 frames the system decides that the eyes are closed and give a fatigue alert.

6 Experimental Results

All the codes were written in MATLAB [9]. The experimental results are shown in the figure.

(a) (b) (c)

Fig. 4. Face Detection (a) Original image (b) Gray scale image (c) Detected face

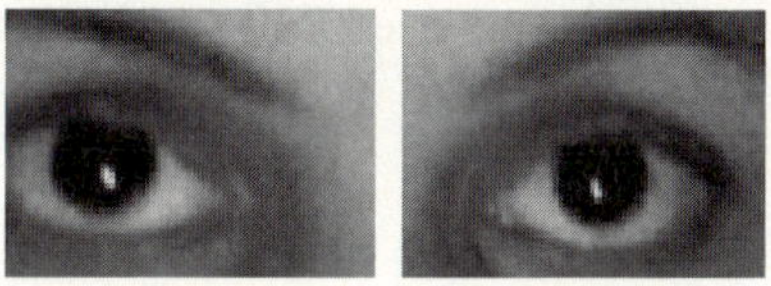

(a) (b)

Fig. 5. Eye Detection, (a) Left eye (b) Right eye

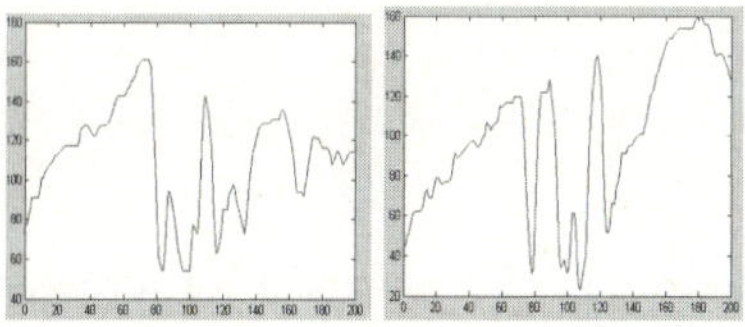

(a) (b)

Fig. 6. Average intensity variation on the face when eyes are open, (a) Left side (b) Right side

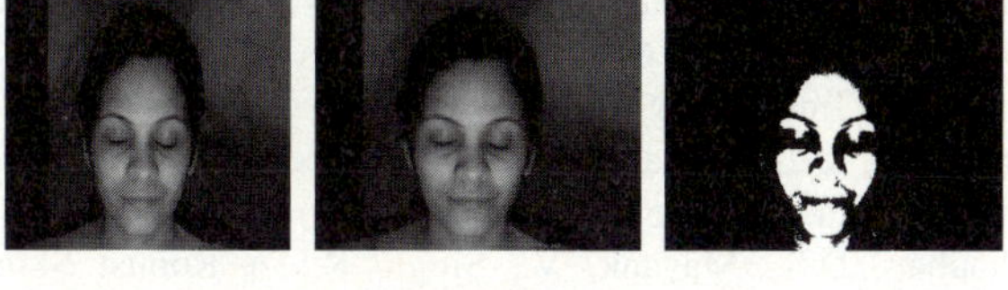

(a) (b) (c)

Fig. 7. Face Detection, (a) Original image (b) Gray scale image (c) Detected face

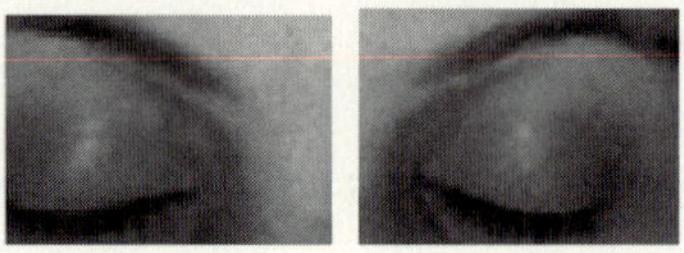

(a) (b)

Fig. 8. Eye Detection, (a) Left eye (b) Right eye

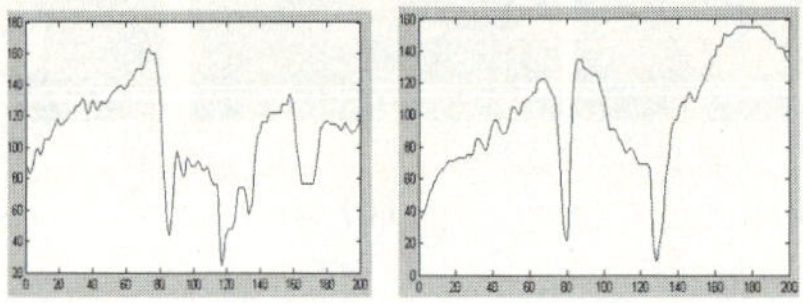

(a) (b)

Fig. 9. Average intensity variation on the face when eyes are close, (a) Left side (b) Right side

7 Conclusion

A driver monitoring system is proposed, designed and implemented which detects the fatigued state of the driver through continuously monitoring the eyes of the driver. The basis of the method used is based on the intensity variation on the face. Based on the fact that the eyebrows are significantly different from the skin in intensity, the eyes are located on the face and the microsleep is detected. The system is very effective to alert the driver.

References

1. Yammamoto, K., Higuchi, S.: Development of drowsiness warning System. J.Soc Automotive Eng.Japan, 127–133
2. Fakuda, J., Adachi, K., Nishida, M.: Development of driver's drowsiness detection technology. Toyota Tech.Rev 45, 34–40 (1995)
3. Singh, Sarbjit, Papanikolopoulos.: Monitoring Driver Fatigue Using Facial Analysis Techniques. IEEE Intelligent Transport System Proceedings, pp. 314- 318 (1999)
4. Ji, Q., Yang, X.: Real time cues extraction for monitoring driver vigilance. In: Proc. Of international workshop on computer vision systems, Vancouver, Canada, (July 7-8, 2001)
5. Hus, R.L., Mottaleb, M.A., Jain, A.K.: Face detection in color images. IEEE Trans. Pattern Analysis and Machine Intell 24, 696–706 (2003)
6. Yang, M., Kriegman, D.J., Ahuja, N.: Detecting faces in images: A survey. IEEE Trans. Pattern Analysis and Machine Intell. 24, 34–58 (2002)
7. Singh, S., Chauhan, D.S., Mayank, V., Singh, R.: A Robust Skin Color Based Face Detection Algorithm. Tamkang Journal of Science and Engineering 6(4), 227–234 (2003)
8. Parmar, N.: Drowsy Driver Detection System. Engineering Design Project Thesis, Ryerson University (2002)
9. Gonzalez, R.C., Woods, R.E., Eddins, S.V.: Digital image processing using MATLAB, Pearson Education, Delhi (2004)

Commentary Describing the Changes Made to the Paper

1 Extremely poor illumination condition

A dash board mounted CCD camera is used which focused at the face of the driver. In driver monitoring systems the camera are used with some optimal light arrangement so that the face is well illuminated and it also does not cause glare to the driver.

2 Real time system

The authors have used vfm files to capture each picture frame at rate 150 frames per second. On every frame the state of the eyes are determined and if the eyes are found closed for 6 consecutive frames then warning signal is issued to the driver.

3 Future work

Driver monitoring can be done by various techniques. Frequent yawning is also result of driver's fatigue. The authors are now working on determining frequency of yawning in order to determine the state of the driver.

JPEG2000 Low Complexity Allocation Method of Quality Layers

Francesc Aulí-Llinàs[1], Joan Serra-Sagristà[1],
Carles Rúbies-Feijoo[2], and Lluís Donoso-Bach[2]

[1] Department of Information and Communications Engineering
Universitat Autònoma de Barcelona
ETSE-UAB, Cerdanyola del Vallès 08290, Spain
fauli@deic.uab.es
[2] UDIAT Centre de Diagnòstic, Corporació Sanitària Parc Taulí
Sabadell 08208, Spain

Abstract. An important issue of JPEG2000 implementations is the allocation of quality layers, since it determines the optimality of the code-stream in terms of rate-distortion. Common strategies of quality layers allocation use both rate-distortion optimization and rate allocation methods, requiring the user to specify the number of quality layers and a distribution function for their rate allocation. This paper presents a new allocation method of quality layers that, neither needing rate-distortion optimization nor requiring user specifications, constructs a near-optimal code-stream in terms of rate-distortion. Besides, the computational cost of the proposed method is in practice negligible, and its application helps to reduce the computational load of the JPEG2000 encoder.

Keywords: JPEG2000 standard, rate allocation, rate-distortion optimization.

1 Introduction

JPEG2000 is the last standard developed by the Joint Photographic Experts Group, constituted by 12 parts that address the coding, transmission, security, and manipulation of images and video. The Part 1 [1] of the standard contains the core coding system and it is the basis of the other parts.

The coding scheme described in JPEG2000 Part 1 is wavelet based with a two-tiered coding strategy built on an Embedded Block Coding with Optimized Truncation (EBCOT) [2]. A key-feature of the JPEG2000 coding system is that each subband is divided in small blocks of coefficients that are encoded independently by the tier-1 coding stage. This stage uses a fractional bit-plane coder and the MQ arithmetic coder, carrying out three coding passes per bit-plane that are called Significance Propagation Pass (SPP), Magnitude Refinement Pass (MRP) and Cleanup Pass (CP). For each code-block, tier-1 produces a quality embedded code-stream that can be truncated at the end of the coding passes.

B. Apolloni et al. (Eds.): KES 2007/WIRN 2007, Part I, LNAI 4692, pp. 478–484, 2007.

In order to supply quality progression and quality scalability, JPEG2000 introduces the concept of quality layers. Quality layers collect incremental contributions of code-blocks, each one forming an optimal rate-distortion representation of the image. The selection of the code-stream segments included in each quality layer is carried out by the rate-distortion optimization method of the encoder and, once quality layers are formed, the tier-2 coding stage compress the auxiliary information of each layer constructing the final code-stream.

Quality layers are an important mechanism of JPEG2000. They provide quality scalability and quality progression, two fundamental features of the coding system. The quality scalability, for example, is needed in interactive image transmissions to allow the delivery of windows of interest at increasing qualities. The quality progression, for example, allows the truncation of the code-stream at different bit-rates without penalizing the coding performance. When the truncation of the JPEG2000 code-stream is at the quality layer boundary, the decoded image yields and optimal rate-distortion representation, but note that truncations at intermediate bit-rates yield approximately optimal representations of the image, depending on the number and rate allocation of quality layers.

It is precisely the number and rate allocation of quality layers what determines the overall rate-distortion optimality of the code-stream [3]. The rate allocation method is, therefore, a fundamental issue for implementations.

This paper is structured as follows: Section 2 reviews the common strategy of quality layers allocation in JPEG2000 and Section 3 introduces our allocation method of quality layers. In order to assess the performance of the proposed method, Section 4 presents several experimental results. Last section points out some noteworthy conclusions.

2 Common Allocation Strategies

Common allocation strategies of quality layers use two methods of the JPEG2000 encoding process: the rate allocation method and the rate-distortion optimization method. The former determines the number of quality layers included in the final code-stream and the bit-rate of each one. The latter selects the code-stream segments included in each quality layer, attaining the bit-rates determined by the rate allocation method.

2.1 Rate-Distortion Optimization

The first rate-distortion optimization method proposed for JPEG2000 was the Post Compression Rate-Distortion optimization (PCRD), introduced in EBCOT. The main idea behind this method is to approach the rate-distortion optimization problem through a generalized Lagrange multiplier for a discrete set of points. Let $R_i^{n_j}$ and $D_i^{n_j}$ denote, respectively, the bit-rate and distortion of the truncation points n_j of the code-block $\mathcal{B}_i$, satisfying $R_i^{n_j} \leq R_i^{n_{j+1}}$ and $D_i^{n_j} > D_i^{n_{j+1}}$. Considering the total distortion of the image and the total bit-rate of the code-stream respectively given by

$$D = \sum_i D_i^{n_j}, \ R = \sum_i R_i^{n_j},$$

then the Lagrange multiplier λ is approached as

$$(D(\lambda) + \lambda R(\lambda)) = \sum_i (D_i^{n_j^\lambda} + \lambda R_i^{n_j^\lambda})$$

to find the set of truncation points $\{n_j^\lambda\}$ which minimizes this expression yielding $R(\lambda) = R^{max}$, where R^{max} is the target bit-rate.

Note that this method needs to collect some information during the encoding of each coding pass, in particular the distortion and bit-rate of every truncation point of all code-blocks. Besides, the Lagrange multiplier only considers those truncation points that lie on the convex hull. Although the results obtained by PCRD are optimal, in its original formulation it compels to encode the complete image even when few coding passes are included in the final code-stream.

In the last five years more than 24 different rate-distortion optimization methods have been proposed in the literature, reducing the computational load of the encoder and supplying other interesting features. An extensive review and comparison among them can be found in [4].

2.2 Rate Allocation

On the other hand, JPEG2000 allocation of quality layers has not been properly addressed until December 2005, giving just some recommendations, based on experience, on the number and bit-rate of quality layers [5, Chapter 8.4.1]. From the point of view of the uneven error protection of embedded code-streams, the rate-distortion optimality of code-streams has been studied under some Expected Multi-Rate Distortion measure (EMRD) in [6] and, in the JPEG2000 framework, this has been studied in [7]. The EMRD measure is extremely useful to evaluate the optimality of a JPEG2000 code-stream in terms of rate-distortion and, consequently, the same authors continued their study, which finalized in December 2005 with the development of a new allocation method of quality layers for JPEG2000 [3].

The main idea used in this study is to weight the distortion of the image recovered at some bit-rates by the probability to recover the image at those bit-rates. In other words, EMRD defines a function that reflects the probability $p(R)$ of the code-stream $\mathcal{X}$ to be decoded at bit-rate R, $R \in [0, length(\mathcal{X})]$. The averaged EMRD over the complete bit-rate of $\mathcal{X}$ is defined as

$$\int_0^{length(\mathcal{X})} D(R)p(R)dR$$

where $D(R)$ represents the distortion of the recovered image at bit-rate R. Under this EMRD measure and considering uniform, exponential and Laplacian distributions, the authors propose a smart allocation algorithm that uses dynamic programming, achieving a near-optimal solution with reasonably computational costs. This is the first study that evaluates the optimality of JPEG2000 code-streams considering the allocation of quality layers but, although the research is outstanding, the degree of improvement when using the proposed method is usually small. The authors explain this poor improvement due to the already good approach of the fractional bit-plane coding of JPEG2000, which already generates code-stream segments with decreasing rate-distortion slopes.

Therefore, the use of a rate-distortion optimization method jointly with a distribution function to determine the rate allocation of quality layers, can already construct a near-optimal code-stream in terms of rate-distortion. This is the common approach in most applications. For instance, the default allocation mode of Kakadu [8], the most optimized implementation of JPEG2000, uses the PCRD method and a logarithmic distribution for the rate allocation. With this approach, however, the user must specify the number of quality layers included in the final code-stream, and this may become, in some cases, a non-obvious task.

3 Proposed Allocation Method

Neither needing a rate-distortion optimization method nor the specification of the number and bit-rates of quality layers, the method proposed in this research is able to determine the allocation of quality layers constructing a near-optimal code-stream in terms of rate-distortion. Besides, the proposed method skips the collecting of measures during the encoding process and avoids the use of the Lagrange multiplier, reducing the computational load of the encoding process.

Our allocation strategy has been conceived from the results obtained by a rate control method that we have recently presented, called Coding Passes Interleaving (CPI) [9]. CPI uses a simple interleaving strategy that selects the coding passes included in the final code-stream through a fixed scanning order. This scanning order is based on the spread belief that coding passes situated at the highest bit-planes have higher rate-distortion contributions than coding passes situated at the lowest bit-planes. Although CPI has been outperformed by the introduction of slight variations on the scanning order [10], or by the characterization of the rate-distortion slope [4], CPI disclosed an important issue for the allocation of quality layers. Let us explain further.

The scanning order followed by CPI uses coding levels. One coding level, referred to as c, identifies unequivocally a bit-plane and coding pass, and it is computed as

$$c = (p \cdot 3) + cp,$$

where cp stands for the coding pass type, computed as

$$cp = \{2 \text{ for SPP},\ 1 \text{ for MRP},\ 0 \text{ for CP}\},$$

and p stands for the bit-plane ($p = 0$ denotes the lowest bit-plane). The highest and lowest coding level of the image are referred to as $C_{max} = max(c)$ and $C_{min} = 0$ respectively. CPI encodes the coding passes of code-blocks belonging to the same coding level, from C_{max} to C_{min} until the target bit-rate is achieved.

In each coding level, the coding passes are selected from the lowest resolution level (LL subband) to the highest resolution level.

The main drawback of CPI is that the coding performance it obtains fluctuates continuously from 0.001 to 0.5 dB worse than the optimal coding performance achieved by PCRD. However, the important issue that we want to stress here is that, at several bit-rates, the coding performance obtained by CPI and PCRD is exactly the same. This evidence is given in Figure 1, that depicts the PSNR difference (in dB) achieved between CPI and the optimal PCRD method when encoding the Cafeteria image of the ISO 12640-1 corpus [11].

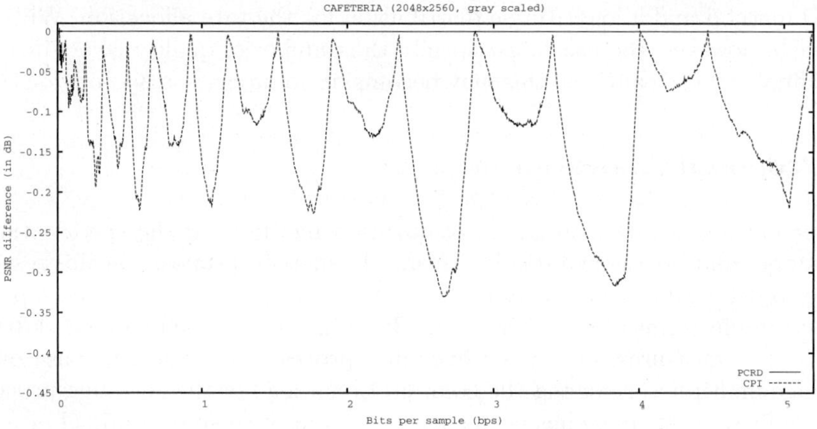

Fig. 1. Coding performance evaluation between CPI and PCRD [4]

The proposed allocation method uses these rate-distortion points achieved by the CPI scanning order to allocate the quality layers. This assures that the truncation of the code-stream at quality layer boundaries achieves an optimal representation and, since CPI does not collect measures during the encoding process nor uses a Lagrange multiplier, the proposed allocation method avoids the use of a specific rate-distortion optimization method. The problem here is to identify when CPI achieves optimal coding performance.

We have accurately observed the execution of both CPI and PCRD, paying special attention on the bit-rates where the coding performance of both methods coincides. This observation has disclosed that CPI obtains optimal results when it finishes the scanning of a coding level containing coding passes of type SPP, or when it finishes the scanning of a coding level containing coding passes of type CP.

The proposed allocation method of quality layers uses the rate-distortion optimal points disclosed by the previous observation to allocate quality layers. The proposed method allocates coding passes c in quality layers l by

$$l = L - 1 - ((c \; div \; 3) * 2) - ((c \; mod \; 3) \; div \; 2).$$

The proposed method neither requires user specifications nor distribution functions, determining the number of quality layers and the bit-rate of each one. The computational costs of our method are negligible; the algorithm just needs to identify the code-stream segments belonging to the coding passes of each code-block, which can be carried out in tier-1.

4 Experimental Results

To evaluate the optimality of code-streams constructed by our method, we compare it to a rate allocation method that uses a logarithmic distribution. Our method is programmed in our JPEG2000 Part 1 implementation, called BOI [12], and the construction of code-streams using the logarithmic rate allocation has been carried out with Kakadu. Coding options are: lossy mode, 5 DWT levels, derived quantization, 64x64 code-blocks, no precincts, restart coding variation.

The tests have been performed for the eight images of the ISO 12640-1 image corpus. For the logarithmic rate allocation method, Kakadu has constructed code-streams containing 10, 20 and 40 quality layers logarithmically spaced, in terms of bit-rate, along 0.001 to 3 bps. Then, the code-streams have been decoded at 300 uniformly distributed bit-rates and the PSNR difference obtained when encoding with PCRD at that particular target bit-rate has been computed. In order to ease the visual interpretation, figures below only depicts the best results obtained by the logarithmic rate allocation method (i.e., code-streams containing 20 logarithmically spaced quality layers).

Figure 2 depicts the average results for all images of the corpus. From 0.001 to 1 bps, the average coding performance obtained by the logarithmic rate allocation method of quality layers is 0.12 dB worse than PCRD. Our methods

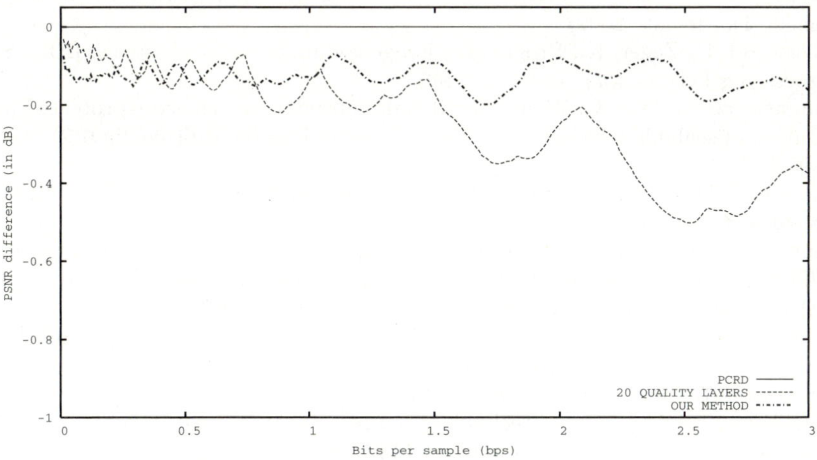

Fig. 2. Average coding performance evaluation between a logarithmic rate allocation, our method and the PCRD method

obtains practically the same average. However, for the overall bit-rate range, our method obtains a coding performance of 0.16 dB worse than the optimal PCRD method, whereas logarithmic rate allocation obtains a coding performance 0.59 dB worse than PCRD.

5 Conclusion

This paper introduces a new allocation method of quality layers. This method is the first one in the literature that, using negligible computational costs, determines automatically the number and rate allocation of quality layers. The strategy followed by our method is based on an important rate-distortion issue of JPEG2000. Besides, our method has negligible computational costs.

Acknowledgments. This work has been partially supported by the Spanish Government (MEC), by FEDER, and by the Catalan Government, under Grants TSI2006-14005-C02-01 and SGR2005-00319.

References

1. ISO/IEC 15444-1, Information technology - JPEG2000 image coding system - Part 1: Core coding system (December 2000)
2. Taubman, D.: High performance scalable image compression with EBCOT. IEEE Transactions on Image Processing 9(7), 1158–1170 (2000)
3. Wu, X., Dumitrescu, S., Zhang, N.: On multirate optimality of JPEG2000 code stream. IEEE Transactions on Image Processing 14(12), 2012–2023 (2005)
4. Auli-Llinas, F.: Model-based JPEG2000, rate control methods. Ph.D. dissertation, Univ. Autònoma de Barcelona (December 2006), Available: `http://www.deic.uab.cat/~francesc/docs/auli-phd.pdf`
5. Taubman, D., Marcellin, M.: JPEG2000 Image compression fundamentals, standards and practice. Norwell, Massachusetts 02061 USA. Kluwer Academic Publishers, Dordrecht (2002)
6. Sherwood, P., Zeger, K.: Progressive image coding on noisy channels. IEEE Signal Processing Letters 4(7), 189–191 (1997)
7. Dumitrescu, S., Wu, X., Wang, Z.: Globally optimal uneven error-protected packetization of scalable code streams. IEEE Transactions on Multimedia 6(2), 230–239 (2004)
8. Taubman, D.: Kakadu software [Online] (February 2007), Available `http://www.kakadusoftware.com`
9. Auli-Llinas, F., Serra-Sagrista, J., Monteagudo-Pereira, J., Bartrina-Rapesta, J.: Efficient rate control for JPEG2000 coder and decoder. presented at the Proc. IEEE Data Compression Conference, 282–291 (March, 2006)
10. Auli-Llinas, F., Serra-Sagrista, J.: Low complexity JPEG-2000 rate control through reverse subband scanning order and coding passes concatenation. IEEE Signal Processing Letters (in press) (April 2007)
11. ISO/IEC 12640-1, Graphic technology - Prepress digital data exchange - Part 1: CMYK standard colour image data (CMYK/SCID) (1997)
12. Group on Interactive Coding of Images. BOI software [Online] (February 2007), Available `http://www.gici.uab.cat/BOI`

Motion Estimation Algorithm in Video Coding

Vibha Bafna[1] and M.M. Mushrif[2]

[1] ECE Dept, G.H.Raisoni College of Engineering, Nagpur, India
vibhabora@rediff.com
[2] ECE Dept, Yaswantrao College of Engineering, Nagpur, India
milindmushrif@yahoo.com

Abstract. This paper is a review of the block matching algorithms used for the motion estimation in video compression to remove the temporal redundancy (i.e. interprediction). It implements and compares three different types of block matching algorithms that range from the very basic Exhaustive Search to the fast adaptive algorithms like Adaptive Rood Pattern Search. It can be used with common video coding standards such as H.263 and H.264.

1 Introduction

DATA compression is the process of reducing the redundancy in data representation in order to achieve savings in storage and communication costs. Due to limited channel bandwidth and stringent requirements of real-time video playback, video coding is an indispensable process, for many visual communication applications and always requires a very high compression ratio. The large amount of temporal correlation called *temporal redundancy* from the compression viewpoint, between adjacent frames in a video sequence, requires to be properly identified and eliminated to achieve this objective. An effective and popular method to reduce the temporal redundancy, called *block-matching motion estimation* (BMME), has been widely adopted in various video coding standards, such as H.261, H.263, H.264, and in any motion-compensated video coding technique. Therefore, fast and accurate block-based search technique is highly desirable to assure much reduced processing delay while maintaining good reconstructed image quality. In either standard the basic flow of the compression decompression process is largely the same and is shown in Fig. 1.

In Fig.1 the encoding side estimates the motion in the current frame with respect to a previous frame. A motion compensated image for the current frame is then created that is built of blocks of image from the previous frame. The motion vectors for blocks used for motion estimation are transmitted, as well as the difference of the compensated image with the current frame (residue) is also JPEG encoded and sent. The encoded image that is sent is then decoded at the encoder and used as a reference frame for the subsequent frames.

The decoder reverses the process and creates a full frame. The whole idea behind motion estimation based video compression is to save on bits by sending JPEG encoded difference images which have less energy and can be highly compressed as compared to sending a full frame that is JPEG encoded. It should be noted that the

B. Apolloni et al. (Eds.): KES 2007/WIRN 2007, Part I, LNAI 4692, pp. 485–492, 2007.

first frame is always sent full, and so are some other frames that might occur at some regular interval (like every 7th frame). The standards do not specify this and this might change with every video being sent based on the dynamics of the video.

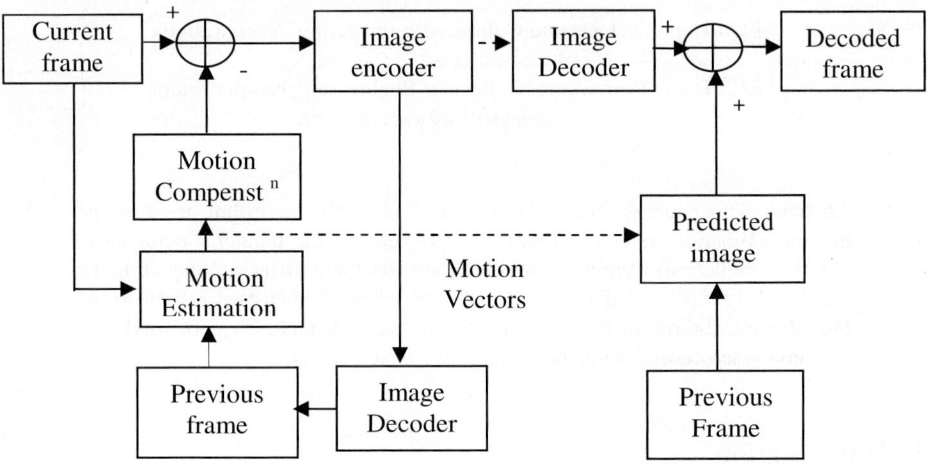

Fig. 1. Video compression process flow in H.26x

This paper implements and evaluates the fundamental block matching algorithms. The algorithms that have been implemented are Exhaustive Search (ES), Three Step Search (TSS), and Adaptive Rood Pattern Search (ARPS).

2 Methodology

Motion estimation[1](ME) is the most time-consuming process. The supposition behind motion estimation is that the patterns corresponding to objects and background in a frame of video sequence move within the frame to form corresponding objects on the subsequent frame. For block matching the current frame is divided into a'macro blocks' that are then compared with corresponding block and its adjacent neighbors in the previous frame to create a vector that stipulates the movement of a macro block from one location to another in the previous frame. The search area for a good macro block match is p pixels on all fours sides of the corresponding macro block in previous frame. Larger motions require a larger p, and the larger the p (search parameter) the more computationally expensive the process of motion estimation becomes. The idea is represented in Fig 2.

The matching of one macro block with another is based on the output of a cost function. The macro block that results in the least cost is the one that matches the closest to current block. There are various cost functions, of which the most popular and less computationally expensive is Mean Squared Error (MSE) given by equation (1). Another cost function is Mean Absolute Difference (MAD) given by equation (2).Sum of absolute error is given by equation (3).

$$MSE = \frac{1}{N^2} \sum_{i=0}^{N-1} \sum_{j=0}^{N-1} (C_{ij}-R_{ij})^2 \quad (1)$$

$$MAD = \frac{1}{N^2} \sum_{i=0}^{N-1} \sum_{j=0}^{N-1} |(C_{ij}-R_{ij})| \quad (2)$$

$$SAE = \sum_{i=0}^{N-1} \sum_{j=0}^{N-1} |(C_{ij}-R_{ij})| \quad (3)$$

where N is the side of the macro bock, Cij and Rij are the pixels being compared in current macro block and reference. Peak-Signal-to-Noise-Ratio (PSNR) given by equation (4) characterizes the motion compensated image that is created by using motion vectors and macro clocks from the reference frame.

$$PSNR = 10 \, Log_{10} \left[\frac{(\text{Peak to peak value of original data})^2}{MSE}\right] \quad (4)$$

3 Design Approach: Algorithms

3.1 Exhaustive Search (ES)

ES[1&2] algorithm, also known as Full Search (raster scan), is the most computationally expensive block matching algorithm of all. This algorithm calculates the cost function at each possible location in the search window. As a result of which it finds the best possible match and gives the highest PSNR amongst any block matching algorithm. The obvious disadvantage to ES is that the larger the search window gets the more computations it requires.

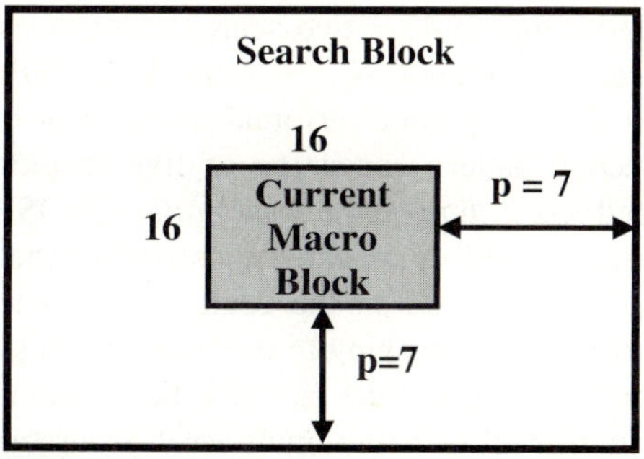

Fig. 2. Block Matching a macro block of side 16 pixels and a search parameter p of size 7 pixels

3.2 Three Step Search (TSS)

TSS[2&5] is one of the earliest fast block matching algorithms. The general idea is represented in Figure 3. It starts with the search location at the center and sets the 'step size' S = 4, for a usual search parameter value of 7. It then searches at eight locations +/- S pixels around location (0, 0). From these nine locations searched so far it picks the one giving least cost and makes it the new search origin. It then sets the new step size S = S/2, and repeats similar search for two more iterations until S = 1. At that point it finds the location with the least cost function and the macro block at that location is the best match. The calculated motion vector is then saved for transmission. It gives a flat reduction in computation by a factor of 9. So that for p = 7, ES will compute cost for 225 macro blocks whereas TSS computes cost for 25 macro blocks.

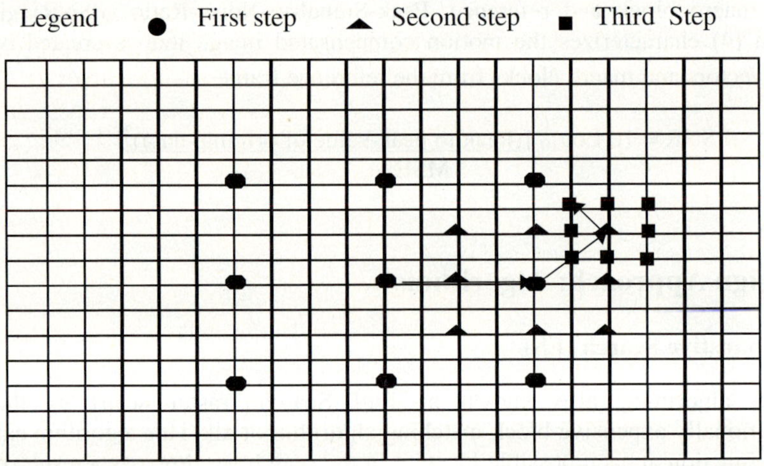

Fig. 3. Three Step Search procedure. The motion vector is (5, -3).

3.3 Diamond Search (DS)

The Diamond Search[4] algorithm employs two search patterns as shown in Fig. 4. The first pattern, called *large diamond search pattern* (LDSP), comprises nine checking points from which eight points surround the center one to compose a diamond shape (◆). The second pattern consisting of five checking points forms a smaller diamond shape,called *small diamond search pattern* (SDSP).In the searching procedure of the DS algorithm, LDSP is repeatedly used until the step in which the minimum block distortion (MBD) occurs at the center point. The search pattern is then switched from LDSP to SDSP as reaching to the final search stage. As the search pattern is neither too small nor too big and the fact that there is no limit to the number of steps, this algorithm can find global minimum very accurately. Among the five checking points in SDSP, the position yielding the MBD provides the motion vector of the best matching block.

The DS algorithm is summarized as follows.

Step 1) The initial LDSP is centered at the origin of the search window, and the 9 checking points of LDSP are tested. If the MBD point calculated is located at the center position, go to **Step 3**; otherwise, go to **Step 2**.

Step 2) The MBD point found in the previous search step is re-positioned as the center point to form a new LDSP. If the new MBD point obtained is located at the center position, go to Step 3; otherwise, recursively repeat this step.

Step 3) Switch the search pattern from LDSP to SDSP. The MBD point found in this step is the final solution of the motion vector which points to the best matching block.

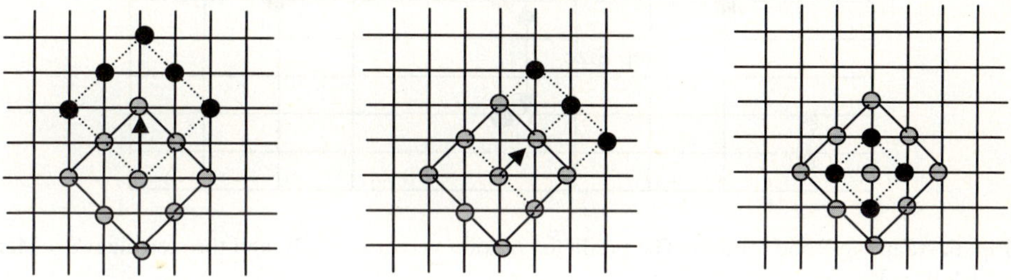

(a) (b) (c)

Fig. 4. (a) The corner point LDSP->LDSP. (b) The edge point LDSP->LDSP. (c) The centre point LDSP->SDSP.

3.4 Adaptive Rood Pattern Search (ARPS)

ARPS[2] algorithm makes use of the fact that the general motion in a frame is usually coherent, i.e. if the macro blocks around the current macro block moved in a particular direction then there is a high probability that the current macro block will also have a similar motion vector [3]. This algorithm uses the motion vector of the macro block to its immediate left to predict its own motion vector. An example is shown in Fig. 5. The predicted motion vector points to (3, -2). In addition to checking the location pointed by the predicted motion vector, it also checks at a rood pattern distributed points, as shown in Fig 5 where they are at a step size of $S = Max(|X|, |Y|)$. It directly puts the search in an area where there is a high probability of finding a good matching block. The point that has the least weight becomes the origin for subsequent search steps, and the search pattern is changed to SDSP. The procedure keeps on doing SDSP until least weighted point is found to be at the center of the SDSP. The main advantage of this algorithm over Diamond Search is if the predicted motion vector is (0, 0), it does not waste computational time in doing LDSP, it rather directly starts using SDSP. Furthermore, if the predicted motion vector is far away from the center, then again ARPS save on computations by directly jumping to that vicinity and using SDSP, where as DS[4] takes its time doing LDSP. Thus ARPS about two to three times faster than that of the *diamond search* (DS), and our method even achieves higher *peak signal-to-noise ratio* (PSNR) particularly for those video sequences containing large and/or complex motion contents. Care has to be taken to not repeat the computations at points that were checked earlier. For macro blocks in the first column of the frame, rood pattern step size is fixed at 2 pixels.

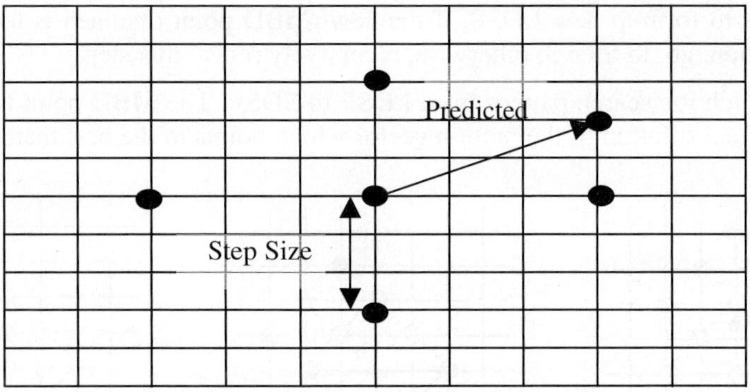

Fig. 5. Adaptive Rood Pattern: The predicted motion vector is (3,-2), and the step size S = Max (|3|, |-2|) = 3

4 Results

'Foreman' video sequence with a distance of 2 between current frame and reference frame was used to generate the frame-by-frame results of the algorithms. Fig.6 shows reference frame, current frame, compensated frame & residue after applying TSS

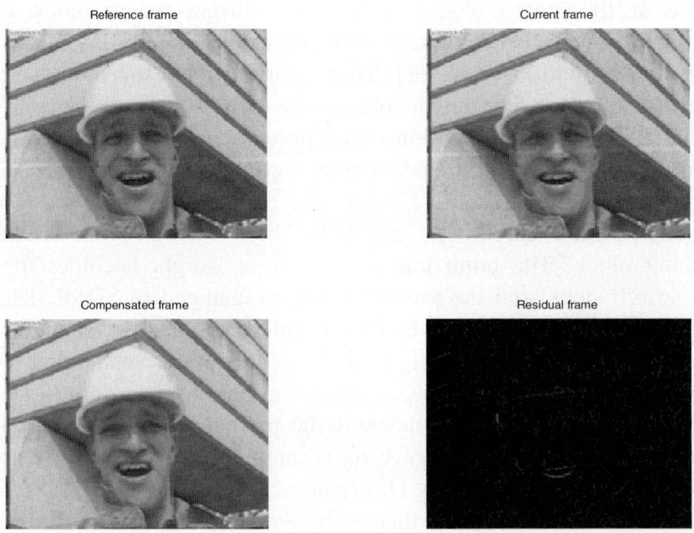

Fig. 6. Three Step search algorithm to foreman.avi

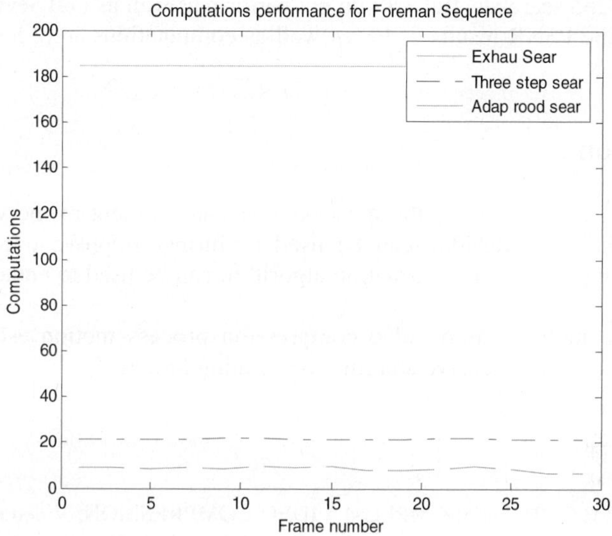

Fig. 7. Computation performance of foreman.avi for three algorithm

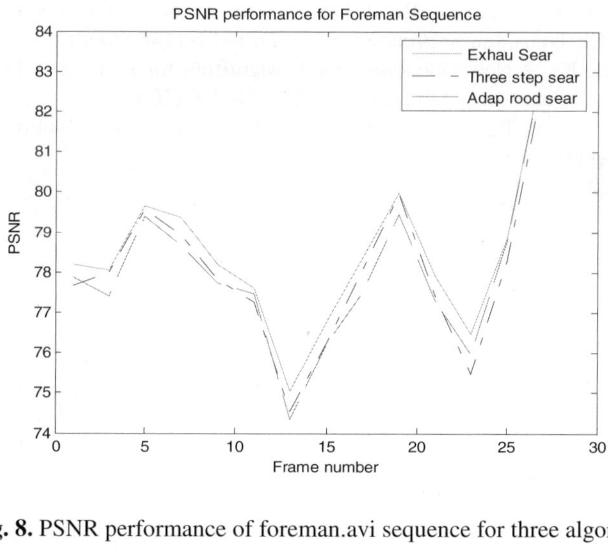

Fig. 8. PSNR performance of foreman.avi sequence for three algorithm

algorithm. Fig.7 shows a plot of the average number of searches required per macro block for the Foreman sequence using the 3 fast block matching algorithms. The PSNR comparison of the compensated images generated using the algorithms is shown in Fig 8. The results are extremely similar to the results of [2] and [3].

Full Search (ES) is guaranteed to find minimum MAD in search window but it is computationally intensive since the energy measure is calculated at every one of the 255 $(2p+1)^2$ locations. The TSS is simpler than Exhaustive search as only 25 searches compared with 255 searches. But TSS do not perform as well as Full Search(ES). ARPS come close to the PSNR results of ES as well as computations are 2 less computation compared to TSS.

5 Conclusion

Block matching techniques are the most popular and efficient of the various motion estimation techniques. Residue can be used to further improve our algorithm. In future fractional pixel motion estimation algorithm can be used to enhance the coding efficiency.

In the entire motion based video compression process motion estimation is the most computationally expensive and time-consuming process.

References

1. Richardson, I.E.G.: H.264 and MPEG-4 VIDEO COMPRESSION, Video Coding for Next Generation Multimedia. Ch.3,5, & 6. John Wiley & Sons Ltd, West Sussex, England (2003)
2. Block Matching Algorithms For Motion Estimation Aroh Barjatya, Student Member, IEEE, DIP 6620 Spring, Paper (2004)
3. Nie, Y., Ma, K.-K.: Adaptive Rood Pattern Search for Fast Block-Matching Motion Estimation. IEEE Trans. Image Processing 11(12), 1442–1448 (2002)
4. Zhu, S., Ma, K.-K.: A New Diamond Search Algorithm for Fast Block-Matching Motion Estimation. IEEE Trans. Image Processing 9(2), 287–290 (2000)
5. Wang, H., Mersereau, R.: Fast Algorithms for the Estimation of Motion Vectors. IEEE Transactions on Image Processing 8(3), 435–438 (1999)

Real-Time Vision Based Gesture Recognition for Human-Robot Interaction

Seok-ju Hong, Nurul Arif Setiawan, and Chil-woo Lee*

Intelligent Image Media & Interface Lab, Department of Computer Engineering, Chonnam National University, Gwangju, Korea
Tel.: 82-62-530-1803
seokju@image.chonnam.ac.kr, arif@image.chonnam.ac.kr, leecw@chonnam.ac.kr

Abstract. In this paper, we propose gesture recognition in multiple people environment. Our system is divided into two modules: Segmentation and Recognition. In segmentation part, we extract foreground area from input image, and we decide the closest person as a recognition subject. In recognition part, firstly we extract feature point of subject's both hands using contour based method and skin based method. Extracted points are tracked using Kalman filter. We use trajectories of both hands for recognizing gesture. In this paper, we use the simple queue matching method as a recognition method. We also apply our system as an animation system. Our method can select subject effectively and recognize gesture in multiple people environment. Therefore, proposed method can be used for real world application such as home appliance and humanoid robot.

Keywords: Context Aware, Gesture Recognition, Multiple People.

1 Introduction

Recently, People prefer new input method such as eye blinks, head motions, or other gestures to traditional computer input devices such as mouse, joystick, or keyboard. Gesture recognition technology is more important than any method since it support instinctive input method. Also it is useful in multiple people environment for home appliance. Currently, there are no researches which focused on gesture recognition in multiple people situation. Most researches are focusing on gesture recognition in single person and multiple people tracking.

First we describe multiple people tracking technology. Multiple people tracking research consist of deterministic and stochastic method. In deterministic method, objects are modeled by color histogram representation, texture, appearance and objects shape such as edgelet. And then tracking is performed by matching process in hypothesized search area [1-4]. This method has a disadvantage that object's movement is fast or discontinuous. Stochastic method use probability to estimate new position of objects based on certain feature [5-7]. But this method needs a lot of computational cost so the numbers of people tracked is limited.

---

* Corresponding author.

B. Apolloni et al. (Eds.): KES 2007/WIRN 2007, Part I, LNAI 4692, pp. 493–500, 2007.

Next we describe gesture recognition technology. Skin color based method use only skin information [8]. But it has a disadvantage that skin extraction will fail in case of complex background and illumination change. Contour based method use distance from body center point to both hand points for recognizing gesture [9]. This method is limited to recognition's number since it only uses distance information. 3D based method use 3d model of human body [10]. But it has a disadvantage such as complicated calculation cost and large database construction. Most of these works only focused on single person gesture recognition.

In this paper we deal with gesture recognition in multiple people. First of all we will define gesture and context. In segmentation part we process multiple people tracking and subject decision. In recognition part we extract feature point of body in decided subject. For extracting feature point we use two methods such as contour based method and skin based method. For recognizing gesture we use queue matching method. We also introduce animation system as an application. Finally we will show experimental result and conclusion. Our system architecture is shown in Fig. 1.

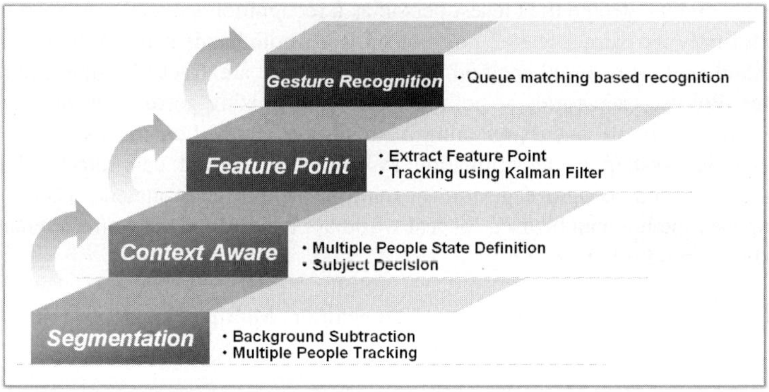

Fig. 1. System architecture (Segmentation module and Recognition module)

2 Context Awareness

In this section we define gesture which is used in our system. Next, we define each individual person's state from input image. Finally we describe state transition model for selecting subject.

2.1 Definition of Gesture

Mankind expresses his mind using eye blink, body movement, or sound. Specifically both hands' movement is used for expressing gesture. So gesture can be analyzed by using movement of both hands. We can not define all gestures used by people. Therefore, we define five gestures for human-robot interaction as shown in Fig. 2. Each gesture meaningfully separated into each other.

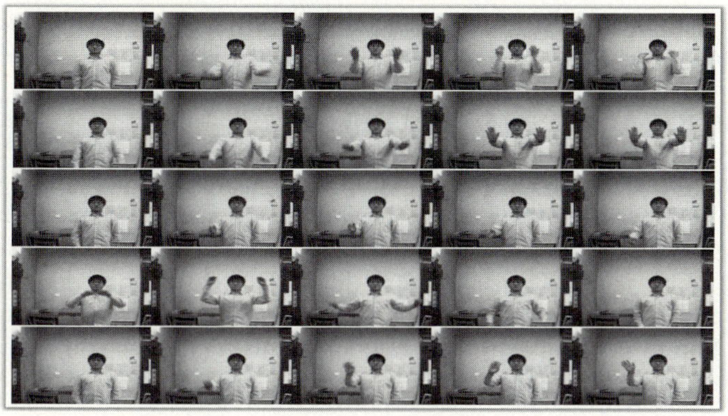

Fig. 2. Definition of Gestures (come here, stops, shake hands, heart, bye bye)

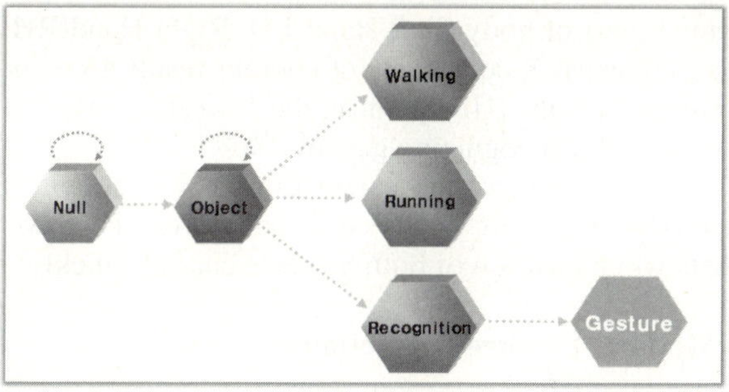

Fig. 3. State transition model of our system

2.2 Definition of States

"Context" is consisting of many situations such as illumination, number of people, temperature, noise and so on. In this paper, we define the "context" as intention state between human and computer. We selected speed and distance as intention of behaviors. According to these factors, state will be decided as shown in Fig. 3. The speed decides [Walking] or [Running] and the distance of behavior is most important factor since it decides to apply gesture recognition algorithm.

Each person has an only one state per every frame. Each state change using state transition model as shown Fig. 3. We assume that there are 3~4 people in input image. If one person going closer, we decide the person as a subject. If a subject decided, we extract feature point from subject's area. In next section, we describe how to extract feature point and how to recognize gesture.

3 Feature Extraction and Tracking

In this paper, we extract area of both hands and head. Segmentation process use Gaussian mixture model in improved HLS space [11]. We use two methods for extracting feature area. First method is contour based method. Second method is skin based method. In this section, we describe these methods and tracking method.

3.1 Contour Based Method (Feature Extraction)

In segmentation process, we extract subject's silhouette from input image. We must eliminate noise since silhouette image has a many noise. To remove this noise we apply dilation operation as shown equation 1. Contour line data is easily extracted from binary image data. We use OpenCV library for extracting contour. It retrieves contours from the binary image and returns the number of retrieved contours. We can get contour line to connect retrieved contour points. Contours can be also used for shape analysis and object recognition.

$$A \oplus B = \{ z \mid [(\hat{B})_z \cap A] \subseteq A \} \tag{1}$$

After extracting contour, we extract feature point for using contour based method. First we define three points of body (Left Hand-LH, Right Hand-RH, and Head Point-HP). [LH] point is the lowest X coordinate of contour result. [RH] point is the highest X coordinate of contour result. [HP] point is the lowest Y between [LH] and [RH]. Extracted points will use for recognizing gesture.

This method has an advantage that calculation cost is simple. But these extract wrong points since both hands are occluded in body area. To solve this problem we must estimate points when position of both hands is change quickly.

3.2 Skin Based Method (Feature Extraction)

Skin is an important factor for extracting both hands and head. There are many methods how to extract skin from image. In this paper we use to extract skin from YCBCR image. First of all, we apply mask in segmentation silhouette image and then we can get only subject area. And then we convert masked RGB image into YCbCr image. If we apply defined threshold in YCbCr image, we can get skin result image.

For recognizing gesture we must decide both hands position from skin result image. Both hands position can get x-y coordinate from x-y projection. Intersection of x projection and y projection is position of both hands and head.

Skin based method has lower calculation cost than contour based method. Also this method can detect both hands points when both hand s occluded. But this method arise problem when illumination change. Also this must apply another skin threshold for different human race.

3.3 Feature Tracking Using Kalman Filter

In this paper, we use a Kalman filter for tracking feature points. The Kalman filter is a set of mathematical equations that provides an efficient computational (recursive) solution of the least-squares method. The filter is very powerful in several aspects: it supports estimations of past, present, and even future states, and it can do so even when the precise nature of the modeled system is unknown. The Kalman filter estimates a process by using a form of feedback control. As such, the equations for the Kalman filter fall into two groups: time update equations and measurement update equations. The time update equations can also be thought of as predictor equations, while the measurement update equations can be thought of as corrector equations. Indeed the final estimation algorithm resembles that of a predictor-corrector algorithm for solving numerical problems.

In this paper, we used Kalman Filter for estimating 2D coordinate of extracted feature points. Feature points means 2D coordinate head and both hands in previous section. Extracted feature points entered by measurement value, and this is used to estimate 2D position in next frame. This process is estimated coordinate of head and both hands. If feature points extraction failed, it uses prediction value. In next section, we will describe gesture recognition using these feature points.

4 Gesture Recognition Using Queue Matching

The gesture contains user's intentions in motions of whole body. Especially, trajectories of hands include more intentions. So, we adopt the different recognition method which uses trajectories of hands as features. Many researchers have tried to develop the matching algorithm for the trajectories in a number of ways. Generally, the methods are used for recognition of handwritten character. But, it is not effective to apply into the gesture recognition, because it is difficult to decide the start and end point of meaningful gestures. Therefore, many researchers are continuing to study about the problems, Gesture Spotting [9].

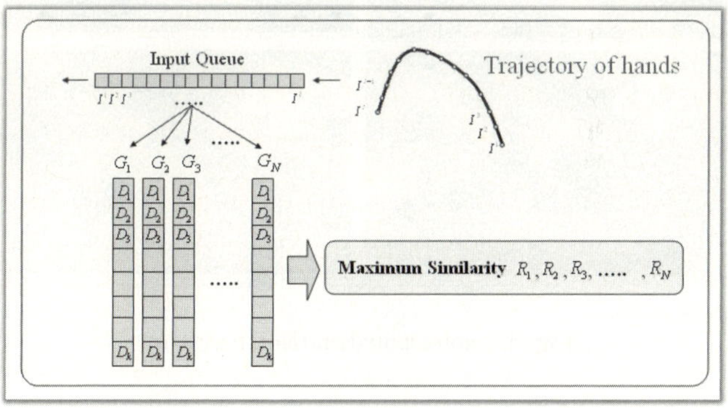

Fig. 4. Queue matching method for recognizing gesture

In this paper, we propose the simple queue matching method instead of gesture spotting algorithm if the gestures are not complicated. And this method has the advantage in fast to process and easy to implement.

The basic concept of this algorithm is as follows. Assume that the model set M has N models. Also, direction vectors represent the trajectories of hands, and these vectors are stored continuously in each gesture models.

We can get directional vectors from each frame. And, input queue with the length I is a set of these vectors. If the meaningful gesture of subject is in the input queue, it can be assumed that this queue includes the subject's intention. And then, input queue is compared with each model gesture. Finally, we can decide the gesture, as a recognition result. In next section we introduce our system as an application.

5 Application Program: 3D Animation System

In this paper we use our system as an animation generation system. From input image we construct 3D body model in virtual space. 3D body model has a similar appearance with subject. Also this model has a similar action with subject's action.

To construct animation system, we use feature point from gesture recognition system. These points used for estimating human body point. Extracted feature points have many noises from general environment. We use NURB algorithm for eliminating noise. And we estimate each body joint position using Inverse Kinematics. To estimate correctly, we use information such as human anatomy, previous frame information and collision process. Finally, we estimate body point using extracted feature point and end-effector.

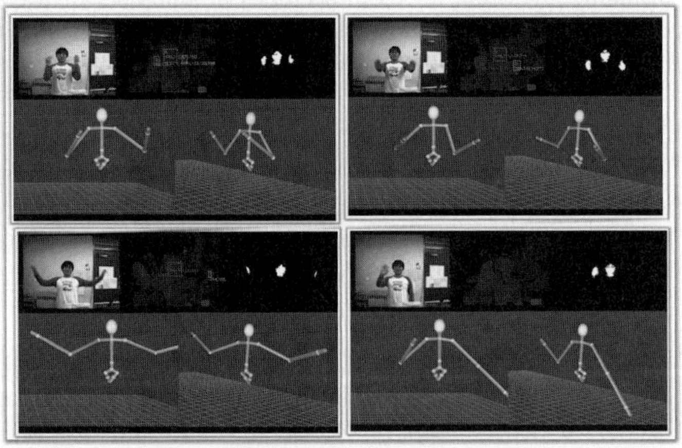

Fig. 5. Implemented animation system

To represent 3D model, first we construct 3D virtual space in animation system. Gesture recognition system send to animation system feature point information. We can get animation system similar doing input gesture.

6 Future Work

The experiment was taken on 2 PCs with 3.0 GHz Intel Pentium 4 CPU and 512MB RAM. We used Bumblebee of Point Grey for extracting stereo information. The system is written in Visual C++ 6.0 based on OpenCV 1.0. Fig. 6 shows results of extracted feature points and gesture recognition result.

Contour based method has a problem when both hands are occluded in body area. For example, both hands position go wrong when [heart] and [bye bye]. Skin based method extract good position in every gesture. It is shown robust result when both hands are occluded in body area. But skin is failed when illumination change.

We have a problem since we use only 2 dimensional data for recognizing gesture. For example, we can not recognize both hands upward and both hands upward in round fashion. This can recognize gestures if we use 3 dimensional data instead of 2 dimensional data. And our system can not make trajectory information when subject doing [shake hands gesture] and subject doing [bye bye gesture]. To solve this problem, we must use time information and movement information of specific area. If we use a convex hull algorithm for extracting feature point, we can have a simple calculation cost and accurate feature points.

Fig. 6. Contour based gesture recognition result (come here, stop, shake hands, heart, bye bye)

Fig. 7. Skin based gesture recognition result (come here, stop, shake hands, heart, bye bye)

Also we have a problem when subject is changed. Subjects have a little different trajectory information of gesture. To solve this problem, we assign a personal ID. Our system recognizes a personal ID, and it uses a model gesture of ID as a model gesture.

In this paper, we proposed gesture recognition in multiple people environment. Our system is divided into two modules – segmentation module and gesture recognition module. Also our system can change subject if subject entered. And then our system tracked feature points using Kalman filter. Finally, our system can recognize gesture using simple queue matching.

In this paper, we propose animation system using implemented gesture system. This system can make 3D information of human. We can get automated animation in future. Our method can use general interface of robot. If it solve previous problem, intelligent robot can communicate with mankind naturally.

Acknowledgments. This work was partly supported by the IT R&D program of MIC/IITA. [2006-S-028-01, Development of Cooperative Network-based Humanoids Technology] and by KOCCA as the result of the research project for 2007 C.N.U culture technology development.

References

1. Zhao, T., Nevatia, R.: Tracking Multiple Humans in Crowded Environment. In: Proceedings of CVPR 2004, pp. 1063–6919 (2004)
2. Wu, B., Nevatia, R.: Detection of Multiple, Partially Occluded Humans in a Single Image by Bayesian Combination of Edgelet Part Detectors. In: Proceedings of ICCV, vol. 1, pp. 90–97 (2005)
3. Haritaoglu, I., Harwood, D., Davis, L.S.: W4: Real-Time Surveillance of People and Their Activities. IEEE Transactions on Pattern Analysis and Machine Intelligence 22(8), 809–830 (2000)
4. Siebel, N.T, Maybank, S.: Fusion of Multiple Tracking Algorithms for Robust People Tracking. In: Proceeding of ECCV 2002, pp. 373–387 (2002)
5. Franc, J.B., Fleuret, o., Fu, P.: Robust People Tracking with Global Trajectory Optimization. In: Proceedings of CVPR 2006, vol. 1, pp. 744–750 (2006)
6. Nguyen, H.T., Ji, Q., Smeulders, A.W.M.: Robust multi-target tracking using spatio-temporal context. In: Proceedings of CVPR 2006, vol. 1, pp. 578–585 (2006)
7. Han, J., Award, G.M., Sutherland, A., Wu, H.: Automatic Skin Segmentation for Gesture Recognition Combining Region and Support Vector Machine Active Learning. In: Proceedings of FGR 2006, pp. 237–242 (2006)
8. Li, H., Greenspan, M.: Multi-scale Gesture Recognition from Time-Varying Contours. In: Proceedings of ICCV 2005, vol. 1, pp. 236–224 (2005)
9. Lee, S.-W.: Automatic Gesture Recognition for Intelligent Human-Robot Interaction. In: Proceedings of FGR 2006, pp. 645–650 (2006)
10. Setiawan, N.A., Hong, S.-j., Lee, C.-w.: Gaussian Mixture Model in Improved HLS Color Space for Human Silhouette Extraction. In: Pan, Z., Cheok, A., Haller, M., Lau, R.W.H., Saito, H., Liang, R. (eds.) ICAT 2006. LNCS, vol. 4282, Springer, Heidelberg (2006)
11. http://www.sourceforge.net/projects/opencvlibrary

Reference Independent Moving Object Detection: An Edge Segment Based Approach

M. Ali Akber Dewan, M. Julius Hossain, and Oksam Chae*

Department of Computer Engineering, Kyung Hee University,
1 Seochun-ri, Kiheung-eup, Yongin-si, Kyunggi-do, South Korea, 449-701
dewankhu@gmail.com, mdjulius@yahoo.com, oschae@khu.ac.kr

Abstract. Reference update to adapt with the dynamism of environment is one of the most challenging tasks in moving object detection for video surveillance. Different background modeling techniques have been proposed. However, most of these methods suffer from high computational cost and difficulties in determining the appropriate location as well as pixel values to update the background. In this paper, we present a new algorithm which utilizes three most recent successive frames to isolate moving edges for moving object detection. It does not require any background model. Hence, it is computationally faster and applicable for real time processing. We also introduce segment based representation of edges in the proposed method instead of traditional pixel based representation which facilitates to incorporate an efficient edge-matching algorithm to solve edge localization problem. It provides robustness against the random noise, illumination variation and quantization error. Experimental results of the proposed method are included in this paper to compare with some other standard methods that are frequently used in video surveillance.

Keywords: Video surveillance, reference independent, chamfer matching, distance image, motion detection.

1 Introduction

Automatic detection of moving objects is a challenging and essential task in video surveillance. It has many applications in diverse discipline such as automatic video monitoring system, intelligent transportation system, airport security system and so on. Detail review on moving object detection algorithms can be found in [1] and [2]. Background subtraction based methods are the most common approaches that are used for moving object detection. In these methods, background modeling is an important and unavoidable part to accumulate the illumination and other changes in the background scene for proper detection [3]. However, most of the background-modeling methods are complex in computation and time-consuming for real time processing [4]. Moreover, most of the time it suffers from poor performance due to lack of compensation with the dynamism of background scene [5].

Edge based methods are robust against illumination change. In [6] and [7], edge based methods are proposed for moving object detection which utilizes double edge

---

* Corresponding author.

B. Apolloni et al. (Eds.): KES 2007/WIRN 2007, Part I, LNAI 4692, pp. 501–509, 2007.

maps. In [6], one edge map is generated from difference image of background and current frame, I_n. Another edge map is generated from difference image of I_n and I_{n+1}. Finally, moving edge points are detected by applying logical *OR* operation on these two edge maps. However, due to illumination change and random noise [6] in background scene, false edge may appear in the first edge map and hence causes false detection in the final detection result. In [7], first edge map is computed from the difference image of I_{n-1} and I_n, and similarly second map is obtained from I_n, and I_{n+1}. Finally, moving edges of I_n are extracted by applying logical *AND* operation on these two edge maps. However, because of noise and illumination change, edge pixels of an edge map may be displaced little bit as compared to previous one. So, exact matching through *AND* operation extracts scattered edge pixels, which fails to represent reliable shape of moving objects. Moreover, pixel based processing for moving edge detection is not feasible in terms of computation. A pseudo-gradient based moving edge extraction method is proposed in [8]. Though this method is computationally faster but its background is not updated to take care of the situation when a moving object stops its movement in the scene. In this situation, stopped object is continuously detected as moving object. As no background update method is adopted in this method, it is not much robust against illumination change. Additionally, this method also suffers from scattered edge pixels of moving objects.

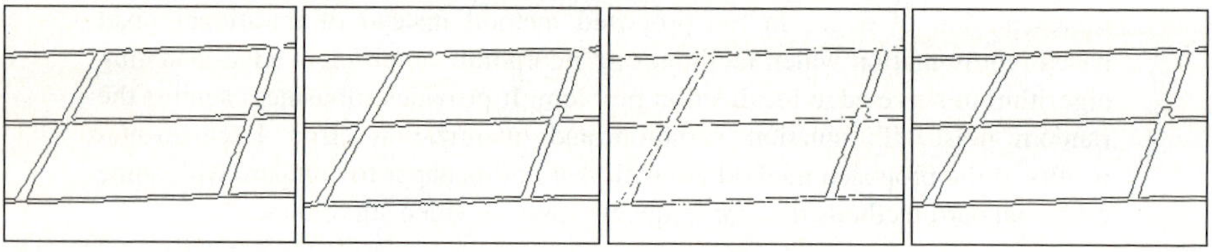

(a) (b) (c) (d)

Fig. 1. Difference between pixel based and segment based matching. (a) Edge image at time *t*; (b) Edge image of same scene at time *t+1*; (c) Result obtained by pixel based matching; (d) Result obtained by segment based matching.

Considering the above-mentioned problems, we present an edge segment based approach which utilizes three successive frames for moving object detection. In our proposed method, two difference image edge maps of three successive frames are utilized to extract moving edges instead of using edge differencing approach. It makes the system robust against random noise as well as illumination variation. Since the proposed method does not require any background model for detection, it is computationally faster and efficient. Moreover, use of most recent frames, embodying the updated information helps to reduce false detection effectively. In our proposed method, the difference image edge maps are represented as segments instead of pixels using an efficiently designed edge class [9]. An edge segment consists of a number of consecutive edge pixels. This novel representation helps to make the decision on matching or in any other operations based on entire edge segment rather than an individual pixel. This representation of edge provides the following benefits:

a) It facilitates to incorporate an efficient and flexible edge-matching algorithm [10] in our proposed method which reduces the computation time significantly.
b) This type of representation facilitates our method to take decision about a complete edge segment at a time instead of an individual edge pixel to keep or discard it from the edge list during matching. Fig. 1 illustrates the advantages of segment based matching over pixel based matching. Here, pixel based matching missed 20% edge pixels due to variation of edge localization in different frames. Segment based matching does not suffer from this problem as it consider all the points of a segment together. As a result, it reduces the occurrence of scattered edge pixels in the detection result.

Since moving object segmentation is a separate problem from detection in video surveillance, we have not considered it in our proposed method. However, because of segment based representation of edges, our proposed method is able to extract reliable shape information of moving objects. Incorporating this shape information with image segmentation algorithm, it is possible to segment out moving objects from current image efficiently. Segment based representation also makes it possible to incorporate knowledge to edge segments which can facilitate the higher level processing of video surveillance such as tracking, recognition, human activity recognition and so on.

2 Description of the Proposed Method

The overall procedure of the proposed method is illustrated in Fig. 2. Detail description of our method is given in the following subsections.

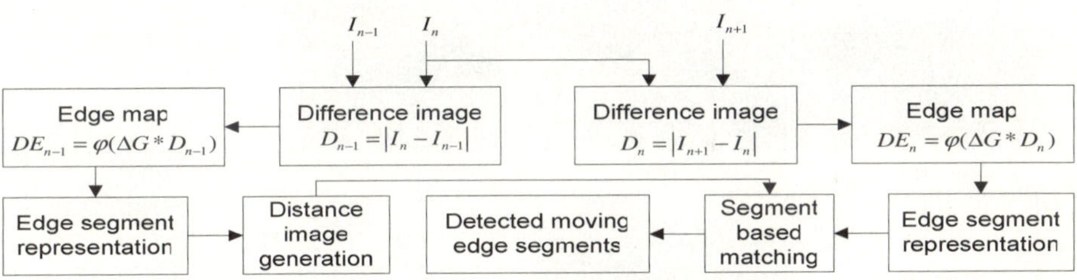

Fig. 2. Flow diagram of the proposed method

2.1 Computation of Difference Image Edge Maps

Simple edge differencing approach suffers a lot with random noise. This is due to the fact that the appearance of noise created in one frame is different from its successive frames. This results in change of edge locations to some extent in successive frames. Hence, instead of using simple edge differencing approach, we utilize difference image for moving edge detection. Edges extracted from difference image are noise robust, comparatively stable and hence partially solve the edge localization problem. Two difference image edge maps are utilized in our proposed method for moving object detection. To compute difference image edge maps, we compute two difference images, D_{n-1}, and D_n utilizing three successive frames I_{n-1}, I_n, and I_{n+1} as follows:

$$D_n = |I_n - I_{n+1}| \tag{1}$$

After computing D_{n-1} and D_n, canny edge detection algorithm [11] is applied and generates difference image edge maps, DE_{n-1} and DE_n, respectively. In the difference image edge maps, edge pixels are grouped together to represent as segments using an efficiently designed edge class [9]. To make the edge segments more efficient for moving edge detection procedure, we maintain the following constrains during edge segment generation:

a) If the edge segment contains multiple branches, then the braches are broken into multiple edge segments from its branching point.
b) If the edge segment bends more than a certain limit at an edge point, the edge is broken into two edge segments from that particular position.
c) If the length of a particular edge segment exceeds a certain limit, then the edge segment is divided into a number of small edge segments of its permitted length.

Segment based representation helps the proposed system to use the geometric shape of edges during matching for moving edge detection. It also helps to extract solid edge segments of moving objects instead of extracting scattered or significantly small edges. In this case no edge pixels are processed independently; rather all the edge pixels in an edge segment are processed together for matching or in any other operations. Fig. 3(d) shows the difference image edge maps generated from Fig. 3(a) and Fig. 3(b). Similarly edge map in Fig. 3(e) is obtained from Fig. 3(b) and Fig. 3(c).

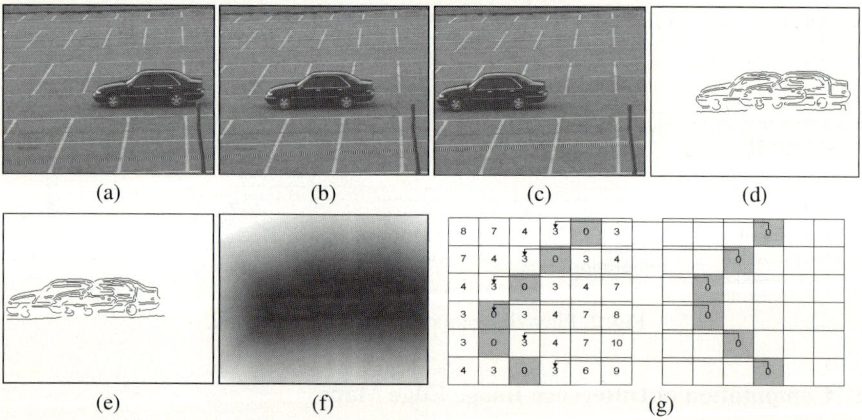

Fig. 3. *DT* image generation and matching. (a) I_{n-1}; (b) I_n; (c) I_{n+1}; (d) DE_{n-1}; (e) DE_n; (f) *DT* image of DE_{n-1}; (g) Edge matching using *DT* image. Here, *Matching_confidence = 0.91287*.

2.2 Moving Object Detection

Edge maps, DE_{n-1} and DE_n are used in this step to extract moving edges for moving object detection in video sequence. DE_{n-1} contains the moving edges of I_{n-1} and I_n, and DE_n contains the moving edges of I_n and I_{n+1}, respectively. Thus, the moving edges of

I_n is common in both of the edge maps. Therefore, to find out moving edges, we superimpose one edge map on another one and compute matching between them. Hence, if two edge segments are of almost similar in size and shape, and situated almost in same positions in the edge maps, then they are considered as moving edges of I_n. However, appearance of noise may cause slightly change of these parameters as well. Hence, instead of exact matching, introducing some variability reduces localization problem to obtain better results. Considering these issues, we have adopted an efficient edge-matching algorithm in this proposed method, which is known as chamfer ¾ matching [10]. According to the procedure of chamfer matching, distance transform (*DT*) image is generated from one difference image edge map and then edge segments from another one are superimposed on it and compute matching confidence. If the matching confidence is less than a certain threshold then the edge segment is enlisted as moving edge. This threshold value gives the variability during matching. In our method, we utilize DE_{n-1} to generate *DT* image and thereafter, edge segments of DE_n are superimposed on it to compute the matching confidence.

To compute *DT* image, we use integer approximation of exact Euclidean distance to minimize the computation time [10]. Each pixel in *DT* represents the corresponding distance to the nearest edge pixel in the edge map. In *DT* image generation, a two-pass algorithm is used to calculate the distance values sequentially. Initially the edge pixels are set to zero and rest of the position is set to infinity. The first pass (forward) modifies the distance image as follows:

$$v_{i,j} = \min(v_{i-1,j-1}+4, v_{i-1,j}+3, v_{i-1,j+1}+4, v_{i,j-1}+3, v_{i,j}) \tag{2}$$

and thereafter, the second pass (backward) works as follows:

$$v_{i,j} = \min(v_{i,j}, v_{i,j+1}+3, v_{i+1,j-1}+4, v_{i+1,j-1}+3, v_{i+1,j+1}+4) \tag{3}$$

where $v_{i,j}$ is the distance at pixel position (i, j). Fig. 3(f) illustrates a *DT* image which is computed from difference image edge map shown in Fig. 3(d). In Fig. 3(f), distance values of *DT* image are normalized into 0 to 255 for better visualization.

During matching, an edge segment of DE_n is superimposed on *DT* image of DE_{n-1} to accumulate the corresponding distance values. A normalized average of these values (root mean square) is the measure of matching confidence of the edge segment in DE_n, shown in following equation:

$$Matching\_confidence[l] = \frac{1}{3}\sqrt{\frac{1}{k}\sum_{i=1}^{k}\{dist(l_i)\}^2} \tag{4}$$

where k is the number of edge points in l^{th} edge segment of DE_n; $dist(l_i)$ is the distance value at position i of edge segment l. The average is divided by 3 to compensate for the unit distance *3* in the chamfer ¾-distance transformation. Edge segments are removed from DE_n if matching confidence is comparatively higher. Existence of a similar edge segments in DE_{n-1} and DE_n produces a low *Matching_confidence* value for that segment. We allow some flexibility by introducing a disparity threshold, τ and empirically we set $\tau = 1.3$ in our implementation. We consider a matching occurs between edge segments, if $Matching\_confidence[l] \le \tau$. The corresponding edge segment is considered as moving edge and consequently enlisted to the moving edge list. Finally, the resultant edge list contains the edge segments of ME_n that belong to moving objects in I_n. Fig. 3(g) illustrates the procedure of computing matching confidence using *DT* image.

3 Experimental Results

Experiments have been carried out with several video sequences captured from indoor as well as outdoor environment to verify the effectiveness of the proposed method. We have applied our proposed method on video formats of size 640x520 and used Intel Pentium IV 1.5 GHz processor and 512 MB of RAM. Visual C++ 6.0 and MTES [12] have been used as of our working tools for implementation.

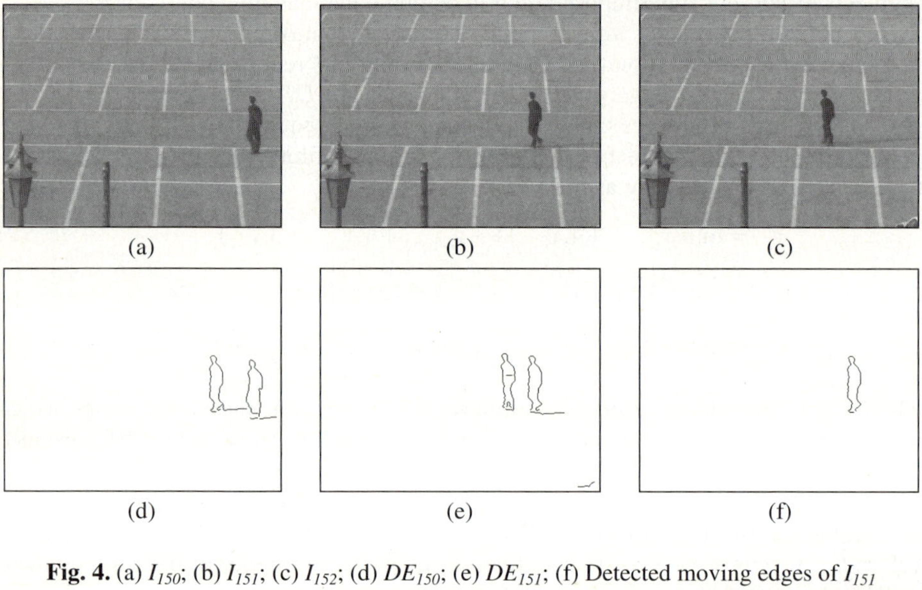

Fig. 4. (a) I_{150}; (b) I_{151}; (c) I_{152}; (d) DE_{150}; (e) DE_{151}; (f) Detected moving edges of I_{151}

Fig. 4 shows experimental result for moving object detection in an outdoor environment. In this case, three consecutive frames, I_{150}, I_{151} and I_{152} shown in Fig. 4(a), Fig. 4(b) and Fig. 4(c), respectively, are used to compute two difference images D_{150} and D_{151}. Thereafter, difference image edge maps, DE_{150} and DE_{151} shown in Fig. 4(d) and Fig. 4(e), are computed using D_{150} and D_{151}, respectively. Fig. 4(f) shows the detected moving edges of I_{151} which were common in both of the difference image edge maps, DE_{150} and DE_{151}.

Fig. 5 shows another experimental result obtained with an indoor video sequence. Fig. 5(a), Fig. 5(b), Fig. 5(c) and Fig. 5(d) show the background and three successive frames I_{272}, I_{273}, and I_{274}, respectively, with different illumination condition and quantization error. Result obtained from the method of Kim and Hwang [6] is shown in Fig. 5(e), where double edge maps have been utilized to detect moving edges. In their method, the difference between background and current frame incorporates most of the noise pixels. Fig. 5(f) shows the result applying the method proposed by Dailey and Cathey [7]. Result obtained from this method is much robust against illumination changes as it uses most recent successive frame differences for moving edge detection. However, it suffers from scattered edge pixels as it uses logical *AND* operation in difference image edge maps for matching. Illumination variation and quantization error induces edge localization problem in difference image edge maps. As a result, some portions of the same edge segment are matched and some are not, and produce scattered edges in final detection result. Our method does not experience this problem because of applying flexible matching between difference image edge maps containing edge segments. The result obtained from our proposed method is shown in Fig. 5(g).

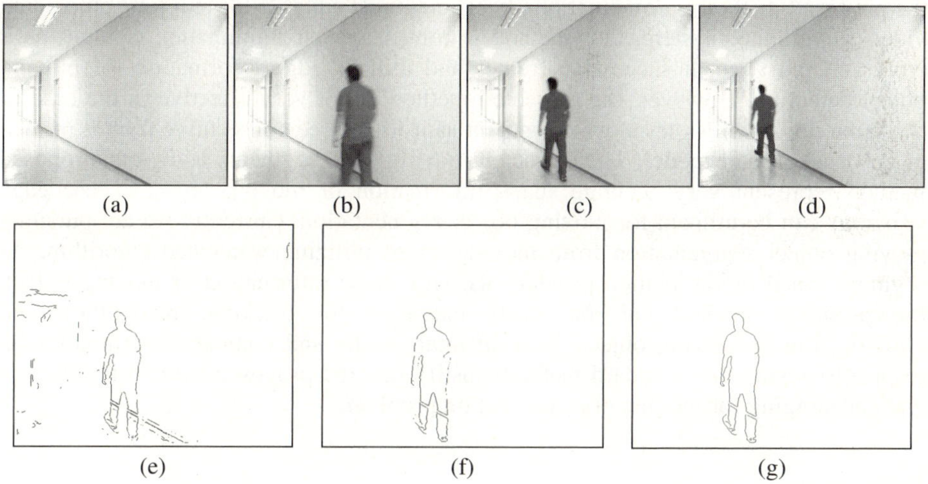

Fig. 5. (a) Background; (b) I_{172}; (c) I_{173}; (d) I_{174}; (e) Detected moving edges of I_{173} using Kim and Hwang method; (f) Detected moving edges of I_{173} using Dailey and Cathey method; (g) Detected moving edges of I_{173} using our proposed method

Table 1. Mean processing time in (ms) for each of the module

Processing steps	Mean time (ms)
Computation of difference images	5
Edge map generation from difference images	39
DT image generation	11
Computation of matching confidence and moving edge detection	19
Total time required	74

In order to comprehend the computational efficiency of the algorithm, it should be mentioned that with the processing power and the processing steps described above, execution time for the moving object detection on grayscale images was approximately 74 ms. Therefore, the processing speed was around 13 frames per second. However, using computers with higher CPU speeds which are available this day and in future as well, this frame rate can be improved. Table 1 depicts approximate times required to execute different modules of the proposed method.

4 Conclusions and Future Works

This paper presents a robust method for moving object detection which does not require any background model. Representation of edges as segments helps to reduce the effect of noise and, incorporates a fast and flexible method for edge matching. So, the proposed method is computationally efficient and suitable for real time automated video surveillance system. Our method is robust against illumination changes as it works on most recent successive frames and utilizes edge information for moving object detection. However, the presented method is not very effective in the case of detecting object with very slow movement as it uses three consecutive frames instead of any background model. The extracted moving edge segments using our proposed method represent very accurate shape information of moving object. These edge segments can be utilized for moving object segmentation. Currently we are pursuing moving object segmentation from moving edges utilizing watershed algorithm. As segment based representation provides us with shape information of moving object, the proposed method can be easily extended for tracking, recognition and classification of moving object. Experimental results and comparative studies with respect to some other standard methods justify that the proposed method is effective and encouraging for moving object detection problem.

References

1. Radke, R., Andra, S., Al-Kohafi, O., Roysam, B.: Image Change Detection Algorithms: A Systematic Survey. IEEE Trans. on Image Processing 14(3), 294–307 (2005)
2. Kastrinaki, V., Zervakis, M., Kalaitzakis, K.: A Survey of Video Processing Techniques for Traffic Applications. Image and Vision Computing 21(4), 359–381 (2003)
3. Chien, S.Y., Ma, S.Y., Chen, L.: Efficient Moving Object Segmentation Algorithm Using Background Registration Technique. IEEE Transactions on Circuits and Systems for Video Technology 12(7), 577–586 (2002)
4. Sappa, A.D., Dornaika, F.: An Edge-Based Approach to Motion Detection. In: Alexandrov, V.N., van Albada, G.D., Sloot, P.M.A., Dongarra, J.J. (eds.) ICCS 2006. LNCS, vol. 3991, pp. 563–570. Springer, Heidelberg (2006)
5. Gutchess, D., Trajkovics, M., Cohen-Solal, E., Lyons, D., Jain, A.K.: A Background Model Initialization Algorithm for Video Surveillance. Proc. of IEEE Intl. Conf. on Computer Vision 1, 733–740 (2001)
6. Kim, C., Hwang, J.N.: Fast and Automatic Video Object Segmentation and Tracking for Content-based Applications. IEEE Trans. on Circuits and Systems for Video Tech. 12, 122–129 (2002)

7. Dailey, D.J., Cathey, F.W., Pumrin, S.: An Algorithm to Estimate Mean Traffic Speed using Un-calibrated Cameras. IEEE Trans. on Intelligent Transportation Sys. 1(2), 98–107 (2000)
8. Makarov, A., Vesin, J.M., Kunt, M.: Intrusion Detection Using Extraction of Moving Edges, International Conf. on. Pattern Recognition 1, 804–807 (1994)
9. Ahn, K.O., Hwang, H.J., Chae, O.S.: Design and Implementation of Edge Class for Image Analysis Algorithm Development based on Standard Edge. In: Proc. of KISS Autumn Conference, pp. 589–591 (2003)
10. Borgefors, G.: Hierarchical Chamfer Matching: A Parametric Edge Matching Algorithm. IEEE Trans. on PAMI 10(6), 849–865 (1988)
11. Canny, J.: A Computational Approach to Edge Detection. IEEE Trans. on PAMI 8(6), 679–698 (1986)
12. Lee, J., Cho, Y.K., Heo, H., Chae, O.S.: MTES: Visual Programming for Teaching and Research in Image Processing. In: Sunderam, V.S., van Albada, G.D., Sloot, P.M.A., Dongarra, J.J. (eds.) ICCS 2005. LNCS, vol. 3514, pp. 1035–1042. Springer, Heidelberg (2005)

Search for a Computationally Efficient Image Super-Resolution Algorithm

Vivek Bannore[1] and Leszek Swierkowski[2]

[1] School of Electrical and Information Engineering, University of South Australia, Mawson Lakes, Adelaide, Australia
[2] Defence Science and Technology Organisation, Edinburgh, Adelaide, Australia
Vivek.Bannore@postgrads.unisa.edu.au,
Leszek.Swierkowski@dsto.defence.gov.au

Abstract. Super-resolution estimates a high-resolution image from a set of observed low-resolution images of the same scene. We formulate the estimation process as a regularized minimization problem and compare its solution, in terms of effectiveness and accuracy, with a fast super-resolution method developed recently in [1]. Results of numerical simulations are presented.

1 Introduction

Image super-resolution refers to an image processing technique that reconstructs a high-resolution image from a sequence of under-sampled and aliased images of the same scene. Due to the relative motion between the sensor and the scene, each low-resolution frame contains a slightly different view of the captured scene. The super-resolution technique fuses these partial views during the reconstruction process, generating an enhanced high-resolution image. The technique can be useful in many visual applications like medical imaging, surveillance, target detection and astronomical imaging, which require high-quality imagery.

Most of the research into image resolution enhancement has been directed towards developing techniques that deliver the highest possible fidelity of the reconstruction process. The computational efficiency issues and the feasibility of developing realistic applications based on super-resolution algorithms have attracted much less attention. Although in some applications like, for example, astronomical imaging or text recognition the computational time constrains are less important, in many other civilian and military applications maintaining low computational time is essential. Super-resolution is a computationally intensive process. Most of the algorithms are based on some kind of optimization that involves minimization of a cost function. The number of unknown variables is then equal to the number of pixels in the reconstructed high-resolution image and is of the order of hundreds of thousands. Moreover, the problem itself is an inverse problem that is underdetermined and ill-conditioned. Clearly, the fidelity of the reconstruction has to be traded-off by performance. For a more extensive overview on super-resolution refer to [2-5].

Maintaining a proper balance between improving spatial resolution and keeping the computational time low is, therefore, an important issue. Recently, we reported a

B. Apolloni et al. (Eds.): KES 2007/WIRN 2007, Part I, LNAI 4692, pp. 510–517, 2007.

hybrid reconstruction scheme for super-resolution restoration [1]. The method makes use of an interpolation technique to produce the first approximation for the reconstructed high-resolution image and then employs an iterative improvement approach to generate the final solution. Numerical simulations showed that the algorithm is efficient and reasonably accurate.

In this paper, we are primarily interested in computational issues of super-resolution restoration of images and in numerical validation of our reconstruction scheme developed in [1]. We adopt Tikhonov regularized optimization formulation of the super-resolution problem and implement a conjugate gradient method of solving it. We then compare the full solution of the optimization problem with calculations based on our fast iterative-interpolation super-resolution (IISR) method [1]. We also propose to speed up the optimization process by initiating it with the high-resolution approximation generated by the IISR hybrid reconstruction algorithm. Finally, we present results of reconstructions from several test image sequences to illustrate the effectiveness of the reconstruction process.

2 The Model

The super-resolution model, in its generic form, assumes that a sequence of N low-resolution (LR) images represent snapshots of the same scene, taken from slightly different directions. The objective of the reconstruction procedure is to combine partial information from all LR frames and to construct a high-resolution representation of the scene. The real scene is represented by a single high-resolution (HR) reference image $\underline{\boldsymbol{X}}$ that we want to reconstruct. We model each LR frame $\underline{b}_k$ as a noisy, down-sampled version of the reference image that is subjected to various imaging conditions like camera and atmospheric blur, motion effects, and geometric warping. It is convenient to represent the observation model in matrix notation:

$$\underline{b}_K = A_K \underline{X} + \underline{E} \qquad \text{for } 1 \le \mathrm{k} \le \mathrm{N}. \tag{1}$$

In the above equation, linear operator $\boldsymbol{A}_{\mathbf{k}}$ represents the process of down-sampling and all the other imaging factors, whereas the additive Gaussian noise is represented by $\underline{\boldsymbol{E}}$. The images are represented in equation (1) as vectors, shown by an underscore, that are ordered column-wise lexicographically.

3 Regularization

In general, for a given sequence of LR images $\underline{b}_k$, the set of equations $\underline{b}_k = A_k \underline{X}$ (k = 1…N) has many solutions or, due to noise, it may have no solution whatsoever. An approximate least square solution might be obtained by minimization of the error $\boldsymbol{\tau}$ between the actually observed and the predicted LR images. Thus, the cost function that has to be minimized is given by the following equation:

$$\tau = \sum_{K=1}^{N} [\underline{b}_K - A_K \underline{X}]^2 \cdot \tag{2}$$

In practice, however, it is well known that the process of estimating the HR image X directly from the equation (2) is very sensitive to even very small changes in $\underline{b}_k$. Thus, the super-resolution reconstruction is ill-conditioned and intrinsically unstable.

The critical part of super-resolution process is, therefore, reformulating the problem in such a way that its solution will constitute a stable and meaningful estimate of the original scene. A commonly used procedure is adding a r*egularization* term to equation (2). The modified equation is given by the following expression:

$$\tau = \sum_{K=1}^{N} [\underline{b}_K - A_K \underline{X}]^2 + \lambda [Q \underline{X}]^2 \tag{3}$$

where the last term is the regularization mechanism, which ensures the uniqueness and the stability of the solution. In the above equation, $\boldsymbol{Q}$ is the regularization or stabilization matrix and $\lambda > 0$ is the regularization parameter.

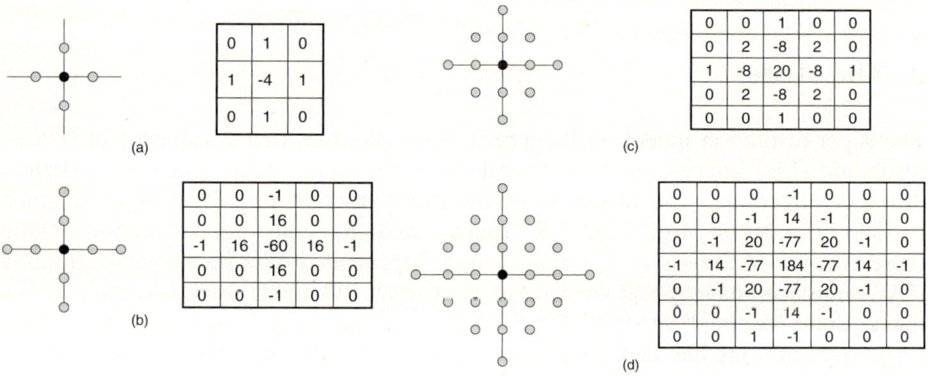

(a)

0	1	0
1	-4	1
0	1	0

(b)

0	0	-1	0	0
0	0	16	0	0
-1	16	-60	16	-1
0	0	16	0	0
0	0	-1	0	0

(c)

0	0	1	0	0
0	2	-8	2	0
1	-8	20	-8	1
0	2	-8	2	0
0	0	1	0	0

(d)

0	0	0	-1	0	0	0
0	0	-1	14	-1	0	0
0	-1	20	-77	20	-1	0
-1	14	-77	184	-77	14	-1
0	-1	20	-77	20	-1	0
0	0	-1	14	-1	0	0
0	0	1	-1	0	0	0

Fig. 1. Regularization/Stabilization Matrix ($\boldsymbol{Q}$): (a) 2D Laplacian Operator ($\boldsymbol{Q}$ = 4-neighbor). (b) 2D Laplacian Operator ($\boldsymbol{Q}$ = 8-neighbor). (c) 2D Biharmonic Operator ($\boldsymbol{Q}$ = 12-neighbor). (d) 2D Biharmonic Operator ($\boldsymbol{Q}$ = 24-neighbor).

Although there is no unique procedure for constructing the regularization term, it is usually chosen to incorporate some prior knowledge of the real HR scene, like degree of smoothness, for example. A popular choice for the matrix $\boldsymbol{Q}$ is a discrete approximation of the Laplacian operator that penalizes large variations in the estimated image. We have implemented four different forms of regularization matrix that are based on various discrete representations of Laplacian and Biharmonic operators. They are shown in Figure 1 as convolution kernels.

The strength of the regularization term in (3) is controlled by the parameter λ. If λ is large, the regularization term will have a dominating effect on the final solution making it smoother but also farther away from the original scene. The estimate will be blurred and, consequently, some information will have been lost. On the other hand, too small value of the regularization parameter λ brings back the risk for the solution being unstable and susceptible to noise amplification. Clearly, the choice of the optimal value of the regularization parameter may strongly influence the fidelity of the reconstruction process. Several estimation techniques for the regularization parameter have been discussed in the literature [6-10], the Discrepancy Principle, Generalized Cross-Validation and the L-curve being the most commonly used in various applications. The choice of the estimation technique depends of the particular application.

4 Optimization Procedure

We adopted the conjugate gradient iterative method for minimizing the cost function, equation (3). The convergence rate for the method is quite rapid and, for most cases, the method is faster than the steepest descent, for example. For the simulations presented in this paper the matrices A_k contain a relatively small amount of blurring and their main role is down-sampling. We assume that the relative motion between frames is well approximated by a single shift vector for each frame. Unlike many other papers investigating super-resolution reconstruction, the down-sampling ratios used in our simulation are large (usually 12 or 16), which we believe brings our calculations closer to reality. Note that the physical enhancement of the resolution achieved during the reconstruction process is usually smaller than the reconstruction magnification ratio. Figure 2 shows an example of super-resolution reconstruction, where the reconstructed HR image is compared to the one of the seventy LR images.

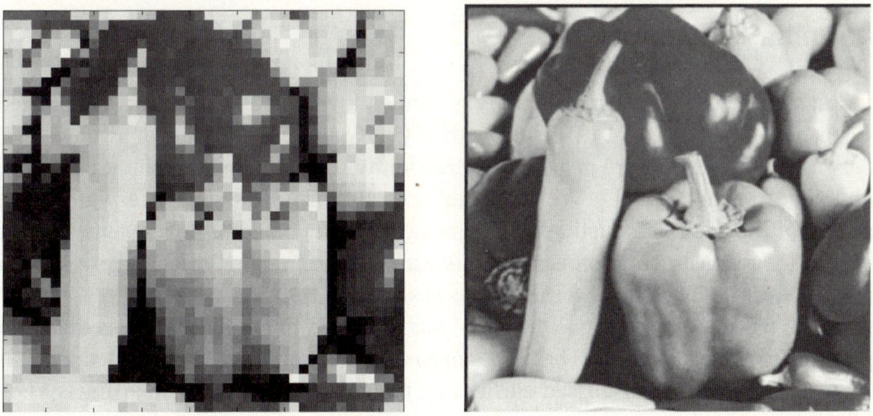

Fig. 2. Low-resolution image (left panel) and reconstructed high-resolution image generated by the optimization procedure (right panel)

5 Iterative-Interpolation Super-Resolution

As we mentioned earlier, maintaining a proper balance between improving spatial resolution and keeping the computational time low is an important issue. Optimization based super-resolution, as described in the previous section, is precise but computationally intensive. On the other hand, the iterative-interpolation super-resolution method (IISR) developed in [1] is relatively fast. We intend to use the full solution of the optimization method as a benchmark for assessing the accuracy of

IISR results. We also propose to speed up the optimization process by initiating it with the high-resolution approximation generated by our IISR hybrid reconstruction algorithm.

The iterative-interpolation method consists of several steps. At the initial stage the sequence of LR images of the scene we want to super-resolve is registered precisely relative to the reference LR frame [11-12]. Once this is achieved, a high-resolution image grid is populated with pixels from low-resolution images by placing them at the appropriate grid-points according to the registration information. Since the number of low-resolution images is limited, the whole composite grid template is not completely filled. The first approximation, X^1, of the high-resolution image is then estimated by interpolating the sparse grid to populate the empty pixels. In this paper, we use cubic spline interpolation as the best tradeoff between accuracy and computational speed [1].

Once the approximate HR image X^1 has been generated it is, then, iteratively improved according to process that is described by the following equation:

$$X^{n+1} = X^n + R_0(b - A \cdot X^n), \rightarrow n = 1,2,3..... \quad (4)$$

where, $\boldsymbol{A}$ is the imaging operator, $\boldsymbol{b}$ is the set of observed LR images, $\boldsymbol{X^n}$ is the $\boldsymbol{n^{th}}$ approximation of the true scene and $\boldsymbol{R_0}$ is the interpolation-based reconstruction operator that was described above. See reference [1] for more detailed description of the procedure.

6 Simulation Results

The test sequences consist of artificially generated low-resolution images. We blurred a test image of size 512 x 512 pixels with a Gaussian kernel of standard deviation 2 pixels. The LR images were generated by randomly sub-sampling the blurred test image with the decimation ratio 12. We generated several random LR sequences with varying numbers of frames. For a qualitative comparison, figures 3 and 4 show results at several stages of the reconstruction process. In this particular example 10 LR images were used for the reconstruction. One of these images is shown in Fig. 3(a).

In the first experiment we applied our iterative-interpolation super-resolution algorithm to this sequence. The first approximation for the reconstructed image is shown in Fig. 3(b). As discussed earlier [1], the existence of periodic artifacts visible in the image is the result of irregular sampling of the scene caused by random movements between LR frames, and inability of the interpolation process to cope with the randomness of the data. The result of reconstruction after 20 iterations of the IISR algorithm is shown in the left panel of Fig. 4. The improvement is evident.

In the second experiment, we applied the optimization procedure described earlier to the same sequence of LR images. In an attempt to increase the convergence of the conjugate gradient solver, we initiated the optimization procedure with the final results of our IISR method presented in the left panel of Fig. 4. The right panel of Fig. 4 shows the final optimized HR image. The additional improvement over the IISR result is rather modest, although quite apparent.

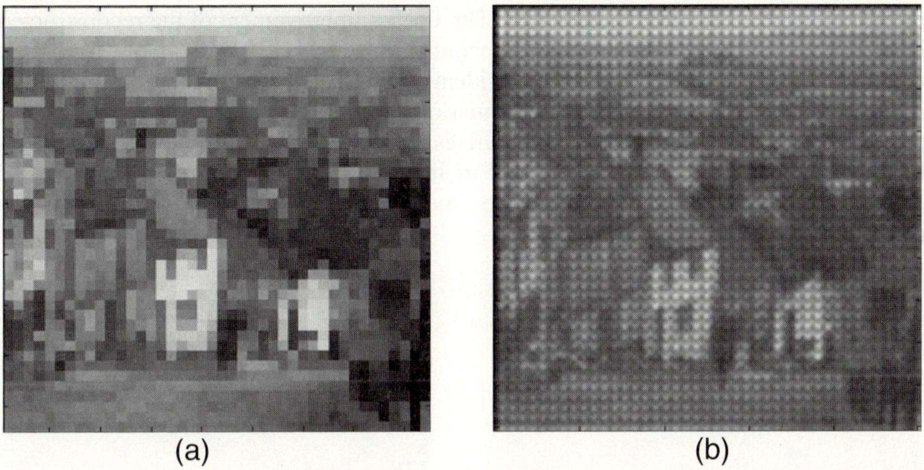

Fig. 3. (a) One of the low-resolution images with a sampling ratio of 12. (b) The first approximation of the HR image generated by the IISR algorithm.

Fig. 4. Reconstructed high-resolution images: generated by our fast IISR algorithm (left panel), and generated by the optimization procedure with the image from the left panel used as a starting point for minimization (right panel)

We tested both algorithms on several test images with various numbers of LR images. In an attempt to quantify the results of our simulations we calculated the Root Mean Square Error (RMSE) between the reconstructed HR images and the original test images. The results are summarized in figures 5 and 6. In Fig. 5, we compare convergence rates of the optimization procedure for three different initialization methods: (1) blank image, (2) interpolated HR image generated by the first approximation of the IISR method and (3) IISR reconstructed image after 20 iterations. Our experiments show that, for both initializations (1) and (2) the optimization routine requires over 80 iterations to reduce the reconstruction error to the initial level of the IISR image error. The further improvement of the reconstructed image is, however, rather small, reconfirming good accuracy of the IISR reconstruction algorithm. It is also evident from the graphs, that this additional improvement over IISR result is computationally expensive since the convergence rate becomes rather slow. Fig. 6 summarizes the accuracy of the reconstruction as a function of the number of LR images included in the reconstruction process. As expected, the more LR frames the more efficient is the super-resolution reconstruction.

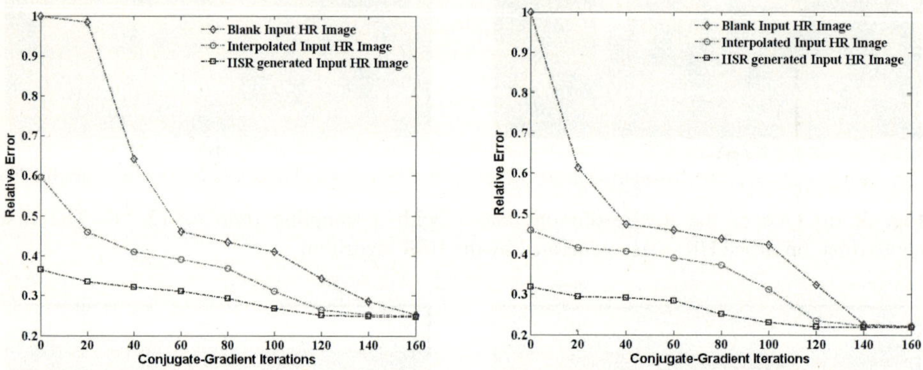

Fig. 5. Convergence plots for the optimization procedure for different initialization methods and for two different numbers of LR images: 20 (left panel) and 70 (right panel)

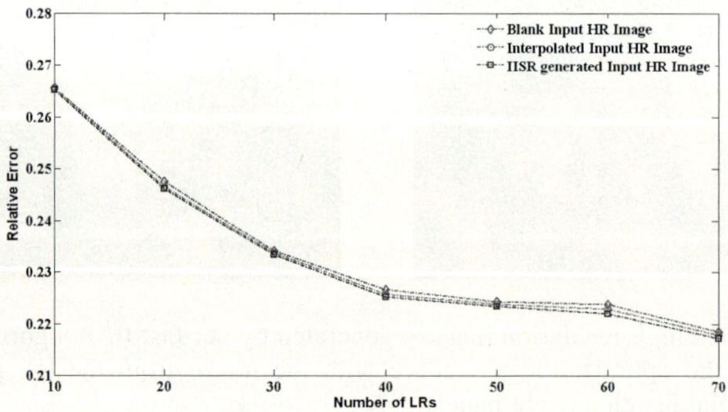

Fig. 6. Accuracy of super-resolution reconstruction as a function of the number of LR images

7 Summary and Conclusions

In this paper, we adopted a Tikhonov regularized minimization formulation of the super-resolution problem and implemented a conjugate gradient method of solving it. We compared the full solution of the optimization problem with calculations based on our fast iterative-interpolation super-resolution (IISR) method. We found that the

IISR reconstruction is reasonably accurate and that further improvement by the optimization procedure is relatively small and computationally expensive.

We proposed to accelerate the optimization process by initializing it with the IISR solution. Further improvement in the rate of convergence can be achieved by preconditioning the minimization procedure. The work in this direction is in progress. Further work is also required to test the robustness of the reconstruction techniques to noise and registration errors.

Acknowledgements. This work is partially supported by Defence Science & Technology Organisation. V. Bannore would like to thank L. Jain and N. Martin for supporting this project. L. Swierkowski acknowledges valuable discussions with B. Smith.

References

[1] Bannore, V., Swierkowski, L.: An Iterative Approach to Image Super-Resolution. In: Shi, Z., S.K.,, F.D. (eds.) Intelligent Information Processing III, Boston, pp. 473–482. Springer, Heidelberg (2006)

[2] Kang, M.G., Chaudhuri, S.: Super-Resolution Image Reconstruction. IEEE Signal Processing Magazine 20, 19–20 (2003)

[3] Super-Resolution Imaging, 1st ed: Kluwer Academic Publishers (2001)

[4] Elad, M., Feuer, A.: Restoration of a Single SR Image from Several Blurred, Noisy, & Under-sampled Measured Images. IEEE Trans. on Image Processing 6, 1646–1658 (1997)

[5] Alam, M.S., Bognar, J.G., Hardie, R.C., Yasuda, B.J.: High-Resolution Infrared Image Reconstruction Using Multiple Randomly Shifted Low-Resolution Aliased Frame. Infrared Imaging Systems: Design, Analysis, Modelling, and Testing VIII, SPIE Proceedings, 3063 (April, 1997)

[6] Tikhonov, A.N., Arsenin, V.Y.: Solutions of Ill-Posed Problems. John Wiley & Sons, Washington, DC (1977)

[7] Hansen, P.C.: Analysis of Discrete Ill-Posed Problems by means of the L-Curve. Siam Review 34(4), 561–580 (1992)

[8] Hanke, M., Hansen, P.C.: Regularization Methods For Large-Scale Problems. Surveys on Mathematics for Industries 3, 253–315 (1993)

[9] Bannore, V.: Regularization for Super-Resolution Image Reconstruction. In: Gabrys, B., Howlett, R.J., Jain, L.C. (eds.) KES 2006. LNCS (LNAI), vol. 4252, pp. 36–46. Springer, Heidelberg (2006)

[10] Tikhonov, A.N.: Regularization of Incorrectly Posed Problems. Soviet Math. Dokl. 4, 1624–1627 (1963)

[11] Sheikh, Y.: Direct Registration of Two Images, http://www.cs.ucf.edu/ yaser/

[12] Bergen, J.R., Anandan, P., Hanna, K.J., Hingorani, R.: Hierarchical Model-Based Motion Estimation. In: Proceedings of the Second European Conference on Computer Vision, pp. 237–252. Springer, Heidelberg (1992)

Step-by-Step Description of Lateral Interaction in Accumulative Computation

Antonio Fernández-Caballero, Miguel A. Fernández, María T. López, and Francisco J. Gómez

Departmento de Sistemas Informáticos, Universidad de Castilla-La Mancha
Escuela Politécnica Superior de Albacete, Albacete, Spain
caballer@dsi.uclm.es

Abstract. In this paper we present a method for moving objects detection and labeling denominated Lateral Interaction in Accumulative Computation (LIAC). The LIAC method usefulness in the general task of motion detection may be appreciated by means of some step-by-step descriptions of significant examples of object detection in video sequences of synthetic and real images.

Keywords: Motion detection, Lateral interaction in accumulative computation method, Video sequences.

1 Introduction

Image segmentation refers to the process of partitioning an image into a set of coherent regions. The segmentation methods lie in (or between) two groups; those detecting flow discontinuities (local operations) and those detecting patches of self-consistent motion according to set criteria (global measurements). Segmentation of an image sequence into moving regions belongs to the most difficult and important problems in computer vision [6]. Spatiotemporal segmentation techniques attempt to identify the objects present in a scene based on spatial and temporal (motion) information [5]. As in [7], we define spatial information as being the brightness information and temporal information as being the motion information. The scene is partitioned into regions such that each region (except the background) represents a moving object. The resulting regions can be identified as moving objects composing the scene [2]. Some approaches rely on a region-merging procedure to identify meaningful objects. First, a set of initial regions is derived. Usually these regions do not represent meaningful objects. These regions are then merged based on some measure of spatiotemporal similarity, so as to obtain meaningful moving objects [1]. We believe that motion from intensity changes is rich enough to warrant precise segmentation.

2 The Lateral Interaction in Accumulative Computation Method

The problem we are putting forward is the detection of the objects moving in a scene. These objects are detected from the motion of any of their parts. Present

B. Apolloni et al. (Eds.): KES 2007/WIRN 2007, Part I, LNAI 4692, pp. 518–525, 2007.

in a video sequence of images, motion allows obtaining the silhouettes of all moving elements. The proposed system is able to detect and even to associate all moving parts of the objects present in the scene [4]. The subtasks implemented in neural network layers, and explained in the following subsections, are (a) LIAC Temporal Motion Detecting, (b) LIAC Spatial-Temporal Recharging, and, (c) LIAC Spatial-Temporal Homogenization.

2.1 LIAC Temporal Motion Detection

This subtask firstly covers the need to segment each input image I into a preset group of gray level bands (n), according to equation 1.

$$x_k(i,j;t) = \begin{cases} 1, \text{if } I(i,j;t) \in [\frac{256}{n} \cdot k, \frac{256}{n} \cdot (k+1) - 1] \\ 0, \text{otherwise} \end{cases} \tag{1}$$

This formula assigns pixel (i,j) to gray level band k. Then, the accumulated charge value related to motion detection at each input image pixel is obtained, as shown in formula 2:

$$y_k(i,j;t) = \begin{cases} v_{dis}, \text{if } x_k(i,j;t) = 0 \\ v_{sat}, \text{if } (x_k(i,j;t) = 1) \cap (x_k(i,j;t-\Delta t) = 0) \\ \max[x_k(i,j;t-\Delta t) - v_{dm}, v_{dis}], \\ \qquad \text{if } (x_k(i,j;t) = 1) \cap (x_k(i,j;t-\Delta t) = 1) \end{cases} \tag{2}$$

The charge value at pixel (i,j) is discharged down to v_{dis} when no motion is detected, is saturated to v_{sat} when motion is detected at t, and, is decremented by a value v_{dm} when motion goes on being detected in consecutive intervals t and $t - \Delta t$ [3].

2.2 LIAC Spatial-Temporal Recharging

This subtask is thought to reactivate the charge values of those pixels partially loaded (charge different from v_{dis} and v_{sat}) and that are directly or indirectly connected to saturated pixels (whose charge is equal to v_{sat}). Formula 3 explains these issues, where v_{rv} is precisely the recharge value.

$$y_k(i,j;t+l\cdot\Delta\tau) = \begin{cases} v_{dis}, & \text{if } y_k(i,j;t+(l-1)\cdot\Delta\tau) = v_{dis} \\ v_{sat}, & \text{if } y_k(i,j;t+(l-1)\cdot\Delta\tau) = v_{sat} \\ \min[y_k(i,j;t+(l-1)\cdot\Delta\tau) + v_{rv}, v_{sat}], \\ \qquad \text{if } v_{dis} < y_k(i,j;t+(l-1)\cdot\Delta\tau) < v_{sat} \end{cases} \tag{3}$$

This step occurs in an iterative way in a different space of time $\tau \ll t$. The value of $\Delta\tau$ will determine the number of times the mean value is calculated.

2.3 LIAC Spatial-Temporal Homogenization

In this subtask the charge is distributed among all connected neighbors holding a minimum charge (greater than θ_{min}), according to equation (4).

$$y_k(i,j;t+m\cdot\Delta\tau) = \frac{1}{1+\delta_{i-1,j}+\delta_{i+1,j}+\delta_{i,j-1}+\delta_{i,j+1}} \times [y_k(i,j;t+(m-1)\cdot\Delta\tau) + \delta_{i-1,j}\cdot y_k(i-1,j;t+(m-1)\cdot\Delta\tau) + \delta_{i+1,j}\cdot y_k(i+1,j;t+(m-1)\cdot\Delta\tau) + \delta_{i,j-1}\cdot y_k(i,j-1;t+(m-1)\cdot\Delta\tau) + \delta_{i,j+1}\cdot y_k(i,j+1;t+(m-1)\cdot\Delta\tau)] \quad (4)$$

where

$$\forall(\alpha,\beta)\in[i\pm 1, j\pm 1], \delta_{\alpha,\beta} = \begin{cases} 1, & \text{if } y_k(\alpha,\beta;t+(m-1)\cdot\Delta\tau) > \theta_{min} \\ 0, & \text{otherwise} \end{cases} \quad (5)$$

Lastly, we take the maximum value of all outputs of the k gray level bands to show the silhouette of a moving object. The result is filtered with a second threshold, namely θ_{max}, eliminating noisy pixels pertaining to non-moving objects:

$$O(i,j;t) = \arg\max_k z_k(i,j;t) \quad (6)$$

$$O(i,j;t) = v_{dis}, \quad \text{if } (O(i,j;t) = \theta_{min}) \cup O(i,j;t) > \theta_{max}) \quad (7)$$

3 Step-by-Step Description

The performance of the method applied to motion detection is demonstrated on a step-by-step description basis of two sets of image sequences. The first set includes synthetic scenes to describe the method's behavior. The second set shows natural images with a real scene from a traffic control system.

3.1 Black over White Motion Detection

In the first sequence a black rectangular region of $8*16$ pixels is moving one pixel per frame rightward on a white $32*32$ pixel background. In this first experiment, motion is detected only on those pixels that pass from black to white at a given frame. General formula (1) is instantiated as $x(i,j;t) = 1$, if $I(i,j;t) = 1$. Fig. 1 (a) to (c) shows the method's output after permanency values calculation on pixels $(16,16)$, $(16,17)$ and $(16,18)$, respectively. Parameters used in this experiment are $v_{sat} = 255$, $v_{dm} = 32$, $v_{rv} = 16$ and $v_{dis} = 0$, whilst $t = 16$. Firstly, this very simple example permits to focus on total recharge, partial discharge, partial recharge, and total discharge. Total recharge occurs at $t = 3$ ($\tau = 48$) at pixel $(16,16)$, $t = 4$ ($\tau = 64$) at pixel $(17,16)$, and $t = 5$ ($\tau = 80$) at pixel $(18,16)$, respectively, just as the black box hits for the first time the pixel in white. From that moment on, you may also appreciate a partial discharge at each new instant t. This is clearly what was expected to occur: a totally or partially charged pixel is partially discharged when no variation is detected in its black level from one frame to another. And, this is true until the black rectangle

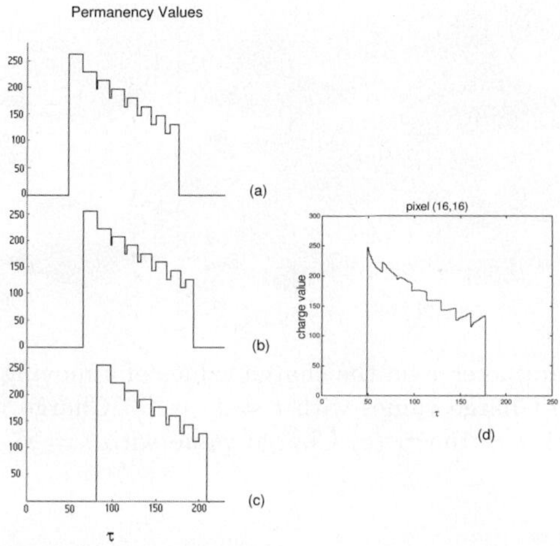

Fig. 1. LIAC permanency and charge values. (a) Permanency values for pixel (16, 16). (b) Permanency values for pixel (16, 17). (c) Permanency values for pixel (16, 18). (d) Charge values for pixel (16, 16).

completely passes the observed pixel. In our special case, the width of the box is eight pixels. Thus, a complete discharge occurs after eight time instants t, that is to say, at $t = 11$ ($\tau = 176$) at pixel (16, 16), $t = 12$ ($\tau = 192$) at pixel (17, 16), and $t = 13$ ($\tau = 208$) at pixel (18, 16), respectively.

Now that the complete recharge, the partial discharge and the complete discharge have been explained from Fig. 1, let us center on the partial recharge notion. Remember, once again, that a partial recharge is the result of being informed by a totally recharged neighbor to sum up some charge. Fig. 1 allows noticing the spatial precedence of this information. In fact, if we consider pixel (16, 16), it may be appreciated that at time instant $t = 4$, it is informed by a neighbor 1 pixel away (pixel (16, 17) in this case); at $t = 5$, it is informed by a neighbor 2 pixels away (pixel (16, 18) in this case; and so on. This simple example offered at Fig. 1 has led us to consider the most relevant ideas in permanency value calculation. Now, Fig. 1d shows the output after charge value calculation on pixel (16, 16). In this figure, you may only notice a quick descent of the charge value until reaching a more stable value at the end.

The moving element (black rectangle) is composed of several charge values due to motion detection. The last step in algorithmic lateral inhibition is the calculation of a common mean charge value. Fig. 2 offers the opportunity to explain the influence of time scale τ. Note that by incrementing τ, the initial ramp is softened. But, in this example, where τ has been fixed with a low value ($t = 16 \cdot \tau$), it is impossible to obtain the mean value desired. We show that, however, by increasing τ, we get the desired solution. Fig. 2e shows the minimum value required in this example for τ to be able to offer a common mean charge value for the moving element. Any greater value for τ gets the same result (Fig. 2f). Compare also the charge value on pixel (16, 16) with $t = 16 \cdot \tau$ and $t = 127 \cdot \tau$ on Fig. 3.

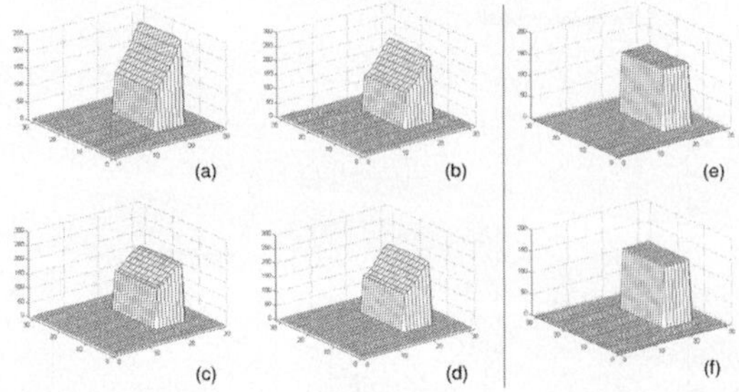

Fig. 2. Influence of parameter τ on the charge values of a moving element. (a) Charge values with $t = \tau$. (b) Charge values with $t = 4 \cdot \tau$. (c) Charge values with $t = 8 \cdot \tau$. (d) Charge values with $t = 16 \cdot \tau$. (e) Charge value with $t = 87 \cdot \tau$. (f) Charge value with $t = 127 \cdot \tau$.

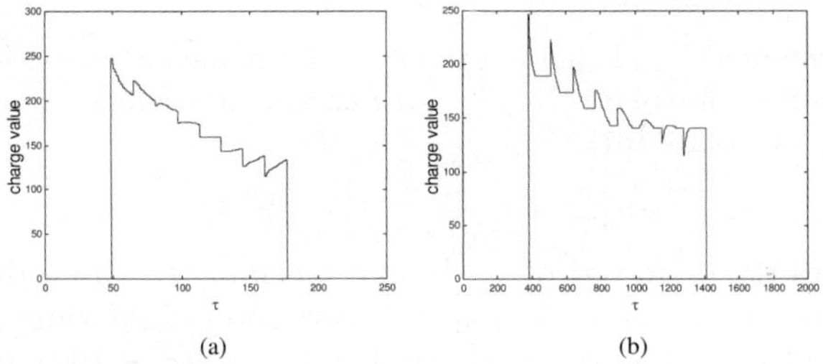

Fig. 3. Influence of parameter τ on the charge values of pixel (16, 16). (a) Charge values with $t = 16 \cdot \tau$. (b) Charge values with $t = 127 \cdot \tau$.

3.2 Noise over Noise Motion Detection

In this second example we consider the synthetic scene shown in Fig. 4, where two random-dot rectangular regions (Fig. 4b1 and 4b2) are moving horizontally one pixel per frame in opposite directions (Fig. 4c) on a random-dot noise background (Fig. 4a). During this motion sequence, there is an overlapping area where both motions are simultaneously perceived.

In this case we shall segment motion of black dots over white background ($x(i,j;t) = 1$, if $I(i,j;t) = 255$), as well as white dots over black background ($x(i,j;t) = 1$, if $I(i,j;t) = 0$), and merge both segmentations. This way, our method perfectly segments moving regions. Fig. 4d shows the result of segmenting from motion of white dots over black background, whereas Fig. 4e shows the result of segmenting from motion of white dots over black background. And, finally, Fig. 4f shows the result of merging both segmentations.

3.3 Gray Level Difference Motion Detection in Real Scenes

We have to highlight that our method applied to motion detection is highly useful when used in real scenes. Let us remember again that the number of

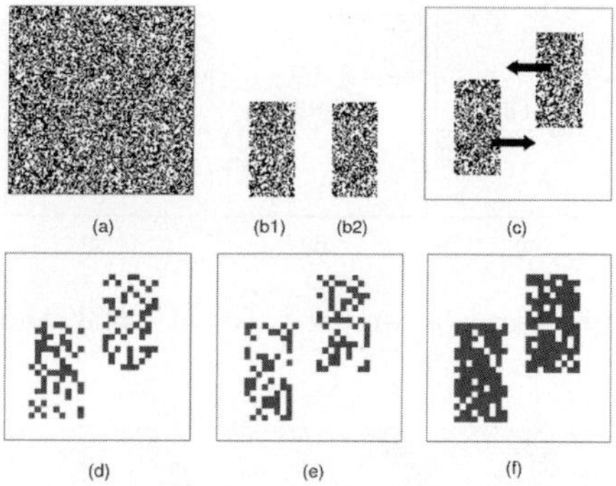

Fig. 4. (a) Random-dot noise background. (b) Random-dot rectangular regions. (c) Motion directions. (d) Segmentation from white dots over black background. (e) Segmentation from black dots over white background. (f) Final result.

Fig. 5. Image segmented into 8 gray level bands (a) at $t = 0$, (b) at $t = 15$, with a frame rate of $\Delta t = 0.04$ seconds

images in a sequence is unlimited. In order to show all these advantages of the neuronal method for lateral interaction in accumulative computation for motion detection we have used a series of real scene test images. This sequence shows a surveillance scene, used with permission from the *PETS2001 dataset 1* (The University of Reading, UK). In this example, we have generalized the method in order to segment from motion due to the change in the current gray level of a pixel.

In this case, we have used $n = 8$ gray level bands. We show in Fig. 5 a little window of the entire scene where images have been segmented in $n = 8$ gray level bands at $t = 0$ and $t = 15$, and where $\Delta t = 0.04$ seconds (image frame rate). The rest of the values were $0 \leq k < n = 8$, $v_{dis} = 0$, $v_{sat} = 255$, and $v_{dm} = 32$ in this case.

Fig. 6 shows some of the outputs of this first part of the whole algorithm after $t = 1$, $t = 2$, $t = 3$, $t = 5$, $t = 11$ and $t = 15$. The implementation of the LIAC Spatial-Temporal Recharging algorithm takes the following values introduced in formula (3): $v_{rv} = 32$ and $1 \leq l \leq 128$, as $t = 128 \cdot \tau$ in this case. Fig. 7 shows, for $t = 12$, the evolution of the LIAC Spatial-Temporal Recharging from $\tau = 1$ up to $\tau = 128$. Notice the effect of fusing pixels to obtain more accurate parts of the vehicle in movement.

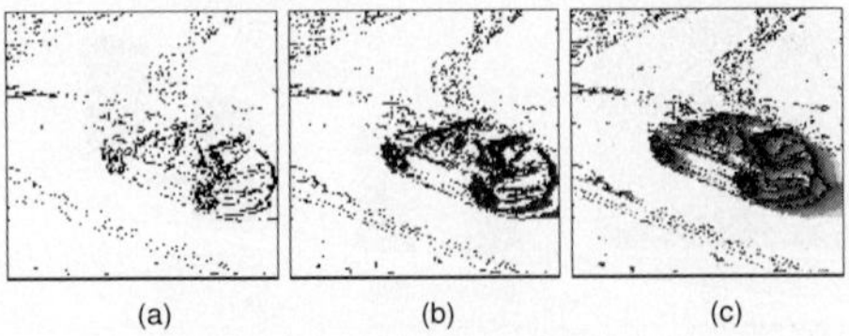

Fig. 6. Image processed (a) at $t = 1$, (b) at $t = 3$, and (c) at $t = 11$

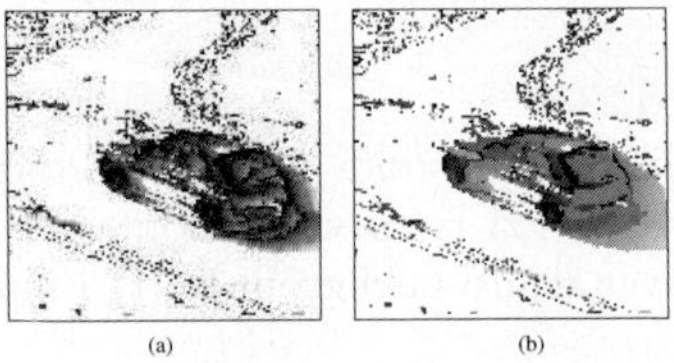

Fig. 7. Image processed at $t = 12$, after (a) $\tau = 1$, and, (b) $\tau = 128$

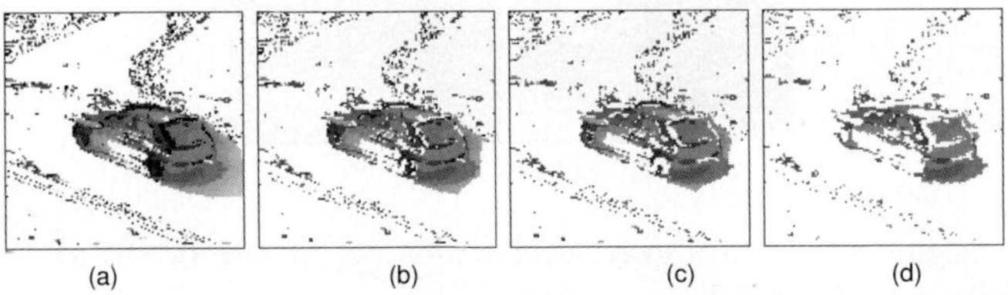

Fig. 8. Result of application of ALI Spatial Homogenization at $t = 12$. (a) Input image. (b) $\theta_{min} = 90$ and $\theta_{max} = 254$. (c) $\theta_{min} = 100$ and $\theta_{max} = 230$. (d) $\theta_{min} = 120$ and $\theta_{max} = 200$.

Lastly, step ALI Spatial-Temporal Homogenization is shown by means of the results offered applying the original formulas (4) and (5), where θ_{min} ranges from 90 to 120 and θ_{max} ranges from 254 down to 200. The results after $t = 12$ are shown in Fig. 8. Obviously there has to be a compromise in the threshold values applied in order to eliminate noise without erasing parts of the moving objects.

4 Conclusion

We have presented a method for motion-based segmentation of images with moving objects. Our approach uses easy local calculation mechanisms. Nevertheless, the global results obtained from these local calculations through the cooperation and propagation mechanisms presented (lateral interaction in accumulative computation mechanisms) may be compared to much more complex methods. Up to some extent, our method can be generically classified into the models based on image difference. The gradient-based estimates have become the main approach in the applications of computer vision. These methods are computationally efficient and satisfactory motion estimates of the motion field are obtained. The disadvantages common to all methods based on the gradient arise from the logical changes in illumination. The intensity of the image along the motion trajectory must be constant; that is to say, any change through time in the intensity of a pixel is only due to motion. This restriction does not affect our model at all. Lastly, region-based approaches work with image regions instead of pixels. In general, these methods are less sensitive to noise than gradient-based methods. Our particular approach takes advantage of this fact and uses all available neighborhood state information as well as the proper motion information. On the other hand, our method is not affected by the greatest disadvantage of region-based methods. Our model does not depend on the pattern of translation motion. In effect, in region-based methods, regions have to remain quite small so that the translation pattern remains valid. The most important limitation of the method applied to motion detection is the impossibility to differentiate among objects that are seen as a whole during occlusions.

Acknowledgements

This work is supported in part by the Spanish CICYT TIN2004-07661-C02-02 grant, and the Junta de Comunidades de Castilla-La Mancha PBI06-0099 grant.

References

1. Ayer, S., Sawhney, H.S.: Layered representation of motion video using robust maximum-likelihood estimation of mixture models and MDL encoding. In: Proceedings of Fifth International Conference on Computer Vision, pp. 777–784 (1995)
2. Dufaux, F., Moscheni, F., Lippman, A.: Spatiotemporal segmentation based on motion and static segmentation. Proceedings of ICIP'95. 1, 306–309 (1995)
3. Fernández, M.A., Fernández-Caballero, A., López, M.T., Mira, J.: Length-Speed Ratio (LSR) as a characteristic for moving elements real-time classification. Real-Time Imaging 9(1), 49–59 (2003)
4. Fernández-Caballero, A., Mira, J., Fernández, M.A., Delgado, A.E.: On motion detection through a multi-layer neural network architecture. Neural Networks 16(2), 205–222 (2003)
5. Goldberger, J., Greenspan, H.: Context-based segmentation of image sequences. IEEE Transactions on Pattern Analysis and Machine Intelligence 28(3), 463–468 (2006)
6. Mansouri, A.R., Konrad, J.: Multiple motion segmentation with level sets. IEEE Transactions on Image Processing 12(2), 201–220 (2003)
7. Vázquez, C., Mitiche, A., Laganiére, R.: Joint multiregion segmentation and parametric estimation of image motion by basis function representation and level set evolution. IEEE Transactions on Pattern Analysis and Machine Intelligence 28(5), 782–793 (2006)

Suitability of Edge Segment Based Moving Object Detection for Real Time Video Surveillance

M. Julius Hossain, M. Ali Akber Dewan, and Oksam Chae*

Department of Computer Engineering, Kyung Hee University,
1 Seochun-ri, Kiheung-eup, Yongin-si, Kyunggi-do, Korea, 449-701
mdjulius@yahoo.com, dewankhu@gmail.com, oschae@khu.ac.kr

Abstract. This paper investigates the suitability of the proposed edge segment based moving object detection for real time video surveillance. Traditional edge pixel based methods handle each edge pixel individually that is not suitable for robust matching, incorporating knowledge with edges, and tracking it. In the proposed method, extracted edges are represented as segments using an efficiently designed edge class and all the pixels belonging to a segment are processed together. This representation helps us to use the geometric information of edges to speed up detection process and enables incorporating knowledge into edge segments for robust matching and tracking. Experiments with real image sequences and comparisons with some existing methods illustrate the suitability of the proposed approach in moving object detection.

Keywords: Video surveillance, reference initialization, segment matching, chamfer distance, dynamic background.

1 Introduction

Detection of moving objects is an important research area for its widespread interest in diverse disciplines such as video surveillance for detection of intruders and traffic flow analysis. Here, the key challenges include variations in illumination, camera motion, calibration error and dynamic background [1], [2]. However, edge-based features are more robust to noise and illumination [2]. Extraction of edge from an image significantly reduces the amount of data to be processed while preserving the important structural properties. Thus, it facilitates to detect moving objects faster than traditional region based methods do.

In the proposed method, we extract the edge information from video frame and represent them as segments using an efficiently designed edge class [3]. We do not work with each edge pixel independently rather all the points belonging to a segment are considered as a unit and are processed together. Once, we construct the edge segment from the edge pixels, we have the location and structural information of each edge segment. It reduces matching time drastically as we do not need to search for edge pixels in the image unlike the traditional edge pixel based methods do. So, our method utilizes the robustness of edge information and also facilitates to incorporate

* Corresponding author.

B. Apolloni et al. (Eds.): KES 2007/WIRN 2007, Part I, LNAI 4692, pp. 526–533, 2007.

fast and flexible matching for background modeling and detection. Representation of edge segment reduces the affect of noises as noises are found sparse and in a small group of points [4], [5]. These scattered pixels are simply ignored in edge extraction step. The proposed method for background modeling generates a robust initial reference that leads to overcome part of the problem caused due to the change in illumination. Reference edges are updated to adapt with the change in background scene. It takes care of dynamic background where foreground acts like background for some period. The proposed matching method tolerates the fluctuation of camera focus or calibration error in a limited scale and thus reduces the false alarm rate.

2 Background of the Research

The research approaches regarding moving object detection can be classified into two categories: the region-based approach and the boundary-based approach. A popular region-based approach is background subtraction followed by thresholding operation. Researchers surveyed and reported experiments on many different criteria for choosing the threshold value and achieve application-specific requirements for false alarms and misses [6]. However, determination of an optimal threshold value for different conditions and applications is very difficult. Some region based motion detection techniques utilize statistical hypothesis test to determine significant change in a particular region [7]. Some researchers focus on optical flow based approach [8], [9]. However, intensity changes in time and space are not unique because temporal changes can be generated by noise or other external factors like illumination drift. Moreover, computational cost is very high in the case of optical flow based methods.

In the case of boundary-based approaches, many researchers use difference of edge pixels, edge-based optical flow, level sets, and active contours. In [5], authors use the difference in edge pixels between a reference image and an input image to adapt the system with changes in illumination [5]. Moving objects are also detected from a combination of edge maps [10], [11]. However, with real image sequence, these methods detect edge pixels belonging to non-moving objects generated by noisy data and variation of illuminations. We propose a new robust edge segment based detection approach to reduce these drawbacks by representing edges as segments along with a flexible scheme for edge matching. We have extended our previous work [4], specially by improving matching scheme to achieve better performance.

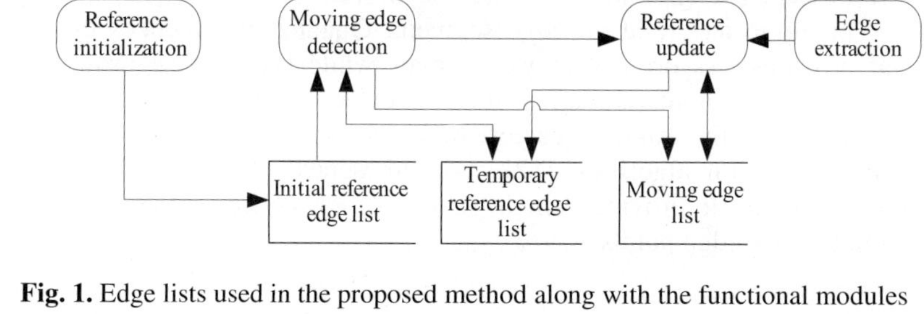

Fig. 1. Edge lists used in the proposed method along with the functional modules

3 Data Structures

The proposed algorithm maintains three different edge lists: initial reference, temporary reference and moving edge, shown in Fig. 1. Initial reference edge list is obtained by accumulating the training set of background images. Extracted edges from current image are searched in the reference edge list and similar edges are eliminated to obtain moving edge list. Initial reference edges are static and no weight value is associated with them for update. Temporary reference edge list is formed by including edge segments from moving edge list having weight value higher than the moving threshold T_M. So, moving edge segments staying in a fixed position for long period of time are considered as temporary reference, also known as dynamic background. Moving edge list is formed by including the moving edges, detected in the current frame. A weight value is associated with each edge segment of the temporary reference and moving edge lists and is updated according to its availability in successive frames. So, the weight value for each edge segment reflects the stability of the edge segment in a particular location. However, there are similarities between temporary reference edge list and moving edge list. Moving edge list can be considered as premature state of temporary reference edge list. The maximum weight or threshold for temporary reference edge list is T_R where, $T_R \geq T_M$. An edge segment in temporary reference or moving edge lists is discarded, if its weight value is zero.

4 Reference Initialization

We generate initial reference edge list from a set of training images. If background scene is free i.e. there is no moving object, a set of frames can be easily selected for background modeling. However, proposed method is able to initialize the reference when moving objects are also present in the scene. In this case training frames are obtained by combining the temporal histogram along with optical flow information [9]. This is very useful especially in the public area where controlling over the monitoring vicinity is difficult or impossible. In the case of reference generation gradient magnitude is extracted from each of the frame in the training set. These values are quantized to n levels and are added to an accumulation array. Quantization is performed by analyzing the cumulative distribution of the gradient image and n quantization levels are selected utilizing the respective n-1 threshold values in CDF. The significant valleys in the histogram are selected as intermediate thresholds. Fig. 2 depicts the CDF, where gradient values are quantized into 8 gray-levels. The lowest level 0 represents background pixel and the highest level, 7 represents the most prominent pixels to a part of an edge segment. Quantization reduces the effect of noise and provides less priority to weak edges while keeping the prominent edge information. The accumulation array is normalized to generate a gradient image having impact of all the training images. Reference edges are extracted by applying canny edge extraction algorithm [12] and represented as segments. Canny edge detector is used, as it provides single response for an edge and achieves good localization to mark edge points.

5 Moving Object Detection

Edge map is generated from current frame and represented as segments. Before extracting the segment, vertices are inserted to the points having more than two branches or belonging to a sharp corner [13]. A vertex divides a connected ridge into more than one edge segment. It helps to break a segment which is part of both background and foreground. In the proposed method, matching between edge segments is performed with a Distance Transform (DT) image rather than computing distance from two edge images. Equation (1) depicts the mathematical representation of generating DT image

$$DT^{(E)}(i, j) = \min_{e \in E} \left\| (i, j) - e \right\| \tag{1}$$

where E is the edge map. DT provides a smooth distance measures between edge segments by allowing more variability between the edges of a template and an object of interest. As we are working for real time detection, we need to apply a very fast edge matching scheme. DT can be computed with a very fast algorithm and subsequently, matching can also be performed by simply counting the distance score of the corresponding pixels of edge of interest. As there are a small deviation between extracted locations of edge points and the actual locations in continuous domain, it is not reasonable to employ an expensive method to calculate the exact Euclidian distances. We utilize chamfer ¾ distance [14], a popular integer approximation of Euclidian distance for computing distance image and edge matching. Matching procedure is performed in two steps. In first phase, distance transformation is performed. Here, all the edge pixels are initialized with zero and all the non-edge pixels with infinity (a very high value) in the distance image D. A forward pass from left to right and top to bottom modifies the distance vector in the following way:

$$D_{i,j} = \min(D_{i-1,j-1} + 4, D_{i-1,j} + 3, D_{i-1,j+1} + 4, D_{i,j-1} + 3, D_{i,j}) \tag{2}$$

Similarly, a backward pass from right to left and bottom to top works as follows:

$$D_{i,j} = \min(D_{i,j}, D_{i,j+1} + 3, D_{i+1,j-1} + 4, D_{i+1,j} + 3, D_{i+1,j+1} + 4) \tag{3}$$

For finding matching confidence, sample edge segments are superimposed on the distance image to calculate the distance between two edge segments. The normalized value, NR is calculated by taking root mean square of all the distances, as equation 4.

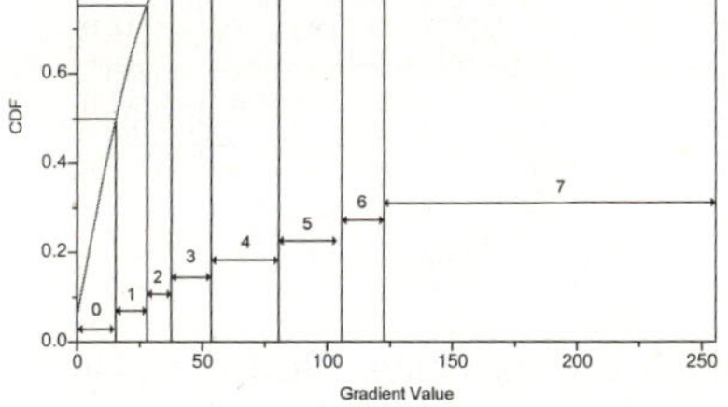

Fig. 2. Quantization of gradient value using the cumulative distribution of the gradient image

$$NR = \frac{1}{3}\sqrt{\frac{1}{n}\sum_{i=1}^{n}\{D(v_i)\}^2} \tag{4}$$

where n is the number of edge points, $D(v_i)$ is the distance value at i^{th} edge point v_i. The average is divided by 3 to compensate for the unit distance *3* in the chamfer 3/4 distance transformation. Fig. 3 depicts the computation of matching confidence.

DT is obtained from reference edge lists. For matching, each edge point is visited in the DT to compute *NR*. If the perfect matching happens, *NR* will be zero. Existence of a similar edge segment in the reference lists produces a low *NR* value. We allow some flexibility by introducing a disparity threshold, τ. We consider a matching if $NR \le \tau$. In this case, the corresponding input edge segment is removed from current edge list. The weight of the reference edge is increased, if it is a temporary reference edge and its weight is less than T_R. An input edge segment that does not match is registered to moving edge list. Flexibility in matching confidence allows little bit of disparity between two edge segments, thus tolerates edge localization problem and minor movement of camera focus. Newly registered edge segments in moving edge list represent the moving object in current frame. However, this process may detect some background edge as moving edge. So, moving edge segments are grouped by analyzing the inter distance information and gray level homogeneity with the neighboring pixels. This process successfully eliminates the scattered edge segments, if any, that are falsely detected as moving edge segments. At this stage each group of moving edges (if any) represent a moving object.

In the detection step, edge segments that are already registered in the moving edge lists are updated by increasing their associated weight value. In this process, segments having the weight value greater than T_M are moved from moving edge list to temporary reference edge list. The proposed method maintains two lists to incorporate dynamic background. Moving edge list is constructed by including the edge segments of moving objects detected in current frame. Temporary reference edge list is constructed by including the edge segments from moving edge list. If a moving edge is found in next frame at same position, the weight of that segment is incremented else it is decremented. If weight of any edge segment reaches T_M, it is moved to the temporary reference edge list. An edge segment is eliminated if the weight of the segment is zero. Temporary reference edges are also updated in similar fashion.

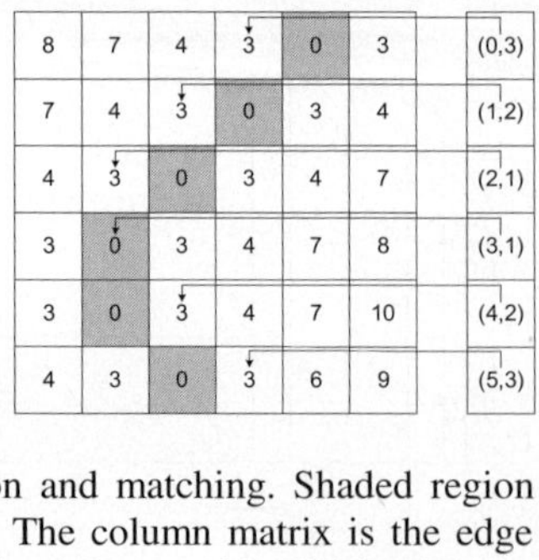

Fig. 3. Distance transformation and matching. Shaded region in left matrix shows the edge points in the template pattern. The column matrix is the edge of interest to be matched. The r.m.s average of the pixel values that are hit divided by three is the edge distance. In this example the computed distance is 0.91287.

6 Results and Analysis

We applied the proposed method on images of size 720x576 that were captured from a corridor and an outdoor parking lot with various changes in constituents and illumination. We used a system, which includes processor of Intel Pentium IV, RAM of 512MB. Visual C++ 6.0 and MTES [15], an image processing environment tool were used for our experiment. The above system processes 7 frames per second. The values of τ , T_M and T_R are set to 2.5, 16 and 32, respectively.

Fig. 4 illustrates moving object detection by the proposed method in different situation. Fig. 4(a) shows a sample background frame, where Fig. 4(b) contains the edge image of the accumulated reference edge list. Fig. 4(c) shows presence of a car at frame 330. The car is detected with respect to initial reference edge list where result is found in Fig. 4(d). Fig. 4(e) and Fig. 4(f) represents the frame 410 and the detected moving object, respectively. The car is parked at this stage for long period of time. At frame number 450 shown in Fig. 4(g), the edge segments of car are registered to the reference as dynamic background and the updated background reference is shown in Fig. 4(h). An intruder is found at frame 509, shown in Fig. 4(i). Fig. 4(j) shows the edge image of the detected moving object for this frame. Here updated reference edge list is used. In many algorithms, a critical situation occurs whenever moving objects are stopped for a long period of time and become part of the background. In this case, when these objects start moving again, a ghost is detected in the area where they stopped. However, as we do not update the initial reference edge list and update only temporary reference edge list, the proposed method does not suffer any problem in this situation.

The proposed method is robust against the change in illumination. Fig. 5 illustrates the results of a separate experiment in indoor environment. Fig. 5(a) and Fig. 5(b) show background and current frame, respectively, having different illuminations. Fig. 5(c) shows the result obtained by the method of Kim and Hwang [11]. In their method the difference between background and current frame incorporates most of the noise pixels. Our proposed accumulation method for generating reference edge and maintaining dynamic background adapt with the change in illumination. The proposed edge matching method detects the edge segments of the moving object successfully, as shown in Fig. 5(d).

Fig. 6 shows that the proposed method is robust against slight movement of camera. In Fig. 6(a), frame 621 of a separate experiment, camera is moved a bit with respect to the background. Fig. 6(b) shows result obtained by the method of Kim and Hwang [10]. Fig. 6(c) shows the result obtained by the method of Dailey and Cathey [11]. However, both of the above approaches detect lots of background edge pixels as foreground due to the movement of camera. This problem is also inherent to most of the image differencing approaches. To solve this problem many of these approaches utilize costly methods to analyze the structure of the detected moving object regions and filter the falsely detected parts. Our method does not suffer from this problem as we applied a flexible matching between edge segments. The result of the proposed method is given in Fig. 6(d). Extracted moving edges along with the accumulated gradient value can be used to segment the moving object. However, moving object segmentation procedure is not included in this paper.

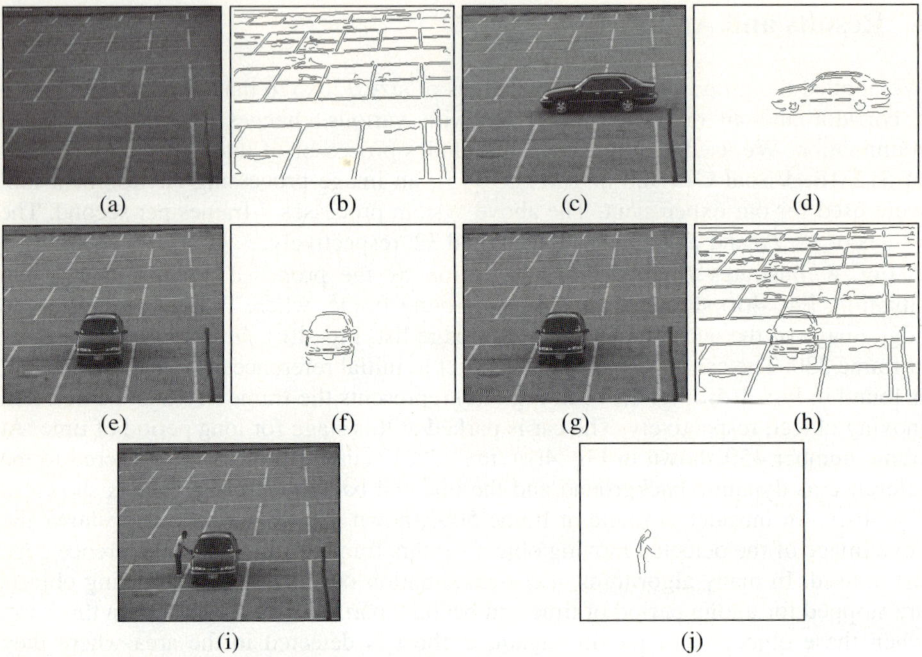

Fig. 4. (a) Sample of background; (b) Edge image of accumulated reference edge list; (c) Frame 330; (d) Edge image of detected moving object at frame 330; (e) Frame 410; (f) Edge image of detected moving object at frame 410; (g) Frame 450; (h) Edge image of updated reference edge at frame 450; (i) Frame 509; (j) Edge image of detected moving object at frame 509

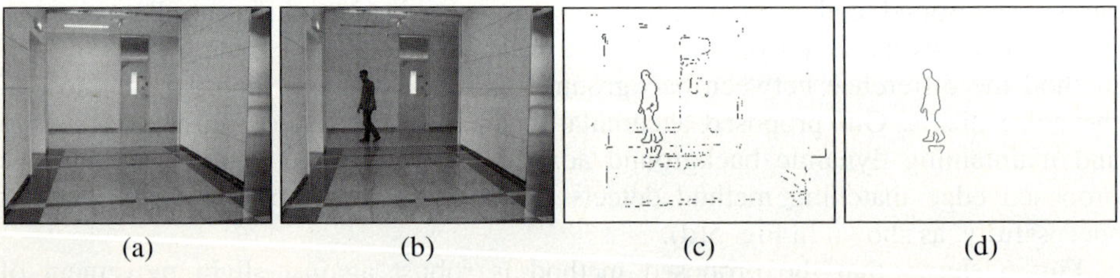

Fig. 5. (a) Sample of background; (b) Frame 205; (c) Edge image of detected moving object by the method of Kim and Hwang; (d) Edge image of detected moving object by the proposed method

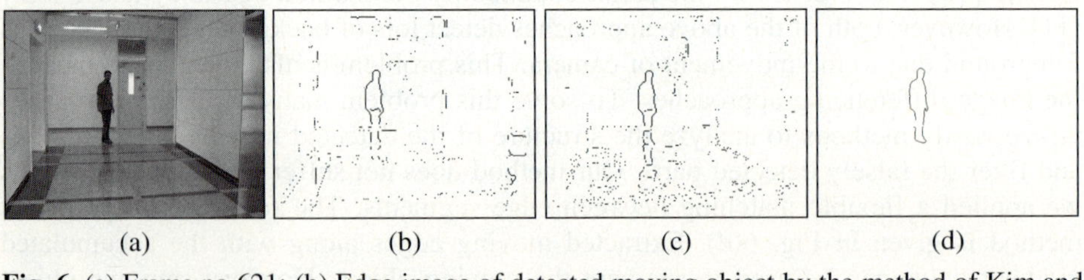

Fig. 6. (a) Frame no 621; (b) Edge image of detected moving object by the method of Kim and Hwang; (c) Edge image of detected moving object by the method of Dailey and Cathey; (d) Edge image of detected moving object by the proposed method

7 Conclusions and Future Works

In this paper, we have presented the suitability of proposed edge segment based moving object detection method for intrusion detection as well as video surveillance. Our intuition is to design a dynamic detection method that will be also robust in case of moving object segmentation, tracking and classification. We have designed edge class to achieve these goals. In the detection part the proposed method proves to perform well by reducing the risk of false alarm due to noise, change of illumination and contents of background. Numerous test results on real scenes and comparisons with some existing approaches justify the suitability of the proposed method. As a future work, our project pursues to segment, track and recognize extracted moving objects.

References

1. Radke, R., Andra, S., Al-Kohafi, O., Roysam, B.: Image Change Detection Algorithms: A Systematic Survey. IEEE Trans. on Image Processing 14(3), 294–307 (2005)
2. Yokoyama, M., Poggio, T.: A Contour-Based Moving Object Detection and Tracking, IEEE Int'l Work. on Visual Surv. and Perfor. Eval. of Track. and Surv., pp. 271–276 (2005)
3. Ahn, K.O., Hwang, H.J., Chae, O.S.: Design and Implementation of Edge Class for Image Analysis Algorithm Development based on Standard Edge. In: Proc. of KISS Autumn Conference, pp. 589–591 (2003)
4. Hossain, M.J., Ahn, K., Lee, J.H., Chae, O.S.: Moving Object Detection in Dynamic Environment. In: Khosla, R., Howlett, R.J., Jain, L.C. (eds.) KES 2005. LNCS (LNAI), vol. 3684, pp. 359–365. Springer, Heidelberg (2005)
5. Makarov, A., Vesin, J.M., Kunt, M.: Intrusion Detection Using Extraction of Moving Edges. Int'l Conf. on Computer Vision & Image Processing 1, 804–807 (1994)
6. Rosin, P.: Thresholding for Change Detection. Computer Vision and Image Understandin 86, 79–95 (2002)
7. Jain, R., Nagel, H.H.: On the Analysis of Accumulative Difference Pictures from Image Sequences of Real World Scenes. IEEE Trans. on PAMI 1, 206–214 (1979)
8. Barron, J.L., Fleet, D.J., Beauchemin, S.S.: Performance of Optical Flow Techniques. Int'l J. Computer Vision 12(1), 43–77 (1994)
9. Gutchess, D., Trajkovics, M., Cohen-Solal, E., Lyons, D., Jain, A.K.: A Background Model Initialization Algorithm for Video Surveillance. In: Proceedings IEEE International Conference on Computer Vision, vol. 1, pp. 733–740 (2001)
10. Kim, C., Hwang, N.J.: Fast and Automatic Video Object Segmentation and Tracking for Content-based Applications. IEEE Trans. on Circuits and Systems for Video Technology 12, 122–129 (2002)
11. Dailey, D.J., Cathey, F.W., Pumrin, S.: An Algorithm to Estimate Mean Traffic Speed using Un-calibrated Cameras. IEEE Trans. on Intell. Trans. Sys. 1(2), 98–107 (2000)
12. Canny, J.: A Computational Approach to Edge Detection. IEEE Transactions on Pattern Analysis and Machine Intelligence 8(6), 679–698 (1986)
13. Smith, S.M., Brady, J.M.: SUSAN - A New Approach to Low Level Image Processing. Int'l J. of Computer Vision, 23(1), 45–78 (1997)
14. Borgefors, G.: Hierarchical Chamfer Matching: A Parametric Edge Matching Algorithm. IEEE Trans. on Pattern Anal. and Machine Intel. 10(6), 849–865 (1988)
15. Lee, J.H., Cho, Y.T., Heo, H., Chae, O.S.: MTES: Visual Programming for Teaching and Research in Image Processing. In: Sunderam, V.S., van Albada, G.D., Sloot, P.M.A., Dongarra, J.J. (eds.) ICCS 2005. LNCS, vol. 3514, pp. 1035–1042. Springer, Heidelberg (2005)

An Ontology for Modelling Human Resources Management Based on Standards

Asunción Gómez-Pérez, Jaime Ramírez, and Boris Villazón-Terrazas

Facultad de Informática, Universidad Politécnica de Madrid,
Campus Montegancedo s/n 28860, Boadilla del Monte, Madrid, Spain
{asun,jramirez,bvillazon}@fi.upm.es

Abstract. Employment Services (ES) are becoming more and more important for Public Administrations where their social implications on sustainability, workforce mobility and equal opportunities play a fundamental strategic importance for any central or local Government. The EU SEEMP (Single European Employment Market-Place) project aims at improving facilitate workers mobility in Europe. Ontologies are used to model descriptions of job offers and curricula; and for facilitating the process of exchanging job offer data and CV data between ES. In this paper we present the methodological approach we followed for reusing existing human resources management standards in the SEEMP project, in order to build a common "language" called Reference Ontology.

Keywords: Human Resources Management Standard, Human Resources Ontologies.

1 Introduction

Nowadays there is an important amount of investment in human capital for economic development. Human resources management refers to the effective use of human resources in order to enhance organisational performance [8]. The human resources management function consists in tracking innumerable data points of each employee, from personal records (data, skills, capabilities) and experiences to payroll records [8]. Human resources management has discovered the Web as an effective communication channel. Although most businesses rely on recruiting channels such as newspaper advertisements, online job exchange services, trade fairs, co-worker recommendations and human resources advisors, online personnel marketing is increasingly used with cost cutting results and efficacy.

Employment Services are becoming more and more important for Public Administrations where their social implications on sustainability, workforce mobility and equal opportunities play a fundamental, strategic importance for any central or local Government. The goal of the SEEMP[1] (Single European Employment Market-Place) project is to design and implement an interoperability architecture for

[1] http://www.seemp.org/

B. Apolloni et al. (Eds.): KES 2007/WIRN 2007, Part I, LNAI 4692, pp. 534–541, 2007.

e-Employment services which encompasses cross-governmental business and decisional processes, interoperability and reconciliation of local professional profiles and taxonomies, semantically enabled web services for distributed knowledge access and sharing. The resultant architecture will consist of: a Reference Ontology, the core component of the system, that acts as a common "language" in the form of a set of controlled vocabularies to describe the details of a job posting or a CV (Curriculum Vitae); a set of local ontologies, so that each ES (E-Employment Services) uses its own local ontology, which describes the employment market in its own terms; a set of mappings between each local ontology and the Reference Ontology; and a set of mappings between the ES schema sources and the local ontologies [4].

A major bottleneck towards e-Employment applications of Semantic Web technology and machine reasoning is the lack of industry-strength ontologies that go beyond academic prototypes. The design of such ontologies from scratch in a textbook-style ontology engineering process is in many cases unattractive for two reasons. First, it would require significant effort. Second, because the resulting ontologies could not build on top of existing community commitment. Since there are several human resources management standards, our goal is not to design human resources ontologies from scratch, but to reuse the most appropriate ones for e-Employment services developed on the framework of the SEEMP project. In this paper we present the methodological approach we followed for reusing existing human resources management standards such as NACE[2], ISCO-88 (COM)[2] and FOET[2], among others.

This paper is organized as follows: Firstly, some related works are briefly explained in section 2. Then, section 3 explains the adopted methodological approach to build the SEEMP Reference Ontology from standards and already existing ontologies. Next, section 4 describes the resultant SEEMP Reference Ontology. Finally, section 5 offers some final conclusions.

2 Related Work

Currently the Human Resource Semantic Web applications are still in an experimental phase, but their potential impact over social, economical and political issues is extremely significant. Bizer et al presents in [2] a scenario for supporting recruitment process with Semantic Web technologies but just within German Government. Mochol et al gives in [9] a brief overview of a Semantic Web application scenario in the Human Resources sector by way of describing the process of ontology development, but its final goal is to merge ontologies. In [3] it is described a competency model and a process dedicated to the management of the competencies underlying a resource related to e-recruitment (mainly CV or a Job Offer). L. Razmerita et al propose in [10] a generic ontology-based user modeling architecture, applied in the context of a Knowledge Management System. E. Biesalski et al explains in [1] some dependencies between Human Resources Management and Knowledge Management in a concrete scenario. Finally, there is an effort described

[2] Available through RAMON Eurostat's Classifications Server at http://ec.europa.eu/comm/eurostat/ramon/

in [1] whose mission is to promote semantic web technology into HR/e-learning standards and applications. Its current focus topics includes: semantic interoperability, semantic of HR-XML[3], etc.

3 Methodological Approach for Reusing Human Resources Management Standards

In this section we describe the adopted approach to build the SEEMP Reference Ontology. This methodological approach follows and extends some of the identified tasks of the ontology development methodology METHONTOLOGY [5]. This methodological approach consists of: specifying, using competency questions, the necessities that the ontology has to satisfy in the new application; selecting the standards and existing ontologies that cover most of the identified necessities; semantic enrichment of the chosen standard; and finally evaluating the ontology content. The steps of this methodology will be explained briefly below:

3.1 Specifying, Using Competency Questions, the Necessities That the Ontology Has to Satisfy in the New Application

This activity states why the ontology is being built, what its intended users are, and who the end-users are. For specifying the ontology requirements we used the competency questions techniques proposed in [6]. These questions and their answers are both used to extract the main concepts and their properties, relations and formal axioms. We have identified sixty competency questions. From the competency questions, we extracted the terminology that will be formally represented in the ontology by means of concepts, attributes and relations. We have identified the terms and the objects in the universe of discourse (instances).

3.2 Selecting the Standards and Existing Ontologies That Cover Most of the Identified Necessities

In order to choose the most suitable human resources management standards for modeling CVs and job offers, the following aspects have been considered: *The degree of coverage of the objects identified in the previous task*, this aspect has been evaluated taking into account the scope and size of the standard. However, a too wide coverage may move us further away the European reality, therefore we have tried to find a tradeoff between this aspect and the following one: *the current european needs*, it is important that standard focuses on the current European reality, because the user partners involved in SEEMP are European, and the outcoming prototype will be validated in European scenarios; and the *user partners recommendations,* in order to asses the quality of the standards, the opinion of the user partners is crucial since they have a deep knowledge of the employment market.

[3] http://www.hr-xml.org

When specifying job offers and CVs, it is also necessary to refer to general purpose international codes such as country codes, currency codes, etc. For this aim, the chosen codes have been the ISO codes, enriched in some cases with user partners classification.

Finally, the representation of job offers and CVs also require temporal concepts such as interval or instant. So, in order to represent these concepts in the final Reference Ontology, the DAML time ontology[4] was chosen.

3.3 Semantic Enrichment of the Chosen Standard

This activity states how we enrich the human resources management standards, the time ontology, the currency classification, the geographic location classification and language classification. For that, all the concept taxonomies were verified; then, ad hoc relationships among concepts of different taxonomies were established; next, concept attributes for describing concept features needed were specified; and finally some formal axioms were defined.

3.4 Evaluating the Ontology Content

The evaluation activity makes a technical judgment of the ontology, of its associated software environments, and of the documentation. We will evaluate the Reference Ontology using the competency questions identified in the first task.

4 SEEMP Reference Ontology

The Reference Ontology described in this section will act as a common "language" in the form of a set of controlled vocabularies to describe the details of a job posting and the CV of a job seeker. The Reference Ontology was developed following the process described in detail in section 2 and with the ontology engineering tool WebODE [5]. The Reference Ontology is composed of thirteen modular ontologies: *Competence, Compensation, Driving License, Economic Activity, Education, Geography, Job Offer, Job Seeker, Labour Regulatory, Language, Occupation, Skill* and *Time*. Figure 1 presents:

- These thirteen modular ontologies (each ontology is represented by a triangle). Ten of them were obtained after wrapping the original format of the standard/ classification, using *ad hoc* translator or wrapper for each standard/classification that transformed all the data stored in external resources into WebODE's knowledge model.
- The connections between the ontologies by means of *ad hoc* relationships. These relationships are defined between specific concepts inside these ontologies.

[4] http://cs.yale.edu/homes/dvm/daml/time-page.html

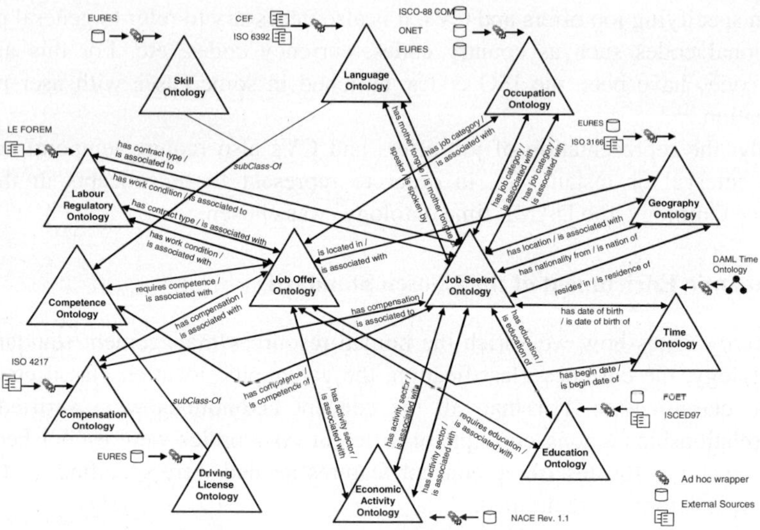

Fig. 1. Main ad-hoc relationships between the modular ontologies

4.1 Wrapping Human Resources Management Standards

As it was mentioned before, these ontologies have been developed following existing human resources management standards and systems classifications, and they are:

- *Compensation Ontology* which is based on the ISO 4217[5]. The ISO 4217 is expressed in HTML format. It is a list of 254 currency names and codes. The resultant Compensation Ontology has 2 concepts: `Currency` and `Salary`. For every currency element specified in the ISO 4217 a different instance of the `Currency` concept is defined. So, the `Currency` concept has 254 instances. An example of instance of the `Currency` concept is `UNITED STATES - US Dollar`.
- *Driving License Ontology* which is based on the levels recognized by the European Legislation[6]. This classification is expressed in HTML format and it is a list of 12 kinds of driving licenses. The resultant Driving License Ontology just has the `Driving License` concept; and for every kind of driving license specified in the European Legislation a different instance of the `Driving License` concept is defined. An example of instance of the `Driving License` concept is `A1 - Light weight motorcycle.`
- *Economic Activity Ontology* is based on the NACE Rev. 1.1[7]. This standard is expressed in MS Access database format and it is a classification of 849 economic activities. The resultant Economic Activity Ontology has 849 concepts. In this case

[5] http://www.iso.org/iso/en/prods-services/popstds/currencycodeslist.html

[6] http://ec.europa.eu/transport/home/drivinglicence/

[7] Available through RAMON Eurostat's Classifications Server at http://ec.europa.eu/comm/eurostat/ramon/

we have defined a concept for every element of the NACE taxonomy in order to preserve the hierarchy.

- *Occupation Ontology* is based on the ISCO-88 (COM)[8], ONET[9] and European Dynamics classification of occupations. ISCO-88 (COM) and ONET are expressed in MS Access database format; European Dynamics classification of occupations is stored in an ORACLE database table. ISCO-88 (COM) is a classification of 520 occupations; ONET is a classification of 1167 occupations and the European Dynamics classification has 84 occupations. The resultant Occupation Ontology has 609 concepts.
- *Education Ontology,* the education fields are based on the FOET[8] and the education levels are based on the ISCED97[8]; both of them are expressed in MS Access database format. FOET has 127 education fields and ISCED97 has 7 education levels. The resultant Education Ontology has 130 concepts. For the education levels we have defined the `Education Level` concept; and for every education level specified in ISCED97 a different instance of the `Education Level` concept is defined. For the education fields we have defined a concept for every element of the FOET taxonomy in order to preserve the hierarchy.
- *Geography Ontology* is based on the ISO 3166[10] country codes and the European Dynamics classifications: Continent and Region. The ISO 3166 is expressed in XML format; Continent and Region classifications are stored in ORACLE database tables. The ISO 3166 has 244 country codes and names; Region classification has 367 regions and Continent classification has 9 continents. The resultant Geography Ontology has four concepts, a `Location` as main concept, which is split into three subclasses: `Continent`, `Region` and `Country`. *Labour Regulatory Ontology* is based on the LE FOREM[11] classifications ContracTypes and WorkRuleTypes, both of them expressed in XML format. ContractTypes classification has ten contract types and WorkRuleTypes has 9 work rule types. The resultant Labour Regulatory Ontology has 2 concepts. For every type of work condition or contract type considered by LE FOREM, a different instance of one of these two concepts (`Contract Type` or `Work Condition`) is included in the ontology. An example of instance of the `Contract Type` concept is `Autonomous`. An example of instance of the `Work Condition` concept is `Partial time`.
- *Language Ontology* is based on the ISO 6392[12] and the Common European Framework of Reference (CEF)[13]. The ISO 6392 is expressed in HTML format and CEF is a description in PDF format. The ISO 6392 has 490 language codes and CEF has 6 language levels. The resultant Language Ontology has 3 concepts: `Language`, `Language Level` and `Language Proficiency`. For every language element specified in the ISO 6392 a different instance of the `Language`

---

[8] Available through RAMON Eurostat's Classifications Server at http://ec.europa.eu/comm/eurostat/ramon/

[9] http://online.onetcenter.org/

[10] http://www.iso.org/iso/en/prods-services/iso3166ma/index.html

[11] LE FOREM is an user partner of the SEEMP project, http://www.leforem.be/

[12] http://www.iso.org/iso/en/prods-services/popstds/languagecodes.html

[13] http://www.cambridgeesol.org/exams/cef.htm

concept is defined, so the `Language` concept has 490 instances. For every language level element specified in the CEF a different instance of the `Language Level` concept is defined, so the `Language Level` concept has 6 instances. An example of instance of the `Language` concept is `eng - English`. An example of instance of the `Language Level` concept is `A2 - Basic User`.
- *Skill Ontology* is based on European Dynamics Skill classification. This classification has 291 skills and it is stored in an ORACLE database table. The resultant Skill Ontology has 2 concepts: `Skill` concept with its subclass `ICT Skill`. For every skill element specified in the European Dynamic classification a different instance of the `ICT Skill` concept is defined. An example of instance of the `ICT Skill` concept is `Hardware programming`.
- *Competence Ontology* defines a concept called Competence as a superclass of the imported concepts `Skill`, `Language Proficiency` and `Driving License`.
- *Time Ontology* is based on DAML ontology[14] and it is expressed in OWL format.

In order to make possible the enrichment of the standards/classifications, it was necessary to import them into the ontology engineering tool WebODE [5]. This process consisted in implementing the necessary conversions mechanisms for transforming the standards/classifications into WebODE's knowledge model.

4.2 Enriching the Ontologies

Once we transformed the standards/classifications into ontologies, the next step is to enrich them introducing concept attributes and *ad hoc* relationships between ontology concepts of the same or different taxonomies. We perform this task by doing the following

- We created from scratch the Job Seeker Ontology and the Job Offer Ontology, which models the job seeker and his/her CV information, and the job offer and employer information, respectively.
- We defined relationships between the concepts of the Job Seeker and Job Offer Ontologies and the concepts defined on the standard (classification) based ontologies.

5 Conclusion

In this paper we have presented the methodological approach we followed for reusing existing human resources management standards in the SEEMP Project. We also described the resultant Reference Ontology which acts as a common "language" in the form of a set of controlled vocabularies to describe the details of a job posting and the CV of a job seeker. The Reference Ontology was developed with the proposed methodology and with the ontology engineering tool WebODE.

---

[14] http://cs.yale.edu/homes/dvm/daml/time-page.html

An important conclusion of the work that we have carried out is that we can reuse human resource management standards in new applications following a systematic approach. Moreover, it is clear such a reuse can save time during the development of the whole system. However, it is not always possible to reuse a standard in a straightforward way, because sometimes the ideal standard does not exist for different reasons (different scope, outdated, etc.), and it is necessary to extend some "imperfect" standard with additional terminology coming from other standards or *ad hoc* classifications.

Acknowledgments. This work has been partially supported by the FP6 EU SEEMP Project (FP6-027347).

References

1. Biesalski, E., Abecker, A.: Human Resource Management with Ontologies. Wissensmanagement. LNCS Volume, pp. 499–507 (2005)
2. Bizer, C., Heese, R., Mochol, M., Oldakowski, R., Tolksdorf, R., Eckstein, R.: The Impact of Semantic Web Technologies on Job Recruitment Processes; 7th International Conference Wirtschaftsinformatik (2005)
3. Bourse, M., Leclère, M., Morin, E., Trichet, F.: Human Resource Management and Semantic Web Technologies;1st International Conference on Information Communication Technologies: from Theory to Applications (ICTTA) (2004)
4. FOREM, UniMiB, Cefriel, ARL, SOC, MAR, PEP: User Requirement Definition D.1.SEEMP Deliverable (2006)
5. Gómez-Pérez, A., Fernández-López, M., Corcho, O.: Ontological Engineering. Springer, Heidelberg (2003)
6. Grüninger, M., Fox, M.: Methodology for the design and evaluation of ontologies. In: Skuce, D (ed.) IJCAI95 Workshop on Basic Ontological Issues in Knowledge Sharing, pp. 1–6 (1995)
7. Jarrar, M.: Ontology Outreach Advisory - The Human Resources and Employment Domain Chapter, http://www.starlab.vub.ac.be/OOA/OOA-HR/OOA-HR.html
8. Legge, K.: Human Resource Management: Rhetorics and Realities. Anniversary ed. Macmillan, NYC (2005)
9. Mochol, M., Paslaru, E.: Simperl: Practical Guidelines for Building Semantic eRecruitment Applications, International Conference on Knowledge Management (iKnow'06), Special Track: Advanced Semantic Technologies (2006)
10. Razmerita, L., Albert, A., Angehrn, A.: Ontology-Based User Modeling for Knowledge Management Systems, pp. 213–217 (2003)

Corpus Building for Corporate Knowledge Discovery and Management: A Case Study of Manufacturing

Ying Liu[1] and Han Tong Loh[2]

[1] Department of Industrial and Systems Engineering
The Hong Kong Polytechnic University, Hung Hom, Kowloon, Hong Kong SAR, China
[2] Department of Mechanical Engineering
National University of Singapore, 21 Lower Kent Ridge Road, Singapore 119077

Abstract. Building a collection of electronic documents, e.g. corpus, is a cornerstone for the research in information retrieval, text mining and knowledge management. In literature, very few papers have discussed the necessary concerns for building a corpus and explained the building process systematically. In this paper, we explain our work of building an enterprise corpus called manufacturing corpus version 1 (MCV1) for corporate knowledge management purpose. Relevant issues, e.g. input texts, category labels and policies, as well as its parallel coding process and quality measurements are discussed. The real-world automated text classification experiments based on MCV1 show the soundness of its coding process. Finally, suggestions are made on how the proposed approach can be implemented in a more economical manner.

1 Introduction

Intensive global competition is pushing manufacturing companies ever harder in their strife for constant profit. As the world is evolving into a knowledge based economy, manufacturing companies are increasingly concerned about the acquisition, management and utilization of advanced R&D information and knowledge from both internal and external resources, e.g. design documents, customer feedbacks, e-journals and digital libraries. For example, product design engineers are concerned about the past design experience, technical tips and solutions of the early models, which have been written down by previous engineers. Successfully handling of such textual information can enrich the company's understanding of market, save development cost and shorten the time-to-market, bring better products to satisfy her customers and in turn lead the company to a prosperous future.

Current studies in information retrieval, knowledge discovery in databases or data mining, text mining and knowledge management are starting to provide feasible solutions to such scenarios. One specific area we focus on is to study a domain specific knowledge management and knowledge retrieval by applying the aforementioned techniques, e.g. automated text classification (TC) and summarization. Since the state-of-the-art techniques are more machine learning (ML) based [2, 3, 7], a collection of domain specific documents is always needed. This has motivated us to study

B. Apolloni et al. (Eds.): KES 2007/WIRN 2007, Part I, LNAI 4692, pp. 542–550, 2007.

the methodology of building such document collections for industry companies who intend to undertake such initiatives. We also intend to see how this building approach will affect the performance of ML based techniques, like TC.

Building a collection of electronic documents, i.e. a corpus, is a cornerstone for the research in information retrieval (IR), text mining (TM) and knowledge management. There are several well-known corpora available for research and benchmarking, such as OSHUMED [4], Reuters21578 [11] and 20 Newsgroups [8] and so on. While these corpora have been extensively tested by IR and TM community, there is lack of documentation about their building process. Reuters Corpus Volume 1 (RCV1) is a recently available corpus [6, 9]. It is an archive of over 800K manually classified news articles between 20/08/1996 and 19/08/1997. While the inputs of RCV1, its building process, quality and many other concerns have been explained, we note the limitation with its serial coding process as well as more than 90 well trained editors were involved at its peak. This imposes more difficulty for industrial companies to create a corpus for their research and application purpose.

This paper aims to give an example of how an industry company can create a quality corpus for their research and application of knowledge discovery and management. All relevant issues, e.g. text inputs, coding process, quality measurements, are discussed and documented. While we present a case study using manufacturing related texts, our approach is intended for general purpose.

Manufacturing Corpus Version 1 (MCV1) is an archive of 1434 English language manufacturing related engineering papers which we gathered by courtesy of the Society of Manufacturing Engineers (SME). It combines all engineering technical papers from SME between 1998 and 2000. All documents have been manually classified. The final output of each document has been formatted as XML files.

Having described the motivation, the rest of this paper is organized as follows. We describe the inputs, coding policies and present a parallel coding process in Section 2. In Section 3, we explain how the coding quality can be measured. The coding performance of human operators is discussed in Section 4. We end this paper with the results of TC experiments using MCV1 in Section 5 and Section 6 concludes.

2 Coding MCV1

2.1 Input Sources: Documents and Coding Labels

The Society of Manufacturing Engineers (SME) provided us with their technical papers from 1998 to 2000. These papers were utilized as the input documents for MCV1. As for the coding labels, basically we adopted the taxonomy implemented by SME for the manufacturing industry. It is called Manufacturing Knowledge Architecture (MKA) in our research. In order to facilitate data processing, all MKA items are coded as shown in Table 1. There are two more levels of subcategories, in the form as CXXYY and CXXYYZZ, under these 18 main categories. Therefore, including manufacturing as the root, there are totally four levels of category labels in MKA. CXXs, which are the 18 major categories, are in the second level. In total, there are 334 category labels.

Table 1. 18 major categories of MKA

C01. Assembly & Joining	C10. Materials
C02. Composites Manufacturing	C11. Measurement, Inspection & Testing
C03. Electronics Manufacturing	C12. Plastics Molding & Manufacturing
C04. Finishing & Coating	C13. Product Design Management
C05. Forming & Fabrication	C14. Quality
C06. Lean Manufacturing, Supply Chain Mgt	C15. Rapid Prototyping
C07. Machining & Material Removal Processes	C16. Research & Development / New Technologies
C08. Manufacturing Engineering & Management	C17. Robotics & Machine Vision
C09. Manufacturing Systems, Automation & IT	C18. Welding

2.2 Coding Policy

The coding policies serve as rules to guide coding operators during the coding process. These need to be explained explicitly at the beginning of the coding process, since it will help to reduce the errors and maintain the coding quality. Based on the coding policies mentioned in literature [6, 9], we had grouped them into two main policies and adopted in our work.

- Lower Bound Policy: Each article has to be assigned with at least one topic label. If none of the suitable labels can be identified, then label <Others> will be chosen. There is no upper limit on the number of labels assigned to each article.
- Hierarchy Policy: Coding operators are required to assign the most specific and suitable labels, may or may not be end leaf labels, to the articles. All ancestors of one specific label are not required to be assigned by coding operators. They can be obtained automatically.

2.3 The Coding Process

Usually, a serial coding process which involves a large number of specialized people is adopted, e.g. RCV1. In Reuters, editors are actually the operators taking care of the coding. In RCV1, a document was coded by one editor first with the results checked by another editor later, and at its peak altogether around 90 editors were involved. As we understood from [9], the second editor could change, delete the labels assigned by the first editor or simply assigned others according to his\her understanding without communicating to the first editor. In the end, the second editor's labels were applied as final ones. We noted that due to this serial process adopted by RCV1, the statistical analysis of coding performance was only applied to the data where the editors engaged in the final coding. In other words, only when they were the second editors [9, 10]. Obviously, there is a lack of information to tell whether these editors have subjective preference with respect to certain labels and whether the second editors have truly enhanced the coding quality, rather than proofreading only. This has motivated us to establish a different process to better examine the coding quality while a large number of operators may not be called upon, which is deemed as more realistic to industry companies.

We developed a parallel process that aims to maximize the output coding quality from human operators. The idea is inspired by the process of customer understanding and conceptual design in a product design and development process [12]. It shows that designers and customers sitting together can enhance understanding greatly and trigger more ideas. We believe that communication among operators will encourage understanding and consequently promote agreement.

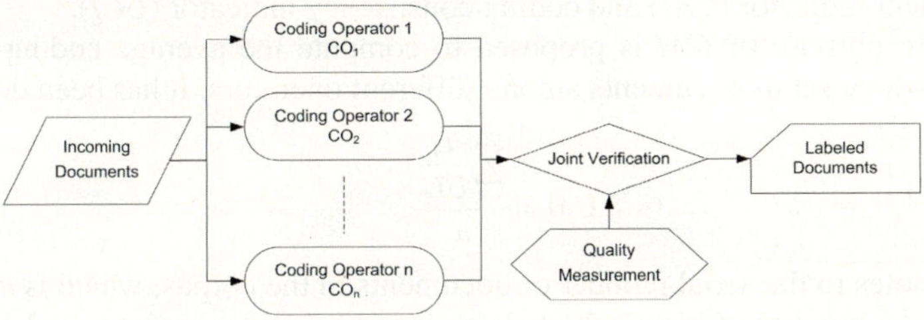

Fig. 1. The parallel coding process used by MCV1

The idea of a parallel coding process is shown in Fig. 1. As indicated, operators are arranged in a parallel way to label documents, where the number of operators needed can be a handful only. After that, a joint verification step is designed to examine and improve the overall coding quality. Basically, four phases are available during the step of joint verification shown in Fig. 2.

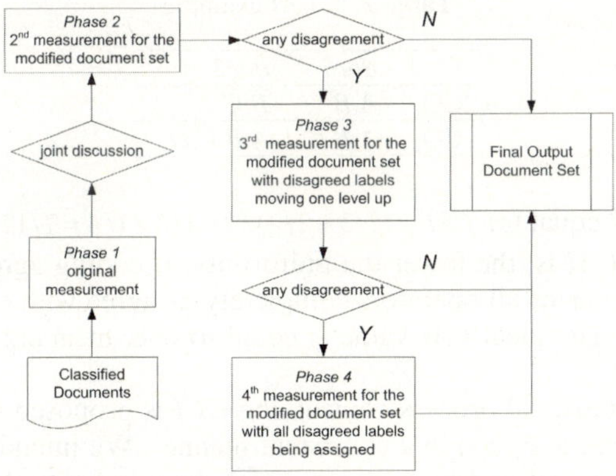

Fig. 2. The illustration of joint verification process

- Phase 1: Rightly after all operators have finished the coding, we can examine the initial performance of operators and investigate their coding patterns.
- Phase 2: If any disagreement exists regarding the labels assigned, operators have to sit down, exchange opinions and try to persuade each other.
- Phase 3: After phases 2, if for some documents, the disagreement still has not been resolved, then the labels are moved one level up, e.g. from the fourth level to the third level. However, this action is only allowed for the disagreement regarding the labels in the 4th level. In other words, moving labels one level up cannot be applied to those beyond the third level. This prevents the labels assigned being too general.
- Phase 4: Finally, for the documents with which operators still do not fully agree, all labels will be assigned. This is mainly to maximize the information coverage.

3 Coding Quality Measurement

In order to ensure a good coding quality of the corpus, quality measurement is a must. Therefore, two quality indicators are utilized for the parallel coding process, i.e. coding agreement indicator (*CAI*) and coding consistency indicator (*CCI*).

The main purpose of *CAI* is proposed to compute the average coding agreement across the whole set of documents among different operators. It has been defined as:

$$CAI = \frac{\sum_{i=1}^{n} \frac{L_i}{UL_i}}{n} \quad (1)$$

where i denotes to the serial number of documents in the corpus, which is n in total. L_i denotes to the number of identical labels assigned by every operator and UL_i denotes to the unique labels assigned by all operators with respect to document d_i.

Here is an example how *CAI* works. Suppose that coding operator 1 (CO_1) and coding operator 2 (CO_2) are classifying a corpus with two documents only, i.e. n=2. The outcome is shown in Table 2, *A*, *B*, *C* and *D* are four labels assigned. For document 1, only two labels, i.e. *A* and *B*, out of three unique labels, i.e. *A*, *B*, and *C*, are agreed with both CO_1 and CO_2. For document 2, only two labels, i.e. *B* and *C*, out of four unique labels, i.e. *A*, *B*, *C*, and *D*, are agreed with both CO_1 and CO_2.

Table 2. A *CAI* example

	doc1	*doc2*
CQ_1	*A,B*	*B,C*
CQ_2	*A,B,C*	*A,B,C,D*

Therefore, *CAI* is equal to: $CAI = (2/3 + 2/4)/2 = 1/3 + 1/4 = 7/12 = 0.5833$

The smaller the *CAI* is, the lower the uniformity of coding agreement. The lowest value is zero which means all operators completely disagree with each other regarding the labels assigned. The ideal *CAI* value is equal to one, meaning operators are completely in agreement.

Inspired by the statistical process control, the *CCI* is proposed to examine whether the category labels are assigned in a consistent manner. We intend to find out if some operators possess unusual coding patterns which may indicate their subjective bias, preference and insufficient knowledge. These measurements include, but not limited to, exams of mean and standard deviation, correlation test, significance test and so on. We skip the details here as they can be found in almost every statistics textbook.

4 Coding Performance of Human Operators

The final coding process was carried out by four full-time graduate students who acted as coding operators. These students were either working on their doctoral or master degrees in engineering at the National University of Singapore. It took about 40 working days averaging four to six hours per day, exclusively of Saturdays and Sundays, for them to accomplish this task. Their daily reading rates generally ranged from 40 to 60 documents. The peak rate could reach 70 documents per day on some days. Each coding operator had to read all 1434 documents.

Table 3. *CAI* values of the first round results

	$CO_{1,2,3,4}$	$CO_{1,2,3}$	$CO_{1,2,4}$	$CO_{1,3,4}$	$CO_{2,3,4}$
CAI (%)	12.17	20.68	17.84	17.64	17.01
Increase (%)	0	+8.51	+5.67	+5.47	+4.84

Table 3 shows the *CAI* values of all four operators and different combinations of any three operators after the first round. As note, the *CAI* value of the combined four operators is only 12.17%. However, if we try the different combinations of any three operators and check their *CAI* values, we note that the combination of operators without CO_4 leads to the highest increase which is over 8.5%. This implies that CO_4 might be an operator who introduced many labels with which others did not agree. Due to this observation, further investigations via *CCI* over all operators have been conducted with more attention on CO_4.

Table 4. Standard deviation analysis of labels assignment for all categories

	CO_1	CO_2	CO_3	CO_4
Within ±1Sigma	13	15	14	6
Within ±2Sigma	19	19	19	19

Table 4 shows CO_4 has an unusual pattern of labeling. We observe that the number of major category labels, i.e. CXX, assigned by each operator is all controlled within ±2 Sigma. However, only six main categories are within ±1 Sigma for CO_4, which is 53% to 60% less compared to others. This unusual pattern basically implies that CO_4 has subjective preferences or incomplete understanding regarding either the labels or their potentially related documents or both. Later, a close look at the coding details of CO_4 reveals that the operator understood many domain concepts incompletely and sometimes wrongly. Therefore, it is very difficult for the person to link up the category labels and their related documents. This led to an unusual coding pattern.

Table 5. CAI values of different phases

	1st Round	2nd Round Joint Discussion	3rd Round Label Upgrading
CAI (%)	20.68	89.44	90.67

Because the coding consistency is always desirable, we rejected the coding results of CO_4 after the first round. Only those from CO_1, CO_2 and CO_3 were moved to the second phrase and beyond. Table 5 shows the final *CAI* value after the joint discussion and label upgrading as mentioned in Fig. 2. Please note only the contributions of CO_1, CO_2 and CO_3 are taken into account. Finally, the *CAI* value of MCV1 stops at 90.67%. Through the analysis, it shows that using *CAI* and *CCI* together is effective to monitor the coding quality.

5 Automated Text Classification Using MCV1

One immediate application based on MCV1 is TC, and subsequently, use the trained classifiers for knowledge discovery and management purposes, such as the classification of product R&D documents and customer service records for marketing research. TC aims to classify documents into a set of predefined categories without human intervention. It has more interests among researchers in the last decade partly due to the dramatically increased availability of digital documents [11].

In our work, we are interested to use TC as a tool to validate the soundness of MCV1 coding process. The conjecture is if the disagreement reduction is effective to enhance the coding integrity of MCV1, we should be able to detect the improvement in TC experiment results. In our experiments, only titles and abstracts were used. The standard text processing procedures were applied, including stop words and punctuations removal, stemming and *tfidf* weighting [1, 11]. We used the state-of-the-art algorithm SVM [13], in particular its implementation SVM^{Light} [5]. The dataset for each category was formed in a typical one-against-all manner. The classic F_1 metric defined as $F = 2pr/(p+r)$ was adopted to measure the TC performance based fivefold cross validation, where p and r are precision and recall respectively [1]. For the sake of simplicity, we upgraded all document labels to CXX, i.e. 18 major categories.

Each operator's results, before and after the coding discussion, were applied separately as the final labels for MCV1. For the convenience of reading, we name F_1CO_X-bfr as the F_1 value of coding operator X before the discussion, and F_1CO_X-aft as the F_1 value of coding operator X after the discussion. Table 6 shows the details of performance difference of TC experiment before and after the coding discussion.

It is now clear that before the coding discussion begins, the results of CO_4 have led to the worst performance compared to other three colleagues. Furthermore, we consider two F_1 values are significantly different if their difference is more than 0.1, i.e. 10%; there are totally nine categories where the CO_4's performance is then significantly lower than the average of rest three operators. This finding supports our exclusion of CO_4's results, and hence, F_1CO_4-aft was not reported. After the coding discussion, the performance of CO_1, CO_2, and CO_3 have been largely improved. For CO_2, the joint discussion has greatly boosted the performance by 8%, while others still managed to achieve marginally 2% increase.

Table 6. Details of coding quality improvement, in terms of macro-averaged F_1

F_1CO_1-bfr	F_1CO_2-bfr	F_1CO_3-bfr	F_1CO_4-bfr
0.6158	0.5585	0.6159	0.4807
F_1CO_1-aft	F_1CO_2-aft	F_1CO_3-aft	F_1CO_4-aft
0.6376	0.6387	0.6358	NA

Table 7. *p*-value of each operator against their own performance before coding discussion

Operator	*n*	*k*	*p*-value
CO_1	17	12	$p(Z>=12) = 0.0717$
CO_2	17	16	$p(Z>=16) = 0.137\text{-E3}$
CO_3	18	13	$p(Z>=13) = 0.0481$

In order to determine whether the performance gained after the coding discussion is significant, we performed the macro sign test (S-test) on the paired F_1 values for individual categories of CO_1, CO_2, and CO_3 [14]. We considered two F_1 values were the same if their difference was not more than 0.01, i.e. 1%. As shown in Table 7, it confirms the improvement.

Table 8 reports the experimental study that any two operators' results and all three operators' results were applied as the final labels respectively and their corresponding TC performances. In both, CO_4 was excluded. We name F_1CO_{XY} as the F_1 value by applying the results of coding operator X and Y, where XY = 12, 13, and 23, and F_1CO_{XYZ} as the F_1 value by applying three operators' results. For the convenience of reading, we also include the results from Table 6, named as F_1CO_X and X = 1 to 3.

Table 8. The details of coding economy experiment

One Operator	F_1CO_1	F_1CO_2	F_1CO_3
	0.6376	0.6387	0.6358
Two Operators	F_1CO_{12}	F_1CO_{13}	F_1CO_{23}
	0.6379	0.6557	0.6359
Three Operators	F_1CO_{123}		
	0.6729		

We observe that the combination of three operators' results has helped SVM generate the best performance so far. Its overall performance, i.e. average F_1 across 18 categories, has been increased with another 3.4% compared to the best performance of each individual. This supports that combining quality results from various operators is an effective step to further boost the coding quality, though each has already delivered a fairly good work. Meanwhile, we also noted that the combination of CO_1 and CO_3 can lead to a performance which is very close to F_1CO_{123}. This implies that combining two best operators' results is able to achieve close-to-best results.

Our experimental study reflects that a joint discussion with at least one more colleague from the same domain is always beneficial and has thus been suggested as an effective means to promote the coding agreement. We also realize that if we can track the coding history of different operators, or be able to tell their coding quality before we start the project, it is possible to code the corpus in a more efficient way without sacrificing the coding quality. One solution is to test operators first over a trial set and then assess their performance. By doing so, we can detect and exclude those operators with abnormal patterns. Subsequently, we can form *k* groups with at least two selected operators each. Each pair of operators can take care of one *k*-th of the collection. As a result, the coding rate can be greatly accelerated.

6 Conclusion

In this paper, we have reviewed the necessity of creating a domain centric corpus for industry companies to undertake knowledge discovery and management initiatives. An example approach using manufacturing texts is demonstrated with all relevant concerns, e.g. inputs, coding policies, the process and quality measurement etc, explained and discussed. The real-world classification experiments conducted using the corpus built show the soundness of its coding process. This study has also made suggestions on building document collections in a more cost-effective way.

References

1. Baeza-Yates, R., Ribeiro-Neto, B.: Modern information retrieval. Addison-Wesley Longman Publishing Co., Inc, Boston, MA, USA (1999)
2. Fayyad, U.M., Piatetsky-Shapiro, G., Smyth, P.: From data mining to knowledge discovery: an overview. In: Fayyad, U.M., Piatetsky-Shapiro, G., Smyth, P. (eds.) Advances in knowledge discovery and data mining. American Association for Artificial Intelligence, Menlo Park, CA, USA (1996)
3. Hearst, M.A.: Untangling Text Data Mining. In: Proceedings of ACL'99, the 37th Annual Meeting of the Association for Computational Linguistics, invited paper (1999)
4. Hersh, W., Buckley, C., Leone, T.J., Hickam, D.: OHSUMED: an interactive retrieval evaluation and new large test collection for research. 17th annual international ACM SIGIR conference on Research and development in information retrieval (SIGIR'94) (1994)
5. Joachims, T.: Text categorization with Support Vector Machines: Learning with many relevant features. In: Nédellec, C., Rouveirol, C. (eds.) Machine Learning: ECML-98. LNCS, vol. 1398, Springer, Heidelberg (1998)
6. Lewis, D.D., Yang, Y., Rose, T.G., Li, F.: RCV1: a new benchmark collection for text categorization research. Journal of Machine Learning Research 5, 361–397 (2004)
7. Mitchell, T.M.: Machine learning and data mining. Communications of the ACM 42, 30–36 (1999)
8. Rennie, J.D.M., Shih, L., Teevan, J., Karger, D.R.: Tackling the Poor Assumptions of Naive Bayes Text Classifiers. In: Proceedings of the Twentieth International Conference on Machine Learning (ICML) (2003)
9. Rose, T., Stevenson, M., Whitehead, M.: The Reuters Corpus Volume 1 - from Yesterday's News to Tomorrow's Language Resources. the third international conference on language resource and evaluation (2002)
10. Rose, T., Whitehead, M.: Private communication: RCV1 building (2003)
11. Sebastiani, F.: Machine Learning in Automated Text Categorization. ACM Computing Surveys (CSUR) 34, 1–47 (2002)
12. Ulrich, K.T., Eppinger, S.D.: Product Design and Development, 2nd edn. McGraw-Hill, New York, USA (2000)
13. Vapnik, V.N.: The Nature of Statistical Learning Theory, 2nd edn. Springer, New York (1999)
14. Yang, Y., Liu, X.: A re-examination of text categorization methods. In: Proceedings of the 22nd annual international ACM SIGIR conference on Research and development in information retrieval (1999)

Intelligent Decision Support System for Evaluation of Ship Designers

Sylvia Encheva[1], Sharil Tumin[2], and Maryna Z. Solesvik[1]

[1] Stord/Haugesund University College, Bjørnsonsg. 45, 5528 Haugesund, Norway
sbe@hsh.no
[2] University of Bergen, IT-Dept., P.O. Box 7800, 5020 Bergen, Norway
edpst@it.uib.no

Abstract. In this paper we propose application of non-classical logic in an intelligent decision support system. Decision making rules an intelligent agent is applying for evaluating a ship designer's reliability are discussed in particular.

Keywords: Shipbuilding, non-classical logic, intelligent agents.

1 Introduction

A very important initial stage of a shipbuilding process is the design of a merchant ship, where the vessel design has always been of collaborative nature. A ship owner and a ship designer are the two parties with the most serious influence on the design development.

The shipping company announces first a tender for making project documentation and sends it to several design agents. Interested naval architects prepare outline specification for the vessel, general arrangement plan, the quotation for classification and drawings along with a rough delivery schedule. The shipping company chooses a suitable design agent, and negotiates a contract to produce a design.

In order to further improve the efficiency of a shipbuilding process we propose use of an automated decision support system. Most automated decision support systems are based on binary logic, i.e. a responce is either positive or negative. One of their disadvantages is that they do not treat incomplete or inconsistent information. Application of many-valued logic allows the system to handle situations with inconsistent and/or incomplete input. In this paper we present decision making rules an intelligent agent is applying for evaluating a ship designer's reliability.

The rest of the paper is organized as follows. Related work and statements from many-valued logic may be found in Section 2 and Section 3 respectively. The main results of the paper are placed in Section 4. Use of a bilattice for two sources ordered reconciliation is presented in Section 5. The system architecture is described in Section 6. The conclusion is placed in Section 7.

B. Apolloni et al. (Eds.): KES 2007/WIRN 2007, Part I, LNAI 4692, pp. 551–557, 2007.

2 Related Work

Inspired by the Aristotle writing on propositions about the future - namely those about events that are not already predetermined, Lukasiewicz has devised a three-valued calculus whose third value, $\frac{1}{2}$, is attached to propositions referring to future contingencies [13]. The third truth value can be construed as 'intermediate' or 'neutral' or 'indeterminate' [16], [14], and [15].

The semantic characterization of a four-valued logic for expressing practical deductive processes is presented in [2]. In most information systems the management of databases is not considered to include neither explicit nor hidden inconsistencies. In real life situation information often come from different contradicting sources. Thus different sources can provide inconsistent data while deductive reasoning may result in hidden inconsistencies. The idea in Belnap's approach is to develop a logic that is not that dependable of inconsistencies. The Belnap's logic has four truth values 'T, F, Both, None'. The meaning of these values can be described as follows:

- an atomic sentence is stated to be true only (T),
- an atomic sentence is stated to be false only (F),
- an atomic sentence is stated to be both true and false, for instance, by different sources, or in different points of time (Both), and
- an atomic sentences status is unknown. That is, neither true, nor false (None).

Extensions of Belnap's logic are discussed in [5] and [12].

Two kinds of negation, weak and strong negation are discussed in [17]. Weak negation or negation-as-failure refers to cases when it cannot be proved that a sentence is true. Strong negation or constructable falsity is used when the falsity of a sentence is directly established.

Logic in preference modeling is discussed in [3], [11], and [14]. In [6] it is shown that additional reasoning power can be obtained without sacrificing performance, by building a prototype software model-checker using Belnap logic.

Python applications are known for increasing overall efficiency in the maritime industry [9].

LAMP is a collective name for the tools of Linux, Apache web server, MySQL database application, PHP scripting language, Perl programming language, and Python programming language. They have the advantage of being freely available, easily configured, and robust. They are a subject of constant development and improvement and are well known to be easily deployed, fully configured, and maintained with minimal efforts. The LAMP tools assist developers to do creative work without being bothered by administrative details.

3 Preliminaries

Contradicting information in classical logic, modal logic, intuitionistic logic entails any arbitrary sentence. The principle is known as 'ex falsum quod libet'(from the false whatever you like). When real application is concerned it is quite common to end up in situations where information are obtained from various inconsistent sources. Solutions for such applications are proposed by a number of alternative systems.

A lattice is a partially ordered set, closed under least upper and greatest lower bounds:

- the least upper bound of α and β is called the join of α and β, and is sometimes written as $\alpha + \beta$,
- the greatest lower bound is called the meet and is sometimes written as $\alpha\dot{\beta}$.

A billatice is a set equipped with two partial orderings $\leq_t$ and $\leq_k$.

- The t partial ordering $\leq_t$ means that if two truth values ϕ, ψ are related as $\phi \leq_t \psi$ then ψ is at least as true ϕ.
- The k partial ordering $\leq_k$ means that if two truth values ϕ, ψ are related as $\phi \leq_k \psi$ then ψ labels a sentence about which we have more knowledge than a sentence labeled with ϕ.

4 Application

An intelligent agent is sending quarries to two independent databases about the reliability of a ship designer via Web services. What should the agent recommend if the responses are for example `'reliable, unreliable'` or `'reliable, no answer'` ?

We propose the following:

a) The responses are {reliable, reliable}. The ship designer is recommended.
b) The responses are {reliable, no answer}. The agent should ask the opinion of a third database.
c) The responses are {reliable, unreliable}. The agent should inquire about the reasons in the database with a negative response and then asks the opinion of a third database.
d) No response from any of the two companies. The agent should find two new databases and consider their responces.
e) The responses are {unreliable, no answer}. The agent should ask the opinion of a third database and inquire about the reasons in the database with a negative response.
f) The responses are {unreliable, unreliable}. The agent should recommend another ship designer.

If at least one of the responces in the second round is of the type {unreliable} or {no answer} the agent then starts sending inquiries about a new ship designer.

5 Ordered Sources

Suppose a shipowner wants to hire a ship designer chosen among five candidates (see Table 2). The shipowner's preferences include

- regular practice in the maritime business, or
- design of modern vessels and interest in platform supply vessels.

Table 1. Notations

Notations	Meaning
1	reliable
O	unreliable
U	no information is available
$\top$	contradiction

The notations used in Table. 2 are described in Table. 1.

Table 2. Designer's reliability

	Regular practice in the maritime business	Design of modern vessels	Interest in platform supply vessels
Designer 1	1U	O1	0U
Designer 2	11	11	1U
Designer 3	0U	UO	U1
Designer 4	UO	U1	11
Designer 5	UU	UU	UU

The results for every tuple in Table 2 are presented in Table 3, where $\oslash$ denotes 'or' and $\odot$ denotes 'and'.

So far we have been working with unordered sources of information. If the opinion of the first source has more weight than the one of the second source we propose use of a bilattice (see Fig. 1) for two sources ordered reconciliation. By more weight we mean that if the first source says *reliable*, this opinion has more value than the same opinion expressed by the second source.

Table 3. Summary of designers' qualifications

	Regular practice in the maritime business (*)	(*) ⊘ (**)
	⊘	⊙
	Design of modern vessels (**)	Interest in platform supply vessels
Designer 1	11	0U
Designer 2	11	1U
Designer 3	UU	UU
Designer 4	U1	U1
Designer 5	UU	UU

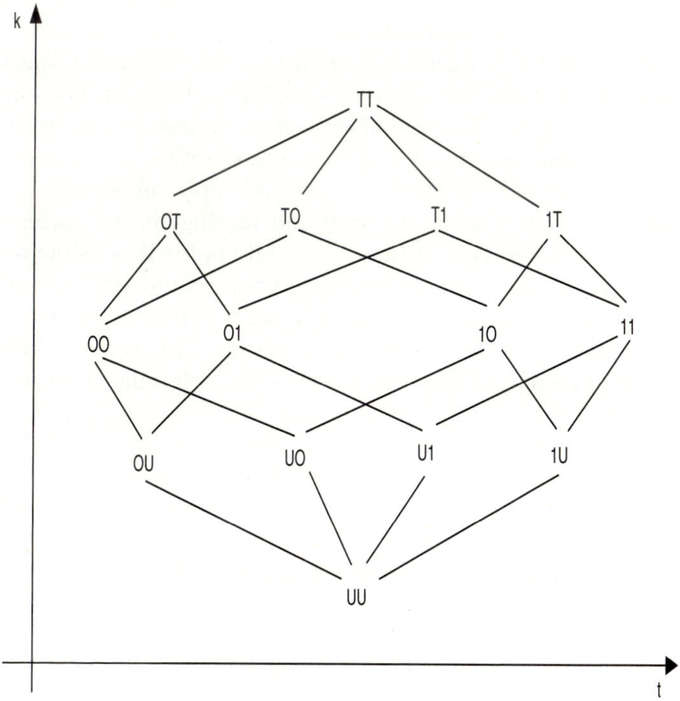

Fig. 1. Bilattice for two sources ordered reconciliation

6 System Architecture

The system implementation uses the so-called LAMP Web application infrastructure and deployment paradigm. It is a combination of free software tools on a Linux operating system of an Apache Web server, a database server and a programming environment using scripting language.

Implementers can choose and mix these tools freely. This in contrast to commercial Web application platforms like for example, WebSphere from IBM [10], JavaServer from Sun [8], and ASP.net from Microsoft [7].

The Web deployment in our system is

- an Apache front end Web server,
- an application middleware for dynamic content, data integration and users' administration written in Python scripting language, and
- a backend database based on lightweight SQLite database engine.

Apache Web server is a robust and extendable Web server. In our implementation, the Web server is extended with a Python interpreter by using 'mod_python' module. A SQLite database engine is a capable relational database engine. It is comparable to MySql and PostgreSQL, but more lightweight and zero administration cost. SQLite does not administer its own user and access control, it uses an operating system file protection mechanism.

This traditional three-tiers Web deployment is joined together with a service support sub-system. A communication framework based on JSON remote procedure call (JSON-RPC) written in Python is used to connect the Web server middleware and the Web application server together. JSON stands for JavaScript Object Notation and it is a lightweight data-interchange format. It is more compact then XML without sacrificing expressiveness. JSON structure is perfect for packaging and sending data in RPC request and reply messages.

The application server provides search and intelligent evaluation services to the Web server. The separation of these two units made it possible to modularly design and implement the system as loosely coupled independent sub-systems. The purpose of the search agent is to search for different reviews from independent reviewers about a particular ship designer. This process will eventually build a database of ship designers' capabilities and reliability reviewed by different reviewers.

By providing a client Web interface, the system invites reviewers to submit their reviews of ship designers they have had experience working with. The user authenticator and user profiler modules play an important role in controlling every particular user, client or administrator authenticity. Only valid reviewers can submit reviews. The administrator can approve the results of a search agent before the data is submitted to the database.

The purpose of the intelligent evaluator is to rank the ship designers' capability and reliability at any one time in response to users' queries.

The Web server's middleware and the application server's software agents can run in parallel, independently of each other. As such, they can be situated on different servers. The middleware implements the Web user interface side of the system while the software agents implement the evaluation side of decision process.

7 Conclusion

This paper presents an intelligent system assessing reliability of a ship designer. The decision making process is based on a many-valued logic. Similar sub-systems can be build up to evaluate the key points in the process of shipbuilding - dimensions, hydrodynamic performance, speed, stability, seakeeping, cargo carrying capacity, propulsion systems, passengers and environment safety standards, and fuel consumption. These sub-systems can be used as building blocks of a complete system that will considerably speed up the process of shipbuilding.

References

1. Belnap, N.J.: How a computer should think. In: Contemporary Aspects of Philosophy. Proceedings of the Oxford International Symposia, Oxford, GB, pp. 30–56 (1975)
2. Belnap, N.J.: A useful four.valued logic. In: Dunn, J.M., Epstain, G. (eds.) Modern uses of multiple-valued logic, pp. 8–37. D. Reidel Publishing, Dordrecht (1977)
3. Briges, D.S., Mehta, G.B.: Representations of preference orderings. Springer, Berlin (1995)
4. Davey, B.A., Priestley, H.A.: Introduction to lattices and order. Cambridge University Press, Cambridge (2005)
5. Font, J.M., Moussavi, M.: Note on a six valued extension of three valued logics. Journal of Applied Non-Classical Logics 3, 173–187 (1993)
6. Gurfinkel, A., Chechik, M.: Model-Checking Software with Belnap Logic. Technical Report 470. University of Toronto (April, 2005)
7. `http://msdn2.microsoft.com/en-us/asp.net/default.aspx`
8. `http://java.sun.com/products/jsp/`
9. `http://www.python.org/about/success/tribon/`
10. `http://www-306.ibm.com/software/websphere/`
11. Kacprzyk, J., Roubens, M.: Non Conventional Preference Relations in Decision Making. Lecture Notes in Economics and Mathematical Systems (LNEMS), vol. 301. Springer, Berlin (1988)
12. Kaluzhny, Y., Muravitsky, A.Y.: A knowledge representation based on the Belnap's four valued logic. Journal of Applied Non-Classical Logics 3, 189–203 (1993)
13. Lukasiewicz, J.: On Three-Valued Logic. Ruch Filozoficzny, 5 (1920), English translation in Borkowski, L. (ed.) 1970. Jan Lukasiewicz: Selected Works. Amsterdam: North Holland (1920)
14. Perny, P., A. Tsoukias, A.: On the continuous extension of a four valued logic for preference modelling. In: Proceedings of the Information Processing and Management of Uncertainty (IPMU) conference, Paris, pp. 302–309 (1998)
15. Priest, G.: An Introduction to Non-Classical Logic. Cambridge (2001)
16. Sim, K.M.: Bilattices and Reasoning in Artificial Intelligence: Concepts and Foundations. Artificial Intelligence Review 15(3), 219–240 (2001)
17. Wagner, G.: Vivid Logic: Knowledge Based reasoning with two kinds of negation. In: Wagner, G. (ed.) Vivid Logic. LNCS (LNAI), vol. 764, Springer, Heidelberg (1994)

Philosophy Ontology for Learning the Contents of Texts

Jungmin Kim[1] and Hyunsook Chung[2,*]

[1] School of Computer Engineering, Seoul National University, Korea
jmkim@idb.snu.ac.kr
[2] Department of Computer Engineering, Chosun University, Korea
hsch@chosun.ac.kr

Abstract. In this paper, we develop a large ontology, which is an explicit formal specification of concepts and semantic relations among them in philosophy. Our philosophy ontology is a formal specification of philosophical knowledge including knowledge of contents of classical texts of philosophy. The philosophy ontology includes contents knowledge of 72 texts of philosophy but will be expanded to almost all the texts of oriental and western philosophy which are considered as essential philosophy texts by experts. It organizes a text-based knowledge base, which includes electronic texts and philosophy articles. The philosophy ontology provides not only easy concepts to novices and students but also more specialized detailed concepts from the contents of texts to domain experts.

Keywords: Philosophy Ontology, Text-based Ontology, Topic Maps Application.

1 Introduction

In this paper, we present a large ontology which conceptualizes knowledge in the philosophy domain. We call it a philosophy ontology. Our philosophy ontology is composed of many small ontologies and each of them represents pieces of philosophical knowledge, such as philosophers, texts of philosophy, terms of philosophy, doctrines of philosophy, schools of philosophy, and so on. The outstanding characteristics of the philosophy ontology are externalization, formalization, and specification knowledge existing within the contents of texts. The philosophy ontology enables people to understand the meaning of main concepts within the contents of texts and semantic relations between them without reading the whole texts.

Ontology building is a labor-intensive work, thus, it requires an engineering method to formalize the building process and activities in each of steps[1]. In this paper, we present a methodology for building the philosophy ontology, which provides a standard building process, detailed activities, and the products to be output at each step for maintaining consistency and validation of ontologies within and between development teams. The building process of the philosophy ontology has 3 major steps like planning, conceptualization, and implementation and 14 minor steps within them.

* Corresponding author.

B. Apolloni et al. (Eds.): KES 2007/WIRN 2007, Part I, LNAI 4692, pp. 558–566, 2007.

We design a three-layered architecture for implementing the philosophy ontology, which is composed of Philosophy Reference Ontology(PRO), Philosophy Domain Ontology(PDO), and Philosophy Text Ontology(PTO). Because the philosophy ontology conceptualizes not only philosophical knowledge, but also textual information, PDO presents philosophical domain knowledge and PTO presents textual philosophical knowledge. Philosophical domain knowledge is general knowledge such as who Immanuel Kant is, who the author of "*German Ideology*" is, and what the relationships between Kant and Hegel are. Textual philosophical knowledge is implicit knowledge that can be learned through reading the whole texts including what practical reason means, and the relationships between practical reason and free will. PRO is the upper-level ontology, which provides a schema and several templates for maintaining consistency of PDO and PTO.

For convenience in creating ontologies with a formal language, we develop a semi-automatic translator for creating XTM(XML Topic Maps)[5] documents from the ontology specification, which is created during conceptualization of the philosophy ontology. XTM is the standard formal language that defines XML syntax for Topic Maps[2] and gives a specific Topic Map data model. A Topic Map is defined as a collection of Topic Map documents, thus, we should write XTM documents which correspond with philosophy ontology specifications.

The contributions of the philosophy ontology built in this research are as follows. Firstly, it conceptualizes and externalizes knowledge of text contents to catch main concepts without reading the texts. Secondly, it provides guidelines for developing text-based ontologies in other learning domains such as, history, language, art, and so on. Thirdly, it can be used as a knowledge map of a digital library, knowledge portal, or document management system used to retrieve and explore semantic information.

2 Related Work

Philosophy web pages, which can be accessed through the Internet, can be classified into the following four types; (1) The philosophy text sites[6], which provide contents of texts written in the original language or translated into English. These websites are useful to people who read computer texts and extract some paragraphs from them. (2) Philosophy journal sites[4], which provide papers and philosophy articles directly or as hyperlinks. These websites support publication and reference of philosophy papers on the web for easy access. (3) Philosophy reference sites[8], which provide philosophy dictionaries or encyclopedias. These websites are useful in obtaining descriptions and explanations of philosophical terms, philosophers, texts of philosophy, philosophical subjects, and so on. (4) Philosophy meta sites[7], which provide directories for accessing philosophy-related websites. These websites introduce information and metadata of texts regarding philosophy, journals of philosophy, organizations of philosophy, and topics of philosophy.

The philosophy ontology here provides not only full contents of texts but also semantic knowledge such as concept thesaurus and semantic-based associations among philosophy concepts. This method also constructs a semantic network through conceptualizing and interconnecting concepts.

3 Knowledge Acquisition

For achieving two levels of granularity of the philosophy ontology knowledge acquisition, two techniques, domain knowledge analysis and formal text analysis, are used. Domain knowledge analysis is used to extract general knowledge from philosophy resources on the Internet, philosophy dictionaries, and encyclopedias. Formal text analysis is used to find and externalize semantic knowledge from the contents of texts.

3.1 Domain Knowledge Analysis

General knowledge in the philosophy ontology can be classified into the following six categories; philosophers, texts of philosophy, terms of philosophy, branches of philosophy, schools of philosophy and doctrines of philosophy. For each category we selected the essential instances from collection of philosophical resources because philosophy consists of a vast knowledge resource from which we cannot extract all knowledge of philosophy. For example, we selected only well-known and influential philosophers according to the period and the geographical criteria, such as Yulkok, Wonhyo, and Yi Hwang as Korean philosopher, Confucius, Laozi, and Mencius as Chinese philosopher, Plato, Socrates, and Aristotle as Ancient western philosopher, Immanuel Kant and Hegel as Modern western philosopher, and so on. But we will expand the philosophy ontology to include more instances continuously because building ontology is long-term project.

For each instance we acquire objective facts rather than subjective arguments from the collection of philosophical resources because knowledge in the philosophy ontology should be acceptable to most of domain experts. For example, we examine philosophical resources to acquire knowledge of philosophers with the following questions.

- What are his original name, English name, and Korean name?
- What is his biography?
- What are his main ideas?
- What are his active fields, schools, branches?
- What are his writings?
- Who are philosophers related with him?

3.2 Text Analysis

Domain experts analyze texts related to their major research fields. First of all, they look for basic information of the allocated texts, such as original title, Korean title, English title, author(s), published date, and so on. Afterward, they analyze their texts to answer the following questions.

- What are the main philosophical subjects(or issues)?
- What are the arguments of author(s)?
- What are the important philosophical terms included in the texts?
- What are the philosophers related with the contents of the texts?
- What are the texts of philosophy related with the contents of the texts?
- What are the branches of philosophy related with the contents of the texts?

- What are the schools of philosophy related with the contents of the texts?
- What are the doctrines of philosophy related with the contents of the texts?

The main philosophical subjects are used for identifying the central concepts, which are importantly described through the whole text by author, of the text. These concepts are specialized with more specific concepts during conceptualization. The arguments of author(s) are also used to identify the central concepts of the text. Domain experts should identify not only philosophers, texts, terms, branches, and schools related with the contents of the texts, but also what types of relations are described in the contents of them.

4 Conceptualization

Our conceptualization process starts by extracting and naming of concepts from texts. After naming of concepts, definition of concept hierarchy, property, and associations follow.

4.1 Concept Naming

It is important to get a comprehensive list of terms that correctly represent the intended meaning of concepts without any ambiguities. Domain experts are responsible for selecting or inventing adequate terms for concepts. They need a naming convention to make consistent naming among them. Table 1 shows our naming convention for concepts in the philosophy domain ontology. During concept naming step, we build a glossary of terms including all terms of the philosophy ontology and their descriptions.

Table 1. Naming convention for philosophy domain concepts

Rule	Rule description	Example
Naming form	Noun, proper noun or complex noun is used	Philosopher
Singular or plural	Both can be used. But it should be consistent	Doctrines of philosophy
Capitalization	First character should be capitalized	Korean Philosopher
Delimiter	Space is used rather than underscore or dash	Pure Reason,
Uniqueness	All terms should be unique	Reason in the oriental
Length	Basically length of a term is not limited but it needs to decide max length of a term.	
Abbreviation	Using abbreviated term is not recommended. Use full name as possible	

4.2 Hierarchy Structure Definition

There are three methods in developing concept hierarchy. These are defined as the top-down approach, bottom-up approach, and middle-out approach[3]. According to the top-down approach the top-most concept is *philosophy* in the philosophy ontology. Then, the philosophy concept is specialized by creating six sub-concepts: philosopher, text of philosophy, term of philosophy, branch of philosophy, doctrine of philosophy, and school of philosophy. We categorize philosophy domain concepts by these, because these sub-concepts are parts of the philosophy concept. Subsequently, the philosopher concept has seven sub concepts, identified in terms of period and geographic viewpoint: Korean philosopher, Chinese philosopher, Indian philosopher, Ancient western philosopher, Medieval western philosopher, Modern western philosopher, and Contemporary western philosopher. There is an "is-a" relationship, between the philosopher concept and its seven sub-concepts.

A middle-out approach should be used to develop a concept hierarchy of the text ontology because a text describes particular subjects and includes specific concepts rather than general concepts. For example, "*Critique of Practical Reason*" written by Kant describes several philosophical issues, practical reason, autonomy, moral law, free will, and so on. These issues are the main concepts that are emphasized in the text. After acquiring the main concepts, more general concepts are derived from the main concepts using a bottom-up approach. General concepts must be used to collect semantically related main concepts. For example, "*reason*" concept aggregates specific concepts, which describe the issues related with reason. These specific concepts are described in the different texts of philosophy, such as "*pure reason*" and "*practical reason*" of Kant, "*social reason*" of Hegel, and so on.

Concept Hierarchy of Philosophy Text Ontology

Text : The Critique of Practical Reason(Immanuel Kant)

CID	Concepts Hierarchy	Resource ID
c1	Practical Reason	
c11	Speculative reason and practical reason	P1.1.2.1
c111	Equivalence of speculative reason and practical	P1.1.3.1
c112	reason	
c113	Use of reason	P1.1.3.5
c1131	Speculative use of reason	P1.1.3.8
c1132	Practical use of reason	P1.1.4.1
c1133	Comparing speculative use to practical use	P1.1.4.3
c12	Superiotiry of practical reason	P1.1.5.1
c121	Semantic of superiority	P1.1.5.2
c122	When speculative reason has superiority	P1.1.5.6
c123	When practical reason has superiority	P1.1.6.5
c2	Moral Law	
c21	Origin of moral law	P1.2.1.3
c22	Moral law and reason	P1.2.2.1
c23	Phase of moral law	P1.2.3.3
c24	Moral law and happiness	P1.2.4.1
c241	Critique of moral sensibility	P1.2.4.4
c242	Principles of self-happiness	P1.2.4.9
c243	Difference between principles of self-happiness	P1.2.5.11
c244	and moral law	P1.2.4.15

Fig. 1. Hierarchy structure of philosophy text ontology

We identify general concepts through analyzing the structure of a main concept name. If a main concept's name was a complex noun or noun phrase, we try to find a noun where it's meaning was restricted by other adjectives or nouns. For example, we know that "*practical*" restricts the meaning of "*reason*" from "*practical reason*", and "*practical reason*" is a specific meaning of the "*reason*" concept. In addition, more specific concepts are derived from the main concepts using a top-down approach. These specific concepts organize sub trees rooted by each of the main concepts. Specialized concepts are extracted from analyzing the contents of those pages, because a main concept is explained over many pages. The leaves of text ontology are specific concepts, representing the meaning included in one or more paragraphs because a paragraph is a meaningful unit that can be summarized and conceptualized by domain experts.

Fig.1 shows an example of a hierarchical relation of a main concept and its specialized concepts. In this figure, CID represents the concept identifier, which may be manually specified by an ontology implementer or automatically generated by an ontology management system. Resource ID indicates one or more paragraphs in a text resource, referenced by a concept.

4.3 Property Definition

A concept must consist of one or more properties describing the meaningful attributes which belong to it. The properties of concepts are defined and documented in the concept dictionary, depicted in Fig.2. In this figure, the philosopher concept has the following properties: English name, original name, biography, biographical sketch, figure, and so on. Similar to a class, all sub-concepts of a concept inherit the properties of that concept. For example, all the properties of the concept philosopher will be inherited to all sub-concepts of philosopher, including Korean philosopher, Chinese philosopher, and Indian philosopher. We will add additional properties, hanja (Chinese character) name, to the Korean philosopher and Chinese philosopher.

CID	philosopher			**LEVEL**	1.1
Concept Type	Class				
Concept Name	Korean name: 철학자		English name: Philsopher		
Super Class	philosophy				
Sub Class / Instance	kor_philosophy anc_philosophy				
Properties	Name	Data / Reference	Cardinality	Domain	Attribute Type
	Korean name	Data	1	String	Person name
	English name	Data	1	String	Person name
	Web page	Reference	0..n	String	Web resource
Relationships	Name	Origin	Ori_Role	Destination	Dest_Role
	author of	philsopher	author	phil_text	book
	teacher of	philsopher	teacher	philsopher	student
	has main idea	philsopher	creator	phil_doctrines	doctrine

Fig. 2. A concept's definition in the concept dictionary

All concepts of the philosophy text ontology have two different kinds of properties, explanation and quotation. A concept's explanation property is a description of the intended meaning of the concept and the quotation property is a reference to particular paragraphs within philosophy texts. An explanation property is divided into an internal explanation and external explanation. An internal explanation is a short description, which exists inside the text ontology, and the external explanation is a reference to certain paragraphs of explanation articles, which are written by domain experts through analyzing philosophy texts. Explanation articles are used to provide more understandable information to novices in the philosophy domain.

4.4 Association Relationship Definition

An association is a binary relation between two concepts, such as "*synonym*", "*disjoint of*", "*author of*", "*contribute to*", "*pupil of*", and so on. An association is a kind of semantic information-based relation that explains how a concept would be related to another concept. An association can be classified into two types, explicit and implicit. An explicit association is an obvious relationship that domain experts can identify and specify with ease. For example, the association, "*author of*", is established between a philosopher and a text and the association, "*pupil of*", is established between one philosopher and another philosopher. An implicit association is a hidden relationship that can be identified from analyzing revealed explicit associations. For example, the association, "*author of*" has an inverted relationship, "*written by*" and if the association, "*opponent of*", exists between text A and text B, the same relationship is established between philosopher A and philosopher B, who are the authors of both text A and B.

In the philosophy text ontology, an association is defined to represent the semantic relationship between concepts in a text or different texts. We define semantic relations which exist between concepts in the text ontology, such as "*be identical to*", "*be opposed to*", "*complementary to*", "*sequence to*", "*cause and result*", and so on. Fig.3 shows the concept association table of the text ontology.

Association of Philosophy Text Ontology

Association	Concept1	Concept2
be identical to	Socialism theory	Communism theory
be identical to	Historic materialism	Materialistic view of history
be opposed to	Historic materialism	Ideological view of history
assertion and foundation	Importance of materialistic production activity	Foundation of societary life
be identical to	Doctrine of science	Positivism
prescription of	Foundation	Social structure
mutual action	Productivity	Production

Fig. 3. Association table of the philosophy text ontology

In contrast to associations of domain ontology, text ontology's associations need to explain the meaning of semantic relationships. For example, we can describe why historic materialism is opposed to the ideological view of history, from fig.3. These associations are called reified associations, which posses explanation and quotation properties to describe the meaning of semantic relations.

5 Implementation

After conceptualization, ontology implementers should design ontology structure and specify the conceptualized ontology using a machine-understandable formal language such as RDF, OWL, or Topic Maps. We use Topic Maps to specify our philosophy ontology. Topic Maps are more appropriate to represent ontologies for knowledge and information management because RDF/S is URI resource-oriented description scheme but Topic Maps are subject-oriented description scheme. Topic Maps can describe semantic information of subjects without their resources.

All philosophy ontology concepts are translated into topics of Topic Map documents. Properties are translated into topics and occurrences because a property itself is a

Table 2. The statistics of the philosophy ontology

# of Texts	# of Topics	# of Topic types	# of Occurrences	# of Assoc. types	# of Associations
72	12055	4315	50406	21	11819

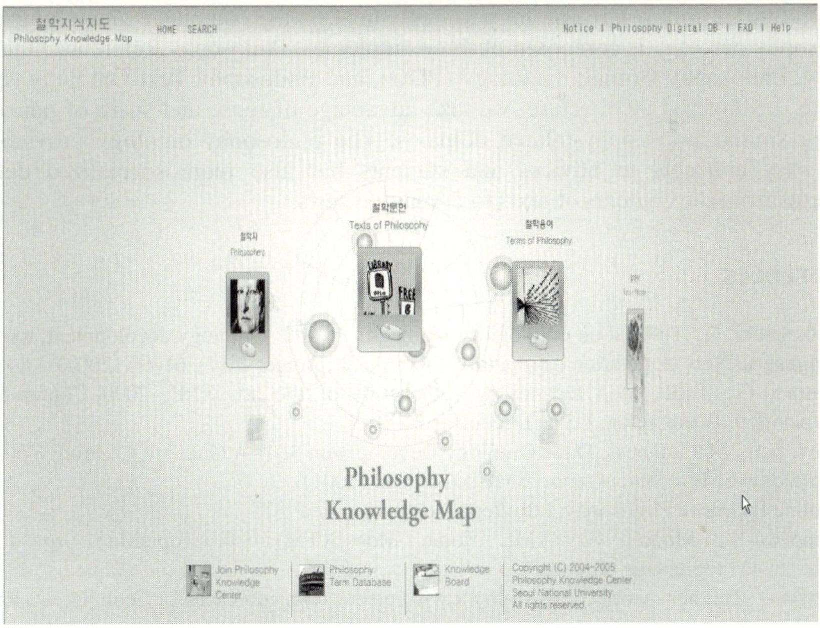

Fig. 4. Philosophy knowledge portal based on the philosophy ontology

concept and specifies a certain value to attribute of a concept. We developed a semi-automatic translator for creating XTM documents from an ontology specification, which is created during conceptualization of the philosophy ontology using the above matching rules. Input data for translation is ontology specification and templates.

Templates are composed of the following seven files: *philosopher_template, phil_text_template, phil_term_template, phil_branch_template, phil_doctrine_template, phil_school_template,* and *phil_textcontent_template*. First, our translator creates uncompleted XTM documents, which have empty properties from input data. Second, domain experts fill empty properties with identified values and manually correct the errors occurring in translated documents. Table 2 shows the statistics of the philosophy ontology.

Fig.4 shows our philosophy knowledge portal based on the philosophy ontology. Users can explore philosophy knowledge through selecting one main category displayed in the main flash.

6 Conclusion

In this paper, we described the philosophy ontology, which conceptualizes knowledge of the philosophy domain and contents of philosophy texts. The philosophy ontology includes contents knowledge of 72 texts of philosophy but will be expanded to almost all the texts of oriental and western philosophy which are considered as essential philosophy texts by experts. It organizes a text-based knowledge base, which includes electronic texts and philosophy articles. We developed a semi-automatic translator for creating XTM documents from the ontology specification, which is created during conceptualization of the philosophy ontology. We designed a three-layered architecture of philosophy ontology to implement on the computer. The implemented philosophy ontology is composed of three ontologies, Philosophy Reference Ontology (PRO), Philosophy Domain Ontology (PDO), and Philosophy Text Ontology (PTO). Due to this layered architecture, we take advantage of reuse and share of ontologies among similar philosophy-related domains. The philosophy ontology provides not only easy concepts to novices and students but also more specialized detailed concepts from the contents of texts to domain experts.

References

1. Mizoguchi, R.: Tutorial on ontological engineering-Part2: Ontology development, tools and languages, New Generation Computing. OhmSha&Springer 22(1), 61–96 (2004)
2. Moore, G.: Topic Map technology - the state of the art. XML 2000 Conference & Exposition, Washington, USA (December 2000)
3. Noy, N.F., McGuinness, D.L.: Ontology Development 101: A Guide to Creating Your First Ontology, SMI technical report SMI-2001-0880 (2001)
4. Online Papers in Philosophy, http://opp.weatherson.net/
5. Pepper, S., Moore, G.: XML Topic Maps(XTM) 1.0, TopicMaps.Org. (2001), http://www.topicmaps.org/xtm/
6. Perseus Classics Collection, http://www.perseus.tufts.edu/cache/perscoll_Greco-Roman.html
7. Philosophy in Cyberspace, http://www-personal.monash.edu.au/~dey/phil/
8. Stanford Encyclopedia of Philosophy, http://plato.stanford.edu/

Recent Advances in Intelligent Decision Technologies

Gloria Phillips-Wren[1] and Lakhmi Jain[2]

[1] The Sellinger School of Business and Management, Loyola College in Maryland,
4501 N. Charles Street, Baltimore, MD 21210 USA
gwren@loyola.edu

[2] University of South Australia, School of Electrical and Information Engineering, Adelaide,
Mawson Lakes Campus, South Australia SA 5095
Lakhmi.Jain@unisa.edu.au

Abstract. Intelligent decision technologies (IDTs) combine artificial intelligence (AI) based in computer science, decision support based in information technology, and systems development based in engineering science. IDTs integrate these fields with a goal of enhancing and improving individual and organizational decision making. This session of the 11th International Conference on Knowledge-Based & Intelligent Information & Engineering Systems (KES) presents current research in IDTs and their growing impact on decision making.

1 Introduction

Intelligent decision technologies (IDTs) combine artificial intelligence (AI) based in computer science, decision support based in information technology, and systems development based in engineering science. IDTs integrate these fields with a goal of enhancing and improving individual and organizational decision making. Rapid expansion of networks and the Internet has made increasing amounts of data available to a decision maker, often in real-time and from many different sources. A decision maker may, for example, face information overload, difficulty interpreting the information presented, lack the needed expertise to put the information into context, require collaboration or agreement with disparate parties, be time-pressured, lose perspective due to the potential impact of the decision, need to assess a problem with many uncertainties, deal with inaccurate information, require information retrieval from datasets that are difficult to access, or need to assess long-range impacts. IDTs can assist the decision maker to overcome these types of problems and improve the outcomes from the decision.

The decision making process was described by the Nobel laureate Simon [1] as consisting of three phases: intelligence, design, and choice. A fourth phase of implementation was added by later researchers. As shown in Figure 1 [2], during the intelligence phase the decision maker acquires information and develops an understanding of the problem. The design phase is the process of identifying criteria, developing the decision model, and investigating alternatives. The user selects an alternative during choice, and acts on the decision during the implementation phase. Intelligent techniques can assist the decision maker during this process. A similar decision making process is recognized by defense tactics and is called the Observe, Orient, Decide, Act (OODA) loop.

B. Apolloni et al. (Eds.): KES 2007/WIRN 2007, Part I, LNAI 4692, pp. 567–571, 2007.

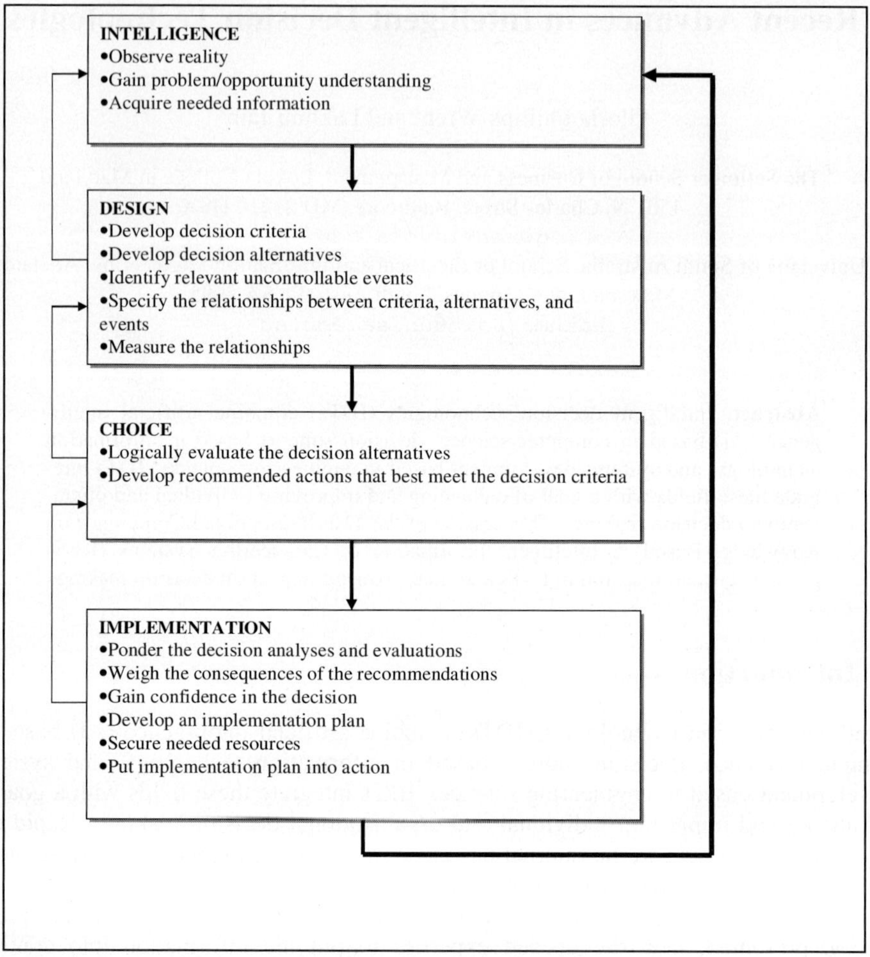

Fig. 1. The decision making process [2]

Decisions are often characterized in terms of the degree of uncertainty in the decision as structured, semi-structured, or unstructured. Structured decisions are algorithmic in nature; the decision data, criteria, and processing are generally agreed upon. At the other end of the spectrum, unstructured decisions have no agreed upon data, criteria, and processing. Semi-structured decisions fall in between these two types, and most decision support systems (DSSs) are designed to support semi-structured decisions. By supporting intelligence, the DSS helps the user acquire data. During the design phase the DSS may assist the user in developing the decision criteria and evaluating potential scenarios. The DSS may help the user select an alternative during choice. Intelligent technologies can enhance and extend the support offered by DSSs. Even if not implemented within a DSS, artificial intelligence can help the decision maker by performing tasks or reducing the cognitive load.

Papers in this session explore applications or advances in methods such as intelligent multi-agent systems that can communicate and collaborate in both competitive and cooperative situations, a classification method to classify overlapping patterns, and fuzzy queries in relational databases to retrieve needed data when the reasoning and decision making process is not well defined. Applications include improving ship design, monitoring the safety and airworthiness of airborne platforms, classification in medical applications, and fuzzy search of a DNA chain.

1.1 AI in Decision Making

The idea that computers could be programmed to rival human intelligence was made famous by Alan Turning who opened his 1950 paper with: "I propose to consider the question, 'Can machines think?'" and ends with "We can only see a short distance ahead, but we can see plenty there that needs to be done." [3]. To answer his initial question and point the way forward, Turing proposed a game in which a human interrogator would query both a human and a machine. If the interrogator could not tell the difference between answers from the two respondents, the machine was said to pass the Turing Test. John McCarthy coined the term "artificial intelligence" in 1956, and a new research field was born [4]. Although there have been many advances in and applications of AI, our interest in the topic is the support of human decision making processes and outcomes.

Recent advances in AI methods and computer technology have increased the accessibility of the techniques and need for intelligent decision making support as seen by the sharp increase in the number of applications, particularly for agent-based systems [5]. For example, *ComputerWeekly* writes, "Agent-based computing has already transformed processes such as automated financial markets trading, logistics, and industrial robotics. Now it is moving into the mainstream commercial sector as more complex systems with many different components are used by a wider range of businesses. Organisations that have successfully implemented agent technologies include DaimlerChrysler, IBM and the Ministry of Defence" [6].

AI is being used in decision support for tasks such as assessing uncertainty and risk, providing up-to-date information for decision making, enabling collaborative decisions, handling routine decisions, monitoring and alerting the decision maker as problems arise, and expanding the knowledge set to enable better decisions. The AI community is "shifting from inward-looking to outward-looking" [7], and intelligent decision support is poised for significant advancement.

1.2 Applications of AI in Decision Making

Recent published research illustrates the growing impact of AI to support decision making tasks. Several examples of practical applications from the literature are given below.

AI has been used to assist decision makers while they are developing designs such as buildings. While early research focused on AI as a revolutionary approach to aiding design processes, the most-promising current efforts use AI as "the glue that holds larger systems together using reasoning systems that represent or manage processes, information, and interaction devices that use conventional procedural programming; effectively blurring the boundaries between AI and non-AI" [8].

Research in intelligent agents has led to pragmatic systems that help the decision maker with strategic choices. For example, the multi-agent distributed goal satisfaction (MADGS) system assistes the commander on the battlefield with mission planning and execution [9]. The environment is complex, distributed, collaborative, and dynamic with competing goals and user preferences. MADGS retrieves, analyzes, synthesizes and distributes information to the decision maker in order to assist the commander with all phases of decision making – intelligence, design and choice.

Intelligent decision systems have been particularly helpful in medical applications. One such system attempts to bring best-practice in oncology to develop an individual patient's treatment [10]. The system examines objective research in order to suggest treatment.

An interesting application involves power providers [11]. The intelligent decision system is based on fuzzy set theory and includes uncertain parameters. Scenario analysis is used to compare alternatives and provide the user with alternatives as well as sensitivity analysis

Additional examples can be found in [12]. The area of intelligent decision technologies is a growth area for future research. Research is needed on artificial intelligent methods that can aid decision making, complex real-world applications, the addition of trust and reputation, collaboration, and systems development.

2 Session Papers

The following section introduces the papers presented in KES 2007. There are 2 papers in this session as described below.

2.1 Description of Session Papers

The paper by Encheva, Solesvik and Tumin [13] entitled "Intelligent Decision Support System for Evaluation of Ship Designers" describes a system to aid the design of a merchant ship by automating collaborations between the ship owner and ship designer. Rather than classical binary logic, the authors propose logic with many values to handle situations with inconsistent and incomplete input. The system aids a ship owner select a designer from among candidates.

Leng, Fyfe and Jain [14] provide a description of "Reinforcement Learning of Competitive Skills with Soccer Agents." Real-time, dynamic, uncertain environments require systems with ability to reason, learn, adapt, and possibly act autonomously. Such characteristics can be found in MASs that can communicate and collaborate in both competitive and cooperative situations. The authors investigate the efficiency of a model-free reinforcement algorithm that uses a function approximation technique known as tile coding to generate value functions. A simulation testbed is applied to test the learning algorithms in the specified scenarios.

3 Conclusion

The papers presented in KES 2007 indicate the diversity of DSS applications and the increasing maturity of artificial intelligence to support decision making. Intelligent decision technologies are an emerging research area with the potential to make significant contributions to difficult applied problems.

Acknowledgements

We appreciate the excellent contribution of the authors. The efforts of the reviewers greatly contributed to the quality of the papers and are gratefully acknowledged.

References

1. Simon, H.: Administrative Behavior, 4th edn. The Free Press, New York (1997)
2. Forgionne, G.: Decision-making support systems effectiveness: the process to outcome link. Information Knowledge-Systems Management 2, 169–188 (2000)
3. Turing, A.M.: Computing machinery and intelligence. Mind 59, 433-460 (October 1950)
4. Stanford Engineering Annual Report 2004-2005, accessed from http://soe.stanford.edu/AR04-05/profiles_mccarthy.html
5. Wooldridge, M.: An Introduction to MultiAgent Systems. John Wiley & Sons, LTD, West Sussex, England (2005)
6. Sedacca, B.: Best-kept secret agent revealed. ComputerWeekly Accessed (12 October, 2006), from http://www.computerweekly.com/Articles/2006/10/12/219087/best-kept-secret-agent-revealed.htm.
7. Mackworth, A.: The Coevolution of AI and AAAI. AI Magazine 26, 51–52 (2005)
8. Maher, M.L.: Blurring the boundaries. Artificial Intelligence for Engineering Design, Analysis and Manufacturing 21, 7–10 (2007)
9. Santos, E., DeLoach, S.A., Cox, M.T.: Achieving dynamic, multi-commander, multi-mission planning and execution. The International Journal of Artificial Intelligence, Neural Networks, and Complex Problem-Solving Technologies 25, 335–357 (2006)
10. Sissons, B., Gray, W.A., Bater, A., Morrey, D.: Using artificial intelligence to bring evidence-based medicine a step closer to making the individual difference. Medical Informatics and the Internet in Medicine 32, 11–18 (2007)
11. Gustave, N., Finger, M.: A fuzzy-based approach for strategic choices in electric energy supply. The case of a Swiss power provider on the eve of electricity market opening. Engineering Applications of Artificial Intelligence 20, 37–48 (2007)
12. Phillips-Wren, G., Jain, L.C. (eds.): Intelligent Decision Support Systems in Agent-Mediated Environments. IOS Press, The Netherlands (2005)
13. Encheva, S., Solesvik, M., Tumin, S.: Intelligent Decision Support System for Evaluation of Ship Designers. In: Proceedings of the 11th International Conference on Knowledge-Based & Intelligent Information & Engineering Systems (2007)
14. Leng, J., Fyfe, C., Jain, L.: Reinforcement learning of Competitive Skills with Soccer Agents. In: Proceedings of the 11th International Conference on Knowledge-Based & Intelligent Information & Engineering Systems (2007)

Reinforcement Learning of Competitive Skills with Soccer Agents

Jinsong Leng[1], Colin Fyfe[2], and Lakhmi Jain[1]

[1] School of Electrical and Information Engineering,
Knowledge Based Intelligent Engineering Systems Centre,
University of South Australia, Mawson Lakes SA 5095, Australia
Jinsong.Leng@postgrads.unisa.edu.au, Lakhmi.Jain@unisa.edu.au
[2] Applied Computational Intelligence Research Unit,
The University of Paisley, Scotland
Colin.Fyfe@paisley.ac.uk

Abstract. Reinforcement learning plays an important role in Multi-Agent Systems. The reasoning and learning ability of agents is the key for autonomous agents. Autonomous agents are required to be able to adapt and learn in uncertain environments via communication and collaboration (in both competitive and cooperative situations). For real-time, non-deterministic and dynamic systems, it is often extremely complex and difficult to formally verify their properties a priori. In this paper, we adopt the reinforcement learning algorithms to verify goal-oriented agents' competitive and cooperative learning abilities for decision making. In doing so, a simulation testbed is applied to test the learning algorithms in the specified scenarios. In addition, the function approximation technique known as tile coding (TC), is used to generate value functions, which can avoid the value function growing exponentially with the number of the state values.

Keywords: Agents, Reinforcement Learning, Decision Making.

1 Introduction

Multi-agent systems (MASs) are composed of several agents that can interact among themselves and with the environment. An agent can be defined as a hardware and/or software-based computer system displaying the properties of autonomy, social adeptness, reactivity, and proactivity [18]. Acting under uncertainty is the key feature for real-time, non-deterministic and dynamic MASs. Due to the inherent complexity of MASs, it is often extremely complex and difficult to formally verify their properties [5]. In addition, it is impossible to use algorithms or other formal methods to predict the whole state space in advance in dynamic environments.

Intelligence of agents can be described as the degree of reasoning and learned behavior with respect to perceiving, reasoning, planning, learning, and communication. MASs inherit many distributed artificial intelligence (AI) motivations, goals and potential benefits, and extend those of AI technologies that may have

B. Apolloni et al. (Eds.): KES 2007/WIRN 2007, Part I, LNAI 4692, pp. 572–579, 2007.

the ability to deal with incomplete information, or the capability for each agent to exhibit distributed control with decentralised data and asynchronous computation [6]. Learning is fundamental to intelligent behavior, and is motivated by the insight that it is impossible to determine all situations a-priori. Fortunately, reinforcement learning [11,16] provides a way to learn control strategies for autonomous agents in uncertain environments.

The autonomous agents are able to learn and adapt themselves in the environment through experience, rather than be controlled externally. Simply speaking, the goal of reinforcement learning is to compute a value function so as to find the optimal or near optimal action when in a given state. Although reinforcement learning is a powerful and effective methodology with underlying theoretical foundations, learning in dynamic and distributed environments is a difficult task due to the large, continuous state-action spaces. Multi-agent teamwork and learning are most attractive fields for AI researchers. Most research has focused on the issues of large state space representation, algorithms' stability and convergence, and agent teaming architectures.

Learning competitive/cooperative behaviors has been widely investigated in computer games such as Soccer [2,3] and Unreal Tournament (UT) [1]. Reinforcement learning has been used to learn both competitive and cooperative skills in RoboCup simulation system using different kinds of learning algorithms and state space representation [7,9,12,13,15]. The success of reinforcement learning critically depends on effective function approximation, a facility for representing the value function concisely, and parameter choices [10].

In this paper, a computer game called SoccerBots is used as a simulation environment for investigating goal-oriented agents' individual and cooperative learning abilities. SoccerBots provides a real-time, dynamic, and uncertain environment with continuous state-action spaces. We adopt Sarsa (λ) to learn competitive/cooperative skills in the SoccerBots simulation environment. In addition, the linear function approximation known as tile coding (TC) [16] is used to approximate value functions, which avoids the state space growing exponentially with the number of dimensions. The contribution of this paper is to build an agent system with competitive and cooperative learning abilities in a dynamic and uncertain environment. The efficiency of Sarsa (λ) and function approximation method can be investigated by varying the value of parameters.

The rest of the paper is organised as follows: Section 2 discusses the reinforcement learning algorithms and state space representation. The major properties of simulation are introduced in section 3. Section 4 presents the experimental results. Finally, we discuss future work and conclude the paper.

2 Reinforcement Learning and State Space Representation

2.1 Reinforcement Learning Techniques

Reinforcement learning is the learning of a mapping from situations to actions so as to maximize a scalar reward or reinforcement signal [16]. The goal of reinforcement learning is to generate a policy that will maximise the observed rewards over the lifetime of the agent.

The approach of reinforcement learning is based on a trial-and-error search, delayed reward, and exploration versus exploitation. A learning algorithm has to balance the trade-off between exploration and exploitation. For reinforcement learning, the value function is the key to find optimal sequential behavior. Most reinforcement learning algorithms are about learning, computing, or estimating values, eventually generating a value function.

The core elements in reinforcement learning can be described as follows [16]:

- A value function – generating the long-term accumulated value of the state.
- A policy – defining the learning agents' way of behaving at a given time.
- A reward function – defining the goal of the system by providing an immediate reward.
- A model (optional)– defining the transition probabilities from one state to next state.

Dynamic programing (DP) works well to compute the value function using an iterative scheme. A prerequisite of dynamic programing is that a model of the environment and state transition functions are known. However, such a model cannot be built for uncertain, dynamic environments. If no explicit model is available (we have no complete knowledge of the environment), we can use model-free reinforcement learning techniques to probe the environment, thereby computing the value functions. The Monte Carlo method generates the value from sampling episodes by averaging the sample returns. Temporal difference (TD) method such as Sarsa and Q-learning learns directly from experience like the Monte Carlo approach (model-free) but employs bootstrapping like DP.

The Sarah(λ) control algorithm with replacing traces modified from [16] is given in Fig. 1:

```
Initialise Q(s,a) arbitrarily and e(s, a) = 0, for all s, a.
Repeat (for each episode):
   Initialise s, a;
   Repeat ( for each step of episode):
      Take action a, observe r, s';
      Choose a' from s' using policy derived from Q (e.g., ε-greedy);
      δ ← r + γQ(s', a') - Q(s, a);
      e(s, a) ← 1;
      For all s, a:
         Q(s, a) ← Q(s, a) + αδe(s, a);
         e(s, a) ← γλe(s, a);
      s ← s'; a ← a';
   until s is terminal.
```

Fig. 1. Sarsa(λ) control algorithm with replacing traces

The temporal difference algorithms fall in two main classes: (1). On-policy learning (Sarsa) – action selection based on the learned policy; (2). Off-policy learning (Q-learning) – action selection based on greedy policy. In Fig. 1, α is a learning rate, γ is a discount rate. The ϵ-greedy policy is the exploration strategy, i.e. the agent takes a random action with probability ϵ and takes best action with probability (1 - ϵ).

TD(λ) represents that the eligibility traces are combined with the algorithms. Eligibility traces are the basic mechanism for improving the speed of learning. From the theoretical view, eligibility traces build a bridge from TD to Monte Carlo methods [16]. From the mechanic view, an eligibility trace is a temporary log to record the occurrence of an event, for example, the visiting of a state or the taking of an action. The purpose of eligibility traces is to assign the credit or punishment for only the eligible states or actions. Traces $e_t(s,a)$ can be accumulated (accumulating traces) by $e_t(s,a) = e_{t-1}(s,a) + 1$ or replaced by 1 (replacing traces).

The replacing traces can be defined as (where as accumulating traces use $e_t(s,a) = e_{t-1}(s,a) + 1$ for the second update):

$$e_t(s,a) = \begin{cases} \gamma\lambda e_{t-1}(s,a), & \text{if } s \neq s_t \\ 1, & \text{if } s = s_t \end{cases} \tag{1}$$

Algorithms combining with replacing traces are known as replacing traces method. Even slightly modified from accumulating traces, the replacing traces method can give a significant improvement in learning rate.

The stability and convergence is the key criteria to algorithms. The convergence of temporal difference algorithms such as Q-learning and Sarsa has been proved to converge with look-up table or sparse representations if the underlying environment is Markovian [4,17].

2.2 State Space Representation

A look-up table is a popular representation to store Q(s,a) values. However, it is impractical for dynamic, and uncertain environments with continuous state-action spaces. The state space explodes exponentially in terms of dimensionality. To make a proper approximate generalisation, two ways can be used to simplify state space: splitting by dividing state space into regions of interest, and aggregation by merging states with similar values.

The function approximation has been widely used to deal with large or continuous state space. The purpose of function approximation is to generate the state/action relationship by using the far fewer parameters. Some important function approximation methods are linear, including Coarse Coding, Tile Coding, and Radial Basis Functions [16]. The approximate function, V_t is represented as a parameterised function, thereby updating parameters instead of entries in a table. The formula is represented as:

$$V_t = \overrightarrow{\theta}_t^T \overrightarrow{\phi}_s = \sum_{i=1}^{n} \theta_t(i)\phi_s(i) \tag{2}$$

In the linear function, $\overrightarrow{\theta}_t$ is the parameter vector, and $\overrightarrow{\phi}_s$ is a corresponding column vector of features for each state. The complexity is related to the size of feature θ rather than the size of state space.

The tile coding method is to split the state space into tilings. In tile coding, tiling partition the state space into cells. Each tiling has only one tile being activated for every input of state, and the active cells of a state is called feature. The receptive fields of the features are grouped into partitions (also called tiling). The value of a state is calculated as $\sum_{i=1}^{n} \theta(i)\phi(i)$. The specifics of partition and the number of tilings may affect the performance.

3 Simulation Domain: SoccerBots

Due to the complex nature of the environment, it is difficult to scale reinforcement learning to soccer games. For example, RoboCup soccer, which includes 23 objects (1 ball, 22 players), 68×105 dimensions, could produce state space $(68 \times 105)^{23}$ without considering players' and ball's velocities, accelerations, and directions.

The soccer game is a real-time, noisy, adversarial domain, which has been regarded as a suitable environment for multi-agent system simulations [14]:

- Real-time: its acts in a dynamically changing environment.
- Noisy: which affects both sensors and effectors.
- Collaborative: the agents must cooperate (play as a team) to achieve the jointly desired goal.
- Adversarial: it involves competing with opponents.

The research areas include (but are not limited to):

- Competitive skills: can the individual agents learn to intercept, shoot, dribble, clear;
- Communication and cooperative skills: when and how to pass; when and where to move to receive a pass etc.
- Coordination: forecasting where and when other individual players are liable to move.
- Learning and Agent team architecture: is there an interaction between the individual learning and that necessary for the team to perform optimally.

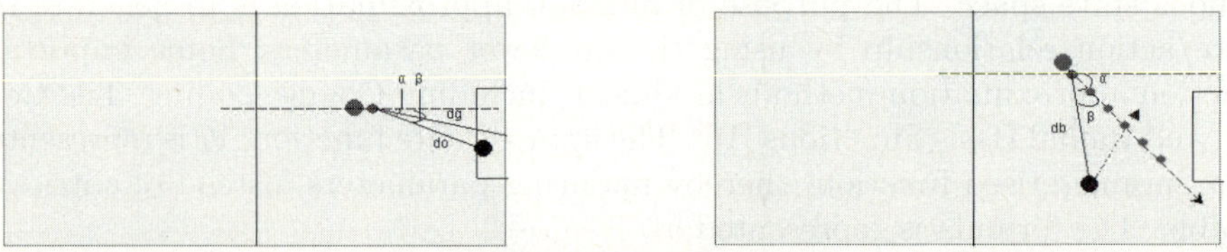

(a) (b)

Fig. 2. (a) Shooting. (b) Intercepting.

The shooting and intercepting behaviors are the most basic individual skills in the soccer game, shown in Fig 2. The shooting problem is to find the optimal kicking direction toward the goal. The intercepting problem is to compute a direction in order to intercept an approaching ball in the shortest time.

In order to evaluate the most realistic performance of reinforcement learning, we adopt the small size soccer league SoccerBots, which is one of a collection of application of TeamBots [2]. Each soccer team can have 2-5 players. Each player can observe the behaviors of other objects such as the ball, a teammate, an opponent, and their locations and motion via a sensor [8]. The ball's direction is influenced by environment noise (at random). In the real-time situation environment, performance is a high priority to be considered to follow the changes of motion of the ball and the players.

4 Experimental Results

To map the reinforcement learning algorithm Sarsa to SoccerBots, we define the scenario to learn shooting behavior for player in Fig 2 (a). An attacker with a ball is placed in front of goal, and a goalie is at the goalmouth moving north or south along the vertical axis. The player kicks the ball the goal ensuring the ball cannot be intercepted by the goalie. Noise influences the ball's velocity and direction at random.

Four parameters are considered: (1). The distance from the ball to the goal; (2) The distance from ball to goalie; (3) The angel between the ball to the goal; (4). The angel between the ball to goalie.

We develop the learning algorithm Sarsa(λ) with linear function approximation, tile coding, and replacing traces. We use four dimensions tiling for the continuous variables, dividing the space into $8 \times 8 \times 10 \times 10$ tiles. All tilings are offset at random variables.

The reward function is defined as follows:

$$Reward(s) = \begin{cases} 100, & \text{if the ball goes into the goal;} \\ -1, & \text{for each step;} \\ 0, & \text{if the ball misses the goal.} \end{cases} \tag{3}$$

The kicking direction is defined as a set of angles in degrees:

$$\{ \ 27, 24, \cdots, 3, 0, \text{-}3, \cdots, 24, 27 \ \}.$$

In Sarsa (λ) (Fig. 1), there are four parameters which can be adjusted: the learning rate α, the discount rate γ, the eligibility trace λ, and the probability of taking random action ϵ. The convergence time may be affected by tuning those parameters. In this case, we run the shooting episodes 1000 times by tuning the learning rate α to 0.05, 0.10, 0.15, and the noise to 0, 0.05, 0.10. The results are shown in Fig 3. Fig 3(a) illustrates that the convergence is quicker for bigger learning rate α, but smoother for smaller learning rate α. Fig 3 (b) shows that the convergence is heavily influenced by the noise.

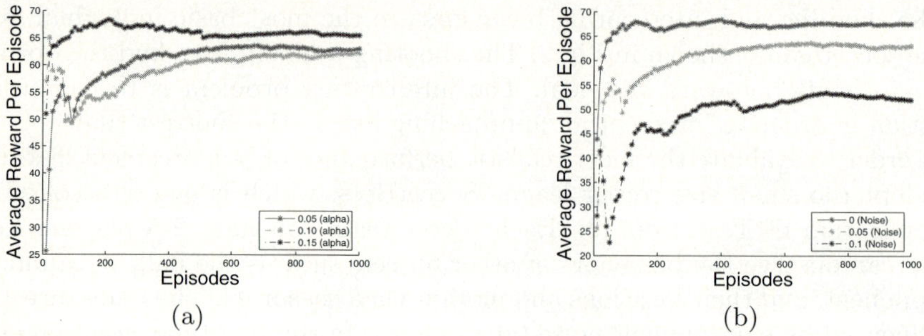

Fig. 3. The diagram of reward and episodes (a) α: 0.05, 0.10, 0.15. (b) Noise: 0, 0.05, 0.1.

5 Conclusion and Future Work

As described above, the efficiency of the linear function approximation like TC, depends on the balance of dimensions and parameter choices. Further work will compare performance by adjusting those parameters, in order to find the good parameters and tile coding dimensions. This paper has demonstrated that the agents are able to learn individual skills using reinforcement learning within a dynamic multi-agent environment.

Our ultimate goal is to develop an effective agent-teaming architecture with competitive and cooperative learning algorithms. Communication and dynamic agents role assignment will be considered for agent teaming. Due to the similar underlying theory to other stochastic techniques such as Bayesian learning, we are able to estimate the transition probabilities that may be possibly combined with reinforcement learning for improving knowledge reconstruction and decision making.

References

1. InfoGrames Epic Games and Digital Entertainment. Technical report, Unreal tournament manual (2000)
2. Teambots (2000), `http://www.cs.cmu.edu/trb/Teambots/Domains/SoccerBots`
3. Humaniod Kid and Medium Size League, Rules and Setup for Osaka 2005. Technical report, Robocup (2005)
4. Dayan, P., Sejnowski, T.J.: TD(λ) Converges with Probability 1. Machine Learning 14(1), 295–301 (1994)
5. Jennings, N.R., Wooldridge, M.: Applications of Intelligent Agents. Agent Technology: Foundations, Applications, and Markets, 3–28 (1998)
6. Jennings, N.R., Sycara, K., Wooldridge, M.: A Roadmap of Agent Research and Development. Autonomous Agents and Multi-Agent Systems 1(1), 7–38 (1998)
7. Kuhlmann, G., Stone, P., Mooney, R., Shavlik, J.: Guiding a Reinforcement Learner with Natural Language Advice: Initial Results in RoboCup Soccer. In: Bredenfeld, A., Jacoff, A., Noda, I., Takahashi, Y. (eds.) RoboCup 2005. LNCS (LNAI), vol. 4020, pp. 30–35. Springer, Heidelberg (2006)

8. Leng, J., Fyfe, C., Jain, L.: Teamwork and Simulation in Hybrid Cognitive Architecture. In: Gabrys, B., Howlett, R.J., Jain, L.C. (eds.) KES 2006. LNCS (LNAI), vol. 4252, pp. 472–478. Springer, Heidelberg (2006)
9. Riedmiller, M.A., Merke, A., Meier, D., Hoffman, A., Sinner, A., Thate, O., Ehrmann, R.: Karlsruhe Brainstormers - A Reinforcement Learning Approach to Robotic Soccer. In: Martin, A. (ed.) RoboCup 2000: Robot Soccer World Cup IV, London, UK, pp. 367–372. Springer, Heidelberg (2001)
10. Sherstov, A.A., Stone, P.: Function Approximation via Tile Coding: Automating Parameter Choice. In: Zucker, J.-D., Saitta, L. (eds.) SARA 2005. LNCS (LNAI), vol. 3607, pp. 194–205. Springer, Heidelberg (2005)
11. Singh, S.P., Sutton, R.S.: Reinforcement Learning with Replacing Eligibility Traces. Machine Learning 22(1–3), 123–158 (1996)
12. Stankevich, L., Serebryakov, S., Ivanov, A.: Data Mining Techniques for RoboCup Soccer Agents. In: Gorodetsky, V., Liu, J., Skormin, V.A. (eds.) AIS-ADM 2005. LNCS (LNAI), vol. 3505, pp. 289–301. Springer, Heidelberg (2005)
13. Stone, P., Veloso, M.: TPOT-RL: Team-partitioned, opaque-transition reinforcement learning. In: RoboCup 98: Robot Soccer World Cup II, p. 221. Springer, Berlin (1998)
14. Stone, P.: Layered Learning in Multiagent Systems: A Winning Approach to Robotic Soccer. MIT Press, Cambridge (2000)
15. Stone, P., Kuhlmann, G., Taylor, M.E., Liu, Y.: Keepaway Soccer: From Machine Learning Testbed to Benchmark. In: RoboCup-2005: Robot Soccer World Cup IX, vol. 4020, pp. 93–105. Springer, Heidelberg (2006)
16. Sutton, R.S., Barto, A.G.: Reinforcement Learning: An Introduction. MIT Press, Cambridge (1998)
17. Tsitsiklis, J.N.: Asynchronous Stochastic Approximation and Q-learning. Machine Learning 16(3), 185–202 (1994)
18. Wooldridge, M., Jennings, N.: Intelligent Agents: Theory and Practice. Knowledge Engineering Review 10(2), 115–152 (1995)

A Bootstrapping Approach for Chinese Main Verb Identification*

Chunxia Zhang[1], Cungen Cao[2], and Zhendong Niu[1]

[1] School of Computer Science and Technology, Beijing Institute of Technology, Beijing 100081, China
[2] Institute of Computing Technology, Chinese Academy of Sciences, Beijing 100080, China
cxzhang@bit.edu.cn, cgcao@ict.ac.cn, zniu@bit.edu.cn

Abstract. The task of main verb identification is to recognize the predicate-verb in a sentence. This task plays a crucial role in various areas such as knowledge acquisition, text mining, and question answering, and is also an important preprocessing for many applications including sentence pattern analysis and semantic roles identification. This paper proposes a domain-independent bootstrapping method to automatically identify main verbs of sentences from unannotated domain-specific Chinese unstructured texts. Experimental results in two domains show that the algorithm is promising. As applications of the main verb identification, we have developed a main verb driven approach of extracting domain-specific terms from unstructured text corpus.

Keywords: Main verb identification, bootstrapping method, domain-specific texts, term extraction.

1 Introduction

A main verb is a predicate-verb of a sentence, and is the most important verb in the sentence. Without it, the sentence would not be complete. Our task is to identify main verbs of sentences, which is a critical problem in various areas such as knowledge acquisition, text mining, automatic summarization and question answering. It is also preconditions for many applications such as sentence pattern analysis, dependency parsing, and semantic roles identification.

Unlike English, Chinese lacks of morphological varieties, and has no variational signs of gender, number, case, and part-of-speech. Moreover, there are no obvious morphological delimiters to separate words in sentences [1]. In addition, Chinese verbs have two features. One is that Chinese verbs appear in the same form no matter whether they are used as nouns, adjectives, or adverbs in sentences. The other is that

* The first and third authors are supported by the Program for New Century Excellent Talents in Universities of China and the IPv6 based National Foundation Education Grid (the Model Project of China Next Generation Internet) and Beijing Institute of Technology Basic Research Foundation (grant no.411002). The second author is supported by the Natural Science Foundation (grant no.60273019, 60496326, 60573063, and 60573064), and the National 973 Program (grant no. 2003CB317008 and G1999032701).

B. Apolloni et al. (Eds.): KES 2007/WIRN 2007, Part I, LNAI 4692, pp. 580–587, 2007.

Chinese verbs have no specific syntactic function, and they can be acted as subjects, predicates, objects, adverbial modifiers, or complements in sentences. Therefore, the essence of the main verb identification problem is to identify main verbs in sentences that have no inflections at all.

There are three kinds of approaches of automatic or semiautomatic main verb identification: heuristic methods or rule-based methods [2-5], statistics-oriented methods [6,7], and hybrid methods [8,9]. Heuristic methods mainly depend on linguistic knowledge. Koong [2] proposed a quantitative model to determine whether a verb candidate really acts as a predicate- verb in the sentence. The information used in this model includes theta grids of verbs, syntactic categories of words, and the animated property of agents. Statistics-oriented methods were introduced by Chen [6] and Sui [7]. Sui proposed a decision tree-based method to identify predicate heads of Chinese sentences. The features include verbs sub-categorization information and the lexicalized context information. Ding [9] employed a support vector machine model to realize the main verb identification based on chunk information.

Most existing works have been developed for detecting predicates only in simple sentences [3,4,7,8]. Furthermore, some works use correct sub-categorization or chunk information as input [6-9]. Although two tasks of sub-categorization and chunk identification have long been studied in the Chinese linguistics community, the results have not been satisfactory. In our work, we make use of more reliable part-of-speech information rather than sub-categorization and chunk information in general.

The purpose of this paper is to automatically identify main verbs of Chinese sentences from un-annotated domain-specific free texts. We propose a domain-independent bootstrapping approach to recognizing Chinese main verbs, which combines discourse context features and self-features of main verbs. Experimental results on two example domains show that the algorithm is promising. As applications of the main verb identification, a main verb driven method has been developed to extract domain-specific terms from unstructured text corpus, extracting terms that other statistic-oriented methods and linguistics-oriented methods could not acquire.

The rest of this paper is organized as follows. Section 2 describes how to identify main verbs of Chinese sentences using the bootstrapping method. Experiments of the algorithm and an application of the main verb identification are given in section 3. Section 4 concludes this paper.

2 A Bootstrapping Method of Main Verb Identification

In this paper, we refer to the definition of the main verb in the work of Ding [9]. A simple sentence is a sentence with only one predicate-verb. The predicate of a simple sentence can be a verb, an adjective, a noun or a subject-predicate in Chinese. A complex sentence is made up of two or more simple sentences. The main verb is the predicate-verb in a simple sentence. It corresponds to a tensed verb in English.

There are two points about our main verb identification. First, in pivotal sentences, series-verb sentences, and sentences with a verb-coordination predicate, the first predicate-verb is defined as the main verb. Second, auxiliary verbs should not be used as main verbs. Features of domain texts [10] and the work of Lu [11] show that the sentences with verb-predicates make up the most part in the corpus. Therefore, we focus on simple and complex sentences with verb-predicates in this paper.

2.1 Model for Chinese Main Verb Identification

Our bootstrapping approach begins with just a few of seed main verbs, and then automatically identifies main verbs of sentences in domain-specific text corpus. This model is called MVB-model (main verb bootstrapping). Seed main verbs are selected from main verbs of sentences in the training corpus.

Our algorithm was motivated by a preliminary experiment on the distribution of verbs in the domain corpus of Archaeological Volume of Chinese Encyclopedia [12], which contains about 3 million characters. The verbs used in the experiment come from the dictionary of Contemporary Chinese Grammatical Information [13]. Table 1 indicates that verbs whose frequency is more than 10 occupy 99.27% of the sum of the frequencies of verbs; and that verbs whose frequency is more than 2 reach about 75.41% in all 1,549 verbs. Based on the distribution of frequencies of verbs, we can conclude that most of the main verbs of sentences appear more than two times, since main verbs of sentences must be verbs. During the first iteration, our algorithm identifies a list of new main verbs of sentences based on the seed main verbs, and those new main verbs were then dynamically added to the learned main verb list. The enhanced learned main verb list was used to determine the main verbs of subsequent sentences during the next iteration. The process repeats for a fixed number of iterations, or until all corpora are used.

Table 1. Distribution of Frequency of Verbs in the Domain-Specific Corpus

Frequency of Verbs	1~2	3~5	6~10	11~50	51~ 100	101~ 500	501~ 1000	1001~ 2000	2001~ 5000	5001~ 10000	10001~ 16382
Number of Verbs	381	216	165	332	120	198	53	51	19	9	5
Frequency	529	847	1255	8416	8613	47587	36099	70830	61406	62811	64720
Percent of Frequency	0.15	0.23	0.35	2.32	2.37	13.11	9.94	19.51	16.91	17.30	17.82

In the following, we focus on the algorithm. Candidate main verbs of sentences are identified by a forward maximum matching method based on three verb sets: *a seed main verb set* S_{smv}, *a verb set* S_v, and *a learned verb set* S_{lv}.

$$S_v = \{v \mid v \in D_v \wedge Freq(v,Cp) > t\},\ S_{lv} = \{v' \mid \exists v\ ((v \in S_{smv} \cup S_v) \wedge SR(v, v'))\}$$

where D_v is a verb dictionary, function Freq(v,Cp) indicates the occurring frequency of v in Cp, predicate SR(v, v′) means that there is a synonymous relation between words v and v′, and t is a threshold. The construction of the learned verb set is based on the dictionary of synonyms [14]. The reason that we build the learned verb set is that the number of verbs in the verb dictionary is finite, and we use synonymies of these verbs to cover verbs in corpus as more as possible. In order to distinguish the candidate main verbs in a sentence, the function Level(cmv) is introduced to denote the level of the candidate main verb cmv, (a) Level(cmv)=3, if $cmv \in S_{smv}$; (b) Level(cmv)=2, when $cmv \in S_v \wedge cmv \notin S_{smv}$; (c) Level(cmv)=1, if $cmv \in S_{lv}$. In addition, the function Num(cmv, L) expresses the number of candidate main verbs with the L.

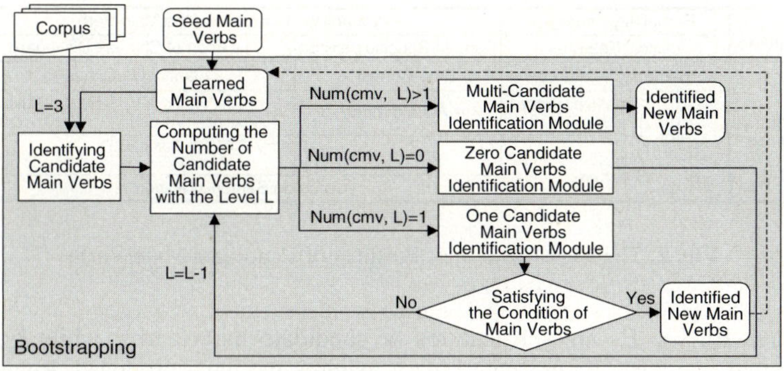

Fig. 1. The Process of the Bootstrapping Algorithm of Main Verb Identification

The process of the domain-independent bootstrapping algorithm to identify main verbs is illustrated with Fig. 1.

Algorithm 2.1.1 A bootstrapping algorithm of main verb identification
Input: Domain-specific Chinese free text corpus Cp;
Output: Texts annotated main verbs of sentences.
Step1: Build the set of seed main verbs S_{smv}, the set of learned main verbs $S_{lmv}(=\phi)$;
Step2: Select features used to determine main verbs of sentences;
Step3: Compute the weight of each feature;
Step4: Read a part of Cp, and set the initial level of candidate main verbs L=3;
Step5: Identify candidate main verbs of a sentence S;
Step6: Compute Num(cmv, L) of S. (a) Num(cmv, L)>1, go to step 7; (b) Num(cmv, L)=1, go to step 8; (c) Num(cmv, L)=0, go to step 9;
Step7: Enter into the multi-candidate main verbs estimation module, and add identified new main verbs of sentences to S_{lmv};
Step8: Go to the one candidate main verb estimation module. If this candidate main verb satisfies the condition of main verbs, then add identified new main verbs of sentences to S_{lmv}; else set L = L–1 and go to step 6;
Step9: Set L=L–1. If L>0, go to step 6; else go to step10;
Step10: If there are sentences to be processed in Cp, go to step4, else exit.

2.2 Candidate Main Verbs Estimation

This section will explicate features selection, feature weights computing, and one or multi-candidate main verbs estimation. Whether a candidate main verb cmv of a sentence acts as a main verb is determined by itself and its contextual characteristics. Our MVB-model captures self and contextual features of cmv, as shown in Fig.2.

Definition 1. The feature vector of a candidate main verb cmv is a quintuple FV=(F_p, F_{dl}, F_{cl}, F_{pp}, F_{np}). (a) F_p is the probability of cmv being the main verb of a sentence; (b) F_{dl} means whether cmv is or is contained by a domain-specific lexicon; (c) F_{cl} indicates whether cmv is or is included by a lexicon of the Chinese dictionary whose part-of-speech is not verb; (d) F_{pp} means whether cmv satisfies positive pattern features increasing possibility of cmv as a main verb; (e) F_{np} shows whether cmv satisfies negative pattern features reducing probability of cmv as a main verb.

<table>
<tr><td rowspan="2">Self-Features</td><td colspan="2">Probability Features</td><td colspan="3">Probability of cmv Being the Main Verb</td></tr>
<tr><td colspan="2">Lexical Features</td><td>Domain-Specific Lexicon</td><td colspan="2">Lexicon of Chinese Dictionary</td></tr>
<tr><td rowspan="4">Contextual Features</td><td colspan="2">Semantic Features</td><td>Time Cohesion</td><td colspan="2">Quantitative Cohesion</td></tr>
<tr><td rowspan="3">Pattern Features</td><td>Positive Pattern Features</td><td>Adverb+cmv</td><td>cmv+ "了/着/过 (Auxiliary Words)"</td><td>"地(An Auxiliary Word)"+cmv</td></tr>
<tr><td rowspan="2">Negative Pattern Features</td><td>cmv+series_nouns+"的"</td><td>cmv+" 的(An Auxiliary Word)"</td><td>"的" + cmv</td></tr>
<tr><td colspan="3">Common Sentence Patterns</td></tr>
</table>

Fig. 2. The Classification of Features of Candidate Main Verbs

The features F_p, F_{dl}, F_{cl} are self-features of candidate main verbs, while F_{pp} and F_{np} are contextual ones. F_{dl}, F_{cl} are new binary features we have proposed. Related works [7-9] have shown that contextual features are helpful to identify main verbs. One difference is that we do not use chunk and sub-categorization information. F_p (=NumMv(cmv)/NumSent(Cp)) is a real value feature, which is estimated in each iteration as follows, based on the increasing text corpus annotated main verbs of sentences. Here, NumMv(cmv) is the number of occurrences of cmv as main verbs of sentences in Cp, NumSent(Cp) is the sum of the number of simple sentences and the number of simple sentences which complex sentences contain. F_{pp} and F_{np} are pattern features about contextual lexicons and contextual part-of-speech. For example, 'Adverb+cmv' is a positive pattern feature; 'cmv+series_nouns(a series of nouns)+'的 (an auxiliary word)'' is a negative pattern feature; '和(and)…+cmv+…+一样(same)' is a common sentence pattern. It is noticed that each pattern feature, as a binary feature, corresponds to one dimension of the feature vector of cmv.

We will discuss how to compute feature weights. $FV=(F_1,F_2,...,F_n)$ is used for the feature vector of cmv. $E_i(cmv)$ is introduced to denote the event that the feature F_i happens when cmv occurs, and $\overline{E_i}(cmv)$ indicates the event that F_i does not occur. The weight, Weight(F_p), of F_p is set to 1. Weight(F_i) ($F_i \neq F_p$) is computed as follows, where MV is the set of main verbs in the training corpus.

$$\text{Weight}(F_i) = \log \frac{P(cmv \in MV \mid E_i(cmv))}{P(cmv \in MV \mid \overline{E_i}(cmv))}$$

$$P(cmv \in MV \mid E_i(cmv)) = \frac{P(cmv \in MV \wedge E_i(cmv))}{P(E_i(cmv))}, \quad P(cmv \in MV \mid \overline{E_i}(cmv)) = \frac{P(cmv \in MV \wedge \overline{E_i}(cmv))}{P(\overline{E_i}(cmv))}$$

For a sentence S, we assume there are m candidate main verbs $cmv_1,cmv_2,...,cmv_m$. The credibility degree, CreDeg(cmv_k), of cmv_k is computed as follows.

$$\text{CreDeg}(cmv_k) = \sum_{i=1}^{n} \text{Weight}(F_i) \times \alpha_i, \quad \alpha_i = \begin{cases} 1, & F_i \in \text{the activated feature set of } cmv_k \\ 0, & F_i \in \text{the inactivated feature set of } cmv_k \end{cases}$$

The feature set which cmv_k satisfies is usually a proper subset of the features set. If m>1, then go to the multi-candidate main verbs estimation module. The candidate main verb with the greatest credibility degree will be identified as the main verb of S. If m=1, then go to the one candidate main verb estimation module. When CreDeg(cmv_1)>β, β is a threshold, then cmv_1 is identified as the main verb of S.

3 Experiments and Application

In our experiments, three counts N, N_1 and N_2 are used to evaluate experimental results obtained by our main verb identification algorithm. Here, N is the total number of sentences in the Corpus Cp, N_1 is the total number of identified main verbs, and N_2 is the total number of correctly identified main verbs. We measure the performance of identification algorithms using recall(R), precision (P), and F-measure (F), where recall=N_2/N, precision=N_2/N_1, F=2RP/(R+P), and F-measure is the harmonic mean of precision and recall. Two example domains of archaeological cultures and sites, and two corpora are used to test the performance of our MVB-model. One corpus is cultures and sites texts of Archaeological Volume of Chinese Encyclopedia, which include 836 thousands characters, about 26,230 sentences. The other corpus is web pages from web sites[1,2], which contain more than 200 items and about 260 thousands characters. 500 sentences as the training corpus are used to evaluate weights of features. For the sentence '1950年河南省文物工作队发现，同年起由安金槐主持发掘(The cultural relic working team of Henan Province found this site in 1950, and Anjinhuai presided over excavation from that year)', the output result is '1950年河南省文物工作队<mv>发现(Find)</mv>, 同年起由安金槐<mv>主持(Preside)</mv>发掘', where '<mv>' and '</mv>' are the starting and ending tags of main verbs.

Three human judges were asked to evaluate experimental results, and the averages of three precisions and recalls are the final precision and recall. The score of all corpora are 91.6% precision, 88.07% recall, and 89.80% F-measure. The reasons that our bootstrapping approach attains high precision are: (a) main verb identification of subsequent corpus continually profit from identified main verbs of frontal corpus during the iteration process. The possibility of a verb as a main verb dynamically increases with the growth of the times of this verb as a main verb. (b) Self-features and contextual features of candidate main verbs are used to determine main verbs of sentences. The main causes of incorrect main verbs are: (a) there are more than one candidate main verbs with the greatest credibility degree; (b) the ambiguity of part-of-speech and errors of word segmentation lead to build incorrect candidate main verbs.

Fig.3 gives the comparison between our work and the works of Tan [5], Sui [7] and Gong [8] in the following aspects: word segmentation and part-of-speech, chunk parsing, sentences type, corpus, number of trained sentences, number of tested sentences, and precision. The first "√" in the volume of our work denotes that we do word segmentation and part-of-speech tagging, while the second "×" means that we do not rely on chunk parsing. Tan does not mention the precision of his results. Sui, Gong, and our works reach 79.95%, 86.5%, 91.6%, respectively.

As an application of the main verb identification, we have developed a main verb driven method to extract domain-specific terms from unstructured Chinese text corpus. The extraction process consists of the following steps: (a) identify main verbs of sentences; (b) label semantic roles of sentences including agents, patients, dative and locative; (c) build domain terms acting as semantic roles; (d) extract terms based

[1] http://www.chinacraft.com/zggy/06wwgys/ej-wwgys-d.htm
[2] http://www.chinaculture.org/gb/cn_zgwh/node-1499.htm

on rules which are composed of time cohesion, quantitative cohesion, part-of-speech and lexicon information. We use the same corpus on cultures and sites to test the performance of our term acquisition algorithm and obtain about 216,000 domain-specific terms of archaeology.

		Sui	Tan	Gong	Our Work
Word segmentation and Part-of-speech		√	√	√	√
Chunk Parsing		×	√	√	×
Sentence Type	Simple Sentence	√	√	√	√
	Complex Sentence	×	√	√	√
Corpus		Unknown	unknown	News texts from www.sina.com.cn	Texts and Web Pages on Archaeological Cultures and Sites
Number of Trained Sentences		3000 simple sentences	unknown	1131 sentences	500 sentences
Number of Tested Sentences		4000 simple sentences	unknown	820 sentences	About 33600 sentences
Precision		79.95%	unknown	86.5%	91.6%

Fig. 3. Comparison between Our Work and Related Works

Sentences 发现人类化石和文化遗物的第4、5、6层，伴出有三门马、中国缟鬣狗、肿骨大角鹿等华北中更新世典型动物，地质时代为中更新世晚期，铀系法断代及古地磁断代为距今40万至14万年。(In the accumulation of the forth, fifth, and sixth layers, there existed the typical animals of Huabei Middle Pleistocene such as equus sanmeniensises, striped hyenas, and megaloceros pachyosteuses. The geologic age of these animals was the Late Middle Pleistocene. Their age was before 400~140 thousands year, based on the uranium-series dating and archaeomagnetic dating.)
The Result of Word Segmentation 发现/v 人类/n 化石/n 和/c 文化/n 遗物/n 的/u 第4/m 、/w 5/m 、/w 6/m 层/q，/w 伴/v 出/v 有/m 三门/ns 马/n 、/w 中国/ns 缟/Ng 鬣狗/n 、/w 肿/v 骨/Ng 大/a 角/n 鹿/n 等/v 华北/ns 中/f 更新/v 世/Ng 典型/a 动物/n ，/w 地质/n 时代/n 为/p 中/f 更新/v 世/Ng 晚期/f ，/w 铀/n 系法/n 断代/v 及/c 古/a 地磁/n 断代/v 为/u 距/v 今/Rg 40/m 万/d 至/v 14/m 万年/m 。/w。
Our Result of Term Extraction 人类化石(Human Fossil), 文化遗物(Artifact), 三门马(Equus Sanmeniensises), 中国缟鬣狗(Striped Hyenas), 肿骨大角鹿(Megaloceros Pachyosteuses), 中更新世(The Middle Pleistocene), 华北中更新世典型动物 (The Typical Animals of Huabei Middle Pleistocene), 中更新世典型动物(The Typical Animals of the Middle Pleistocene), 地质时代(The Geologic Age), 动物(Animal), 铀系法断代(The Uranium-Series Dating), 古地磁断代(Archaeomagnetic Dating), 中更新世晚期(The Late Middle Pleistocene)

Fig. 4. Examples of Our Term Extraction Results

Currently, there are two major approaches in terminology extraction: statistical and linguistic. Statistical methods can cope with high frequency terms but tend to miss low frequency terms [15]. Linguistic techniques rely on the assumption that terms present specific syntactic structures or patterns [16]. Our term extraction method depends on main verbs and semantic roles of sentences to acquire terms. Therefore, this approach not only extracts the terms that occur once or more times in the corpus, but also extracts the terms that act as semantic roles and may not fit well to the specific syntactic patterns. Comparisons between the results of the word segmentation system [17] and our results show that our algorithm acquires about 57,450 new terms, which occupy 26.54% of all extracted terms. Fig. 4 gives the word segmentation result of one sentence and the terms extracted from the sentence by our method.

4 Conclusion

Main verb identification plays an increasingly important role in knowledge acquisition and many natural language processing tasks. In this paper, we present a bootstrapping technique that is used to identify main verbs from un-annotated domain-specific Chinese free texts, and have shown its effectiveness through experimental results. A main verb driven approach has been proposed to extract domain-specific terms, and it obtains a large amount of terms, which could not be acquired via statistic-oriented and linguistics-oriented methods. In our bootstrapping approach, the verb sets used to build candidate main verbs are independent of domains. In addition, self-features and the contextual features of main verbs utilized to determine main verbs are also independent of domains. These two advantages indicate that our method can be applied to any domain. In future, we would like to add the resolution of anaphora and co-reference in order to improve the performance of main verbs identification. We will also use main verbs information to extract relationships between domain-specific terms.

References

1. Zhang, C., Hao, T.: The State of the Art and Difficulties in Automatic Chinese Word Segmentation. Journal of System and Simulation 1, 138–143 (2005)
2. Koong, H., Soo, V.: Hypothesis Scoring over Theta Grids Information in Parsing Chinese Sentences with Serial Verb Constructions. International Conference on Computational Linguistics, Kyoto, Japan, pp. 942–948 (1994)
3. Luo, Z., et al.: An Approach to the Recognition of Predicates in the Automatic Analysis of Chinese Sentence Patterns. In: Proceedings of 3rd National Computational Linguistics, Beijing, China, pp. 159–164 (1995)
4. Sui, Z., Yu, S.: The Research on Recognizing the Predicate Head of a Chinese Simple Sentence in EBMT. Journal of Chinese Information Processing 4, 39–46 (1998)
5. Tan, H.: Center Predicate Recognization for Scientific Article. Journal of Wuhan University 6, 1–3 (2000)
6. Chen, X., Shi, D.: To Mark Topic and Subject in Chinese Sentences. In: Proceedings of the Fourth National Conference on Computational Linguistics, pp. 102–108 (1997)
7. Sui, Z., et al.: The Acquisition and Application of the Knowledge for Recognizing the Predicate Head of a Chinese Simple Sentence. Journal of Peking University 223, 221–230 (1998)
8. Gong, X., Luo, Z., Luo, W.: Recognizing the Predicate Head of Chinese Sentences. Journal of Chinese Information Processing 2, 7–13 (2003)
9. Ding, B., Huang, C., Huang, D.: Chinese Main Verb Identification: from Specification to Realization. Computational Linguistics and Chinese Language Processing 1, 53–94 (2005)
10. Huang, B.R., Liao, X.D.: Modern Chinese. High Education Publisher, Beijing (2002)
11. Lu, S., et al.: Elementary Study of Chinese Grammar. The Commercial Press, Beijing (1999)
12. Chinese Encyclopedia. Encyclopedia of China Publishing House, Beijing (1998)
13. Yu, S., et al.: A Dictionary of Contemporary Chinese Grammatical Information. The Tsinghua University Press, Beijing (1998)
14. Mei, J., et al.: A Dictionary of Synonyms. Shanghai Thesaurus Press, Shanghai (1983)
15. Bourigault, D.: Lexter: A Natural Language Processing Tool for Terminology Extraction. In: Proceedings of the 7th Euralex International Congress (1996)
16. Evans, D., et al.: Noun-Phrase Analysis in Unrestricted Text for Information Retrieval. In: Proceedings of the 34th Association for Computational Linguistics, pp. 17–24 (1996)
17. Chinese Text Segmentation and POS Tagging, http://www.icl.pku.edu.cn

A Novel Method of Extracting and Rendering News Web Sites on Mobile Devices

Harshit Kumar[1], Sungjoon Park[2], and Sanggil Kang[3]

[1] University of Suwon, South Korea
harshitkumar@suwon.ac.kr
[2] Kongju Communication and Arts College, South Korea
sjpark@kcac.ac.kr
[3] Inha University, South Korea
sgkang@inha.ac.kr

Abstract. In this paper, we focus on the problem of displaying news web sites on mobile devices. Our middleware acts as an interface between the existing news web site and user. News items are extracted using spatial location of HTML elements and a DOM based page segmentation algorithm. The nature of the proposed algorithm is dynamic; it learns from the desktop user trails on news items, and correspondingly displays news item of the highest rank on mobile devices. The rank is calculated based on event and access frequency of news item. The experimental results show that the proposed algorithm saves highly on the bandwidth usage.

Keywords: Information retrieval, Page Segmentation, HCI, Web Mining.

1 Introduction

Most of the news websites are developed keeping desktop users in mind. However, the numbers of mobile Internet users are increasing due to the rapid proliferation of ubiquitous Internet access using mobile phones, PDAs, and other digital devices. Therefore the need for presenting information effectively and efficiently on mobile devices is increasing. The problem of rendering news item on mobile device can be broadly divided into two parts web page segmentation and rendering information content.

Recently, websites are being increasingly developed using CSS [1], which has not been considered in any of the previous page segmentation algorithm. We implemented support for CSS in the page segmentation module.

This paper makes the following contributions:

- We introduce an algorithm that uses spatial clues to extract menu items and DOM to segment news web site pages into blocks, called 'news items'. Note that spatial clues are different from vision based segmentation algorithms.
- As the size of the image takes up a major chunk of web page size, images and advertisements are eliminated during segmentation.
- The proposed algorithm can distinguish between requests from desktop computer and mobile devices; hence render corresponding related content.

B. Apolloni et al. (Eds.): KES 2007/WIRN 2007, Part I, LNAI 4692, pp. 588–595, 2007.

- A mobile device on placing a request is rendered with the highest ranked news item in a particular category. Ranking of news items are calculated as a function of *EventId* and *Access Frequency*. An Event is defined as something that happens once in a few years or every year, for instance, the Soccer World Cup, Cricket World Cup, elections, conference etc. Note that, a league match is not an event.

The rest of the paper is organized as follows: Section 2 discusses related work. Section 3 introduces middleware architecture and its modules. Experimental results and conclusions are presented in section 4 and 5 respectively.

2 Related Work

The problem of displaying news websites on mobile devices can be segregated into two parts; web page segmentation and rendering information content. For web page segmentation, many algorithms were proposed by researchers in the past, which fall into either of two categories; DOM-based [2] and Vision-based [3]. The DOM-based approach explores the DOM tree and extracts contents from web pages. The Vision-based approach uses visual cues to extract contents. Here visual cues means, difference of color or font size between paragraphs. Recently, web pages are increasingly being made using <div>, <p> tags and CSS, which makes most of the proposed approaches obsolete. The second part of the problem deals with rendering of extracted information on mobile devices. There are four general approaches for rendering web pages on mobile devices: Device Specific Authoring, Multiple Device Authoring, Client Side Navigation, and Automatic Re-authoring. Device Specific Authoring implies creating a new version for each device, i.e. creating different web pages for desktop users and mobile devices, which is tedious. Multiple Device Authoring [4] implies creating web pages in a fashion so that it is compatible for display on multiple devices. In Client Side Navigation [5, 6], the approach is to modify the presentation aspects of web page before rendering them on a user screen. The Automatic Re-authoring method automatically converts web pages to WML pages as proposed in [7, 8]. The approach proposed in [9] is partly similar to our approach, which segments a web page and splits a web page into blocks. Since this is a general approach, it is not suitable for news web sites. The approach proposed in [10, 11] is a precursor to our approach. Both of them fail to explain segmentation algorithm as well as supporting event based information retrieval. Also, there is no support for advertisement removal and rendering long pages in parts.

3 Middleware Architecture

Fig. 1 shows the overall middleware architecture. The main modules of the middleware architecture are Device Detection Module (*DDM*), Page Segmentation Module (*PSM*), and Log Module (*LM*). *DDM* is an interface between the user device and news web server. It redirects user request to the appropriate place depending upon the type of browser. If the request comes from a desktop computer (1.1), the requested web page is rendered on the desktop computer (1.2).

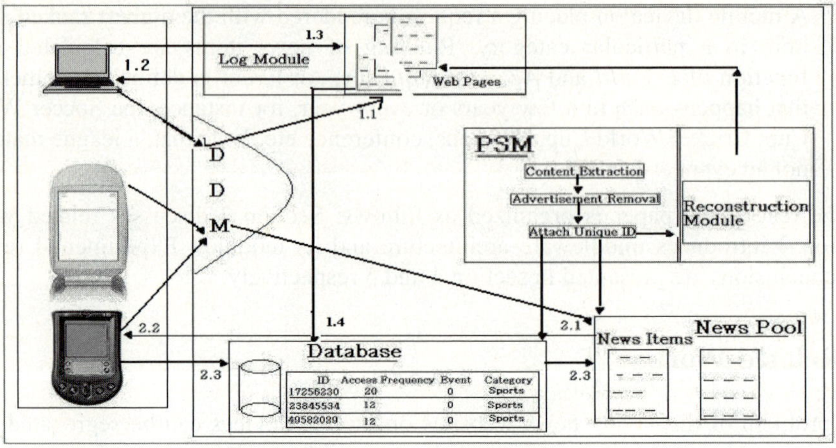

Fig. 1. System Architecture

For each news item clicked by the user (1.3), LM module (1.4) increments the *access frequency* of the corresponding news item in the database. This process doesn't hinder the user interaction as it is implemented through AJAX. If the request comes from a mobile device (2.1), index.wml is rendered on the mobile device (2.2). The user chooses either one of the menu items or "Top Stories". If the user clicks one of the menu items, a query executes which locates the highest ranked news item from the database within that category (2.3). The corresponding news item is fetched from news pool and is rendered on the mobile browser (2.2). The '*rank*' is defined as a function of two parameters, A*ccess Frequency*, which is the number of times rendered news item is accessed, and *EventID*.

3.1 Device Detection Module

The *DDM* detects whether the client device is a desktop computer or a mobile device.

```
HttpServletRequest req;
String userAgent = req.getHeader("user-agent");
String httpUserAgent = Request.ServerVariable("HTTP_USER_AGENT");
IF (userAgent.indexOf("WAP") || userAgent.indexOf("PALM") || httpUserAgent.indexOf("Windows CE") THEN
        Load index.wml
   ELSE
        Load index.html
   ENDIF
```

Fig. 2. An algorithm that detects type of client device

On detecting the type of device, *DDM* loads the relevant web page, i.e. index.html or index.wml. The algorithm for a device detection module is shown in Fig. 2. The parameter "user-agent" and HTTP_USER_AGENT are two properties of a

HTTPServletRequest object that can provide information about a client device. These parameter values are checked for either of following strings "WAP" or "PALM" or "Windows CE". If there is any match, then the request is redirected to index.wml, otherwise to the index.html.

3.2 Page Segmentation Module

This section is the central module of our research work. The *PSM* takes index.html as input, extract menu items using spatial clues and constructs index.wml. Fig. 3 presents an algorithm for extraction of menu items.

```
FOR EACH pid either of <table> or <div> or <p>
        FOR EACH cid of <td>
                IF majority of <td>'s have either of same left or top value THEN
                        Extract text inside the tag using innerText property.
                ENDIF
        NEXT cid
        FOR EACH cid of <div>
                IF majority of <div> tags have either of same left or top value THEN
                        Extract text inside the tag using innerText property.
                ENDIF
        NEXT cid
        FOR EACH cid of <li>
                IF majority of <div> tags have either of same left or top value THEN
                        Extract text inside the tag using innerText property.
                ENDIF
        NEXT cid
Compare extracted text with common words like politics, entertainment, sports. If words match, then result is correct.
NEXT fid
```

Fig. 3. An algorithm for extraction of menu items

To extract menu items, the value of the top attribute or left attribute for <div> or <td> or <li> tags are investigated. Note that menu items are generally aligned as a single row or single column. If menu items are aligned as a single row (column), they must have the same value for top (left) attribute.

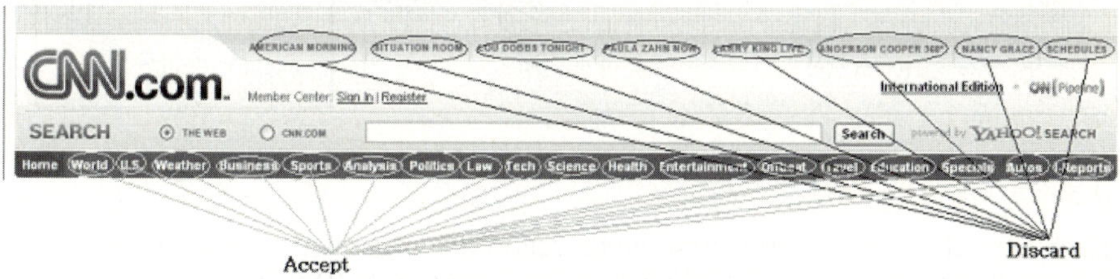

Fig. 4. An example that shows extraction of menu items from www.cnn.com

Fig. 4 shows the menu item for news website www.cnn.com. The topmost menu items shown in blue circles are all aligned in one row, however, they are discarded because none of the enclosed text string matches the expected menu item strings. On the other hand, the 2nd row of menu items shown in white circles are accepted because they are all aligned in one row and some of the enclosed text string matches the expected menu item. After extracting menu items, *PSM* takes the .html file which is associated with the first menu item. Using HTML parser, we developed a java program that extracts news item on the basis of semantic structure of web pages. Fig. 5 shows the algorithm used. Each <table> or <div> tag encountered is pushed into the stack. Each closing </div> or </table> tag results in a pop operation. If the length of text enclosed in tag is greater than 100 words, it is considered as a news item.

```
FOR EACH tag of <td> or <li> or <div> tag
IF (tag == "<td>" or "<li>" or "<div>") THEN
        Push (tag) into the stack.
ENDIF
IF (tag == </td> or "</li>" or "</div>") THEN
        Set tag= pop();
        String data = tag.innerText;
        IF (length (data) > 100) THEN
                Generate a random ID. Attach ID with the news item.
                Place the news item in news pool.
                Make an entry in the database with ID as primary key.
                Assign the same ID to the tag in .html file, reconstruct the html file.
        ENDIF
    ENDIF
    NEXT tag
```

Fig. 5. An Algorithm for extraction of news items from news web site

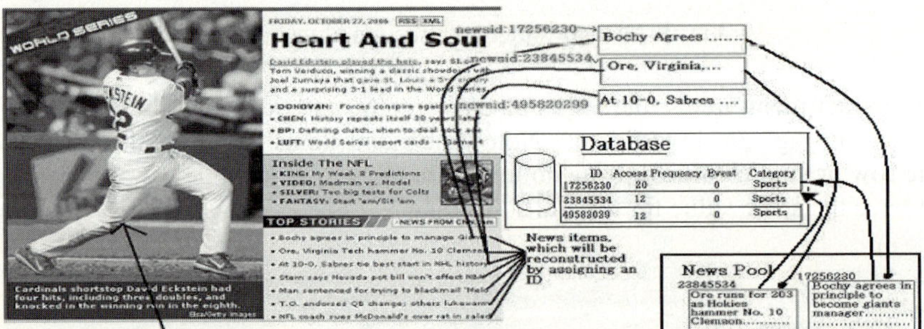

Fig. 6. Sports web page from CNN sports menu item

Each extracted news item is assigned a randomly generated unique *ID*. A record is created in the database for each extracted news item with *ID* as its primary key. Each record inserted in the database has an *Access Frequency* 0 and if it is some event (refer section 3.3), then the *EventId* column is assigned a value 10 otherwise 1. Manual assignment is tedious but it is possible, as there are very few events (Refer definition of event). Also, same *ID* value is attached correspondingly to the new item on the web page, which is the reconstruction process shown in fig 6. During the extraction process, the values of the "src" and "href" attributes throughout the page are surveyed. These attribute values determine which link they refer to. If an address matches against a list of common servers that particular tag containing these attributes is ignored, and is not included in the news pool.

3.3 Log Module

Log Module (*LM*) executes every time a user accesses a news item from a desktop computer. Consequently JavaScript code executes, which makes an update (increment *access frequency* by 0.2) in the database corresponding to the *ID* of the news item clicked. The value of *access frequency* and *event* is used for calculating the rank when a request comes from a mobile device. An event is something that happens once in few years like a World Cup or Grand Slam. It may sometimes happen that a user wants to read about a new event which has just started or is about to start. Because this is a new event, which is just added, the *Access Frequency* of this news item is low, which prohibits it from showing up on the browser. Assigning a high value (refer section 3.2) to the *EventId* column helps event based news items to get a higher rank. The rank of a news item is calculated as follows

$$R(n_i) = AC(n_i) * Event(e_i) / \sum_j (AC(n_j) * Event(e_j)) \tag{1}$$

where n_i is news item *i*. $AC(n_i)$ is the access frequency for news item *i*. $Event(e_i)$ is the *eventId* for the news item n_i. The denominator calculates the summation of all news items in that particular category. This is done to normalize the rank. Equation 1 returns rank for news item *i*.

After the user receives a news item, there are two options at the bottom of the news item. These options are used for receiving a user response. The options are; "Like" or "Dislike." If user chooses the first option, his/her mobile device is rendered with the next news item in the same category. If the user chooses the second option, he/she does not like this news, which can be interpreted in two ways; either the user wants to read different news or wants to go to a different menu. For both cases, the proposed algorithm renders menu items on mobile devices.

4 Experimental Results and Setup

In this section, we present the details of the experimental setup and results obtained. We created a simulated environment wherein there are downloaded sample web pages of yahoo news service, CNN news service, and Korea Herald news. The PalmOS Emulator serves as a mobile device. Downloaded sample web pages were stored on the Apache web server. The server Operating System is Windows XP and Microsoft Access was used as the database. For this experiment, web pages were accessed using a desktop computer or using a Palm OS Emulator. The experiment was performed at Suwon University Computer Science Lab, and twelve students were selected. Six among them were accessing sample web pages using desktop computers and the rest were accessing sample web pages using Palm OS Emulators.

4.1 Experimental Results

We show two types of results here; one is success rate of news items extraction and the other is percentage saving in bandwidth. The middleware installed on the web server starts segmenting sample web pages and creates a news pool, which consists of news items. Students reported that the extracted news items were well legible understandable. Table 1 show that *PSM* is 100% successful in extracting menu items and 78% successful in extracting news items from the sample web pages.

Table 1. PSM performance

	% of menu items extracted	**% news item extracted**
www.cnn.com	100%	85%
www.koreaherald.co.kr	100%	75%
News.yahoo.com	100%	74%

Furthermore, we investigated the reasons for the 22% failure rate and found that some news items were not extracted because the constructed web pages had the wrong sequence of tags, which is quite common in web pages. HTML, being a poorly authored language, does not report errors if tags are not used properly. If a web developer adheres to HTML guidelines, it is certain that success rate will reach 99%. To prove our point, we identified news items that were not extracted. The tags surrounding them were tweaked and we finally submitted the reconstructed web page to *PSM*. This time, all the news items were extracted successfully.

Table 2. Results showing saving in bandwidth

	No. of news items.	**Without Middleware**	**With Middleware**	**% Saving**
www.koreaherald.co.kr (Business)	5	3.9 MB	50 KB	98.76%
www.cnn.com (Politics)	10	5.5 MB	100 Kb	98.22%
news.yahoo.com (Politics)	22	7.4 MB	250 KB	96.7%

To understand how bandwidth is saved; for example, if user views a web page using his/her mobile device, he/she will first open a home page followed by news item of interest. On an average, 700 KB of data will be downloaded in the whole process. In addition, a user will face problems in finding news item of interest. When the same news web site is accessed using middleware, the user first receives menu items. Downloading menu items will take 1 KB of bandwidth. Assume that the user selects sports, the highest rank news item in sports category will be rendered on his/her mobile device, and the average size of the news item is 10 Kb. In this whole sequence of operations, approximately 11 Kb of data was downloaded, which means 98.4% bandwidth is saved. Because images and advertisements were filtered, only relevant text corresponding to the selected news item is rendered on the mobile device. Hence, there is an enormous saving in bandwidth. Table 2 shows bandwidth utilization results for specific menu items like business and politics from different web sites when accessed with, and without middleware.

5 Conclusion and Future Work

In this paper, we introduced an algorithm, which extracts news items from news web sites using spatial logic and DOM. The extracted news items are then rendered on a mobile device using combination of the Automatic Re-authoring and Client Side Navigation approach. The results as shown in the experimental section, extracted 100% of menu items and 78% of news items. However, the algorithm can perform better if web developers adhere to HTML guidelines. The advertisements and images were eliminated during the news item extraction; hence, we are able to save the bandwidth. Major events are given preferential treatment, thus bringing event based news items on par during the calculation of rank.

Currently, this method covers only specific domain and is applicable for news web sites. We have not evaluated the method for ranking news item. This will form future research.

References

1. http://www.w3.org/Style/CSS
2. Lin, S.-H., Ho., J.-M.: Discovering Informative Content Blocks from Web Documents. In: Proceedings of ACM SIGKDD'02 (2002)
3. Cai, D., Yu, S., Wen, J.-R., Ma, W.-Y.: Extracting Content Structure for Web Pages based on Visual Representation. Fifth Asia Pacific Web Conference (APWeb-03) (2003)
4. Nylander, S., Bylund, M., Waern, A.: The Ubiquitous Interactor - Device Independent Access to Mobile Services. CADUI, 269–280 (2004)
5. http://www.sourceforge.net/projects/wpar
6. http://www.webwiper.com
7. Buyukkokten, O., Garcia-Molina, H., Paepcke, A., Winograd, T.: Power Browser: Efficient Web Browsing for PDAs. In: proc. of the Conf. on Human Factors in Computing Systems, CHI'00, pp. 430–437 (2000)
8. Chen, Y., Ma, W.Y., Zhang, H.J.: Detecting webpage structure for adaptive viewing on small form factor devices. In: proceedings of the 12th International Conference on World Wide Web, pp. 225–233 (2003)
9. Kaasinen, E., Aaltonen, M., Kolari, J., Melakoski, S., Laakko, T.: Two Approaches to Bringing Internet Services to WAP Devices. In: proc. of 9th World-Wide-Web Conf. WWW'9, Amsterdam, pp. 231–246 (15-19 May, 2000)
10. Kang, S., Park, W., Kim, Y.: Dynamically Personalized Web Service System to Mobile Devices. In: Larsen, H.L., Pasi, G., Ortiz-Arroyo, D., Andreasen, T., Christiansen, H. (eds.) FQAS 2006. LNCS (LNAI), vol. 4027, pp. 416–426. Springer, Heidelberg (2006)
11. Jeon, Y., Hwang, E.: Automatically customizing service pages on the Web for mobile devices. In: Bianchi-Berthouze, N. (ed.) DNIS 2003. LNCS, vol. 2822, pp. 53–65. Springer, Heidelberg (2003)

An Adaptation Framework for QBH-Based Music Retrieval*

Seungmin Rho[1], Byeong-jun Han[2], Eenjun Hwang[2], and Minkoo Kim[1]

[1] Graudate School of Information and Communication, Ajou University, Suwon, Korea
{anycall,minkoo}@ajou.ac.kr
[2] Department of Electronics and Computer Engineering, Korea University, Seoul, Korea
{hbj1147,ehwang04}@korea.ac.kr

Abstract. In this paper, we present a new music query transcription and refinement scheme for efficient music retrieval. For the accurate music query transcription into symbolic representation, we propose a method called WAE for note onset detection, and DTC for ADF onset detection. Also, in order to improve the retrieval performance, we propose a new relevance feedback scheme using genetic algorithm. We have built a prototype system based on this scheme and performed various experiments. Experimental results show that our proposed scheme achieves a good performance.

Keywords: Genetic Algorithm, Music Retrieval, Relevance Feedback.

1 Introduction

For decades, most traditional approaches for retrieving multimedia data were based on the textual information. Compared with this traditional method, content-based music retrieval provides more flexibility and expressive power to the users. Content-based music retrieval is usually based on a set of extracted music features such as pitch, duration and rhythm. One common approach for developing a content-based music retrieval system is to represent music into a string of characters using three possible values for the pitch change: U(p), D(own), and S(ame) or R(epeat). In our previous work [1], we described the limitation of a pure UDR notation based on the pitch contour and proposed new notations such as uUdDr and LSR to overcome such limitations. So far, most music information retrieval systems have used symbolic representation, with input ranging from note sequences using CMN (Common Music Notation) to user-hummed tunes.

There are some researches for symbolic MIR which is based on the ideas from the field of traditional text IR. Examples of using traditional IR techniques such as probabilistic modeling and approximate string matching are described in [2] and [3],

* This research was partially supported by the MIC, Korea, under the ITRC support program supervised by the IITA. (IITA-2006-(C1090-0603-0002)) and by the Ubiquitous Computing and Network (UCN) Project, the Ministry of Information and Communication (MIC) 21st Century Frontier R&D Program in Korea.

respectively. Some work addresses other IR issues such as ranking and relevance. Hoashi et al. [4] used relevance feedback for music retrieval based on the tree-structured vector quantization method (TreeQ), developed by Foote [5]. Their approach of the TreeQ method is to train a vector quantizer instead of modeling the sound data directly. Lopez-Pujalte [6] implemented a genetic algorithm for relevance feedback in textual information retrieval and running it with different order-based fitness functions. Among the fitness functions present in literature, the ones that yield the best result are those that take into account not only when documents are retrieved, but also the order in which they are retrieved.

QBH (Query by humming) is one of the most popular content-based querying methods for a large-scale music database. QBH systems can take a user's acoustic input (a short clip of singing, whistling or humming) through a microphone, extract useful features from it and then retrieve matched songs from a music database by comparing their similarity. This is very useful when the user does not know detailed information about the music such as its title or singer, but just remembers a small segment of the music. However, the quality of QBH is strictly dependent on the accuracy of the audio transcription such as duration or pitch of each note. Thus, an efficient algorithm to transcribe an audio signal into a note-like representation is one of the critical components in a QBH-based music retrieval system.

There are many techniques to extract pitch contour, pitch interval and duration from a voice humming query. In general, methods for detecting pitches can be divided roughly into two categories: time-domain based and frequency-domain based. In the time-domain, ZCR (Zero Crossing Rate) and ACF (Auto Correlation Function) are two popular methods. The basic idea is that ZCR gives information about the spectral content waveform cross zero per unit time [7]. In the frequency-domain, FFT (Fast Fourier Transformation) is one of the most popular methods. This is based on the property that every waveform can be divided into simple sine waves. But, a low spectrum rate for longer window may increase the frequency resolution while decreasing the time resolution.

In this paper, we propose a new music retrieval scheme based on two prominent features: (i) Voice humming signal can be transcribed into notes. Especially, in this stage, we improved the transcription accuracy a lot using two new methods: WAE (Windowed Average Energy) and DTC (Dynamic Threshold Curve)-based ADF which is an improved version of the ADF (Amplitude-based Difference Function) which was proposed in our previous work [8]. (ii) We propose a new GA-based relevance feedback scheme to improve the quality of query results by reformulating a user query.

The remainder of this paper is organized as follows. We describe our overall system architecture in Section 2. Details on our transcription and query refinement scheme are presented in Section 3. In Section 4, we report some of the experimental results. In Section 5, we conclude this paper and describe our future directions.

2 System Architecture

In this section, we describe the overall architecture of our music retrieval system. The system consists of two main parts: Transcriber Module for music query transcription into symbolic representation and GA Engine Module for GA-based relevance feedback for user query reformulation.

Fig. 1 shows main components of our prototype system and their interactions. Typical query processing scenario is as follows: User first makes an initial query using one of four different user query interfaces: QBE (Query by Example), QBH (Query by Humming), QBMN (Query by Music Notation) and QBC (Query by Contour). When a user query is given, Transcriber Module interprets the query as a signal or a sequence of notes, and extracts audio features such as pitch and time contour. Next, it transcribes those extracted features into a symbolic representation such as uUdDr and LSR string.

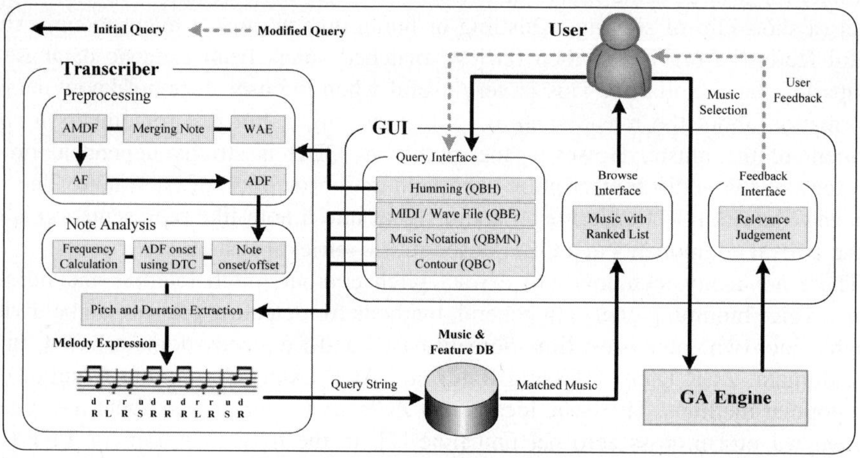

Fig. 1. System Architecture for Music Retrieval

In order to transcribe voice queries using humming interface, we first preprocessed them using WAE and ADF. After that, notes are analyzed and their pitch and duration features are extracted. We also used the AMDF (Average Magnitude Difference Function) to get a fundamental frequency of each frame. Furthermore, we applied the ADF to each frame in order to get ADF onset from WAE information. This makes it a lot easier to extract note information. For clipping ADF onset efficiently, we propose a new dynamic threshold method based on DTC.

Matched melodies are displayed according to their rank on the browse interface. When the user selects a melody or its segment as relevant one, GA engine generates new music segments and evaluates fitness of each segment using our genetic algorithm. A modified query is generated by the user's relevance judgment via feedback interface, and then the whole query process will be repeated until the user is satisfied.

3 Algorithms

In this section, we describe all the details and algorithms on the main components of our system including the voice humming transcription and query refinement.

3.1 Voice Humming Transcription

As we mentioned in Section 2, Transcriber Module transcribes voice humming input into the melody expression such as uUdDr and LSR notation. It is often necessary to segment the human voice signal into several windowed frames in order to analyze and extract features including the fundamental frequency and notes. It is already known that a framing size in the range of 20~50ms is appropriate for processing. Also, human voice can be changed in any interval length. However, if the interval is too short, then analyzable frequency range is also too narrow. Thus, we assumed the framing length of 20ms as the minimum analyzable length. Furthermore, we used the frame overlapping ratio of 50% for the continuity between frames.

In order to detect the note onset and offset, we applied a new feature called WAE. WAE is an improved version of the Average Energy (AE) which is a traditional energy estimation method. AE itself indicates the average amount of energy of some signal range. However, we observed several limitations of this traditional approach. First, its classification is solely dependent on one global threshold value. Hence, it is not robust to the variation of amplitudes when the strength of human voice changes. Rather, it might be more appropriate to define multiple local thresholds according to the variation. For this reason, we revised the traditional AE and propose a method called WAE by augmenting AE with local thresholds. In WAE, local thresholds are defined for each window to classify silent/voiced frame. This could improve the accuracy of frame discrimination.

In our work, we define the ‘meaningful segment’ as one having minimal length for further analysis. With this definition, we removed tiny segments by merging tiny voiced/silent segments to the neighboring silent/voice segments, respectively. After this, AMDF (Average Magnitude Difference Function) was applied to compute the fundamental frequency of each classified frame for further usage.

In order to detect ADF onset, AF (Amplitude Function) is first calculated by summing the absolute values of amplitudes within human voice's frequency range in each frame [9]. ADF onset occurs when energy goes up rapidly from a low level or long silence. In [8], a global threshold was used in the ADF onset detection. However, for better accuracy, we propose a dynamic threshold method for ADF onset detection.

In general, when human starts singing or humming, the energy of voice contour increases suddenly. On the other hand, as singing or humming approaches to the offset, the energy of voice contour decreases gradually by a certain ratio. Based on this observation, we defined a DTC (Dynamic Threshold Curve) for classifying useful and useless ADF values. DTC can be defined as follows:

$$\mathrm{DTC}(x) = A \cdot r^{x-k}, \tag{1}$$

where x is the input element, r represents the reduction ratio, A is the amplitude element, and the last element k represents the parallel translation element.

Assuming s and e are the first and last position of a note segment, it can be shown that the following two conditions are satisfied.

$$\begin{cases} \mathrm{DTC}(s) = A \cdot r^{s-k} = \mathrm{ADF}(s) \\ \mathrm{DTC}(e) = A \cdot r^{e-k} = \hat{r} \cdot \mathrm{ADF}(s) \end{cases}, \tag{2}$$

where $\hat{r}$ is the voice contour reduction ratio. From these two equations, we can compute the value r to represent the DTC within the note segment.

$$r = \hat{r}^{\frac{k-s}{e-s}} \tag{3}$$

Also, A can be computed from the initial condition.

$$A = \mathrm{ADF}(s) \cdot \hat{r}^{\frac{k-s}{e-s}} \tag{4}$$

From these, we can get the general expression for the DTC:

$$\mathrm{DTC}(x) = \mathrm{ADF}(s) \cdot \hat{r}^{\frac{x-s}{e-s}} \tag{5}$$

Fig. 2 shows the algorithm for calculating ADF onsets using DTC. After detecting the note onset/offset and ADF onset, we integrated them with the Query Representation Algorithm based on [8]. ADF onset was applied to note onset/offset result. After this, we can obtain more accurate note segment information.

```
Procedure ADFOnsetByDTC (ADF, r, MinTh, MaxLength)
Step 1: Initialize s by 1
Step 2: For each position i ∈ [s, s+size of ADF],
          find ADF onset where ADF(i) > MinTh
          If found, store the new ADF onset into ADFOnsetList
          Otherwise, exit.
Step 3: s ← i; e ← i + MaxLength
Step 4: Make new DTC using the following equation:
          DTC(x) = ADF(s) · r^((x-s)/(e-s))
Step 5: For each position i ∈ [s+1,e],
          find another ADF onset where ADF(i)>DTC(i) and ADF(i)>MinTh
          If found, store the new ADF onset into ADFOnsetList,
          and go to Step 3
Step 6: s ← e+1, and go to Step 2
```

Fig. 2. Dynamic threshold method

Finally, it is essential to recalculate the frequency of note, because the fundamental frequencies of each frame are different within a note segment. Thus we applied the k-means algorithm for clustering the result of AMDF.

3.2 Query Refinement

As we mentioned before, in this paper, we implemented a GA-based relevance feedback scheme to improve retrieval performance. Our GA starts with initializing the population and then evaluates this population using a fitness function that returns a value for each chromosome indicating its quality as a solution to the problem.

We calculate the fitness value of the chromosome with the following formula:

$$Fitness = \frac{1}{\sum_{i=1}^{N} relevance(M_i)} \sum_{i=1}^{N} \frac{1}{i} \sum_{j=1}^{i} relevance(M_j)\text{'} \tag{6}$$

where N is the total number of music objects retrieved in population P and *relevance*(M_i) is a function that returns the relevance of the music M_i. The equation for calculating the relevance is:

$$relevance(M_i) = \frac{QueryLength - LD(M_i, Query)}{QueryLength}, \tag{7}$$

where LD is a function that calculates the lowest operating cost. Each relevance value of music object Mi ranges from 0 to 1, where '1' represents the case where the music is relevant to the user's query with full confidence and '0' indicates the opposite case. The Levenshtein Distance (LD) [10] function represents the distance between 3wo strings by the number of operations such as deletions, insertions, or substitutions required to transform one string into the other.

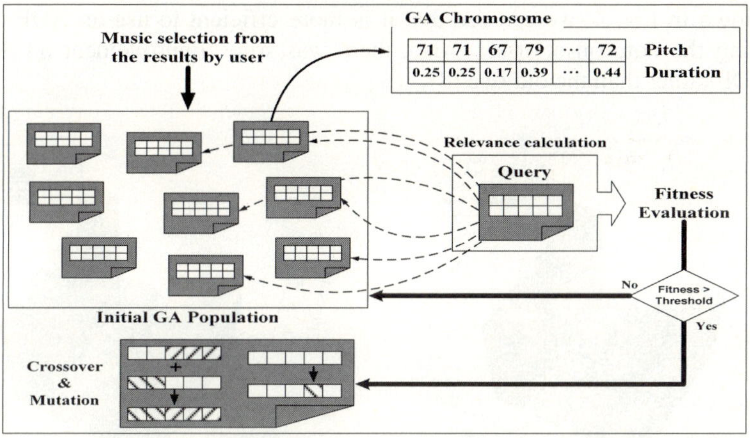

Fig. 3. Overall process of genetic algorithm

Fig. 3 shows the overall process of how the population is generated and genetic operation is performed in our system. Given the chromosomes, the genetic algorithm often requires a fitness function that returns a numeric value that represents its fitness score. This score will be used in the process of selection of the parents in current population, so that the fittest chromosome will have a greater chance of being selected. In our experiment, for the initial query result using the approximate matching, we got an initial population, for which we applied the fitness function. If the fitness score is below some threshold value, we filter the initial query result with a higher success rate and then apply the fitness function again until the fitness score gets higher than the threshold value. The chromosomes evolve in generations by means of genetic operators such as crossover and mutation. At each generation, chromosomes are selected by a tournament selection method with tournament size of 5. We used the classical single-point crossover and mutation which is implemented as a random process. A random number is generated in a given interval, in our case [-10, 10], and that number is taken as the new value for the gene that has to mutate.

4 Experiments

The voice signals were captured from a microphone, and stored as PCM wave files with 8-bit, 22.05kHz, mono. In our experiment, we considered four different types of errors. The drop error indicates when a note was lost or merged to an adjacent one during the transcription. The add error is related to the appearance of not-existing notes and the pitch error indicates that adjacent pitch changes are wrong. The duration error occurs when the difference between the detected duration and the original duration was larger than the smaller one. Each error type was represented by the following error estimation equation.

$$Error\ rate = \frac{\#\,of\ notes\ where\ errors\ occured}{\#\,of\ notes} \tag{8}$$

As shown in Fig. 4, we can see that it is more efficient to use an ADF onset for rearranging the note onset/offset. Also, there was some improvement on the DTC-based ADF rather than the old one [8].

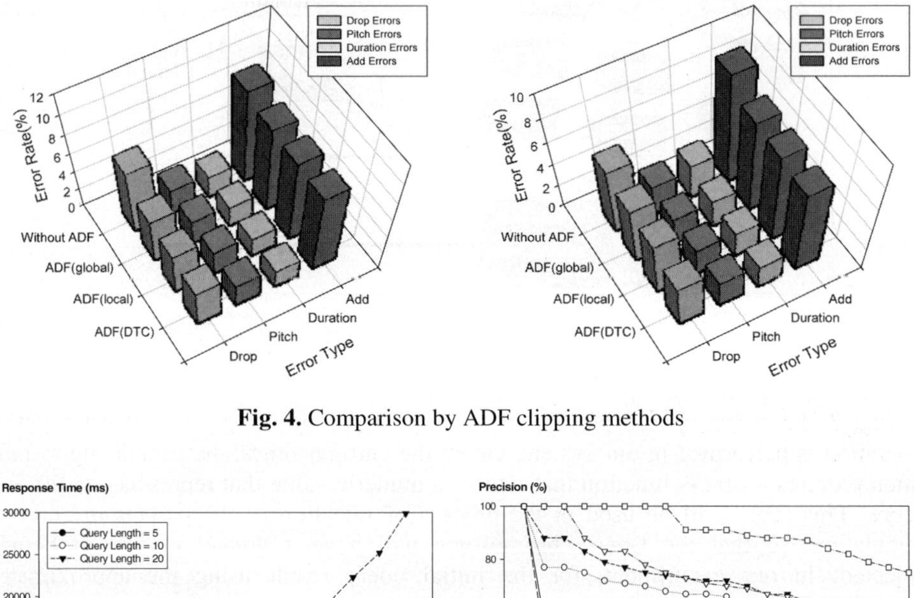

Fig. 4. Comparison by ADF clipping methods

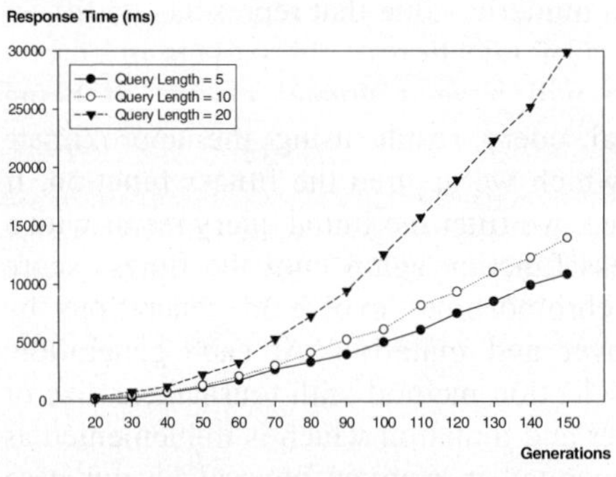

Fig. 5. Efficiency of our algorithm

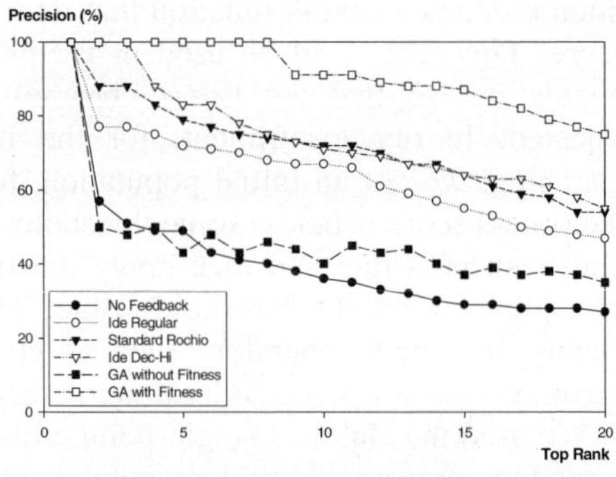

Fig. 6. Comparison with other RF methods

In Fig 5, we measured the relationship between the query length and the response time in each generation of our genetic algorithm. As we expected, a small number of generations with a few notes query such as 5 or 10 notes gives much better results than large number of generations with longer notes. We empirically found out that the optimal number of generation is 20 and reasonable query size is about 10.

In order to evaluate the effectiveness of our retrieval scheme, we ranked the retrieved music objects based on their score, and calculated the precision of the top n songs in each of the methods we considered. As shown in Fig. 6, our GA-based feedback method improved the retrieval accuracy up to 20~40%.

5 Conclusion

In this paper, we have presented a new music query transcription and refinement scheme for music retrieval. For the query transcription, we proposed two new methods, WAE (Windowed Average Energy) and DTC (Dynamic Threshold Curve)-based ADF in order to improve the transcription accuracy. For the query refinement, we proposed a new GA-based relevance feedback scheme to improve the quality of query results by reformulating a user query. We have built a prototype music retrieval system and demonstrated its effectiveness and efficiency through various experiments.

References

1. Rho, S., Hwang, E.: FMF: Query adaptive melody retrieval system. Journal of Systems and Software (JSS) 79(1), 43–56 (2006)
2. Pickens, J.: A Comparison of Language Modeling and Probabilistic Text Information Retrieval Approaches to Monophonic Music Retrieval. In: Proceedings of the International Symposium on Information Retrieval (ISMIR), Plymouth, Massachusetts, (October 2000)
3. Lemstrom, K., Wiggins, G.A., Meredith, D.: A three layer approach for music retrieval in large databases. In: 2nd International Symposium on Music Information Retrieval, Bloomington, IN, USA, pp. 13–14 (2001)
4. Hoashi, Matsumoto, Inoue.: Personalization of User Profiles for Content-based Music Retrieval Based on Relevance Feedback. ACM Multimedia, 110–119 (2003)
5. Foote, The TreeQ Package, ftp://svr-ftp.eng.cam.ac.uk/pub/comp.speech/tools/treeq1.3.tar.gz
6. Lopez-Pujalte, C., Guerrero-Bote, V., Moya-Anegon, F.: Order-based Fitness functions for genetic algorithms applied to relevance feedback. Journal of the American Society for Information Science 54(2), 152–160 (2003)
7. Gerhard, D.: Pitch Extraction and Fundamental Frequency: History and Current Techniques. In: Technical Report TR-CS 2003-06 (November 2003)
8. Park, S., Kim, S., Byeon, K., Hwang, E.: Automatic Voice Query Transformation for Query-by-Humming Systems. In: Proc. of the IMSA 2005, pp. 197–202 (August 2005)
9. Chai, W.: Melody Retrieval On the Web. In: Requirements of the degree of Master of Science in Media Arts and Sciences at the Massachusetts Institute of Technology (2001)
10. Baeza-Yates, R., Ribeiro-Neto, B.: Modern Information Retrieval. Addison Wesley, London, UK (1999)
11. Han, B., Rho, S., Hwang, E.: An Efficient Voice Transcription Scheme for Music Retrieval. In: Proc. of the IEEE MUE 2007, Seoul, Korea, pp. 366–371 (2007)

An Association Method Using Concept-Base

Noriyuki Okumura, Eriko Yoshimura, Hirokazu Watabe, and Tsukasa Kawaoka

Dept. of Knowledge Engineering & Computer Sciences
Graduate School of Engineering, Doshisha University
Kyo-Tanabe, Kyoto, 610-0394, Japan
{nokumura,eyoshimura}@indy.doshisha.ac.jp
{hwatabe,tkawaoka}@mail.doshisha.ac.jp

Abstract. This paper proposes an association method using Concept-base. Concept-base is a large-scale relational database constructed by some electronic dictionaries and some newspapers. Each concept which is defined in Concept-base has attributes which characterize the concept and weights which show each attribute's importance. Concept-base enables to calculate the depth of relationship (Degree of Association) of two any concepts using attributes and weights. This paper proposes One Word Association Method which associates one word (concept) with some relative words (concepts). Also, a threshold of Degree of Association is defined using the evaluation result of One Word Association Method. Finally, Two Words Association Method is proposed based on the threshold of Degree of Association.

Keywords: Concept-base, Degree of Association, Association Method.

1 Introduction

We human beings have achieved a wide conversation and communications by various associations in daily life. This paper proposes an Association Method using Concept-base to achieve human beings' association ability on computer[1,2,3].

An Association Method is a function which enables to associate some input words with another words on computers. For example, if Human input is 'Apple', then Association Method outputs 'Red, Sweet, Fruit $\cdots$'.

The Method is constructed with Concept-base and Calculation Method of Degree of Association[4,5]. Calculation Method of Degree of Association is the method to evaluate depth of relationship between any concepts defined in Concept-base. For example, not only similar words (Ex: Red, Purple), but also relative words (Ex: Red, Blood), Calculation Method of Degree of Association is able to quantify the relation. Using Association Mechanism, it is able to associate relative concepts from some words or some sentences, and to use knowledge flexibility.

This paper proposes One Word Association Method which associates one word (concept) with some relative words (concepts). Also, a threshold of Degree of Association is defined using the evaluation result of One Word Association Method. Finally, Two Words Association Method is proposed based on the threshold of Degree of Association.

B. Apolloni et al. (Eds.): KES 2007/WIRN 2007, Part I, LNAI 4692, pp. 604–611, 2007.

2 Concept-Base

Concept-base is constructed from Japanese Electronic Dictionary and Newspaper mechanically. It is composed, the headword is assumed to be a concept, the content words included in the explanation are assumed to be attributes. It is the database of important factor to Association Mechanism for Calculation Method of Degree of Association. Concept-base is set of concepts (words), and concept (A) is sets of pairs of some attributes (a_i) characterizing concept (A) and weights (w_i) of importance of each attributes. (Eq. 1)

$$A = (a_1, w_1), (a_2, w_2), \cdots, (a_i, w_i), \cdots, (a_{znum}, w_{znum}) \quad (1)$$

Attributes are also defined in Concept-base as concepts; so one concept is defined as n-th-order attributes chain model. In this paper, Concept-base (thirty thousand concepts) made form electronic dictionary is extended to wide Concept-base (ninety thousand concepts) added from electronic newspaper. Calculation Method of Degree of Association is evaluated with extended Concept-base.

3 Calculation Method of Degree of Association

For measure method between concepts, Calculation Method of Degree of Association [4,5] is proposed. In Calculation Method of Degree of Association, concepts are defined as attributes sets with weight, and Degree of Association is calculated in consideration of Degree of Match. For the calculation of Degree of Match, pairs of attributes sets which are Match or Similar are defined between the concepts.After the process, for the pairs of attributes, it is calculated in consideration of weights of the attributes.

3.1 Degree of Match

For two concepts: A and B, the first-order attributes and the weights are defined as follows (Eq. 2).

$$A = \{(a_i, u_i) | i = 1 \sim L\} \quad (2)$$
$$B = \{(b_j, v_j) | j = 1 \sim M\}$$

Then, it is defined of Degree of Match ($MatchWR(A, B)$) between concepts: A and B. In addition, each attributes' weights are normalized to the total of sum of weights equal 1.0.

$$MatchWR(A, B) = \sum_{a_i = b_j} Min(u_i, v_j) \quad (3)$$
$$Min(u_i, v_j) = \begin{cases} u_i \ (u_i \leq v_j) \\ v_j \ (v_j < u_i) \end{cases}$$

Degree of Match is defined by Eq. 3, because if common attributes are detected, then effective weight is smaller weight of both common attributes. In addition, Degree of Match is from 0.0 to 1.0.

3.2 Calculation Method of Degree of Association

Refined Eq. 2, lower number of attributes is A $(A(L \leq M))$. Rank of first-order attributes of concept A draw in order.

$$A = (a_1, w_1), (a_2, w_2), \cdots, (a_L, w_L) \quad (4)$$

First, it searches attributes which become $a_i = b_j$ $(MatchWR(a_i, b_j) = 1.0)$. The case is searched which becomes $a_i = b_j$ (Completely Match), if $u_i > v_j$, then $u'_i = u_i - v_j$, if $v_j \leq u_i$, then $v'_j = v_j - u_i$, then using u'_i or v'_j, it is made to correspond to other attribute. The relation is more correctly appreciable.

Second, rank of first-order attributes of concept B is decided with concept B as constructed with concept A by Eq. 2 considered with Completely Match. The number of Completely Match is α. In addition, $b_{L+\alpha+1}, b_{L+\alpha+2}, \cdots, b_M$is not used.

$$B_x = \{(b_{x1}, v_{x1}), (b_{x2}, v_{x2}), \cdots, (b_{xL+\alpha}, v_{xL+\alpha})\} \quad (5)$$
$$x_k = \{Number of attributes to which a_i is decided correspondence\}$$

Third, $ChainWR(A, B)$ is defined as follows, because $L + \alpha$ attributes correspond.

$$ChainWR(A, B) = \sum_{i=1}^{L+\alpha} MatchWR(a_i, b_{x_i}) \times \frac{u_i + v_{x_i}}{2} \times \frac{min(u_i, v_{x_i})}{max(u_i, v_{x_i})} \quad (6)$$

$$Min(u_i, v_{x_i}) = \begin{cases} u_i \ (u_i \leq v_{x_i}) \\ v_j \ (v_{x_i} < u_i) \end{cases} \quad Max(u_i, v_{x_i}) = \begin{cases} u_i \ (v_{x_i} \leq u_i) \\ v_j \ (u_i < v_{x_i}) \end{cases}$$

$ChainWR(A, B)$ corrects Degree of Association by multiplying average $((u_i + v_{x_i})/2)$ of both attribute's weight and ratio $(Min(u_i, v_{x_i})/Max(u_i, v_{x_i}))$ of both attributes' weights to $MatchWR(A, B)$. Degree of Association is from 0.0 to 1.0.

4 Association Method Using Concept-Base

This paper proposes One Word Association Method which associates one word (concept) with some relative words (concepts). Also, a threshold of Degree of Association is defined using the evaluation result of One Word Association Method. Finally, Two Words Association Method is proposed based on the threshold of Degree of Association.

4.1 One Word Association Method

One Word Association Method is a function which enables to associate a word (concept) with other words (concepts) on computers. One Word Association Method is constructed by Concept-base. For example, attributes (first-order attributes) to the concept (Driver) are extracted. Concepts which have 'Driver'

Table 1. First-order attributes and concepts to 'Driver'

Concept : Driver				
First-order attributes	Automobile	Drive	Rest Stop	Train
First-order concepts	Automobile	Drive	Train	Sunday Driver

as an attribute are gotten, because Concept-base is defined by n-th-order attributes. Table 1 shows attributes to the concept (Driver) and concepts which have 'Driver' as an attribute.

These words are assumed to be association words about "Driver". One Word Association Method outputs these words in order with high Degree of Association.

4.2 Evaluation of One Word Association Method

For evaluation of One Word Association Method, a hundred words which are defined in Concept-base are readied. About these words, first-order attributes and concepts are gotten as outputs in order with high Degree of Association. 10 persons visually evaluate the outputs. The words judged that 8 persons or grater are commonsense are assumed to be correct answer words.

Fig. 1 shows Precision and Recall about first-order attributes. Fig. 2 shows Precision and Recall about first-order concepts. Precision is the average of (the

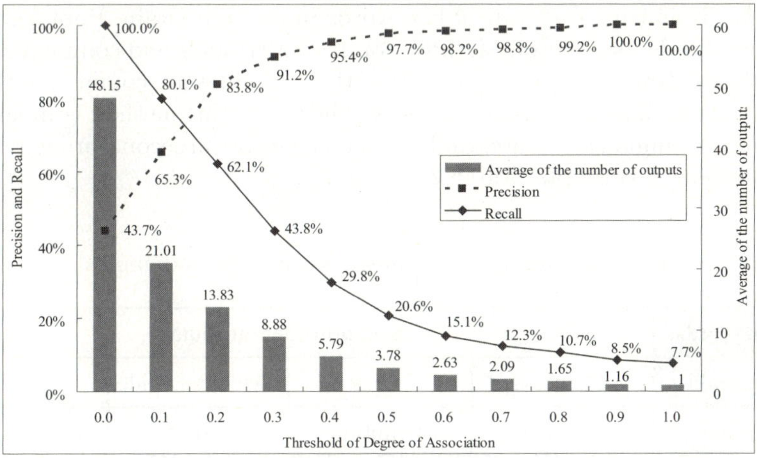

Fig. 1. Evaluation of first-order attributes

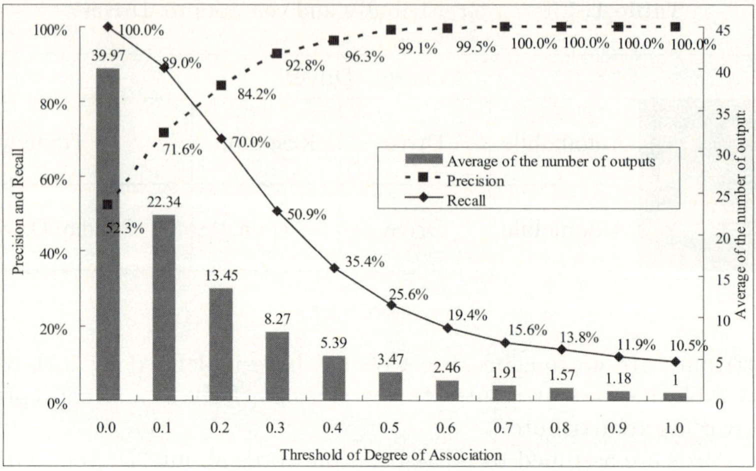

Fig. 2. Evaluation of first-order concepts

number of common outputs) / (all of outputs) for all words. Recall is the average of (the number of common outputs) / (the number of common outputs included in all outputs) for all words.

Fig. 1 and Fig. 2 show a similar tendency. Therefore, the threshold of Degree of Association is thought that 0.2 is excellent from both sides of Precision and Recall.

4.3 Two Words Association Method

'Two Words Association Method' is constructed that corresponds outputs (Ex: (Umbrella, Rain cape ···)) out of two words input (Ex: (Rain, Boots)).

In Two Words Association Method, common attributes and common concepts are presented. However, it is rare case that the attributes are common in the first-order attributes and first-order concepts. Therefore, this method is mounted by presenting a common attributes and common concepts in second-order attributes and second-order concepts (Table 2, Table 3).

Table 2. Example of common second-order attributes

Input words	Second-order attributes			
Mountain	Climbing	Peak	Mountain side	...
Climbing	Ascent	Climbing	Apex	...

Table 3. Example of common second-order concepts

Input words	Second-order concepts			
Morning	Early-morning	Breakfast	Wake up	...
Food	Eat	Breakfast	Meat	...

Degree of Association of Outputs and each input word are calculated. The average is assumed to be Degree of Association of input words and outputs. The output words are assumed to be words that became more than the threshold (0.2) which is provided by paragraph4.2.

4.4 Evaluation of Two Words Association Method

A hundred test sets (Noun, Noun) is readied for evaluation of 'Two Words Association Method'. As the evaluation method, three persons measure the outputs. If all of three judged correct answer words, then the outputs are correct. Average of (common outputs)/ (number of outputs) with each test sets is Precision. Recall is the average of (the number of common outputs) / (the number of common outputs included in all outputs) for all test sets.

Though ten high-ranking words were assumed to be an output word in 'Dynamic Calculation Method of Degree of Association between Concepts', the method and the proposal method are compared in this paper so far because the threshold of Degree of Association was set.(Fig. 3)

The performance has improved the proposal method by fig. 3 in all items (Precision, Recall and Average number of outputs). This is because proposal method sets the threshold of Degree of Association. And it is a big cause that the association number of words has expanded by having output all words more than the threshold of Degree of Association. Table 4 shows a result sample of

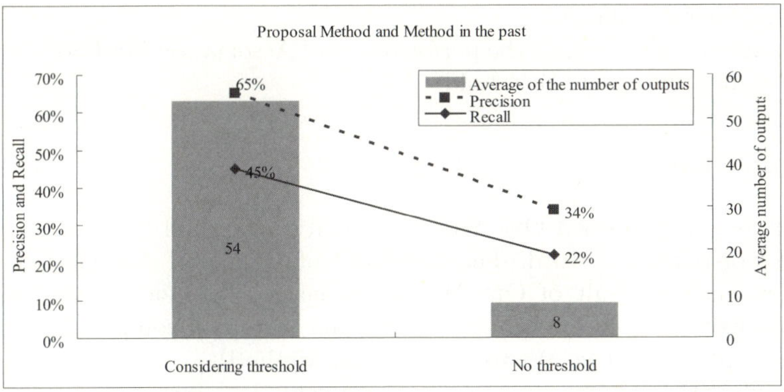

Fig. 3. Comparing proposal method and method in the past

Table 4. A result of Two Words Association Method

Input words (Capital, Company)			
Output Words	Enterprise	Company	The capital
	Business	Our company	Capital increase
	Economy	Employee	Going to the office
	Principal	Limited Company	Company house
	Investment	Joining a company	Joint management
	Shi	Property	Corporation
	Visit	Quotient association	The stocks
	Collateral company	Parent company	Stockholder
	Subsidiary company	In-house	Head quarters
	Leaving the office	Yield	Branch Office
	Fund	Family company	Joint-stock company
	Production	Foreign capital	Unlimited partnership

Two Words Association Method. The cells of the net multiplication are correct answer words.

Proposal method uses the threshold of Degree of Association defined by One Word Association Method. So in the case of Two Words Association Method, it is necessary to set the threshold of Degree of Association again. This is the reason why Precision of Two Words Association Method is lower than Precision of One Word Association Method. Two Words Association Method uses second-order attributes and concepts, so these words are not most-relative words to based input word (concept).

This paper confirmed that the performance of Association Method is improved by setting an appropriate threshold.

5 Conclusion

In this paper, it proposed One Word Association Method that associated two or more words from a word. The threshold of Degree of Association was set by the evaluation result of One Word Association Method. It was shown to have improved the performance by setting the threshold compared with a past association Method (Two Words Association Method).

Expanding the number of evaluation objects, and being construct a more flexible Association Method becomes targets. Moreover, the application to the conversation system etc. that use Association Method is expected.

Acknowledgements. This work was supported with the Aid of Doshisha University's Research Promotion Fund.

References

1. Tsuchiya, S., Watabe, H., Kawaoka, T.: A time judgement system based of an association mechanism. In: Proc. of KES2005 (Knowledge-Based Intelligent Information and Engineering System) Part III, pp. 742–748 (2005)
2. Horiguchi, A., Watabe, H., Kawaoka, T.: Constructing a sensuous judgement system based on conceptual processing. In: Gelbukh, A. (ed.) CICLing 2002. LNCS, vol. 2276, pp. 86–95. Springer, Heidelberg (2002)
3. Tsuchiya, S., Watabe, H., Kawaoka, T.: A sensuous association method using an association mechanism for natural machine conversation. In: Gabrys, B., Howlett, R.J., Jain, L.C. (eds.) KES 2006. LNCS (LNAI), vol. 4251, pp. 1001–1010. Springer, Heidelberg (2006)
4. Watabe, H., Kawaoka, T.: The degree of association between concepts using the chain of concepts. In: Proc. of SMC2001(IEEE International Conference on Systems Man and Cybernetics), pp. 877–881 (2001)
5. Okumura, N., Araki, T., Watabe, H., Kawaoka, T.: Dynamic calculation method of degree of association between concepts. In: Gabrys, B., Howlett, R.J., Jain, L.C. (eds.) KES 2006. LNCS (LNAI), vol. 4251, pp. 506–514. Springer, Heidelberg (2006)

Fair News Reader: Recommending News Articles with Different Sentiments Based on User Preference

Yukiko Kawai[1], Tadahiko Kumamoto[2], and Katsumi Tanaka[3]

[1] Undergraduate School of Science, Kyoto Sangyo University
Motoyama, Kamigamo, Kita-Ku, Kyoto-City 603-8555, Japan
Tel.: +81-75-705-2958; Fax: +81-75-705-1495
kawai@cc.kyoto-su.ac.jp

[2] Faculty of Information and Computer Science, Chiba Institute of Technology
2-17-1, Tsudanuma, Narashino, Chiba 275-0016, Japan
kumamoto@net.it-chiba.ac.jp

[3] Graduate School of Informatics, Kyoto University
Yoshida-Honmachi, Sakyo-ku, Kyoto 606-8501, Japan
Tel.: +81-75-753-5979; Fax: +81-75-753-4957
ktanaka@i.kyoto-u.ac.jp

Abstract. We have developed a news portal site called Fair News Reader (FNR) that recommends news articles with different sentiments for a user in each of the topics in which the user is interested. FNR can detect various sentiments of news articles, and determine the sentimetal preferences of a user based on the sentiments of previously read articles by the user. While there are many news portal sites on the Web, such as GoogleNews, Yahoo!, and MSN News, they can not recommend and present news articles based on the sentiments they are likely to create since they simply select articles based on whether they contain user-specified keywords. FNR collects and recommends news articles based on the topics in which the user is interested and the sentiments the articles are likely to create. Eight of the sentiments each article is likely to create are represented by an "article vector" with four elements. Each element corresponds to a measure consisting of two symmetrical sentiments. The sentiments of the articles previously read with respect to a topic are then extracted and represented as a "user vector". Finally, based on a comparison between the user and article vectors in each topic, FNR recommends articles that have symmetric sentiments against the sentiments of read articles by the user for fair reading about the topic. Evaluation of FNR using two experiments showed that the user vectors can be determined by FNR based on the sentiments of the read articles about a topic and that it can provide a unique interface with categories containing the recommended articles.

1 Introduction

As the amount of Web content continues to increase, users are more strongly demanding novel Web sites that provide better quality content. Web portal sites

B. Apolloni et al. (Eds.): KES 2007/WIRN 2007, Part I, LNAI 4692, pp. 612–622, 2007.

now gather high-quality content from many Web sites and provide integrated pages. Users can then access various kinds of information from these integrated pages without accessing many Web sites. With a news portal site on the Web such as GoogleNews and Yahoo! , users search for articles of potential interest from many collected articles by using keywords or selecting a category. Retrieval using aspects other than keywords (e.g. sentimental aspect of articles) is not supported.

Conventional news portal sites have two basic ideas behind article search and classification: one is the frequency at which keywords occur in an article[1][2][3], and the other is the structure of the links between pages[4][5]. While the user may be able to browse articles of potential interest listed in the search results, if he or she wants to read articles based on their sentimental aspects, he or she has to judge the sentimental aspects based only on the contents of the search results. For example, a search using the keywords "Iraq" and "terror" with existing news portal sites will return articles about "Iraq" with such topics as "suicide bombing in Iraq", which are likely to create a sad sentiment. A system that can recommend, for example, articles that create a happy sentiment, such as "released hostage in Iraq", should thus be useful.

We have developed a novel news portal system called Fair News Reader (FNR) that recommends articles that have symmetric sentiments against the sentiments of read articles by a user for balanced reading about each of the topics in which the user is interested. When the number of news articles about a topic that create the same or similar sentiments for a user is small, the articles are defined as "non mainstream articles" for the user. When the number is large, the articles are defined as "mainstream articles". The new technical contributions of FNR are as follows:

- mining sentiments of a news article,
- determining sentimental preferences for a user based on his or her browsing history, and
- discovering non mainstream articles for the user based on sentiments of articles and his or her sentimental preferences.

FNR recommends non mainstream articles to users in the following way. It first estimates the eight sentiments each article is likely to create: happy, unhappy, acceptance, rejection, relaxation, strain, fear, and anger. "An article vector" is created from each article; it has four elements based on the eight sentiments, and each element has a value ranging from 0 to 1, and is calculated based on the sentiments of words in the article. For example, when an element has the sentiment of "happy $\Leftrightarrow$ unhappy" from 1 to 0, if many of the words in an article create a happy sentiment, this element represented by a happy sentiment to be generated from the article has a large value. Next, the sentiments the user has about the topic of articles previously read are extracted by the article vectors and represented as "a user vector". The user vector for each topic is created from the article vectors of read articles, and calculated using the standard deviation represented by the fluctuations of each element of the vector by using read articles. Finally, for balanced reading about a topic, FNR recommends non mainstream articles that have symmetric sentiments against the sentiments of read articles by the user. To find such non mainstream articles for the topic, FNR compares the article vectors with the user vector, and selects the articles for which each element of article vector has symmetric values of the user vectors' elements. Naturally, the recommendation algorithm can change the selections from non mainstream to mainstream articles (enabling the user to read articles more in line with his or her sentiments).

2 Overview of FNR

The FNR is implemented on the system which we have developed called "My Portal Viewer (MPV)" [2], which gathers and integrates articles from many news sites based on the interests of the user. In this section, we introduce MPV and then describe the concept of FNR on MPV.

2.1 News Portal System (MPV)

MPV collects articles from various news sites by crawling through them, stores the articles in a database, integrates the content as needed, and presents them on one page, the "MPV page". This involves two unique concepts: the user's interests and knowledge based on his or her access history and the MPV page, which emulates the "look and feel" of the user's favorite news page, as shown in Fig.1.

The MPV layout mirrors that of the users' favorite news page, and parts of the original content are replaced by the integrated content. In the example shown in Fig.1, the user had specified the CNN top page as his or her favorite news page, and some of the original content have been replaced by integrated

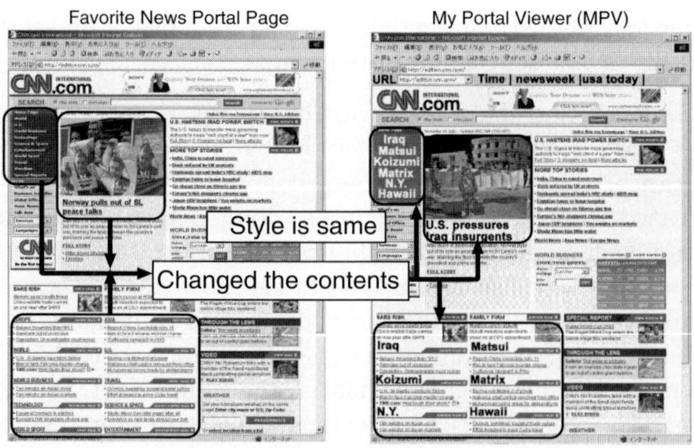

Fig. 1. Original information on favorite page is replaced with gathered information, and integrated information is shown on MPV page

content in the MPV page. Because the layout of the user's favorite news page is retained, he or she can easily locate particular information. The replaced content, categories, top news, and articles in each category, on the other hand, is created based on the frequency of term occurrence, as determined from the history of articles previously read by the user. The new category names are taken from the "interest keywords", which represent the topics in which a user is potentially interested based on the his or her access history. In the example shown in Fig.1, the categories "Iraq", "Koizumi", and "Matsui" were presented based on these keywords. Using the interest keywords as category names enables the user to easily grasp the content of articles in each category.

2.2 Concept of FNR

FNR uses the following three techniques developed for MPV. The first technique is that the gathered articles are categorized by interest keywords and "co-occurrence keywords", which are extracted from the user's browsing history. The second point is the category names are taken from the interest keywords. And the last point is that the layout of the user's favorite news portal page is used for the integrated portal page, and only part of the content is replaced by integrated content. For example, some category names are replaced with ones based on the interest keywords.

FNR offers the following three new techniques.

- It introduces new method of measure such as sentiments to news integration system.
- It recommends fair news articles for a user to him or her after it modeled the sentiments of the user about a topic.
- It integrates two or more categories that contain many of the same articles.

FNR is thus able to model a user's interests and sentiments, and the constructed model can be used to create a recommendation system.

The sentiments of the user about each article are represented by a vector with four elements, i.e. sentiments measure. They are based on Plutchik's eight basic emotions: joy, acceptance, fear, surprise, sadness, disgust, anger, and anticipation. Each scale represents two contrasting basic emotions; "happy ⇔ unhappy", "acceptance ⇔ rejection", "relaxation ⇔ strain", and "fear ⇔ anger". The value for each scale a real number between 1 and 0. For example, if an article has a value is 0.1 for "acceptance ⇔ rejection", the article should create a strong sentiment of "rejection". FNR calculates the average value and standard deviation for each scale. The four calculated standard deviations reflect the fluctuations in the interest keyword. If the standard deviation for a scale is larger than a threshold, FNR assumes that the user is not interested in news articles with a variety of sentiments in that category, and the value for that scale is defined as "*don't care*". If the standard deviation is smaller than the threshold, FNR assumes that the user is interested in news articles with biased sentiments in that category, and the value for that scale is set to the average value. As a result, the four scale have a value of "*don't care*" or the "average value", each of which is determined by the standard deviation. The determined vector of the interest keyword and the interest keyword are defined as the user's preferences. FNR categorizes articles by the interest keyword, and it selects and recommends articles in each category by using this vector.

3 User's Sentimental Preferences for Fair Reading

In this section, we describe how the user's preferences are determined based on the user's interests and the sentiments for the user about previously read articles. FNR creates a user vector after extracting the interest keywords based on the user's browsing history.

3.1 Generating Vectors from News Articles

A user sentiment vector is generated for each input article.

1. Words whose parts of speech are action nouns, adjectives, or verbs are extracted from the information obtained from Web page P_i in step 1 for interest word extraction to be described in 3.2.
2. Scale value S_{je} and weight M_{je} for $e(e = 1, 2, 3, 4)$ of sentiment scale from 0 to 1 of each word j are obtained by consulting a sentiment dictionary, as described below.
3. Scale value O_{ie} for sentiment scale e of P_i is calculated:

$$O = \sum^{j} S_{je} \times |2S_{je} - 1| \times M_{je} \Big/ \sum^{j} |2S_{je} - 1| \times M_{je}, \qquad (1)$$

   where the $|2S - 1|$ term denotes an inclined distribution depending on scale value S. When scale value S is 0.5, it is 0. When scale value S is 0 or 1, it is 1. Many of the words that appear in articles are independent of the feelings created by the articles. The inclined distribution described here was been introduced to remove the adverse effect such general words can cause in the calculations.
4. An sentiment vector for P_i is generated in the form of "$(O_{i1}, O_{i2}, O_{i3}, O_{i4})$".

The sentiment dictionary used in step 2 was automatically constructed by analyzing the Nikkei Newspaper Full Text Database[1] [6] using an extended version of the method proposed in Ref. [7]. The original method creates a sentiment scale from a pair of sentiment words, while our extended version creates an sentiment scale from two or more sentiment words. That is, we formulated which of two groups of sentiment words that composed an sentiment scale each of the words extracted from an input article would co-occur with more often

The groups of sentiment words used in constructing our sentiment dictionary are listed in Table 1, and part of the sentiment dictionary is shown in Table 2. The upper lines of each entry show the scale values, and the lower lines show the weights.

[1] This database has two million news articles accumulated over a 12-year period, from 1990 to 2001. Each edition consists of about 170,000 articles (about 200 MB).

Table 1. Impression scales designed for MPV Plus

Impression Scale	Impression Words
1. Approval – Disgust	Shonin (acceptance), Shonin-suru (approve), Aikou (love), Aikou-suru (love), Suki-da (like), Kyohi (rejection), Kyohi-suru (reject)
	Ken'o (aversion), Ken'o-suru (take an aversion), Kirai-da, (dislike)
2. Grad – Sad	Akarui (bright, encouraging), Ureshii (glad), Tanoshii (happy)
	Kurai (dark), Kanashii (sad), Kurushii (painful)
3. Relaxation – Strain	Yuttari (comfortable), Yuttari-suru (feel easy), Nonbiri (peaceful)
	Nonbiri-suru (feel relieved), Yukkuri (slowly), Yukkuri-suru (take one's time)
	Kincho (tension), Kincho-suru (become tense), Kinkyuu (emergency)
4. Anger – Fear	Okoru (get angry), Dogou (roar), Osoreru (dread), Kowai (scary), Kyofu (fear)

Table 2. Examples of entries in impression dictionary

Entry word	No. 1	No. 2	No. 3	No. 4	Entry word	No. 1	No. 2	No. 3	No. 4
Sosei (revival)	0.91	0.521	0.429	0.000	Shototsu-suru (collide)	0.344	0.353	0.315	0.529
	0.464	0.582	0.732	0.328		1.004	1.016	1.099	0.948
Shukkoku	0.596	0.209	0.762	0.201	Kenen-suru (worry)	0.373	0.319	0.246	0.293
(departure from a country)	0.975	1.049	1.065	0.701		1.447	1.440	1.521	1.275
Shibou (death)	0.28	0.358	0.260	0.364	Hofu-da (rich)	0.597	0.676	0.761	0.466
	1.132	1.272	1.306	1.112		1.416	1.352	1.299	1.109
Dassen (derailment)	0.31	0.546	0.403	0.291	Saiteki-da (optimum)	0.622	0.671	0.743	0.192
	0.514	0.603	0.737	0.549		1.185	1.164	1.145	0.899
Dekakeru (go out)	0.639	0.754	0.887	0.590	Konnan-da (difficult)	0.318	0.305	0.307	0.317
	1.430	1.394	1.304	1.114		1.451	1.526	1.528	1.274
Chosen-suru (challenge)	0.618	0.687	0.752	0.500	Fumei-da (unknown)	0.359	0.367	0.336	0.359
	1.399	1.330	1.251	1.090		1.241	1.337	1.364	1.18

3.2 Extracting Interest Keywords

1. FNR extracts and stores the metadata (title, description, URL, etc.) after it downloads pages $P_1 to P_n$ from several news sites.
2. The description and title are morphologically analyzed, and the proper nouns and general nouns are extracted.
3. The weight of each word is calculated using the term frequency and weight of the three parts of speech as in the following equation: $w_{ij} = tf \cdot idf = (log(F_j + 1)/log(F_{all}))\ (cdotlog(N/N_j))$, where F_j is the frequency of the appearance of word j in page P_i, and F_j is the frequency of appearance of all words in P_i. N is the number of all pages gathered, and N_j is the number of pages with appearance of a word j.
4. When a user reads articles on M pages, the weight W_j of word j on M pages is the summation of w_{ij}: $W_j = \sum_{i=1}^{M} w_{ij}$.
5. If W_j is larger than a certain threshold, j is identified as an interest keyword.

The detected interest keywords are used as new category names. FNR does not replace all the original category names because the number of categories on the original news portal page is limited. Instead, an "others" category is created, and the remaining interest keywords are placed there. When the user selects this category, the remaining interest keywords are displayed.

3.3 User Vector

The vector for a user is determined from the user's interest keywords and the article vectors as follows.

1. $R_1, R_2, \cdots, R_m$ are article pages read by the user, and these pages have interest keyword j.
2. The vector for the article page R_i is defined by $v_i = (v_{i1}, v_{i2}, v_{i3}, v_{i4})$.
3. μ_{je} is the average value for each element $e(e = 1, 2, 3, 4)$, and σ_{je} is the standard deviation for each element.

$$\mu_{je} = \sum_{i=1}^{m} v_{ie} \Big/ m, \quad \sigma_{je} = \sqrt{\sum_{i=1}^{m} (v_{ie} - \mu_{je})^2 \Big/ (m-1)} \qquad (2)$$

4. When σ_{je} is less than the threshold, the fluctuations in the element of the vector are small, and the value for that element for interest keyword j is defined as μ_{je}. When σ_{je} is more than the threshold, the fluctuations in the element are large, and the value of that element is defined by a "*don't care*". For example, when σ_{j2} and σ_{j3} are less than the threshold and others are more than the threshold, the user vector of a topic j is determined by (*don't care*, μ_{j2}, μ_{j3}, *don't care*).

4 Selection and Ranking of Articles Based on User Preferences

FNR selects articles from pages gathered using the user's interest keywords, and ranks the selected articles by using co-occurrence keywords and the user vectors.

1. Co-occurrence keyword k is extracted from the pages that have interest keyword j.
2. Value c_{jk} of the co-occurrence is calculated using c_{jk} = {(the number of cooccurrences of j & k) + 1 } / { (frequency of j) + (frequency of k)}
3. Article page P_i, which includes interest keyword j, is selected from m accessed articles by the user.
4. Cosine similarity is computed based on the distance between P_i and c_{jk} based on interest keyword j.
5. If P_i is more than the threshold, P_i is selected
6. The cosine similarity of sentiment D_i is computed based on the distance between article vector $v_i = (v_{i1}, v_{i2}, v_{i3}, v_{i4})$ of P_i and the $v_j = (v_{j1}, v_{j2}, v_{j3}, v_{j4})$ user vector for interest keyword j, which is calculated as described in Section 3.3, and D_i is calculated using

$$D_i = \sum_{e=1}^{4} (v_{ie} \times v_{je}) \Big/ \sqrt{\sum_{e=1}^{4} v_{ie}^2 \times \sum_{e=1}^{4} v_{je}^2}. \qquad (3)$$

However, if v_{je} is more than the threshold, the calculation of v_{je} is excluded because v_{je} is a "*don't care*" term".
7. P_i is displayed if D_i is larger than a threshold.

5 Evaluation

We have developed a prototype FNR.It was developed using Microsoft Visual Studio .Net C# and Perl. The morphological analysis was done using Mecab[8]. This section presents the experimental results obtained with the prototype system and discusses the user vector changes based on the user's browsing history. The articles were collected on April 28, 2005, between 9:00 and 9:30 a.m. from six news web sites: there were a total of 255 articles with metadata. The MPV site categorized and integrated the metadata based on user preferences. The extraction threshold for keywords of interest was set at 0.06 because at least one interest keyword must be extracted from each article; the extraction threshold for words such as proper nouns and general nouns in each article was set at 0.1 because at least 11 words must be extracted from each article.

5.1 User Vector

Figure 2 shows the changes in the average and the standard deviations for our first experiment. Both graphs show the values for each vector element, *e2* to *e4*. In this case, the user selected an article about "topic of *a country*" which has the opinion of agreement or opposition. The user initially selected the article at random and then gradually selected articles expressing opposing opinions. The standard deviations for *e2*, "grad ⇔ sad", and *e3*, "relaxation ⇔ strain" were initially relatively high, as shown in (Fig.2 (a)). Both gradually decreased as the user selected articles. The average value of *e2* was smaller than 0.5(Fig.2 (b)), indicating that the user had selected an article that should create an sad (unhappy) sentiment and that the sentiment was modeled correctly. The other average values of *e1*, *e3* and *e4* showed that the user also selected articles that created "rejection", "strain", and "anger" sentiments.

These results show that FNR can model the sentiments for a user about topics based on the article vectors by using his or her browsing behavior.

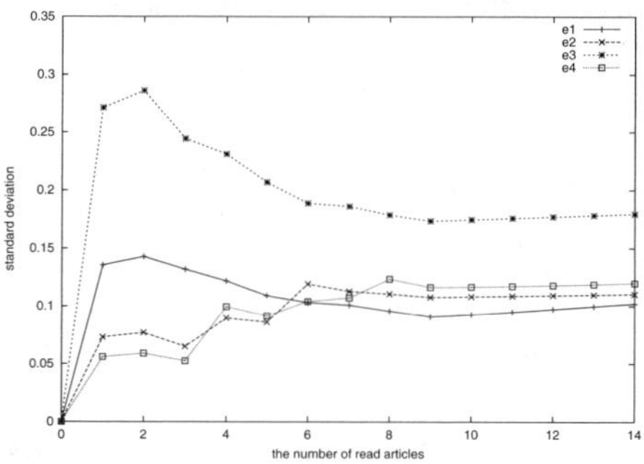

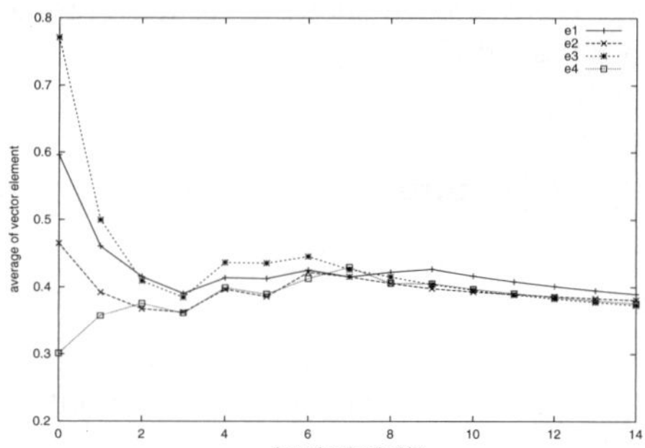

Fig. 2. Changes in average and standard deviation of impression vector based on user's browsing history (keyword of interest is a country name): Changes in standard deviation value (left), Changes in average value (right)

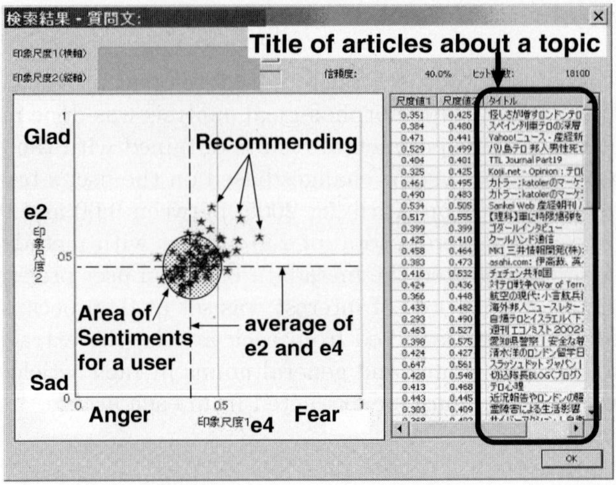

Fig. 3. Plots of sentiments for user and article vector for a topic

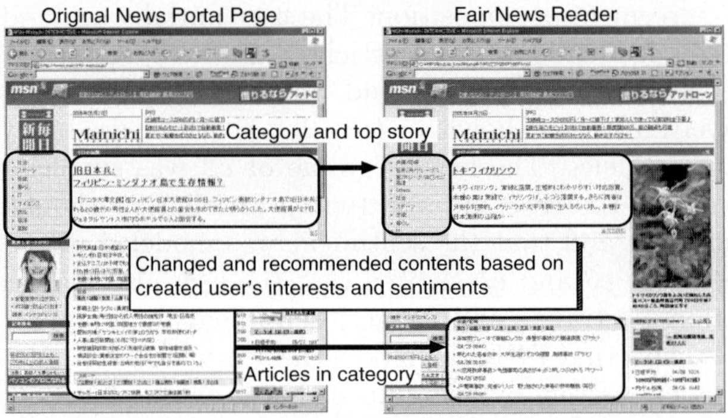

Fig. 4. FNR page showing results obtained using different original news portal page from that in Fig. 2

5.2 Recommendation Based on Sentiments for User

Figure 3 shows a plot of the article vectors and an area of sentiments for a user about a topic of "*a country*" the same as in Fig.2. The dotted lines are the average of *e2* and *e4* of the user's sentiment vector, and the stars of dots represent the articles collected about a topic.

The area of sentiments for the user is represented by circle of center *e2* and *e4*. When the value of the standard deviation was under 0.2, *e2* was 0.47 and *e4* was 0.39. The user had read articles that created "sad" and "anger" sentiments, and FNR recommended articles that created opposite sentiments, such as "glad" and "fear", for balanced reading.

5.3 FNR Prototype

Figure 4 shows an example original news portal page and the FNR page. FNR changed the content in three areas based on the user's browsing of several news pages. First, the original category keyword area was mapped onto the user's interest keywords. Next, the top news article with an image was selected based on the interest keywords and was displayed as a title that has not been read yet. Then, the titles of the article in each category were replaced with those of articles containing the interest keywords and having higher valued vector.

6 Related Work

There has been considerable investigation of portal site technology for gathering, categorizing, integrating, and recommending information.

Columbia's Newsblaster [9] is an online news summarization system in which collected news articles are categorized by event using a topic detection and tracking (TDT) method and TF · IDF. After each news article has been assigned to one of the six categories, each category is summarized using language processing technology. The user can then read a brief summary of an event based on information collected several Web pages. However, the method of article's summarization can not consider about the various sentiments or aspects.

MSN Newsbot[3] uses not only collection and classification technology but also personalization technology. A user's preferred articles are selected using personalized information based on his browsing history. However, the method of article selection is not good enough because the system only adapts to user interest, and the selected articles are not categorized.

Methods of extracting information about writers from movie reviews, book reviews, and production evaluation questionnaires also have been studied. Turney [10] proposed a method of classifying various genres of reviews into "recommended" or "not recommended". His method extracts specific patterns of phrases from input text, calculates mutual information, and takes the difference, where the two reference words were heuristically determined by him. However, using this method it is difficult to satisfy multiple impressions because the two reference words were designed only for a specific impression scale; "recommended – not recommended". We hit on the idea of classifying input documents into two or more impression classifications using a text classification method.

Many researchers have previously tackled the problem of creating more accurate classifiers using less accurate answer data [11,12]. However, they were not successful because their methods required a large amount of correct answer data. Our method, by contrast, can classify documents using only a little correct data, which can be reused as correct answer data, and which may contribute to the creation of classifiers that are sufficiently accurate.

7 Conclusion

We have developed a news portal system called Fair News Reader (FNR) which unbiasly recommends news articles with different sentiments for a user. The algorithm dynamically determines the sentiments of news articles and the sentimental preferences for a user based on previously read articles, and discovering symmetric articles against the read articles.

We will evaluate FNR more fully by monitoring the browsing behavior of many users, and will adapt the user preferences' modeling to other types of Web sites, such as shop sites.

References

1. NewsCrawler, http://www.newzcrawler.com/
2. Kawai, Y., Kanjo, D., Tanaka, K.: My Portal Viewer: Integration System based on User Preferences for News Web Sites. In: Andersen, K.V., Debenham, J., Wagner, R. (eds.) DEXA 2005. LNCS, vol. 3588, pp. 156–165. Springer, Heidelberg (2005)
3. Newsbot, http://uk.newsbot.msn.com
4. GoogleNews, http://news.google.co.jp
5. Cai, D., Yu, S., Wen, J.-R., Ma, W.-Y.: Extracting Content Structure for Web Pages Based on Visual Representation. In: Zhou, X., Zhang, Y., Orlowska, M.E. (eds.) APWeb 2003. LNCS, vol. 2642, pp. 406–417. Springer, Heidelberg (2003)
6. Nihon Keizai Shimbun, Inc.: Nikkei Newspaper Full Text Database DVD-ROM, 1990 to 1995 editions, 1996 to 2000 editions, 2001 edition, Nihon Keizai Shimbun, Inc.
7. Kumamoto, T., Tanaka, K.: Proposal of Impression Mining from News Articles. In: Khosla, R., Howlett, R.J., Jain, L.C. (eds.) KES 2005. LNCS (LNAI), vol. 3681, pp. 901–910. Springer, Heidelberg (2005)
8. MeCab (2004), http://chasen.org/~taku/software/mecab/
9. McKeown, K., Barzilay, R., Evans, D., Hatzivassiloglou, V., Klavans, J., Sable, C., Schiffman, B., Sigelman, S.: Tracking and Summarizing News on a Daily Basis with Columbia's Newsblaster (2002)
10. Peter, D.: Thumbs up or thumbs down? Semantic orientation applied to unsupervised classification of reviews. In: Proc. of Conference on Association for Computational Linguistics (2002)
11. Nagata, M., Taira, H.: Text classification — Trade fair of learning theories. In: IPSJ Magazine, vol. 42 (2001)
12. Tsukamoto, K., Sassano, M.: Text categorization using active learning with AdaBoost. In: IPSJ SIG Notes, NL126-13 (2001)

Location Name Extraction for User Created Digital Content Services

Dragan Jevtic, Zeljka Car, and Marin Vukovic

Faculty of Electrical Engineering, University of Zagreb, Unska 3, Zagreb, Croatia
{dragan.jevtic,zeljka.car,marin.vukovic}@fer.hr

Abstract. The increase in the amount of electronically stored textual data over the past decade has opened new communication possibilities in which users significantly participate in services. The result is a variety of digital content services in which new formats, such as free text form, have become particularly accepted and attractive. To support user service interaction, services must be able to obtain specific information from user defined texts. This paper proposes a model for location name extraction, constructed using a neural network trained with a Backpropagation learning algorithm. An analysis of location name interpretation, the semantic and binding problem in Croatian, as well as language specific variations are given.

Keywords: Digital content, free text, backpropagation.

1 Introduction

The telecommunications infrastructure allows interactivity, participation, and participant diversities through online connectivity, opening up new possibilities for user-created contents. In such environments, digital content has become an increasingly important part of communication services enabling significant user-participation. Digital content services include three major activities, i.e. development, distribution and access [4]. Including knowledge-based and intelligent information technologies in service domains to support user-service interaction has become an imperative. This interaction could additionally be supported by introducing intelligent methods for network and resource coordination [1][2][3]. Users become important participants in service delivery, taking a major role in content creation, marketing and distribution. In such services, free text forms for user correspondences and user-service control are in expansion.[1]

Location information provides vital service information with different forms of accessibility and various features necessary for service control. In general, a communication service can detect the source, destination and throughput of calls and messages using data from connected devices. On the other hand, documents and message content have become important new factors for service control. To

[1] This work was partially supported by Croatian ministery of science, education and sports under the project Knowledge-based network and service management.

B. Apolloni et al. (Eds.): KES 2007/WIRN 2007, Part I, LNAI 4692, pp. 623–630, 2007.

extract particular information from builtin textual content, the system must discover the syntax and semantics required to find the desired control information. In this paper, we present a method for free text analysis in Croatian, based on the application of neural networks.

This paper is organized as follows. First, the essentials of free text form are described in Section 2. A detailed description of the location name structure, the interpretation and binding problem for the Croatian language, and the model requirements for name extraction are given in Section 3. In section 4, the general characteristics and learning features of suitable neural network families and the proposed model for name extraction are introduced and described. The results of the location name extraction process applied to a data base from a Croatian dictionary are given in Section 5, followed by some concluding remarks.

2 Free Text Form in Digital Content

Digital content frequently operates with free text forms, i.e. it typically includes data which is entered into a field without any formal or predefined structure other than the normal use of grammar and punctuation. From such content, control information could be extracted and utilized to help users with their connections and to help connections performing services. Consequently, processing system have to recognize the semantic from symbol sequences and/or digits, i.e. strings. We selected the location name as an important discrimination parameter for service control, but hidden in written variants depending on text form. Service control based on location name could be used for:

- document and message content related reorganization, i.e. arranging a document in user-expected form based on the requested location,
- location-based service coordination, (for example, in user interaction with the service for information retrieval)
- travelling information services for location-based tour planning within corresponding destinations.

2.1 Location Name Structure and Semantic

A *location name* is a spoken and/or written language entity used to identify a geographical region, or a natural or man-made object. For the services that include free text form, the location name can have the following properties:

- it can consist of one or a collection of words called location name attributes,
- these attributes can be nouns, adjectives, pronouns, verbs and adverbs, combined in different forms of usage,
- it can be part of specific peoples names or surnames,
- it can include word variants which appear as a declension of nouns, adjectives or pronouns, combined with plural forms, together with verb conjugation, and variants of adverbial modifications of verbs and adverbs (typical in Slavonic languages),

– semantically, the location name can be identical to phrases used in the surrounding text,
– there are identical location names for geographically different locations,
– each name attribute can be coloured using some local modifications.

2.2 The Binding Problem and Semantic Representation

Binding representation is an important part of the general problem of language processing. A binding is the relation of an attribute and an object. In this case, the attributes are one or more words, i.e. components of location name, while the object is the location. For a single location name, with a certain construction of name attributes, several meanings could exist, each of them related to the written form. The exact meaning can be deciphered by inspecting the location name attributes and/or relations of the name and position inside the surrounding text. For example, the regular written form of the location name Sveti Martin pod Okicem (Saint Martin under Okic) could infer different meanings depending on the attribute order and options:

– "Sveti Martin pod Okicem" - meaning: the real location is inside the named place,
– "Okic Sv Martin" - could have two meanings:
  - a place in the mountain named "Okic", but the actual location is in the vicinity of place Sv Martin, or
  - exactly inside the place Sv Martin.

The exact meaning could be additionally inspected by considering surrounding text. Similar consideration can be applied for the location name Sveti Martin na Muri (Saint Martin on Mura), where the word Mura is the name of a river. On the other hand, for the location name Sveti Petar u Sumi (Saint Peter in the Forest) with a similar structure of attributes, these variations do not exist as a consequence of the semantics of the attribute sumi (forest).

2.3 Name Interpretation and Semantic Connotation

In general, there exist two types of location names. The first type includes locations whose official name contains exactly one word and can be uniquely identified from that attribute. However, there is a possibility that a user will write two or more consecutive simple location names, each consisting of one attribute, in order to emphasize the district in which the location he is referring to is, for example Pula Istra (Pula Istria). The second type of location name contains those whose regular name consists of two or more attributes. Name interpretation should consider the following characteristics:

– *Similarity* - similarity between two or more different location names, which differ only by a letter, or in the position of a letter in the word (for example, Barci and Baric; Marina, Marine, Marini and Marin, etc.). In this case, it is crucial to avoid generalization, i.e. to classify these typical noun declensions as different location names, which is quite in contrast to the property required for other textual parts.

- *Abbreviated forms* - the user may try to shorten the name of a location and write an acronym instead, for example V. Gorica, or Vel. Gorica instead of Velika Gorica. However, since there are a limited number of acronyms commonly used, the system can include all, or most, human expected textual forms.
- *Prefix and suffix* - another issue concerns location names which contain name attributes representing other actual location names. For example, it is common that part of a town has the same name as the town itself, but with a certain prefix. For example, Novi Zagreb is a part of Zagreb, where "novi" means "new". The same can hold true for smaller neighbouring towns. Since these prefixes (and suffixes) are typically not just location-related words, in some circumstances is necessary to carefully inspect the surrounding words.

In addition, some location names consist of attributes which typically arise in other textual slogans. For instance, the rural community named "Dobra Kuca" means "Good House", but neither "Dobra" nor "Kuca" are authentic location names. Additionally, the name "Dobra" is the name of river. Therefore, in some circumstances, to find a specified location it is necessary to inspect the words, even if they are not adjacent to actual location names.

3 Short Review of Backpropagation

Backpropagation learning is an essential algorithm for learning neural networks, with certain modifications to fit the exact network structure and its specifics. A neural network learns from examples, trying to separate the input vector space [6]. The task of learning is to make decision regions for a set of input patterns. Ordinary, the network structure consists of more than one processing layer (Fig. 1 shows an example). During the learning process, weights are adjusted by an error backpropagation algorithm which minimizes the sum of the squared errors. The weights in the next (n+1) step are changed by the amount,

$$\Delta W_{pq,k}(n+1) = \eta(\delta_{q,k} \cdot OUT_{p,j}) + \alpha[\Delta W_{pq,k}(n)] \tag{1}$$

$$W_{pq,k}(n+1) = W_{pq,k}(n) + \Delta W_{pq,k}(n+1) \tag{2}$$

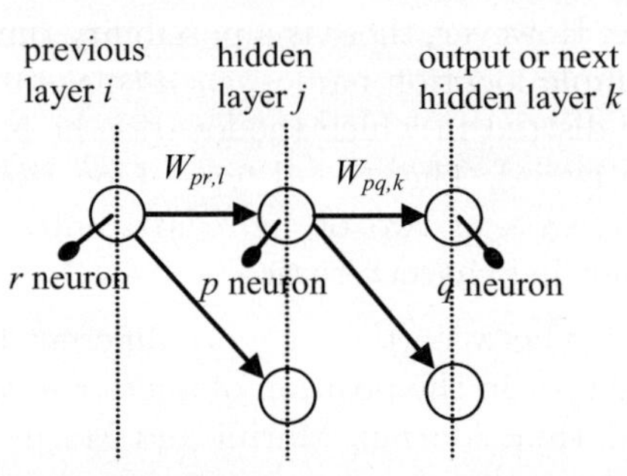

Fig. 1. General structure of neural network elements, units, links and their weights in multilayer feed-forward neural network

where $W_{pq,k}$(n) is the weight from the neuron p in the j layer to the neuron q in the next layer, the layer k; the n represents the n-th step, the step before adjustement and $n+1$, the step after adjustment; ΔW is the amount of weights adjustment; $OUT_{p,j}$ is the output of the neuron p in the j-th layer; $\delta_{q,k}$ is the factor in gradient descent of the sum of squared errors and, finally, η and α are learning rate and momentum, respectively. Factor δ for the output layer neurons has a value of

$$\delta_{q,k} = OUT_{q,k}(1 - OUT_{q,k}) \cdot (T_k - OUT_{q,k}) \tag{3}$$

where T_k is the q-th component of the teaching vector corresponding to the q-th component of the output vector $OUT_{q,k}$. For hidden layer, neurons factor $\delta$is given by the expression:

$$\delta_{q,k} = OUT_{q,k}(1 - OUT_{q,k}) \cdot (\sum_r \delta_{r,l} W_{pr,l}) \tag{4}$$

where $W_{pr,l}$ is the weight from the neuron p in the j-th layer to the neuron r in the previous (i-th) layer.

The equations (1) to (3) are the standard error backpropagation function for training in addition with equation (4) which is used only for sigmoidal feed-forward neural networks. Hence, depending on the input vector space characteristics, in order to improve its learning properties, a combination of activation functions for the hidden layer can be applied [5]. The width of input is constant, and re-learning includes repeating the learning process with old and new patterns included.

4 The Proposed Model for Location Name Extraction

A system with two classification modules was used for location name extraction. These modules were previously learned[2] by error backpropagation. The interface contained control logic and monitored the outcomes of both modules. The system architecture is depicted in Fig. 2. Strings of free text served as input to the word module which discovered the corresponding semantics and assigned a classification index to each identified word. The output was recognized and the classified words were ordered in the same array as in the free text. Unrecognized strings were cancelled. Recognized strings were categorized as words with labels which represent their semantics and classification. Three classes were used, i.e. the word is a single attribute location name, the word is an attribute of a location name (composite name), and the word is not an attribute of a location name. Subsequent words indexed as the attributes of location names could have other uses in the free text and were therefore additionally inspected by the binding module. Thus, the input to the binding module contained only groups of adjacent words, all of which were classified as location name attributes.

[2] Learning was performed by the Stutgart Neural Network Simulator.

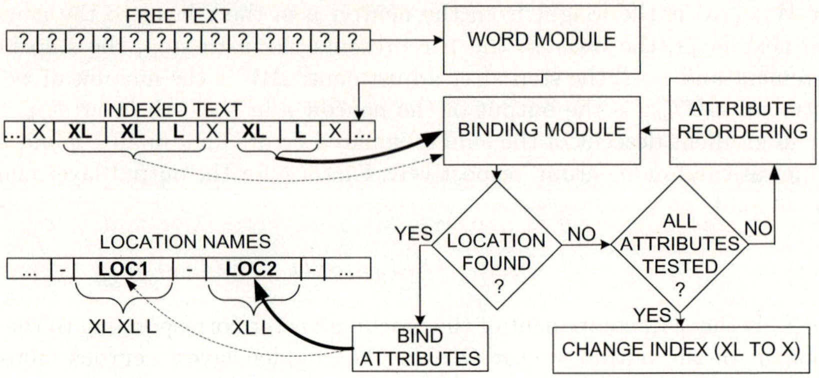

Fig. 2. Flow chart of program logic which employs two modules built by neural networks for classification and binding

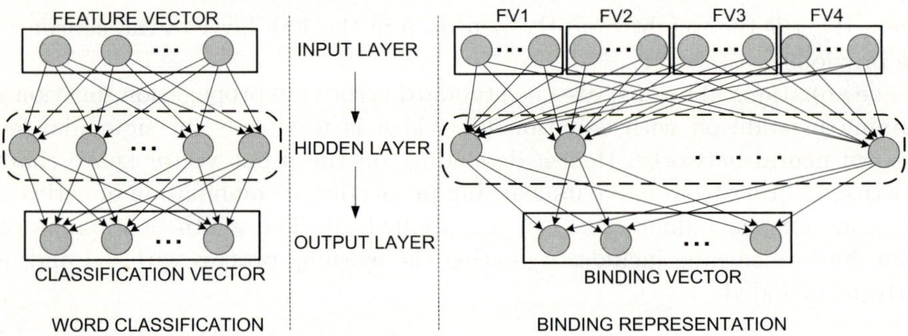

Fig. 3. The internal structure of the word classification and binding modules

The word module was used to recognize and classify the words that appeared in free text form of a particular category. The task of the word module was to identify the words, make syntax corrections, and classify them into three groups using indexes: X-not location name, XL - location name attribute and L single attribute location name (fig. 2). This task was performed by neural networks built inside the module. The word module consisted of 25 neural networks, each of them with 25 input, 500 hidden, and 750 output units. The inputs for each network were feature vectors (FV) with 25 dimensions which computed each word separately on the basis of letter positioning. To train and test the word set, it was divided into subsets based on the first letter of each word. Simple program logic was used to transmit an actual string to the input of the appropriate network, corresponding to the right subset. All known variations of a single word may be included in the training set. The corresponding network is trained to activate a single output unit which was assigned a label to identify, and a label to classify the word.

The binding module consisted of one neural network with 100 inputs, 475 hidden, and 360 output units. Up to four feature vectors were presented to the network input. These feature vectors were selected outputs of the word module, i.e. these words were previously inspected (fig. 3). Further, each feature vector was created in the same way as the input for the word module. Bindings were selected manually, by a human expert. Up to four subsequent words, indexed by XL or XL and L, were inspected by the binding module. The task of the binding module was to detect the existence of composite location names consisting of a group of attributes.

Binding discovery was designed as a recursive process in which name attributes, represented as FV, were presented multiple times to the network input by reordering attributes. Reordering attributes is the process of eliminating end words from the array, i.e. removing the first word on the list, then returning the first and removing the last, then removing both and so on. Each of these combinations was inspected by a neural network (fig. 3). The elimination process stopped when one word remained, or a group was recognized and classified by group index $LOCn$ (fig. 2). Eliminated words previously classified by XL changed its class to X.

5 Results

The neural networks in the word module in fig. 2 were taught to recognize 2000 location names and tested with the first 60,000 words from a Croatian dictionary data base with a total of 300,000 words. Letters with diacritical marks were substituted. Two experiments were performed. In the first one, the learning data base consisted of only 2000 location names chosen from one region with one neural network for all patterns inside the word module. In the second, the word module consisted of 25 neural networks as described in section 4. The training set was created manually by a human expert. Independent learning and testing was performed for each neural network separately. The results are shown in table 1. The first column of the table represents the selected location names. The outcomes of the first experiment are placed in columns 2 and 3. The numbers inside the parenthesis of column 2 show the total number of variations

Table 1. Outcomes of classification module trained for location names

Location name	(1st experiment) Classified as location	(1st experiment) Replacement errors	(2nd experiment) Classified as location	(2nd experiment) Replacement errors
Ajduci	1(6)	627	1(6)	0
Benkovac	3(5)	681	4(5)	0
Biocic	1(6)	514	2(6)	1
Delnice	1(5)	829	2(5)	9
Biograd	3(5)	75	3(5)	5

of location name and generalization features. For example, 3 from 5 variants of town Benkovac were recognized, but 681 replacement errors appeared (see Column 3). The second experiment in which 25 neural networks were used, shows a small generalization improvement in column 4 and a considerable decrease of replacement errors (see column 5). On the other hand, bindings were inspected for 2187 composite location names consisting of up to 4 attributes, and chosen from a total of 6000 location names in Croatian. The proposed configuration of neural networks in the binding module was able to carry out the required bindings and mainly eliminate undesired binding combinations without the need to extend the training set with undesired bindings.

6 Conclusion

This paper deals with the application of artificial neural network techniques for location name detection. The use of text intensive services, such as mail systems, has grown significantly in flexible information retrieval systems, aiming to help users find relevant information. Location name was selected as a vital control parameter for messaging services independent of message category. As expected, results indicate the need for extra carefully selected training patterns, to bring the classification boundary to the required level due to the conflicting features described in section 3.

References

[1] Jevtic, D., Kunstic, M., Matijasevic, S.: Load Protection Model based on Intelligent Agent Regulation. In: Gabrys, B., Howlett, R.J., Jain, L.C. (eds.) KES 2006. LNCS (LNAI), vol. 4252, pp. 284–291. Springer, Heidelberg (2006)
[2] Jevtic, D., Kunstic, M., Ouzecki, D.: The Effect of Alteration in Service Environments with Distributed Intelligent Agents. In: Khosla, R., Howlett, R.J., Jain, L.C. (eds.) KES 2005. LNCS (LNAI), vol. 3683, pp. 16–22. Springer, Heidelberg (2005)
[3] Kusek, M., Lovrek, I., Sinkovic, V.: Teamwork Coordination in Large-scale Mobile Agent Networks. In: Gabrys, B., Howlett, R.J., Jain, L.C. (eds.) KES 2006. LNCS (LNAI), vol. 4251, pp. 236–243. Springer, Heidelberg (2006)
[4] Organisation for Economic Co-operation and Development (OECD): Digital Brodband Content – Digital content strategies and policies DSTI/ICCP/IE (2005)
[5] Weymaere, N., Martens, J.P.: On the Initialization and Optimization of Multilayer Perceptrons. IEEE Transactions on Neural Networks 5(5), 738–751 (1994)
[6] Rumelhart, D.E., McClelland, J.L., Wiliams, R.J.: Parallel Distributed processing Explorations in the Microstructure of Cognition. MIT Press, Cambridge (1987)

Understanding Support Method of Unknown Words Using Robot Type Search Engine

Kazuto Goto, Noriyuki Okumura, Hirokazu Watabe, and Tsukasa Kawaoka

Dept. of Knowledge Engineering & Computer Sciences,
Doshisha University, Kyo-Tanabe,
Kyoto, 610-0394, Japan
{kgotou,Nokumura,Hwatabe,Tkawaoka}@indy.doshisha.ac.jp

Abstract. In the daily conversation, new words and proper nouns (unknown words) appear in some cases. This paper proposes the method to paraphrase unknown words to common words using Thesaurus in which knowledge concerning noun is defined by a tree structure. This method uses the search results of robot type search engine (Google) as a source concerning unknown words. The relation to each node of Thesaurus is evaluated based of the search results, and unknown words are classified as an appropriate node. This method can support the user's understanding of unknown words.

Keywords: Thesaurus, Concept-base, Degree of Association, Unknown Words.

1 Introduction

In the daily conversation, new words and proper nouns (unknown words) appear in some cases. Recently, we can get various information about unknown words by the spread of the Internet. However, it is not necessarily the case that the page which is described about unknown words appears upper level of the search results.

This paper proposes the method to paraphrase unknown words to common words using Thesaurus[1] in which knowledge concerning noun is defined by tree structure. This method uses the search results of robot type search engine (Google[2]) as a source concerning unknown words. The relation to each node of Thesaurus is evaluated based of the search results, and unknown words are classified as an appropriate node. This method can support the user's understanding of unknown words.

2 Thesaurus

Figure 1 shows Thesaurus. Thesaurus is a database in which the relation of general nouns is defined by a hierarchical tree structure. Number of general nouns (node) is 2710. There are super-sub relation and whole-part relation between nodes. Also, a concrete noun is allocated in each node as leaf. The Leaf of about 130000 words is defined in entire Thesaurus. This paper uses 370 nodes out of 2710 nodes.

B. Apolloni et al. (Eds.): KES 2007/WIRN 2007, Part I, LNAI 4692, pp. 631–638, 2007.

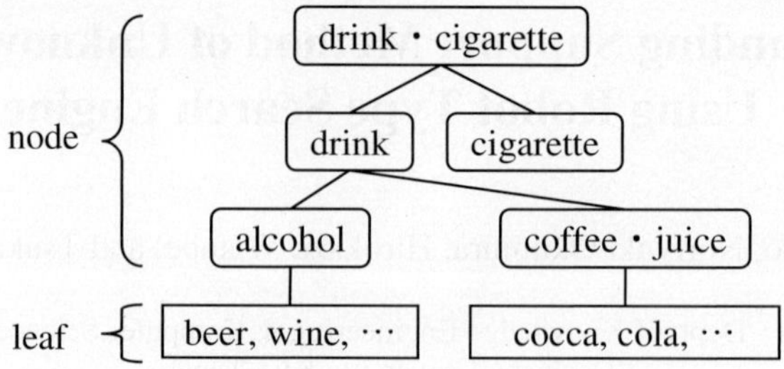

Fig. 1. Thesaurus

3 Concept-Base[3,4]

The concept-base is a large, automatically generated database that consists of keywords from multiple electronic dictionaries as the concepts and independent words in the descriptions for the keywords as concept attributes. In this paper, the concept-base was used, which has approximately 90000 concepts after the refining process, that is, after automatic generation, the attributes that were not appropriate as human senses were removed and the necessary attributes were added.

In the concept-base, a given concept A is expressed with the pair of the attribute a_i that expresses the semantic characteristics of the concept and the weight w_i that indicates how important the attribute a_i is to express the concept A. When the number of attributes of the concept A is m, the concept A can be expressed as Eq.1. The attribute a_i here is called the primary attribute of the concept A.

$$A = \{(a_1, w_1), (a_2, w_2), \cdots, (a_m, w_m)\} \tag{1}$$

Since the primary attribute a_i of the concept A is the concept defined in the concept-base, a_i can also have its own attributes. This is called the secondary attribute of the concept A. Table 1 shows the expansion of the Concept "train" to secondary attributes.

Table 1. The Concept "train" expanded to secondary attribute

	attribute	weight	attribute	weight	attribute	weight
private attribute	train	0.36	locomotive	0.21	railroad	0.10
secondary attribute	train	0.36	locomotive	0.42	railroad	0.38
	locomotive	0.21	train	0.23	railway	0.25
	railroad	0.10	chuffer	0.11	train	0.18

In this paper, concept-base is used for Acquiring Attributes of Unknown Words from Web(ref. Section 5.1) and Acquiring Attributes of Nodes of Thesaurus (ref. Section 5.2).

4 Calculation of Degree of Association[5]

For concepts A and B with primary attributes a_i and b_i and weight u_i and v_i , if the number of attributes are L and M , respectively $(L \le M)$, the concepts can be expressed as follows.

$$A = \{(a_1, u_1), (a_2, u_2), \cdots, (a_L, u_L)\},\ B = \{(b_1, v_1), (b_2, v_2), \cdots, (b_L, v_L)\}$$

The Degree of Identity $I(A, B)$ between concepts A and B is defined as Eq.2 (the sum of the weight of the various concepts is normalized to 1).

$$I(A, B) = \sum_{a_i = b_j} \min(u_i, v_j) \tag{2}$$

The Degree of Association is calculated by calculating the Degree of Identity for all of the targeted primary attribute combinations and then determining the correspondence between primary attributes. Specifically, priority is given to determining the correspondence between matching primary attributes. For primary attributes that do not match, the correspondence between primary attributes is determined so as to maximize the total Degree of Matching. Using the Degree of Matching, it is possible to give consideration to the Degree of Association even for primary attributes that do not match perfectly.

When the correspondences are thus determined, the Degree of Association $R(A, B)$ between concepts A and B is as Eq.3.

$$R(A, B) = \sum_{i=1}^{L} I(a_i, b_{xi})(u_i + v_{xi}) \times \{\min(u_i, v_{xi}) / \max(u_i, v_{xi})\} / 2 \tag{3}$$

In other words, the Degree of Association is proportional to the Degree of Identity of the corresponding primary attributes, and the average of the weight of those attributes and the weight ratios.

This paper uses calculation method of Degree of Association to judge the strength of relation between the attribute of undefined words and the attribute of nodes of Thesaurus.

5 Understanding Support Method of Unknown Words

Understanding support method of unknown words acquires attributes of unknown words and nodes after inputting an unknown word. These attributes are words which exist in the concept-base. Then, the unknown word is classified as an appropriate node using these attributes.

As the processing flow, first, attributes of an unknown word and nodes are acquired to compare the character of the unknown word and nodes. Next, calculation of Degree of Association is performed using acquired attribute group, and the belonging candidate nodes are narrowed. In addition, the node which the unknown word belongs is defined based on Thesaurus-specific information. Figure 2 shows the flow which an unknown word is classified as the node of Thesaurus.

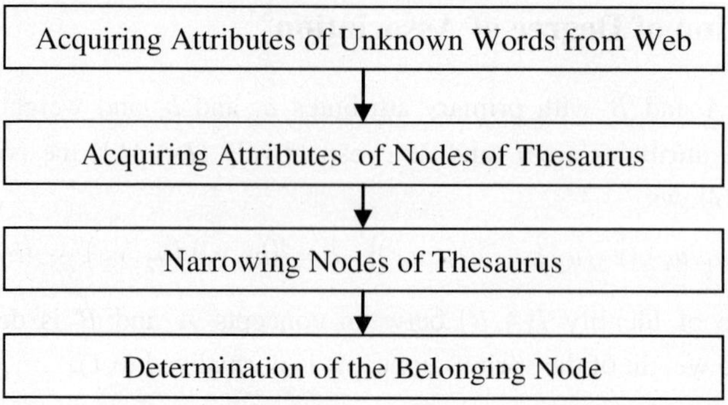

Fig. 2. The flow of determining the node

5.1 Acquiring Attributes of Unknown Words from Web

An unknown word is input, and the documents which are described about the unknown word are acquired from the search result pages of Google. The words included in the search result pages are attributes of unknown words, and attributes of unknown words is concepts which are defined in concept-base. The weight of attributes is granted by tf and idf. Tf is the frequency which unknown words appear on the search result pages. Idf is calculated from the number of the search result pages and the number of all pages of Google. Table 2 shows some examples of the acquired attributes of unknown words.

Table 2. The attributes of unknown words "Harrison Ford" and "FinePix"

Harrison Ford		FinePix	
attribute	weight	Attribute	weight
movie	225.16	Digital	331.21
actor	120.77	Camera	326.95
appearance	87.46	Pixel	301.11

5.2 Acquiring Attributes of Nodes of Thesaurus[6]

The attributes and weight of all leafs which are allocated in each node are acquired from the concept-base. The assembly of these attributes and weight is attributes and weight of the node. A similar operation is performed all nodes, and the attributes of nodes of Thesaurus is determined. The weight of each attribute is a value which acquired weight is multiplied by idf. Table 3 shows the attributes of node "clock" as an example.

Table 3. The attributes of node "clock"

attribute	weight
pocket watch	4733.49
Wall clock	3476.39
bracket clock	2791.44

5.3 Narrowing Nodes of Thesaurus

Calculation of Degree of Association is performed using acquired attributes of unknown words in Section 5.1 and acquired attributes of nodes in Section 5.2. And then, nodes which Degree of Association is 0.02 or more are belonging candidate nodes. Degree of Association of 0.02 is experimentally-determined.

5.4 Determination of the Belonging Node

5.4.1 Node-Verb

Node-verb is a verb which is right as the meaning of sentence, when sentences have a node as a subject or an object. This paper provides each node with one node-verb. Table 4 shows an example of node-verb.

Table 4. An example of node-verb

node	node-verb
tobacco	smoke
snack	eat
camera	film

To assist the determination of the belonging node, the number of hits of AND search in Google is acquired by using node-verb and unknown words. For example, when an unknown word is "Marlboro" and belonging candidate nodes are "tobacco" and "snack", AND search of "Marlboro, smoke" and "Marlboro, eat" is performed in Google. Then, the number of hits of this AND search is acquired (Table 5).

Table 5. The number of hits by node-verb

Keyword	The number of hits
Marlboro smoke	1220000
Marlboro eat	659000

5.4.2 Coincidence-Hit

It is highly possible that two related words appear in documents of Web at the same instant. And so, to assist the determination of the belonging node, the number of hits of AND search in Google is acquired by using belonging candidate nodes and unknown words. For example, when an unknown word is "Marlboro" and belonging candidate nodes are "tobacco" and "snack", AND search of "Marlboro, tobacco" and "Marlboro, snack" is performed in Google. Then, the number of hits of this AND search is acquired (Table 6).

Table 6. The number of hits by coincidence-hit

Keyword	The number of hits
Marlboro tobacco	1210000
Marlboro snack	217000

5.4.3 Determination Method of the Belonging Node

The following four methods are proposed as determination method of the belonging node (Table 7). Eq.4, Eq.5, Eq.6, and Eq.7 are computation expression which is used by each method. The node which has highest node-score $NodeValue$ among belonging candidate nodes $node_i$ is the belonging node of the unknown word.

Table 7. The list of Determination method of the belonging node

method-id	determination method of the belonging node	computation expression
(1)	calculation Degree of Association between acquired attributes of unknown words and node	Eq.4
(2)	(1) + node-verb	Eq.5
(3)	(2) + coincidence-hit	Eq.6
(4)	(3) + node-verb + coincidence-hit	Eq.7

$$NodeValue(node_i) = RelValue(node_i) \tag{4}$$

$$NodeValue(node_i) = RelValue(node_i) \cdot \log(VerbHit(node_i)) \tag{5}$$

$$NodeValue(node_i) = RelValue(node_i) \cdot \log(CoincidenceHit(node_i)) \tag{6}$$

$$NodeValue(node_i) = RelValue(node_i) \cdot \log(VerbHit(node_i)) \cdot \log(CoincidenceHit(node_i)) \tag{7}$$

$RelValue$ is Degree of Association between the unknown word and $node_i$, $VerbHit(node_i)$ is the number of hits when AND search of the unknown word and node-verb is performed, and $CoincidenceHit(node_i)$ is the number of hits when AND search of the unknown word and node is performed.

6 Evaluation of Understanding Support Method of Unknown Words

To evaluate understanding support method of unknown words, this paper prepares test set which is the pair of 200 words and belonging nodes of these words. Table 8 shows an example of test set which is used for evaluation.

Table 8. Testset

unknown word	belonging node
Harrison Ford	Actor
FinePix	camera
Marlboro	tobacco

Words of test set are allocated to the node by understanding support method of unknown words. The methods which are shown in Table 7 were tried. Precision rate are defined as follows.

Precision rate = number of correct answers / total number of words (200)
Correct answer = the word which is allocated to an appropriate node
Incorrect answer = the word which is allocated to an inappropriate node

Figure 3 shows the evaluation result for each method, and Figure 4 shows the average calculation time of 200 words.

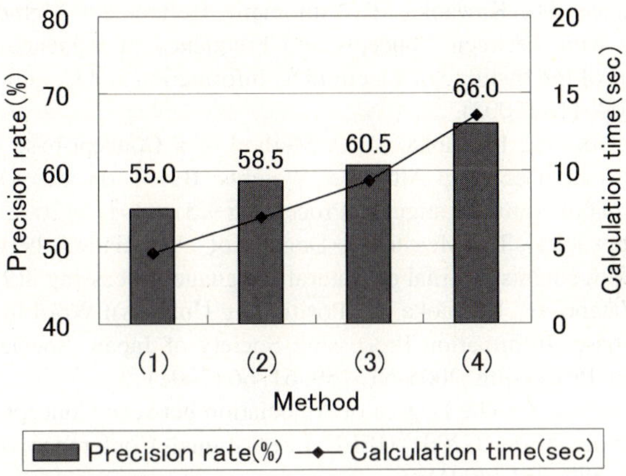

Fig. 3. Precision rate and the average calculation of each method

As shown Figure 3, the method for determining the appropriate node only by Degree of Association (method (1)) had 55.0% precision rate. In contrast, the method for determining the appropriate node considering node-verb and coincidence-hit (method (4)) had 66.0% precision rate. Also, the average calculation time lengthens from method to method. Because communication time with Web happens to find the number of hits, the method which uses node-verb or coincidence-hit needs more calculation time.

7 Conclusion

This paper proposed the method to paraphrase unknown words which appear in the daily conversation to common words by allocating node of Thesaurus. This method can translate unknown words into one word, and it can be concluded that the smooth conversation can materialize. In addition, the calculation time is only about 10 seconds, and users themselves can reduce time and labour to acquire information about unknown words from the search result page.

This proposed method has a problem that can not handle ambiguity, i.e. that a word can have different meanings. As challenges for the future, it is necessary to consider eliminating ambiguity over words from anteroposterior contexts.

Acknowledgements

This work was supported with the Aid of Doshisha University's Research Promotion Fund.

References

1. NTT Communication Science Laboratory: NIHONGOGOITAIKEI, Iwanami Shoten (1997)
2. Google, http://www.google.co.jp
3. Hirose, T., Watabe, H., Kawaoka, T.: Automatic Refinement Method of Concept-base Considering the Rule between Concepts and Frequency of Appearance as an Attribute. Technical Report of the Institute of Electronics, Information and Communication Engineers, NLC2001-93, 109–116 (2002)
4. Kojima, K., Watabe, H., Kawaoka, T.: A Method of a Concept-base Construction for an Association System: Deciding Attribute Weights Based on the Degree of Attribute Reliability. Journal of Natural Language Processing 9(5), 93–110 (2002)
5. Watabe, H., Kawaoka, T.: Measuring Degree of Association between Concepts for Commonsense Judgements. Journal of Natural Language Processing 8(2), 39–54 (2001)
6. Shunsuke, I., Watabe, H., Kawaoka, T.: Positioning Unknown Word in Existing Thesaurus Using Concept-base. Information Processing Society of Japan, Special Interest Group of Natural Language Processing, 2004-NL-159, 61–66 (2004)
7. Watabe, H., Kawaoka, T.: The Degree of Association between Concepts using the Chain of Concepts. In: Proc. of SMC2001 (IEEE International Conference on Systems Man & Cybernetics), pp. 877–881 (2001)

AI Techniques for Waste Water Treatment Plant Control Case Study: Denitrification in a Pilot-Scale SBR

Davide Sottara, Luca Luccarini, and Paola Mello

[1] DEIS, Facoltà di Ingegneria, Viale Risorgimento 2 Bologna
[2] ACS-PROT-IDR – Gestione Risorse Idriche - ENEA Bologna
dsottara@deis.unibo.it, luca.luccarini@bologna.enea.it
pmello@deis.unibo.it

Abstract. We propose to show how different AI techniques might be used in the development of a modular expert system, acting as a manager and advisor for the operation of a pilot-scale SBR urban wastewater treatment plant, fed with real sewage. The plant's depurative effectiveness and global biomass' health depend on the reactions of nitrification and denitrification, with the former taking place as soon as the latter is complete. Since the duration of the reaction cannot be predicted, we have trained an intelligent software to recognize the event analyzing the profiles of some available signals, namely pH, orp and dissolved oxygen, thus allowing us to optimize the process' yield and detect possible failures. Using a SOM neural network, the system has been trained to remember an adequate set of reference signals, which have been given meaning using Bayesian belief techniques. Eventually, using the formalism provided by logical languages, reasoning capabilities have been imparted to the system, allowing the real-time, online deduction of new pieces of needed information. Thanks to the integration of these techniques the system is able to assess the status of the plant and act according to the adequate known policies.

1 Introduction

In Italy, the largest WasteWater Treatment Plants are usually equipped with Software Control And Data Acquisition (SCADA) and/or Programmable Logic Controller (PLC) systems, but the large amount of data acquired from the sensors is seldom actually used in plant management, but simply plotted and stored for later use, typically by human operators in case of faults. An intelligent control software, instead, could act as a "virtual operator", monitoring the processes continuously, in order to optimize the yield and detect and possibly correct faults at an early stage.

Sequencing Batch Reactors are natural candidate targets for this kind of technology, since such plants allow highly customizable, different operational conditions without the need of structural modifications.

The treatment of urban wastewater requires a sequence of two chemical reactions, denitrification in an anoxic environment and nitrification in aerobic conditions. Research in the field has focused on the detection of the end of such reactions in order to optimize the operational cycles duration and reduce costs. The former, in fact, may be identified analyzing the profiles of the observed signals pH and orp, when the first reaches a maximum and the second shows a sort of flex or an abrupt change in trend.

B. Apolloni et al. (Eds.): KES 2007/WIRN 2007, Part I, LNAI 4692, pp. 639–646, 2007.

Actually, this is but one of many possible control policies, which could be handled by a knowledge-based system capable of learning and interpreting the sensor data's meaning in real-time. Such a system should be able to learn, in order to adapt to seasonal and process fluctuations, to recognize known situations, extracting features and patterns from the signals, and apply domain knowledge to assess the plant's health status and advise or choose itself the most appropriate control actions, effectively acting as a Decision Support System (DSS) with Fault Detection and Isolation (FDI) capabilities.

Since we have to deal with problems requiring both recognition and interpretation, we have tried to integrate different AI techniques: neural networks are particularly suited for learning and recognition tasks, but their inference capabilities are quite limited unless they are specifically designed. On the contrary, logic-based tools can be used to formalize and elaborate facts and rules, but must be integrated with some kind of external interface to become context-aware: fuzzy logic, in particular, is used because of its high degree of expressiveness in describing the "vagueness" of many real world situations. Hence, we tried to complement the inferential weaknesses of a sub-symbolic approach with the formalism of a symbolic one.

2 Intelligent Monitoring

According to [1], an intelligent DSS estimates the process status analyzing the features of the measured signals, comparing and assigning them to known classes according to the detection of known patterns, then chooses the optimal control actions according to the known policies, eventually diagnosing and explaining the causes that led to the present conditions.

In terms of intelligent behaviour, we should assume that the system possesses a long term memory containing general facts and rules about process and plant control, sensorial capabilities to acquire stimuli from the external world, and interpretation and inference capabilities to match the outer inputs to the inner knowledge.

2.1 Bayesian Rational Pavelka Logic

Traditionally, state knowledge is formalized using first-order predicate logic, but since we deal with a high degree of variability and uncertainty, a fuzzy-logic based approach has turned out to be preferable.

Fuzzy logic is widely used in control applications, mostly because of its ability to describe local approximations of unknown functions in a human-readable way. The expressiveness of fuzzy logic, however, has been thoroughly studied from a theoretical point of view under the name of fuzzy logic "in a narrow sense" or "rational logic" [2], of which we use a belief-oriented variant we can only introduce here.

Given:

- an n-ary predicate P
- its argument set $\mathbf{X} = \{X_1, \dots, X_n\}$
- its absolute membership function $\mu_P{}^0(\mathbf{X}) : \mathbf{C}^n \rightarrow [0,1]$
- the truth interval $I = [0,1]$ divided into $N = 10$ intervals $\varepsilon_{j:1..N} = [\ (j\text{-}1)/N\ ,\ j/N\ )$

we consider the event set $\{E_{j:0..N-1} : \mu_P{}^0(\mathbf{X}) \in \varepsilon_j\}$.

Since sometimes information about **X** may be noisy or incomplete, or since it may be useful to add subjectiveness to a predicate, and events E_j may be considered random and mutually exclusive, a Dirichlet distribution can be used to estimate the probability distribution $\{ p_{j:0..N-1} = p(E_j) \}$.

The distribution shape is determined by its parameter vector $\beta = \{\beta_1..\beta_N\}$. Given its L1 norm, which is directly correlated to confidence or support, $B = \Sigma_{j:1..N}\, \beta_j$, and according to the principle of maximum likelihood, we may set:

$$p_j = \max\{\mathrm{dir}(\mathbf{p},\boldsymbol{\beta})_j \mid \boldsymbol{\beta}\} = \beta_j / B \qquad (1)$$

Under these assumptions, the subjective membership function μ_P fuzzifies the input arguments **X** by computing the corresponding parameter vector β.

$$\mu_P : \mathbf{X} \rightarrow \boldsymbol{\beta}$$

The example in Fig.1 shows clearly how this approach makes vagueness and uncertainty orthogonal concepts.

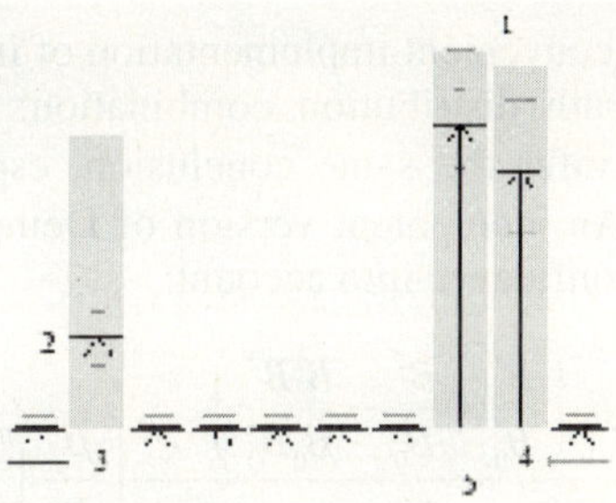

Fig. 1. Truth distribution for parameter values *{0,150,0,0,0,0,0,325,275,0}*/750, describing a predicate which is very likely to be almost true (1) but might also be nearly false (2)

This notion of "truth value" is particularly useful to deal with narrow fuzzy logic. As stated in [2], fuzzy modus ponens relies on the notion of T-norm **&** to compute a conclusion's truth value from a combination of the truth values of the premise and implication. This value, however, is only a lower bound and not necessarily the greatest.

$$\varepsilon(P)\ \&\ \varepsilon(P \rightarrow C) \Rightarrow \tau(C) \leq \varepsilon(C)$$

If no other information is available, ε(C) may be assumed to be uniformly distributed in [τ(C),1].

The choice and definition of the T-norm (and its dual, the S-norm) leads to the definition of the logical operators. In the following, we assume Pavelka's choice for T-norm, Lukasiewicz's "strict and", and negation, thus obtaining the following constructs. Given operand predicates P_i with essence ε^i:

- ¬ NOT $\varepsilon(\neg P) = 1 - \varepsilon(P)$
- ∨ OR $\varepsilon(\vee P_i) = \max\{ \varepsilon^i \}$
- ∧ AND $\varepsilon(\wedge P_i) = \min\{ \varepsilon^i \}$

- $\otimes$ AND Strict $\varepsilon(\otimes P_i) = \max\{0, \Sigma\varepsilon^i - (N-1)\}$
- $\oplus$ OR Strong $\varepsilon(\oplus P_i) = \min\{1, \Sigma\varepsilon^i\}$
- $\lozenge$ XOR $\varepsilon(P \lozenge Q) = |\varepsilon(P) - \varepsilon(Q)|$
- $!^n$ POW $\varepsilon(!^n P) = \varepsilon(P)^{\otimes n} = P \otimes P \otimes ... \otimes P$ n times ($n \in \Re^+$)
- $\rightarrow$ IMPLY $\varepsilon(A \rightarrow B) = \min\{1, 1 - \varepsilon(A) + \varepsilon(B)\}$

Since for a generic n-ary operator $*$ there are N^n possible combinations of truth values $\{\varepsilon^1..\varepsilon^n\}$, each one will mapped to a slot $\varepsilon^* = *(\varepsilon^1,...,\varepsilon^n)$ and contribute with probability $\Pi_{k:1..n}\, p(\varepsilon^k)$. The confidence parameter B is then computed as the geometric average of the confidences $B^{k:1..n}$ of the operands.

The operators are also useful in defining the quantifiers. Given a predicate with a free variable *P*(..,X,..) and a context **C** for X, the truth distribution of the predicate varies as X is bound to values $x \in$ **C**.

- $\forall$X "All X" : $\varepsilon(\forall X : P(X)) = \wedge_{x \in C}\, \varepsilon(P(x))$
- $\exists$X "One X": $\varepsilon(\exists X : P(X)) = \vee_{x \in C}\, \varepsilon(P(x))$
- $\int$X "Any X": $\beta(\int X : P(X)) = \Sigma_{x \in C}\, \beta(P(x))$

The last quantifier shows a convenient implementation of induction.

The last operation is truth distribution combination: in fuzzy inference, many premises may be used to infer the same conclusion, especially if both P→C and Q→¬C can be exploited. An idempotent version of Dempster-Shafer's combination rule has been used to take confidence into account:

$$\beta_j \left| \overline{\beta}^1, \overline{\beta}^2 \right. = \frac{B_0\left(\frac{B^1}{B_0} + \frac{B^2}{B_0} - \frac{B^1 B^2}{B_0 B_0}\right)}{1 - \sum_{k=1}^{N} \left(\frac{\beta_k^1}{B^1}\right)^{\frac{B^1}{B^1+B^2}} \cdot \left(\frac{\beta_k^2}{B^2}\right)^{\frac{B^2}{B^1+B^2}}} \cdot \left(\frac{\beta_j^1}{B^1}\right)^{\frac{B^1}{B^1+B^2}} \cdot \left(\frac{\beta_j^2}{B^2}\right)^{\frac{B^2}{B^1+B^2}}$$

The following sections show how this formalism can be used to evaluate a simple reasoning applied to signal interpretation.

2.2 An Architecture Outline

First of all, the system has been trained to learn the correct plant operating conditions: the data relative to the anoxic phase $[pH(T) ; orp(T) ; DO(T)]_{T:1..60/90}$ have been extracted from the optimal cycles acquired in summer-fall 2004, then they have been denoised, first derivatives have been computed and the resulting 6-dimensional vectors have been clustered using a nearest neighbour algorithm and DTW distance function. The 56 series from the 5 largest clusters have been chosen as a training set for a SOM neural network [3]. We have used 25x5 neurons on a rectangular lattice, assuming that, on average, every neuron would have covered a 3-sample span for a given signal class.

Whenever the network tries to recognize a new sample S, the Malahanobis distance $\alpha_N = (N^T\Sigma^{-1}S)^{½}$ is computed for every neuron N, where Σ is the input space's covariance matrix. Every neuron then activates with intensity which depends on a "scope" parameter, chosen such that:

$$y = e^{-\left(\frac{2\alpha_N}{D}\right)^3} \qquad \begin{cases} y(D/4) \approx 0.9 \\ y(D/2) \approx 0.35 \\ y(D) \approx 0 \end{cases}$$

This allows network activity to be formalized by writing the predicate *active*(N,T), meaning "Neuron N is active at time T". Conversely, it can be stated that the network evaluates the truth degree of a class of predicates for different N and T:

$$\mu_{\text{active(N,T)}} = B_0\, \delta(j - j^*)\ ,\ j^* = \texttt{floor}(10y) \tag{2}$$

δ is Dirac's delta, meaning that the truth value is certainly determined by the activation level. Notice that ∃N: *active*(N,t) describes the "winner neuron" at time t, and ∃N ∫T : *active*(N,T) depends on the net's ability to have at least one of its neurons recognize every given input, thus estimating the net's coverage. Other quantifications give different meanings and results.

After the first phase, a second training has been carried out to give the activations meaning. For every time serie in the training set, a human expert has identified the "denitrification complete" event within a ~15 minute interval $[T_s,T_e]$, so the samples have then been colored in red $[0,T_s)$, green $[T_s,T_e]$ and blue $(T_e,T_{max}]$. This form of implicit training-by-example has been used for two main reasons: it does not require the system to learn complete explicit knowledge about the event and does not require the human expert, who might not possess it himself, to express it.

The rational predicate *det*(N,C) states whether a neuron activates only in front of signals of a given color C:

$$\mu_{\text{dett(N,C)}} = \Sigma_T\, K\, y(T)\, \delta(j - j^*), \quad j^* = \texttt{floor}(10\varphi_C(T)) \tag{3}$$

Scaling constant K apart, belief is induced over ∫T : *color*(C,T) the more a neuron activates, where the phase-membership function φ evaluates the truth degree: the recognition of a red sample by neuron n will make *det*(n,red) truer, but *det*(n,grn) and *det*(n,blu) falser since $\varphi_R = 1$ and $\varphi_G = \varphi_B = 0$. Notice that the use of φ_C allows also to define fuzzy transitions from one phase to the other.

These predicates may be used to define a sequence pattern matching reasoning:

∃N : {*active*(N,T) ⊗ *det*(N,C)} ↔ *type*(T,C).

The activation of a neuron N which activates only in sub-phase C allows to deduct that the input sample has the features of phase C.

type(T,red) ⊕ *type*(T,grn) → ¬*type*(T,blu).
type(T,red) ⊕ *type*(T,blu) → ¬*type*(T,grn).
type(T,blu) ⊕ *type*(T,grn) → ¬*type*(T,red).

Obviously, phases are mutually exclusive (or, more generally, for a given time instant t : $\varphi_R(t)+\varphi_G(t)+\varphi_B(t) = 1$). This implication gives the upper bounds for *type*(T,C).

ack(red,T-1) ⊕ $!^2$ *type*(red,T) → *ack*(red,T).
(*ack*(grn,T-1) ⊕ $!^2$ *type*(grn,T)) ⊗ ack(red,T) → *ack*(grn,T).
(*ack*(blu,T-1) ⊕ $!^2$ *type*(blu,T)) ⊗ ack(grn,T) → *ack*(blu,T).

If *type*(C,T) becomes sufficiently true for a sufficient number of times, the system acknowledges sub-phase C. Notice that later phases cannot be actually acknowledged until the earlier ones have been recognized.

$$ack(\text{red},T) \otimes ack(\text{grn},T) \otimes ack(\text{blu},T) \rightarrow goal.$$

Eventually, as soon as the post-event conditions have been recognized the reaction is assumed complete.

3 Discussion and Results

The left chart in Figure 2 shows the signals pH and orp during the anoxic phase, with the denitrification reaction being completed around T_{den}. The coloured rows show the temporal evolution of the truth distribution of several key predicates, with an highlighted upper part meaning "true" and lower part "false". The neurons recognize the three phases correctly and with a high degree of certainty, except, as should be expected, during transitions around T_{start} and T_{end}. In the example, the system actually acknowledges the green phase some minutes before T_{start}, but a human expert confirmed that these signals do have "green" features earlier than stated, crediting to the net's generalization capabilities.

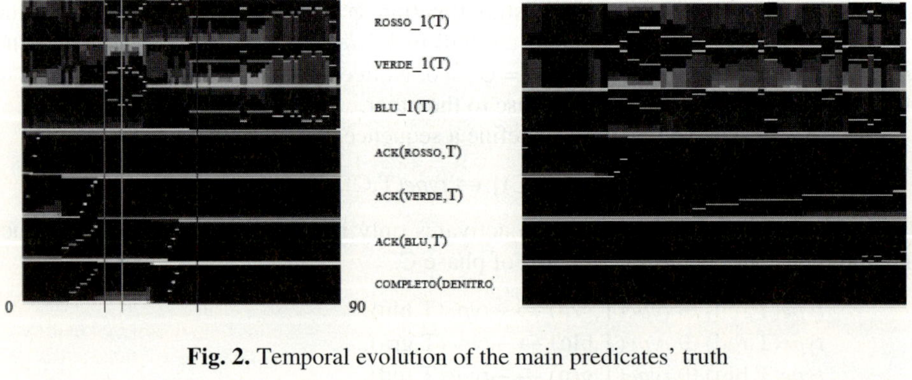

Fig. 2. Temporal evolution of the main predicates' truth

As shown in the last row, the goal becomes certainly true near T_{end}, meaning that the detection system would have saved about 40 minutes over a total of 90.

The software has not yet been installed on the plant, but it has been tested on the training set, with a 95% accuracy and only a 5% of false negative mistakes due to overly noisy signals. The software has also been validated on 50 time series extracted from other clusters, only half of which presented evidence of a correct denitrification:

this time, it achieved 100% accuracy, not answering at all when it could not isolate the required event. The right chart in Fig. 2 is an example of this last class: the reaction starts so the system acknowledges the red phase, but the signals show that it does not end within the 60 min. deadline and no green neuron activates.

4 Related and Future Works

The monitoring and optimization of SBR cycles is a well-known problem in literature and has been dealt with using several techniques. In [4], for example, we see how optimal fuzzy pattern recognition can give comparable results in assessing end-phase criteria. Our training process, instead, has been based purely on example learning, classification and recall, but not on feature isolation: in particular, our solution resembles a subjective experience-induced Learning Vector Quantization of the reference signals memorized by the SOM. Hence, a similar approach is suitable for situations where explicit knowledge is not available. Furthermore, from a theoretical point of view it tries to emphasize the connections and differences between the two levels of recognition and inference.

The problem of system control by signal analysis and interpretation is a primary issue in many critical applicative fields, from bioprocesses [6] to health care [7] to nuclear systems [8]. Often, fuzzy logic is used in such applications because of its ability to describe (linear) local function approximations, as shown by [2]. Since many types of neural networks share the same property, it is both common practice [9] to extract rules from nets and to use nets for predicate calculus evaluation [10], [11].

Thus, we have tried to show that narrow fuzzy logic may be as effectively used to describe a network's output or, dually, that a neural network is capable of assessing rational fuzzy predicates by computing the membership function of its arguments. However, we agree with [5] in stating that fuzzy control may be viewed as a specific case of direct "cause-effect" fuzzy reasoning, so we also tried to show that the expressiveness of rational logics may be used to formalize and compute one of a broader class of reasoning.

This preliminary work aimed at verifying the feasibility of a hybrid, context-aware adaptive tool which can rely on the full expressive capabilities of a narrow-fuzzy rule-based system. Our future works will improve and expand the set of available policies, while developing a customizable inferential infrastructure.

Acknowledgments

Part of this work was funded by the SPINNER consortium. We are also grateful to the PROT-IDR division of the Ente Nazionale per le Nuove Tecnologie e l'Ambiente (ENEA) for allowing access to the plants and supporting this research.

References

1. Olsson, G., Newell, B.: Wastewater Treatment Systems: Modelling, Diagnosis and Control. IWA Publishing (1999)
2. Hajek, P.: Metamathematics of Fuzzy Logic. Kluwer Academic Publishers, Dordrecht (1998)
3. Haykin, S.: Neural Networks: A comprehensive foundation. Pearson (1999)

4. Marsili-Libelli, S.: Control of SBR switching by fuzzy pattern recognition. Water Research 40, 1095–1107 (2006)
5. Gerla, G.: Fuzzy control as a fuzzy deduction system. Fuzzy Sets and Systems 121, 409–425 (2001)
6. Feng, E., Yang, H., Rao, M.: Fuzzy expert system for real-time process condition monitoring and incident prevention. Expert Systems with Applications 15, 383–390 (1998)
7. Choi, J., Min, B.G.: An intelligent remote monitoring system for artificial heart. IEEE Transactions on Information Technology in Biomedicine 9(4) (2005)
8. Guimaraes, A.C.F., Lapa, C.M.F.: Fuzzy inference to risk assessment on nuclear engineering systems. Applied Soft Computing 7, 17–28 (2007)
9. Tan, S.C., Lim, C.P., Rao, M.V.C.: A hybrid neutral network model for rule generation and its application to process fault detection and diagnosis. Engineering Applications of Artificial Intelligence 20, 203–213 (2007)
10. Gobi, A.F., Pedrycz, W.: The potential of fuzzy neural networks in the realization of approximate reasoning engines. Fuzzy Sets and Systems 157, 2954–2973 (2006)
11. Murata, T., Zhang, D.: A Predicate-Transition Net Model for Parallel Interpretation of Logic Programs. IEEE Transactions on Software Engineering 14(4), 481–497 (1988)

An Embedded Real-Time Automatic Lane-Keeping System

Salvatore Vitabile[1], Salvatore Bono[2], and Filippo Sorbello[2]

[1] Dipartimento di Biotecnologie Mediche e Medicina Legale,
Università degli Studi di Palermo
Via del Vespro, 90127 Palermo, Italy
vitabile@unipa.it

[2] Dipartimento di Ingegneria Informatica
Università degli Studi di Palermo
Viale delle Scienze, Ed. 6, 90128 Palermo, Italy
sorbello@unipa.it

Abstract. Advanced Driver Assistance Systems (ADAS) are safety systems capable of identifying an unavoidable dangerous scenario and reacting coherently. An automatic lane-keeping system is designed to prevent dangerous events when the driver left inadvertently his/her own lane. In this paper a real-time automatic lane-keeping system is proposed. The main features of the system are the lane markers location process as well as the generation of the vehicle's steering angle. The system has been prototyped using the Celoxica RC203 board, equipped with a Xilinx Virtex II FPGA, for real-time processing of real motorway scenes coming from a CCD camera on a moving vehicle. The required processing time is 25,521 ms per frame, so that the system could process 39 frames/sec. Vehicle's steering angle is computed with good approximation.

1 Introduction

The trend towards Intelligent Transportation Systems (ITS) paradigm is shifting the philosophy of vehicle system design from specific, closed devices to heterogeneous, intelligent, cooperating systems. Advanced monitoring systems will soon lead to efficient, automatic traffic signal recognition, pedestrian detection, automatic collision-avoidance systems, and drowsiness detection systems [4, 6]. The vehicle will have to adapt to a potential situation of danger and behave differently from normal, shifting control modes, commands, reconfiguring in real-time vehicle resources (sensors, logic, actuators) and obtaining an unified action.

A real-time automatic lane-keeping system could be used in drowsiness detection systems as well as in automatic driving systems. A typical lane-keeping system is essentially composed of two subsystems: the first one deals with lane marker identification, based on a segmentation method, whereas the second one deals with the calculation of the vehicle's steering angle. Clearly, it is necessary to identify appropriate criteria which allow a reliable and effective result. A segmentation method could be able to distinguish the road surface from other elements in all the frames, with good generalization capabilities. The identification of lane marker

B. Apolloni et al. (Eds.): KES 2007/WIRN 2007, Part I, LNAI 4692, pp. 647–654, 2007.

position is necessary to characterize road geometry for detecting the vehicle's position and orientation. A steering angle calculation method is necessary to calculate the steering angle of the vehicle, beginning with the analytic formulation of the lane marker to simulate a person driving.

In the literature, various approaches are available to manage the above issues. In [1], the authors propose a method which combines an image segmentation technique based on maximum entropy with a bi-normalized adjustable template. This method first produces the best road image segmentation based on maximized entropy. Then, lane mark parameters are acquired based on the bi-normalized adjustable template. In [2], the authors propose a histogram image based approach. This system exploits the characteristics of the grey level histogram of road images to detect lane markers. Each lane marker is then analyzed using a decision tree, and finally the relations between lane markers are analyzed to create structures defining the lane boundaries. In [3], Arkin describes a schema-based reactive control system that allows remote operators to direct a mobile platform without undue concern for local obstacle avoidance. By using an analogue of the potential field methodology, repulsive forces are constructed around nearby obstacles allowing the reactive system to avoid them.

In this paper, a system which takes its cue from these results, modifying and adapting them to an operating context (embedded system) in which it is often difficult or inefficient to use certain mathematical elements, such as decimal numbers or high computational cost operations, is proposed. With more details, a segmentation method based on the entropy function and the grey image level modal value has been adapted and used. As far as calculation of steering angle is concerned, we have adopted an approach which is similar to the Arkin reactive systems, having partially modified it in order to adapt some aspects to the context of a road vehicle. The system has been prototyped using the Celoxica RC203 board, equipped with a Xilinx Virtex II FPGA, for real-time processing of real motorway scenes coming from a CCD camera on a moving vehicle. The system has been described using the algorithm-like Handel-C language and the Celoxica DK4 design environment.

The paper is organized as follows. Section II deals with segmentation algorithms used for lane marker identification, while Section III presents and describes the proposed system. In Section IV the experimental results are provided and discussed. Finally, in Section V, the conclusion of the current work are reported.

2 Segmentation Algorithms Overview and Integration

Image segmentation is fundamental because it provides essential information to the identification of the kind of road and, therefore, to the development of the whole system. In the literature, various methods and various solutions have been proposed for extracting the regions of interest from processed images.

One of the most used methods in the literature [2] is able to extract a threshold from the histogram gray levels and use that for objects segmentation. In each acquired image, the areas of larger size will represent the road surface with medium brightness, whereas the smaller areas which indicate the lane markers will be nearly white, and the region which indicate other objects in the image will be of variable brightness.

The histogram therefore has a pseudo-Gaussian structure, with a modal value close to average value which we can consider as the characteristic value of the asphalt brightness. So, we can decide to take the threshold as having an intermediary value, between the modal value and the maximum value, using a medium weighed between these two quantities. After some preliminary studies, however, we became aware that system segmentation results improve when a composed threshold, depending on the whole image (global feature) and 8 local sub-images (local features), is taken.

Another interesting approach is based on the entropy measure of an image which is computed after reaching the threshold [1]. This method is used to calculate the entropy which has the maximum contrast. We can consider this as a measure of the entropy. The algorithm works as follows. For an 8 bit grey image, set *T* as the supposed segmentation threshold. Pixels whose grey values are less than *T* can be specified as the background area B and greater than *T* as target area O. For each area, the frequency of pixels is calculated by the number of each area based on a certain grayscale step. This frequency is defined as the relative grayscale probability $p(i)$. In this way, target area grayscale probability p_o and background grayscale probability p_b are described in Equation 1:

$$\begin{cases} p_o = \sum_{i=T}^{255} p(i) \\ \\ p_b = \sum_{i=0}^{T-1} p(i) \end{cases} \tag{1}$$

In this paper the grayscale step is 1 for 0-255 grey level. The one-dimensional entropies of the target area and background area are defined in Equation 2:

$$\begin{cases} H_o(T) = -\sum_{i=T}^{255}\left(\frac{p(i)}{p_o}\log\left(\frac{p(i)}{p_o}\right)\right) \\ \\ H_b(T) = -\sum_{i=0}^{T-1}\left(\frac{p(i)}{p_b}\log\left(\frac{p(i)}{p_b}\right)\right) \end{cases} \tag{2}$$

whereas the one-dimensional entropy of the global image can be defined as:

$$H(t) = H_o(t) + H_b(t) \tag{3}$$

The threshold of T, which we are looking for, will maximize the entropy function. As a result, *t* is the global threshold *T*.

The previous described algorithms produce excellent results, even if, as it stands, it is not possible to replicate them in embedded devices as a result of the heavy emphasis on floating points arithmetic and the need for calculating computationally expensive operations, such as logarithms. Therefore a modification and adaptation of those algorithms is required for an efficient implementation on hardware devices and for a fast, real-time execution. The testing of the algorithms was undertaken by using the software environment MatLab®, which allows for fast and useful prototyping. So, a various number of tests were performed using different sample images performing histogram calculation and, successively, entropy calculation.

Before algorithms implementation in a FPGA platform, some modifications are necessary in order to use entire arithmetic operations. Algorithms have been profiled and weighted adopting the index proposed in [5]. In addition, algorithms implementation on hardware devices has been performed following the implementation guidelines proposed in [5]. As example, elimination of logarithm calculations (see Equation 2) has been made by rounding up p(i), pb(i) and po(i) into discrete quantities and implementing the (x/y)*log(x/y) function values with look-up tables.

3 The Proposed System

The system has been prototyped using the Celoxica RC203 board, equipped with a Xilinx Virtex II FPGA, for real-time processing of real motorway scenes coming from a CCD camera on a moving vehicle. The system has been described using the algorithm-like Handel-C language and the Celoxica DK4 design environment. The automatic real-time lane-keeping system is essentially composed by three pipelined subsystems: i) Acquisition; ii) Elaboration; iii) Visualization. The Visualization subsystem function is only to show each processed frame through a VGA output. The block scheme of the whole system is depicted in Figure 1.

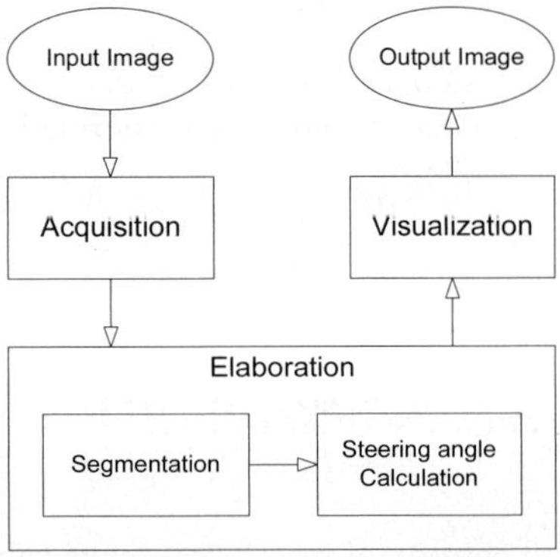

Fig. 1. The overall scheme of the realized lane-keeping system

3.1 The Acquisition Subsystem

Image acquisition is carried out by a CCD camera on a moving vehicle. Image pixels have been represented through the YCbCr space format, so that it is necessary to convert it into monochromatic format. Such conversion is made using the 0.299, 0.587, 0.114 coefficients for the Y, Cb, Cr channels, respectively.

After that, pixels pass through a FIFO buffer to ensure that none of the acquired pixels are lost. System processing is pipelined, i.e. the first frame is acquired, the first frame is processed while the second frame is acquired, and so on. If *Tf* is the total clock cycles required by all filters, the FIFO buffer length must be dimensioned to store the pixels acquired in $2Tf$. The acquisition phase ends with the image storing on the two FPGA's external RAM. The acquisition phase and the process flow of the pipeline operations, are depicted in Figure 2 and Figure 3, respectively.

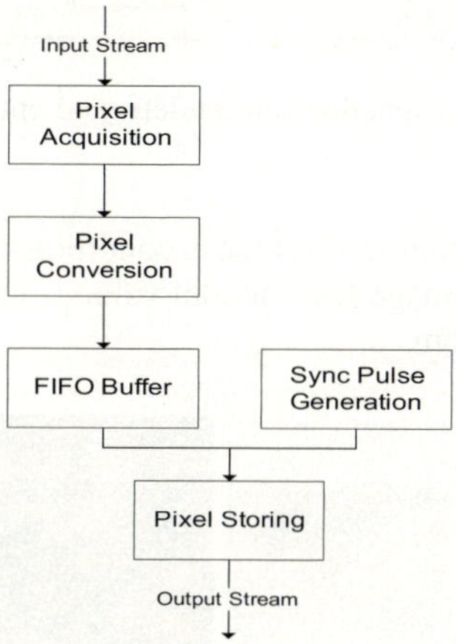

Fig. 2. The pixel acquisition and storing process

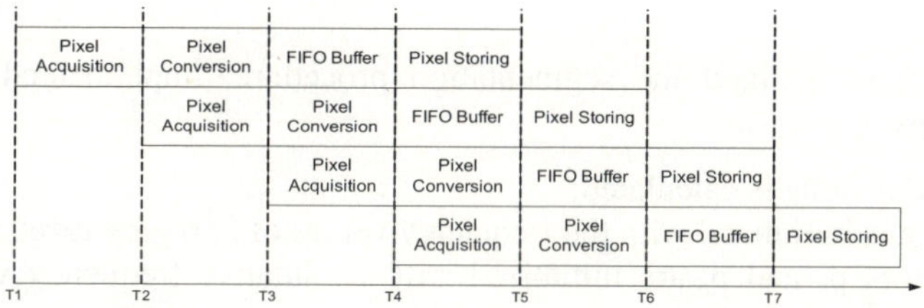

Fig. 3. The process flow of the pipeline operations

3.2 The Elaboration Subsystem

The Segmentation Module

The elaboration phase begins with the image clipping operation which eliminates pixels that are beyond the vanishing-lines. It proceeds with the calculation of the image histogram. For all frames, information is collected about pixels in the input stream active region and is stored in a particular structure. At the end of the frame, such values are available through a set of macro.

As showed in Figure 4, in our trials entropy is a function which has two very closed relative maximums. It is probable that a small variation can trigger a jump from second maximum (which is the result on the right) to the first and, consequently, trigger an outstanding change in the results.

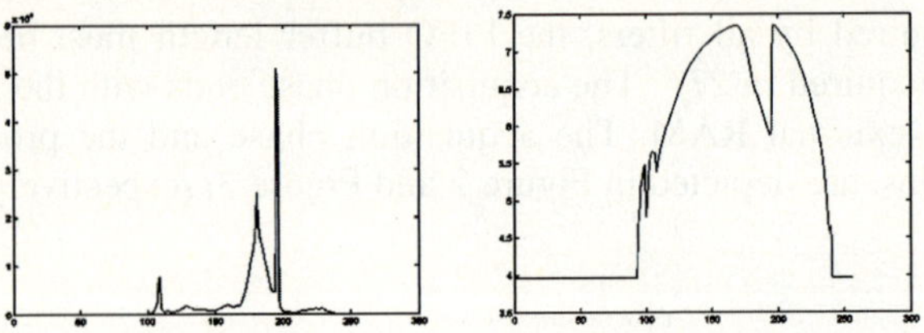

Fig. 4. An example of histogram function (on the left) and entropy function (on the right) for our road scenes

A robust behavior with different brightness conditions is reached when a maximum values greater than the grey image level modal value is considered. Figure 5 shows an elaboration result of our system.

Fig. 5. An acquired road scene (on the left) and a segmented road scene (on the right)

In Figure 6 is outlined the segmentation procedure, implemented through the following steps:

1. Pixel modal value is calculated;
2. A vector p is initialized with the frequency values of 256 grey levels.
3. Two vectors p_b and p_o are initialized with cumulative frequency values of 256 grey levels, lower and upper respectively;
4. p, p_b and p_o values are quantified and stored in vectors p_l, p_{bl} and p_{ol};
5. Entropy value is calculated for all grey levels and is compared with the previous found maximum in order do find threshold which has the greatest entropy value at the end of cycle. However, such research is only performed on values which are greater than modal value;
6. Threshold operation is performed.

The Steering Angle Computation Module

For steering angle calculation, lane marker analytic formulation is the first operation to carry out. An interpolation of lane marker characteristic points is performed in order to make the system more reactive. Once such information is obtained, it is necessary to calculate the steering angle. The implemented method is similar to the method proposed for robotic reactive architectures [3]. Lane marker base points are considered as they have an electromagnetic field with oppose charge in comparison with vehicle charge. Consequently, a repulsive force will be generated according to the inverse square of the distance.

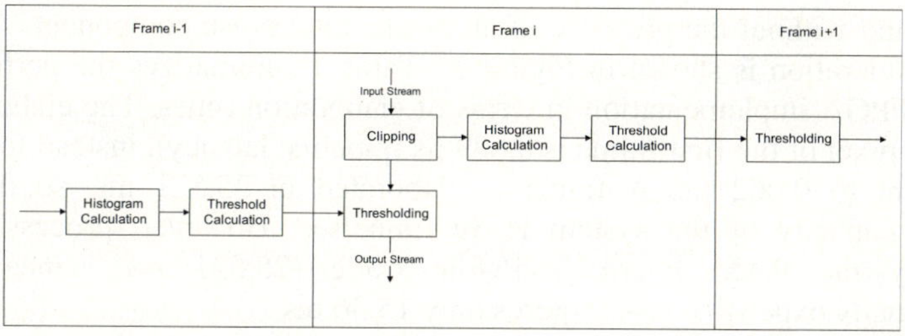

Fig. 6. The steps for the segmentation procedure and the inter-frame interaction

4 Experimental Results

System behavior denotes good performance with different brightness conditions and kind of streets and motorways. The elaborated image emphasizes lane marker, in

Fig. 7. An acquired image (on the left) and the system at work (on the right)

Table 1. The performance of the system in terms of elaboration times

Routine	*Clock cycles*	*Latency for pixel (μs)*	*Latency for frame (ms)*	*Work Freq. (MHz)*
Convert	1	0.015	6.221	65
FIFO	1	0.015	6.221	65
VGASyncGen	4	0.062	25.521	65
PALPL1RAMFrameBufferDB	4	0.062	25.521	65
FIFO	1	0.015	4.608	65
Send	1	0.015	4.608	65
Receive	1	0.025	7.68	40
Clip_Rectangle	2	0.05	15.36	40
Analyze	1	0.025	7.68	40
Threshold	1	0.025	7.68	40
Find_Lane	2	0.05	15.36	40
Rectangle	1	0.025	7.68	40
FIFO	1	0.025	7.68	40
Send	1	0.025	7.68	40
Receive	1	0.04	12.203	25.175
Convert	1	0.04	12.203	25.175

clean way and without the presence of an outstanding noise component. An example of this consideration is shown in Figure 7. Table 1 summarizes the performance of the digital FPGA implementation in terms of elaboration times. The elaboration time for the first pixel of the first frame is 0.689 μs (pipeline latency), instead for the others is equivalent to 0.062 μs. A frame is elaborated in 25.521 ms, so that the top processing capacity of the system is 39 frame/sec. However, processing time is affected by the RAM board read/write cycle (25.521 ms), since the most computationally expensive task requires only 15.36 ms.

The Steering Angle Computation Module has a robust behavior with an average error of 6% on real motorway sequences with straight stretches and curves.

5 Conclusion

In this paper a real-time automatic lane-keeping system has been proposed. The required processing time is 25,521 ms per frame, so that the system could process 39 frames/sec, even if potential speed execution can raise to 65 frames/sec. Vehicle's steering angle is computed with good approximation. Some aspect of the system can be modified in order to increase its reliability. A possible system enhancement deals with a better analytics formulation of the lane markers using the Hough transform.

References

[1] Tianhong, Y., Rongben, W., Lisheng, J., Jiangwei, C., Lie, G.: Lane Mark Segmentation Method Based on Maximum Entropy. In: Proceedings of the 8th International IEEE Conference on Intelligent Transportation Systems, Vienna, Austria, September 13-16, pp. 722–726 (2005)

[2] Gonzalez, J.P., Özgüner, Ü.: Lane Detection Using Histogram-Based Segmentation and Decision Trees. 2000. In: IEEE Intelligent Transportation Systems – Conference Proceedings Dearborn (MI), USA, October l-3, pp. 346–351 (2002)

[3] Arkin, R.C.: Reactive control as a substrate for telerobotic systems. Aerospace and Electronic Systems Magazine. IEEE 6(6), 24–31 (1991)

[4] Aparicio, F.: EEVC WG19 Activities on Primary and Secondary Safety Interaction, European Enhanced Vehicle-safety Committee (EEVC) (2005), http://www.eevc.org/publicdocs/ESV2005_WG19_05-0462-0.pdf

[5] Vitabile, S., Gentile, A., Siniscalchi, S.M., Sorbello, F.: Efficient Rapid Prototyping of Image and Video Processing Algorithms. In: Proc. of DSD 2004, pp. 452–457. IEEE Computer Society Press, Los Alamitos (2004)

[6] Vitabile, S., Gentile, A., Sorbello, F.: A Neural Network based Automatic Road Signs Recognizer. In: Proc. of 2002 IJCNN, pp. 2315–2320. IEEE Computer Society Press, Los Alamitos (2002)

Effects of Kinematics Design on Tracking Performance of Model-Based Adaptive Control

Serdar Kucuk

Kocaeli University, Electronics and Computer Education,
Umuttepe Campus, Kocaeli, Turkey
skucuk@kou.edu.tr

Abstract. In this paper, the relationships between the kinematics design and tracking performance of the model-based adaptive control are studied. For this purpose, the position tracking error convergences of three serial manipulators with joint types of RR, RP and PP are considered. The physical parameters and desired trajectories of these manipulators are assumed same for the proper comparison. Since the model-based adaptive control can completely account for nonlinear structure of robot dynamics, it has been preferred as control method.

Keywords: Kinematics design, Model-based adaptive control, Robot dynamics.

1 Introduction

Robot manipulators are composed of rigid links connected to each other with revolute (R) or prismatic (P) joints. Combination of prismatic and revolute joints constructs the different types of kinematics designs such as Cartesian (Epson Cartesian robot), cylindrical (Seiko RT3300) and the spherical (The Standford Arm) [1]. In general, high manipulator performance for a manipulator is a challenging kinematics design problem. Therefore, kinematics-related performance indices have been used in order to find the optimal kinematics design for the manipulators [2], [3], [4], [5]. Performance indices can also be used for comparing the efficiency and usefulness of one kinematics design to another. Recently, Kucuk and Bingul [5] have compared the performance indices for the serial manipulators. In this work, it has been concluded that manipulators with revolute joints have the best kinematics designs.

Generally, the design procedures involve in integration of a kinematics design and a control method. The kinematics design influences directly some important criteria such as accuracy, high mechanical stiffness and lower positioning error. The control method has significant effects on the performance of the manipulators such as stability, speed convergence rate and good disturbance rejection. Both the kinematics design and the control method are strongly relevant in the manipulator performance. For instance, control method used for the Cartesian and the spherical manipulators are fundamentally different from each other.

As the robotic manipulators have the highly nonlinear dynamic structure with uncertainties, various control strategies have been the interests of many researchers [6], [7], [8], [9], [10]. Basically, the control strategies can fall into the linear and nonlinear

B. Apolloni et al. (Eds.): KES 2007/WIRN 2007, Part I, LNAI 4692, pp. 655–663, 2007.

control of manipulators. The linear control of robotics manipulators involves traditional feedback controls which are proportional–integral–derivative (PID) control or proportional-derivative (PD) control [11], [12]. The stability of the system depends on the selection of the suitable control gains in feedback controllers. The difficulties in adjusting suitable gains under possible rapid changes in robot dynamics can not guarantee the system stability. The nonlinear control of robotics manipulators involves model-based control, switching control, adaptive control or combination of these. Since model-based adaptive control algorithm [13] gives response to a wide range of parameter uncertainties, such as the mass, inertia, and frictional nonlinearity, it is chosen as the control method.

Several issues mentioned above aim at improving manipulator performance employing either control strategies or kinematics designs. In this paper, the correlations between the kinematics design and tracking performance of the model-based adaptive control are presented.

2 Model-Based Adaptive Control Design

When the Lagrangian formulation is evaluated symbolically, the dynamic model of a robot manipulator can be written in the form

$$\tau = M(q)\ddot{q} + C(q,\dot{q}) + G(q) + F(q,\dot{q}) \tag{1}$$

where, τ is the nx1 generalized torque vector applied at joints. q, $\dot{q}$ and $\ddot{q}$ are the nx1 joint position, velocity and acceleration vectors, respectively. M(q) is the nxn mass matrix, $C(q,\dot{q})$ is an nx1 vector of centrifugal and Coriolis terms, G(q) is an nx1 vector of gravity and $F(q,\dot{q})$ is nx1 vector of friction terms of actual mechanism.

If the parameters in the model are known, the model-based control works accurately. In this method the controller is partitioned into two parts. The system parameters appear only in the model-based portion and the servo portion is independent of these parameters. To develop the error equation, control law can be written as follows

$$\tau = \alpha\tau' + \beta \tag{2}$$

where,

$$\alpha = \hat{M}(q), \quad \beta = \hat{C}(q,\dot{q}) + \hat{G}(q) + \hat{F}(q,\dot{q}) \tag{3}$$

where, $\hat{M}(q)$, $\hat{C}(q,\dot{q})$, $\hat{G}(q)$ and $\hat{F}(q,\dot{q})$ denote the estimated terms in the dynamic model in (1). Equation (2) can be written in the more compact form for simplicity as

$$\tau = \hat{M}\tau' + \hat{C} + \hat{G} + \hat{F} \tag{4}$$

Let's equate (1) and (4).

$$M\ddot{q} + C + G + F = \hat{M}\tau' + \hat{C} + \hat{G} + \hat{F} \tag{5}$$

τ' is obtained from model-based controller as follows

$$\tau' = \ddot{q}_d + K_v\dot{E} + K_pE \tag{6}$$

where E and $\dot{E}$ are nx1 vectors of position and velocity errors, respectively. Substituting (6) into (5) and subtracting $\hat{M}\ddot{q}$ from both sides of the obtained equation, then the resulting equation is given by

$$\ddot{E} + K_v\dot{E} + K_pE = \hat{M}^{-1}[(M-\hat{M})\ddot{q} + (C-\hat{C}) + (G-\hat{G}) + (F-\hat{F})] \tag{7}$$

The error equation results from errors in the model parameters given by (7) can be rewritten as follows

$$\ddot{E} + K_v\dot{E} + K_pE = \hat{M}^{-1}[\tilde{M}\ddot{q} + \tilde{Z}] \tag{8}$$

where, $\ddot{E} = \ddot{q}_d - \ddot{q}$, $\dot{E} = \dot{q}_d - \dot{q}$, $E = q_d - q$, $\tilde{Z} = (C-\hat{C}) + (G-\hat{G}) + (F-\hat{F})$ $\tilde{M} = M - \hat{M}$, and K_v and K_p are nxn positive define diagonal gain matrices. Because the robot manipulators have parametric uncertainties and disturbances, the right hand side of the (8) would not be zero and the errors occur. To overcome this problem an adaptation shame given in Fig. 1 can be used in the model-based reference system. It updates the values of the model parameters until the errors disappear. The right hand side of the error equation given by (8) can be written in the form

$$\ddot{E} + K_v\dot{E} + K_pE = \hat{M}^{-1}(q)W(q,\dot{q},\ddot{q})\Phi \tag{9}$$

where W is a nxi matrix which includes known parameters of the dynamic model and Φ is ix1 vector of parameter errors stated as

$$\begin{aligned}\Phi = H - \hat{H} &= [h_1\ h_2\ \cdots\ h_i]^T - [\hat{h}_1\ \hat{h}_2\ \cdots\ \hat{h}_i]^T \\ &= [m_1 \cdots m_n\ k_1 \cdots k_n\ v_1 \cdots v_n]^T - [\hat{m}_1 \cdots \hat{m}_n\ \hat{k}_1 \cdots \hat{k}_n\ \hat{v}_1\ \cdots\ \hat{v}_n]^T\end{aligned} \tag{10}$$

where 3n=i and $m_1, m_2 \ldots m_n$ represent link masses, $k_1, k_2 \ldots k_n$ and $v_1, v_2 \ldots v_n$ are the elements of F. ^ stands for the estimate of a term. In order to obtain adaptation law, let's write the full error state representation.

$$\dot{X} = AX + B\hat{M}^{-1}W\Phi \tag{11}$$

and

$$E_1 = CX \tag{12}$$

where A,B,C are block diagonal matrices X is vector of state variables and E_1 is filtered error. The Lyapunov function candidate is defined as

$$v(X,\Phi) = X^TPX + \Phi^T\Gamma^{-1}\Phi \tag{13}$$

where P and Γ are any symmetric positive definite matrices with proper dimensions. Differentiating the Lyapunov function yields the following equation.

$$\dot{v}(X,\Phi)=\dot{X}^T P X + X^T P \dot{X} + \dot{\Phi}^T \Gamma^{-1}\Phi + \Phi^T \Gamma^{-1}\dot{\Phi} \tag{14}$$

Since P and are diagonal matrices, $\dot{X}^T P X + X^T P \dot{X}$ can be shortened as $2\dot{X}^T P X$. Using this simplification and substituting $\dot{X}$ into (14) gives

$$\dot{v}(X,\Phi)=2X^T A^T P X + 2\Phi^T W^T (\hat{M}^{-1})^T B^T P X + 2\Phi^T \Gamma^{-1}\dot{\Phi} \tag{15}$$

The strictly positive real lemma Q is stated as

$$A^T P + PA = -Q \tag{16}$$

and

$$PB = C^T \tag{17}$$

Since A, B and P are block diagonal matrices $2A^T P = -Q$ and $B^T P = C$. Using these simplifications in (15) yields

$$\dot{v}(X,\Phi)=-X^T Q X + 2\Phi^T W^T (\hat{M}^{-1})^T C X + 2\Phi^T \Gamma^{-1}\dot{\Phi} \tag{18}$$

As a result, substituting (12) into (18) yields

$$\dot{v}(X,\Phi)=-X^T Q X + 2\Phi^T (W^T \hat{M}^{-1} E_1 + \Gamma^{-1}\dot{\Phi}) \tag{19}$$

If $\dot{\Phi}$ is selected as

$$\dot{\Phi}=-\Gamma W^T \hat{M}^{-1} E_1 \tag{20}$$

Equation (19) would be negative definite. Since $\Phi = H - \hat{H}$ and $\dot{\Phi}=\dot{\hat{H}}$ then the adaptation law can be obtained as

$$\dot{\hat{H}}=\Gamma W^T \hat{M}^{-1} E_1$$

$$=\begin{bmatrix}\gamma_1 & \cdots & 0\\ \vdots & \ddots & \vdots\\ 0 & \cdots & \gamma_n\end{bmatrix}_{ixi}\begin{bmatrix}w_{11} & \cdots & w_{n1}\\ w_{12} & \cdots & w_{n2}\\ \vdots & \vdots & \vdots\\ w_{1i} & \cdots & w_{ni}\end{bmatrix}_{ixn}\begin{bmatrix}\hat{M}_{11}^{-1} & \cdots & \hat{M}_{1n}^{-1}\\ \vdots & \ddots & \vdots\\ \hat{M}_{n1}^{-1} & \cdots & \hat{M}_{nn}^{-1}\end{bmatrix}_{nxn}\begin{bmatrix}e_{11}\\ \vdots\\ e_{nn}\end{bmatrix}_{nx1} \tag{21}$$

where Γ is a gain matrix whose diagonal elements are $\gamma_1, \gamma_2 \ldots \gamma_n$, w_{in} are the elements of W^T matrix and $e_{11} \ldots e_{ii}$ are filtered errors stated as

$$\begin{bmatrix}e_{11}\\ \vdots\\ e_{nn}\end{bmatrix}=\begin{bmatrix}\dot{e}_1+\Psi_1 e_1\\ \vdots\\ \dot{e}_n+\Psi_n e_n\end{bmatrix} \tag{22}$$

where $\psi_1, \psi_1 \ldots \psi_n$ are positive constants. The detailed theory of model-based adaptive control can be found in [14].

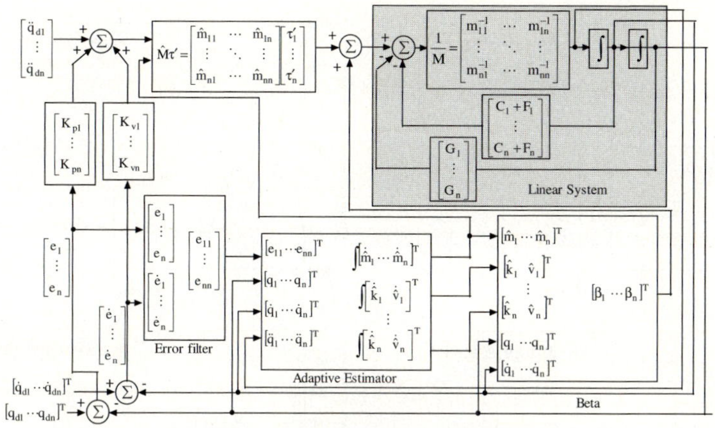

Fig. 1. A model-based adaptive control scheme

3 Simulation Case Study

A simulation study has been conducted in order to demonstrate the correlations between the tracking performances of the model-based adaptive control and kinematics designs. The physical parameters and desired trajectories of RR, RP and PP manipulators illustrated in Fig. 2 are assumed same for the proper comparisons. The following physical parameters are used for the manipulators: l_1=l_2=1 m, lengths of link 1 and 2;

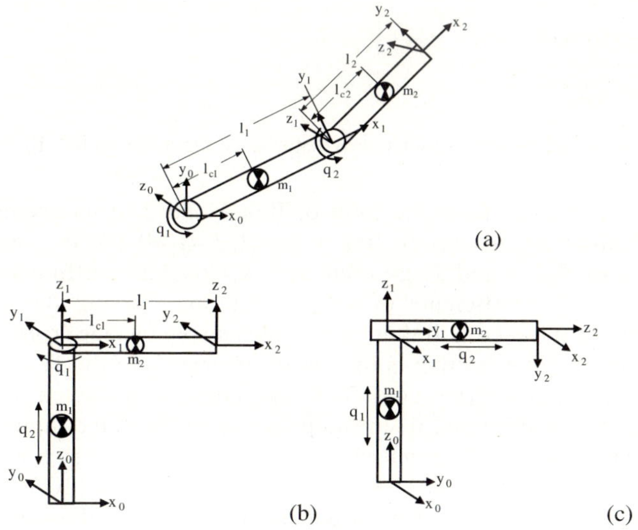

Fig. 2. The configurations of (a)RR, (b) RP and (c) PP robot manipulators

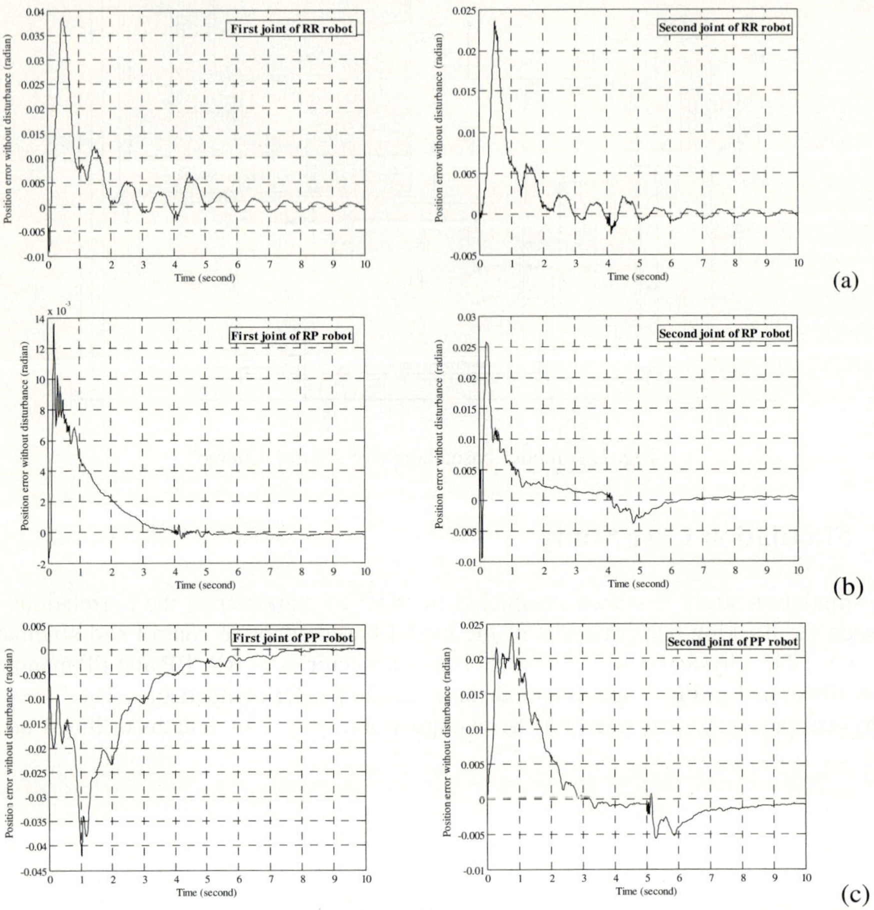

Fig. 3. Position tracking errors under zero disturbance inputs for a) RR, b) RP and c) PP

$lc_1=lc_2=0.5$ m, distances from the joint of link 1 and 2 to its centers of gravity; $m_1=2$ kg and $m_2=1$ kg, masses of link 1 and 2; $I_1=I_2=0.2$ kgm^2, lengthwise centraidal inertias of link 1 and 2; g=9,8m/sec^2; $k_1=k_2=0.2$, coefficients of dynamic friction; $v_1=v_2=0.2$, coefficients of viscous friction. During the operation, the mass at the end of the second link m_2 undergoes a step change from 1 kg to 1.5 kg at t=4 sec. The control objective is to control the state q of the system to track the desired trajectory q_d = [cos(t) sin(t)]. The simulation is conducted in two stages, i) under zero disturbance inputs, ii) in the presence of the same input disturbances. The control gains and the diagonal elements of the gain matrices were also set to be the same as K_p=diag{16,16}, K_v=diag{8,8} and γ=diag{10,10,10,10,10,10}. The model-based adaptive control for RR has been presented as an example. The simulation results of the tracking performances of the RR, RP and PP are also presented for comparisons.

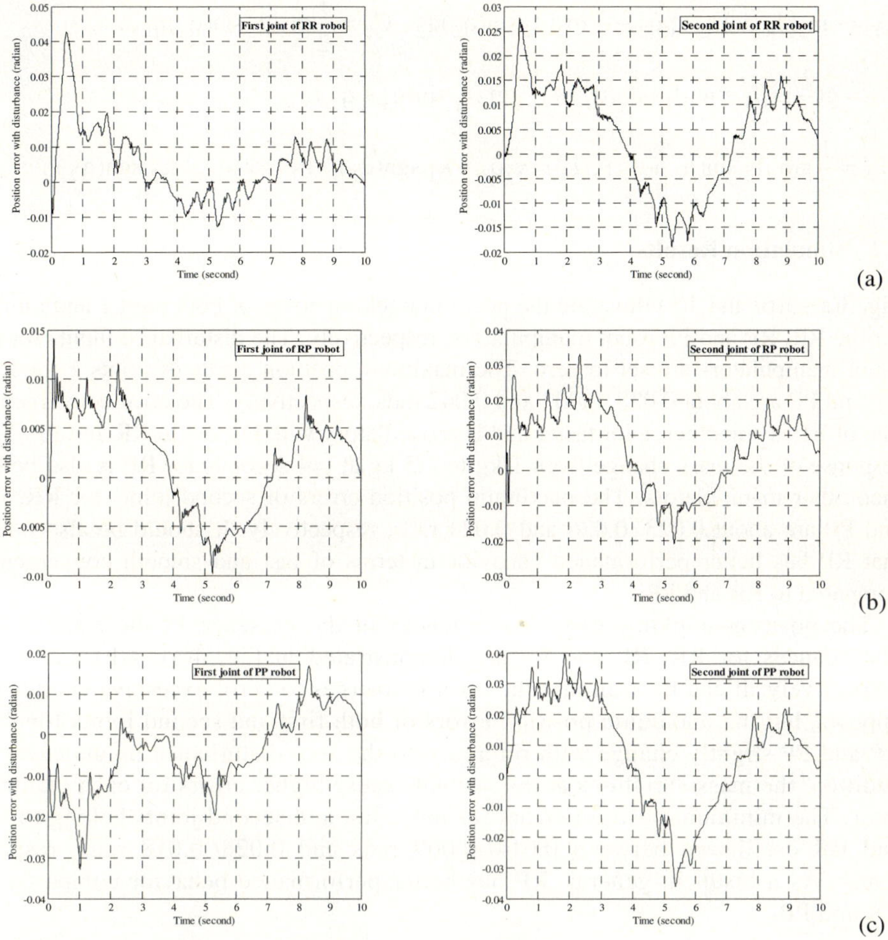

Fig. 4. Position tracking errors with the same input disturbances for a) RR, b) RP and c) PP

3.1 Adaptive Design Example for RR Manipulator

The configuration of the RR manipulator is illustrated in Fig. 2(a). The dynamic equation of the RR two-link robot manipulator is

$$\begin{bmatrix} \tau_1 \\ \tau_2 \end{bmatrix} = \begin{bmatrix} m_{11} & m_{12} \\ m_{21} & m_{22} \end{bmatrix} \begin{bmatrix} \ddot{q}_1 \\ \ddot{q}_2 \end{bmatrix} + \begin{bmatrix} C_1 \\ C_2 \end{bmatrix} + \begin{bmatrix} G_1 \\ G_2 \end{bmatrix} + \begin{bmatrix} F_1 \\ F_2 \end{bmatrix} \tag{23}$$

where

$$m_{11} = \frac{1}{4} m_1 l_1^2 + m_2 [l_1^2 + \frac{1}{4} l_2^2 + l_1 l_2 \cos(q_2)] + I_1 + I_2 ,$$

$$m_{12} = m_{21} = m_2 [\frac{1}{4} l_2^2 + \frac{1}{2} l_1 l_2 \cos(q_2)] + I_2 , \; m_{22} = \frac{1}{4} m_2 l_2^2 + I_2 ,$$

$$C_1 = -m_2 l_1 l_2 \sin(q_2)\dot{q}_1\dot{q}_2 - \frac{1}{2} m_2 l_1 l_2 \sin(q_2)\dot{q}_2^2, \; C_2 = \frac{1}{2} m_2 l_1 l_2 \sin(q_2)\dot{q}_1^2,$$

$$G_1 = g(\frac{1}{2} m_1 l_1 + m_2 l_1)\sin(q_1) + \frac{1}{2} g m_2 l_2 \sin(q_1 + q_2),$$

$$G_2 = \frac{1}{2} g m_2 l_2 \sin(q_1 + q_2), \; F_1 = v_1\dot{q}_1 + k_1 \operatorname{sgn}(\dot{q}_1), \; F_2 = v_2\dot{q}_2 + k_2 \operatorname{sgn}(\dot{q}_2).$$

3.2 Simulation Results

Fig. 3(a), 3(b) and 3(c) illustrate the position tracking errors of both joint 1 and joint 2 for the RR, RP and PP robot manipulators, respectively. The disturbance inputs for all robot manipulators are set to zero. The maximum position errors of joints 1 for RR, RP and PP are about 0.038, 0.002 and 0.042 rads, respectively. The error convergence rate of RP is smoother, a lot faster and less oscillatory with respect to RR and PP. The response of the mass change from 1 kg to 1.5 kg at t=4 seconds for RP is also better than other manipulators. The maximum position errors of second joints for RR, RP and PP are about 0.023, 0.026 and 0.024 rads, respectively. It should be also noted that RP has better performance behavior in terms of fast and smooth convergence compared to RR and PP.

The position tracking error convergences in the presence of the same input disturbances for RR, RP and PP are demonstrated in Fig. 4(a), 4(b) and 4(c), respectively. It can be seen that the error converges to zero; excessive chattering appears, but the maximum position errors of both first and second joints for RR, RP and PP slightly change with reference to the zero disturbance inputs case. In addition, the mass variations at t=4 seconds cause higher errors for each manipulator. The minimum position errors for both first and second joints belong to RP and RR oscillating between 0.014/-0.009 rads and 0.028/-0.018 rads, respectively. As a result in general, RP has better performance behavior compared to RR and PP.

4 Conclusion

In this paper, the effects of the kinematics design on tracking performance of the adaptive control are considered. The position tracking error convergences for the RR, RP and PP manipulators are obtained and compared. Although manipulators with revolute joints have the best kinematics designs, under similar conditions, such as physical parameters, desired trajectories and input disturbances, the tracking performance of the RR is not the best one amongst the others. This is mainly resulted from that complex kinematics designs require more sophisticated control approaches. Conclusively, the results achieved in the simulation case studies show that the RP has outperformed the others. This study can be furthered to determine what kind of kinematics design is appropriate for a particular control scheme.

References

1. Spong, M.W., Hutchinson, S., Vidyasagar, M.: Robot Modeling and Control, pp. 15–17. John Wiley & Sons, Inc, Chichester (2006)
2. Gosselin, C., Angeles, J.: A Global Performance Index for the Kinematics Optimization of Robotic Manipulators. Journal of Mechanical Design 113, 220–223 (1991)
3. Zhigang, Y., Huijun, Z., Shengze, W., Weizhong, G., Song, H.: The establishment and reasoning of knowledge base system for mechanism kinematic schemes. The International Journal of Advanced Manufacturing Technology 23, 295–300 (2004)
4. Mayorga, R.V., Carrera, J., Oritz, M.M.: A Kinematics Performance Index Based on the Rate of Change of a Standard Isotropy Condition for Robot Design Optimization. Robotics and Autonomous Systems 53, 153–163 (2005)
5. Kucuk, S., Bingul, Z.: Comparative Study of Performance Indices for Fundamental Robots. Robotics and Autonomous Systems 54(7), 567–573 (2006)
6. Bezinea, H., Derbelb, N., Alimi, A.M.: Fuzzy Control of Robot Manipulators: Some Issues on Design and Rule Base Size Reduction. Engineering Applications of Artificial Intelligence 15(5), 401–416 (2002)
7. Kelly, R.: A Tuning Procedure for Stable PID Control of Robot Manipulators. Robotica 13, 141–148 (1995)
8. Zha, X., Cui, P.: Sliding Mode Control for Uncertain Nonlinear Systems Using RBF Neural Networks. In: Wang, J., Liao, X.-F., Yi, Z. (eds.) ISNN 2005. LNCS, vol. 3498, pp. 21–29. Springer, Heidelberg (2005)
9. Li, Q., Tso, S.K., Zhang, W.J.: Trajectory Tracking Control of Robot Manipulators Using a Neural-Network-Based Torque-Compensator. J Sys. Contr 212(5), 361–372 (1998)
10. Ouyang, P.R., Zhang, W.J., Gupta, M.: An Adaptive Switching Learning Control Method for Trajectory Tracking of Robot Manipulators. Mechatronics 16, 51–61 (2006)
11. Kerry, R.: PD Control with Desired Gravity Compensation of Robotic Manipulators: a Review. Int. J. Robot Res. 16(5), 660–672 (1997)
12. Chen, Q.J., Chen, H.T., Wang, Y.J., Woo, P.Y.: Global Stability Analysis for Some Trajectory Tracking Control Schemes of Robotic Manipulators. J. Robot Syst. 18(2), 69–75 (2001)
13. Craig, J.J., Hsu, P., Sastry, S.S.: Adaptive Control of Mechanical Manipulators. Int. J. Robotics Res. 6, 16–28 (1987)
14. Craig, J.J.: Introduction to Robotics: Mechanics and Control. Addison-Wesley publishing company, London, UK (1989)

Fault Detection with Evolution Strategies Based Particle Filter and Backward Sequential Probability Ratio Test

Katsuji Uosaki[1] and Toshiharu Hatanaka[2]

[1] Department of Management and Information Sciences,
Fukui University of Technology,
Gakuen, Fukui 910-8505, Japan
uosaki@fukui-ut.ac.jp

[2] Department of Information and Physical Sciences,
Graduate School of Information Science and Technology, Osaka University
Suita, Osaka 565-0871, Japan
hatanaka@ist.osaka-u.ac.jp

Abstract. Fault detection in dynamic systems has attracted considerable attention in designing systems with safety and reliability. Though a large number of methods have been proposed for solving the fault detection problem, it is hardly apply to nonlinear stochastic state space models. A novel filter called the Evolution Strategies based particle filter (ESP) proposed by recognizing the similarities and the difference of the processes between the particle filters and Evolution Strategies is applied here to fault detection of nonlinear stochastic state space models. Numerical simulation studies have been conducted to exemplify the applicability of this approach.

Keywords: Particle filter, evolutioary strategies, fault detection, probability rartio test, nonlinear stochastic systems.

1 Introduction

The problem of fault detection in dynamic systems has attracted considerable attention in designing systems with safety and reliability. In the past two decades, a large number of methods have been proposed for solving the fault detection problem, see the survey papers [4], [9], [16] and the books [5], [6], and references therein. Among these, the model-based approaches using the quantitative analytical model of the system to be monitored are by nature the most powerful ones. For all model-based approaches, the decision of a fault is based on the innovations based on the state estimate obtained from the observed input-output data and a mathematical model of the system. Though the fault detection method can be easily constructed in linear/Gaussian state space models where the well-known Kalman filters [10], [1], [13] can employ to evaluate the state estimate, the idea is generally difficult to apply to nonlinear systems with non-Gaussian noises. In this paper, a new fault detection method is proposed for nonlinear/non-Gaussian state space models using the idea of the backward sequential probability ratio test (BSPRT) and the evolution strategies based particle filter (ESP).

B. Apolloni et al. (Eds.): KES 2007/WIRN 2007, Part I, LNAI 4692, pp. 664–671, 2007.

2 Fault Detection

Consider the following set of nonlinear state space models indexed by $m = 0, 1$.

$$x_{t+1} = f^{(m)}(x_t, u_t) + v_t \tag{1}$$

$$y_t = g^{(m)}(x_t) + w_t, \qquad m = 0, 1 \tag{2}$$

where x_t, u_t, y_t are the state variable, input and observation, respectively, $f^{(m)}(\cdot)$ and $g^{(m)}(\cdot)$ are known possibly nonlinear functions, and v_t and w_t are independently identically distributed (i.i.d.) system noise and observation noise sequences with probability density function (pdf) $p_v(\cdot)$ and $p_w(\cdot)$, respectively. It is assumed that, the system initially operates normally with the given normal mode model (1) and (2) indexed by $m = 0$, and then, fault may occur at unknown time $t = \tau$ where the model changes to the given fault mode model indexed by $m = 1$. The problem here is to develop a detection procedure to find a model change as quickly as possible.

The fault detection problem can be reduced to perform a hypothesis testing for the hypotheses:

$$H_0 \text{ (Normal mode)} : \text{System models are indexed by } m = 0$$
$$H_1 \text{ (Fault mode)} : \text{System models are indexed by } m = 1$$

Wald's sequential probability ratio test (SPRT) (Wald ([15])) is a common procedure for hypothesis testing, where the logarithm of likelihood ratio function (LLR)

$$\lambda_t = \log \frac{p(y_{1:t}|H_1)}{p(y_{1:t}|H_0)} \tag{3}$$

is evaluated and compared with two threshold values $B^* < 0 < A^*$ until it exceeds these thresholds. It is known that the fault detection system based on the above mentioned Wald's SPRT formulation minimizes, on the average, the time to reach a decision for specified error probabilities if the system is either in the normal mode or the fault mode from the beginning of the test and no transition of the models occurs during the test. However, the characteristics of the fault process differs from it; the system initially operates in normal mode and then transition occurs to the fault mode at time instant τ during observations. This difference causes the extra time delay in failure detection. To fit this situation, the idea of the backward SPRT (BSPRT) is introduced.

Rewriting the hypotheses representing normal and fault modes in fault detection process as

$$H_0 \text{ (Normal mode)} : \text{System models at time } t-k+1 \text{ are indexed by } m = 0$$
$$H_1 \text{ (Fault mode)} : \text{System models at time } t-k+1 \text{ are indexed by } m = 1,$$
$$t > \tau,\ k = 1, \ldots, t-\tau+1$$

we can introduce a backward LLR (BLLR), where LLR is computed in reverse (*backward*) from the current observation to the past observations:

$$\lambda^B_{t,k} = \log \frac{p(y_t, y_{t-1}, \ldots, y_{t-k+1}|H_1)}{p(y_t, y_{t-1}, \ldots, y_{t-k+1}|H_0)} \tag{4}$$

Assuming $p(y_{1:t}) = p(y_{1:k})p(y_{k+1:t}|y_{1:k}) \approx p(y_{1:k})p(y_{k+1:t})$, i.e., $y_{1:k}$ and $y_{k+1:t}$ are independent, we can express the BLLR approximately with the conventional LLR as

$$\lambda^B_{t,k} = \lambda_t - \lambda_{t-k}, \quad k = 1, 2, \ldots, n \tag{5}$$

with $\lambda_0 = 0$. Then the decision rule for acceptance of the hypothesis that the system is in the fault mode can be restated as

$$\lambda^B_{t,k} = \lambda_t - \lambda_{t-k} > K \quad \text{for some } k = 1, 2, \ldots, t, \tag{6}$$

or

$$\lambda_t - \min_{1 \le k \le t} \lambda_k > K \tag{7}$$

Introducing the statistics called the maximum BLLR,

$$\begin{aligned} S_t &= \max[0, S_{t-1} + \ell_t], \quad t = 1, 2, \ldots \\ S_0 &= 0 \end{aligned} \tag{8}$$

with

$$\ell_t = \log \frac{p(y_t|y_{1:t-1}, H_1)}{p(y_t|y_{1:t-1}, H_0)}, \quad t = 1, 2, \ldots \tag{9}$$

where $p(y_t|y_{1:t-1}, H_m)$ is the density of one-step-ahead output prediction y_t under the hypothesis H_m, $(m = 0, 1)$. Then the decision rule based on the BLLR can be expressed as

"*If $S_t > K$, where K is a suitable constant, one terminates observation with acceptance of the hypothesis that the system is in the fault mode. Otherwise, one continue observations as the system is likely not in the fault mode.*"

In order to apply this failure detection approach, we have to evaluate the density $p(y_t|y_{1:t-1}, H_m)$ of one-step-ahead output prediction y_t. It is hard to evaluate it for nonlinear/non-Gaussian models and some approximations such as extended Kalman filter [1] should be introduced. It is known that the EKF estimate is often divergent due to linearization error for severe nonlinear system, and then the fault detection may fails, and the evolution strategies based particle filter (ESP) [14] is applied.

3 Fault Detection by Evolution Strategies Based Particle Filters

It is known that particle filter, a simulation-based filter based on Bayesian sequential analysis, is quite effective for state estimation of nonlinear/non-Gaussian systemsĄ-Cand a lot of modifications have been developed [7], [2]. One of the most important modification is introduction of resampling process [11] to avoid the degeneracy phenomenon in the conventional particle filter, where all but one of the normalized importance weights are very close to zero after a few iterations and a large computational effort is wasted to updating trajectories whose contribution to the final estimate is almost zero. Resampling is carried out by selecting new grid points $x^*_{t|t}$ randomly with probability proportional to the importance weight $w^{(i)}_{t|t}$ and the weights are reset to $1/n$, and hence the grid points with larger weight are likely to be selected. Particle filter with the resampling process is called as the Sampling importance resampling (SIR) particle filter (SIR). On the other hand, the selection process of Evolution Strategies (ES) [12], one of so-called Evolutionary Computation approaches [8], [3], chooses the individuals with higher fitness from the union of parents and offspring or offspring only. We can recognize that this selection process corresponds to resampling processes in SIR by considering the fitness as the weight $w_{t|t}^{(i)}$ in SIR. The main difference is resampling in SIR is carried out probabilistically and the weights are reset as $1/n$, while the selection in ES is deterministic and the fitness function is never reset. Hence, by replacing the resampling process in SIR by the deterministic selection process in ES, we can derive a novel particle filter called Evolution strategies based particle filter comma (ESP(,)) and Evolution strategies based particle filter plus (ESP(+)) [14]. As shown in [14], the ESP filters behave more stable than SIR both in squared estimation errors and processing time by their deterministic selection process, we will develop fault detection methods using the ESP.

In order to apply the BLLR to fault detection, we need the information of $p(y_t|y_{1:t-1}, H_m)$ in (9). Here we use the grid approximation of the pdf $p(x_t|y_{1:t}, H_m)$ as

$$p(x_t|y_{1:t}, H_m) \approx \sum_{i=1}^{n} w_{t|t}^{(i,m)} \delta(x_t - x_{t|t}^{(i,m)}), \quad (m = 0, 1) \tag{10}$$

where the second superscript m is corresponding to the models, and evaluate it by using the particle filter. We can evaluate

$$\begin{aligned} p(x_t|y_{1:t-1}, H_m) &= \int p(x_t|x_{t-1}, H_m) p(x_{t-1}|y_{1:t-1}, H_m) dx_{t-1} \\ &= \int p_v(x_t - f^(m)(x_{t-1})) p(x_{t-1}|y_{1:t-1}, H_m) dx_{t-1} \\ &\approx \sum_{i=1}^{n} w_{t-1|t-1}^{(i,m)} p_v(x_t - f^{(m)}(x_{t-1|t-1}^{(i,m)})) \end{aligned} \tag{11}$$

Then, we can approximate the pdf $p(y_t|y_{1:t-1}, H_m)$ in (9) by

$$\begin{aligned} p(y_t|y_{1:t-1}, H_m) &= \int p(y_t|x_t, y_{1:t-1}, H_m) p(x_t|y_{1:t-1}, H_m) dx_t \\ &= \int p_w(y_t - g^m(x_t) p(x_t|y_{1:t-1}, H_m) dx_t \\ &\approx \frac{1}{n} \sum_{i=1}^{n} p_w(y_t - g^{(m)}(x_{t|t-1}^{(i,m)})) \end{aligned} \tag{12}$$

where $x_{t|t-1}^{(i,m)}$, $(i = 1, \ldots, n)$ are samples from the pdf $p(x_t|y_{1:t-1}, H_m)$ evaluated by (11). Thus, we can conduct the fault detection by evaluating the BLLR S_t with the pdf estimates obtained by two ESP filters corresponding to the system models H_0 and H_1, and compare S_t with suitable threshold K.

4 Numerical Examples

To exemplify the applicability of the proposed ESP filters, we carried out a numerical simulation. We consider the following nonlinear state space model with known parameters.

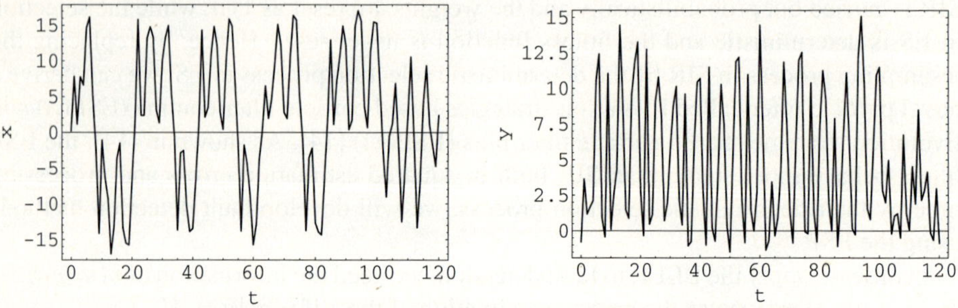

(a) State (b) Observation

Fig. 1. Sample behavior of state and observation processes

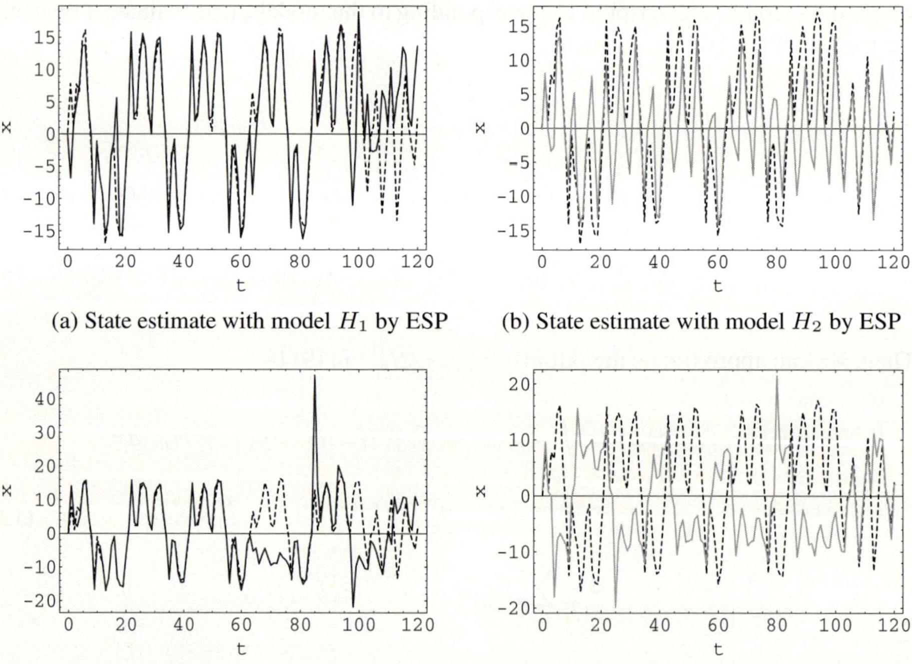

(a) State estimate with model H_1 by ESP (b) State estimate with model H_2 by ESP

(c) State estimate with model H_1 by EKF (d) State estimate with model H_2 by EKF

Fig. 2. Sample behaviors of state estimates (solid line : estimate; dotted line : true state)

$$x_t = \frac{x_{t-1}}{2} + \frac{a^{(m)} x_{t-1}}{1 + x_{t-1}^2} + 8\cos(1.2t) + v_t,$$
$$= f^{(m)}(x_{t-1}) + v_t, \qquad (13)$$
$$y_t = \frac{x_t^2}{20} + w_t = g^{(m)}(x_t) + w_t, \qquad m = 0, 1$$

with $a^{(0)} = 25$ for normal mode and $a^{(1)} = 12.5$ for fault mode, and v_t and w_t are i.i.d. zero-mean Gaussian random variates with variance 10 and 1, respectively. We assume that the fault occurs at $t = \tau = 101$. A sample behavior of the true state and corresponding observation processes is shown in Fig.1. Here the Gaussian distribution with mean $f(x_{t-1|t-1}^{(i)})$ and variance 10 is chosen as the importance density $q(x_t|x_{t-1|t-1}^{(i)}, y_{1:t})$.

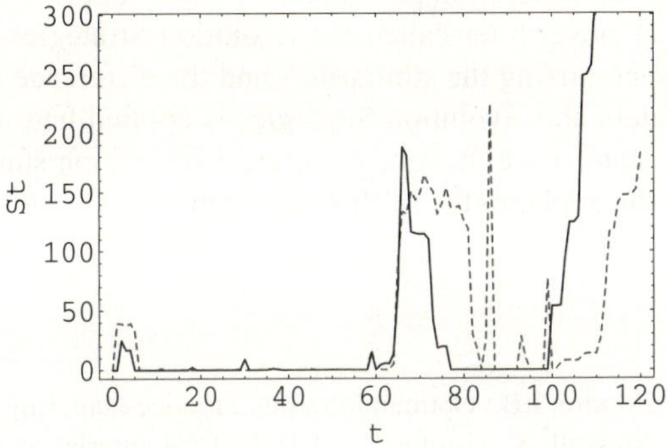

Fig. 3. Sample behaviors of test statistics BLLR by ESP and EKF (solid line : ESP; dotted line : EKF)

Table 1. Fault detection result

	False alarm rate	Miss alarm rate
Fault detection by ESP	2/20	0/20
Fault detection by EKF	6/20	1/20

Sample behaviors of state estimates by ESP(,) with $n = 10$, $r = 2$ based on the model H_m, $(m = 0, 1)$, and the test statistics BLLR S_t are given in Figs. 2 and 3 with corresponding results by EKF as well for comparison. The test statistics BLLR S_t takes positive value and is growing up rapidly after the change point τ both in ESP and EKF, we can detect the model change when the BLLR exceeds the suitable threshold K. Moreover, it should be noted that, as shown in Fig. 2, the state estimate by EKF shows poor behavior and hence the test statistics sometimes exceeds the threshold even where no fault occurs. Eventually, the rate of false alarm[1] and miss alarm[2] are higher by the detection procedure using EKF than by the procedure using ESP as shown in Table 1 that summarizes the result of 20 simulation runs of fault detection with the threshold $K = 250$. These results illustrate the applicability of the proposed approach for fault detection of nonlinear stochastic state space models.

This approach is applicable to general nonlinear/non-Gaussian stochastic models and an application to a simple model of air craft suspension system is now pursued. To apply this approach, preliminary simulation experiments should be carried out to determine suitable choice of design parameters in particle filter and fault detection systems.

5 Conclusion

Fault detection in dynamic systems has attracted considerable attention in designing systems with safety and reliability. Though a large number of methods have been proposed for solving the fault detection problem, it is hardly apply to nonlinear stochastic state space models. A novel filter called the Evolution Strategies based particle filter (ESP) proposed by recognizing the similarities and the difference of the processes between the particle filters and Evolution Strategies is applied here to fault detection of nonlinear stochastic state space models. Numerical simulation studies have been conducted to exemplify the applicability of this approach.

References

1. Anderson, B.D.O., Moore, J.B.: Optimal Filtering. Prentice-Hall, Englewood Cliffs (1979)
2. Arulampalam, S., Maskell, S., Gordon, N., Clapp, T.: A tutorial on particle filters for on-line non-linear/non-Gaussian Bayesian tracking. IEEE Trans. on Signal Processing SP-50, 174–188 (2002)
3. Bäck, T.: Evolutionary Algorithms in Theory and Practice: Evolution Strategies, Evolutionary Programming, Genetic Algorithms. Oxford University Press, London (1996)
4. Basseville, M.: Statistical approaches to industrial monitoring problems — Fault detection and isolation. In: Proc. 11th IFAC System Identification, pp. 413–432 (1997)
5. Basseville, M., Benveniste, A. (eds.) Detection of Abruptly Changes in Signals and Dynamic Systems. Springer, Berlin (1986)
6. Basseville, M., Nikiforov, I.: Detection of Abrupt Changes — Theory and Application. Prentice-Hall, Englewood Cliffs (1993)
7. Doucet, A., de Freitas, N., Gordon, N. (eds.) Sequential Monte Carlo Methods in Practice. Springer, Berlin (2001)
8. Fogel, D.B.: Evolutionary Computation. IEEE Computer Society Press, Los Alamitos (1995)
9. Frank, P.M.: Fault diagnosis in dynamic systems using analytical knowledge based redundancy — A survey and new results. Automatica 26, 459–474 (1990)

---

[1] The decision that the system model has changed is made even when the system model does not change. In this example, the test statistics exceeds the threshold between $t = 0$ and $t = 100 < \tau$.

[2] The decision that the system model does not change is made even when model changes. In this example, the test statistics never exceeds the threshold between $t = 101 = \tau$ and $t = 120$.

10. Kalman, R., Bucy, R.: New results in linear filtering and prediction theory. J. of Basic Engineering, Trans. ASME Series D 83, 95–108 (1961)
11. Rubin, D.B.: Using the SIR algorithm to simulate posterior distribution. In: Bernardo, J.M., DeGroot, M.H., Lindley, D.V., Smith, A.F.M. (eds.) Bayesian Statistics, vol. 3, pp. 395–402. Oxford University Press, Oxford (1988)
12. Schwefel, H.-P.: Evolution and Optimum Seeking. J. Wiley, New York (1995)
13. Sorenson, H.W.: Kalman Filtering: Theory and Application. IEEE Computer Society Press, New York (1985)
14. Uosaki, K., Hatanaka, T.: Nonlinear state estimation by evolution strategies based particle filters. In: Proc. 16th IFAC World Congress on Automatic Control (2005)
15. Wald, A.: Sequential Analysis, New York. J. Wiley, Chichester (1947)
16. Willsky, A.S.: A survey of design methods for failure detection in dynamic systems. Automatica 12, 601–611 (1976)

Infringing Key Authentication of an ID-Based Group Key Exchange Protocol Using Binary Key Trees*

Junghyun Nam[1], Juryon Paik[2], Youngsook Lee[2], Jin Kwak[3], Ung Mo Kim[2], and Dongho Won[2,**]

[1] Department of Computer Science, Konkuk University, Korea
jhnam@kku.ac.kr
[2] Department of Computer Engineering, Sungkyunkwan University, Korea
juryon.paik@gmail.com, {yslee,dhwon}@security.re.kr, umkim@ece.skku.ac.kr
[3] Department of Information Security, Soonchunhyang University, Korea
jkwak@sch.ac.kr

Abstract. Group key exchange protocols allow a group of parties communicating over a public network to come up with a common secret key called a session key. Due to their critical role in building secure multicast channels, a number of group key exchange protocols have been suggested over the years for a variety of settings. Among these is the ID-based group key exchange protocol proposed by Yang and Shieh in 2001. In this paper, we revisit the Yang-Shieh ID-based protocol and conduct a security analysis on the protocol. The consequence of our analysis is that the Yang-Shieh protocol fails to achieve its basic goal of securely establishing a session key among the intended parties. This is shown via a collusion attack on the protocol. We also show how to fix the security problem with the protocol.

1 Introduction

Protocols for group key exchange (GKE) are essential in building secure multicast channels for applications where quite a large number of users are likely to be involved (e.g., digital contents distribution, video conferencing, and massive online gaming). As these group-oriented applications proliferate in modern computing environments, the design of secure and efficient GKE protocols has attracted many researchers' attention over the years [3].

The highest priority in designing a key exchange protocol is placed on ensuring the security of session keys to be established by the protocol. Even if it is computationally infeasible to break the cryptographic algorithms used, the whole system becomes vulnerable to all manner of attacks if the keys are not

* This work was supported by the Korean Ministry of Information and Communication under the Information Technology Research Center (ITRC) support program supervised by the Institute of Information Technology Assessment (IITA).
** Corresponding author.

B. Apolloni et al. (Eds.): KES 2007/WIRN 2007, Part I, LNAI 4692, pp. 672–679, 2007.

securely established. But unfortunately, the experience has shown that the design of secure key exchange protocols is notoriously difficult. In particular, the difficulty is greatly increased in the group setting where a session key is to be established among an arbitrary number of parties. Indeed, there is a long history of protocols for this domain being proposed and years later found to be flawed (e.g., [4,2]). Thus, group key exchange protocols must be subjected to a thorough and systematic scrutiny before they are deployed into a public network, which might be controlled by an adversary.

Secure key establishment requires some form of authentication among protocol participants. Public Key Infrastructure (PKI) is based on asymmetric cryptography and is one of the most common authentication method. In a PKI, both the private key and the corresponding public key are random in appearance. This implies that the authenticity of a public key certificate should be verified using the certifying authority's public key before the public key contained in the certificate can be used. Hence, in practice, the use of public key certificates incurs a significant amount of administrative work and costs. Eventually, it was this drawback that, in 1984, prompted Shamir to introduce a new type of asymmetric cryptography called identity-based cryptography [5]. In ID-based cryptography, publicly verifiable information such as email or IP address instead of digital certificates can be used as user's public key for encryption or signature verification. This obviates the need for certificates and therefore simplifies the task of managing public keys.

In 2001, Yang and Shieh [6] proposed a group key exchange protocol which provides distributed functionality in collaborative and dynamic groups. The Yang-Shieh protocol eliminates the need for public key certificates by employing ID-based cryptography. Moreover by leveraging a binary key tree structure, this ID-based protocol achieves $O(\log m)$ round complexity and $O(\log m)$ computation complexity in a group of size m. But despite its many merits, the Yang-Shieh protocol does not achieve its main goal of securely establishing a session key among the intended parties. In this paper, we demonstrate this by showing that the protocol is vulnerable to a collusion attack in which two active adversaries together can easily violate the security of session keys established by the protocol. In addition, we suggest a simple patch for fixing the security vulnerability.

2 Review of the Yang-Shieh Protocol

The ID-based group key exchange protocol by Yang and Shieh [6] consists of three phases: the system setup phase, the user registration phase, and the group key agreement phase.

System Setup. To set up system parameters, the key generation center (KGC): (1) picks two distinct large primes p and q and sets $n = p \cdot q$, (2) computes the master key d such that $3 \cdot d \mod (p-1)(q-1) = 1$, (3) finds an integer g that is a generator of both $\mathbb{Z}_p^*$ and $\mathbb{Z}_q^*$, and (4) chooses a one-way function f. KGC makes the parameters $\langle n, g, f \rangle$ public while keeping the master key d secret.

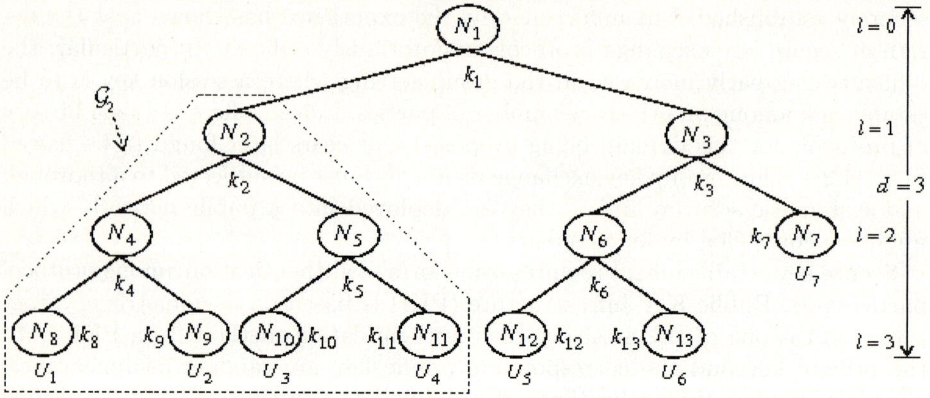

Fig. 1. A binary key tree structure for $\mathcal{G} = \{U_1, \ldots, U_7\}$

User Registration. Let $\mathcal{U}$ denote the set of all potential users. Each user $U_i \in \mathcal{U}$ submits its identity ID_i to KGC for registration. For each given ID_i, KGC performs the following: it first computes an extended identity $EID_i = f(ID_i) \in \mathbb{Z}_n^*$, then calculates U_i's private key s_i as $s_i = EID_i{}^d \mod n$, and finally delivers $\langle n, g, f, s_i \rangle$ to U_i via a secure channel. Notice that the extended identity EID_i and the private key s_i are related by the equation

$$EID_i = s_i{}^3 \mod n.$$

Group Key Agreement. Group key agreement procedure makes use of a binary key tree structure, where every node is either a leaf or a parent of two nodes. The root is located at level 0 and all leaves are at level d or $d-1$, with d being the height of the key tree. Nodes of the tree are numbered in such a way that node N_j has N_{2j} as its left child and N_{2j+1} as its right child. Fig. 1 shows an example key tree of height 3 with 7 leaves and 6 internal nodes.

Let $\mathcal{G} = \{U_1, U_2, \ldots, U_m\}$ be a set of users wishing to agree on a common secret key. The users in $\mathcal{G}$ are arranged at leaves of the key tree; all interior nodes are logical nodes hosting no users. We denote by $\mathcal{G}_j$ the subgroup consisting of all users located in the subtree rooted at node N_j. For each subgroup $\mathcal{G}_j$, there is a representative user called a *designated negotiator* and denoted by DN_j. Any user in $\mathcal{G}_j$ could be elected as the designated negotiator DN_j.

Each node N_j in the key tree is associated with a node key k_j. In the protocol, the node key k_j is shared as the subgroup key among the users in $\mathcal{G}_j$. Consequently, the root node key k_1 serves as the group key (i.e., the session key) shared by all users in $\mathcal{G}$. The process of session key generation corresponds to establishing all node keys of the tree. As the protocol proceeds, the node keys become generated in a bottom-up manner, starting with the leaf nodes and proceeding towards the root node. This process is quite straightforward to see. For each leaf node N_j, the node key k_j is just a random number chosen (uniformly from $\mathbb{Z}_n^*$)

by the user positioned at node N_j. Each internal node key k_j is computed from two child node keys k_{2j} and k_{2j+1}, and is established by two designated negotiators DN_{2j} and DN_{2j+1}. For ease of exposition, assume without loss of generality that U_α and U_β play the role of DN_{2j} and DN_{2j+1}, respectively. Then, U_α and U_β agree upon the internal node key k_j by performing the following two-party key agreement protocol:

1. U_α chooses a random number $r_{2j} \in \mathbb{Z}_n^*$ and computes

$$x_\alpha = g^{3h(r_{2j}\|k_{2j})} \bmod n,$$
$$y_\alpha = s_\alpha \cdot g^{2h(r_{2j}\|k_{2j})} \bmod n,$$

   where h is a one-way hash function. (Recall that s_α is the private key of U_α). U_α then sends the message $\langle ID_\alpha, x_\alpha, y_\alpha \rangle$ to U_β.
2. Upon receiving $\langle ID_\alpha, x_\alpha, y_\alpha \rangle$, U_β computes $EID_\alpha = f(ID_\alpha)$ and checks whether the following equation holds:

$$EID_\alpha \stackrel{?}{=} \frac{{y_\alpha}^3}{{x_\alpha}^2} \bmod n. \tag{1}$$

   If it is untrue, U_β aborts the protocol. Otherwise, U_β selects a random number $r_{2j+1} \in \mathbb{Z}_n^*$ and computes the node key k_j as

$$k_j = {x_\alpha}^{h(r_{2j+1}\|k_{2j+1})} \bmod n$$
$$= g^{3h(r_{2j}\|k_{2j})h(r_{2j+1}\|k_{2j+1})} \bmod n.$$

   After that, U_β computes

$$x_\beta = g^{3h(r_{2j+1}\|k_{2j+1})} \bmod n,$$
$$y_\beta = s_\beta \cdot g^{2h(r_{2j+1}\|k_{2j+1})} \bmod n,$$
$$z_\beta = \mathcal{E}_{k_j}(x_\beta),$$

   where $\mathcal{E}_{k_j}(x_\beta)$ is a symmetric encryption of x_β under the key k_j. U_β then sends the message $\langle ID_\beta, x_\beta, y_\beta, z_\beta \rangle$ to U_α.
3. After receiving $\langle ID_\beta, x_\beta, y_\beta, z_\beta \rangle$, U_α computes $EID_\beta = f(ID_\beta)$ and checks if the following equation holds:

$$EID_\beta \stackrel{?}{=} \frac{{y_\beta}^3}{{x_\beta}^2} \bmod n. \tag{2}$$

   If it is not the case, U_α aborts the protocol. Otherwise, U_α derives the node key k_j as $k_j = {x_\beta}^{h(r_{2j}\|k_{2j})} \bmod n$ and verifies the correctness of k_j by decrypting z_β.

After establishing k_j as above, the designated negotiator U_α (resp. U_β) should send $\mathcal{E}_{k_{2j}}(k_j)$ (resp. $\mathcal{E}_{k_{2j+1}}(k_j)$) to the rest of subgroup $\mathcal{G}_{2j}$ (resp. $\mathcal{G}_{2j+1}$) so that k_j can be shared within the subgroup.

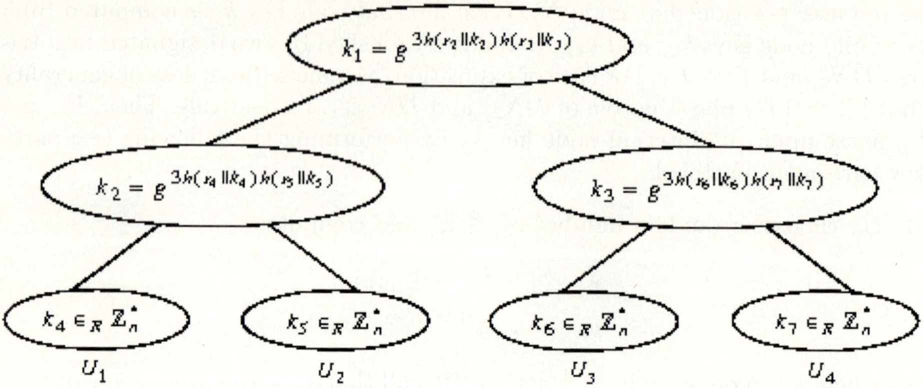

Fig. 2. Node keys established among the users in $\mathcal{G} = \{U_1, U_2, U_3, U_4\}$

The process above is repeated recursively until the root node key k_1 is shared among the users in $\mathcal{G}$. Fig. 2 shows all node keys established for the case of $\mathcal{G} = \{U_1, U_2, U_3, U_4\}$. When the root node key k_1 has been shared, the session key sk is simply set equal to k_1. In this manner, the Yang-Shieh protocol bounds the maximum amount of computation performed by any single user by $O(\log m)$ in a group of size m. This becomes obvious once we notice that in the protocol, each user needs to compute the node keys only on the path from its leaf node to the root node.

3 Security Analysis

Implicit key authentication is the fundamental security property that any given key exchange protocol is expected to possess [4].

Definition 1 (implicit key authentication). *Let $\mathcal{G}$ be a set of parties who wish to share a secret key by running a key exchange protocol* KEP. *Let K_i be the key computed by a party $U_i \in \mathcal{G}$ as a result of an execution of protocol* KEP. *We say that* KEP *satisfies implicit key authentication if each $U_i \in \mathcal{G}$ is assured that no $U_a \notin \mathcal{G}$ can learn the key K_i unless helped by a dishonest $U_j \in \mathcal{G}$ or by any other trusted party.*

In many real world applications, it is typical that a communication party establishes several concurrent sessions possibly with different parties. Hence, implicit key authentication has to be satisfied even when multiple instances of the protocol are run concurrently, of course, in the presence of active adversaries who may read, modify, insert, delete, replay and delay messages. A protocol achieving implicit key authentication is called an authenticated key exchange protocol, and is of fundamental importance in much of modern cryptography and network security.

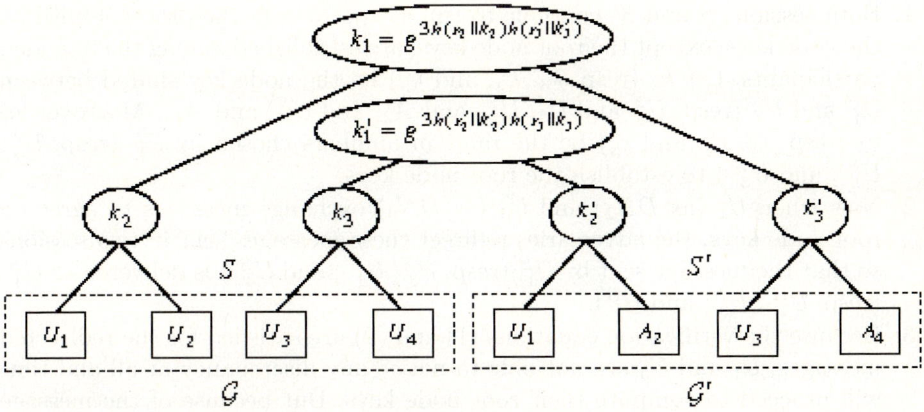

Fig. 3. A high-level depiction of a collusion attack on the Yang-Shieh protocol

Unfortunately, the Yang-Shieh protocol fails to satisfy implicit key authentication, unlike the claim that it is an authenticated key exchange protocol. We demonstrate this by revealing that the Yang-Shieh protocol is vulnerable to an active attack mounted by two colluding adversaries A_2 and A_4. For the collusion attack, we assume that the adversaries are legitimate users registered with KGC, and so are able to set up normal protocol sessions with other users.

We start off discussing the setting for the attack. First, consider a protocol session S to be conducted by the users of group $\mathcal{G} = \{U_1, U_2, U_3, U_4\}$, where U_1 and U_3 play the roles of DN_2 and DN_3, respectively. Now assume that U_1 and U_3 accept the invitation by the adversaries A_2 and A_4 to participate in a new concurrent session S' and to serve respectively as DN_2 and DN_3 for the new session. In summary our attack involves the following two sessions running concurrently:

$$\begin{aligned} \text{Session } S: &\quad \mathcal{G} = \{U_1, U_2, U_3, U_4\}, \\ \text{Session } S': &\quad \mathcal{G}' = \{U_1, A_2, U_3, A_4\}. \end{aligned}$$

We note however that this setting is used only for simplicity and clarity of the attack scenario. Indeed, our attack also holds for more general situations, as will become evident from the attack description below.

The collusion attack leads to a serious consequence as depicted in Fig. 3. At the end of the attack, every user in $\mathcal{G}$ computes a session key as per protocol specification and thinks that the session is finished successfully, when, in fact, the computed key is available also to the adversaries A_2 and A_4. The adversaries' strategy towards this end is simple and clever. Let U_i^S (resp. $U_i^{S'}$) be the instance of U_i participating in S (resp. S'). Then, all that the adversaries need to do is to have U_1^S establish its root node key with $U_3^{S'}$ instead of U_3^S and to have $U_1^{S'}$ establish its root node key with U_3^S instead of $U_3^{S'}$. A more detailed explanation of this is as follows:

1. Both sessions S and S' continue to run as specified in the protocol until all the node keys, except the root node keys, are established among the intended participants. Let k_2 (resp. k_3, k_2', and k_3') be the node key shared between U_1^S and U_2 (resp. U_3^S and U_4, $U_1^{S'}$ and A_2, and $U_3^{S'}$ and A_4). Moreover let r_2 (resp. r_3, r_2', and r_3') be the random numbers chosen by U_1^S (resp. U_3^S, $U_1^{S'}$, and $U_3^{S'}$) to establish the root node keys.
2. Now when U_1 (as DN_2) and U_3 (as DN_3) exchange messages to agree on root node keys, the adversaries redirect these messages sent in two sessions so that the message sent by U_1^S (resp. U_3^S, $U_1^{S'}$, and $U_3^{S'}$) is delivered to $U_3^{S'}$ (resp. $U_1^{S'}$, U_3^S, and U_1^S).
3. Because the verification equations (1) and (2) are satisfied for the redirected messages, U_1 and U_3 are not able to notice any discrepancy at all and thus will proceed to compute their root node keys. But because of the message redirections, root node keys are shared between U_1^S and $U_3^{S'}$ and between U_3^S and $U_1^{S'}$. More precisely, U_1^S and $U_3^{S'}$ compute the same root node key $k_1 = g^{3h(r_2\|k_2)h(r_3'\|k_3')} \bmod n$, while U_3^S and $U_1^{S'}$ derive their root node key as $k_1' = g^{3h(r_2'\|k_2')h(r_3\|k_3)} \bmod n$.
4. As the final step, the DNs U_1 and U_3 make the keys k_1 and k_1' available to appropriate subgroup users. Therefore, at the end of two sessions, U_1^S, U_2, $U_3^{S'}$ and A_4 share the same session key $sk = k_1$, while U_3^S, U_4, $U_1^{S'}$ and A_2 hold the common session key $sk' = k_1'$. Consequently, the implicit key authentication property of the protocol is violated.

Through the collusion attack, the authentication mechanism of the Yang-Shieh protocol has been completely compromised. Indeed, the effect of the attack is much the same as that of a man-in-the-middle attack. The users of group $\mathcal{G}$ believe that they have established a secure session among them, while in fact they have shared the session keys sk and sk' with the adversaries. As a result, the adversaries can not only access and relay any confidential communications among the users of $\mathcal{G}$, but can also send arbitrary messages for their own benefit impersonating any of the users to the others. In conclusion, it can be stated that the Yang-Shieh protocol becomes very hazardous as soon as two protocol sessions run concurrently with the same DNs DN_2 and DN_3.

4 Preventing the Attack

We now figure out what is wrong with the Yang-Shieh protocol and how to fix it. The main flaw of the protocol lies in the process of two-party key agreement (TKA) between the designated negotiators U_α and U_β. The problem with the TKA protocol is that it does not provide U_α and U_β with any means of checking whether they are members of the same group or not. All that U_α (resp. U_β) can be sure of by verifying the equality (2) (resp. equality (1)) is that the received message is sent by U_β (resp. U_α). This oversight allows the adversaries in our attack to switch the messages between two different sessions without being detected.

Hence to prevent the attack, the TKA protocol should be fixed so that it can provide U_α and U_β with a way of verifying that they are communicating with a same-group user. A simple yet effective fix is to incorporate mutual key confirmation into the TKA protocol. Let $k_{j,\alpha}$ and $k_{j,\beta}$ denote k_j computed respectively by U_α and U_β as a result of the original TKA protocol. Moreover let us denote by $\mathcal{ID}$ the identities of all users in $\mathcal{G}$. Then the key confirmation technique used for the fix is the widely accepted one [1] with a slight modification. Specifically, it works as follows:

- For key confirmation from U_β to U_α, U_β sends $z_\beta = h(ID_\beta \| \mathcal{ID} \| x_\alpha \| k_{j,\beta})$ to U_α. Upon receiving z_β, U_α checks if z_β equals $h(ID_\beta \| \mathcal{ID} \| x_\alpha \| k_{j,\alpha})$. If so, then U_α computes the final node key k_j as $k_j = h(ID_\alpha \| ID_\beta \| \mathcal{ID} \| x_\alpha \| x_\beta \| k_{j,\alpha})$. Otherwise, it aborts the protocol.
- Key confirmation from U_α to U_β is symmetric to the above case. U_α sends $z_\alpha = h(ID_\alpha \| \mathcal{ID} \| x_\beta \| k_{j,\alpha})$ to U_β, and U_β verifies that z_α equals $h(ID_\alpha \| \mathcal{ID} \| x_\beta \| k_{j,\beta})$. U_β computes the node key $k_j = h(ID_\alpha \| ID_\beta \| \mathcal{ID} \| x_\alpha \| x_\beta \| k_{j,\beta})$ if the verification succeeds, and otherwise aborts the protocol.

In the modified TKA protocol, the key confirmation values z_α and z_β become tightly linked to all user identities $\mathcal{ID}$. Thus if U_α and U_β are users of different groups, they will stop executing the protocol because the key confirmation process will fail. Consequently, the collusion attack does not work for the fixed protocol.

References

1. Bresson, E., Chevassut, O., Pointcheval, D., Quisquater, J.-J.: Provably authenticated group Diffie-Hellman key exchange. 8th ACM Conference on Computer and Communications Security (CCS '01), pp. 255–264 (2001)
2. Choo, K.-K., Boyd, C., Hitchcock, Y.: Errors in computational complexity proofs for protocols. In: Roy, B. (ed.) ASIACRYPT 2005. LNCS, vol. 3788, pp. 624–643. Springer, Heidelberg (2005)
3. Katz, J., Yung, M.: Scalable protocols for authenticated group key exchange. In: Boneh, D. (ed.) CRYPTO 2003. LNCS, vol. 2729, pp. 110–125. Springer, Heidelberg (2003)
4. Pereira, O., Quisquater, J.-J.: A security analysis of the Cliques protocols suites. 14th IEEE Computer Security Foundations Workshop (CSFW '01), pp. 73–81 (2001)
5. Shamir, A.: Identity-based cryptosystems and signature schemes. In: Blakely, G.R., Chaum, D. (eds.) CRYPTO 1984. LNCS, vol. 196, pp. 47–53. Springer, Heidelberg (1985)
6. Yang, W.-H., Shieh, S.-P.: Secure key agreement for group communications. International Journal of Network Management 11(6), 365–374 (2001)

Multiresolution ICA for Artifact Identification from Electroencephalographic Recordings

Nadia Mammone, Giuseppina Inuso, Fabio La Foresta, and Francesco Carlo Morabito

Neurolab, DIMET, University of Reggio Calabria
Via Graziella, Feo di Vito, Reggio Calabria, Italy
nadia.mammone@unirc.it

Abstract. This paper addresses the issue of artifact extraction from Electroencephalographic (EEG) signals and introduces a new technique for EEG artifact removal, based on the joint use of Wavelet transform and Independent Component Analysis (WICA). In fact, EEG recordings are often contaminated by the artifacts, signals that have non-cerebral origin and that might mimic cognitive or pathologic activity and therefore distort the analysis of EEG. The proposed technique extracts the artifacts taking into account the frequencies of the four major EEG rhythms. An artificial artifact-laden EEG dataset was created mixing a real EEG with a set of synthesized artifacts and the performance of WICA was measured. WICA had the best artifact separation performance for every kind of artifact with respect to other techniques and allowed for minimum information loss.

Keywords: Artifacts, EEG, Independent Component Analysis, Wavelet Transform.

1 Introduction

Electroencephalography is the neurophysiologic measurement of the electrical activity of the brain recorded through electrodes placed on the scalp. The neurons, in fact, communicate by means of electrical impulses and generate a bio-electromagnetic field that propagates through the brain tissues, the skull and the scalp. By means of some electrodes placed on the scalp over multiple areas of the brain, we can detect and record the electric field that reaches the scalp and that, even though attenuated, reflects the underlying brain activity: the resulting traces are called electroencephalogram (EEG). Historically, four major types of continuous rhythmic sinusoidal EEG activity are recognized (*alpha*, *beta*, *delta* and *theta*) although there is no precise agreement on the frequency ranges for each type.: **Delta** is the frequency range up to 4 Hz; **Theta** is the frequency range from 4 Hz to 8 Hz; **Alpha** (Berger's wave) is the frequency range from 8 to 12 Hz; **Beta** is the frequency range above 12 Hz. Unfortunately, EEG if often contaminated by signals that have non-cerebral origin and are called *artifacts*.

Artifacts can be clustered in two categories: internal (biological) artifacts and external artifacts. The former have biological but not-cognitive origin, like for example electro-oculogram (EOG) and electromyographic signals (EMG). The latter have external origin, most of times linked to the electrical line noise, to the impedance or settling of the electrodes.

B. Apolloni et al. (Eds.): KES 2007/WIRN 2007, Part I, LNAI 4692, pp. 680–687, 2007.

This is one of the reasons why it takes considerable experience to interpret EEG clinically, because artifacts might mimic cognitive or pathologic activity and therefore distort the analysis or completely overwhelm the EEG waves.

A very powerful approach, but still not common in clinical practice, was proposed in literature some years ago: it consists in artifactual signals extraction, detection and cancellation. This approach concentrates the artifactual content of the EEG dataset in a few signals to be rejected, so that we do not have to cancel the entire affected data segment. This approach is based on ICA because the hypothesis that an artifact is statistically independent from the rest of the signals is widely accepted and therefore it is also well accepted that it can be extracted by Independent Component Analysis (ICA) [1], [2], [3], [4], [5], [6], [7], [8]. Because of this, ICA has been widely exploited as a tool for artifactual signals extraction, ever since it was proposed for the first time by Makeig et al. [8], Vigario et al. [7] independently.

However, artifacts often involve a very narrow frequency range and exploiting these features in the frequency domain would help, but ICA operates in time domain. Sometimes this means that, even when the separation is good, some useful EEG information content can be seen in the component accounting for the artifact, thus cancelling it would cause information loss.

On the contrary, Wavelet analysis is a time-frequency technique and might help to overcome the limitations of ICA. So far, Wavelet analysis has been exploited in EEG artifact extraction jointly with ICA, but only as a pre- or post-denoising tool [9] [10]. In our opinion, the joint use of Wavelets and ICA holds a much wider potential, thus in this paper we propose the new technique WICA for EEG artifact extraction. WICA is compared to other techniques based on the joint use of wavelet denoising and ICA, showing to outperform them in separation and in saving useful information.

2 Wavelet Analysis and Independent Component Analysis

2.1 Wavelet Analysis

Wavelet analysis is a multiresolution analysis (MRA) method and tool, which has been widely used in signal processing and data analysis.

The wavelet expansion relative to the *wavelet mother* $\psi(t)$ and the *scaling function* $\varphi(t)$ of a signal $x(t)$ has the following expression [11]:

$$x(t) = \sum_{k} c_{j_0k}\varphi_{j_0k}(t) + \sum_{j=j_0} \sum_{k} d_{jk}\psi_{jk}(t) \quad (1)$$

where j_0 is an arbitrary starting scale. The first term in (1) is the *approximation* at scale j_0 and the second term represents the sum of the *details*.

2.2 ICA for Artifact Extraction

Independent component analysis (ICA) is a statistical method for transforming an observed multidimensional random vector into components that are statistically as independent from each other as possible.

The independent components (ICs) extracted from the multidimensional random vector **x** have the following expression:

$$\mathbf{u} = \mathbf{W}\mathbf{x}$$

where $\mathbf{x}$ represents a n-dimensional vector to be processed in order to extracts the ICs. In this case, $\mathbf{x}$ represents the dataset of selected WCs. The rows of $\mathbf{u}$ are an estimation of the ICs.

The unmixing matrix $\mathbf{W}$ is computed according to the extended-INFOMAX learning rule [1]:

$$\Delta\mathbf{W} \propto [\mathbf{I} - \mathbf{K}tanh(\mathbf{u})\mathbf{u}^T - \mathbf{u}\mathbf{u}^T]\mathbf{W}$$

where $\mathbf{I}$ is the n-dimensional identity matrix and $\mathbf{K}$ is a $n \times n$ diagonal matrix whose elements k_i are equal to 1 for supergaussian ICs and -1 for subgaussian ones.

As introduced above, it is widely accepted that an artifact is statistically independent from the rest of the signals. This hypothesis is clear when the artifact is external but it is also well accepted if it is internal because even though the triggering event starts from a brain area (for example the motor cortex) the time course of the artifact carries no information about the triggering event signal. This is why ICA has been widely exploited for artifactual signals extraction. ICA is usually able to concentrate the artifactual information in a single component, anyway, most of times this component carries non-artifactual information as well, thus rejecting it may cause information loss. Moreover, ICA performance depends on the size of the dataset (number of samples): the larger the dataset processed, the higher the probability that the effective number of sources will overcome the number of channels (*overcomplete ICA*), because the number of channels is fixed over time, but the number of contributions from different neuronal sources is likely to increase with the length of the recording. In this case redundancy is not sufficient to estimate the sources and an independent component might account for more than one contribution, in other words, the algorithm might not be able to separate the artifactual signals from the rest. On the contrary, the smaller the number of samples, the more difficult the estimation of the parameters and thus the performance of ICA suffers. The best choice is a trade-off between a small dataset and a high performance. In order to overcome this limitation, the proposed methodology includes a step prior to ICA that increases the redundancy of the dataset, thanks to wavelet decomposition, bypassing the possible problem of overcomplete ICA.

Another limitation is that ICA cannot take advantage of the features of the artifacts in frequency domain: artifacts have a typical frequency range and their spectrum is overlapped to the spectrum of the EEG, thus filtering the dataset is not an optimal solution because this would lead to a great information loss. But we can make the most of this limitation in frequency domain performing ICA in the range where the artifact is concentrated.

2.3 Wavelet Denoising and ICA

As cited in the Introduction, other efficacious techniques, that exploit jointly ICA and Wavelet transform for EEG artifact rejection, were proposed in Literature, but wavelet analysis was just proposed as a denoising tool, prior or after ICA processing. The strong point of WICA methodology is that wavelets are not used as a denoising or filtering tool but wavelet decomposition is an integral part of the separation process. The first method (Method A) performs wavelet denoising over the EEG dataset and then applies ICA to reject artifacts [9]. The second method (Method B) performs ICA and then passes the ICs through wavelet denoising [10]. Method A, Method B and W-ICA were applied and their performances in artifact extraction were compared in this paper.

The wavelet denoising procedure involves a wavelet decomposition step, depending on the chosen wavelet function and the number of levels. A thresholding rule is applied to the the details and then the denoised signal is reconstructed by the approximation and the processed details. In particular the details were processed by soft Stein's unbiased risk estimate (SURE) thresholding rule. The same wavelet family (Daubechies) and the same level of decomposition were used either in WICA and in the wavelet denoising step of Method A and B. The level of decomposition will be detailed in Section 4 because it depends on the bandwidth of the dataset.

3 WICA Methodology

A WICA technique was first proposed for the processing of electromyographic signals (EMG) [12], [13]. In this paper a new WICA technique optimized for EEG processing is presented: according to the limitations of ICA in artifact extraction discussed in Section 2.2. This Section will explain why the proposed WICA technique allows for either increasing redundancy and exploiting the features of EEG artifacts in frequency domain.

If we suppose to detect an artifactual event in one channel and to decompose the channel into a certain number of wavelet components (WCs), if the spectral content of the artifactual signal is not strictly concentrated in one of the levels of the decomposition, applying ICA to these wavelet components we will take advantage of the increased redundancy, because the event that was visible in one signal (the original channel) is now visible in more that one signal (the wavelet components). In case the spectral content of the artifactual signal is concentrated only in a certain level of the wavelet decomposition, ICA could focus on the same level wavelet components, working much better than processing the entire dataset that includes also frequencies not involved in the artifactual event. This way, even in case of non-optimal separation, the independent component accounting for the artifact might account also for non-artifactual information only in the frequency range where the artifact was concentrated, limiting the information loss in the other ranges.

The first stage of WICA for EEG artifact extraction is a wavelet decomposition that partitions the original dataset into the four major ranges of brain activity described in the introduction, projecting the data into a n-dimensional space where ICA is performed. The basis of this new space consists in the scaling function and the wavelet functions, where n-1 is the number of *levels* of the decomposition, and the scaling and wavelet functions depends on the chosen *wavelet family*. In this paper *Daubechies* family is adopted [11]. Once the raw data have been so projected into the n-dimensional space, the WCs linked to artifactual events are selected for ICA processing.

ICA will concentrate the artifactual information in one or more ICs, these ICs will be rejected, obtaining an artifact-free set of ICs ($\mathbf{u_{af}}$), and inverse ICA performed multiplying $\mathbf{u_{af}}$ by the inverse of the estimated unmixing matrix $\mathbf{W}$, so that the dataset of corrupted WCs is reconstructed without the artifact ($\mathbf{x_{rec}}$).

$$\mathbf{x_{rec}} = \mathbf{W^{-1}u_{af}}$$

The artifact-free wavelet components $\mathbf{x_{rec}}$ are restored into the original dataset of WCs. Then wavelet reconstruction is performed in order to reconstruct the clean EEG recording.

4 EEG Artifact Extraction

4.1 Data Simulation

The dataset is an apparently artifact-free EEG with eight channels recorded simultaneously (F3, F4, C3, C4, P3, P4, O1, O2) according to the standard 10/20 System. The sampling rate is 100Hz and time duration is 7.5 sec (the dataset is available at ftp://ftp.ieee.org/uploads/press/rangayyan/).

We simulated four types of artifacts and mixed them to two channels of the EEG (with a 0 dB signal/noise) in order to simulate a real artifact-laden EEG. We modelled electrical shift artifacts by implementing discontinuities. The synthesized artifact is shown on top of Figure 2.a whereas the mixed EEG is shown in Figure 1.a. We modelled linear trends with selected a slope of 100 V per second. The synthesized artifact is shown on top of Figure 2.b whereas the EEG mixed to the artifact is shown in Figure 1.b. We modelled EOG time courses using random noise band-pass filtered (FIR) between 1 and 3 Hz. The synthesized artifact is shown on top of Figure 2.c. The mixed EEG is shown in Figure 1.c. We modelled temporal muscle artifacts using random noise

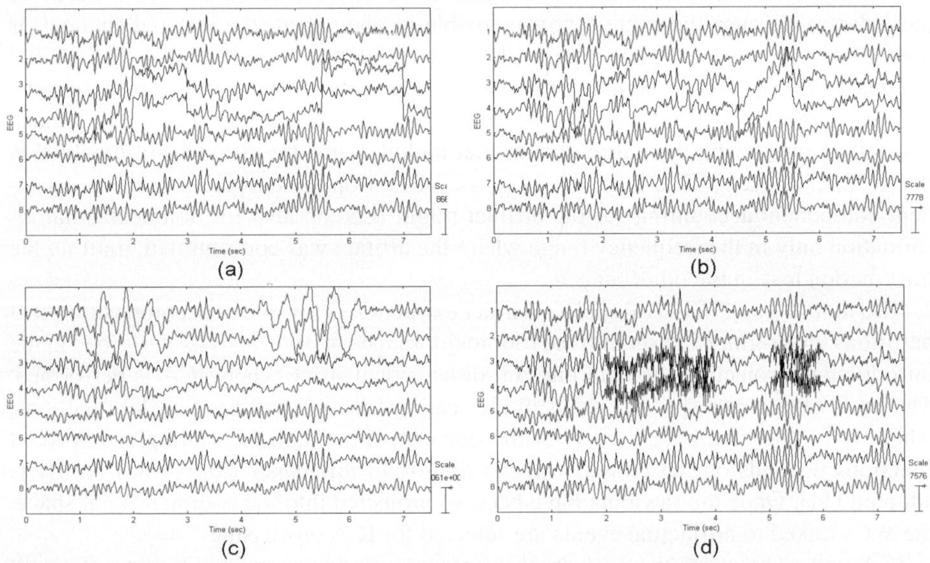

Fig. 1. Semi-simulated data obtained by adding simulated artifacts to the apparently artifact-free EEG data shown in Figure .b. The synthesized artifacts are plotted on top of each picture in Figure 2.

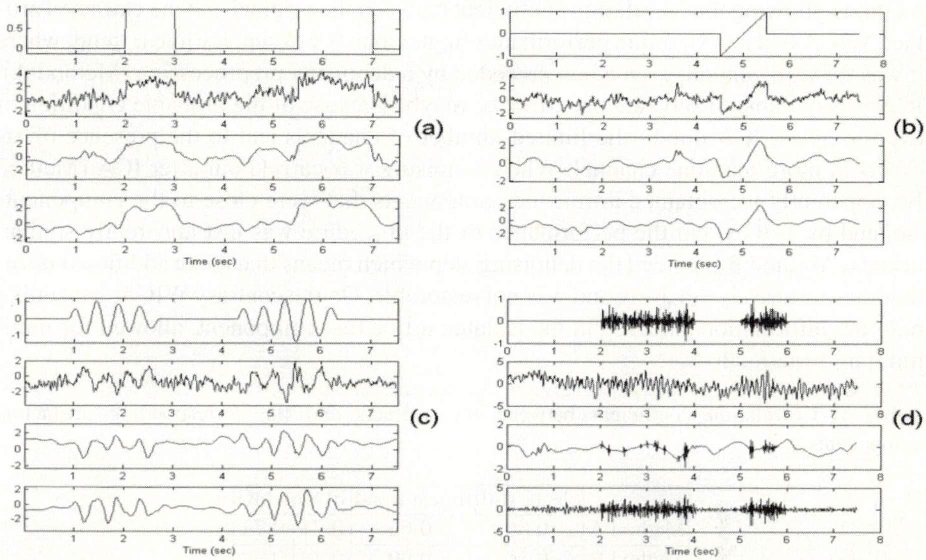

Fig. 2. Comparison of the performance of artifact separation of the three methods for each synthesized dataset: (a) EEG with electrical shift; (b) EEG with linear trend; (3) EEG with EOG; (d) EEG with EMG. The original artifact and the artifactual signals isolated by Method A, B and W-ICA, respectively, are shown.

band-pass filtered (FIR) between 20 and 60 Hz. The synthesized artifact is shown on top of Figure 2.d. The mixed EEG is shown in Figure 1.d.

4.2 Results

The three techniques Method A, Method B and WICA, described in Section 2, were implemented in MATLAB® environment and were applied to the four synthesized artifact-laden dataset shown in Figure 1. Either for WICA and for the denoising step of Method A and B, the chosen wavelet family is Daubechies, *db4* in particular. Being the band of the EEG 0-50Hz, a four levels wavelet decomposition was sufficient to partition the frequency range 0-50Hz into the ranges of the major EEG rhythms (delta: 0-4Hz, theta: 4-8Hz, alpha: 8-12Hz, beta: 12-50Hz).

Figure 2 shows the comparison of the performance in artifact separation of the three methods for each synthesized dataset. On top of each picture in Figure 2 the original simulated artifact is plotted, then the signals accounting for that artifact isolated by Method A, B and W-ICA respectively, are shown. Looking at Figure 2, we can see that too high information loss would be caused by the rejection of the components isolated by Method A: it always accounted for more than the artifactual information. The components isolated by Method B were close enough to the original artifact in case of discontinuity, linear trend and EOG artifacts, but different for the EMG one.

WICA had the best performance and showed to be suitable for the extraction of any kind of artifact. Table 1 summarizes the comparison of the performance of the three methods, showing the correlation coefficient between the original and the extracted artifact. WICA had an extraction performance higher than 0.9 except for linear trend, where it was 0.73. In general, even when preceded by a denoising preprocessing (Method A), ICA was not able to separate the artifacts, maybe because of the probable condition of over-complete ICA due to the limited number of channels and to the presence of artifacts in more than one channel. When denoising was carried out after ICA (Method B), apparently we obtained artifactual components that were close to the components isolated by WICA, but the performance of the procedure was just apparently similar, because Method B included the denoising step which means that some additional information was already cut away and was not restorable. On the contrary WICA, cancelling only the information gathered in the isolated artifactual component, allowed for minimum information loss.

Table 1. Correlation coefficient between the artifacts and the corresponding artifactual components

	electr. shift	linear trend	EOG	EMG
Method A	0.86	0.65	0.71	0.74
Method B	0.86	0.69	0.73	0.22
W-ICA	0.92	0.73	0.9	0.91

5 Conclusion

In this paper, a new W-ICA technique optimized for EEG artifact extraction is presented. This technique integrates wavelet analysis and ICA and separate the artifacts exploiting the knowledge about the major rhythms in the EEG. A real EEG and a set of synthesized artifacts were mixed in order to obtain an artificial artifact-laden EEG dataset. This dataset was processed by WICA and two other methods and the performance of the artifact separation was compared. The proposed technique had the best artifact separation performance enforcing the hypothesis that WICA technique improves the performance of ICA because it projects the data into a new space where the redundancy is higher and the features of the signal in frequency domain are fully exploited. Future efforts will be devoted to the automation and optimization of the intermediate step of the procedure consisting in the selection of the critical wavelet components to be processed by ICA.

References

1. Lee, T.-W.: Independent Component Analysis - Theory and Applications. Kluwer Academic Publishers, Dordrecht (1998)
2. Jung, T., Humphries, C., Lee, T.-W., Makeig, S., McKeown, M.: Extended ica removes artifacts from electroencephalographic recordings. Advances in Neural Information Processing System 10, 894–900 (1998)
3. Jung, T., Makeig, S., Humphries, C., Lee, T.-W., McKeown, M.: Extended ica removes artifacts from electroencephalographic recordings. Advances in Neural Information Processing System 37(2), 163–178 (2000)

4. Jung, T.P., Makeig, S., Westerfield, M., Townsend, J., Courchesne, E., Sejnowski, T.J.: Removal of eye activity artifacts from visual event-related potentials in normal and clinical subjects. Clinical Neurophysiology 111(10), 1745–1758 (2000)
5. Cichocki, A., Vorobyov, S.A.: Application of ica for automatic noise and interference cancellation in multisensory biomedical signals. In: Second International Workshop on Independent Component Analysis and Blind Signal Separation, pp. 187–193 (2000)
6. Karhunen, J., Hyvaerinen, A., Vigario, R., Hurri, J., Oja, E.: Application of neural blind separation to signal and image processing. In: International Conference on Acoustics, Speech, and Signal Processing, pp. 187–193 (1997)
7. Vigario, R., Jousmaki, V., Hamalainen, M., Hari, R., Oja, E.: Independent component analysis for identification of artifacts in magnetoencephalographic recordings. In: Jordan, M.I., Kearns, M.J., Solla, S.A. (eds.) Advances in neural information processing systems, vol. (10), pp. 229–235. MIT Press, Cambridge, MA (1997)
8. Makeig, S., Bell, A.J., Jung, T., Sejnowski, T.J.: Independent component analysis of electroencephalographic data. Advances in Neural Information Processing Systems 8, 145–151 (1996)
9. Zhou, W., Gotman, J.: Removal of emg and ecg artifacts from eeg based on wavelet transform and ica. In: 26th Annual International Conference of the IEEE EMBS (EMBC), pp. 392–395 (2004)
10. Castellanos, N.P., Makarov, V.A.: Recovering eeg brain signals: artifact suppression with wavelet enhanced independent component analysis. J Neurosci Methods 158(2), 300–312 (2006)
11. Daubechies, I.: Ten Lectures on Wavelets. Society for Industrial and Applied Mathematics (1992)
12. Azzerboni, B., Finocchio, G., Ipsale, M., La Foresta, F., Morabito, F.C.: A new approach to detection of muscle activation by independent component analysis and wavelet transform. In: Marinaro, M., Tagliaferri, R. (eds.) Neural Nets. LNCS, vol. 2486, pp. 109–116. Springer, Heidelberg (2002)
13. Azzerboni, B., Finocchio, G., Ipsale, M., La Foresta, F., Morabito, F.C.: Neural-ica and wavelet transform for artifacts removal in surface emg. In: International Joint Conference on Neural Networks (IJCNN), pp. 3223–3228 (2004)

Neural Networks for Matching in Computer Vision

Giansalvo Cirrincione[1] and Maurizio Cirrincione[2]

[1] Department of Electrical Engineering, Lab. CREA
University of Picardie-Jules Verne
33, rue Saint Leu, 80039 Amiens - France
exin@u-picardie.fr

[2] Universite de Technologie de Belfort-Montbeliard (UTBM)
Rue Thierry MIEG, Belfort Cedex 90010, France
maurizio.cirricione@utbm.fr

Abstract. A very important problem in computer vision is the matching of features extracted from pairs of images. At this proposal, a new neural network, the Double Asynchronous Competitor (DAC) is presented. It exploits the self-organization for solving the matching as a pattern recognition problem. As a consequence, a set of attributes is required for each image feature. The network is able to find the variety of the input space. DAC exploits two intercoupled neural networks and outputs the matches together with the occlusion maps of the pair of frames taken in consideration. DAC can also solve other matching problems.

1 Introduction

In computer vision, structure from motion (SFM) algorithms recover the motion and scene parameters by using a sequence of images (very often only a pair of images is needed). Several SFM techniques require the extraction of features (corners, lines and so on) from each frame. Then, it is necessary to find certain types of correspondences between images, i.e. to identify the image elements in different frames that correspond to the same element in the scene. This paper addresses this specific problem, also known as *matching*. The techniques for image matching can be continuous or discrete. The latter are feature-based methods (FBM) and treat the images as samples of the scene taken at discrete times. They allow accurate estimation of motion parameters and structure of the scene even under a relatively large motion, do not suffer much from the problem of varying image intensity and do not need intensity smoothness. However, they suffer from several problems, like occlusion, depth discontinuities, repetitive patterns in the scene. As a consequence, image matching belongs to the class of the so called inverse problems, which are known to be ill-posed. It can be converted to a well-posed problem by introducing constraints in the correspondence formulation. The commonly used constraints are the similarity (or compatibility) constraint (matching features must have similar attribute values), the uniqueness constraint (almost always, a given pixel or feature from one view can match no

B. Apolloni et al. (Eds.): KES 2007/WIRN 2007, Part I, LNAI 4692, pp. 688–695, 2007.

more than one pixel or feature from the other view), the epipolar constraint (for calibrated stereo rig), the continuity (or disparity) constraint (correspondences for neighboring points in one image must have similar disparities), the disparity gradient constraint (condition on the tangent to the object surface), the ordering constraint, the smooth motion constraint and several geometric and kinematic constraints (see [1] for reference).

The formulation of the correspondence problem is well-suited to computation by a neural network. Indeed, it can be formulated as an optimization task where a cost function (energy equation), representing the constraints, is minimized. The minimization can be mapped into a Hopfield neural network [2] such that the energy function is the same as the Lyapunov function of the network, with the synaptic interconnection weights between the neurons representing the constraints imposed by the corresponding problem. These applications yield bad results for real images because of their dimensionality and high computational demand, suffer from the local minima and the possible improvements (e.g. simulated annealing) are too time consuming for real time applications. The most interesting applications are the Zhou and Chellappa neural network [3] for static stereo, the Sarigianidis and Pycock neural network [4] which uses a motion correspondence method and the Nasrabadi and Choo neural network [5] for stereo vision correspondence (the cost function is devised by implementing the continuity, uniqueness and 2D rigidity). Unlike the previous ones, the Bellando and Kothari neural network [6] uses topology preserving maps (see after), but works well only for toy problems.

In the next section a novel neural network is proposed. It considers matching as a pattern recognition problem. Two examples of matching are then given, the first by using a pair of synthetic images, the second by using a road sequence.

2 The Double Asynchronous Competitor (DAC)

The Double Asynchronous competitor (DAC) exploits the features of the self organizing feature maps (SOFM, [7]) for solving the correspondence problem. The SOFM defines a topologically preserving mapping from a high-dimensional input data space into a low dimensional output space: the latter is usually a 2D lattice of neurons. The neurons are usually arranged on a rectangular grid and connected according to a given neighborhood relation. Each neuron possesses a weight vector whose dimensionality is the same of the input vector. The learning is unsupervised and its updating is *winner-take-most*: after presentation of the input vector x to the network, the neuron i whose weight w_i is the nearest to the input, according to a predefined metric, is selected (winning neuron with index i^*). The winner-take-most approach updates both the winning neuron and its lattice neighbors according to:

$$w_i \longleftarrow w_i + \alpha(t)G(i, i^*, t)(x - w_i) \qquad \forall i \tag{1}$$

where $\alpha(t)$ is the learning rate, usully decreasing with time, and $G(i, i^*, t)$ is a weighting function (*neighborhood kernel*) depending on the grid distance around the winning neuron. The kernel can be Gaussian:

$$G(i, i^*, t) = e^{-\frac{\|w_i - w_{i^*}\|_2^2}{\rho^2(t)}} \qquad (2)$$

where the Euclidean distance is used and $\rho(t)$ (decreasing with time) is the neighborhood radius. The SOFM quantizes the input distribution and is able to capture the underlying variety of the input space. The vector quantization is constrained by the grid neighborhood structure of the neurons. These units project the input distribution toward the neuron grid by respecting locally the concept of proximity: in this sense, SOFM is topologically preserving.

DAC works for sequences of images without need of epipolar constraints. It is able to solve the matching problem even in the case of several moving objects and mobile camera. It also yields an occlusion map for each image. The following description considers the two-view problem, but can be easily extended to multiple views. DAC is composed of two SOFMs (NN12 and NN21) which are coupled. Every SOFM has a particular lattice: the neurons are placed in one image (2D lattice) at the geometrical position of the feature points and the training set is composed of the attribute vectors of the other image. An advantage of this approach is the possibility of using as many attributes as desired, without any care of eventual correlations: in every case, the network is able to detect the unknown variety of the input space. The choice of the attributes depends on the kind of image sequence. In the simulations, the SUSAN corner detector [8] has been used. The advantage of this choice is given by the fact that no explicit image derivatives are needed and the feature extraction is very quick. According to the value of its scalar parameter t (brightness difference threshold), corners or edge points (edgels) can be extracted. A lot of attributes can be attached to these features, e.g., the intensity (gray value) of the feature point, the edgeness as defined in [9], the positive and the negative cornerness as defined in [9], the USAN area computed by the SUSAN detector and the USAN centre of gravity w.r.t. the feature point [8], the measure of interest of the Moravec operator (computed ad hoc for the feature point, [10]), the Laplacian using 3x3 or 5x5 masks for the second derivatives [11], the second derivative in the direction of the gradient and in the direction orthogonal to the gradient [11] and the coordinates of the feature points, weighted far less than the other attributes in the training set.

NN12 has the first image (feature points) as grid (neurons), the attribute vectors of feature points of the second image as training set (TS) and the attribute vectors of feature points of the first image as initial weight vectors. NN21 is the opposite: the second image as grid, the first image attribute vectors as training set and the second image attribute vectors as initial weight vectors. This choice for the initial conditions helps implementing the search for the correspondence similarity by means of the SOFM quantization process. The choice for the grid implies the continuity and 2D rigidity constraints thanks to the topologically preserving behavior of the SOFM. Indeed, the network seeks to 2D project the one image attribute vectors in order to respect the relative other image feature point positions. In a certain sense, it is a way of implementing the 2D rigidity constraint without need of introducing an a priori shift parameter as in the existing matching techniques. The coupling between NN12 and NN21 implements the uniqueness constraint. The SOFMs used in DAC have Gaussian neighborhood kernels. An important difference with the SOFM learning is the possibility for an input vector of not having a winning neuron. In fact, a neuron is accepted as winner only if its weight vector stays into a hypersphere (search window in the attribute/input space) centered at the input vector, whose radius is given by the threshold $\delta(t)$ which decreases with time for increasing resolution. The association input vector-winning neuron is considered as a possible match.

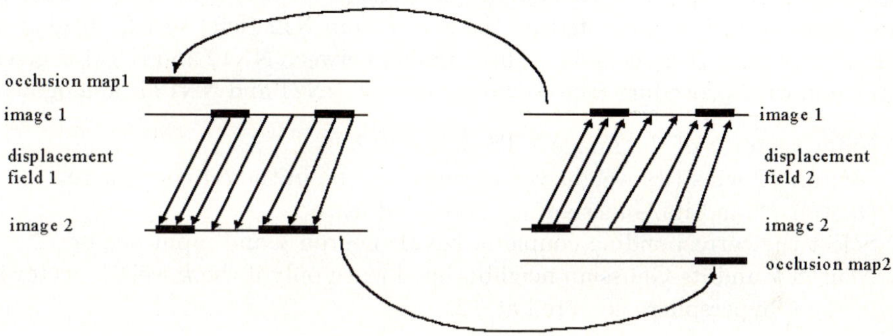

Fig. 1. 1D illustration of the determination of the occlusion maps. The images are one-dimensional. The displacement fields illustrate correspondences between two 1D images.

There are several possible architectures for DAC. The basic arrangement is made of the series of NN21 and NN12 and a feedback from NN12 to NN21. One DAC epoch is equal to the sum of one NN21 epoch (half DAC epoch) and one NN12 epoch (half DAC epoch). Every half DAC epoch, a test is done for the setting of the first and second image occlusion maps. Fig.1 shows the testing for 1D images. The first half epoch (presentation of the whole image 1 training set) is given by the NN21 learning (right side of fig.1). This learning is coupled to the other SOFM. At the end of the epoch, a first approximated disparity map $2 \rightarrow 1$ is given by the input data and associated winning neurons. This first matching does not use occlusion information. Hence, this matching may *jam* the occluded parts of the second image into parts of the first image. The objective of this first half DAC epoch is to compute the occlusion map 1 (first image). The jamming generally will not affect the computation of the occlusion map 1, since the occluded regions of image 1 to be detected may only occur on the opposite side across the jammed region [9]. Those regions (input data/features) in the first image that have not been matched are occluded in the second image and are marked as occluded: every neuron/input vector has a state variable which is set to zero in case of occlusion, i.e. when no neuron wins. If the input vector (feature) is not occluded, its state variable is set to the number of the winning neuron. Hence, a test is done on the input data state variables. If an input vector is occluded, it is excluded from the TS for the successive N21 learnings and the corresponding (i.e. attached to the same feature point of image 1) neuron in NN12 is inactivated and does not enter the N12 learning. This is recorded by a state variable set to zero for the corresponding neuron. This inactivation is represented by the arrow from N21 (right) to N12 (left) in fig.1. Once this inactivation is done, NN12 learns for one half DAC epoch with the remaining neurons (presentation of the whole N12 training set). A first approximated disparity map $1 \rightarrow 2$ is output. The results of this learning determine occlusion map 2 analogously. The same test is done on the NN12 input data. As a consequence, the occluded data are ruled out the TS and the corresponding neurons in the N21 grid are inactivated. This inactivation is represented by the arrow from N12 (left) to N21 (right) in fig.1. This procedure is recursive. The coupling between NN12 and NN21 is given by the following procedure (the coupling between NN21 and NN12 is analogous):

1. Enter vector k of the image 2 TS into NN12.
2. Neuron $f2$ wins (its weight vector must also be in the δ-hypersphere).
3. Neuron $f2$ and its Gaussian neighborhood learn.
4. Select the corresponding couple in NN21: neuron k and input vector $f2$.
5. Neuron k and its Gaussian neighborhood learn only if the k weight vector is in the δ-hypersphere centered at $f2$.

As a consequence of this coupling, one half epoch in a SOFM induces learning in the other one before the corresponding half epoch. This coupling implements the uniqueness constraint as shown in fig.2. Indeed, the asymmetry of DAC (one feature has only one winning neuron, but one neuron may win for more than one feature) and the interchange of the function of neuron and feature between the SOFMs imply the uniqueness.

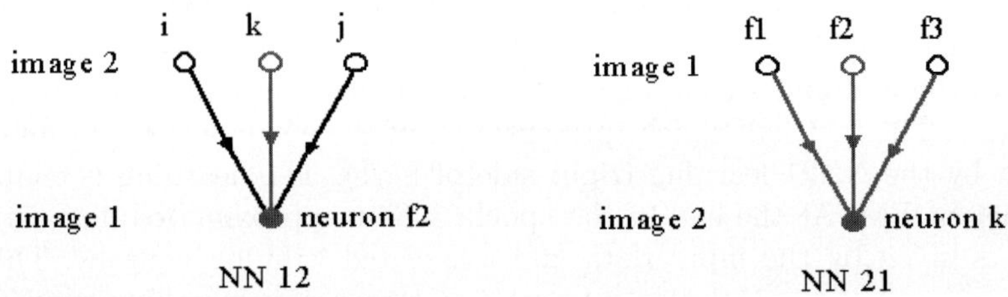

Fig. 2. DAC uniqueness coupling property

Resuming, the only DAC parameters are the neighborhood radius $\rho(t)$, the learning rate $\alpha(t)$ and the threshold $\delta(t)$ for the acceptance of the winner. Experience has suggested the following heuristics for setting a constant value for δ: merge the two training sets and compute all the possible Euclidean distances D between couples of vectors; then, set σ to $(1.1 \div 1.2)\ d_{\min}$, where $d_{\min}$ is equal to $\bar{D} - \sigma_D$ and $\bar{D}$ and σ_D are the average and standard deviation of the set of distances D.

After every DAC epoch, a control is done on the DAC results as a stop criterion:

1. Every non occluded input vector/feature point from image 1 has a nonzero state variable representing the number of the associated winning neuron for image 2.
2. The winning neuron represents also the input vector for the feature point in the same image (i.e. image 2).
3. If the state variable of this input vector (image 2) coincides with the state variable of the input vector of image 1 (first step), then the match is accepted.
4. If all matches are accepted, then DAC stops and outputs the list of the correspondences and the lists of the two occlusion maps.

3 Simulations

The first simulation deals with a pair of synthetic images. Given a world reference frame xyz, centered at the center of projection of the normalized camera (unit focal distance) and of z-axis parallel to the optical axis, a certain number of points $P_i = (x_i, y_i, z_i)$, where $x_i, y_i \in [-5, 5]$ and $z_i \in [3, 10]$, has been selected randomly. These points are then translated of the vector $[2, 3, 1]$ and rotated around the axis $[0, 0, 1]$ of 0.1 rad. Then, the initial and final points are projected (perspective projection) into an image plane whose coordinates are parallel to the x- and y-axes of the world reference frame and whose centre is the intersection with the optical axis. Hence, the correspondences are known a priori. A white Gaussian noise ($\mu = 0$, $\sigma = 1$) is then added to all image points. A comparison is made between the network of Nasrabadi and Choo and DAC. In the first network, the vertical disparity is also implemented in the compatibility measure and the selected parameters (used for the comparison) are $\lambda = 1$ and $\vartheta = 10$ (for an explanation of these terms, see [5]). In the second network, the input vectors are only composed of the coordinates of the image points. The learning rate $\alpha(t)$ is a decreasing function (inversely proportional to the iteration number) from 0.3 to 0.03; after reaching the minimum, the function remains constant. at 0.03. The same function is used both for the Gaussian standard deviation $\rho(t)$ and for the hypersphere ray $\delta(t)$, with initial values, respectively, 1.5 and 3 and same final value 0.03. Tab.1 shows the results, averaged over five experiments for every choice of the number of points. DAC is more accurate, both for noiseless and, above all, for noisy images, even if only two attributes can be used. Furthermore, it is by far faster than the other network and does not suffer from the problem of local minima. In fact, only two DAC epochs are sufficient for all the experiments.

The second simulation considered here is fully presented in [12]. Two image frames are extracted from a road sequence of 100 frames, given by the CSTV CNR Computer Vision Laboratory of Turin (Italy). It has been captured by a camera mounted on the roof of a van (mobile laboratory MOBLAB) which slowly turns to left. The frames have 256 grey levels and pixel dimension 720 × 280. In the camera model (frame XYZ) for the test road image pair, the Y-axis has the vertical direction and is oriented upward. A certain number of feature points

Table 1. Correct match percentages for the synthetic pair

	DAC		Nasrabadi & Choo	
points	noiseless	noisy	noiseless	noisy
10	100 %	100 %	98.2 %	96.4 %
30	100 %	100 %	95.2 %	90.6 %
70	100 %	98.3 %	92.1 %	87.5 %
100	100 %	96.9 %	89.1 %	77.2 %

Fig. 3. DAC results (arrows) for the test road image sequence, superimposed on the first frame. The black arrows are excluded by the CASEDEL EXIN neural network.

(corners) are extracted by the SUSAN corner finder. The image matching has been performed by using DAC, which uses all the attributes described before. In fact DAC is able to automatically capture the variety of the input space. The learning rate $\alpha(t)$ is constant and equal to 0.01; the Gaussian standard deviation $\rho(t)$ is a decreasing function (inversely proportional to the iteration number) with initial value 150; the hypersphere ray $\delta(t)$ is constant and equal to 1. DAC converges after only two epochs and finds 31 correspondences. These correspondences are visualized as superimposed on the first frame in fig.3. In the neural SFM module presented in [12], the neural network called CASEDEL EXIN is able to characterize the cluster of black arrows in the upper left of fig.3 as cluster of mismatches (outliers) and so it refines the DAC matching. The presence of this group of black arrows is explained by the absence of detected corners for the advertising poster in the first frame [12]. Despite the claims of Stephen Smith, SUSAN does not show a good stability in this image sequence. This explains the inaccuracy of some matches.

4 Conclusion

This paper has presented a novel kind of neural network, DAC, which exploits the self-organization for solving the matching problem in SFM as a pattern recognition problem. As a consequence, a set of attributes is required for each feature point. The network is able to find the variety of the input space. DAC is faster and more accurate than the Hopfield based neural techniques and is by far better than the BPN based neural techniques which would require nearly unlimited training sets. DAC is an open architecture in the sense that different schemas are possible and other self-organizing learning laws can be chosen. In [12] DAC is integrated with the EXIN SNN neural network, which is able to segment the image and has a similar learning law. Hence, it can be implemented directly in the DAC learning law and can refine the matching.

Future work will deal firstly with the choice of other feature detectors, better if neural, which must be less unstable than SUSAN. A more detailed analysis of the limits and the convergence properties of DAC is still needed. Its coupling and temporal scheduling must be further investigated. Furthermore, the DAC matching technique can be extended to other problems as the template matching, the range point matching and the graph matching. The parallelization of the DAC architecture must be still analysed.

References

1. Faugeras, O.: Three-Dimensional Computer Vision: a Geometric Viewpoint. MIT Press, Cambridge, Massachusetts (USA) (1993)
2. Hopfield, J.: Neural networks and physical systems with emergent collective computational abilities. Proc. Nat. Acad. Sci. 79, 2554–2558 (1982)
3. Zhou, Y., Chellappa, R.: Artificial Neural Networks for Computer Vision. Research Notes in Neural Computing, vol. 5. Springer, Heidelberg (1992)
4. Sarigianidis, G., Pycock, D.: Motion correspondence using a neural network. British Machine Vision Conference, 649–658 (1993)
5. Nasrabadi, N., Choo, C.: Hopfield network for stereo vision correspondence. IEEE Trans. on Neural Networks 3(1), 5–12 (1992)
6. Bellando, J., Kothari, R.: On image correspondence using topology preserving mappings. In: Proc. Int. Conf. On Neural Networks, Washington, D.C (USA), pp. 1784–1789 (June 1996)
7. Kohonen, T.: Self-Organization and Associative Memory. Springer, Berlin (1989)
8. Smith, S., Brady, J.: Susan-a new approach to low level image processing. International Journal of Computer Vision 23(1), 45–78 (1997)
9. Weng, J., Huang, T., Ahuja, N.: Motion and Structure from Image Sequences. Springer Series in Information Sciences. Springer, Heidelberg (1993)
10. Moravec, H.: Visual mapping by a robot rover. In: Proc. Of the 6th Int. Joint Conf. On Artificial Intelligence, pp. 598–600 (1979)
11. Beaudet, P.: Rotational invariant image operators. In: Proc. Of the Int. Conf. On Pattern Recognition, pp. 579–583 (1978)
12. Cirrincione, G.: A Neural Approach to the Structure from Motion Problem. PhD thesis, LIS INPG Grenoble (December, 1998)

SNNR-Based Improved Multi-modal Fusion and Fission Using Fuzzy Value Based on WPS and Web

Jung-Hyun Kim and Kwang-Seok Hong

School of Information and Communication Engineering, Sungkyunkwan University, 300, Chunchun-dong, Jangan-gu, Suwon, KyungKi-do, 440-746, Korea
kjh0328@skku.edu, kshong@skku.ac.kr
http://hci.skku.ac.kr

Abstract. This paper implements the Multi-Modal Instruction Agent (hereinafter, MMIA) including a synchronization between audio-gesture modalities, and suggests improved fusion and fission rules depending on SNNR (Signal Plus Noise to Noise Ratio) and fuzzy value for simultaneous multi-modality, based on the embedded KSSL (Korean Standard Sign Language) recognizer using the WPS (Wearable Personal Station) and Voice-XML. Our approach fuses and recognizes the sentence and word-based instruction models that are represented by speech and KSSL, and then translates recognition result that is fissioned according to a weight decision rule into synthetic speech and visual illustration (graphical display by HMD-Head Mounted Display) in real-time. The experimental results, average recognition rates of the MMIA for the prescribed 62 sentential and 152 word instruction models were 94.33% and 96.85% in clean environments, and 92.29% and 92.91% were shown in noisy environments.

1 Introduction

Human-Computer Interaction (HCI), alternatively Man-Machine Interaction (MMI) or Computer-Human Interaction (CHI), is one of the most extensive research areas that are concerned with the design, evaluation and implementation of interactive computing systems for human use and with the study of major phenomena surrounding them [1], and a basic goal of HCI is to improve the interaction more user-friendly and receptive to the user's needs between users and computers (or machines). Namely, in the desktop PC and wire communications net-based traditional computer science and HCI, according as the user interface (of a computer program) referred to the graphical, textual, auditory information and the control sequences (such as movements of the computer mouse and selections with the touch-screen), generally they have some restrictions and problems such as conditionality on the space, limitation of motion and so on. However, the next generation HCI for more advanced and personalized PC system such as wearable computer and PDA based on wireless network and wearable computing, may require and allow new interfaces and interaction techniques such as tactile interfaces with haptic feedback methods, and gesture interfaces based on hand gestures, or mouse gestures sketched with a computer mouse or a stylus, to serve different kinds of users. Namely, for perceptual experience and behavior to benefit from the simultaneous stimulation of multiple sensory modalities that are concerned with a human's the (five) senses, fusion and fission technologies of the

B. Apolloni et al. (Eds.): KES 2007/WIRN 2007, Part I, LNAI 4692, pp. 696–703, 2007.

information from these modalities are very important and positively necessary. Consequently, we implement MMIA including synchronization between audio-gesture modalities by coupling the WPS-based embedded KSSL recognizer with a remote Voice-XML user, for improved multi-modal HCI in noisy environments. In contrast to other proposed multi-modal interaction approaches, our approach is unique in two aspects: First, because the MMIA provides different weight and a feed-back function in individual (speech or gesture) recognizer, according to SNNR and fuzzy value, it may select an optimal instruction processing interface under a given situation or noisy environment, and can allow more interactive communication functions in noisy environment. Second, according as the MMIA fuses and the sentence and word-based instruction models that are represented by speech and KSSL, and then translates recognition result, which is fissioned according to a weight decision rule into synthetic speech and graphical display by HMD-Head Mounted Display in real-time, it provides a wider range of personalized information more effectively.

2 WPS-Based Embedded KSSL Recognizer

For the WPS-based embedded KSSL recognizer, we used 5DT company's wireless data gloves and Fastrak® which are popular input devices in the haptic application field, and utilized blue-tooth module for the wireless sensor network [2]. And the i.MX21 test board was selected, which is developed the next-generation wearable PC platform in the Rep. of Korea. The i.MX21 test board consists of an ARM926EJ-S (16KB I-Cache, 16KB D-Cache) CPU, and includes ARM Jazelle technology for Java acceleration and MPEG-4 and H.263 encode/decode acceleration [3].

2.1 Feature Extraction and Instruction Recognition Models Using RDBMS

We constructed 62 sentential and 152 word instruction models by coupling KSSL hand gestures with motion gestures that are referred to "Korean Standard Sign Language Tutor (KSSLT) [4]". In addition, for a clustering method to achieve efficient feature extraction and construction of training / recognition models based on distributed computing, we utilize and introduce an improved RDBMS (Relational Data-Base Management System) clustering module [5].

2.2 Fuzzy Logic: Max-Min Composition of Fuzzy Relation

As the fuzzy logic for KSSL recognition, we applied trapezoidal shaped membership functions for representation of fuzzy numbers-sets, and utilized the fuzzy max-min composition.

$$\textit{For}\ (x, y) \in A \times B,\ (y, z) \in B \times C,$$
$$\mu_{S \bullet R}(x, z) = \underset{y}{Max}\ [Min\ (\mu_R(x, y), \mu_S(y, z))] \tag{1}$$

Two fuzzy relations *R* and *S* are defined in sets *A*, *B* and *C* (we prescribed the accuracy of hand gestures and basic KSSL gestures, object KSSL recognition models as the sets of events that occur in KSSL recognition with the sets *A*, *B* and *C*). That is, $R \subseteq A \times B$, $S \subseteq B \times C$. The composition $S \cdot R = SR$ of two relations *R* and *S* is expressed by the relation from A to C, and this composition is defined in Eq. (1) [6], [7]. $S{\cdot}R$ from this elaboration is a subset of $A{\times}C$. That is, $S{\cdot}R \subseteq A{\times}C$. If the relations R and S are represented by matrices M_R and M_S, the matrix $M_{S\bullet R}$ corresponding to $S{\cdot}R$ is obtained from the product of M_R and M_S; $M_{S\bullet R} = M_R{\cdot}M_S$. The matrix $M_{S\bullet R}$ represents max-min composition that reason and analyze the possibility of C when A occurs, and it is also given in Fig. 1.

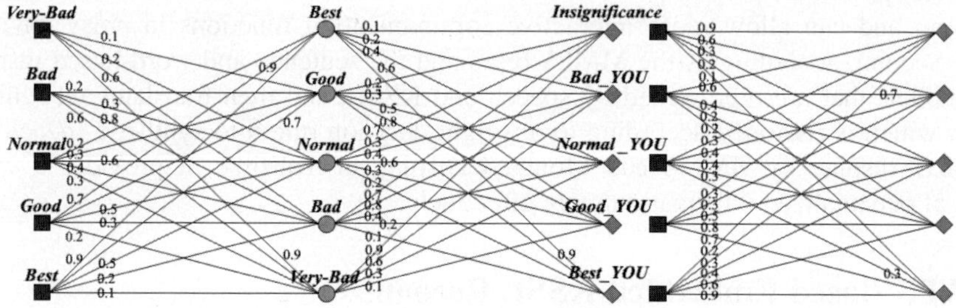

Fig. 1. Composition of fuzzy relation

In addition, WPS-based embedded KSSL recognizer calculates and produces a fuzzy value from the user's dynamic KSSL via a fuzzy reasoning and composition process, and then decides and recognize user's various KSSL according to produced fuzzy value. The flowchart of KSSL recognizer is shown in Fig. 4 (in section 4) together with an outline and flow-chart of the MMIA.

3 Speech Recognition and Synthesis Based on Voice-XML

Voice-XML is the W3C's standard XML format for specifying interactive voice dialogues between a human and a computer. For ASR-engine in architecture [8] of W3C's VXML 2.0, we used the HUVOIS solution that is Voice-XML-based voice

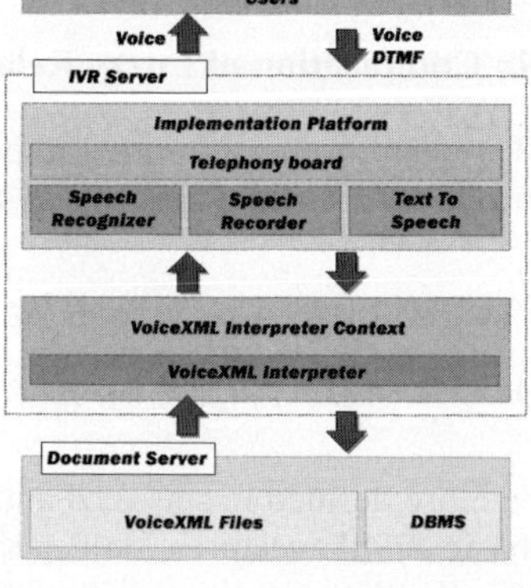

Fig. 2. The Voice-XML's architecture

software developed by KT Corp. in Korea for those with impaired sight that converts online text into voice and reads out the letters and words punched in through the computer keyboard, thus enabling them to use computers and the internet. The HUVOIS solution consist of HUVOIS-ARS based on HMM, TTS using tri-phone unit and HUVOIS Voice-XML, and supports client-sever network, LSS(Load Share Server) and modular structure. The Voice-XML's architecture is shown in Fig. 2. A document server (e.g. a web server) processes requests from a client application, the Voice-XML interpreter, through the VXML interpreter context. The server produces Voice-XML documents in reply, which are processed by the Voice-XML interpreter. The Voice-XML interpreter context may monitor user inputs in parallel with the Voice-XML interpreter. For example, one Voice-XML interpreter context may always listen for a special escape phrase that takes the user to a high-level personal assistant, and another may listen for escape phrases that alter user preferences like volume or text-to-speech characteristics. The implementation platform is controlled by the Voice-XML interpreter context and by the Voice-XML interpreter.

4 Simultaneous Multi-modality-Fusion and Fission Between Modalities

4.1 Fusion Scheme Between Audio-Gesture Modalities

The integration scheme consists of seven major steps: 1) the user connects to Voice-XML server via PSTN and internet using telephone terminal and WPS based on wireless networks (including middleware), and then inputs prescribed speech and KSSL, 2) the user's speech data, which are inputted into telephone terminal, is transmitted to ASR-engine in Voice-XML, then ASR results are saved to the MMDS (Multi-Modal Database Server; The MMDS is the database responsible for synchronizing data between speech and KSSL gesture), 3) user's KSSL data, which are inputted into WPS, are recognized by embedded KSSL recognizer, then the WPS transmits and saves recognition results to the MMDS, using middleware over TCP/IP protocol and wireless networks(blue-tooth module), 4) at this point, the user's KSSL and speech data run the synchronization session using internal SQL logic of the MMDS, 5) while suggested the MMIA runs comparison arithmetic (validity check) on ASR and KSSL recognition results with pre-scribed instruction models by internal SQL logic, the

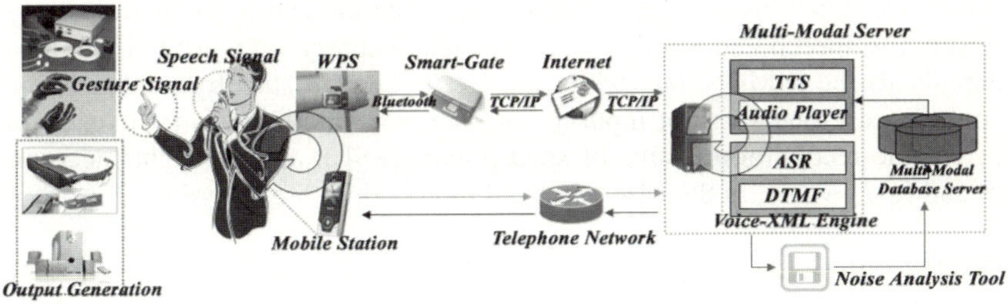

Fig. 3. The components and fusion architecture

NAT(Noise Analysis Tool) analyzes noise for user's speech data (wave file) which is recorded by Voice-XML, 6) According to analyzed noise and arithmetic result, the MMIA gives weight into an individual (gesture or speech) recognizer, 7) finally, user's intention is provided to the user through TTS and visualization. The suggested fusion architecture and flowchart of MMIA are shown in Fig. 3 and Fig. 4.

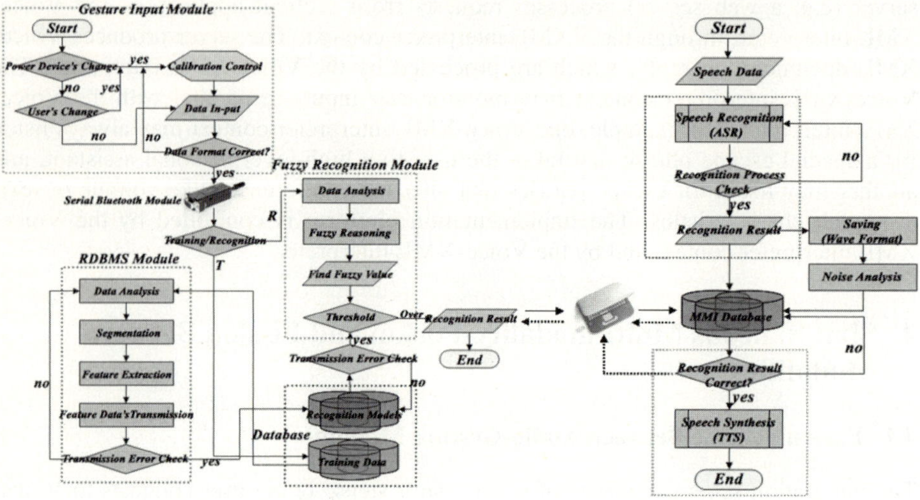

Fig. 4. The flowchart of the MMIA integrating 2 sensory channels with speech and gesture

4.2 Synchronization Between Audio-Gesture Modalities

All multi-modal inputs are synchronized, because while speech recognizer generates absolute times for words, gesture movements generate *{x, y, t}* triples, and initial work identifies an object or a location from gesture inputs accordingly, as speech understanding constrains gesture interpretation. This paper solves the asynchronous control problems between speech and gesture signals using a web-logic and word-unit input method based on the MMDS in section 4.1. For synchronization between speech and gesture signals, after individual speech and KSSL recognizer recognizes inputted speech and the KSSL recognition models, they transmit recognition results into the MMDS for weight application. However, the transmission time of recognition results has some time delay because of asynchronous communication of two input signals. As a result, the speech and KSSL recognition results based on word-unit are recorded sequentially to the MMDS, and while the DB is kept in standby mode via internal web-logic in case one was not input among the two input signals (where, two input signals are the recognition results of speech and KSSL), apply weights according to SNNR and fuzzy value, in the case where all input values are recorded.

4.3 Noise Analysis

In noisy environments, speech quality is severely degraded by noises from the surrounding environment and speech recognition systems fail to produce high recognition rates

[9], [10]. Consequently, we designed and implemented Noise Analysis Tool (NAT) for weight decision in individual (gesture or speech) recognizer. The NAT calculates average energy (mean power; [dB]) for a speech signal that recorded by wave-format in the Voice-XML; and then computes SNNR by Eq. (2), where, P is average energy (mean power; [dB]).

$$SNNR(dB) = 10\log_{10} \frac{P_{signal + noise}}{P_{noise}} \tag{2}$$

4.4 Fusion and Fission Rules Depending on SNNR and Fuzzy Value

Speech recognition rate does not usually change to a SNNR of 25 dB, but if the rate lowers, the speech recognition rate falls rapidly. Therefore, the MMIA provides feed-back function according to SNNR critical value. In case SNNR critical value for weight decision is ambiguous, according as a feed-back function requests re-input (speech and KSSL) to user for clear a declaration of intention, more improved instruction processing is available. In addition, we utilized an average speech recognition rate as speech probability value for weight decision, and to define speech probability value depending on SNNR, we repeatedly achieved speech recognition experiments 10 times with the 20 test speech recognition models in noisy and clean environments, for every 5 reagents. The average speech recognition rates are given in Table 1.

Table 1. Weight value according to the SNNR and critical value for the feed-back function

SNNR Critical value	Weight value (%)		Average speech recognition rates for the 20 test recognition models (%)						
	Speech (W_S)	KSSL(W_G)	Reagent 1	Reagent 2	Reagent 3	Reagent 4	Reagent 5	Average(S)	Difference
more than 40 [dB]	99.0	1.0	98.2	98.4	97.9	98.5	98.2	98.2	0.9
35 [dB] ≤ SNNR < 40 [dB]	98.0	2.0	97.8	97.3	96.6	97.1	97.5	97.3	0.3
30 [dB] ≤ SNNR < 35 [dB]	96.0	4.0	97.5	96.5	96.6	97.0	97.4	97.0	0.2
25 [dB] ≤ SNNR < 30 [dB]	94.0	6.0	97.2	96.5	96.5	96.9	96.9	96.8	0.2
20 [dB] ≤ SNNR < 25 [dB]	92.0	8.0	96.9	95.9	96.4	96.8	96.8	96.6	2.2
15 [dB] ≤ SNNR < 20 [dB]	*Feed-Back*		*92.4*	*96.2*	*93.8*	*95.2*	*94.1*	*94.3*	*11.1*
10 [dB] ≤ SNNR < 15 [dB]	6.0	94.0	83.6	83.4	83.5	82.6	83.2	83.3	8.8
5 [dB] ≤ SNNR < 10 [dB]	4.0	96.0	71.9.	72.5	70.2	79.5	75.6	74.5	22.4
0 [dB] ≤ SNNR < 5 [dB]	2.0	98.0	53.4	51.3	52.6	51.6	51.3	52.0	14.0
less than 0 [dB]	1.0	99.0	38.5	37.6	37.5	38.2	38.5	38.1	-

$$P\_W = W_S \times S + W_G \times G \tag{3}$$

- *P_W : a probability value after weight application*
- *W_S : Defined Weight for Speech recognition mode in Table 1.*
- *W_G : Defined Weight for KSSL recognition mode in Table 1.*
- *S : speech probability (an average speech recognition rate)*
- *G : KSSL probability (the critical value depending on normalized fuzzy value)*

$$G = \frac{Fuzzy\ Value\_Current}{Fuzzy\ Value\_Max} = \frac{Fuzzy\ Value\_Current}{3.5} \tag{4}$$

- *Fuzzy Value_Current : Fuzzy value to recognize current gesture(KSSL)*
- *Fuzzy Value_Max = 3.5 : The maximum fuzzy value for KSSL recognition*

For fusion and fission rules depending on SNNR and fuzzy value, we defined *P_W* that is a probability value after weight application and the *KSSL probability* (*G*) of the embedded KSSL recognizer in Eq. (3) and (4). This *P_W* value depending on SNNR and fuzzy value gives standard by which to apply weights, and because *KSSL probability (G)* is changed according to *Fuzzy Value_Current*, the *P_W* is changed justly. Where, the maximum fuzzy value for KSSL recognition is 3.5, and the minimum critical value is 3.2, in our application. As a result, if *P_W* value is over than 0.917, the MMIA fissions and returns recognition result of speech recognizer based on Voice-XML, while the MMIA fissions the embedded KSSL recognizer in case *P_W* value is less than 0.909. The *P_W* values depending on SNNR and fuzzy value are given in Table 2.

Table 2. In case Fuzzy Value_Current is 3.2, P_W values using the Eq. (3) and (4)

SNNR	Speech		KSSL		P_W
	W_S	S	W_G	G	
more than 40 [dB]	0.99	0.982	0.01	0.914	0.981
35 [dB] ≤ SNNR < 40 [dB]	0.98	0.973	0.02	0.914	0.972
30 [dB] ≤ SNNR < 35 [dB]	0.96	0.970	0.04	0.914	0.968
25 [dB] ≤ SNNR < 30 [dB]	0.94	0.968	0.06	0.914	0.965
20 [dB] ≤ SNNR < 25 [dB]	0.92	0.966	0.08	0.914	0.917
15 [dB] ≤ SNNR < 20 [dB]	*Feed-Back*				
10 [dB] ≤ SNNR < 15 [dB]	0.06	0.833	0.94	0.914	0.909
5 [dB] ≤ SNNR < 10 [dB]	0.04	0.745	0.96	0.914	0.907
0 [dB] ≤ SNNR < 5 [dB]	0.02	0.520	0.98	0.914	0.906
less than 0 [dB]	0.01	0.381	0.99	0.914	0.909

5 Experiments and Results

The experimental set-up is as follows. The distance between the KSSL input module and the WPS with a built-in KSSL recognizer approximates radius 10M's ellipse

Table 3. MMIA's recognition results for the 62 sentential and 152 word instruction models

Evaluation (R:Recognition) / Reagent	Uni-modal Instruction Processing Interface								The MMIA			
	KSSL (%)				Speech (%)				KSSL + Speech (%)			
	Noise		Clean		R-Noise		Clean		Noise		Clean	
	sentence	word	sentence	word	sentence	word	sentence	word	sentence	word	sentence	word
Reagent 1	92.7	92.8	92.7	92.7	85.7	83.6	94.2	98.1	92.5	92.7	94.2	98.1
Reagent 2	91.8	93.8	91.7	93.9	84.6	83.5	94.9	95.4	91.5	93.3	94.8	95.5
Reagent 3	92.9	94.1	92.9	94.1	79.8	82.4	93.2	95.6	92.3	93.7	93.2	95.7
Reagent 4	93.1	92.9	93.2	92.8	82.8	85.1	93.8	96.3	92.9	92.3	93.8	96.3
Reagent 5	93.2	93.1	93.3	93.2	85.9	85.6	93.9	96.7	92.7	93.5	93.8	96.9
Reagent 6	92.1	91.8	92.2	91.8	84.2	84.6	95.5	95.9	91.5	91.1	95.5	96.1
Reagent 7	91.8	92.7	91.7	92.4	79.9	84.3	92.7	95.7	91.2	92.5	92.7	95.8
Reagent 8	91.5	94.6	91.4	94.5	78.2	82.6	93.2	96.8	90.9	94.5	93.3	97.1
Reagent 9	93.1	93.4	93.1	93.5	85.3	83.4	94.6	97.5	93.0	93.3	94.6	97.6
Reagent10	93.3	93.1	93.1	93.2	82.1	84.9	93.8	97.3	92.8	92.9	94.1	97.4
Reagent11	92.2	93.7	92.2	93.6	84.6	83.7	96.7	97.8	91.8	93.6	96.7	97.9
Reagent12	93.3	92.4	93.1	92.5	84.1	82.9	93.9	96.4	93.0	92.2	93.7	96.3
Reagent13	92.8	92.6	92.5	92.6	83.6	83.6	93.4	96.6	92.1	92.1	93.3	96.9
Reagent14	92.8	93.3	92.8	93.2	82.5	84.1	95.2	97.1	92.5	92.8	95.3	97.3
Reagent15	94.0	93.3	93.9	93.2	84.7	84.1	94.2	97.1	93.7	93.4	94.2	97.3
Average	92.71	93.17	92.65	93.15	83.20	83.85	94.21	96.71	92.29	92.91	94.33	96.85

form. In KSSL gesture and speech, we move the wireless data gloves and the motion tracker to the prescribed position. For every 15 reagents, we repeat this action 10 times in noisy and clean environments. While the user inputs KSSL using data gloves and a motion tracker, and speak using the blue-tooth headset in a telephone terminal. Experimental results, the uni-modal and the MMIA's average recognition rates in noisy and clean environment, are shown in Table 3.

6 Conclusion

This study combines natural language and artificial intelligence techniques to allow human computer interaction with an intuitive mix of speech, gesture and sign language based on the WPS and Voice-XML. In addition, our experiments were achieved on below experimental conditions and weights.

- *As a noisy environment, the average SNNR using actual waveform data is recorded in laboratory space, including the music and the mechanical noise, was about 13.59[dB].*
- *Also, the average SNNR using actual waveform data that remove noise elements for a clean environment was about 38.37[dB].*
- *if the SNNR changes by experimental conditions such as the music and the mechanical noise, because weight also is changed, experiment result can be changed.*

The MMIA supports two major types (simultaneous multi-modality and sequential multi-modality) of multi-modality for instruction recognition based on embedded and ubiquitous computing. Finally, we clarify that this study is a fundamental study for implementation of an advanced multi modal recognizer integrating the human's five senses such as sight, hearing, touch, smell, and taste, to take the place of the traditional uni-modal recognizer for natural speech and sign language processing.

References

1. Gary Perlman. et al.: HCI Bibliography.: Human-Computer Interaction Resources, http://www.hcibib.org/
2. Kim, J.-H., et al.: Hand Gesture Recognition System using Fuzzy Algorithm and RDBMS for Post PC. In: Wang, L., Jin, Y. (eds.) FSKD 2005. LNCS (LNAI), vol. 3614, pp. 170–175. Springer, Heidelberg (2005)
3. MX21 Processor Data-sheet, http://www.freescale.com/
4. Kim, S.-G.: Korean Standard Sign Language Tutor, 1st edn. Osung Publishing Company, Seoul (2000)
5. Kim, J.-H., et al.: An Implementation of KSSL Recognizer for HCI Based on Post Wearable PC and Wireless Networks KES 2006, Lecture Notes in Artificial Intelligence. In: Gabrys, B., Howlett, R.J., Jain, L.C. (eds.) KES 2006. LNCS (LNAI), vol. 4251, pp. 788–797. Springer, Heidelberg (2006)
6. Chen, C.H.: Fuzzy Logic and Neural Network Handbook. McGraw-Hill, New York (1992)
7. kandasamy, W.B.V.: Smaranda Fuzzy Algebra. American Research Press, Seattle (2003)
8. McGlashan, S., et al.: Voice Extensible Markup Language (VoiceXML) Version 2.0. W3C Recommendation (1992), http://www.w3.org
9. Martin, W. H.: DeciBel – The New Name for the Transmission Unit, Bell System Technical Journal (January 1929)
10. NIOSH working group.: STRESS. AT WORK NIOSH, Publication No. 99-101,U.S. National Institutes of Occupational Health (2006)

Vision Technologies for Intelligent Vehicles

Massimo Bertozzi, Alberto Broggi, L. Bombini, C. Caraffi, S. Cattani, Pietro Cerri, Alessandra Fascioli, M. Felisa, R.I. Fedriga, S. Ghidoni, Paolo Grisleri, P. Medici, M. Paterlini, P.P. Porta, M. Posterli, and P. Zani

VisLab, Università di Parma, Italy
www.vislab.it

Abstract. This paper surveys the different technologies involved in the development of computer vision applications for Intelligent Vehicles derived from the 15 year experience of VisLab. Some illustrative examples are also discussed.

Keywords: Machine vision, intelligent vehicles, intelligent transportation systems.

Introduction

The use of vision for Intelligent Transportation Systems developed in the last 20 years and, after the first experiments, rapidly evolved and is now widely considered as one of the most convenient perception technologies. The first experiments were performed in the second half of the eighties but it took some years before different prototypes of vision-based autonomous vehicles were tested in real situations. In the last few years, initiatives like the Grand Challenge and Urban Challenge caused a novel flourishing of research activities in the autonomous vehicles field [2].

In order to effectively develop vision based systems for intelligent vehicles both hardware and software issues have to be considered.

The different technologies involved in the development of these systems at VisLab, University of Parma, are described in the following sections: hardware and software.

The hardware side affects both the sensing technologies, namely the cameras, and the computing engine. Different camera technologies can be exploited: daylight, far infrared (thermal), near infrared, or even range cameras; the choice mainly depends on the specific application and on a cost vs benefit analysis. Also in the case of the computing engine, the potential choices highly depend on the application and on computational power constraints. In fact, in the intelligent vehicles field, real time performance is generally mandatory. Few years ago, dedicated hardware was designed to obtain enough computational power to fulfill this constraint. Nowadays, standard off-the-shelf components, namely standard PCs, can be used since they deliver enough power. Anyway, the technology transfer of these systems towards a mass market production often requires to minimize size and power supply requirements of the computing engine, therefore also embedded systems like DSPs or FPGAs have to be considered.

B. Apolloni et al. (Eds.): KES 2007/WIRN 2007, Part I, LNAI 4692, pp. 704–711, 2007.

On the software side, two different issues have to be examined: the development phase and the final system software. In order to enable the programmers to focus on the specific vision problem without having to care about other details, a rapid application development tool have to be used. In the last 10 years, VisLab developed such a tool (called GOLD) which provides programmers with a rich set of functionalities and APIs for commonly required tasks like I/O operations (acquisition from cameras and other sensors as well, sensors synchronization, local net and file management, data graphical display...) or low-level image processing functions. A convenient graphical user interface is also used to support programming activities. Anyway, for some applications, such a software environment is not always suited to be used in the final system especially in the case of small or embedded architectures. Therefore, particular care has been used in designing the software internal structure: in fact, GOLD is structured in different logical layers, each one devoted to a specific task. Thanks to a great degree of independence between the different layers, the software can be easily shrunk removing functionalities and layers not necessary to the specific application.

In order to show how these technologies dwell in the real world, this paper also includes the description of some real applications characterized by different requirements. In the case of the Grand and Urban Challenges the huge number of functionalities and sensors require to use a powerful computing system made of multiple PCs. Conversely, in the case of the Start Inhibit functionality, after an initial development phase in which a standard PC was used, embedding constraints require to port the system towards a DSP architecture. A third example (Boat Speed Monitor) completes the picture by showing the capabilities of the GOLD software: GOLD has been adapted removing unnecessaries functionalities and the porting towards an ARM-based architecture embedded into an intelligent camera is under development.

This paper is organized as follows: the next section illustrates the different camera technologies and computing engines that can be used for artificial vision while section 2 details the functionalities and structure of the software development framework used at VisLab. Section 3 ends the paper discussing examples of three different systems developed by VisLab researchers.

1 Hardware for Vision-Based Intelligent Vehicles

In the following, camera technologies and computing architectures used on board of vision-based intelligent vehicles are discussed.

Camera Technologies. Images acquired in different spectral domains have different characteristics and bear different costs.

In the far infrared (FIR) domain the image of an object relates to its temperature, namely on its emitted heat and barely depends on illumination. In addition, FIR images lacks of textures and colors. Therefore, FIR cameras are suitable for the detection of objects warmer (or colder) than the background and ease the initial steps of a detection process thanks to the absence of small details.

Fig. 1. Comparison amongst color, grey level, and NIR images

Conversely, object detection in the visible domain is often more difficult due to the presence of details, shadows, and changes in the luminance or sharpness of images. Anyway, small details are of paramount importance for object classification. Moreover, the use of color can be important in some applications, such as traffic signs detection.

Another interesting technology is near infrared (NIR). Objects reflect incident NIR radiation very similarly to visible light and the appearance of objects depends on how the material reflects the NIR radiation. Clothings generally reflect well the NIR light, therefore exhibit less grey-level variability in the image, although some fabric absorb all NIR light. A comparison between the appearance of clothings in color visible images, grey level visible images, and NIR images is shown in figure 1.

Daylight and NIR cameras generally provide larger resolutions than FIR sensors and are cheaper and smaller.

Recently, alternative sensing technologies are being considered. For example range cameras can acquire depth information. This technology is quite new and, unfortunately, current sensors feature a very low resolution and a limited working range. Anyway, many research activities (like in the EU project PReVENT) are trying to overcome the current limits of this technology.

Computing Engines. At the beginning of the 90s, dedicated and expensive hardware were still mandatory for artificial vision. Just a few years later COTS (Commercial Off-The-Shelf components, like standard PCs) started to be used as a computing engine for vision system, since they were able to deliver enough computational power.

As an example, in 1996 lane and obstacle detection took approximately 75 ms to be executed on dedicated system (the PAPRICA multi-processor architecture). Only three years after, in 1999, a 200 MHz Pentium Processor was capable to complete the same task using the available MMX instructions [1].

Nowadays, PCs are able to handle much more complex tasks. As an example, in the Grand and Urban Challenges a huge list of vision based functionalities are performed by PCs exploiting modern CPUs features; multithreading oriented algorithms are able to use multi core CPUs and small portion of codes are optimized using assembly multimedia instructions. Moreover, also the modern graphics processing units, specifically designed to perform image processing, can be exploited as a coprocessor for image processing tasks.

Unfortunately, PCs are not suited to fit into small embedded systems; in such a case devices like FPGAs and DSPs are used. They require the porting and adaptation of algorithms.

2 The GOLD Software Framework

GOLD is a framework aimed at providing a complete set of tools to allow fast development of computer vision applications. Moreover, GOLD is also used to run the on-board final system.

Functionalities. GOLD allows to build vision-based applications as independent plugins. Each application can take advantage from an increasingly rich set of tools and debugging facilities. These components are described in the following.

GOLD frees the developer from using the real hardware, offering an abstraction layer over real devices. Various classes of devices are currently supported, several cameras, laserscanners, radars, and inertial systems. Data can be acquired from multiple sensors with or without synchronization. This feature allows to easily develop data fusion oriented algorithms. Moreover, virtual devices are also implemented to use pre recorded data. GOLD also allows to preprocess acquired data performing common operations (i.e. image stabilization, distortion removal...) and sharing the results among all the active plugin applications.

Developers can also easily produce highly interactive and consistent user interfaces, through the use of a set of widgets specifically designed to seamlessly integrate with the processing code. A windows subsystem provides a set of simple and powerful drawing primitives. These can be used to display intermediate and final results during the debugging phases or as the final result. This system hides the complexity of the underlying graphics API, and thus ensures portability and ease of use. The window subsystem itself supports many target platforms (OpenGL, X11, SVG...). Hardware acceleration is used to reduce the CPU load and boost the performance of the visualization system.

Developers may use GOLD to record data sequences from multiple data streams in order to allow a later playback for in-lab development. Captured data can be stored in different formats, depending on their nature and on the application constraints. GOLD allows advanced playback capabilities (such as looping, stepping, jumping, moving forward or backwards at a given frame rate and setting bookmarks), which guarantee high flexibility for the development and testing of algorithms. During playback, acquired data are fed to the application through the same APIs used for acquiring from real devices. Recorded data contain an index file with a timestamps for each data. This additional file is used to cope with problems involving time, such as tracking and motion stereo, allowing a time based mode of replay. In this case, playback is performed considering the actual processing time and therefore emulating the system behavior in real working conditions.

GOLD Architecture. The GOLD architecture has been conceived to ensure scalability and independence among its components. The whole system is organized into several subsystem layers, which provide the different functionalities to the developers. Thanks to a cleanly defined interface, subsystems can be easily added or removed depending on the specific needs of the target project. Also the vision-based applications are seen as plug-ins by the framework, and can be easily removed or ported to different platforms.

Each software layer has been developed to achieve the maximum reusability and performance, often using machine-code level optimizations; multithreaded code has also proven very effective in boosting speed (up to 40%) of CPU-intensive tasks on multi-core machines and coping with I/O latencies.

The hardware abstraction layer masks the complexity of input devices to the programmer. Sensors are organized in a tree structure: the first level contains the set of different device classes (cameras, laserscanners or inertial measurement units) while the second level manages the connected devices of each set. Applications can traverse the tree to the available devices. The tree can also be dynamically reconfigured to take in account device hotplug or malfunctions.

A second software layer for image processing enables programmers to use primitives that cover many fields of computer vision: low-level image processing and stabilization, features tracking, perspective mappings, stereovision...

A graphical user interface layer has been developed to provide graphic widgets and controls for interacting with applications allowing programmers to forget about available graphic backends. A rich set of APIs allow programmers to output final or partial results, producing graphics with very low resource consumption, thanks to its multithreaded client-server architecture.

Porting to Embedded Architectures. DSPs and FPGAs feature different constraints and capabilities than a modern PC and the code must be rewritten and/or best-managed to run on such architectures. As an example of major steps to be performed, floating point must be converted to fixed-point and conditional loops need to be converted to operations that can better exploit DSP pipeline. External RAM accesses are slow and algorithms have to efficiently manage the small internal RAM and L2 cache.

Proprietary compiler for embedded systems often provide an efficient way to develop software, regardless of compatibility to language standard. Anyway, even such compilers are not able to reach the same code optimization of a skilled assembly programmer. Therefore, critical portions of the code have to be manually implemented.

3 Examples of Vision-Based Systems

Vision technologies for intelligent vehicles can be employed in a number of different applications that present different constraints. In the following, two different examples are given: the Grand and Urban Challenges sensing architecture and a Start Inhibit system. In the first case, several sensing tasks had to be performed on a prototype vehicle; therefore, a very powerful hardware architecture and a complete version of the GOLD platform were used. Conversely, the Start Inhibit system was conceived to be installed on commercial vehicles; in this case, embedding constraints required to port the system on small architectures.

In addition, also the application of the same technologies in a different field is presented in order to show the versatility of such an approach: a surveillance system for boats speed detection.

Grand & Urban Challenges. The DARPA Grand Challenge was a robotic vehicle competition intended to energize the engineering community to tackle the major issues in autonomous vehicle development. Vehicles had to operate with full autonomy as they maneuvered around obstacles on a 132 mile route in an off-road environment.

Oshkosh Truck Corporation, Rockwell Collins, and the University of Parma partnered together to form Team TerraMax^{TM}in 2005. The developed vehicle (Fig. 2.a) is an MTVR truck equipped with electronic actuators for steering, brake, throttle, and transmission control. Team TerraMax^{TM}is one of the only five participants that completed the course, and the only one that used vision as its primary sensor [3].

Fig. 2. The TerraMax vehicle (*a*) and its on-board PCs (*b*) fitted under the passengers seat

VisLab is now developing the artificial vision systems to sense the environment for the 2007 *Urban Challenge*: a new competition in which fully autonomous vehicles have to complete a 60-mile long race in an urban environment in less than 6 hours, obeying to road rules and negotiating traffic.

The vision software is now spread over 4 high performance dual core PCs, used to sense the environment thanks to 11 cameras, and to provide a wide range of information: drivable path, lane position, far and close frontal obstacle, close rear obstacles and overtaking vehicles, and traffic at junctions.

In such complex scenarios, the design efforts are totally focused on reaching the needed levels of accuracy, robustness, speed, and completeness. Issues regarding power consumption, hardware size, etc. are not taken into account. The PCs (fig. 2.b) were chosen among the most powerful ones. Each one is hosting a full-featured copy of GOLD for both acquisition and processing.

Start Inhibit for Heavy Good Vehicles. Many accidents involving heavy goods vehicles are related to the limited field of view of the driver (see fig. 3.a).

Although, some of these blind areas can be partly covered by additional mirrors, an electronic system provides a better and safer solution.

VisLab developed a system that monitors the driver's frontal blind area; it warns the driver and prevents the vehicle from taking off when any obstacle or pedestrian is present, namely a Start-Inhibit functionality.

The system has been tested on several prototypes: a Volvo truck, an Iveco truck, and it is also used as close obstacle detector on the TerraMax vehicle. Anyway, for commercial vehicle installation, the use of an embedded architecture is mandatory. Initially, the system has been implemented on a barebone PC (see fig. 3.b) with the output directly sent on the CAN-bus. Thanks to the low computational power needed by this algorithm, it was also successfully ported to a DSP-based architecture. In particular, a Texas Instruments DM642 DSP running at 720 MHz has been chosen: on this platform the system runs up to a 30 Hz rate. This allows the system to be easily integrated on-board real vehicles and become a viable solution for mass-production.

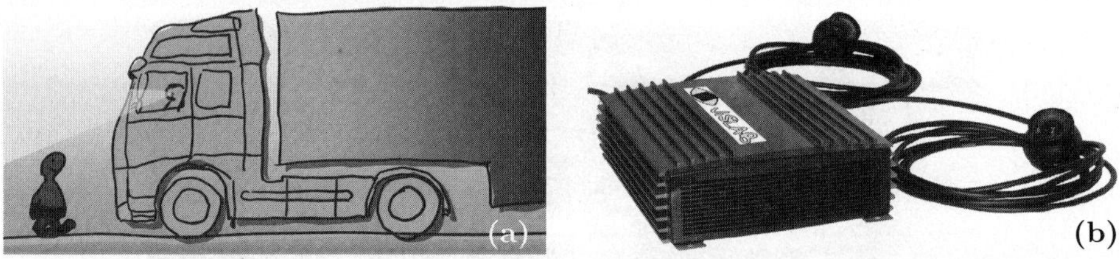

Fig. 3. (*a*) the driver's field of view does not allow to detect pedestrians and (*b*) the barebone PC on which the whole system has been installed and tested

Boat Speed Monitor. The vision technologies specifically developed for intelligent vehicles can also be successfully exploited in other research fields. As an example, in the following a surveillance system for boats speed estimation is presented. Despite the different issues and scenario involved in developing this system, the usual GOLD platform was used to develop and run this system.

In the city of Venice, a lot of watercrafts pass too fast through water channels, generating waves that erode seacoasts. The preservation of this coast is important to save the unique ecosystem of Venice.

To tackle such a problem, a system that can detect and display the boat speed with the intent to deter excessive speeds has been developed.

A high resolution camera (Tattile TAG 1600) is mounted on the top of a 10 m pole and used to detect boats and evaluate their speed. The camera is connected via Ethernet to an off-the-shelf PC (Pentium 4, 3 GHz, 1 GB RAM) used for the whole processing. The PC is connected by a RS232 link to two large displays that show the speed to the boat's driver.

For this application only a reduced portion of GOLD is needed: only low level vision filters, hardware abstraction layer, and real time management are used. The final commitment is to port the whole system on a smart camera (Tattile

TAG PLUS) with an ARM processor and a Linux embedded operating system. Therefore, the GOLD platform will be pruned, keeping its low-level structure and the processing plug-in only.

4 Conclusion

In this paper the most important technologies for the development of on-board vision-based systems for Intelligent Vehicles have been presented, examining both software and hardware issues.

On the hardware side, the different camera technologies and the potential choices for the processing engine have been discussed.

For the software issues, the GOLD platform for the rapid development of vision-based systems has been presented, and its structure and functionalities detailed. GOLD can be used both as a development system and as the engine of the final system, enabling a fast software deployment towards the final system.

Two different case studies were presented: the perception system for the Grand and Urban Challenges and a Start Inhibit system. Moreover, to demonstrate the adaptability of these vision technologies in other situations, also a boats speed monitor system has been presented.

References

1. Bertozzi, M., Broggi, A., Conte, G., Fascioli, A.: The Experience of the ARGO Autonomous Vehicle. In: Procs. SPIE - Enhanced and Synthetic Vision 1998, Orlando, FL, vol. 3364, pp. 218–229 (April 1998)
2. Bertozzi, M., Broggi, A., Fascioli, A.: VisLab and the Evolution of Vision-Based UGVs. IEEE Computer 39(12), 31–38 (2006)
3. Braid, D., Broggi, A., Schmiedel, G.: The TerraMax Autonomous Vehicle concludes the 2005 DARPA Grand Challenge. In: Procs. IEEE Intelligent Vehicles Symposium 2006, Tokyo, Japan, pp. 534–539 (June 2006)

A Geographic Event Management, Based on Set Operation Among Geographic Objects

Masakazu Ikezaki[1], Toyohide Watanabe[1], and Taketoshi Ushiama[2]

[1] Department of Systems and Social Informatics,
Graduate School of Information Science, Nagoya University,
Furo-cho, Chikusa-ku, Nagoya, 464-8603, Japan
{mikezaki,watanabe}@watanabe.ss.is.nagoya-u.ac.jp
[2] Faculty of Design, Kyushu University,
9-1 Shiobaru 4-chome, Minami-ku, Fukuoka, 815-0032, Japan
ushiama@design.kyushu-u.ac.jp

Abstract. Recently, a lot of researches on geographic information models for handling events have been proposed. However, many of researches are application-oriented. Namely, they are not suitable to represent an event spreading broadly in the geographic space, such as earthquake, typhoon, and so on. In this paper, we propose a new model for geographic information system. In our model, event information and geographic information are represented independently, and are associated flexibly from a viewpoint of users for an event.

Keywords: GIS, Event, Spatio-temporal Information.

1 Introduction

Recently, Geographic Information Systems (GISs) are rapidly spreading in various fields such as city administration, economics, and so on, and take an important role as a fundamental social infrastructure system. In order to manage the historical information about geographic changes, traditional researches on GIS have focused on the historical versions of geographic objects or sequential changes of attribute values based on the cell-based representation method. However, the meta-information about geographic changes like circumstances or situations is not sufficiently represented in the traditional GIS, although such information is useful in various application domains of GIS such as statistics processing of spatial information, spatio-temporal data mining and so on.

We have worked on the framework for treating event information. In [1], we defined an event as a phenomenon that induces geographic changes, and we represented an event as an object with properties aggregated from changes of geographic objects. In addition, in order to represent the relationships among events, we defined three relations: *is-a relation* between event concepts, *part-of relation* and *causal relation* between events. There are other researches to represent situations for geographic changes [2][3][4]. In [3], Worboy et al. represented phenomena with three concepts: objects, events, and situations. A geographic event involves some objects which have applicable roles in the event such as terminator, initiator, etc. in a geographic situation. In [4], Galton et al. modeled the event in fully four-dimensional space. They represented a

B. Apolloni et al. (Eds.): KES 2007/WIRN 2007, Part I, LNAI 4692, pp. 712–719, 2007.

phenomenon as a collection of object's coordinated behaviors, and analyzed properties of the collection from the viewpoint of object-like and event-like aspects.

The models in traditional researches can provide the correct representation of phenomena such as traffic jams, tour trips and so on. In other words, these models can represent when, where, and what happens in the space and time. However, they are unsuitable to represent phenomena spreading in the geographic space broadly such as earthquakes, typhoons, and so on. This is because these models need to manage the relation between events and geographic objects explicitly. The cost for managing these relation between events and a large number of geographic objects is too high. This problem become prominent specifically in integrated GISs which are wide-spreading as our fundamental information infrastructure systems. In integrated GISs, geographic information is managed in distributed databases, from one data domain to another. Therfore, by traditional model, it is difficult to manage the associations between an event and geographic objects explicitly in integrated GIS. In this paper, we propose a new framework for representing events which supports the following properties:

- to handle relations between events and a large number of geographic objects,
- to represent various aspects of an event.

This paper is organized as follows. In Section 2, we show a framework. In Section 3, the formal definition of our conceptual model and the event handling mechanism are provided. In Section 4, we describe a concrete example and results of handling event information in our prototype system. The discussion for our model is set in Section 5. Finally, in Section 6 we state our conclusion and future works.

2 Approach

In representing event information on a map, traditionally event information are divided into each geographic domain such as roads, buildings and etc., and represented as changes of geographic objects concerning to the event;destruction of roads, reparation of buildings, etc. are associated to the event. If changes caused by events in all the geographic domains, it is possible to handle all the effects of the event by collecting these geographic changes over the domains(Fig.1). The region of geographic objects that have the collected geographic changes is corresponding to the region of the event. Reversely, it seem to be possible to collect the geographic objects that are changed by the event from the region of the event, even if the geographic changes caused by the event are not represented explicitly.

In our model, based on this prospect, the event region, which is the existing region of the event in the space and time, is managed, and also separated management of event information and geographic objects are achieved. The association between events and geographic objects is not represented explicitly, but is computed based on the space and time. Namely, if geographic objects and events, which have the existing region and the lifespan, have the shared region in the space and time, the geographic objects are treated as participants of the event. In addition, the machanism for treating events with a certain perspective is needed, because the operations in GIS, in general, are performed on the interesting geographic domain. As an example of earthquake events, some focus

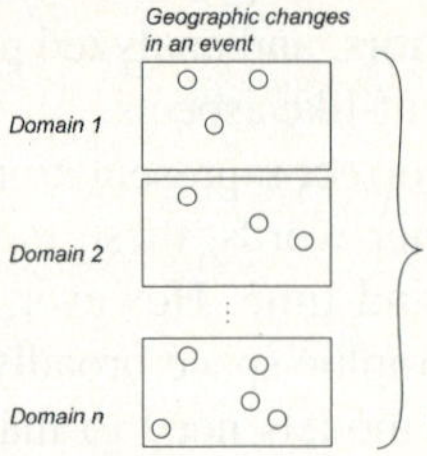

Fig. 1. Domain and event in geographic change

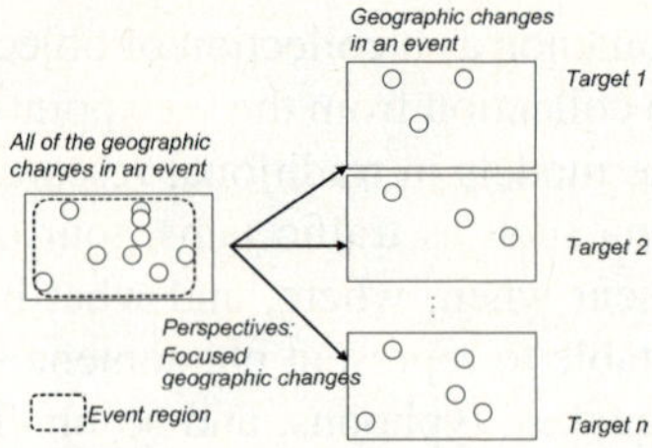

Fig. 2. Perspectives for events

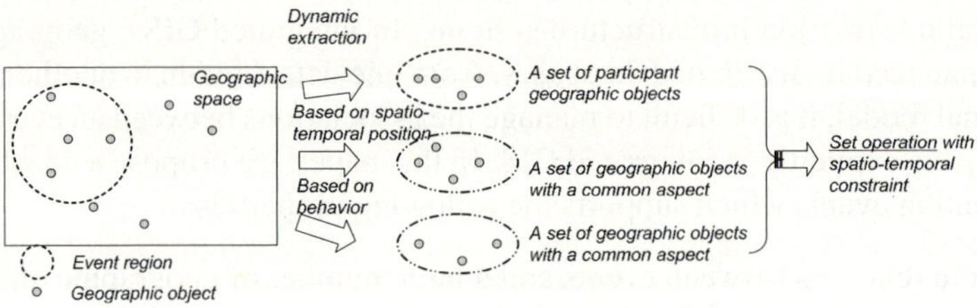

Fig. 3. Conceptual diagram of proposal model

on the buildings destructed by the earthquake, and others focus on continental plates which induce the earthquake. From this example, it can be said that the perspective for the events can be indicated by focused geographic changes(Fig.2). In order to treat geographic changes, a concept of geographic object's behaviors is introduced. A geographic object is an entity in the geographic space such as a river, a building, and so on. Geographic objects are changeable over time and have their own lifespan. In addition, each attribute value also has the valid time. We define the behavior of geographic objects as an ordered set of attribute values in geographic objects. By indicating the behavior of geographic objects, a set of geographic objects can be obtained from all of the geographic objects.

The event handling is performed based on set operations among a set of participant objects or a set of geographic objects that have an interesting behavior. However, the interesting time interval about a geographic object is a temporal segment of the geographic object's lifespan. For example of a set of participant objects of an event, the interesting time interval for the element object is a time interval while the geographic object participates in the event. And also, the interesting time interval for an element objects of a set of geographic objects with an interesting behavior is a time interval while the geographic object have the interesting behavior. In order to treat these sets of geographic objects, a concept of geographic object's aspect is introduced. An aspect represents a geographic object in an interesting time interval. Namely, an aspect indicates a geographic object with a time interval while the geographic object have a common property for a set of geographic objects.

Our framework is applicable to the integrated GISs, because the event information does not need to be associated with geographic objects directly. Our conceptual diagram is depicted in Fig.3. In our model, event information and geographic objects are managed seperatedly, and participant relation is computed based on their spatio-temporal positions. In addition, by indicating the interesting behavior of geographic objects, a set of geographic objects with a common behavior are extracted. An element of these geographic objects set is represented as an aspect of geographic object. Handling events is performed by set operation among these set of aspects.

3 Formal Description

3.1 Aspect of Geographic Object

A geographic object is a entity with the shape and the lifespan in the geographic space. Attributes of a geographic object are changeable during the lifespan. The attributes values are managed with the valid time. A geographic object o is described as follows.

$$\begin{aligned} o &= (t_s, t_e, ATTR_o, POS_o), \\ ATTR_o &= \{\{(val_{i,0}, t_s, t_{i,1}), ..., (val_{i,m_i-1}, t_{i,m_i-1}, t_e)\} | i = 0, \ldots, n\}, \\ POS_o &= \{(pos_0, t_s, t_1), ..., (pos_{l-1}, t_{l-1}, t_e)\}. \end{aligned} \quad (1)$$

Here, t_s and t_e are the onset time and termination time, respectively. $ATTR_o$ is a set of attribute values of o. Each attribute is corresponding to the type of o, and represented as a set of tuples: the attribute values and the valid time intervals. POS_o represents the existence region, and each pos_i represents the existence space with the valid time.

We call an ordered set of attribute values in an object as geographic object's behavior. By indicating a pattern of an interesting geographic object's behavior, a set of geographic objects which have the interesting behavior can be specified. The behavioral pattern *bpt* is described as follows.

$$bpt = \{(classname_i, state+) | i = 0, \ldots n\}, state = (atname, cond). \quad (2)$$

Here, $classname$ is a name of geographic object class, and $state$ represents the state of a geographic object. The signature $+$ means more than one iteration. $state$ is represented as a tuple of an attribute name $atname$ and a condition of the attribute value $cond$. As an example of destruction of buildings and roads, if the class names and attributes of buildings and roads are $Buildings$ and $strength$, $Roads$ and $state$, respectively, then the behavioral pattern is described as

$$\begin{aligned} bpt_{collapse} &= \{(Buildings, < st_1, st_2 >), (Roads, < st_3, st_4 >)\}, \\ st_1 &= (strength, 6 \leq val < 10), st_2 = (strength, 1 \leq val < 5), \\ st_3 &= (state, val = normal), st_4 = (state, val = destructed). \end{aligned} \quad (3)$$

Here, the notation $<>$ represents an ordered list on the time axis and val represents the attribute value corresponding to $atname$. In this example, we treat the changes of buildings from a state, whose $strength$ value is 6 to 10, to a state, whose $strength$ value is 1 to 5, and the changes of roads from a $normal$ state to a $destructed$ state as the "collapsed" behavior. All the buildings and roads which behave like $bpt_{collapse}$ have the destructed aspect.

The aspect of a geographic object is represented by a tuple of a geographic object and a time interval corresponding to the behavior.

$$aspect = (obj, T), start(obj) < t \in T < end(obj). \quad (4)$$

Here, obj represents the geographic object. T is a set of instant times, and represents the focused time interval. Accurately, T represents an instant time if the corresponding behavioral pattern represents a state transition of o, or T represents a time interval if the behavioral pattern represents a sequence of state transitions or a state of an attribute. $start$ and end represents the start time and termination time of the geographic object. The function ex_{aspect} that extracts a set of the geographic objects with a common aspect indicated by a behavioral pattern *bpt* is defined as follows.

$$\begin{aligned} ex_{aspect} &: BehavioralPattern- > \{Aspect\}, \\ ex_{aspect}(bpt) &= \{(obj, T) | obj \text{ behaves like } bpt \text{ at } t \in T.\}. \end{aligned} \quad (5)$$

Here, *bpt* is a behavioral pattern that specifies the interesting aspect. obj is a geographic object that have a behavior like *bpt* during the time interval represented by T. Concerning to the concept of the aspect, a function $period$ which returns the time interval attented with a geographic object for an aspect $as = (obj, T)$ is defined.

$$period : Aspect- > \{Time\}, period(as) = T. \quad (6)$$

3.2 Event

An event is represented by properties and occurrence fields in space and time. Properties of an event are corresponding to the event type such as typhoons, earthquakes and so on. The occurrence fields represent the histories about the existence space of an event.

$$\begin{aligned} ev &= (ATTR_e, t_s, t_e, POS_2), \\ ATTR_e &= (val_0, ..., val_{n-1}), POS_e = \{(pos_0, t_s, t_1) \ldots (pos_{m-1}, t_{m-1}, t_e)\}. \end{aligned} \quad (7)$$

Here, $ATTR_e$ is a tuple of invariant attributes, corresponding to the event type. t_s and t_e are the onset time and the termination time of the event, respectively. POS_e represents the occurrence field of the event, and each pos_i is a spatial region of the event with the valid time. Let an event type "Earthquake" have attributes ($mainsource$, $name$). As an example, an event "$earthquake_1$" can be described as follows.

$$earthquake_1 = (((x_0, y_0), \text{"TokaiQuake"}), t_3, t_4, \{(region_1, t_3, t_4)\}) \quad (8)$$

Here, $earthquake_1$ has the main source at (x_0, y_0) and the name is "TokaiQuake". This earthquake has the lifespan from t_3 to t_4, and exists at $region_1$ between t_3 and t_4.

An event has its spatio-temporal field, and each geographic object has its positions during the lifespan. Geographic objects which are placed in an event's field are regarded as participants of the event. The notation of event's participant is as follows.

$$\begin{aligned} ex_{participant} &: Event- > \{Aspect\}, \\ ex_{participant}(ev) &= \{(obj, T) | t \in T.contain(pos(t, ev), pos(t, obj)), \\ & start(obj) < t < end(obj)\}. \end{aligned} \quad (9)$$

pos is a function which represents the spatial region of a geographic object obj or an event ev (indicated by the second argument) at time t (indicated by the first argument). $start$ and end represent the onset time and the end time of a geographic object or an event (indicated by an argument), respectively. Consequently, the function $ex_{participant}$ returns a set of geographic objects with valid time intervals while the set of geographic objects exists in the event field.

3.3 Data Handling

Spatio-Temporal Constraint. Participant geographic objects for an event and geographic objects with a similar behavior can be represented as a set of aspects by functions $ex_{participant}$ and ex_{aspect}, respectively. Therefore, we can perform the set operation such as $union$, $intersection$, etc. among these sets of geographic objects. However, general set operations cannot reflect the users' intention properly, since these operations are regardless of the spatio-temporal relation between elements in two sets of aspects. In this section, we introduce the three constraints for the set operations; spatial constraints, temporal constraints and spatio-temporal constraints.

The spatial constraint is represented by spatial relations between the elements in the sets, such as neighboring, overlapping and so on. For example, if two building objects obj_1 and obj_2 are placed enough closely at time t, this relation is described as $neighboring(pos(t, obj_1), pos(t, obj_2))$. The topological relation and distant relation are employed as the spatial relation. The temporal constraint is represented by temporal relations between time intervals, while geographic objects in two sets have a common feature: i.e. having a common behavior or participating in an event. Representation of the temporal relations is based on Allen's interval logic [5]. For example, if a geographic object obj_1 behaves like bpt_1 after bpt_2 and these behavior are represented as $as_1 \in ex_{aspect}(bpt_1)$ and $as_2 \in ex_{aspect}(bpt_2)$, respectively, then the temporal relation $before(period(as_1), period(as_2))$ is specified. There are six relations in Allen's interval logic; $before$, $meets$, $overlaps$, $starts$, $contains$, and $finishes$. The spatio-temporal constraints is a combination of spatial relation and temporal relation.

Set Operation. Spatial, temporal, and spatio-temporal constraints restrict a couple of elements in two sets of aspects. In this section, set operations among aspects with these constraints are introduced. Here, these constraints are alternated as $constraint$.

$$\begin{gathered}
union(AS_1, AS_2) = \{as|as \in AS_1 \cup AS_2\}, \\
intersection_c(AS_1, AS_2, constraint) = \\
\{as|as \in AS_1, as_2 \in AS_2.constraint(as, as_2) \wedge as = as_2\}, \\
production_c(AS_1, AS_2, constraint) = \\
\\
\{(as_1, as_2)|as_1 \in AS_1, as_2 \in AS_2.constraint(as_1, as_2)\}, \\
selection_c(AS_1, AS_2, constraint) = \\
\{as|as \in AS_1, as_2 \in AS_2.constraint(as, as_2)\}.
\end{gathered} \tag{10}$$

AS_1 and AS_2 are aspect sets of participant geographic objects or geographic objects that have a common behavior. The notation $union(AS_1, AS_2)$ is a general union set of AS_1 and AS_2. The notation $intersection_c(AS_1, AS_2, constraint)$ is a set of common elements of AS_1 and AS_2. Moreover, the elements must satisfy $constraint$.

$production_c(AS_1, AS_2, constraint)$ represents a set of couples in $AS_1 \times AS_2$ that satisfy $constraint$. $selection_c(AS_1, AS_2, constraint)$ extracts the elements from AS_1 that satisfy $constraint$ for elements in AS_2. The examples of these operations are provided in the next section.

4 Prototype System and Example

We developed a prototype system based on our proposed model. We made a data set by simulation on the map of Japan (Map 2500 of Geographical Survey Institute in Japan). Typical behavioral patterns and handling method for aspect sets are implemented by Java. We simulated the trajectories of several typhoons and changes of geographic objects. Examples of computational results are presented below.

In Fig.4, a set of geographic objects that participate in a typhoon event (ev_1) is visualized. Namely, $ex_{participant}(ev_1)$ is displayed. By using the event information as medium, these geographic objects that are managed separately in each domain can be handled simultaneously, although the relations among these geographic objects are not represented explicitly in databases.

Fig.5 describes the aspect set of building and road objects with destructed behavior(Eq.3): $ex_{aspect}(bpt_{collapse})$. These are caused by several typhoons or other factors.

Fig.6 shows the result of the set operation computing intersections of two aspect sets in Fig.4 and Fig.5, with the time constraint that the relation between two time intervals, participating in the event and behaving like $bpt_{collapse}$, is $overlaps$. This computation is represented formally as $intersection_c(ex_{aspect}(bpt_{collapse}), ex_{participant}(ev_1), overlaps)$. Consequently, the result described in Fig.6 can be treated as the roads and buildings destructed by the typhoon ev_1.

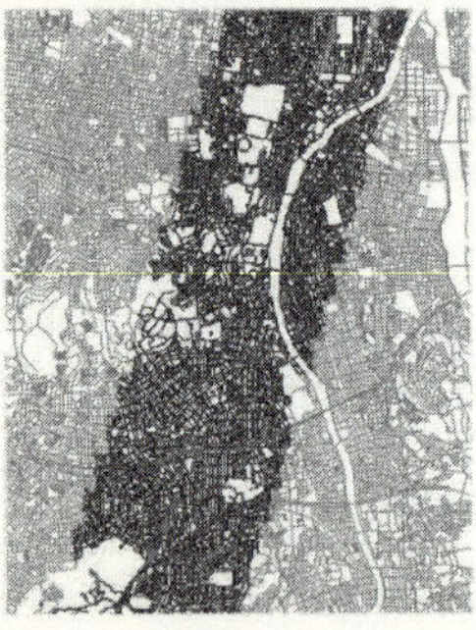

Fig. 4. Participant Object Set

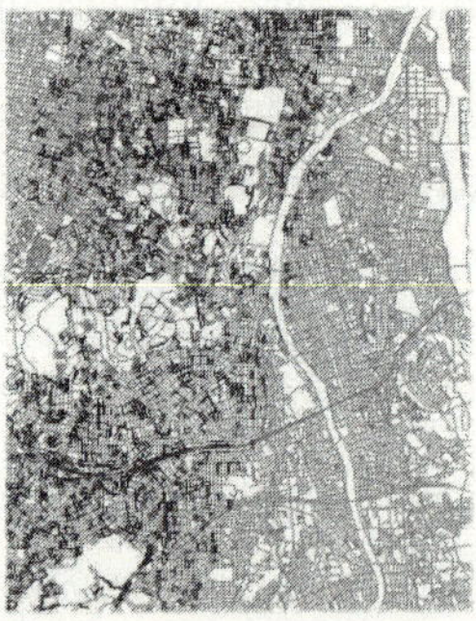

Fig. 5. Aspect set with destructed behaviour

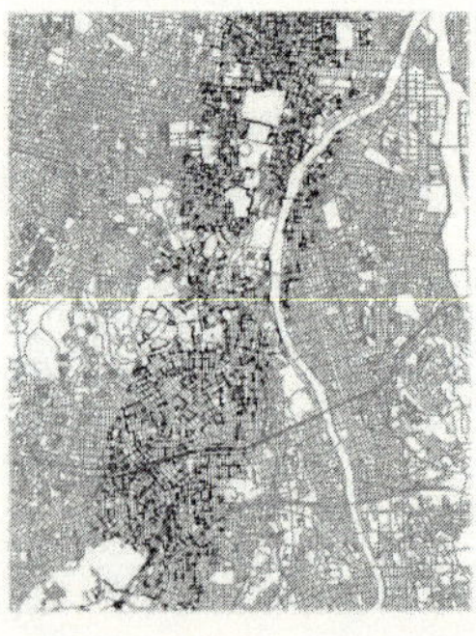

Fig. 6. The result of intersection operation

5 Consideration

In this paper, we proposed a framework for integrating a large number of geographic changes related to events by representing event information and geographic information independently. In our model, a set of geographic objects can be specified by indicating interesting geographic changes and associating event information and geographic information dynamically. This mechanism make it possible to handle event information in an integrated GIS. The process of this mechanism is the definition of viewpoint for events. Namely, this mechanism works as the procedure for defining what geographic changes are related to events. Therefore, our model can supports the constitution process in the application domain. Consequently, the position of our proposed model is that it provides the views for geographic data at each application domain in integrated GIS in which various kinds of geographic information are managed.

6 Conclusion

In this paper, we addressed the data model for representing events. Our model can represent the relation between events and a large number of geographic objects, since event information and geographic information are represented independently and the relation between an event and geographic objects is calculated dynamically. In addition, by indicating interesting behavior of geographic objects, a set of geographic object that user intends in an event can be obtained. This mechanism is performed based on set operations so that we can specify the way to obtain the target information intuitively.

In the future work, we must develop the indexing structure for retrieving geographic objects effectively based on the behavior of geographic objects. In order to specify an aspect, it is needed to retrieve the geographic object based on the geographic objects' behaviors. However, the retrieval of geographic objects based on behaviors is too expensive to handle in real-time. Therefore we need to consider the indexing mechanism in particular to the behavior of geographic objects.

Acknowledgement

This work was supported in part by Research grant of Daiko Financial Group.

References

1. Ikezaki, M., Mukai, N., Watanabe, T.: Event handling mechanism for retrieving spatio-temporal changes at various detailed level. In: IEA/AIE, pp. 353–356 (2005)
2. Peuquet, D., Duan, N.: An event-based spatiotemporal data model (estdm) for temporal analysis of geographical data. International Journal of Geographical Information Systems 9(1), 7–24 (1995)
3. Worboys, M.F., Hornsby, K.: From objects to events: Gem, the geospatial event model. In: GIScience, pp. 327–344 (2004)
4. Galton, A.: Dynamic collectives and their collective dynamics. In: Cohn, A.G., Mark, D.M. (eds.) COSIT 2005. LNCS, vol. 3693, pp. 300–315. Springer, Heidelberg (2005)
5. Allen, J.F.: Towards a general theory of action and time. Artif. Intell. 23(2), 123–154 (1984)

A Method for Judging Illogical Discourse Based on Concept Association and Common-Sense Judgment

Eriko Yoshimura, Noriyuki Okumura, Hirokazu Watabe, and Tsukasa Kawaoka

Dept. of Knowledge Engineering & Computer Sciences, Doshisha University
Kyo-Tanabe, Kyoto, 610-0394, Japan
{eyoshimura,nokumura,watabe,kawaoka}@indy.doshisha.ac.jp

Abstract. In recent years, research into natural language processing for implementing natural discourse has attracted considerable attention. However, up to now, research into natural language processing has focused mainly on discourse processing for certain limited objectives and under specific conditions, with emphasis on the superficial formation of sentences. The understanding of the user-generated words is constrained by the size of the constructed knowledge base and the quality of actual data collected. Overcoming this limitation requires the capability to autonomously generate discourse sentences using a computer. However, there is a risk that the generation of response sentences by a computer results in illogical word combinations caused by the mechanical combination of words. To avoid this risk, the capability of detecting illogical word combinations is needed. This paper proposes a technique for judging illogical word combinations by creating a knowledge model of human discourse and words.

Keywords: Illogical discourse, Natural language, Common-sense, Association.

1 Introduction

A great deal of attention is being paid to research on natural language processing for enabling natural discourse. However, up to now, research into natural language processing has focused mainly on discourse processing for certain limited objectives and under specific conditions, with emphasis on the superficial formation of sentences. Also, with the advances in computer technology, there has been a strong tendency to collect large quantities of response examples to create a knowledge base, although the understanding of such user-generated words is constrained by the size of the constructed knowledge base and the quality of data collected. Because this system does not include common sense and is unable to understand information from the surrounding environment or discourse partner, discourses tend to follow a fixed pattern, resulting in illogical responses.

Overcoming this limitation requires the capability to autonomously generate discourse sentences using a computer. However, there is a risk that the generation of response sentences [1] by a computer results in illogical word combinations caused by the mechanical combination of words. To avoid this risk, the capability of detecting illogical word combinations is needed.

B. Apolloni et al. (Eds.): KES 2007/WIRN 2007, Part I, LNAI 4692, pp. 720–728, 2007.

Therefore, in order to create discourses by machine, it is necessary to have "common sense" knowledge in addition to word knowledge and grammatical knowledge. When humans converse, their discourse develops by natural association from the discourse sentences. For this, it is necessary not only to accumulate past data, but also to perceive the associations between words and carry out common-sense recollection. Without the ability to perform common-sense recollection, discourses will fall into a fixed pattern, based on the accumulation of past data, and will result in illogical responses due to the inability to think logically. This paper discusses a method for detecting illogical combinations of words.

2 Purpose

The occurrence of logical responses indicates to the user that the machine has understood the meaning and is able to converse with common sense. It is expected that this ability will enable very free, flexible discourse with machines. In order to show the possession of common sense, we are conducting research on the development of techniques to prevent the generation of illogical discourse.

In this paper, "illogical discourse" is used to mean "strange expressions." These include expressions that make listeners feel that something is wrong and expressions that seem unnatural. For example, the following sentences are called "illogical discourse" for various reasons.

Sensory factor: "I ate a black apple."
Time factor: "I went skiing in August."
Quantity factor: "I put a house in my desk."

If someone hears the sentence, "I ate a black apple," they will obviously feel that something is wrong because of the association of "apple" with "black." The expression, "I went skiing in August," also creates an odd impression, because of the association of "skiing" with "August," which is a summer month. To understand these kinds of sentences, it needs to possess not only grammatical knowledge, but also common sense about words. This common sense is accumulated through experience.

The above sentences provoke a sense of oddness in the human listener. This sense of oddness then becomes a subject of conversation that conditions the development of the discourse. In a natural discourse, if someone feels that the discourse is unnatural, then he or she will typically respond by expressing doubt to the discourse partner.

Humans can immediately identify what is unnatural about the discourse and why it is unnatural. This is because humans possess common sense in relation to words. Thus, in order for computers to be aware that "this is unnatural," or "this is a strange expression," they, too, must possess common sense in relation to words. In other words, systems that can deal with such sentences must also be capable of understanding their meaning and responding to them, based on common sense.

Therefore, in this study, we propose a method for recognizing illogical discourse. Our system is constructed using the Japanese language. For this reason, it is created to match the characteristics of the Japanese language and the Japanese environment.

3 Association Judgment Mechanism

Humans possess a common knowledge of words and can conduct a discourse based on the common sense relating to words. For example, humans possess certain knowledge of "airplanes"—that they are flying machines, they have wings, and they are related to airports. They also possess the common-sense knowledge that airplanes are bigger than people and chairs, and faster than trains and people. Assuming this common knowledge, humans can converse with each other on this subject.

Thus, by modeling human knowledge of discourse and words and making this understandable by machines, we believe it is possible to construct a discourse mechanism similar to the human discourse system. We call a mechanism that is constructed based on this concept an association judgment mechanism.

An association judgment mechanism is composed of the Concept Association Mechanism and a Common-Sense Judgment mechanism. The Concept Association Mechanism defines the common meanings relating to words, while a Common-Sense Judgment mechanism defines the common sense relating to words. Using this association judgment mechanism, we are developing an illogical discourse system, as explained in Section 6 and subsequent sections.

4 Concept Association Mechanism

The Concept Association Mechanism incorporates word-to-word relationships as common knowledge. It is a structure that includes a mechanism for capturing various word relationships. In this section, we describe the Concept Base and a method of calculating the degree of association using this base.

4.1 Concept Base [2]

The Concept Base is a knowledge base consisting of words (concepts) and word clusters (attributes) that express the meaning of these words, mechanically and automatically constructed from multiple sources, such as Japanese dictionaries and newspaper texts. The concept base that we used in this study contains approximately 90,000 registered words organized in sets of concepts and attributes.

4.2 Calculation the Degree of Association [3]

The degree of association is a parameter that quantitatively evaluates the strength of the association between a concept and another. The method for calculating the degree of association involves developing each concept up to second-order attributes, determining the optimum combination of first-order attributes by a process of calculation using weights, and evaluating the number of these matching attributes. The value of the degree of association is a real number between 0 and 1. The higher the number, the higher the association of the word.

5 Common-Sense Judgment Mechanism

The Common-Sense Judgment Mechanism derives common-sense associations from words in terms of various factors (e.g., quantity, time and physical sense). This "physical sense" indicates a sense that can be acquired by stimulation through any of the five physical senses (vision, hearing, smell, taste, and touch). These associations are constructed using the Concept Association Mechanism. In this section, from the various logic factors, we focus particularly on physical sense, time, and quantity.

5.1 Sensory Judgment System [4]

A sensory judgment system obtains a physical sense that humans can recollect and characterize using common sense. A sensory judgment system clarifies the relationship between a noun (e.g., apple) and the "physical sense" that characterizes it through the construction of a knowledge base of everyday nouns from the sensory judgment knowledge base (e.g., red), and extracts the necessary sensory words.

Also, by using the Concept Base and thesaurus of common knowledge, the system can recollect the "physical sense" of words that are not contained in the sensory judgment KB (unknown words). Table 1 is examples of the sensory judgment system.

Table 1. Examples of the sensory judgment system

Input	Output
Apple	Red, sweet, round
Sunset	Beautiful, red

5.2 Time Judgment System [5]

A time judgment system judge time such as seasons and times of day from nouns, using a knowledge base(Time Judgment KB) of words indicating the time (time word). The system sorts out the relationships between a noun and time through the construction of the Time Judgment KB and extracts the necessary time words.

We identified a set of basic representative time words—"spring, summer, autumn, winter, rainy season," and "morning, daytime, evening, night"—and applied these to all of the time words registered in the system.

The system can also handle time words not contained in the time judgment knowledge base (unknown words) through use of the concept base of common knowledge. Table 2 shows examples of this system.

Table 2. Examples of the time judgment system

Input	Output		
	Time word	Start time	End time
Sunset	Evening	4 p.m.	5 p.m.
Snow	Winter	December	February

5.3 Quantity Judgment System [6]

A quantity judgment system is something for capturing multiple things from quantitative aspects and comparing their size, length, weight, and so on. This system performs comparisons based on 10 quantitative factors: “size,” “length,” “width,” “height,” “depth,” “weight,” “fatness,” “thickness,” “speed,” and “value.”Table 3 shows examples of the quantity judgment system.

Table 3. Examples of the quantity judgment system

Input		Output
Comparison object	Comparison factor	
Apple, watermelon	Size (which is bigger)	Watermelon
Pencil, chimney	Shortness(which is shorter)	Pencil

6 Illogical Discourse Processing System

The illogical discourse processing system proposed in this paper is a discourse processing system that aims to handle “strange expressions.” Thus, the system needs to possess common sense to understand that an expression is unnatural. Common sense has various aspects, but in this paper we propose a system that is capable of handling expressions relating to physical sense, time, and quantity.

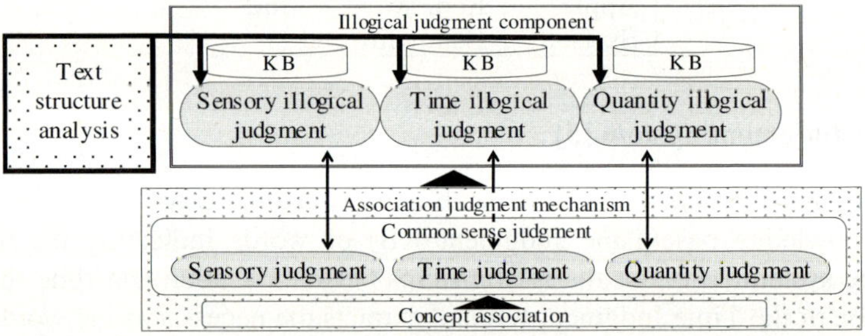

Fig. 1. Illogical discourse judgment

For detecting illogical discourse, there are methods of using the database made by statistical value of words and method of using the database made by human. First, using the database made by statistical value of words, it can detect illogical discourse by searching a set of object words. However, if an expression appears even once, it is judged a general expression. Therefore, it can’t capture an illogical discourse on purpose to attract public notice; a phrase of ad, a title of novel and so on. Second, using the database [7] made by human, it can detect logical discourse because the human arrange the database logically and generally. However, it is impossible to store data all-inclusive and the data might be different depending on a manufacturer.

The Commonsense Judgment system also associates on the word that doesn't exist in the database made by human. Therefore, using the method described in this report, the covered range can be expanded more than the database only made by human.

Illogical discourse processing involves a judgment component and a response component, but the explanation in this paper focuses on the judgment component. Figure 1 shows the structure of an illogical judgment.

In order to perform illogical discourse judgments, it is first necessary to extract words from a text to serve as judgment objects. For the object words to appear, a fixed pattern must exist in the text structure. Thus, we created a database of fixed patterns. Then, by performing a text structure analysis of input sentences and determining whether or not these matched the patterns, we extracted judgment object words. We called this database of collected fixed patterns "text structure patterns." We prepared text structure patterns for each aspect of physical sense, time, and quantity.

In order to analyze the text structure, we use a meaning understanding system. A meaning understanding system performs sorting in order to enable machines to grasp the content of a text. The system stores input text (single sentences) by dividing them into 6W1H (what, who, whom, why, where, how) + verb frames. Figure 2 shows an example of the meaning understanding system.

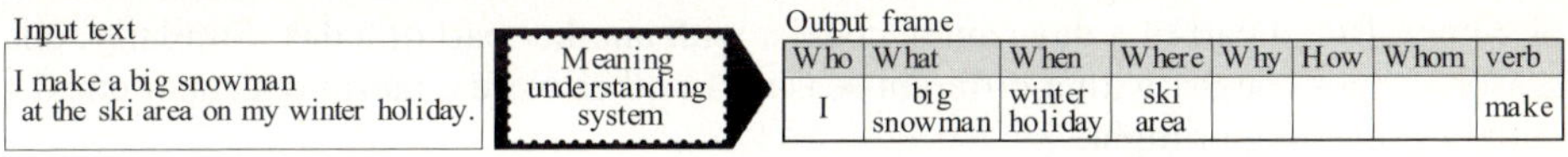

Fig. 2. Example of using a meaning understanding system

When using the meaning understanding system, the combination of frames containing object words has a fixed pattern. For this reason, text structure patterns are selected according to the presence or absence of words in the frames and part-of-speech patterns. The text structure patterns include five patterns for physical sense, eight patterns for time, and four patterns for quantity. This database also stores the relationships between two extracted words for each pattern.

By comparing the frames resulting from inputting the text into the meaning understanding system with the text structure patterns, it is possible to extract words for use as judgment objects. This method makes it easier to extend the system by adding similar rules, even when introducing new common-sense factors.

The extracted object words are judged for each factor to determine whether they are illogical. With the physical sense factor, the object words are nouns and descriptive words. For judging their relationship, it needs to have a knowledge structure relating to the common-sense qualities and characteristics of the object nouns. For example, "apple" has the qualities of color, taste, and smell, and the physical characteristics red, round, and sweet. But, it does not possess the qualities of brightness or sound.

These characteristics and qualities can be efficiently expressed using a thesaurus structure. The qualities of object nouns are derived from parent nodes. For this reason, we created a thesaurus-structured database expressing the qualities of object words, and we call it the Sensory Judgment Knowledge Base. In order to grasp the relationship between object nouns and descriptive words, the qualities of objects are expressed using the sensory judgment knowledge base, and the characteristics of objects are expressed using the sensory judgment system. Figure 3 shows a flow chart of an illogical judgment process relating to physical sense.

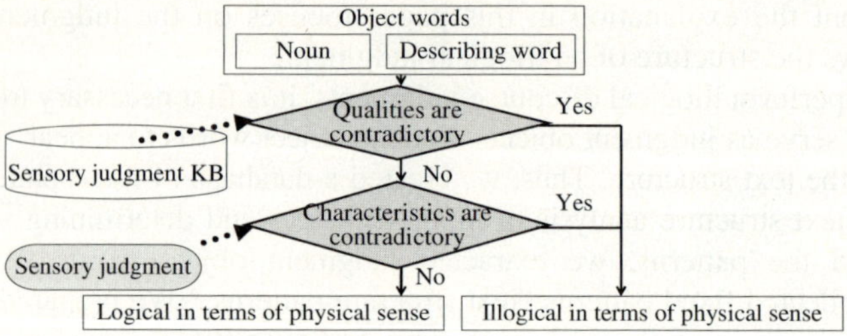

Fig. 3. Flow chart of illogical judgment of physical sense

Using time factors, the system can judge time-based contradictions when there are object words that recollect time. By the way, the concept of time is interconnected. A day is a part of a season. Changes of days coexist with seasons. That is, morning, daytime, evening, and night can exist in any of the seasons—spring, summer, autumn, or winter. But, a part of a day cannot coexist with another part of a day. Similarly, one season cannot coexist with a different season. In other words, morning cannot coexist with day, evening or night.

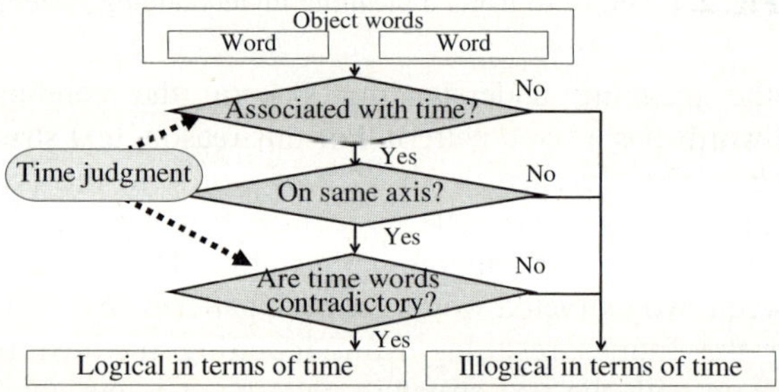

Fig. 4. Flow chart of illogical judgment of time

Thus, the system can detect these kinds of contradictions using the time judgment system. When object words recollect time, the system judges whether the time words are on the same axis. In the case of two words on the same axis, the criterion for judgment is whether the time words are different. Figure 4 shows a flow chart for illogical judgment relating to time.

Using the quantity factors, the system checks the size relationships of the object words. Based on the text structure and verbs, it checks whether the size relationship that should exist between the object words actually exists. For example, even if it is possible to have "a goldfish in a glass," the phrase "a whale in a glass" is unnatural because of the size relationship. In this text structure (A in B), there is a size relationship in which B is bigger than A. Such a text-structure size relationship is stored in the text structure pattern of quantity. To make an illogical judgment relating to quantity, the system checks the size relationship of the object words using the quantity judgment system (glass > goldfish, and glass < whale). A judgment is then made by checking if the size relationship of the object words matches the size relationship of the text structure pattern. Figure 5 shows a flow chart of the illogical judgment process relating to quantity.

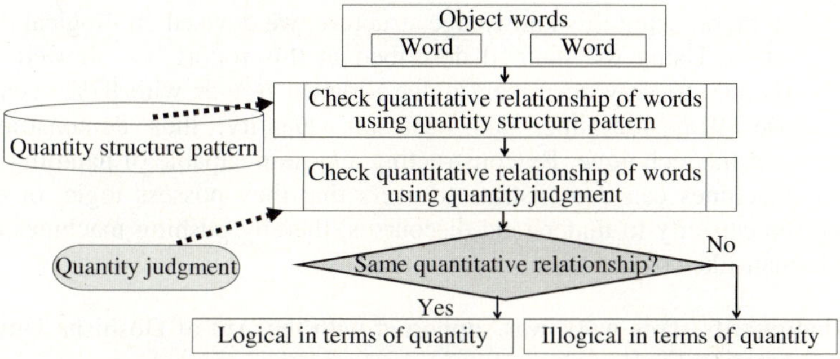

Fig. 5. Flow chart of illogical judgment of quantity

7 Evaluation

We evaluated the proposed illogical discourse judgment technique. The evaluation was performed according to three different factors—time, physical sense, and quantity. For each factor, we manually prepared 50 illogical discourse texts and 50 logical discourse texts. For each of these texts, we evaluated the proportion of correctly classified texts using the illogical discourse judgment technique.

Table 4 shows the evaluation of the illogical discourse judgment technique for the factors of time, physical sense, and quantity. In Table 4, "F-F" indicates the judgment of illogical texts as illogical; "T-T" indicates the judgment of logical texts as logical; "F-T" indicates the judgment of illogical texts as logical; and "T-F" indicates the judgment of logical texts as illogical. In this study, we calculate the accuracy of the technique as the total of "F-F" and "T-T" results as a proportion of all results.

Table 4. Evaluation results

	F-F	T-T	F-T	T-F
Sense	43%	44%	6%	7%
Time	44%	47%	6%	3%
Quantity	40%	48%	10%	2%

As shown in Table 4, the accuracy was 87% for physical sense, 91% for time, and 88% for quantity. For physical sense, "tasting spicy honey" was judged as illogical, and "drinking cold beer" as logical. For time, "dandelions blossom in autumn" was judged as illogical, and "stars twinkling in the night sky" as logical. For quantity, "planting a tree in a vase" was judged as illogical, and "putting coins in a purse" as logical. Thus, through this method we showed that this judgment system is effective.

8 Conclusion

In this study, we propose an illogical discourse judgment technique using a concept association system and a Common-Sense Judgment system, with the aim of enabling computer-generated logical discourse. By sorting out the relationships between the object words of a text, we created a knowledge structure model for detecting illogical words. Furthermore, using this knowledge structure, we devised an illogical discourse judgment system. Using the method described in this report, we showed that this technique was able to very accurately judge illogical usages with 87% accuracy for physical sense, 91% for time, and 88% for quantity, thus demonstrating the effectiveness of the technique. By constructing a system capable of handling illogical discourses, machines can demonstrate to users that they possess logic, or common sense, and the capacity to understand discourses, thereby pushing machines one step closer to human-like conversation.

Acknowledgments. This work was supported with the Aid of Doshisha University's Research Promotion Fund.

References

1. Yoshimura, E., Watabe, H., Kawaoka, T.: An Automatic Enhancing Method of Greeting Sentences Using Association Knowledge Mechanism. Journal of Natural Language Processing 13(1), 117–141 (2006)
2. Kojima, K., Watabe, H., Kawaoka, T.: A Method of a Concept-base Construction for an Association System: Deciding Attribute Weights Based on the Degree of Attribute Reliability. Journal of Natural Language Processing 9(5), 93–110 (2002)
3. Watabe, H., Kawaoka, T.: Measuring Degree of Association between Concepts for Commonsense Judgements. Journal of Natural Language Processing 8(2), 39–54 (2001)
4. Tsuchiya, S., Watabe, H., Kawaoka, T.: A Sensuous Association Method Using an Association Mechanism for Natural Machine Conversation. In: Proc. of KES2006 Knowledge-Based Intelligent Information and Engineering Systems, pp. 1001–1010. Springer, Heidelberg (2006)
5. Tsuchiya, S., Watabe, H., Kawaoka, T.: A Time Judgement System Based on an Association Mechanism. In: Proc. of KES2005 (Knowledge-Based Intelligent Information and Engineering Systems), pp. 742–748. Springer, Heidelberg (2005)
6. Kikuyama, Y., Watabe, H., Kawaoka, T.: Construction of Quantitative Judgement Mechanism with scale of common knowledge. The Science Engineering Review of Doshisha University 41(1), 7–15 (2000)
7. Kawahara, D., Kurohashi, S.: Case Frame Compilation from the Web using High-Performance Computing. In: Proceedings of the 5th International Conference on Language Resources and Evaluation (2006)

A Query-Strategy-Focused Taxonomy and a Customizable Benchmarking Framework for Peer-to-Peer Information Retrieval Techniques

Alfredo Cuzzocrea

DEIS Department, University of Calabria, Italy
cuzzocrea@si.deis.unical.it

Abstract. P2P IR techniques are gaining momentum in both academic and industrial research communities, mainly due to the fact that they are extensively used-in-practice in a wide set of advanced applications ranging from e-business to e-government and e-procurement systems. P2P IR research is devoted to design innovative search strategies over P2P networks, whit the goal of making these strategies as more efficient and sophisticated as possible. In this respect, benchmarking P2P IR techniques is a leading aspect, and, at the same time, a non-trivial engagement as modeling the strongly decentralized nature and the rapidly-evolving dynamics of real-life P2P systems is still an open and un-completely solved research challenge. Starting from the proposal of a taxonomy of P2P IR techniques, which emphasizes the query strategy used to retrieve information and knowledge from peers, this paper focuses on a customizable benchmarking framework that allows us to study, analyze, and benchmark P2P IR techniques according to several useful metrics, and under the ranging of a number of input parameters. Finally, a comparative analysis of some state-of-the-art P2P IR techniques developed on top of the proposed frame-work is presented and discussed in detail. This analysis further confirms the effectiveness and the reliability of our benchmarking framework for P2P IR techniques.

Keywords: P2P Information Retrieval Techniques.

1 Introduction

P2P networks are natively built on top of a very-large repository of data objects (e.g., files), which is intrinsically distributed, fragmented, and partitioned among participant peers. P2P users are usually involved in (*i*) retrieving data objects containing information of interest, like video and audio files, and (*ii*) sharing information with other (participant) users/peers. From the *Information Retrieval* (IR) perspective, P2P users (*i*) typically submit short, loose queries by means of keywords derived from natural-language-style questions (e.g., "*find all the music files containing Mozart's compositions*" is posed through the keywords "*compositions*" and "*Mozart*"), and (*ii*), due to resource-sharing purposes, are usually interested in retrieving as result a set of data objects rather than only one. As a consequence, well-founded IR methodologies (e.g., rank), which have already reached a relevant degree of maturity, can be successfully applied in the context of P2P systems in order to improve the capabilities

B. Apolloni et al. (Eds.): KES 2007/WIRN 2007, Part I, LNAI 4692, pp. 729–739, 2007.

of these systems in retrieving useful information and knowledge, and achieve performance better than that of more traditional database-like query schemes. Furthermore, the consolidate IR mechanism naturally supports the "self-alimenting" nature of P2P systems, as in such a mechanism intermediate results can be then (re-)used to share new information, or to set and specialize new search activities. As regards schemas, from the database perspective, P2P users typically adopt a semi-structured (data) model for querying data objects rather than a structured (data) model. This feature poses unrecognized problems concerning the issue of integrating heterogeneous data sources over P2P networks. In addition to this, efficiently access data in P2P systems, which is an aspect directly related to both issues above, is still a relevant research challenge.

Basically, P2P IR techniques extend traditional functionalities of P2P systems (i.e., file sharing primitives and simple look-up mechanisms based on partial- or exact-match of strings), by enhancing the latter via useful (and more complex) knowledge representation and information extraction features. Accomplishment of the definition and the development of innovative knowledge delivery paradigms over P2P networks is the goal that underlies the idea of integrating IR techniques inside core layers of P2P networks. Therefore, in general, we can assert that P2P IR techniques can be effectively used to support even-complex processes like knowledge representation, discovery, and management over P2P networks, being the retrieval of information in the vest of appropriate sets of data objects the basic issue to be faced-off.

Nevertheless, several characteristics of P2P networks pose important limitations to the accomplishment of this goal. Among these, we recall (*i*) the completely decentralized nature of P2P networks, which enable peers and data objects to come and go at will, (*ii*) the absence of global or mediate schemas of data sources, which is very common in real-life P2P networks, and (*iii*) excessive computational overheads that could be introduced when traditional IR methodologies (such as those developed in the context of distribute databases) are applied as-they-are to the context of P2P systems. To overcome these limitations, P2P IR research is devoted to design innovative search strategies over P2P networks, whit the goal of making these strategies as more efficient and sophisticated as possible. In this respect, benchmarking P2P IR techniques is a leading aspect, and, at the same time, a non-trivial engagement as modeling the strongly decentralized nature and the rapidly-evolving dynamics of real-life P2P systems is still an open and un-completely solved research challenge. Therefore, with respect to the probing problem of modeling virtual P2P environments devoted to test the reliability of IR techniques, the most important requirement is to capture the main features of real-life P2P systems, as very often, due to the same nature of P2P networks, real-life P2P settings and dynamics do not match in-laboratory-built analytical models. Starting from these considerations, in this paper we first propose a taxonomy of state-of-the-art P2P IR techniques, which emphasizes the query strategy used to retrieve information and knowledge from peers (Sect. 2), and, then, we present a customizable benchmarking framework that allows us to (*i*) rigorously model critical factors of P2P systems such as distribution of peers and data-loads of the network, and (*ii*) meaningfully define several useful metrics mainly devoted to study the accuracy and the efficiency of the target P2P IR technique under the ranging of a number of input parameters (Sect. 3). Then, we complete our analytical contribution with an experimental study of some state-of-the-art P2P IR techniques developed on top of the proposed framework, which puts in evidence the effectiveness and the reliability of our framework (Sect. 4). Finally, we state conclusions and future work (Sect. 5).

2 A Query-Strategy-Focused Taxonomy of P2P IR Techniques

It is well established (e.g., [9]) that P2P systems are mainly characterized by their proper query strategies allowing us to retrieve useful information and knowledge in form of data objects (e.g., files). According to this vision, in this Section, we propose a query-strategy-focused taxonomy of state-of-the-art P2P IR techniques.

The first general classification distinguishes between *Keyword-based P2P* (KbP2PS) and *Object Identifier-based P2P* (OIDbP2PS) *Systems*, by looking at the atomic construct they use to drive the search mechanisms. In KbP2PS, traditional keywords are used to drive the search across peers, whereas, in OIDbP2PS, object identifiers are implemented on peers to enhance IR performance by biasing the search towards specific sub-sets of peers (in particular, due to the decentralized nature of P2P systems, object identifiers are usually embedded into distributed indexing data structures such as *Distributed Hash Tables* (DHTs)).

Among the *Basic Search Techniques* (BST – e.g., *Gnutella* [5]), the *Breadth First Search* (BFS) is one of the most popular ways of supporting IR over P2P networks. In BFS, a peer p_i receiving a query message q from a sender peer p_j first forwards q to all its neighboring peers, other than p_j, and then searches its local repository for relevant matches. Furthermore, if a peer p_k reached by q finds a match in its repository, it sends across the network the hit message along with (*i*) the identifiers needed to download from it the data objects of interest, and (*ii*) the state of its network connectivity. Finally, if p_j receives hit messages from more than one peer, it may decide to download the retrieved documents from peers on the basis of their network connectivity states. BFS performance is usually very poor due to an inefficient network resources utilization that generates a lot of service messages across the P2P network, and peers with low bandwidth can become serious bottlenecks for such search mechanism. However, equipping query messages with the *Time-To-Live* (TTL) parameter, which determines the maximum number of hops allowed for any query message, can limit network flooding and sensitively increase performance.

Random Search Techniques (RST) represent a simple yet effective derivation from BST. Kalogeraki *et al.* [6] propose a significant extension to the naïve version of BFS, called *Random Breadth First Search* (RBFS), which consists in propagating the query message from a peer to a randomly determined sub-set of its neighboring peers rather than all of them. A setting parameter establishes how many neighboring peers must be involved by the propagation of the query message (e.g., if the parameter is equal to 0.5, then the query message is propagated to half of all the neighboring peers, chosen at random). The major benefit of the RBFS approach consists in the fact that performance of the BFS approach is dramatically improved yet ensuring low computational overheads because of the random choice does not require any global knowledge. On the other hand, being RBFS a probabilistic technique, it could happen that large segments of the P2P network are neglected by the random choice, thus reducing the efficiency of the query task. Lv *et al.* [8] present the *Random Walkers Algorithm* (RWA), according to which each peer forwards the query message (called, in this context, *walker*) to another of its neighboring peers at random. To improve performance and reduce query time, the original idea of using one walker only is extended to the usage of $k > 1$ walkers, which are consecutively sent from the sender peer. RWA resembles RBFS but, indeed, in RBFS the query message is propagated to a sub-set of peers instead that at only one for time (as in RWA): as a consequence, in RBFS the number of service messages across the P2P network can become exponential, whereas in RWA such a number is bounded by a linear complexity [8].

Beyond the previous "basic" approaches, another line of P2P IR research aims at integrating intelligent techniques, perhaps inherited from similar experiences in related-but-different scientific disciplines, into P2P middleware as to enforce the quality of the search task. We name such a class of proposals as *Intelligent Search Techniques* (IST). The *Intelligent Search Mechanism* (ISM), proposed by Zeinalipour-Yazti *et al.* [11], belongs to the latter technique class, and represents a novel approach for supporting IR over P2P networks by (*i*) minimizing the number of messages sent among the peers, and (*ii*) minimizing the number of peers that are involved for each search request. To this end, ISM is composed by: (*i*) a *Profile Mechanism*, according to which each peer builds a "profile" for each of its neighboring peer; (*ii*) a *Query Similarity* function, which calculates the similarity queries to a new query; (*iii*) a *Relevance Rank*, which is a ranking technique for peers that takes as input the (neighboring) peer profiles, and produces as output a ranked list of (neighboring) peers used to bias the search towards the most relevant peers; and (*iv*) a *Search Mechanism*, which implements the ISM search policy. In [11], authors show how ISM works well (*i*) when peers hold some specialized knowledge about the P2P environment, and (*ii*) over P2P networks having high degrees of query locality; in these particular conditions, ISM outperforms BFS as well as RBFS techniques.

Statistics-based Search Techniques (SST) are another important result for IR over P2P networks. These techniques use some aggregated statistics to forward queries to a particular sub-set of peers, and, usually, the same statistics are maintained by mining results of past queries. Example techniques belonging to such a class are: (*i*) the *Most Results in Past* (>RES) heuristic, proposed by Yang and Garcia-Molina [10], where query messages are routed to those peers that returned the most results for the last *m* queries, being *m* a technique parameter (it should be noted that, in this case, the statistics employed is very simple being based on a "quantitative" approach); (*ii*) Galanis *et al.*'s data summaries and histograms [4], which are built on each peer by means of data replication techniques, in order to exploit such information at query time to bias the search towards the most relevant peers.

Index-based Search Techniques (IndST) efficiently exploit the hierarchical nature of structured P2P networks, and extensively use and take advantages from well-known data indexing solutions coming from the RDBMS technology (e.g., B^+-trees and *R*-trees). Among IndST proposals, we recall *XP2P*, proposed by us in [1], where lightweight XPath expressions are encoded in few KB by means of Rabin's fingerprints in order to build a (lightweight) distributed index for efficiently supporting look-up queries over structured P2P networks. Just like SST, even if query capabilities are improved and well-supported, also including new paradigms that were missing in first-generation P2P systems (such as range queries), IndST mainly suffer from scalability limitations, and updating distributed indexes over large P2P networks is still an open problem.

Semantics-based Search Techniques (SemST) are the new frontier for IR over P2P networks. Such techniques aim at adopting formal semantics to both model and query distributed resources over P2P networks, in order to improve the capabilities of traditional resource-sharing P2P systems. The first advantage of SemST is the amenity of re-using and re-adopting well-founded results coming from semantic models and query languages. Another advantage consists in the possibility of meaningfully integrating IR techniques in P2P networks with leading new research trends like *Ontologies* and *Semantic Web*. Since most papers are still focused on query performance of unstructured and structured P2P systems, in literature, there are very few proposals addressing the described research challenges. However, it is expected that integrating semantics in P2P networks will be one of the most relevant research topic for next-generation P2P applications.

In this context, Crespo and Garcia-Molina propose the notion of *Semantics Overlay Networks* (SON) [2], which are an efficient way of grouping together peers sharing the same schema information. Thus, peers having one or more topics on the same *thematic hierarchy* belong to the same SON. This approach well-supports query routing as every peer p_i can quickly identify peers containing relevant information, namely the set $N(p_i)$, by avoiding network flooding. Here, "relevant" means that a certain semantic relation exists between information held in p_i and information held in peers belonging to $N(p_i)$. Such semantic hierarchies are naturally represented (and processed) via the *Resource Description Framework* (RDF), by also taking advantages from several declarative languages for querying and defining views over RDF bases. In [2], authors demonstrate that SON can significantly improve query performance while at the same time allow users to decide what content to publish in their (peer) hosts, i.e. how to form a SON. A possible limitation of SON is represented by the overwhelming volume of messages that can be generated for supporting data object replications on peers, as required by SON design guidelines [2]. Thus, P2P applications running on top of SON-based model for query route incur excessive overheads on network traffic. An interesting solution to this problem has been proposed by Li *et al.* [7]: they suggest using *signatures on neighboring peers* for directing searches along selected *network paths*, and introduce some schemes to facilitate efficient searching of data objects. Signatures are a way of adding semantics to data, by building a bit vector V; V is generated according to the following two steps: (*i*) hashing the content of a data object into bit strings, said BS, and (*ii*) applying a bitwise OR operator on BS. The so-built bit strings are used at query time by performing a bitwise AND operation on the *search signature* (i.e., the signature of the term used as search key) and the *data signature* (i.e., the signature stored on the *current* peer). In [7], authors show how some proposed flooding-based search algorithms allow the signatures of the neighboring peers to be efficiently exploited for enhancing search results, and, moreover, an extensive experimental part clearly confirms the effectiveness of the neighborhood signature technique.

Finally, in [3], we propose an innovative semantics-based framework for supporting KD- and IR-style resource querying on large scale P2P XML repositories (e.g., those that one can find in corporate B2B and B2C *e*-commerce systems). In more detail, such a framework is based on (*i*) modeling both XML repositories/ documents and queries in terms of the concepts they express by means of formal reasoning flat models, like lists, and hierarchical models, like graphs, and (*ii*) applying ad-hoc knowledge extraction algorithms that efficiently exploit such models. Specifically, these algorithms enhance the semantic expressiveness of the reasoning task by exploiting the *local knowledge* given by mining, according to some meaningful two-dimensional abstractions, past (successful) query results flooded through neighboring peers [3].

3 A Customizable Benchmarking Framework for P2P IR Techniques

As discussed in Sect. 1, another goal of this paper is presenting our customizable benchmarking framework for P2P IR techniques, where several aspects of real-life P2P systems are captured and modeled, along with an experimental study of some state-of-the-art proposals which allows us to show the effectiveness and the reliability of our proposed framework. The main feature of this framework is the capability of capturing several aspects of real-life P2P systems. Furthermore, as we better discuss next, our framework also allows any query strategy to be implemented and integrated inside the core information and knowledge extraction layer, thus being completely "customizable" in dependence on the needs of the target P2P IR technique.

In our framework (see Fig. 1), we implement a P2P environment simulator composed by (*i*) a *P2P Layer*, where peers are modeled, and (*ii*) a *Data Layer*, where, without any loss of generality, Internet-like XML repositories are modeled. We highlight that, being our framework designed and used to measure the *accuracy* and the *efficiency* of P2P IR techniques rather than the *performance* of the techniques (e.g., measuring the absolute time needed for evaluating a given query), in our setting we do not take into consideration other even-important parameters for P2P simulators such as signal propagation delay etc. It should be noted that the kind of analysis supported by our benchmarking framework is consistent with a plethora of research experiences on P2P IR of the literature (e.g., [9]).

As regards the data support of our framework, we employ a modified version of the popular synthetic XML data set generator *XMark*, where an ad-hoc built thesaurus, called *ContentThesaurus*, replaces the original XMark Shakespeare's literature contents with context-oriented contents (e.g., *e*-insurance context, or *e*-government context etc). Furthermore, in such a data support solution, concepts expressed by contents stored in *ContentThesaurus* are materialized within another specific thesaurus, called *ConceptThesaurus*, in the form required by the specific P2P IR technique (e.g., flat, hierarchical etc). Besides, a very simple cluster-based indexing data structure is in charge of maintaining pointers between concepts in *ConceptThesaurus* and related contents in *ContentThesaurus*. Among other well-understood and accepted advantages, usage of XMark inside our framework allows us to generate synthetic XML repositories having different structure and size, thus making experiments built upon the framework close to scenarios drawn by real-life P2P applications.

One of the first problems we addressed in modeling the P2P environment of our framework was about how to distribute peers over the network. Regarding this aspect, we chose a network topology such that peers are uniformly distributed over the network. This choice was driven by the literature evidence claiming that several previous experiences focused on similar experimental methodologies (e.g., [7,8,10]) have shown that this configuration is the "best-in-laboratory" one. In doing this, a two-dimensional representation of the whole P2P network is employed, in order to identify each peer p by means of a unique Cartesian coordinate $\langle X_p,Y_p\rangle$ with respect to a two-dimensional reference scheme P related to the target network. Then, the Cartesian coordinates are obtained by means of a Uniform distribution u, thus generating a (virtual) P2P network with uniformly distributed peers. To give an example, if u is defined on the range $[a, b]$, such that $a < b$, then each Cartesian coordinate $\langle X^u{}_p,Y^u{}_p\rangle$ generated by u is such that $a \leq X^u{}_p \leq b$ and $a \leq Y^u{}_p \leq b$. In our benchmarking framework, this strategy is implemented inside the component *PeerUnDist* (see Fig. 1).

Since the Data Layer is another critical component of P2P systems, in our framework we also implement the component *RepDist*, which is able to populate the (virtual) P2P environment (i.e., to distribute the XML repositories across the target P2P network) according to an input, customizable distribution, in a similar way to what happens with the modeling of the network topology. In more detail, given a domain of XMark-generated XML repositories D, and a distribution f, *RepDist* "puts" XML repositories in D over peers in the target (virtual) P2P network by extracting via f synthetic Cartesian coordinates $\langle X^f{}_p,Y^f{}_p\rangle$ from the two-dimensional reference scheme P. At the end of this process, we obtain that the XML repositories in D are distributed over the P2P network according to f, i.e. according to the data-loads shaped by f. As regards distributions used to populate the P2P environment, we exploit the Uniform (similarly to the previous case) and the Zipf ones because of, as widely-established, they meaningfully represent the two opposite situations where, in the first case, data are uniformly distributed over the P2P network, and, in the second case, data are non-uniformly distributed over the P2P network, and asymmetric data-loads are present. However, any distribution can be used inside our proposed framework, as we make available a distribution repository (see Fig. 1) to enhance the experimental validation of the target P2P IR technique by generating different kinds of data-loads, and observing the behavior of the technique under different application scenarios, thus taking advantages from a cross-analysis-oriented experimental methodology.

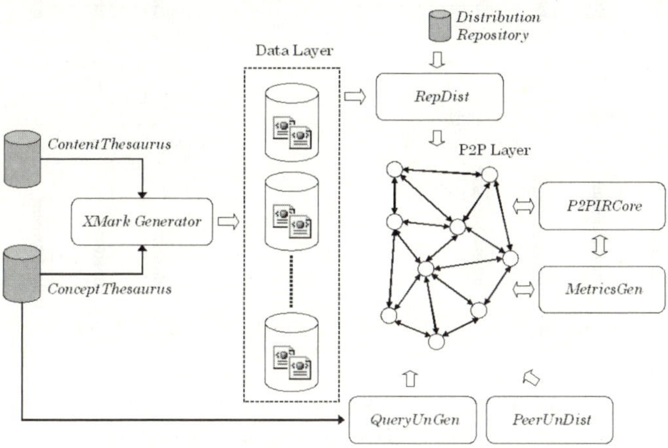

Fig. 1. A customizable benchmarking framework for P2P IR techniques

As regards modeling the input, in our framework, queries are simply modeled by means of uniform extractions over *ConceptThesaurus*, implemented inside the component *QueryUnGen* (see Fig. 1), thus obtaining random lists of concepts. However, this component can be further specialized to capture more complex class of queries arising from the dynamics of real-life P2P systems (e.g., queries including complex search predicates, such as Boolean queries).

P2P IR technique-wise, to ensure flexibility, the P2P IR policy is also maintained completely customizable within our benchmarking framework; to this end, the component *P2PIRCore* defines a neutral software (abstract) interface codifying the most popular methods which are common to any P2P IR technique (such as, for instance, `selectNextPeer`, `rank` etc), and, as a consequence, the comparison P2P IR techniques are codified as implementing this interface, plus, by exploiting the amenities of object-oriented-languages' inheritance, other possible (more specific) methods. Finally, the component *MetricsGen* (see Fig. 1) is in-charge-of generating the metrics of the target experiments, according to the specific goals of the actual analysis; just like *P2PIRCore*, *MetricsGen* is also totally customizable for different purposes.

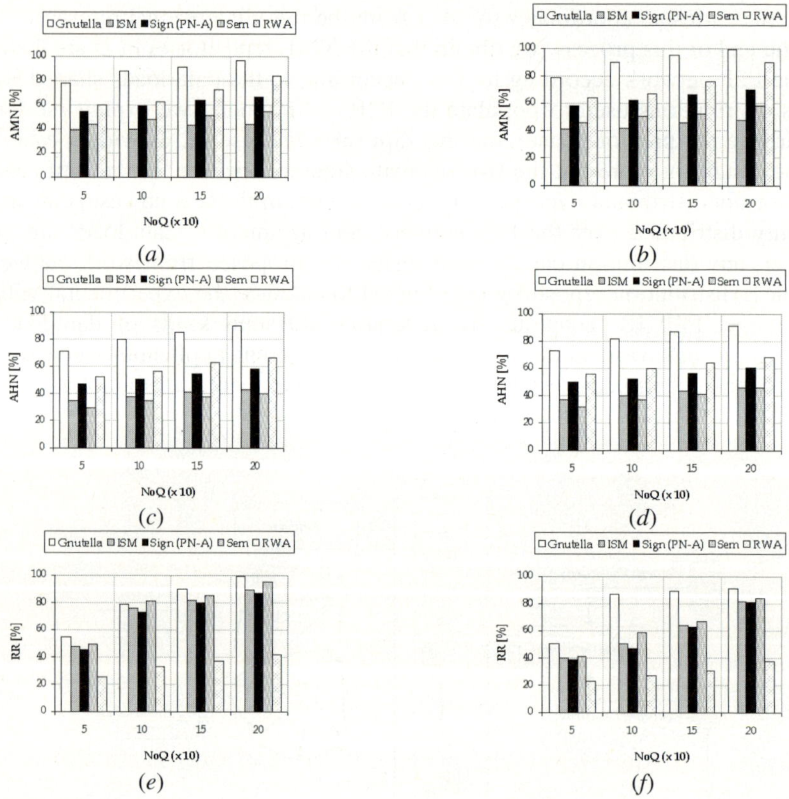

Fig. 2. Experimental results (TTL = 5): AMN w.r.t. NoQ for Uniform data-loads (*a*); AMN w.r.t. NoQ for Zipfian data-loads (*b*); AHN w.r.t. NoQ for Uniform data-loads (*c*); AHN w.r.t. NoQ for Zipfian data-loads (*d*); RR w.r.t. NoQ for Uniform data-loads (*e*); RR w.r.t. NoQ for Zipfian data-loads (*f*)

4 Experimental Study

In order to test the effectiveness and the reliability of our proposed benchmarking framework, we performed a comparative analysis of some state-of-the-art P2P IR techniques developed on top of the framework. In our experiments, the hardware infrastructure was composed by 10 workstations interconnected with a 10/100 LAN, each of them equipped with a processor *AMD Athlon XP 2600+* at 2.133 GHz and 1 GB RAM, and running *SUSE Linux 9.1*. On the top of our P2P benchmarking framework, we deployed 200 peers, each of them running like a single peer on a single host, thanks to the multi-threading programming environment offered by the underlying software platform. The data layer of our experimental setting was obtained by producing synthetic contents and concepts concerning with *e*-tourism P2P XML repositories; these repositories contained related data from which useful knowledge such as information for reaching tourist places, locations of hotels and restaurants, train and bus timetables, information on historical/archeological sites, information on gastronomic tours, information on events and movies etc can be extracted. By exploiting our modified version of the XML data set generator XMark, we obtained several (synthetic) XML repositories having various structure (in particular, the maximum depth and the maximum width of the XML documents ranged in the intervals [5, 15] and [50, 650] respectively) and size (in particular, the size of the XML documents ranged in the interval [1, 45] MB). To model the data-loads of the P2P network, we used the Uniform distribution defined over the range [0, 199], and the Zipf distribution with characteristic parameter z equal to 0.5. Synthetic queries were generated by means of uniform extractions over *ConceptThesaurus*, by making use of concepts coming from the tourism context like as "destination", "holiday", "social event" etc.

As comparison P2P IR techniques, we chose the following ones: Gnutella [5], RWA [8], ISM [11], the neighborhood signature technique [7] (particularly, in this case we used the PN-A scheme – see [7] for further details), and our semantics-based framework for P2P KD and IR [3] (which we name as "*Sem*" in this paper).

In our experiments, we considered the following metrics: (*i*) the *Average Message Number* (AMN), which measures the average number of messages used to retrieve documents of interest; (*ii*) the *Average Hop Number* (AHN), which measures the average number of hops needed to retrieve documents of interest; (*iii*) the *Recall Rate* (RR), which is the fraction of retrieved documents with respect to the collection of all the documents involved by a given search task. It should be noted that AMN and AHN mainly test the efficiency of the target P2P IR technique, whereas RR is used to investigate on the accuracy of the target P2P IR technique. For what instead regards to the independent parameters of the experimental analysis (i.e., the parameters with respect to which the previous metrics are generated and observed), we considered: (*i*) the *Number of Queries* (NoQ) used to retrieve documents of interest; (*ii*) the *Number of Peers* (NoP) populating the network; (*iii*) the "classical" TTL. It should be noted that the parameter NoQ influences, from a side, the quality of the search results as, in general, the higher is the number of queries employed during the search task, the higher is the quality of the retrieved results; from another side, such as parameter is also a way of evaluating the *reliability* of the investigated techniques, as, in general, the higher is the number of queries employed in the search task, the higher are the computational overheads introduced in the P2P network, as a high volume of query (route) messages is needed. In addition to this, the parameter NoP allows us to study the performance of the target P2P IR technique under the ranging of the number of peers populating the network, i.e. to probe the *scalability* of the technique. Finally, the parameter TTL gives us the opportunity of studying how to efficiently bound the network flooding phenomenon that can occur in the target P2P IR technique. All these metrics and independent parameters were codified inside the component *MetricsGen* (see Fig. 1), by effectively taking advantages from the "extendibility" of our proposed framework.

Fig. 2 shows the results of our experimental study for both Uniform and Zipfian data-loads. All considering, from the analysis of such results, it follows that ISM and *Sem* present the best behaviors, with different performance on the AMN (where ISM is better) and AHN (where *Sem* is better) metrics. Therefore, our conclusion about the experimental study is that, depending on the particular application context, and by trade-offing the two benefits (i.e., low AMN, or low AHN), one can decide to choose between the two different solutions. Nevertheless, as regards the definitive goal of both the techniques (i.e., retrieving as more documents of interest as possible), the RR-based metrics confirms the goodness of *Sem* against ISM.

5 Conclusions and Future Work

In this paper, we have proposed a query-strategy-focused taxonomy of P2P IR techniques, and a customizable benchmarking framework that allows us to experimentally study the features of any P2P IR technique by capturing several aspects of real-life P2P systems. This framework is also capable of modeling the data layer of the virtual P2P environment according to customizable input distributions (e.g., Uniform, Zipf etc), thus stressing the query capabilities of the target P2P IR technique under different data-loads. Another benefit coming from the proposed framework consists in supporting various useful metrics, mainly devoted to test the accuracy and the efficiency of the target P2P IR technique, which can also be personalized in dependence on the actual analysis. Finally, an experimental study of some state-of-the-art P2P IR techniques based on the proposed framework has been presented, along with its experimental results, which have further confirmed the effectiveness and the reliability of our framework. Future work is oriented at making our framework able to deal with the problem of studying how updates that can occur in the data layer of P2P networks influence the performance of the target P2P IR technique, which is a leading aspect to be captured when a P2P IR technique is benchmarked.

References

[1] Bonifati, A., Cuzzocrea, A.: Storing and Retrieving XPath Fragments in Structured P2P Networks. Data & Knowledge Engineering 59(2), 247–269 (2007)

[2] Crespo, A., et al.: Semantic Overlay Networks for P2P Systems. Stanford TR, Stanford University (2003)

[3] Cuzzocrea, A.: On Semantically-Augmented XML-based P2P Information Systems. In: Larsen, H.L., Pasi, G., Ortiz-Arroyo, D., Andreasen, T., Christiansen, H. (eds.) FQAS 2006. LNCS (LNAI), vol. 4027, pp. 441–457. Springer, Heidelberg (2006)
[4] Galanis, L., et al.: Locating Data Sources in Large Distributed Systems. In: Aberer, K., Koubarakis, M., Kalogeraki, V. (eds.) Databases, Information Systems, and Peer-to-Peer Computing. LNCS, vol. 2944, pp. 874–885. Springer, Heidelberg (2004)
[5] The Gnutella File Sharing System, http://gnutella.wego.com
[6] Kalogeraki, V., et al.: A Local Search Mechanism for Peer-to-Peer Networks. In: Proc. of CIKM, pp. 300–307 (2002)
[7] Li, M., et al.: Neighborhood Signatures for Searching P2P Networks. In: Proc. of IDEAS, pp. 149–159 (2003)
[8] Lv, Q., et al.: Search and Replication in Unstructured Peer-to-Peer Networks. In: Proc. of ICS, pp. 84–95 (2002)
[9] Tsoumakos, D., et al.: A Comparison of Peer-to-Peer Search Methods. In: Proc. of WebDB, pp. 61–66 (2003)
[10] Yang, B., et al.: Efficient Search in Peer-to-Peer Networks. In: Proc. of ICDCS, pp. 5–14 (2002)
[11] Zeinalipour-Yazti, D., et al.: Exploiting Locality for Scalable Information Retrieval in Peer-to-Peer Systems. Information Systems 30(4), 277–298 (2005)

An Approach for Four Way Set Associative Multilevel CMOS Cache Memory

Prasanna Palsodkar[1], Amol Deshmukh[2], Preeti Bajaj[3], and A.G. Keskar[4]

[1] Research Associate, ETRX Dept, G.H.Raisoni College of Engineering, Nagpur, India
prasanna_palsodkar@yahoo.co.in
[2] Asst.Professor & Research Associate, ETRX Dept, G.H.Raisoni College of Engineering, Nagpur, India
amolydeshmukh@yahoo.com
[3] Professor & Head, ETRX Dept, G.H.Raisoni College of Engineering, Nagpur, India
preetib123@yahoo.com
[4] Dean R & D, Visvesvaraya National Inst. Of Technology, Nagpur, India

Abstract. The approach for design of four way set associative multilevel CMOS cache memory is discussed here. The cache hierarchy, organization and cache structure has been discussed. Apart from these the two levels i.e. level 1 & level 2 of cache memory design approach are discussed. The type of cache memory organization to be used is the major part design. The approach for the design of CMOS cache memory uses set associative mapping over the other cache organization as set associative mapping uses several direct-mapped caches which is referred as set. This four way set associative cache memory can be used for VLSI systems in computer and wireless communication systems.

1 Introduction

The increased demand for the high data transfer rate in recent times leads to the success of these systems in many applications. The main reason for the success is the high data transfer rate which can be achieved by using cache memory in between the two systems which communicates with each other very oftenly.

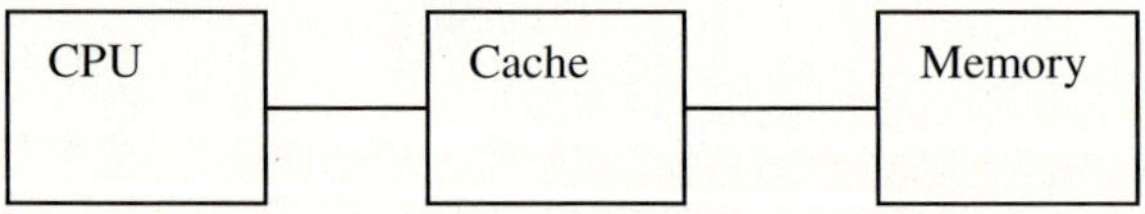

Fig. 1. Cache based memory system

Figure 1 shows the cache based memory system where a cache is placed between CPU and memory. The system first copies the data needed by the CPU from memory into the cache, and then from the cache into a register in the CPU. If an application needs the same data again, data access time is reduced significantly if the data is still in the cache [1].

Figure 2 shows the four way set associative cache memory system. A set-associative cache design uses several direct-mapped caches. Each cache is often referred to as a set. On an incoming request, the cache controller decides which set the line will go into. Within the set, a direct-mapped scheme is used to allocate a slot in the cache [2].

B. Apolloni et al. (Eds.): KES 2007/WIRN 2007, Part I, LNAI 4692, pp. 740–746, 2007.

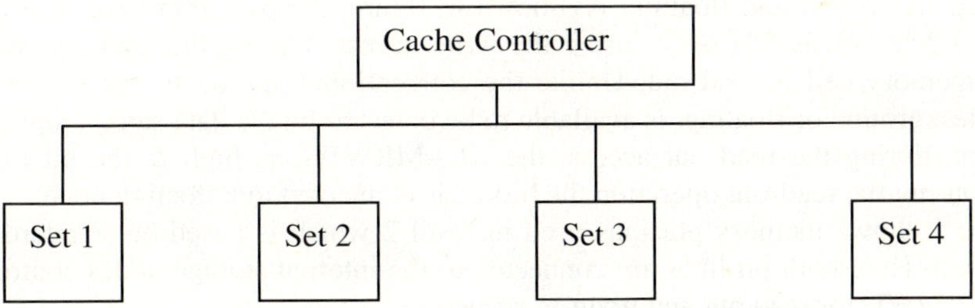

Fig. 2. Four Way Set Associative Cache memory system

2 Methodology

Figure 3 shows the logical organization of set associative cache. This system is called as set associative because the system is partitioned into distinct set of blocks. The sets are represented by the rows. In this the cache has N sets and each set has four blocks. When an access occurs to this cache, the cache controller does not search the entire cache looking for a match instead, the controller maps the address to a particular set of the cache and searches only the set for a match. Because the search is conducted over four blocks, the cache is said to be four-way set associative [5].

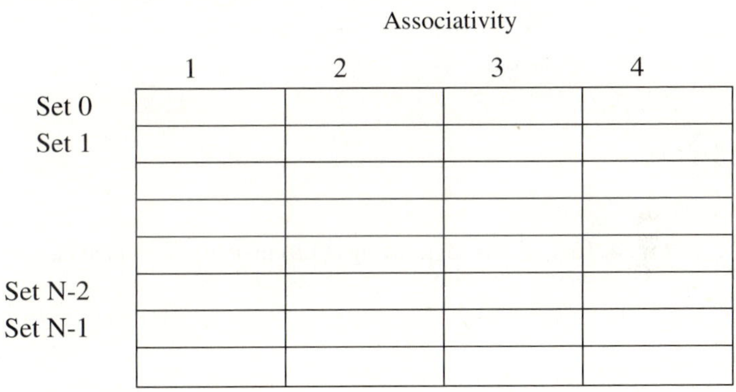

Fig. 3. A Logical organization of four way set associative cache

3 Design Approach

This paper reports a four way set-associative two-level CMOS cache memory. This cache memory is designed to have the memory portions, the tag portions in two levels.

3.1 Memory Portion

Figure 4 shows the 10-T shrunk logic swing (SLS) memory cell in the L1 As shown in figure 4, the L1 10-T SLS memory cell is targeted for G/F logic operation. In order to reduce the switching time, the SLS memory cell is designed for operation with logic state of ground and floating. As shown in figure, the pass transistor controlled by the L1-MWWL is ON only during the write access. During the read access, the data in memory cell is read out. Unlike the conventional approach, one of the two logic gates, ground or floating, is available to be detected by the data sense amplifiers. Therefore, during the read out access, the L1 –MRWWL is high & the bit line is raised. During the read-out operation the bitline is connected to a floating node.

Figure 5 shows memory portions used in level 2 which is based on standard 6-T SRAM cell. Here both bit lines are connected to the internal storage nodes controlled by the L2-MWL for read out and write-in accesses.

A 6-T SRAM cell uses a pair of cross-coupled inverters as its bi-stable storage element with two additional NMOS devices for read and writes access. The bit lines (BL and /BL) have been precharged to a reference voltage, and the cell drives a differential current onto the bit lines according to the stored value [4].

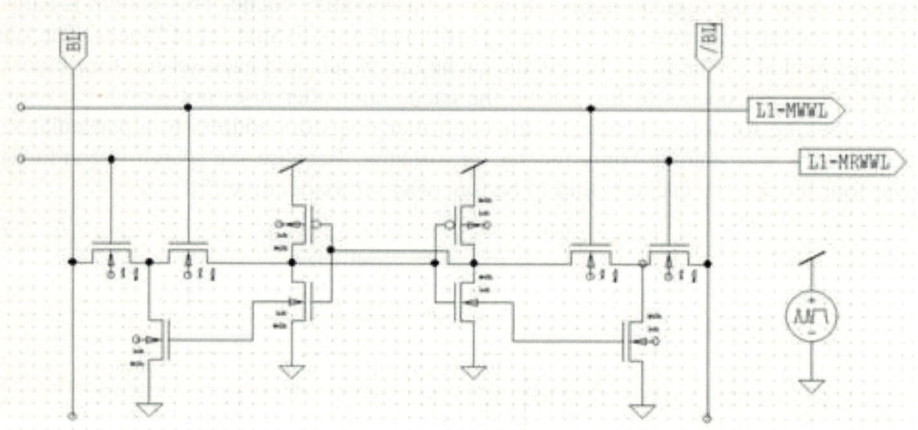

Fig. 4. 10-T shrunk logic swing (SLS) memory cell in the L1

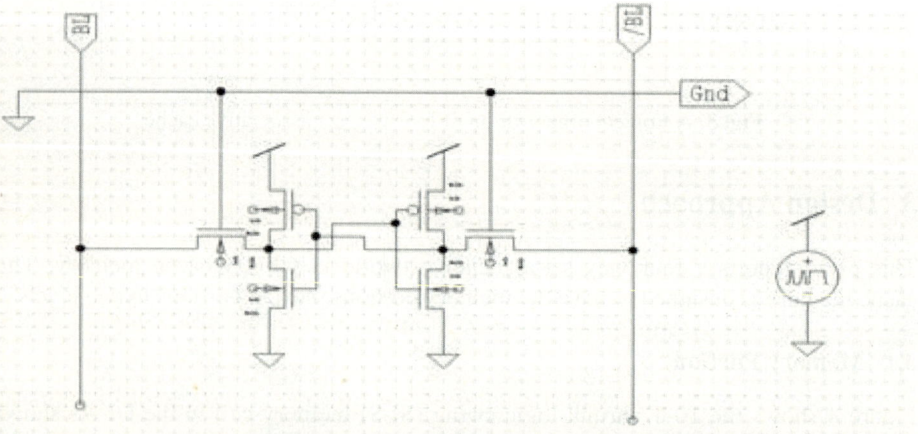

Fig. 5. 6-T shrunk logic swing (SLS) memory cell in the L2

3.2 Tag Portion

Tag portion of level 1 consists of decoder circuit, WLOTC tag compare scheme and memory driver circuit.

Figure 6 shows the decoder circuit used to generate the write-enable word line (WEWL) and read-enable word line (REWL).

The other important tag related circuit is the WLOTC tag sense amplifier as shown in figure 7 [3].

It is used to provide the miss signal and the TRWL, which are based on the output of decoder circuit.

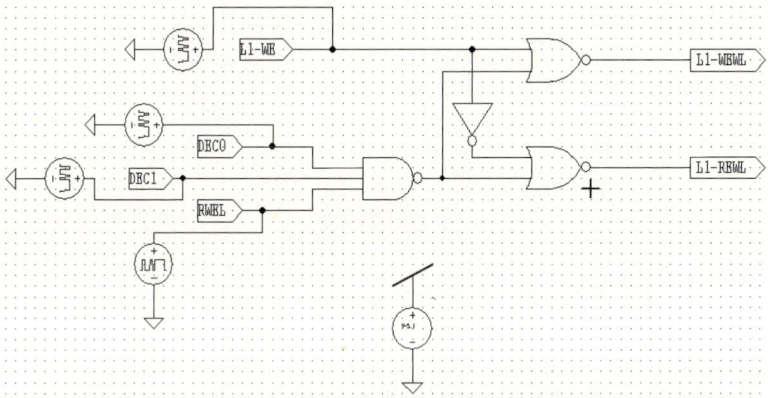

Fig. 6. Decoder circuit used in level 1

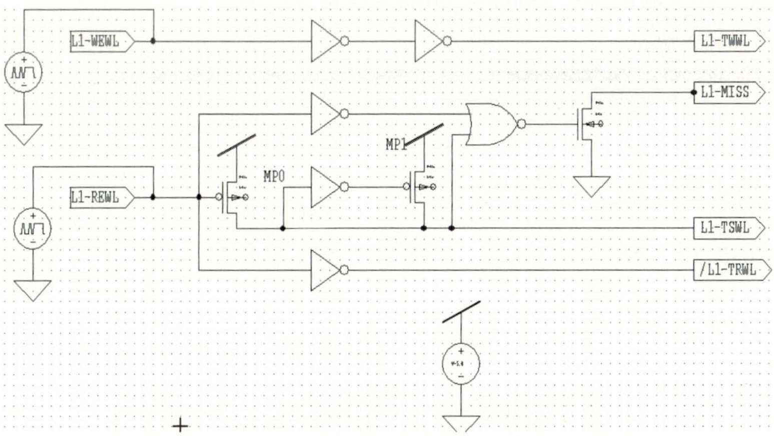

Fig. 7. Word line oriented tag compare of L1

4 Simulation Results

Figure 8 shows the simulation result of 10- T shrunk logic swing memory cell used in level 1.This shows that when writing in the memory cell is in process, the read access in not allowed in that particular memory portion. This 10-T SLS memory cell with a G/F data sense amp in Level 1 can be used for high-speed operation for low-voltage low-power VLSI system applications.

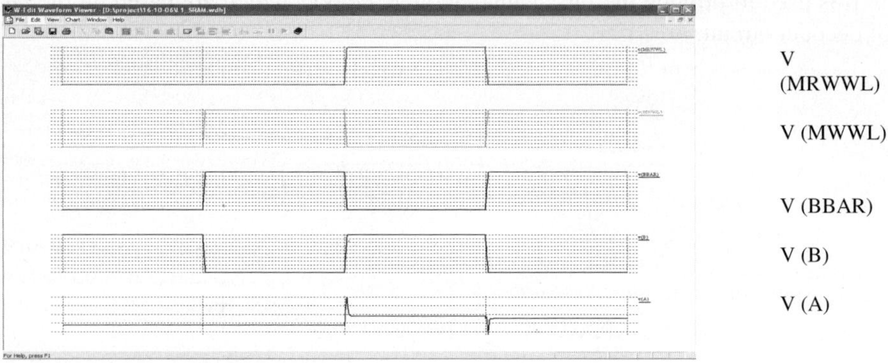

Fig. 8. Result of 10-T SLS memory cell used in Level 1

Figure 9 shows the simulation result of 6-T shrunk logic swing memory cell used in level 2. By using 6-T SLS memory cell in level 2, reduces parasitic capacitances. Here both bit lines V (A) & V (B) are connected to the internal storage node which continuously checks the status of the memory whether a hit signal or a miss signal has been generated and it allows the operation accordingly [8].

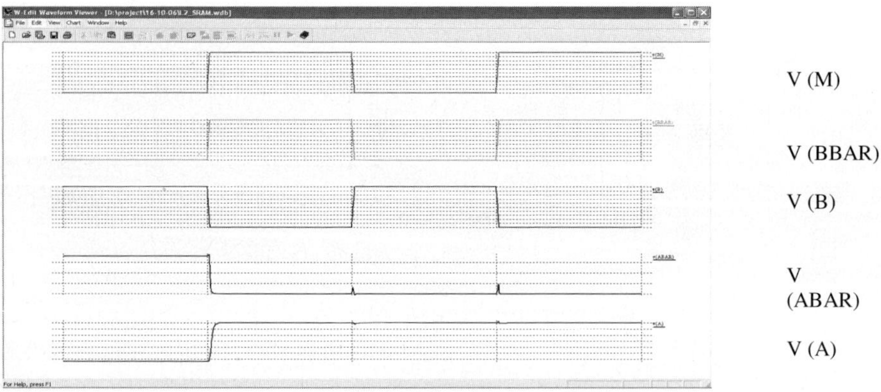

Fig. 9. Result of 6-T SLS memory cell used in Level 2

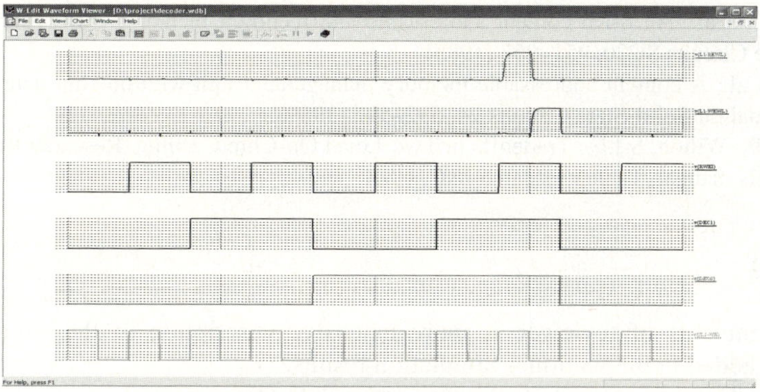

Fig. 10. Result of Decoder circuit used in Level 1

Figure 10 shows the simulation result of decoder circuit used in level 1. This decoder is used to generate write- enable word line (L1-WEWL) and read-enable word line signal (L1-REWL). The L1-WEWL and L1-REWL signal is generated depending on the status of the read-write enable line [9].

5 Conclusion

This approach elaborates the design issues related with various circuits of memory portion, tag portion and shrunk logic swing memory cell used in level 1 and level 2. The outputs of the decoder circuit acts as an input to the WLOTC tag sense amplifier. The memory driver circuit is used to generate memory read and memory write signal depending on the output of the WLOTC tag sense amplifier. This cache memory reduces the accessing time of memory portion as indicated in the simulation results.

References

[1] Kuo, J.B., et al.: Low-Voltage CMOS VLSI Circuits. John Wiley, New York (1999)

[2] Lin, P.-F.: A 0.8-V 128Kb Four-Way Set-Associative Two-Level CMOS Cache Memory Using Two-Stage Wordline/Bitline-oriented Tag-Compare (WLOTC/BLOTC) Scheme. IEEE Journal Of Solid-State Circuits 37(10) (October 2002)

[3] Mizuno, H.: A 1-V, 100-MHz, 10-mW Cache Using a Separated Bit-Line Memory Hierarchy Architecture and Domino Tag Comparators. IEEE Journal Of Solid-State Circuits 31(11) (1996)

[4] Kuo, J.B.: A 1-V 128-Kb Four-Way Set-Associate CMOS Cache Memory Using Wordline-Oriented Tag-Compare (WLOTC) Structure with the Content-Addressable-Memory (CAM) 10-Transistor Tag Cell. IEEE Journal Of Solid-State Circuits 36(04) (2001)

[5] Sirisantana, N., Wei, L., Roy, K.: School of Electrical and Computer Engineering, Purdue University *Intel Corp., Hillsboro. In: The Proceedings of the 2000 IEEE International Conference on Computer Design (2000)

[6] kadota, H., et al.: An 8-Kb content addressable and reentrant memory. IEEE Journal Of Solid-State Circuits SC20, 951–957 (1985)
[7] Godsel., et al.: A content addressable memory management unit with on chip data cache. IEEE Journal Of Solid-State Circuits 24, 592–596 (1989)
[8] Jouppi, N.P., Wilton, S.J.E.: Tradeoffs in Two-Level On-Chip Caching. Research Report
[9] Tanner tools manual, Tanner research INC, USA

Comments:

1. As per comments of reviewer 1, more extensive description of the simulation results are added in the section 4 Simulation results.
2. The more references are included in the section 6 References.
3. As per reviewer 2, experimental section has been improved and added in the section 4 Simulation results.

An Intelligent Typhoon Damage Prediction System from Aerial Photographs*

Chien-Chang Hsu and Zhi-Yu Hong

Department of Computer Science and Information Engineering,
Fu-Jen Catholic Universit,
510 Chung Cheng Rd., Hsinchuang, Taipei, Taiwan 242

Abstract. Damage prediction of typhoons or hurricanes is an intractable task. It needs complete meteorological information and disaster condition evaluation. It also depends heavily on the experiences of the government and evaluator. Different angles and characteristics of typhoons may cause different degrees of damage. It is a difficult task for the government to predict the possible damages correctly. This paper proposes an intelligent typhoon damage prediction system from aerial photographs. The system uses wavelet transformation, support vector machines, and fuzzy neural networks for image compression, classification, and error correction. The system then uses case-based reasoning and fuzzy damage measurement to compute possible damages to agricultural products and inhabitants. Jaiosi, Taiwan is a place used by the system as an example to illustrate its functionality. The experiment shows that the system can find important surface features from aerial photographs as well as predict possible typhoon damage correctly.

Keywords: Typhoon damage prediction, Case-based reasoning, Support vector machines, Fuzzy neural networks.

1 Introduction

Typhoons are the most fearful disasters in island countries especially in near-sea nations. Heavy rains, strong winds, mudflows, landslides, and floodwaters usually cause serious damages to inhabitants as well as large amounts of damage to agricultural products every year. For example, Hurricane Katrina devastated Southern America in 2005. It caused serious damages to America as well as world economy. Incorrect damage prediction and rescue strategies cause the additional infliction to inhabitants. Reducing the damage to the wealth and life of inhabits is an urgent issue during typhoon impact. Correct estimation of typhoon damage is a difficult task for government officials. Finding and predicting possible damages earlier is very important problem for the government so the right rescue strategy can be taken. Many meteorologists engaged in the prediction task of typhoon track and influenced region. There are many systems proposed to analyze and forecast typhoon information, such

* This work is partly supported by National Science Council of ROC under grants NSC 95-2745-E-030-004- URD.

B. Apolloni et al. (Eds.): KES 2007/WIRN 2007, Part I, LNAI 4692, pp. 747–756, 2007.

as track analysis and numeric model construction. For example, R. Sakai uses a numeric prediction model for tropical cyclone prediction [14]. K. S. Chen uses microwave observation from different satellites and ground radar data to construct the typhoon model [2]. T. L. Lee uses back propagation neural networks to predict the storm surge [10]. A. Kitamoto uses data mining and principle component analysis to analyze and predict typhoon track according to satellite image [7].

However, typhoon impacts usually cause serious damage in the affected areas. Different angles and energy characteristics of the typhoon may cause different influences. Few of the systems have been proposed to predict damages to man-made architectures and farm products by typhoons. Case-based reasoning (CBR) is a problem solving methodology that uses past experiences to solve similar problems [8], [11]. It can help to predict possible impact by incorporating different sources of information. Topography, meteorology, and agriculture can be used to predict typhoon disaster. Surface feature recognition from aerial photographs can find useful information on the ground. However, aerial photographs contain too many surface features in the photo. It needs to decompose the aerial photograph into different sub-images for feature recognition. It also needs to compress the image into a more compact format without losing information. Wavelet transformation and SVM are considered as the newest and the better approach to analyzing high dimension photographs. Wavelet transformation can extract local information layer by layer from the raw image for image compression and noise filtering. SVM can find global minimum in geometric space and classify patterns. It can be used as a feature classifier to recognize surface features in the photograph. For example, F. Melgani uses SVM in hyperspectral dataset to classify the data of a scene taken over northwest Indian pines by the AVIRIS sensor [12]. L. Cui uses wavelet transformation to compress the audio image [3]. B. Wang uses wavelet transformation to decompose an image into multi-channel images [16]. S. Li integrates wavelet transformation into SVM to classify texture images. Moreover, feature smoothing is also an important problem in the post-processing of segmented image analysis. It needs to eliminate misclassified features and smooth the features in the recombined photograph. The fuzzy neural filter can be used as post-processor to filter noise information [17]. The agricultural database can be used as the information source for agricultural damage prediction in agriculture. The workers in meteorology can provide typhoon forecast information.

This paper proposes an intelligent typhoon damage prediction system from aerial photographs. The system contains two modules, feature extractor and damage predictor. Feature extractor uses wavelet transformation, SVM, and fuzzy neural network for image reduction and feature extraction from the aerial photographs. Wavelet transformation is used to filter noisy data and decompose images of aerial photographs. SVM is used as a feature classifier to recognize surface features of imagines into sketched graphs. Fuzzy neural network is used to filtering and smoothing sketched graphs for CBR. Damage predictor uses CBR to predict human and crop damage. The typhoon damage prediction system can use agricultural data, surface features of the ground, and past typhoon cases to predict possible damage.

The rest of this paper is organized as follows. Section 2 introduces the architecture of the intelligent typhoon damage prediction system. Section 3 uses the Jaiosi, a city in Taiwan, as a model application to illustrate the functionality of the system. Finally, section 4 concludes the work.

2 System Architecture

Figure 1 shows the architecture of the case-based typhoon damage prediction system. The system contains two components, namely, feature extractor and damage predictor. Feature extractor uses wavelet transformation to reduce the dimension of aerial photographs. SVM is used to classify surface features of the earth. Fuzzy neural networks are used as noise filters to enhance damage prediction accuracy for CBR. Damage predictor uses typhoon case library, crop data, and typhoon forecast to predict typhoon damage by CBR. Moreover, it uses fuzzy damage measure to compute possible damage. The following section provides more details on each module.

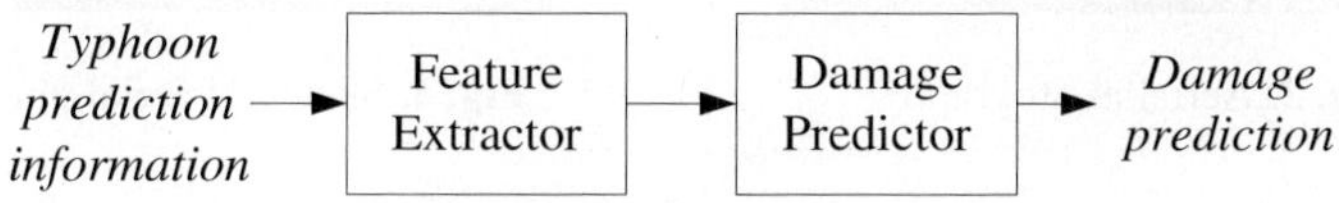

Fig. 1. System architecture

2.1 Feature Extractor

Feature extractor conducts three main tasks, which are, image decomposition and compression, surface feature recognition, and noise filtering. Basically, image decomposition and compression divides the aerial photograph into many sub-images with 16-by-16 gray levels of color. Wavelet transformation is used by the sub images of aerial photographs to conduct image compression. It uses image decomposition and coefficient elimination to extract important features and reduce image size [13], [15]. It is achieved by successive high-pass and low-pass filtering of the image (Figure 2). The original signal can be restructured by discrete wavelet transformation (DWT) for image compression.

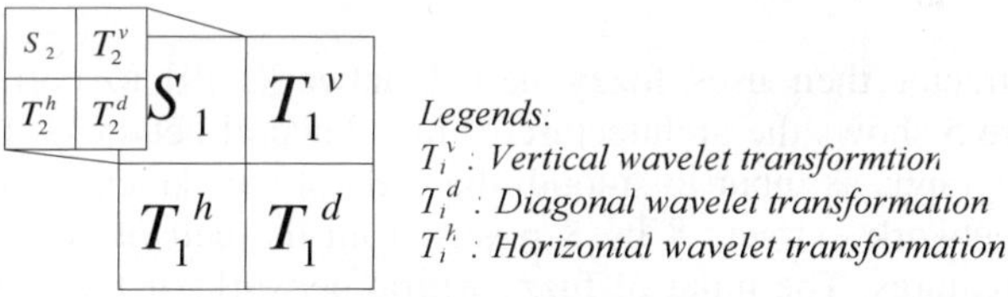

Fig. 2. Wavelet transformation of image S

Feature extractor then uses SVM to classify surface features of aerial photographs. The hyperplane separates the data samples into two classes [1], [5], [6].

$$f(x) = sign(\sum y_i \alpha_i^0 K(x_i, x) - b^0) \tag{1}$$

where *x, y, b, α,* and *sign* are input data, class, bias, Lagrange multiplier, and sign symbol. One-Against-All strategy of SVM is used to recognize surface features. SVM uses sketched graphs with 256 colors to represent different surface features of the aerial photograph. Figures 3 and 4 display the example of the aerial photograph and recognized features. Notably, the colors pink, gray, blue, red, green, white, orange, and yellow represent the farm, house, river, fruit farm, wood, sand, sea, and miscellaneous correspondingly. Each cell of Figure 3 is represented by 16 by 16 pixels of colors.

Fig. 3. Aerial photograph

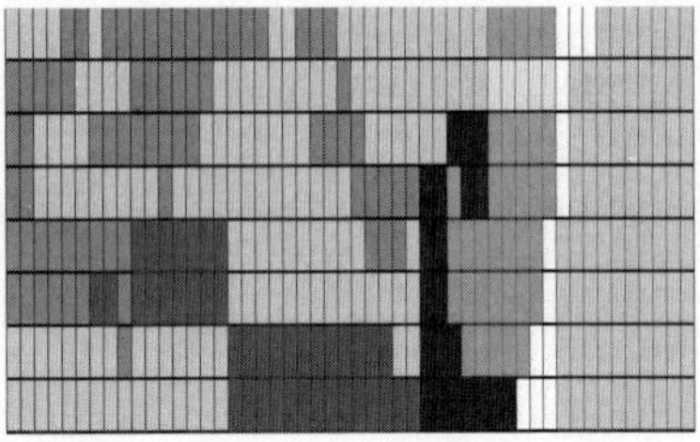

Fig. 4. Sketch of aerial photograph

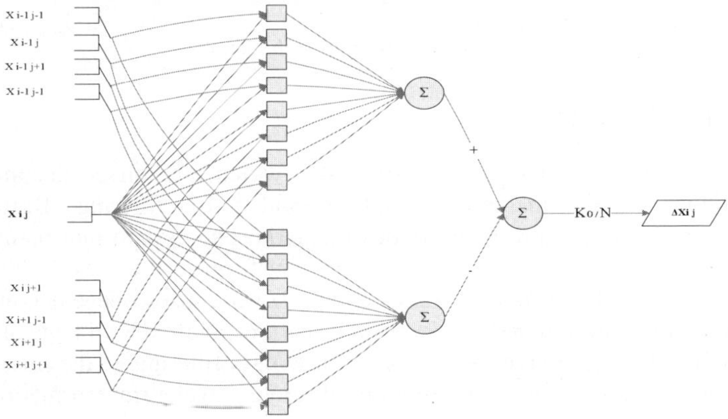

Fig. 5. Fuzzy neural networks architecture

Feature extractor then uses fuzzy neural networks [9] to correct misclassified features. Figure 5 shows the architecture of fuzzy neural networks. It uses the sketch of aerial photographs as input to smooth the cells of the sketch image. Each node of fuzzy neural network extracts 8 by 8 pixels from the cell of sketch graph to filter misclassified features. The input of fuzzy neural networks is the central pattern with 64 pixels and eight neighbor patterns of sketch graph. Each pattern is affected by its eight neighboring patterns from the following equations.

$$\Delta X_{i,j} = \frac{\alpha}{N}[\sum_{X_{m,n} \in A} \mu R_1(X_{i,j}, X_{m,n}, \alpha) - \sum_{X_{m,n} \in A} \mu R_2(X_{i,j}, X_{m,n}, \alpha)] \tag{2}$$

$$\mu R_q(u, v, \alpha) = \begin{cases} MAX\left\{\left(1 - \frac{|u - v - \alpha|}{2\alpha}\right), 0\right\} & q = 1 \\ MAX\left\{\left(1 - \frac{|u - v + \alpha|}{2\alpha}\right), 0\right\} & q = 2 \end{cases} \tag{3}$$

where $X_{i,j}$, $X_{m,n}$, N, u, v, and α are central pattern, neighbor pattern, number of neighbors, value of central pattern, value of neighbor pattern, and scaling factor. The value range ofαis between 0 and 8. Patterns of sketch graph are evaluated from left to right and top to bottom manner sequentially.

2.2 Damage Predictor

Damage predictor uses CBR to predict possible damage of the typhoon. CBR uses typhoon weather forecasts of government weather departments to find a similar case from the typhoon case library. The typhoon case library contains past typhoon cases to the analyzed regions. Each case of the typhoon case library contains basic typhoon characteristics and damage. The former includes track, speed, degree, month, coordinates, affected hours, average amount of rain, and atmospheric pressure of typhoon. The latter is the influenced region, number of person casualties, amount of crop damage, and percentage of damage.

Damage predictor uses feature similarity of typhoon information to select a similar case.

$$F(T, C_j) = \sum_{i=1}^{n} sim(f_{T_i}, f_{C_i}) \times w_i \tag{4}$$

where T, C_j, f_T, f_C, and w are the typhoon forecasting information, case j, typhoon feature, case feature, and feature weight.

Damage predictor then uses selected case, sketch of aerial photographs, and crop statistical database to predict damage. Crop statistical database stores farm product history of the plant area in every season. It can be used to measure the damage plant area and amount of crop. Damage predictor uses the features of the similar case to compute possible damage. It uses fuzzy damage measure, FDA, to evaluate possible damage.

$$FAD = IA(x) \times PR(x) \times DD * CR \tag{5}$$

where *IA, x, PR, DD* and *CR* are influenced area, damage object, damage amount, damage degree, and amount of planted farm. *IA, PR,* and *DD* are the feature value of the similar case. CR is the amount of planted farm or inhabitants from the crop statistical database and county government. Table 1 lists the linguistic values and representative intervals of each variable. The triangle fuzzy number is used as the shape of fuzzy set (Fig. 6). It then uses the centre of gravity (COG) method to estimate the possible damage.

Table 1. Linguistic variable and representative interval

Linguistic variable & value	IA(crop) (ha.)	IA(resident) (m^2)	PR(crop) (t.)	PR(resident) (Person)	DD (Percentage)
Very Low	[0, 25]	[0, 3]	[0, 20]	[0, 5]	[0, 20]
Low	[10, 35]	[1, 7]	[10, 40]	[3, 9]	[10, 40]
Medium	[30, 45]	[5, 11]	[30, 60]	[7, 13]	[30, 60]
High	[40, 55]	[9, 15]	[50, 85]	[11, 17]	[50, 85]
Very High	[50, 70]	[13, 20]	[70, 100]	[15, 20]	[70, 100]

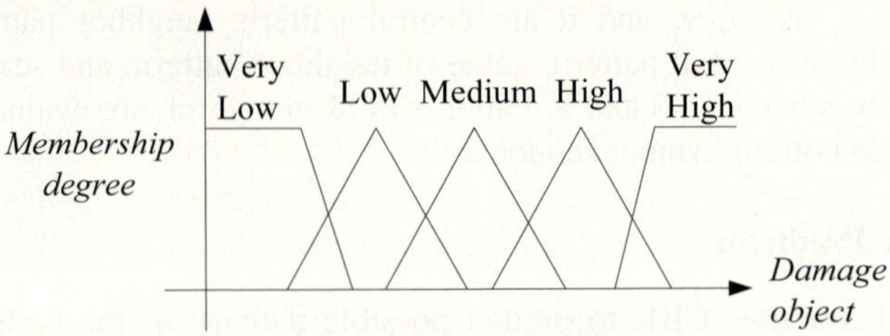

Fig. 6. Fuzzy sets of Table 1

3 Example Application of Jaiosi in Taiwan

Jasosi is a famous mineral spring locale in northeastern of Taiwan. It attracts tens of thousands of sightseers every year. However, typhoons usually threat Jasosi from July to November. Beaches, buildings, and farms are overwhelmed and hammered by ruthless winds and rain. Numerous human beings, buildings, and crops are damaged after typhoon. Predicting possible damages to select and adopt the right prevention policy is an important mission.

3.1 Feature Extractor

Figure 7 shows the input screen of typhoon information. The screen contains the fields of movement speed, central air pressure, maximal wind velocity, strength,

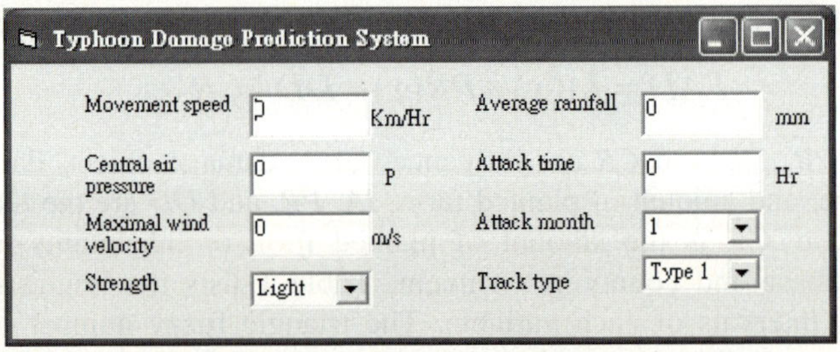

Fig. 7. Input screen of typhoon information

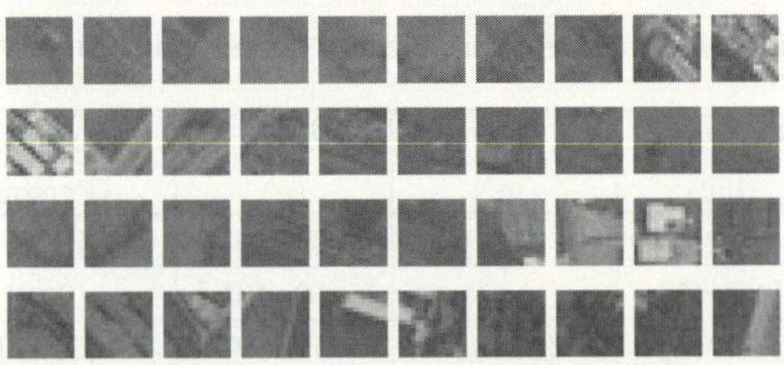

Fig. 8. Example sub-graph of aerial photograph

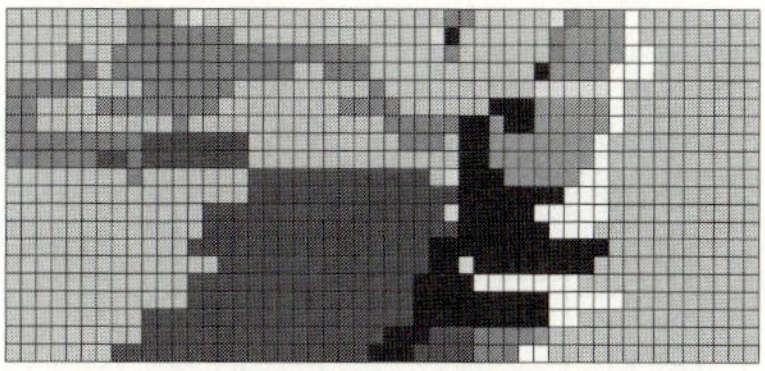

Fig. 9. Input image of fuzzy neural network

Fig. 10. output of example sketch graph

Table 2. Training and test data of each class

Class	Sub-image #	Training #	Test #	Accuracy
Sea	502	402	104	100%
House	313	250	63	100%
Sand	64	51	13	100%
River	87	70	17	100%
Farm-1	695	555	140	100%
Farm-2	531	425	106	100%
Wood	319	255	64	100%

average rainfall, attack time, attack month, and track type. Notably, track type is classified into nine classes according to the weather reports from the Taiwan Central Weather Bureau [4]. Feature extractor then uses the above typhoon information and aerial photographs to recognize surface features on the earth. First, feature extractor divides the aerial photograph into 2600 sub-images with 16*16 pixels for image compression. Figure 8 shows example sub-images. The image uses wavelet transformation to reduce the size for pattern recognition. Wavelet transformation uses DWT and harr function with two decomposition levels to compress the image. Feature extractor then uses SVM to recognize the surface features. Each compressed sub-image is represented by 256 numerical attributes by transforming each pixel of the sub-image into the gray level value. SVM then extracts the numeric value in row-major order as the value of the sub-image for surface feature classification. Each value of the image is considered as an attribute. SVM classifies the sub-image into one of the following classes, which are, sea, house, sand, river, farm_1, farm_2, and wood. SVM uses 2080 and 520 sub-images as the training data and test data (Table 2). The accuracy of image recognition is 100%. The kernel function of SVM is a radial basis function. Feature extractor then uses fuzzy neural networks to filter misclassified surface features for CBR reasoning. Figures 9 and 10 display the input and output of example sketch graph by fuzzy neural networks. Eight sub-images are changed after image filtering and smoothing.

3.2 Damage Predictor

Damage predictor uses feature sketch, agricultural data, and past typhoon cases to predict possible damage. Tables 3 and 4 show the example agricultural data and

Table 3. Example agricultural data

Season	Banana	Balsam pear	Cucumber	Taros	Green onion	Peanut	Sweet potato	Cabbage
1	0	7200	1427	0	5000	0	171400	75833
2	1860	8082	0	1856	30223	1895	265320	32177
3	0	12000	5143	0	9000	0	72000	38333
4	17000	3204	2259	1406	6813	687	12327	7346

Table 4. Example typhoon data

Name	Year	Month	Area	Stay hours	Average speed (km/hr)	Central pressure	Rain(mm)	Central speed (m/s) (degree)
HERB	1996	7	Jaiosi	72	19.3	920	350.84	53(16)
ALEX	1987	7	Jaiosi	46	12.06	970	119.772	35(12)
OFELIA	1990	6	Jaiosi	55	20.24	968	153.752	38(13)
YANCY	1990	8	Jaiosi	48	9.30	960	191.268	40(13)
ELLIE	1991	8	Jaiosi	49	20.90	965	44.828	33(12)

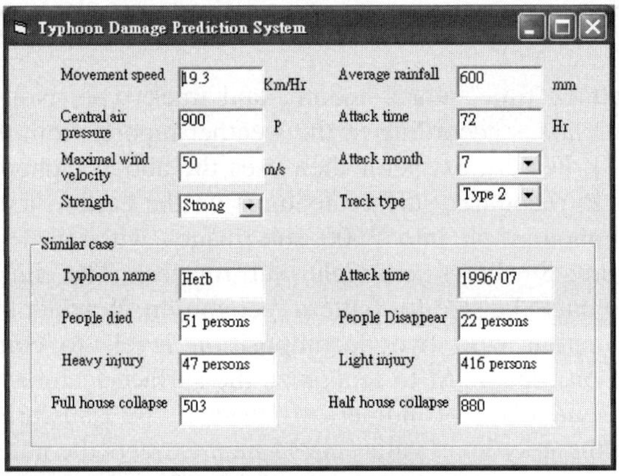

Fig. 11. Similar cases

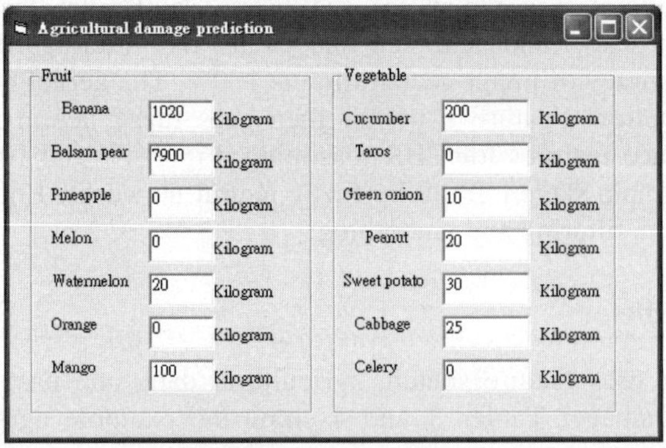

Fig. 12. Agricultural damage prediction

typhoon information of Jaiosi, Taiwan. Figure 11 shows the forecasted typhoon information and selected similar typhoon case. Finally, damage predictor computes the value of FDA to find the possible agricultural damage (Figure 12).

4 Conclusion

This work proposes a case-based typhoon damage prediction system from aerial photograph with the help of hybrid feature analysis. The system contains two components, feature extractor and damage predictor. Feature extractor uses image decomposition and compression, classification, and error correction to analyze the surface features of the aerial photograph. Wavelet transformation and SVM are used as image preprocessor to reduce the size of the photograph and recognize surface features. Fuzzy neural network is used to smooth the sketch graph by filtering misclassified features. The hybrid feature analysis provides the capability to process image data. Damage predictor uses case-based reasoning and fuzzy damage measurement to compute possible damages of a typhoon. Jaiosi in Taiwan as an example used by the system explains its efficiency. The experiment shows that the system can find important surface features from aerial photographs.

However, the case library only contains 30 practical cases of agriculture damage in Taiwan. The real similar cases of Jaiosi and accuracy of damage prediction are 5 and 70%. The lack of similar cases in the case library may cause low prediction accuracy by case-based reasoning. It is a difficult task to evaluate the accuracy of the damage prediction system. We will try to enrich the case library from different regions in order to improve the accuracy of the intelligent typhoon prediction in the future.

References

1. Burges, C.J.C.: A Tutorial on Support Vector Machines for Pattern Recognition. Data Mining and Knowledge Discovery 2(1), 121–167 (1998)
2. Chen, K.S., Wang, J.T., Mitnik, L.M.: Satellite and Ground Observations of the Evolution of Typhoon Herb near Taiwan. Remote Sensing of Environment 75(3), 397–411 (2001)
3. Cui, L., Wang, S., Sun, X.: The Application of Wavelet Analysis and Audio Compression Technology in Digital Audio Watermarking. In: Proceedings of the IEEE International Conference on Neural Networks and Signal Processing, pp. 1533–1537 (2003)
4. Central Weather Bureau, Meteorological Bulletin, Central Weather Bureau, Ministry of Transportation and Communications (2006), Available at, http://photino.cwb.gov.tw/rdcweb/lib/cd/cd07mb/index.htm
5. Foody, G.M., Mathur, A.: A Relative Evaluation of Multiclass Image Classification by Support Vector Machines. IEEE Transaction on Geosciences and Remote Sensing 42(6), 1335–1343 (2004)
6. Huang, C., Davis, L.S., Townshend, J.R.G.: An Assessment of Support Vector Machines for Land Cover Classification, International. Journal of Remote Sensing 23(4), 725–749 (2002)
7. Kitamoto, A.: Spatio-temporal Data Mining for Typhoon Image Collection. Journal of Intelligent Information Systems 19(1), 25–41 (2002)

8. Kolodner, J.: Case-Based Reasoning, Morgan Kaufmann, San Mateo (1993)
9. Lee, C.-S., Guo, S.-M., Hsu, C.-Y.: Genetic-based Fuzzy Image Filter and Its Application to Image Processing. IEEE Transactions on Man and Cybernetic 35(4), 694–711 (2005)
10. Lee, T.L.: Neural Network Prediction of a Storm Surge. Ocean Engineering 33(4), 483–494 (2006)
11. Lenz, M., Bartsch, S.B., Wess, S.: Case-Based Reasoning Technology, from Foundations to Applications. Springer, Berlin (1998)
12. Melgani, F., Bruzzone, L.: Classification of Hyperspectral Remote Sensing Images With Support Vector Machines. IEEE Transactions on Geosciences and Remote Sensing 42(8), 1778–1790 (2004)
13. Qu, Y., Adam, B.-L., Thornquist, M., Potter, J.D., Thompson, M.L., Yasui, Y., Davis, J., Schellhammer, P.F., Cazares, L., Clements, M.A., Wright, G.L., Feng, Z.: Data Reduction using a Discrete Wavelet Transform in Discriminate Analysis of Very High Dimensionality Data. Biometrics 59(1), 143–151 (2003)
14. Sakai, R.: A History and Future Plan for Numerical Typhoon Prediction at JMA, (2006), Available at, http://www.bom.gov.au/bmrc/basic/wksp16/ papers/Sakai.pdf
15. Thuillard, M.: Wavelets in Soft Computing. World Scientific Publishing, NJ (2001)
16. Wang, B., Zhang, L.M.: Supervised Texture Segmentation using Wavelet Transformation. In: Proceedings of the. IEEE International Conference on Neural Networks and Signal Processing, pp. 1078–1082 (2003)
17. Zhang, J., Dong, J., Shi, M.: An Adaptive Method for Image Filtering with Pulsed-Coupled Neural Networks. Proceedings of the IEEE International Conference on Image Processing 2, 11–14 (2005)

Analysis and Research of Predictive Algorithm in NCS with Time Delay

Zaiping Chen[1], Rui Lou[2], Xunlei Yin[1], Nan Yang[1], and Gang Shao[1]

[1] Tianjin University of Technology, Tianjin, 300191, China
[2] Tianjin vocational College of Mechanics and Electricity, Tianjin, China
zpchen@tjut.edu.cn, lourui2033@163.com, sdwhyxl@sina.com, yangnan19811003@163.com, clickto@163.com

Abstract. In this paper, grey predication theory and grey calamity predication model are explained. An approach to design of a full-state estimator is discussed with single packet transmission and network-induced random time delay less than one sample period. Then a scheme of compensation for network-induced delay τ_{ca} in NCSs is proposed with the grey calamity predication theory. Finally, the simulation has been carried out, and the effectiveness of the compensation scheme proposed in this paper is verified.

Keywords: Grey predication, Intelligent system, Networked control.

1 Introduction

Networked Control Systems(NCSs) is a type of closed-loop control system with real-time communication networks imported into the control channel and feedback channel. NCSs usually includes of DCS、Fieldbus and Industrial Ethernet, etc., which show the trends of control system towards network, integration,distribution and node intelligent has become a hot research spot in control domain.

The network-induced delay is the key problem in NCSs[1]. There exist three conventional treatments with network-induced delay. Firstly, stochastic delay time is fixed with bumper[2,3],but delay time has been artificially lengthened. Secondly, delay time is compensated by designing observer with time sampling[4-6], but which usually applied to delay time between sensor and controller. Thirdly, using jump markov predicts delay time only when the distribution of time delay is confirmed[7-9].

In this paper, the relation between sampling rate and time delay is studied and the necessity of precise compensation is discussed first. Then grey predictive theory and grey calamity model are explained. An model of NCSs, in which the setup with single-packet transmission and random time delay less than one sample period is taken into account is introduced. The compensation scheme of delay time including τ_{ca} and τ_{sc} based on grey calamity prediction is proposed, then refinement is carried out by the sampling rate's effect on delay time. Finally, the simulation is carried out and the effectiveness of the compensation scheme presented here is verified.

B. Apolloni et al. (Eds.): KES 2007/WIRN 2007, Part I, LNAI 4692, pp. 757–764, 2007.

2 Relation Between Sampling Rate and Networked Time Delay and Impact on Stability of NCSs

Usually fast sampling rate is expected, but too fast sampling rate can make network overloading so that increasing the delay time. So it is very important to study the relation between sampling and delay time in NCSs.

According a model of time delay less than one sampling period T, system model normally can be given as follows.

$$\begin{aligned} \dot{x}(t) &= Ax(t) + Bu(t) \quad t \in [kT+\tau, (k+1)T+\tau] \\ y(t) &= Cx(t) \\ u(t^+) &= -Kx(t-\tau) \quad t \in \{kT+\tau, k = 0,1,2\cdots\} \end{aligned} \tag{1}$$

where $u(t^+)$ segmented continues, and at kh+ τ_k point value of function is refreshed.

When sampling period is T, derive model as in Eq.(2).

$$\begin{aligned} x[(k+1)T] &= \Phi x(kT) + \Gamma_0(\tau)u(kT) + \Gamma_1(\tau)u[(k-1)T] \\ y(kT) &= Cx(kT) \end{aligned} \tag{2}$$

$$\Phi = \ell^{AT}, \Gamma_0(\tau) = \int_0^{T-\tau} \ell^{As} B ds \quad , \Gamma_1(\tau) = \int_{T-\tau}^{T} \ell^{As} B ds$$

Define

$$z(kT) = \left[x^T(kT), u^T((k-1)T)\right]^T,$$

Corresponding closed loop system is given in Eq. (3).

$$z((k+1)T) = \tilde{\Phi}(k) z(kT). \tag{3}$$

Where, $\tilde{\Phi}(k) = \begin{bmatrix} \Phi - \Gamma_0(\tau)K & \Gamma_1(\tau) \\ -K & 0 \end{bmatrix}$

To keep a NCSs steady, the relation between sampling rate and delay time must meet the following inequality:

$$\begin{aligned} \max\left\{\frac{1}{2}T - \frac{1}{K}, 0\right\} &< \tau < \min\left\{\frac{1}{K}, T\right\} \\ \max\left\{\frac{1}{2} - \frac{1}{KT}, 0\right\} &< \frac{\tau}{T} < \min\left\{\frac{1}{KT}, 1\right\}. \end{aligned} \tag{4}$$

From Eq. (4), when sampling period is small, the upper bound of time delay can reach to sampling period; while sampling period is longer, the upper bound of τ/T becomes smaller.

Giving a system model for simulation as followings:

$$\dot{x} = \begin{bmatrix} -2 & -3 \\ 4 & -9 \end{bmatrix} x + \begin{bmatrix} 0 \\ 1 \end{bmatrix} u, \; y = \begin{bmatrix} 1 & 0 \end{bmatrix} x \quad (5)$$

when sampling period T=0.02s,T=0.08s, and delay time $\tau = 0.02s$ are utilized, the simulation is carried out, and the simulation results are given in Fig.1-2 respectively.

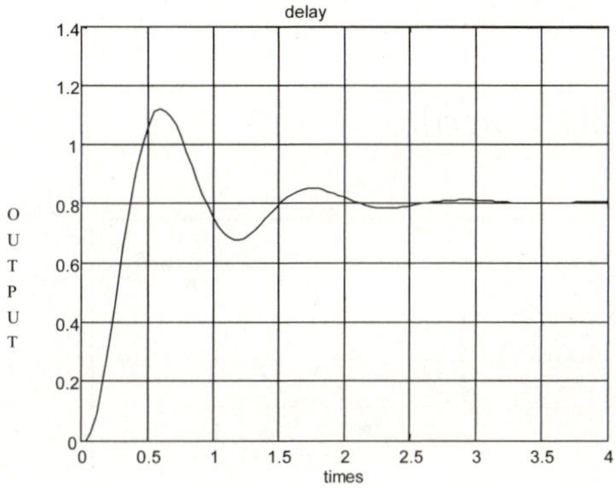

Fig. 1. The system simulation with T=0.02s $\tau = 0.02s$

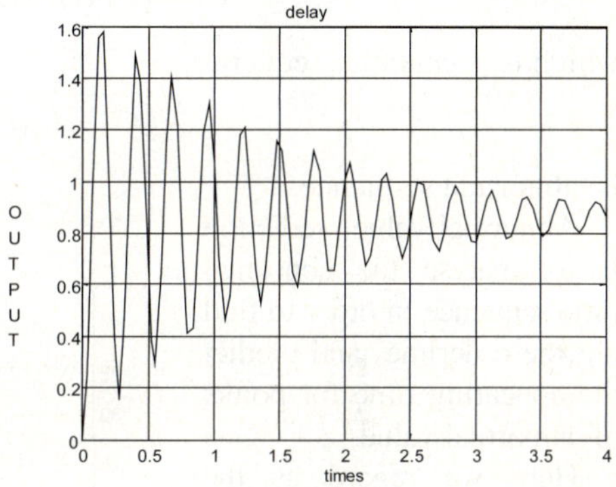

Fig. 2. The system simulation with T=0.08s $\tau = 0.02s$

From the analysis of simulation result, a conclusion can be reached that the result of compensation may get worse if compensation isn't precise when sampling period is too big. So refination of time delay is necessary.

3 Grey Calamity Prediction

(1) Grey prediction

In grey theory, observed data sequence is treated as time-varying grey process, and the data is accumulated so that the inner ordinal exponential distribution rule is excavated, furthermore relevant predictive model is built.

The model reflects the trend change of actual observed value, which is proved well in prediction of increasing or drop sequences but can't get results well in more stochastic sequences.

(2) Grey calamity prediction

In fact grey calamity is prediction for abnormal value, in which we can determine the abnormal value by experience. The duty of grey calamity prediction is to figure out the appearing time for points of abnormal value.

Definition one:

Original sequence X= $(x(1), x(2), \ldots, x(n))$, the upper bound of abnormal value is ξ, calling $X_\xi = (x[q(1)], x[q(2)], \ldots x[q(m)]) = \{x[q(i)] \mid x[q(i)] \geq \xi, i = 1,2,\ldots,m\}$ the upper bound abnormal sequence.

Definiation two: Original sequence X= $(x(1), x(2), \ldots, x(n))$, the lower bound abnormal is ξ, calling $X_\xi = (x[q(1)], x[q(2)], \ldots x[q(l)]) = \{x[q(i)] \mid x[q(i)] \le \xi, i = 1,2,\ldots,l\}$ the lower bound abnormal sequence.

Defination three:
Giving original sequence as follow:

$$X_\xi = (x[q(1)], x[q(2)], \ldots x[q(m)]) \subset X$$

which is is calamity sequence, so

$$Q^{(0)} = (q(1), q(2), \ldots, q(m))$$

is abnormal sequence.

Abnormal value prediction is to analyse the abnormal time sequence in order to find out the orderline and predict the appearing time for points of abnormal value.

Here we regard as the delay time as data sequence. Because of stochastic change and bound limitation, Grey model can't be directly used to predict delay time, the grey calamity prediction is employed for predicting time for points of time delay. In the prediction time delay is regarded as calamity data, the appearing time point of data sequence is treated time sequence, which is increasing and has no boundary limition, and grey model can be used to predict and get one sequence:

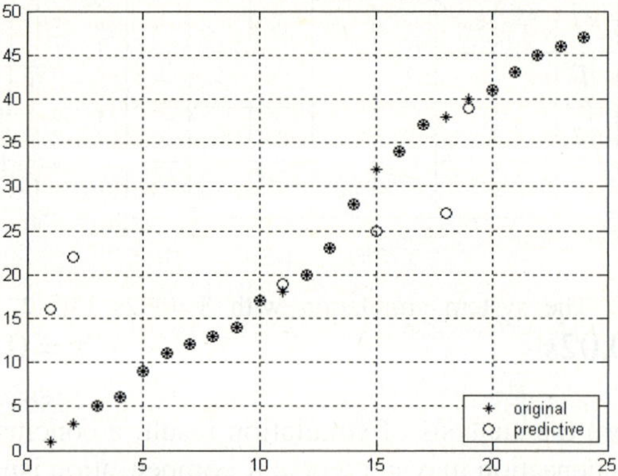

Fig. 3. Grey predictive result with grey calamity theory

$$\{q(i) \mid x(q(i)) \ge 0.01, i = 1,2,\ldots,n\}$$

Prediction steps are following:

(1) Define stochastic sequence, sampling period is h, delay time $\tau \le h$;

(2) select upper bound sequence, abnormal value is $\xi = h/n$,n=1,2,....

(3) to lower randomness, use second-order sequence operator to soften calamity sequence, operator can be taken as follows:

$$x(i)d = \frac{1}{m-i+1}(x(i) + x(i+1) + \ldots + x(m)),$$

$$x(i)d^2 = \frac{1}{m-i+1}(x(i)d + x(i+1)d + \ldots x(m)d)$$

where m is the upper bound of sequence.

(4) using grey model to soften the time sequence.
(5) carry inverse soft operation to get estimated time point sequence.

Giving the stochastic sequence:

{1,3,5,6,9,11,12,13,14,17,18,20,23,28,32,34,37,38,40,41,43,45,46,47} the predictive result by grey calamity theory is in Fig. 3.

The predictive sequence is as follow.

{5,6,9,11,12,13,14,16,17,19,20,22,23,25,27,28,34,37,39,41,43,45,46,47}, the result shows the predictive method is effective.

4 Compensation of NCS Time Delay Based on Grey Calamity Prediction

Previous research works mainly focus on τ_{sc} to predict and compensate, because τ_{sc} can be got at controller through synchronization method, the compensation using state feedback based on state observer can be carried out, while the τ_{ca} between controller and actuator can't be got beforehand at controller, so the compensation to τ_{ca} will not be obtained.

In this paper we use grey calamity predictive model to estimate the time point at time delay in different level. Then τ_{ca} can be predicted, and after that the time delay compensation based on state observer will be utilized.

For a NCS model which delay time is less than one sampling period,we can give the state equation model as follows:

$$\dot{x}(t) = Ax(t) + Bu(t) + \varpi \qquad y(t) = Cx(t) \tag{6}$$

discrete controller is

$$u(kT) = -Kx(kh) + r(kh), \tag{7}$$

where r is reference input, k = 0, 1, 2 ⋯

$x \in R^n, u \in R^m, y \in R^p$, A is system matrix, B is control matrix, C is observersation matrix, K is state feedback matrix, h is sampling period ω, v is white noise, $E(\varpi) = 0$.

The discretization of (1) is:

$$\begin{aligned} x(k+1) &= \Phi x(k) + \Gamma_0(\tau_k)u(k) + \Gamma_1 u(k-1) \\ y(k) &= Cx(k) \end{aligned} \tag{8}$$

$$\Phi = \ell^{Ah}$$
$$\Gamma_0(\tau_k) = \int_0^{h-\tau_k} \ell^{As} B ds$$
$$\Gamma_1(\tau_k) = \int_{h-\tau_k}^{h} \ell^{As} B ds$$

Its full order observation dynamic equation is:

$$\dot{\hat{x}}(t) = (A - GC)\hat{x}(t) + Bu(t) + Gy(t)$$
$$\hat{y}(t) = C\hat{x} \tag{9}$$

$(A - GC)$ is observer matrix, the design of observer must meet the limitation under any initial condition:

$$\lim_{t\to\infty}(\hat{x}(t) - x(t)) = 0 \tag{10}$$

by（4）and（7）:

$$\dot{x}(t) - \dot{\hat{x}}(t) = (A - GC)(x(t) - \hat{x}(t)) \tag{11}$$

get the equation (8) resolution:

$$\hat{x}(t) = x(t) - e^{(A-GC)(t-t_0)}(x(t_0) - \hat{x}(t_0)) \tag{12}$$

The descretization of equation（9）and taking into account equation（5）,can get the observer model:

$$\begin{aligned}\hat{x}((k+1)h) &= x((k+1)h) - e^{(A-GC)h}[x(kh) - \hat{x}(kh)] \\ &= \Phi x(kh) + \Gamma_0(\tau_k)u(kh) + \Gamma_1(\tau_k)u((k-1)h) \\ &\quad - e^{(A-GC)h}[x(kh) - \hat{x}(kh)],\end{aligned} \tag{13}$$

Thc state space equation of full order state observer is:

$$\begin{aligned}\begin{bmatrix} x((k+1)h) \\ \hat{x}((k+1)h) \end{bmatrix} &= \begin{bmatrix} \Phi & -\Gamma_0 K \\ \Phi - \Phi_e & h \end{bmatrix}\begin{bmatrix} x(kh) \\ \hat{x}(kh) \end{bmatrix} + \\ \begin{bmatrix} 0 & -\Gamma_1 K \\ 0 & -\Gamma_1 K \end{bmatrix}&\begin{bmatrix} x((k-1)h) \\ \hat{x}((k-1)h) \end{bmatrix} + \begin{bmatrix} \Gamma_0 K_r & \Gamma_1 K_r \\ \Gamma_0 K_r & \Gamma_1 K_r \end{bmatrix}\begin{bmatrix} r(kh) \\ r((k-1)h) \end{bmatrix}\end{aligned} \tag{14}$$

Here delay time is regarded as continuous sequence and first divided into two levels to use grey calamity to predict it. Then more level is employed.

Giving a system model for simulation analysis as follows.

$$\dot{x} = \begin{bmatrix} -2 & -3 \\ 4 & -9 \end{bmatrix}x + \begin{bmatrix} 0 \\ 1 \end{bmatrix}u, \quad y = [1 \quad 0]x$$

the pole is P=[-5-10j -5+10j]; feedback coefficient is K=[-32.333 -1]; state feedback gain coefficient is

G=[-1;-34.6667]; the reference input is r=1.

The simulation results are shown in Fig. 5-6, in which delay time $\tau = 0.02s$, sampling period h=0.02s, h=0.04s are utilized respectively, two levels divided is used, and the abnormal value is $\xi = h/2$.

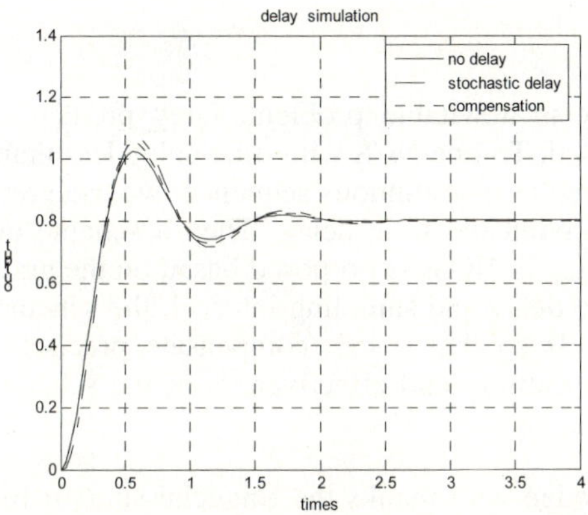

Fig. 4. System simulation with two levels compensation, (=0.02 and h=0.02s

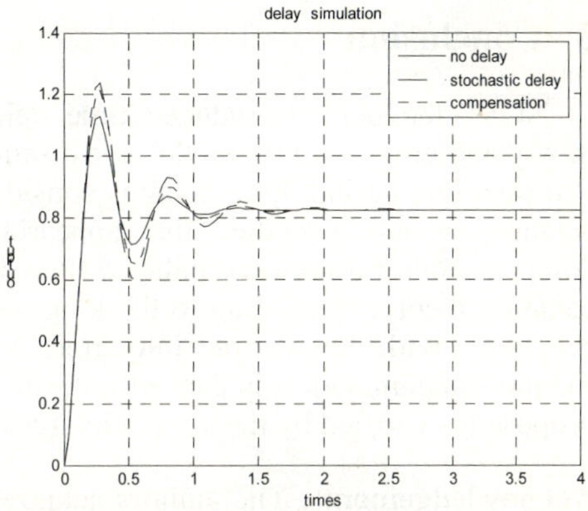

Fig. 5. System simulation with two levels compensation, (=0.02 and h=0.04s

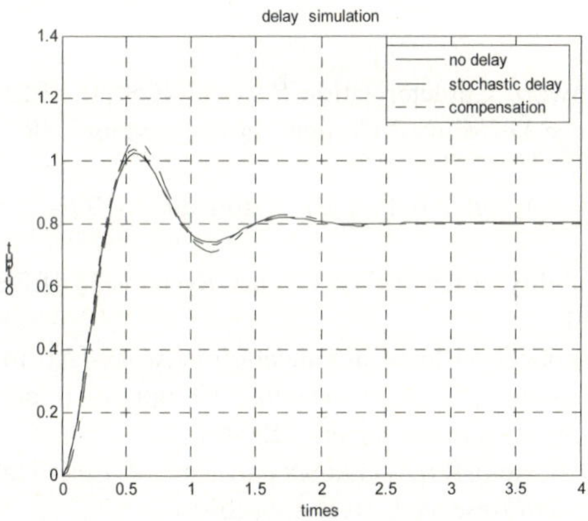

Fig. 6. System simulation with four levels compensation, τ=0.02 and h=0.02s

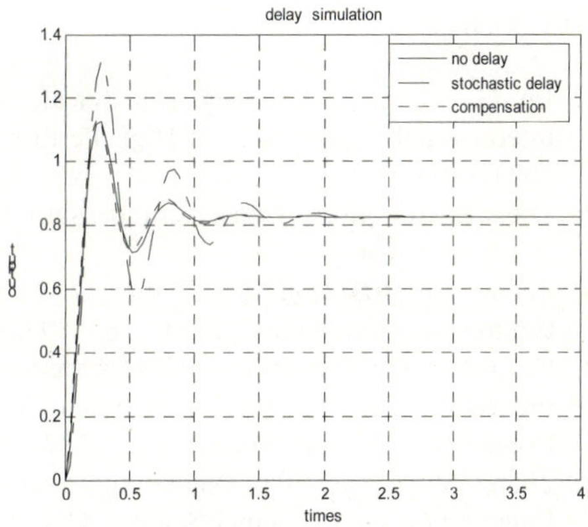

Fig. 7. System simulation with four levels compensation, τ=0.02 and h=0.04s

The simulation results in Fig.4-5 show that the compensation scheme proposed can improve the system performance, but when sampling period is longer(h=0.04s), the delay time needs to be divided more levels.

Fig. 6 and Fig. 7 show that the effectiveness of the compensation with four levels scheme, in which delay time $\tau = 0.02s$, sampling period h=0.02s, h=0.04s are utilized respectively, four levels divided is used, and the abnormal value is $\xi = h/4$.

Comparing simulation result in Fig. 5 with result in Fig.7, we can see that time delay compensation scheme based on grey calamity prediction with four levels can effectively improve system performance. The scheme based on grey calamity prediction can also further divide delay time into more levels when delay time interval is bigger, which can't be carried out with other compensation schemes.

5 Conclusion

In NCSs, the network-induced time delay is inevitable problem. Grey prediction theory and grey calamity model are discussed. To one NCS with time delay less than one sampling period, time delay is considered as a continuous sequence, we use grey calamity theory to predict the appearing point of time delay. Then a scheme of compensation for network-induced time delay in NCSs is proposed based on the gray calamity theory. According to the long time delay and sampling interval, the scheme can also divide delay time into more levels so that it can compensate precisely. Finally, the simulation is carried out, the feasibility and effectiveness of the scheme proposed is verified by the simulation results.

Acknowledgements. The authors acknowledge with thanks the financial support by Tianjin Natural Science Foundation (Grant: 043601511), Tianjin Natural Science Foundation key project(Grant: 07JCZDJC09600), and Tianjin Education Committee (Key Discipline Grant: 2004BA08)

References

1. Jin, G., Yang, G., Crowley, B.R., et al.: Network Chracterization Service(NCS)[J]. IEEE International Symposium on High Performance Distributed Computing Proceedings, 289–302 (2001)
2. Luck, R.: An observer-based compensator for distributed delays [J]. Automatic 26, 903–908 (1990)
3. Zhang, W., Branicky, M S, Philips, S M.: Stability of networked Control Systems[J]. IEEE Control Systems Magazine 21(1), 85–89 (2001)
4. Hespanha, J P.: Stochastic hybrid systems: application to communication networks[A]. In: Proceedings of the 7th International Workshop on hybrid systems: Computation and Control [C], Philadelphia, USA, pp. 387–401. Springer, Heidelberg (2004)
5. Zhang, W.: Analysis of Networked Control Systems[D]. USA: Department of Electrical Engineering and Computer Science. Case Western Reserve University (2001)
6. Nilsson, J., Bemhardson, B., Wittenmark, B.: Stochastic analysis and control of real-time systems with random time delays[J].Automatica 34(1), 57–64 (1998)
7. Lineoln, B., Bernhardsson, B.: Optional Control over networks with long random delays[A]. In: Proceedings of the Fourteenth International Symposium on Mathematical Theory of Networks and Systems[C], France, pp. 1–7. University of Perpignan (2000)
8. Xiao, L., Hassibi, A., How, J P.: Control with random communication delays via a discrete time jump systems approach[A]. In: Proceedings of the 2000 American Control Conference[C], Chicago, USA, pp. 2199–2204. IEEE Computer Society Press, Los Alamitos (2000)
9. Zhixun, Y., Huitang, C., Yuejuan, W.: The research of closed loop control system based on Markov[J]. Control Theory and Application 19(2), 263–267 (2002)

Automated Planning and Replanning in an Intelligent Virtual Environments for Training

Jaime Ramírez and Angélica de Antonio

Universidad Politécnica de Madrid, Madrid, Spain
{jramirez,angelica}@fi.upm.es
http://decoroso.ls.fi.upm.es

Abstract. The application of Artificial Intelligence (AI) planning techniques to the development of Intelligent Tutoring Systems (ITS) has focused mainly on instructional planning, in settings where the initiative is taken primarily by the system. 3D Virtual Environments (VE) have emerged in the last years as a good means to apply a case-based training approach, placing a more active role on the student. Here AI planning turns out to be an interesting solution for the dynamic resolution of the problems (cases) that are posed to the student. These environments allow the students to navigate through and interact with a virtual representation. This paper describes MAEVIF, a platform for the development of intelligent virtual environments for training (IVETs) whose architecture is based on a collection of cooperative software agents. The role of AI planning in the teaching-learning approach followed by MAEVIF is described, with two main planning services: the generation of a plan as an ideal solution to the case, and the evaluation of the effect of the student's actions during their resolution of the case.

Keywords: Intelligent Tutoring Systems, planning, replanning, multi-agent system.

1 Introduction

ITSs appeared back in the 70s as a result of combining AI methods with Computer Aided Instruction. One of the most remarkable consequences of this evolution was the possibility for the instructional system to adapt the tutoring process to the characteristics and progress of each individual student thanks to a specific process devoted to student modeling and to a tutoring strategy that bases decisions on the current state of the student's knowledge and the instructional goals. Classical AI planning techniques, when applied to ITSs, have been used mainly to implement the instructional planning process, that is responsible for the selection of the content to be presented (content planning) and the selection of the best way to present the material (delivery planning) [1]. An early example of the use of classical AI planning for this purpose was [2], and many systems have followed this path later on [3]. Other systems use planners not to make the instructional decisions, but to try to infer the student's mental plan. One example in this category is PLAIT [4]. Yet a third possibility is to use AI planning to

B. Apolloni et al. (Eds.): KES 2007/WIRN 2007, Part I, LNAI 4692, pp. 765–772, 2007.

find a solution to the proposed activities thanks to an expert module containing knowledge about the state of the domain. Classical ITSs based their expert model on other AI paradigms such as production systems or semantic networks. However, these representational approaches are not well suited for IVETs. In this kind of systems the student is faced to a situation in which a defined goal must be accomplished, starting from a given initial state of the world, and the student must find a proper sequence of actions to transform the initial state into the final desired one. The system plays the role of a coach that observes the student's behavior and intervenes whenever it is necessary to correct or guide the student. The use of AI planning for this purpose in ITSs is scarce. Maybe the most significant representative is Steve (Soar Training Expert for Virtual Environments) [5] an animated agent that cohabits the virtual world with one or more students and can demonstrate procedures, answer questions and monitor the students' performance. However, the language for specifiying the task model of Steve is little expressive, since it does not support the use of variables in the preconditions or the postconditions.

In this paper we present a generic platform for the development of IVETs called MAEVIF (Model for the Application of Intelligent Virtual Environments to Education). We propose an agent-based architecture in which the use of AI planning is central for the tutoring module. Next the general agent-based architecture of MAEVIF will be described and how it has been designed to allow the integration of planning capabilities (section 2). In section 3, the structure proposed for the plan is described. Section 4 will present the way in which MAEVIF conducts the tracking of the student's behavior during the resolution of a case, and the important role that the planner plays in this process. Finally, in section 5, some conclusions and future work lines will be shown.

2 The Planning Agent in MAEVIF's Architecture

The MAEVIF platform is the result of several previous experiences integrating virtual environments and intelligent tutoring systems. The objectives are the definition of a generic model for intelligent training and learning environments based on the use of virtual worlds and the definition of an open and flexible agent-based software architecture to support this generic model. MAEVIF's agent-based architecture is inspired by the separation into four modules classical of ITSs: student, expert, tutoring and communication modules. Figure 1 shows a detailed view of the agent-based server part, which has been implemented using JADE (Java Agent DEvelopment Framework), while the client, responsible for the visualization of the 3D virtual environment and for managing the interaction with the user, has been built using C++ and OpenSceneGraph. The communication between the agents and the clients is made through a Message Center via sockets.

In this architecture, the Communication Agent acts as the interface between the agent-based ITS and the 3D graphical environment with which the student interacts. Therefore, every significant event provoked by the student (for instance

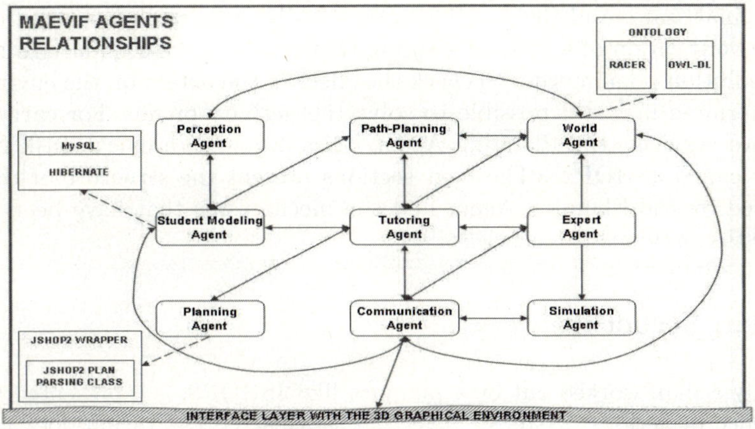

Fig. 1. Agent-based server architecture

by moving around, or by interacting with virtual objects) will be notified to the server through the communication agent, and every information or command issued by the tutoring part (for instance a recommendation to the student, or a movement of the virtual representation of the tutor) will be transmitted to the 3D graphical environment in the same way. The Tutoring Agent is responsible for the instructional decisions. Therefore, it will select the activity or case to be posed to the student and it will act as the coach during the execution of the activity by the student or team of students. The Student Modeling Agent is in charge of building a model of the student's knowledge and behavior. The Simulation Agent is the one that encapsulates the simulation of the systems behind the 3D world. As an example, if the 3D environment reproduces a control room for a nuclear power plant, the Simulation Agent will contain the model and simulation of the plant. The World Agent is a very important part in MAEVIF's architecture. It maintains an ontology that contains both geometrical and semantic information about virtual objects and spaces, and it provides some inference capabilities that are able to reason on this information.

The solution plan for an activity that has been selected by the tutoring agent is worked out by the Planning Agent. The Path-Planning Agent will complete the plans with respect to those actions involving movement along the 3D environment, since it is able to calculate a trajectory from a certain point of the world to another one. Once the solution plan has been computed the Tutoring Agent will start monitoring the students' actions in the virtual environment, checking if they are valid or not with respect to the plan worked out by the Planning Agent, and making tutoring decisions.

Since one of the most likely modifications in the system is a change in the planning algorithm, the planning agent acts as a wrapper for the planner, hiding details of its functioning to the rest of the system. The services provided by the Planning Agent to the Tutoring Agent are: the construction of a solution plan for an activity, given the initial state of the environment and simulation, the desired final state, and the possible actions to be applied in that environment; and the verification of a student's action that is out of the sequence established by the solution plan in order to check the effect of the action on the environment and determine if is still possible to solve the activity or not. For carrying out these two services, the Planning Agent relies on a Hierarchical Task Network planner called JSHOP2[1]. The next sections present the structure of the plans generated by the Planning Agent and the mechanisms that have been used to provide the previous two services.

3 Plan Structure

Normally a plan worked out by a planner, like JSHOP2, is represented through a sequence of actions or steps. Nevertheless, this representation does not cope with the possibility of some actions (or even subsequences) in the plan being executed in any order. For example, if we have to put two garments B1 and B2 in a washing machine, the order of introduction is not important. In order to cope with this possibility, a more sophisticated plan structure has been adopted. This plan structure is outlined in figure 2.

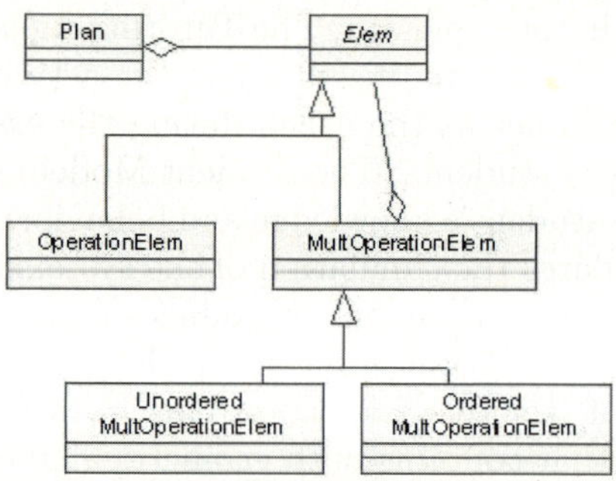

Fig. 2. Plan structure

A plan is formed by a sequence of elements (class Elem). In turn, a element is either an operation element or a multiple operation element. While an operation element represents a unique action, a multiple operation element comprises a sequence of elements (recursive definition) where the sequence of elements may specify a fixed order of execution (ordered multiple operation element), or any permutation of the elements in the sequence (unordered multiple operation element).

However, as it was mentioned above, JSHOP2 provides its plans as plain sequences of actions, and therefore nothing to do with the plan structure shown

[1] http://www.cs.umd.edu/projects/shop/description.html

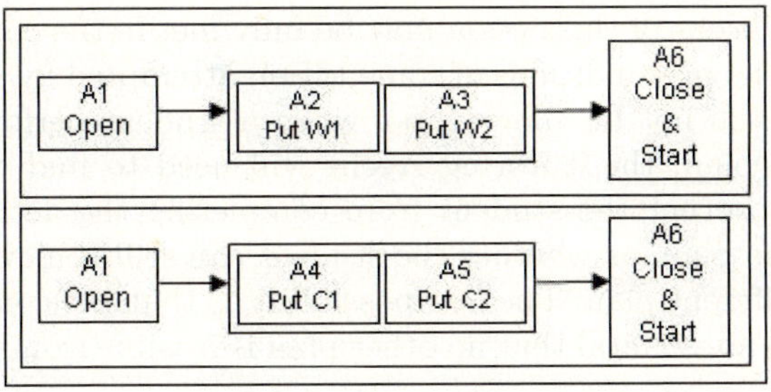

Fig. 3. Plan example

above. On the other hand, when designing the plan domains for JSHOP2, it is not difficult for the designer to identify the unordered blocks of tasks, and therefore to delimit them by using artificial operators like brackets in an arithmetic expression. In this way, a parser integrated in the planning service of the JSHOP2 wrapper may later build the plan structure outlined above from the plan in JSHOP2 format enriched with the artificial operators.

Next, an example of a plan represented through the enriched structure will be shown. In order to wash some garments, two white ones and two colored ones, by using a washing machine, it is necessary to perform two washing cycles, one for the white garments and another for the colored garments. Each cycle requires opening the machine, then putting the garments in it, and finally closing the machine and starting the washing. Lets assume that the washing machine is smart enough to detect whether there are both white and colored garments inside. In this case, the washing machine will not start washing to avoid damaging the garments. For the sake of simplicity, the actions for programming the machine and taking out the garments have been skipped. The plan for washing the garments is depicted in figure 3. Clearly, there are two ordered multiple operations (the elements are joined with arrows), and three unordered multiple operations elements, one in the first level representing that the order of the washing cycles is irrelevant, and another in each cycle representing that the order for putting in the garments of the same type (white or colored) does not matter. The example of figure 3 specifies four different sequences for achieving the same goals, in this case, having all the garments clean. The plan structure can be seen conceptually as a tree in which each path from the root to each leaf represents one of the aforementioned four sequences.

Thanks to the designed plan structure, the Tutoring Agent will need the replanning service less often, because it will be able to find out by itself whether an action is valid w.r.t. the plan associated with the activity.

4 Plan Tracking with Replanning

At some moment of the learning session, a student may perform an action that is not a correct action according to the plan. Even if an action is not correct w.r.t. the plan, this does not mean that the action cannot not be performed in the VE, since the preconditions of this action may be fully met in the current state of the VE. However, if the preconditions of some action attempted by a student are not met, this action will not be allowed. So, when a student performs an incorrect action w.r.t. the plan, the Tutoring Agent will need to find out whether this action does not prevent the student from completing the activity by carrying out the rest of the plan, or whether the student can still achieve the same goals by following a different plan. The last possibility is that the consequences of the performed action are so fatal that no other plan is feasible from the current state of the VE.

4.1 Is the Last Performed Action Relevant?

In order to obtain an answer to these questions, the Tutoring Agent needs to resort to the Planning Agent. When the Planning agent receives the request including the last performed action, first of all, taking into account the current state of the world, finds out whether the remaining plan is still applicable, and whether after executing the remaining plan, all the goals will be met. That is, the Planning agent must check whether the last action is *relevant* w.r.t. the current plan. Following with the example of the washing machine, a relevant action would be "putting a color garment (A4) in the machine" after "having put a white garment in (A2)", since the washing machine will not start washing (A6) because there are garments of different colors in the washing machine (one precondition of A6 is not met). In addition, an irrelevant action would be "going down on his knees" after "having opened the washing machine (A1)", because it is supposed that a person can load or unload the machine on kneeled or standing up.

For achieving this, the Planning agent will need to know the current state of the world, so it must have been informed of all the actions performed by the student in the VE so far. Only this way, the Planning agent can check whether the remaining steps of the current plan may still be carried out. It must be noticed that the remaining plan may represent more than one valid sequence of steps. Hence, the Planning agent must check whether one of these sequences is still valid, because we assume if one sequence is applicable, then the others will be also applicable. Therefore, the plan will be valid as a whole, and it must not be discarded.

In order to check whether a sequence of steps is still valid, the Planning agent must traverse the sequence applying each traversed action to the state obtained as a result of applying the previous action in the sequence. The first action must be applied to the current state of the VE expressed in terms of the JSHOP2 language. It may be happen that some action of the sequence cannot be applied, since its preconditions cannot be satisfied in the corresponding state. In this case, the last action performed by the student will be considered as relevant, and it will require a further treatment, as it will be shown below. Nevertheless, if the sequence is completely applied and all the goals are met in the final state, then the action will be considered as irrelevant, and this will be notified to the Tutoring Agent.

4.2 Is the Last Performed Action Reversible?

Once the Planning Agent has classified the action as relevant, the next step consists in finding out whether the last performed action is reversible or irreversible. An action is *reversible* if there is still some other plan that can allow the student to reach the final desired state from the current state. The previous example of relevant action ("putting a color garment (A4) in the machine" after "having put a white garment in (A2)") would be also an example of reversible action, because there would still exist a plan to reach the desired final state (all the garments clean) comprising the actions "taking out one of the garments", and then "putting a garment of the same color as the garments inside the machine" and "starting the washing". On the other hand, an example of irreversible action would be "starting the washing" after having put just one garment in the machine, because it is not possible to stop the washing to put another garment in so that we can get the desired final state in the end.

For finding out whether the action is relevant, the Planning Agent runs the JSHOP2 planning service through a JSHOP2 wrapper service providing as initial state the current state of the VE, and as planning domain a tailored version of the domain employed for working out the initial plan. Basically, this tailored version of the domain is able, unlike the normal version, to deal with initial states that for the other domain can only be intermediate states. As a result of this execution of JSHOP2, two possible outputs may be obtained. First, JSHOP2 may return a plan, which means that the action at hand is reversible, because its effects will not prevent the student from reaching a desired final state. Otherwise, that is, if JSHOP2 fails when working out a plan, the action will be considered irreversible from the point of view of JSHOP2. However, this does not mean that the action is irreversible necessarily, as there is no reason why the tailored planning domain has to be complete in the sense of being able to return a plan for each possible initial state, if this plan exists. In general, when designing the tailored planning domain, it is very difficult to prove its completeness. Hence, when designing the Planning Agent, it is sensible to leave open the possibility of JSHOP2 failing to work out a plan, even though it exists. Thus, if JSHOP2 fails to work out a plan, the Planning Agent will resort to a plan repair technique. Basically, the proposed technique relies on defining a direct correspondence between actions and reverse actions. In this way, the Planning Agent may undo the last performed action by creating a plan whose first part is in charge of undoing the effects of this action, so that the Planning agent can reuse the remaining steps of the initial plan.

Finally, if the action is reversible, the Planning Agent will notify it to the Tutoring Agent, and it will send the new plan for the learning activity. Otherwise, if the Planning Agent is not able to work out a plan neither by using the tailored planning domain, nor by using the plan repair technique, then the last performed action will be classified as irreversible, and this will be notified to the Tutoring Agent. From this point, the Tutoring agent may follow different tutoring strategies like, for example, stopping the learning session and explaining to the student his/her mistake, or leaving the student going on with the learning session so that he/she can discover by himself/herself that he/she has made a mistake.

5 Conclusions and Future Work

This paper presents how a planner can enrich the tutoring process performed by an IVET. This enrichment can be appreciated in two phases of the tutoring process: the initial generation of a plan as an ideal solution to an exercise, and the evaluation of the effect of the student's actions during their resolution of the exercise. The use of automated planning in ITSs, as it has been presented in this work, permits to give to the student a greater freedom of action during the resolution of an exercise. Moreover, we think this greater freedom improves the learning experience, therefore, in our opinion, this work represents a remarkable contribution to Computer Aided Instruction. In addition, the planner (JSHOP2) employed by MAEVIF supports a more expressive language for defining the task model than the planner employed by previous contribution such as Steve. Hence, MAEVIF will permit to learn more complex activities than Steve.

As future work, we will tackle the planning, tracking and replanning for co-operative activities, in which more than one student will be involved. This will require some extensions in the MAEVIF architecture, and more particulary, in the Planning Agent, so that it can deal with temporal planning and planning of concurrent actions.

References

1. Wasson, B.: Determining the Focus of Instruction: Content Planning for Intelligent Tutoring Systems. PhD thesis, Department of Computational Science, University of Saskatchewan (1990)
2. Peachey, D., MacCalla, G.: Using planning techniques in intelligent tutoring systems. International Journal of Man-Machine Studies 24, 77–98 (1986)
3. Vassileva, J.: Reactive instructional planning to support interacting teaching strategies. In: Proceedings of the 7-thWorld Conference on AI and Education, Washington, USA, pp. 334–342 (1995)
4. Martens, A., Uhrmacher, A.M.: Adaptive tutoring processes and mental plan. In: Cerri, S.A., Gouardéres, G., Paraguaçu, F. (eds.) ITS 2002. LNCS, vol. 2363, pp. 71–80. Springer, Heidelberg (2002)
5. Rickel, J., Johnson, W.L.: Animated agents for procedural training in virtual reality: Perception, cognition and motor control. Applied Artificial Intelligence 13, 343–382 (1999)

Determination of Illuminance Level Using ANN Model

Vedat Topuz[1], Selcuk Atis[2], Sureyya Kocabey[3], and Mehmet Tektas[4]

[1, 2, 4] Marmara University, Vocational School of Technical Sciences,
34722 Istanbul, Turkey
[3] Marmara University, Technical Education Faculty,
34722 Istanbul, Turkey
{vtopuz,satis,sureyyakocabey,tektas}@marmara.edu.tr

Abstract. In this study, an illuminance determining method, using an artificial neural network (ANN) model, has been designed. The model was realized as an alternative to existing simulation programs to determine the illuminance of a working place. In the model, maintenance factor (MF), working plane (WP), suspension height (SH) of luminaries were selected as input parameters. average illuminance (Eav), minimum illuminance (Emin) and maximum illuminance (Emax) of working plane were selected as output parameters that are the effective parameters in establishment and maintenance of luminance. Comparison between the real time measurements, illuminance simulation program (ISP) and ANN model results has shown that designed ANN model is satisfied.

Keywords: Illuminance Level, ANN Model.

1 Introduction

Illumination affects the mood and motivation level of people. Its systemic effect on mood has been expressed through experimental studies [1]. In this context, many studies have been made to research suitable illumination conditions, particularly for working places. While some of these studies have used classical model approaches, the others have used artificial intelligence modeling approaches [2-10].

One of the most important criteria implemented in these studies is to ensure that the illuminance must be in desired level depending on the tasks in working places. For this aim, illuminance level can be determined either by real time measurements or illumination simulator programs (ISP).

An alternative illuminance determining method using an ANN model has been designed in this study. The ANN model has been implemented by ANN simulator developed by research team. Illuminance data related to a working plane has been obtained by designed ANN model as well as the real time measurements and a commercial ISP. The obtained illuminance results have been compared with real time measurements both for ISP results and ANN model results.

In the following sections, mathematical and ANN modeling of illuminance have been outlined and result of the studies have been explained.

B. Apolloni et al. (Eds.): KES 2007/WIRN 2007, Part I, LNAI 4692, pp. 773–780, 2007.

2 Mathematical Model

Illuminance (E) is defined in SI system as follows

$$E = \frac{d\Phi}{dA} \quad (1)$$

where, E is illuminance (lux), A is area receiving the flux (m2), Φ is luminous flux (lumen). It can be treated as a vector quantity. It can be calculated as at a specific point (point-by-point method) illuminance or in an average uniform horizontal illuminance (lumen method) across the working plane.Consider a point source illuminating a surface at an angle θ to the normal as in Fig. 1

$$E_p = \frac{I_\theta . Cos\theta}{r^2} = E_{max} . \cos\theta \quad (2)$$

where Ep is illuminance value at point P (lux) Emax is the maximum illuminance that the source could produce at point P, when θ=00 (lux), Iθ is the luminous intensity of the source in the direction of the illuminated point (Iθ =dΦ/dω) (candela [cd]), r is distance of the light source to the object (meter), θ is angle of light source as to normal.

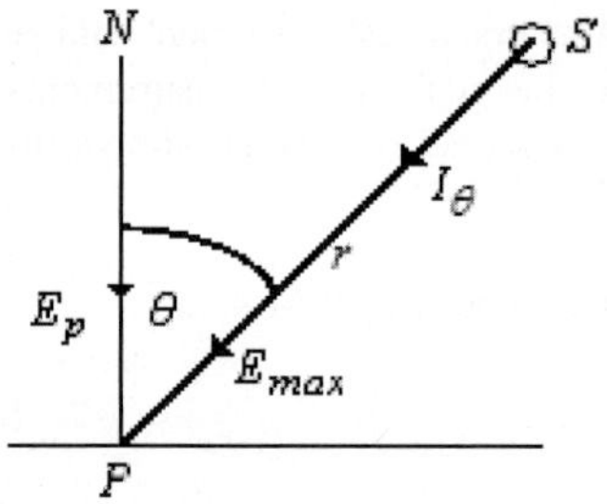

Fig. 1. A point source illuminating a surface at an angle θ to the normal

Specifications often require the lighting professional to know or design for average uniform horizontal illuminance. To do this with the Inverse Square Law for a large number of points would be both tedious and expensive. In addition, a second set of calculations would have to be made to determine the interreflected components. The lumen method is used to calculate the number of luminaries required for a uniform or general lighting layout. The lumen method calculates the average, uniform, horizontal maintained illuminance throughout a room. The average surface illuminance is calculated from the following equation

$$E_{av} = \frac{N \times \Phi_{in} \times n \times UF \times MF}{A} \quad (3)$$

where Eav is average illuminance(lux), N is the number of luminaries, Φin is initial luminous flux of the light source (lumen), n is number of lamps per luminaire, UF is utilisation factor. UF is the ratio of the total flux received by surface to the total lamp flux of the installation. It depends on the dimensions of room, the ceiling colour, the wall colour and the floor colour, A is area to be illuminated (m2), MF is the maintenance factor [11-12].

3 ANN Model

There are multitudes of different types of ANN models. Some of the more popular of them include the multilayer perceptron, which is generally trained with the back-propagation algorithm. Realized three layers perceptrons is shown in Fig.2.

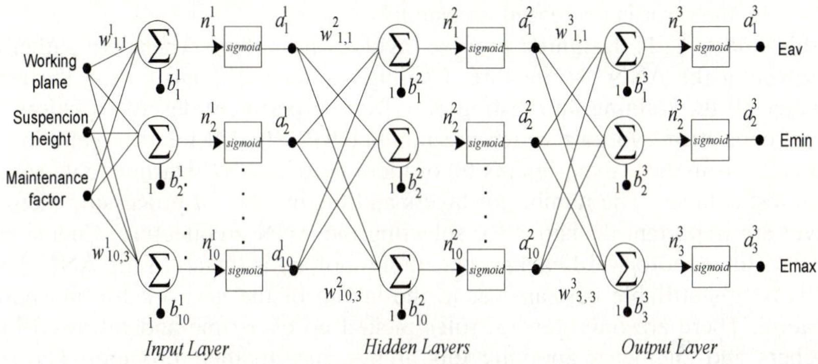

Fig. 2. Three-layer feedforward network

ANN must be trained before it becomes useful.. The training method tries to minimize the current errors for all processing elements. The training continues until the ANN reaches user defined performance level. Test is critical to insure that the network has not simply memorized a given set of data, but has learned the general patterns involved within an application. For train to multilayer perceptron with backpropagation, the first step is propagating the inputs towards the forward layers through the network. For a three-layer feedforward network, training process is initiated from the input layer [13]:

$$\begin{aligned} &a^0 = p \\ &a^{m+1} = f^{m+1}(W^{m+1}a^m + b^{m+1}), \qquad m = 0,1,2 \\ &a = a^3 \end{aligned} \tag{4}$$

where a output vector, p is input vector, f(.) is the activation function, W is weighting coefficients matrices, b is bias factor vector and m is the layer index.

Second step is propagating the sensibilities (s) from the last layer to the first layer through the network: s^3, s^2, s^1. The error calculated for output neurons is propagated to the backward through the weighting factors of the network. It can be expressed in matrix form as follows:

$$\begin{aligned} &s^3 = -2\dot{F}^3(n^3)(t-a) \\ &s^m = \dot{F}^m(n^m)(W^{m+1})^T s^{m+1}, \quad \text{for } m = 2,1 \end{aligned} \tag{5}$$

where t is target vector and $\dot{F}^m(n^m)$ is Jacobian matrix. The last step in backpropagation is updating the weighting coefficients. The state of the network always changes in such a way that the output follows the error curve of the network towards down.

$$W^m(k+1) = W^m(k) - \alpha s^m (a^{m-1})^T \quad (6)$$

where α represents the training rate, k represents the epoch number. By the algorithmic approach known as gradient descent algorithm using approximate steepest descent rule, the error is decreased repeatedly.

In this study, ANN designing process involves four steps. These are gathering the data, selecting the ANN architecture, training the network, and testing the network. We gather all the training and testing data from experiment described below. These data sets consist of two part which are inputs (WP, SH, MF) and target (Eav, Emin and Emax). From these experiments 60 of them used as ANN learning data set and 30 of them test data set. The number of layers and the number of processing elements in per layer are important decisions for selecting the ANN architecture. Choosing these parameters to a feedforward backpropagation topology is the art of the ANN designer. There is no quantifiable best answer to the layout of the network for any particular application. There are only general rules picked up over time and followed by most researchers and engineers applying this architecture to their problems.The optimal numbers of neurons in the first and second layers have been chosen as 10. Also, the activation function has been chosen as a sigmoid function for all of the layers. Realized ANN model is shown in Fig. 2.

4 Results

The experimental space is a classroom in sizes of 10.33 x 6.98 x 3.72 m at the University of Marmara, Technical Education Faculty. The classroom is illuminated by six without reflector luminaires which have 2x36W T8 fluorescent lamp and each lamp luminous flux is 3350 lumen. The values of wall, ceiling and floor reflectance are respectively 60%, 83% and 33%. In Real time measurement, Lutron LX-1102 luxmeter is used.The experimental measures were made in September at night. Firstly, while luminaires were mounted on ceiling, illuminance measurement was made at 150 points on the working plane in height of 0.5, 0.6, 0.7, 0.8, 0.83 meters. Then, suspension height of luminaire was changed. 0, 0.3, 0.5, 0.6, 0.7, 0.9 meters and every measurements were repeated for each suspension height as shown in Fig. 3.

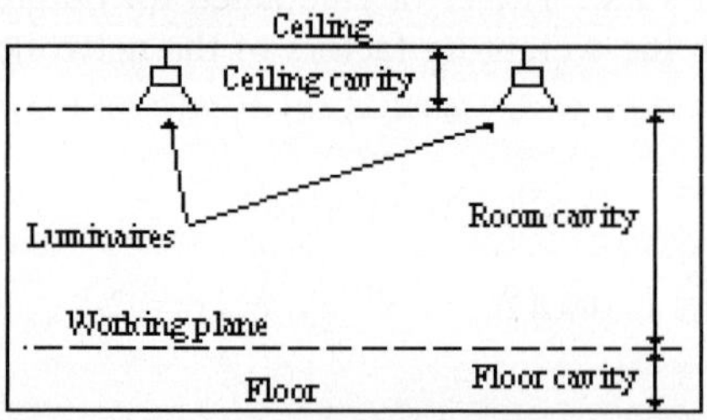

Fig. 3. Working plane height (floor cavity) and luminaries suspension height (ceiling cavity)

According to EN12464-1:2002 standard, the maintenance coefficient of measurement class is determined as 0.81. Besides, by considering the different maintenance conditions of environment, 0.7, 0.6, 0.5, 0.4 maintenance factor coefficients were determined. As you seen in equation (3) maintenance factor (MF) is direct proportional to illuminance (E). For this reason, the Eav, Emin and Emax for 0.7, 0.6, 0.5, 0.4 maintenance coefficients were calculated.

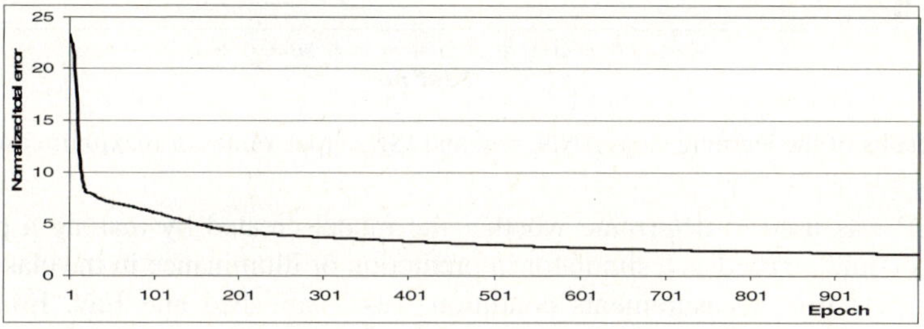

Fig. 4. Variation of the total training error through the one thousand epochs

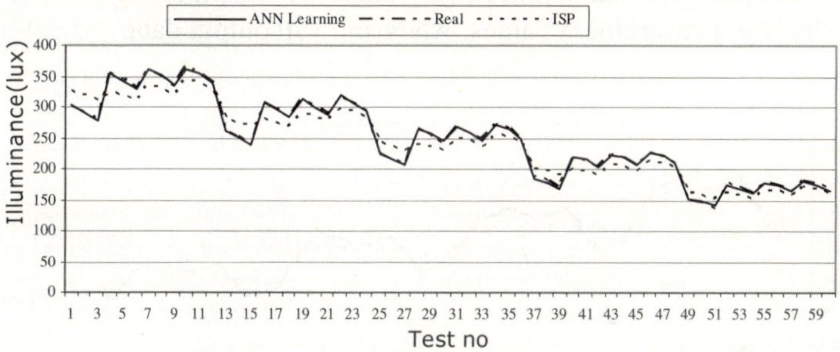

Fig. 5. Results of the learning step; ANN, real and ISP output values of average illuminance

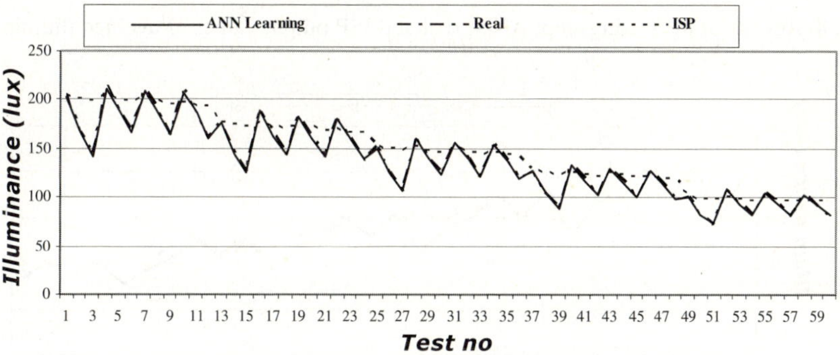

Fig. 6. Results of the learning step; ANN, real and ISP output values of minimum illuminance

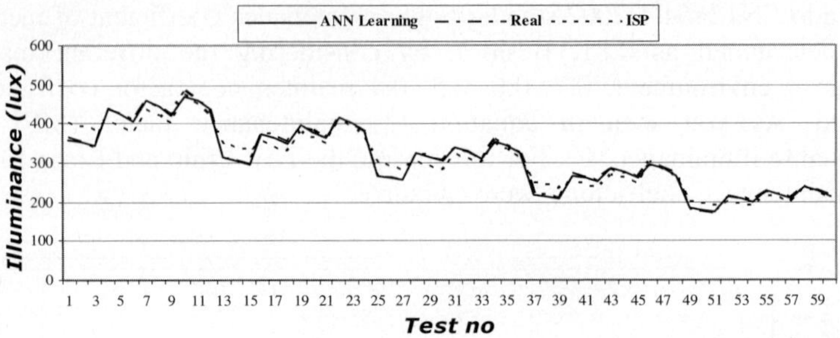

Fig. 7. Results of the learning step; ANN, real and ISP output values of maximum illuminance

The ISP was used to determine whether the model created by making a prediction with ANN could be used as a simulator in prediction of illuminance in the classroom. In the ISP a real time measurements condition was established and Eav, Emin, Emax values were recorded. Then ANN simulator has been trained through the one thousand epochs. The training process has been stopped when the error has become stable. Variation of the total absolute error through the epochs is shown in Figure 4. End of the training step, founded Eav, Emin and Emax values are shown in Fig. 5, 6, 7. Each figure also shows the real measurement values, ANN and ISP output data.

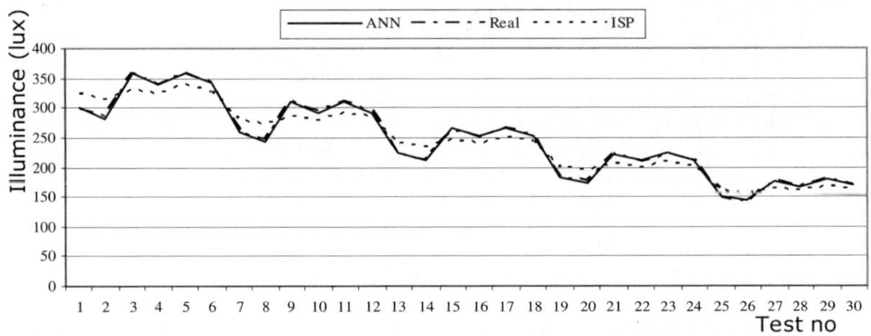

Fig. 8. Results of the testing step; ANN, real and ISP output values of average illuminance

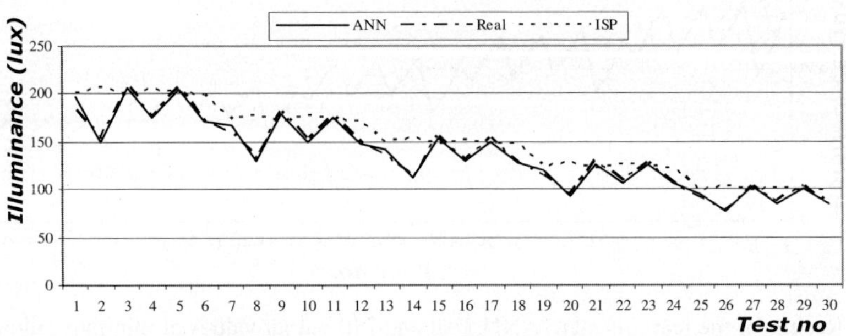

Fig. 9. Results of the testing step; ANN, real and ISP output values of minimum illuminance

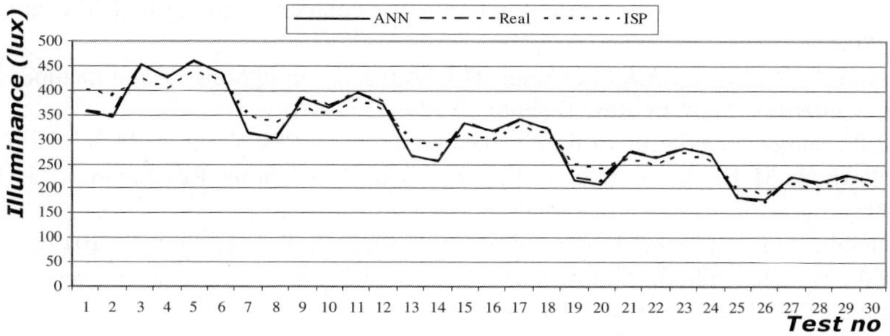

Fig. 10. Results of the testing step; ANN, real and ISP output values of maximum illuminance

After the training step testing step was realized. In the testing step, the data which were not used in learning step were applied to ANN, to prove the performance of realized ANN. model. Also end of the testing step, founded Eav, Emin and Emax values are shown in Fig. 8, 9, 10. Each figure also shows the real measurement values, ANN and ISP output data.

5 Conclusion

In this study, an illuminance determining method using an ANN model has been designed. This model was realized as an alternative to existing ISP to determine the illuminance of a working place.

The root mean square error (RMS) between the real time measurements and ANN model results are shown in Table.1. The RMS errors between the real time measurements and ISP results are also shown in Table.1, depending on both the learning data and test data.

Table 1. RMS error obtained by ANN and ISP

	Eav		Emin		Emax	
	ANN	ISP	ANN	ISP	ANN	ISP
Learning Data	0,21	2,31	0,14	2,51	0,44	2,75
Test Data	0,16	1,61	0,31	1,96	0,25	2,05

Referring to Table 1, Eav error obtained by ANN model has been reduced comparing to error obtained by ISP, both learning and test step. In the same manner, Emin and Emax error obtained by ANN model have been reduced comparing to ISP. So, the designed model can be used to develop an illumination control system reducing the operation and maintenance cost. Such a control system study will be planned in the future.

References

1. McCloughan, C.L.B., Aspinall, P.A., Webb, R.S.: The impact of lighting on mood. Lighting Res.Technol. 31, 81–88 (1999)
2. Newsham, G.R, Richardson, C., Blanchet, C., Veitch, J.A.: Lighting quality research using rendered images of offices, Lighting Res. Technol. 37, 93–115 (2005)

3. Moeck, M.: Lighting design based on luminance contrast, Lighting Res. Technol. 32, 55–63 (2000)
4. Lupton, M.Y., Leung, A.S.M., Carter, D.J.: Advances in lighting design methods for non-empty interiors, Lighting Res. Technol. 28, 29–41 (1996)
5. Jay, P.: Subjective criteria for lighting design, Lighting Res. Technol. 34, 87–99 (2002)
6. Aizlewood, M E.: Innovative daylighting systems. Lighting Res.Technol. 25, 141–152 (1993)
7. Littlefair, P.: Daylighting design and research, Lighting Res. Technol 32, 101 (2000)
8. Çolak, N., Onaygil, S.: Prediction of the artificial illuminance using neural network, Lighting Res. Technol 31(2), 63–66 (1999)
9. Kurian, C.P., Kuriachan, S., Bhat, J.: An adaptive neuro-fuzzy model for the prediction and control of light in integrated lighting schemes, Lighting Res. Tech. 37, 343–352 (2005)
10. Corcione, M., Fontana, L.: Optimal design of outdoor lighting systems by genetic algorithms Lighting Res. Technol. 35, 261–280 (2003)
11. Simons, R.H., Bean, A.R.: Lighting Engineering. Architectural Press (2001)
12. Chen, K.: Energy Management in Illuminating Systems. CRC press, Boca Raton, USA (1999)
13. Hagan, T.M., Demuth, H.B., Beale, M.: Neural Network Design, PWS Publishing Company, Boston, MA vol. 4, 11, pp. 1–13, 1–14 (1996)

Efficient Content Distribution Method Based on Location and Similarity in Unstructured P2P System

Suhong Min[1], Byong Lee[2], and Dongsub Cho[1]

[1] Department of Computer Science and Engineering,
Ewha Women's University, Seoul, Korea
shmin@ewhain.net, dscho@ewha.ac.kr
[2] Department of Computer Science,
Seoul Women's University, Seoul, Korea
byongl@swu.ac.kr

Abstract. The peer-to-peer (P2P) systems have grown significantly over the last few years due to their high potential of sharing various resources. However, identifying the location and content similarity between peers is very costly and time-consuming as it involves the cooperation of many peers. In this study, we propose an ISP2P (Intelligent Super-Peer based P2P) system, which classifies the peers into three types, ASP (Agent Super-Peer), SP (Super-Peer) and OP (Ordinary-Peer), and utilizes the location information and content types to provide efficient content distribution and retrieval service. The proposed system can improve the performance of average message response time and reduce the message traffic by using the similarity of the network distance and contents.

Keywords: Peer-to-Peer (P2P), Super-Peer (SP), Network distance, Content Similarity.

1 Introduction

For the last few years, there has been a large volume of research on Peer-to-Peer (P2P) system, resulting in many hybrid P2P models. Many researches have shown that P2P systems become very effective, especially in query processing, when dividing the peers into two layers, SP (Super-Peer) and OP (Ordinary-Peer). With this layer separation, OP should select only one SP for sharing resources and can participate in the network only through the chosen SP. Compared with pure P2P systems, Super-peer based P2P systems have to deal with a large number of queries from OPs. As a result, SP's workload can be overloaded due to the number of query processing. In this paper, we propose ISP2P system which classifies peers into three type, ASP (Agent Super-peer), SP (Super-peer), and OP (Ordinary-peer). The aim of this classification of peer is to equitably distribute their workload and contents based on their locations and content types. We suggest the following approaches: First, we consider the network distance between peers so that it allows us to find content's location. To do this, we compute the network distance with modified GNP (Global Network Positioning) algorithm which estimates the distance between peers without direct measurement with Cartesian coordinate system. Second, we analyze the similarity of

B. Apolloni et al. (Eds.): KES 2007/WIRN 2007, Part I, LNAI 4692, pp. 781–788, 2007.

sharing content between peers to retrieve content efficiently. To do this, we compute the similarity of content between peers with the information retrieval algorithm such as VSM (Vector Space Model) and ISM (Intelligent Search Mechanism) which are used to evaluate the similarity between query and content. The main contributions of this paper are: We can improve the performance of average message response time and reduce the network bandwidth by consideration of the location and content similarity between peers.

The rest of the paper is organized as follows: Section 2 reviews some related works briefly. Section 3 propose the ISP2P system which computes the network distance between peers and evaluates the similarity of content with VSM and ISM algorithm. Section 4 shows the simulation results of the proposed mechanism; finally, the conclusion and the future work are added in Section 5.

2 Related Works

2.1 GNP (Global Network Positioning)

Ng et al., proposed a Cartesian coordinate-based approach, called Global Networking Positioning (GNP) [2]. Instead of using the original network distances, GNP represents the location of each host in N-dimensional Cartesian coordinate system, where N is the number of beacon nodes. The coordinate of a host is the distances from itself to the beacon nodes, and the distance between two hosts is calculated as the Euclidean distance in the Cartesian coordinate. The major advantage of representing network distances in a coordinate system is to extract topological information from the measured network distances. As a result, the accuracy in estimating the distance between two arbitrary hosts will be improved especially in the case that the number of beacon nodes is small [3].

2.2 VSM (Vector Space Model)

This algorithm is for evaluation of the similarity between query and document. In this model, a document is represented as a term vector [4]. Each component of the vector represents the importance of the corresponding term within the document. Components are calculated using TFxIDF scheme, where TF (term frequency) is the frequency of the term within the document and IDF (inverse document frequency) is the inverse of the number of documents in which the term appears. The idea behind this scheme is that terms that appear more frequently in a document are good candidates for representing this document, whereas terms that appear in too many documents should be penalized. A document set with d documents can be represented by a t £ d term document matrix, which is obtained by combining the vector representations of all documents, where t is the number of distinct terms appearing in the set. Each query is also represented as a vector, which is the weighted sum of the vectors of the terms appearing in the query. A common similarity measure is cosine similarity, which is the cosine of the angle between the vectors [5]. The cosine similarity between two n-dimensional vectors A and B is expressed as formula (1).

2.3 ISM (Intelligent Search Mechanism)

The ISM is a new mechanism for information retrieval in P2P networks. The objective of the algorithm, which was proposed in [7], is to help the querying peer to find the most relevant answers to its query quickly and efficiently. Keys to improving the speed and efficiency of the information retrieval mechanism is to minimize the communication costs, that is, the number of messages sent between the peers, and to minimize the number of peers that are queried for each search request. To achieve this, a peer estimates for each query, which of its peers are more likely to reply to this query, and propagates the query message to those peers only. The Intelligent Search mechanism consists of two components: A Profile Mechanism, which a peer q uses to build a profile for each of its neighboring peers. The profile keeps the most recent replies of each peer. A Relevance Rank, which is a peer ranking mechanism that uses the peer's profiles to select the neighbors that will lead a query to the most relevant answers [8].

3 ISP2P System

3.1 System Architecture

We have extended ISP2P (Intelligent Super-peer based P2P system) [1, 6]. Proposed system's functions are divided into two kinds as follows. First, we consider the network distance between peers so that it helps us to efficiently distribute contents according to peer's location. Second, we calculate the similarity of content sharing between peers to improve content retrieval. In this proposed system, peers are divided into three types, ASP (Agent Super-peer), SP (Super-peer), and OP (Ordinary-peer). The aim of classification of peers is to distribute workloads among peers by grouping them with specific domains such as their locations and content types. ASPs are linked with each other and they are organized as a virtual network on top of the P2P network. When OPs or SPs join the network, they first register with an ASP. ASP's main function is to provide the connection management of peers that join or leave the network and distribute peers according to their location and the category of content. We calculate the network distance between peers with modified GNP (Global Network Positioning) algorithm. SP's primary function is to retrieve contents on behalf of OPs. SPs explore common features indicating the similarity of content sharing between peers and compute the similarity of content corresponding to queries with VSM (Vector Space Model). And also they compute the similarity of current query through comparing to past queries with ISM algorithm. It applies the Nearest Neighbor classification technique and cosine similarity. OPs provide contents to other peers to enable them to download the contents. Also OPs generate CDF (Content Description File) to allow SPs to gain their content information. The CDF includes peer's content information such as content ID, name, type, and size etc. When OPs connect to one SP, they provide the CDF to its SP. The architecture of ISP2P is shown in Fig 1.

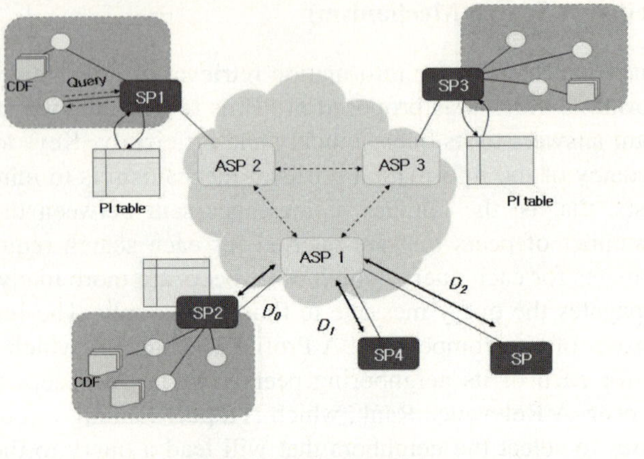

Fig. 1. ISP2P Overview

3.2 Content Distribution Based on Network Distance

In our proposed system, OPs should choose a SP to request queries. The network distance between SP and OP affects response time needed to process OP's queries. Therefore, OPs should select the closest SP to look for contents in a short time. The primary goal of estimating network distance is to enable measurement of the network distance between arbitrary peers without direct measurement between SPs and OPs. To do this, we calculate the network distance between OP and SP using GNP algorithm. Several approaches have been proposed among which GNP may have received the most attention. GNP transforms the original distance data space into a Cartesian coordinate system and uses coordinates in the coordinate system to represent the location [3]. In this paper, we modify the existing GNP to apply to our system. In our ISP2P system, first, OPs should choose ASPs to join the network. OPs selects ASPs based on RTT (Round Trip Time). To do this, we classify the range of possible latency values into a number of levels. For example, we separate the range of possible latency values into 3 levels; level 0 (L0) for latencies in the range [0,100] ms, level 1 (L1) for latencies between [100,200] ms and level 2 (L2) for latencies greater than 200ms. OP selects ASP with the best level vector, and requests SP's information from ASP. Second, ASPs provide SP's location information to OPs. They should calculate the network distance between SP and ASP with Euclidean distance in the Cartesian coordinate. We divide the range of possible Euclidean distance values into 3 levels; level 0 (l0) for distance in the range [0,2], level 1 (l1) for distance between [2,4] and level 2 (l2) for distance greater than 4. Through this procedure, we can calculate the network distance without direct measurement between OP and SP. If ASP = (Ai, Aj) and SP = (Si, Sj) for two-dimension, the distance is computed as:

$$D_{euclidean} = \sqrt{(A_i - S_i)^2 + (A_i - S_i)^2} \quad (1)$$

3.3 Content Discovery

Our proposed system classifies peer's role to share contents or provide contents retrieval. OPs extract the feature of contents and generate the CDF (Content Description File) information to provide content to other peers. On the other hand, SPs provide content retrieval service to OPs. To provide this, SPs compute the similarity between requested query and CDF information of OPs and the similarity between new query and past queries with VSM and ISM algorithm. The process of content discovery is shown in Fig 2.

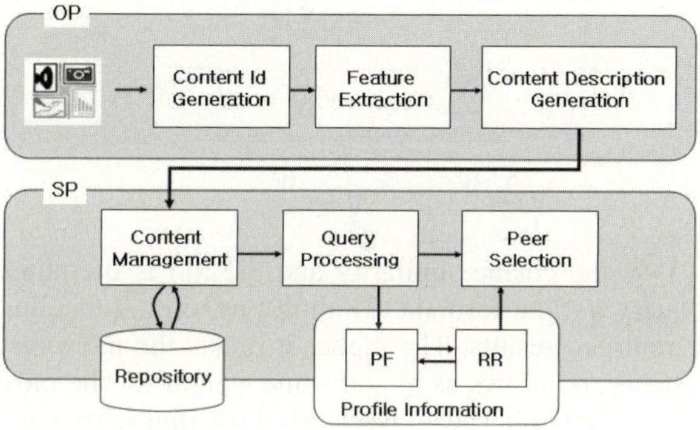

Fig. 2. Content Discovery Process

First, an OP generates a unique identifier to create the content ID using a hashing function. This step ensures that the same content will have the same identifier no matter where they are stored. The 32-bit cyclic redundancy checksum (CRC-32) algorithm is used to generate a unique identifier (ID). The CRC-32 is counted based on resources, and each resource is mapped to a single 32-bit number. Second, OP extracts common features from contents to make a content description. To extract file features, we should consider which resource type is used to share between peers in P2P system. If the aim of P2P system is to share specific resource type such as music file, image data, or document, we should focus on the resource type to extract features of resources. For example, if peers want to search music information using P2P system, we should consider MIR (Music Information Retrieval) [9]. In this paper, we deal with various media files and do not depend on specific resource types. Therefore, we focus on common resource features such as content ID, name, type and size. After extracting features, OPs make CDF (Content Description File) indicating sharing files. And then each OP transmits its CDF to SP connected. After receiving the CDF, a SP stores them at its repository. Initially, the SP computes the similarity between query and CDF with VSM algorithm when OPs request content retrieval to SPs. As time goes by, the SP generates PIT (Profile Information Table) which consists of PF (ProFile list) and RR (Relevance Rank) information. They are created by modified ISM algorithm that uses Nearest Neighbor classification technique and cosine similarity. We modify this to apply to our proposed system because it is originally for pure P2P system. The PF contains the list of the most recent past queries that provided the answer for. The SP monitors and records the list of past queries with respect to query hit message in the PF. The relevance rank information provides the peer's list with the most appropriate resources to answer to the query efficiently. To compute the ranking, the SP compares the query to previously seen queries and finds the most similar ones. The Relevance-Rank (RR) is calculated as follows:

$$RR_{Pl}(P_i, q) = \sum Qsim(q_j, q)^{\alpha} \times S(P_i, q_j) \tag{2}$$

$$Qsim(q_j, q) = \frac{q_j \cdot q}{|\vec{q_j}| \times |\vec{q}|} = \frac{\sum_{t=1}^{t} w_{i,j} \times w_{i,q}}{\sqrt{\sum_{i=1}^{t} W^2_{i,j}} \times \sqrt{\sum_{t=1}^{t} W^2_{i,q}}} \tag{3}$$

The formula (3) is the cosine similarity and $S(P_i,q_j)$ is the number of results returned by Pi for query q_j. The formula (2) allows us to rank the clusters with respect to the number of returned results. The higher it ranks, the more results it returns. In addition, the parameter, α, allows us to add some weight for the most similar queries. In this paper, we use PIT like a cache. Because of its limit table size, once PIT is full, the SP uses a Least Recently Used (LRU) policy to keep the most recent queries in the table.

As a result, when OPs request content discovery, we can provide it to OPs in a short time if there is content information in PIT. But, if not, we should find it in the SP's repository with CDF lists by comparing the query with the contents.

4 Experiment Evaluations

In this section, we present the simulation model used to evaluate the performance of ISP2P system and discuss simulation results. The simulation model is implemented in C++ using CSIM [10]. It consists of a number of OPs, SPs and ASPs. During simulation, OPs join and leave the network following a Poisson process with an arrival rate of λ and departure rate of μ. All peers repeat these processes during the simulation time. In our simulation, we verify that the proposed system can improve response time of query processing and bandwidth cost by considering the location and similarity between peers. To do this, we assume the performance of system as follows. All OPs send messages at the same frequency to SPs to connect to a SP and request queries. We also assume that the ratio of number of SPs is 10% of OPs. Hence, the number of SPs also increases as the number of OPs augments. We set the group size of ASPs to 10, and it doesn't change during the simulation time.

First, we compare the performance of content discovery algorithm as the number of peer changes. We measure the average message response time by each algorithm with VSM and combining of ISM and VSM. First, we experiment that SPs process queries based on CDF lists with VSM algorithm whenever OPs request content retrieval. We do not use ISM to look for content. Second, we measure the average response time based on combining VSM and ISM algorithm. In Fig. 3, comparing to two algorithms, we can improve the average message response time when we use proposed scheme.

Second, based on the distance and the similarity, we examine the influence of message traffic on various ratios of peers. Fig. 4 shows the message traffic size as peers increase. We compare the performance of proposed system with a conventional approach. Fig. 4 shows that the message traffic can be reduced in a network when we consider the similarity of distance and content. We demonstrate the effectiveness of proposed scheme which can show good performance with considering the location and the content similarity of peers.

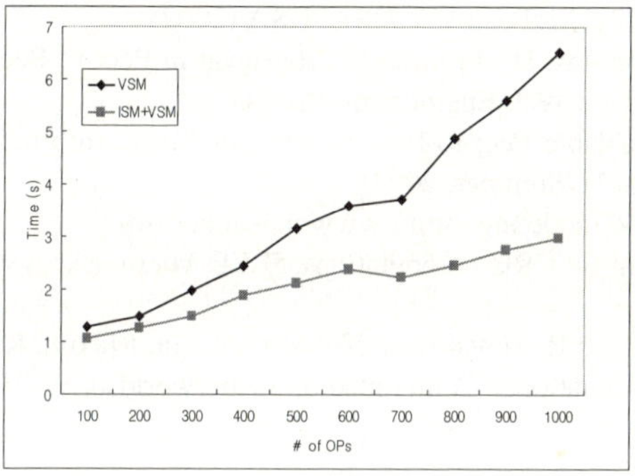

Fig. 3. The number of peer vs. Response Time

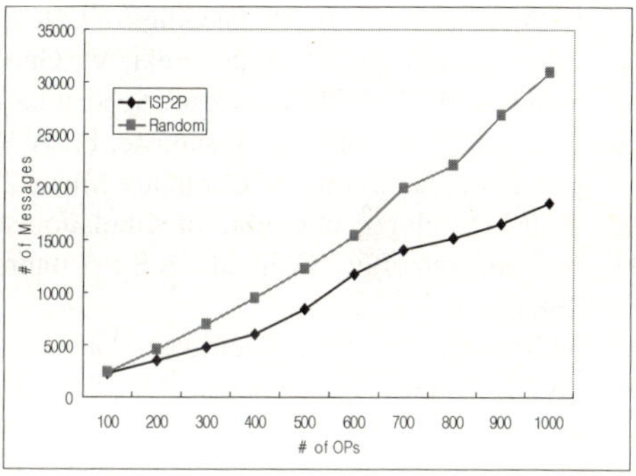

Fig. 4. The number of peer vs. Message Traffic

5 Conclusion and Future Work

We have presented the ISP2P system which classifies peers into three types to distribute workload and content of peers. In this paper, we can provide content distribution and information retrieval based on the location and the similarity of contents being shared between peers in P2P network. Especially, we focus on 1) recognizing the peer's location without direct measurement of distance between peers with GNP algorithm, and 2) providing the efficient content retrieval to peers with combining VSM and ISM algorithm. As you can see, we successfully demonstrated the performance of the proposed scheme using a number of simulations. In our experiment, we show that our proposed system can improve the performance of average message response time and reduce the message traffic by the similarity of the network distance and content in the network. We plan to extend ISP2P system to include additional functions in the future work.

References

1. Min, S.H., Cho, D.S.: Super-peer Selection based Framework using Dynamic Capacity and Similarity. In: Levi, A., Savaş, E., Yenigün, H., Balcısoy, S., Saygın, Y. (eds.) ISCIS 2006. LNCS, vol. 4263, Springer, Heidelberg (2006)

2. Ng, E., Zhang, H.: Predicting Internet network distance with coordinates-based approaches. In: Proceedings of Infocom (2002)
3. Lim, H., Hou, J., Choi, C.: Constructing Internet coordinate system based on delay measurement. In: Proceedings of the ACM/SIGCOMM Internet Measurement Conference (IMC-03) (October, 2003)
4. Lee, D.L., Huei, C., Seamons, K.: Document Ranking and the Vector Space Model. the journal of IEEE software 14(2) (March/ April, 1997)
5. Sahin, O.D., Emekci, F., Agrawal, D., Abbadi, A.: Content-Based Similarity Search over Peer-to-Peer Systems. The 1st international workshop on Databases, Information System and Peer-to-Peer computing (2003)
6. Min, S.H, Cho, D.: An Intelligent Performance based Hybrid P2P System. Journal of Korea Electrical Engineering and Technology 5(2) (February, 2006)
7. Kalogeraki, V., Gnuopulos, D., Zeinalipour-Yazti, D.: A Local Search Mechanism for Peer-to-Peer Networks. In: Proceedings of CIKM'02, McLean VA, USA, USA (2002)
8. Zeinalipour-Yazti, D., Kalogeraki, V., Gunopulos, D.: Information Retrieval in Peer-to-Peer Systems. IEEE CiSE Magazine, Special Issue on Web Engineering (2004)
9. Tzanetakis, G., Gao, J., Steenkiste, P.: A Scalable Peer-to-Peer System for Music Information Retreival. journal of Computer Music 28(2) (Summer, 2004)
10. CSIM Development toolkit for simulation and modeling, http://www.mesquite.com
11. Ramabadran, T.V., Gaitonde, S.S.: A tutorial on CRC computations. IEEE Micro (August, 1988)
12. Maxim, R., Hui, S.C.: Intelligent Content-Based Retrieval for P2P Networks. In: Maxim, R., Hui, S.C. (eds.) Proceedings of the 2003 International Conference on Cyberworlds(CW'03) (2003)

GIGISim – The Intelligent Telehealth System: Computer Aided Diabetes Management – A New Review

Joanna Koleszynska

AGH University of Science and Technology al. Mickiewicza 30, 30-059 Kraków, Poland
koleszyn@galaxy.agh.edu.pl

Abstract. The projected epidemic of diabetes will emerge as a major threat to the public health. This paper concludes a review of the contribution of the IT to the diabetes management. As the design and development of a computer-aided system for the education of diabetes, have been described previously in various papers, this one presents the analysis of clinicians and diabetes requirements. Software implementation of newly proposed innovations is presented together with effectiveness and suitability rate of systems, prior to the identified requirements. The second aim of this paper was to present a complex web system: the GIGISim developed to facilitate the management of Diabetes mellitus through the co-ordination of multiple healthcare tools available directly in Internet.

Keywords: Telehealth system, Teleeducation, web tools, glucose-insulin model, interactive simulation, diabetes therapy, evolutionary computing.

1 Introduction

This paper presents the review of the computer aided diabetes education introducing GIGISim (Glucose-Insulin and Glycemic Index Web Simulator) e-learning tool and a complex approach to the therapy management. Together with the system, recently reported solutions were summarized and their functionality compliance with the broad spectrum of diabetes therapy requirements, presented briefly in section 2, checked. The third section highlights the application of computers as teaching tools and ways in which such educational approaches might be applied to improve patients' management and health awareness. The fourth chapter presents interactive, diabetes-dedicated simulators which have a great deal of educational potential for patients and their families, and may also offer a means of training for health-care professionals. The role of decision support tools for planning insulin therapy, using various clinical algorithms, together with certain knowledge-based approaches was also presented. The novel features and potential benefits of such methodologies are highlighted and some of the limitations of currently available software are discussed.

The proposed new software – GIGISim with newly designed additional components is presented in section 5 together with observations and recommendations surrounding the use of the current version of the developed web system. A detailed description of the simulation algorithms and adaptation procedure provided to assure adequate results for

B. Apolloni et al. (Eds.): KES 2007/WIRN 2007, Part I, LNAI 4692, pp. 789–796, 2007.

each patient was reported in previous papers on GIGISim [1][2]. This one focuses in particular on the analysis of the patient needs and following component system architecture. Close attention is devoted to the practical application of each of the proposed tools included in GIGISim, with the possible effects on the therapy, conclusion and comments also detailed in section 6.

2 Diabetes Therapy

2.1 Clinician and Patients' Requirements

Our research on medical requirements in diabetes management follows the principles of Canadian and the American Diabetes Association. The basics of therapy are:

- Self-monitoring of capillary blood glucose, blood pressure, body weight and the level of cholesterol as a means of assessing metabolic control.
- Keeping records of above values to adapt insulin and drug doses.
- Making healthy food choices.
- Exercising and applying nutrition correlated to physical activity.
- Administrating insulin and drugs if necessary.
- Visiting physician and finding support end education in local diabetes centers.

The analysis of diabetes patients needs has been established through a series of intermediate research surveys and literature studies. Clinical survey was carried among diabetes patients in Krakow's Hospital and among members of the local Diabetes Association, to identify additional requirements and possible application of diabetes oriented software. Suggestions of patients are listed below:

- All individuals would like to receive nutritional counseling from registered dietitian as often as possible, preferably online via Internet.
- Automatically generated dietary recommendations should be individualized to accommodate person's preferences and lifestyle.
- Diet diaries and statistical analysis of improvements in body weight control should be provided.
- User graphical interface should be facilitated to improve the process of data entering and available online to assure help on therapy.

The implementation of those requirements in GIGISim and other popular systems will be discussed in the following chapters.

2.2 New Nutritional Requirements - The GI

Nutrition management is a key component of diabetes. Many factors should be taken into account, like the individual's micro - and macronutrient needs, physical activity, lifestyle and finally insulin dosages. Despite the importance of healthy diet in diabetes, patients often neglect nutrition recommendations like the role of the Glycemic Index (GI). In previous papers on GIGISim we proved that the exact effect of the GI on daily glucose variation is difficult to identify without dedicated methods.

The Glycemic Index is a scale that ranks carbohydrate-rich foods by how much they raise blood glucose levels and lowering the total GI of the diet may improve blood glucose control significantly [4][5].

According to our surveys, patients often ignore the importance of basic nutritional meal assessment, despite the necessity of the properly balanced diet. Calculations of fats, proteins, carbohydrates and calories are long and tedious and if the glycemic index should be included, as the additional nutrition factor, the process itself becomes even more complicated. The dedicated software should therefore be designed to facilitate this procedure, and additional nutritional recommendations cited from the Canadian Diabetes Association [3] should also be considered to provide a sufficient meal description:

- Eat at regular times.
- Limit sugar, sweets and products with high GI.
- Include foods high in fiber.

GIGISim was preliminary designed to educate patients on the GI impact on their diet using visualizations of the postprandial glucose profiles correlated to user's diet. Additional tools assessing all aspects of proper diabetes diet have been lately added.

3 Is There a Need for Computer Aided Diabetes Management?

The overall goal of computer-aided diabetes education is to help individuals with Diabetes mellitus gain the necessary knowledge about physiological processes, control dietetic habits and offer support needed on daily-basis. Despite the introduction of many therapeutic aids such as intelligent insulin pumps and injectors, blood glucose tests, insulin dose algorithms and new pharmacological formulas the metabolic care of most patients is still insufficient, mainly because of the number of therapy recommendations that should be taken under consideration and lack of the education. A meal and insulin dose planning requires experience from patients, their families and physicians - patients should correlate their insulin regime with blood glucose variations and daily diet. Type II diabetes should often modify eating habits to loose weight and improve their body's sensitivity to insulin. This all could be successfully managed with computer help as long as the software complies with user requirements and assures that given help is medically reviewed.

Although the impressive number of diabetes dedicated software is available in WWW, there are very few accessible solutions which would enable virtual, online patient-clinician contact, and could be use both by clinicians and patients in everyday life. Telehealth system equipped with graphical analysis and diaries would be used by individuals in home to monitor blood glucose, cholesterol and blood pressure results while registered physician would gain the access to the reports and statistical analysis on the patient's therapy progress. Finally diet-oriented software would help user identify the effect of the novel, worldwide approved nutritional trend - the Glycemic Index. One of the solutions complying with the vision above is GIGISim.

4 Existing Systems Review

4.1 Approaches

Diabetes-dedicated computer programs fall into various categories:

- *Educational programs* - diabetes is particularly amenable to computer aided education because, using dedicated metabolism simulators, developed recently on a large scale, patient is allowed to experiment with various hypothetical simulated cases, like insulin overdosing or hypoglycemia without the consequences of the wrong decision in real life. Implemented metabolism models cover the principles of human carbohydrates metabolism and often allow interactive mode.
- *Clinical systems* are designed to help the physicians, diabetes educators and patients detect patterns and trends in the patient's blood glucose readings, perform statistical analysis for the need of scientific research. Those systems require medical supervision and are dedicated for clinical use only.
- *Intelligent insulin dosage programs*, devices and algorithms have been used by patients to advise them on insulin dosage adjustment on a daily basis. This example of software is especially interesting as the insulin dose adjustment is considered the most difficult skill to develop for a new diabetes patient.

Although limited number of e-learning diabetes tools complies with a full variety of requirements, discussed in previous sections, selected group of diabetes dedicated systems will be presented in this chapter to present the overview of interesting facilities and functionalities available.

4.2 Educational and Assistant Systems

Computer educational systems are created to complement and enhance, rather than replace the functions of the diabetes educator. Since now the only way to provide such an education is during clinic or physician visits and alternative ways to educate people about diabetes need to be identified.

Several investigators have pointed out the possibility of using the glucose and insulin plasma levels simulation models, to help diabetes interpret human carbohydrate metabolism. This approach to the diabetes education was proved to be effective and numbers of mathematical models of the diabetes mellitus metabolism have been previously reported in literature [6][7]. AIDA or The DiasNet software [8][9] had a clear clinical functional scope which has been extended to an educational. Both solutions are dedicated for education via the Internet. Similarly to mentioned systems DIABLOG [10] is able to simulate glucose and insulin profiles of a 24 h period and display them graphically as curves however, opposite to the DiasNet, this system doesn't support clinicians supervising so the results of the simulation must be carefully interpreted by potential users.

When it comes to the classic educational systems diabCAL [11] may be presented as a comprehensive learning tools accessible both as a modular course or a desktop quick-reference, dedicated mainly for diabetes patients whereas expert system for diabetes management called DIABETES may be used for the education of medical personnel as well as students of medicine in the field of diabetes management.

When it comes to assistant systems commercial software named DiabetesWorksTM should be mentioned. A combination of five programs developed to assist diabetic management integrates tools for blood glucose monitoring, an interpreter module used to analyze the adequacy of diet and insulin treatment for diabetics, and an advisory module suggesting alterations in diet and/or insulin regimen in order to improve glycemic control. The analysis is based on models of insulin effect and carbohydrate absorption using meal-time.

Mentioned requirement about clear and useful diet diaries are met in Librae - a diabetes simulator and in the UTOPIA - complex home management system proposed to support home data analysis comparing them with clinical measurements. Learning relationships between insulin adjustments and glucose variations is performed by time series analysis and parametric, linear systems model [12].

4.3 Clinical Systems

In The Center of Diabetes and Metabolic Disorders of Berlin, a computer-aided care system has been used since 1974, [13] improving medical staff routine tasks and analyzing data from approximately 55,000, signaling possible epidemiological research on an unselected diabetes population. This is only one of the various clinical systems aiming at prevalence, incidence, duration of diabetes, and modes of treatment. Mentioned AIDA, designed for the clinical use or the CADMO with an intelligent automated analysis and interpretation of data relevant to glycemic control - are intended to assist health care professionals. Rule-based logic, statistical methods, and a physiologic model of insulin and glucose dynamics are used to help detect meaningful patterns and trends in glucose and insulin data and to suggest approaches for optimizing insulin regimens [14].

5 The GIGISim

5.1 The System Architecture

We believe the computer aided-diabetes therapy has a promising perspective if only the designed system will be able to assure and facilitate contact between the patient and his physician via web application. The GIGISim is a clinician supervised system designed for home use. A basic functionality of the GIGISim is to improve or maintain the quality of life for people with diabetes by assisting decision-making and enhancing the patient's personal sense of control. All suggestions generated by AI algorithms (expert base and evolutionary algorithms) will be presented to the user but also sent to the physician or dietitian. Therapeutic interventions will be based on the retrospective analysis of glycemic response to various factors. For example new insulin treatment schema proposed by the system, will be sent in a form of a report to the registered physician and, if approved, presented to the patient via web-portal.

GIGISim is an ASP.NET 2.0 web application equipped with user friendly interface, and many additional web-components providing help, tips and hints about diabetes therapy. User may adapt the Graphical User Interface individually thanks to a new .NET 2.0 features: Themes and Web Parts. GIGISim is deployed on a web server and registered users' accounts are correlated with their physicians account. Depending on the type of the diabetes, physician controlling individuals' therapy may recommend and enable access to additional features:

- The system of trackers and diaries to control patient's progress in weight, blood glucose, and blood pressure control. The patient enters the data and simple recommendations are presented warning against possible hyperglycemia or high blood pressure. When necessary, diabetes-related complications symptoms are highlighted. Meanwhile a complex report is generated and sent to an assisting clinician.
- Diet assessment feature controls saturated fat, cholesterol, sodium, proteins, fiber, vitamins and minerals of the patient's diet. The role of the Glycemic Index on the daily glucose variation was presented briefly in the section 2. Only few systems available on the market, including GIGISim, actually include GI as a new factor, evaluating high-quality carbohydrates for diabetes diet together with meal time dependencies. Glucose variations are presented as colorful plots rather than as calculated figures - graphical method illustrates metabolism reaction more clearly and intuitively and patient may identify patterns of blood glucose responses, which are strongly correlated to GI of the meal.
- Adaptation components have been added to personalize simulation – both the algorithm and adaptation method where reported in previous papers about GIGISim. Each user, registered to the GIGISim web-portal, might obtain, after completing a short form, appropriate values of model parameters corresponding to his carbohydrate metabolism.
- Genetic algorithms (AI) in an intelligent Meal Planner individualize nutritional recommendations, to satisfy the person's taste preferences and lifestyle and to assure optimal blood glucose control. The tool analyzes the data base containing 1000 products to find and suggest optimal diet basing on nutrient requirements.
- Graphical simulator may educate on physiological processes and human carbohydrate metabolism.
- Insulin and drug planner advice on insulin and medicaments doses and timing, always assuring medical personnel approval before presenting advices to the user.

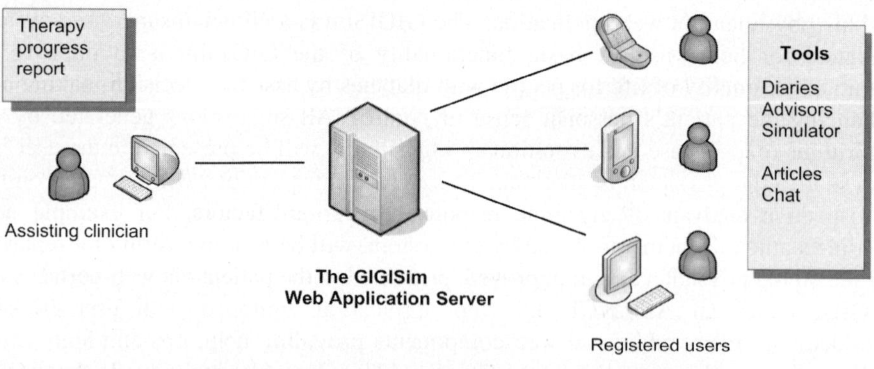

Fig. 1. The GIGISim system web architecture

System will be broadly and freely accessible in WWW providing demo-simulations and facilities for online users, news on diabetes care, newsletters or discussions. Registered users and physicians will have the access to the patient-profiled simulations and data, report generator and statistical analysis. Online communication between patient and his physicians will be also handled.

At this stage of development we plan also to implement the simple GIGISim simulation software designed for mobile devices. That would give patients the unique chance to simulate glucose behavior pattern offline, after a meal and take results in to account while planning everyday activities.

6 Conclusion

The GIGISim visualizes the postprandial glucose profiles correlated to user's diet, allowing patients to practice, and gain experience with insulin dosage adjustment and diet planning. The system was tested on non-diabetes group (10 healthy students age 20-25) as well as on diabetes patients in clinical tests made in Krakow's Hospital (11 women, age 65-78, all treated with insulin). Results of the simulation were satisfactory and educational goal was reached. All subjects show high interest on presented visualization, many of them wished to carry simulation on their own. After the graphical visualization, that demonstrated the effect of GI on the blood glucose, both groups declared changing eating habits, for example switch to whole grain products.

For the evaluation of the educational quality of the designed tool one group was presented a classic lecture about GI whereas the second group worked with interactive, graphic simulator in GIGISim. The results proved the second method to be far more meaningful and educative. The WWW-accessible architecture and interface was acceptable to most users, even to those with little computer experience.

The value of computer aided therapy management is undisputed but the challenge is not only to develop innovative, validated algorithms and tools to aid patient decision-making and optimize glycemic control but also to assure high usability, accessibility and compliance with patients' requirements.

References

1. Izworski, A., Koleszynska, J., Tadeusiewicz, R., Bulka, J., Wochlik, I.: GIGISM (Glucose-Insulin and Glycemic Index Web Simulator) - The Online System Supporting Diabetes Therapy. The IASTED International Conference on Telehealth, Banff, Canada (2005)
2. Izworski, A., Koleszynska, J., Tadeusiewicz, R., Bulka, J., Wochlik, I. (eds.): Internet Tools and Computer-Aided Diabetes Education: Introducing GIGISim Online. Proceeding of Communications, Internet, and Information Technology, St. Thomas, USVI, USA (2006)
3. Guidelines, C.D.A.: for the Nutritional Management of Diabetes Mellitus in the New Millennium. 23(3), 56–69 (1999)
4. Sheard, N.F., Clark, N.G., Brand-Miller, J.C., Franz, M.J., Mayer-Davis, F.E.: Dietary Carbohydrate (Amount and Type) in the Prevention and Management of Diabetes: A statement by the American Diabetes Association. Diabetes Care 27(9), 2266–2271 (2004)

5. Frost, G., Wilding, J., Beecham, J.: Dietary advice based on the glycemic index improves dietary profile and metabolic control in type 2 diabetic patients. Diabetic Med. 11, 397–401 (1994)
6. Lehmann E.D: Interactive educational simulators in diabetes care. Med. Inform. 22 (1997)
7. Worthington, D.R.L.: The use of models in the self-management of insulin-dependent diabetes mellitus. Computer Methods & Programs in Biomedicine 32, 3–4 (1999)
8. Lehmann, E.D.: British Diabetic Association review of the AIDA v4 diabetes software simulator program. Diabetes Technol. Ther. 6(1), 87–96 (2004)
9. Plougmann, S., Hejlesen, O.K., Cavan, D.: DiasNet-a diabetes advisory system for communication and education via the internet. Int. J Med. Inform. 64(2-3), 319–330 (2001)
10. Biermann, E., Mehnert, H.: DIABLOG: a simulation program of insulin-glucose dynamics for education of diabetics. Computer Methods and Programs in Biomedicine 32(3-4), 311–318 (1990)
11. Emma-Jane Berridge, A.V.: Roudsari: diabCAL: evaluating computer-aided learning for diabetes patient education. Aslib Proceedings 6, 367–378 (2003)
12. Deutsch, T., Roudsari, A.V., Leicester, H.J., Theodorou, T., Carson, E.R., Sonksen, P.: UTOPIA: a consultation system for visit-by-visit diabetes management. Med. Inform. 21(4), 327–344 (1996)
13. Thoelke, H., Meusel, K., Ratzmann, K.: Computer-aided system for diabetes care in Berlin, G.D.R. Comput. Methods. Programs Biomed. 32(3-4), 339–343 (1990)
14. Berger, M.P., Gelfand, R.A., Miller, P.L.: Combining statistical, rule-based, and physiologic model-based methods to assist in the management of diabetes mellitus. Comput. Biomed. Res. 23(4), 346–357 (1990)

Image Mining Using Wavelet Transform

Sanjay T. Gandhe[1], K.T. Talele[2], and Avinash G. Keskar[1]

[1] Visvesvaraya National Institute of Technology, Nagpur, India
[2] S.P. College of Engineering, Andheri (W), Mumbai, India

Abstract. In this paper, we propose an image mining using wavelet transform. It uses general pattern matching, pattern recognition and data mining concepts so that a real life scene/ image can be related to a particular category, helping in different prediction and forecasting mechanisms. It is a three-step process i.e. image gathering, learning and classification. As wavelet transform uses time frequency relation, it can be used for image mining instead of Fourier transform. Wavelet transform is used to decompose an image into different frequency sub bands and a low frequency sub band is used for Principal Component Analysis. Classification relates to identifying the category to which an image belongs. We have developed prototype system for recognition using DWT + PCA system. The concept of image mining thus can be efficiently used for weather forecasting so that we can know the natural disasters that may occur in advance.

Keywords: Image mining, wavelet transform, pattern recognition, Eigen values, data mining, image learning, and image classification.

1 Introduction

Image mining [1] is a three-step process of processing an image. These steps are image gathering, learning and classification. The image -gathering step uses general search techniques used on the web that are keyword based and result into a collection of different images related to a particular search keyword entered by the user. The second step is image learning, which relates to learning about a particular category depending upon particular features of an image. This is important because there is a big difference between images captured by professional photographers and amateur photographers.

The image mining is needed because many times the real life scenes are to be classified to a particular category. The third step is of image classification where some real life scene/ image is given as an input to the image-mining tool and then depending upon the features of the image, the image is matched with a particular category and is then classified under a particular category. The concept of image mining thus can be efficiently used for weather forecasting so that we can know the natural disasters that may occur in advance. General data mining techniques can be used for image mining. The concept of image mining refers to relating a real life scene/ image to some category of images in the database that helps in identifying a general image as a specific image of a particular category.

B. Apolloni et al. (Eds.): KES 2007/WIRN 2007, Part I, LNAI 4692, pp. 797–803, 2007.

2 Data Mining and Image Mining

Any mining terminology associated with computers relates to knowledge discovery techniques. The concept of image mining has come from data mining because everything in computers is nothing but data- the information. Knowledge discovery [3] makes use of information existing in databases to give more precise information for future use. The knowledge discovery is a four-step process of data management, data preprocessing data mining and post processing as shown in fig 1.

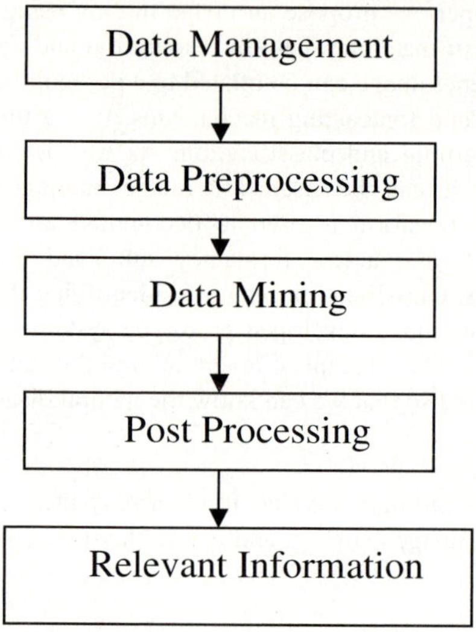

Fig. 1. Knowledge Discovery using Data Mining

The data management step defines different structures for storing and accessing the data efficiently. The second step of data preprocessing is the heart of knowledge discovery. Here main intention is given on improvement of efficiency and ease of data mining process. Data preprocessing is necessary because the real world data is noisy, incomplete and many times irrelevant. The data mining is the last stage of knowledge discovery that uses various algorithms for data mining which when applied result in knowledge discovery. Sometimes post processing is also necessary after data mining.

The image mining process is somewhat different from the general data mining approach. The image mining is a three-step process as shown in Fig 2 where the very first task is image gathering. In this step, a search is carried out for a particular keyword on the Internet. The result of this search a large volume of images related to a particular keyword. This collection can be used to create an image database where images of a particular type can be separated. The second step of knowledge discovery using image mining is image learning. In this step, images related to a particular keyword are studied thoroughly.

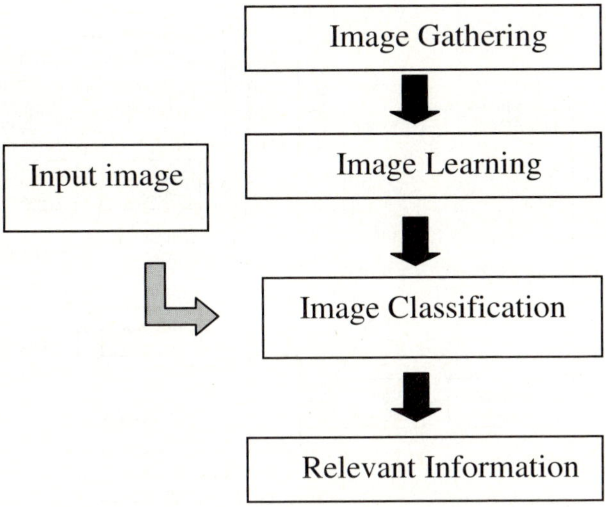

Fig. 2. Knowledge Discovery using Image Mining

This step analyses different features of a particular image and names these features after a particular category. The third step is image classification. In this step any real life image or real life scene is given as an input so that the computer can decide the category of the image just by matching features of the image with the features stored already in the database. The reason why basic data mining concepts are not used for image mining is twofold. The first reason is the way images are stored and represented in databases and the second reason is the pattern matching of real life scenes to the existing images that are in the database.

3 Wavelet Transform and Image Mining

Wavelet transform is used for data mining because wavelet transform is capable of providing time and frequency localizations simultaneously whereas Fourier transform provides only frequency representations. Moreover the Fourier transform is not suitable for non stationary waves. The Short Time Fourier Transform (STFT) was developed to support both time and frequency representation. In STFT, the signal is divided into small segments and each segment is considered as stationery. Thus longer segments give better frequency resolutions and poorer time resolutions whereas shorter segments give better time resolutions and poorer frequency resolutions. Now considering the wavelet transform where, Wavelet is designed to give good time resolution and poor frequency resolution at high frequencies and vice versa.

A wavelet transformation converts data from an original domain to a wavelet domain by expanding the raw data in an orthonormal basis generated by dilation and translation and mother wavelet. For example, in image processing the original domain is spatial domain, and the wavelet domain is frequency domain. An inverse wavelet

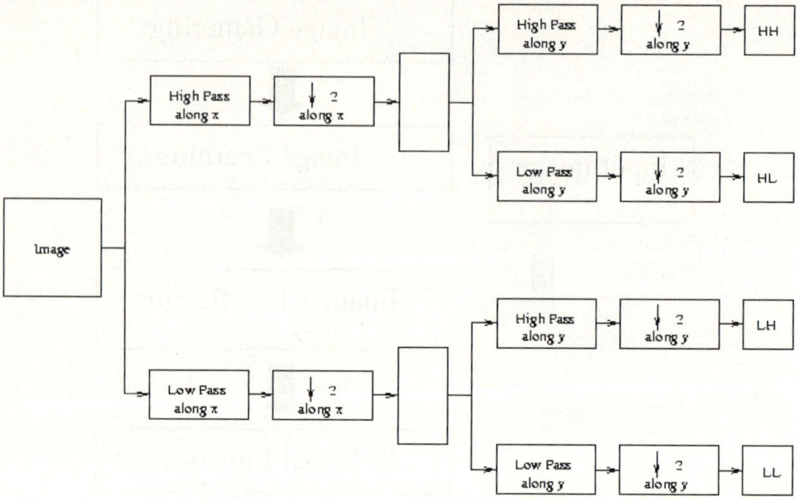

Fig. 3. Forward DWT Algorithm

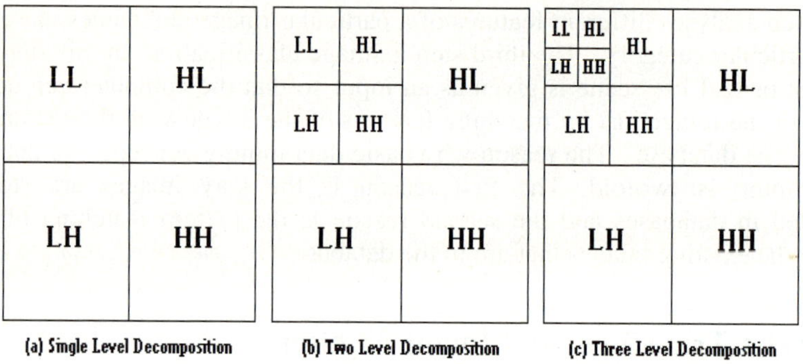

Fig. 4. Image decomposition

transformation converts data back from the wavelet domain to the original domain. The wavelet transformations preserve the structure of data. Computation complexity (space complexity is linear), vanishing moments (where noisy data can be eliminated), compact support (processing the data inside the wavelet region without affecting the data outside its region), and de-correlated coefficients (used to reduce complex process in time domain to simple process in wavelet domain) are some of the properties of the wavelets that can be used for the data mining purpose.

The wavelet transform can be used for image mining too. The first step of knowledge discovery using image mining is the image gathering. In the second step, which is heart of the total process, large contents are captured that describe and distinguish image. These contents are color, texture and shape features that form an image vector for an image in the collection. A visual dictionary is created for each

(a) Input images

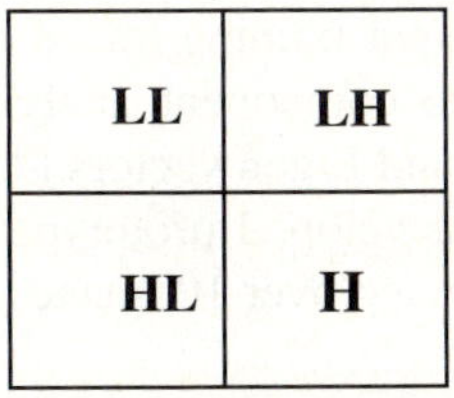

(b) Frequency bands

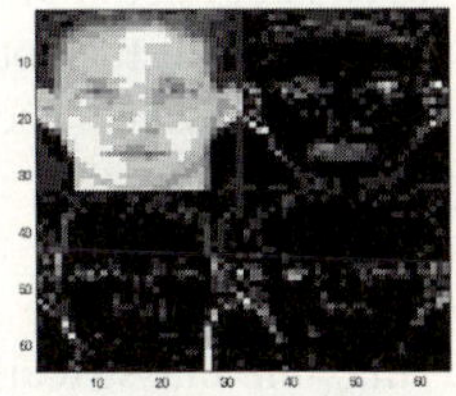

(c) DWT images

Fig. 5. One level decomposition

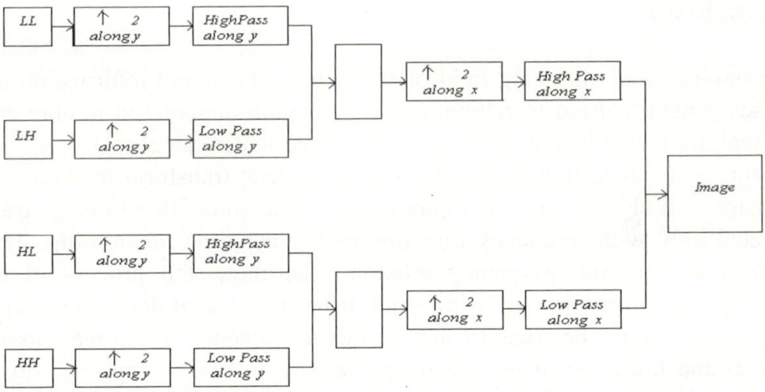

Fig. 6. Inverse DWT Algorithm

feature attribute (color, texture and shape). This leads to discovery of hidden relationships in different images.

Wavelet transform is used to decompose an image into different frequency sub bands and a low frequency sub band is used for Principal Component Analysis. Classification relates to identifying the category to which an image belongs.

Discrete wavelet transform algorithm and one level decomposition of input image are shown in fig 3,4 and 5. The inverse DWT algorithm is as shown in fig 6.

DWT of input training image as well as test image is calculated and it is cascaded with principle component analysis for recognition. We have selected LL band of DWT image and Eigen vectors [4,5,6] of LL image is calculated.

We have developed prototype system for recognition using DWT + PCA system and experimented over 10 images. The efficiency of the proposed method is 100 %.

4 Issues and Future Work

There are different issues related to image mining using wavelet transform. The very first issue is the image management. Since the first step is image gathering, large number of images is collected at the source machine after a keyword search. Handling large number of images at the source machine is a crucial task and hence gives rise to memory management issue. The database should be so efficient to handle images efficiently. By handling images means storing and accessing images efficiently. The next issue is applying wavelet transform over images for content analysis so that different features of the image can be separately studied that will help in image learning and thus in classification. Applying inverse wavelet transform converts images in wavelet domain to the original domain again is one more issue.

Applying wavelet transform in distributed environment is an area of research where image mining is a topic of main focus.

5 Conclusion

Image mining is an emerging field of computer science and information technology that uses general pattern matching, pattern recognition and data mining concepts so that a real life scene/ image can be related to a particular category, helping in different prediction and forecasting mechanisms. The wavelet transform method can be used for image mining instead of Fourier transform since the Fourier transform is associated only with frequency relationship but not time relationship. The wavelet transform uses a time-frequency relation. The three-step process of knowledge discovery using image mining is different from the general data mining approach. A robust system could be used to index video-documents (video-mail messages, for example) and image archives. An image archive indexed in such a way would be useful for criminal identification by the investigation department.

References

[1] Michael, C., Burel, M.C., Fowlkes, C., Roden, J., Stechert, A., Muktar, S.: Dimond eye: A distributed Architecture for image data mining. at SPIE DMKD, Orlando (April 1999)

[2] Burel, M.C., Fowlkes, C., Roden, J.: Mining for image content. Jet Propulsion lab Jornal, 126–347

[3] Chu, W.W, Hsu, C.-C., Cardinas, A., K, R.: Knowledge Based Image Rewith spatial and temporial Constructs. University of California

[4] Gandhe, S.T., Keskar, D.A.G.: Image analysis and inference using soft computing tools: A comparison of algorithms. In: proceedings of National Conference on High Performance Computing (VISION-05), Government College of Engineering, Tirunelveli, Tamilnadu, pp. 331–335 (8th April, 2005)

[5] Gandhe, S.T., Talele, K.T., Keskar, D.A.G.: Face Recognition Using Isodensity Maps: A neural network approach. In: Gandhe, S.T., Talele, K.T. (eds.) 1st international conference on Emerging Applications of Information technology (EAIT 06) organised by Computer Society of India, Kolkata, 11th and 12th February, pp. 129–132 (2006)

[6] Turk, M., Pentland, A.: Eigenfaces for recognition. Journal of cognitive neuroscience (1993)

[7] Yao, Y.: Face Feature Extraction with PCA and Kernel PCA. School of Computer Engineering (17 September 2003)

Implementation of Intelligent Active Fault Tolerant Control System

Seda Postalcıoğlu, Kadir Erkan, and Emine Doğru Bolat

Kocaeli University, Technical Education Faculty
Department of Electronic & Computer Education, Umuttepe/Kocaeli, Turkiye
postalcioglu@ieee.org, {erkan,ebolat}@kou.edu.tr

Abstract. This paper addresses implementation of intelligent active fault tolerant control for experiment set which is a FODPT (First Order Plus Dead Time) system. Temperature control has been done using fuzzy logic controller (FLC). Faults often cause undesired reactions, so to keep the system stable and acceptable control performance is an important problem for control system design. In this paper, multiplicative, additive types of sensor faults have been examined and disturbance has been applied for temperature sensor as a fault. Feature vectors of the sensor faults have been constructed using wavelet analysis, sliding window and a statistical analysis. Classifier of the feature vectors has been done using The Self Organizing Map (SOM). Switch mode control with fuzzy logic controller chosen by supervisor has been used for reconfiguration. When a fault occurs in the system, a suitable controller has been selected online to maintain the closed-loop performance of the system.

Keywords: Fault detection, fault identification, fault tolerant control, temperature control.

1 Introduction

In automatic processes, faults will often cause undesired reactions and shut-down of a controlled plant and the consequences could be damage to technical parts of the plant, to personnel or the environment [1]. To keep the system stable and acceptable control performance is an important problem for control system design. Control system stability and reliability are critical for nuclear power stations, passenger airplanes, but also essential for large-scale productions. Most of modern industrial plants are complex and often include a number of subsystems which may compensate for the effects of sensor faults [2]. Ideally, when faults happen, the closed-loop system should be capable of maintain its present operation. This leads to the recently studied area of fault tolerant control (FTC). FTC has three levels for an automatic response to a fault [3]. These levels are:

(a) Detect and identify the presented failures,
(b) Determine the operating conditions,
(c) Stabilize the closed-loop system within the predetermined operation conditions.

FTC combines fault detection and identification with control methods to handle faults in an intelligent way [4].

B. Apolloni et al. (Eds.): KES 2007/WIRN 2007, Part I, LNAI 4692, pp. 804–812, 2007.

Fault detection consists of decision as to whether a fault occurred or not. Fault identification imposes a stronger condition. When one or more faults occur, this method identify, which faults have occurred. [5]. Fault tolerant control system can be established by adding a supervision system on top of the control system. FTC is realized by hardware redundancy in most real industrial systems. For example, the majority-voting scheme is used with redundant sensors to cope with sensor faults [2]. But hardware redundancy has two major restrictions. These are, high cost and taking more space .Over the last two decades analytical redundancy has been investigated as a solution for hardware redundancy. There are generally two different approaches using analytical redundancy. These are passive and active approaches.

Passive fault tolerant control use robust control techniques to design closed-loop systems so that it is insensitive to certain faults. Active fault tolerant control use online fault detection and identification information and reconfigurable controllers.

According to FTC, the occurrence of faults in sensors or actuators is modeled by local linear models. It is assumed that for each set of operating conditions there is a local controller which stabilizes it while embodying matching constraints. FTC is made up by these local controllers which alter among them selves according to the decision of a supervisor.

This paper addresses on-line intelligent FTC for temperature control. Sensor faults for temperature control have been examined on the oven. Switch mode controller approach has been applied.

2 Design of Intelligent Active Fault Tolerant Control

The structural design of online intelligent fault tolerant control is presented in Fig. 1. As shown in Fig. 1, the structural design has three layers: a bottom layer with the control loop, a second layer with online fault detection and identification and a third layer with supervisor unit [6].

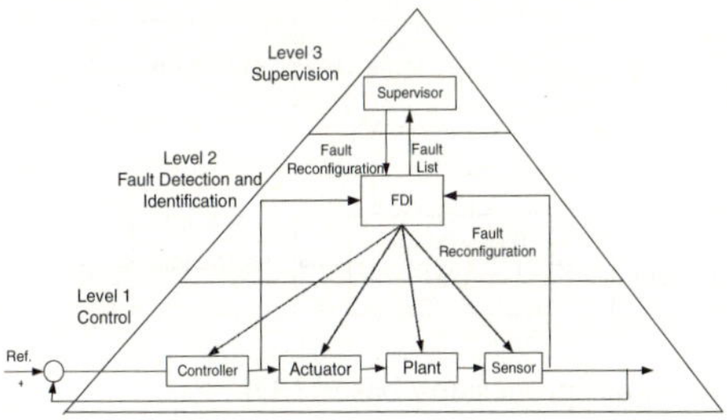

Fig. 1. Structural design of intelligent active fault-tolerant control

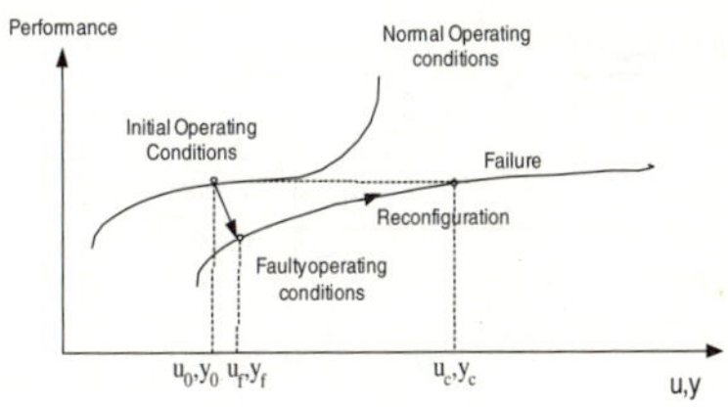

Fig. 2. The principle of FTC approaches

Separating the structural design of FTC into three layers is provides a clear development structure, independent design and development of each layer. Level 1 comprises a traditional control loop with sensor and actuator interface and the controller. Level 2 comprises the functions of online fault detection and identification module. Level 3 comprises the supervisor functionality. The supervisor activates reconfiguration action in response, which can be pre-determined for each fault or obtained from real-time analysis and optimization [7,6]. When a fault occurs in the system, an appropriate controller can be selected on-line to maintain the closed-loop performance of the system. The principle of FTC approaches is illustrated in Fig. 2 [8]. The fault occasion, the system deviates from its nominal operating point defined by its input-output variables (u_0,y_0) to a faulty one (u_f ,y_f,). The goal of FTC is to determine a new controller that takes the corrupted system parameters into account and drives the system to a new operating point (u_c,y_c,) to main performance of the system.

2.1 Design of Fault Detection and Identification

The monitoring of faults in feedback control system components has come to be known as fault detection and identification (FDI) [9]. The FDI methods can he classified into two major groups: model-free method and model-based method [10].

In the last two decades the main effort has been made in the research of analytical approaches based on quantitative models. The main idea is the generation of residuals, signals that reflect the difference between nominal and faulty system. The residuals are usually generated using analytical approaches, such as observers, parity equations or parameter estimation, based on analytical redundancy. Since most of the real processes are non-linear, FDI model-based methods require a precise and accurate model. The traditional model-based FDI methods cannot guarantee satisfactory performance. For this reason, knowledge-based methods have been developed combination of the analytical approaches with the artificial intelligence methods e.g. neural networks, fuzzy logic, neuro-fuzzy, evolutionary programming, etc. [10]. Knowledge-based methods are sometimes called model-free or qualitative methods.

In this study model-free method has been used with artificial intelligence and signal processing techniques. Fig. 3 shows the scheme of artificial intelligence and signal processing based FDI [12].

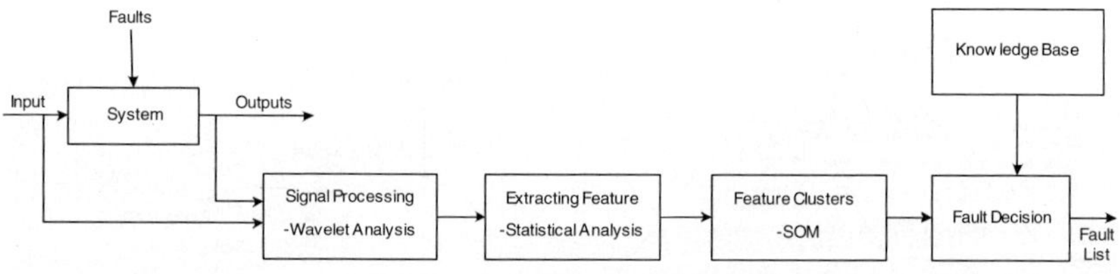

Fig. 3. Artificial intelligence and signal processing based FDI

2.1.1 Wavelet Analysis

Faults in the system cause certain changes in the response of measured signals, changes in time response and in frequency response. These changes would result in transient behavior of system variables and transient analysis becomes critical for fast and accurate fault detection. Wavelet analysis is capable of detecting the change or transition in the signal [11]. For this reason wavelet analysis has been used in this study.

In signal processing stage of FDI, decomposition for discrete signal is computed using a series of low-pass and high-pass filters. Filters computation has shown in equations (1-5) [15].

$$y_{high}[k] = \sum_{n} x[n].g[2k - n] \tag{1}$$

$$y_{low}[k] = \sum_{n} x[n].h[2k - n] \tag{2}$$

$$x[n] = \sum_{k} (y_{high}[k].g[2k - n]) + (y_{low}[k].h[2k - n]) \tag{3}$$

As shown in equations (4-5), c[k] is called approximation coefficients and dj[k] is called detail coefficients. Parameter j determines the scale or the frequency range of each wavelet basis function Ψ. Parameter k determines the time translations.

$$c[k] = \int_{-\infty}^{\infty} f(t)\varphi(t - k)dt \tag{4}$$

$$d_j[k] = \int_{-\infty}^{\infty} f(t)2^{j/2}\psi(2^j t - k)dt \tag{5}$$

Discrete wavelet transform (DWT) is a linear transform that is very suitable to represent the non-stationary events in signals. DWT has good localization properties of high frequency components.

2.1.2 Feature Extraction

For real time processing, a moving time window technique has been used. In moving time window, the latest several samples carry the most up to date information on any changes in the signal. Sliding window is needed technique to track dynamic data and detect the transient state of faults.

Detecting the variations along with the time using wavelet analysis, equation 6-7 have been used to calculate the information of the signal. Maximum absolute value changing ratio (ΔS) and variance changing ratio (ΔV) has been used for feature extraction of the signal [12]. Feature vector contains the changing ratio values for more than one system measurements.

$$\Delta S_{max} = \frac{|S|_{max\_now} - |S|_{max\_past}}{|S|_{max\_past}} \tag{6}$$

$$\Delta V = \frac{var_{now} - var_{past}}{var_{past}} \tag{7}$$

3 The Experiment Set

The experiment set is shown in Fig. 4. It has two volumes, internal and external. There is a 2.5 cm space between two volumes there. To isolate the internal side of the set from outside, 2 cm thick isolation material is fixed in this space. External volume is (25x25x30) cm and internal volume is (20x20x25) cm. There are two holes on the top of the set to be used for disturbances. A 800 W resistance is put on the internal bottom of the experiment set for heating the internal volume. A thermocouple temperature sensor is fixed into the middle point of the front side of the set. The experiment set is a FODPT (First Order Plus Dead Time) system [13,14].

Fig. 4. The experiment set

Units for the oven are shown in Table 1. PIC based card's main hardware capabilities are listed in Table 2.

Table 1. Units for the oven

(1) Oven
(2)Termocouple (Temperature sensor)
(3)Fan
(4)Disturbances (Two holes on the top of the oven)

Table 2. PIC based card's main hardware capabilities

Three input for K type temperature sensor (0-1000 ^{0}C)
Two feed resistance output with phase controller (40+40 Ampere-rms),
Three relay output (10 Ampere-rms),
RS232 output (as much as 56Kbit)
Input is AC mains voltage

PIC (PIC 18F4585) based card gets the temperature data from the experiment set by using a temperature sensor and makes the data appropriate for the computer. Then, this unit transmits the data to the computer by using an RS-232 protocol. Read temperature is evaluated in Matlab-Simulink and desired controller signal is generated. Afterwards, this control data is transmitted to PIC based card again. Finally, the phase angle of the triac BTA 41 determines the energy level of the resistance. So, the control is achieved by applying necessary amount of energy to the resistance.

Different sensor fault types are presented for the experiment set. Control of the experiment set is provided by fuzzy logic controller is shown in Fig. 5. The key benefit of fuzzy logic controller is that it lets the operator describe the system behavior relationship with simple if-then rules [9]. 3-D map of the input – output relationships for fuzzy rules are shown in Fig. 6.

Disturbance, Additive and multiplicative faults has been implemented for temperature sensor. Sensor faults for the experiment set are shown in Table 3.

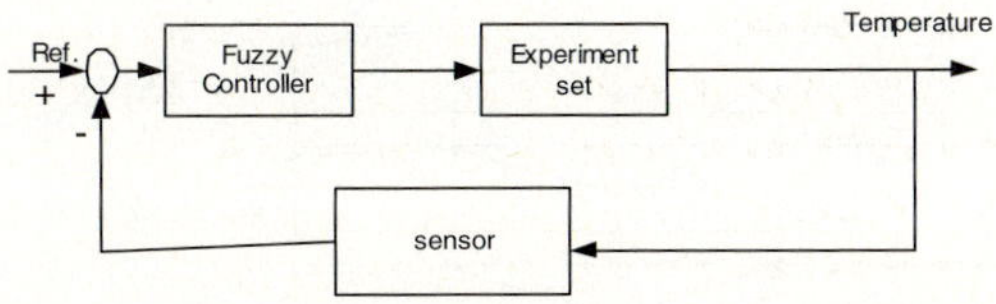

Fig. 5. The experiment set with fuzzy logic controller

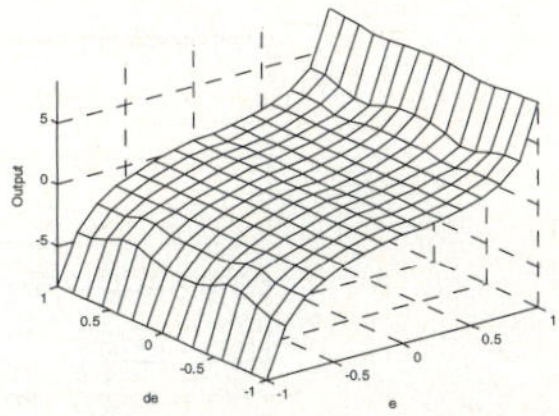

Fig. 6. 3-D map of the input – output relationships

Table 3. Sensor faults for the experiment set

Status No	Status
0	Normal operating condition
1	Additive fault (+30)
2	Multiplicative fault (x0.8)
3	Additive fault (-10)
4	Disturbance
5	Multiplicative fault (x1.2)

4 Implementation Results

Faults are applied to temperature sensor. Sensor fault start time is 1500 sec for each fault. Sensor faults are shown in Fig. 7 for nominal controller. Feature vectors have been constructed using wavelet analysis, maximum absolute value changing ratio (ΔS) and variance changing ratio (ΔV). Feature vectors of sensor faults are shown in Fig. 8. Classification is supplied by SOM which is subtype of artificial neural network.

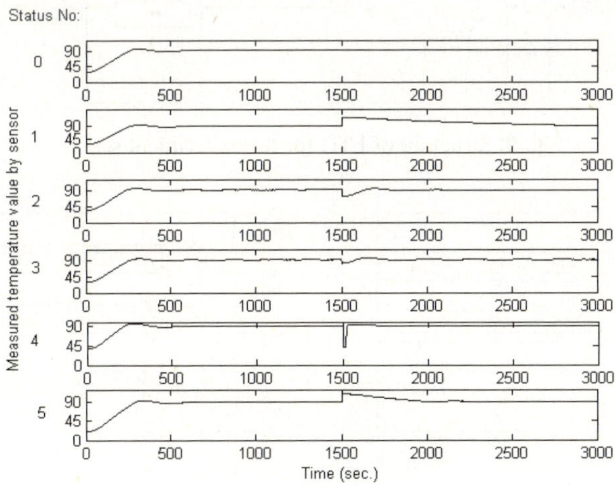

Fig. 7. Sensor faults

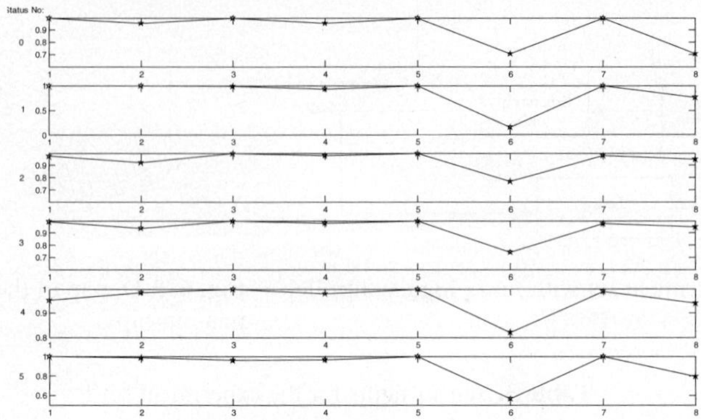

Fig. 8. Feature vectors of sensor faults

Intelligent active FTC has been used. Switch mode control is obtained by supervision using fuzzy logic controller. Fig. 9 shows the structure of FTC for the experiment set. When a sensor fault occurs in the system, a suitable controller can be selected on-line to maintain the closed-loop performance of the system. All the sensor faults have been detected and identified for 1502 sec. afterwards suitable controller is selected by supervisor. Reconfiguration has been done using switch mode controller.

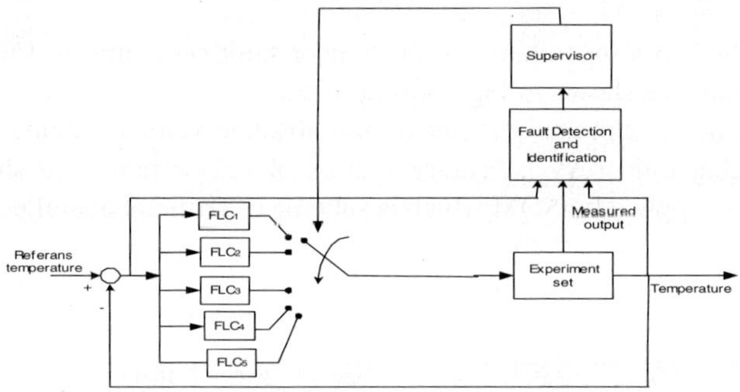

Fig. 9. Structure of FTC for the experiment set

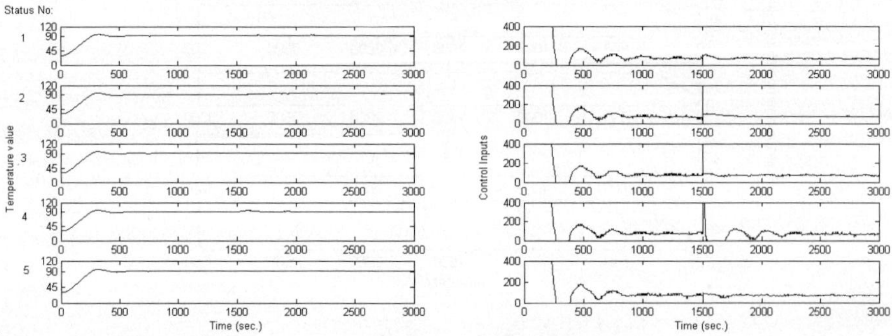

Fig. 10. Temperature values with sensor faults using FTC

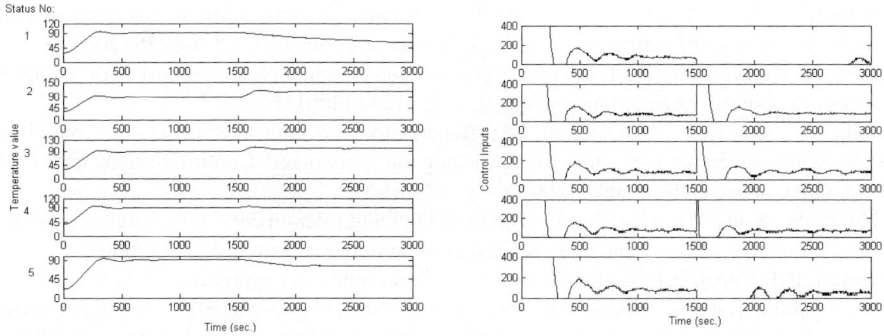

Fig. 11. Temperature values with sensor faults not using FTC

Fig. 10 shows FTC results and control inputs with sensor faults. Desired temperature value was determined as 90 ^{0}C. With FTC, the temperature value was nearly 90 ^{0}C despite of the sensor faults. Fig. 11 shows the effects of the sensor faults and control inputs with out FTC. As shown in Fig. 11, temperature value deviated from the desired temperature value with out FTC.

5 Conclusion

Most of modern industrial plants are complex and often include a number of subsystems which may compensate for the effects of sensor faults [2]. To keep the system stable and acceptable control performance is an important problem for control system design. For this reason, this paper addresses intelligent active fault tolerant control. The experimental set which is a FODPT (First Order Plus Dead Time) system has been used for temperature control. Multiplicative and additive types of sensor faults have been examined and disturbance has been applied for temperature sensor. Feature vectors of the sensor faults have been constructed using wavelet analysis, sliding window and statistical analysis. Feature vector has been classified using Self Organizing Map. Fault tolerant control has been achieved by switch mode control using fuzzy logic controller chosen by supervisor. When a fault occurs in the system, a suitable controller has been selected on-line to maintain the closed-loop performance of the system.

References

1. Blanke, M., Marcel, S., Wu, E.N.: Concepts and methods in fault tolerant control. Proceedings of the american control conference, pp. 2606–2620 (2001)
2. Yu, D., L., Chang, T., Yu, D.W.: Fault tolerant control of multivariable process using auto-tuning PID controller. IEEE transactions on systems, man, and cybernetics 35(1), 32–43 (2005)
3. Wang, H., Wang, Y.: Neural network –based fault tolerant control of unknown nonlinear systems. IEE Proe-Control Theory Appl. 146(5) (1999)
4. Niemann, H.: Editorial, Int. journal of robust and nonlınear control 10, 1153–1154 (2000)

5. Saberi, A., Stoorvogel, A.A., Sannuti, P., Niemann, H.: Fundamental problems in fault detection and identification. Int. J. Robust Nonlinear Control 10, 1209–1236 (2000)
6. Puig, V., Quevedo, J.: Fault tolerant PID controllers using a passive robust fault diagnosis approach. Control engineering practice 9, 1221–1234 (2001)
7. Puig, V., Quevedo, J., Ocampo, C.: Benchmark for fault-tolerant control based on Barcelona sewer network. In: 1st Workshop on Networked Control System and Fault Tolerant Control conf., France (2005)n
8. Noura, H., Sauter, D., Hamelin, F., Theilliol, D.: Fault tolerant control in dynamic systems: application to a winding machine. IEEE control system magazine , 33–49 (2000)
9. Patton, R.J, Uppal, F.J., Lopez-Toribio, C.J.: Soft computing approaches to fault diagnosis for dynamic systems: survey, http://66.102.9.104/search?q=cache:fB8U-mhpELMJ:www.eng.hull.ac.uk/research/control/softcomp.ps+Soft+computing+approaches+to+fault+diagnosis+for+dynamic+systems:+survey&hl=tr&gl=tr&ct=clnk&cd=1
10. Palma, L.B., Coito, F.V., Silva, R.N.: Fault Diagnosis based on Black-Box Models with Application to a Liquid-Level System, Emerging Technologies and Factory Automation, 2003. Proceedings. ETFA '03. IEEE Conference 2, 739–746 (2003)
11. Postalcıoğlu, S., Erkan, K., Bolat, E.: Discrete Wavelet Analysıs Based Fault Detectıon. Wseas Transactıons on Systems 5(10), 2391–2398 (2006)
12. Xu. Z.: Design of Knowledge-based Fault Detection and Identification for Dynamical Systems. Master of Science, University of Alberta. (2002).
13. Bolat, E., Erkan, D., Postalcıoğlu, K.: Microcontroller Based Temperature Control of Oven Using Different Kinds of Autotuning PID Methods. In: Zhang, S., Jarvis, R. (eds.) AI 2005. LNCS (LNAI), vol. 3809, pp. 1295–1300. Springer, Heidelberg (2005)
14. Bolat, E.D., Erkan, K., Postalcıoğlu, S.: Experimental Autotuning PID Control of Temperature Using Microcontroller. EUROCON 2005, Serbia & Montenegro, Belgrade, pp. 266-269 (2005)
15. Postalcıoğlu, S., Erkan, K., Doğru Bolat, E.: Applicatıon of Discrete Wavelet Transform to Fault Detection. 10th WSEAS Int. Conf. on Systems, Greece (2006)

Natural Language Understanding for Generating Grasp Actions

Hirokazu Watabe, Seiji Tsuchiya, Yasutaka Masuda, and Tsukasa Kawaoka

Dept. of Knowledge Engineering & Computer Sciences, Graduate School of Engineering, Doshisha University, Kyo-Tanabe, Kyoto, 610-0394, Japan

Abstract. For achieving smooth communications between human and an intelligent robot, it is important that the robot understands the meaning of the sentences of natural language. Most of the instructions given to the robot are related to some actions. This paper reports the method to generate the action of the humanoid robot, especially grasp actions, by understanding natural language (Japanese) conversation. For this purpose, comprehension of action instruction written by natural language is necessary. The system that understands natural language and generates proper grasp action is constructed. This system consists of a semantic understanding method to arrange input sentences, the knowledge base of vocabulary related to objects to grasp, the knowledge base of the parameter for robots to act, and association mechanism to handle a word that is unknown. The system is capable of understanding many input sentences by the association mechanism using Concept-Base and a degree of association.

1 Introduction

Intelligent humanoid robots, which can become a partner to us humans, need ability to act naturally. In actions of the robot, it is very important to understand natural language from humans.

Conventionally, the knowledge of action for the robot was mostly constructed by humans for the fixed number of input sentences. Therefore, the robot cannot act anymore if the given sentence is not acceptable one. This paper proposes the method to generate the robot's action that grasps or holds some object, appropriately by understanding the meaning of the human's instructions. In this method, the robot analyzes the input text of Japanese natural language and understands the meaning, then gets the information necessary for a series of action by referring to the word knowledge base about actions. Proposed system consists of a semantic understanding method to arrange input sentences, the knowledge base of vocabulary related to particular actions, the knowledge base of the parameter for the robot to act, and the word association system to handle unknown words. The system is capable of understanding many input sentences by the word association system using Concept-Base and the degree of association among word concepts.

B. Apolloni et al. (Eds.): KES 2007/WIRN 2007, Part I, LNAI 4692, pp. 813–820, 2007.

2 Virtual Humanoid Robot

In this paper, it is used the humanoid robot called "Robovie" [1] shown in fig. 1(a). The robot that has a human-like appearance has various sensors, such as vision, sense of touch and so on. It has two arms (4 degrees of freedom x 2), a head (3 degrees of freedom), and a mobile platform (2 driving wheels).

The virtual robot (VRobovie) is constructed based on Robovie. Each arm has 6 degrees of freedom with fingers in order to simulate grasp actions.

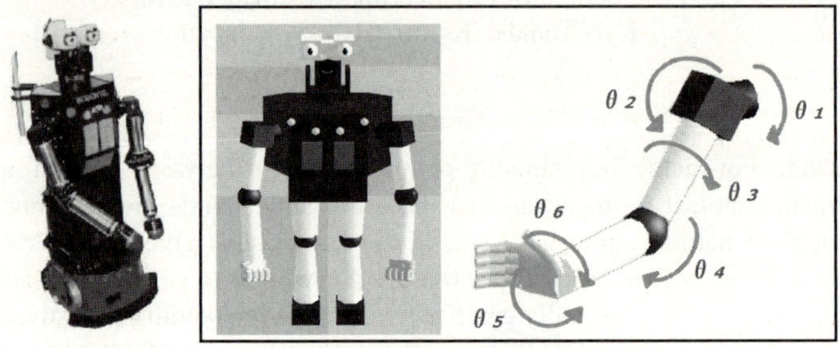

(a) Robovie (b) VRobovie

Fig. 1. Robovie and Virtual Robovie

3 Structure of the Proposed System

Fig. 2 shows the structure of the proposed system. The following is the algorithm to generate action of the robot from input sentence.

Given input sentence (Japanese typed text) is stored into semantic frame as separated words by semantic meaning understanding system.

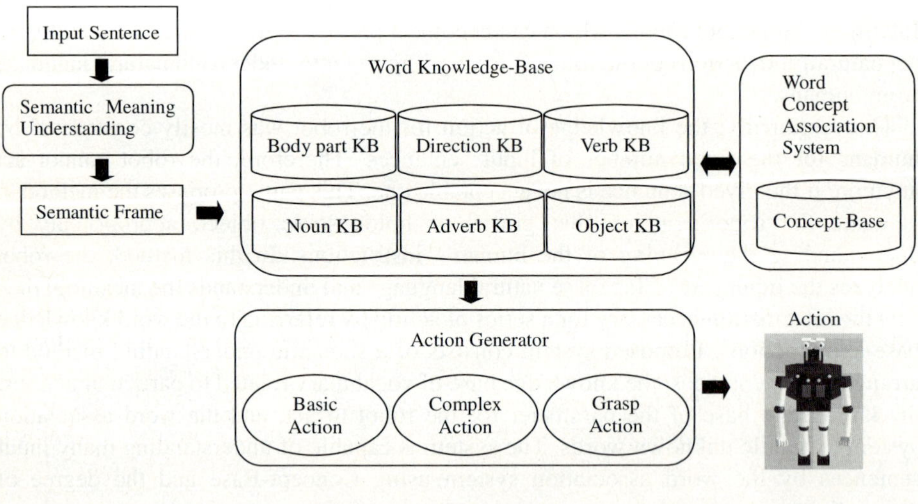

Fig. 2. System structure

Each word in the semantic frame is compared with Word Knowledge-Base (WKB) for action. When there are no matched words in WKB, search semantic similar word using Word Concept Association System (WCAS). WCAS consists of Concept-Base (CB) and the calculation module of the degree of association between concepts [2][3].

Action Generator generates each joint parameter of the robot and action data using each module.

4 Semantic Meaning Understanding

It is used Semantic Meaning Understanding System (SMUS)[4] in order to get words about actions. SMUS divides the input sentence into semantic frame like shown in fig. 3. And action frame is derived from semantic frame. These words in the action frame are candidates of action words.

Input sentence: 右手で林檎を持つ.
(I have an apple with my right hand.)

Semantic frame

Who	What	when	Where	why	How	Whom	Pred.
私	林檎				右手		持つ
(I)	(an apple)				(right hand)		(have)

Action frame

Part	Object	Pred.
右手	林檎	持つ
(right hand)	(an apple)	(have)

Fig. 3. Semantic frame (Action frame)

5 Word Knowledge-Base About Action

Word knowledge-base about action (WKB) is constructed in order to generate actions from action words. WKB consists of six KB as shown in fig. 2. Table 1 shows the part of WKB. Body part KB has nouns representing human body parts with verb about possible actions. There are 128 nouns of human body parts and 263 verbs in body part KB. Verb KB has 275 verbs. Each verb in verb KB represents some action by itself, such as 'stand up', 'have' and so on. Noun KB has 167 nouns. Each noun in noun KB represents some action by itself, such as 'straightening one's back', 'deep breathing' and so on. Direction KB has 106 words representing directions or positions including body parts, such as 'up', 'down', 'eye', 'head' and so on. Adverb KB has 54 adverbs representing the degree of actions, such as 'a little', 'pretty', 'well enough' and so on.

Table 2 shows the part of object KB. It has the field of object name, kana reading, and ID. Each object is classified by its shape (sphere, box, Cylinder) and size (least, small, medium, large). These shape and size is decided by not the object itself but the grasp part of the object. Object KB has 39 parameters according to the combination of shape and size. One parameter (One or Two) is decided by whether the robot uses one hand or both hands when he grasps the object. 18 parameters decide the position and orientation (two perpendicular vectors) of the right hand and the left hand in the object coordinate (Fig. 4). The other 20 parameters are rotation values of 10 fingers when the robot grasps the object.

Table 1. Word knowledge base about action

Body part KB

Part	Verb	ID
頭	曲げる	52
頭	振る	10
手	上げる	53

Verb KB

Verb	kana reading	ID
起き上がる	おきあがる	22
立つ	たつ	27
持つ	もつ	100

Noun KB

Noun	kana reading	ID
背伸び	せのび	22
握手	あくしゅ	3
拍手	はくしゅ	9

Table 2. Object knowledge base

Object KB

Object	kana reading	ID
林檎	りんご	3
辞書	じしょ	13
段ボール	だんぼーる	14
ペン	ぺん	21
鞄	かばん	22
缶	かん	23

ID	Shape	Size	One or Two	Right position	Right vector 1	Right vector 2	...	Right Thumb 1	Right Thumb 2	...
1	Sphere	least	1	(-.6,0,.8)	(0,0,-1)	(0,-1,0)	...	-30	70	...
3	Sphere	medium	1	(-.5,0,0)	(1,0,0)	(0,-1,0)	...	-60	90	...
4	Sphere	large	2	(-1.4,0,-.1)	(.9,0,.43)	(0,-1,0)	...	-45	30	...
12	Box	small	1	(0,1,0)	(0,-1,0)	(-1,0,0)	...	-40	90	...
13	Box	medium	1	(0,1,0)	(0,-1,0)	(-1,0,0)	...	-60	90	...
14	Box	large	2	(-2,0,0)	(1,0,0)	(0,-1,0)	...	-45	0	...
21	Cylinder	least	1	(-.6,0,.8)	(0,0,-1)	(0,-1,0)	...	-30	70	...
22	Cylinder	small	1	(0,0,.2)	(0,0,1)	(0,-1,0)	...	-60	90	...
23	Cylinder	medium	1	(-.5,0,0)	(1,0,0)	(0,-1,0)	...	-60	90	...

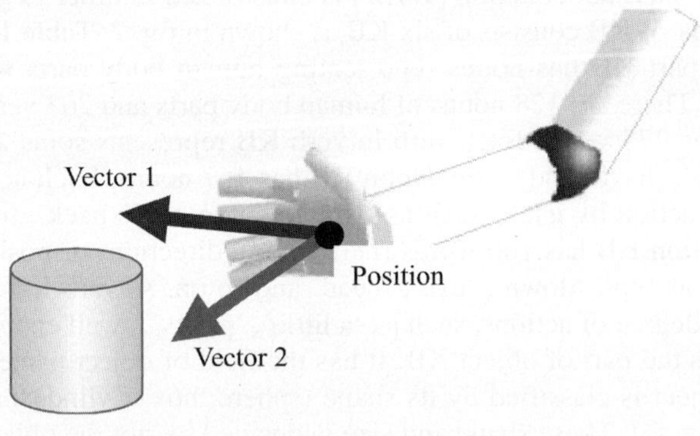

Fig. 4. Position and orientation of hand

6 Unknown Word Processing

Since the input sentence is arbitrary, there are many words, which are not stored in the WKB. These words are called unknown words. If there are unknown words in the input sentence, unknown word processing is executed and these unknown words are substituted with the words in WKB.

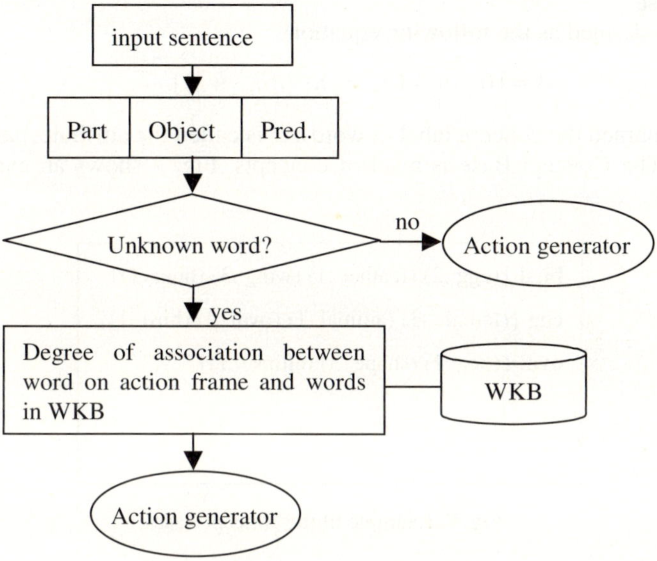

Fig. 5. Unknown word processing

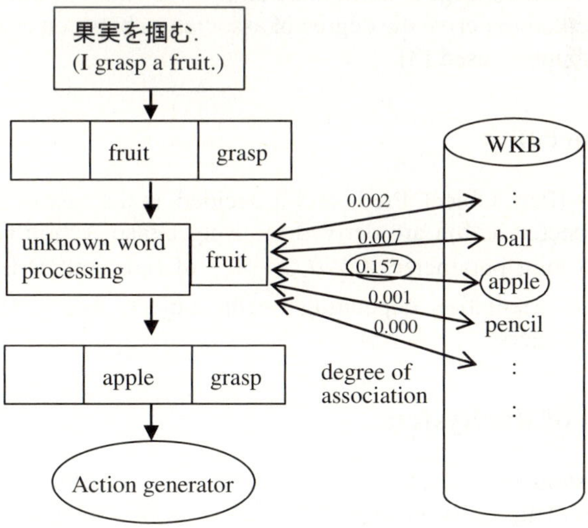

Fig. 6. Example of the unknown word processing

Fig. 5 shows the flow of the unknown word processing, and fig. 6 shows an example of the unknown word processing. If each word on action frame is not the word in the WKB, the degree of association [3] between the unknown word and words in WKB are calculated, and the word, which has the highest degree of association that is greater than threshold (0.03), is selected. The degree of association derives the semantic distance between two words.

Concept-Base

A Concept is defined as the following equation.

$$A = \{(a_1, w_1), (a_2, w_2), \cdots, (a_m, w_m)\} \quad (1)$$

Where, A is named the concept label (a word), a_i is called the attribute, and w_i is called the weight. The Concept-Base is a set of concepts. Fig. 7 shows an example of the concept -base.

bird {(egg .2) (feather .1) (wing .3) (hunt .1)}

egg {(female .3) (animal .1) (oval .2) (bird .1)}

oval {(egg .2) (shape .3) (ellipse .1)}

.
.
.

Fig. 7. Example of the concept-base

Degree of Association

Each concept is defined as a set of attributes and each attribute is also a concept as described above. Each concept is defined by an infinite chain of attributes (concepts). In this paper, a method to derive the degree of association between concepts using up to second order attributes is used [3].

7 Action Generator

After action frame (Part, Object, Predicate) is decided, in the case of grasp action, joint parameters of expected action are derived by using Grasp Action Generator (GAG). GAG generates 6 joint parameters $(\theta_1, \theta_2, \cdots, \theta_6)$ of right and/or left hand from the hand position and orientation depending on the object. And GAG also generates rotation angles of fingers.

8 Evaluation of the System

The proposed system is evaluated by using 120 input sentences, which correspond some grasp actions (fig. 8, 9). In these 120 sentences, 60 sentences contain unknown words. The robot acts using action data that the system generates, and humans judge that the action is correct or wrong. As a result, the rate of correct actions is 81%.

Fig. 8. Experiment environment

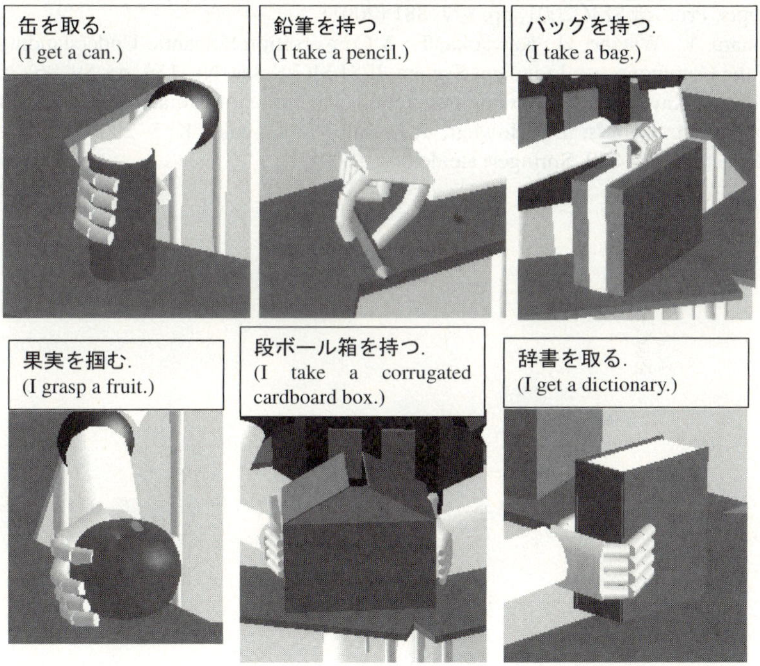

Fig. 9. Experimental results

9 Conclusion

In this paper, the method to generate the grasp action of the humanoid robot from the natural language input was proposed. Proposed system consists of a semantic meaning understanding module that arrange input sentence, the knowledge base of word related to actions, the action generator module to generate parameters for the robot to act, and association mechanism to handle a word which is unknown. In the word knowledge base about objects, each object is defined with class of shape and size, and from the view of grasping. By the evaluation, from 81% of input sentences, the robot made correct grasp actions.

Acknowledgements. This work was supported with the Aid of Doshisha University's Research Promotion Fund.

References

1. Ishiguro, H., Ono, T., Imai, M., Kanda, T.: Development of an Interactive Humanoid Robot Robovie – An interdisciplinary approach, Robotics Research, STAR 6, pp. 179–191. Springer, Heidelberg (2003)
2. Kojima, K., Watabe, H., Kawaoka, T.: Concept-base Refining with Thesaurus and Logical Relations for a Word Association-system, Proc. of KES2001, Part 2, pp. 1590–1594 (2001)
3. Watabe, H., Kawaoka, T.: The Degree of Association between Concepts using the Chain of Concepts, Proc. of SMC2001, pp. 877–881 (2001)
4. Shinohara, Y., Watabe, H., Kawaoka, T.: A Conversation Semantic Understanding Method using the Commonsense Judgment System, IPSJ SIG Notes, NL-153, pp. 89–96 (2003)
5. Watabe, H., Kawaoka, T.: Autonomous Action Generation of Humanoid Robot from Natural Language. In: Gabrys, B., Howlett, R.J., Jain, L.C. (eds.) KES 2006. LNCS (LNAI), vol. 4251, pp. 882–889. Springer, Heidelberg (2006)

New Machine Scores and Their Combinations for Automatic Mandarin Phonetic Pronunciation Quality Assessment

Fuping Pan, Qingwei Zhao, and Yonghong Yan

ThinkIT Laboratory, Institute of Acoustics, Chinese Academy of Sciences, Beijing, China
{fpan,qzhao,yyan}@hccl.ioa.ac.cn

Abstract. This paper discusses Mandarin vowel pronunciation quality assessment. The phonetic pronunciation quality is traditionally evaluated under the speech recognition framework by the phonetic posterior probability score, which may be computed by normalizing the frame-based posterior probability or be calculated on the phone segment directly. By the first method, we can achieve a human-machine scoring correlation coefficient (CC) of 0.832 for vowel; and by the second, the CC can be up to 0.847. This paper proposes a novel kind of formant feature and applies the feature to the evaluation of vowel: we transform the formant plots on the time-frequency plane to a bitmap and extract its Gabor feature for pattern classification; when use the classification probability for pronunciation assessment, we can get a CC of 0.842. Finally we combine the three scores with various linear or nonlinear methods; the best CC of 0.913 is gotten by using neural network.

Keywords: Computer Assisted Language Learning, Speech Recognition, Formant, Gabor Feature, Neural Network.

1 Introduction

Over the last decades many research groups have started to investigate on automatic pronunciation quality assessment by using speech recognition techniques. These works are mainly focused on the assessment at sentence level or speaker level[1-4]. The assessment at phone level, which is usually measured by the phonetic posterior probability score under the speech recognition framework, is only used as a reference of high level's evaluation and not investigated much. There are primarily two algorithms to compute the phonetic pronunciation score: one is to calculate the average of the logarithm of the frame based posterior probability (AFBPP)[1-3]; the other is to calculate the phone log-posterior probability (PLPP)[5]. However, these two algorithms are not very accurate in some cases due to the limited discriminating ability of the acoustic model.

This paper introduces our phonetic pronunciation quality assessment algorithm, which is also mainly under the speech recognition framework, and special attention is paid to some improvements of the vowels' evaluation. Our algorithm of phonetic assessment is originally based on the AFBPP score. In order to improve the vowel

B. Apolloni et al. (Eds.): KES 2007/WIRN 2007, Part I, LNAI 4692, pp. 821–830, 2007.

assessment accuracy, we suppose to employ the long-term information of speech, which is critical to vowel perception, by utilizing formant. Considering the difficulty of accurate formant tracking, a novel kind of formant feature is suggested. That is to convert the formant candidate plots on the time-frequency plane to a bitmap and then extract its Gabor feature to represent the formant trajectory. We use Gaussian Mixture Model (GMM) to classify the formant patterns and calculate the formant classification posterior probability (FCPP) score to assess the pronunciation quality. Such the formant classification score is complementary to AFBPP and PLPP, so we further investigate to combine them with various linear or nonlinear methods, and the best result is obtained by using neural network to combine the three scores.

The rest of this paper is organized as the follows: section 2 introduces the system structure and traditional phonetic scores; section 3 discusses the assessment method of formant classification; section 4 is dedicated to the combination of scores; some experiments and results are given in section 5; and finally the conclusion is drawn.

2 Traditional Phonetic Pronunciation Assessment Method

2.1 System Overview

Phone is the most fundamental assessment unit of our evaluation system. The phonetic evaluation is carried out by using speech recognition techniques of HMM and Viterbi searching. Block diagram of the system is shown in Fig. 1.

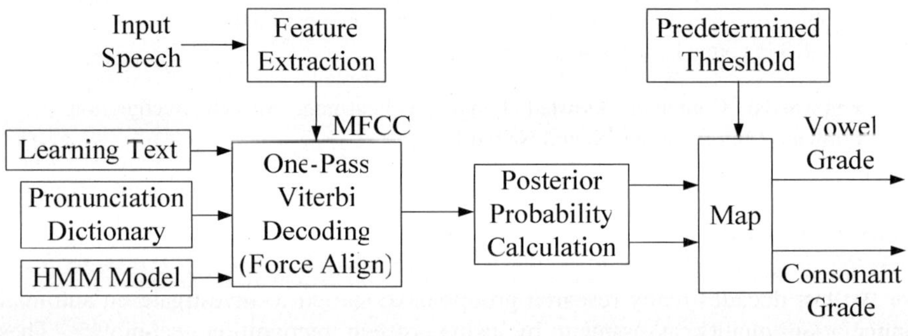

Fig. 1. The front-end feature extraction converts the speech waveform to MFCC coefficients and these are fed into HMM model net to do Viterbi decoding. The force alignment results are then used to calculate the phonetic posterior scores. And the final stage of evaluation uses predetermined thresholds to map the scores to evaluation grades.

The front-end feature extraction converts the speech waveform to a sequence of mel-frequency cepstral coefficients (MFCC) and these are fed into HMM model net to do one-pass Viterbi decoding. The HMM model net only consists of the models of the learning text, and the Viterbi decoding is only a force alignment between the speech frames and the HMM models in the net. The force alignment results are used to calculate the phonetic posterior scores, which will be discussed in the following. The final stage of evaluation uses predetermined thresholds, which are trained by the training data set and its human grades, to map the scores to evaluation grades.

2.2 Phonetic Posterior Probability Score

The phonetic posterior probability score is a good measure of the similarity between the native speech and the students' speech of each phone q. There are mainly two algorithms to calculate it.

The first one is the average of logarithm of the frame based posterior probabilities (AFBPP) belonging to q [1-3]:

$$\rho = \frac{1}{t_e - t_b + 1} \sum_{t=t_b}^{t_e} \log P(s_t \mid x_t) . \tag{1}$$

Where $P(s_t \mid x_t)$ is the frame based posterior probability of the force-aligned state s_t given the observation vector x_t ; t_b is the start frame and t_e is the end frame of q. The second one is phone log-posterior probability (PLPP)[5]:

$$\rho = \frac{1}{\tau} \log[P(q \mid O^{(q)})] = \frac{1}{\tau} \log \frac{p(O^{(q)} \mid q)}{\sum_{p \in Q} p(O^{(q)} \mid p)} . \tag{2}$$

Where τ is the number of frames in the acoustic segment $O^{(q)}$ of phone q ; Q is the set of Mandarin consonants when q is consonant, otherwise is the set of Mandarin vowels when q is vowel.

3 Vowel Pronunciation Assessment by Formant Classification

Formants have long been regarded as the dominant parameters to describe vowels. Formant trajectory of the vowel segment is supposed to convey the long-term information that is critical to vowel identification. We are to use the Gaussian Mixture Model to classify the pattern of the formant trajectory and use the classification posterior probability score for pronunciation quality assessment.

Numerous experiments have been carried out to classify a set of monophthong vowels of a specific language with formants[6-8]. It appeared that the most difficult problem is accurate automatic formant tracking. Errors tend to occur in highly transient phone boundaries[9]. We suggest a novel kind of formant feature to avoid the tracking errors and can better represent dynamic properties of formant trajectory.

3.1 Feature Extraction

2-D Gabor function was pioneered by Daugman to model the spatial summation properties of simple cells in the visual cortex[10]. Local image features extracted by Gabor filter bank are widely used in face recognition, fingerprint identification, contour detection and many other image processing or computer vision applications. We propose to convert the formant candidate plots on the time-frequency plane to a bitmap and calculate its Gabor feature for formant pattern classification. Block diagram of the process of feature extraction is shown in Fig. 2.

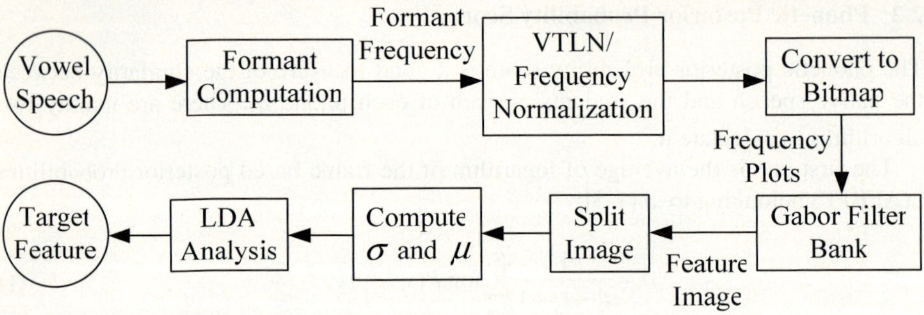

Fig. 2. The first step is to estimate the formants of the vowel portion; in the second step, the formant frequencies are normalized by using VTLN; after converting the formant plots to a bitmap, its Gabor feature is extracted; and finally use LDA to reduce the feature dimension

After the vowel portion of the utterance is segmented by the force-alignment, its formant candidates are estimated at first. This is automatically done by using PRAAT with the LPC-analysis method, and for each frame, five formants are estimated.

Differences of vocal tract length among different speakers result in an apparent expansion or compression of the frequency axis of formant trajectories, which will damage the performance of classification. In order to make formant locations stationary across speakers, we seek vocal tract length normalization (VTLN) to compensate for the variation of formant location by a warp of frequency axis, as the second step shown in Fig. 2.

After plotting the formant candidates of each frame on a time-frequency plane, we only convert the plots in the region up to 4000 Hz to a bitmap, because such the region can safely contain all the first three formant candidates. And to be uniform, we resize the bitmaps to 200×100 pixels.

Typically, an input image $I(x, y), (x, y)\in \Omega$ (Ω -the set of image points), is convolved with a 2-D Gabor function, $g(x, y), (x, y)\in \Omega$, to obtain a Gabor feature image $r(x, y)$ as the following[11]:

$$r(x, y) = \iint_{\Omega} I(\xi,\eta)g(x-\xi, y-\eta)d\xi d\eta \,. \tag{3}$$

We use the following family of Gabor functions:

$$g_{\lambda,\Theta,\varphi}(x, y) = \exp(-\frac{x'^2+\gamma^2 y'^2}{2\sigma^2})\cos(2\pi\frac{x'}{\lambda}+\varphi)\,. \tag{4}$$

Where $x' = x\cdot\cos\Theta + y\cdot\sin\Theta$, $y' = -x\cdot\sin\Theta + y\cdot\cos\Theta$, $\sigma = 0.56\lambda$ and $\gamma = 0.5$. In our experiments, the phase offset is set as $\varphi = 0$; the wavelength λ and the orientations Θ are selected according to experiment results.

Suppose the filter bank has I wavelengths and J orientations, it will generate $I\times J$ Gabor feature images. We equidistantly split each image into M rows and N columns to produce $M\times N$ blocks, and calculate the mean $\mu^{(l)}_{mn}$ and standard deviation $\sigma^{(l)}_{mn}$ of the magnitude of each block as feature elements:

$$\mu^{(l)}_{mn} = \iint \left| r^{(l)}_{mn}(x, y) \right| dxdy \ , \ \sigma^{(l)}_{mn} = \sqrt{\iint (\left| r^{(l)}_{mn}(x, y) \right| - \mu^{(l)}_{mn})^2 dxdy} \ . \tag{5}$$

According to[6-8], inclusion of vowel duration D will result in a consistent classification improvement. So the feature vector $\overline{f}$ is constructed as:

$$\overline{f} = [\mu_{11}{}^{(1)}, \sigma_{11}{}^{(1)}, ..., \mu_{MN}{}^{(1)}, \sigma_{MN}{}^{(1)}, ..., \mu_{mn}{}^{(l)}, \sigma_{mn}{}^{(l)}, ..., \mu_{MN}{}^{(I\times J)}, \sigma_{MN}{}^{(I\times J)}, D] \ . \tag{6}$$

Where l is the index of Gabor feature image, and m, n are the block indexes.

$\overline{f}$ usually has a very high dimension. However, a low-dimensional representation of the vector is especially important for machine learning. We use linear discriminant analysis (LDA) to project the high dimensional feature to a lower dimensional space, and exploit the projected vector as the final classification feature.

3.2 Classification Posterior Probability Score

We use GMM to classify the Gabor feature for vowel pronunciation evaluation. One model is trained for each Mandarin vowel. For each coming testing feature x, a formant classification posterior probability (FCPP) score is calculated as:

$$P(Vowel_{ref} \mid x) = \frac{p(x \mid Vowel_{ref})}{\sum_{k\in V} p(x \mid Vowel_k)} \ . \tag{7}$$

Where $Vowel_{ref}$ is model of the answer vowel; V is the Mandarin Chinese vowel set. $P(Vowel_{ref} \mid x)$ is mapped to evaluation grades by using predetermined thresholds.

4 Score Combination

Scores of AFBPP, PLPP and FCPP contain information of different time scale and so are complementary to each other. We investigate to combine them to improve the assessment performance by using various linear or nonlinear methods. This is a problem of predicting the human subjective evaluations by the machine scores.

4.1 Linear Regression

We suggest the expected value of human grade $\tilde{h}$ is a linear combination of two or more machine scores m_n plus a bias term b.

$$\tilde{h} = a_1 m_1 + a_2 m_2 + ... + a_n m_n + b \ . \tag{8}$$

The linear coefficients $a_1, a_2, ..., a_n$ and b are estimated on the training corpus by minimizing the mean square error between the predicted and the actual human grades.

4.2 Nonlinear Regression

In general, the expected human grade is a nonlinear function of the machine scores. Two nonlinear methods are tried to predict the human grades.

4.2.1 Probability Distribution Estimation

In this approach we compute the expected human grades by using estimates of the necessary conditional probabilities. The predicted human grade $\tilde{h}$ is computed as:

$$\tilde{h} = \arg\max_{h_i}[P(h_i \mid m_1, m_2, \ldots m_n)] . \tag{9}$$

where $P(h_i \mid m_1, m_2, \ldots m_n)$ is the estimated conditional probability of the human grade h_i given the machine scores $[m_1, m_2, \ldots m_n]$. Suppose $P(h_i) = P(h_j)\ (i \neq j)$, by using Bayes rule, the predicted human grade $\tilde{h}$ can be deduced as:

$$\tilde{h} = \arg\max_{h_i}[P(m_1, m_2, \ldots m_n \mid h_i)] . \tag{10}$$

In this work we model $P(m_1, m_2, \ldots m_n \mid h_i)$ by using Gaussian mixture model and the model parameters are estimated on the training data set.

4.2.2 Neural Network

A neural network can be capable of implementing arbitrary maps between input and output spaces. With this approach, the machine scores to be combined are the input to a neural network; the predicted grades are the output values of the network; the actual human grades provide the targets for the training of the network. Neural network parameters, the weights, are adjusted by the training algorithm to minimize the error criterion.

After some preliminary experiments with different network architectures, we choose the two-layer back propagation network with a single linear output unit and a hidden layer of log-sigmoid units. We vary the number of hidden layer units; the best performance is obtained with 10 hidden units. The number of input units corresponds to the number of machine scores combined. The network is trained by using the mean square error criterion. A momentum term is used in the weight update rule to accelerate the training speed. To avoid over fitting to the training data and to obtain good generalization, we use a cross-validation set formed with 15% of the training data. Prediction performance is assessed after each training iteration on this set; the training is stopped when performance do not improve on the cross-validation set[3].

5 Experiments and Results

5.1 Corpus

Because our system is used to assist the Hong Kong Putonghua-Shuiping-Kaoshi (PSK) test, the following experiments are performed on the PSK test samples. A PSK test set has 75 utterances to be pronounced, including 50 mono-syllable words and 25 double-syllable words. We only focus on vowel assessment in this paper. Each vowel in the test is graded on a 0-2 scale. A rating of 2 indicates excellent pronunciation, and a rating of 0 indicates completely wrong pronunciation. We collect 195 sets of test samples of the same content from 195 test attendees, among whom half are male and half are female. The test attendees are all Hong Kong native residents. They have very strong South China accent and often mispronounce Mandarin vowels. 80% of the collected samples are used as training set and the other 20% are used as testing set.

We use speech from a native Mandarin mono-syllable database to train the GMM model for formant classification. About 5000 utterances are collected for every Mandarin vowel. They are averagely spoken by 286 native Chinese speakers, among whom half are male and half are female.

5.2 Correlation Coefficient

The popular way to evaluate the performance of one pronunciation assessment system is to calculate the correlation between machine grades and human expert's subjective grades. The correlation coefficient (CC) of a specific vowel v is defined as:

$$CC = \frac{\langle \overline{H}_v, \overline{M}_v \rangle}{\|\overline{H}_v\| \cdot \|\overline{M}_v\|} = \frac{\sum_i h_{vi} m_{vi}}{\sqrt{\sum_i h_{vi}^2} \sqrt{\sum_i m_{vi}^2}} . \tag{11}$$

Where $\overline{M}_v$ is the machine grade vector of v constituted by grades of different speakers i, and $\overline{H}_v$ is the corresponding human rating vector. Both the testing and the training corpora have been graded by 5 human experts, whose average inter-rater correlation is 0.94. We use mean of the human experts' grades to calculate the human-machine grading correlation coefficients.

5.3 Methods

In the first experiment we use AFBPP and PLPP to grade the vowel pronunciation quality. Two thresholds are set to map those scores to grades of 0, 1 and 2. The thresholds are trained on the training data set by maximizing the human-machine correlations and then applied to the testing set. The vowels' average correlation coefficients of the testing data are shown in Table 1. The posterior probability score directly computed on the phone segment shows better correlation with human grades.

Table 1. Corelation coefficients of traditional phonetic posterior probability score

Machine score	Average CC
AFBPP (Baseline)	0.832
PLPP	0.847

Then the same experiment procedure is followed to exam the FCPP score. We compare performances of four different kinds of formant features. The first one is Gabor feature. Extensive experiments are done to determine the optimal Gabor feature parameters. We set $\lambda = 10$ and $\Theta = [0^\circ, 45^\circ, 90^\circ, 135^\circ]$ to generate the Gabor filter bank and equidistantly split each Gabor feature image into 10 rows and 4 columns, that is 40 blocks. Mean and standard deviation of each block together with the vowel segment duration form the feature $\overline{f}$, whose dimension is

$1\times4\times10\times4\times2+1=321$. By LDA analysis, we reduce the vector size to 50. In order to demonstrate the profit of Gabor transformation, we compose the second kind of feature with means and standard deviations of the original untransformed image blocks (Non-Gabor M&SD). The original image is also split into 10 rows and 4 columns that lead to a vector size of 81 (including vowel duration). After LDA analysis, the dimension is reduced to 50, too. The third kind of feature is constituted by the vowel segment duration and direct formant measurements, which are sampled at 20%, 50% and 80% of the vowel continuance[6]. And the last one utilizes third-order Legendre polynomials of the formant trajectories[7] together with duration of the vowel segment. Evaluation results of the four features are compared in Table 2. It can be seen that because the Gabor feature can exactly detect the shape of formant trajectories and is robust to inaccurate formant measurements, it results in the best correlation coefficient of 0.842, which is better than the baseline system.

Table 2. Corelation coefficients of formant classification posterior probability score

Formant feature	Feature dimension	Average CC	CC improvement to baseline
Gabor feature	50	0.842	0.01
Non-Gabor M&SD	50	0.818	-0.014
Formant samples	10	0.819	-0.013
Legendre polynomials	13	0.821	-0.011

At last we evaluate the three different types of predictors, including linear regression, probability distribution estimation, and neural network, in mapping and combining different types of machine scores to increase the correlations. The parameters of the regression and estimation models are trained on the training set and evaluated on the testing set.

The evaluation results are shown in Table 3. The FCPP score contains long-term information which is critical to vowel identification, so its linear combination with AFBPP increases the correlation by 3.6% with respect to AFBPP alone and its linear combination with PLPP increases the correlation by 1.5% with respect to PLPP alone. The correlation of PLPP+FCPP is close to that of AFBPP+FCPP but not as good as it, because the computation of PLPP on the whole vowel segment is likely to add some long-term information but lose some short-term information, which makes it less complementary to FCPP. The linear combination of AFBPP PLPP and FCPP can further increase the correlation, but the extent of improvement is small. The nonlinear combination (mapping) methods are more effective than the linear one. Correlations of AFBPP+PLPP combined by distribution estimation and neural network are largely improved due to the effect of nonlinearity. The correlation changing tendencies of nonlinear methods are like that of linear method. When using distribution estimation, the combination of AFBPP, PLPP and FCPP increases the correlation by 5.6% with respect to the baseline. And the best case is combining the three scores with neural network, which increase the correlation by 9.7% with respect to the baseline.

Table 3. Corelation coefficients of different mapping methods and combinations of scores

Combination method	Machine scores	Average CC	CC improvement to baseline
Linear regression	AFBPP+PLPP	0.852	0.02
Linear regression	AFBPP+FCPP	0.862	0.03
Linear regression	PLPP+FCPP	0.860	0.028
Linear regression	AFBPP+PLPP+FCPP	0.864	0.032
Distribution estimation	AFBPP+PLPP	0.859	0.027
Distribution estimation	AFBPP+FCPP	0.876	0.044
Distribution estimation	PLPP+FCPP	0.876	0.044
Distribution estimation	AFBPP+PLPP+FCPP	0.879	0.047
Neural network	AFBPP+PLPP	0.885	0.053
Neural network	AFBPP+FCPP	0.910	0.078
Neural network	PLPP+FCPP	0.909	0.077
Neural network	AFBPP+PLPP+FCPP	0.913	0.081

6 Conclusion

Accurate formant tracking is a very difficult problem, which limits its application in speech recognition and pronunciation assessment. This paper bypasses the problem by using Gabor feature to represent the formant trajectory and gets very good results. The formant trajectory contains long-term information of speech, which is critical to vowel identification. After the formant classification score is combined with the traditional two kinds of phonetic posterior probability scores, the correlation between machine and human grades is greatly improved. It should be noticed that the mapping from machine scores to evaluation grades is more likely to be a nonlinear function than a linear one. The combination method of neural network gets the best evaluation result.

Acknowledgments. This work is partially supported by MOST (973 program, 2004CB318106), National Natural Science Foundation of China (10574140, 60535030), The National High Technology Research and Development Program of China (863 program, 2006AA010102, 2006AA01Z195).

References

1. Franco, H., Neumeyer, L., et al.: Automatic pronunciation Scoring for Language Instruction. ICASSP, Munich, pp. 1471–1474. Munich (1997)
2. Neumeyer, L., Franco, H.: Automatic Scoring of Pronunciation Quality. Speech Communication 30, 83–93 (2000)
3. Franco, H., Neumeyer, L., Digalakis, V., Ronen, V.: Combination of machine scores for automatic grading of pronunciation quality. Speech Communication 30, 121–130 (2000)

4. Yasushi, T., Masatake, D., Tatsuya, K.: Practical use of English pronunciation system for Japanese students in the CALL classroom. INTERSPEECH, pp. 1689–1692 (2004)
5. Witt, S.M., Young, S.J.: Phone-level pronunciation scoring and assessment for interactive language learning. Speech communication 30, 95–108 (2000)
6. Hillenbrand, J., Getty, L.A., Clark, M.J.: etc: Acoustic characteristics of American English vowels. Journal of the Acoustical Society of America 97, 3099–3111 (1995)
7. Schmid, P., Barnard, E.: Explicit, n-best formant features for vowel classification. ICASSP, pp. 21–24 (1997)
8. Nearey, T.M., Assmann, P.F.: Modeling the role of inherent spectral change in vowel identification. Jorunal of the Acoustical Society of America 80, 1297–1308 (1986)
9. Lee, M., VanSanten, J., Mobius, B., Olive, J.: Formant Tracking Using Context-Dependent Phonemic Information. IEEE Transactions on Speech and Audio Processing 13, 741–750 (2005)
10. Petkov, N.: Biologically motivated computationally intensive approaches to image pattern recognition. Future Generation Computer Systems 11, 451–465 (1995)
11. Grigorescu, S.E., Petkov, N., Kruizinga, P.: Comparison of texture features based on Gabor filters. IEEE Transactions on Image Processing 11, 1160–1167 (2002)

Particle Swarm Optimization Applied to Vertical Traffic Scheduling in Buildings

Zhonghua Li, Hong-Zhou Tan, and Yunong Zhang

Department of Electronics and Communication Engineering,
Sun Yat-sen University, Guangzhou 510275, China
{lizhongh,issthz,zhynong}@mail.sysu.edu.cn

Abstract. Vertical traffic scheduling is significant in evaluating the quality of property management. An excellent vertical traffic scheduling approach aims at enhancing the system handling capacity and reducing the waiting time, journey time and energy consumption, especially in up-peak traffic pattern. To balance workloads of elevator cars in the system is a good choice for any peak traffic. This paper proposed a novel PSO-based dynamic scheduling algorithm for vertical traffic in buildings. The service zones for elevators are determined by considering their expected round-trip time. Our preliminary simulation results show that the proposed algorithm balances the round-trip time over the elevator system and further improves the service quality of elevator system in buildings.

Keywords: Particle swarm optimization, vertical traffic scheduling, elevator group control system, dynamic zoning, round-trip time.

1 Introduction

With the development of economy and the increase of land-value, more and more buildings have been established and more elevators have been installed, so that elevator group control systems are usually excellent choices to significantly enhance such performance indices of elevator systems as handling capacity, waiting time, journey time and energy consumption [1,3-4,8-9,12-13,17]. Even the collaborative control of multiple elevator groups would be conceptually formed. The conception of elevator group control system is emphasized not only in building design but also after lift installation [11]. How elevator group control systems function, becomes one of key factors which determine the performance of elevator systems.

Excellent elevator group control systems should satisfy the requirements of passenger traffic during each traffic pattern, especially during peak traffic pattern. A series of elevator group control policies have been proposed in previous twenty-five years [3-4,9,12-13]. Even though the implementations of these polices are relatively different, their primary targets is unchanged, i.e., to minimize passengers' waiting time and to avoid from long waiting time. For peak traffic pattern, more attentions are also paid to the handling capacity of elevator systems. Zoning is one of effective methods, where one elevator corresponds to a single zone. In earlier days, vertical traffic zoning is mostly static or time-scheduling. It is apparent that these two methods are both inadaptable to variable vertical traffic in buildings [1,3-4,17].

B. Apolloni et al. (Eds.): KES 2007/WIRN 2007, Part I, LNAI 4692, pp. 831–838, 2007.

Later, dynamic zoning for up-peak traffic is conceptually proposed by Powell [2], where the waiting time of passengers at the main terminal is reduced. Furthermore, Albert proposed the concept of uniform round-trip time (uRTT), which aims at maximizing the total handling capacity of elevator system. Algebraic algorithm is ever employed to find the optimal zoning solution. Some heuristic approaches, e.g. genetic algorithm and immune algorithm, have been applied to optimize the elevator dispatch during lunch-peak traffic [12-13].

Particle swarm optimization (PSO) is a population-based search algorithm, which has been successfully applied in many areas: function optimization, artificial neural network, fuzzy system, and other industrial areas [4-6]. PSO is easy to implement and there are few parameters to adjust. Therefore, this paper will apply a standard PSO to solve vertical traffic scheduling (also called PSO-VTS) in buildings. With the elevator traffic changes, PSO will periodically optimize the floor zones served by elevators in order to keep the whole elevator system higher handling capacity.

The remainder of this paper is organized as follows. Firstly, vertical traffic scheduling (e.g. dynamic zoning) is reviewed. Secondly, a novel PSO-based vertical traffic scheduling is presented. Thirdly, some numerical experiments are arranged and the corresponding results are given. Comparisons are conducted between PSO and other conventional policies. Finally, the conclusions are made.

2 Vertical Traffic Scheduling

A building with N floors excluding the *Main Terminal* (MT) is considered. If the total demand within 5-minute duration of peak traffic is U and the demand of the kth floor is u_i, $i = 1, 2, ..., N$, then the relation between U and u_k is defined by:

$$U = \sum_{i=1}^{N} u_k \tag{1}$$

According to the spirit of the mentioned zoning, let Z_j and n_j are zones served by the jth car and the highest floor in Z_j, respectively, thus we obtain the following sets:

$$\begin{aligned} Z_1 &= \{0, n_0, ..., n_1\} \\ Z_j &= \{0, n_{j-1}+1, ... n_j\} \\ Z_M &= \{0, n_{M-1}+1, ... n_M\} \\ n_M &= N, n_0 = 1, j = 1, \hbar\ , M \end{aligned} \tag{2}$$

Clearly, n_j is constrained by $0<n_1<...<n_j<...<N$. Let $t_{rt}(j)$ be the round-trip time of the jth car and $\bar{t}_{rt}$ be the average round-trip time of all cars. By utilizing the concept of uRTT, the objective function to be optimized is defined as below:

$$\min f(n_1, n_2, ..., n_{M-1}) = \left(\frac{1}{M} \sum_{j=1}^{M} (t_{rt}(j) - \bar{t}_{rt})^2 \right)^{1/2}, \quad \bar{t}_{rt} = \frac{1}{M} \sum_{j=1}^{M} t_{rt}(j) \cdot \tag{3}$$

And $t_{rt}(j)$ is the sum of flight time, stop time and passenger transfer time, which is determined by the following formula:

$$t_{rt}(j) = 2H(j)t_v + (S(j)+1)t_s + 2P(j)t_p, \tag{4}$$

where t_v is the interfloor flight time, t_s is defined as the stop time and t_p is the average passenger transfer time. These three parameters are determined by the elevator system. The remainder 3 parameters in Eq.(4): *H*(*j*) is the highest reversal floor of the *j*th car, *S*(*j*) is the expected number of floor stops of the *j*th car and *P*(*j*) is the possible number of passengers in the *j*th car. With the real-time traffic changes, *H*(*j*), *S*(*j*) and *P*(*j*) are constantly adjusted. In this paper, PSO will be employed to dynamically find the optimal service zones for elevator cars during peak traffic duration.

The simulation platform of elevator group control system is described in document [18]. There is a function interface for a group controller to read information about elevator cars, e.g. position, direction, car-call and hall call, etc, and to write the best hall call assignment to elevator cars.

3 PSO-Based Vertical Traffic Scheduling

A. Particle Swarm Optimization

Assume an *l*-dimensional (*l*-D) search space, $S \subset \Re^l$, and a swarm consisting of *L* particles. The i th particle is in fact an *l*-D vector, $X_i = (x_{i1}, x_{i2}, \dots, x_{il})^{\mathrm{T}} \in S$. The velocity of this particle is also a *l*-D vector, $V_i = (v_{i1}, v_{i2}, \dots, v_{il})^{\mathrm{T}}$. The best position encountered by the i th particle is a point in S, denoted as $P_i = (p_{i1}, p_{i2}, \dots, p_{il})^{\mathrm{T}}$.

```
procedure particle swarm optimization;
    begin
        set time t:=0;
        initialize the population X(t);
        while the termination condition is not met do
            begin
                evaluate fitness of each member of the population X(t);
                update the best position encountered by the ithparticle;
                update the global position among all the swarm
                mutate members of X(t) according to Eq.(5) and Eq.(6);
                set time t:=t+1;
            end
    end
```

Fig. 1. The pseudocode of particle swarm optimization

Assume g to be the index of the particle that attained the best previous position among all the individuals of the swarm, and t to be the iteration counter. Then, according to the standard version of PSO, the swarm is evolved as follows:

$$V_i(t+1) = wV_i(t) + c_1 r_1 (P_i(t) - X_i(t)) + c_2 r_2 (P_g(t) - X_i(t)) \tag{5}$$

$$X_i(t+1) = X_i(t) + V_i(t+1) \,. \tag{6}$$

Where w is the weight factor to adjust the inertia of particles' velocity, c_1 and c_2 are positive constants, called the acceleration constants, r_1 and r_2 are random numbers, uniformly distributed within the interval [0,1]. Compared to $V_i(t)$, $V_i(t+1)$ is added two items: $c_1 r_1 (P_i(t) - X_i(t))$ and $c_2 r_2 (P_g(t) - X_i(t))$, as shown in equation (6). The termination condition of PSO is commonly that the evolutionary process exceeds the maximum epochs predefined. Fig.1 gives the pseudocode of standard PSO.

B. PSO-based Vertical Traffic Scheduling (PSO-VTS)

During the course of elevator system running, a feasible traffic predictor can constantly predict the traffic flow demand in the building [10]. When the predicted vertical traffic demand exceeds 80% of the total capacity of the elevator system, the vertical traffic is easily recognized as peak traffic pattern by neural network or support vector machine [8,14-16]. Thus the dynamic zoning will be triggered. Similarly the dynamic zoning will be cancelled when the predicted vertical traffic demand falls off below 80% of the total capacity of the elevator system [7]. Outside peak traffic pattern, some conventional control policies, e.g. queue-based control algorithm (QCA) are executed out of the requirement of elevator group control.

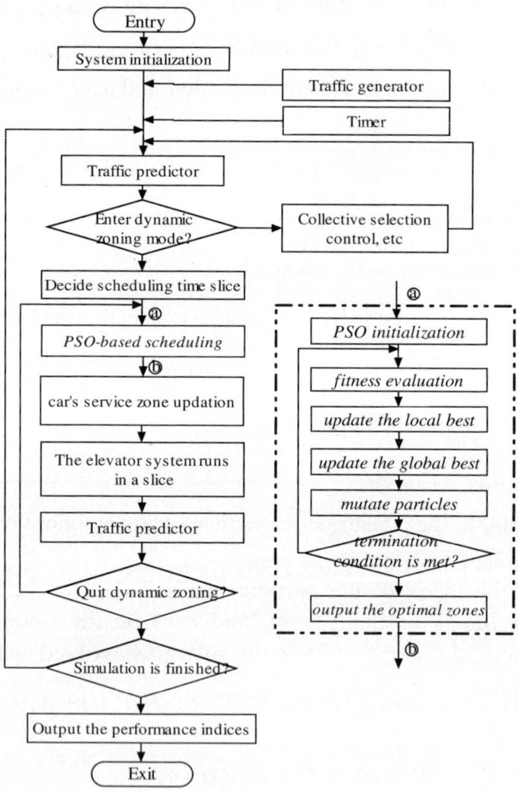

Fig. 2. The flow chart of PSO-based vertical traffic scheduling

The flowchart of PSO-VTS is shown in Fig.2. During each scheduling time period, PSO is used to determine the service zone of elevator cars. While terminating, PSO will output an optimal solution $Z^*=(n_1,n_2,\ldots,n_M)$, which consists of new service zones of elevator car. For technical details, the PSO-based scheduling algorithm is expanded as in the right part of Fig.2, denoted by blue dash-dotted rectangle.

4 Simulation Experiments

4.1 Parameter Settings

In this section, there are 4 elevators installed in an office building, which has 3 meters in interfloor height and 20 floors in total excluding the terminal. Each elevator car has the same electrical and mechanical specifications, e.g., the rated velocity is 3 meters/ second, the rated capacity is 1000 kilograms, the door-opening time, the door-closing time, the car-accelerating time and the car-decelerating time are all 3s, that is t_s=12s, the passenger-transfer-in/out time is 1.2s. During the pear traffic, the default traffic demand is re-defined as 100 persons/ 5 minutes.

For PSO-VTS, there are such parameters to give in advance as w=0.6, c_1=0.5, c_2=0.5 by trial and error method, and the maximum number of iterations is 1000 epochs.

The simulations run on the Fujitsu computer system (LifeBook S2110) with Windows XP, AMD Athlon 1.65GHz CPU and 512Mbytes memory (video-shared).

4.2 Experiment Results

Within experiments, the examined indices are average waiting time, average journey time and average in-system time. In reality, the in-system time is equal to the waiting time plus the journey time. The simulation experiments will compared among *queue-based control algorithm* (QCA), *static zoning-based control algorithm* (SZCA) and PSO-VTS. Under QCA, each car in the elevator group runs upward and downward in the light of queue theory, while under SZCA the service zones of elevator cars are intuitively configured as Z_1={0,1-6}, Z_2={0,7-11}, Z_3= {0,12-16}, Z_4={0,17-20}.

For the requiremtns of comparison among three algorithms, the following two scenarios are to be considered: (1) even floor-traffic, where each floor has the identical traffic demand; and (2) uneven floor-traffic, where any floor may have different traffic demand from each other. Accordingly, two attribution variables, the floor-traffic intensity λ and the distribution ζ of floor-traffic demand, are considered, respectively. Therefore, the total intensity is equated by $\lambda = \lambda_1\zeta_1 + \ldots + \lambda_N\zeta_N$. The traffic flow within 15 minutes is generated under peak traffic pattern for simulation.

A. Even Floor-Traffic

Under even floor-traffic, the distribution ζ of floor-traffic demand and the floor-traffic intensity is identical. For a building with 20 floors, as an instance, each floor has 5% of the total traffic demand under even floor-traffic. When λ ranges from 40 persons/5 minutes to 150 persons/5 minutes, the responding performance indices of elevator system are listed as in Table 1 through comparisons among three algorithms.

Table 1. Performance comparisons among three algorithms under even floor-traffic

λ (persons /5min)	average waiting time (s)			average journey time (s)			average in-system time (s)		
	QCA	SZCA	PSO-VTS	QCA	SZCA	PSO-VTS	QCA	SZCA	PSO-VTS
40	102.09	47.81	54.66	105.45	58.21	58.91	207.54	106.02	113.57
60	93.04	46.94	55.09	103.90	59.03	57.98	196.94	105.97	113.06
80	100.43	47.01	54.69	107.65	59.00	57.73	208.08	106.01	112.42
100	86.15	50.21	49.89	102.17	58.62	57.94	188.31	108.83	107.83
120	81.09	49.42	55.93	100.50	57.53	58.49	181.59	106.96	114.42
150	98.15	49.42	53.84	106.28	57.92	59.78	204.43	107.34	113.62

Seen from Table 1, SZCA and PSO-VTS both have lower value than QCA in all three indices. It is apparently concluded that SZCA and PSO-VTS outperform QCA under peak traffic pattern. In other words, zoning control is more suitable to peak traffic scheduling than the conventional queue control. However, PSO-VTS doesn't behave better than SZCA. Apart from the average journey time, PSO-VTS has larger value in other two indices than SZCA. It is because that the predefined zoning solution of SZCA is relative suitable to this kind of even floor-traffic. After all, the traffic demands in real buildings constantly waves, forming into uneven floor-traffic. PSO-VTS is focused on how to deal with this uneven floor-traffic.

B. Uneven Floor-Traffic

Under uneven floor-traffic, λ is kept fixed, that is 100 persons/5 minutes, and ζ_i is defined by the percentage of the building population. In this experiment, 6 scenarios will be considered: (a) u_i =5%U, the arrival time of passengers leaving for 1F-5F, 6F-10F, 11F-15F and 16F-20F has a latency with 150 seconds in sequence; (b) u_i =5%U, the arrival time of passengers leaving for 1F-5F, 6F-10F, 11F-15F and 16F-20F has an advance with 150 seconds in sequence; (c) u_i={3.0%/1F, 6.8%/2F, 4.3%/3F, 3.7%/4F, 5.1%/5F, 2.4%/6F, 3.5%/7F, 6.0%/8F, 4.8%/9F, 4.0%/10F, 5.8%/11F, 6.2%/12F, 6.7%/13F, 4.5%/14F, 3.9%/15F, 8.1%/16F, 7.4%/17F, 6.1%/18F, 3.4%/19F, 4.3%/20F}, the arrival time of passengers towards each floor is the same as

Table 2. Performance comparisons among three algorithms under uneven floor-traffic

No. of scenario	average waiting time (s)			average journey time (s)			average in-system time (s)		
	QCA	SZCA	PSO-VTS	QCA	SZCA	PSO-VTS	QCA	SZCA	PSO-VTS
(a)	199.19	153.90	85.49	94.52	63.61	56.75	293.71	217.51	142.24
(b)	162.04	164.53	125.61	90.08	64.27	62.01	252.12	228.79	187.63
(c)	166.13	98.62	79.05	103.19	62.77	60.06	269.32	161.40	139.11
(d)	164.01	74.61	71.38	103.95	61.81	58.23	267.96	136.42	129.60
(e)	166.91	84.52	73.46	102.85	62.79	61.51	269.76	147.31	134.97
(f)	207.30	107.55	71.17	100.82	62.22	61.42	308.12	169.77	132.59

(b); (d) u_i is set the same as (c), and the arrival time of passengers to 1F-5F and 16-20F is earlier 300 seconds than that of 6F-15F; (e) $u_1=u_2=\ldots=u_{10}=4\%U$, $u_{11}=u_{12}=\ldots=u_{20}=6\%U$, and each floor has the identical distribution of passenger arrival time; (f) $u_1=u_2=\ldots=u_5=4\%U$, $u_{16}=u_{17}=\ldots=u_{20}=4\%U$, $u_6=u_7=\ldots=u_{15}=6\%U$, and the floor-traffic arrival distribution is the same as (d). By comparisons between three algorithms, the performance indices of elevator system are illustrated as in Table 2.

Observed from Table 2, QCA has the largest average waiting time, the largest average journey time and the largest average in-system time, SZCA has the medium value in all three indices, while PSO-VTS has the smallest value in average waiting time, average journey time and average in-system time. Among 6 scenarios, compared to SZCA, PSO-VTS decreases at most 68.41s and at least 3.23s in average waiting time, decreases at most 6.86s and at least 0.8s in average journey time, and depresses at most 75.27s and least 6.82s in average in-system time. It is clear that PSO-VTS gains a great improvement in performance of elevator system. Hence PSO can effectively handle the real uneven peak traffic of office buildings.

4.3 Discussions

Based on the spirit of zoning control, SZCA and PSO-VTS service fewer floors during a round-trip period, and thus encounter fewer stops along the movement trajectory. Therefore, passengers can arrive at his/her destination floor within shorter time, including waiting time and journey time.

For some special traffic scenarios, e.g. even floor-traffic, PSO-VTS and SZCA both perform well, even SZCA is better than PSO-VTS in three indices related with time. It is a pity that even floor-traffic doesn't occur in practice.

Among uneven floor-traffic scenarios, PSO-VTS outperforms SZCA in all three performance indices. By applying PSO-VTS to elevator system, many improvements are harvested in average waiting time, average journey time and average in-system time, which interest passengers. Through the experiments under 6 uneven floor-traffic scenarios, it is concluded that PSO-VTS is more suitable to practical vertical traffic scheduling during peak traffic pattern. Down-going peak traffic is to be considered in our future investigation work.

5 Conclusion

This paper employs PSO to dynamically determine service zones of elevator cars during peak traffic in the morning. As an instance, a typical office building is considered and the model of dynamic zoning is described in detail. Then how to apply PSO to vertical traffic scheduling is focused on and a formed procedure is deduction. Simulation experiments are executed under even-floor traffic and uneven floor-traffic, respectively. Compared to SZCA and QCA, PSO-VTS has more robust and more adaptable to real peak traffic scheduling, which indicated PSO-VTS is an effective approach to elevator group control system.

Acknowledgements. This work is supported by the National Natural Science Foundation of China under the grant No.60643004 and No.60575006, and the Youth Foundation of Sun Yat-sen University under the grant No. 1131100.

References

1. Barney, G.C., dos, S.M.: Elevator Traffic Analysis, Design and Control. IEE Peter Peregrinus, London (1985)
2. Powell, B.A: Important Issues in Up-peak Traffic Handling. In: Proceedings of the 1992 International Conference on Elevator Technologies, Amsterdam, Holland, pp. 207–218 (1992)
3. Chan, W.L., So, A.T.P., Lam, K.C.: Dynamic Zoning in Elevator Traffic Control. In: Proceedings of the 1995 International Conference on Elevator Technologies, Hong Kong, China, pp. 132–140 (1995)
4. So, A.T.P., Chan, W.L.: Comprehensive Dynamic Zoning Algorithms. Elevator World, XLV(8), 899–109 (1997)
5. Zeng, J.C., Jie, J., Cui, Z.H.: Particle Swarm Optimization Algorithms. Chinese Press of Science (2004)
6. Bergh, F., Engelbrecht, A.P.: A Cooperative Approach to Particle Swarm Optimization. IEEE Transactions on Evolutionary Computation 8(3), 225–239 (2004)
7. The Chartered Institute of Building Services Engineers' (CIBSE), CIBSE Guide D - Transportation System in Buildings, London (1993)
8. So, A.T.P., Beebe, J.R., Chan, W.L., et al.: An Artificial Neural-Network Based Traffic Patterns Recognition System. International Journal of Elevator Engineering 1, 35–46 (1996)
9. Chiang, T.C., Fu, L.C.: Design of Modem Elevator Group Control Systems, Proc. of IEEE Int. Conf. on Robotics and Automation, Washington, DC, USA. Proc. of IEEE Int. Conf. on Robotics and Automation, Washington, DC, USA 470, 1465–1470 (2002)
10. Kameli, Nader.: Predictor Elevator for Traffic during Peak Conditions, United States (Patent, No. 5276295) (1994)
11. Zhang, Y., Li, Z.H., Mao, Z.Y.: Elevator Traffic Analysis and Elevator Scheduling, Chinese. Journal of China Elevator 17(3), 10–15 (2006)
12. Cortes, P., Larraneta, J., Onieva, L.: Genetic Algorithm for Controllers in Elevator Groups. Analysis and Simulation during Lunchpeak Traffic, Applied Soft Computing 4, 159–174 (2004)
13. Li, Z.H., Tan, H.-Z., Zhang, Y.N., Mao, Z.Y.: Dynamic Optimization of Elevator Group Control Based on Artificial Immune Algorithm under Inter-Floor Peak Traffic During Lunch-time. Chinese Journal of Control Theory and Applications 24(2), 177–182 (2007)
14. Li, Z.H., Chen, S.B., Zheng, R.R., Wu, J.P., Mao, Z.Y.: A Novel Approach to Clustering Analysis Based on Support Vector Machine. In: Yin, F.-L., Wang, J., Guo, C. (eds.) ISNN 2004. LNCS, vol. 3173, pp. 565–571. Springer, Heidelberg (2004)
15. Li, Z.H., Tan, H.Z.: A Combinational Clustering Method Based on Artificial Immune System and Support Vector Machine, Lecture Notes in Computer Science, Vol. In: Gabrys, B., Howlett, R.J., Jain, L.C. (eds.) KES 2006. LNCS (LNAI), vol. 4251, pp. 153–162. Springer, Heidelberg (2006)
16. Xu, Y.G, Luo, F.: Pattern Recognition of Traffic Flows in Elevator Group Control Systems Based on SVM, Chinese. Journal of South China University of Technology (Natural Science) 33(6), 32–35 (2005)
17. Li, Z.H., Wu, J.P., Mao, Z.Y.: Research on Dynamic Zoning of Elevator Traffic Based on Artificial Immune Algorithm, Proceedings of the 8th International Conference on Control, Automation, Robotics and Vision, Kunming, China, pp. 2169–2174 (2004)
18. Li, Z.H., Mao, Z.Y.: Matlab-based Simulation of Elevator Group Control System with Multi-level Velocity Mode. Journal of System Simulation 17(8), 1997–2000 (2005)

Person Identification Using Lip Motion Sequence

Salina Abdul Samad, Dzati Athiar Ramli, and Aini Hussain

Department of Electrical, Electronic and Systems Engineering, Faculty of Engineering, Universiti Kebangsaan Malaysia, 43600 Bangi, Selangor, Malaysia
Tel.: +603-89216331; Fax: +603-89216146
salina@vlsi.eng.ukm.my

Abstract. In this paper, we propose a new method based on lip motion sequence and the Unconstrained Minimum Average Correlation Energy (UMACE) filter as a classifier for person identification. UMACE filter has advantages in terms of their characteristics such as distortion-tolerant, shift-invariant and high discrimination ability. Therefore, executing UMACE filter provides an ideal application for person identification because we deal with the variations of the lip appearance in each frame sequence. The results obtained using a Digit Database show that the use of lip motion sequence and UMACE filter offer a good potential and can be an alternative technique for a person identification system.

Keywords: Unconstrained minimum average correlation energy (UMACE), lip motion, person identification.

1 Introduction

Automatic recognition of a person's identity based on biometrics has received a lot of attention in recent years. Biometric identification is the task of identifying a user from a set of subjects based on specific information either on his/her physiological or behavioral characteristics such as fingerprint, face, voice, iris and gait. Identification of a person's identity based on biometric information is important for many security and surveillance applications where access to services, buildings or files are restricted to authorized individuals [1][2][3].

The use of visual information specifically a person's mouth region as a feature to recognition system has been reported. Wark and Sridharan investigated the use of shape and intensity information from a person's lips in speaker identification systems [4]. Then, Wark et al extended this work by using these features in the presence of background noise for a robust speaker verification system [5]. Broun et al developed a feature extraction algorithm for the purpose of segmentation of lip region and extraction of geometric dimensions using MRF which is claimed to be highly effective in reducing both false acceptance and false rejection rates in speaker recognition tasks [6]. Multi-modal speaker authentication systems developed by combining speech and dynamic lip features have been investigated by other researchers [7][8]. Person verification by lip-motion using velocity estimation by orientation detection has been reported in [9].

B. Apolloni et al. (Eds.): KES 2007/WIRN 2007, Part I, LNAI 4692, pp. 839–846, 2007.

In summary, lip features for recognition system can be categorized into two main groups: lip contour (shape) based features and pixel (appearance) based features [10]. For the lip contour based features, inner and outer lip contour are extracted or geometric features such as mouth height and width are used. In pixel based group, the entire region containing the speaker's mouth (region of interest – ROI) is considered as informative for lipreading [10][11][12].

So far, many classifiers have been used for person identification systems. The familiar approaches are Dynamic Time Warping, Hidden Markov Model (HMM) and Gaussian Mixture Model (GMM) [13]. A support vector machine (SVM) for speaker recognition has been investigated by Campbell [14].

We propose a novel approach by implementing Unconstrained Minimum Average Correlation Energy (UMACE) filter in this study because of its ability to trade-off between discrimination and distortion tolerance e.g. variations in pose, illumination and facial expressions. The UMACE filter has been successfully applied in the field of automatic target recognition as well as in biometric recognition [15][16]. Face verification and fingerprint verification using correlation filters have been investigated in [17] and [18], respectively. Savvides et al performed face authentication and identification using correlation filters based on illumination variation [19].

In this paper, we develop a lip recognition system based on lip images which are the frame sequence of the speaker mouth region, captured from video while the person utters the speech. The motivation for implementing lip motion images instead of static image is that we shall use the information from each frame to develop a more powerful person recognition system which is robust under image variation. Furthermore, the changes of the speaker's lips appear in each frame sequence containing personal information that can lead to the improvement of a person recognition system performance. In this study, each consecutive frame containing the ROI from the training sequence is synthesized by the UMACE filter before the testing process takes place. Because the UMACE filter use the whole image of the lips, therefore some of the tedious feature extraction process may be reduced such as extraction of the height, width and angle of the lips for each frame.

2 Unconstrained Minimum Average Correlation Energy (UMACE) Filters

Correlation filter evolves from matched filters which are optimal for detecting a known reference image in the presence of additive white Gaussian noise. Correlation filters are attractive because of their characteristics such as shift-invariant, close form expression and high discrimination ability [15]. Therefore, executing correlation filters provides an ideal application for the recognition process.

Unconstrained Minimum Average Correlation Energy (UMACE) filter is synthesized in the Fourier domain using closed form equations. The aim of UMACE filter is to yield sharp peaks when the input image is authentic. In order to achieve this target, the filter strives to minimize the average correlation energy while maximizing the correlation output in the origin [15][19].

This optimization leads to the following filter equation

$$\mathbf{U}_{mace} = \mathbf{D}^{-1}\mathbf{m} \tag{1}$$

where $\boldsymbol{D}$ is a diagonal matrix with the average power spectrum of the training images placed along the diagonal elements and $\boldsymbol{m}$ is a column vector containing the mean of the Fourier transforms of the training images [19].

Peak-to-Sidelobe ratio (PSR) metric is used to measure the sharpness of the peak. The PSR is given by

$$\mathrm{PSR} = \frac{\mathrm{peak} - \mathrm{mean}}{\sigma} \tag{2}$$

Here, the peak is the largest value of the test image yield from the correlation output. Mean and standard deviation are calculated from the 20x20 sidelobe region by excluding a 5x5 central mask [15]. The correlation filter for each person is designed by using several training images of his/her own lip images. The number of training images used to synthesize the filter depends on the variations among the training images. The test image is then cross-correlated for each designed filter. Fig. 1 illustrates the process for person 1 using lip motion images.

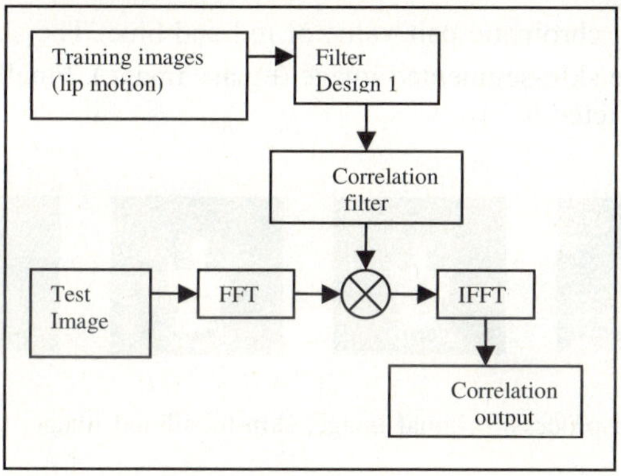

Fig. 1. Illustration of the correlation filter process

Fig. 2 shows the sample of the correlation plane for the test image for high (left) and low (right) PSR values.

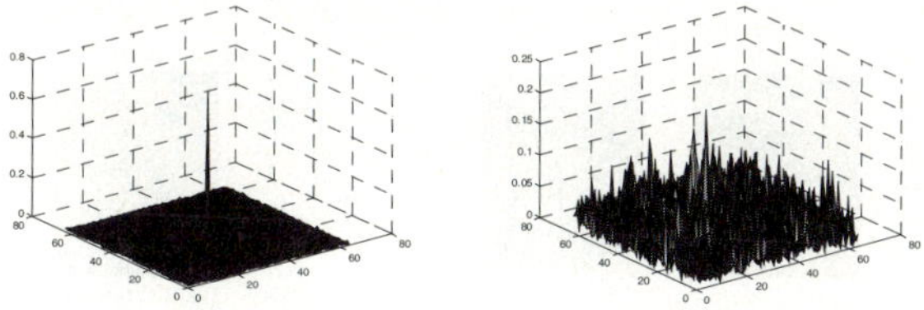

Fig. 2. Example of the correlation plane for high (left) and low (right) PSR values

3 Methodology

In this section, we shall describe in detail the lip identification process. The process can be divided into 2 main phases, feature extraction and the lip sequence identification process. The audio-visual Digit database developed by Sanderson was used for the experiments described in this paper [20]. The database consists of video and corresponding audio reciting digits zero to nine. The video of each person is stored as a sequence of JPEG images with a resolution of 512 x 384 pixels while the corresponding audio provided as a monophonic, 16 bit, 32 kHz WAV file.

3.1 Feature Extraction

In this phase, face detection and lip localization techniques are employed in order to locate the lips in a face area for each frame sequence. To segment human skin regions from non-skin color, i.e. for face detection algorithm, we apply a color-based and template matching technique. Summarily, the Gaussian model $N(\mu,C)$ with mean vector $\mu = E[x]$ and covariance matrix $C = E[(x-\mu)(x-\mu)^T]$ which are obtained from different skin color images have been used to find out the skin likelihood for any pixel of an image. The skin likelihood is calculated as $P(r,b) = \exp[-0.5(x-m)^T C^{-1}(x-m)]$ where $x = (r,b)^T$ i.e. chromatic pair value of red and blue. The skin-likelihood image is transformed to the skin-segmented image (binary image). Finally, the face region is then detected as depicted in Fig. 3.

Fig. 3. Face detection process: original image, skin-likelihood image, skin-segmented image, detected face region

For the lip localization task, we implement hue/saturation color thresholding to differentiate the lips from the skin. The detection of the lip in hue/saturation color is much easier owing to its robustness under wide range of lip colors and varying illumination conditions [21]. From the hue-saturation image, a binary image is then obtained by setting the threshold values, $H_0 = 0.04$ and $S_0 = 0.1$. Here, by employing

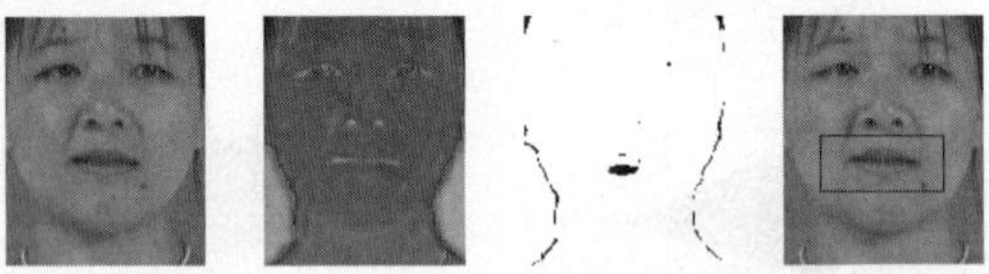

Fig. 4. Lip localization process: face detection region, hue-saturation image, binary image, localized lip region

morphological image processing, the lip region can be localized by finding the largest blob in the binary image. The lip region of 90 x 60 pixels is then extracted. Fig. 4 shows the lip localization process.

3.2 Lip Sequence Identification

Our lip motion database consists of 20 x 11 x 10 =2200 localized lip images. We use only the first 20 frames of each sequence in the video from 10 people in the database. 11 sequences of frames for each person have been utilized for the purpose of this study. One of the 11 sequences (20 images) for each person is chosen as a training sequence to be synthesized for the UMACE filter. Fig. 5 illustrates the details of the identification process in this experiment.

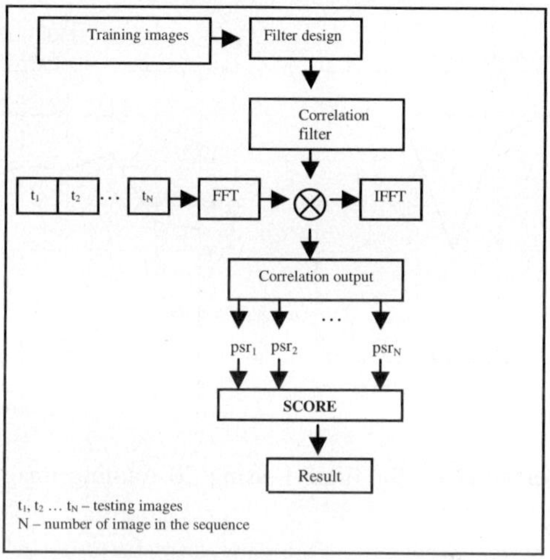

Fig. 5. Illustration of lip sequence identification

In our case, we have 10 filters. During the testing stage, for each filter, we performed cross correlations with 20 lip motion images from the person we want to verify. Because we use 20 lip images during the testing phase, the correlation output will produce 20 PSR values for each filter. First, we sort all the PSR values in descending order and then only the first 8 values are considered to be computed as the SCORE value. The SCORE value is then calculated as in equation 3. The person is then identified by finding the highest SCORE value produced by all 10 filters which represent the person.

$$\text{SCORE} = \sum_{i=1}^{n} \frac{psr_i}{n}, \quad n\text{ - number of image} \tag{3}$$

4 Results and Discussion

In the experiments, the identification process was executed by cross-correlating all the images for the person which we want to identify, for each of the 10 filters.

Then, their corresponding PSR and SCORE values were computed and recorded. Fig. 6 (left) illustrates person 1 SCORE performance using the entire images in the training sequence. Solid line refers to the SCORE of the filter 1 (the filter that has been trained by person 1 images) while dotted lines refer to the remaining 9 filters. For filter 1, SCORE values are above 30 whereas for the others filters the values are below 10.

We also attempt to adjust the system by reducing the number of training images to be synthesized by the filter. Instead of utilizing the entire images in the sequence as training images, we select only 4 images based on the largest variation among the 20 images for each training sequence. This is to ensure that the filter will be completely design for the purpose of superior identification. Fig. 6 (right) illustrates person 1 SCORE performance with 4 training images.

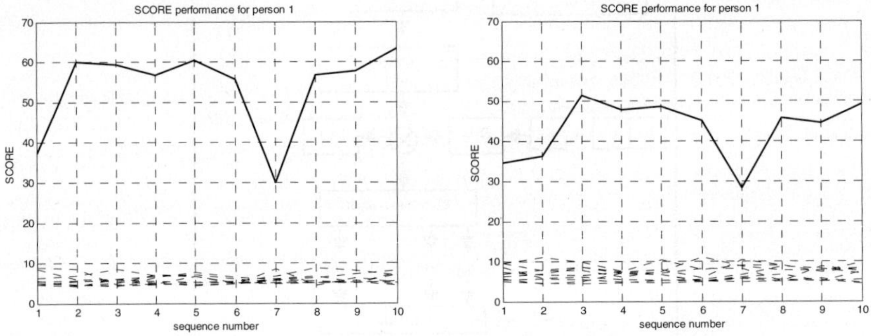

Fig. 6. The SCORE performance for filter 1 using 20 training images (left) and 4 training images (right)

We observed that, the SCORE value tends to drop when we cut the number of training images. However, the system still produces 100% identification performance accept for person 4. Table 1 and 2 summarize the percentages of identification performance using 20 and 4 training images, respectively.

Table 1. Percentages of identification performance for all 10 persons using 20 training images

Person	1	2	3	4	5	6	7	8	9	10
Identification rate (%)	100	100	100	100	100	100	100	100	100	100

Table 2. Percentages of the identification performance for all 10 persons using 4 training images

Person	1	2	3	4	5	6	7	8	9	10
Identification rate (%)	100	100	100	90	100	100	100	100	100	100

5 Conclusion

This paper has shown that lip motion sequence and UMACE filter can be an alternative technique for performing person identification. UMACE filter is also capable in performing well even with image variations. In addition, the system also performs well although a small number of training images are employed for the purpose of filter synthesizing.

References

1. Brunelli, R., Falavigna, D.: Personal Identification using Multiple Cue. IEEE Trans. on Pattern Analysis and Machine Intelligence 17(10), 955–966 (1995)
2. Zhang, D., Ghobakhlou, A., Kasabov, N.: An Adaptive Model of Person Identification Combining Speech and Image Information. IEEE International Conference on Control, Automation, Robotic and Vision, 413–418 (2004)
3. Fox, N.A., Reilly, R.B.: Robust Multi-Modal Person Identification with Tolerance of Facial Expression. IEEE International Conference on System, Man and Cybernetics, 580–585 (2004)
4. Wark, T., Sridharan, S.: A Syntactic Approach to Automatic Lip Feature Extraction for Speaker Identification. IEEE International Conference on Acoustics Speech and Signal Processing 6, 3693–3696 (1998)
5. Wark, T., Sridharan, S., Chandran, V.: The use of Speech and Lip Modalities for Robust Speaker Verification under Adverse Conditions. IEEE International Conference on Acoustics Speech and Signal Processing 6, 3061–3064 (1999)
6. Broun, C.C., Zhang, X., Mersereau, R.M., Clements, M.: Automatic Speechreading with Application to Speaker Verification. IEEE International Conference on Acoustics Speech and Signal Processing 1, 685–688 (2002)
7. Chetty, G., Wagner, M.: Liveness Verification in Audio-Video Speaker Authentication. In: Proceeding of International Conference on Spoken Language Processing ICSLP, pp. 2509–2512 (2004)
8. Chetty, G., Wagner, M.: Automated Lip Feature Extraction for Liveness Verification in Audio-Video Authentication. In: Proceeding of Image and Vision Computing, pp. 17–22 (2004)
9. Faraj, M.I., Bigun, J.: Person Verification by Lip-Motion. IEEE International Conference on Computer Vision and Pattern Recognition Workshop, 37–43 (2006)
10. Hennecke, M.E., Stork, D.G., Prasad, K.V.: Visionary speech: Looking Ahead to Practical Speechreading Systems. Speechreading by Humans and Machines, pp. 331–349. Springer, Heidelberg (1996)
11. Potamianos, G., Neti, C., Gravier, G., Garg, A., Senior, A.W.: Recent Advances in Automatic Recognition of Audio-Visual Speech. Proceeding of the IEEE 91(9), 1306–1326 (2003)
12. Potamianos, G., Neti, C.: Improved ROI and Within Frame Discriminant Features for Lipreading, pp. 250–253 (2001)
13. Reynolds, D.A.: An Overview of Automatic Speaker Recognition Technology. Proceeding of IEEE on Acoustics Speech and Signal Processing 4, 4072–4075 (2002)
14. Campbell, W.M.: A SVM/HMM System for Speaker Recognition. Proceeding of IEEE on Acoustics Speech and Signal Processing 2, 209–212 (2003)

15. Savvides, M., Venkataramani, K., Vijaya Kumar, B.V.K.: Incremental Updating of Advanced Correlation Filters for Biometric Authentication System. Proceeding of ICME 3, 229–232 (2003)
16. Vijaya Kumar, B.V.K., Savvides, M., Venkataramani, K., Xie, C.: Spatial Frequency Domain Image Processing for Biometric Recognition. Proceeding of International. Conference on Image Processing (ICIP) 1, 53–56 (2002)
17. Savvides, M., Vijaya Kumar, B.V.K., Khosla, P.: Face Verification using Correlation Filters. Proceeding of Third IEEE Automatic Identification Advanced Technologies, pp. 56–61 (2002)
18. Venkataramani, K., Vijaya Kumar, B.V.K.: Fingerprint Verification using Correlation Filters. In: Kittler, J., Nixon, M.S. (eds.) AVBPA 2003. LNCS, vol. 2688, pp. 886–894. Springer, Heidelberg (2003)
19. Savvides, M., Vijaya Kumar, B.V.K.: Efficient Design of Advanced Correlation Filters for Robust Distortion-Tolerant Face Recognition. In: Proceeding of the IEEE Conference on Advanced Video and Signal Based Surveillance (AVSS'03), pp. 229–232 (2003)
20. Sanderson, C., Paliwal, K.K.: Noise Compensation in a Multi-Modal Verification System. Proceedings of International Conference on Acoustics, Speech and Signal Processing, 157–160 (2001)
21. Matthews, I., Cootes, J., Bangham, J., Cox, S., Harvey, R.: Extraction of visual features for Lipreading. IEEE Trans. on Pattern Analysis and Machine Intelligence 24(2), 198–213 (2002)

Proposal of Method to Judge Speaker's Emotion Based on Association Mechanism

Seiji Tsuchiya[1], Eriko Yoshimura[2], Hirokazu Watabe[3], and Tsukasa Kawaoka[3]

[1]Institute of Technology and Science, The University of Tokushima
Minami Josanjima, Tokushima, 770-8506, Japan
tsuchiya@is.tokushima-u.ac.jp
[2,3] Dept. of Knowledge Engineering & Computer Sciences, Doshisha University
Kyo-Tanabe, Kyoto, 610-0394, Japan
eyoshimura@indy.doshisha.ac.jp
{hwatabe,tkawaoka}@mail.doshisha.ac.jp

Abstract. We are conducting research aiming to develop new interfaces that follow the mechanism of human communication, particularly focusing on human common sense. In this paper, the authors focus on the emotion of common sense and attempt to establish a method to judge the user's feelings based on what the user says. We proposed a method to judge a user's emotion, categorized into 10 types, from a sentence the user utters, based on the four components of the sentence. We built an actual system that uses this method and evaluated its performance. The system gave the "common sense" answer in 76.5% of the cases examined. When the "not out-of-common-sense" answers were included, the correct-answer ratio reached 89.0%.

Keywords: Emotion, Common Sense, Concept Base, Degree of Association.

1 Introduction

We are conducting research aiming to develop new interfaces that follow the mechanism of human communication, particularly focusing on human common sense. Humans, in their communication, are able to appropriately interpret ambiguous information that they receive and carry on a smooth conversation.

In this paper, however, the authors focus on the emotion of such common sense and attempt to establish a method to judge the user's feelings based on what the user says. The effectiveness of this method is examined using an actual system. We expect that use of this system can, for instance, select an appropriate expression if the content that the system tries to provide the user contains expressions that are unpleasant or remind the user of unhappy events.

The method proposed in this paper uses an Association Mechanism to evaluate the meaning and relevance of unknown words with the vocabulary already learned by the system. Thus, we believe that the proposed method has a unique superiority, in that it is able to appropriately process words that the system has yet to learn.

B. Apolloni et al. (Eds.): KES 2007/WIRN 2007, Part I, LNAI 4692, pp. 847–854, 2007.

2 The Emotion Judgment System

As this study is still in its early stages, the components of uttered sentences to be used to judge speaker feelings were limited to four ("subjects", "modifiers", "objects" and "action words"). Table 1 shows examples of the uttered sentences processed.

A "subject" is a noun that refers to the agent of the uttered sentence. The study herein is restricted to cover only those sentences with "I" as the subject, which denotes the speaker him/herself, as the use of "I" provides the foundation for future development of this study.

A "modifier" is an adjective or "adjectival verb" that modifies the "object" which follows the modifier. We permitted the omission of modifiers, as they are not always necessary in textual expression.

An "object" is a noun that denotes the object of the subject's action, behavior, or state. Hereafter, we collectively refer to "modifiers", explained above, and "objects" together as "object words".

An "action word" is a verb, adjective, or "adjectival verb" that describes the subject's action, behavior, or state.

In this paper, we attempt to judge a speaker's emotion based on the "subject", "object words", and "action words". We registered some knowledge related to the "generation of emotions", the "action words", and the "modifiers" of the "object words" in the Emotion Judgment Knowledge Base. Based on this, the system associates words and expands its knowledge within the range of common sense, making it possible to handle many expressions. The word association is realized by using the huge Concept Base[1,2] that was automatically built from multiple digital dictionaries, and a method to calculate the Degree of Association[3] that evaluates the relationship between words. Hereafter, this Concept Base and the calculation method are called the "Association Mechanism". "Objects" of the "object words" are processed using the Sense Judgment System that is able to extract the sense and perception features of a noun.

Many psychologists, philosophers, etc., have studied human emotions [4,5,6]. However, these researchers have had different interpretations of emotions and devised different models for emotions, as emotions have no substance and are quite ambiguous. However, the authors of this paper have defined emotions as "something one feels instantaneously when an action takes place" and has defined the following ten emotions to judge: "joy", "sadness", "anger", "ease", "fear", "disappointment", "shame", "regret", "sense of guilt", and "no emotion". Because we are attempting to develop a system that makes common sense judgments,

Table 1. Examples of uttered sentences covered by this study

Uttered sentence	Subject	Modifie	Object	Action word
I got some pretty	I	pretty	jewelr	got
I am afraid of ghosts.	I	-	ghosts	am afraid of

we have decided to exclude "hatred", as like/dislike of something depends on individual taste.

Because this study uses a machine system to handle and express human common sense, we have built and evaluated databases based on human common-sense thinking and feeling. In terms of system performance, we are aiming for correct judgments at least 80% of the time.

3 Concept Base and Degree of Association Algorithm

3.1 Concept Base

The Concept Base is a large-scale database that is constructed both manually and automatically using words from multiple electronic dictionaries as concepts and independent words in the explanations under the entry words as concept attributes. In the present research, a Concept Base containing approximately 90,000 concepts was used, in which auto-refining processing was carried out after the base had been manually constructed. In this processing, attributes considered inappropriate from the standpoint of human sensibility were deleted and necessary attributes were added.

In the Concept Base, Concept A is expressed by Attributes a_i indicating the features and meaning of the concept in relation to a Weight w_i denoting how important an Attribute a_i is in expressing the meaning of Concept A. Assuming that the number of attributes of Concept A is N, Concept A is expressed as indicated below. Here, the Attributes a_i are called Primary Attributes of Concept A.

$$A = \{(a_1, w_1), (a_2, w_2), \cdots, (a_N, w_N)\}$$

Because Primary Attributes a_i of Concept A are taken as the concepts defined in the Concept Base, attributes can be similarly elucidated from a_i. The Attributes a_{ij} of a_i are called Secondary Attributes of Concept A. Figure 1 shows the elements of the Concept "train" expanded as far as Secondary Attributes.

3.2 Degree of Assosiation Algorithm

For Concepts A and B with Primary Attributes a_i and b_i and Weights u_i and v_j, if the numbers of attributes are L and M, respectively ($L \leq M$), the concepts can be expressed as follows:

Concept						
train	train, 0.36	locomotive, 0.21	railroad, 0.10	...	a_i, w_i	Primary Attributes
	train, 0.36	locomotive, 0.21	railroad, 0.10	...	a_{i1}, w_{i1}	Secondary Attributes
	locomotive, 0.21	streetcar, 0.23	subway, 0.25	...	a_{i2}, w_{i2}	
	:	:	:	:	:	
	a_{1j}, w_{1j}	a_{2j}, w_{2j}	a_{3j}, w_{3j}	...	a_{ij}, w_{ij}	

Fig. 1. Example of the Concept "train" expanded as far as Secondary Attributes

$$A = ((a_1, u_1), (a_2, u_2), \cdots, (a_L, u_L))$$
$$B = ((b_1, v_1), (b_2, v_2), \cdots, (b_M, v_M))$$

The Degree of Identity $I(A, B)$ between Concepts A and B is defined as follows (the sum of the weights of the various concepts is normalized to 1):

$$I(A, B) = \sum_{a_i = b_j} \min(u_i, v_j)$$

The Degree of Association is calculated by calculating the Degree of Identity for all of the targeted Primary Attribute combinations and then determining the correspondence between Primary Attributes. Specifically, priority is given to determining the correspondence between matching Primary Attributes. For Primary Attributes that do not match, the correspondence between Primary Attributes is determined so as to maximize the total degree of matching. Using the degree of matching, it is possible to give consideration to the Degree of Association even for Primary Attributes that do not match perfectly.

When the correspondences are thus determined, the Degree of Association $R(A, B)$ between Concepts A and B is as follows:

$$R(A, B) = \sum_{i=1}^{L} I(a_i, b_{xi})(u_i + v_{xi}) \times \{\min(u_i, v_{xi}) / \max(u_i, v_{xi})\}/2$$

In other words, the Degree of Association is proportional to the Degree of Identity of the corresponding Primary Attributes, and the average of the weights of those attributes and the weight ratios.

4 Processing of Object Words

4.1 Processing of Modifiers

As a relatively small number of modifiers are used in everyday life (6,358 words), we herein classify all of these into the two categories, depending on their modifying method. The direct modification and dependent modification types are further divided into different groups having similar meaning according to the adjectives describing the modifiers (203 groups of "sense words" [meaning classification of objects] judged by the Sense System, to be described later), and they then are registered in the knowledge base for emotion judgment.

(1) Direct modification type (785 words): Modifiers that are directly involved in the generation of emotion and that convert the classification of the meaning of the noun following the modifiers into that of the modifiers.

(2) Dependent modification type (54 words): Modifiers with meaning categories that change depending on the noun that they modify.

4.2 Processing of Objects

The meaning categorization of objects is made by using the Sense Judgment System[7,8], which was developed in another study.

The Sense Judgment System judges the features people commonly think of when they hear or see a noun, in the form of adjectives and/or adjectival verbs. The "five-sense words" refer to those words referring to features directly felt by the human senses, such as "painful" and "smelly". The "perception words" refer to those words denoting features other than those directly felt by the five senses, such as "happy", "unhappy", etc. We refer to the words in both of these two groups as "sense words" and have defined 203 such words. To process the Sense Judgment System, the words and typical knowledge related to their relationship with corresponding senses and perceptions are registered in the Sense Knowledge Base. Based on this registered knowledge, the system makes precise judgments of corresponding sense words using the Association Mechanism described in Chapter 3.

Objects are also classified according to their meanings using the 203 sense words that the Sense Judgment System can judge. These 203 sense words share the common meaning categories with the modifiers discussed earlier. In short, the 203 sense words are used to categorize the meanings of the object words.

Sense Judgment System[7,8]. The knowledge base for the sense and perception judgments has a structure like a thesaurus, and it contains sense and perception words that are associated with typical nouns, which have been entered manually. In cases when an unknown word not registered in the Sense Knowledge Base needs to be processed, the system calculates the Degree of Association with those known words registered in the knowledge base for the sense and perception judgments and chooses the one with the highest Degree of Association for processing. This lets the system obtain the rough corresponding sense and perception. In addition, the system refers to the attributes registered in the Concept Base to find the sense and perception particular to that word. Due to its structure, these attributes in the Concept Base contain some inappropriate words as senses and perceptions to be associated, and thus the system is carefully designed so that the correct sense and perception is selected using the Degree of Association.

5 Processing of Action Words

Action words denote the speaker's action, behavior, and state, and include verbs, adjectives and adjectival verbs. For instance, in the sentence "My stomach aches", the action word is "aches". An action word converts the feature related to the sense and perception that associated with an object word. Features expressed in terms of senses and perceptions can be roughly divided into positive and negative expressions. For example, positive word expressions include "is beautiful", "is important", etc., and negative ones include "ache" and "is dirty". Likewise, emotions can also be categorized into two groups, positive and

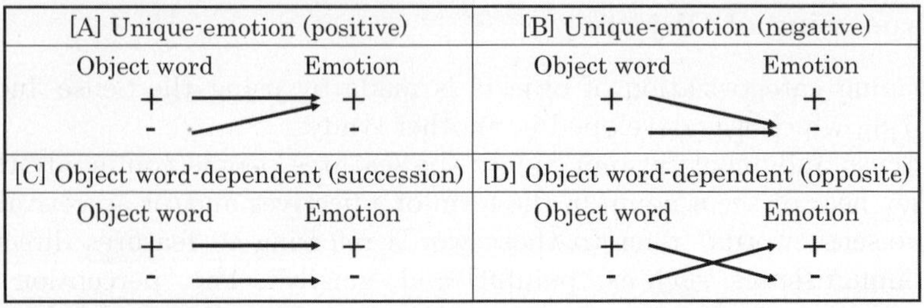

Fig. 2. Concepts of the four types of action word interaction

negative. For instance, "joy" and "ease" belong to the positive feelings, while "sadness" and "anger" belong to the negative feelings. Therefore, four types of effect can be found in the action words (Figure 2).

The action words in [A] and [B] shown in Figure 2 denote the unique emotions themselves without any dependence on object words. Examples of words in [A] include "happy" and "enjoy", while those in [B] include "grieve" and "be afraid of". Words in [C] and [D] have different feelings depending on the meaning category of the object word. A word in the [C] category generates an emotion that "succeeds" the meaning category of the object word. In contrast, a word in the [D] category creates an emotion "opposite to" the meaning of the object word. Examples of words in [C] are "see" and "get", while the [D] group includes "lose" and "throw away". Hereafter, the action words in [A] and [B] will be referred to as "unique-emotion" words and those in [C] and [D] as "object word-dependent". The "unique-emotion" action words are further divided into 10 sub-categories of emotions for the system to judge. For instance, words that indicate "happiness" are categorized in the "happy" sub-category and those that indicate "sadness" in the "sad" sub-category.

This system's knowledge of action words was built with a thesaurus that systematically arranges verbs according to the actions and states they represent. Thus, the system's Emotion Judgment Knowledge Base has registered all the self-sufficient verbs, as well as all 17,676 nouns indicating actions and states (those Japanese nouns that are frequently and easily turned into verbs with a suffix). This thesaurus of verbs enables the system to classify large numbers of verbs with ease. In order to handle adjectives and adjectival verbs, the system uses its knowledge of modifiers.

6 Processing of the Emotions Generated

As described in Chapter 2, there are a great variety of opinions regarding the definitions of the emotions that humans have. Therefore, we have defined an emotion as "something one feels instantaneously when an action takes place" and defined the following ten emotions as emotions to judge: "joy", "sadness", "anger", "ease", "fear", "disappointment", "shame", "regret", "sense of guilt", and "no emotion".

With respect to the emotions that were generated, those associated with a total of 406 pairs of the meaning categories of object words (203 categories) and action words (2 categories of "succession" and "opposite") were manually defined and registered in the system's Emotion Judgment Knowledge Base. The 10 sub-categories of change words were not registered in the Emotion Judgment Knowledge Base as rules to generate emotions, as these sub-categories directly indicate "joy", "anger", or other emotions.

7 Performance Evaluation of the Emotion Judgment System

In order to evaluate how valid the emotions generated by this Emotion Judgment System were, pairs of object and action words that are associated with an emotion were collected from 40 college students, and 200 pairs were randomly selected to be used as data for evaluation. Five test subjects were then asked to judge whether the emotions generated by the system were common sense. If four or more of judges said an emotion generated was a common sense emotion, the emotion was considered as "common sense" (correct answer). In cases when two or three subjects said the emotion was "common sense", the generated emotion was considered as being "not out-of-common-sense". If only one subject or no subjects considered the generated emotion as "common sense", the emotion was thought of as "out of common sense" (an error). For cases when multiple emotions were generated, they were considered as being "common sense" (correct answer) if all of the generated emotions were common sense ones. If any one of the emotions was considered to be "out of common sense", then the particular emotion generated was considered as being "out of common sense" (an error). All others were regarded as being "not out-of-common-sense".

Table 2 shows the results of the system's emotion judgments. Table 3 shows the evaluation data and the evaluation results. The system gave the correct answer

Table 2. Result of the System's Emotion Judgement

	Common sense answer (correct answer)	Not out-of-common-sense answer	Out of common sense answer (an error)
Rate of result	76.5%	12.5%	11.0%

Table 3. Examples of data for evaluation and the emotions generated

Data for evaluation	Generate	Evaluation result
I saw a boring movie.	Disappointment	Common sense answer
I live alone.	Sadness	Common sense answer
I saw a ghost.	Fear	Common sense answer
I gave birth to a daughter	Joy, Sadness	Not out-of-common-sense answer
I overlooked an error.	Ease, Joy	Out of common sense answer

in 76.5% of cases. If the "not out-of-common-sense" answers were counted as part of the "correct answers", the correct-answer ratio rose to 89.0%. Because these ratios were both high, we believe that this system is effective.

8 Conclusion

In this paper, we focused on emotions, which are part of the common sense judgments humans make in everyday communication. We proposed a method to judge a user's emotion, categorized into 10 types, from a sentence the user utters, based on the four components of the sentence; "subjects", "modifiers", "object words", and "action words". We built an actual system that uses this method and evaluated its performance. The system gave the "common sense" answer in 76.5% of the cases examined. When the "not out-of-common-sense" answers were included, the correct-answer ratio reached 89.0%. This demonstrates that the Emotion Judgment System we have in this study performs very well, and that the method we have proposed is effective.

Acknowledgment

This work was supported with the Aid of Doshisha University's Research Promotion Fund.

References

1. Hirose, T., Watabe, H., Kawaoka, T.: Automatic Refinement Method of Concept-base Considering the Rule between Concepts and Frequency of Appearance as an Attribute. Technical Report of the Institute of Electronics, Information and Communication Engineers. NLC2001-93, pp. 109–116 (2002)
2. Kojima, K., Watabe, H., Kawaoka, T.: A Method of a Concept-base Construction for an Association System: Deciding Attribute Weights Based on the Degree of Attribute Reliability. Journal of Natural Language Processing 9(5), 93–110 (2002)
3. Watabe, H., Kawaoka, T.: Measuring Degree of Association between Concepts for Commonsense Judgements. Journal of Natural Language Processing. 8(2), 39–54 (2001)
4. Hukui, Y.: KANJO-NO-SHINRIGAKU (Japanese). Kawashima Shoten (1990)
5. Rita, C.: NO-TO-KOKORO-NO-CHIZU (Japanese). Hara Shobo (1999)
6. Susan, G.: NO-NO-TANKYU (Japanise). Mumeisha (2001)
7. Horiguchi, A., Tsuchiya, S., Kojima, K., Watabe, H., Kawaoka, T.: Constructing a Sensuous Judgement System Based on Conceptual Processing. Computational Linguistics and Intelligent Text Processing (Proc. of CICLing-2002), pp. 86–95 (2002)
8. Watabe, H., Horiguchi, A., Kawaoka, T.: A Sense Retrieving Method from a Noun for the Commonsense Feeling Judgement System. Journal of Artificial Intelligence 19(2), 73–82 (2004)

The Automatic Peer-to-Peer Signature for Source Address Validation

Yan Shen, Jun Bi, Jianping Wu, and Qiang Liu

Network Research Center, Tsinghua University,
Beijing 100084, China

Abstract. SPM (Spoofing Prevention Method) proposed a peer-to-peer anti-spoofing method which could effectively filter spoofed packets and support incremental deployment. However, mechanism of SPM key updating has some serious problems: (1) heavy management cost; (2) risk of becoming the target of DOS/DDOS attacks during the key updating; (3) limitation of the scale of the SPM union; (4) impossibility of tackling the threat of real-time packet sniffing. This paper proposes an Automatic Peer-to-Peer Anti-spoofing Method (APPA). APPA resolves the key management problem and could effectively prevent source address spoofing.

Keywords: Source Address Spoofing, Spoofing Prevention, Key Management.

1 Introduction

Spoofing of source IP address is often used by malefactors to launch DOS/DDOS (Denial of Service) attacks [1] [2] [3]. Attackers take advantages of spoofing to make it hard for the victim destination to categorize malicious traffic by source. Nowadays economic and trade services are in fast development in internet, such as e-commerce, e-bank, IP spoofing are much more commonly used by criminals. Spoofing gives rise to a number of challenges for law enforcement. If not deterred, spoofing would make much more serious threat to the internet and even the social life in the foreseeable future.

Great efforts have been made to prevent IP address spoofing. uRPF (Unicast Reverse Path Forwarding) [4] and Ingress filtering [6] are famous of them. Ingress filtering filters out packets containing unexpected source IP address at the source of the packets' route, while uRPF filters out packets with source IP address which is not expected to exist in a route. Both of them are effective and fast. IETF OPSEC (Operation Security) work group suggests that the combination of Ingress filtering and uRPF can solve the spoofing problem. However, Ingress filtering and uRPF are not designed for stepwise deployment and won't work well if it is not deployed in large scale. Ingress filtering also can't recognize spoofing in the same AS (Autonomous System) and uRPF can't filter packet spoofing in the same route. SPM (Spoofing Prevention Method) [5] presents an anti-spoofing method which is effective and also supports stepwise deployment. However, SPM is not DOS resilient in the key-update process. It also suffers from sniffing and can't be used for anti-spoofing of smaller granularity.

This paper present a new anti-spoofing method called Automatic Peer-to-Peer Anti-spoofing Method (APPA), which derives from and improves upon SPM. APPA

B. Apolloni et al. (Eds.): KES 2007/WIRN 2007, Part I, LNAI 4692, pp. 855–863, 2007.

uses the same tagging and verifying key mechanism as SPM and makes the key-update process automatically. It is effective, DOS resilient, applicable in large scale, anti-sniffing and designed for stepwise deployment.

The rest of this paper is as follows. Section 2 introduces the related work of anti-spoofing. Section 3 introduces the basic rationale and the implementation choices of APPA. In section 4, we proposed a detailed solution of APPA used for inter-AS anti-spoofing. Section 5 is the comparison of APPA and SPM. Section 6 is a brief introduction of future work of APPA. Section 7 concludes the whole paper.

2 Related Work

Methods for IP address anti-spoofing could be categorized into 3 types: traceback, filter in route, and filter at destination. Traceback is used to trace the real source of the malicious packets when attack happens. It is of forensic value of law enforcement and will help catch and prosecute the attackers. According the way of tracing, Traceback has three different types. (1) The routers on the route add information into the packets. [7][8][9][15] are typical traceback of this type. (2) The routers on the packets' route send ICMP messages to the destinations of the packets. [10] [11] belong to this type. (3) The routers on the packets' route deposit the digest of the packets. [12][13] are of this type. Traceback does not work well against spoofing. Even it could exactly trace back to the attacker it will not work well, since many hosts in the internet are comprised by attacks, traceback may find out the comprised host instead of the real attackers.

Filtering in the route, the second category of anti-spoofing methods is more effective. The typical two are Ingress filtering [6][14] and uRPF[4]. Ingress filtering brings benefits to the network that deploys it only if it is deployed in large scale, hence it is lack of incentive for stepwise deployment. uRPF is based on that a packet destined to a given AS would traverse a specific route. uRPF permits attackers to spoof any units on the same route to a victim. It also suffers from asymmetry route. Methods mentioned in [16] to solve the problem of uRPF in asymmetry route is rough and would even weaken the original effect of uRPF. uRPF also does not support stepwise deployment.

Hop count [16] and SPM [5] do filter at the destination. Hop count uses TTL value as a key for each (source, destination) pair. Hop count could filter out 90% of the forged packet if the attackers do nothing to the TTL value. However, the TTL value is small and the attackers could easily learn the right TTL value he should put in packets. It does not work to prevent sophisticated attacks.

SPM proposed a novel peer-to-peer anti-spoofing method. It is effective and light-weighted, also brings great incentive for internet owner to deploy it and directly benefits them by conquering the deficiency of trace back and route filtering. If adequately deployed, SPM could filter out 99.999999% of the forged packet and it is much safer than Hop count. Compared with AH [17] of IPSec, it is light-weighted and would not become a target of DOS/DDOS attack.

In SPM, a packet is tagged with a key at the border of the source AS, and the key is verified and removed at the border of the destination AS. The key is associated with the (source, destination) pair and is kept secret from any AS else. SPM suggests to change keys every two 2 hours for safety. It provides two ways to update keys (one active way and one passive way), both needs safe interactions between every pairs. As SPM thought it is difficult to sniff in bone network, keys are used in plain text.

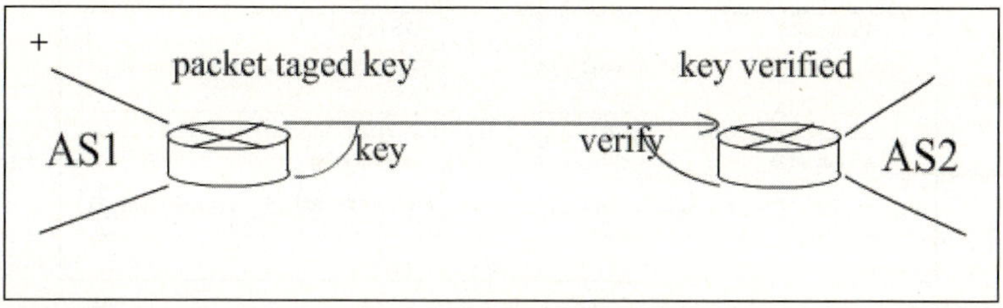

Fig. 1. SPM, packet is tagged with a key and the key is verified at the destination

We developed an implement of SPM and found out SPM have the flowing flaws:

(1) SPM could not be used in large scale. Keys need be changed to keep safe. SPM AS needs to interact with every ASes else in the SPM league. When the network is busy, it is hard to finish all the interactions in half an hour when 2000 ASes take part in the SPM league.

(2)The interaction could easily become the target of DOS/DDOS attack. This makes the key updating became much more difficult.

(3) SPM has no idea to deal with sniffing so could not be used where sniffing is not so hard.

3 APPA

3.1 Basic Rationale

As a signature-and-verification method, APPA tags a key into each packet at the source and verifies the key at the destination. Each key is used only in one packet and will change in the next packet. This makes eavesdropping and replaying the key or the whole packet useless. The key updating is finished automatically by special mechanism rather than interaction. This improves its safety and efficiency.

3.2 Updating Keys

The most important thing for APPA is updating key safely and efficiently without interaction, in spite of any states of the network. To implement APPA, the following issues must be satisfied. 1. The key change can not be inferred from the known serial keys by anyone except the source and destination. The only opportunity left for attackers is brute-force guess. 2. Updating key should be very fast. 3. The choice of algorithms should be pretty much, so APPA would keep robust and safe.

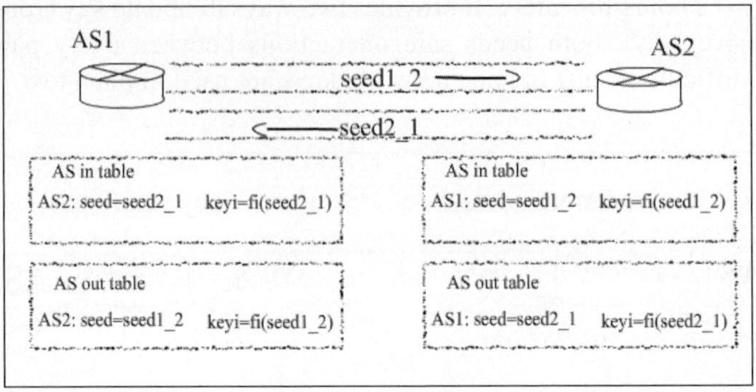

Fig. 2. The seed exchange process uses special method (e.g. Diffle Hellman protocol) to exchange seeds between peers. Keys are generated from the seed.

Random Number Generator(RNG) fits all the request of APPA. RNG generates a series of numbers that satisfies several tests of statistics in mathematics. Good RNGs have several characteristics: 1. Fast. 2. Well distributed, pass all statistical tests known. 3. Independent. 4. Portable and reproducible. 5. Long periods. 6. Large seed space. 7. Unpredictable and irreproducible.

The last one is an additional request for cryptology. [18][19] describe RNGs in detail.

APPA needs the request of 1,6,7 of RNGs, that is fast, large seed spaces and unpredictable .There are many RNGs fit for APPA such as [22][24]. Every source and destination pair only needs to exchange a seed for RNG, then all the sources and destinations could easily update and synchronism keys.

To enhance safety, we could make the result of f(seed_a) ⊕ f(seed_b) the key, where f is a RNG. This is similar to the One-Time-Pad (OTP) cipher mechanism [18] in cryptology.

Countless good RNGs are fit for APPA such as [20] [21] [22].The progress in RNG theory may cause some RNGs unsafe but we still have choices. The long period of RNGs omits any more interactions in pairs except the first seed exchange.

4 Inter-AS Solution of APPA

4.1 Overall

In this section we give one Inter-ASes anti-spoofing solution in detail.

The phase of tagging keys and verifying keys in packet is similar as SPM. Packet is tagged with a key at the border of the AS and the key is verified at the border of destination AS. Keys are changed periodically to enhance safety. APPA updates keys automatically rather than by interaction in SPM, it can generate millions of keys in 1 second.

4.2 Initialization

The initialization is very important to APPA because it decides all the keys. In our solution we choose The KISS Generator as the key generator. The task of initialization is to exchange seeds between pairs. Suppose there are N ASes in the APPA league and 1 new AS now takes part in. The fresh APPA AS needs to exchange seeds with all the N ASes. This is called the Initialization.

To accomplish the initialization of seeds exchange safely, the SA (Security Association) process of IPSec and the Diffie Hellman exchange are practical ways. We use a combination of TCP Interception and Diffle Hellmann protocol [26] to exchange seeds. The KISS Generator needs a seed of 128 bits and we use two seeds to generate one key, so the seed's length is 256 bits:

$$key = KISS(seed\_1) \oplus KISS(seed\_2)$$

The KISS Generator is safe, fast and has a period of 2^{124} .No more interactions is needed after the initialization of seeds exchange, unless one AS thinks it necessary to change the seed.256 bits seed make it safe to brute force guess. The KISS can generate millions of keys per second with normal hardware nowadays, so keys can change as fast as possible and the APPA league can be very large.

4.3 The Length, Position, and Period

The length of the key is directly associated with the safety while under attack. SPM suggested 32bits-length key. In IPV6, the length is not the problem because it can be put in expansion head, in [24] the key is even up to hundreds of bytes .In IPV4, the length is a problem. Considering the efficiency of padding and verifying, the key should be placed in the IP head. The IP option field is 32 bits and enough in length, but based on the report of [23], more than 70% of the connection requests would be failed and 44% of the connection would break if the packets contain unknown IP option field. The IP identification could work but is only 16bits long. So in IPV6 the key could be placed in expansion head, while in IPV4 in expansion head but keys need faster change.

Obviously, the faster the key changes, the safer is the key. Shannon proved one-time key is the most safe in [25]. SPM suggested the 32-bit key changes every two 2 hours that is not safe under brute-force attack, we'll give an analysis for that in the next section. APPA suggests the 32-bit key period 10 minutes and could provide even one-time key for each packet.

4.4 Failure Recovery

When a router reboots by spontaneity or accident, its keys for other ASes is out of time and it also does not know what keys other ASes would use, unless it knows how much time it has shut down. An AS often has more that one border router and the rebooted router can easily get the AS-in-table and AS-out-table from other routers. Only if all the routers which have the tables reboot at the same time period, we have to learn the tables from other ways. One possible way to learn from the in-bound packets, compute keys use the table before rebooting to match the keys gets in the packets. That is fast and easy .Another way is to use series number as part of the key, or use the same keep-alive scheme in TCP. The failure recovery is not difficult for APPA.

4.5 Anti-sniffing

SPM mentioned that it is hard to sniff in backbone network so the key can be used in plain text. We agree that it is feasible in inter-AS solution and also think it is useful to develop ways for anti-sniffing, as sniffing possibly exists in backbone network and anti-spoofing should finally be deployed for end-to-end systems.

The main threat of sniffing is that attackers could launch replay attack, which makes many secure mechanisms brittle. If attacker could get the key by sniffing, he could launch spoofing attack easily. The key point is anti-replay for anti-sniffing. Many methods have been proposed for anti-replay, such as time-stamp, sequence number, bloom filters [27]. Time-stamp needs global clock synchronization, while bloom filters need large cache spaces. IPSec AH [17] uses sequence number and a 32-size slide window to performance anti-replay, it is feasible because AH is end-system oriented, while the packets between two end systems' session are less possible in disorder. The slider window needed in inter-AS solution is large, and every peer has to keep n-1 windows while the league size is n. Nowadays the flux of packets between ASes is very huge and fast. For the reasons above, anti-replay is a difficult task in inter-AS solution, unless we don't care the huge cost which even becomes a bottle-neck.

APPA could achieve anti-replay by two ways. 1) To change keys as fast as possible, the limit of that is to change key for each packet and every key is used only in one packet. If most of the packets arrive the destination in order as they leave the source, keep a small slide-window would achieve anti-replay effectively as APPA generates keys very fast. 2) To use a combination of sequence number and Chaos Theory[28], and also take advantage of APPA's fast changing key. One possible solution is as following steps:

1) APPA generates a key, called *k0*.
2) Each packet from *S* to *D* contains an increasing sequence number, which are 1,2,3,4...
3) The key tagged in the packet is calculated from *L(k0, sequence number)*, where *L* is a Chaos-based function, e.g., *L* is a four-times call of Logistic mapping, that is,

$$X_{n+1} = ax_n(1-x_n), 0 < a \leq 4, 0 < x_n < 1, \text{(Logistic mapping of Chaos Theory)}$$

Logistic mapping has the butterfly effect after 3 or 4 times iteration. Hence *L(k0, sequence number)* produces safe and unpredictable keys. The source and destination could calculate and verify the keys easily but the attackers not. The *k0* is changing fast so it is safe. A bit map is kept to denote which sequence number has been used. If *k0* is updated, the sequence is reset to 0 and the bit map too.

Anti-replay is not necessary in inter-AS solution and could be an option function. However, APPA is easy to against replay attacks.

5 Analysis and Comparison

The main purpose of this section is to give analysis and comparison to evaluate the advantage that APPA over SPM and other relative methods. The management cost, safety and expansibility is evaluated in relative-comparative manner with SPM, and the efficiency and safety with other methods.

The first and foremost deficiency of SPM is the high-frequency interaction for update keys, which is not DOS resilient and even not busy-network resilient. That is, the interaction would become hard while DOS attack exists or the network is in busy state. In contrast, APPA only needs the initial interaction to exchange the seeds, the trait of long period and unpredictable makes the key safe, and the large seed space makes it safe under brute force attack.

The total number of key-update interactions during a short key period in SPM is increased by $O(n^2)$, which is zero in APPA after the initial interaction. Attackers could take advantage of this to launch attacks to block part of key-update process or DOS at the router/server. On the other hand, though each participant only does $O(n)$ interactions but it is still difficult to finish in short time. Our experiments show that a SPM league with 2,000 participants is hard to update keys in half an hour while the network is just normally busy. That means neither could SPM update keys fast nor could SPM have many participants.

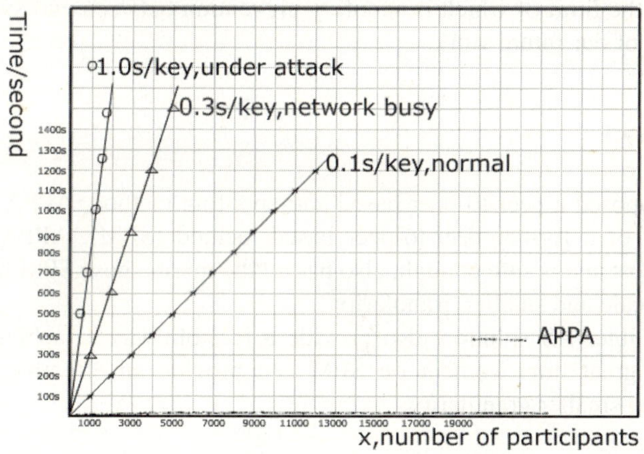

Fig. 3. This figure shows the time each participant spends to update keys with other participants. The x axis denotes the size of the league, while the y axis denotes the time needed to update keys. SPM needs much more time to update keys, while APPA needs only few time. APPA could update millions of keys per second, even billions of participants won't bring problem.

Another aspect should be considered is the length of the key.The key should be as short as possible to improve the effective rate payloads. This is contradicting with the safety request. If the attackers control 1GB bandwidth, a 32-bit key would be doped out in $\frac{2^{32}\times 64\times 8}{2^{30}}$ seconds. Updating the key faster provide more safety and reduce the length of the key. APPA could easily make the key period a few seconds even reach one key per packet, which provide perfect safety for anti-spoofing.

6 Future Work

As described in section 3, anti-replay is difficult in inter-AS solution, because the inter-AS flux of packets in huge and fast, any known anti-replay methods would bring great loss of network performance. The APPA could change keys very fast, that brings a glimpse of hope to solve anti-replay in inter-AS solution. That is, each key is used only in one packet. This makes it impossible to launch replay attack because each key is used only once. However, because of the packet would arrive in disorder or lost in the route, a slide window is needed for each peer, to keep memory that which key has been used. When the league size is large, this needs large memory spaces to deposit information. The usage of Logistic mapping also needs a bit map for checking used sequence number for each peer, which also needs much space. One of our future works is to perfect one-time key.

APPA's low management cost makes it easy to be used in smaller granularity anti-spoofing, the limit of which is spoofing prevention between end-systems. That is another aspect of our future work.

7 Conclusion

This paper proposed a peer-to-peer based anti-spoofing method, APPA, which solves the deficiency of a former peer-to-peer based method (SPM). APPA requests low management cost and is safe enough, also has all the merits those SPM has. It could be used in large scale and smaller granularity.

The APPA inter-AS solution also provide a feasible way to achieve anti-replay. APPA brings great improvement on SPM and provide an effective way to solve anti-spoofing in smaller granularity.

References

[1] Moore, D., Voelker, G., Savage, S.: Inferring internet Denial-of-Service activity. In: in Proc. USENIX Security Symposium, pp. 9–22 (2001)
[2] CERT Advisory CA-, 01:Denial-of-service Development (January 2000), http://www.cert.org/advisories/CA-2000-01.html
[3] Mazu, Networks. Enforcer, 2002, http://www.mazunetworks.comproducts
[4] Cisco, I.O.S.: Unicast reverse path forwarding (1999)
[5] Anat, B.B., Hanoch, L.: Spoofing Prevention Method. In: Proceedings IEEE Infocomm (2005)
[6] Baker, F., Savola, P.: Ingress Filtering for Multihomed Networks. RFC 3704 (March 2004)
[7] Dawn, X.iao,D.S., Adrian, P.: Advanced and authenticated marking schemes for IP traceback. In: Proceedings IEEE Infocomm 2001 (April 2001)

[8] Kihong, P., Heejo, L.: On the effectiveness of probabilistic packet marking for IP traceback under denial of service attack. Tech. Rep. CSD-00-013, Department of Computer Sciences. Purdue University (June 2000)
[9] Micah, A.: Tradeoffs in probabilistic packet marking for IP traceback. In: Proceedings of 34th ACM Symposium on Theory of Computing (STOC), New York (2002)
[10] Belenky, A., Ansari, N.: On IP Traceback. IEEE Communications Magazine 41(7) (July 2003)
[11] Bellovin, S., Leech, M., Taylor, T.: Icmp traceback messages (February 2003), http://www.ietf.org/internet-drafts/draft-ietf-itrace-04.txt
[12] Alex, C.S., Craig, P., Luis, A.S., Christine, E.J., Fabrice, T., Beverly, S., Stephen, K., Strayer, W.: Single-packet IP traceback. ACM/IEEE Transactions on Networking (December 2002)
[13] Timothy, W.S., Christine, E.J., Fabrice, T., Regina, R.H.: SPIE-IPv6: Single IPv6 Packet Traceback, Local Computer Networks, 29th Annual, IEEE International Conference on (16-18 November 2004), pp. 118 – 125 (2004)
[14] Ferguson, P., Senie, D.: Network ingress filtering: Defeating denial of service attacks which employ ip source address spoofing (May 2000), RFC (2827)
[15] Amin, S.O., Kang, M.S., Hong, C.S.: A Lightweight IP Traceback Mechanism on IPv6. In: Zhou, X., Sokolsky, O., Yan, L., Jung, E.-S., Shao, Z., Mu, Y., Lee, D.C., Kim, D., Jeong, Y.-S., Xu, C.-Z. (eds.) Emerging Directions in Embedded and Ubiquitous Computing. LNCS, vol. 4097, pp. 671–680. Springer, Heidelberg (2006)
[16] Cheng, J., Wang, H.N., Kang, G.S.: Hop-count filtering: An effective defense against spoofed DDoS traffic. In: Proceedings of ACM CCS ' (2003)
[17] Kent, S.: IP Authentication Header. RFC 4302 (December 2005)
[18] Mathew, S., Jacob, K.P.: A New Fast Stream Cipher: MAJE4. IEEE INDICON 2005.
[19] Stephen, K.P., Keith, W.M.: Random number generators: good ones are hard to find. Communications of the ACM archive 31(10) (October 1988)
[20] Matsumoto, M., Nishimura, T.: Mersenne twister: A 623-dimensionally equidistributed uniform pseudo-random number generator. ACM Trans. Model. Comput. Simul. 8(1), 3–30 (1998)
[21] George, M., Wai, W.T.: The 64-bit universal RNG, Letters in Statistics and Probability. 6(2), 183–187 (January 2004)
[22] George, M.: The KISS generator. http://oldmill.uchicago.edu/~wilder/Code/random/Papers/Marsaglia_2003.html
[23] Alberto, M., Mark, A., Sally, F.: Measuring the Evolution of Transport Protocols in the Internet. ACM Computer Communication Review, 35(2), April (2005), http://www.icir.org/mallman/papers/tcp-evo-ccr05.ps
[24] Li, X., Yang, X.W.: Efficient and Secure Source Authentication with Packet Passports. In: Proc. of USENIX SRUTI, San Jose, CA (2006)
[25] Shannon, C.E.: Communication Theory of Secure Systems. Bell System Technical journal 28(4), 656–715 (1949)
[26] Whitfield, D., Martin, E.H.: New directions in cryptography. IEEE Transactions on Information Theory 22(6), 644–654 (1976)
[27] Li, F., Pei, C., Almeida, J., Broder, A.Z.: A scalable wide-area web cache sharing protocol. Technical Report 1361, Department of Computer Science. University of Wisconsin-Madison (1998)
[28] Baptista, M.S.: Cryptography with chaos[J]. Physics Letters A, 50–54 (1998)

Traffic Demand Prediction Using ANN Simulator

Vedat Topuz

Marmara University, Vocational School of Technical Sciences, 34722 Istanbul, Turkey
vtopuz@marmara.edu.tr

Abstract. The prediction of the traffic data is a vital requirement for advanced traffic management and traffic information systems, which aim to influence the traveler behaviors, reducing the traffic congestion, improving the mobility and enhancing the air quality. Both the stochastic time series (TS) techniques and artificial intelligent (AI) techniques can be used for this aim. Daily traffic demand in Second Tolled Bridge of Bosphorus, which has an important role in urban traffic networks of Istanbul has been predicted by both a TS approach using an autoregressive (AR) model, and an AI approach using an artificial neural network (ANN) model. The results have shown that the prediction error obtained by ANN model is smaller than the error obtained by AR model. The results have also pointed out that many other transportation data prediction studies can be implemented easily and successfully by using the developed ANN simulator.

Keywords: Traffic demand, Prediction, Artificial Neural Network, Time series analysis.

1 Introduction

Detecting the trends and patterns in transportation data is a popular area of research in the business world to support the decision-making processes. The primary means of detecting trends and patterns has involved statistical methods such as clustering and regression analysis most of which use linear models. The goal is to predict a future value of data by using the past data set [1, 2].

On the other hand, non-linear approaches offer more powerful models, since they give more possibilities in the choice of the input-output relation. ANN models have been applied to a large number of problems because of their non-linear system modeling capability by learning ability using collected data. They offer highly parallel, adaptive models that can be trained by the experience.

During the last decade ANN models have been applied widely to prediction of the traffic data [3,4]. Such a prediction study has been completed in this paper, to compare the effectiveness of ANN and TS models.

2 Prediction by ANN Model

The daily traffic demand in Second Tolled Bridge of Bosphorus (FSM Bridge) which has an important role in the urban traffic networks of Istanbul has been predicted by

B. Apolloni et al. (Eds.): KES 2007/WIRN 2007, Part I, LNAI 4692, pp. 864–870, 2007.

both an AR(4) model and an ANN model implemented by ANN simulator developed by research team. The developed ANN simulator and test studies completed by using this simulator are introduced in this section.

2.1 ANN Simulator

The ANN simulator has different two choices of ANN configurations: single layer and three layers feedforward networks. In the single layer configuration, the hebbian, perceptron, delta, and widrow-hoff algorithms can be chosen for training. In the three layers configuration, the backpropagation algorithm is used for training. The screen structure of backpropagation training module with three layers perceptron model is shown in Fig. 1.

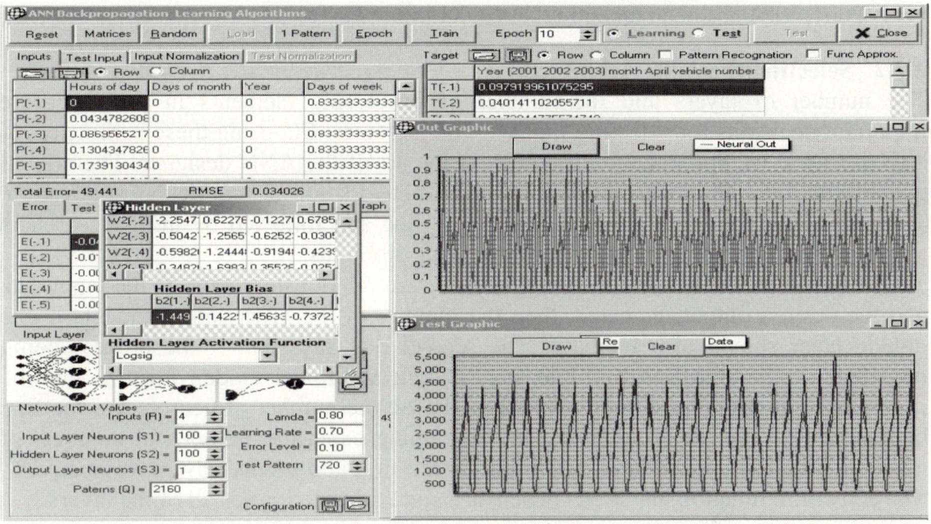

Fig. 1. The backpropagation screen of the ANN simulator

In order to use the ANN simulator for any application, first the number of neurons in the layers, type of activation function (purelin, tansig, logsig), the number of patterns, and the training rate must be chosen. Then, the input and target data must be loaded as a text file. If the user wants to normalize the input and target data, the normalization screen should be run.

At the beginning of training phase, the weighting coefficient matrix is assigned with random numbers. For each epoch, it is possible to observe the error, weighting coefficient matrix, and total error on the screen. In addition to this, the variation of the error, target and output values can be seen in a graphic format. All errors during the training and test can be calculated and the results will be saved in the data files.

2.2 Designing Process

ANN designing process involves four steps. These are gathering the traffic data, selecting the ANN architecture, training the network, and testing the network. In the training step, four input variables have been used. These are, hours of the day: 0-23; days of the month: 1-30; days of the week: 1-7; and years: 2001-2003. Selecting the daily traffic demands as distinct variables both for days of the month and days of the week are mandatory situation, because of the weekly and monthly variance of the daily traffic demands. In the training phase, the target variables have been used as 2160 data related to the volumes, or hourly passing number of the vehicles (demand) of 4th months (April) of the years 2001-2003. In the testing phase, 720 data related to the hourly demand of 4th month (April) of the year 2004, have been used to compare the actual and neural output of the system performance.

2.2.1 Gathering the Traffic Data

Data recorded by the tolls has been used for training, testing and analysis. Only 2880 number of traffic demand data related to the hours of the days of 4th months of the years 2001-2004 has been taken into consideration [3].

2.2.2 Selecting the ANN Architecture

The number of layers and the number of processing elements in per layer are important decisions for selecting the ANN architecture. Choosing these parameters to a feedforward backpropagation topology is the art of the ANN designer. There is no quantifiable best answer to the layout of the network for any particular application. There are only general rules picked up over time and followed by most researchers and engineers applying this architecture to their problems.

The first rule states that if the complexity in the relationship between the input data and the desired output increases, then the number of the processing elements in the hidden layer should also increase. The second rule says that if the process being modeled is separable into multiple stages, then additional hidden layer(s) may be required. Some tests have been applied to choose the number of neurons for the first and second layers. So, following a lot of training experiments, optimal numbers of neurons have been founded as 100 both for the first and second layers, respectively. The activation function has been chosen as a sigmoid function for all of the layers.

2.2.3 Training the Network

ANN simulator has been trained through the four hundred epochs. The training data have been used as the traffic demand related to the hours of the days of 4th months of

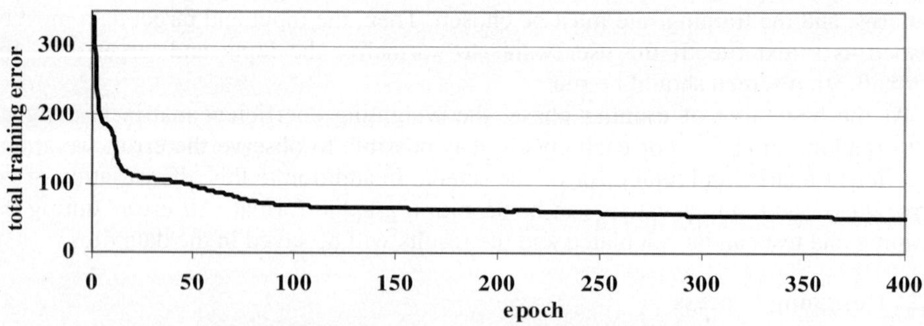

Fig. 2. Variation of the total training error through 400 epochs

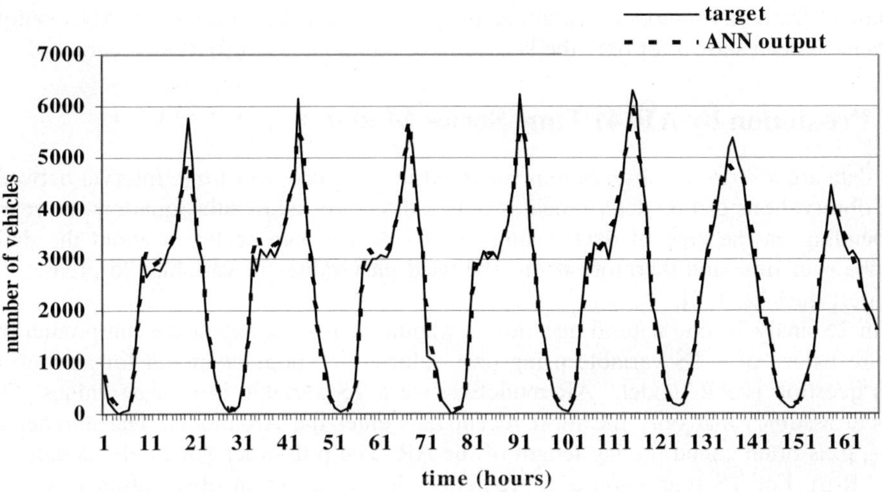

Fig. 3. Variation of the ANN output data versus the target data (first analysis)

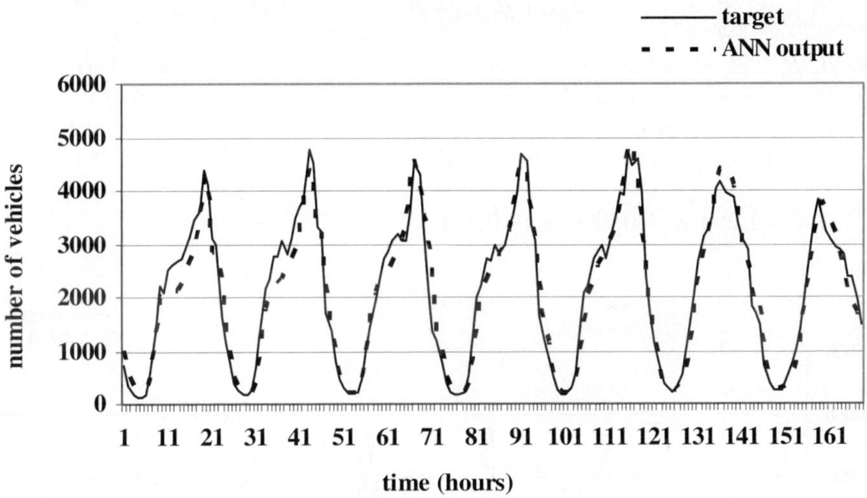

Fig. 4. Variation of the ANN output data versus the target data (2nd analysis)

the years 2001-2003. The training process has been ended when the error has become stable. Variation of the total absolute error through the epochs is shown in Fig. 2.

An example of the daily traffic demand variation, obtained from ANN is shown in Fig. 3 (first analysis). Here, ANN output data and target data have been shown for only the last week of 4th month of the year 2003.

2.2.4 Testing the Network

In the testing phase, an unknown input pattern has been applied to the ANN, and the output has been calculated. Fig. 4 shows an example of the daily traffic demand, obtained from ANN model (second analysis). Here, the target data and ANN output data have been shown for only the last week of 4th month (April) of the year 2004.

3 Prediction by AR(4) Time Series Model

TS data are a series of observations recorded in sequence over time. Interval between the observations can be determined as hours, days, weeks, months, quarters, or years, depending on the type of data. Using the TS data, it can be learnt about the daily traffic over time and therefore it may be used past values of variables to predict the future values [2, 4, 5].

In TS analysis, one natural question is whether and what extent we can predict the future values of a TS variable using past values. One popular model for answering this question is AR model. AR models relate a TS variable to its past values. The model assumes that only the most recent lags enter the AR model. The number of lags, p, is often called the lag length of the AR. The p-th order AR model is denoted as AR(p). For TS regression and prediction, let y_t denote an observation in date t. Then, the AR(p) model has the form as follows:

$$y_t = \beta_0 + \beta_1 y_{t-1} + \beta_2 y_{t-2} + \ldots\ldots + \beta_p y_{t-p} + u_t \quad (1)$$

where $\beta_0, \beta_1, \ldots\beta_p$ are unknown regression coefficients, and ut is unobserved error term.

For the comparison, an AR(4) model has been used. After the search of the best fit with SPSS package program [6], the regression coefficients and unobserved error term have been found. Then the AR(4) model has been established as follows:

$$y_t = 1{,}294\, y_{t-1} - 0{,}0232\, y_{t-2} - 0{,}704\, y_{t-3} - 0{,}137\, y_{t-4} + 379{,}076 \quad (2)$$

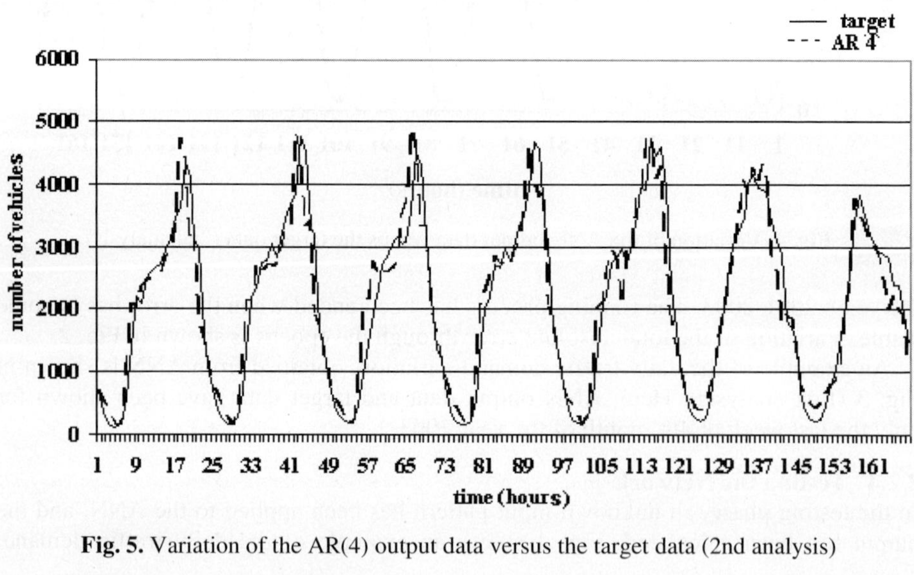

Fig. 5. Variation of the AR(4) output data versus the target data (2nd analysis)

Two different analysis have been completed, using this model. In the first group of analysis, the traffic demand, yt , related to the hours of the days of 4th months of the years 2001-2003 has been used. In the second group of analysis, the traffic demand related to the hours of the days of 4th month of the year 2004 has been used.

An example of the daily traffic demand variation obtained by AR(4) model, related to the result of the second analysis is shown in Fig. 5. Here, AR(4) model output data and target data have been shown for only the last week of 4th month of the year 2004.

4 Conclusion

The daily traffic demand in Second Tolled Bridge of Bosphorus which has an important role in urban traffic networks of Istanbul has been predicted by an AR(4) model and an ANN model implemented by a developed ANN simulator. Two group of analysis have been completed using both ANN model and AR(4) model, to compare the obtained errors. First group of the analysis have been completed using the traffic demand data related to the hours of the days of 4th months of the years 2001-2003. Second group of the analysis have been completed using the traffic demand data related to the hours of the days of 4th month of the year 2004. Obtained errors, both normalized and actual root mean square errors (RMSE), are shown in Table 1.

Referring to Table 1, in the first analyze, the error obtained by ANN model has been reduced 40,35 % comparing to AR(4) model. Also, in the second analyze, the error obtained by ANN model has been reduced 36,92 %. The results have shown that ANN model approach can be more powerful than AR model in this application.

Table 1. Obtained errors for two groups of the analysis

	Analysis by ANN		Analysis by AR (4)	
Related years of the data	2001-2003	2004	2001-2003	2004
RMSE(normalized)	0,034	0,041	0,057	0,065
RMSE(actual)	49,44	59,62	82,88	94,52

It is also concluded that, many other transportation data prediction studies can be implemented easily and successfully by using the developed ANN simulator. So, new studies are being planned in this context.

References

1. Faraway, J.: Time series forecasting with neural networks, a comparative study using the airline data. Applied Statistics, Royal Statistical Society, UK, 231–250 (1998)
2. Box, G.E.P., Jenkins, G.: Time Series Analysis, Forecasting and Control, Golden-Day Publishing Company, San Francisco, CA, pp. 20–35 (1970)
3. General Directorate Of Highways. Traffic and Transportation Survey 2000-2004. GDOH, Ankara, Turkey (2004)

4. Hipel, K.W., Mcleod, A.I.: Time Series Modeling of Water Resources and Environmental Systems, pp. 14–25. Elsevier, Amsterdam (1994)
5. De Gooijer, J.G., Kumar, K.: Some recent developments in non-linear time series modeling, testing, and forecasting. International Journal of Forecasting 8, 135–156 (1992)
6. SPSS For WINDOWS ver.10. Web site: http://www.spssscience.com

Author Index

Printing: Mercedes-Druck, Berlin
Binding: Stein+Lehmann, Berlin

Lecture Notes in Artificial Intelligence (LNAI)

Vol. 4694: B. Apolloni, R.J. Howlett, L. Jain (Eds.), Knowledge-Based Intelligent Information and Engineering Systems: KES 2007 - WIRN 2007, Part III. XXIX, 1126 pages. 2007.

Vol. 4693: B. Apolloni, R.J. Howlett, L. Jain (Eds.), Knowledge-Based Intelligent Information and Engineering Systems: KES 2007 - WIRN 2007, Part II. XXXII, 1380 pages. 2007.

Vol. 4692: B. Apolloni, R.J. Howlett, L. Jain (Eds.), Knowledge-Based Intelligent Information and Engineering Systems: KES 2007 - WIRN 2007, Part I. LV, 882 pages. 2007.

Vol. 4682: D.-S. Huang, L. Heutte, M. Loog (Eds.), Advanced Intelligent Computing Theories and Applications. XXVII, 1373 pages. 2007.

Vol. 4660: S. Džeroski, J. Todorovski (Eds.), Computational Discovery of Scientific Knowledge. X, 327 pages. 2007.

Vol. 4659: V. Mařík, V. Vyatkin, A.W. Colombo (Eds.), Holonic and Multi-Agent Systems for Manufacturing. VIII, 456 pages. 2007.

Vol. 4651: F. Azevedo, P. Barahona, F. Fages, F. Rossi (Eds.), Recent Advances in Constraints. VIII, 185 pages. 2007.

Vol. 4635: B. Kokinov, D.C. Richardson, T.R. Roth-Berghofer, L. Vieu (Eds.), Modeling and Using Context. XIV, 574 pages. 2007.

Vol. 4632: R. Alhajj, H. Gao, X. Li, J. Li, O.R. Zaïane (Eds.), Advanced Data Mining and Applications. XV, 634 pages. 2007.

Vol. 4629: V. Matoušek, P. Mautner (Eds.), Text, Speech and Dialogue. XVII, 663 pages. 2007.

Vol. 4626: R.O. Weber, M.M. Richter (Eds.), Case-Based Reasoning Research and Development. XIII, 534 pages. 2007.

Vol. 4617: V. Torra, Y. Narukawa, Y. Yoshida (Eds.), Modeling Decisions for Artificial Intelligence. XII, 502 pages. 2007.

Vol. 4612: I. Miguel, W. Ruml (Eds.), Abstraction, Reformulation, and Approximation. XI, 418 pages. 2007.

Vol. 4604: U. Priss, S. Polovina, R. Hill (Eds.), Conceptual Structures: Knowledge Architectures for Smart Applications. XII, 514 pages. 2007.

Vol. 4603: F. Pfenning (Ed.), Automated Deduction – CADE-21. XII, 522 pages. 2007.

Vol. 4597: P. Perner (Ed.), Advances in Data Mining. XI, 353 pages. 2007.

Vol. 4594: R. Bellazzi, A. Abu-Hanna, J. Hunter (Eds.), Artificial Intelligence in Medicine. XVI, 509 pages. 2007.

Vol. 4585: M. Kryszkiewicz, J.F. Peters, H. Rybinski, A. Skowron (Eds.), Rough Sets and Intelligent Systems Paradigms. XIX, 836 pages. 2007.

Vol. 4578: F. Masulli, S. Mitra, G. Pasi (Eds.), Applications of Fuzzy Sets Theory. XVIII, 693 pages. 2007.

Vol. 4573: M. Kauers, M. Kerber, R. Miner, W. Windsteiger (Eds.), Towards Mechanized Mathematical Assistants. XIII, 407 pages. 2007.

Vol. 4571: P. Perner (Ed.), Machine Learning and Data Mining in Pattern Recognition. XIV, 913 pages. 2007.

Vol. 4570: H.G. Okuno, M. Ali (Eds.), New Trends in Applied Artificial Intelligence. XXI, 1194 pages. 2007.

Vol. 4565: D.D. Schmorrow, L.M. Reeves (Eds.), Foundations of Augmented Cognition. XIX, 450 pages. 2007.

Vol. 4562: D. Harris (Ed.), Engineering Psychology and Cognitive Ergonomics. XXIII, 879 pages. 2007.

Vol. 4548: N. Olivetti (Ed.), Automated Reasoning with Analytic Tableaux and Related Methods. X, 245 pages. 2007.

Vol. 4539: N.H. Bshouty, C. Gentile (Eds.), Learning Theory. XII, 634 pages. 2007.

Vol. 4529: P. Melin, O. Castillo, L.T. Aguilar, J. Kacprzyk, W. Pedrycz (Eds.), Foundations of Fuzzy Logic and Soft Computing. XIX, 830 pages. 2007.

Vol. 4520: M.V. Butz, O. Sigaud, G. Pezzulo, G. Baldassarre (Eds.), Anticipatory Behavior in Adaptive Learning Systems. X, 379 pages. 2007.

Vol. 4511: C. Conati, K. McCoy, G. Paliouras (Eds.), User Modeling 2007. XVI, 487 pages. 2007.

Vol. 4509: Z. Kobti, D. Wu (Eds.), Advances in Artificial Intelligence. XII, 552 pages. 2007.

Vol. 4496: N.T. Nguyen, A. Grzech, R.J. Howlett, L.C. Jain (Eds.), Agent and Multi-Agent Systems: Technologies and Applications. XXI, 1046 pages. 2007.

Vol. 4483: C. Baral, G. Brewka, J. Schlipf (Eds.), Logic Programming and Nonmonotonic Reasoning. IX, 327 pages. 2007.

Vol. 4482: A. An, J. Stefanowski, S. Ramanna, C.J. Butz, W. Pedrycz, G. Wang (Eds.), Rough Sets, Fuzzy Sets, Data Mining and Granular Computing. XIV, 585 pages. 2007.

Vol. 4481: J. Yao, P. Lingras, W.-Z. Wu, M. Szczuka, N.J. Cercone, D. Ślęzak (Eds.), Rough Sets and Knowledge Technology. XIV, 576 pages. 2007.

Vol. 4476: V. Gorodetsky, C. Zhang, V.A. Skormin, L. Cao (Eds.), Autonomous Intelligent Systems: Multi-Agents and Data Mining. XIII, 323 pages. 2007.

Vol. 4456: Y. Wang, Y.-m. Cheung, H. Liu (Eds.), Computational Intelligence and Security. XXIII, 1118 pages. 2007.

Vol. 4455: S. Muggleton, R. Otero, A. Tamaddoni-Nezhad (Eds.), Inductive Logic Programming. XII, 456 pages. 2007.

Vol. 4452: M. Fasli, O. Shehory (Eds.), Agent-Mediated Electronic Commerce. VIII, 249 pages. 2007.

Vol. 4451: T.S. Huang, A. Nijholt, M. Pantic, A. Pentland (Eds.), Artifical Intelligence for Human Computing. XVI, 359 pages. 2007.

Vol. 4441: C. Müller (Ed.), Speaker Classification. X, 309 pages. 2007.

Vol. 4438: L. Maicher, A. Sigel, L.M. Garshol (Eds.), Leveraging the Semantics of Topic Maps. X, 257 pages. 2007.

Vol. 4434: G. Lakemeyer, E. Sklar, D.G. Sorrenti, T. Takahashi (Eds.), RoboCup 2006: Robot Soccer World Cup X. XIII, 566 pages. 2007.

Vol. 4429: R. Lu, J.H. Siekmann, C. Ullrich (Eds.), Cognitive Systems. X, 161 pages. 2007.

Vol. 4428: S. Edelkamp, A. Lomuscio (Eds.), Model Checking and Artificial Intelligence. IX, 185 pages. 2007.

Vol. 4426: Z.-H. Zhou, H. Li, Q. Yang (Eds.), Advances in Knowledge Discovery and Data Mining. XXV, 1161 pages. 2007.

Vol. 4411: R.H. Bordini, M. Dastani, J. Dix, A.E.F. Seghrouchni (Eds.), Programming Multi-Agent Systems. XIV, 249 pages. 2007.

Vol. 4410: A. Branco (Ed.), Anaphora: Analysis, Algorithms and Applications. X, 191 pages. 2007.

Vol. 4399: T. Kovacs, X. Llorà, K. Takadama, P.L. Lanzi, W. Stolzmann, S.W. Wilson (Eds.), Learning Classifier Systems. XII, 345 pages. 2007.

Vol. 4390: S.O. Kuznetsov, S. Schmidt (Eds.), Formal Concept Analysis. X, 329 pages. 2007.

Vol. 4389: D. Weyns, H.V.D. Parunak, F. Michel (Eds.), Environments for Multi-Agent Systems III. X, 273 pages. 2007.

Vol. 4386: P. Noriega, J. Vázquez-Salceda, G. Boella, O. Boissier, V. Dignum, N. Fornara, E. Matson (Eds.), Coordination, Organizations, Institutions, and Norms in Agent Systems II. XI, 373 pages. 2007.

Vol. 4384: T. Washio, K. Satoh, H. Takeda, A. Inokuchi (Eds.), New Frontiers in Artificial Intelligence. IX, 401 pages. 2007.

Vol. 4371: K. Inoue, K. Satoh, F. Toni (Eds.), Computational Logic in Multi-Agent Systems. X, 315 pages. 2007.

Vol. 4369: M. Umeda, A. Wolf, O. Bartenstein, U. Geske, D. Seipel, O. Takata (Eds.), Declarative Programming for Knowledge Management. X, 229 pages. 2006.

Vol. 4343: C. Müller (Ed.), Speaker Classification I. X, 355 pages. 2007.

Vol. 4342: H. de Swart, E. Orłowska, G. Schmidt, M. Roubens (Eds.), Theory and Applications of Relational Structures as Knowledge Instruments II. X, 373 pages. 2006.

Vol. 4335: S.A. Brueckner, S. Hassas, M. Jelasity, D. Yamins (Eds.), Engineering Self-Organising Systems. XII, 212 pages. 2007.

Vol. 4334: B. Beckert, R. Hähnle, P.H. Schmitt (Eds.), Verification of Object-Oriented Software. XXIX, 658 pages. 2007.

Vol. 4333: U. Reimer, D. Karagiannis (Eds.), Practical Aspects of Knowledge Management. XII, 338 pages. 2006.

Vol. 4327: M. Baldoni, U. Endriss (Eds.), Declarative Agent Languages and Technologies IV. VIII, 257 pages. 2006.

Vol. 4314: C. Freksa, M. Kohlhase, K. Schill (Eds.), KI 2006: Advances in Artificial Intelligence. XII, 458 pages. 2007.

Vol. 4304: A. Sattar, B.-h. Kang (Eds.), AI 2006: Advances in Artificial Intelligence. XXVII, 1303 pages. 2006.

Vol. 4303: A. Hoffmann, B.-h. Kang, D. Richards, S. Tsumoto (Eds.), Advances in Knowledge Acquisition and Management. XI, 259 pages. 2006.

Vol. 4293: A. Gelbukh, C.A. Reyes-Garcia (Eds.), MICAI 2006: Advances in Artificial Intelligence. XXVIII, 1232 pages. 2006.

Vol. 4289: M. Ackermann, B. Berendt, M. Grobelnik, A. Hotho, D. Mladenič, G. Semeraro, M. Spiliopoulou, G. Stumme, V. Svátek, M. van Someren (Eds.), Semantics, Web and Mining. X, 197 pages. 2006.

Vol. 4285: Y. Matsumoto, R.W. Sproat, K.-F. Wong, M. Zhang (Eds.), Computer Processing of Oriental Languages. XVII, 544 pages. 2006.

Vol. 4274: Q. Huo, B. Ma, E.-S. Chng, H. Li (Eds.), Chinese Spoken Language Processing. XXIV, 805 pages. 2006.

Vol. 4265: L. Todorovski, N. Lavrač, K.P. Jantke (Eds.), Discovery Science. XIV, 384 pages. 2006.

Vol. 4264: J.L. Balcázar, P.M. Long, F. Stephan (Eds.), Algorithmic Learning Theory. XIII, 393 pages. 2006.

Vol. 4259: S. Greco, Y. Hata, S. Hirano, M. Inuiguchi, S. Miyamoto, H.S. Nguyen, R. Słowiński (Eds.), Rough Sets and Current Trends in Computing. XXII, 951 pages. 2006.

Vol. 4253: B. Gabrys, R.J. Howlett, L.C. Jain (Eds.), Knowledge-Based Intelligent Information and Engineering Systems, Part III. XXXII, 1301 pages. 2006.

Vol. 4252: B. Gabrys, R.J. Howlett, L.C. Jain (Eds.), Knowledge-Based Intelligent Information and Engineering Systems, Part II. XXXIII, 1335 pages. 2006.

Vol. 4251: B. Gabrys, R.J. Howlett, L.C. Jain (Eds.), Knowledge-Based Intelligent Information and Engineering Systems, Part I. LXVI, 1297 pages. 2006.

Vol. 4248: S. Staab, V. Svátek (Eds.), Managing Knowledge in a World of Networks. XIV, 400 pages. 2006.

Vol. 4246: M. Hermann, A. Voronkov (Eds.), Logic for Programming, Artificial Intelligence, and Reasoning. XIII, 588 pages. 2006.

Vol. 4223: L. Wang, L. Jiao, G. Shi, X. Li, J. Liu (Eds.), Fuzzy Systems and Knowledge Discovery. XXVIII, 1335 pages. 2006.

Vol. 4213: J. Fürnkranz, T. Scheffer, M. Spiliopoulou (Eds.), Knowledge Discovery in Databases: PKDD 2006. XXII, 660 pages. 2006.

Vol. 4212: J. Fürnkranz, T. Scheffer, M. Spiliopoulou (Eds.), Machine Learning: ECML 2006. XXIII, 851 pages. 2006.